PRINTED
CIRCUITS
HANDBOOK

PRINTED CIRCUITS HANDBOOK

Clyde F. Coombs, Jr. Editor-in-Chief

Fifth Edition

McGRAW-HILL

New York   Chicago   San Francisco   Lisbon
London   Madrid   Mexico City   Milan
New Delhi   San Juan   Seoul
Singapore   Sydney   Toronto

To Ann

Cataloging-in-Publication Data is on file with the Library of Congress

McGraw-Hill

A Division of The McGraw·Hill Companies

1 2 3 4 5 6 7 8 9 0 DOC/DOC 0 7 6 5 4 3 2 1

P/N 0-07-138144-9
ISBN 0-07-135016-0

The sponsoring editor for this book was Steven Chapman, the production supervisor was Pamela Pelton. It was set in Times Roman by North Market Street Graphics.

Printed and bound by R.R. Donnelley & Sons Company.

McGraw-Hill books are available at special quantity discounts to use as premiums and sales promotions, or for use in corporate training programs. For more information, please write to the Director of Special Sales, McGraw-Hill, Inc., Two Penn Plaza, New York, NY 10121-2298. Or contact your local bookstore.

CONTENTS

Part 2 Materials

Chapter 5. Introduction to Base Materials   5.3

Chapter 6. Base Material Components   6.1

Part 3   Engineering and Design

Chapter 13. Physical Characteristics of PCB   13.3

Chapter 14. The PCB Design Process **14.1**

Chapter 15. Electrical and Mechanical Design Parameters **15.1**

Chapter 16. Controlled Impedance   16.1

Chapter 17. Multilayer Design Issues   17.1

Chapter 18. Planning for Design, Fabrication, and Assembly   18.1

Chapter 19. Manufacturing Information Documentation and Transfer   **19.1**

Chapter 20. Electronic Contract Manufacturing Supplier Selection and Management   **20.1**

Part 4   High-Density Interconnect

Chapter 21. Introduction to High-Density Interconnection Technology   21.3

Chapter 22. High-Density Interconnect-Build-up Technologies   22.1

Chapter 23. Microvia Hole Technologies   23.1

Part 5   Fabrication Processes

Chapter 24. Drilling Processes   24.3

Chapter 25. High-Density Interconnect Drilling

25.1

Chapter 26. Imaging

26.1

Chapter 27. Multilayer Materials and Processing   **27.1**

Chapter 28. Preparing Boards for Plating   **28.1**

Chapter 29. Electroplating   29.1

Chapter 30. Direct Plating   **30.1**

Chapter 31. PWB Manufacture Using Fully Electroless Copper   **31.1**

Chapter 32. Surface Finishes   **32.1**

Chapter 33. Etching Process and Technologies 33.1

Chapter 34. Solder Resist Material and Processes   **34.1**

Chapter 35. Machining and Routing   **35.1**

Chapter 39. Bare Board Test Equipment   39.1

Chapter 40. HDI Bare Board Special Testing Methods   40.1

Part 6   Assembly

Chapter 41. Assembly Processes   41.3

Part 7   Soldering

Chapter 42. Design for Soldering and Solderability   **42.3**

Chapter 43. Solder Materials and Processes 43.1

Chapter 44. No-Clean Assembly Process   **44.1**

Chapter 45. Lead-Free Soldering   **45.1**

Chapter 46. Fluxes and Cleaning   46.1

Chapter 47. Press-Fit Connections   47.1

Part 8   Quality Control and Reliability

Chapter 48. Acceptability of Fabricated Boards   48.3

Chapter 49. Acceptability of Printed Circuit Board Assemblies

49.1

Chapter 50. Assembly Inspection   50.1

Chapter 51. Design for Testing

51.1

Chapter 52. Loaded Board Testing

52.1

Chapter 53. Reliability of Printed Circuit Assemblies   **53.1**

Chapter 54. Component-to-PWB Reliability   **54.1**

Part 9 Environmental Issues and Waste Treatment

Part 10 Flexible Circuits

Chapter 59. Termination of Flexible Circuits   59.1

Chapter 60. Special Constructions of Flexible Circuits   60.1

Chapter 61. Quality Assurance of Flexible Circuits   **61.1**

LIST OF CONTRIBUTORS

Steven M. Allen *AVEX Electronics, Fremont, CA* (CHAP. 20)

A. D. Andrade *Sandia National Laboratories, Livermore, CA (retired)* (CHAP. 48)

Simon Ang *University of Arkansas, Fayetteville, AR* (CHAP. 3)

Joyce M. Avery *Avery Environmental Services, Saratoga, CA* (CHAP. 55)

Fred Barlow *University of Arkansas, Fayetteville, AR* (CHAP. 3)

Todd A. Barnett *Excellon Automotive, Torrence, CA* (CHAP. 25)

David W. Bergman *IPC, Northbrook, IL* (APPENDIX)

Bruce Bolliger *Agilent Technologies, Singapore* (CHAP. 50)

Richard Boulanger *Universal Instruments, Binghamton, NY* (CHAP. 41)

Mark Brillhart *Cisco Systems, San Jose, CA* (CHAP. 54)

William Brown *University of Arkansas, Fayetteville, AR* (CHAP. 3)

James Cadile *NOVA Drilling Service, Santa Clara, CA* (CHAP. 35)

Hugh Cole (CHAP. 42)

Brian F. Conaghan *Parelec, Rocky Hill, NJ* (CHAP. 26)

Edward F. Duffek *Adion Engineering, Cupertino, CA* (CHAP. 28)

C. D. DuPriest *Lockheed Martin, Dallas, TX* (CHAP. 27)

Sylvia Ehrler *Multek Europe, Böblingen, Germany* (CHAP. 12)

Timothy A. Estes *Conductor Analysis Technologies, Albuquerque, NM* (CHAP. 36)

Gary M. Freedman *Compaq, Alpha High Performance Systems, Marlboro, MA* (CHAPS. 43, 45, 47)

Judith Glazer *Hewlett-Packard, Palo Alto, CA* (CHAP. 53)

Ceferino G. Gonzalez *Dupont, Research Triangle Park, NC* (CHAP. 11)

Marshall I. Gurian *Marshall Gurian Consulting, Tempe, AZ* (CHAP. 33)

Paul W. Henderson *Hewlett-Packard, Palo Alto, CA* (CHAP. 44)

Charles G. Henningsen *Insulectro, Mountain View, CA* (CHAP. 35)

Ralph J. Hersey, Jr. *Lawrence Livermore National Laboratory, Livermore, CA* (CHAP. 15)

Happy T. Holden *Westwood Associates, Loveland, CO* (CHAPS. 1, 18, 21, 22)

Robert R. Holmes *AT&T, Richmond, VA* (CHAP. 17)

Edward J. Kelley *Sanmina, Derry, NH* (CHAPS. 5–10)

George Milad *Shipley Ronal, Marlborough, MA* (CHAPS. 29, 32)

Peter G. Moleux *Peter Moleux and Associates, Newton Centre, MA* (CHAP. 55)

Hayao Nakahara *N.T. Information, Huntington, NY* (CHAPS. 4, 23, 30, 31, 34)

Dominique K. Numakura *DKN Research, Haverhill, MA* (CHAPS. 56–61)

Stig Oresjo   *Agilent Technologies, Loveland, CO* (CHAP. 50)

Kenneth P. Parker   *Agilent Technologies, Loveland, CO* (CHAPS. 51, 52)

Clyde Parrish   *PC World, Toronto, Canada* (CHAP. 19)

Tarak Railkar   *Conexant Systems, Newport Beach, CA* (CHAP. 3)

Ronald J. Rhodes   *Conductor Analysis Technologies, Green Brook, NJ* (CHAP. 36)

Lee W. Ritchey   *3Com, Santa Clara, CA* (CHAPS. 13, 14)

Michael Roesch   *Hewlett-Packard, Palo Alto, CA* (CHAP. 12)

John W. Stafford   *JWS Consulting PLC, Phoenix, AZ* (CHAP. 2)

Ken Taylor   *Polar Instruments, Garenne Park, Guernsey, UK* (CHAP. 16)

Leland E. Tull   *Contouring Technology, Mountain View, CA* (CHAP. 35)

Laura J. Turbini   *University of Toronto, Canada* (CHAP. 46)

Hans Vandervelde   *Laminating Company of America, Garden Grove, CA* (CHAP. 24)

David J. Wilkie   *Everett Charles Technologies, Pomona, CA* (CHAPS. 37–40)

Warren Wong   *Excellon Automotive, Torrence, CA* (CHAP. 25)

Bruce Wooldridge   *DSC Communications, Plano, TX* (CHAP. 49)

PREFACE

There has been a "density revolution" in the technology of printed circuits. The need for ever smaller board feature sizes has resulted in a situation in which we can no longer do the same things in the same ways, only smaller. For geometries that meet or exceed the definition of high-density interconnect (HDI), new approaches to design, fabrication, assembly, and testing have been developed, and entirely new materials, tools, and processes have emerged. This would seem to be the very definition of a technical revolution.

This edition of *Printed Circuits Handbook* addresses these new elements of the printed circuit processes, while still maintaining its foundation on the basics of the technology. As a result, almost one-quarter of the chapters in this book are new to this edition, while half have been revised and expanded to include HDI-related information in the traditional process areas. The result is a new book clearly founded on the basics, but looking to the future.

To achieve this, we have included new chapters on materials, design, microvia hole creation, sequential build-up of multilayer boards, buried and blind via construction, small-geometry imaging and plating, special assembly, and soldering, as well as testing of both fabricated and assembled boards made with HDI feature size, and the reliability of these assemblies.

No matter how sophisticated the leading edge of the technology becomes, however, at the core of all printed circuits is the plated through-hole in its various forms. This remains one of the most important technical achievements of the twentieth century, and certainly one of the least appreciated. It provides the means to interconnect, and therefore make useful, the complex components that make up all the electronic products that are so much a part of daily life. This represents the process elements that form the enduring foundation of printed circuit technology. Printed circuits have evolved over the years to be more reliable, efficient, and reproducible, but the process described in the first edition of this book is still recognizable in the fifth. Therefore, those new to the technology will still find introductory information, while experienced practitioners will find the industry standard methods and best practices that help with the most recent developments. All will find a comprehensive discussion of HDI technologies, materials, and methods to help them function in that complex field as well.

As the industry has grown, it has become more specialized. This has created the need to standardize documentation and communication techniques as well as to understand the specific capabilities of all suppliers in the overall value delivery chain. The result is that process capabilities and limitations, at each step, must be known, the board must be designed with these clearly in mind, and consistent acceptability criteria must be agreed to in advance, before the responsibility for the board passes from designer to fabricator to assembler to end user. This has created a community of people who have not been intimately involved in printed circuit issues before, and who now find a working knowledge of printed circuits critical for their performance. This book provides information for these people as well. They will not only find the basic information useful in understanding the issues, but there are also specific guidelines on the development and management of the value chain for the success of all.

While the industry's preferred term for the subject of this book is *printed wiring,* the term *printed circuits* has passed into the world's languages as representing the process and products described. As a result, we will use the terms interchangeably.

I wish to thank the staff of the IPC, especially Dieter Bergman, David Bergman, and Tony Hilvers, for their efforts in support of this book. While not directly involved here, Ray Pritchard, retired executive director of the IPC, remains a strong influence on all of us in the industry.

Clyde F. Coombs, Jr.

INTRODUCTION TO PRINTED CIRCUITS

CHAPTER 1
ELECTRONIC PACKAGING AND HIGH-DENSITY INTERCONNECTIVITY*

Clyde F. Coombs Jr.
Editor-In-Chief, Los Altos, California

Happy T. Holden
Westwood Associates, Loveland, Colorado

1.1 INTRODUCTION

All electronic components must be interconnected and assembled to form a functional and operating system. The design and the manufacture of these interconnections have evolved into a separate discipline called *electronic packaging*. Since the early 1950s, the basic building block of electronic packaging is the printed wiring board (PWB), and it will remain that into the foreseeable future. This book outlines the basic design approaches and manufacturing processes needed to produce these PWBs.

This chapter outlines the basic considerations, the main choices, and the potential trade-offs that must be accounted for in the selection of the interconnection methods for electronic systems. Its main emphasis is on the analysis of potential effects that the selection of various printed wiring board types and design alternatives could have on the cost and performance of the complete electronic product.

1.2 MEASURING THE INTERCONNECTIVITY REVOLUTION (HDI)

The continuing increase in component performance and lead density, along with the reduction in package sizes, has required that PWB technology find corresponding ways to increase the interconnection density of the substrate. With the introduction and continued refinement of such packaging techniques as the ball grid array (BGA), chip-scale packaging (CSP), and chip-on-board (COB), traditional PWB technology has approached a point where alternative ways of providing high-density interconnection have had to be developed. (See Chaps. 2 and

---

* Adapted from Coombs, Clyde F. Jr., *Printed Circuits Handbook* (4th ed.), chap. 1, "Electronic Packaging and Interconnectivity," (McGraw-Hill, New York, 1996.)

3 for detailed discussions of component and packaging technologies.) This has been called at times high-density interconnects (HDI), the interconnection revolution, or the density revolution, because doing the same things in the same way, only smaller, was no longer sufficient.

1.2.1 Interconnect Density Elements

The extent of these interconnect density issues is not always observable, but the chart[1] in Fig. 1.1 can help one define and understand it. The chart portrays the interrelationship between component packaging, surface-mount technology (SMT) assembly, and PWB density. As can be seen, these three elements are interlinked. A change in one has a significant effect on the overall interconnection density. The metrics are as follows:

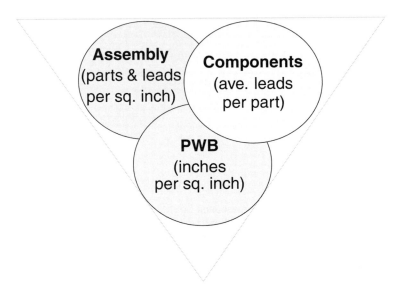

FIGURE 1.1 Representation of the metrics of assembly, component, and PWB technologies and their general relationship to each other.

- *Assembly complexity:* the measure of the difficulty of assembling surface-mounted components in parts per square inch and leads per square inch.
- *Component packaging complexity:* the degree of sophistication of a component, measured by its average leads (I/Os) per part.
- *Printed wiring board density:* the amount of wiring a PWB has as measured by the average length of traces per square inch or the area of that board, including all signal layers. The metric is inches per square inch.

1.2.2 Interconnect Technology Map

To visualize the interrelationships of the three elements, see Fig. 1.2. It shows these elements as axes of a three-dimensional technology map that defines the passage from conventional PWB structures to advanced technologies and shows how changes in just one of the elements can increase or decrease the total density of the entire electronic package.

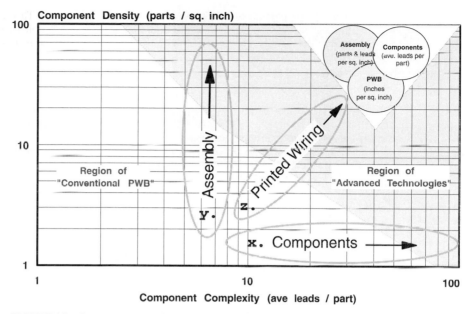

FIGURE 1.2   Component technology map, showing the relationship of assembly, PWB, and component technologies on overall package density and technology.

To describe the component complexity of an assembly, the total component connections (I/Os) include both sides of an assembly, as well as edge fingers, or contacts, which are divided by the total number of components on the assembly. The resulting average leads (I/Os) per part provides the x axis of Fig. 1.2. The horizontal oval shape shows how the component complexity can vary from two leads per part in discrete circuit elements to the very large numbers seen on BGA and application-specific integrated circuits.

When Fig. 1.2 is used to describe surface-mount assemblies, the vertical (y-axis) dimension (shown as a vertical oval) indicates how complex it is to assemble the board by number of components per square inch or square centimeter for the surface area of the PWB. This vertical oval can vary from 1 to over 100 parts per square inch. As the parts become smaller and closer together, this number naturally goes up. A second assembly measure is average leads (I/Os) per square inch or square centimeter. This is the x-axis value multiplied by the y-axis value. (For a further description of these issues, and equations for quantifying them, see Chap. 18.)

The z-axis oval in Fig. 1.2 describes the printed wiring board's density. This is the wiring required to connect all the I/Os of the components at the size of the assembly specified, assuming three nodes per net. This axis has the units inches per square inch, or centimeters per square centimeter. A further description of this metric is provided in this chapter and in more detail in Chap. 18.

1.2.3   An Example of the Interconnect Revolution

By charting products of a particular type over time, an analysis will show how the interconnect technology has changed and continues to change, its rate of change, and the direction of these changes. An example is given in Fig. 1.3. This shows how component technology, assembly technology, and PWB technology have led to the evolution of the same computer CPU from:

a. First RISC processor (1986)
(8"x16" 14 layer-TH)

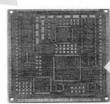

b. Same RISC processor as MCM-L (1991)
(4"x4"   10 layer-TH/Bv seq. lam)

c. Same RISC processor as HDI (1995)
(2"x2"   2+2+2 HDI-microvia)

FIGURE 1.3   Example of the same computer CPU board as it used alternative component, assembly, and PWB technologies. (*a*) Size and appearance of each generation.
(*b*) Movement of total board density from traditional to HDI. (*c*) HDI.

A 14-layer board through-hole with a surface area of 128 in^2 in 1986 (Fig. 1.3[*a*]) to

A 10-layer surface-mount technology board with a surface area of 16 in^2 in 1991 (Fig.
1.3[*b*]) to

A high-density interconnect board with sequential build-up microvias, buried and blind
vias, and a surface area of 4 in^2 in 1995 (Fig. 1.3[*c*])

1.2.4   Region of Advanced Technologies

A second valuable feature of the chart in Fig. 1.2 is the area identified as the region of
advanced technologies. This is where calculations and data have shown that it is necessary to
have an HDI structure. Therefore, this is the barrier, or wall of HDI: on one side it is most cost
effective to use traditional PWB technologies; on the other side it becomes cost effective to
use HDI technologies. Continuing beyond this point, HDI becomes necessary.

1.3   *HIERARCHY OF INTERCONNECTIONS*

To have the proper perspective on where PWBs fit into electronic systems, it will be helpful to
describe briefly the packaging hierarchy of electronic systems. Some time ago, the Institute
for Interconnecting and Packaging Electronic Circuits (IPC)[2] proposed eight categories of

system elements in ascending order of size and complexity, which will be used here to illustrate typical electronic packaging structures. These are as follows:

Category A consists of fully processed active and passive devices. Bare or uncased chips and discrete capacitors, resistors, or their networks are typical examples of this category.

Category B comprises all packaged devices (active and passive) in plastic packages, such as DIPs, TSOPs, QFPs, and BGAs, as well as those in ceramic packages, such as PGAs, and connectors, sockets, and switches. All are ready to be connected to an interconnecting structure.

Category C is substrates that interconnect uncased or bare chips (i.e., the components of category A) into a separable package. Included here are all types of multichip modules (MCMs), chip-on-boards (COBs), and hybrids.

Category D covers all kinds of substrates that interconnect and form assemblies of already packaged components, i.e., those from categories B and C. This category includes all types of rigid PWBs, flexible and rigid-flexible, and discrete-wiring boards.

Category E covers the back planes made by printed wiring and discrete-wiring methods or with flexible circuits, which interconnect PWBs, but not components, from the preceding groups.

Category F covers all intraenclosure connections. Included in this category are harnesses, ground and power distribution buses, RF plumbing, and co-ax or fiberoptic wiring.

Category G includes the system assembly hardware, card racks, mechanical structures, and thermal control components.

Category H encompasses the entire integrated system with all its bays, racks, boxes, and enclosures and all auxiliary and support subsystems.

As seen from the preceding list, PWBs are exactly in the center of the hierarchy and are the most important and universally used element of electronic packaging.

The packaging categories F, G, and H are used mainly in large mainframes, supercomputers, central office switching, and some military systems. Since there is a strong trend toward the use of miniaturized and portable electronic products for the majority of electronic packaging designs, trade-offs are made in the judicious application and selection among the elements of the first five categories. These are discussed in this chapter.

1.4   *FACTORS AFFECTING SELECTION OF INTERCONNECTIONS*

Selection of the packaging approaches among the various aforementioned elements is dictated not only by the system function, but also by the component types selected and by the operating parameters of the system, such as the clock speeds, power consumption, and heat management methods, and the environment in which the system will operate. This section provides a brief overview of these basic constraints that must be considered for proper packaging design of the electronic system.

1.4.1   Speed of Operation

The speed at which the electronic system operates is a very important technical factor in the design of interconnections. Many digital systems operate at close to 100 MHz and are already reaching beyond that level. The increasing system speed is placing great demands on the ingenuity of packaging engineers and on the properties of materials used for PWB substrates.

The speed of signal propagation is inversely proportional to the square root of the dielectric constant of the substrate materials, requiring designers to be aware of the dielectric properties of the substrate materials they intend to use. The signal propagation on the substrate between chips, the so-called *time of flight,* is directly proportional to the length of the conductors and must be kept short to ensure the optimal electrical performance of a system operating at high speeds.

For systems operating at speeds above 25 MHz, the interconnections must have transmission line characteristics to minimize signal losses and distortion. Proper design of such transmission lines requires careful calculation of the conductor and dielectric separation dimensions and their precise manufacture to ensure the expected accuracy of performance. For PWBs, there are two basic transmission line types:

1. Stripline
2. Microstrip (for details, see Chap. 16)

1.4.2 Power Consumption

As the clock rates of the chips increase and as the number of gates per chip grows, there is a corresponding increase in their power consumption. Some chips require up to 30 W of power for their operation. With that, more and more terminals are required to bring power in and to accommodate the return flow on the ground planes. About 20 to 30 percent of chip terminals are used for power and ground connections. With the need for electrical isolation of signals in high-speed systems operation, the count may go to 50 percent.

Design engineers must provide adequate power and ground distribution planes within the multilayer boards (MLBs) to ensure efficient, low-resistance flow of currents, which may be substantial in boards interconnecting high-speed chips consuming tens of watts and operating at 5 V, 3.3 V, or lower. Proper power and ground distribution in the system is essential for reducing *di/dt* switching interference in high-speed systems, as well as for reducing undesirable heat concentrations. In some cases, separate bus-bar structures have been required to meet such high power demands.

1.4.3 Thermal Management

All the energy that has been delivered to power integrated circuits (ICs) must be efficiently removed from the system to ensure its proper operation and long life. The removal of the heat from a system is one of the most difficult tasks of electronic packaging. In large systems, huge heat-sink structures, dwarfing the individual ICs, are required to air-cool them, and some computer companies have built giant superstructures for liquid cooling of their computer modules. Some computer designs use liquid immersion cooling. Sill, the cooling needs of large systems tax the capabilities of existing cooling methods.

The situation is not that severe in smaller, tabletop or portable electronic equipment, but it still requires packaging engineers to ameliorate the hot spots and ensure longevity of operation. Since PWBs are notoriously poor heat conductors, designers must carefully evaluate the method of heat conduction through the board, using such techniques as heat vias, embedded metal slugs, and conductive planes.

1.4.4 Electronic Interference

As the frequency of operation of electronic equipment increases, many ICs, modules, or assemblies can act as generators of radio frequency (RF) signals. Such electromagnetic interference (EMI) emanations can seriously jeopardize the operation of neighboring electronics or even of other elements of the same equipment, causing failures, mistakes, and errors, and must be prevented. There are specific EMI standards defining the permissible levels of such radiation, and these levels are very low.

Packaging engineers, and especially PWB designers, must be familiar with the methods of reducing or canceling this EMI radiation to ensure that their equipment will not exceed the permissible limits of this interference.

1.4.5   System Operating Environment

The selection of a particular packaging approach for an electronic product is also dictated by its end use and by the market segment for which that product is designed. The packaging designer has to understand the major driving force behind the product use. Is it cost driven, performance driven, or somewhere in between? Where will it be used—for instance, under the hood of a car, where environmental conditions are severe, or in the office, where the operating conditions are benign? The IPC[2] has established a set of equipment operating conditions classified by the degree of severity, which are listed in Table 1.1.

1.4.5.1   Cost.   The universal digitization of most electronic functions led to the merger of consumer, computer, and communication technologies. This development resulted in the increased appeal of electronics and the need for mass production of many electronic products. Thus, product cost has become the most important criterion in any design of electronic systems. While complying with all the aforementioned design and operation conditions, the design engineer must keep cost as the dominant criterion, and must analyze all potential trade-offs in light of the best cost/performance solution for the product.

TABLE 1.1   Realistic Representative-Use Environments, Service Lives, and Acceptable Cumulative-Failure Probabilities for Surface-Mounted Electronics by Use Categories

Use category	Worst-case use environment					Years of service	Acceptable failure risk, %
	T_{min}, °C	T_{max}, °C	ΔT,* °C	t_D, h	Cycles/ year		
1—Consumer	0	+60	35	12	365	1–3	~1
2—Computers	+15	+60	20	2	1460	~5	~0.1
3—Telecomm	−40	+85	35	12	365	7–20	~0.01
4—Commercial aircraft	−55	+95	20	12	365	~20	~0.001
5—Industrial & automotive	−55	+95	20	12	185		
(passenger compartment)			&40	12	100	~10	~0.1
			&60	12	60		
			&80	12	20		
6—Military	−55	+95	40	12	100		
ground & ship			&60	12	265	~5	~0.1
7—LEO	−40	+85	35	1	8760	5–20	~0.001
Space GEO				12	365		
8—Military *b*	*a*			40	2	365	
avionics *c*	−55	+95	60	2	365	~10	~0.01
			80	2	365		
			&20	1	365		
9—Automotive			60	1	1000		
(under hood)	−55	+125	&100	1	300	~5	~0.1
			&140	2	40		

& = in addition.
* ΔT represents the maximum temperature swing, but does not include power dissipation effects; for power dissipation, calculate ΔT_e.

The importance of the rigorous cost trade-off analysis during the design of electronic products is underscored by the fact that about 60 percent of the manufacturing costs are determined in the first stages of the design process, when only 35 percent of the total design effort has been expended.

Attention to manufacturing and assembly requirements and capabilities (so-called design for manufacturability and assembly [DFM/A]) during product design can reduce assembly costs by up to 35 percent and PWB costs by up to 25 percent.

The elements that must be considered for the most cost-effective electronic packaging designs are:

- Optimization of the PWB design and layout to reduce its manufacturing cost
- Optimization of the PWB design to reduce its assembly cost
- Optimization of the PWB design to reduce testing and repair costs

The following sections provide some guidelines on how to approach such optimization of PWB designs. Basically, the costs of the electronic assemblies are directly related to their complexity and there are a number of measurements relating the effects of various PWB design elements to their costs to guide the design engineer in selection of the most cost-effective approach.

1.5 ICS AND PACKAGES

The most important factors influencing PWB design and layout are the component terminal patterns and their pitches, especially those of ICs and their packages, since these dictate the density of the interconnecting substrates. Thus, this element will be considered first.

Driven by the need for improved cost and performance, the complexity of ICs is constantly increasing. Due to relentless progress in IC technology, the gate density on a chip is increasing by about 75 percent per year, resulting in the growth of IC chip I/O terminals by 40 percent per year, which places ever increasing demands on the methods of their packaging and interconnection.

As a result, the physical size of electronic gears keeps shrinking by 10 to 20 percent per year, while the surface area of substrates is being reduced by about 7 percent per year. This is accomplished by continuously increasing wiring densities and reducing linewidths, which has severely stressed PWB manufacturing methods, reduced processing yields, and increased the costs of the boards.

1.5.1 IC Packages

Since their inception, IC chips have been placed within ceramic or plastic packages. Until about 1980, all IC packages had terminal leads that were soldered into plated through-holes (PTHs) of the PWBs. Since then, an increasing number of IC packages have their terminals made in a form suitable for surface-mounting technology (SMT), which has become the prevailing method of component mounting.

There has been a proliferation of IC package types, both for through-hole assembly as well as for surface mounting, varying in their lead configurations, placement, and pitches. (See Chap. 2 for details.) Also, IPC-SM-782[3] provides a good catalog of the available SMT packages and of the PWB footprint formats they require for their assembly.

Basic I/O termination methods of IC packages include the following:

- *Peripheral,* where the terminations are located around the edges of the chip or package
- *Grid-array,* where the terminations are located on the bottom surface of the chip or package

Most IC packages have peripheral terminations at their edges. The practical limit on the peripheral lead pitches on packages is about 0.3 mm, which permits locating, at most, 500 I/Os on a large IC package, as shown in Table 1.2. It has also become evident that, in typical board assembly operations, the yields plummet as the lead pitches go below 0.5 mm.

TABLE 1.2 Various Array Package Body Sizes, Configurations, and Lead Pitches

Body size (mm)	Number of I/Os	Minimum pitch (mm)
8 × 8	24	0.5
9 × 9	68	0.5
10 × 10	144	0.5
13 × 13	154	0.65
23 × 23	168	1.27
23 × 23	208	1.27
23 × 23	217	1.27
23 × 23	240	1.27
23 × 23	249	1.27
27 × 27	225	1.27
27 × 27	256	1.27
27 × 27	272	1.27
27 × 27	292	1.27
27 × 27	300	1.27
27 × 27	316	1.27
31 × 31	304	1.50
31 × 31	329	1.27
31 × 31	360	1.27
31 × 31	385	1.27
35 × 35	313	1.27
35 × 35	352	1.27
35 × 35	388	1.27
35 × 35	420	1.27
35 × 35	456	1.27
37 × 37	676	0.8
42.5 × 52.5	1247	1.0
52.5 × 52.5	2577	1.0

Various area array components come with a large variety of body sizes, numbers of I/Os, and I/O pitches. These components are called chip-scale packages (CSPs), plastic ball grid arrays (PBGAs), ceramic ball grid arrays (CBGAs), plastic pin grid arrays (PPGAs), and ceramic column grid arrays (CCGAs).

It is expected that chips with terminal counts below 150 to 200 will continue to use packages with peripheral leads, if these can be soldered within practical assembly yields. But for IC packages with over 150 to 200 I/Os, it is very attractive to use the grid-array terminations, since in such a case the entire bottom surface area can be utilized for terminations, which makes it possible to place large numbers of I/Os within a limited area.

This consideration has led to the development of a number of area array solder-bumping termination methods for IC and multichip module (MCM) packages, variously called pad grid, land grid, or ball grid arrays (BGAs) with terminal grids set at 1 mm (0.040 in), 1.27 mm (0.050 in), and 1.50 mm (0.060 in), respectively.

Use of grid arrays provides a number of benefits. The most important is the minimal footprint area on the interconnecting substrate, but grid arrays also offer better electrical performance due to low electrical parasitics in high-speed operation, simplified adaptation into SMT component placement lines, and better assembly yields, despite the impossibility of direct visual inspection of the joints.

Due to continuous decrease of the terminal pitches on packages, it is important that PWB designers carefully assess the manufacturing and assembly capabilities of PWB substrates requiring such fine-pitch terminations, to ensure the greatest yields and lowest cost of the product.

1.5.2 Direct Chip Attach

The relentless pressures of size, weight, and volume reduction of electronic products have resulted in a growth in interest in direct chip attach (DCA) methods, where bare IC chips are mounted directly to the substrate. These methods are extensively used on chip-on-board (COB) and multichip module (MCM) assemblies, as shown in Fig. 1.4.

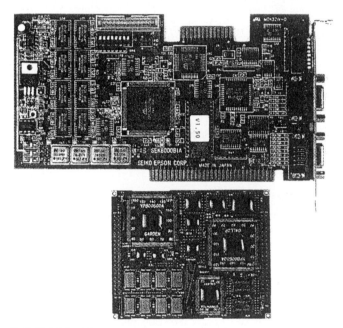

FIGURE 1.4 COB and MCM assemblies.

Three methods of bare chip attachment to the substrate are as follows:

1. *Wire-bonding* is the oldest and the most flexible and widely used method. (More than 96 percent of all chips today are wire-bonded.)
2. *Tape-automated bonding* (TAB) is useful with small I/O pitches and provides the ability to pretest the chips before assembly.
3. *Flip-chipping* is used for its compactness and improved electricals, typical of which is the C4 process of IBM.

The problems of thermal coefficient of expansion (TCE) mismatch between silicon chips that are directly flip-chipped onto a laminate substrate have been effectively eliminated by using a filled epoxy underfill encapsulation technique between the chip and the substrate (see

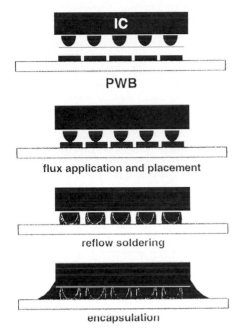

flux application and placement

reflow soldering

encapsulation

FIGURE 1.5 Underfill between the chip and PWB on flip-chip on board.

Fig. 1.5). This method distributes stresses over the entire area of the chip and thus significantly improves the reliability performance of this assembly method.

While the chips that need area array or flip-chip terminations—i.e., those with high pin counts—are the fastest-growing category of ICs, they still represent only a very small percent of all ICs used. Designers must, therefore, ascertain which of the DCA methods will be the most cost beneficial for a particular application.

1.5.3   Chip-Scale Packages (CSPs)

When mounting unpackaged chips on these interconnecting substrates, it is not always possible to ascertain that only properly operating chips have been assembled. By now, there are a number of methods proposed to solve this known good die (KGD) problem.

As one of the ways to resolve this problem, a number of manufacturers have developed a set of miniature packages, only slightly larger than the chip itself, which protect the chip and redistribute the chip termination to a grid array. These miniature packages permit testing and burning in of chips prior to their final assembly. A typical example of such chip scale packages is shown in Fig. 1.6. There are a number of such packages on the market.

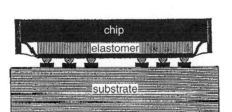

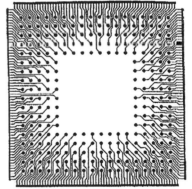

FIGURE 1.6   CSP by Tessera, Inc., San Jose, California.

The designer, however, must analyze the termination pitches of these new CSPs because some use very tight grids, such as 0.5 mm (0.020 in) or less, which need special PWB techniques to permit signal redistribution from these packages to the rest of the board.

In general, the current PWB technology is adequate to provide direct chip terminations if wire-bonding or TAB techniques are used for interconnecting bare chips to the substrate. It requires placing suitable bonding pads spaced by the required pitch in one or two rows around the chip site. While this somewhat reduces the packaging efficiency of the board, it is still an effective method for DCA assemblies.

With grid array, the situation is more difficult because the signals from internal rows of area grid terminations must be routed between the terminals located closer to the edge, which do not permit more than one, or at most two, conductors to pass through. In most cases, these signals from internal rows are brought down into internal layers of MLBs.

The conventional PCB constructions today cannot handle any grid arrays with pitches below 0.020 in, while some flip-chip ball grid arrays go below 0.010-in pitch. In cases when grid distances of the area terminations are below 0.50 mm (0.040 in), special redistribution layers are frequently used, which distribute signals to the conventionally made PTHs in supporting MLB.

Such layers consist of unsupported dielectric layers where small vias or blind holes are formed by laser or plasma etching or are photoformed and then plated using additive or semi-additive metallization processes. While this approach requires some extra area beyond the chip perimeter to complete the signal transfer and increases the costs of substrates, it permits the mounting of flip-chips and CSPs on PWBs. A typical method for forming such redistribution layers, called surface laminar circuit (SLC),[4] has been developed at IBM's Yasu plant.

1.6 DENSITY EVALUATIONS

1.6.1 Component Density Analysis

Because the components and their terminations exert a major effect on the design of the PWB, a number of metrics have been developed to establish the relationships between component density and PWB density. A major analysis of these relationships has been made by H. Holden[5] and some of his charts and derivations are provided here to guide the design engineers during the development of a rational PWB design.

This information is very useful in determining where the designed product will fit in the component density spectrum and what, therefore, is to be expected for PWB density.

Figure 1.7 provides a generalized view of the relationships among the component density, their terminal density, and the necessary wiring density that will be required to accommodate the selected degree of component complexity. The definition of the wiring connectivity W_f is provided.

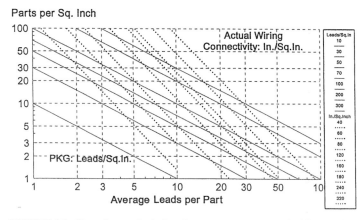

FIGURE 1.7   Plot of general relations between component and wiring density.

1.6.2 PWB Density Metrics

It is essential for the proper design of PWBs to determine the density requirements and then analyze alternative methods of board construction for the most cost-effective design. There are a number of basic terms and equations used for the calculation and analysis of PWB wiring density.

$$W_c = \frac{T*L}{G} \text{ in/in}^2 \tag{1.1}$$

where W_c = wiring capacity
T = tracks per channel
L = number of signal layers
G = channel width

But it is more important to determine the required wiring density that will be sufficient to interconnect all the components on the desired board size. There have been a number of empirically developed equations that permit the calculation of such a wiring demand. The simplest has been developed by Dr. D. Seraphim:[6]

$$W_d = 2.25 \, N_t*P \tag{1.2}$$

where W_d = wiring demand
N_t = number of I/Os
P = pitch between packages

1.6.3 Special Metrics for Direct Chip Attach (DCA)

The assembly of uncased or bare chips on substrates has become popular mostly due to the ability of such assemblies to reduce the area of interconnections. The ideal limit for such assembly would be to place all the chips tightly together, without any space in between. This would result in 100 percent packaging efficiency, a metric measuring the ratio of silicon area to the substrate area. Naturally, such 100 percent efficiency is not achievable, but this metric is still useful in ranking various substrate construction or bare chip attachment methods, as shown in Fig. 1.8.

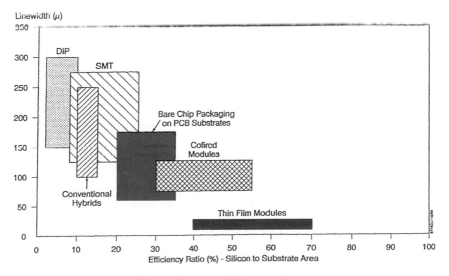

FIGURE 1.8 Packaging efficiency. *(Courtesy of BPA, used with permission.)*

Packaging efficiency of 100 percent is impossible to achieve because all chip-mounting methods require some space around the chips. Even with flip-chips, there must be a distance left between the chips to permit room for the placement tool.

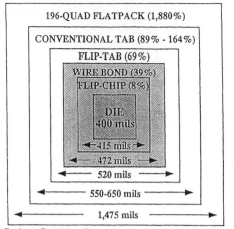

Package Penalties: Excess Interconnect Area

FIGURE 1.9   Chip area required to accommodate bonding methods.

Dr. H. Charles[7] of Johns Hopkins University has listed the dimensions in Table 1.3 for the necessary spacing between the chips (or the total width of the frame around the chips) for various chip attachment methods. These or very similar distances have also been cited by a number of other sources.

Even with the flip-chip mounting, packaging efficiency must be derated to about 90 percent, for wire-bonding to 70 percent and for TABs to about 50 percent, and in some cases much more. A very similar situation is shown graphically in Fig. 1.9. The packaging efficiency deratings shown in Fig. 1.9 are required to accommodate only the wiring bond pads on the substrates. But the mounting of bare chips on PWBs requires additional signal redistribution area to permit placement of larger-diameter PTHs farther out for communication with internal layers. It is evident that the packaging efficiencies on PWBs could be reduced to the range of only 20 to 30 percent, unless special surface signal redistribution layers (as previously mentioned) are used, which are made of unreinforced dielectric material. In such cases, packaging efficiency and the chip-to-chip distances will again be similar to the values cited in Table 1.3.

TABLE 1.3   Spacing Required Between Chips

Attachment method	Chip spacing, mils
Flip-chip	15–20
Wire-bonding	70–80
Flip TAB	100–120
Regular TAB	150–400

It is apparent that direct chip attach on PWBs will result in the significant reduction of the packaging efficiency of such assemblies, except for the fact that components can be mounted on both sides of the PWB substrate. It has been shown that wire-bonding can be done on both sides of a PWB with some special fixturing; also, outer lead bonding (OLB) of TABs can be done on both sides of the PWB substrate. Thus, while single-sided bare chip assembly on PWBs reduces its packaging efficiency to about half that of other types of substrate constructions, the ability to place components on both sides of PWBs brings it back to the same packaging efficiency level as others.

1.7   METHODS TO INCREASE PWB DENSITY

There are three basic ways to increase the connectivity or available conductor capacity of PWBs:[8]

- Reduce hole and pad diameters
- Increase the number of conductive channels between pads by reducing the widths of the conductors
- Increase the number of signal planes

The effect of each approach on manufacturing yields, and thus on board costs, will be discussed in sequence. It should be noted that the last option is the simplest but the most costly solution, and thus should be used only after the methods suitable for resolving the first two conditions have been proven inadequate for achieving the desirable board density.

1.7.1 Effect of Pads on Wiring Density

The major obstacles preventing increase of conductor channel capacity are large pad diameters around the plated through-holes (PTHs), since, at the present state of technology, PWBs still require pads wider than the conductors at their location. These pads reduce the obtainable connectivity of PWB boards and must be accounted for in a proper analysis of interconnection density I_d. For instance, in one design, the reduction of pad diameters from 55 to 25 mils (by 55 percent) doubled the interconnection density, while the reduction in conductor pitch C_p from 18 to 7 mils (by 61 percent) increased it only by 50 percent. It is obvious that the reduction of pad diameters, or their total elimination, could be a more efficient way to increase the wiring capacity of complex PWBs.

The purpose of copper pads surrounding the drilled holes in PWBs is to accommodate any potential layer-to-layer or pattern-to-hole misregistrations and thus prevent any hole breakout outside the copper area of the pads. This misregistration is caused mainly by the instability and movement of the base laminate during its processing through the PWB or multilayer board (MLB) manufacturing steps.

The base material standards specify that such movement be limited to a maximum of 300 ppm, but the actual base material excursions are closer to 500 ppm, producing 10 mils of layer movement within a 20-in distance. For many applications this tolerance is too wide, as it requires at least a 10-mil-wide annular ring around drilled holes, resulting in considerable conductor channel blockage.

Another cause of material instability in MLBs is the excessive material movement that occurs if the laminating temperature exceeds the glass transition temperature T_g of the laminate resin. On the other hand, if the laminating temperature remains below the T_g of the resin, there is minimal dimensional variation of the base material, as the resin is still in its linear expansion phase. This explains the need for use of high-T_g resins in the PWB industry.

The data obtained from the performance of new, more stable unidirectional laminates indicate that the base material movement is reduced, for instance to 200 ppm from 500 ppm, and the requirements for the annular ring width will be reduced to 4 mils from 10 mils.

Table 1.4 illustrates the connectivity gains made possible when a more stable laminate material is used, permitting a reduction in the initial diameters of the pads (as given in the first column) spaced at 2.5 mm (0.100 in) while keeping the conductor pitches constant. The most effective use of the signal plane area is achieved when the pads are eliminated and the z-axis interconnects are confined within the width of the conductors forming the invisible vias.

This derivation is based on actual data obtained from the performance of new, more stable, unidirectional laminates. While MLBs using these new, more dimensionally stable, unidirectional laminates with reduced pad diameters could be manufactured by conventional manufacturing methods, the production of MLBs with invisible vias requires the use of a sequential manufacturing process similar to the SLC process previously described.

PWB manufacturers are reasonably comfortable with the production of boards with 4- or 5-mil-wide conductors, but they still require large pads around plated holes to ensure against hole breakout. This limits the currently available wiring density to about 40 to 60 in/in^2 per plane, as seen from Table 1.4. A technology that will permit PWB manufacturers to fabricate invisible vias could increase the connectivity per PWB signal plane from this current range to the level of 100 to 140 in/in^2. Conductor widths of 0.002 in will offer a PWB of 200 to 250 in/in^2 per signal plane.

TABLE 1.4   Effect of Pad Diameters on Interconnectivity Density

Pad dia, in	Cond pitch, in	I_d @ 500 ppm, in/in^2	I_d @ 200 ppm, in/in^2	I_d @ invisible via, in/in^2
0.055	0.010	20	37	55
0.036	0.018	30	48	55
0.025	0.009	40	96	100
0.025	0.007	60	130	143

TABLE 1.5 Effect of Increased Connectivity on Reduction of Layers

Pad dia, in	Cond pitch, in	I_d @ 500 ppm, layers	I_d @ 200 ppm, layers	I_d @ invisible via, layers
0.055	0.010	10	6	4
0.036	0.018	7	4	4
0.025	0.009	5	2	2
0.025	0.007	4	2	2

Table 1.5 illustrates the most important result of increased connectivity per layer: a reduction in the number of signal layers needed to provide the same wiring density W_d. Table 1.5 was constructed by applying connectivity data from Table 1.4 to a 50-in² MLB with total wiring length of 10,000 in. Note also that the layer count in Table 1.4 has been brought up to the next higher full-layer value, i.e., the calculated 1.4 layers have been recorded as 2 layers.

The major benefit of such a reduction in the layer count is that it can result in a significant reduction of the manufacturing cost while providing the same total interconnection length.

1.7.2 Reduction of Conductor Width

An obvious method of increasing the connectivity of PWBs is to reduce the widths of conductors and spaces and thus increase the number of available wiring channels on each signal plane, as described previously. This is the direction that has been used in the IC and PWB industries for many years. However, it is impossible to decrease conductor widths or spaces indefinitely. The reduction of the conductor width is limited by the current-carrying capacity of thin, small conductors, especially when these conductors are long, as they frequently are on PWBs. There are processing limits to this conductor reduction, since manufacturing yields may plummet if the reduction stretches the process capabilities beyond their normal limits.

There is also a limit to the reduction of the spaces between the conductors, governed mainly by electrical considerations, i.e., by the need to prevent excessive cross talk, to minimize noise, and to provide proper signal propagation conditions and characteristic impedance.

Still, such conductor reductions, if achieved within the described limits, can be an effective path for increasing the PWB density and the reduction of PWB manufacturing costs. As seen from Table 1.6, constructed from cost data derived from the Columbus program of BPA, the reduction of conductor widths from 6 to 3 mils halves the number of signal layers necessary to

TABLE 1.6 Effect of Conductor Widths on Number of Layers and Board Cost for a 6-in × 8-in MLB, with $I_d = 450$ in/in², 65 to 68 Percent Yields

Line-space	Total no. of layers	No. of signal layers	Board cost, %
3–3	8	4	55
4–4	10	6	64
5–6	12	7	77
5–7	14	8	87
6–6	16	8	90
7–8	20	10	100

ensure the same connectivity (while their yields, interconnection density, and board area were kept constant). This reduction in the number of layers can significantly reduce the manufacturing costs of PWB boards.

1.7.3 Effect of Conductor Widths on Board Yields

It is obvious that any successful increase of conductor density I_d in PWBs would be effective only if the processes exist that permit manufacture with reasonable yields. Unfortunately, the yields of thin conductors in PWBs fall rapidly as their widths are reduced below 5 mils, as shown in Fig. 1.10. Therefore, the understanding of manufacturing yields is very important for analysis of the most cost-effective manufacturing process, because the process yields have a major effect on the cost of interconnection substrates.

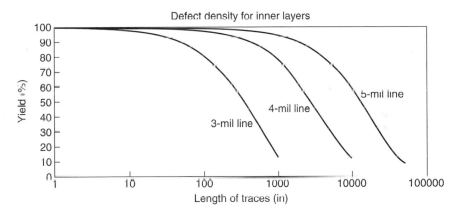

FIGURE 1.10 Board yields vs. conductor width.

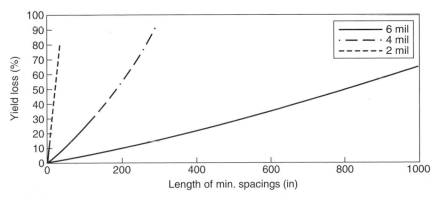

FIGURE 1.11 Yield loss from shorts.

A useful empirical equation for calculating the manufacturing cost is:

$$\text{Cost } C = \frac{(\text{material} + \text{process costs})}{\text{Yield } Y} \tag{1.3}$$

To establish the effect of the interconnection density I_d on the final yield of substrates, the total processing yield can be split into two components: one that depends on the conductor density, i.e., Y_{Id}, and the second, which is a function of the combined yields of the rest of the manufacturing processes:

$$Y_{\text{total}} = Y_{Id} * Y_{\text{proc}} \tag{1.4}$$

In a well controlled manufacturing operation, the process-dependent yields (such as plating) remain fairly constant for a given technology, permitting the yield function to be based solely on the changes in the conductor widths.

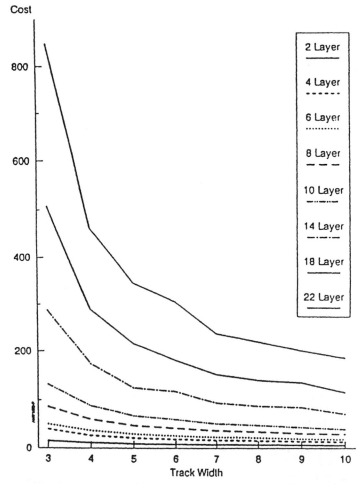

FIGURE 1.12   Cost relationships between number of layers and conductor widths.

As seen from Fig. 1.11, the defects that affect this density-dependent yield function Y_{ld} are conductor opens and shorts between them. It would be reasonable to assume that such defects have a Poisson distribution over the total length TL of conductors of a substrate, with an average defect frequency of v. The yield is the probability of zero defects ($n = 0$) in the total conductor length TL. Thus,

$$Y = (\text{at } n = 0) = e^{(-v * \text{TL})} \quad \text{(Poisson distribution)} \tag{1.5}$$

As seen from Figs. 1.10 and 1.11, the defect frequency v depends also on the widths of lines and spaces, i.e., on the conductor pitch C_p. With a decrease of C_p, v will increase, but for very large C_p, v should be 0, since Y_{ld} will be 100 percent.

For instance, in the case of a design using invisible pads, where $C_p = 2w$, the interconnection density I_d can be expressed as $I_d = \text{TL}/A$, and I_d is proportional to C_p, i.e., $I_d * C_p = 1$, and TL $= A/C_p$. Therefore, v in this equation can be empirically expressed as:

$$v = -\ln \frac{Y_0}{\text{TL}_0} * \left(\frac{C_{p0}}{C_p} \right)^b \tag{1.6}$$

where b is an exponent dependent on the technology or process used to form the conductors. This exponent b varies considerably from facility to facility and among various pattern formation methods, and must be empirically determined for each case.

1.7.4 Increase in Number of Conductor Layers

This is the simplest and most straightforward solution: when there is insufficient room on existing layers to place all the necessary interconnecting paths, add a layer. This approach has been widely practiced in the past, but when cost effectiveness of the substrates is of paramount importance, a very careful design analysis must be made to minimize layer counts in MLBs, because there is a significant cost increase with every additional layer in the board. As seen from Table 1.6, calculated for 6- × 8-in MLBs produced in large quantity with yields and conductor density kept constant, there is almost a linear relationship between board costs and layer count.

Table 1.6 also shows that any increase in the number of signal layers in boards operating at frequencies requiring transmission line characteristics will double the total number of layers, due to the need to interleave ground or DC power planes between signal planes.

A typical example of the effect of layer count on the finished MLB yield can be seen from Fig. 1.12, prepared some years ago by BPA. We can see that there is a definite decrease in the manufacturing yields with an increased number of layers in any of the linewidth categories. This is rather a typical situation in board manufacturing because increased complexity and thickness of MLB with a higher number of layers usually leads to a larger number of problems on the production floor.

REFERENCES

1. Toshiba, "New Polymeric Multilayer and Packaging," *Proceedings of the Printed Circuit World Conference V,* Glasgow, Scotland, January 1991.

2. The Institute for Interconnecting and Packaging Electronic Circuits, 7380 N. Lincoln Ave, Lincolnwood, IL 60646.

3. IPC-SM-782, "Surface Mount Design and Land Pattern Standard," The Institute for Interconnecting and Packaging Electronic Circuits.

4. Y. Tsukada et al., "A Novel Solution for MCM-L Utilizing Surface Laminar Circuit and Flip Chip Attach Technology," *Proceedings of the 2d International Conference on Multichip Modules,* Denver, CO, April 1993, pp. 252–259.

5. H. Holden, "Metrics for MCM-L Design," *Proceedings of the IPC National Conference on MCM-L,* Minneapolis, MN, May 1994.

6. D. Seraphim, "Chip-Module-Package Interface," *Proceedings of Insulation Conference,* Chicago, IL, September 1977, pp. 90–93.

7. H. Charles, "Design Rules for Advanced Packaging," *Proceedings of ISHM* 1993, pp. 301–307.

8. G. Messner, "Analysis of the Density and Yield Relationships Leading Toward the Optimal Interconnection Methods," *Proceedings of Printed Circuits World Conference VI,* San Francisco, CA, May 1993, pp. M 19 1–20.

CHAPTER 2
SEMICONDUCTOR PACKAGING TECHNOLOGY

John W. Stafford
JWS Consulting PLC, Phoenix, Arizona

2.1 INTRODUCTION

A revolution has occurred in the electronics industry due to advances in semiconductor design and manufacturing, the packaging of semiconductor die, the packaging of electronics systems, and product physical design.

The driving force in this revolution originates with the advances that have occurred in integrated circuit (IC) technology and the levels of integration obtained. The initial driver to this was the development of a micrometer-level lithographic capability.[1] Semiconductor packaging and printed circuit board (PCB) evolution tracks the advances in IC technology. Figures 2.1 and 2.2 show the trend of chip transistor density and the increase in chip frequency vs. time. There is concern that, with the current semiconductor technology advances, by about 2010 we may begin to approach significant process and technology obstacles. As a consequence of being able to put more circuitry on a silicon IC, its package size and the number of package input/output (I/O) pins have increased, as has the wiring density required of the medium that interconnects the packaged ICs.

This continuing thrust for higher levels of integration has forced an ongoing effort for smaller and cheaper means of packaging these ICs so they can be interconnected in a cost-effective manner that does not degrade the electrical performance of the assembled circuit. As a result, high-performance systems require consideration of both the IC design and its packaged format and the design of the interconnect that connects the ICs.

Table 2.1 compares the computing capability available over the years. What has occurred is that there has been a constant increase in functionality (i.e., instructions per second) along with continuous decreases in the cost per instruction. The results of this trend are the ever increasing functionality of portable wireless communications, such as pagers and cellular phones, and their reduction in size to minimum ergonomic standards.

Worldwide in 1998 there were some 308 million cell phone subscribers; the number grew to 475 million in 1999. It is estimated that in 2001 there were 1000 million (i.e., 1 billion) cell phone subscribers. From 2000 to 2001, more than half a billion cellular phones were manufactured, and these are now considered a commodity product. This same technology trend is spawning new wireless products for local area networks. These applications will allow cellular

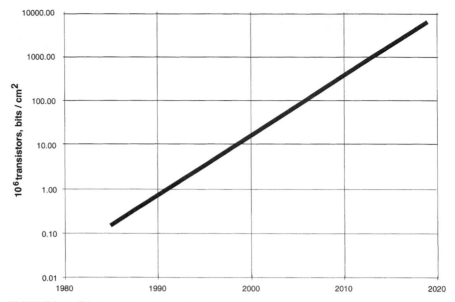

FIGURE 2.1 Chip density trend for logic and DRAM components.

phones, personal computers, etc. to exchange data (including video) using appropriate radio frequency (RF) protocols (Bluetooth operating in the 2.4-GHz frequency range is one such protocol) if they are in proximity to one another.

The requirements of the Internet for increasing functionality and bandwidth and lower cost will require new concepts for electronic physical design as well as the incorporation of optical components and interconnects. These trends will continue indefinitely.

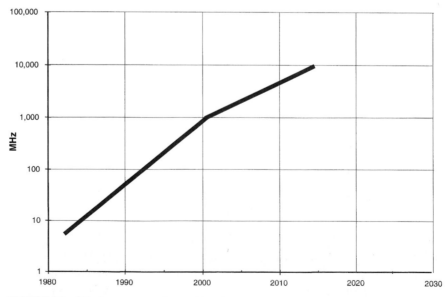

FIGURE 2.2 Chip frequency requirement trend.

tion is required in the packaging of RF components to control the electrical parasitics, the thermal resistance for components such as power amplifiers, etc.

2.1.3 Requirements for Electronic Systems

The requirements for electronics systems and products driven by semiconductor technology developments are as follows:

1. The advances in integrated semiconductor technology mean products operate at higher speeds and have higher performance and greater functionality.
2. The reliability and quality of products are givens and are expected to be built in at no cost premium.
3. The volume (i.e., size) of the electronics products is diminishing, and is constrained only by ergonomic requirements and the ability to dissipate heat (i.e., power).
4. The costs of the components and assembly are expected to continuously decrease with time.
5. The time to market impacts all of the preceding items.

2.2 SINGLE-CHIP PACKAGING

Prior to 1980, the semiconductor package predominately used was the dual inline package (DIP). The package is rectangular in shape with leads on an 0.100-in pitch along the long sides of the package.

Figure 2.4 shows the various semiconductor formats and package trends. The packages on the left side of Fig. 2.4 are essentially perimeter I/O packages—i.e., DIPs, quad flat packages (QFPs), plastic leaded chip carriers (PLCCs, tape automated bonding (TAB), etc. The package types on the lower right side of Fig. 2.4 represent area array packages such as pin grid array (PGA) packages (an array of pins attached to the package base for electrical connection), land grid array packages (an array of conducting pads on the package base for electrical connection, sometimes called pad array carriers [PACs], or, when the lands have reflowed solder balls attached, ball grid array [BGA] packages), and multichip modules. The pitch of the I/Os of DIPs and PGA packages is 0.1 in, while the I/O pitches of the balance of the parts

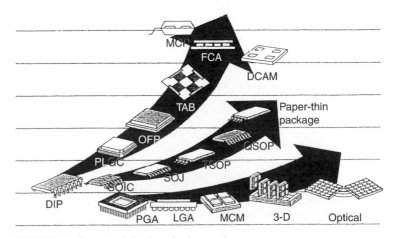

FIGURE 2.4 Integrated circuit packaging trends.

are 0.060 in or less (i.e., 0.050 in, 0.5 mm, 0.4 mm, 0.3 mm). Array-type packaging concepts have emerged to provide higher package electrical performance and/or lower packaging densities (package area/die area).

2.2.1   Dual Inline Packages (DIPs)

Figure 2.5 shows the configuration of a DIP. DIPs are available with a cofired ceramic body with the leads brazed along the long edges or in a postmolded construction where the die is bonded to a lead frame and gold wires interconnect the chip to the lead frame leads prior to molding of a plastic body around the lead frame. DIPs are limited to 64 or fewer I/Os.

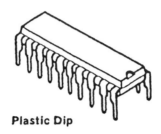

Plastic Dip

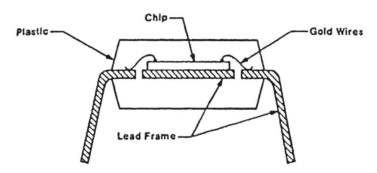

FIGURE 2.5   Dual inline package.

2.2.2   Leadless Ceramic Chip Carriers

To improve the form factor, packages for commercial and military applications were developed called leadless ceramic chip carriers, consisting essentially of the cavity portion of the ceramic hermetic DIP (see Fig. 2.6) with solderable lands printed onto the bottom of the leadless ceramic chip carrier package. These parts were assembled on ceramic substrates and used in both military and telecommunications products. Almost concurrently, leaded versions of the leadless packages begin to appear. The pitches of the leadless parts were 0.040 and 0.050 in, while the leaded parts were on a 0.050-in pitch. Reference 3 discusses these developments in detail. By 1980, thrust-to-quad surface-mount packaging had begun, with the emphasis on leaded plastic quad packages.

Reference 4 shows that by 1993 there was a dramatic swing away from through-hole-mounted parts (i.e., DIPs) to surface-mount packages. In 1993, 50 percent of the semiconduc-

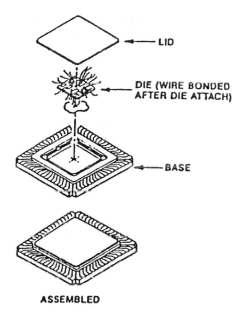

FIGURE 2.6   Leadless ceramic chip carrier.

tor packages fabricated were through-hole-mounted DIPs. By 2000, this percentage dropped to 30 percent, and by 2005 it is projected to drop to only 15 percent.

2.2.3   Plastic Quad Flat Package (PQFP)

The driver for the surface-mount plastic packages has been the development of the plastic quad flat package (PQFP), which consists of a metal leadframe with leads emanating from all four sides. The leadframe is usually copper, to which the semiconductor die is "die bonded" (usually epoxy die bonded). The I/Os of the die are connected by wire bonds to the leadframe leads. The conventional method of wire-bonding is thermosonic gold ball wedge bonding. A plastic body is then molded around the die and the leads are trimmed and formed. Figure 2.7(a) shows a cross-sectional view of a PQFP.

PQFPs have their leads formed in a gull-wing fashion (see Fig. 2.7), while PLCCs have their leads in the shape of a J, which are formed (i.e., folded) underneath the package.

Figure 2.8 shows the lead pitch and pin count limit vs. QFP size and lead pitch. QFPs are in production and readily used in the assembly of product with 0.5-mm pitch. Based on molding capability and impact of lead length on electrical performance,

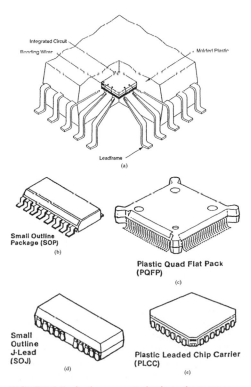

FIGURE 2.7   Surface-mount plastic package types.

a molded body 30 mm on a side is thought to be the practical limit. QFPs with 0.5-mm pitch based on the preceding are limited to around 200 I/Qs. QFPs with 0.4-mm pitch have been implemented.

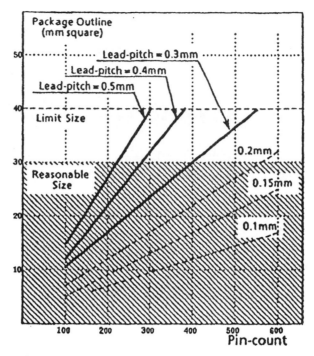

FIGURE 2.8 Lead pitch and pin count limit vs. QFP size and lead pitch.

Figure 2.7 shows the variety of surface-mount plastic packages that have been developed based on PQFP technology.

Ceramic and plastic QFPs, as well as PLCCs, are used to package gate array and standard cell logic and microprocessors. Small-outline IC and small-outline J-lead packages are used to package memory (SRAM and DRAM) as well as linear semiconductors. Pin count for all package types is limited only by molding capability and the demand for ever thinner molded packages.

2.2.4 Pin Grid Array (PGA) and Pad Array Carrier (PAC)

Consider the impact of using a perimeter I/O package vs. an area array package. Figure 2.9 illustrates the differences between a perimeter array package (leadless chip carrier) and an area array package (PAC). Figure 2.10 shows the relation between the package area vs. I/O for perimeter and area array packages. It is clear from Fig. 2.10 that for semiconductors with more than 100 I/Os, PGA and PAC packages have become increasingly attractive for packaging very-large-scale semiconductor ICs and ultra-large-scale semiconductor ICs. The scalable limit is determined only by fatigue issues of the solder connection or joint.

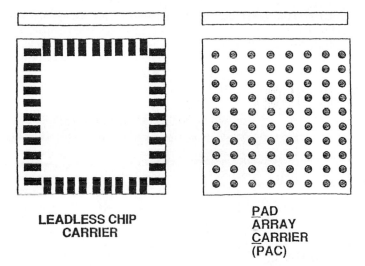

**LEADLESS CHIP
CARRIER**

**PAD
ARRAY
CARRIER
(PAC)**

FIGURE 2.9   Perimeter I/O package vs. area array package.

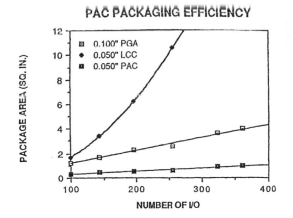

FIGURE 2.10   PAC packaging efficiency

2.2.4.1   *Perimeter Array I/O Package Advantages and Disadvantages*

- Perimeter array I/O packages at 0.050-in pitch can be readily surface-mount assembled. Both 0.4-mm pitch perimeter I/O QFPs and 0.3-mm pitch perimeter I/O QFPs are in now in manufacture. Reviewing Fig. 2.8, which shows QFP size vs. lead pitch and pin count, it is apparent that the usefulness of QFPs is limited to around 400 I/Os.

- There still is a controversy in the packaging and assembly community as to whether 0.3-mm pitch QFPs, particularly with a large number of I/Os, are a high-throughput, high-assembly-yield part due to the fragility of the lead (0.015 mm wide) and the possibility of solder shorts between leads.

2.2.4.2 *Pad Array and BGA Packages.* This disconnect is being addressed first by pad array and BGA package technology. References 5 and 6 provide details of this emerging technology, particularly as it relates to low-cost plastic array packages with reflowed solder balls attached to the package array I/Os. Figure 2.11 shows a ceramic BGA (CBGA) and Fig. 2.12 shows a cross section of the plastic BGA.

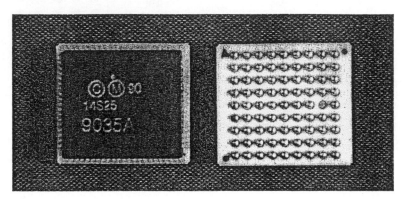

FIGURE 2.11 Ceramic ball grid array.

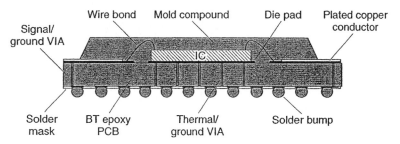

FIGURE 2.12 Ball grid array cross section.

The advantage of ball grid array packages (ceramic or plastic) include the following:

- The package offers a high-density interconnect. Pad array pitches of 1.27, 1.0, 0.8, and 0.5 mm are now commercially available.
- The packages have achieved six-sigma soldering (demonstrated for 1.27- and 1.00-mm I/O pitches) because of the large volume of solder on the I/O pad.
- The package is a low-profile part (package thicknesses as low as 1 mm are now available).
- The package has superior electrical performance in that the total lead length is short, controlled impedance interconnects can be designed in, and low–dielectric constant and low-loss materials can be used for the substrate.
- The package depending on the design has the potential for superior thermal performance.
- The package concept is extendable to multichip packages (MCPs).

Table 2.2 gives the expected ranges of I/O for BGA packages for high-end microprocessors and for dies used in portable products. In 2000, approximately 60 percent of all BGA packages had 1.27- or 1.00-mm pitch and the balance had 0.8- or 0.5-mm pitch. By 2004, 60 percent of all BGA packages will have 0.8- or 0.5-mm pitch, with a small number having less than 0.5-mm pitch.

TABLE 2.2 BGA Pin Counts

Year	2001	2005	2010
Pin count, high-end logic	700–1000	1000–1900	1200–4000
Pin count, portable products	256	312	360

The area efficiency of BGA packages that have a relatively low pad count (i.e., less than 140) has increased in the last several years. This new form of the BGA package, called chip-scale package (CSP), has an area efficiency of around 80 percent (i.e., 80 percent of the package area comprises the silicon die area). The pad pitch on CSPs range from 0.8 to 1 mm. CSPs should be considered a variant of the BGA package technology. Figure 2.13 shows a typical CSP package.

Amkor Tape CSP

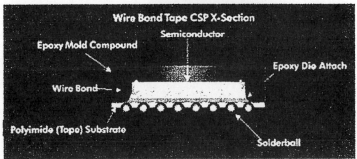

FIGURE 2.13 Typical CSP technology. (*Courtesy of Amkor Inc.*)

A new and smaller package called a wafer-scale package (WSP) has been developed. Figure 2.14 shows such a package. The area efficiency of this package is 100 percent (i.e., 100 percent of the package area comprises the silicon die area). A WSP is a package where all the packaging processes are completed on the silicon wafer, including application of the I/O solder balls. The WSP has relatively low I/O count and has pad pitches ranging from 0.8 to 1 mm. Generally, a redistribution of the chip I/Os uniformly over the die area is required to obtain these pitches. The WSP requires no chip under encapsulation, as is required by direct chip attach (DCA), discussed in Sec. 2.2.5.

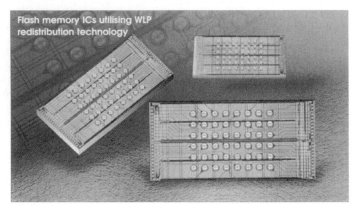

FIGURE 2.14 Wafer-scale package. (*Courtesy of Flipchip Technologies.*)

2.2.5 Direct Chip Attach (DCA)

The next step down from area array packages is DCA. The methods of attaching a semiconductor die directly to an interconnect board (PCB, multilayer ceramic, etc.) are:

- Die bond/wire-bond
- TAB
- Flip-chip bonding

An exhaustive discussion of wire-bond, TAB and controlled-collapse chip connection (C4) or solder-bumped flip-chip-to-board interconnect technology can be found in Chap. 6 of Ref. 7. Figure 2.15 illustrates these chip interconnect methodologies, which will be discussed in the following text.

2.2.5.1 Die Bonding to Printed Circuit. The preferred method of die bonding and wire-bonding is epoxy die bonding to the interconnect (i.e., PCB or multilayer ceramic) and gold ball-wedge wire-bonding. One of the advantages of gold ball-wedge wire-bonding is that a wedge bond can be performed on an arc around the ball bond. This is not true for wedge-wedge wire-bonding.

Wire bonds can be made with wire diameters as small as 0.8 mil. Thermosonic ball-wedge bonding of a gold wire, shown in Fig. 2.15(*e*), is performed in the following manner:

1. A gold wire protrudes through a capillary.
2. A ball is formed over the end of the wire by capacitance discharge or by passing a hydrogen torch over the end of the gold wire.

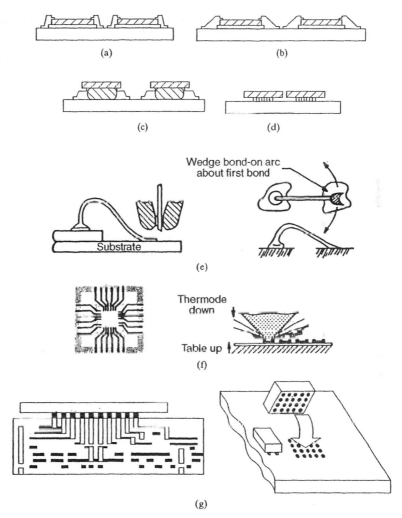

FIGURE 2.15 DCA interconnect methodologies: (*a*) die bond/wire-bond module; (*b*) TAB module; (*c*) flip TAB module; (*d*) flip-chip module; (*e*) thermosonic gold wire-bonding; (*f*) TAB bonding; (*g*) flip-chip bonding.

3. Bonding of the ball is accomplished by simultaneously applying a vertical load to the ball bottomed out on the die bond pad while ultrasonically exciting the capillary (the die and substrate are usually heated to a nominal temperature).

4. The capillary is moved up and over to the substrate or lead bond pad, creating a loop, and, under load and ultrasonic excitation, a bond is made.

5. The wire is clamped relative to the capillary and the capillary moves up, breaking the wire at the bond.

Die bond and wire-bond attach suffer from the problem that this method of chip attach is difficult to repair, particularly if the chip is encapsulated.

2.2.5.2 Tape-Automated Bonding (TAB). TAB, shown in Fig. 2.15(*f*), is more expensive than wire bonding and may require a substantial fan-out from the die to make the outer lead bond. TAB is a process in which chemically etched, prefabricated copper fingers, in the form of a continuously etched tape consisting of repetitive sites, are simultaneously bonded using temperature and pressure to gold or gold-tin eutectic bumps that are fabricated on the I/Os of the die. The outer leads of the TAB-bonded die are excised and simultaneously bonded to tinned pads on the interconnect, using temperature and pressure.

2.2.5.3 Solder-Bumped Dies. The use of solder-bumped dies for packaging electronic systems was pioneered by IBM and was called controlled-collapse chip connection (C4) by IBM. The solder bump composition of the C4 die is approximately 95Pb/5Sn. The C4 dies in the IBM application were attached to multilayer ceramic substrates by reflow soldering, using a flux that required cleaning after reflow. The initial IBM application of C4 technology was for high-end computer packaging. Flip-chip attach shown in Fig. 2.15(*g*), where the die I/Os are solder bumped (usually with 95Pb/5Sn or eutectic Pb/Sn solder) and the chip reflow is attached to its interconnect, has now emerged as a viable packageless technology for consumer commercial products. For the DCA technology, the PCB flip-chip lands are usually solder-finished with a eutectic Pb/Sn solder. A no-clean flux is used for DCA soldering. Once the chip has been solder-attached, it is underencapsulated to provide moisture protection and to enhance the thermal cycling performance of the assembled die. References 8 to 10 discuss some of the emerging developments for direct attach of solder-bumped die to PCBs for commercial product applications. Figure 2.16 shows the DCA application discussed in Ref. 10. In this application, the CBGA packaged microprocessor was solder-bumped and direct chip attached to illustrate the potential savings in PCB real estate.

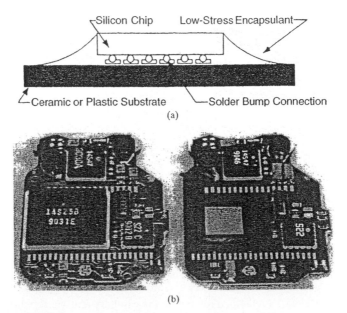

FIGURE 2.16 DCA applications. The microprocessor on the left is in BGA, while the microprocessor on the right is attached by solder bumps to show potential area savings.

2.3   MULTICHIP PACKAGES

MCPs have emerged as a packaging option where performance is an issue or where interconnect area for connecting packaged ICs is at a premium. The performance issue is one of increasing semiconductor performance and the impact of the interconnect medium on the performance of the assembled die. The consequence of this is that the interconnects can no longer be considered to provide an instantaneous electrical connection. It can be shown that for copper conductors on polyimide at a clock frequency of 200 MHz, the maximum allowable length of interconnect is 170 mm, or about 6.8 in. As a reference point, the clock speed for the DEC ALPHA RISC microprocessor was 200 MHz or better in 1995. The clock speed of the Pentium III microprocessor is 500 MHz, and microprocessors with 1 GHz clock speeds are commercially available.

2.3.1   Multichip vs. Single-Chip Packages

MCPs for memory have found broad application in personal computer and laptop computer products. Multiple memory dies (packaged or unpackaged) are assembled on a rectangular interconnect (PCB or multilayer ceramic) with I/Os along one rectangular edge. Such a package is called a single inline package (SIP). The I/Os of the SIP can have DIP leads assembled on them for through-hole attachment onto PCB, or they can have lands on the package to mate with a suitable connector. For through-hole-mounted SIPs, the leads are usually on a 0.100-in pitch. Memory packaged in system-on-a-package or thin system-on-a-package or small-outline J-lead packages are assembled onto SIPs whose substrate is a PCB in appropriate multiples to offer enhanced memory capability. DCA or memory die is also possible. Figure 2.17 shows several SIP configurations. Figure 2.17(a) shows an early version of a SIP using a ceramic substrate and leadless ceramic chip carriers with soldered-on leads. Figure 2.17(b) shows a typical version of a SIP to mate with a SIP-type connector.

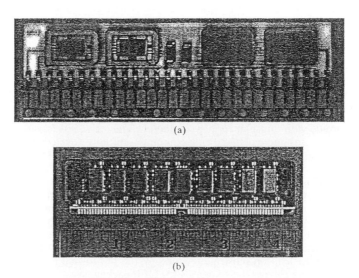

(a)

(b)

FIGURE 2.17   SIP configurations.

2.3.2 MCPs Using Printed Circuit Technology

MCPs using PCB technology, but using leads on the substrate, are also being implemented to provide improved product functionality and board space savings. Figure 2.18 shows the concept of the lead-on-substrate MCP. The leads to the substrate can be fabricated integrally with the board during the PCB fabrication or can be soldered on using a suitable high-temperature solder. The lead-on-substrate MCPs can be postmolded or cover-coated with a suitable encapsulant.

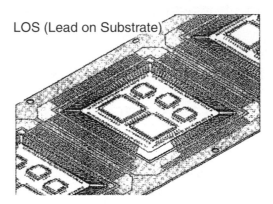

FIGURE 2.18   LOS printed circuit MCP.

The lead-on-substrate MCP could also be implemented using a cofired ceramic package with the leads brazed on and finished by cover coating with a suitable encapsulant or hermetic sealing of the assembled ICs. Chapter 7 of Ref. 7 gives a detailed discussion of the cofired multilayer ceramic package manufacturing processes.

2.3.3 MCPs Using Organic Substrates

The more prevalent form of MCPs uses a multilayer organic interconnect built on a substrate such as silicon, alumina ceramic, or metal composite. The dielectric films are patterned serially one on top of another to produce a multilayer interconnect. The dielectric of choice is polyimide with thin-film copper conductors. The technology for the interconnect in question uses processes and manufacturing equipment initially developed for semiconductor manufacture. This technology has a 1-mil line and space capability. The packages can have the cavity facing up (facing away from the PCB) or down (facing the PCB) if a heat sink is required for heat dissipation.

Figure 2.19 shows wire-bonded die on a silicon substrate packaged in a premolded QFP and a ceramic QFP. A premolded QFP is a lead frame about which a plastic body has been molded and which contains a cavity for a die or a substrate. The method of interconnect from the silicon substrate to the package is wire-bonding. As shown in Fig. 2.15, bare die can be attached to the silicon substrate by die/wire-bonding, TAB bonding, or flip-chip bonding. As with the lead-on-substrate technology, the I/O format for the QFPs that house the multichip substrate follows the standards for QFP packages.

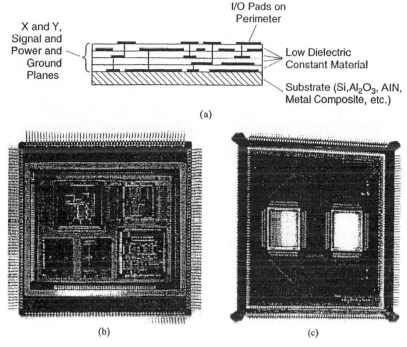

FIGURE 2.19   MCP example.

2.3.4   MCP and PGA

PGA packages can also be used as MCPs containing the type of substrate shown in Fig. 2.19. The PGA packages can have their cavity facing up (facing away from the PCB) or down (facing the PCB) if the package requires a heat sink for power dissipation. The pitch of the pins of the PGA is usually 0.1 in.

2.3.5   Multichip Stacked-Die Packages

A new form of MCP is emerging at the time of this writing. The packaged dies are stacked one on top of each other and are interconnected generally by wire bonds. The dies are thinned in wafer format to thicknesses of 6 mil or less. Figure 2.20 shows the various formats. The substrates for the multichip stacked-die packages can be tape, multilayer organic (FR-4, FR-5, etc.), multilayer organic with high-density interconnect layers, or multilayer ceramic. The preferred package format is a BGA. The initial application of stacked die has been to incorporate memory with logic dies rather than integrate logic and memory on a single silicon die.

2.3.6   MCP and Known Good Die

One of the issues in the application of MCP technology is the availability of known good die. Clearly considering that, for most dies, the package die test yield after wafer test runs

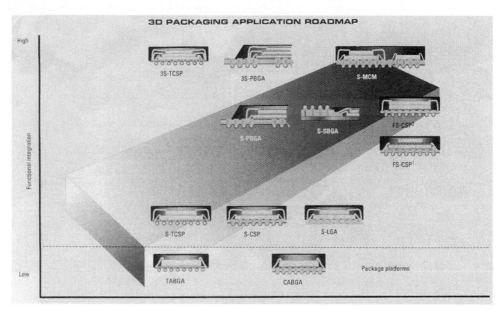

FIGURE 2.20   Formats for multichip stacked die packages.

somewhere around 95 percent, the MCPs are limited to perhaps five IC dies per package at best. The limitation is the resultant yield loss of a finished multichip module package. As an example, for a two-die MCP where each die has an assembled yield of 90 percent the MCP yield would be 81 percent (i.e., 0.9^2) and for a four-chip MCP the yield would be 65.6 percent (i.e., 0.9^4).

2.3.7 System-in-a-Package

An enhanced form of MCP called system-on-a-package or system-in-a-package has emerged. This package incorporates some or all of the multichip packaging technologies previously discussed. Figure 2.21 shows all of the various packaging features that make up a system-on-a-package or system-in-a-package. One feature of this package concept is that it includes not only silicon die but all the passives, etc., required to provide a system function. Passives (i.e., resistors, capacitors, and inductors) can be assembled on the substrate or can be integrated into the substrate itself (i.e., embedded passive substrate technology). Reference 11 provides a detailed overview of the system-on-a-package or system-in-a-package concept. Another example of this concept is the RF front-end HiperLAN module shown in Fig. 2.22.

2.4 OPTICAL INTERCONNECTS

As the performance of semiconductors increases, as measured by clock frequency, the allowable length of interconnect that does not degrade the device performance decreases. The emergence of the Internet and the need for ever increasing bandwidth has become the

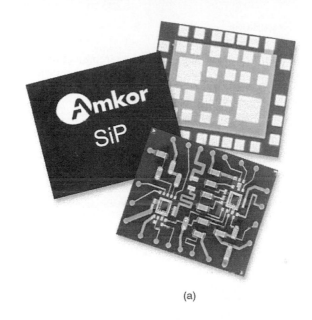

(a)

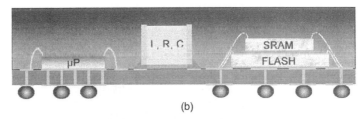

(b)

FIGURE 2.21 (*a*) Packaging technologies features that make up system-in-a-package; (*b*) cross section of system-in-a-package.

INTARSIA RF FRONT-END

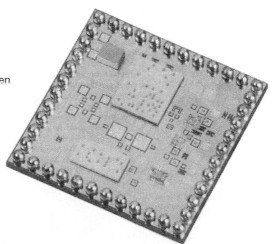

- RF front-end for HiperLAN
  - For 5.2 GHz wireless LAN module
  - 12.2 x 12.2 mm wafer-level CSP

- Provides all functions to convert between first IF and 5.2 GHz RF
  - GaAs transceiver
  - GaAs power amplifier
  - T/R
  - LNA transistors
  - Prescaler
  - System RF filters

- Silicon carrier with integrated passives
  - 2 GaAs flip chip
  - 1 Si flip chip
  - 2 wirebonded GaAs discretes

FIGURE 2.22 Example of system-in-a-package.

major driver for optical component and interconnect technology development. For wave-division multiplexing and Ethernet systems we have reached the copper bandwidth limit of 5 Gbs.

2.4.1   Components and Packages

The demand for increased bandwidth is being driven closer to the end user (i.e., residence or desktop). The issues facing low-cost optoelectronic physical designs are (1) low-cost assembly/processes and low-cost packages for optoelectronic components and (2) the capability for standard automated board assembly of all optoelectronic parts. Figure 2.23 shows a variety of package types used to package light emitters and detectors. The packages are generally of the TO header type or the butterfly DIP or through-hole-mounted DIP type. In general the packages are hermetic. The through-hole-mounted DIP and the butterfly DIP packages are most often used to package edge-emitting lasers. An example of a fiber pigtailed butterfly DIP packaged edge-emitter laser is shown in Fig. 2.24.

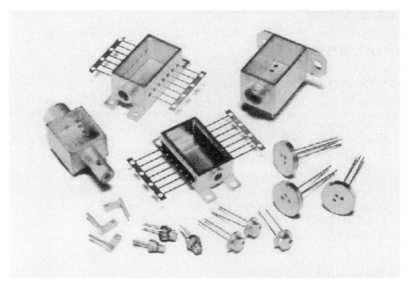

FIGURE 2.23   Package types used to package light emitters and detectors.

FIGURE 2.24   DIP laser 0/7.

2.4.2 Advantages of Optical Interconnects

The advantage of optical interconnects are as follows:

- The signal propagation is independent of the bit rate (up to 20,000 Gbits/s)
- The component is immune to electromagnetic interference and cross talk
- The optical signals can pass through one another (optical noninteraction)
- The components lend themselves readily to multiplexing

Advances in optoelectronic devices are leading to active consideration of local optical signal distribution to:

- Reduce the pin count of packages for complex very-large-scale semiconductor IC chips by replacing I/O complexity with bandwidth
- Carry very-high-bit-rate signals in hybrid circuits and PCB assemblies
- Allow highly complex interconnections to be achieved without the cost and difficulties of a metal-based backplane

2.5 HIGH-DENSITY/HIGH-PERFORMANCE PACKAGING SUMMARY

Where high performance and a high level of interconnects are required, MCPs and multichip modules have emerged as one packaging idea that can provide the performance required. In addition, fiberoptic transmitters and receivers are now commercially available to provide high-performance offboard optical interconnects in place of a hard-wired backplane interconnect using standard PCB technology and cabling. These fiberoptic transmitters and receivers generally use hermetic and nonhermetic custom DIP package formats for packaging. With advances in optoelectronic semiconductors, onboard optical interconnects are also being investigated and demonstrated. Reference 12 discusses one such demonstration of onboard optical interconnects where a polymeric optical waveguide, including branch and cross-circuits, was fabricated in the polyimide dielectric of a multilayer copper polyimide silicon substrate. Optical waveguide technology that is embedded in PCBs is now being developed.

2.6 ROADMAP INFORMATION

The NEMI *Year 2000 Roadmap*[13] is a good source of detailed roadmap information for IC packaging, PCB technology, optoelectronics, and so on.

REFERENCES

1. D. R. Herriott, R. J. Collier, D. S. Alles, and J. W. Stafford "EBES: A Practical Lithographic System," *IEEE Transactions on Electron Devices,* vol. ED-22, no. 7, July 1975.
2. Ira Deyhimy, Vitesse Semiconductor Corp., "Gallium Arsenide Joins the Giants," *IEEE Spectrum,* February 1995.
3. J. W. Stafford, "Chip Carriers—Their Application and Future Direction," *Proceedings of the International Microelectronics Conference,* Anaheim, CA, February 26–28, 1980, New York, June 17–19, 1980. Also published in *Electronics Packaging and Production,* vol. 20, no. 7, July 1980.
4. Ron Iscoff, "Costs to Package Die Will Continue to Rise," *Semiconductor International,* December 1994, p. 32.

5. Bruce Freyman and Robert Pennisi, Motorola, Inc., "Overmolded Plastic Pad Array Carriers (OMPAC): A Low Cost, High Interconnect Density IC Packaging," *Proceedings of the 41st Electronics Components Technology Conference,* Atlanta, GA, May 1991.

6. Howard Markstein, "Pad Array Improves Density," *Electronics Packaging and Production,* May 1992.

7. Rao R. Tummala and Eugene J. Rymaszewski, *Microelectronics Packaging Handbook,* Van Nostrand Reinhold, New York, 1989.

8. Yutaka Tsukada, Dyuhei Tsuchia, and Yohko Machimoto, IBM Yasu Laboratory, Japan, "Surface Laminar Circuit Packaging," *Proceedings of the 42nd Electronics Components and Technology Conference,* San Diego, CA, May 18–20, 1992.

9. Akiteru Rai, Yoshihisa Dotta, Takashi Nukii, and Tetsuga Ohnishi, Sharp Corporation, "Flip Chip COB Technology on PWB," *Proceedings of the 7th International Microelectronics Conference,* Yokohama, Japan, June 3–5, 1992.

10. C. Becker, R. Brooks, T. Kirby, K. Moore, C. Raleigh, J. Stafford, and K. Wasko, Motorola, Inc., "Direct Chip Attach (DCA), the Introduction of a New Packaging Concept for Portable Electronics," *Proceedings of the 1993 International Electronics Packaging Conference,* San Diego, CA, September 12–15, 1993.

11. William F. Shutler, Alberto Parolo, Stefano Orggioni, and Claudio Dall'Ara, "Examining Technology Options for System On a Package," *Electronics Packaging and Production,* September, 2000.

12. K. W. Jelley, G. T. Valliath, and J. W. Stafford, Motorola, Inc., "1 Gbit/s NRZ Chip to Chip Optical Interconnect," *IEEE Photonics Technology Letters,* vol. 4, no. 10, October 1992.

13. National Electronic Manufacturing Initiative, Inc. (NEMI) *Year 2000 Roadmap,* NEMI, Herndon, VA, 2000.

CHAPTER 3
ADVANCED PACKAGING

William Brown, Simon Ang, and Fred Barlow
University of Arkansas, Fayetteville, Arkansas

Tarak Railkar
Conexant Systems Inc., Newport Beach, California

3.1 INTRODUCTION

Packages that contain electronic devices such as integrated circuits (ICs) must, at a minimum, perform four basic functions:

1. The package must provide for electrical interconnection (signal, power, and ground) between the various components in the package.
2. The package must offer mechanical, electrical, and environmental protection for the devices it contains.
3. The package must provide sufficient input/output connections to allow fan-out between it and other elements of the electronic system of which it is a part.
4. The package must provide for dissipation of heat if any significant amount is generated by the electronic devices contained within the package when the system is powered.

In the simplest terms, an electronic package must provide for circuit support and protection, power distribution, signal distribution, and heat dissipation. These basic requirements must be met if a package contains a single IC, multiple ICs, or a combination of ICs and passive devices.[1]

3.1.1 Package Drivers

Driven by the desire, and need, to make electronic systems smaller, faster, cheaper, and less power hungry, semiconductor technology continues its relentless assault on increasing component and interconnect density, input/output (I/O) capability, and power dissipation, which in turn places increasing demands on packaging technology.[2] These advances in ICs drive packaging technology by increasing the number of I/O connections, the operating speeds, and the thermal dissipation requirements. Furthermore, systems packaging interconnections must accommodate passive components required for electromagnetic interference (EMI) reduction, filter circuits, terminations, and impedance matching.[3] Consequently, for high-performance ICs, the package will continue to play a larger and larger role in determining performance and cost of the IC. As a result, packaging technologies must continue to improve on the protection provided for the IC, the handling of thermal dissipation, and the routing of more and more signal interconnections, as well as power and ground distribution, through smaller and smaller spaces.[4] It is also anticipated that the package will eventually contain most, if not all, of the passive devices in the form of integral passives. The advent of optoelectronic packages and applications has further added to the already high expectations for packaging. Optoelectronic packages contain not only semiconductor devices, but also optical components, such as optical fibers, lens assemblies, and, depending on the application, elements such as optical multiplexers/demultiplexers. In view of the increased scope, *advanced packaging* is probably a more appropriate term than simply *electronic packaging*.

3.1.2 Packaging as a Basic Design Element

Essentially, semiconductor technology has progressed to the point where packaging is not a mundane activity. The package plays an integral and increasingly more important role in the performance of the semiconductor device or devices it contains. The 1999 *ITRS Roadmap* addresses this fact,[5] stating in the beginning of its assembly and packaging section: "There is an increased awareness in the industry that assembly and packaging has become a differentiator in product development. Package design and fabrication are increasingly important to system applications. It is no longer just a means of protecting the integrated circuit (IC), but also a way for the systems designer to ensure form fit and function for today's product—spanning consumer products to high-end workstations." Although the definition of high-density packaging varies all the way from the I/O or interconnect pitch to a package that has to be co-designed with the chip, it is understood that high-density packaging is and will continue to be a requirement for high-performance ICs and systems.[4] Figure 3.1 shows the evolution of what have been termed advanced packaging technologies in the past in terms of system-level packaging efficiency. System-on-a-chip (SOC) and system-on-a-package (SOP) technologies will continue to emerge and may dominate advanced packaging in the future.

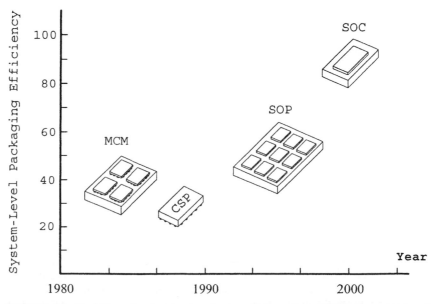

FIGURE 3.1 Evolution of system-level packaging efficiency from 1980 to the present and beyond.

3.2   *SYSTEM-ON-A-CHIP (SOC) VS. SYSTEM-ON-A-PACKAGE (SOP)*

For decades, the least expensive way to add more functions to an electronic system has been to integrate more functions on a chip either by reducing minimum linewidths or by increasing the size of the chip or both. Recently, however, it has been recognized that on-chip interconnects are the limiting factor in high-performance ICs. To complicate the problem further, off-chip interconnects are beginning to contribute to the challenge. What this means is that interconnects, and not the ICs, are dominating, and often limiting, the system performance and increasing its cost.[6] Consequently, there is mounting support for moving interconnects off the chip and onto the package.[7] As a result of opposing views, there is an ongoing debate pitting supporters of SOC against those who support SOP, also referred to as system-in-a-package (SIP).[8–10]

3.2.1   SOC

There are many reasons why semiconductor manufacturers continue to push on-chip system integration.[11] The reason most often cited is that on-chip integration allows faster interconnection between circuit components simply because of the shorter distances involved. However, it has been pointed out that this may not be the case for all designs. For example, on a large digital IC running at a high clock frequency, a signal traveling on a global interconnection trace may take dozens of clock cycles to reach its destination. In such a case, dividing the device into smaller dies and using high-density interconnects on a package substrate can actually be faster.[12]

Another reason for continuing with on-chip system integration is to avoid the configuration of a multichip package (MCP) because, historically, packages containing a system composed of multiple chips have been larger than those containing a single chip system.

Even though SOC integration is the best choice in some situations, it may not be the best choice in every situation. Multichip packaging solutions can be the right approach when cost and/or performance is an issue. Furthermore, intellectual property issues and technology incompatibilities can make multichip packaging an attractive approach to system integration. In fact, one of the driving forces for SOP is the use of predesigned blocks of circuitry, referred to as intellectual property (IP), in SOC design. In this approach, blocks of IP are integrated into a single chip solution. This may involve combining IP blocks from different sources (vendors) that use different processes. Thus, combining a microprocessor core designed for a 0.25-µm process with a memory core designed for a 0.18-µm process may offer quite a challenge.[13]

Mixed semiconductor technologies, such as gallium arsenide (GaAs) and silicon germanium (SiGe), also present a challenge when they are incorporated into an SOC solution. An alternative to the SOC approach is the use of few-chip packaging (FCP), in which the various technologies, in bare die format, are interconnected on a high performance substrate.[13] Advances in laminate substrate technology, such as sequential build-up, low-*k* dielectrics, and microvias, have made FCP commercially viable.[13] Thus, while SOC may deliver high performance, FCP may offer a practical and cost-effective approach to combining IP from different sources or incorporating mixed semiconductor technologies into a system.

Although designers will continue to be pressured to integrate as much functionality onto a single chip as possible, there are many factors that must be considered when deciding if SOC is the right approach to the design of an electronic system. Furthermore, as minimum geometries continue to shrink and chip sizes continue to increase due to the incorporation of additional functionality, a point may be reached where it becomes economically impractical, if not impossible, to fabricate SOC. At this point, SOP may provide the only practical alternative to SOC. Now that chip and package are starting to be co-designed, an opportunity exists to develop capabilities for designing the SOP solutions that would cost less and perform better than SOC solutions.[9] Figure 3.2 provides a list of some of the challenges which must be addressed if SOC is to become the standard approach to electronic system integration.

SOC challenges	SOP challenges
Fundamental: latency, SiO_2 insulation	Design: high-speed digital, optical analog, RF
Process complexity	Large-area intelligent manufacturing; cost/yield
SOC design and test	Thermal management
Wafer fab costs and yields	Testing and reliability
Intellectual property for integrated functions	High performance for low cost fab

FIGURE 3.2   Issues that must be addressed prior to full-scale implementation of SOC and SOP.

To date, SOC has not been the paradigm shift that was originally envisioned, because of myriad problems faced by SOC developers. For example, SOCs are often much too expensive. In addition, SOC developers are faced with the difficulty of integrating analog functions without degrading signal-to-noise ratios or output, inadequate design rules and inexperienced designers, the cost of masks, competition from chip-scale packaging, the length of time it takes to bring an SOC to market, etc.[10] Finally, many people feel that SOC is not really an advanced packaging technology, but merely an advanced design strategy. In fact, the package containing an SOC may be optimized for high performance, but the end result is still a single chip in a package. Consequently, many companies are using both SOC and SOP, with the approach used to design a given system based on performance and cost.[10]

3.2.2 SOP

For many years, systems (or subsystems) have been assembled on printed circuit boards (PCBs), more recently referred to as printed wiring boards (PWBs).[14] Whatever they are called, the boards contain ICs and other components assembled on the board and interconnected to perform a specific electronic function. In some cases, a single board may contain an entire system. For most PWBs, the board only provided mechanical support for the components and interconnects and had little impact on the electrical performance of the system it contained. Unlike traditional PWBs, SOP requires a packaging technology that is optimized to take full advantage of the performance capability of the ICs and components it contains.[15] In other words, the objectives of SOP technology are to reduce the physical size, increase the electrical performance, and reduce the cost of an electronic system through the use of advanced technologies and materials. Thus, SOP consists of assembling the required ICs and passive components, either as discrete devices or integrated within an interconnection, onto a single interconnection substrate (or package) with the specific intent of producing a functional high-density, high-performance electronic subsystem or system.

What actually constitutes an SOP depends on what is considered to be a system. In fact, much of what is referred to today as an electronic system is, in reality, a subsystem, meaning that it does not perform an autonomous electronic function. Consequently, SOP is defined differently by different people. For example, some people might consider a multiple number of logic and memory ICs in a single package as being an SOP. On the other hand, if the SOP did not contain analog peripheral device drivers, one might argue that the package really does not contain a system. Additionally, complete systems usually contain passive devices in addition to the ICs. Passive devices are required to make systems work, and if an entire system is to be placed in a package, then the required passives must go in as well.[3,9]

Reduction in the substrate (board) area required for the total system is perhaps the biggest advantage of SOP. Performance is enhanced because of shorter interconnection distance between dies and the ability to control impedance levels. Regions of very dense wiring permit interconnection of ICs with high lead counts. This means that the second-level assemblies require less space, allowing room for more components at that level.[8] Although SOP is an extremely attractive packaging technology, for it to be really cost effective, large-panel pro-

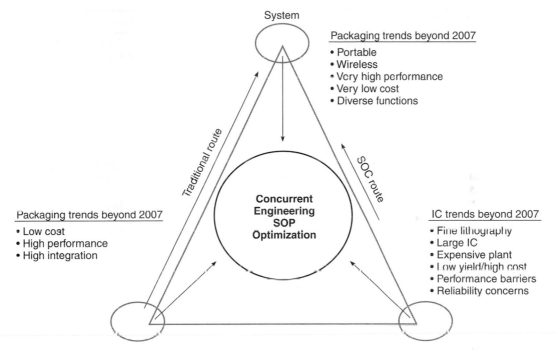

System

Packaging trends beyond 2007
- Portable
- Wireless
- Very high performance
- Very low cost
- Diverse functions

Traditional route

SOC route

Concurrent
Engineering
SOP
Optimization

Packaging trends beyond 2007
- Low cost
- High performance
- High integration

IC trends beyond 2007
- Fine lithography
- Large IC
- Expensive plant
- Low yield/high cost
- Performance barriers
- Reliability concerns

FIGURE 3.3 Future trends in IC, packaging, and system technologies as envisioned by personnel of the Packaging Research Center at The Georgia Institute of Technology. *(Drawing provided courtesy of the Packaging Research Center, The Georgia Institute of Technology.)*

cessing is a requirement. Additionally, challenges still exist in the interfacing of mixed technology components, such as optical devices. Figure 3.2 lists some of the challenges that must be considered for the successful development and implementation of SOP.

Area array packaging has become an enabling technology for addressing the needs of customized SOP solutions in order to reduce size, weight, and pin count at the second level of interconnection.[16] This type of packaging technology includes pin grid array (PGA), ball grid array (BGA), and chip-scale packages (CSPs), with BGA playing the largest role, primarily because of its versatility. Compared to conventional leaded packages, BGA packages exhibit improved electrical performance due to a shorter distance between the IC and the solder balls, improved thermal performance due to the use of thermal vias or heat dissipation through power and ground planes, reduced handling-related lead damage, and increased manufacturing yields due to their self-alignment feature.[17] In fact, BGA has been one of the biggest contributing factors to the proliferation of mobile phones and other wireless communication systems. However, even with the advantages created by area array packaging, performance, reliability, and cost requirements are challenging currently available area array electronic package design, primarily because of materials issues.[18] Figure 3.3 provides an overview of IC, packaging, and system trends that are anticipated for the year 2007 and beyond.

3.3   MULTICHIP MODULES

Just as the name implies, multichip module (MCM) technology mounts multiple, unpackaged ICs (bare dies), along with signal conditioning or support circuitry such as capacitors and resistors, to form a system, or a subsystem on a single substrate.[1] In fact, in the late 1980s and early 1990s, MCMs were considered to be the ultimate interconnect and packaging solution, capable of meeting every challenge of the electronics industry. This technology placed ICs in close proximity to each other, thereby enhancing system performance by reducing interconnect delay.[13]

A typical MCM from this time period contained anywhere from 2 to 20 or more bare dies, either hermetically sealed on a ceramic or laminate substrate or placed in a conventional semiconductor package. Surface-mount components, such as capacitors, resistors, and inductors, were usually assembled on such packages. In the late 1980s, predictions were that the thin-film MCM market would reach $10 billion per year by the year 2000. Obviously, this did not happen.[19] While companies like IBM and Fujitsu have had a successful history of using MCMs in their mainframe computers, MCMs never attained the commercial success that was predicted for them. In other words, they were never adopted by the lucrative, high-volume, PC/workstation market, primarily due to bare die testing and substrate rework issues.[13] The liquidation of MicroModule Systems (MMS), the last major North American firm trying to make a go of the MCM business, has left people wondering if all MCM technology might soon disappear from the electronics scene. However, there are still people in the MCM business who feel that MCMs may one day become more mainstream.[19] In fact, there are quite a few primarily semiconductor companies that offer MCMs as a standard packaging option for their products, primarily on laminate substrates.[20]

Size is often the primary driver for MCM-based systems. The typical multicomponent discrete assembly provides a silicon-to-board efficiency of <10 percent (actual total die area vs. total printed circuit board area). MCM technology can often increase the silicon-to-board efficiency to 35 or 40 percent with chip-and-wire assembly processes, and to 50 percent or higher with some of the higher-density processes.[21] Thus, with reduced size and weight, MCMs offer a practical approach to reducing overall system size while, at the same time, providing enhanced performance due to a reduction in the interconnect distance between chips.[1] Multichip modules typically use three to five times less board area than their equivalent discrete solution.[21]

Since a multichip module is, by definition, a single substrate containing two or more ICs, it is unlikely that they are a thing of the past. In fact, they are alive and well, and have experienced a kind of rebirth over the past few years. They have just been renamed. Instead of referring to them as MCMs, we now have multichip packages (MCPs) and few-chip packages (FCPs). These lower-cost approaches usually contain only two to six ICs, which are packed onto an inexpensive laminate substrate. Although they are multichip modules, now they are intentionally called something else in order to avoid the MCM tag. Intel refers to its two-chip package as a chip-scale package (CSP) although in fact it is a multichip package (MCP) or MCM.[19] Interestingly, there are still those who believe that when you must reach market quickly with a low-cost, reliable product, and area is not an issue, multichip solutions win. On the other hand, there are those who believe that it is only a matter of time before SOCs take over completely and multichip packaging will not be necessary. They view multichip solutions only as stopgaps.[10]

The IPC, an industry association, has established formal definitions of MCMs (Publication IPC-MC-790), "Guideline for Multichip Technology Utilization," IPC, Lincolnwood, Illinois, 1990.[1] Depending on the nature of the carrier and the composition of the interlevel dielectric separating the electrical conductors in the package, an MCM can be categorized into one of three types:

1. MCM-L (laminate)
2. MCM-C (ceramic)
3. MCM-D (deposited)

For the interconnection of the chips, MCM-L uses printed wiring board wiring, while MCM-C uses cofired ceramic or glass-ceramic with thick-film metallization. MCM-D uses thin

TABLE 3.1 Properties and Parameters of Several Multichip Module Technologies

Technology →	MCM-D		MCM-C		MCM-L	MCM-D/C	MCM-D/L
Parameter/ property ↓	Inorganic	Organic	Conventional ceramic	Low-K ceramic	Micro-PWB	Deposited thin film on ceramic	Deposited thin film on laminate
Maturity	Good	Good	Excellent	Very good	Very good	Good	Good
Cost	High	High	Medium	Medium	Low	High	Medium
Substrate material	SiO_2	Polyimide BCB	Al_2O_3	Glass-ceramic	Epoxics, polyimides, etc.	Glass-ceramic, glass, etc.	Epoxies polyimides, etc.
Number of metal layers	5	5	70+	>50	>20	>50	>10
Line width (μm)	15	10	100	100	25	15	20
Line pitch (μm)	25	25	200	200	150	25	150
Via diameter (μm)	25	25	100	100	200	25	25
Conductor resistivity	2–40 mΩ/□		2–20 mΩ/□		0.1–3 mΩ/□	10 mΩ/□	1 mΩ/□
Thermal conductivity	0.1–1 W/m·C		1–20 W/k·C		0.1–0.3 W/k·C	1–20 W/k·C	0.1–0.3 W/k·C
Dielectric constant	2–4		5–10		2.5–5	2–4	2.5–5
CTE (ppm/°C)	3–8		3–9		4–15	3–8	3.8

metal films and an organic dielectric over a rigid base substrate such as silicon or aluminum nitride.[22] Table 3.1 compares some properties and parameters of these three conventional MCM technologies.

3.3.1 MCM-D

MCM-Ds are formed by the sequential deposition of thin-film conductors and dielectric layers, which may be polymers or inorganic dielectrics, on a substrate base. MCM-Ds are some-

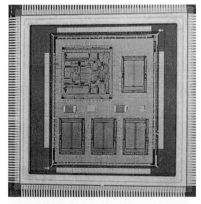

what unique compared to MCM-Ls and MCM-Cs because their fabrication is significantly different. MCM-D fabrication makes use of conventional semiconductor processing technologies. Thus, the fabrication steps and materials used to construct MCM-Ds are identical or very similar to those developed and used to manufacture ICs. Consequently, MCM-D offers a very high reproducibility of very small line dimensions and is therefore an excellent technology for the integration of RF and microwave circuits. Additionally, various types of high-performance integrated passives (spiral inductors, TaN resistors, and Ta_2O_5 capacitors, among others) can be placed directly on the substrate, providing a size and cost reduction and increasing the packaging density.[23] Because of the fabrication technologies used, MCM-D provides the highest interconnect density, but it is also the most expensive. Figure 3.4 is a photograph of a packaged MCM-D fabricated on silicon in the early 1990s.

FIGURE 3.4 A packaged, silicon-based, high-performance MCM-D. *(Photograph provided courtesy of HiDEC, The University of Arkansas.)*

3.3.1.1 Substrates. Possible substrate candidate materials are silicon; ceramics such as alumina, aluminum nitride, mullite, silicon carbide, glass, glass-ceramics, and beryllia; and metals such as aluminum, copper, steel, tungsten, etc. The dimensional stability of this type of

TABLE 3.2 Properties of Some Materials Used for MCM-D Substrates

Property → Material ↓	Dielectric constant	Dissipation factor (tan δ)	Electrical resistivity (Ω/cm)	Thermal conductivity W/m·C	CTE (ppm/°C)
Silicon	12	—	10^5	125–150	2.5–4.5
Alumina	9–10	0.001	$>10^{14}$	20–40	6.5–7.2
AlN	8.5–10	—	$>10^{14}$	150–260	3.0–4.5
BeO	7–9	<0.001	$>10^{15}$	250–300	6.8–8.5
SiC	20–40	0.06	$>10^{14}$	100–270	3.0–4.6
Glass	6–7	0.005	$>10^{14}$	1.6–2.0	5–9
Glass-ceramic	5–9	0.0025	$>10^{13}$	1.0–2.5	3–7

module is determined by the underlying substrate. The substrates are generally round or square. Since the substrate of an MCM-D only provides a platform for independently applied dielectric and interconnect layers, its required properties are not as stringent as those for MCM-L and MCM-C. However, silicon is often used because of its extremely smooth surface and because, when a silicon substrate is used with silicon chips, there is no CTE mismatch. Table 3.2 provides some properties of materials used as MCM-D substrates.

3.3.1.2 Dielectrics. The dielectric materials normally have a dielectric constant of <5 and consist of chemically vapor-deposited silicon dioxide; a liquid polymer, such as polyimide; benzocyclobutene (BCB); or a fluoropolymer deposited by conventional spin coating. Silicon dioxide is attractive because it is well understood and easy to deposit. Polyimides are the most widely used because of their low dielectric constant. Benzocyclobutene is a good alternative to polyimides because it has a lower dielectric constant, lower moisture absorption, and excellent planarization properties, and does not react with copper, which is attractive from a reliability point of view. Table 3.3 gives some properties of commonly used dielectric materials.

3.3.1.3 Interconnects. Interconnects on MCM-Ds are formed from high-conductivity metals by sputtering or plating. The most common metals are aluminum, gold, and copper. Often, barrier layer metals are required to prevent diffusion of the interconnect metal into the dielectric or intermixing of different metals in a structure such as a solder ball. Barrier metals that are commonly used include chromium, titanium, and nickel. Vias may be formed during the conductor deposition/plating processes. When deposition and plating are used in combination, the vias are plated after conductor deposition. The choice of metal is usually made in conjunction with the choice of dielectric because metals and dielectric materials can interact with each other, resulting in serious problems with reliability. Table 3.4 gives some properties of metals used in electronic packaging, including MCM-Ds.

TABLE 3.3 Properties of Some Dielectric Materials Used in MCM-Ds

Property → Material ↓	Dielectric constant	Dissipation factor (tan δ)	Electrical resistivity (Ω/cm)	CTE (ppm/°C)	Thermal conductivity (W/m·K)	Degree of planarization (%)
Silicon dioxide	3.9	0.004	$>10^{15}$	0.5	1.6	Conformal
Silicon nitride	8	—	$>10^{15}$	2.4–3.2	25–35	Conformal
Polyimides	2.5–4.0	0.01–0.002	$>10^{17}$	3–50	6.5–7	10–35
Benzocyclobutene (BCB)	2.6–2.8	0.001	$>10^{18}$	35–70	—	90–95
Polyphenylquinoxaline (PPQ)	2.7–3.0	—	—	50–60	—	20–70

TABLE 3.4 Properties of Some Metals Used in Packaging

Property → Metal ↓	Resistivity (μΩ/cm)	Thermal conductivity (W/m·K)	CTE (ppm/°C)	Melting temperature (°C)
Tungsten	5.5	160–200	4.5	3415
Molybdenum	5.2	146	5.0	2620
Platinum	10.6	71	9.0	1775
Palladium	10.8	70–92	12–13.3	1551
Chromium	13–20	66	6.3–6.5	1885
Titanium	5.5	22	9.0	1665
Tantalum	15.6	58	6.5	2980
Aluminum	2.65–4.3	240–247	23–25	660
Gold	2.2–2.35	295–297	14–14.2	1065
Copper	1.67–1.75	390–420	17–20	1070
Silver	1.6	420	20	961
Nickel	6.8–10.8	92	13.3–13.5	1455

As practiced by the IBM Microelectronics Division, the copper wiring in MCM-D packages is deposited by electroplating using the plate-through-mask technique. A number of factors led to the selection of plating over a dry process such as sputtering. Some of these factors are as follows: plated copper has a more desirable metallurgical structure so that it has low stress, is equiaxial, and is ductile; plating provides improved filling of trenches and vias (less tendency for voids to form); the tools and processes used in plating are more scalable to large-format substrates; the processing time for plating is faster, thus providing higher throughput; and, factoring in tooling, raw materials, and maintenance, plating is a relatively low-cost manufacturing process.[22]

3.3.2 MCM-C

For years, ceramic (an alumina/silica combination) substrates have been used as a platform to mount multiple bare (unpackaged) dies in combination with thick/thin-film conductors, resistors, and capacitors to form what are commonly referred to as "hybrid" circuits. Hybrid circuits are fabricated using thick-film multilayer (TFM) technology in which multiple layers of inks (or pastes), formed one layer at a time, are screen-printed onto a substrate. Thus, screen-printing is used to form the conductors, resistors, capacitors, and inductors, depending on the properties of the ink. Following application to the substrate, the inks must be properly dried and fired in order to realize their final property values.

The latest evolution of the hybrid circuit is called a multichip module–ceramic (MCM-C).[14] MCM-C technology utilizes cofired ceramics as the basic building block to form the substrate. Because of the higher interconnect density requirement for MCMs, multiple substrates containing screen-printed conductors and vias must be formed, one substrate at a time, and then laminated together. Several layers (up to 100 have been used) containing the interconnection patterns and vias are cofired under pressure to form a completed MCM substrate. ICs and passive devices are then attached to the substrate to complete the module. The term *cofired* implies that multiple ceramic and conductor layers are heated at the same time. Despite consistent predictions of its demise, ceramic usage continues to grow and the technology continues to offer advantages in both high-performance (digital, wireless, microwave) and cost-sensitive (automotive) applications. There are three ceramic-based technologies that can be classified as MCM-Cs. These are thick-film multilayer (TFM), high-temperature cofired ceramic (HTCC), and low-temperature cofired ceramic (LTCC).[1]

HTCC MCMs, as noted earlier, are merely a variation of TFM technology and were developed to increase packaging density by building individual conductor/dielectric layers and then laminating them by firing at high temperatures (in the range of 1400 to 1600°C) under pressure. The individual layers are formed by screen-printing conductor material onto sheets of dielectric tape, called green tape. Vias are formed by punching and filling with a conductive material prior to screen-printing the conductor traces. After the requisite number of sheets are prepared to provide for the circuit interconnect, they are stacked, aligned, and laminated together under pressure and temperature.

LTCC packaging combines the advantages of the cofired process (dielectric tape) with standard thick-film conductor materials that have significantly higher electrical conductivities. In this case, dielectric and conductor materials are borrowed from TFM technology so that the lamination firing temperature is reduced to around 750 to 850°C. This lower temperature allows the use of much higher conductivity interconnects, such as gold, silver, and copper.[1] The technology is based on green tape dielectric thick-film tape-casting instead of screen-printed dielectrics.[24] Green (unfired) dielectric tape is blanked to size, and registration holes are punched. After punching or drilling to form vias, conductor lines are screen-printed on the tape. When all layers have been punched and printed, they are registered, laminated, and cofired. Since the cofiring process requires fewer firing steps and allows for the inspection of punched printed layers prior to lamination, it can produce a greater number of layers with higher final yield than other MCM-C processes. Figure 3.5 shows a photograph of a conventional MCM-C built on eight layers of LTCC.

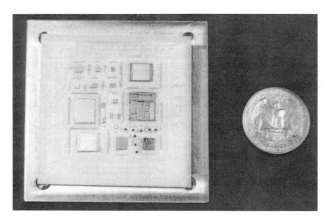

FIGURE 3.5   A conventional MCM-C. *(Photograph provided courtesy of HiDEC, The University of Arkansas.)*

3.3.2.1 Substrates.   A number of different materials can be used for an MCM-C substrate. These include, but are not limited to, alumina (the most common), aluminum nitride, beryllia, and mullite. Of these, alumina containing various amounts of glass is the one most commonly used. The substrates are formed by pressing or casting. Casting produces green tape from a slurry of ceramic powder, organic binders, and solvents. For proper electrical performance, the substrate must be an insulating material with a low dielectric constant in order to provide electrical isolation for the interconnections and passive components. Thus, the substrate material acts as both the dielectric interlayer and the supporting plane for attached electronic components. Table 3.5 provides the properties of some MCM-C substrate materials.

3.3.2.2 Dielectrics.   The material that provides the mechanical support for the circuit (i.e., the substrate) also provides for electrical isolation of the various interconnect metal layers and passive devices. Thus, the discussion in the previous section is also appropriate for the interlayer dielectric material in an MCM-C. The materials usually have a dielectric constant ≥5 (see Table 3.5).

TABLE 3.5 Properties of Some MCM-C Substrate Materials

Property → Material ↓	Dielectric constant	Dissipation factor (tan δ)	Electrical resistivity (Ω/cm)	Thermal conductivity (W/m·K)	CTE (ppm/°C)	Flex strength (MPa)	Density (g/cm³)
Alumina							
96%	9–9.5	0.0003–0.001	>10¹⁴	20–25	6.6–6.7	350–360	3.75–4.0
99.5%	9.6–9.8	0.0001–0.0004	>10¹⁴	25–35	6.9–7.1	380–390	3.75–4.0
99.9%	9.9–10.2	0.0001–0.0004	>10¹⁵	35–45	7.0–7.2	390–400	3.75–4.0
AlN	8.5–10	0.001	>10¹⁴	150–260	2.7–4.5	271–304	3.2
BN	4.0–4.5	0.005–0.1	>10¹⁴	55–60	2.5–6.5	109–117	2.1–3.1
BeO	6.5–8.9	0.0004–0.001	>10¹⁵	250–300	6.5–9.0	176–253	3.0
Glasses	4.0–7.2	0.006	>10¹⁴	2–3	8.5–9.5	50–56	2.7–3.0
Glass-ceramics	4.5–8.5	0.002	>10¹³	1–2	2.5–6.5	152–248	2.8–3.2
Mullite	6.2–6.8	0.005–0.02	>10¹⁴	5–9.8	4.0–5.0	141–165	2.85–3.1
Cordierite	4.5–6.0	0.004	10⁶–10¹³	1.3–3.9	1.5–2.5	72–100	2.6–2.75
Forsterite	6.2	0.005	10¹⁰–10¹²	2.2–4.2	9.5–10.0	170–181	2.82–2.98
Steatite	5.5–7.5	0.001	>10¹²	2.1–3	4.0–4.5	165–188	2.7–2.84

3.3.2.3 Interconnects.   Conductors are usually a fireable metal material such as tungsten, molybdenum, nickel, molymanganese, or the screenable frit metal thick-film conductors gold, silver, platinum, palladium, and copper. The metal chosen for a particular type of MCM-C depends on a number of factors although the primary factor is the cofiring temperature, which differs by several hundred degrees Celsius for LTCC and HTCC.[1] To form the interconnects, inks (or pastes) consisting of some combination of metal or dielectric powder, vitreous or oxide powder, and organic binders are screen-printed onto the ceramic substrate. Vias necessary during the interconnecting process are formed during the conductor screen-printing. The vias are usually the same materials as the conductors. The properties of some metals used in MCM-Cs are given in Table 3.4

3.3.3 MCM-L

In one sense, MCM-L has been around for a long time, because it is essentially a miniature version of PWBs utilizing ICs that are directly attached to the module using chip-on-board (COB) technology. Thus, MCM-L structures are laminated PWBs that have been scaled to meet the requirements and dimensions of multichip modules. They are miniaturized in terms of interconnect linewidth and spacing. Simply, then, MCM-L refers to a multichip module that utilizes unpackaged chips on a laminate substrate. If the laminate is not reinforced, it is quite flexible and is referred to as a flexible substrate. MCM-L is the least dense in terms of interconnect density, but it is also the least expensive technology.

MCM-Ls are expected to be the most commercially dominant MCM because of prior investments and the existing infrastructure. Furthermore, the PWB base is being enhanced by new materials, such as aramid fiber, BT resin, maleimide styryl, COPNA resin/E-glass fabrics, and photosensitive epoxy; new processes, such as additive plate and laser/photovias; and large-area, low-cost processing in not-so-clean facilities. The chip mounting process used to complete the multichip module has already been practiced by direct wire bonding, TAB, and flip-chip bonding of bare dies.[25] Figure 3.6 is a photograph of an MCM-L prototype engine controller.

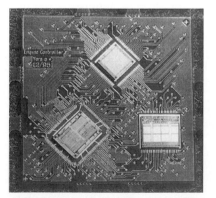

FIGURE 3.6 A prototype engine controller MCM-L. *(Photograph provided courtesy of The Center for Advanced Vehicle Electronics, Auburn University.)*

3.3.3.1 Substrates. The substrate (base) material of MCM-Ls is a resin—often an epoxy, cyanate ester, phenolic, polyimide, Teflon, or acrylic. The resin is usually reinforced with woven fibers of E-glass, F-glass, S-glass, D-glass, quartz, Kevlar, etc., to provide mechanical strength. Once metallic interconnects have been formed on one or both sides of some number of the single substrates using photolithographic definition, they are laminated together to form the module on which the ICs are mounted and through which they are interconnected. The number of these substrates needed for a particular module is determined by the required chip interconnect density. Electrical contact between the various layers of interconnections is provided by vias that are drilled prior to final lamination. Following lamination, the vias are plated to provide electrical connection between the different layers of the MCM-L. The MCM is then populated with bare ICs mounted directly onto the MCM (commonly referred to as chip-on-board [COB]), instead of individually packaged devices as is generally done in conventional PWB technology.[1] FR-4, which consists of epoxy and E-glass, is a very common laminate material. Table 3.6 gives some commonly used rigid laminate materials and their important properties. Table 3.7 provides properties of some flexible substrate materials.

TABLE 3.6   Properties of Some MCM-L Substrate Materials

Property → Material ↓	Dielectric constant	Dissipation factor (tan δ)	Thermal conductivity (W/m·K)	CTE (ppm/°C) xy/z
E-glass/epoxy	4.65	0.018	0.33	12–15/65–80
E-glass/PTFE	2.3	0.005	0.25	24–26/255–260
E-glass/polyimide	4.45	0.021	0.35	10–14/60–85
Kevlar/Quatrex	3.68	0.032	0.15	3–10/90–105
Kevlar/polyimide	3.63	0.008	0.12	3–8/83–87
Quartz/Quartrex	3.55	0.025	0.14	5–12/62–65
Quartz/polyimide	3.6	0.01	0.12	6–12/34–40

TABLE 3.7   Properties of Some Flexible Substrate Materials

Material → Property ↓	Polyimide (Kapton)	Polyester (Mylar)	Fluoropolymers
Dielectric constant	3.4–4.0	3.3	2.0–2.3
Dissipation factor (tan δ)	0.003–0.01	0.005–0.015	0.00025
Dielectric strength (V/mil)	7650	7500	5000
Tensile strength (Mpa)	230	160–175	20–30
Dimensional stability (%)	0.18–1.25	1.4–1.5	0.3–0.5
CTE (ppm/°C)	20–45	28–31	10–15
Moisture absorption (wt. %)	3–5	0.8–1.0	0.1–0.5
Typical thickness (mils)	0.5–5	1.0–7.0	1–7

3.3.3.2 Dielectrics.   In this technology, the same material that forms the substrate also provides for dielectric isolation of the interconnects. Tables 3.6 and 3.7 provide performance parameters for materials used as the dielectric in these structures.

3.3.3.3 Interconnects.   The most commonly used interconnect material is copper. The substrate material is usually purchased as copper-clad laminates. The interconnects are then deposited additively or are generated as part of the printed board subtractive process by electroplating. Vias are copper, electrolessly plated initially, followed by additional electrodeposition. When MCM-L requires control of the coefficient of thermal expansion (CTE), lamination techniques use materials that have a low CTE, such as copper-Invar-copper. Properties for metals used in the fabrication of MCM-Ls are given in Table 3.4.

3.3.4   MCM-D/C

MCM-D/C combines deposited dielectric layers on top of a multilayer cofired ceramic substrate, yielding a product with the high-frequency and high density attributes of MCM-D and the physical attributes of MCM-C. Table 3.1 provides some properties and parameters of this technology. In this structure, cross talk is minimized by embedding the conductors in low-dielectric-constant material, while power and ground planes can be surrounded by high-dielectric-constant ceramic. Additionally, the cofired substrate contains the substrate base, signal interconnect, package body, and module-level I/O, negating the need to place the module in a conventional package. By bringing vias out the bottom of the module and mating them with pins, the I/O simulates a standard pin grid array (PGA) package.[1] Figure 3.7 is a photograph of a 16-chip memory MCM-D/C built on 12 layers of LTCC using copper/BCB.

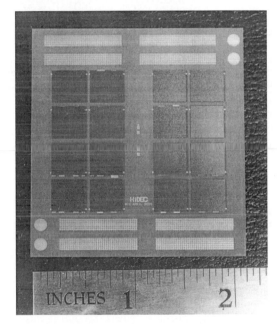

FIGURE 3.7 An MCM-D/C. *(Photograph provided courtesy of HiDEC, The University of Arkansas.)*

3.3.5   MCM-D/L

The fabrication of MCM-L using PWB techniques places limits on the fanout available to the MCM-L as a result of minimum interconnect width and spacing and the cost associated with drilling vias. Consequently, conventional MCM-L cannot accommodate the very dense area array I/O patterns encountered in fully populated ball grid array (BGA) packages, chip-scale packages, and flip-chip-on-board (FCOB) assemblies. These problems led to the development of special conductor redistribution layers placed between the ICs and the conventional MCM-L module. Table 3.1 gives some properties and parameters for this technology. Because these redistribution layers can consist of photodefinable materials, they allow for much smaller minimum geometries (similar to MCM-D) and the creation of all vias at the same time. Other materials that lend themselves to laser "drilling" also provide for small vias. However, as is the case for MCM-D/C, thermal issues, planarization, and warping of the substrate are all challenges to the realization of MCM-D/L. Row *b* of Fig. 3.8 shows an MCM-D/L.

FIGURE 3.8 Three different multichip packages (or modules): (*a*) Pentium™ CPU module; (*b*) 320-pin MCM-D/L (1.0-mm pitch), (*c*) 48-pin MCP (1.0-mm pitch). *(Photograph provided courtesy of Fujitsu Microelectronics, Inc.)*

3.4   MULTICHIP PACKAGING

In general, a multichip package (MCP) is any package that contains more than one die. As noted previously, multichip modules have been around for decades, mostly for low-volume, high-reliability applications. Because of this, the term *multichip* has become synonymous with *expensive*. However, new, lower-cost options for multichip module packaging are available for high-volume applications.[9]

Advanced electronic packaging technologies, such as MCMs, have historically been limited to applications in high-performance products such as supercomputers. Today, they are being moved into lower-cost products through the use of MCM with chip-on-board (COB), ball grid array (BGA), or land grid array (LGA) for second-level interconnect assembly, and SMT mass production equipment. Improvement in the silicon efficiency of MCM designs can be accomplished by packaging ICs onto very thin, compact, and lightweight microcarriers using BGA or LGA connections, such as those shown in Fig. 3.9. Because of the development of lower-cost materials (i.e., low-cost photosensitive materials), large-area processing (as practiced in the fabrication of displays), large-area lithography, and low-cost metallization processes (as practiced in PWB fabrication), economically viable multichip packaging is realistic.[25]

3.4.1   Few-Chip Packaging (FCP)

An increasing number of companies are embracing FCP for technical and business reasons. While these FCPs are virtually indistinguishable from the single-chip packages, they are a radical departure from the MCMs of the early 1990s. Instead of 10 to 20 dies, today's FCPs typically contain 2 to 5 dies mounted on a laminate substrate in a ball grid array (BGA) package (see the multichip package in row *c* of Fig. 3.8). This rebirth can be attributed, at least in

FIGURE 3.9 Three different BGA packages and 1 LGA package: (*a*) 1600-pin FC-BGA (1.0-mm pitch, ceramic substrate), (*b*) 1140-pin FC-LGA (1.0-mm pitch, build-up substrate), (*c*) 768-pin TAB-BGA (1.0-mm pitch), (*d*) 672-pin EBGA (1.27-mm pitch). (*Photograph provided courtesy of Fujitsu Microelectronics, Inc.*)

part, to improved bare die testing and handling, along with the availability of low-cost, high-performance laminate substrates. Furthermore, there is a growing trend toward the use of FCP as an alternative to system on-a-chip (SOC), resulting in a system-in-a-package (SIP). This approach, also referred to as system-on-a-package (SOP), is discussed in a previous section of this chapter.[13]

3.4.2 Partitioned Silicon (Tiling)

As noted in Sec. 3.2.1, IC manufacturing has reached the point in system integration where the delays of some long on-chip interconnects are starting to have a significant impact on the ICs' performance.[9] Thus, in the not-too-distant future, SOC may encounter serious problems in the form of long lossy lines and unacceptable yield if IC size continues to increase, because, as the die gets larger, on-chip wiring becomes longer and slower. Although additional on-chip drivers can be added to reduce the long global interconnect delays, this practice leads to increased power dissipation and, sometimes, complications in signal routing. Using larger on-chip wiring traces to compensate for the length increases of larger ICs increases the required number of metal layers. This adds to the number of masking levels with the accompanying potential yield problems. On the other hand, wiring traces between dies can have lower resistance than long on-chip traces. If an IC is laid out to minimize long traces, separated into smaller dies, and the dies are then mounted on an interconnect substrate (module), net performance can improve.[7, 26] Simulation results suggest that once on-chip interconnects get beyond 5 or 6 mm in length, it is better to go off the chip than to try to run long interconnects on-chip with very thin metal.[26]

The long lossy line problem is not the only difficulty in making larger devices. As the die becomes larger, a single killer defect can disrupt more functions in disabling a single die. A larger die also can result in a net loss of functionality created per wafer, due to the area loss problem at the edge of the wafer. The maximum die size, according to present technology, seems to be about 15 to 20 mm per side. Once the dimensions exceed about 15 mm on a side, the yield starts dropping off to the point where the feasibility of a decision to manufacture such large chips becomes questionable.[6]

It has been suggested that separating system functions into more than one die can be preferred, so multiple dies are used by choice rather than necessity.[27] However, partitioned silicon has not been widely used to date. This approach involves the intentional separation of a single technology system (i.e., a large silicon die) into more than one die and then interconnecting the smaller die through the package substrate. In addition to the fact that smaller dies have a better yield, moving long global interconnects off the die and onto the package substrate can shorten their signal propagation delay. This requires adjustments to the IC's architecture, appropriately designed chip-to-chip drivers, and high-density packaging technology to be successful. Also, packaging a chip set within a package can lead to simpler chip designs. Additionally, the smaller ICs can be altered in some way to optimize the overall system that would have been contained on a single die. While this allows performance and cost gains, it also is the main drawback. In addition to known good die (KGD) issues, which are still considerable, the dies are not really complete until they are integrated into the multichip package.[9] Figure 3.10 is an illustration of silicon tiling.

System Board

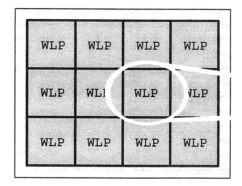

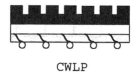

CWLP

FIGURE 3.10   The concept of tiling.

3.4.3   Chip-Scale Packaging (CSP)

Chip-scale packaging (CSP) was introduced in Japan in the early 1990s and in the United States in 1994 as a less expensive alternative to multichip modules (MCMs).[28] The accepted definition of a CSP is a package that has a perimeter no more than 20 percent larger than that of the die it houses.

CSPs offer the same space and material savings and short signal paths that direct chip attach (DCA) methods such as chip-on-board (COB) and flip-chip-on-board (FCOB) offer. The advantages of using a chip-scale package over DCA are easier handling, more protection for the chip, and simpler board assembly. Figure 3.11 shows a selection of commercially available CSPs. CSPs are limited to use in moderate-I/O ICs.[28] However, it is anticipated that full–area array BGA packages will attain I/Os in the 1000 to 2000 range in the not too distant future.

Recently, blending has started to occur between chip and package. In Tessera's "tiles" concept, the chip is no longer the finished device. The topmost interconnection layers on the chip that handle power, ground, and some clock signals are moved out to the package.[28] In these cases, tape or flex substrates may be used in chip-scale BGA packages.[29]

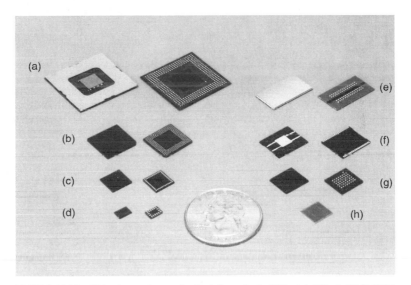

FIGURE 3.11   Chip-size packages. On the left are logic CSPs: (*a*) 352-pin TAB-BGA (0.80-mm pitch); (*b*) 288-pin FBGA (0.5-mm pitch); (*c*) 48-pin BCC (0.5-mm pitch); (*d*) 16-pin BCC (0.65-mm pitch). On the right are memory CSPs: (*a*) 60-pin μBGA (0.8 mm pitch); (*b*) 46 pin SON (0.5-mm pitch); (*c*) 48-pin FBGA (0.8-mm pitch); (*d*) 16-Mbit Flash Memory Chip. *(Photograph provided courtesy of Fujitsu Microelectronics, Inc.)*

3.4.4   Wafer-Scale Integration (WSI)

In wafer-scale integration, an entire electronic system, composed of a multiple number of different ICs (subsystems), is fabricated on a single wafer. Because of yield considerations, redundancy is used to actually create more than one system on the wafer. Once the wafer is fabricated, it is statically and dynamically tested to determine which of the ICs are good, but they are not separated from each other. Instead, the good ICs, which are required to form the electronic system, are interconnected. The yield of WSI wafers after testing is a concern. This complicates processing requirements because testing often creates damage to the wafers.

The attractiveness of WSI lies in its promise of greatly reduced cost, high performance, high levels of integration, greatly increased reliability, and significant application potential. However, there are still major problems with WSI technology, such as redundancy and yield, that are unlikely to be solved in the near future.[30]

3.4.5   Three-Dimensional (3-D) Packaging

Another type of multichip packaging gaining wider acceptance is stacked-die packaging.[31–33] At the present time, most packaging of this type is in the form of stacked memory in CSPs and the stacking is done only to save space. However, it is a growing trend that can be applied to any electronic system where volume density is of concern. Stacked-die packaging is illustrated in the Fig. 3.12(*a*), while Fig. 3.12(*b*) shows how packaged devices can be stacked.

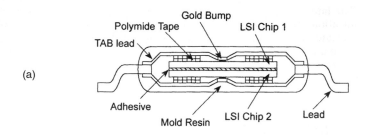

(a)

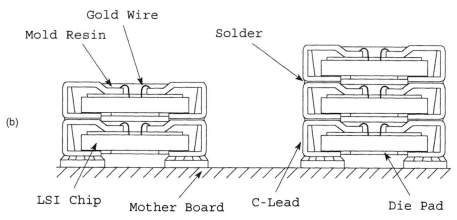

(b)

FIGURE 3.12 Three-dimensional (3-D) packaging: (*a*) die (or chip) stack packaging; (*b*) package stacking.

The advantages of stacked die packaging are further enhanced by using thinned dies. Thinning dies has the potential of enhancing first-level interconnect reliability while minimizing the increase in vertical profile. A silicon die with a thickness <100 µm is quite flexible and can relieve stresses induced by packaging and CTE mismatch.[34]

The simpler, wire-bonded CSPs are suitable for stacking memory, where the main concern is saving space. Where high die-to-die I/O density is needed, flip-chip-on-chip (FCOC) is used and is denoted as FS-CSP.[9]

Many companies, particularly in Asia, have been stacking dies physically without connecting them to each other. A three-die stacked memory device has been demonstrated. The bottom die is the largest and the top die is the smallest. They are all wire-bonded to the same package. Many companies use this scheme to put two dies in a single package. Since the dies are not connected to each other, this scheme simply saves board space.[9] Although this type of packaging does not meet the definition of a multichip module, it does qualify as a multichip packaging technology.

Other approaches use such techniques as memory devices in the form of interconnected stacks using metal traces on the side of the stack to form the interconnects. Multiple dies are placed in a wafer-shaped carrier, and, after singulation, interconnects are completed on the sides of the stack, producing stacked wafer-level packaging technology based on through-silicon-contact technology. In the latter case, contacts are imbedded in the wafer from the front side and, after wafer processing, the wafer is thinned to reveal the contacts through the back of the wafer. Additional specifics are found in Refs. 6 and 9.

The latest in IC packaging is a stacked ball grid array (SDBGA) package for the communications market. It is used to stack various ICs in one package, with resulting savings in real estate on the motherboard and manufacturing cost. Compared to conventional packages, an area savings of up to 70 percent can be realized. Popular SDBGA sizes range from 8×8 mm to 14×14 mm and pin counts from 80 to 140. The total package height is typically 1.4 mm.

3.5 ENABLING TECHNOLOGIES

Once a multichip substrate, or module, has been fabricated, the ICs and passive devices must be attached. This section addresses the issues of known good die (KGD), chip attach techniques, chip-on-board (COB), and passive devices.

3.5.1 Known Good Die (KGD)

Anytime an electronic system is fabricated using bare (unpackaged) dies, KGD issues must be considered.[35,36] In particular, IC manufacturers perform only limited performance testing on dies prior to placing them in a package. Since bare ICs are used in multichip packaging, some effort must be expended to ensure that an IC is "good" prior to placing it onto a substrate (module). At the present time, a few IC manufacturers have implemented extended testing of their bare die products; not all have done so, however.[37] Consequently, there are still concerns about reworking substrates or modules (i.e., replacing bad dies) after they have been assembled and final testing reveals that they are nonfunctional.

3.5.2 Chip Thinning

As noted in Sec. 3.4.5, one approach to increasing the functional density of an electronic system is to stack ICs vertically. Reducing the thickness of the chips allows for more chips to be stacked within a package, thereby substantially increasing the functional density. In addition to the obvious increase in functionality per unit area of substrate (module), thinning chips also improves their thermal performance, results in a more mechanically reliable device, allows for flexibility so the chips can conform to a curved surface, and relieves stress induced on chips due to packaging.[34,38] A disadvantage of thinning chips is the resulting back surface effects in the form of stresses induced by the thinning process, the formation of microcracks, and the creation of crystal dislocations fairly deep into the chip. Additionally, thinned chips are more vulnerable to cracking during handling and attachment procedures. For single-chip packaging, the package can be made thinner, which results in increased functionality per volume of space. Such packages are already under development and are referred to as ultrathin or paper-thin packages.

The two primary methods of chip thinning are plasma etching and backside grinding/polishing. Other techniques being used are laser chemical, wet chemical, and focused ion beam etching. Thicknesses routinely achieved vary from 50 to 100 μm, although thicknesses as small as 15 to 25 μm (silicon is flexible in this thickness range) have been demonstrated.[39] Figure 3.13 is a photograph of a silicon wafer that has been thinned to approximately 50 μm using a plasma process that causes the bending of the wafer.

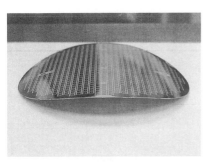

FIGURE 3.13 Silicon wafer that has been thinned to approximately 50 μm by plasma etching. Warpage was caused by the thinning process. *(Photograph provided courtesy of the Materials and Manufacturing Research Laboratories, The University of Arkansas.)*

3.5.3 Chip Attach

Packaged ICs required a large amount of real estate because of the requirement for a lead-frame that is used to provide electrical connections from the IC bond pads to the module (or substrate). Bare die attachment eliminates the leadframe, thereby saving valuable module space and reducing interconnect distance, which enhances system performance. Direct die attachment can be accomplished by conventional die attach/wire bonding, tape automated bonding (TAB), and flip-chip processes.

3.5.3.1 Wire Bond.   Conventional die attach and wire-bonding are the oldest and the best-understood processes for attaching dies and providing electrical interconnection to the rest of the circuit. Wire-bonding can be classified as ultrasonic, thermocompression, or thermosonic, depending on whether the bonding process involves ultrasonic energy, heat, pressure, or some combination of the three. Conventional wire-bonding is depicted in Fig. 3.14, along with TAB and flip-chip bonding for comparison purposes. Aluminum, gold, and copper wire is used to create wedge or ball bonds, depending on the metal and the bonding process used. This process has the advantage of being adaptable to almost any size die and bond pad pattern via computer control. However, there is an upper limit on the number of bond pads per chip that can be wire-bonded. Furthermore, wire-bonds introduce a significant amount of parasitic resistance and inductance, making the process less attractive for extremely high-performance systems. Some properties of the three metals used for bonding wire are given in Table 3.4.

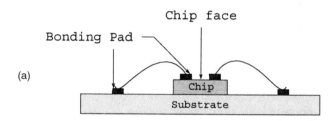

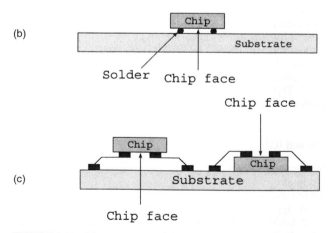

FIGURE 3.14   (*a*) Wire bonding; (*b*) flip-chip bonding; (*c*) TAB.

3.5.3.2 *Tape Automated Bonding (TAB).* Tape automated bonding (TAB) is a method of achieving die attach/electrical connection in a single bonding operation. ICs are usually mounted onto a polymeric film (referred to as the TAB tape) that contains etched metal (usually electroplated copper) patterns that physically connect the IC to the module and electrically connect the bond pads on the IC to the module using thermocompression bonding. Attachment of the IC to the copper interconnect pattern on the TAB tape is accomplished by solder bumps either on the IC or on the tape. Attachment of the IC to the substrate (module) is accomplished using the same technology.

TAB is rather expensive because it requires a custom tape for each different IC and its construction requires a significant amount of processing. Furthermore, the process is equipment intensive because of testing of the IC while it is mounted on the tape, excising the IC from the tape, and shaping (forming) its leads prior to mounting it onto a substrate, requiring a custom die for gang bonding. However, it is attractive because it provides the opportunity to perform burn-in and die testing prior to final assembly. TAB is illustrated in Fig. 3.14.

3.5.3.3 *Flip-Chip Bonding.* Another bare die assembly process is the flip-chip technique, in which the IC is attached face down directly onto either a substrate or a carrier, usually employing either solder bumps or z-axis conductive polymers. Another popular mode of attaching silicon dies in the flip-chip configuration is called stud-bumping, where a gold wire is used in a similar fashion as in wire-bonding, but the wire is truncated after establishing contact to the die pads. The resultant stud is then flattened using a *coining* process. The original flip-chip assembly, pioneered by IBM, eliminates all the packaging area that is not occupied by the die itself, generally resulting in the smallest footprint and the lowest profile. However, wire-bonded devices can be back ground to reduce the overall package thickness. It is not trivial to perform a similar back-grinding operation on flip-chip devices for two reasons. First, the bumping process in its present form cannot accept back-ground wafers due to breakage concerns, and second, it is extremely difficult to back-grind bumped wafers.

In general, flip-chips can use either grid arrays or peripheral pads for I/O connection at pitches between 0.1 and 0.65 mm, but most are used at an area array pitch of 0.2 to 0.3 mm. Of all the techniques available, flip chip assembly offers the highest packaging density and there is increased use of flip-chip attachment. Consequently, many bumping contractors also offer redistribution services. Although redistribution is done on the chip—actually on the wafer—it is considered a packaging operation. Unitive Electronics (Research Triangle Park, North Carolina) and Pac Tech (Nauen, Germany) offer redistribution with benzocyclobutene (BCB) as the dielectric, while Flip Chip Technologies (FCT, Phoenix, Arizona) uses polyimide as its dielectric material. Many other companies, such as MicroFab Technology of Singapore, Advanced Interconnect Technology of Hong Kong, ChipBond Technology of Taiwan, PacTecH of Nauen, Germany, Elcoteq of Helsinki, Finland, etc., offer merchant bumping and redistribution services.[40] With the shortest interconnection length, flip-chips have the lowest interconnect inductance and resistance, resulting in the highest performance among all packaging technologies.[41] Typical design rules for and photographs of solder bumps for flip-chip bonding are shown in Fig. 3.15.

3.5.4 Chip-on-Board (COB)

In chip-on-board (COB) assembly, the back of a bare (unpackaged) IC is attached directly onto a printed wiring board (PWB), wire-bonded, and then encapsulated with a polymer. The die bond pad pitch is generally around 0.25 mm and IC placement must be very accurate. For many applications requiring miniaturization, and especially those where space is limited, COB assembly can be the most cost-effective packaging option. It is a mature technology and offers high packaging density, low packaging cost, and fast signal speed because the dies are

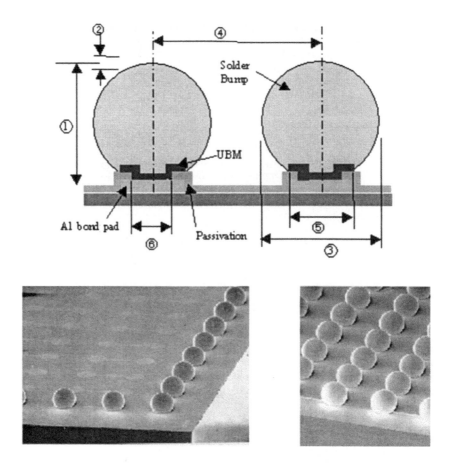

100-μm solder bump on 250-μm pitch

Typical Design Rules for Plated Solder Bump (200mm Wafer)	
Bump height (average) (1)	100 μm
Bump height allowance (2)	±12% (wafer), ±5% (die)
Bump size (3)	150 μm
Bump pitch (4)	150 μm
UBM size (5)	Passivation opening + 15 μm
Passivation opening (6)	Bond pad width 15 μm
Invalid area	5 mm from edge of wafer
Wafer thickness	600 μm
Solder selection	37Pb/63Sn, 95Pb/5Sn

FIGURE 3.15 Photographs of and typical design rules for plated solder bumps for flip-chip attachment. *(Photographs provided courtesy of MicroFAB, Singapore.)*

FIGURE 3.16 Chip-on-board (COB).

wire bonded directly onto a board.[41] Figure 3.16 shows details of a COB mounted power diode. Usually, as a final step, a glob top is deposited onto the chip to serve as passivation and protection.

3.5.5 Passive Devices

Since the invention of the IC, almost continuous progress has been made in creating a given functionality in a smaller area. Unfortunately, surface-mount (SM), discrete passive (DP) component size, which has seen continuing size reductions from 0805 components (80×50 mils or 2×1.25 mm) to 0201 components (20×10 mils or 0.5×0.25 mm) with preliminary work being done on 0105 components (10×5 mils or 0.25×0.125 mm), has not shrunk accordingly. Furthermore, increasing numbers of DPs are required in newer electronic systems. Consequently, DPs can occupy significant substrate area.

3.5.5.1 *Discrete Passives (DPs).* In fact, system substrates are now dominated by individually placed DP components. The key challenges presented by DPs are their fairly constant physical size, their increasing numbers, the cost of individual handling and attachment, the parasitic effects of their associated electrical connections, the reliability of solder joints, and the wide range of required values.[3,42,43] Thus, they continue to be a major hurdle to the miniaturization of electronic systems, particularly for the consumer market.

3.5.5.2 *Integrated Passives (IPs).* Much attention has been directed toward the use of integrated passive (IP) devices (also referred to as embedded or integral passive devices). IPs are located within the substrate and are formed during the fabrication of the substrate. Consequently, they require processing steps in addition to those that provide only interconnection. There are no external leads to be individually attached because all IPs are connected to metallizations within the substrate, whether it is laminate, ceramic, or any other material. They do not compete with ICs for surface space, but may complicate routing within inner substrate layers. They have no casing or packaging of their own to contribute weight and volume to the system. The lead length between the IP and an interconnect is usually only mils. Also, since the IP is typically a planar structure, there is much less parasitic inductance, capacitance, and resistance compared to DPs. Furthermore, IPs eliminate pick-and-place operations and two solder joints. A shortcoming of IPs is the fact that they cannot be tested individually, so the yield of the IPs will directly impact the yield of the substrates. The arguments for changing to IPs include reduced cost, smaller system size, better electrical performance, and improved system reliability. Figure 3.17 shows schematically the difference between discrete passives, located on the surface of a substrate, and integrated passives, which are located below the surface of a substrate.

3.5.5.3 *Resistors.* Thin-film passive components rely on deposition and photolithography to physically define conductive, resistive, and dielectric materials that produce a desired electrical response.[3] Integrated resistors are fabricated by depositing and patterning a layer of thin-film or thick-film resistive material in series with an interconnect line on an insulating substrate. Most IP resistor applications can be realized using only two ranges of film sheet resistance; 10 to 50 Ω/square and 500 to 1000 Ω/square. The lower range can be realized using any one of several materials, including TaN_x, CrSi, NiCr, and TiN_x. TaN_x is popular because its resistivity varies from 0.1 to 1 mΩ/cm, depending on deposition conditions and nitrogen stoi-

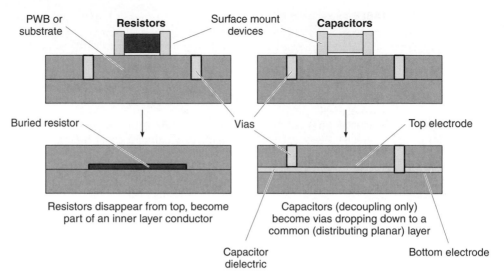

FIGURE 3.17   The physical transition from discrete passives (*top*) to integrated passives (*bottom*).

chiometry. The higher resistance range can be realized using cermet materials containing a small percentage ($\approx$0.5%) of a metal such as Ta, Cr, Ti, W, or Mo. By cosputtering one of these metals and SiO_x, films can be realized with resistivities of 10 to 40 mΩ/cm or a sheet resistivity of 1 to 4 kΩ/square. Table 3.8 gives some characteristics of typical tantalum nitride and nichrome resistors.

3.5.5.4 Integrated Capacitors.   The integration of capacitors into interconnect substrates is largely driven by three applications: decoupling of ICs from power supplies, analog functions such as those in RF/wireless, and termination of transmission lines. There is a wide variety of dielectric materials that can be used to build integrated capacitors. There are two broad classifications of dielectrics—paraelectrics and ferroelectrics—and they have very distinct electrical properties, the most significant being the dielectric constant. Some of the dielectric materials that can be used to fabricate integrated capacitors are listed in Table 3.9, along with their composition and dielectric constant. The physical size of capacitors should be as small as possible to maximize the number of components per unit substrate area, but if the specific

TABLE 3.8   Properties of Two Materials Used to Create Integrated Resistors

Resistor material → Characteristic ↓	Tantalum nitride	Nichrome
Sheet resistance	20–150 Ω/□ 100 Ω/□ (typical)	25–300 Ω/□ 100–200 Ω/□ (typical)
Sheet resistance tolerance	±10% of nominal value	±10% of nominal value
Temperature coefficient of resistance	−75 ± 50 ppm/°C (typ.) 0 ± 25 ppm/°C (anneal)	0 ± 50 ppm/°C 0 ± 25 ppm/°(anneal)
Resistance drift (1000 h at 150° in air)	<1000 ppm	<1000 ppm (anneal) <200 ppm (sputter/ann.)
Resistor tolerance after anneal and laser trim	±0.10% standard ±0.03% (bridge trim)	±0.10%

TABLE 3.9   Dielectric Material Candidates for Application in Integrated Capacitors

Material	Composition	Dielectric constant	Paraelectric (P) or ferroelectric (F)
Silicon monoxide	SiO	4.5–6.8	F
Silicon dioxide	SiO_2	4–5	F
Silicon nitride	Si_3N_4	6–7	F
Silicon carbide	SiC	20–45	F
Aluminum oxide	Al_2O_3	6–10	F
Aluminum nitride	AlN	8–10	F
Tantalum oxide (a)	Ta_2O_5	25	F
Tantalum oxide (h)	Ta_2O_5	50	F
Titanium oxide	TiO_2	10–100	F
Diamond-like-carbon	sp^2 and sp^3 carbon	4–6	F
BCB	Organic	2.7	F
Polyimide	Organic	3–4.5	F
Barium strontium titanate	$BaSrTiO_4$	up to 1000	P
Lead zirconate titanate	$PbZr_xTi_{1-x}O_3$	up to 2000	P
BPZT	$Ba_{0.8}Pb_{0.2}(Zr_{0.12}Ti_{0.88})O_3$	up to 3000	P
Barium titanate	$BaTiO_3$	up to 5000	P

capacitance of the films is too high, it may be difficult to fabricate a low-value capacitor in a size that is large enough to be easily or reproducibly defined by standard board-level photolithography.

3.5.5.5   Integrated Inductors.   Integrated inductors are the easiest of the IPs to fabricate because they are usually spirals of the conductor material. Also, since inductors are magnetic devices, they pose an integration problem not shared by resistors and capacitors; they perform best when there is a sufficient volume of space to allow their magnetic fields to be unimpeded by other structures. As a result, there is a "keep-away" distance required for inductors in order to avoid loss of inductance relative to an isolated structure and prevent interference with nearby signal lines and ground planes from the inductor's field.

The easiest way to fabricate integrated inductors is to simply form a spiral out of the interconnect material, typically having one to eight turns with a total outer diameter of about 0.2 mm, which will provide an inductance of around 1 to 40 mH.[3]

3.5.5.6   Thin-Film Integrated Passives.   Since thin-film IPs are not currently in widespread usage, it is difficult to find vendors that can supply the necessary materials, chemicals, and patterning supplies for large-scale production. However, as system integration pushes toward smaller, cheaper, and faster technologies, there is a high probability that IPs will soon become an important aspect of conventional packaging.

3.6   DRIVERS FOR ADVANCED PACKAGING

Electronic systems are complex assemblies of ICs, discrete and passive components, wiring boards, power supplies, and interfaces to people, information, and communication. In fact, they must interface with mechanical, chemical, biological, optical, and electromagnetic functions. Unfortunately, the rate of progress in size, weight, performance, and cost of systems lags that of ICs for several reasons, including the fact that design tools are far less comprehensive and integrated than those for ICs. Also, the functional density of systems has progressed slowly compared with that of ICs because of lagging progress in interconnect density, passive components, and other support hardware. A major part of the problem is that there is no industry-accepted standard for advanced packaging. Each company develops its own package.

3.6.1 Materials

It is clear that some of the materials (e.g., lead-free solder, laminate, polymers, etc.) and assembly processes (e.g., flip-chip, BGA, thin chip, etc.) being used in advanced packaging at the present time will continue to play a major role in the high-density, high-performance packaging of the future. However, advanced electronic packaging must continue to evolve with the objectives of providing the materials and processes that will allow electronic systems to operate at their absolute optimum.

3.6.2 Lead-Free

The choice of materials for advanced packaging applications depends on the application, as well as the cost-performance expectations from the device. A detailed discussion of materials is presented elsewhere in this book. However, advanced (high-performance) packaging can be impacted either positively or negatively by the materials that are available to the technology. For example, eutectic tin-lead (63 percent tin with 37 percent lead, also called 63Sn37Pb) is one of the most popular solder materials used in the assembly and packaging industry because it has a relatively low eutectic/melting point, it is readily available, and it is low in cost. Unfortunately, lead has been identified as a health hazard. Consequently, a worldwide effort, primarily led by the Japanese Ministry of International Trade and the European electronics community, has been focused on reducing, and eventually eliminating, the lead content in electronic assembly.[44,45] The eventual elimination of lead-based solders has far-reaching implications for the processing, assembly, reliability, and cost aspects of electronic packaging, as can be seen from Table 3.10.

3.6.3 Micro-Electro-Mechanical Systems and Micro-Opto-Electro-Mechanical Systems

Another challenge for the future is the packaging of micro-electro-mechanical systems (MEMS) and micro-opto-electro-mechanical (MOEMS).[46] Many MEMS devices must inter-

TABLE 3.10   Lead-Free Candidate Solder Alloys

Property → Alloy ↓	Melting range (°C)	Metal cost (per lb.)*	Density at 25°C (lbs/in³)	Metal cost (per in³)	Patented alloy?
63Sn/37Pb (standard)	183	$ 2.67	0.318	$0.85	No
42Sn/58Bi	138	$ 3.54	0.316	$1.12	No
77.2Sn/20In/2.8Ag	179–189	$23.47	0.267	$6.27	Yes
85Sn/10Bi/5Zn	168–190	$ 3.70	0.273	$1.01	No
91Sn/9Zn	199	$ 3.63	0.263	$0.95	No
95Sn/5Sb	232–240	$ 3.80	0.263	$1.00	No
90Sn/7.5Bi/2Ag/0.5Cu	186–212	$ 5.19	0.273	$1.42	No
96.5Sn/3.5Ag	221	$ 6.24	0.368	$1.67	No
98Sn/2Ag	221–226	$ 5.25	0.266	$1.40	No
99.3Sn/0.7Cu	227	$ 3.92	0.264	$1.03	No

* Metal cost only—does not include fabrication costs, margins, etc.

face with the environment in order to perform their intended function, and the package must be able to facilitate access to the environment while protecting the device. Furthermore, the package cannot interfere with the proper operation of the device. Because many MEMS devices are application specific, they require custom packaging. In fact, many need media-compatible packaging that protects the devices from the environment in which they are intended to operate. Thus, packaging of MEMS devices is as critical to their performance and reliability as their design and fabrication.[47]

3.6.4 Technology Developments and Trends

Figure 3.18 shows a packaging concept proposed by researchers of the High Density Electronics Center (HiDEC) at the University of Arkansas, Fayetteville, Arkansas, to realize a 50- to 100-fold improvement in volumetric efficiency of the packaging of parts per cubic inch, in what may be the next revolution in packaging. All ICs are thinned and flip-chip attached in, not just on, a build-up of flex layers with copper conductors. The flex layers contain thin-film passive devices. Integrated heat pipes are added where necessary in the glue layers. Thin-film batteries would also be included in the structure. Designing a system-in-a-package such as the one described here will require more sophisticated design tools than are available at the present time.[9] Another pressing challenge is in high-density interconnection. Flip-chip pitches are hovering around the 200- to 250-μm area. Filling vias with diameters below 100 μm also will be a challenge.[29]

In order to realize the system packaging approach of Figure 3.18 in the next few years, work on many technologies must be accelerated. For example, optics have some inherent advantages over their electronic counterparts, such as a very high bandwidth, excellent noise reduction, minimal cross talk, and high operating frequencies that far exceed the current capabilities of electronic devices. However, the assembly operation for optoelectronic devices requires an extremely high degree of precision. This requirement translates into the need for more expensive placement and assembly equipment. Further, high-accuracy placement equipment generally has a lower throughput, which further contributes to the increased cost of assembly and thus to the overall cost. Thus, effort must be directed at reducing the complexity and cost of incorporating optoelectronic technologies into systems.

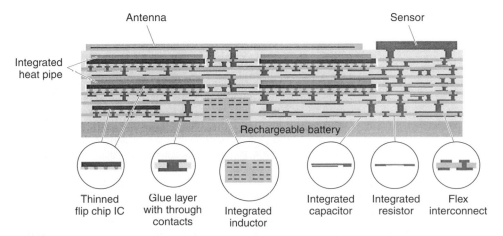

FIGURE 3.18 Conceptual mockup of high-density, high-performance electronic system packaging of the future.

REFERENCES

1. William D. Brown (ed.), *Advanced Electronic Packaging,* chap. 1, IEEE, Piscataway, NJ, 1999.

2. Terrence Thompson, "Top 10+ Critical Issues for HDI Manufacturing," *HDI,* August 1999, pp. 34–43.

3. R. K. Ulrich, W. D. Brown, S. S. Ang, F. D. Barlow, A. Elshabini, T. G. Lenihan, H. A. Naseem, D. M. Nelms, J. Parkerson, L. W. Schaper, and G. Morcan, "Getting Aggressive with Passive Devices," *Circuits and Devices,* vol. 16, no. 5, 2000, pp. 17–25.

4. John Baliga, "High-Density Packaging: The Next Interconnect Challenge," *Semiconductor International,* February 2000, pp. 91–100.

5. *International Technology Roadmap for Semiconductors,* 1999 Edition.

6. J. Fjelstad and T. DiStefano, "Where Are We Headed?" *Electronic Packaging and Production,* December 1999, pp. 16–20.

7. K. U. Maner, E. V. Porter, L. W. Schaper, S. S. Ang, and W. D. Brown, "The Seamless High Off-Chip Connectivity (SHOCC) Packaging Technology," *Electrochemical Society Proceedings,* vol. 99-7, 1999, pp. 188–194.

8. William F. Shutler, "Examining Technology Options for System on a Package," *Electronic Packaging and Production,* September 2000, pp. 32–36.

9. John Baliga, "Packaging Provides Viable Alternatives to SOC," *Semiconductor International,* July 2000, pp. 169–181.

10. James B. Brinton, "SOC Isn't Cutting It Yet. Is Multi-Chip Package a Better Answer Today?" *Semiconductor Business News,* January 28, 2000.

11. Jack Woida, "The Challenge of System on a Chip," *Electronic Packaging and Production,* May 2000, pp. 18–26.

12. E. Davidson, "Large Chip vs. MCM for a High-Performance System," *IEEE Micro,* July/August 1998, pp. 33–41.

13. Kevin Rinebold, "Few-Chip Packaging—An MCM Renaissance," *HDI,* October 2000, pp. 18–21.

14. Thomas L. Landers, William D. Brown, Earnest W. Fant, Eric M. Malstrom, and Neil M. Schmitt, *Electronic Manufacturing Processes,* Prentice Hall, Upper Saddle River, NJ, 1994.

15. William F. Shutler, Alberto Parolo, Stefano Oggioni, and Claudio Dall' Ara, "Examining Technology Options for System on a Package," *Electronic Packaging and Production,* September, 2000, pp. 32–36.

16. Claudio Truzzi and Steve Lerner, "Broadening the Platforms for System-in-Package Solutions," *Solid-State Technology,* November 2000, pp. 115–120.

17. Analog Devices, Technology Background, Web site: www.analog.com/industry/dsp/bga/background.html.

18. Darrel R. Frear, "Materials Issues in Area-Array Microelectronic Packaging," *Journal of Materials,* vol. 51, no. 3, 1999, pp. 22–27.

19. Kelly Barrett, "A Rose and an MCM by Any Other Name . . . ," *Electronic News Online,* May 17, 1999, Web site: www.electronicnews.com/enews/Issue/1999/05171999/20mcmbl.asp.

20. Karen Carpenter, "IBM Packaging Technology Keeping Pace with Semiconductor Roadmap," *IBM Microelectronics MicroNews,* vol. 4, no. 4, 1998, Web site: www.chips.ibm.com/micronews/vol4_no4/packaging.html.

21. Larry Barnes, "Multichip Modules Increase Functionality in Small Places," *Analog Devices Products and Datasheets Magazines,* Web site: analog.com/magazines/Briefings/vol10no2/mcms.html.

22. K. K. H. Wong, S. Kaja, and P. W. DeHaven, "Metallization by Plating for High-Performance Multichip Modules," *IBM Journal of Research and Development,* vol. 42, no. 5, 8 pages, Web site: www.research.ibm.com/journal/rd/425/wong.html.

23. G. Carchon, S. Brebels, K. Vaesen, P. Pieters, B. Nauwelaers, and E. Beyne, "Design of Microwave MCM-D CPW Quadrature Couplers and Power Dividers in X-, Ku- and Ka-band," *International Journal of Microcircuits and Electronic Packaging,* vol. 23, no. 3, 2000, pp. 257–264.

24. R. R. Tummala and E. J. Rymaszewski, *Microelectronic Packaging Handbook,* chap. 7, Van Nostrand, New York, 1989, pp. 501–511.

25. *Advanced Multichip Packaging Developments,* chap. 4, sec. 6, Web site: http://itri.loyola.edu/ep/c4s6.htm.

26. Sergio Afonso, Leonard W. Schaper, James P. Parkerson, William D. Brown, Simon S. Ang, and Hameed A. Naseem, "Modeling and Electrical Analysis of Seamless High Off-Chip Connectivity (SHOCC) Interconnects," *IEEE Transactions on Advanced Packaging,* vol. 22, no. 3, 1999, pp. 309–320.

27. E. Davidson, "Large Chip vs. MCM for a High Performance System," *IEEE Micro,* July/August 1998, pp. 33–41.

28. John Baliga, "Making Room for More Performance with Chip Scale Packages," *Semiconductor International,* vol. 21, no. 11, October 1998, pp. 85–92.

29. John Baliga, "Ball Grid Arrays: The High Pincount Workhorse," *Semiconductor International,* 6 pages, September 1999, Web site: www.semiconductor-intl.com/semiconductor/issues/issues/1999/sep99/docs/feature2.asp.

30. WMU VLSI, Part 5: *System Packaging Styles,* Web site: www.cs.wmich.edu/ vlsi2/Pages/what/part_5 .html.

31. Keith D. Gann, "Neo-Stacking Technology," *HDI,* December 1999, pp. 2022.

32. Daniel Chen, Steffen Hellmold, Mike Yee, and Kevin Kilbuck, "Stacked Multi-Chip CSP Standards—A Push Forward," *Advanced Packaging,* June/July 2000, pp. 49–53.

33. David Francis and Linda Jardine, "Stackable 3-D Chip-Scale Package Uses Silicon as the Substrate," *Chip Scale Review,* May/June 2000, p. 79.

34. S. Leseduarte, S. Marco, E. Beyne, R. Van Hoof, A. Marty, S. Pinel, O. Vendier, and A. Coello-Vera, "Residual Thermomechanical Stresses in Thinned-Chip Assemblies," *IEEE Transactions on Components and Packaging Technologies,* vol. 23, no. 4, 2000, pp. 673–679.

35. Jim Roton, "KGD: A State of the Art Report," *Advanced Packaging,* September 1999, pp. 50–56.

36. Craig Beddingfield, Walid Ballouli, Frank Carney, and Raj Nair, "Wafer Level KGD," *Advanced Packaging,* September 1999, pp. 26–30.

37. Michael A. Fry, Jerry D. Kline, John L. Prince, and Gary A. Tanel, "In-Line KGD Test Speeds Flip Chip Assembly," *Electronic Packaging and Production,* February 2001, pp. 36–41.

38. Dave Francis, "Thinning Wafers for Flip Chip Applications," *HDI,* May 1999, pp. 18–21.

39. Jedediah J. Young, Ajay P. Malshe, W. D. Brown, and Timothy Lenihan, "Modeling and Analysis of Very Thin Silicon Chips for Conformal Electronics Systems," *Proceedings of the International Conference and Exhibition on High-Density Interconnect and Systems Packaging,* Santa Clara, CA, April 18–20, 2001.

40. J. Baliga, "Contract Bumping is Increasing," *Semiconductor International,* March 1999, p. 42.

41. Nicholas Brathwaite and Kangsen Huey, "Advanced Electronic Packaging Techniques Enhance Function, Performance, and Portability," *Advanced Electronic Packaging,* 5 pages, 1997, Web site: www.devicelink.com/mem/archive/97/10/009.html.

42. L. Nappi, J. Rapelo, H. Pohjonen, P. Holloway, and E. Ristolainen, "Integrating Passive Components with Thin-Film Deposition," *HDI,* July 2000, pp. 38–48.

43. Elizabeth A. Logan and James L. Young, "Integration Ends Passives' Domination of Wireless Circuits," *Electronic Packaging and Production,* August 2000, pp. 26–30.

44. Zhenwei Hou, R. Wayne Johnson, Erin Yaeger, Mark Konarski, and Larry Crane, "Lead-Free Solder for Flip Chip," *HDI,* September 2000, pp. 38–47.

45. Susan Crum, "The Green Revolution: The Search for Green Products Goes Beyond Lead-Free," *Electronic Packaging and Production,* April 2000, pp. 20–24.

46. Ajay P. Malshe, Chad O'Neal, Sushila B. Singh, William D. Brown, William P. Eaton, and William M. Miller, "Challenges in the Packaging of MEMS," *The International Journal of Microcircuits and Electronic Packaging,* vol. 22, no. 3, 1999, pp. 233–241.

47. Chad B. O'Neal, Ajay P. Malshe, William F. Schmidt, Matthew H. Gordon, Robert R. Reynolds, William D. Brown, William P. Eaton, and William M. Miller, "A Study of the Effects of Packaging Induced Stress on the Reliability of the Sandia MEMS Microengine," *Proceedings of the Pacific Rim/ASME International Electronic Packaging Technical Conference and Exhibition,* Kauai, HI, July 8–13, 2001.

CHAPTER 4
TYPES OF PRINTED WIRING BOARDS

Dr. Hayao Nakahara
N.T. Information Ltd., Huntington, New York

4.1 INTRODUCTION

Since the invention of printed wiring technology by Dr. Paul Eisner in 1936 several methods and processes have been developed for manufacturing printed wiring board (PWBs) of various types. Most of these have not changed significantly over the years; however, some specific trends continue to exert major influences on the types of PWBs required and the processes that create them:

1. Computers and portable telecommunications equipment require higher-frequency circuits, boards, and materials, and also use more functional components that generate considerable amounts of heat that need to be extracted.
2. Consumer products have incorporated digital products into their design, requiring more functionality at ever-lower total cost.
3. Products for all uses continue to get smaller and more functional, driving the total circuit package itself to become more dense, causing the PWBs to evolve to meet these needs.

These trends have led to the larger use of nonorganic base substrates, such as aluminum and soft iron. In addition, alternate ways to create boards have been developed. These will be discussed in this chapter, along with the traditional board structures and processes. The terms *printed wiring board, PWB,* and *board* will be used synonymously. Also, the words *laminate, substrate,* and *panel* will be used interchangeably.

4.2 CLASSIFICATION OF PRINTED WIRING BOARDS

PWBs may be classified in many different ways according to their various attributes. One fundamental structure common to all of them is that they must provide electrical conductor paths which interconnect components to be mounted on them.

4.2.1 Basic PWB Classifications

There are two basic ways to form these conductors:

1. *Subtractive:* In the subtractive process, the unwanted portion of the copper foil on the base substrate is etched away, leaving the desired conductor pattern in place.
2. *Additive:* In the additive process, formation of the conductor pattern is accomplished by adding copper to a bare (no copper foil) substrate in the pattern and places desired. This can be done by plating copper, screening conductive paste, or laying down insulating wire onto the substrate on the predetermined conductor paths.

The PWB classifications given in Fig. 4.1 take into consideration all these factors, i.e., fabrication processes as well as substrate material. The use of this figure is as follows:

- Column 1 shows the classification of PWBs by the nature of their substrate.
- Column 2 shows the classification of PWBs by the way the conductor pattern is imaged.
- Column 3 shows the classification of PWBs by their physical nature.
- Column 4 shows the classification of PWBs by the method of actual conductor formation.

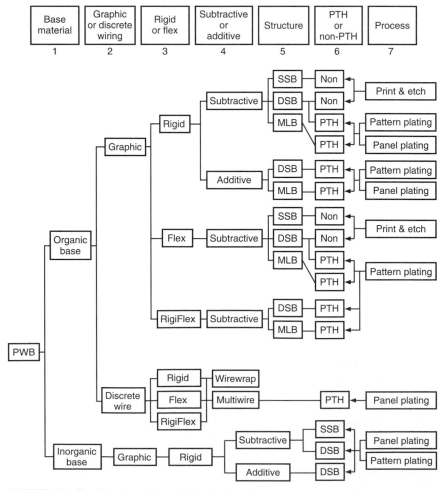

FIGURE 4.1 Classification of printed wiringing boards.

- Column 5 shows the classification of PWBs by the number of conductor layers.
- Column 6 shows the classification of PWBs by the existence or absence of plated-through-holes (PTHs).
- Column 7 shows the classification of PWBs by production method.

4.3 ORGANIC AND NONORGANIC SUBSTRATES

One of the major issues that has arisen with the ever-higher speed and functionality of components used in computers and telecommunications is the availability of materials for the PWB substrate that are compatible with these product and process needs. This includes the stresses on substrate material created by more and longer exposure to soldering temperatures during the assembly process, as well as the need to match the coefficient of thermal expansion for components and substrate. The resultant search has found new materials, both organic and nonorganic based. The details of these materials are explained in Chaps. 5 through 10, but this outlines the basic character of the two types of substrate.

4.3.1 Organic Substrates

Organic substrates consist of layers of paper impregnated with phenolic resin or layers of woven or nonwoven glass cloth impregnated with epoxy resin, polyimide, cyanate ester, BT resin, etc. The usage of these substrates depends on the physical characteristics required by the application of the PWB, such as operating temperature, frequency, or mechanical strength.

4.3.2 Nonorganic Substrates

Nonorganic substrates consist mainly of ceramic and metallic materials such as aluminum, soft iron, and copper-invar-copper. The usage of these substrates is usually dictated by the need of heat dissipation, except for the case of soft iron, which provides the flux path for flexible disk motor drives.

4.4 GRAPHICAL AND DISCRETE-WIRE BOARDS

Printed wiring boards may be classified into two basic categories, based on the way they are manufactured:

1. Graphical
2. Discrete-wire

4.4.1 Graphical Interconnection Board

A *graphical PWB* is the standard PWB and the type that is usually thought of when PWBs are discussed. In this case, the image of the master circuit pattern is formed photographically on a photosensitive material, such as treated glass plate or plastic film. The image is then transferred to the circuit board by screening or photoprinting the artwork generated from the master. Due to the speed and economy of making master artwork by laser plotters, this master can also be the working artwork.

Direct laser imaging of the resist on the PWB can also be used. In this case, the conductor image is made by the laser plotter, on the photoresistive material, which is laminated to the board, without going through the intermediate step of creating a phototool. This tends to be somewhat slower than using working artwork as the tool and is not generally applied to mass production. Work continues on faster resists, as well as exposure systems, and this method will undoubtedly continue to emerge.

4.4.2   Discrete-Wire Boards

Discrete-wire boards do not involve an imaging process for the formation of signal conductors. Rather, conductors are formed directly onto the wiring board with insulated copper wire. Wire-wrap® and Multiwire® are the best known discrete-wire interconnection technologies. Because of the allowance of wire crossings, a single layer of wiring can match multiple conductor layers in the graphically produced boards, thus offering very high wiring density. However, the wiring process is sequential in nature and the productivity of discrete-wiring technology is not suitable for mass production. Despite this weakness, discrete-wiring boards are in use for some very high density packaging applications. See Fig. 4.2 for an example of a discrete-wiring board.

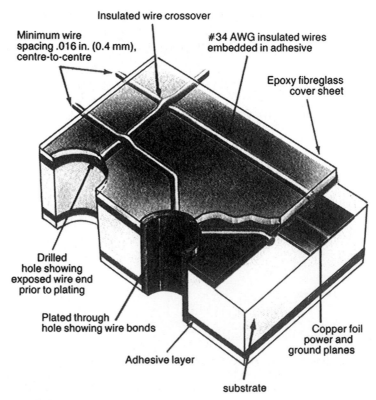

FIGURE 4.2   Example of discrete-wiring board.

4.5 RIGID AND FLEXIBLE BOARDS

Another class of boards is made up of the *rigid* and *flexible* PWBs. Whereas boards are made of a variety of materials, flexible boards generally are made of polyester and polyimide bases. *Rigi-flex boards,* a combination of rigid and flexible boards usually bonded together, have gained wide use in electronic packaging (see Fig. 4.3). Most rigi-flex boards are three-

FIGURE 4.3   Rigi-flex printed wiring board.

dimensional structures that have flexible parts connecting the rigid boards, which usually support components; this packaging is thus volumetrically efficient.

4.6 GRAPHICALLY PRODUCED BOARDS

The majority of boards produced in the world are graphically produced. There are three alternative types:

1. Single-sided boards
2. Double-sided boards
3. Multilayer boards

4.6.1 Single-Sided Boards (SSBs)

Single-sided boards (SSBs) have circuitry on only one side of the board and are often referred to as print-and-etch boards because the etch resist is usually printed on by screen-printing techniques and the conductor pattern is them formed by chemically etching the exposed, and unwanted, copper foil.

4.6.1.1 Typical Single-Sided Board Materials. This method of board fabrication is generally used for low-cost, high-volume, and relatively low functionality boards. In the Far East, for example, the majority of SSBs are made of paper-based substrates for lowest cost, with the most popular grade of paper-based laminate being XPC-FR, which is a flame-retardant phenolic material that is also highly punchable. In Europe, FR-2 grade paper laminate is the most popular substrate for SSBs because it emits less odor than XPC-FR when placed in high-voltage, high-temperature environments, such as inside a television set chassis. In the United States, CEM-1 material, which is a composite of paper and glass impregnated with epoxy resin, is the most popular substrate for SSBs. While not as low cost as XPC-FR or FR-2, CEM-1 has gained popularity because of its mechanical strength and also because of the relative unavailability of paper phenolic laminates.

4.6.1.2 Single-Sided Board Fabrication Process. Given the emphasis on cost and low complexity, SSBs are generally produced in highly automated, conveyorized print-and-etch lines, using the following basic process flow.

Step 1: Cut substrate into appropriate panel size by either sawing or shearing.

Step 2: Place panel in loader which feeds them into the line.

Step 3: Clean panels.

Step 4: Screen panel with ultraviolet curable etch-resist ink.

Step 5: Cure the etch-resist ink.

Step 6: Etch exposed copper.

Step 7: Strip the resist.

Step 8: Apply solder resist.

Step 9: Screen legend.

Step 10: Form holes by drilling or punching.

Step 11: Test for shorts and opens.

The conveyor speed of automated print-and-etch lines ranges from 30 to 45 ft/min. Some lines are equipped with an on-line optical inspection which enables the elimination of the final electrical open/short test.

As previously noted, after the conductor pattern is generated in the print-and-etch line, holes for component insertion are formed on the panel by punching when the panel is made of paper-based substrate, but must be formed by drilling when the panel is made of glass-based substrate.

4.6.1.3 Process Variations.   In some variations, the conductor surface of the PWB gets insulated, exposing only pads, and then conductive paste is screened to form additional conductors on the same side of the board, thus forming double conductive layers on a single side.

Most metal-core PWB consumer applications are made of aluminum substrate, which comes as a copper-clad material. PWBs made of such material do not have through-holes, and components are usually surface-mount types. These circuits are frequently formed into three-dimensional shapes.

4.6.2 Double-Sided Boards

By definition, double-sided boards (DBs) have circuitry on both sides of the boards. They can be classified into two categories:

1. Without through-hole metallization
2. With through-hole metallization

The category of through-hole metallization can be further broken into two types:

1. Plated-through-hole (PTH)
2. Silver-through-hole (STH)

4.6.2.1 Plated-Through-Hole Technology.   PTH technology is discussed in some detail in Sec. 4.7; however, some comments are appropriate here.

Metallization of holes by copper plating has been practiced since the mid-1950s. Since PWB substrate is an insulating material, and therefore nonconductive, holes must be metallized first before subsequent copper plating can take place. The usual metallization procedure is to catalyze the holes with palladium catalyst followed by electroless copper plating. Then, thicker plating is done by galvanic plating. Alternately, electroless plating can be used to plate all the way to the desired thickness, which is called *additive plating*.

The biggest change in the manufacturing process of double-sided PTH boards, and also of multilayer boards (MLBs), is the use of *direct metallization* technologies. (See Chap. 30 for full discussion of electroless and direct metallization for through-hole boards.) Here, simply, it eliminates the electroless copper process. The hole wall is made conductive by palladium catalyst, carbon, or polymer conductive film, then copper is deposited by galvanic plating. The elimination of electroless copper, in turn, allows the elimination of environmentally hazardous chemicals, such as formaldehyde, and EDTA, which are two main components of electroless copper-plating solutions.

4.6.2.2 Silver-Through-Hole Technology.   STH boards are usually made of paper phenolic materials or composite epoxy paper and glass materials, such as CE-1 or CE-3. After double-sided copper-clad materials are etched to form conductor patterns on both sides of the panel, holes are formed by drilling. Then the panel is screened with silver-filled conductive paste. Instead of silver, copper paste can also be used.

Since STHs have a relatively high electrical resistance compared with PTHs, the application of STH boards is limited. However, because of their economic advantage (the cost of

STH boards is usually one-half to two-thirds that of functionally equivalent PTH boards), their application has spread to high-volume, low-cost products such as audio equipment, floppy disk controllers, car radios, remote controls, etc.

4.6.3 Multilayer Boards (MLBs)

By definition, MLBs have three or more circuit layers (see Fig. 4.4). Main applications of MLBs used to be confined to sophisticated industrial electronic products. Now, however, they are the mainstream of all electronic devices, including consumer products such as portable video cameras, cellular phones and audio discs.

4.6.2.3 Layer Count. As personal computers and workstations become more powerful, mainframe computers and supercomputers are being replaced in many applications by these smaller machines. As a result, the use of highly sophisticated MLBs, which have layer counts over 70, are being reduced, but the technology to produce them is proven. At the other end of the layer-count spectrum, thin and high-density MLBs with layer counts between 4 and 8 are mainstream. The drive towards thinner MLBs will continue and is made possible by the continuing concurrent advancement of materials and equipment to handle thin core materials.

4.6.2.4 Via and Via Production Technologies. As PWBs have had to address the issues of higher speed, higher density, and the rise of surface-mount components that use both sides, the need to communicate between layers has increased dramatically. At the same time, the space available for vias has decreased, causing a continuing trend towards smaller holes, more holes on the board, and the decline of the use of holes that penetrate the entire board, which use space on all layers. As a result, the use of buried and blind vias has become a standard part of multilayer board technology, driven by the need for this increased package density (Fig. 4.4).

One of the immediate issues that arise from these trends is the problems of drilling and the associated cost of this fabrication step. Printed wiring boards, which once were stacked three high on a drilling machine, must be drilled individually, and the number of holes per board has risen, to accommodate the need for vias. This has caused a major problem for fabricators, who find that a lack of drilling capacity is creating a big demand on funds for additional machines, while the cost of drilling continues to increase dramatically. Therefore, alternate methods for creating vias are being developed. These pressures will be ongoing, and therefore the process listed here, or some equivalent, will undoubtedly become more important as the drive to miniaturization continues and drilling individual holes becomes less and less practical.

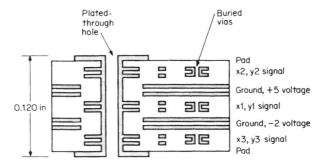

FIGURE 4.4 Cross-section multilayer board with buried via holes. Buried vias are built into each of the double-sided boards that make up the final multilayer structure.

These processes have been developed to mass-produce vias without drills.

Surface Laminar Circuits (SLCs).   The most notable MLB technology developed to form vias is the *sequential* fabrication of multilayers without press operations. This is particularly important for surface blind via holes.

The process for fabricating a board using surface laminar circuits is as follows (see Fig. 4.5):

1. Innerlayer ground and power distribution patterns are formed.
2. Panel receives an oxide treatment.
3. Insulating photosensitive resin is coated over the panel by curtain or screen-coating methods.
4. Holes are formed by photoexposure and development.
5. Panel is metallized by usual copper reduction process (consisting of catalyzing and electroless copper plating or by direct metallization process).
6. Thicker deposition of copper is made by continuation of electroless copper plating or galvanic plating.
7. Circuit patterns are formed by dry film tenting process.

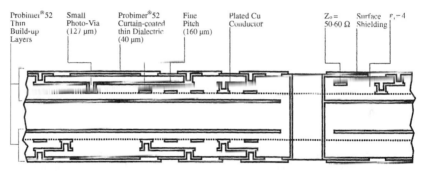

FIGURE 4.5 Example of surface laminar circuit (SLC) board cross section. (*Courtesy IBM Yasu and Ciba-Geigy Limited.*)

DYCOstrate®.   A different approach to small via creation has been taken by Dyconex AG of Switzerland. After ground and power patterns are formed on the panel, and the panel is oxide-treated, polyimide-backed copper foil is laminated on the panel. Holes in the copper are formed by a chemical etching process, and the insulating polyimide material underneath the holes is removed by plasma etching. PWBs made in such a way are called DYCOstrate. In other, similar technologies, different dielectric materials are used, and they are removed by alkaline solutions. The rest of the process is similar to that for SLC; that is, holes are metallized and a thick copper deposition is made by electroless or galvanic plating, and the circuit pattern is formed by a tent-and-etch process (see Fig. 4.5).

Drilled Vias.   In both SLC and DYCOstrate cases, through-holes can also be made by conventional drilling and plating processes, in addition to the surface blind via holes.

Cost Impact.   The manufacturing cost of these sequential technologies is not necessarily directly cheaper than conventional MLB technology, which depends on a laminating press operation. However, since the cost of making standard holes in a board can be as high as 30 percent of the total manufacturing cost, and the creation of holes in these processes is comparatively inexpensive, the overall cost for equivalent functionality can be less. In addition, the fine pattern capability for this process is excellent. For example, an eight-layer conventional structure can often be reduced to a four-layer structure, reducing the total cost for the same packaging density.

4.7 MOLDED INTERCONNECTION DEVICES

Three-dimensional circuit technology was of great interest in the early to mid-1980s. The proponents for this technology, however, realized the mistake of trying to make it directly competitive with conventional flat circuits and have developed a niche where the substrate also offers other functional uses, such as structural support for the product.

Manufacturers of three-dimensional circuits prefer to call them *molded interconnection devices* (MIDs). In many applications of MIDs, the number of components to interconnect the electronic and electrical components can be reduced, thus making the total assembly cost cheaper and the final structure more reliable.

4.8 PLATED-THROUGH-HOLE (PTH) TECHNOLOGIES

In 1953, the Motorola Corporation developed a PTH process called the *Placir* method,[1] in which the entire surface and hole walls of an unclad panel are sensitized with $SnCl_2$ and metallized by spraying on silver with a two-gun spray. Next, the panel is screened with a reverse conductor pattern, using a plating resist ink, leaving metallized conductor traces uncovered. The panel is then plated with copper by an electroplating method. Finally, the resist ink is stripped and the base silver removed to complete the PTH board. One problem associated with the use of silver is the migration caused by silver traces underneath the copper conductors.

The Placir method was the forerunner of the semiadditive process, which is discussed in Chap. 31.

In 1955, Fred Pearlstein[2] published a process involving electroless nickel plating for metallizing nonconductive materials. This catalyzer consists of two steps. First the panel is sensitized in $SnCl_2$ solution, and then it is activated in $PdCl_2$ solution. This process presented no problem for metallizing nonconductive materials.

At the same time that Pearlstein's paper was published, copper-clad laminates were starting to become popular. Manufacturers of PWBs applied this two-step catalyzing process to making PTHs using copper-clad laminates. This process, however, turned out to be incompatible with the copper surface. A myriad of black palladium particles called *smads* were generated between copper foil and electrolessly deposited copper, resulting in poor adhesion between the electroless copper and the copper foil. These smads and electroless copper had to be brushed off with strong abrasive action before the secondary electroplating process could begin. To overcome this smad problem, around 1960 researchers began attempting to develop better catalysts; the products of their research were the predecessors of modern palladium catalysts.[3]

The mid-1950s was a busy time in the area of electroless copper-plating solutions. Electrolessly deposited nickel is difficult to etch. But since it adheres somewhat better to the base than does electroless copper, research for the development of stable electroless copper-plating solutions was quite natural. Many patent applications for these solutions were filled in the mid-1950s. Among the applicants were P. B. Atkinson, Sam Wein, and a team of General Electric engineers, Luke, Cahill, and Agens. Atkinson won the case, and a patent[4] teaching the use of Cu-EDTA as a complexing agent was issued in January 1964 (the application had been filed in September 1956).

4.8.1 Subtractive and Additive Processes

Photocircuits Corporation was another company engaged throughout the 1950s in the development of chemicals for PTH processes. Copper-clad laminates were expensive, and a major portion of expensive copper foil had to be etched (*subtracted*) to form the desired conductor pattern. The engineers at Photocircuits, therefore, concerned themselves with plating (*adding*) copper conductors wherever necessary on unclad materials for the sake of economy. Their efforts paid off. They were successful in developing not only the essential chemicals for reliable PTH processes but also the fully additive PWB manufacturing technology known as the CC-4* process. This process is discussed in detail in Chap. 31.

With the use of $SnCl_2$-$PdCl_2$ catalysts and EDTA-base electroless copper-plating solutions, the modern PTH processes became firmly established in the 1960s. The process of metallizing hole walls with these chemicals for the subsequent formation of PTHs is commonly called the *copper reduction process*. In the subtractive method, which begins with copper-clad laminates, pattern plating and panel plating are the two most widely practiced methods of making PTH boards. These methods are discussed in the following subsections.

4.8.2 Pattern Plating

In the pattern-plating method, after the copper reduction process, plating resist layers of the reverse conductor image are formed on both sides of the panel by screening resist inks. In most fineline boards, photosensitive dry film is used instead. There are some minor variations in the pattern-plating method (see Fig. 4.6):

1. Catalyzing (preparing the nonconductive surface to cause copper to come out of solution onto that surface)
2. Thin electroless copper (0.00001 in) followed by primary copper electroplating; thick electroless copper (0.0001 in)
3. Imaging (application of a plating resist in the negative of the desired finished circuit)
4. Final electroplating copper
5. Solder plating (as etching resist) 0.0002 or 0.0006 in
6. Stripping plating resist
7. Etching of base copper
8. Solder etching (0.0002-in case); solder reflow (0.0006-in case)
9. Solder mask followed by hot-air solder coater leveler if solder etching is used
10. Final fabrication and inspection

Most manufacturers of DSBs with relatively wide conductors employ thick electroless copper plating. However, thin electroless copper followed by primary electroplating is preferred for boards having fine-line conductors, because a considerable amount of surface is brushed off for better adhesion of dry film. This provides a higher reliability for PTHs. Solder reflow boards had been preferred by many customers, particularly in military and telecommunications applications, until the emergence of hot-air solder coater levelers. Although the solder-over-copper conductors protect the copper from oxidization, solder reflow boards have some limitations. Solder mask is hard to apply over reflowed solder, and it tends to wrinkle and peel off in some areas when the boards go through component soldering. A more serious problem is the solder bridging that occurs when the conductor width and clearance become very small.

In step 9, the entire surface of the board except for the pads is covered by solder mask, and then the board is immersed into the hot-air solder coater leveler, resulting in a thin coating of

---

* CC-4 is a registered trademark of Kollmorgen Corporation.

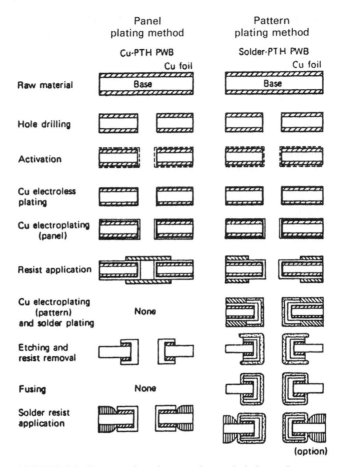

FIGURE 4.6 Key manufacturing steps in panel plating and pattern-plating methods.

solder over the pads and the hole walls. The operation sounds simple, but it requires constant fine-tuning and maintenance of the hot-air solder coater leveler; otherwise, some holes may become heavily clogged with solder and are then useless for component insertion.

One advantage of the pattern-plating method over the panel-plating method is in etching. The pattern-plating method needs to etch only the base copper. The use of ultrathin copper foil (UTC), which is usually ⅛ or ¼ oz thick, offers a real advantage. However, as long as electroplating is used, the pattern-plating method cannot escape from a current distribution problem, regardless of the thickness of the base foils. The panel-plating method by electroplating suffers from the same problem but to a lesser degree. Good current distribution is very difficult to achieve when the boards are not of the same size or type, and particularly if some have large ground planes on the outer faces being plated. When the board has a few holes in an isolated area remote from the bulk of the circuitry, these tend to become overplated, making component lead insertion difficult during assembly. To minimize this current distribution problem, various countermeasures are practiced, such as special anode position, anode masking, agitation, and plating thieves. But none of these offers a decisive solution to the distribution problem, and they are extremely difficult to implement flexibly and effectively in a large plating operation, where a large number of product mixes have to be handled all the time.

Another advantage of the pattern-plating method is its ability to form padless micro-via holes of a diameter ranging from 0.012 to 0.016 in. Micro-via holes enable better usage of conductor channels, thereby increasing the connective capacity of the board.

4.8.3 Panel Plating

In the panel-plating method, there are two variations for finishing the board after the panel is plated with electrolytic copper to the desired thickness. In the *hole-plugging* method, the holes are filled with alkaline-etchable ink to protect the hole walls from being etched; this is used in conjunction with screened etch resist. In the other method, called *tent-and-etch* or simply *tenting,* the copper in the hole is protected from etching by covering the hole or tenting with dry film, which is also used as an etch resist for conductors on the panel surface. The simplified sequence of the panel-plating method is as follows (see Fig. 4.6):

1. Catalyzing
2. Thin electroless copper deposition (0.0001 in)
3. Electroplating copper (0.001 to 0.0012 in)
4. Hole plugging with alkaline-resolvable ink; tenting (dry film lamination)
5. Screen-print etching resist (conductor pattern); photoexpose the panel for conductor pattern
6. Etching copper
7. Stripping etching resist
8. Solder mask
9. Solder coater leveler (optional)
10. Final fabrication and inspection

The panel-plating method is ideal for bare copper board. However, it is a difficult way to make padless via holes, which are becoming more popular. Generally, the conductor width of 0.004 in is considered to be the minimum realizable by this method for mass production.

Although the use of the panel-plating method in the United States and western Europe is limited, nearly 60 percent of the PTH boards in Japan are manufactured by this method.

4.8.4 Additive Plating

Plated-through holes can be formed by additive (electroless) copper deposition, of which there are three basic methods: fully additive, semiadditive, and partially additive. Of these, semiadditive involves pattern electroplating for PTHs with very thin surface copper, but the other two form PTHs solely by electroless copper deposition. The additive process has various advantages over the subtractive process in forming fineline conductors and PTHs of high aspect ratio. A detailed account of the additive process is given in Chap. 31.

4.9 SUMMARY

Modern electronic packaging has become very complex. Interconnections are pushed more into lower levels of packaging. The choice of which packaging technology to use is governed by many factors: cost, electrical requirements, thermal requirements, density requirements, and so on. Material also plays a very important role. All things considered, PWBs still play important roles in electronic packaging.

REFERENCES

1. Robert L. Swiggett, *Introduction to Printed Circuits,* John F. Rider Publisher, Inc., New York, 1956.
2. Private communication with John McCormack, PCK Technology, Division of Kollmorgen Corporation.
3. C. R. Shipley, Jr., U.S. Patent 3,011,920, Dec. 5, 1961.
4. R. J. Zebliski, U.S. Patent 3,672,938, June 27, 1972.

P · A · R · T · 2

MATERIALS

CHAPTER 5
INTRODUCTION TO BASE MATERIALS

Edward J. Kelley
Sanmina Corporation, Derry, New Hampshire

5.1 INTRODUCTION

The field of base materials can appear deceptively simple. In simple terms, base materials are made up of just three components: the resin system, the reinforcement, and the conductive foil. However, the variants in each of these components and the many possible combinations of these components make the discussion of base materials much more complex. One of the primary reasons for this complexity is the fact that printed circuits are used in so many differing applications. This results in many different sets of requirements in terms of cost and performance, and therefore many grades of base materials.

Also, because base materials are the most fundamental components of the printed circuit itself, they interact with virtually every other printed circuit manufacturing process. Therefore, not only are the physical and electrical properties of the materials critical, but their compatibility with manufacturing processes is also of great importance. Many of the key interactions between base materials and printed circuit manufacturing processes are discussed in subsequent chapters. This chapter discusses grades and specifications of base materials.

5.2 GRADES AND SPECIFICATIONS

The various types of base materials can be classified by the reinforcement type, the resin system used, and the glass transition temperature T_g of the resin system, as well as many other properties of the material.

5.2.1 NEMA and IPC Grades

Two commonly used classification schemes are:

- National Electrical Manufacturers Association (NEMA) grades
- IPC-4101 specification for base materials for rigid and multilayer boards

5.2.2   Glass Transition Temperature

Since it is also a common practice to classify base materials by their glass transition temperature, an understanding of this property is important before proceeding.

The T_g (pronounced "T sub g") of a resin system is the temperature at which the material
transforms from a relatively rigid or "glassy" state to a more deformable or softened state. A
material is not in a liquid state when it is above the T_g, as some discussions of T_g imply. It is a
temperature at which physical changes take place due to the weakening of molecular bonds
within the material. It is important to understand T_g because the properties of base materials
are different above the T_g than below T_g. While the T_g is typically described as being a very
precise temperature, this is somewhat misleading, because the physical properties of the
material can begin to change as the T_g is approached and some of the molecular bonds are
affected. As the temperature increases, more of the bonds become weakened until, for all
practical purposes, all relevant bonds are affected. This explains the curved line in Fig. 5.1,
which is discussed in the following text.

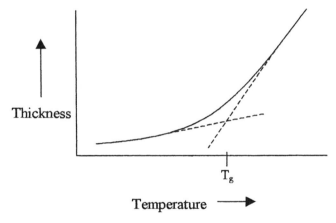

FIGURE 5.1   Glass transition temperature by TMA.

The T_g of a resin system has several important implications. These include:

- Thermal expansion
- A measure of the degree of cure of the resin system

5.2.2.1   Thermal Expansion.   All materials undergo changes in physical dimensions in
response to changes in temperature. The rate at which the material expands is much lower
below the T_g than above. Thermomechanical analysis (TMA) is a procedure used to measure
dimensional changes vs. temperature. Extrapolating the linear portions of the curve to the
point at which they intersect provides a measure of the T_g (see Fig. 5.1). The slopes of the linear portions of the curve above and below the T_g represent the respective rates of thermal
expansion, or as they are typically called, the coefficients of thermal expansion (CTEs). CTE
values are important because they influence the reliability of the finished circuit. Other things
being equal, less thermal expansion will result in greater circuit reliability as less stress is
applied to plated holes.

5.2.2.2   Degree of Cure.   The resin systems used in base materials begin with subcompo-
nents that contain reactive sites on their molecular structures. The application of heat to the
resin system and curing agents causes the reactive sites to *cross-link* or bond together. This

process of curing the resin system brings about physical changes in the material in proportion to the degree to which the cross-linking occurs, including increases in the T_g. When most of the reactive sites have cross-linked, the material is said to be fully cured and its ultimate physical properties are established.

Besides TMA, two other thermal analysis techniques are also commonly used for measuring T_g and degree of cure: differential scanning calorimetry (DSC) and dynamic mechanical analysis (DMA).

- DSC measures heat flow vs. temperature rather than the dimensional changes measured by TMA. The heat absorbed or given off also changes as the temperature increases through the T_g of the resin system. T_g measured by DSC is often somewhat higher than that measured by TMA.

- DMA measures the modulus of the material vs. changes in temperature.

Materials that are not fully cured can cause problems in circuit manufacturing processes and with finished circuit reliability. For example, an undercured multilayer circuit at the drilling process can result in excessive resin smear across the internal circuit connections in the hole formed. This happens because an undercured resin system, and the resulting lower than normal T_g, result in excessive softening of the resin when it is subjected to the heat generated during the drilling process. If this smeared resin is not completely removed, it can prevent electrical connection to other layers of circuitry when the hole is plated with copper. In addition, finished circuits in which the resin system is not fully cured exhibit greater amounts of z axis thermal expansion. This adversely affects circuit board reliability due to the increased stress on the plated through-holes as the circuit expands.

Because increased cross-linking requires greater amounts of energy (heat) to weaken the bonds in the molecular structure of the resin system, measurements of T_g can be used to determine the degree of cure of the resin system. For example, a measure of the degree of cure can be obtained by performing two thermal analyses, such as TMA, on the same sample. After the first analysis, the sample is subjected to a thermal cycle designed to promote any additional cross-linking available within the resin system, and then a second thermal analysis is performed. The degree of cure is measured by comparing the difference in measured T_g between the two analyses (see Fig. 5.2). If the material is fully cured, the difference between T_{g2} and T_{g1} will be very small, typically within a few degrees Celsius. Negative differences in these measured values or negative "delta T_g" values where T_{g1} is greater than T_{g2} are also indicative of full cure.

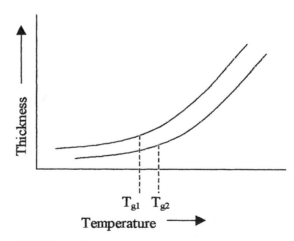

FIGURE 5.2   Measuring degree of cure by measuring T_g values.

5.2.2.3 Advantages and Disadvantages of High **T**$_g$ *Values.*

Implicit in many discussions of T_g is the assumption that higher values of T_g are always better. This is not always the case. While it is certainly true that higher values of T_g delay the onset of high rates of thermal expansion for a given resin system, total expansion can differ from material to material. A material with a lower T_g could exhibit less total expansion than a material with a higher T_g, due to differing resin CTE values or to the incorporation into the resin system of fillers that lower the CTE of the composite material. This is illustrated in Fig. 5.3. Material C exhibits a higher T_g than material A, but material C exhibits more total thermal expansion because its CTE value above T_g is much higher. On the other hand, with the same CTE values above and below the T_g, the higher-T_g material B exhibits less total thermal expansion than material A. Finally, although the T_g values are the same, material B exhibits less total expansion than material C due to a lower CTE value above T_g.

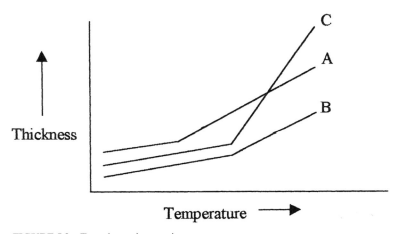

FIGURE 5.3 T_g vs. thermal expansion.

In addition, other material properties are also correlated with T_g. For example, higher-T_g resin systems are generally harder and more brittle than lower-T_g systems. This can adversely affect productivity in the printed circuit manufacturing process. In particular, the drilling process can be affected because slower drilling, reduced drill bit life, or reduced stack heights can be required for high-T_g materials.

Lower copper peel strength values and shorter times to delamination can also be correlated with higher T_g values, though other factors influence these properties as well. The time to delamination is a measure of the time it takes for the resin and copper, or resin and reinforcement, to separate or delaminate. While higher T_g does not always imply shorter times to delamination, it occurs frequently enough to be a concern. This test utilizes TMA equipment to bring a sample to a specified temperature and then measures the time it takes for failure to occur. Temperatures of 260°C (T260) and 288°C (T288) are commonly used for this testing.

In conclusion, T_g is just one property that must be considered when choosing a resin system. Balancing this property with others is important in achieving good circuit manufacturability and reliability.

5.3 NEMA INDUSTRIAL LAMINATING THERMOSETTING PRODUCTS

One of the first classifications of base materials for printed circuits (and other electrical components) was completed by the National Electrical Manufacturers Association (NEMA). NEMA's industrial laminating thermosetting products standard documents many of the materials used in printed circuits as well as specifications for some of their properties. Historical NEMA grades are outlined in Table 5.1 and cross-referenced with IPC-4101 in Table 5.2. Some of the commonly used materials have been FR-2, CEM-1, CEM-3, and, of course, FR-4.

TABLE 5.1   NEMA Base Material Grades

Grade	Resin	Reinforcement	Flame retardant
XXXPC	Phenolic	Cotton paper	No
FR-2	Phenolic	Cotton paper	Yes
FR-3	Epoxy	Cotton paper	Yes
FR-4	Epoxy	Woven glass	Yes
FR-5	Epoxy	Woven glass	Yes
FR-6	Polyester	Matte glass	Yes
G-10	Epoxy	Woven glass	No
CEM-1	Epoxy	Cotton paper/woven glass	Yes
CEM-2	Epoxy	Cotton paper/woven glass	No
CEM-3	Epoxy	Woven glass/matte glass	Yes
CEM-4	Epoxy	Woven glass/matte glass	No
CRM-5	Polyester	Woven glass/matte glass	Yes
CRM-6	Polyester	Woven glass/matte glass	No
CRM-7	Polyester	Matte glass/glass veil	Yes
CRM-8	Polyester	Matte glass/glass veil	No

FR-2 is composed of multiple plies of paper impregnated with a flame-resistant phenolic resin. It possesses good punching characteristics and is relatively low in cost. It has typically been used in simple applications like radios, calculators, toys, and television games where dimensional stability and high performance are not required. FR-3 is also paper-based, but uses an epoxy resin system.

CEM-1 uses a paper-based core with woven glass cloth on the surfaces, both impregnated with an epoxy resin. This enables the material to be punched while realizing improved electrical and physical properties. CEM-1 has been used in both consumer and industrial electronics.

CEM-3, a composite of dissimilar core materials, uses an epoxy resin–impregnated nonwoven fiberglass core with epoxy resin–impregnated woven fiberglass cloth surface sheets. It is higher in cost than CEM-1, but is more suitable for plated through-holes. CEM-3 has been used in home computers, automobiles, and home entertainment products.

FR-4, by far the most commonly used material for printed circuits, is constructed of woven fiberglass cloths impregnated with an epoxy resin or epoxy resin blend. The outstanding electrical, mechanical, and thermal properties of FR-4 have made it an excellent material for a wide range of applications including computers and peripherals, servers, telecommunications, aerospace, industrial controls, and automotive applications. FR-4 is discussed in more detail in the following text.

TABLE 5.2 IPC-4101 Base Material Summary

Spec. Sheet #	Reinforcement type	Resin system	ID reference*	T_g range
00	Cellulose paper	Phenolic	NEMA XPC	N/A
01	Cellulose paper	Modified phenolic	NEMA XXXPC	N/A
02	Cellulose paper	Phenolic, flame resistant	NEMA FR-1	N/A
03	Cellulose paper	Modified phenolic, flame resistant	NEMA FR-2	N/A
04	Cellulose paper	Modified epoxy, flame resistant	NEMA FR-3	N/A
10	Woven E-glass surface/ cellulose paper core	Modified epoxy, flame resistant	NEMA CEM-1	N/A
11	Woven E-glass face sheets/E-glass felt core	Polyester, flame resistant	NEMA CRM-5	≥80°C
12	Woven E-glass face/ nonwoven glass core	Epoxy, flame resistant	NEMA CEM-3	N/A
20	Woven E-glass fabric	Epoxy, non-flame resistant	NEMA G-10, MIL-S-13949/ 03—GEN	≥100°C
21	Woven E-glass fabric	Epoxy, flame resistant	NEMA FR-4, MIL-S-13949/ 04—GF/GFN/GFK	≥110°C
22	Woven E-glass fabric	Epoxy, hot strength retention, non-flame resistant	NEMA G-11—GB	135–175°C
23	Woven E-glass fabric	Epoxy, hot strength retention, flame resistant	NEMA FR-5, MIL-S-13949/ 05—GH	135–175°C
24	Woven E-glass fabric	Majority epoxy modified or unmodified, flame resistant	NEMA FR-4, MIL-S-13949/ 04—GF/GFG	150–200°C
25	Woven glass	Epoxy/PPO, flame resistant	NEMA FR-4, MIL-S-13949/ 04—GF/GFG	150–200°C
26	Woven glass	Majority epoxy modified or unmodified, flame resistant	NEMA FR-4, MIL-S-13949/ 04—GF/GFT	170–220°C
27	Unidirectional, cross-plied fiberglass	Majority epoxy, flame resistant	None	≥110°C
30	Woven E-glass	Triazine and/or bismaleimide modified epoxy, flame resistant	MIL-S-13949/26—GFT	170–220°C
40	Woven E-glass	Polyimide	MIL-S-13949/10—GI/GIN	≥200°C
41	Woven E-glass	Polyimide	MIL-S-13949/10—GI/GIL	≥250°C
42	Woven E-glass	Polyimide	MIL-S-13949/10—GI/GIJ	200–250°C
50	Woven aramid fabric	Modified epoxy	MIL-S-13949/15—AF	135–190°C
53	Nonwoven aramid	Polyimide	MIL-S-13949/31—BI	≥220°C
54	Unidirectional, cross-plied, aramid fiber	Cyanate ester resin	None	≥230°C
55	Nonwoven aramid	Modified epoxy	MIL-S-13949/22—BF	135–190°C
60	Woven quartz fabric	Polyimide	MIL-S-13949/19—QIL	≥250°C
70	Woven S-2 glass	Cyanate ester	MIL-S-13949/27	≥230°C
71	Woven E-glass	Cyanate ester	MIL-S-13949/29—GC	≥230°C
80	Woven E-glass surface/ cellulose paper core	Modified epoxy (catalyzed for additive process), flame resistant	NEMA CEM-1	N/A
81	Woven E-glass face/ nonwoven glass core	Epoxy resin (catalyzed for additive process), flame resistant	NEMA CEM-3	N/A
82	Woven E-glass	Epoxy resin (catalyzed for additive process), flame resistant	NEMA FR-4	110°C

* MIL-S-13949 is no longer an active standard but is indicated for reference.

5.4   IPC-4101 SPECIFICATION FOR BASE MATERIALS FOR RIGID AND MULTILAYER PRINTED BOARDS

The common standard that applies to base materials for printed circuits is IPC-4101. This standard presents a classification scheme and specification sheets for the various materials in use. Table 5.2 summarizes the various materials by specification sheet number. Each specification sheet in IPC-4101 includes property requirements for that particular material type.

IPC-4101 also provides a classification scheme for identification of laminate and prepreg materials; this is provided in the following sections.

5.5   FR-4 ISSUES

As can be seen from Table 5.2, there are several materials listed that are considered a NEMA FR-4 grade. This begs the question, What is FR-4? Directly from Table 5.2, it can be seen that each material considered NEMA grade FR-4 is flame resistant, uses a woven fiberglass reinforcement, and is primarily an epoxy-based resin system, with one type specifically designated as an epoxy-polyphenylene oxide (PPO) blend.

5.5.1   The Many Faces of FR-4

So what can be different about these materials? The most obvious difference between the commercially available FR-4 materials is in T_g values. Specification sheet 21 refers to an FR-4 with a minimum T_g of 110°C. More common are FR-4 materials with T_g values between 130 and 140°C and between 170 and 180°C. Therefore, it is obvious that FR-4 materials can consist of different types of epoxy resins. Indeed, there are many types of epoxy resins that provide differing T_g values as well as differences in other properties. In addition, as the epoxy-PPO blend makes obvious, the resin system can include resins other than epoxies. Finally, specification sheet 82 covers an FR-4 material catalyzed for an additive copper plating process. A catalyst used for this material is a kaolin clay filler coated with palladium. Obviously, materials with fillers can also be classified as FR-4.

5.5.2   The Longevity of FR-4

FR-4 materials have been the most successful, most commonly used materials in printed circuit manufacturing for many years. Why? Well, as we just discussed, FR-4 actually encompasses a range of material types, even though they share certain properties and are primarily epoxy based. The result is that there is typically an FR-4 material available for the most common end use applications. For relatively simple applications, the 130 to 140°C T_g FR-4s have become the material of choice. In higher-layer-count multilayer circuits and in very thick circuits, as well as in circuits requiring improved thermal properties, the 170 to 180°C T_g FR-4s have become the material of choice. In circuits requiring improved electrical properties, FR-4 materials with low dielectric constants and low loss properties are also available. In other words, as the range of end use applications has grown, so has the range of available FR-4 materials.

In addition, the components used in FR-4 materials, particularly woven fiberglass cloth and epoxy resins, provide a very good combination of performance, processability, and cost characteristics. The range of woven fiberglass cloth styles available makes it easy to control dielectric and overall circuit thicknesses. As noted earlier, the range of epoxy resin types also makes it relatively simple to tailor material properties to end use applications. The combination of good electrical, thermal, and mechanical properties of epoxies also explains why they have become the primary type of resin used in printed circuits. Epoxy-based materials are also relatively easy to process through conventional printed circuit manufacturing processes, at least in comparison to some of the other types of materials available. The processability of these materials has helped control the cost of FR-4-based printed circuits.

Last, the development of FR-4 materials took advantage of established manufacturing processes and materials. Weaving processes have been in place for many, many years. The fundamental process of weaving fiberglass yarns into glass cloth is not so different from weaving textile yarns into cloth fabrics. Using the same basic manufacturing technology avoids additional up-front research and development costs, but, more importantly, it avoids to some extent highly specialized capital assets and aids in achieving economies of scope for the suppliers of fiberglass cloth. The result is good control of the costs of these materials. Likewise, epoxy resins have been in use in a wide range of applications outside of printed circuits, resulting in a very large installed manufacturing base for these types of materials. This also results in good cost competitiveness for epoxy resins. In summary, the range of FR-4 types and their unique combination of performance, processability, and cost characteristics have made them the workhorse of the printed circuit industry.

5.6   LAMINATE IDENTIFICATION SCHEME

An example of a laminate designation in IPC-4101 is:

L	25	1500	C1/C1	A	A
Material designator	Specification sheet number	Nominal laminate thickness	Metal cladding type and nominal weight/thickness	Thickness tolerance class	Surface quality class

- *Material designator:* L refers to laminate.
- *Specification sheet number:* references the specification sheet number as in Table 5.2.
- *Nominal laminate thickness:* is identified by four digits. May be specified either over the cladding or over the dielectric. For metric specification, the first digit represents whole millimeters, the second represents tenths of millimeters, etc. For orders requiring English units, the four digits indicate the thickness in ten thousandths of an inch (tenths of mils). In the example shown, 1500 is designated for the English usage of 0590.
- *Metal cladding type and nominal weight/thickness:* five designators are used to specify cladding type and thickness. The first and fourth indicate the type, the second and fifth indicate thickness, and the third is a slash mark to differentiate sides of the base material. Table 5.3 lists the types of metal cladding. Table 5.4 lists copper foil weights and thicknesses.
- *Thickness tolerance class:* references the thickness tolerance as agreed between user and supplier (see Table 5.5). Classes A, B, and C refer to measurement by micrometer of the base material without cladding. Class D requires measurement by microsection (see Fig. 5.4). Classes K, L, and M refer to measurement by micrometer of the base material with the metal cladding.
- *Surface quality class:* identified by either A, B, C, D, or X as agreed upon between user and supplier. Indentations are located by 20/20 vision with the longest dimension measured with a reticle on a minimum 4× magnifier, with referee inspections at 10×. A point value system is used to determine point count for any 300 × 300-mm area (see Table 5.6). There should be no adherent material in an indentation nor exposure of base laminate. Table 5.7 shows the allowable point counts for each class.

TABLE 5.3   Metal Cladding Types from IPC-4101

Designation	Metal cladding type
A	Copper, wrought, rolled (IPC-MF-150, grade 5)
B	Copper, rolled (treated)
C	Copper, electrodeposited (IPC-4562, grade 1)
D	Copper, electrodeposited, double treat (IPC-4562, grade 1)
G	Copper, electrodeposited, high ductility (IPC-4562, grade 2)
H	Copper, electrodeposited, high-temperature elongation (IPC-4562, grade 3)
J	Copper, electrodeposited, annealed (IPC-4562, grade 4)
K	Copper, wrought, light colled rolled (IPC-4562, grade 6)
L	Copper, wrought, annealed (IPC-4562, grade 7)
M	Copper, wrought, rolled, low-temperature annealable (IPC-4562, grade 8)
P	Copper, electrodeposited, high-temperature elongation, double treat (IPC-4562, grade 3)
R	Copper, reverse-treated electrodeposited (IPC-4562, grade 1)
S	Copper, reverse-treated electrodeposited, high-temperature elongation (IPC-4562, grade 3)
T	Copper, copper foil parameters as dictated by contract or purchase order
U	Aluminum
Y	Copper Invar copper
N	Nickel
O	Unclad
X	Other, as agreed between user and supplier

TABLE 5.4   Foil Weights and Thicknesses from IPC-4562

Foil designator	Common industry terminology	Area weight (g/m^2)	Nominal thickness (μm)	Area weight (oz/ft^2)	Area weight (g/254 in^2)	Nominal thickness (mils)
E	5 μm	45.1	5.0	0.148	7.4	0.20
Q	9 μm	75.9	9.0	0.249	12.5	0.34
T	12 μm	106.8	12.0	0.350	17.5	0.47
H	½ oz	152.5	17.2	0.500	25.0	0.68
M	¾ oz	228.8	25.7	0.750	37.5	1.01
1	1 oz	305.0	34.3	1	50.0	1.35
2	2 oz	610.0	68.6	2	100.0	2.70
3	3 oz	915.0	103.0	3	150.0	4.05
4	4 oz	1220.0	137.0	4	200.0	5.40
5	5 oz	1525.0	172.0	5	250.0	6.75
6	6 oz	1830.0	206.0	6	300.0	8.10
7	7 oz	2135.0	240.0	7	350.0	9.45
10	10 oz	3050.0	343.0	10	500.0	13.50
14	14 oz	4270.0	480.0	14	700.0	18.90

TABLE 5.5 Base Laminate Thickness Tolerances from IPC-4101

Nominal thickness of laminate	Class A/K	Class B/L	Class C/M	Class D
0.025–0.119 mm	±0.025 mm	±0.018 mm	±0.013 mm	−0.013 + 0.025 mm
0.120–0.164 mm	±0.038 mm	±0.025 mm	±0.018 mm	−0.018 + 0.030 mm
0.165–0.299 mm	±0.050 mm	±0.038 mm	±0.025 mm	−0.025 + 0.038 mm
0.300–0.499 mm	±0.064 mm	±0.050 mm	±0.038 mm	−0.038 + 0.051 mm
0.500–0.785 mm	±0.075 mm	±0.064 mm	±0.050 mm	−0.051 + 0.064 mm
0.786–1.039 mm	±0.165 mm	±0.10 mm	±0.075 mm	Not applicable
1.040–1.674 mm	±0.190 mm	±0.13 mm	±0.075 mm	Not applicable
1.675–2.564 mm	±0.23 mm	±0.18 mm	±0.10 mm	Not applicable
2.565–3.579 mm	±0.30 mm	±0.23 mm	±0.14 mm	Not applicable
3.580–6.35 mm	±0.56 mm	±0.30 mm	±0.15 mm	Not applicable

TABLE 5.6 Pit and Dent Point Values from IPC-4101

Longest dimension	Point value
0.13–0.25 mm	1
0.26–0.50 mm	2
0.51–0.75 mm	4
0.76–1.00 mm	7
>1.00 mm	30

TABLE 5.7 Surface Quality Grades from IPC-4101

Surface quality grade	Requirement
Class A	The total point count shall be 29 maximum for any 300 mm × 300 mm area.
Class B	The total point count shall be 5 maximum for any 300 mm × 300 mm area. There shall be no foil indentations with a maximum dimension >0.38 mm.
Class C	The total point count shall be 17 maximum for any 300 mm × 300 mm area.
Class D	The total point count shall be 0 (zero) for any 300 mm × 300 mm area. Foil indentations >/= 125 microns shall not be acceptable. Resin spots shall be 0 (zero) as inspected with 20/20 vision. If Class D is specified, other quality related features are also required of this quality class per IPC-4562.
Class X	Requirements shall be agreed upon between user and supplier.

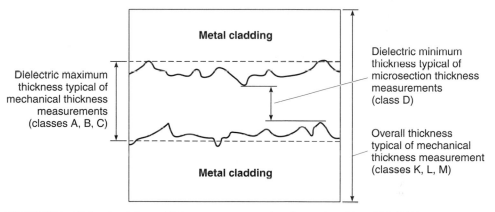

FIGURE 5.4 Dielectric minimum thickness measurements by microsection. *(From IPC-4101.)*

5.7   *PREPREG IDENTIFICATION SCHEME*

An example of a prepreg identification in IPC-4101 is:

P	25	E7628	TW	RE	VC
Material designator	Specification sheet number	Reinforcement style	Resin content method	Flow parameter method	Optional prepreg method

- *Material designator:* P refers to prepreg.
- *Specification sheet number:* references the specification sheet number as in Table 5.2.
- *Reinforcement style:* indicates reinforcement type and style of prepreg by five digits based on the chemical type and style. In the example, E refers to E-glass and 7628 is the glass fabric style.
- *Resin content method:* refers to the method used to specify resin content of a prepreg. The two methods are RC, which refers to percent resin content of the prepreg, and TW, which refers to the treated weight of the prepreg. One of the two methods must be designated.
- *Flow parameter method:* indicates how much the resin in the prepreg will flow under specified conditions—a critical property. The options are:
  1. MF (percent resin flow)
  2. SC (scaled flow)
  3. NF (no flow)
  4. RE (rheology flow)
  5. PC (percent cure)
  6. DH (delta H)
- *Optional prepreg method:* other test methods may also be designated. These include:
  1. VC (volatile content)
  2. DY (dicy inspection)

3. GT (gel time)
4. 00 (none specified)

The choice of test methods and their required nominal values and tolerances are normally agreed upon between user and supplier.

REFERENCES

1. NEMA, Industrial Laminating Thermosetting Products Standard, 1998.
2. IPC-4101, Specification for Base Materials for Rigid and Multilayer Printed Boards, 1997.
3. IPC-4562, Metal Foil for Printed Wiring Applications, 2000.

CHAPTER 6
BASE MATERIAL COMPONENTS

Edward J. Kelley
Sanmina Corporation, Derry, New Hampshire

6.1 INTRODUCTION

While there are many types of base materials, they all contain three components:

1. The resin system, including additives
2. The reinforcement(s)
3. The conductor type

Each of these components is important in its own right, and in combination they determine the properties of the base material as well as the relative cost of the material.

6.2 EPOXY RESIN SYSTEMS

The most successful and widely used resin systems for printed circuit applications are epoxy resin systems. There are many types of epoxy resins, and this class of resin system continues to be the workhorse among printed circuit board materials. This is a result of the combination of good mechanical, electrical, and physical properties and the relatively low cost of epoxies in comparison to the higher-performance resins. In addition, epoxy systems are relatively easy to process, which aids in keeping manufacturing costs down.

6.2.1 Definition of Epoxy

One of the most common versions of epoxy resin for printed circuit applications is manufactured from reacting epichlorohydrin and bisphenol A. This reaction is shown in Fig. 6.1. The bromination of the bisphenol A provides flame retardancy to the finished resin system (Fig. 6.2). The triangular rings on either end of the difunctional epoxy are the epoxide functional groups. In subsequent resin polymerization, these groups react and result in the curing of the resin system.

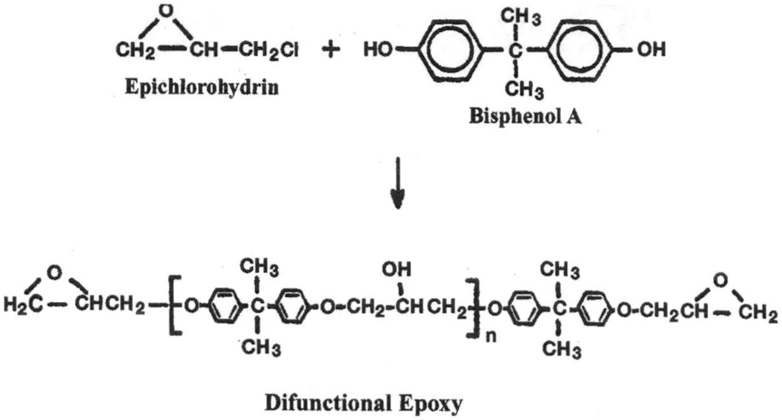

FIGURE 6.1   Reaction to form difunctional epoxy resin.

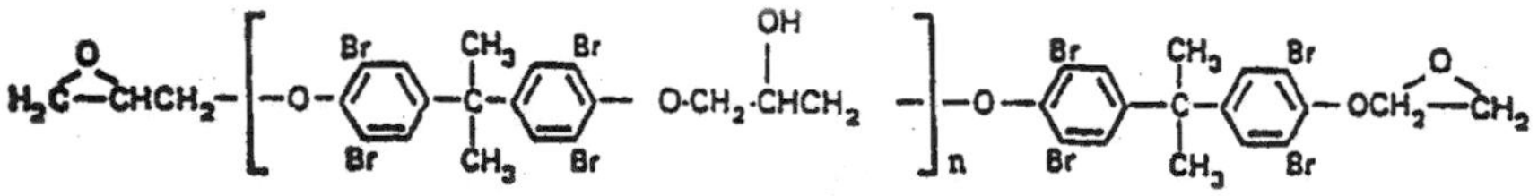

FIGURE 6.2   Brominated difunctional epoxy resin.

6.2.2   Difunctional Epoxies

The epoxy resins in Figs. 6.1 and 6.2 are difunctional epoxies. The molecular weight of the epoxy can be varied based on the number of repeating groups shown in the center of the molecule. On either end of the molecule are the epoxide functional groups. The name *difunctional epoxy* is derived from the fact that there are two epoxide groups, one on either end of the molecule. The molecular weight, the curing agents used to react the resin, and other factors result in the cured resin having a certain glass transition temperature T_g. The T_g is the temperature at which the resin turns from a rigid or glassy state to a softer, more deformable state. The T_g is important because it affects the thermal and physical properties of the base material and finished circuit board. In general, higher T_gs result in better thermal performance and reduced z-axis expansion.

Difunctional epoxies can have a range of T_gs, but values are typically below 120°C. These epoxies are typically used in relatively unsophisticated products such as simple double-sided printed circuits and are commonly blended with other epoxies in higher-performance systems.

6.2.3   Tetrafunctional and Multifunctional Epoxies

The use of epoxy compounds with more than two epoxide functional groups per molecule results in greater cross-linking when the resin is cured. Among other things, this results in higher T_g levels. Resin systems with these types of epoxies exhibit improved thermal and physical properties. Common commercially available epoxies of this type can be segmented into a few T_g ranges: 125 to 145°C, 150 to 165°C, and 170 to 185°C (there are epoxy systems with T_gs above 185°C, but they are not common). These resin systems are normally blends of difunctional, tetrafunctional, and multifunctional epoxy resins. Figures 6.3 and 6.4 are examples of tetrafunctional and multifunctional epoxies.

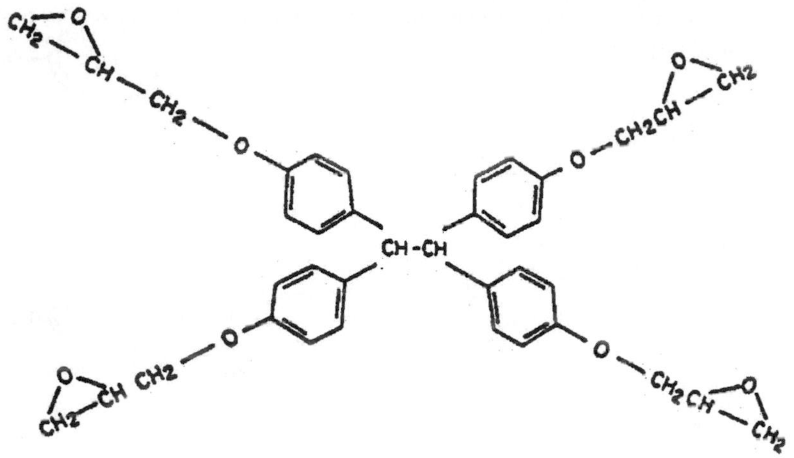

FIGURE 6.3 A tetrafunctional epoxy resin.

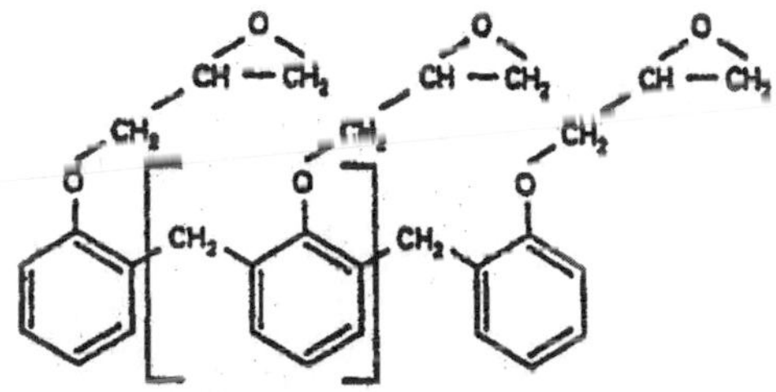

FIGURE 6.4 A multifunctional epoxy novolac resin.

While the differences in cost among these ranges are shrinking, a cost-performance trade-off still exists. The base cost of the high-T_g systems is still somewhat higher than that for lower-T_g systems. In addition, the high-T_g systems can incur increased circuit manufacturing costs, due primarily to increased multilayer lamination cycle times and decreases in drilling productivity.

Tetrafunctional and multifunctional epoxy resin systems, blended with difunctional epoxy resins, are the most common material type used in multilayer printed circuits. These are used in a wide range of applications, including automotive, industrial, computer, and telecommunications products.

6.3 OTHER RESIN SYSTEMS

Many other resin systems are also available. In choosing a resin system, the circuit board designer and fabricator have to consider what level of performance is needed, as there is typically a strong cost-performance relationship. This cost-performance relationship is driven not only by the base price of the material itself, but also by the impact the material has on processing costs, during laminate and prepreg manufacturing as well as printed circuit manufacturing processes.

6.3.1 Epoxy Blends

Blends of epoxy resins with other types of resins have also been developed. These are used when performance demands exceed the capabilities of even the high-T_g epoxies, but when the costs of the highest-performance materials cannot be justified. In many cases, the driving force behind these materials is the need for improved electrical properties vs. the standard epoxy offerings. Specifically, improvements in the dielectric constant (permittivity) and dissipation factor (loss tangent) are the properties of interest. Materials with lower values for these properties are needed for circuits that operate at high frequencies.

These blends include epoxy-polyphenylene oxide (PPO) and epoxy–cyanate ester. While these materials have been developed to minimize impacts to common printed circuit manufacturing processes, they can impact productivity in multilayer lamination and drilling and can require special desmear and hole wall conditioning processes. On the other hand, they typically have less of an impact on these processes when compared to the even higher-performance materials.

These epoxy blends are often used in high-frequency applications including cellular base stations, wireless communication interconnects, and high-speed computing applications.

6.3.2 Bismaleimide Triazine (BT)/Epoxy

Typically, epoxy is added to BT resins in order to modify the properties of pure BT. Therefore these materials are also considered epoxy blends. BT/epoxies normally have T_g values near 180°C and exhibit a good combination of electrical, thermal, and chemical resistance properties. BT/epoxy is commonly used in BGA substrates and chip-scale packages because it can meet the requirements of specifications for use in semiconductor chip packaging. It is also suitable for high-density multilayers requiring good thermal, electrical, and chemical performance.

6.3.3 Cyanate Ester

Cyanate ester systems possess very high T_g values, typically in the neighborhood of 250°C. The thermal performance of cyanate ester combined with its very good electrical performance has made it a good choice in high-speed circuits and thick backpanels. However, cyanate esters are relatively expensive and require special processing that can add additional cost to the finished circuit. For these reasons, cyanate ester materials are still used only in niche applications.

6.3.4 Polyimide

When extreme heat resistance is required, polyimide resins offer the best performance. With a T_g of 260°C for the purest systems to 220°C for modified or toughened systems, and very high decomposition temperatures, circuits manufactured with polyimide materials exhibit very high levels of reliability. These materials have commonly been used in burn-in boards and military applications where thermal resistance is vital. Again, however, these resin systems are relatively expensive and difficult to process, making them suitable for only limited applications.

6.3.5 Polyester

Polyester-based resins have also been used in printed circuits. These materials have been designed primarily for high-frequency applications such as microstrip antennas and wireless communications products where low loss is required. Processing of polyester materials can be comparable to that for FR-4. However, thermal performance is typically not as good as that of other resin systems.

6.3.6 Polytetrafluoroethylene (PTFE, Teflon®)

A variety of PTFE-based products are also available for applications in which extremely good electrical properties are required. These materials typically require very specialized processing and are relatively expensive. Hybrids of PTFE-based materials and other material types are often used to gain the performance benefits of the PTFE-based materials where needed, while controlling the total circuit cost by using other, less costly materials in other layers of the multilayer circuit.

6.3.7 Allylated Polyphenylene Ether (APPE)

APPE-based products offer promise in filling additional gaps in the cost-performance spectrum. This material offers superior electrical properties compared to the epoxy blends just discussed, and also offers excellent thermal performance. This resin system is well suited for wireless communications and high speed computing applications. In addition, APPE can be processed using conventional PCB manufacturing processes, although some adjustments to these processes may be required.

6.4 ADDITIVES

The resin systems just discussed typically contain a variety of additives that either promote curing of the resin system or modify the properties in some way. These additives are as follows.

6.4.1 Curing Agents and Accelerators

Each resin system contains organic components that must be reacted together in order to promote resin cross-linking. Curing agents and accelerators are used to promote these reactions. A common curing agent used in epoxy resin systems has been dicyandiamide or "dicy." However, non-dicy systems have been developed to promote faster curing, reduce moisture sensitivity, and improve reliability. The choice of curing agent and accelerator is also driven by the resin type.

6.4.2 Flame Retardants

In the case of standard epoxy materials, flame retardancy has commonly been achieved by brominating the epoxy resin. This normally involves manufacturing the epoxy resins with tetrabromobisphenol A (TBBPA), which contains bromine within the chemical backbone (see Fig. 6.2). Because TBBPA is reacted into the epoxy resin itself, it is not available for release into the environment. Under excessive exposure to heat, the bromine is released and prevents flammability. TBBPA has been used successfully for many years as a flame retardant and is still used in the overwhelming majority of materials. However, environmental and health concerns over brominated flame retardants in general, though not specifically TBBPA, have spurred interest in alternative flame retardants.

Halogen-free (bromine is a halogen) resin systems are also available. Alternatives include nitrogen-based compounds, phosphorus-based compounds, and hydrated fillers. These can be subdivided into reactive components and additives. A reactive component is one that becomes directly incorporated into the resin system itself, like TBBPA. As with TBBPA, reacted components have the advantage of not being available for release to the environment through leaching or solvent extraction. An example of a reactive, nonhalogen flame retardant would be epoxidized phosphorus compounds. The epoxide groups of these compounds make them reactive and result in chemical bonding into the polymer backbone. In contrast, red phosphorus is an inorganic solid that can be dispersed into an epoxy formulation. Hydrates, such as aluminum hydroxide or magnesium hydroxide, decompose and liberate water, which then suppresses the burning process.

In choosing a flame retardant, consideration must be given to the impact on the performance of the resin system and the finished base material. These materials, at the levels required for flame retardancy, can affect the physical properties of the laminate, change rheological properties, and alter cure kinetics of the resin system. In summary, the reactive compounds are generally preferred because they are bound to the polymer backbone, which prevents release into the environment, and they seem to facilitate obtaining the desired material properties in comparison to additives or fillers.

6.4.3   Ultraviolet (UV) Blockers/Fluorescing Aids

Some resins naturally absorb ultraviolet light. Others absorb very little. There are two reasons why this property is important. First, automatic optical inspection (AOI) equipment is often laser based and relies on the resin system in the base material to fluoresce upon exposure. In this way, the AOI equipment can distinguish between the base laminate and the conductor pattern. The second reason involves photoimaging on very thin circuits—during the solder mask imaging process, for example. This process commonly involves exposure to ultraviolet (UV) light through an image of the desired pattern onto both sides of the circuit, which is coated with a mask or photoresist. The UV light initiates changes to the photoresist, typically polymerization, which makes it less soluble in a developer solution. In very thin circuits, if the base material does not absorb UV light sufficiently, the UV light from one side of the circuit board can pass through to the other side and cause unwanted expose of the photoresist. Therefore, components that absorb UV light are sometimes added to resin systems that do not sufficiently absorb UV light.

6.5   REINFORCEMENTS

While there are also a variety of reinforcements used in base materials, woven fiberglass cloths are by far the most common. Other materials include paper, glass matte, nonwoven aramid fibers, nonwoven fiberglass, and a variety of fillers. The advantages of woven glass cloths include a good combination of mechanical and electrical properties, a wide range of types for achieving various laminate thicknesses, and good economics.

6.5.1   Woven Fiberglass

The process of manufacturing woven fiberglass cloth begins with melting the various inorganic components required for a particular grade of glass. The molten components travel through a furnace and ultimately flow through specialized bushings to form the individual fiberglass filaments and yarns. These yarns are then used in a weaving process in order to manufacture the fiberglass cloth. The relative concentrations of the components used affect the chemical, mechanical, and electrical properties of the fiberglass. The compositions of a few fiberglass types are provided in Table 6.1.

TABLE 6.1 Fiberglass Compositions

Component	E-glass	S-glass	D-glass	Quartz
Silicone dioxide	52–56%	64–66%	72–75%	99.97%
Calcium oxide	16–25%	0–0.3%	0–1%	
Aluminum oxide	12–16%	24–26%	0–1%	
Boron oxide	5–10%		21–24%	
Sodium oxide and potassium oxide	0–2%	0–0.3%	0–4%	
Magnesium oxide	0–5%	9–11%		
Iron oxide	0.05–0.4%	0–0.3%	0.3%	
Titanium oxide	0–0.8%			
Fluorides	0–1.0%			

FIGURE 6.5 Plain weave. *(From Clark-Schwebel Handbook.)*

By far, E-glass is the most commonly used fiberglass for printed circuits. E-glass provides an excellent combination of electrical, mechanical, and chemical properties at a reasonable cost. S-glass provides greater strength but is more difficult to process through the mechanical drilling operation. Other types of glass may be used in specialty applications, including those requiring lower dielectric constants and dissipation factors.

These cloths are also coated with a finish or coupling agent tailored for improving the bond between the glass filaments and the specific resin coated onto the glass. There are several diameters of filaments available (Table 6.2), as well as many different yarn types. Combining the various yarn types in the weaving process leads to many different types of cloth styles. While there are also many different weave types, virtually all cloths for printed circuits use a plain weave (Fig. 6.5). The plain weave consists of yarns interlaced in an alternating fashion, one over and one under every other yarn. This weave pattern provides good fabric stability. Some of the common plain-weave cloth styles used for printed circuits are shown in Table 6.3.

TABLE 6.2 Glass Filament Diameters

Filament designation	Nominal diameter (μm)	Nominal diameter (in)
B	3.5	0.00015
C	4.5	0.00018
D	5	0.00021
DE	6	0.00025
E	7	0.00028
G	9	0.00037
H	10	0.00043
K	13	0.00051

TABLE 6.3   Common Fiberglass Cloth Styles for Printed Circuit Base Materials

Style	Fiberglass thickness (in)	Warp yarn	Fill yarn	Count (ends/in)	Weight (oz/yd.^{2})
106	0.0014	ECD 900-1/0	ECD 900-1/0	56×56	0.73
1080	0.0023	ECD 450-1/0	ECD 450-1/0	60×47	1.42
1500	0.0052	ECE 110-1/0	ECE 110-1/0	49×42	4.95
1652	0.0045	ECDE 150-1/0	ECDE 150-1/0	52×52	4.18
2113	0.0029	ECE 225-1/0	ECD 450-1/0	60×56	2.31
2116	0.0038	ECE 225-1/0	ECE 225-1/0	60×58	3.22
2165	0.0040	ECE 225-1/0	ECG 150-1/0	60×52	3.55
2313	0.0032	ECE 225-1/0	ECD 450-1/0	60×64	2.38
3070	0.0031	ECDE 300-1/0	ECDE 300-1/0	70×70	2.74
3313	0.0033	ECDE 300-1/0	ECDE 300-1/0	60×62	2.40
7628	0.0068	ECG 75-1/0	ECG 75-1/0	44×32	6.00
7629	0.0070	ECG 75-1/0	ECG 75-1/0	44×34	6.25
7635	0.0080	ECG 75-1/0	ECG 50-1/0	44×29	6.90

6.5.2   Yarn Nomenclature

Because there can be so many types of yarn manufactured from the available grades of glass and filament diameters, special nomenclature systems have been developed. The two systems used are the U.S. System and the TEX/Metric System.

6.5.2.1   The U.S. System.

An example of a yarn name in the U.S. system is ECD 450-1/0. This yarn is used in making a 1080-style glass cloth. Each letter and number in the name describes something about the yarn:

- *First letter:* describes the glass composition. Electrical-grade or E-glass is the most common grade used for manufacturing materials for printed circuits.
- *Second letter:* C indicates that the yarn is composed of continuous filaments. S indicates staple filaments. T indicates texturized continuous filaments.
- *Third letter:* indicates the individual filament diameter (Table 6.2).
- *First number:* represents 1/100 the normal bare glass yardage in 1 lb of the basic yarn strand. In this example, multiply 450 by 100, which results in 45,000 yards in 1 lb.
- *Second number:* represents the number of basic strands in the yarn. The first digit represents the original number of twisted strands. The second digit, after the diagonal slash, represents the number of strands plied or twisted together.

The name may also include a designation indicating the number of turns per inch in the twist of the final yarn and the direction of the twist. An example would be 3.0S, or three turns per inch with an S-direction twist. An S twist has spirals that run up and to the left. A Z twist has spirals that run up and to the right.

6.5.2.2   The TEX/Metric System.

An example of a yarn name in the TEX/Metric System is EC9 33 1X2.

- *First letter:* designates the glass composition.
- *Second letter:* C indicates that the yarn is composed of continuous filaments. T indicates textured continuous filaments. D indicates staple filaments.
- *First number:* indicates the individual filament diameter expressed in micrometers.
- *Second number:* represents the nonlinear weight of the bare glass strand expressed in TEX. TEX is the mass in grams per 1000 m of yarn.

- *Third number:* indicates yarn construction or the basic number of strands in the yarn. The first digit represents the original number of twisted strands and the second digit, after the X, indicates the number of these strands twisted or plied together.

6.5.3 Fiberglass Cloths

With the possible combinations of glass compositions, filament diameters, yarn types, and the number of different weave patterns available, the number of possible fiberglass cloths can almost be unlimited. The effects that these glass fabrics have on the base material are driven by these variables. In addition, the fabric count, or the number of warp yarns and fill yarns, also helps determine the properties of the fabric and the base material. Warp yarns are those that lie in the length (machine direction) of the fabric, while the fill yarns lie across the warp direction. The warp direction is also commonly called the grain direction.

As pointed out earlier, glass cloths used in printed circuits are typically, though not always, made from E-glass, and virtually all use what is called a plain weave. A plain weave results in a fabric constructed so that one warp yarn passes over and under one filling yarn (and vice versa). This weave pattern offers good resistance to yarn slippage and fabric distortion. Common fiberglass cloths used in materials for printed circuits are shown in Table 6.3. Figure 6.6 shows three different fabric styles. At a given resin content, each style yields a different nominal thickness. Having the flexibility of many glass styles and thicknesses is important for meeting controlled impedance and overall thickness requirements.

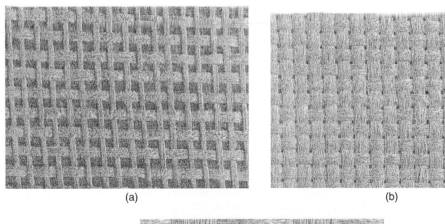

(a)   (b)

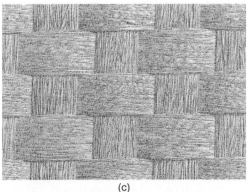

(c)

FIGURE 6.6 (*a*) 1080 cloth; (*b*) 2116 cloth; (*c*) 7628 cloth.

During the manufacture of the filaments and woven fabrics, a variety of surface finishes may be applied to the glass in order to improve manufacturability, to help prevent abrasion and static, and to aid in holding the filaments together. The most important of these surface finishes to the laminate and printed circuit manufacturer is the coupling agent applied to the finished fabric. The coupling agent, typically an organosilane compound, aids in the wetting of and adhesion to the resin applied to the glass. It is important for the reliability of the finished circuit board that this coupling agent improve the resin-to-glass adhesion during both the circuit manufacturing process, e.g., in mechanical drilling as well as during the actual end use environment of the circuit. A variety of these compounds are commercially available, with the specific choice primarily driven by the type of resin to be applied to the fiberglass cloth.

6.5.4 Other Reinforcements

While woven fiberglass cloth makes up the overwhelming majority of the reinforcements used for printed circuit base materials, there are other types that can be used exclusively or in combination with woven glass fabric. These other reinforcements include the following.

6.5.4.1 Glass Matte.   As opposed to woven cloths, glass matte reinforcement exhibits a more random orientation. Chopped-strand matte consists of fiberglass strands that have been chopped into 1- to 6-in lengths and distributed evenly. Continuous-strand matte, as the name implies, consists of continuous strands of fiberglass in a random, spiral orientation. Glass matte is used in the core of CEM-3, which is used in circuit boards for home computers, automobiles, and home entertainment products.

6.5.4.2 Aramid Fiber.   As opposed to the inorganic chemistry of fiberglass reinforcements, aramid fibers consist of aromatic polyamide organic compounds and therefore exhibit different properties. The unique properties of aramid fibers can offer advantages in specific high-performance printed circuits and laminate based multichip modules (MCM-Ls). For example, aramid fiber–reinforced materials are often used in microvia applications because they are easily ablated by plasma or laser. Other interesting properties of aramid fibers are their low weight, high strength, and negative coefficient of thermal expansion (CTE) in the axial direction. When aramid fibers are combined with the resin system, the resulting composite can offer reduced overall CTEs in the *x-y* plane compared to conventional materials.

6.5.4.3 Linear Continuous-Filament Fiberglass.   A unique process for producing fiberglass-reinforced laminates winds continuous yarns in a linear orientation. The resulting laminate possesses three layers of fiberglass filaments, with the outside layers parallel to each other and the middle layer perpendicular to the outside. With an equal number of linear filaments running in each axis, this type of reinforcement results in a laminate with improved dimensional stability.

6.5.4.4 Paper.   Cellulose-based papers can also be used as a reinforcement for base materials. Paper-based reinforcements are also used in conjunction with other reinforcements such as woven glass, and these materials can also allow punching of through holes rather than drilling. This makes them economical in some high-volume, low-technology applications including consumer electronics such as radios, toys, calculators, and television games. Paper is used in FR-2, FR-3, and the core of CEM-1.

6.5.4.5 Fillers.   Fillers are small particles that can be added to a resin system in order to modify the properties of the composite material. These materials range from kaolin clay powders to tiny hollow glass spheres to a variety of other inorganic materials. These materials are typically used to tailor the properties of the base material for specific uses. For example, kaolin clay powders coated with a layer of palladium and dispersed within the base materials have been used as catalysts for electroless copper plating. Hollow glass microspheres have been used to reduce the dielectric constant of materials. Other fillers are being used to reduce thermal expansion properties and improve reliability, to improve the machinability of materials in the drilling process, to alter electrical properties, including increasing the dielectric constant in capacitive applications, and to lower total costs.

6.5.4.6   Expanded Teflon.   Although not typically thought of as a reinforcement, Teflon that has been "expanded" into a spongelike structure on a microscopic scale is currently being used in applications that require prepregs with very low dielectric constants or loss properties. A resin system is applied to the expanded Teflon structure and is made into a B-staged sheet capable of bonding printed circuit layers together. This material can be used to make prepregs for high-frequency applications and is shown in Fig. 6.7.

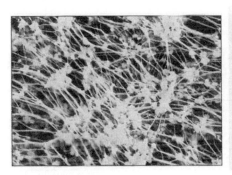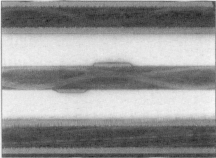

FIGURE 6.7   Expanded Teflon and PCB cross section showing prepreg using expanded Teflon. *(Courtesy of W. L. Gore and Associates.)*

6.6   CONDUCTIVE MATERIALS

The primary conductive material used in printed circuits is copper foil. However, the trend toward increased circuit density has brought about recent developments in copper foil technology as well. In addition, copper foils subsequently plated with other metal alloys are employed to manufacture printed circuits with resistive components buried within a multilayer structure. Copper foil grades are shown in Table 6.4.

TABLE 6.4   Copper Foil Grades from IPC-4562

Grade	Foil description
1	Standard electrodeposited
2	High-ductility electrodeposited
3	High-temperature elongation electrodeposited
4	Annealed electrodeposited
5	As rolled—wrought
6	Light cold rolled—wrought
7	Annealed—wrought
8	As rolled—wrought, low-temperature annealable

6.6.1 Electrodeposited Copper Foil

The foil most commonly used in manufacturing printed circuits is electrodeposited copper foil (ED foil). ED foil is produced through the use of an electrochemical process in which copper feed stock or scrap copper wire is first dissolved in a sulfuric acid solution. The purified copper sulfate–sulfuric acid solution is then used to electroplate copper onto a cylindrical drum typically made from stainless steel or titanium. This is illustrated in Fig. 6.8. This process results in a copper foil with a relatively smooth, shiny side, and a coarser matte side (see Fig. 6.9). The shiny side mirrors the surface of the plating drum, while the microscopically rough matte side is formed by the copper grain structure. By controlling the plating solution chemistry, the surface conditions of the plating drum, and the electroplating parameters, the properties of the copper foil can be modified for various usage environments. For example, mechanical properties such as tensile strength and elongation, as well as the surface profile of the matte side, can all be adjusted through control of these process variables.

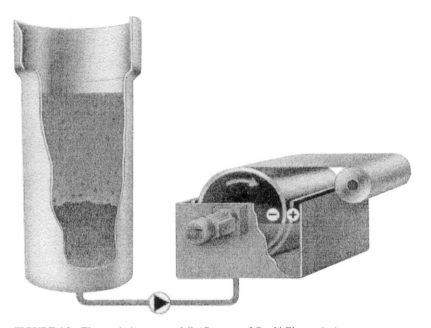

FIGURE 6.8   Electroplating copper foil. *(Courtesy of Gould Electronics.)*

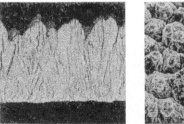 

FIGURE 6.9   Cross section and matte side of standard Grade 1 foil. *(Courtesy of Gould Electronics.)*

The foils most commonly used in printed circuit manufacturing are Grade 1 and Grade 3 foils. Unlike Grade 1 foil, Grade 3 foil must meet specific elongation requirements at elevated temperatures (180°C). Grade 3 foil, also commonly referred to as high-temperature elongation (HTE) foil, is widely used in base laminates used to manufacture multilayer printed circuit boards. Its increased ductility at elevated temperatures provides resistance to copper foil cracks when the multilayer circuit is thermally stressed and expands in the z axis. Changes to the plating bath are made to alter the grain structure of HTE foil. This results in different mechanical properties. Tables 6.5 and 6.6 show the tensile strength and ductility requirements for standard Grade 1 copper foil and high-temperature elongation Grade 3 copper foil.

TABLE 6.5 Tensile and Elongation Properties of Grade 1 Copper Foil

Property @ 23°C	½ oz	1 oz	2 oz
Tensile strength:			
kpsi	30	40	40
MPa	207	276	276
Elongation %	2	3	3

TABLE 6.6 Tensile and Elongation Properties of Grade 3 Copper Foil

Property	½ oz	1 oz	2 oz
Tensile strength @ 23°C:			
kpsi	30	40	40
MPa	207	276	276
Elongation % @ 23°C	2	3	3
Tensile strength @ 180°C:			
kpsi	15	20	20
MPa	103	138	138
Elongation % @ 180°C	2	2	3

The surface profile of the copper foil is also important in printed circuit manufacturing. On the one hand, a relatively rough surface profile aids in the bond strength of the foil to the resin system. On the other hand, a rough profile may require longer etching times, which impact productivity and the geometry of the etched features. With longer etching times, there is an increased propensity toward forming trapezoidal circuit traces because there is more time for lateral etching of the conductor. This has obvious implications for manufacturing fine-line circuits in high yield and controlling impedance properties. Low-profile and very-low-profile characteristics are included in the IPC-4562 specification and are summarized in Table 6.7. Comparing Figs. 6.9 and 6.10 illustrates the profile differences between a standard and low-profile copper foil.

TABLE 6.7 Foil Profile Criteria

Foil profile type	Maximum foil profile (μm)	Maximum foil profile (μin)
Standard (S)	N/A	N/A
Low-profile (L)	10.2	400
Very-low profile (V)	5.1	200
No treatment or roughness (X)	N/A	N/A

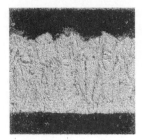

FIGURE 6.10 Cross section and matte side of low-profile Grade 1 foil. *(Courtesy of Gould Electronics.)*

Subsequent to the manufacturing of the base copper foil, a variety of surface treatments are typically applied to the matte side of the foil, and these too vary depending on the usage environments. These treatments fall into four categories.

6.6.1.1 Bonding Treatments or Nodularization.   This treatment increases the surface area of the foil by plating copper or copper oxide nodules to the surface of the foil. The increased surface area results in increased foil-to-resin bond strengths. The thickness of this treatment is relatively small, but can be tailored for adhesion to high-performance resin systems such as polyimides, cyanate esters, BT, etc. The matte side images in Figs. 6.9 and 6.10 include these nodules.

6.6.1.2 Thermal Barriers.   A coating of zinc, nickel, or brass is usually applied over the nodules. This coating can prevent thermal or chemical degradation of the foil-to-resin bond during manufacture of the laminate, the printed circuit, and the circuit assembly. These coatings typically measure several hundred angstroms in thickness and vary in color due to the specific metal alloy used, although most treatments are brown, gray, or a yellow mustard color.

6.6.1.3 Passivation and Antioxidant Coatings.   In contrast to the other coatings, these treatments are virtually always applied to both sides of the foil. Although many of these treatments are chromium based, organic coatings can also be utilized. The primary purpose of these treatments is to prevent oxidation of the copper foil during storage and lamination. These coatings are usually less than 100 Å thick and are typically removed by the cleaning, etching, or scrubbing processes normally used at the start of printed circuit manufacturing processes.

6.6.1.4 Coupling Agents.   Coupling agents, primarily silanes like those used to promote fiberglass-to-resin adhesion, can also be used on copper foils. These coupling agents provide a chemical bond between the foil and the resin system and can also help prevent oxidation.

6.6.2 Reverse-Treated Foils

Reverse-treated foil (RTF) is also an electrodeposited copper foil, but the treatments are coated onto the smooth drum side of the foil rather than the matte side, as with conventional electrodeposited foil. This results in a very low-profile surface bonded to the laminate, with the rough matte side facing out (see Figs. 6.11 and 6.12). The low surface profile against the laminate aids in the production of fine circuit features on the innerlayer, while the matte surface can aid in photoresist adhesion. In addition, in very thin laminates the low surface profile can aid in achieving consistent dielectric thickness with a reduced concern about inadequate dielectric separation between the corresponding points on the tooth profile.

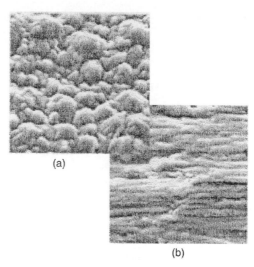

FIGURE 6.11 External sides for (*a*) RTF and (*b*) standard foils. *(Photos courtesy of Gould Electronics.)*

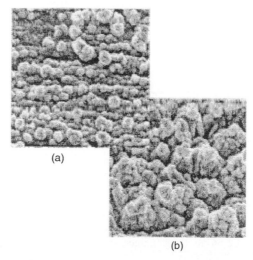

FIGURE 6.12 Sides bonded to laminate for (*a*) RTF and (*b*) standard foils. *(Photos courtesy of Gould Electronics.)*

6.6.3 Wrought Annealed Copper Foils

Wrought annealed foil is typically used in flexible circuit manufacturing because of its superior ductility. In contrast to an electrodeposition process, wrought annealed foil begins with a slab or ingot of copper that is worked through a series of rollers in conjunction with heat cycles to obtain the desired thickness and mechanical properties. The resulting grain structure of wrought annealed copper foil, which is very random compared to the columnar or fine-grain structure of electrodeposited foil, is a significant contributor to the differences in mechanical properties. In addition, wrought annealed foil consists of two very-low-profile sides, so the treatment steps account for any surface roughness. The tensile and elongation requirements for Grade 7 wrought annealed copper are provided in Table 6.8.

TABLE 6.8 Tensile and Elongation Properties of Wrought Annealed Foil

Property	½ oz	1 oz	2 oz
Tensile strength @ 23°C:			
kpsi	15	20	25
MPa	103	138	172
Fatigue ductility % @ 23°C	65	65	65
Elongation % @ 23°C	5	10	20
Tensile strength @ 180°C:			
kpsi	TBD	14	22
MPa		97	152
Elongation % @ 180°C	TBD	6	11

6.6.4 Other Foil Types

6.6.4.1 Double-Treated Copper Foil. As discussed in the preceding text, the foil surface that is bonded to the base laminate is specially treated with coatings designed to improve foil-to-resin bond strength and reliability. In double-treated foils, these coatings are also applied to the foil surface that forms the outside laminate surface. It is also possible to have a "reverse-treated" double-treated foil, meaning that the smooth surface is bonded to the laminate with the matte surface facing out, while both sides have been treated.

The advantage of double-treated foil is that it eliminates the oxide or other surface preparation process typically used to prepare the innerlayer circuitry for multilayer lamination. However, no abrasion of this double-treat coating can be tolerated, and removal of any surface contamination becomes difficult. This also makes double-treated foil more sensitive to handling practices in the circuit manufacturing process.

6.6.4.2 Resistive Foils. Other treatments can also be applied to the base foil for use in manufacturing innerlayer circuits with buried resistors. This technology can enable the creation of resistors on internal layers of a multilayer circuit, with removal of many of the resistors commonly assembled on the outside of the multilayer circuit. This can improve board reliability and free up space on the outside of the board for active components. These foils typically use a resistive metal alloy coated onto the base foil. The laminate made with this foil can then be sequentially imaged and etched to produce the desired circuit pattern along with resistive components.

6.7 REFERENCES

1. Marcia L. Hardy, "Regulatory Status of the Flame Retardant Tetrabromobisphenol A," IPC Printed Circuits Expo, April 2000.

2. Manfred Cygon, "Base Materials and Ecological Aspects," EPC European PCB Convention, November 1999.

3. Manfred Cygon, Marc Hein, and Marty Choate, "Halogen Free Base Materials for PWB Applications," IPC Printed Circuits Expo, April 2000.

4. Nelco product reference materials.

5. Polyclad product reference materials.

6. Clark-Schwebel industrial fabrics guide.

7. BGF Industries, Inc. fiberglass guide.

8. Gould Electronics product reference materials.

9. Edward J. Kelley and Richard A. Micha, "Improved Printed Circuit Manufacturing with Reverse Treated Copper Foils," IPC Printed Circuits Expo, March 1997.

10. Martin W. Jawitz, *Printed Circuit Materials Handbook,* McGraw-Hill, New York, 1997.

11. IPC-4562, Metal Foil for Printed Wiring Applications.

12. Edward Kelley, "Meeting the Needs of the Density Revolution with Non-Woven Fiberglass Reinforced Laminates," EIPC/Productronica Conference, November 1999.

13. W. L. Gore technical literature.

CHAPTER 7
BASE MATERIAL MANUFACTURING PROCESSES

Edward J. Kelley
Sanmina Corporation, Derry, New Hampshire

7.1   LAMINATE AND PREPREG MANUFACTURING PROCESSES

While there are different manufacturing processes that can be used to integrate the components that make up printed circuit base materials, the overwhelming majority of materials are manufactured using a common process. In recent years, however, new techniques have been developed and continue to be developed. These new processes have been designed to lower the cost of manufacturing, to improve material performance, or both. Common to all processes is the need to manufacture a copper-clad laminate and the bonding sheets or *prepregs* used to manufacture multilayer circuits.

7.2   CONVENTIONAL MANUFACTURING PROCESSES

Figure 7.1 illustrates the overall conventional manufacturing process. This process can be subdivided into two processes, prepreg manufacturing and laminate manufacturing. Prepreg is also called B stage or bonding sheets, while laminate is sometimes called C stage. The terms *B stage* and *C stage* refer to the degree to which the resin system is polymerized or cured. B stage refers to a state of partial cure. B stage is designed to melt and continue polymerizing when it is exposed to sufficient temperatures. C stage refers to a state of "full" cure (typically we never realize full cure, in the sense that not all reactive sites on the resin molecules have cross-linked; however, we use the term *full cure* to mean that the overwhelming majority of such sites have reacted and additional exposure to temperature will do little to advance the state of cure).

7.2.1   Prepreg Manufacturing

The first step in most of these processes involves coating a resin system onto the chosen reinforcement, most commonly a woven fiberglass cloth. Rolls of fiberglass cloth or alternative reinforcement type are run through equipment called treaters (see Figs. 7.2 and 7.3). In the treatment process, the fiberglass cloth is first pulled through a pan containing the resin system to be applied to the cloth.

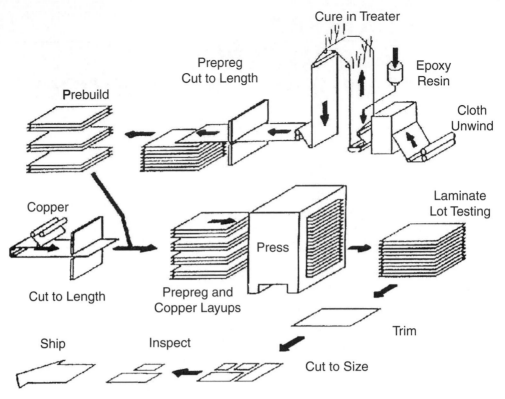

Prebuild

Prepreg
Cut to Length

Cure in Treater

Epoxy
Resin

Cloth
Unwind

Copper

Cut to Length

Prepreg and
Copper Layups

Press

Laminate
Lot Testing

Trim

Cut to Size

Ship   Inspect

FIGURE 7.1 The overall conventional manufacturing process. *(Courtesy of ParkNelco.)*

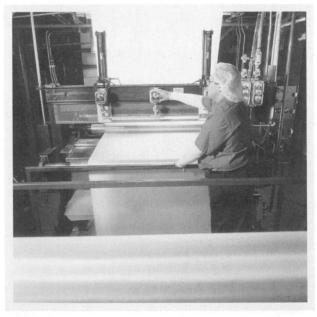

FIGURE 7.2 Fiberglass cloth being fed into a treater. *(Photo courtesy of ParkNelco.)*

FIGURE 7.3   Prepreg after treatment with resin. *(Photo courtesy of ParkNelco.)*

The cloth is then pulled through metering rolls that control how much resin is applied to the cloth. Next the cloth is pulled through a series of heating zones. These heating zones commonly utilize forced-air convection, infrared heating, or a combination of the two. In the first set of zones, any solvent used to carry the resin system components is evaporated off. Subsequent zones are dedicated to partially curing the resin system, or B-staging the resin. Finally, the prepreg is then rewound into rolls or cut into sheets.

As you can imagine, there are a number of process controls required in this operation. The concentrations of the resin system components must be controlled and the viscosity of the resin system must be maintained within acceptable limits. Tension on the cloth as it is pulled through the treater is also important, because, among other reasons, it is important not to distort the weave pattern of the cloth. Control of the resin-to-glass ratio or resin content, the degree of cure of the resin, and cleanliness are also critical.

Because the resin system at this point is only partially cured, prepregs must typically be stored in temperature- and humidity-controlled environments. Temperature, for obvious reasons, could affect the degree of cure of the resin and therefore its performance in laminate or multilayer circuit pressing. Because moisture can affect the performance of many curing agents and accelerators, not to mention the performance of the resin system during lamination or press cycles, control of humidity is also important during prepreg storage. Absorbed moisture that becomes trapped during lamination cycles can also lead to blisters or delaminations within the laminate or multilayer circuit.

7.2.2 Laminate Manufacturing

The process of manufacturing copper-clad laminate begins with the prepreg material discussed in the previous chapter. Prepregs of certain fiberglass cloth styles and specific resin contents are combined with the desired copper foils to make the finished laminate. First, the prepregs and copper foils are sheeted to the desired size. These materials are then laid up in the proper sequence to produce the desired copper-clad laminate (see Figs. 7.4 and 7.5). Several of these individual sandwiches are stacked on top of each other, typically separated by stainless steel plates, although other separator materials, including aluminum, are also used (see Fig. 7.6). These stacks are then loaded into multiopening lamination presses (see Fig. 7.7) where pressure, temperature, and vacuum are applied. The specific press cycle used varies depending upon the particular resin system, the degree of cure of the prepregs, and other factors. The presses themselves have many platens that can be heated by steam or hot oil that flows through the platens. They can also be heated electrically.

Process controls in these operations are just as important as in prepreg manufacturing. Cleanliness in the manufacturing environment and cleanliness of the steel separator plates

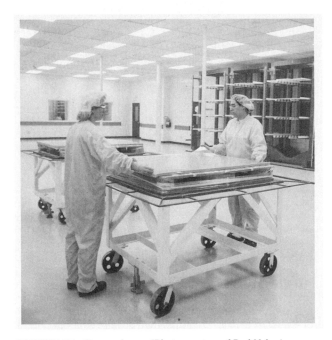

FIGURE 7.4 Prepreg layup. *(Photo courtesy of ParkNelco.)*

are critical in achieving good surface quality and in avoiding embedded foreign material within the laminate. Control of the temperature rise during lamination provides the desired amount of resin flow, while control of the cooldown rate can impact warp and twist. The length of time that the laminate is above the temperature required to initiate the curing reaction determines the degree of cure, as does the degree to which the actual temperature exceeds this temperature.

While these descriptions of prepreg and laminate manufacturing give a simple description of the processes used, it is important to understand that there are many variables that influence the quality and performance of the finished products. In addition, many of these variables interact with each other, meaning that a change in one may influence others and may require adjustments to these other process variables. In summary, prepreg and laminate manufacturing is much more complex than it appears at first glance.

FIGURE 7.5 Laminate setup for pressing. *(Photo courtesy of ParkNelco.)*

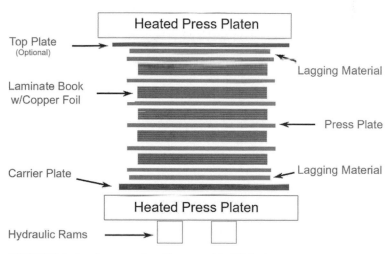

FIGURE 7.6 Laminate pressing. *(Courtesy of ParkNelco.)*

FIGURE 7.7   Laminate press room. *(Photo courtesy of ParkNelco.)*

7.3   DIRECT-CURRENT OR CONTINUOUS-FOIL MANUFACTURING

Continuous-foil or direct-current manufacturing is an alternative method used to manufacture copper-clad laminates (see Fig. 7.8). Prepregs are still used in this process; however, the layup and pressing operations are somewhat different. In this process, the copper foils are not sheeted but are kept in rolls. To start, while still part of a continuous roll, copper foil is positioned with the side to be bonded to the prepreg facing up. The prepregs are laid up in the

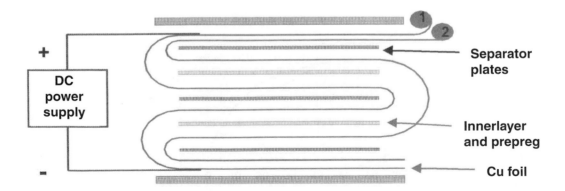

FIGURE 7.8   Laminate pressing with continuous copper foil and direct-current heating. *(Illustration courtesy of Cedal.)*

proper sequence, and then the roll of copper foil is passed over the prepregs so that it is applied to the top of what will become the finished laminate. Typically, two rolls of foil are used to allow dissimilar copper weights or types to be bonded to either side of the laminate (Fig. 7.8). A separator plate, typically made of anodized aluminum, is then placed on top of this sandwich, and the process is repeated so that several laminates are stacked up.

At this point, the stack of laminates is loaded into a press and subjected to heat, pressure, and vacuum. However, as opposed to the conventional process, this technique involves applying direct current to the copper foil that runs throughout the stack. The current heats the foil and therefore the prepregs adjacent to the foils. By controlling the amount of current, the temperature of the materials in the stack is controlled.

This process offers a few potential benefits. In the conventional process, the heat transfer occurs from the platens through the stack, resulting in the laminates nearest the platens being heated first. This creates a temperature lag from the outside to the inside of the stack. Variations in the thermal profile from outside to inside can create variations in resin flow and therefore laminate thickness variations. In this continuous-foil process, each laminate experiences essentially the same thermal profile during heatup. Therefore, improvement in laminate thickness control is possible. This also implies consistent resin content from laminate to laminate. This, coupled with the very low pressures possible in this process, can also improve laminate dimensional stability. Finally, since the copper surface is only briefly exposed to the environment, it is possible to achieve very good surface quality as long as the separator plates are clean and smooth.

7.4 CONTINUOUS MANUFACTURING PROCESSES

Over the years, continuous lamination processes have also been designed. Rather than requiring the prepregs and foils to be cut into sheets, laid up, and pressed as individual laminates, these processes use rolls of prepreg or bare fiberglass cloth and rolls of copper foil that are continuously unwound and fed together into a continuous horizontal press. One process currently in use starts with rolls of prepreg. Another starts with the bare, untreated fiberglass cloth, then applies a resin to the cloth and sandwiches it between the copper foil as it is fed continuously into a horizontal press. At the back end of the press, sections of the continuous laminate can be cut into sheets, or, with thin laminates, rolls of copper-clad laminate can be manufactured.

These processes also offer potential benefits. First, with proper controls in the press, each laminate experiences a consistent thermal profile resulting in good control of resin content and thickness. Since the surface of the copper foil is only briefly exposed to the environment, excellent surface quality can be obtained with this process as well. In addition, since the prepreg or glass cloth is also in roll form, it too is exposed to the environment only briefly, reducing the chance that airborne contaminants could become trapped within the finished laminate. These continuous processes also eliminate some of the steps required in other processes, such as sheeting of the prepregs and foils and the layup of these materials.

REFERENCES

1. Nelco and Nelco-Dielektra product reference materials.
2. Cedal product reference materials.
3. Matsushita electronic materials product reference materials.

CHAPTER 8
PROPERTIES OF BASE MATERIALS

Edward J. Kelley
Sanmina Corporation, Derry, New Hampshire

8.1 INTRODUCTION

There are a variety of properties of base materials that are of interest to the printed circuit manufacturer, the assembler, and the end user. These properties fall into the following categories:

- Thermal
- Physical
- Mechanical
- Electrical

In this section an introduction to some of the important elements of these properties is provided along with some comparisons of material types. Most of these properties are determined by testing that follows standardized procedures that can be found in the IPC test methods manual, IPC-TM-650.

8.2 THERMAL, PHYSICAL, AND MECHANICAL PROPERTIES

Two of the most important of these properties are the glass transition temperature T_g of the resin system used, as discussed in Chap. 5, and the coefficients of thermal expansion (CTEs), as discussed in Chap. 5. Surface quality, in other words a smooth laminate surface free of pits or dents, is important for the imaging of fine-line circuits. Surface quality grades from IPC-4101 are presented in Table 5.7. Additional important properties include peel strength of the conductive foil, moisture absorption, chemical resistance, resin decomposition temperature, flexural strength, and flammability.

8.2.1 Coefficient of Thermal Expansion (CTE)

Materials undergo changes in physical dimensions in response to changes in temperature. The rates of expansion of fiberglass cloth–reinforced materials are different in the respective axes of the material due to the directionality of the reinforcement. The length and width of the laminate or printed circuit is termed the x/y plane, while the axis perpendicular to this plane is the z axis.

Thermal expansion can be measured by thermomechanical analysis (TMA). TMA uses a device that measures a dimension of a sample versus temperature. Depending upon the orientation of the sample in the device, either the x/y CTE or the z-axis CTE can be measured.

8.2.1.1 z-Axis Coefficient of Thermal Expansion.

Thermal expansion in the z axis can significantly affect the reliability of printed circuits. Since plated through-holes run through the z axis of the printed circuit, thermal expansion and contraction in the base materials cause deformation in the plated through-holes and can also cause stress on the copper pads on the surface of the printed circuit. With sufficient stress on the external pads, they can be pulled toward the plated through-hole during thermal stress and subsequently appear lifted from the surface upon cooling. These lifted pads are an indication of excessive thermal expansion. Thermal cycling over time can fatigue the plated through-hole and ultimately cause failure due to separation of the conductor from the hole wall or cracking of the conductor within the barrel of the hole.

8.2.1.2 x/y Coefficient of Thermal Expansion.

Thermal expansion in the x/y axes is of more importance when discussing the attachment of components to the printed circuit. This is of particular importance when chip-scale packages (CSPs) and direct chip attach components are used because the difference in thermal expansion between the printed circuit board and the component can compromise the reliability of the bond between them as they undergo thermal cycles. Some CTE values of various materials are shown in Table 8.1.

TABLE 8.1   Thermal Expansion Values of Some Common Base Materials

Material	T_g (°C)	z-Axis expansion (% from 50 to 260°C)	x/y CTE (ppm/°C from −40 to 125°C)
FR-4 epoxy	140	4.5	12–16
Filled FR-4 epoxy	155	3.7	12–14
High-T_g FR-4 epoxy	180	3.7	10–14
BT/epoxy blend	185	3.75	10–14
Low-D_k epoxy blend	210	3.5	10–14
Cyanate ester	250	2.7	11–13
Polyimide	250	1.75	12–15
APPE	170/210*	3.0	14–18

* 170 by TMA, 210 by DMA.

8.2.1.3 Controlling Thermal Expansion.

The rate of thermal expansion is a function of the components used in the base material and their relative concentrations. The resin system will have a relatively high coefficient of thermal expansion compared to fiberglass cloth or other types of inorganic reinforcements.

In controlling z-axis expansion, the key factors to consider are the choice of resin system, the resin system T_g, and the resin content of the base materials. The use of fillers in the resin system, in addition to the fiberglass cloth, can also be used to lower the CTE of the material. Table 8.1 compares the thermal expansion of several commercially available base materials. Note the general difference in CTE values as T_g increases. Also, notice that the filled 155°C-T_g FR-4 material has the same z-axis expansion as the 180°C FR-4 material, which is unfilled. The filler used in this laminate aids in reducing the CTE values.

8.2.2 Time to Delamination

The time to delamination refers to a specific test procedure used to measure how long a material will resist blistering or delamination at a specific temperature. The procedure utilizes a thermomechanical analyzer in which a sample is heated to the specified temperature. The most common temperature used is 260°C. This test is commonly called a T260 test. Other temperatures are also used, such as 288°C or 300°C. Table 8.2 compares the performance of various material types. When reviewing this table, it is important to note that there can be large variations between similar materials across the manufacturers of these materials and even within manufacturers. T260 values can be affected by the specific resins and curing agents used in a given material and caution should be used in interpreting the results. For example, note the lower T260 values reported for the 180°C epoxy versus the 140°C epoxy. While the T260 value is shorter, long-term reliability tests typically show that the higher-T_g epoxies perform better. As a result, while there is an absolute measure of performance using this test, it is often used in failure analysis of a given material sample by checking to see if the material performs as it was designed to. In addition, there is great variance in T260 values across the many specific epoxy materials available.

TABLE 8.2 Some Thermal Properties of Common Base Material Types

Material	T_g (°C)	T260 (min)	Decomposition temperature (°C 5% wt loss)	Arc resistance (sec, minimum)
FR-4 epoxy	140	8–12	290–310	65
Filled FR-4 epoxy	155	13	317	124
High T_g FR-4 epoxy	180	4–30	300–330	65
BT/epoxy blend	185	30+	334	118
Low-D_k epoxy blend	210	30+	357	123
Cyanate ester	250	30+	376	160
Polyimide	250	30+	389	136
APPE	170/210*	30+	360	

* 170 by TMA, 210 by DMA.

8.2.3 Decomposition Temperature

While the time to delamination measures the bonds between the components of the base materials, the decomposition temperature measures actual physical degradation of the resin system. This test uses thermogravimetric analysis (TGA), which measures the mass of a sample versus temperature. The decomposition temperature is reported as the temperature at which 5 percent of the mass of the sample is lost to decomposition.

Table 8.2 includes decomposition temperatures of some common materials. Again, a note of caution is warranted. There is a fairly wide selection of FR-4 epoxies. Even those with comparable T_g values can have significantly different decomposition temperatures. Therefore, in choosing a material, the various options within a given material class should be investigated.

8.2.4 Arc Resistance

In this test technique, a low-current arc is placed above the surface of a material. Arc resistance describes the time the material resists tracking, or forming a conductive path, under this condition.

8.2.5 Density

The density of various materials is given in Table 8.3.

TABLE 8.3 Densities of Common Base Material Types

Material	Density (g/cm^3)
FR-4 epoxy	1.79
Filled FR-4 epoxy	1.97
High-T_g FR-4 epoxy	1.79
BT/epoxy blend	1.77
Low-D_k epoxy blend	1.77
Cyanate ester	1.71
Polyimide	1.68
APPE	1.51

8.2.6 Copper Peel Strength

Peel strength testing is the most popular method used to measure the bond between the conductor and the substrate. Peel strengths can be measured in the "as-received" condition, after thermal stress, at elevated temperature, and after exposure to processing chemicals. The standard procedure requires that the sample consist of traces or strips of copper foil or other metal to be tested, and that they be imaged onto the sample using standard printed circuit manufacturing processes. The strips should be at least 0.79 mm (0.032 in) in width for testing after exposure to processing chemistries and 3.18 mm (0.125 in) in width for the other tests. Metal cladding thickness can influence the measured peel strength and therefore it is accepted practice to use 1-oz. cladding. If a thinner cladding is used, it may be plated up to 1 oz in thickness.

One end of the strip is peeled back and attached to a load tester such as a tensile strength tester equipped with a load cell. The peel strength is calculated per the formula in Eq. (8.1):

$$\text{lb/in} = L_m/W_s \tag{8.1}$$

where L_m = minimum load
W_s = measured width of the peel strip

For testing after thermal stress, the sample is first floated on solder at 288°C for 10 s. For testing after exposure to processing chemistries, the sample is exposed to a series of conditions. First the sample is exposed to an organic stripper at 23°C for 75 s. Historically methylene chloride has been used, but due to environmental concerns, equivalents are now allowed. After drying, the sample is then immersed in a solution of 10 g/L sodium hydroxide at 90°C for 5 min. The sample is rinsed and then exposed to 10 g/L sulfuric acid and 30 g/L boric acid at 60°C for 30 min. The sample is again rinsed and dried and then immersed in a hot oil bath maintained at 220°C for 40 s. Finally, the sample is immersed in a degreaser at 23°C for 75 s to remove the oil and then dried. Testing at an elevated temperature may be performed by placing the sample in a hot fluid or in hot air while performing the peel strength measurements. For FR-4 materials, 125°C is commonly used.

Table 8.4 shows 1-oz copper peel strength values in pounds per inch at various conditions for several common materials.

Note that polyimide, long used for its thermal reliability characteristics, exhibits the lowest peel strength values. This illustrates that the absolute value of peel strength is not always the

TABLE 8.4   Peel Strength Values for Common Base Material Types

Material	T_g (°C)	Peel strength after solder float	Peel strength at elevated temperature (lb/in)	Peel strength after exposure to chemistries (lb/in)
FR-4 epoxy	140	9.0	7.0	9.0
Filled FR-4 epoxy	155	7.0	6.8	6.8
High-T_g FR-4 epoxy	180	9.0	7.0	9.0
BT/epoxy blend	185	8.9	8.3	9.4
Low-D_k epoxy blend	210	7.5	8.1	9.0
Cyanate ester	250	8.0	7.5	8.0
Polyimide	250	7.5	6.0	7.0
APPE	170/210*	6.1	6.1	7.1

* 170 by TMA, 210 by DMA.

best indicator of quality or reliability. Measuring and monitoring of peel strengths for a given material type for comparison to its normal values can be used as a process control tool or an inspection criterion, or for failure analysis purposes, but in choosing a material, a higher peel strength does not necessarily imply higher reliability.

FIGURE 8.1   Flexural strength test.

0.2.7 Flexural Strength

Flexural strength is a measure of the load that a material will withstand without fracturing when supported at the ends and loaded in the center, as shown in Fig. 8.1. IPC-4101 specifies the minimum flexural strengths of various materials, some of which are summarized in Table 8.5.

TABLE 8.5   Flexural Strength Requirements of Base Materials

Material type	Minimum flexural strength lengthwise (kg/m²)	Minimum flexural strength crosswise (kg/m²)
XXXPC	8.44×10^6	7.39×10^6
CEM-1	2.11×10^7	1.76×10^7
CEM-3	2.32×10^7	1.90×10^7
FR-1	8.44×10^6	7.04×10^6
FR-2	8.44×10^6	7.39×10^6
FR-3	1.41×10^7	1.13×10^7
FR-4	4.23×10^7	3.52×10^7
FR-5	4.23×10^7	3.52×10^7
Polyimide/woven E-glass	4.23×10^7	3.17×10^7
Cyanate ester/woven E-glass	3.52×10^7	3.52×10^7

8.2.8 Water and Moisture Absorption

The ability of a material to resist absorbing water, either from the air or when immersed in water, is important for printed circuit reliability. Besides the obvious concerns of moisture causing defects when a material is subjected to thermal excursions, absorbed moisture also affects the ability of a material to resist conductive anodic filament (CAF) formation when a bias is applied to the circuit.

The test method for measuring water absorption for metal-clad base laminates involves immersing a sample in distilled water at 23°C for 24 h after etching off the metal cladding, drying the sample for 1 h at 105 to 110°C, and cooling in a dessicator. The sample is weighed after drying, immersed in water under the specified conditions, and weighed again. The water absorption is calculated as follows:

Increase in weight, % = (wet weight − conditioned weight)/conditioned weight × 100

An additional moisture absorption test measures weight gain after 60 min at 15 psi. Table 8.6 shows the moisture absorption of some common material types.

TABLE 8.6   Moisture Absorption and Methylene Chloride Resistance of Common Materials

Material	T_g (°C)	Moisture absorption (%)	Methylene chloride resistance (%)
FR-4 epoxy	140	0.1	0.7
Filled FR-4 epoxy	155	0.22	0.42
High-T_g FR-4 epoxy	180	0.1	0.7
BT/epoxy blend	185	<0.5	0.7
Low-D_k epoxy blend	210	0.1	0.7
Cyanate ester	250	<0.5	0.32
Polyimide	250	0.35	0.41

8.2.9 Chemical Resistance

One common method used to evaluate the chemical resistance of base laminates is to measure absorption of methylene chloride. Similarly to water absorption testing, etched samples are exposed to this solvent and the weight gain is measured. The standard procedure starts by etching off the metal cladding of the samples, drying them in an oven for 1 h at 105 to 110°C, and measuring the initial weights. The samples are then soaked in methylene chloride at 23°C for 30 min, allowed to dry for 10 min, and weighed again. The calculation is as follows:

Change in weight, percent = (final weight − initial weight)/initial weight × 100

Table 8.6 lists methylene chloride resistance for some common materials.

8.2.10 Flammability

Flammability properties are classified by Underwriters Laboratories (UL) as 94-V-0, 94-V-1, or 94-V-2. Definitions of these classifications are as follows:

- *94-V-0:* Specimens must extinguish within 10 s after each flame application and a total combustion of less than 50 s after 10 flame applications. No samples are to drip flaming particles or have glowing combustion lasting beyond 30 s after the second flame test.
- *94-V-1:* Specimens must extinguish within 30 s after each flame application and a total combustion of less than 250 s after 10 flame applications. No samples are to drip flaming particles or have glowing combustion lasting beyond 60 s after the second flame test.

- *94-V-2:* Specimens must extinguish within 30 s after each flame application and a total combustion of less than 250 s after 10 flame applications. Samples may drip flaming particles, burning briefly, and no specimen will have glowing combustion beyond 60 s after the second flame test.

8.3 ELECTRICAL PROPERTIES

There are a variety of base material electrical properties that are important to understand when designing and manufacturing printed circuits. Some of the most important properties are discussed in this section. As noted earlier in the chapter, the demand for circuits operating at high frequencies requires materials with good permittivity and loss characteristics.

8.3.1 Dielectric Constant or Permittivity

The dielectric constant can be defined as the ratio of the capacitance of a capacitor with a given dielectric material to the capacitance of the same capacitor with air as a dielectric, as illustrated in Fig. 8.2. In other words, the dielectric constant is a measure of the ability of a material to store an electric charge. There are actually several test methods used to measure dielectric constant or permittivity, and a complete discussion of these methods is beyond the scope of this chapter. However, it should be noted that measuring permittivity at high frequencies can be very difficult and that reported values can vary with the specific test method used. Therefore, when comparing the reported values for different materials, the best comparisons are those that use the same test method.

FIGURE 8.2 Dielectric constant.

Furthermore, the dielectric constant is not really a constant. As just implied, the dielectric constant will vary with frequency. It will also vary with temperature and humidity. Therefore, in addition to the test method, the frequency, temperature, and humidity conditions must also be considered. Last, even with the same material type, variations in resin content (resin-to-reinforcement ratio) also affect the dielectric constant. These variations are further discussed in Chap. 9. Table 8.7 shows dielectric constants for some common fiberglass (E-glass)-reinforced materials at 50 percent resin content.

TABLE 8.7 Dielectric Constants and Dissipation Factors of Common Materials

Material	Dielectric constant		Dissipation factor	
	1 MHz	1 GHz	1 MHz	1 GHz
FR-4 epoxy	4.4	3.9	0.027	0.015
Filled FR-4 epoxy	4.5	3.96	0.023	—
High-T_g FR-4 epoxy	4.4	3.9	0.023	0.012
BT/epoxy blend	4.1	3.8	0.013	0.010
Epoxy/PPO	3.9	3.8	0.010	0.011
Low-D_k epoxy blend	3.9	3.8	0.009	0.010
Cyanate ester	3.8	3.5	0.008	0.006
Polyimide	4.3	3.7	0.013	0.007
APPE	3.7	3.4	0.005	0.007

8.3.2 Dissipation Factor or Loss Tangent (Tan δ)

FIGURE 8.3   Dissipation factor.

The dissipation factor in an insulating material is the ratio of the total power loss in the material to the product of the voltage and current in a capacitor in which the material is a dielectric (Fig. 8.3). Many of the test methods used for measuring dielectric constant also measure the dissipation factor. The dissipation factor also varies with frequency, resin content, temperature, and humidity. The dissipation factor is discussed in more detail in Chap. 9. Table 8.7 lists the dissipation factors of some common fiberglass (E-glass)-reinforced materials.

8.3.3 Insulation Resistance

The insulation resistance between two conductors or plated holes is the ratio of the voltage to the total current between the conductors. Two measures of electrical resistance are volume and surface resistivities. Because these properties can vary with temperature and humidity, testing is normally performed at two standardized environmental conditions, one involving humidity conditioning, the other involving elevated temperature. Humidity conditioning subjects the sample to 90 percent relative humidity and 35°C for 96 h (96/35/90). Elevated temperature conditioning normally subjects the sample to 125°C for 24 h (24/125).

8.3.4 Volume Resistivity

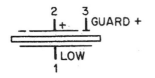

FIGURE 8.4   Circuit for volume resistance.

Volume resistivity is the ratio of the DC potential applied to electrodes embedded in a material to the current between them, typically expressed in megohm-centimeters. The measured current flows between electrodes 1 and 3 while stray current flows between electrodes 2 and 3, as shown in Fig. 8.4. Table 8.8 shows the volume resistivity values of some common fiberglass reinforced material types.

TABLE 8.8   Additional Electrical Properties of Common Base Materials

Material	Volume resistivity		Surface resistivity		Electric strength (V/mil)
	(96/35/90)	(24/125)	(96/35/90)	(24/125)	
FR-4 epoxy	10^8	10^7	10^7	10^7	1250
Filled FR-4 epoxy	10^{11}	10^{10}	10^8	10^9	1250
High-T_g FR-4 epoxy	10^8	10^7	10^7	10^7	1300
BT/epoxy blend	10^7	10^7	10^6	10^7	1200
Low-D_k epoxy blend	10^8	10^7	10^7	10^7	1200
Cyanate ester	10^7	10^7	10^7	10^7	1650
Polyimide	10^7	10^7	10^7	10^7	1350

8.3.5 Surface Resistivity

The surface insulation resistance between two points on the surface on any insulating material is the ratio of the DC potential applied between the two points to the total current between them. For surface resistivity, the measured current flows between electrodes 1 and 2,

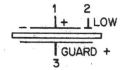

FIGURE 8.5 Circuit for surface resistance.

while stray current flows between electrodes 1 and 3, as shown in Fig. 8.5. Table 8.8 shows the surface resistivity values of some common fiberglass-reinforced materials.

8.3.6 Electric Strength

Electric strength is a measure of the ability of an insulating material to resist electrical breakdown perpendicular to the plane of the material when subjected to short-term high voltages at standard AC power frequencies of 50 to 60 Hz and is reported in volts per mil. Results can be affected by moisture content in the sample, so measurements may vary with different preconditioning environments. Unless otherwise noted, measurements are taken at 23°C, after preconditioning for 48 h in distilled water at 50°C and immersion in ambient-temperature distilled water for 30 min minimum, 4 h maximum. Measurements are performed under an oil medium to prevent flashover on a small specimen. The values may decrease with increasing specimen thickness for an otherwise identical material. Table 8.8 compares the electric strength of some common fiberglass-reinforced materials.

8.3.7 Dielectric Breakdown

Dielectric breakdown measures the ability of rigid insulating materials to resist breakdown parallel to the laminations (or in the plane of the material) when subjected to extremely high voltages at standard AC power frequencies of 50 to 60 Hz. As with electric strength, values obtained on most materials are highly dependent on the moisture content and preconditioning method. Unless otherwise noted, measurements are performed at 23°C after preconditioning for 48 h in distilled water at 50°C followed by immersion in ambient-temperature distilled water for 30 min minimum, 4 h maximum. Dielectric breakdown is also performed in an oil medium. Dielectric breakdown for the materials shown in Table 8.8 is normally above 50 kV.

REFERENCES

1. NEMA, Industrial Laminating Thermosetting Products Standard, 1998.
2. IPC-4101, Specification for Base Materials for Rigid and Multilayer Printed Boards, 1997.
3. Nelco product reference materials.
4. General Electric product reference materials.
5. IPC-TM-650 Test Methods Manual, 2000.

CHAPTER 9
DENSIFICATION ISSUES FOR BASE MATERIALS

Edward J. Kelley
Sanmina Corporation, Derry, New Hampshire

9.1 IMPACT OF TRENDS IN IC TECHNOLOGY AND PCB DESIGN

Besides the variety in end use applications, printed circuit technology is also driven by the trends in integrated circuit (IC) packaging. In the 1970s, surface mount technology was introduced and changed forever the nature of packaging and component interconnection. This in turn drove changes in printed circuit technology in order to accommodate and take advantage of this new development. The shift from dual-sided leadframe packages to the quad flat pack (QFP), with subsequent reductions in lead pitch, resulted in significant increases in interconnect density.

As with most technologies, limitations to further developments were ultimately experienced. The advent of the ball grid array (BGA) package was an approach to solving the problems of the QFP by adapting the array interconnect approach of flip-chip techniques to the package-to-board interconnect problem. Experience with the BGA has generally shown that BGA assembly processes are much simpler. However, design and use of BGA and chip-scale packages (CSPs) is highly dependent on printed circuit board technology. This is due to the fact that it is more difficult to route conductors to the array footprint of the BGA than the inline peripheral footprint of the QFP and that, for most BGAs, a printed circuit is used as the principal component of the package structure and takes the place of the leadframe in a QFP. The design and manufacturing of high-density BGA packages are paced by the interconnect density of available PCB technologies, which are in turn affected by available base material technologies.

The increasing density in BGA devices results in smaller vias and via-to-pad and via-to-clearance ratios, higher circuit layer counts to route conductors, and thinner conductors in order to maximize circuit routing. In turn, these design requirements create a need for base materials that are more dimensionally stable, even as individual laminates get thinner, in order to properly align the layers within a multilayer circuit. As the densification trend continues, the need for improved thermal and electrical performance, flatness, and resistance to warp and twist, and a host of other properties, becomes increasingly important for the printed circuit to support the requirements of IC technology.

In high-frequency circuits, signal speed and signal integrity issues are also critical considerations. The electrical performance required of the printed circuit in these applications places strict demands on the base materials. Specifically, materials with low and consistent dielectric constants and dissipation factors are needed in these applications.

Taken together, these developments are driving the density revolution in printed circuit technology. The implications of this density revolution on base materials is also addressed in this chapter.

9.2   METHODS OF INCREASING CIRCUIT DENSITY

There are basically only three ways to increase printed circuit density:

- Decrease conductor linewidths and the spacings between them
- Increase the number of circuit layers in the PCB
- Reduce via and pad sizes

Decreasing conductor linewidths requires very low-profile copper foils for high yields in circuit etching processes. Increasing circuit layer counts has resulted in both greater overall multilayer thicknesses and thinner individual dielectrics, making thickness control and thermal reliability more important than ever. Reducing via and pad sizes requires improved laminate dimensional stability for registration of high-layer-count circuits.

Circuits operating at high frequencies are driving the use of materials with low dielectric constants, low dissipation factors, and tighter thickness tolerances. IC packaging technologies such as BGA and chip-scale packages require improved PCB flatness.

9.3   COPPER FOIL

One obvious method of increasing printed circuit functionality is to include more circuitry per unit area of the circuit. Printed circuit densification has driven several improvements in copper foil technology. One of the first developments was high-temperature elongation (HTE) foils. Other advances include low-profile and very-low-profile foils, thin foils, and foils for high-performance resin systems.

9.3.1   HTE Foil

HTE, or Class 3, copper foil exhibits improved elongation properties at elevated temperatures as compared to standard electrodeposited or Class 1 copper foil. HTE copper foil typically exhibits elongations of 4 to 10 percent at 180°C.

The growth in multilayer printed circuits has resulted in HTE becoming the most commonly used foil because its excellent ductility at elevated temperatures helps prevent innerlayer copper foil cracking. As a printed circuit experiences a thermal cycle, the base materials expand. The z-axis expansion applies stress to the connection of the innerlayer foil and the plated hole. With HTE foil, the reliability of this connection is improved. This property is particularly important in thicker circuits and high-resin-content constructions where increased z-axis expansion will occur.

9.3.2   Low-Profile and Reverse-Treated Copper Foils

Three classifications describe the profile of the copper foil surface, as shown in Table 9.1.

Copper foil profile is important for etching of fine-line circuits. Figures 6.9 and 6.10 in Chap. 6 illustrate the difference between standard and low-profile foils. As can be seen in those photos, the tooth profile of the standard-profile foil is much more pronounced. As you

TABLE 9.1   Copper Foil Profiles

Foil profile type	Maximum foil profile (μm)	Maximum foil profile (μin)
Standard (S)	N/A	N/A
Low-profile (L)	10.2	400
Very-low-profile (V)	5.1	200
No treatment or roughness (X)	N/A	N/A

can imagine, the etching of the lower-profile foil results in more control of the geometry of the circuit trace. In addition, in very thin laminates the large tooth structure of the standard-profile foil can result in very inconsistent dielectric thickness, making impedance control more difficult, and can even result in electrical failures if the tooth structures from the opposing sides of the laminate protrude sufficiently.

Reverse-treated foils (RTFs), as discussed in Chap. 6, take this concept a step further. As discussed earlier, when copper foil is manufactured, there is a very smooth, shiny side and a rougher matte side. Conventional technology involves treating the matte side and laminating this side to the base material. Reverse-treated foil, as its name implies, involves putting the treatments on the smooth, shiny side of the foil and laminating this side to the base material. This has two important effects. First, the side bonded to the base material has an extremely low profile that aids in etching very fine circuit traces. Second, the rougher matte side, which is now on the surface of the laminate, can improve photoresist adhesion. This enables the removal of surface roughening processes and can also improve innerlayer imaging and etching yields. Figure 9.1 compares laminates made with conventional and RTF foils.

9.3.3   Thin Copper Foils

The capability to etch fine-line circuits is also improved through the use of thinner copper foils. While electrical considerations can limit the use of very thin foils on innerlayer circuits, these thin foils can be used on external layers because the outerlayer process involves plating

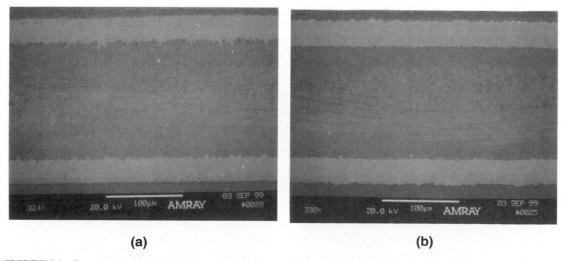

(a)   (b)

FIGURE 9.1   Comparison of laminates with standard (*a*) vs. RTF (*b*) copper foils. *(Photos courtesy of Polyclad Technologies.)*

on top of the foil to the desired overall thickness. For fine-line, dense circuitry, 5.0-μm and 9.0-μm copper foils are sometimes used. Processes that use 3.0-μm copper foil have also been developed.

9.3.4 Foils for High-Performance Resin Systems

Many of the high-performance resin systems such as BT, polyimide, cyanate ester, and even some high-T_g epoxies exhibit lower peel strengths and resistance to undermining of the copper foil when exposed to aggressive chemistries. For these applications, foils with increased nodularization and coupling agents tailored to the resin system are often used. The increased nodularization results in more surface area for mechanical adhesion, while the specific coupling agent aids in chemically bonding the foil to the resin system.

9.4 DIMENSIONAL STABILITY

As circuit layer counts grow and via-to-pad size ratios get tighter, the alignment or registration of the layers of circuitry becomes extremely important. While there are several material and process variables that contribute to the capability to achieve layer-to-layer and via-to-innerlayer feature registration, laminate dimensional stability is one of the most important. This is especially true in high-layer-count circuits that use thin laminates, since thinner laminates are generally not as dimensionally stable as thicker laminates. An example can help illustrate this point.

9.4.1 A Model of Printed Circuit Registration Capability

Table 9.2 lists several of the key variables that influence an overall via-to-innerlayer pad registration capability in the printed circuit manufacturing process. The values in the table represent individual process standard deviations using a specific multilayer circuit design. If we make the assumption that these processes are normally distributed and that they are centered on the desired nominal value, we can use additivity of variance to estimate an overall registration capability. In other words, taking the square root of the sum of the squares of the standard deviation of each individual process gives an overall process standard deviation that can be used to assess capability.

TABLE 9.2   Multilayer Registration Variables

	Process σ	
Process variable	Material type A	Material type B
Artwork plotting	0.33	0.33
Artwork alignment	0.70	0.70
Postetch punch	0.40	0.40
Laminate stability	1.70	0.90
Drill setup	0.80	0.80
Hole location	1.00	1.00
Overall σ	2.30	1.79
Overall ±3σ	6.90	5.37
Drill ⌐ ? capability	13.80	10.75

Doubling the overall $\pm 3\sigma$ value gives an estimate of the added innerlayer feature diameter over the drilled hole diameter necessary to maintain acceptable registration. In other words, if we are trying to maintain at least tangency of a drilled hole to an innerlayer pad with a 13.5-mil hole, the pad needs to be almost 27.5 mils in diameter for material type A (13.5 + 13.8 = 27.3). The same procedure can be used to calculate clearance or antipad diameters with the desired plated through-hole–to–clearance distance, which is also an important design consideration. It is important to note that items such as dimensional stability and hole location are significantly impacted by the specific material type, board design, and manufacturing processes used, and that actual registration systems are more complex than described here. The point of this analysis is to show the impact of laminate dimensional stability on printed circuit design. As you can see in this example, improving laminate stability from a standard deviation of 1.70 to 0.90 improves the overall registration capability from drill size +13.8 mils to drill size +10.75 mils, or a reduction of over 3 mils in the required diameter of the internal feature. Reductions in the required size of internal features allows increased circuit density. Therefore, increased circuit density drives base materials with improved dimensional stability.

9.4.2 Dimensional Stability Test Methods

A common test method used to evaluate dimensional stability starts with a sample of a copper-clad laminate with scribed targets or holes in the four corners of the sample. Baseline measurements of the distances between these holes are taken prior to conditioning of the sample. One conditioning procedure involves etching the copper cladding off and remeasuring and determining the dimensional movement compared to the baseline. A second method subjects the sample to a thermal cycle, commonly a bake at 150°C for 2 h. Again, measurements are taken after conditioning and compared to the baseline dimensions. A third method involves first etching the cladding off, measuring, and then subjecting the sample to the bake cycle and measuring again. Each of these methods can be used as a process control tool in the laminate manufacturing process.

However, these test methods are of limited value to the printed circuit manufacturer, who etches circuit images on these laminates, combines them with prepreg materials and other laminates, and presses them together under temperature and pressure to form a multilayer circuit. It is the predictability and consistency of laminate movement through the printed circuit manufacturing process, across a variety of circuit patterns, and especially through the multilayer lamination cycle that is of concern for the circuit manufacturer and ultimately for the designer who wants to increase circuit density. Multilayer lamination cycles commonly reach 185°C or higher, and normally exceed the T_g of the base material. Above the T_g, the resin softens and allows tension in the laminate to be released and is also subject to stresses from the surrounding materials and lamination pressure. Most of the movement of the laminate in the circuit manufacturing process occurs during this lamination cycle.

9.4.3 Improving Dimensional Stability

While there are many variables in laminate and circuit manufacturing processes that can influence dimensional stability, some common techniques involve laminate press cycle optimization, control of laminate resin contents, and the use of higher-T_g materials. New materials and process techniques have also been developed to improve dimensional stability.

9.4.3.1 Laminate Stabilization Processes. Some printed circuit manufacturers have the laminate manufacturer bake the laminate prior to shipment, or bake it themselves prior to use. The intent is to relieve any stress that may be stored within the laminate prior to use in the circuit manufacturing operation. While this process may help, the added material handling and cycle time do not usually justify the process. Instead, many laminate manufacturers reduce the lamination pressure at a specified point in the lamination cycle in order to minimize the stress that becomes stored in the finished product.

New laminate manufacturing techniques may also offer improved dimensional stability. The direct current and continuous manufacturing processes described in Chap. 7 result in consistent thermal profiles from laminate to laminate and can use low lamination pressures. Optimizing these parameters can lead to consistent dimensional stability.

9.4.3.2   Fiberglass Cloths and Resin-to-Glass Ratios.   Each fiberglass cloth style used in laminates and prepregs has a resin content range that results in sufficient wet-out of the glass cloth, is relatively easy to control, and therefore results in more uniform thickness and more consistent dimensional stability. The specification of laminates and prepregs with resin contents in the desired ranges can result in improvements in dimensional stability and therefore registration capabilities. As a result, having a range of fiberglass cloth styles available is important in order to be able to achieve a wide range of dielectric thicknesses.

9.4.3.3   Nonwoven Reinforcements.   Because of the woven, serpentine geometry of the yarns in fiberglass cloths, they can behave like springs when used to reinforce base materials. During the resin-treating process and the laminate-pressing operation, the fiberglass cloth is subjected to stresses that can be stored in the laminate as the resin is cured. These stresses can then be released during the circuit manufacturing process, causing dimensional changes.

Nonwoven materials can avoid these stresses. In one type of nonwoven material, short, randomly oriented fibers are treated with the resin system. In a second type, linear strands of fiberglass are laid down in a balanced, cross-plied orientation that resists subsequent stresses.

9.4.3.4   Multilayer Press Cycles Below Laminate $\mathbf{T_g}$.   Most of the dimensional changes in the laminate occur during the multilayer press cycle when the temperature exceeds the T_g of the resin in the laminate. Above the T_g the resin system in the laminate softens, allowing any stored stress in the reinforcement to be released and allowing the laminate to be affected by the adjacent materials and the pressure of the lamination cycle. Using a resin system in the prepreg materials that can be cured below the T_g of the resin system in the laminate can avoid the softening of the laminate resin system and therefore prevent much of this movement.

9.5   THERMAL PROPERTIES AND RELIABILITY

The ultimate reliability of the finished printed circuit board can be greatly affected by the base materials. Increased circuit densities make base material reliability even more critical because of the heat generated by dense circuitry and the very thin dielectrics used in high-layer-count multilayers. The use of lead-free solders, which require higher soldering temperatures, also makes the resistance to thermal stress more important. Choosing the right resin system and resin content in the base materials improves circuit reliability. In addition, as conductor spacings, via spacings, and via-to-conductor spacings shrink, the resistance to conductive anodic filament (CAF) growth or electromigration becomes an important base material property.

9.5.1   Reliability Testing

There are a variety of test procedures designed to assess the long-term reliability of printed circuits and the materials used in them. While a full description of these test methods is beyond the scope of this chapter, they all subject samples to extreme environmental conditions in order to accelerate failure. Some of these tests include:

- *Solder shock:* typically involves floating or dipping a sample in molten solder at 288°C. Multiple shocks provide a relatively quick measure of reliability, although they are more qualitative than the other tests. Failures consist of material delamination, separation of the plated copper from the via hole wall, and excessive lifting of copper pads on the circuit board surface.

- *Thermal cycling:* these test methods subject the samples to alternating temperature extremes. Q-1000 test methods, commonly used in the automotive industry, include air-to-air thermal cycles from −40 to 125°C and −50 to 150°C. Other procedures use oil as the medium rather than air in order to accelerate heat transfer. Interconnect stress testing (IST) utilizes current to rapidly heat test samples and cycle them between the desired temperatures. Delamination and other physical defects in the sample also constitute failure in these methods. But, in addition, these procedures typically involve measuring electrical continuity or resistance in a test circuit and using these electrical measures to determine failure as well. The number of cycles the sample withstands before failure provides the measure of reliability.

- *Other environmental conditions:* includes exposure to conditions of temperature and humidity or pressure and humidity.

9.5.2 Conductive Anodic Filament (CAF) Growth

CAF formation, or electromigration, is a term used to describe an electrochemical reaction in which conductive paths are formed through a dielectric material. These paths may form between two circuit traces, between two vias, or between a trace and a via. By definition, as circuit density increases, the space between these features decreases. With shorter paths between features, CAF growth becomes a more critical reliability consideration.

In order for CAF growth to occur, a bias and a path for this filament growth must be present. In fiberglass-reinforced materials, the fiberglass filaments can bridge the gap between traces and vias. If the bond between the resin system and the fiberglass filaments is insufficient or is compromised, this can become such a path. Hollow fiberglass filaments could also be a path. In addition, there must be a medium in which this electromigration can occur. Absorbed moisture within the resin system allows dissolved ionic species to migrate and promote the electrochemical reaction that results in CAF formation.

Test methods for CAF resistance typically include trace-to-trace, trace-to-via, and via-to-via test coupons. A voltage is applied to the coupons while they are exposed to specified temperature and humidity conditions. As expected, higher voltages, higher temperatures, and higher relative humidities tend to accelerate CAF formation.

9.5.3 Choosing a Base Material

For resistance to CAF growth, the resin system and the bonding of the resin system to the fiberglass cloth are key considerations. Coupling agents, typically silanes, are used to promote adhesion between the fiberglass filaments and the resin system. The molecular structure of the coupling agent includes a portion that bonds well to the fiberglass surface and other portions that bond well to the resin system. Specific coupling agents have been developed for specific resin types. Achieving a good bond between the fiberglass filaments and the resin system with an appropriate coupling agent helps prevent this interface from becoming a path for CAF growth.

In addition, low rates of moisture absorption and low levels of ionic contamination within the resin system also aid in resisting CAF growth. Epichlorohydrin is a precursor material used to make epoxy resins. The use of this compound can result in residual hydrolyzable chlorides in the epoxy resin. In the presence of moisture, these hydrolyzable chlorides form ionic species that can promote electromigration. When CAF resistance is a critical consideration, the use of high-purity resins is required.

As discussed earlier, the thermal reliability of printed circuits is affected by the z-axis expansion of the base materials. Specifically, the degree to which the printed circuit expands and contracts in response to temperature cycling determines the amount of stress applied to the plated vias. The overall expansion of the base material is determined by the resin type, the resin content, and the T_g of the resin system.

- *Resin system type:* the coefficients of thermal expansion (CTE) below and above the T_g should be low.
- *Resin content:* lower levels of resin content result in lower overall expansion values.
- T_g: since coefficients of thermal expansion above the T_g are much higher than below the T_g, other things being equal, higher T_g values will result in less overall expansion.

Bear in mind, however, that these parameters interact. For example, it is possible for two different resin types to have the same T_g but differing CTE values. Likewise, a material with a lower relative T_g can have the same overall level of expansion as a material of a higher relative T_g. Table 8.1 in Chap. 8 illustrates this.

In addition to z-axis expansion, the resin decomposition temperature can also affect reliability. The temperatures to which printed circuits can be exposed, either during assembly or in operating environments, can approach or exceed the decomposition temperatures of some resin systems. It is important to understand what temperatures the circuit materials will be exposed to and to choose a resin system with an appropriate decomposition temperature. Table 8.2 in Chap. 8 shows the decomposition temperatures of some commonly available materials.

9.6 ELECTRICAL PROPERTIES

Base material electrical properties are an important consideration in sophisticated printed circuits operating at high frequencies. The dielectric constant (D_k) and dissipation factor (D_f) of the base materials become very important in these applications. Moreover, the consistency of these properties over a large frequency range is also important. These properties are defined in Chap. 8 as follows:

- *Dielectric constant/permittivity:* the ratio of the capacitance of a capacitor with a given dielectric material to the capacitance of the same capacitor with air as a dielectric. It refers to the ability of a material to store an electric charge.
- *Dissipation factor/loss tangent:* the ratio of the total power loss in a material to the product of the voltage and current in a capacitor in which the material is a dielectric.

9.6.1 Importance of These Properties

These properties are important because they affect signal transmission in the printed circuit. At low frequencies, a signal path in a printed circuit can typically be represented electrically as a capacitance in parallel with a resistance. However, as frequencies increase, at some point signal paths must be considered transmission lines where the electrical and dielectric properties of the base materials have a greater effect on signal transmission. A full discussion of capacitive versus transmission line environments is beyond the scope of this chapter, but the premise is to determine, for the transmission of a signal pulse of a given rise time, the acceptable length of a conductor before a significant voltage difference is realized along its length. Conductors longer than this critical value are then regarded as transmission lines. Because the velocity of signal propagation is inversely proportional to the square root of the permittivity of the dielectric, a lower permittivity value results in faster signal speeds and a longer rise distance. With a larger rise distance, larger conductor lengths are acceptable before a significant voltage drop is experienced. However, if the ratio of conductor length to rise distance is large enough, signal reflections from a mismatched load impedance may be received back at the source after the pulse has reached its maximum plateau value, and pulse additions that occur under these circumstances may lead to false triggering of a device.

On the other hand, signal attenuation can result in missed signals. One of the causes of signal attenuation is dielectric loss. As the circuit operates, energy from the signal is absorbed by the dielectric medium. Attenuation of the signal by the dielectric is directly proportional to the square root of permittivity and directly proportional to the loss tangent. In addition, dielectric losses increase as frequencies increase. When a high bandwidth is desired, this results in a larger effect on the higher-frequency components, and the bandwidth of the propagating pulse will decrease and degrade the rise time.

Because the permittivity and loss tangent vary with frequency, and other factors to be discussed later, the degree to which these properties vary is also an important circuit design consideration. Obviously, if these properties vary significantly with frequency, designing a circuit with devices that operate at various frequencies becomes that much more complex. In addition, operating within a given bandwidth becomes that much more difficult as different frequency components experience different dielectric properties that in turn lead to differences in signal propagation and loss.

Therefore, base materials with low permittivity values and low loss factors are desired for high-speed, high-frequency printed circuits. In addition, consistency of these properties across frequencies is also required. Besides frequency dependence, since operating environments can also vary, the consistency of these properties across environmental conditions is also important and will be discussed later.

9.6.2 Choosing a Base Material

The dielectric constant and dissipation factor of the dielectric material are determined by both the resin system and the reinforcement type. Therefore, each of these should be considered when choosing a material.

There are a variety of low-dielectric, low-loss resin systems available for high-speed circuit applications. These include polytetrafluoroethylene (PTFE or Teflon®), cyanate ester, epoxy blends, and allylated polyphenylene ether (APPE). Likewise, there are a few different reinforcements and fillers available that can be used to modify the electrical properties of the base material. While E-glass is still the most commonly used fiberglass reinforcement, it should be noted that others are available. In addition, inorganic fillers are sometimes used to modify electrical properties as well. Table 9.3 provides electrical property data on some of the available fiberglass materials. Table 9.4 provides data on some of the base material composites available.

TABLE 9.3 Dielectric Constants and Dissipation Factors of Common Glass Types

Reinforcement	D_k at		D_f at	
	1 MHz	1 GHz	1 MHz	1 GHz
E-glass	6.6	6.1	0.0025	0.0038
S-glass	5.3	5.2	0.0020	0.0068
D-glass	3.8	4.0	0.0010	0.0026

TABLE 9.4 Dielectric Constants and Dissipation Factors of Common Resin/Reinforcement Composites

Resin system	Reinforcement	D_k at		D_f at	
		MHz	1 GHz	1 MHz	1 GHz
Epoxy	E-glass	4.4	3.9	0.025	0.015
Epoxy	Aramid fiber	3.9	3.8	0.024	0.019
Epoxy/PPO	E-glass	3.9	3.9	0.011	0.009
Epoxy blend	E-glass	3.9	3.8	0.009	0.010
Epoxy blend	SI™-glass	3.6	3.5	0.008	0.007
Cyanate ester	E-glass	3.8	3.5	0.008	0.006
Polyimide	E-glass	4.3	3.7	0.013	0.007
APPE	E-glass	3.7	3.4	0.005	0.007
APPE	SI™-glass	3.4	3.0	0.003	0.004
PTFE	E-glass	2.3	2.3	0.0013	0.0009
Epoxy*	Expanded PTFE	3.0	2.8	0.023	0.019
Cyanate Ester*	Expanded PTFE	2.5	2.6	0.0038	0.0035
Hydrocarbon	E-glass/ceramic	3.4	3.3	0.0025	0.0024

* Prepreg only.

Table 9.4 lists values for approximately 50 percent resin content. These values change as resin content varies, with the magnitude of the change depending on the specific resin system. Also, comparable resin system types from various suppliers could differ. For example, because there are many types of epoxies, with a wide range of electrical properties, specific epoxy formulations from different material suppliers can exhibit somewhat different dielectric and loss properties. The purpose of Table 9.4 is to highlight the relative differences between various resin systems and reinforcement types. Figure 9.2 illustrates the resin content dependence of the dielectric constant for several materials. Figure 9.3 shows the impact of resin content on the dissipation factor of the low-D_k epoxy blend.

Beyond frequency and resin content effects, the dielectric constant and dissipation factor can vary with temperature and moisture absorption. When choosing a material for a specific application, it is important to understand the operating conditions and environment the circuit will be used in. Some resin systems exhibit less sensitivity to these conditions than others. For example, base material suppliers are continually developing resin systems to meet the demanding electrical properties requirements of high-speed and wireless applications, which include consistent properties over frequency ranges and environmental conditions. Figure 9.4

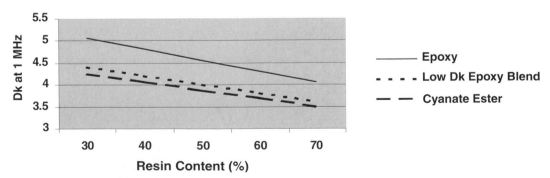

FIGURE 9.2 Dielectric constant vs. resin constant.

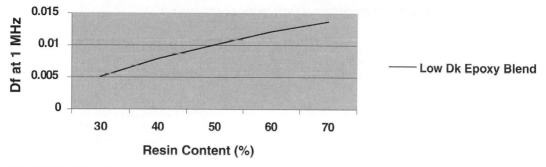

FIGURE 9.3 Dissipation factor vs. resin content for a low-D_k epoxy blend.

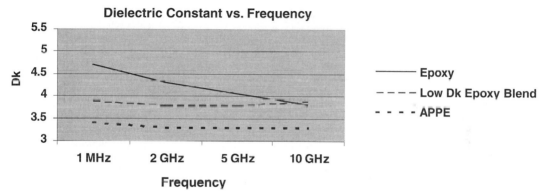

FIGURE 9.4 Dielectric constant vs. frequency.

shows the dielectric constant response to frequency variation of two of these materials in comparison to a standard epoxy FR-4 material. As you can observe from the graph, these new materials exhibit much more consistent values across frequency.

In addition, when selecting a material, it is important not to simply select the material with the lowest dielectric constant or dissipation factor, since there are typically cost-performance trade-offs to be made. In general, the lower the dielectric constant and dissipation factor, the more costly the material, and very often the more difficult it is to process.

In summary, some general relationships include:

- The dielectric constant often decreases with increasing resin content.
- The dissipation factor often increases with increasing resin content.
- The dielectric constant typically drops as the frequency increases.
- The dielectric constant and dissipation factor typically increase as water absorption rises.
- The dielectric constant of E-glass is only mildly frequency dependent and therefore lower resin contents result in less variation across frequencies.
- The dissipation factor typically exhibits maxima at certain frequencies.

9.7 *HIGH-DENSITY INTERCONNECT/MICROVIA MATERIALS*

One method used to increase circuit density is to use blind and buried vias. Rather than extending completely through the printed circuit board, blind and buried vias go only partly through the multilayer circuit, joining only the layers that require connection. Because these vias do not go through the entire multilayer, real estate on the other layers becomes available for additional circuit routing. Buried vias are those that are not visible from the outside of the finished circuit board, and are formed in a subcomposite or copper-clad laminate. Blind vias are those that are visible from the outside of the multilayer circuit but do not go completely through it. By limiting the size of these vias, significantly increased interconnection density is achievable. Microvia or high-density interconnection (HDI) printed circuit designs utilize these technologies to increase circuit density.

While the materials already discussed are employed in blind and buried via applications that use conventional processes, there are additional materials that can be used to increase density in more specialized process techniques. The specialized processes used to form microvias include laser ablation, plasma etching, and photoimaging.

Certain types of lasers are limited in their ability to ablate the fiberglass reinforcement common to most base materials. Also, plasmas are not effective in etching through fiberglass. As a result, materials that use an alternative reinforcement or that do not contain an inorganic reinforcement have been developed for these applications.

For blind-via applications, resin-coated copper foil can be used to form the external circuit layer and dielectrics between layers 1 and 2 and n to $n - 1$, using laser or plasma processes to form the vias. Two basic types of resin-coated copper foil are available, although there are several resin systems available. The first type uses one layer of partially cured resin, as shown in Fig. 9.5. This resin-coated foil is then laminated to the rest of the multilayer circuit. A second type of resin-coated copper foil uses two layers of resin, as shown in Figs. 9.6 and 9.7. The first layer is fully cured, while the second is partially cured. This technique guarantees a minimum dielectric separation between the external foil and the circuitry on the next layer in, because the cured resin layer limits how close the internal circuit layer can get to the external foil.

Another material used in HDI designs utilizes an organic reinforcement that can be laser-ablated or plasma-etched. The most common organic reinforcement used is aramid fiber–based.

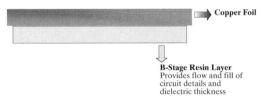

FIGURE 9.5 Resin-coated copper foil with a single layer of B-staged resin. *(Courtesy of Polyclad Technologies.)*

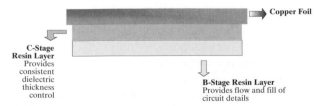

FIGURE 9.6 Resin-coated copper foil with a C-stage and B-stage layer. *(Courtesy of Polyclad Technologies.)*

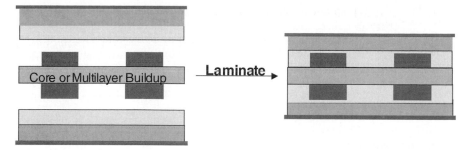

FIGURE 9.7   C-stage plus B-stage resin-coated copper foil laminated to PCB. *(Courtesy of Polyclad Technologies.)*

The aramid fibers are randomly oriented and formed into a sheet that is impregnated with the resin system. In this way, both copper-clad laminates and prepregs can be manufactured and used in multilayer applications. Table 9.5 shows some available thicknesses of Thermount® aramid fiber reinforcement with 50 percent resin content. An additional reinforcement that can be used to make a prepreg material is expanded PTFE. This material has a spongelike structure that can also be impregnated with a resin system and used in HDI applications (see Fig. 6.7 in Chap. 6). Expanded PTFE also has a very low dielectric constant and loss factor.

TABLE 9.5   Commonly Available Thicknesses of Thermount®

Thermount® type	Prepreg thickness	Laminate thickness
E210	0.0018 in/46 µm	0.0020 in/51 µm
E220	0.0030 in/76 µm	0.0032 in/81 µm
E230	0.0037 in/94 µm	0.0039 in/99 µm

The third process technique used in these applications involves photoimaging a permanent dielectric material in order to form the microvias. These photoimageable dielectrics resemble plating resists but must be able to be catalyzed for subsequent plating operations that will form the external circuit image, and must adhere sufficiently to the rest of the multilayer circuit to provide long-term reliability.

9.8   INTEGRATED PASSIVES

Typically the passive components used in the finished printed circuit are assembled onto the surface of the printed circuit board or in through-holes. In some applications these devices can be replaced by integrating the devices or functions within the circuit board itself. Buried capacitance and buried resistors can replace external capacitors and resistors in these applications. The common materials used for integrating these passive components into the printed circuit board itself are described in the following text.

9.8.1   Buried Capacitance

Printed circuit assemblies are designed with bypass capacitors mounted adjacent to the integrated circuit devices on the surface of the PCB. These capacitors take up valuable real estate and contribute to parasitic inductance. Buried capacitance is a technology that utilizes copper-clad laminates with very thin dielectrics in order to achieve capacitance between these layers that can replace many of the surface bypass capacitors. The common form of this technology utilizes 0.0020-in dielectrics with very tight thickness tolerances in order to achieve a specific range of capacitance. Power and ground circuit designs are imaged on either side of this laminate, forming the capacitive planes that are connected to the rest of the printed circuit. In addition, copper foils with very low profiles are required to ensure that the dielectric separation is consistent and that reliability will be sufficient. This buried capacitance technology is currently patented and licensed by the Sanmina Corporation.

9.8.2 Buried Resistors

Just as buried capacitance enables the elimination of surface capacitors, buried resistors enable the elimination of surface resistor components. In this case, the common material used is copper foil that has been coated with a thin film of a nickel-phosphorus alloy. This coated copper foil is then used to manufacture a base laminate, with the alloy side of the foil against the base laminate. The alloy type and thickness, along with the imaged size of the resistor, is designed to provide the desired resistance values in the finished circuit. In the manufacturing process, sequential imaging and etching processes are used to first form the desired copper circuit pattern, and then to image and etch the resistive layer. The result is a circuit pattern with resistors integrated within it.

REFERENCES

1. James Hayward, "The Problem of Interconnection and the Impact of Printed Circuit Board Technology on Integrated Circuit Package Development," *Future Circuits International,* iss. 4, pp. 19–22.
2. IPC-4101, Specification for Base Materials for Rigid and Multilayer Printed Boards.
3. Nelco product reference materials.
4. Polyclad product reference materials.
5. Clark-Schwebel industrial fabrics guide.
6. BGF Industries, Inc. fiberglass guide.
7. Gould Electronics product reference materials.
8. Edward J. Kelley and Richard A. Micha, "Improved Printed Circuit Manufacturing with Reverse Treated Copper Foils," *IPC Printed Circuits Expo,* March 1997.
9. IPC-TM-650 test methods manual.
10. IPC-4562, Metal Foil for Printed Wiring Applications.
11. Edward Kelley and Owen Christofferson, "Multilayer Printed Circuits with Exacting Registration Requirements, Achieving Next Generation Design Capabilities," *IPC Printed Circuits Expo,* May 1998.
12. Edward Kelley, "Meeting the Needs of the Density Revolution with Non-Woven Fiberglass Reinforced Laminates," *EIPC/Productronica Conference,* November 1999.
13. Sanmina Corp. technical literature.
14. Ohmega Corp. technical literature.

CHAPTER 10
INTRODUCING BASE MATERIALS INTO THE PCB MANUFACTURING PROCESS

Edward J. Kelley
Sanmina Corporation, Derry, New Hampshire

10.1 INTRODUCTION

Besides the electrical and physical properties that must be considered for the end use environment of printed circuits, the ease of processing of these materials is also important because it will ultimately affect the cost-performance relationship. While manufacturers of printed circuits may have their own qualification procedures and specific applications may require more specialized qualifications, there are several fairly common procedures used to introduce a new base material into a manufacturing process. Several of these procedures and qualifications are discussed in this chapter.

10.2 VALIDATION OF PHYSICAL, THERMAL, AND ELECTRICAL PROPERTIES

Many of the properties of base materials discussed in Chap. 8 are reported by the material supplier on product data sheets. The data may have been obtained from testing performed by the material supplier, by independent lab testing, or both. To the extent that these data are certified by the supplier or upon availability of test reports, the printed circuit manufacturer may choose not to retest all of these properties. However, when the material is to be used for specific applications, the printed circuit manufacturer may want to have the properties relevant to that application verified. For example, when low–dielectric constant materials are chosen due to the demands of high operating frequencies, the circuit manufacturer may want to have various material constructions tested for dielectric constant vs. frequency, vs. temperature and humidity, etc.

Also, some of the properties are relatively important for the circuit manufacturing process and are fairly easy and inexpensive to validate. These include verification of T_g, copper peel strengths, z-axis CTE, time to delamination, and bond strengths to internal layers. These properties are typically verified when introducing the material into a circuit manufacturing process. Reliability testing, as described in Chap. 9, is also an important part of a new material introduction.

10.3 *LAMINATE CONSTRUCTIONS*

During introduction of a new material is often the best time to reevaluate the preferred laminate constructions used in the printed circuit manufacturing process. Several considerations must be taken into account when choosing laminate constructions, leading to a choice that balances the requirements of cost, processability, and circuit performance. Several examples of laminate constructions are given in Table 10.1.

TABLE 10.1 Common Laminate Constructions

Laminate thickness	Fiberglass cloth(s)	Approximate resin %
0.0020 in	1-106	69–70%
0.0025 in	1-1080	56–57%
0.0030 in	1-1080	62–63%
0.0030 in	1-2313	43–44%
0.0035 in	1-2313	48–49%
0.0040 in	1-2313	53–54%
0.0040 in	1-2116	45–46%
0.0040 in	1-106/1-1080	59–60%
0.0050 in	1-2116	52–53%
0.0050 in	1-2165	47–48%
0.0050 in	2-1080	56–57%
0.0050 in	1-2313/1-106	52–53%
0.0062 in	1-2157	47–48%
0.0062 in	1-1080/1-2313	54–55%
0.0080 in	1-7628	44–45%
0.0080 in	1-2116/1-2313	49–50%
0.0080 in	2-2116	45–46%
0.0090 in	2-2116	49–50%
0.0140 in	2-7628	39–40%

10.3.1 Single-Ply vs. Multiple-Ply Constructions

With dielectrics below 0.0040 in, there is often no choice but to use a single ply of fiberglass cloth in order to achieve the desired thickness. With dielectrics in the 0.0040- to 0.0080-in range, there are both single-ply and multiple-ply options. Above 0.0080 in, multiple plies of cloth are typically required to achieve the desired thickness. Within each range, there are often multiple cloth and resin content combinations that can achieve the desired dielectric thickness. The choice of laminate construction can significantly impact both cost and performance. The choice of single-ply vs. multiple-ply construction, when the option is available, is no exception.

As should be expected, a single-ply construction typically represents a cost savings compared to a multiple-ply construction. The magnitude of this savings depends on the specific glass styles involved and a host of other parameters. Performance can also be affected and should be considered when specifying the constructions to be used. First, single-ply constructions are typically lower in resin content, as can be observed in Table 10.1. Resin content issues are discussed later. The other main benefit of single-ply constructions is dielectric thickness control beyond resin content considerations. Other things being equal, tighter thickness tolerances can be achieved using a single-ply construction than a multiple-ply construction because the variation in thickness control with one ply of prepreg is statistically less than with multiple plies.

10.3.2 Resin Contents

As Table 10.1 illustrates, the same dielectric thickness can be achieved with multiple fiberglass cloth and resin content combinations. Constructions with relatively lower resin contents are often preferred because they result in less z-axis expansion and can therefore improve reliability in many applications. In addition, lower resin contents can also improve dimensional stability, resistance to warpage, and dielectric thickness control. On the other hand, constructions with higher resin contents result in lower dielectric constant values, which are sometimes preferred for electrical performance, as discussed in Chap. 9.

10.3.3 Laminate Flatness and Flexural Strength

During manufacture of the innerlayer circuit patterns, the flatness and flexural strength of the base material is important for successfully transporting these circuit layers through conveyorized equipment. This is particularly important for very thin laminates. If the laminate is curled, it can become caught or damaged inside this equipment. If the thin laminate with the circuit image sags on the conveyors, similar damage can occur.

For these reasons, it is often preferred to use constructions with relatively high glass contents and to use the thickest cloths possible, as this will result in greater flexural strengths. In addition, balanced constructions (or those that use a symmetrical construction) are normally preferred in order to avoid curling. Asymmetrical constructions are more prone to curling and causing problems in conveyorized equipment. Consider a 0.0080-in laminate as an example. Table 10.1 shows three constructions for this thickness: 1-7628, 1-2313/1-2116, and 2-2116. The asymmetrical 1-2313/1-2116 construction is more prone to curling than the other constructions. A word of caution is warranted, however. Curling is but one property that must be considered. Dimensional stability, thickness control, and other properties must also be considered when choosing a construction.

10.4 PREPREG OPTIONS AND YIELD PER PLY VALUES

Just as there are a variety of base laminate constructions, there are a variety of prepreg options. Each fiberglass cloth style can be treated to achieve multiple resin contents and flow values. While the same properties must be considered for prepreg materials as for base laminates, prepregs must also contain sufficient resin to flow and fill in the innerlayer circuitry of a multilayer circuit board. Because the innerlayer circuits can vary in terms of copper thickness and circuit density, a variety of resin content and flow value options are typically required when specifying prepreg styles. Higher resin contents may be needed when filling thick copper weights and signal layers, while lower resin contents can be used against thinner copper weights and power or ground circuit patterns. Table 10.2 shows some common prepreg styles with various resin contents and thickness yields per ply.

10.5 MULTILAYER PRESS CYCLE QUALIFICATION

With materials to be used for multilayer circuits, one of the initial objectives is to establish the optimal multilayer press cycle. Typically the material supplier has performed resin rheology and cure studies and has initial recommendations for the heat rise and pressure required to achieve the proper amount of prepreg resin flow, and for the required time at temperature to achieve the desired level of resin cure. Based on these values, it is common for an initial multilayer press cycle to be established and tested. Product produced in this cycle is sampled for

TABLE 10.2 Common Prepreg Styles

Glass style	Resin content (%)	Thickness (in)*
106	64.1	0.0015
106	66.1	0.00175
106	69.7	0.0020
106	74.6	0.00225
1080	52.8	0.00225
1080	56.5	0.0025
1080	62.5	0.0030
2313	48.9	0.0035
2313	53.7	0.0040
2116	49.1	0.0045
2116	50.9	0.00475
2116	55.6	0.00525
2165	47.2	0.0050
2165	49.1	0.00525
2165	54.0	0.0060
2165	58.0	0.00675
2157	44.4	0.00575
2157	46.2	0.0060
2157	47.8	0.00625
2157	49.3	0.0065
7628	39.4	0.0070
7628	40.9	0.00725
7628	42.3	0.0075
7628	43.7	0.00775

* Assumes a given flow value and represents thickness without filling circuitry.

T_g values and ΔT_g values, or the difference in T_g value between successive measurements with a thermal cycle in between. This intermediate thermal cycle should be sufficient to promote any additional curing of the resin system in the event the initial press cycle is not sufficient for full cure. If the second T_g measurement exceeds the first by more than a few degrees Celsius, additional time or temperature may be required in the press cycle. In addition, dielectric and overall thicknesses should be measured, and samples of circuits from the outside and inside of the stack of circuits pressed should have the copper foil etched off so that inspection for lamination voids, poor resin flow, or other defects can be performed.

It is also important to evaluate the possible variations in product type and equipment performance. Evaluating the number of multilayer circuits that can be pressed in a given stack is one of these variables. A larger stack or a stack of greater heat capacity will require more time to reach the cure temperature than a smaller stack. The chosen press cycle must be able to handle these variations. It is also common for the outside and inside circuits to experience slightly different heat rise profiles. The difference in thermal profile can affect resin flow and is the reason why outside and inside circuits should be etched and inspected in the qualification process. Different resin systems may be more or less sensitive to these heat rise differences. Ongoing process controls typically involve monitoring of T_gs and ΔT_gs, and heat rise and cooldown rates, at both the inside and outside of the stack.

While full cure of the resin system during the press cycle is typically the desired process, there are some cases in which a partial cure in the press cycle followed by a post-cure process is used. This process is sometimes used when the material needs to reach temperatures that exceed the capability of the manufacturer's press equipment, or when the press process limits the capacity of the manufacturer. While this process is still used successfully, it increases mate-

TABLE 10.3   Multilayer Cure Times for Some Common Base
Materials

Resin system	Time @ temperature required for full cure
Epoxy	45–90 min @ 340–360°F
Low-D_k epoxy blends	90–120 min @ 380–395°F
BT/epoxy	60–120 min @ 360–375°F
APPE	90 min @ 360°F
Polyimide	135–240 min @ 425°F
Cyanate ester	240 min @ 425°F

rial handling and can result in increased warp and twist if the postcure process is not well
designed.

The time required for the press cycle is also an important parameter to consider when
introducing a new material. Small changes in press cycle time can have dramatic effects on
press capacity. When evaluating a new material it is important to quantify this value. Table
10.3 shows the time at temperature required in the multilayer press cycle for some commonly
available materials. The range in values for the different resin systems is indicative of the multiple formulations available.

10.6   PREPREG-TO-INNERLAYER CIRCUIT ADHESION

Obviously the bond between the prepreg and the internal circuitry is important for reliability.
When introducing a new material into a printed circuit manufacturing process, it is a fairly
standard procedure to test this adhesion. One of the considerations when evaluating this
property is the type of adhesion promotion used between the copper circuitry and the resin in
the prepreg. These range from double-treated copper to copper oxide to tin oxide or micro-
roughened copper that may also be coated with a coupling agent. Most common FR-4 materials are compatible with all of these treatments. However, some specialty materials may
require specific adhesion promotion types.

Regardless of which type of adhesion promotion is used, it is common to perform a few
basic reliability tests followed by the longer-term reliability tests already discussed in Chap. 9.
The basic reliability tests normally consist of multiple solder shocks, time to delamination tests
such as T260 testing, and peel strengths of the copper to the prepreg material. Multiple solder
shocks are performed to ensure that no delamination between the prepreg and the innerlayer
circuits occurs. Similarly, time to delamination testing also stresses the adhesion between the
innerlayer conductive pattern and the prepreg material. However, time to delamination testing
proceeds until failure occurs, measuring the time it takes for delamination to occur. The greater
the adhesion, the longer it takes until delamination occurs, provided the temperature used does
not exceed the decomposition temperature of the resin system. If the test temperature exceeds
the resin decomposition temperature, the failure may be due to resin degradation rather than
to the bond between the prepreg and innerlayer conductor. Likewise, if the failure occurs
between the resin and the glass within the prepreg, little can be concluded about the prepreg-
to-innerlayer copper bond. For these reasons, evaluation of the failure mode is important
before drawing conclusions on the prepreg to innerlayer conductor adhesion.

Peel strength testing also provides a measure of adhesion between these layers. The procedure used for this purpose differs somewhat from that described earlier when the peel
strength to the base laminate was discussed. While the test method is essentially the same, the
sample preparation is very different. In testing the peel strength of the copper conductor to

the prepreg, we start with samples of copper foil and then treat the surface to be bonded to the prepreg with the appropriate adhesion promotion. This surface of the copper foil is then laminated to sheets of the prepreg we want to evaluate using the qualified press cycle for multilayer production. From this point on, the procedure is the same as that for evaluating base laminate peel strength. This procedure may also be used as a process control tool.

10.7 DIMENSIONAL STABILITY CHARACTERIZATIONS

Understanding the dimensional changes exhibited by thin laminates as they proceed through the printed circuit manufacturing process is critical for registration of the many layers that make up a multilayer circuit. This dimensional change is affected by many factors, both in the laminate manufacturing process and the printed circuit manufacturing process. The construction and resin content of the laminate as well as the prepregs used in the multilayer construction both affect dimensional stability. In addition, the circuit design itself can influence this dimensional movement. As a result, it is difficult for the laminate manufacturer to provide the circuit manufacturer with anything but general information regarding dimensional stability.

Therefore, the printed circuit manufacturer must characterize this movement for the range of material types and laminate constructions used. In doing this, both the average movement and the variability of this movement are critical to understand. The average movement is needed to scale the artworks used in imaging innerlayer circuits, so that when the multilayer circuit is laminated together, the circuits on each layer are aligned and are the correct size. The variability is important as well, since excessive variation can cause misalignment of these layers even if the average movement is reflected in the artwork scaling. This was discussed in Chap. 9.

When introducing a new material, it is common to use test patterns or actual printed circuit designs to assess the dimensional stability characteristics of the material. It is important to evaluate the range of laminate constructions to be used, as well as the range of circuit designs, including those that result in most of the copper foil being etched off the laminate vs. those that involve etching very little copper off the laminate. This is because the amount of copper removed from the laminate can greatly impact the dimensional changes observed, particularly for thin laminates.

10.8 IMPEDANCE CHARACTERIZATIONS

Impedance evaluations may also be done through the use of test patterns or actual printed circuits or both. Laminate thickness and dielectric constant both affect the measured impedance. Since resin contents can vary with different laminate constructions, so can dielectric constants. Typically, the laminate manufacturer can provide dielectric constant values vs. resin content values, and therefore the printed circuit manufacturer can calculate the expected impedance with a specific dielectric thickness and circuit design. Manufacturing test circuits to verify the expected impedance values using various constructions is then a fairly straightforward process.

10.9 DRILLING OPTIMIZATION

The ease with which a material can be drilled is an important characteristic for the printed circuit manufacturer because it impacts both manufacturing costs and circuit board reliability.

Materials that require reduced drilling stack heights, reduced drill bit hit counts, or relatively slow drilling parameters in order to achieve acceptable hole wall quality can cause significant cost increases and can negatively affect the capacity of the drilling process. Materials prone to resin smear during the drilling process can result in separation between the internal circuit layers and the plated copper in the hole if the smeared resin is not removed prior to the plating process.

Therefore it is a standard procedure to evaluate the drilling performance of a new material. One such evaluation procedure starts with drilling a series of holes at a given drill surface speed and infeed rate. When a given quantity of holes is drilled, the infeed rate is increased and another set of holes is drilled. This process of increasing the feed rate is continued until drill breakage occurs. The advantage of this test is that it is much faster than drilling holes at a more standard speed and feed combination until breakage occurs. Multiple trials of this test provide an average and standard deviation for the infeed rate at breakage, which is one measure of the drillability of a material. In assessing performance in this test, it is important to consider both the average and standard deviation obtained. When compared in a production environment, it has been shown that materials that result in a lower average but that are more consistent can outperform materials with a higher average but more variation. It should also be noted that different types of drill bits can also result in differences in performance for a given material.

Beyond the accelerated infeed test just described, it is also common practice to manufacture a series of test vehicles consisting of various printed circuit constructions and designs. A design of experiments (DOE) is often used in conjunction with these test boards in order to determine the drilling parameters that will provide the best combination of hole wall quality and manufacturing process output. Figure 10.1 shows the inside of a drilled hole. The upper half of the photograph shows the resin and fiberglass, while the lower half shows an internal copper layer.

FIGURE 10.1 SEM photo of the inside of a drilled hole. *(Photo courtesy of Shipley.)*

10.10   *DESMEARING AND ELECTROLESS COPPER DEPOSITION CHARACTERISTICS*

In conjunction with the drilling optimization tests, resin desmear, hole wall conditioning, and electroless copper deposition evaluations are normally performed. These tests are typically performed simultaneously because the hole wall quality obtained in the drilling process determines the amount of resin removal required and affects the ability to properly condition and copper-plate the hole wall. For example, a drilling process that results in relatively higher amounts of resin smear requires a more aggressive desmearing process.

While most desmearing processes use a permanganate-based solution to remove this resin smear, some advanced materials may require the use of plasma desmear even with a well-designed drilling process. Plasma desmearing is a process that subjects the drilled multilayer circuits to a partially ionized gaseous environment. The ionized gases are used to remove the resin smear from the internal conductors in the hole wall.

Catalyst absorption and electroless copper plating coverage is also commonly evaluated when introducing a new material. The objective in this process is to ensure complete coverage of the hole wall with electroless copper. Figure 10.2 shows photographs of a drilled hole after a permanganate desmearing and electroless copper plating process.

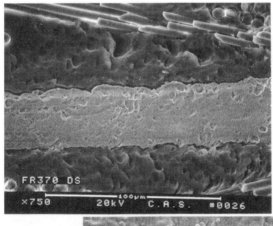

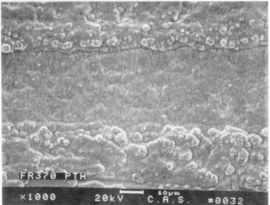

FIGURE 10.2   SEM photographs after desmearing and electroless copper plating. *(Photos courtesy of Shipley.)*

10.11   *ABSORPTION OF ULTRAVIOLET LIGHT*

The ultraviolet (UV) light absorption characteristics of base materials are important when it comes to imaging processes once the external circuit image is formed (or in primary imaging in the case of additive printed circuits). This is typically of concern only in relatively thin printed circuits.

The reason for this is that imaging operations, such as forming the solder mask pattern, normally use photosensitive polymers applied to both sides of the printed circuit. These polymers are selectively exposed to ultraviolet light in order to form the proper image. Many of these photosensitive masks allow much of the ultraviolet light to pass through the film and into the base materials. If the circuit is thin enough and does not sufficiently absorb ultraviolet light, the ultraviolet light exposed to the circuit from one side can pass through to the other side, causing unwanted exposure of the mask, which is commonly termed *shoot-through*.

While the more sophisticated techniques of UV spectroscopy could be used to quantify UV absorption, a simpler technique is commonly used in practice. This technique uses a UV densitometer, which is a simple device normally employed to measure the artwork used in imaging processes and is therefore available to most printed circuit manufacturers. By obtaining base material samples of varying thicknesses, etching off the metal cladding, and measuring the samples with the UV densitometer, an understanding of the UV absorption characteristics vs. material thickness can be obtained. The circuit manufacturer can then determine the minimum thickness required to prevent shoot-through. If necessary, base material suppliers can tailor this property through the use of specific resins or of additives that prevent the transmittance of UV light through the material. Figure 10.3 shows an example of the measurements taken on a commercially available 180°C-T_g FR-4 material. In this particular case, a measurement of 2.3 on the densitometer is required to prevent shoot-through. As such, a thickness of at least 15 mils is required. One final note concerns the resin content of the samples. Higher resin content in the material improves the UV absorption characteristics, and therefore this should be understood when evaluating the thickness required. The use of laminates of a given thickness, but different glass styles and resin contents, also results in different UV absorption levels.

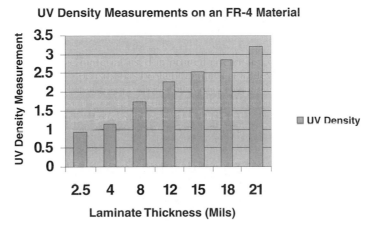

FIGURE 10.3   UV densitometer readings on an FR-4 material.

10.12   *FLUORESCENCE AT AUTOMATIC OPTICAL INSPECTION*

Laser-based automatic optical inspection (AOI) requires that the base material fluoresce in response to the laser in order to detect the base laminate vs. the conductive pattern. When introducing a new material, imaged laminates should be run through this type of AOI equipment to verify that the signal received from the fluorescence of the laminate is sufficient. Similar to UV absorption, any deficiency in this regard can typically be rectified by addition of the appropriate additive or resin type.

REFERENCES

1. Nelco product reference materials.
2. Polyclad product reference materials.
3. Clark-Schwebel industrial fabrics guide.
4. BGF Industries, Inc. fiberglass guide.
5. IPC-TM-650 test methods manual.
6. Edward Kelley and Owen Christofferson, "Multilayer Printed Circuits with Exacting Registration Requirements: Achieving Next Generation Design Capabilities," IPC Printed Circuits Expo, May 1998.

CHAPTER 11
HDI MICROVIA MATERIALS

Ceferino G. Gonzalez
*E. I. du Pont de Nemours & Co., Inc.,
Research Triangle Park, North Carolina*

11.1 INTRODUCTION

This chapter provides an overview of the dielectric and applied conductive materials used in microvia and via filling. Some of these materials can be used in both IC chip carrier and PWB HDI applications. The discussions are focused on the HDI PWB arena (Fig. 11.1) and on materials for which information is readily available.[1] A brief technology review is covered because of the strong tie-in with material selection. In Sec. 11.7, cross-references are made to the relevant material specifications of the IPC/JPCA-4104 specification for HDI and microvia materials. A brief material roadmap discussion is included to illustrate material property trends.

11.2 DEFINITIONS

IPC-2315 and IPC-4104 define microvia (build-up via per JPCA) as blind and buried vias that are <0.15 mm in diameter and target pad (via pad bottom land per JPCA) that is <0.35 mm in diameter. High-density interconnections (HDIs) (build-up PWB per JPCA) are substrates with reduced geometries (Fig. 11.1). HDI is used to reduce size and weight and to enhance electrical performance. Its nature also allows innovation in the three-dimensional packaging.

11.3 TECHNOLOGY CONSIDERATIONS FOR HDI MICROVIA FABRICATION

The four major processes for HDI microvia formation are as follows:

1. Laser via fabrication
2. Dry/wet etch via fabrication
3. Photovias
4. Conductive ink/stacked vias and circuitry

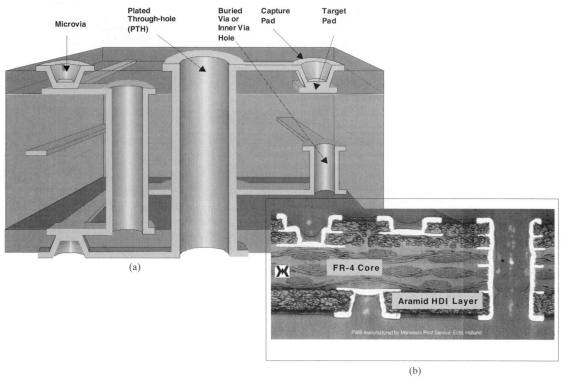

FIGURE 11.1 Sample of generic HDI microvia printed wiring board. (*a*) Schematic; (*b*) actual HDI PWB.

Table 11.1 shows the material and equipment choices. Laser RCF is the dominant technology and material choice because of its relative ease of use, availability of suppliers, and current technical needs. As lines and spaces become even smaller, this is driving the technology toward unclad dielectrics (or thin copper-clad dielectrics).

11.3.1 Laser Via Formation

The laser actually ablates material rather than removing it when creating a hole. However, this creation of a hole in a substrate has become known in the industry as laser drilling. There are two main types of laser drilling methods: UV or CO_2. (See Chap. 23 for a complete discussion of lasers.) The UV type can laser-drill through copper and dielectric, while the CO_2 type can only laser-drill through the organic dielectric. A UV-type laser can define vias as small as 25 μm in most any material. Laser drilling is the most universally adaptable method of via formation. This can be a single or multiple via generation technology that replaces existing mechanical drilling process with a laser drilling process. Laser via creation differs from mechanical drilling in that the focused beam used to create the holes can produce smaller holes than conventional drilling, but the hole wall quality is much rougher. These lasers are generally categorized by the wavelength of light they use. One of the most important advantages of laser drilling is that it is compatible with many copper-clad or unclad dielectric materials, reinforced or nonreinforced. Laser drilling can be used to create both

TABLE 11.1 Equipment and Material for Microvia PWBs

	Photovias	Laser/RCF	Laser/epoxy resin	Stacked via
Required facilities	• Vacuum laminator/ curtain or roller coater • Oven • Exposure/ registration • Clean room	Laser	• Curtain, screen, or roll coater • Oven • Laser • Clean room	Laser and others
Major equipment suppliers	• Furnace coaters: Vantico, Buerkle, Coates, Systronic • Vacuum laminators: Hitachi, DuPont, Meiki • Exposure/ registration machines: ORC, OTEC, Ono-Sokki, Bacher	Laser machines: Hitachi, Mitsubishi, Sumitomo, ESI, GSI Lumonics	• Furnace coaters: Buerkle, Systronic • Laser machines: Hitachi, Mitsubishi, Sumitomo, ESI, Lumonics	Laser machines: Hitachi, Mitsubishi, Sumitomo, ESI, GSI Lumonics
Major materials suppliers	• Vantico, Taiyo	Mitsui, Hitachi, Polyclad, Sumitomo, Bakelite, Isola	• Taiyo, Vantico	DuPont, Teijin

Source: Prismark, Electronics Industry Report 2001.

blind vias and holes. The process follows standard multilayer process steps, but it is capable of resolving smaller features. Low productivity is one of the weaknesses of laser drilling, i.e., one beam produces one via at a time. This means that throughput is affected by the number and placement of the vias on a panel. Via ablation speed is likewise affected by the material set selected. Vias ablated through only dielectric material require a shorter time than vias ablated through copper, resin, and glass. Therefore material selection can affect laser throughput.

11.3.2 Etching-Process-Formed Vias

Microvias can be formed by various etching techniques. The two most common are as follows:

1. Dry etching
2. Wet etching

11.3.2.1 Dry Etching. The most common etching technique is to use a microwave-gas plasma, which is a dry etching process. The plasma approach to microvia production relies on an etched copper foil mask (conformal mask) that only exposes the dielectric that requires removal, in the vias' locations. As a result, plasma processing can only be subtractive and the material set is limited to copper-clad dielectric materials. Other issues such as via shape and the anisotropic nature of the etching are seen as additional detractors. The process has the capacity for mass via formation, but production throughput is low compared to alternative methods. As a result, this method has not been widely adopted.

11.3.2.2 *Wet Etching.* Wet etching by hot KOH has been used for polyimide films. Both dry and wet methods are isotropic, such that they etch inward while they etch down. This results in an undercut beneath the copper layer that makes the subsequent metallization operation very difficult (or requires additional process steps). On the positive side, these formation techniques are mass via generators in that they form all vias at the same time without regard to number or diameter.

11.3.3 Photovia Formation

First commercially used in 1990 by IBM at Yasu, Japan, the photoimageable dielectric was a modified liquid solder mask. Modern photoimageable dielectrics (PIDs) are either liquid or dry film. Microvias are formed in mass by photoimaging. The dielectric is applied over the base substrate. The microvias are imaged, developed, and cured. This via layer is then "roughened" and copper-plated to connect the microvias. A final primary resist is applied to complete the outerlayer imaging; then the circuits are etched or pattern-plated. Conductive paste can also be used to fill the microvias and circuit pattern.

By its nature the photovia process is always semi- or fully additive. Specially formulated photosensitive resin systems are applied onto subcomposite structures. The photosensitive material is exposed to UV radiation using hard contact imaging exposure through a phototool. Mass via generation results from this exposure. The time (and ultimately the cost) required to create 1 or 100,000 vias remains constant.

Early products did not demonstrate ideal dielectric characteristics, robust via formation capability, or good plating adhesion. Later, significant advancements were made in the dielectric materials that can be used for HDI. Many dielectrics have been formulated to accept plating with adhesion values approaching those of standard laminated low-profile copper foil. The photo speeds have improved and resolution capability in the range of 1:1 is possible with some formulations.

Once vias are formed and the dielectric is cured, the next step is plating. Significant gains in copper-to-dielectric adhesion have been demonstrated with unclad dielectrics. Many dielectric materials have adhesion of greater than 6.5 lb/in (1.17 kg/cm) per 1 oz (35.6 μm) of plating. The standard processes used for multilayer board desmearing and metallization are most frequently employed.

11.3.4 Dry Metallization (Conductive Inks/Conductive Paste/ Insulation Displacement)

Three types of dry metallization are covered: conductive ink, conductive paste, and insulation displacement. Conductive ink describes a single-layer dielectric with microvias formed by photoimaging, laser, or insulation displacement. A conductive paste is used to fill the microvias and act as the conductive path between layers. Surface metallization may be accomplished either by laminating copper foil onto the dielectric surface or by chemical deposition.

Standard activities concentrate on the use of photoimaging and screen printing while overcoming the limitations and expense associated with processes such as plating and etching. The concept is built on the use of a photoimageable dielectric to produce both vias and circuit channels that are then filled with a conductive ink. These approaches eliminate the need for a separate dielectric and clad metal layer that requires either etching or a combination of plating and etching to produce a circuit. They also eliminate the resist deposition and stripping previously required to define that circuit. The conductive ink technique of metallization virtually eliminates the generation of metal waste streams.

11.4 ALTERNATIVE HDI CONSTRUCTIONS FOR IMPROVED AREA UTILIZATION

Conventional multilayer printed wiring boards must be subjected to drilling and through-hole plating to create interconnections. These holes represent inefficient use of printed wiring board area. In addition, to connect the printed wiring with part connection lands, some through-holes must be provided in areas other than where the lands are located.

11.4.1 ALIVH®

Any-layer interstitial via hole (ALIVH)[2] has inner via hole (IVH) structure in all innerlayers. It can even provide an innerlayer connection immediately under the component pad. In addition, it has no through-holes at all and any number of layers can be interconnected, and it can be designed more easily than other types. As a result, it has greater potential for significant downsizing than conventional types. In addition, its wiring can be shortened, making it ideal for high-speed applications.

The key technological elements that make this possible are shown in Fig. 11.2. The short design time enabled by ALIVH is due to the fact that vias can be sited on other vias. ALIVH offers a high degree of design freedom and facilitates automatic layout. Other microvia constructions, on the other hand, cannot place vias atop one another readily or position a land over a via, because the copper plate used to fill the via creates a nonflat surface. ALIVH also allows for a thinner and lighter board (improved 50 percent) vs. FR-4 core boards because ALIVH uses aramid prepreg, as discussed in the following text.

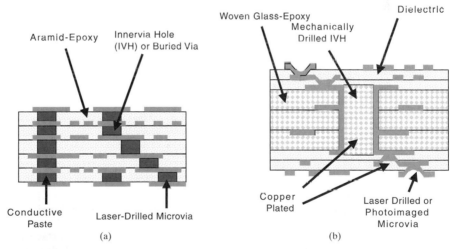

FIGURE 11.2 (*a*) Cross section of ALIVH printed wiring. (*b*) Cross section of traditional HDI printed wiring board.

The main components of ALIVH fabrication are the use of aramid prepreg, pulse-oscillated CO_2 gas laser, and interconnection by conductive paste stacking. Also, all aramid prepregs are processed in parallel, which improves productivity and reduces yield loss.

11.4.1.1 Aramid Base Material. Nonwoven aramid prepreg impregnated with conventional epoxy resin (FR-5 or equivalent) offers greater heat resistance than FR-4, which is used in conventional multilayer printed wiring boards. The other features of aramid fibers are low thermal expansion coefficient, low dielectric constant, high heat resistance, high rigidity, and light weight. Printed wiring boards with aramid fibers used for reinforcement are already being put into practical use, mainly in the military and aerospace industries.

11.4.1.2 Microvia Drilling. ALIVH uses a pulse-oscillated CO_2 gas laser drilling machine for high-speed, microdiameter drilling of interconnection holes. This laser machine employs a galvanic mirror that optically scans the laser beam, which is then focused by an *f*-lens. The machine is capable of drilling 200-µm vias in prepreg developed for ALIVH. The creation of vias using a CO_2 laser drill has been realized by the adoption of an all-organic combined material of aramid fibers and epoxy whose laser ablation rates are close to each other. In short, the introduction of nonwoven aramid fabric has played a highly critical role in realizing low-cost drilling of increased numbers of small-diameter vias.

11.4.1.3 Interconnection Technique. To realize interconnection by a means that supersedes conventional copper plating, a conductive paste consisting of copper powder, epoxy resin, and a hardener has been newly developed. This conductive paste is injected into laser-drilled vias. Copper foil is attached to both surfaces of the aramid-epoxy prepreg, then heated and pressed using a hot press. This hardens the epoxy resin in the aramid-epoxy prepreg and in the conductive paste, achieving simultaneous bonding and electrical connection between the conductive paste and copper foil. This enables a low connection resistance closely equivalent to that of conventional copper-plated through-holes. Further, since the conductor is made up solely of copper foil with no plating, it is of uniform thickness, which is advantageous when forming an even finer conductor.

ALIVH exhibits extremely high reliability, as demonstrated in various storage tests in which it is left to stand under conditions of high temperature, low temperature, high temperature and humidity, or high humidity (pressure cooker test [PCT]). ALIVH was adopted in mobile cellular phones in 1996, and has profoundly influenced the direction taken by printed wiring board technology.

11.4.2 B²it

Figure 11.3 shows a buried bump interconnection technology (B²it)[3] board. B²it is a process patented by Toshiba and has emerged as one of the major microvia production methods. B²it is an alternative to ALIVH for creating buried vias. This technology is based on "piercing" the prepreg with silver bumps on copper plate. Sony, Nokia, Shinko, and Mitsui Toatsu are some of the bigger customers of this technology. B²it offers a selection of materials that provide a range of mechanical and electrical characteristics.

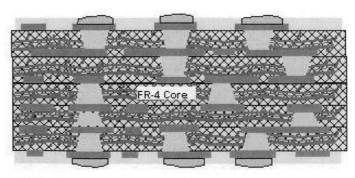

FIGURE 11.3 B²it board. *(Courtesy of Toshiba.)*

11.5   MATERIALS FOR HDI MICROVIA FABRICATION

Figure 11.4 illustrates a material and technology selection flowchart for use when choosing dielectric materials. In using the flowchart, the following questions should be asked about the dielectric being considered:

- Will the dielectric use chemistry compatible with current chemistry used by core substrate material?
- Will the dielectric have acceptable plated copper adhesion? (Many OEMS want ≥6 lb/in [1.08 kgm/cm] per 1 oz [35.6 μm] copper.)
- Will the dielectric provide adequate and reliable dielectric spacing between metal layers?
- Will it meet thermal needs?
- Will it provide a (desirable) "high" T_g for wire-bonding and rework?
- Will it survive thermal shock with multiple SBU layers (i.e., solder floats, accelerated thermal cycles, multiple reflows)?
- Will it have platable/reliable microvias (that is, will it have latitude to ensure good plating to the bottom of the via)?

11.5.1   Copper-Clad Dielectric Materials

Due to relative ease of implementation, copper clad dielectrics are used on a larger scale than unclad dielectrics. Copper-clad dielectrics provide a method that requires the least number of changes in manufacturing flow because they typically use the same dielectric and reinforcements found in standard PWBs. Copper-clad-based materials have a longer history in making blind vias than any other method. This makes many designers, OEMs, and PWB fabricators more comfortable with copper-clad-based materials.

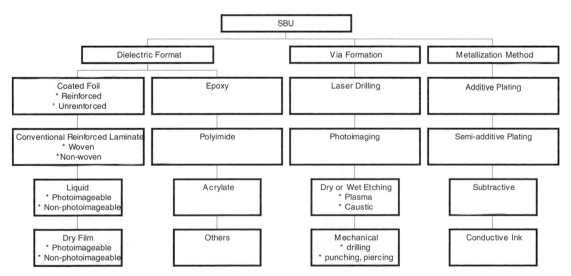

FIGURE 11.4   Material and technology choices for sequential build-up (SBU) fabrication. *(Courtesy of DuPont.)*

These materials can be nonreinforced or reinforced. The reinforcement can be woven or nonwoven and can be aramid, glass, etc. These dielectrics are suitable for via formation by laser or other mechanical removal methods.

Due to its wide availability, and familiarity, FR-4 material is often initially evaluated in laser-drilling microvias. Reinforced with thin 106 or 1080 woven glass (since thicker glass is more difficult to vaporize with lasers), one- or two-ply laminates are selected with a resin content close to 70 percent. The "laser drilling" is done by ablation of the via using a conformal mask or a directly focused beam, with either a UV Nd:YAG or a CO_2 laser. These materials may also be coated on a copper foil. Typical applications use single-side clad material where the copper clad is used as the outer layer and the C stage is bonded to the subcomposite. These materials are suitable for via formation using methods such as plasma or laser. To meet fine circuitry and smaller via needs, very thin copper is available. Another approach is what many Asian PWB fabricators practice when they thin down incoming copper clad by etching and/or planarizing the copper surface precisely.

11.5.2 Unclad Dielectric Materials

If the dielectric is reinforced, microvias are formed by laser drilling or other mechanical means. If it is nonreinforced, it can be photoimeagable in addition to the previous choices. To add conductivity, subtractive processing is the standard manufacturing practice in the United States and Europe. Semi- or fully additive techniques have been practiced in continuous high volume only in the Asian region, most notably Japan. In Japan, which leads the world in microvia production, about 22 percent of manufacturing begins with a non-copper-clad dielectric material. Japan has long accepted the additive manufacturing methodology for creating circuits on the board surface as well as through hole connection.

The trade-off between clad and unclad dielectric materials follows a benefit analysis similar to that used for subtractive versus additive processing. Copper-clad-based material is fabricated using standard practices familiar to most shops. However, it innately generates more waste and has a higher cost due to the subtractive nature of the manufacturing sequence. In addition, copper-clad materials demonstrate limitations in fine-line capability similar to those seen with standard manufacturing methods used today.

Unclad dielectrics require considerable resources for optimized plating of copper to achieve reliable and consistent copper peel strengths. Unclad dielectrics and thin copper-clad dielectrics, however, gain in importance as the requirements for line and space geometries reach ≤75 μm and required via sizes are lower than 75 μm in diameter. Figure 11.5 depicts the

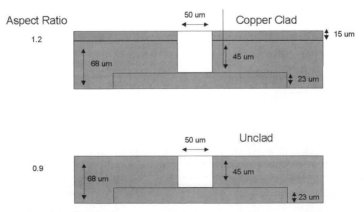

FIGURE 11.5   Impact of ½-oz base copper on the aspect ratio of a 50-μm via. *(Courtesy of Vantico.)*

3 μm

9 μm

12 μm

FIGURE 11.6   Cross section of a 40-μm-wide conductor. *(Courtesy of Mitsui Mining and Smelting Co. Ltd.)*

influence of base copper on the aspect ratio of a via. The base copper will increase the aspect ratio, making it more difficult to plate the via, and can cause issues with bottlenecking whereby the via will plate shut at the top without sufficient plating in the bottom. Figure 11.6 illustrates what happens to fine circuits. In this example, the thickest copper is the farthest from the 40-μm nominal conductor. Other data show that thinner copper will etch more uniformly, which will help the cause for controlled impedance.

11.5.3   Clad vs. Unclad Dielectric Materials

Table 11.2 provides a comparison of the utilization of clad vs. unclad dielectric materials as circuit geometry shrinks. This table shows that as the technology requirements get tougher, the trend toward using unclad dielectrics increases. As line and space requirements get denser, and via diameters get smaller, requirements for chip package substrates to use unclad or thin copper dielectrics increase. In the case of mobile phones and other portable applications, for example, it is foreseen that by the year 2005 lines and spaces will be down to 50 to 75 μm, respectively, and via sizes to 50 μm, with a subsequent high usage of unclad material.

TABLE 11.2   Utilization of Unclad Dielectrics by Key Market Segments[4]

	Chip package substrates	Cellular, camcorder, PDA	Computer/ telecommunication*
HDI penetration	High	High	Low
Line/space and via diameter (year 2000)	50-μm lines and spaces; 50-μm via diameter	75-μm lines and spaces; 100-μm via diameter	100-μm lines and spaces; 125-μm via diameter
Use of unreinforced dielectrics	High	High	Low
Use of unclad dielectrics	High	Low, but increasing	Very low

* Excluding laptop computers

11.6 *MATERIAL AND TECHNOLOGY DRIVERS*

Drivers for the choice of microvia dielectric depend upon the capabilities of the fabricator and the needs of the circuit. The use of copper-clad laminate, laser drilling, and ALIVH technology generally makes use of technologies already installed or readily available. Additive and semi-additive processes, though having distinct advantages in line definition, require a set of different equipment and skills than traditionally needed. In addition, as lasers become faster, with associated higher throughput, the break-even for photovia processing is harder to meet. Thicker dielectrics will be the best way to meet RF and controlled-impedance requirements until fabrication of high-tolerance thinner materials and line formation processes is common, and design rules reflect this. Table 11.3 shows the strengths and weaknesses of the two most common microvia production processes, while Table 11.4 compares dielectric materials. The road map for substrate materials to meet component technology needs is shown in Tables 11.5 and 11.6.

TABLE 11.3   Via Formation Technology Strengths and Weaknesses

	Strength	Weakness
Photovia	• Mass via formation • High productivity • Lower cost for high via count (industry trend) • Capable of producing large cavities	• Yield concern (false vias with negative system) • Limitation in via diameter (resolution capability)
Laser drilling	• Wide choice of dielectric • No false vias • "Turnkey" • Layer 1–3 connection easier	• Capital expenditure • Productivity affected by board design (number of vias and density)

TABLE 11.4   HDI Material Strengths by Type

Copper-clad dielectric	Liquid unclad dielectric	Dry-film unclad dielectric
• Copper adhesion • Drop-in process with 1 SBU construction • Familiar dielectric materials	• Material cost • Available automation • User-determined thickness • Demonstrated capability • High density	• Coating quality • Thickness control • Planarization • Process cost with multi-SBU construction • High density

TABLE 11.5 Roadmap of Substrate Materials for LSI Package

Year	1998	2005	2010
Pin count	600–700	1200–1500	2000–2500
Package technology	BGA	BGA, CSP	CSP, bare chip
T_g (TMA) (°C)	160–180	180–200	200–220
CTE (ppm/°C)	14–15	8–10	6–8
Dielectric constant (1 MHz)	4.4–4.6	3.0–3.5	<3.0
Dielectric loss (1 MHz)	0.02–0.025	0.01–0.015	<0.005
Conductor thickness (μm)	12, 18	9	5
Insulator thickness (μm)	50–60	40–50	30–40
Peel strength (kN/m)	1.0–1.2	1.0–1.2	1.0–1.2
Via diameter (μm)	100–150	60–80	25–50
Resist resolution (μm)	16–65	6–35	5–30

Source: Conference International Packaging Forum, Tokyo, September 1999.

TABLE 11.6 Roadmap of Substrate Materials for Printed Wiring Board

Year	1998	2005	2010
T_g (TMA) (°C)	120–130	140–160	160–180
CTE (ppm/°C)	14–15	12–13	11–12
Dielectric constant (1 MHz)	4.4–4.6	3.0–4.0	<3.5
Dielectric loss (1 MHz)	0.02–0.025	0.013–0.015	<0.01
Conductor thickness (μm)	12, 18	12, 18	12, 18
Insulator thickness (μm)	60–100	60–100	40–60
Peel strength (kN/m)	1.0–1.2	1.0–1.2	1.0–1.2
Via diameter (μm)	150–250	100–150	80–100
Resist resolution (μm)	50–100	30–70	25–50
Flame retardancy	V-0	V-0	V-0

Source: Conference International Packaging Forum, Tokyo, September 1999.

In 1999, "halogen-free" dielectrics and laminates started to get more market attention because of concerns over the long-term effect of bromine as a waste product. The WEEE, JPCA, and IPC are some of the leading bodies addressing these activities. The effect on microvia formation materials is no different than that on the general board design and fabrication.

11.7 EXAMPLES OF HDI MICROVIA ORGANIC SUBSTRATES

From a mechanical standpoint, materials may be grouped as reinforced and nonreinforced laminates and prepregs, as in Fig. 11.7. Reinforced materials are generally better in dimensional stability, lower in coefficient of thermal expansion (CTE), and less sensitive to thermal cracking, while nonreinforced materials often have a lower dielectric constant (D_k) and may be photoimageable.

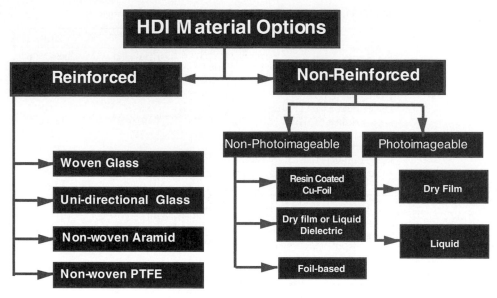

FIGURE 11.7   Material options by reinforcement. *(Source: DuPont.)*

11.7.1 Nonreinforced Dielectric Materials

11.7.1.1   Resin-Coated Copper Foils.   Resin-coated foils (RCFs) are common materials used for build-up multilayer microvia applications. Many product variations are available, and they fit well within the existing multilayer manufacturing infrastructure. Epoxy-based coated foils are the most common and have performance properties similar to those of FR-4 but with no E-glass reinforcement. Peel strengths, thermal performance, and electrical properties are compatible with standard FR-4. Other resin systems are also available. Laminate suppliers should be contacted when special needs are required, as the resulting composite will require understanding of fabrication and design issues.

Resin-coated foils come in two general types:

1. One-pass coated foils have a single B-stage layer designed to flow and fill and to provide thickness control all at the same time.

2. Two-pass coated foils have a C-stage resin layer against the foil and a B-stage layer for flow and fill. The fully cured C-stage layer acts as a "stop" during the lamination process, typically enabling better thickness control.

Resin-coated foils are available in a variety of thicknesses, yielding between 1.0-mil (25-μm) and 3.0-mil (76-μm) finished dielectrics.

The copper foils most commonly used are ½ oz (18 μm) and ⅜ oz (13.34 μm), but there is substantial interest in thinner copper foils for improved laser efficiency and better fine-line circuitry definition.

The basic process used for production of build-up microvia structures with resin-coated foil is illustrated in Fig. 11.8.

Step 1: Starting with a substrate package, the lamination process is identical or very similar to that for traditional multilayer press processes for FR-4 in the case of epoxy-coated foils.

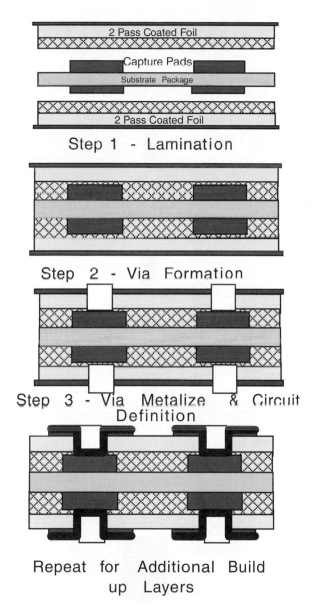

FIGURE 11.8   General coated-foil build-up microvia. *(Courtesy of Polyclad.)*

Step 2: Via formation is most commonly done using lasers to create a "blind" opening between the surface copper and the capture pad on the layer below.

Step 3: Via metallization and circuit definition is done using traditional techniques for plating and outerlayer definition on traditional multilayer PWBs.

Step 4: Repeat for additional build-up layers.

With the exception of the laser process, all the basic technology to produce SBU using resin-coated foil is the same as, and found in, most multilayer PWB production operations.

Because build-up technology is evolving very fast, many different and diverse approaches are being taken to the resins used and variations on the actual via formation process. Low-D_k/D_f resins such as polyphenylene ether (PPE) are being used to address signal speed and integrity demands in resin-coated foil build-up structures. Another approach is to use additively platable resin and use the copper as a sacrificial carrier, eliminating the need to laser through copper. This also gives excellent surface topography for good peel strengths. See IPC-4104 specification sheets 12, 13, 19, 20, 21, and 22 for more complete material performance information.

Another alternative is to use additively platable resins. In this approach, the copper foil acts as a sacrificial carrier. The process is similar to that described earlier, but the first step after lamination is to chemically etch off all the copper. This eliminates the need to laser through copper and makes acquisition of registration targets simpler. The structure left behind by the copper also gives excellent surface topography to the resin for good peel strengths from the subsequent additive plating step. This approach allows for extremely fine feature definition and much greater laser efficiency.

11.7.1.2   *Unclad Nonreinforced Photoimageable Dielectric Materials.*   Material chemistry options for this group include epoxies, epoxy blends, polynorborenes, polyimides, etc. They can be applied as liquid or dry film, negative or positive imaging, and can be solvent or aqueous developable. To improve the dielectric's adhesion to copper, most dielectric suppliers require a copper pretreatment with a black oxide or conversion coating (oxide replacement) prior to application of the dielectric. Most of these materials are either epoxy or epoxy and novolac based to provide high T_g and good plating, and most provide adhesion values for plated copper of at least 1.1 kg/cm at a 25-μm copper thickness. Typically these dielectrics are metallized using conventional solvent swell and permanganate etching techniques. The liquid materials use only safe solvents (those not known to lead to harmful health effects with prolonged exposure).

A unique advantage for photoimageable materials is that the speed used to make small or large vias is the same. With the growing need to make embedded reactives (passives), large rectangular areas need to be opened up to place these devices. Currently, this is only economical with photoimaging. Most photoimageable materials are easy and fast to laser-drill because they are nonreinforced. Figure 11.9 shows a typical microvia appearance before and after desmearing. See IPC-4104 specification sheets 1, 2, 7, 8, 9, 10, and 16 for more complete material performance information.

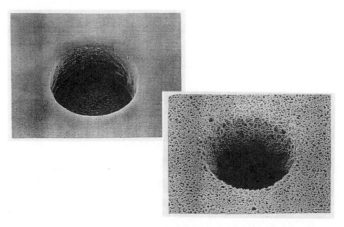

FIGURE 11.9   Typical microvia appearance before and after desmearing. *(Courtesy of Vantico.)*

11.7.1.3 Nonphotoimageable, Nonreinforced Dielectric.   This group can be laser-drilled, plasma-etched, and or mechanically treated to form microvias. As stated earlier, many of the photoimageable dielectrics are laser-drillable by following the process described in Fig. 11.10. As in the photoimageable group, to improve the dielectric's adhesion to copper, most dielectric suppliers require a copper pretreatment with a black oxide or conversion coating (oxide replacement) process. See IPC-4104 specification sheets 6, 11, 17, and 18 for more complete material performance information.

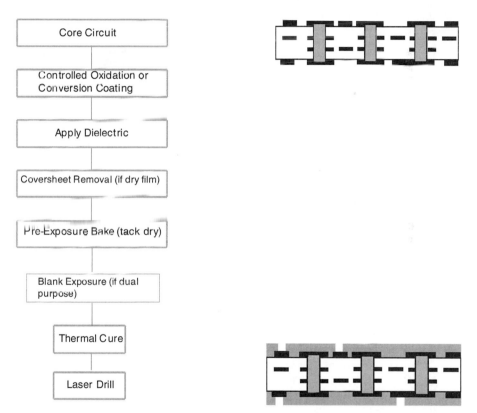

FIGURE 11.10   Typical laser-drilling process of unclad, nonreinforced dielectric material. *(Courtesy of DuPont.)*

11.7.2   Aramid-Reinforced, Nonwoven, Nonglass Laminate

In 1965, scientists at DuPont discovered a method for producing an almost perfect polymer chain extension using the polymer poly-*p*-benzamide. The key structural feature of this molecule is the paraorientation on the benzene ring, which allows it to form rod-like structures with a simple repeating molecular backbone. The term *aramid* now refers generically to organic fibers in the aromatic polyamide family. Kevlar® was the first para-aramid fiber to become familiar, due to its use in bullet-resistant vests and as a lightweight, high-strength structural reinforcement. Aramid-reinforced prepregs and laminates have proven their functionality for a number of years in high-reliability applications and more recently in consumer electronics.

The low coefficient of thermal expansion (CTE) of aramid nonwoven reinforced prepreg and laminate provides a closer match to the CTE of the silicon chip (Fig. 11.11). Depending on the type of resin and the resin and copper content of the laminate and the prepreg, the CTE of the PWB can be tailored to between 10 and 16 ppm/°C (Fig. 11.12). This allows the designer the option of finding a best fit of the CTE of the PWB to the CTE of the components.

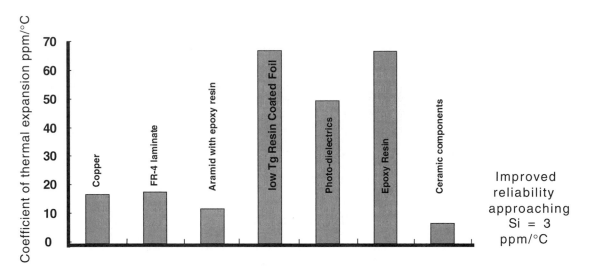

FIGURE 11.11 Typical in-plane CTE of various materials. *(Courtesy of DuPont.)*

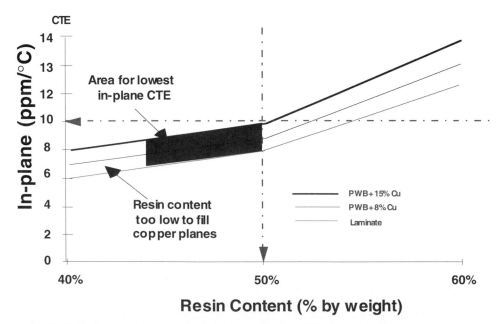

FIGURE 11.12 Impact of copper and resin content on in-plane CTE. *(Source: DuPont.)*

Reliability can be designed in by PWB designers, as they know which component packages are used, what the CTE requirements of these packages are, and what the life expectancy requirements of the electronic equipment are. The ability to tailor in-plane CTE has made nonwoven aramid-reinforced PWBs one of the most favorable material options for avionics, satellites, and telecom applications where long life expectancy and high reliability are needed. In mobile phone applications, where chip scale packages (CSPs) are commonly employed, low-CTE nonwoven aramid reinforcement extends solder joint life as much as three times over FR-4 and resin-coated foil (RCF). After more than 1000 thermal cycles (−40 to +125°C), nonwoven aramid-reinforced epoxy resin does not crack as is common with nonreinforced dielectrics.

Aramid-reinforced laminate and prepreg allow fast microvia hole formation and, at the same time, maintain the performance characteristic of a smooth surface for fine-line conductor imaging. The ablation speed of nonwoven (aramid) laminates and prepregs is close to that achieved when using nonreinforced materials like resin-coated foil, dry film, or liquid dielectrics. Since aramid laminates are very stable, they allow the fabrication of double-sided very thin etched innerlayers, which are then pressed to a multilayer package in a single pressing cycle. Thus, these thin innerlayers can be processed in parallel. Aramid-reinforced material is a cost-effective alternative when laser hole formation is used to form single- and multiple-layer interconnects, and is compatible with FR-4 materials used in the core of aramid–FR-4 mixed constructions. Laser drilling of stepped microvia holes is accomplished in one subsequent operation. This allows the designer to interconnect up to four layers on each side of the inner core without sequential processing—a substantial productivity advantage for the PWB fabricator that results in the lowest possible manufacturing cost. See IPC-4104 specification sheets 5 and 23 for more complete material performance information.

11.8   VIA FILLING

The demand for filling of through-holes with epoxy resin or conductive paste has been rising ever since surface-mount technology became widely adopted in the PWB industry.

11.8.1   Basics

Following are the traditional examples of the types of applications that require via filling processing capability:

- Preventing acid residue from attacking the Cu plating at the through-hole opening, which can in turn cause an open circuit.
- Preventing mishandling due to loss of vacuum during board-level assembly or during vacuum-assisted transport in production.
- Preventing blowout of flux and/or solvent residue during assembly and solder reflow.
- Preventing flux from dripping through hole to opposite side of PWB.
- Preventing solder resist ink from migrating into through-holes during screen printing, which can cause formation of nodules at the through-hole opening during solder plating and/or gold plating.
- Enhancing planarity of solder resist on surface of filled through-hole and for planarity of core layer for SBU PWBs.
- Enhancing stability of solder paste printing volume for via-in-pad designs.
- Preventing migration of solder into through-hole in via-in-pad designs.

The filling of through-holes with nonconductive resin and conductive materials using dispensing, screen-printing, and roll-coating methods has been extensively evaluated and tested.

Of the three basic methods described earlier, screen printing is the most common process that allows for efficient, selective filling of large numbers of through-holes.

11.8.2 Screen-Printed Via-Filling Materials

Selection of the proper resin-based material is the most important objective when using a screen-printing process to fill through-hole vias in build-up multilayer cores. In particular, the primary issues to be considered when choosing an appropriate fill material are:

- Ease of printing
- Ease of grinding (planarizing)
- Adhesion to hole wall and panel surface

The most commonly used materials for permanently filling through-holes include single-cure (thermal) resins, photoimageable dielectrics, conductive pastes, and the dual-cure (UV + thermal) epoxy resin utilized in the Noda screen flat plug process. Following are brief descriptions and comparisons of the characteristics of each of these types of via fill materials.

11.8.2.1 Dual-Cure (UV + Thermal) Epoxy Resin.   The main purpose for using a dual-cure epoxy resin is the stability of the B-cured material and the ease with which it can be removed by grinding during planarization. A homogeneous, semicured state can be reached because the liquid environment within the UV exposure system helps to control the temperature and stabilize the intensity of UV light. Furthermore, because the temperature of the liquid environment is kept low, the expansion of air bubbles that may have been trapped in the fill material at room temperature is prevented.

In formulating epoxy resin via-fill ink, one must consider the adhesion between the fill material and the copper plating in the hole barrel. Normally it may seem suitable to use a material that does not include an acrylic component. However, some suppliers have found it more beneficial to add an acrylic component because of its resistance to moisture absorption. They determined that it would be preferable to improve the ink's adhesion to copper by pretreatment with a black oxide or chemical etching process. This is similar to the approach taken by many dielectric suppliers.

11.8.2.2 Photoimageable Dielectrics.   An advantage of a dielectric being photoimageable is that unwanted material is washed off during development, thus eliminating the planarization step. Typically, one screen pass is sufficient to fill or plug vias, but two passes can also be used to maximize hole filling. Standard screen-printing equipment is needed with a bed-of-nails or dimple plate to maximize air release out of the plugged holes. If using a squeegee, single-sided application is recommended. Two coats are recommended to minimize the effects of material. Cores ≥ 0.030 in (0.76 mm) require a secondary coating application.

11.8.2.3 Conductive Paste.   In addition to the previous advantages of via filling stated earlier, conductive via fills also dissipate heat, increase electrical conductivity in the hole, and allow a via-in-pad design option (Fig. 11.13). The latter advantage is another way to recover real estate.

One example* is a screen-printable paste made of silver, copper, and epoxy. The optimum mix of metal particle shapes and sizes bonded together with an epoxy resin produces essentially zero shrinkage. In general, the plated or unplated through-hole is filled with paste using

---

* DuPont has such a product. It is called CB100 ViaPlug.

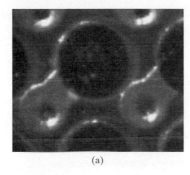 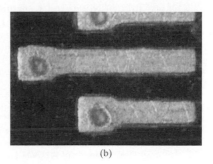

(a)   (b)

FIGURE 11.13   Via-in-pad option with conductive via fill. (*a*) Traditional "dog-bone" design. (*b*) Via-in-pad design. (*Courtesy of DuPont.*)

a stencil-printing process. After drying and curing, the plugged through-holes are planarized and plated to make them solderable. Some typical design rules are:

Aspect ratio	1:1 to 6:1 with vacuum assist
Via size	6 to 25 mils (152 to 635 μm)
Core thickness	6 to 85 mil (152 to 2159 μm)

They are extremely reliable during assembly.

11.8.2.4   Single-Cure (Thermal) Resin.   Typically, thermally cured resins exhibit good adhesion strength to copper. Unfortunately, the rate of heat dissipation throughout the filled core material is unstable, making it difficult to achieve a reliable semicured state prior to planarization. As a result, a rapid and unpredictable rise in temperature is likely to occur; as such, it is imperative to completely fill the holes, leaving no voids, if this type of material is to be used.

11.8.2.5   Solder Mask Ink.   Using this type of material, it is relatively easy to fill the hole. However, the presence of volatile elements (solvents) and the degree of shrinkage during curing have a significant effect on the stability of the manufacturing process. In the case of small-diameter through-holes, the solvent residue may remain in craters or dishes at the mouth of the hole opening. Furthermore, the copper adhesion strength of these materials is typically lower than that of other hole fill materials.

11.8.2.6   Resin-Coated Foil (RCF).   Similar to the potential problems faced with photo-imageable dielectrics, there is a divergence between the material characteristics required for hole filling and the characteristics that are desirable for an innerlayer dielectric-copper material. Nonetheless, it is likely that the RCF foil materials will prove to be acceptable for filling vias in the limited range of specific applications that exhibit a low aspect ratio between hole diameter and core thickness.

11.9   ACKNOWLEDGMENTS

The author would like to thank the following contributing authors: Kathy Nargi-Toth, Enthone-OMI; Ken Ikeda, Hitachi Chemicals America; Torsten Reckert, Vantico; Phil Knudsen, Shipley-Ronal; Dr. Goran Matijesivic, Ormet; Stephen Pierce, SGP Ventures; Hiroshi

Ogawa, Takuya Noda, and Keiichiro Sato, Noda Screen Co., Ltd.; Kazunori Kitamura and Yukihiro Koga, San-ei Kagaku Co., Ltd.; Erik Bergum, Polyclad; and Michael Weinhold, Dave Powell, Neville White, and Rich Wessel, E. I. du Pont de Nemours & Co., Inc.

REFERENCES

1. Ceferino G. Gonzalez, "Materials for Sequential Build-up (SBU) of HDI-Microvia Organic Substrates," *Circuit Tree HDI Materials,* 1999 and 2000.

2. Kazunori Sakamoto, Shingo Yoshida, Kazuyoshi Fukuoka, and Daizo Andô, "The Evolution and Continuing Development of ALIVH High-density Printed Wiring Board," presented at IPC Expo 2000; featured in *CircuiTree,* May 2000.

3. Motoaki Itou, tp://www.nikkeibp.com/nea/nov99/tech/.

4. "Microvia Substrates; An Enabling Technology for Minimalist Packaging 1998–2008," BPA Group Ltd., 1999, pp. 4–3 to 4–17.

CHAPTER 12
LAMINATE QUALIFICATION AND TESTING

Michael Roesch
Hewlett-Packard Co., Palo Alto, CA

Sylvia Ehrler
Multek Europe GmbH, Böblingen, Germany

12.1 INTRODUCTION

Copper-clad laminate materials (single- or double-sided) are the base material for almost all printed circuit boards (PCBs). The properties of these materials have a large influence on the quality of the final product and therefore need to be understood and tested prior to release/manufacturing. In almost all cases the laminate manufacturer will provide a set of data for the specific materials. This data sheet is a good starting point and can certainly be used to evaluate laminates at an early state of a project. Often, though, it is necessary to perform additional tests in the PCB facility to verify or complement the manufacturer's data. This chapter introduces the most important laminate properties and their characterization methods. In addition to the individual test methods, we propose a best-practice qualification guide for copper-clad laminates.

This chapter describes the testing of laminate materials for printed circuit board manufacturing. It serves as a quick reference guide for the printed circuit board specialist as well as an introduction to laminate testing for people who are new to the field. Laminate properties according to industry standards like those of the National Electrical Manufacturers Association (NEMA) and the Institute for Interconnecting and Packaging Electronic Circuits (IPC) are introduced. Then the actual test methods are discussed in the form of a best-practice guide. This will help the PCB specialist make quick decisions during the testing process. Methods of testing mechanical, thermomechanical, and electrical properties of printed circuit board materials are described. The focus is on how the tests are performed, what the results actually mean, and how relevant the obtained test data are. In addition to the raw base material testing, the chapter introduces a best-practice qualification guide to assess interactions between manufacturing processes and laminate properties.

12.2 INDUSTRY STANDARDS

There are a number of different industry standards that can be used to evaluate the properties of copper-clad laminate. Many of the test methods for similar properties are very comparable, though, and in the end it is up to the materials engineer to decide on which standard to base his or her characterization. In some cases it may even be necessary to define a new characterization method to qualify a laminate material for a specific customer application or requirement.

12.2.1   IPC-TM-650

This document is the most comprehensive and most widely used compilation of copper-clad laminate test methods. It is administered by the IPC and is available in print or online at www.ipc.org/html/testmethods.htm.

Most of the test methods described in this chapter are based on the test manual IPC-TM-650. It contains industry-approved test techniques and procedures for chemical, mechanical, electrical, and environmental tests on all forms of printed wiring and connectors. An IPC test method is a procedure by which the properties or constituents of a material, an assembly of materials, or a product can be examined. These test procedures do not contain acceptability levels for specific properties.

12.2.2   IPC Specification Sheets

The IPC publishes a wide variety of technical specification sheets for PCB and PCA products. A few are important in this context and are discussed in the following text.

12.2.2.1   IPC-4101, "Specifications for Base Materials for Rigid and Multilayer Printed Boards."   This document contains specification sheets for laminate materials used to manufacture PCBs for commercial and military applications. The minimum requirements for key material properties can be found in these specification sheets.

12.2.2.2   IPC-L-125, "Specification for Plastic Substrates, Clad or Unclad, for High Speed/High Frequency Interconnections."   This document contains specification sheets for Teflon™ (PTFE) laminate materials used to manufacture PWBs for high-speed/high-frequency applications.

12.2.2.3   IPC-4104, "Specification for High-Density Interconnect (HDI) and Microvia Materials."   This document contains specification sheets for materials used in HDI and microvia applications.

12.2.3   American Society for Testing and Materials (ASTM)

The ASTM develops standard test methods and specifications, which are in some cases related to laminate materials. They will be referenced as appropriate. More information can be found at www.astm.org.

12.2.4   National Electrical Manufacturers Association (NEMA)

NEMA publishes standards on laminated thermosetting materials. The last published version is the LI 1-1998 standard for industrial laminated thermosetting products, which includes the latest information concerning the manufacture, testing, and performance of laminated thermosetting products. Most of today's copper-clad laminate materials are named according to NEMA grades, but it is important to understand that NEMA grades only represent material categories and not one individual material. We often think of FR-4 as a specific laminate material, though it only represents a material class defined by properties that fall into a certain class. This means that the properties of different FR-4 materials from different suppliers may not be the same. More information can be found at www.nema.org.

12.2.5 NEMA Grades

Table 12.1 shows technical and industrial laminate grades as they apply to the most important printed circuit board materials.

TABLE 12.1 NEMA Grade Material Designations and Properties

NEMA grade	Resin system	Reinforcement	Description
XXXPC	Phenolic	Paper	Phenolic paper, punchable
FR-2	Phenolic	Paper	Phenolic paper, punchable, flame resistant
FR-3	Epoxy	Paper	Epoxy paper, cold punchable, high insulation resistance, flame resistant
CEM-1	Epoxy	Paper-glass	Epoxy paper core and glass on laminate surface, flame resistant
CEM-2	Epoxy	Paper-glass	Epoxy paper core and glass on laminate surface
CEM-3	Epoxy	Glass matte	Epoxy nonwoven glass core and woven glass surface, flame resistant
CEM-4	Epoxy	Glass matte	Epoxy nonwoven glass core and woven glass surface
FR-6	Polyester	Glass matte	Polyester nonwoven glass, flame resistant
G-10	Epoxy	Glass (woven)	Epoxy woven glass, not flame resistant
FR-4	Epoxy	Glass (woven)	Epoxy woven glass, flame resistant
G-11	Epoxy	Glass (woven)	High-temperature epoxy, woven glass, not flame resistant
FR-5	Epoxy	Glass (woven)	High-temperature epoxy, woven glass, flame resistant

12.3 LAMINATE TEST STRATEGIES

Laminates contain different reinforcement materials (woven/nonwoven glass/organic fibers, expanded PTFE, etc.), resin types (phenolic, epoxy, cyanate ester, polyimide, APPE, BT, etc.), resin formulations (blended, functionality, etc.), hardeners (dicyanodiamide [dicy], Phenol-Novolak, Cresol-Novolak, P-aminophenol, isocyanurate, etc.), and sometimes filler particles (ceramic or organic). The ratio of all of these individual components can vary widely. In order to define a test strategy for laminates, it is important to understand the different main components of the materials as well as the conditions during manufacturing, as these have a large influence on laminate properties and quality.

The evaluation and qualification of a laminate material can be a quite complex task. There are a number of different qualification test methods available to the materials engineer. However, very often not all of them are relevant or necessary for the final product. At the same time the materials engineer will face pressure to produce quick results driven by increasingly shorter development cycles. The organization of the test methods in this chapter follows a proposed best-practice guide aimed to help in performing full material evaluations efficiently.

The different sections allow quick decisions, especially when a number of materials are being evaluated at the same time.

12.3.1 Data Comparison

The first step in every material evaluation is a data comparison and assessment step. This may already allow the elimination of some candidate materials without the performance of actual tests.

The first and most important thing to understand is the state of the material that is being considered. Is it an established material formulation that has an established history at other manufacturing locations, or is it a newly developed material that has not seen mass production yet? New developments often do not come with complete data sheets that allow an early assessment, while for established materials the laminator will often be able to provide one. Cost is another important factor. The laminator will be able to provide guidance about volume pricing for the new material. Another consideration is the history of the supplier in terms of having met commitments in the past and how much of the material can be made available if it is needed for a high-volume product. If more than one PWB manufacturer is using the material and no second source can be identified, then this may raise a red flag in consideration of the material for use.

12.3.2 Two-Tier Strategy Approach

The qualification procedure proposed in this chapter follows a two-tier strategy approach that will enable quick decisions during a material qualification program. The first set of tests is easy to perform and focuses on the key properties that should be evaluated when testing a new laminate material. Failure to pass any of these tests will likely result in a material not being accepted for manufacturing. If multiple materials are being evaluated, the first tests will allow the easy elimination of the least probable candidates.

The second set of tests is an expanded set that focuses on all key properties that are commonly evaluated for today's laminates. However, the final decision about a qualification test plan for a new laminate should always be driven by product requirements, which means that additional tests may need to be added to the set of tests listed below.

12.4 INITIAL TESTS

This first set of tests is most commonly performed when starting a material evaluation and qualification project. They can be performed easily and give some early indications if the material exhibits any major problems.

12.4.1 Surface and Appearance

This test is one of the very first to be done as the new material arrives in the shop. The laminate is inspected for its appearance and visual quality. Very often the materials engineer will be able to assess the quality of the product without applying specification guidelines. Even packaging and shipping materials should be used to assess the laminator's overall quality control.

IPC-TM-650 method 2.1.2 and method 2.1.5 define procedures for inspecting the surface and appearance of copper-clad laminates. These standards should only be applied to surface areas that are pertinent to the finished board using good engineering judgment. More than 90

TABLE 12.2 Pit and Dent Point Value–Based Rating System

Longest dimension (mm)	Point value
0.13–0.25	1
0.26–0.50	2
0.51–0.75	4
0.76–1.00	7
Over 1.00	30

percent of the copper is usually removed during manufacturing and a new material should not be discarded solely for cosmetic reasons. On the other hand, areas that will contain edge card connector pads or very fine-line traces will require these surface standards to be applied.

The specifications categorize the longest permissible dimensions of any copper pits and dents and define a point value–based rating system as shown in Table 12.2.

The total point value for an area of 300×300 mm of laminate determines the material class as shown in Table 12.3.

TABLE 12.3 Total Point Value Determines Material Class

Class	A	B	C	D	X
Total point count	<30	<5 (All dents < 0.38 mm)	<17	0	Requirements agreed upon

In addition to copper pits and dents, the specifications contain procedures on how to inspect for scratches, wrinkles, and inclusions.

12.4.2 Copper Peel Strength

The copper peel strength of copper-clad laminates is an indicator for a variety of performance characteristics of the laminate. It will give an indication of the bond strength of the copper circuitry of a double-sided printed circuit construction after manufacturing and any successive assembly or repair cycles. In order to assess the copper bond strength of the outerlayer circuitry of a multilayer construction accurately, it is necessary to take a close look at the exact buildup and copper foil used in the construction and to build specific test vehicles that are similar to the original stack-up. Copper peel strength measurements have grown increasingly important as circuit traces and pads are becoming smaller. In addition, this test will allow an assessment of any potential weaknesses in multilayer constructions and identify the interface that might delaminate first after thermal stress. The higher the initial bond strength of a material, the better. For all of the characterizations after thermal stress, it is also important to look at the decrease in the bond strength. This is especially important for applications that require multiple reflow operations and one or more repair cycles.

The test method is specified in IPC-TM-650, method 2.4.8, and the applicable test patterns are specified in IPC-TM-650, method 5.8.3. Figure 12.1 shows an actual test pattern.

The material being tested should be processed using the same fabrication steps that will be applied for the actual product. IPC-TM-650, method 2.4.8, recommends a specimen size of 50.8×50.8 mm (2.0×2.0 in) and a minimum sample size of two for both the warp and fill (x and y) directions of the laminate. The end of the test strip is peeled off so a tensile tester can clamp it. The load cell should be calibrated to account for the weight of the clamp. The copper foil strip is then peeled from the material at a rate of 50.8 mm/min (2 in/min). The measured value is dependent on the copper thickness. The thicker the copper foil, the more force is re-

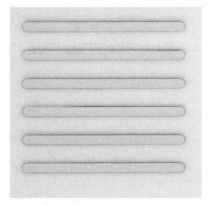

FIGURE 12.1 Test pattern used for copper peel strength test.

quired to deform it plastically. The value is also dependent on the angle at which the strip is being pulled. Therefore, it is always important to specify the thickness of the copper foil tested and to maintain a peeling force radius of 90° during the test. The minimum load is determined as specified and the peel strength is calculated using Eq. (12.1):

$$\text{lb/in} = L_M/W_S \tag{12.1}$$

where L_M = minimum load
$\quad\quad W_S$ = measured width of peel strip

Copper bond retention after exposure to soldering or touch-up temperatures is measured as outlined in IPC-TM-650, method 2.4.8, condition B. The test samples are covered with silicone grease before floating them on the solder pot at 288°C for 5 to 20 s to avoid solder contamination on the test strip. After allowing the samples to cool down to room temperature and removing the grease, the test is performed in the same way as described previously. It is important to also inspect the specimen for any blistering or delaminations prior to testing (see also Sec. 12.5.2.3.).

Another factor that can have an influence on the adhesion strength of the copper foil is exposure to processing chemicals during manufacturing. Details about this test method can be found in IPC-TM-650, method 2.4.8, condition C. This test is usually only performed if there are specific requirements at the end product level that call for it or if the material exhibits low copper bond retention properties during manufacturing.

12.4.3 Solder Shock

The solder shock test is one of several methods of assessing the thermal resistance of copper-clad laminates. It is easy to perform and represents another key test during the early assessment of a material. There are a number of different methods to choose from, which will be described in detail in the following text. During the initial assessment of the material it is important to choose at least one of the described test methods to make certain that the material meets the minimum requirements. If more than one material is being tested, the test will identify potential differences between candidates.

12.4.4 Glass Transition Temperature T_g

The determination of the glass transition temperature T_g completes the initial assessment. Different test methods are described in more detail in Sec. 12.5.2.1. For the first assessment, one method should be selected. When choosing the method, it is important to make sure that the test procedure is compatible with the material being tested (for some materials the glass transition can only be determined using a dynamic mechanical analysis [DMA]). A minimum of two samples should be tested. If a data sheet of the material is available, the measured value should correspond with the manufacturer's data (within the measurement accuracy of the equipment). The values of T_{g1} and T_{g2} (measured on the same sample) should not be more than 7°C apart.

12.5 FULL MATERIAL CHARACTERIZATION

The following set of characterization methods summarizes the most commonly used test methods to qualify copper-clad laminates for manufacturing. It is intended to guide the materials engineer through the laminate qualification process. Yet it can only serve as a best-practice guide because the final set of qualification methods will vary for each material and should always be driven by its specific application.

12.5.1 Mechanical Tests

12.5.1.1. Modulus and Tensile Strength (Tensile Test).   The test method to determine modulus and yield stress for copper-clad laminates described in the following text is based on ASTM-D-882 ("Standard Test Methods for Tensile Properties of Thin Plastic Sheeting"). Other relevant industry standards are IPC-TM-650, method 2.4.19 (for flexible laminate materials) and IPC-TM-650, method 2.4.18.3 (for deposited organic free films). Elastic modulus and yield stress are always determined by the same method.

The elastic modulus is the indicator value for the stiffness and strength of a composite material under load, while the yield stress indicates the maximum load the material can take in the *x/y* direction before it breaks. The modulus is the more important of the two values. It is used to differentiate materials and their actual strength during qualification testing. In addition to this, the elastic modulus is also one of the key properties that need to be understood about a material when performing computational modeling of packages and printed circuit board assemblies. In a lot of these applications materials with different properties are combined, which leads to thermomechanical stresses that cause fatigue or interfacial failures. Computational modeling is often used when there is not sufficient time or resources available to perform actual product test, but fast reliability assessments are needed.

The standard test specimen for this test consists of a strip of laminate material 152.4 mm (6 in) long and 12.7 mm (0.5 in) wide. All copper needs to be removed before testing. A minimum of 10 samples should be prepared. It is important that only samples with identical thickness and glass stackup be compared. The test engineer also needs to make sure that both the warp and fill direction of the laminate composite are measured separately for each laminate being tested.

After the samples are cut, they need to be inspected for any fractures, delamination, or roughness along the edges, and if necessary they need to be sanded. The cross-sectional area of each specimen is calculated from the width and thickness in the measurement area. Then the individual samples need to be conditioned at 23°C (73.4°F) and 50 percent RH for 24 h. While most specifications recommend conditioning of the material prior to cutting and preparation of the samples, the authors recommend first cutting the specimens to size and then conditioning them. This will ensure more uniform properties of the individual samples.

The tests are performed using a standard tension and compression apparatus, which can be operated at a controlled rate of cross-head movement. The load-extension curve should be recorded and the specimen tested until it breaks. If extensometers are used, it is necessary to minimize the stress on the sample at the contact points of specimen and indicator.

The tensile strength is calculated by dividing the maximum load at break by the original minimum cross-sectional area. The unit is megapascals (MPa). Dividing the elongation at break by the initial gauge length and multiplying it by 100 will calculate percent elongation. When extensometers are employed, only the length of the section between them is used; otherwise the distance between the grips represents the initial gauge length. Young's modulus is calculated by drawing a tangent to the linear portion of the stress-strain curve, selecting any point on this tangent, and dividing the tensile stress at that point by the corresponding strain. The unit is gigapascals (GPa). For all calculations, the values for 10 samples should be recorded and the average calculated.

In addition to the absolute values for the different mechanical properties, the stress-strain curve will give the materials engineer information about the elastic and plastic portion of the deformation in the laminate samples. The elastic (linear) part of the deformation is restorable, while plastic changes in the materials are not. Figure 12.2 shows a typical stress-strain curve.

12.5.1.2 Flexural Strength.   The test method for determining the flexural strength of copper-clad laminates is based on ASTM-D-790 ("Flexural Properties for Un-reinforced and Reinforced Plastics and Insulating Material") and IPC-TM-650, method 2.4.4B ("Flexural Strength of Laminates at Ambient Temperatures"). The test method described earlier of determining the modulus in tensile mode does have the inherent problem that the glass rein-

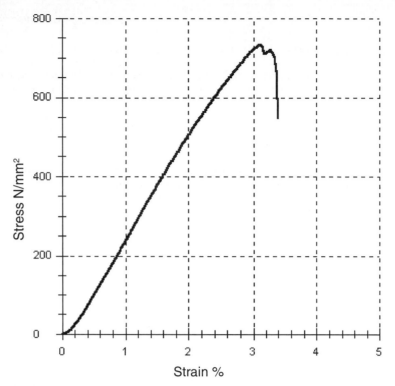

FIGURE 12.2 Typical stress-strain curve.

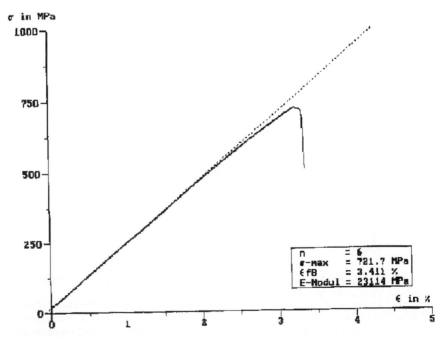

FIGURE 12.3 Typical flexural strength measurement curve.

forcement and its properties have a large influence on the results. This may be avoided when testing copper-clad laminates in flexural mode. Here the attributes of the resin system have a greater influence on the outcome of the test. When comparing the test results of different laminate materials it is important to only compare data that were generated using the same test method.

This test also requires that the copper be completely removed from the laminate material by means of standard etching processes. The specific size (length and width) of the samples and the test parameters (test span and speed) are dependent on the thickness of the laminate and can be found in IPC-TM-650, method 2.4.4.

The samples are conditioned to room temperature. They are then centered on the supports of the standard tension and compression test apparatus with the long axis of the specimen perpendicular to the loading nose and support. The samples are loaded at the defined speed and tested until they break. The load at breakage should be recorded. The flexural strength is calculated by Eq (12.2):

$$S = 3PL \, / \, 2bd^2 \tag{12.2}$$

where S = flexural strength (psi)
  P = Load at breakage (pounds)
  L = Span (inches)
  B − Width of specimen (inches)
  D = Thickness (inches)

Figure 12.3 shows a typical flexural strength measurement curve.

12.5.1.3 Copper Peel Strength (Before and After Thermal Stress). Test methods are described in Sec. 12.4.2 and results can be transferred to this point in the evaluation. The data can be completed by performing additional tests after thermal stress or after exposure to plating solutions if necessary.

12.5.2 Thermomechanical Tests

12.5.2.1 Glass Transition Temperature T$_g$ ***(TMA, DSC, DMA).*** The resin systems used in today's copper-clad laminates are almost all three-dimensionally cross-linked thermoset polymers. This means that the materials are manufactured in a curing reaction into a hard final product and cannot be remelted many times in the way thermoplastic polymers can. In many cases different resin systems are blended to create a material with specific properties. The functionality of the individual resins and their ratio in the blend define the properties of the thermoset. The degree of cure of the hardened thermoset resin system can be described by characteristics such as percent of uncured monomer and heat of final cure or by mechanical properties such as hardness, modulus, or yield stress. But one of the most commonly used characteristics to describe the degree of cure and cross-link density of a thermoset polymer is the glass transition temperature.

The simplest definition of the T_g of a laminate is to call it the temperature at which the mechanical properties of a laminate change (deteriorate) rapidly. It is at this temperature that the polymer changes from a rigid, brittle and glass-like state to a softer and more rubber-like material. However, this should not be confused with the melting point of a crystalline substance or the extreme softening of a thermoplastic polymer. What happens in a thermoset material at T_g is that the relative mobility of the cross-linked polymer molecules changes. Below T_g they are immobile and the material is rigid; above T_g the relative mobility of the molecules is increased and slight movements are possible, which leads to a loss of mechanical stability in the material. While all of the methods described in the following text allow us to determine a specific glass transition temperature for laminate materials, it also needs to be emphasized that the glass transition in thermoset resins is not a single defined temperature. It usually occurs within a temperature range (for some resin systems this range can be >100°C) and the T_g is determined in this temperature range.

There are a number of different measurement techniques to determine the T_g. Those most commonly used are thermal analysis techniques. There are other techniques available (i.e., spectral analysis, electrical characterization, etc.) but their use is limited and therefore they will not be discussed here. The three thermal analysis methods discussed in this chapter are as follows:

- Thermomechanical analysis (TMA)
- Differential scanning calorimetry (DSC)
- Dynamic mechanical analysis (DMA)

All of these techniques measure slightly different property changes in the laminate material and therefore the values that are obtained will be different, though they usually follow this rule of thumb:

$$T_g \, (\text{DMA}) > T_g \, (\text{DSC}) > T_g \, (\text{TMA})$$

T_g *by TMA.* Thermomechanical analysis measures dimensional changes in a material as it is heated from room temperature to a preset final temperature. The change in length, width, or height of the specimen with the change in temperature determines the coefficient of thermal expansion (CTE) of the material (see also Sec. 12.5.2.2). At T_g the expansion coefficient of the material changes and it is this property change that is used to determine the T_g by TMA.

The test procedure is specified in IPC-TM-650, method 2.4.24C. The specimen should have a minimum thickness of 0.51 mm, and at least two samples taken from random locations in the material need to be tested. Any copper cladding needs to be removed by etching. The treatment of the copper foil leaves a negative imprint on the laminate surface, which can lead to problems during measurement right below the actual glass transition temperature. Therefore, the surface of the specimen should be sanded lightly prior to testing. The edges of the sample should also be smooth and free of burrs. Use care to minimize stress or heat on the specimen.

The samples need to be preconditioned for 2 h at 105°C (221°F), then cooled to room temperature in a desiccator. The actual measurement should be started at a temperature no higher than 35°C (95°F); an initial temperature of 23°C is recommended. Unless otherwise specified, a scan rate of 10°C (18°F) per minute is commonly used. The temperature ramp needs to be continued at least 30°C (54°F) above the anticipated transition region. The T_g is defined as the temperature at which the two tangent lines for the thermal expansion coefficient intersect. Figure 12.4 shows an example of a typical TMA scan. The material has a T_g of 136.7°C.

T_g *by DSC.* Differential scanning calorimetry measures the difference in heat absorption or emission from a test specimen in comparison to a reference sample (usually nitrogen gas). This makes the technique applicable to determine a variety of property changes in polymer materials. It can detect the exothermic cure reaction, crystallization energy, and residual reactivity in polymers, as well as endothermic melting points. For epoxy-based resin systems, DSC is a well suited technique as these materials go through a crystalline transition at T_g and this property change can be used to determine the glass transition. For other resin systems that are more amorphous and have a T_g that occurs over a wider range (i.e., polyimides), detection may be more difficult with DSC than with TMA.

The test procedure is specified in IPC-TM-650, method 2.4.25C. The specimen should be a solid piece weighing between 15 and 25 mg. In the case of very thin materials, multiple pieces may be used. The specimen should be of a size and configuration that fits within the sample pan of the DSC equipment. All of the specimen preparation needs to take place before preconditioning to avoid any moisture influence prior to testing. The edges of the sample should be smoothed and burrs should be removed by light sanding or the equivalent to achieve proper thermal conduction. Use care to minimize stress or heating of the specimen. Even

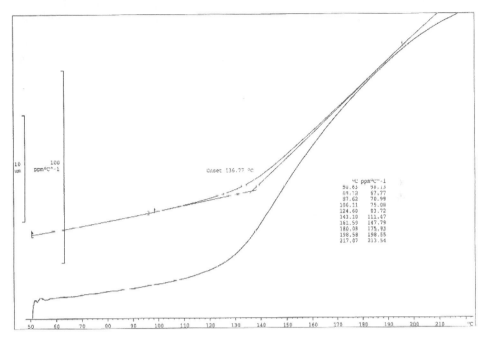

FIGURE 12.4 Typical TMA scan.

though the IPC test method allows samples to be tested with copper cladding, the authors recommend testing only without copper cladding. This increases the relative mass of polymer in the DSC sample pan and leads to a better signal and detection of T_g.

The samples need to be preconditioned for 2 h at 105°C (221°F), then cooled to room temperature in a desiccator for at least ½ h prior to testing. Place the specimen in a standard aluminum sample pan with an aluminum lid. The use of a lid and crimping tool is optional. For reference purposes, a cover lid crimped onto the sample pan should be used. Start the scan at a temperature at least 30°C lower than the anticipated onset of T_g. Unless otherwise specified, a scan at a rate of 20°C/min (36°F/min) is recommended and the temperature ramp needs to be continued at least 30°C (54°F) above the anticipated transition region. The glass transition temperature is determined using the heat flow curve. In many cases the DSC equipment has applicable software installed to determine T_g. A tangent line is fit onto the curve above the transition region and a second tangent line is fit below the transition region. The temperature on the curve halfway between the two tangent lines, or ½ δ C_p, is the glass transition point T_g. Figure 12.5 shows an example of a DSC scan where the material has been scanned twice. The T_g of the first scan was 192.9°C. The second scan, as expected, yielded a slightly higher value of 195.3°C.

T_g *by DMA.*   Dynamic mechanical analysis applies an oscillatory stress or strain on the sample while the temperature is increased during the experiment. The ability of the material to elastically store the mechanical strain energy changes during the heat cycle as it changes its viscoelastic behavior from glassy to rubbery. This property change determines the glass transition temperature by DMA.

The test procedure is specified in IPC-TM-650, method 2.4.24.2. The test specimen should consist of a strip of laminate material compatible with the measuring equipment. For all samples with woven reinforcement, it is necessary to make sure that the specimens are cut paral-

DSC

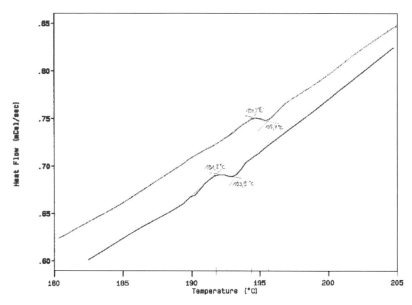

FIGURE 12.5   Example of DSC test result.

lel or perpendicular to the woven structure. The analysis is based on an assumption of constant specimen geometry; therefore the test specimens must be stiff enough not to deform plastically during the experiment. All copper needs to be etched off.

While the IPC method recommends preconditioning of the samples at 23°C and 50 percent RH for at least 24 h prior to testing, the authors recommend preconditioning similar to that employed for TMA or DSC measurements. Samples are baked for 2 h at 105°C (221°F), then cooled to room temperature in a desiccator for at least ½ h prior to testing. The sample is mounted in the fixture, making sure that it is perpendicular to the clamps. Both clamps are tightened using torque screwdrivers to ensure that the sample does not slip during the measurement and that there are no stresses built up around the sample that could negatively influence the results. The sample is loaded at a frequency of 1 Hz (6.28 radians/s) while being heated in dry nitrogen or dry air at a rate of no faster than 2°C/min. The temperature ramp needs to be continued at least 50°C above the transition region. The glass transition temperature is defined as the temperature corresponding to the maximum in the tan δ–vs.–temperature curves at a frequency of 1 Hz. Tan δ is calculated from Eq. (12.3):

$$\tan \delta = E''/E' \tag{12.3}$$

where E'' is the loss modulus
E' is the storage modulus

Figure 12.6 shows an example of a typical DMA scan. The material has a T_g of 193.17°C.
    Comparison of Methods.   Table 12.4 shows a comparison of methods used to determine T_g.

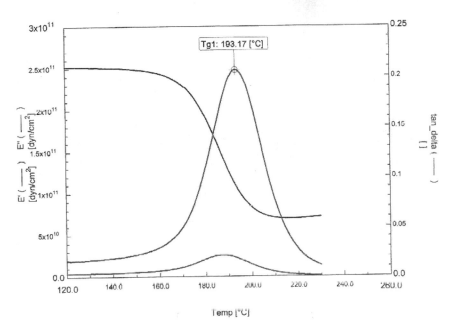

FIGURE 12.6   Typical DMA scan.

TABLE 12.4   Comparison of Methods Used to Determine T_g

		TMA		DSC		DMA
Sample preparation and setup	+	Samples need some care. Easy setup.	+	Easy sample preparation and setup	–	Samples need care. Setup can be complex.
Measurement duration	+	Short.	+	Short	–	Long.
Data analysis	+	Clear and easy.	–	Can be complex and may be difficult to detect	+	Clear and easy.
Signal sensitivity	+	Very sensitive.	–	Less sensitive and not suitable for all materials	+	Very sensitive.
Sensitivity to external disturbance	–	High.	+	Very low	+	Low.

Some typical glass transition temperature ranges for commonly used resin systems are listed below. The list is meant as a guideline and the ranges are not dependent on a specific measurement method.

Polyimides	260 to 270°C
Polyimide blends	240 to 260°C
Epoxy/polyimide blends	240 to 260°C
Cyanate ester	230 to 250°C

BT/bismaleimide blends	225 to 240°C
Multifunctional epoxies	160 to 190°C
Multi-/tetrafunctional epoxy blends	145 to 160°C
FR-5	140 to 160°C
Tetrafunctional epoxies	130 to 140°C
FR-4	115 to 125°C

12.5.2.2 Coefficient of Thermal Expansion (CTE). The CTE describes the property of materials to expand as they are being heated (see also Sec. 12.5.2.1). Because of the reinforced composite construction of most laminates, the CTE in the x and y directions is usually different from the CTE in the z direction. By convention, the x direction corresponds to the warp direction of the reinforcement and the y direction to the fill direction of the fabric. The z direction is the direction perpendicular to the plane of the laminate. The difference between the values is caused by the reinforcement, which severely restrains the expansion in the x and y directions while in the z direction the resin can expand unrestrained.

For copper-clad laminate materials used in printed circuit boards, the CTE in the x, y, and z directions is of great importance. The x and y directions are critical because of all of the components that will be mounted on the final printed circuit board. The greater the mismatch between the laminate material and the soldered components, the higher the risk that temperature changes will lead to solder fatigue and eventually to a reduction in the reliability of the board. The z direction is of equal importance because the expansion of the laminate during thermal cycles will lead to stresses in the copper plating (CTE of 17 ppm/K) of the plated through-holes in the board. A low z-axis CTE and high T_g are therefore generally desirable for increased through-hole reliability.

There are a number of different measurement techniques available to determine the CTE of copper-clad laminates. The most commonly used is TMA following IPC-TM-650, method 2.4.24C, as described in Sec. 12.5.2.1. When determining the CTE of a laminate, the temperature scan must start at a temperature sufficiently lower than the specified temperature range for which the CTE is being determined to allow the heat rate to stabilize. The typical scan rate is also 10°C (18°F) per minute and the scan should be continued to at least 250°C (482°F). Because the expansion properties of the material change at T_g, there are usually two CTE values that are being reported: the CTE below T_g is commonly referred to as α_1 and the CTE above T_g as α_2.

$$\text{CTE: } \alpha = \Delta L / \Delta T \text{ ppm/K} \tag{12.4}$$

where ΔL = change in length
$\quad \Delta T$ = change in temperature
$\quad$ K = degrees Kelvin

In the TMA scan example (Fig. 12.4) the material has a z-axis CTE(α_1) of ~65 ppm/K and a z-axis CTE(α_2) of ~150 ppm/K.

The CTE in the x and y directions can also be measured by TMA, although care must be taken during the preparation of the sample to avoid any influence of reinforcement material on the TMA probe. The results are strongly dependent on the properties of the reinforcement fabric.

Another method to determine the x- and y-axis CTE of laminate materials employs strain gauges. This method is described in IPC-TM-650, method 2.4.41.2. There are a number of details that need to be taken into consideration when making use of this method. The strain gauges need to be calibrated for the specific temperature range and the adhesive used to attach them needs to be stable over the whole range. Special care needs to be taken during specimen preparation and gauge attachment. It is recommended that one heat cycle be run prior to the actual measurement to remove any residual stresses. More details can be found in the IPC test method referenced earlier.

12.5.2.3 Thermal Resistance. The thermal resistance of laminate materials is an important property. For one thing, it is a key indicator of the performance of printed circuit boards during assembly operations. It is not unusual for printed circuit boards to be exposed to more than one reflow process, and, if components need to be repaired afterward, the total number of thermal excursions that the board material sees can often be four or more. Another important factor that can be assessed with thermal resistance tests are printed circuit board applications that expose the board to high operating temperatures. There are a number of different test methods that should be considered when qualifying laminates.

Solder Dip Resistance.   This test method is described in IPC-TM-650, method 2.4.23. The purpose is to assess the ability of laminate materials to withstand the temperatures in a molten solder bath. The samples are tested both with and without copper cladding. Performance factors being considered are resistance to softening, loss of surface resin, scorching, delamination, blistering, and measling.

The material is tested in three different surface configurations: (1) a surface upon which no metal cladding was ever applied (if possible), (2) a surface with the metal cladding removed by standard etching processes, and (3) a surface with metal cladding as received. The sample size for all specimens is 31.75 × 31.75 mm (1¼ × 1¼ in) × thickness, and three samples are required for each surface configuration. All samples are tested using the same procedure.

Preclean the samples by immersing them for 15 s in 10 percent HCl (by volume) and then rinsing in water. The HCl should be at a temperature of 60°C (140°F). Dry the specimen quickly to avoid excess oxidation of the sample. Dip the sample into a flux agent and allow to drain for 60 s before proceeding with the solder dip. Stir, then skim the surface of the molten solder with a clean stainless steel paddle to ensure that the solder is of uniform composition and at a temperature of 245°C (473°F). Immerse the specimen edgewise into the molten solder. The insertion and withdrawal should be at a rate of 25.4 mm/s (1 in/s) and the dwell time in the solder should be 4 s. Upon withdrawal, the solder should be allowed to solidify by air-cooling while the specimen is in a vertical position. Thoroughly remove the flux.

The samples are examined for any evidence of discoloration or surface contaminants, loss of surface resin, softness, delamination, interlaminar blistering, or measles. The specimens having metal cladding also need to be examined for blistering or delamination of the metal foil from the laminate material.

Solder Float Resistance.   Compared to the solder dip resistance test just described, this test addresses the resistance of the laminate material floating on the solder bath. Because this method subjects the sample to a thermal gradient across the *z* axis of the material, similar to an actual wave solder operation, the authors consider the results of this test particularly important.

The test method is described according to IPC-TM-650, method 2.4.13. At least two samples need to be tested for each material and they should be taken from random locations. For double-clad laminate, the copper foil needs to be removed from the backside of each specimen by using standard etching processes. The test samples need to be preconditioned in an air-circulating oven at 135°C (275°F) for 1 h to remove any excess moisture that could lead to premature failures. After preconditioning, the specimens can be held in a desiccator at room temperature. After attaching the sample to the solder float test fixture, float it for 10 s, foil side down, on the surface of the molten solder at a temperature of 260°C (500°F) (for method A) or 288°C (550°F) (for method B). Then remove the sample and tap the edges to remove any excess solder. Thoroughly clean and visually examine the sample for blistering, delamination, or wrinkling.

If more than one material is being qualified and no failures are being observed after testing for 10 s, or if the goal is to determine the weakest point in the laminate and test to failure, the authors recommend increasing the solder float time incrementally or repeating the test with the same sample until failures are observed.

T260 (TMA).   Aside from testing materials in solder baths, there is one other technique of verifying the thermal resistance of laminates. This test method is described in IPC-TM-650, method 2.4.24.1, and is used to determine the time to delamination of laminates and printed boards using a TMA system (see also Sec. 12.5.2.1).

The sample requirements are identical to the ones for determination of the T_g by TMA. A minimum of two specimens should be tested, which can be taken from random locations of the material. Materials are tested as received; the metal cladding is not removed. The samples are preconditioned for 2 h at 105°C (221°F) and then cooled down to room temperature in a desiccator. The TMA temperature ramp should start from an initial temperature no higher than 35°C (95°F) at a scan rate of 10°C/min. After the scan reaches the specified isothermal temperature, hold at that temperature (onset) for 10 min or until failure. The time to delamination is determined as the time from the onset of the isotherm to failure. Failure is defined as any event or deviation of the data plot where the thickness is shown to have irreversibly changed. On occasion some materials will delaminate before the isotherm is reached. In this case, the temperature at the time of failure is recorded. For epoxy laminates and similar materials, the recommended isothermal temperature is 260°C (500°F). For polyimides and other high-temperature materials, increasing the isothermal temperature to 288°C (550°F) is recommended.

12.5.3 Electrical Characterization

12.5.3.1 Dielectric Constant D_c and Dissipation Factor D_f. The Dielectric constant (permittivity, D_c, ε_r) is defined as the ratio of the capacitance of a capacitor with a given dielectric (laminate) to the capacitance of the same capacitor with air as the dielectric. It is a measure of the ability of a material to store electrostatic energy and determines the relative speed that an electrical signal will travel within that material. The higher the D_c, the slower the resulting signal propagation speed. The signal speed is inversely proportionate to the square root of the dielectric constant. D_c is not an easy property to measure or to specify, because it depends not only on the material properties and the resin-to-glass ratio but also on the test method, the test frequency, and the conditioning of the samples before and during the test. It also tends to shift with temperature. Still, D_c values are important for computer simulations that are used to predict the performance of impedance-controlled high-end multilayer constructions, especially when new laminate materials are considered. See Fig. 12.7 for a schematic representation of the test.

FIGURE 12.7 Dielectric constant test.

Related to the dielectric constant is the dissipation factor D_f or loss tangent. This is a measure of the percentage of the total transmitted power that will be lost as electrons dissipate into the laminate material. See Fig. 12.8 for a schematic representation of the test.

There are a number of test methods specified in IPC-TM-650 to determine the dielectric constant and the dissipation factor (methods 2.5.5, 2.5.5.1, and 2.5.5.2). Both methods 2.5.5 and 2.5.5.2 test laminate material with fully removed copper. At least three samples need to be tested because of the large influence of the dielectric thickness on the determined value. Method 2.5.5.1 employs samples with an etched pattern. All of these measurements are based on the capacitance of the corresponding sample. With this capacitance and the area and thickness of the sample capacitor it is possible to calculate the dielectric constant.

FIGURE 12.8 Dissipation factor test.

The dissipation factor is determined with methods 2.5.5.1 and 2.5.5.2. In one case the value can be read from the equipment display (Agilent Technologies Mdl 4271A); in the other it can be calculated from the measured sample conductance and capacitance and the measuring frequency.

12.5.3.2 Surface and Volume Resistivity. Electrical resistivity of laminate materials is differentiated between specific surface resistance and volume resistance. The surface resistance σ characterizes the electrical resistance between two conductors along the surface of the laminate material. The volume resistance ρ describes the electrical resistance between two layers of conducting copper along the z axis of the laminate material. The higher the values for both of these electrical properties, the better, as this will ensure proper isolation of individual copper conductors in the printed circuit board.

Both resistance values are determined according to IPC-TM-650, method 2.5.17. All resistance measurements are done with equipment capable of measuring up to 10^{12} MΩ while applying 500 V direct current to the test specimen (i.e., Agilent Technologies 16008A). The sample size is 101.6 × 101.6 mm for laminates thicker than 0.51 mm and 50.8 × 50.8 mm for laminates less than 0.51 mm thick. The test pattern according to method 2.5.17 should be applied to the samples using standard photo and etch processes.

All measurements are performed by applying 500 V direct current. The voltage needs to be applied to the samples for 60 s before taking the resistance reading to allow the test structures to stabilize. The surface resistance is determined between the outer ring electrode and the inner solid electrode. The volume resistance is determined between the solid front and back electrodes after changing the connecting cables appropriately. The values for volume and surface resistivity can be calculated form the measured resistance values as follows:

$$\text{Volume resistivity } \rho \text{ (MΩ-cm)} \qquad (12.5)$$

$$\rho = (R * A)/T$$

where R = measured resistance, MΩ
$\quad A$ – effective area, square cm
$\quad T$ = average thickness of specimen, cm

$$\text{Surface resistivity } \sigma \text{ (MΩ)} \qquad (12.6)$$

$$\sigma = (R * P)/D$$

where R = measured resistance in MΩ
$\quad P$ = effective perimeter of electrode in cm
$\quad D$ = width of test gap in cm

12.5.3.3 *Dielectric Strength Breakdown.*   The dielectric strength of a laminate material is its ability to resist electrical breakdown. The dielectric strength defines a specific voltage that the laminate resists for a specified time; the dielectric breakdown voltage defines the maximum voltage at which the laminate fails. These properties can be measured perpendicular to the reinforcement (z axis) or parallel to the reinforcement (x-y axis). The more important value is the z-axis strength, because more and more thin prepreg and laminate cores are being used in high-end multilayer applications. The minimum thickness in the laminate is defined as the shortest distance between the copper treatment peaks that needs to resist the desired test voltage. The values for dielectric strength vary with test setup, temperature, humidity, frequency, and wave shape, but are, if tested under controlled conditions, comparable between materials.

The test method (see Fig. 12.9) following IPC-TM-650, method 2.5.6.2, describes the determination of the perpendicular electric strength of laminates. Four specimens should be tested. The sample size is recommended to be 101.6 × 101.6 mm (4 × 4 in) with any copper cladding removed. Unless otherwise specified, the samples need to be conditioned for 48 h in distilled water at 50°C. After that the samples are immersed in ambient-temperature distilled water for a minimum of 30 min to allow the samples to achieve temperature equilibrium without a substantial change in moisture content. The test is performed at ambient temperature (23°C). Relative humidity is not significant as the tests are performed under oil. Samples are inserted into the high-voltage test equipment and tested to failure at a 500-V/s increase. The results are reported in volts per mil.

FIGURE 12.9   Dielectric strength test.

The test method following IPC-TM-650, method 2.5.6, describes the determination of the parallel electric strength of laminates. Four specimens should be tested, two in the machine direction and two in the transverse direction for reinforced materials. The sample size is recommended to be 76.7 mm (3 in) long by 50.8 mm (2 in) wide with any copper cladding removed. Two holes 4.77 mm (0.188 in) in diameter are to be drilled along the centerline of the 76.7-mm (3-in) dimension and midway between the edges in the 50.8-mm (2-in) dimension, with a spacing of 25.4 mm (1 in) from center to center. Conditioning and test setup in oil are identical. Electrodes are inserted in the holes and the samples are tested to failure at a 500-V/s increase. The results are reported in kilovolts.

12.5.4   Other Laminate Properties

12.5.4.1   Flammability.   The flammability of laminate materials is classified according to Underwriters Laboratories (UL) specifications. All of the tests are performed using a standard test setup under an exhaust hood using a Bunsen burner as a source for the flame. The categories follow.

UL-94-V-0.   Specimens must extinguish within 10 s after each flame application and within a total of less than 50 s after a total of 10 flame applications. No samples are to drip flaming particles or exhibit glowing combustion that lasts for more than 30 s after the second flame test.

UL-94-V-1.   Specimens must extinguish within 30 s after each flame application and within a total of less than 250 s after a total of 10 flame applications. No samples are to drip flaming particles or exhibit glowing combustion that lasts for more than 60 s after the second flame test.

UL-94-V-2.   Specimens must extinguish within 30 s after each flame application and within a total of less than 250 s after a total of 10 flame applications. Samples may drip flaming particles or burn briefly; however, no specimen may exhibit glowing combustion that lasts for more than 60 s after the second flame test.

In most cases the laminator will provide test results for these tests routinely. In the case of new materials when the laminator has not supplied any data in this area, it may be a good idea to check for this property. This does not require an elaborate test setup and a first indication can be gained by igniting a sample specimen under an exhaust hood using a lighter.

12.5.4.2   Water Absorption.   Depending on its specific molecular composition, every laminate material will absorb a certain amount of water. This will happen not only during the many wet processing steps in printed circuit board manufacturing but also as a result of exposure to normal environmental conditions. The absorbed moisture may change the properties of the laminate and increase the risk of causing blistering and delaminations during high-temperature processes such as reflow soldering.

The test method according to IPC-TM-650, method 2.6.2.1, determines the amount of water that is absorbed by a laminate material sample when immersed in water for 24 h. The test is easy to perform and the results for different laminates are readily comparable.

The test samples for this test need to be 50.8 mm (2 in) long by 50.8 mm (2 in) wide. The thickness is not specified but should not vary widely when more than one material is characterized. The edges of the samples need to be sanded smooth and copper cladding is to be removed using standard etching processes. The samples are preconditioned in a drying oven for 1 h at 105°C (221°F), cooled down to room temperature in a desiccator, and weighed immediately after removal. Then the samples are placed in distilled water at 23°C (73.4°F). It is important to place the samples on their edge to maximize the laminate area exposed to water. After 24 h the samples are removed, dried with a dry cloth, and immediately weighed. The moisture absorption is reported in percent weight increase.

12.5.4.3   Additional Tests.   In addition to the described test methods, there are many more that can be found in IPC-4101 and in the test methods manual IPC-TM-650. All of these tests address laminate properties that may have a significant impact on the performance of the final product. The final decision as to which qualification tests to include and which not to include always needs to be made on a case-by-case basis depending on the performance requirements for the printed circuit board.

At the same time there are a number of laminate qualification tests that rarely will be performed in the printed circuit manufacturing facility. In many cases the laminate supplier will perform a number of qualification tests and share the results with customers. This may in many cases be sufficient, especially after a close relationship between the supplier and the PCB manufacturer has been established.

12.5.4.4 Prepreg Testing.   During the qualification of a new laminate material it may also become necessary to perform prepreg-specific tests to verify the material's quality. The most commonly tested properties of prepregs are resin content, flow during mutilayer processing, and gel time. Details about additional properties and all corresponding test methods can be found in IPC-4101.

12.6 CHARACTERIZATION TEST PLAN

Table 12.5 summarizes all of the test procedures.

TABLE 12.5   Laminate Characterization Test Plan

Step 1	Data comparison	Unit				Test result
	Material available in production volume	Yes/no				
	Material cost	$/m^2				
	Material data sheet available	Yes/no				

Step 2	First test runs in small volume	Unit	Test method	Conditioning	Data sheet	Test result
	Surface and appearance		IPC-TM-630, methods 2.1.2 and 2.1.5	As received		
	Copper peel strength laminate, condition A	N/mm	IPC-TM-650, method 2.4.8	As received		
	Copper peel strength laminate, condition B	N/mm	IPC-TM-650, method 2.4.8	Solder pot, 288°C/10 s		
	Solder shock laminate, 288°C/10 s	Pass/fail	IPC-TM-650, method 2.4.23	As received		
	Solder shock laminate, 288°C/60 s	Pass/fail	IPC-TM-650, method 2.4.23	As received		
	Glass transition temperature (DSC or TMA or DMA)	°C	IPC-TM-650, as applicable	As received		
	z-axis expansion laminate, TMA	ppm/°C	IPC-TM-650, method 2.4.24	As received		

Step 3	Material characterization	Unit	Test method	Conditioning	Data sheet	Test result
	Dielectric constant (1 MHz)	—	IPC-TM-650, method 2.5.5.2	24 h/23°C/50%		
	Dielectric constant (1 MHz)	—	IPC-TM-650, method 2.5.5.2	96 h/35°C/90%		

TABLE 12.5 Laminate Characterization Test Plan (Continued)

Step 3	Material characterization	Unit	Test method	Conditioning	Data sheet	Test result
	Dissipation factor (1 MHz)	—	IPC-TM-650, method 2.5.5.2	24 h/23°C/50%		
	Dissipation factor (1 MHz)	—	IPC-TM-650, method 2.5.5.2	96 h/35°C/90%		
	Surface resistance	$M\Omega$	IPC-TM-650, method 2.5.17	24 h/23°C/50%		
	Surface resistance	$M\Omega$	IPC-TM-650, method 2.5.17	96 h/35°C/90%		
	Volume resistance	$M\Omega$ cm	IPC-TM-650, method 2.5.17	24 h/23°C/50%		
	Volume resistance	$M\Omega$ cm	IPC-TM-650, method 2.5.17	96 h/35°C/90%		
	Dielectric withstanding voltage	V/mil	IPC-TM-650, method 2.5.6	48 h in 50°C H_2O		
	Glass transition temperature (DSC)	°C	IPC-TM-650, method 2.4.25C			
	Glass transition temperature (TMA)	°C	IPC-TM-650, method 2.4.24C			
	Glass transition temperature (DMA)	°C	IPC-TM-650, method 2.4.24.2			
	CTE x,y (α_1)	ppm/K	IPC-TM-650, method 2.4.41.2			
	CTE x,y (α_2)	ppm/K	IPC-TM-650, method 2.4.41.2			
	CTE z (α_1)	ppm/K	IPC-TM-650, method 2.4.24			
	CTE z (α_2)	ppm/K	IPC-TM-650, method 2.4.24			
	Water absorption	%	IPC-TM-650, method 2.6.2.1			
	Flammability		UL-94			

12.7 MANUFACTURABILITY IN THE SHOP

During the qualification of a new material it is also crucial to run the laminate through the production processes to check if they are compatible. Table 12.6 summarizes all the steps that are necessary to control.

The first process step is innerlayer processing. Here a change in stiffness of the new laminate material can have a significant impact on the processability in horizontal manufacturing lines. This is especially important for thin cores. A change in the copper quality of the new laminate may impact the adhesion of the innerlayer photoresist and can also affect the copper etch rates. During automatic optical inspection (AOI) it is necessary to verify the contrast between copper circuitry and laminate surface and to adjust the AOI settings if necessary. In case of necessary innerlayer repairs, the settings for the welding process may need adjustment to avoid thermal damage of the laminate and ensure reliable interconnects. The last step dur-

TABLE 12.6   Summary of Steps to Control Material
Introduction to Fabrication Process

Compatibility with innerlayer (IL) process steps
Preclean
Resist lamination
Exposure
Develop—etch—strip
Punching
IL automatic optical inspection
IL repair/OL repair
Black oxide/Alternative surface

Compatibility with multilayer process steps
IL drying
Layup
Press cycle
Edge routing
Dimensional stability

Compatibility with process steps from drill to electroless copper
Drilling
Brush/pumice/deburr
Desmear
Electroless Cu

Compatibility with solder mask process step

Compatibility with different finish metallization processes
Hot-air solder leveling (HASL)
Electroless Ni/Au
Immersion Sn
Immersion Ag

Compatibility with routing process

ing innnerlayer processing is copper black oxide. Here it is necessary to verify the compatibility of the black oxide type (reduced/nonreduced or alternative) with the base material and confirm that the innerlayer bond strength of the multilayer product is sufficient.

After the multilayer lamination cycle, the dimensional stability of the innerlayers needs to be assessed and the innerlayer scaling factors adjusted if necessary. The next step after multilayer lamination and edge routing is drilling. A change in the thermomechanical properties of the laminate may impact the quality of the drilled holes, and adjustments of drill speeds or desmear settings may be necessary. After metallization and plating, the adhesion of the copper in the plated through-hole as well as reliable contacts to all innerlayers need to be verified.

The adhesion of solder mask may change, as well, as a new material is introduced. This is especially critical in combination with any metal finishing process steps (electroless Ni/Au, immersion Sn, immersion Ag). In addition to a negative impact on solder mask adhesion, these processes in combination with new laminates can also lead to nonselective metal plating or skip plating. One of the last manufacturing steps that should be evaluated during the introduction of a new laminate material is board routing and scoring. Here the use of a new base material with changed stiffness or different reinforcement may lead to necessary process adjustments.

P · A · R · T · 3

ENGINEERING AND DESIGN

CHAPTER 13
PHYSICAL CHARACTERISTICS OF THE PCB

Lee W. Ritchey
3Com Corp., Santa Clara, California

13.1 CLASSES OF PCB DESIGNS

Printed circuit boards (PCBs) or printed wiring boards (PWBs) can be divided into two general classes which have common characteristics based on their end functions. These two classes have very different materials and design requirements and functions and, as a result, need to be treated differently throughout the design and fabrication processes. The first class contains analog, RF, and microwave PCBs such as are found in stereos, transmitters, receivers, power supplies, automotive controls, microwave ovens, and similar products. The second contains digital-based circuitry such as is found in computers, signal processors, video games, printers, and other products that contain complex digital circuitry. Table 13.1 lists many of the characteristics of each class of PCBs.

TABLE 13.1 Characteristics of RF/Analog vs. Digital-Based PCBs

RF, microwave, analog PCB	Digital-based PCB
Low circuit complexity	Very high circuit complexity
Precise matching of impedance often needed	Tolerant of impedance mismatches
Minimizing signal losses essential	Tolerant of lossy materials
Small circuit element sizes often essential	Small circuit element sizes desirable
Only 1 or 2 layers	Many signal and power layers
High feature accuracy needed	Moderate feature accuracy needed
Low/uniform dielectric constants needed	Dielectric constant secondary

13.1.1 Characteristics of Analog, RF, and Microwave PCBs

As can be seen from Table 13.1, the materials, design, and fabrication needs of this class of PCBs are markedly different from those of PCBs commonly referred to as digital.

- Circuit complexity is low because most components used have two, three, or four leads. This is due to the high usage of resistors, transistors, capacitors, transformers, and inductors.

- Traces, pads, and vias often act as inductors, capacitors, and coupling elements in the actual circuit. Their shapes may have a material effect on overall circuit performance. For example, the lead inductance and capacitance in a transistor collector circuit wire may act as the resonant components for an RF amplifier or it may degrade performance if it is unwanted. Figure 13.1 shows the impedance of traces as a function of their capacitance.

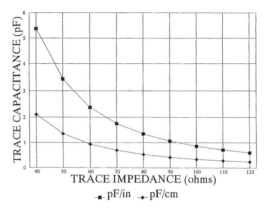

FIGURE 13.1   Trace capacitance vs. trace impedance, based on $L_0 = 8.5$ nH/in. (*Prepared by Ritch Tech, 1992.*)

- Two traces running side by side may be used to couple a signal from one circuit to another as is done in directional couplers of microwave amplifiers. (This same coupling in a digital circuit may result in a signal getting into a neighboring circuit causing a malfunction.)
- A series of conductors running side by side may function as a band-pass filter. Proper performance of filters, as well as most other wideband RF circuits, depends on all frequencies traveling with equal speed through the structures. To the extent that this is not true, frequencies that arrive later distort the signal being processed. This is called *phase distortion.*

Figure 13.2 illustrates the dielectric constant of various PCB materials as a function of frequency. Notice that some materials exhibit a dramatic decline in dielectric constant as frequency increases. The speed with which a signal travels through a dielectric is a function of the dielectric constant. Figure 13.3 illustrates signal velocity as a function of dielectric constant. From these two graphs it can be seen that using a dielectric material with a nonuniform dielectric constant in RF applications may result in severe phase distortion because the higher-frequency components arrive at the output before the lower frequencies.

- A trace in a power supply circuit may be expected to carry several amps without significant heating or voltage drop. Its resistance may even be used as a sense element to detect current flow. Similarly, handling large currents with insufficient copper in a trace may result in a voltage drop that degrades circuit performance. Figure 13.4 illustrates trace resistance of a copper trace as a function of its width and thickness. Figures 13.5 and 13.6 illustrate conductor heating as a function of width, thickness, and current flow.
- PCBs used in consumer electronics tend to share lower circuit complexity with RF and analog PCBs. However, the performance demands are far lower. The need for the lowest possible cost offsets this. To achieve the cost objectives, every effort is made to keep all connections on a single side and to form all holes in a single operation by punching. This

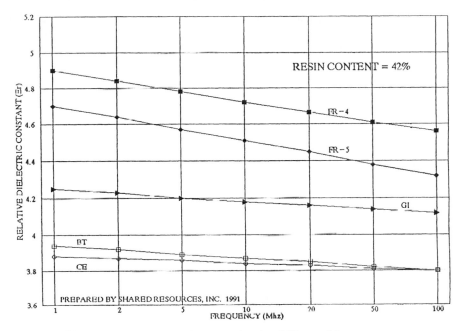

FIGURE 13.2 Dielectric constants vs. frequency of various PCB materials.

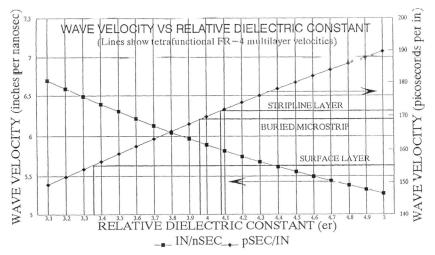

FIGURE 13.3 Signal velocity as a function of dielectric constant. (*Prepared by Shared Resources, Inc., 1991*)

eliminates both drilling and plating. The substrate material system is often resin impregnated paper, the lowest-cost substrate system for electronic packaging.

Summarizing, successful RF and analog design depends heavily on the properties of the materials used and on the physical shapes of the conductors and their proximity to each other

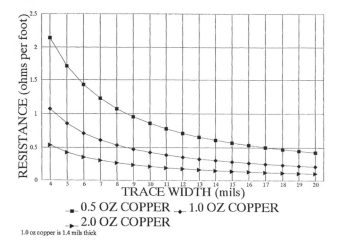

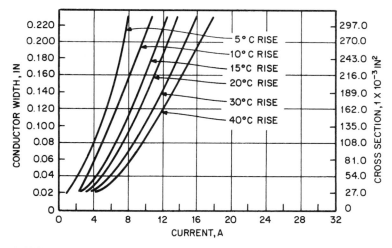

FIGURE 13.4 Trace resistance vs. trace width and thickness. (*Prepared by Ritch Tech.*)

FIGURE 13.5 Temperature rise vs. current for 1-oz copper.

rather than on the ability to handle very large numbers of circuits simultaneously. Hand routing or connecting of the individual parts coupled with manipulating the shapes of individual copper features are essential parts of this design process. For these reasons, the design tools and design team must be chosen to meet these needs. Physical layout tools that provide convenient graphical manipulation of PCB shapes are a must.

13.1.2 Characteristics of Digital-Based PCBs

Compared to RF and analog PCBs, digital-based PCBs have complex interconnections, but are tolerant of rather wide feature size and materials variations.

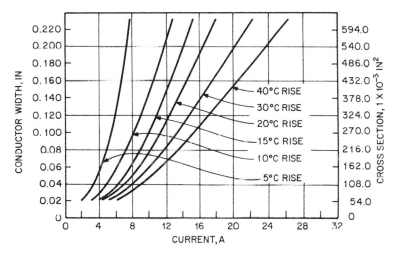

FIGURE 13.6   Temperature rise vs. current for 2-oz copper.

- They are characterized by very large numbers of components, often numbering in the hundreds and sometimes the thousands.

- Digital components often have large numbers of leads, as high as 400 or more. This high lead count stems from logic architectures that have data and address buses as wide as 128 bits or more. To connect PCBs with these wide data buses, digital systems often contain board-to-board connectors with as many as 1000 pins.

- Digital circuits have increasingly fast edges and low propagation delays to achieve faster performance. Edge rates as fast as 1 ns are now encountered in devices destined for products as common as video games. Table 13.2 lists edge speeds of some commonly used logic families, edge rate being the time required for a logic signal to switch from one logic level to the other (switching speed). Propagation delays, the time required for a signal to travel through a device, are decreasing along with edge rates.

TABLE 13.2   Typical Logic Family Switching Speeds

Logic family	Edge speed, ns	Critical length, in
STD TTL	5.0	14.5
ASTTL	1.9	5.45
FTTL	1.2	3.45
HCTTL	1.5	4.5
10KECL	2.5	7.2
BICMOS	0.7	2.0
10KHECL	0.7	2.0
GaAs	0.3	0.86

- These fast edges and short propagation delays lead to transmission line effects such as coupling, ground bounce, and reflections that can result in improper operation of the resulting PCB. Table 13.2 illustrates the degree to which a fast switching signal will couple into a neighboring line as a function of the edge-to-edge separation and the height of the signal pair above the underlying power plane. The critical length listed in Table 13.2 is the length of parallelism between two traces at which the coupling levels in Fig. 13.7 are reached.

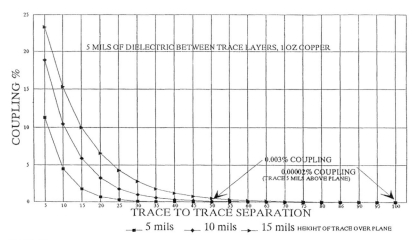

FIGURE 13.7 Trace-to-trace coupling. (*Prepared by Shared Resources, Inc.*)

The digital circuits themselves are designed to function properly with input signals that vary over a relatively wide range of values. Figure 13.8 illustrates the signal levels for a typical logic family, in this case ECL. The smallest output signal from an ECL driver is the difference between VOL_{max} and VOH_{min} or 0.99 V. The smallest input voltage to a device at which the logic part is designed to work properly is the difference between VIL_{max} and VIH_{min} or 0.37 V. The difference between these two levels, the noise margin of 0.62 V is available to counteract losses in the wiring and the dielectric and from other sources such as coupling and reflections. From this it can be seen that digital logic has a high tolerance of losses and higher immunity to noise.

This tolerance of noise and losses makes it possible to have trace features and base materials that introduce substantial losses and distortion while still achieving proper operation. It is this relatively high tolerance of distortion that makes it possible to manufacture economical digital PCBs.

Summary.   The large number of connections in digital PCBs generally requires multiple wiring layers to successfully distribute power and interconnect all the devices. As a result, the design task is heavily weighted on the side of successfully making many connections in a limited number of routing layers while obeying transmission line rules. The base materials need to have characteristics that result in a PCB that is economical to fabricate and able to withstand the soldering processes while preserving high-speed performance. Compared to RF PCBs, losses in the dielectric tend to be of little concern for digital PCBs. The actual shapes of conductors, pads, holes, and other features have little effect on performance. (For detailed treatment of these topics, see Howard W. Johnson and Martin Graham, *High Speed Digital Design: A Handbook of Black Magic,* Prentice Hall, New York, 1993.)

The PCB design system and the design skill set for digital PCBs must be optimized to ensure accuracy in making large numbers of connections while successfully handling the high speed requirements of the system. Achieving this in a reasonable amount of time demands the use of a CAD system that contains an automatic router for use in connecting the wires.

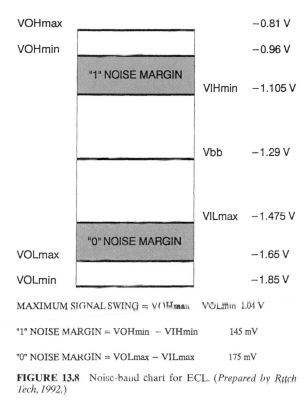

TYPICAL 10K ECL LOGIC VOLTAGE LEVELS

MAXIMUM SIGNAL SWING = VOHmax − VOLmin 1.04 V

"1" NOISE MARGIN = VOHmin − VIHmin   145 mV

"0" NOISE MARGIN = VOLmax − VILmax   175 mV

FIGURE 13.8 Noise-band chart for ECL. (*Prepared by Ritch Tech, 1992.*)

13.2 TYPES OF PCBs OR PACKAGES FOR ELECTRONIC CIRCUITS

The range of choices for packaging electronic circuits is quite broad. Some of the parameters that influence the choices are weight, size, cost, speed, ease of manufacture, repairability, and function of the circuit. The more common types are listed as follows with a brief description of their characteristics.

Levels of packaging are often used when referring to how electronics circuits are packaged. The first level of packaging is the housing of an individual component. This is usually an encapsulating coating, a molded case, or a cavity-type package such as a PGA (pin grid array). The second level of packaging is the PCB or substrate on which individual components are mounted. Third-level packaging is any additional packaging beyond these two. Most often third-level packaging takes the form of a multichip module (MCM) that has bare components mounted in it which is itself mounted on a PCB along with other components.

13.2.1 Single- and Double-Sided PCBs

These PCBs have conductor patterns on one or both sides of a base laminate with or without plated through-holes to interconnect the two sides. These are the workhorses of consumer electronics, automotive electronics, and the RF/microwave industry. They are the lowest-cost choice for consumer products. Laminate materials range from resin-impregnated paper for consumer electronics to blends of low-loss Teflon™ for RF applications.

13.2.2 Multilayer PCBs

These PCBs (see Fig. 13.9) have one or more conductor layers (usually power planes) buried inside in addition to having a conductor layer on each outside surface. The inner layers are connected to each other and to the outer layers by plated through-holes or vias. These are the packages of choice for nearly all digital applications ranging from personal computers to supercomputers. Numbers of layers range from 3 to as many as 50 in special applications. Laminate materials are nearly always some type of woven glass cloth impregnated with one of several resin systems. The resin system is chosen to satisfy requirements such as the ability to withstand high temperatures, cost, dielectric constant, or resistance to chemicals.

13.2.3 Discrete-Wire or Multiwire PCBs

This class of PCBs is a variation of the multilayer package. A circuit board is constructed by etching a pair of power layers back to back on a laminate substrate. A layer of partially cured, still sticky laminate is bonded to each side of this power plane structure. Discrete wiring is rolled into this sticky adhesive in patterns that will connect leads or serve as access to surface-mount component pads.

Once all of the wires have been rolled into place, a second layer of laminate is placed over the wires, followed by copper foil sheets. This sandwich is then laminated, drilled, and processed like any multilayer PCB. The resulting PCB has outer layers and power planes much like any multilayer PCB. The principle difference is the printed wiring signal layers have been replaced by discrete wiring layers. In some cases of very high wiring density, alternating layers of power planes between wiring layers serve as isolation.

Designing a discrete-wiring PCB (see Fig. 13.10) involves adding a special discrete-wiring router to a standard PCB design system to generate the files for the machine that rolls the wire into the dielectric. Discrete wiring once provided a faster prototyping alternative to multilayer fabrication. At the present time, both technologies are equally rapid and cost effective during prototyping. However, for modest to high-volume production, multilayer technology is more cost effective than discrete wiring.

13.2.4 Hybrids

These circuits are usually single- or double-sided ceramic substrates with a collection of surface-mount active components and screened-on resistors made from metallic pastes. They are most often found in hearing aids and other miniature devices.

13.2.5 Flexible Circuits

These circuits are made by laminating copper foil onto a flexible substrate such as Kevlar® or Kapton®. They can range from a single conductor layer up to several layers. They are most often used to replace a wiring harness with a flat circuit to save weight or space. Often, the flexible circuit will contain active and passive components. Common applications are cameras, printers, disk drives, avionics, and video tape recorders.

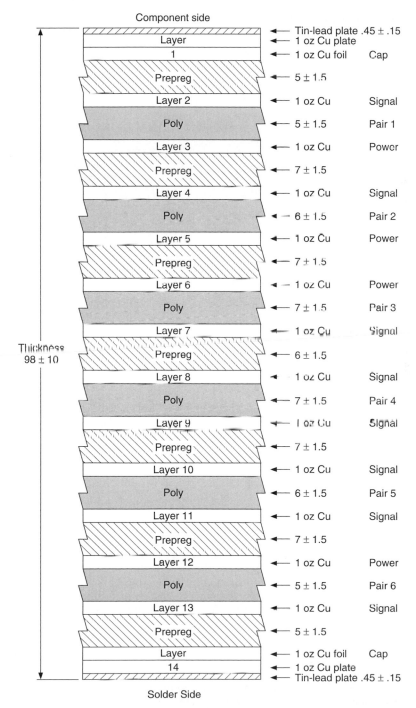

FIGURE 13.9   Cross section of 14-layer multilayer printed wiring board, showing a typical inner layer and prepreg material relationship. In this case, to reduce z-axis expansion, the innerlayers are polyimide, while the prepreg material is semicured polyimide. Typical signal, power, and ground layers are also indicated, as well as the thickness of the copper foil for each layer.

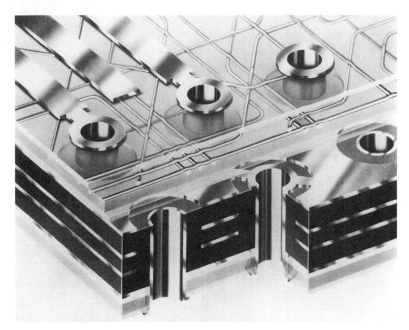

FIGURE 13.10   Cross section of a discrete-wire PCB. (*Courtesy of Icon Industries.*)

13.2.6   Flexible Rigid or Flex-Rigid

As the name suggests, these are combinations of flexible PCBs and rigid PCBs in a single unit. The flexible portion of the circuit is made first and included in the lamination process of the rigid portion of the assembly. This process eliminates wiring harnesses and the associated connectors. Applications include avionics and portable equipment such as laptop computers. As a rule, a flex-rigid assembly is more expensive than an equivalent combination of PCBs and cables.

13.2.7   Backplanes

Backplanes are special cases of multilayer PCBs. They tend to contain large quantities of connectors that have been installed using press fit pins. In addition, backplanes are used to distribute large amounts of dc power to the system. This is accomplished by laminating several power planes inside the backplane and by bolting bus bars onto the outside surfaces. Some applications require that active components, such as surface-mount ICs, be soldered to their surfaces. This greatly increases the difficulty of assembly as a result of the need to solder fine-pitch parts to a large, thick PCB.

13.2.8   MCMs (Multichip Modules)

Multichip modules are essentially miniature PCBs. Miniaturization is achieved by removing components such as ICs from their packages and mounting them directly to the substrate using wire bonds, flip chip, TAB or flip TAB. The motivation for using an MCM is miniaturization, reduction in weight, or a need to get high-speed components as close to each other as possible to achieve high-speed performance goals. MCMs usually represent a third level of packaging in a system between packaged components and the carrier PCB. As a result, this additional level of packaging virtually always results in a more expensive, more complex assembly than the equivalent circuits in standard packages. There are several types of MCM package.

13.2.8.1   MCM-L, Multichip Module, Laminate.   This version of an MCM is manufactured from very thin laminates and metal layers using the same techniques employed in the manufacture of standard PCBs. Features such as holes, lands, and traces are much finer and require tooling similar to that used to manufacture semiconductors. This is the least expensive MCM type to design, tool, and manufacture. The same design tools and methodologies used for PCBs can be used.

13.2.8.2   MCM-C, Multichip Module Ceramic.   This version of an MCM is manufactured by depositing conductor layers on thin layers of uncured ceramic material, punching and backfilling holes for vias, stacking the layers, and firing the total to create a hard ceramic multilayer substrate. This is the second least expensive MCM type to design, tool, and manufacture. It has been the workhorse of IBM's large mainframe computers for at least two decades. The same design tools and methodologies used for PCBs can be used.

13.2.8.3   MCM-D, Multichip Module Deposited.   This version of an MCM is manufactured by depositing alternating thin films of organic insulators and thin films of metal conductors on a substrate of silicon, ceramic, or metal. The design and manufacturing techniques used for this technology resemble that used to create integrated circuit metallization. The thermal conductivity of the substrate is quite good. Design and fabrication support for MCM-D is limited.

13.2.8.4   MCM-D/C, Multichip Module Deposited and Cofired.   This version of an MCM is a combination of a cofired, multilayer ceramic substrate containing the common wiring for a family of modules and deposited conductor and insulation layers containing the personality wiring. It has all of the problems of each technology it uses plus problems related to mismatches in temperature coefficient of the two materials systems.

13.2.8.5   MCM-Si, Multichip Module Silicon.   As the name implies, this MCM technology starts with a silicon substrate like that used to make integrated circuits. Conductor patterns are formed using silicon dioxide (glass) as an insulator and aluminum or a similar metal for the wiring patterns in the same manner as is employed to build an integrated circuit. In fact, the same design tools and fabrication tools used to build ICs are used to build MCM-Si modules.

A significant advantage of MCM Si is the fact that the substrate is the same material as the ICs that will be attached to it. Therefore, it is thermally matched to the ICs, ensuring reliable contacts over extremes of temperature.

13.2.8.6   Summary of MCM Technologies.   MCM packaging may be seen as a way to achieve higher performance from a collection of high-speed ICs than can be accomplished by mounting them onto a PCB or as a way to reduce size and weight. In reality, higher levels of integration nearly always result in a more economical solution. For all but low-volume, specialty applications, such as aerospace electronics and specialty processors for very high performance equipment, this has proven to be true for quite some time. This is likely to continue to be so for some time as semiconductor technology continues to improve the density of functionality that can be placed on a single IC. One need only examine the progression of microprocessor performance to see this phenomenon at work.

When a high-performance product requires integrated circuits made with different processing technologies, such as analog and CMOS or ECL and CMOS, integration does not represent a reasonable alternative to MCMs. Examples of this type of product are high-performance graphics products and video signal processing equipment.

13.3   METHODS OF ATTACHING COMPONENTS

A wide range of methods has evolved for attaching components to PCBs. The methods chosen as well as the combinations of methods chosen for a product have a substantial impact on the final cost, ease of assembly, availability of components, ease of test, and ease of rework. The five basic attachment combinations are: through-hole only, through-hole mixed with surface-mount on one side, surface-mount one side only, surface-mount both sides, and surface-mount both sides with through-hole.

13.3.1   Through-Hole Only

All component leads attach to the PCB by being inserted into holes that pass through the PCB. The components may be secured by wave soldering or by pressing into holes that result in an interference fit (press fit). Assembly involves a component placement operation followed by a wave-soldering operation. This method is still the workhorse of the low-cost consumer electronics industry.

13.3.2   Through-Hole Mixed with Surface-Mount

Components such as connectors and PGAs are attached to the PCB with through-hole technology. All other components are mounted using surface-mount packages. This is the most common method used to assemble electronics products. Assembly is a two-step process that involves placing all surface-mount parts and soldering them in place with a solder reflow system, then inserting all through-hole parts and soldering them in place in a wave-soldering operation. Alternatively, the through-holes may be hand-soldered if the quantity is small.

13.3.3   Surface-Mount, One Side Only

This type of package is made up of only surface-mount parts all mounted on the same side of the PCB. Assembly is a one-step process that involves placing all components and soldering them in place using some form of solder reflow.

13.3.4   Surface-Mount, Both Sides

This type of package contains surface-mount components on both sides. Assembly is a two-step process that involves placing all components on one side and reflow soldering them, followed by placing all components on the other side and reflow soldering them. In addition to more complex assembly operations, designing and testing this type of assembly is much more complex, as parts often share the same area on opposite sides of the PCB, causing conflicts when trying to locate vias and test points.

13.3.5   Surface-Mount, Both Sides with Through-Hole

As the name implies, this package type contains surface-mount parts on both sides, as well as through-hole parts such as connectors. In most cases, the surface-mount components on one side are passives, such as bypass capacitors and resistors that can withstand being passed through wave soldering. Assembly is a three-step process that involves placing the surface-mount components on the primary side and reflow soldering them. Once this is complete, the secondary-side surface-mount components are glued in place, the through-hole components are inserted, and the PCB is sent through wave soldering.

Because of the extra operations and exposure of components on the secondary side to molten solder, this type of assembly is prone to many assembly defects. It is important to note that soldering fine-pitch parts on the secondary side using wave soldering results in excessive solder bridging and should be avoided. In addition, active components, such as ICs, may be damaged by exposure to excessive heating.

13.4 COMPONENT PACKAGE TYPES

Over time, a wide assortment of packages has been developed to house components. Selecting the correct package type for each component is one of the most important parts of the design process. Package types chosen affect ease of design, assembly, test, and rework, as well as product cost and component availability.

13.4.1 Through-Hole

This class of component is characterized by parts that have wire or formed leads. These leads pass through holes drilled or punched in the PCB and are soldered to lands on the back side or to plating in the holes. This is the original package type used for electronic components. A major benefit of through-hole components is the fact that every component lead passes all the way through the PCB. Because of this, there is automatic access to any PCB layer to make connections. Further, every lead is available on the bottom of the PCB, so test tooling is easy to construct. With the advent of surface-mount components, through-hole is used primarily for connectors and pluggable devices such as microprocessors mounted in PGA packages.

Through-hole packages are often preferred for ICs and other components that dissipate large amounts of heat, because of the relative ease with which heat-sinking devices can be fitted to them. In addition, it is much easier to provide a socket for a through-hole device. This eases the task of changing programmable parts and microprocessors when it is necessary to upgrade a system.

Caution: Integrated circuits in through-hole packages are becoming difficult to find as they are displaced by surface-mount equivalents and should be avoided in new designs unless a secure supply of components is available for the production life of the design.

13.4.2 Surface Mount

This package type is the mainstream choice for packaging electronic components of every type, including connectors. Its principle characteristic is that all connections between a component lead and the PCB or substrate is made with a lap joint to a pad on the surface of the PCB. This has both advantages and disadvantages. On the advantage side, since there are no holes piercing the PCB, wiring space on inner layers and on the reverse side is not consumed with component lead holes. Because of this, it is usually possible to wire a PCB in fewer layers than would be true with through-hole parts. Another and larger benefit is the fact that surface-mount components are always smaller than their through-hole equivalent, making it possible to fit more parts in a given area.

The main disadvantages of surface-mount components stem from the fact that there are no leads to easily grip with instrumentation probes and that there may not be access to the leads from the reverse side for purposes of production testing. This gives rise to the need to add a test pad to most nets on the back side in order to perform production test. It also gives rise to the need for very expensive, complex adapters in order to provide access to leads on processors and other complex devices to probe their inputs and outputs when performing diagnostic work.

Yet another disadvantage of surface-mount parts stems from their small size. It is more difficult to remove heat from SMT packages than it is for their through-hole equivalent. In some cases, such as high-performance processors, the heat generated by the IC is too high to permit proper operation in an SMT package.

13.4.3 Fine Pitch

Fine pitch is a special class of surface-mount components. This class is characterized by lead pitches lower than 0.65 mm (25 mils). These fine lead pitches are usually driven by very high lead count ASICs (160 pins and up) or by the extreme miniaturization requirements of PCMCIA (Personal Computer Memory Card Industry Association) cards, PDAs (personal digital assistants), and other small, high-performance products. The motivation for designating a special fine-pitch-component class of surface-mount parts is the extra difficulty of successfully testing, assembling, and reworking these parts on PCBs, as well as in building PCBs with accurately formed patterns and solder masks to mate with the leads of fine-pitch parts. Fine-pitch parts are the source of most manufacturing defects in a well-run SMT assembly line. The defects stem from lack of coplanarity of the leads, bent leads, insufficient solder on the joints, and poor alignment of the leads to the patterns on the PCB.

Successful manufacture using fine-pitch components involves very tight cooperation among the PCB designer, the PCB fabricator, the component manufacturer, and the PCB assembler/tester. It almost always involves specialized assembly, test, and rework tooling. Design is most often done by convening a series of meetings of the engineering personnel of all these groups to evolve a set of rules, processes, equipment, tooling, and components. These meetings need to start at the product development stage and continue to be held until the proper pad shapes and sizes have been established and the production process is stable.

13.4.4 Press Fit

Press fit is a special form of through-hole technology. Components are fastened to the PCB by deliberately designing an interference fit between the component lead and the plated through-hole in the PCB. The principle application of press-fit technology is the attachment of connectors into backplanes. The reason for this is that early backplanes were built by wire wrapping the signal connections onto the connector pins extending out the back of the backplane. Trying to solder the connector pins to the backplane through this field of pins proved difficult, if not impossible. The solution was press fit.

Successful assembly of a press-fit backplane rests in designing a hole size small enough to create a solid connection with the pins while ensuring that the hole is large enough to permit the insertion of the pin without fracturing the hole barrel.

Caution: Hot-air leveling the solder on a backplane results in a hole with irregular diameter. This irregularity will almost certainly result in damaged hole plating when the insertion operation is done. Be sure to note on the fabrication drawing of a press-fit backplane that hot-air solder leveling is prohibited. Solder must be plated onto the backplane traces and pads and fused using IR reflow or hot-oil reflow to fuse the lead and tin into a solder alloy.

13.4.5 TAB

TAB stands for tape-automated bonding, which is a technique for attaching bare IC die to a printed circuit board. It uses a subminiature lead frame that attaches directly to the bonding pads of an IC at one end, spreads out to a much larger pitch, and attaches to pads on a PCB. The tape in the title describes the method for carrying the parts prior to assembly, which is a tape with the TAB lead frame and IC built into it. The tape is wound onto a reel for handling. The principle application of TAB components is products such as pagers and portable phones that are made in very high volume and can justify the automation involved in attaching TAB parts to substrates.

13.4.6 Flip Chip

Flip-chip technology involves plating raised pillars of metal on the bonding pads of ICs, turning them upside down, and attaching them to a matching pattern on a substrate. The substrates are most often silicon and precision ceramics. From this description, it can be seen that this is a very specialized packaging method. To succeed, a source of tested good bare die with plated-on pillars must be available. This situation occurs almost exclusively in very high performance supercomputers where the ICs have been specially designed for the application or in very high volume applications such as pagers and cellular phones.

13.4.7 BGA

BGA or ball grid array is a relatively new technology that is a cross between pin grid arrays and surface mount. High-pin-count IC die are mounted on a multilayer substrate made from ceramic or organic material. The die are connected to the substrate using standard wire-bond techniques and encapsulated in epoxy or another form of cover. The bottom side of the substrate contains an array of high melting-point solder balls which connect to the wire-bond pads through the multilayer substrate. These balls mate with a matching array of pads on the PCB and are reflow soldered during the same operation as all other surface-mount parts. (See Fig. 13.11.)

The appeal of BGA technology is as an alternative to high pin-count, fine-pitch SMT ICs. As mentioned earlier, assembling high pin-count, fine pitch SMT parts is very difficult, owing to the fragile nature of the component leads. BGAs represent a much more robust package during component-level test and during assembly.

As with most technologies, there are some disadvantages to BGAs. Among these are:

- The solder joints are hidden from view, so inspecting them requires x-rays.
- The solder joints are not accessible for rework, so the soldering process must have a very high success rate.
- Part removal requires special tooling.
- The pattern on the PCB surface is an array of pads that require one via each all the way through the PCB, hampering routing.
- All component leads are concentrated in a much smaller area than the equivalent part in a fine-pitch SMT package. This concentrates the wiring in a small area pierced with many vias. As a result, the BGA PCB will likely have more wiring layers than the equivalent SMT PCB.
- The BGA package is more expensive than the equivalent fine-pitch SMT package, by as much as three times.

13.4.8 Wire-Bonded Bare Die

As the name implies, this method of assembly involves attaching bare IC die directly to a substrate using an adhesive or reflowing solder and connecting the bonding pads to pads on the

FIGURE 13.11 Typical ball grid array package. (*Courtesy of Icon Industries.*)

PCB using wire-bond techniques. Virtually all digital watches and many other similar consumer products use this assembly technique. It is very inexpensive when used in very high volume products with only a single IC to connect.

13.5 MATERIALS CHOICES

A wide variety of materials has been developed for use in packaging electronic circuits. These can be divided into three broad classes: reinforced organics, unreinforced organics, and inorganics. These are used primarily for rigid PCBs, flexible and microwave/RF PCBs, and multichip modules, respectively. The following treatment of the available materials concentrates on the principle materials systems used in PCB design along with the properties that warrant their use. See Chaps. 5 and 8 in this handbook for detailed data provided on loss tangents, temperature of coefficient of expansion, glass transition temperature, and other electrical properties.

IPC, the Institute for Interconnecting and Packaging Electronic Circuits, publishes a comprehensive series of standards that list in detail the properties of all types of laminates, resins, foils, reinforcement cloths, and processes that are candidates for the manufacture of PCBs. These standards start with IPC-L-108B and run up to IPC-CF-152. It is recommended that copies of the applicable standards be obtained at the start of a program in order to ensure a thorough understanding of all important characteristics of the materials being considered for a design.

Properties important to the manufacture of PCBs include (see Table 13.3):

- *Glass transition temperature T_g*—The temperature at which the coefficient of thermal expansion in a resin system makes a sharp change in rate from a slow rate of change to a rapid rate of change. A high T_g is important for PCBs that are very thick to guard against barrel cracking or pad fractures during the soldering operation.

- *Coefficient of thermal expansion T_{CE}*—Surface-mount assembly process subjects the printed wiring assembly to more numerous temperature shocks than typical through-hole processes. At the same time, the increase in lead density has caused the designer to use more and more layers, making the board more susceptible to problems concerned with the base material's coefficient of thermal expansion T_{CE}. This can be a particular problem with regard to the z-axis expansion of the material, as this induces stresses in the copper-plated hole, and becomes a reliability concern. Figure 13.12 shows typical z-axis expansion for a variety of printed circuit base laminate materials.

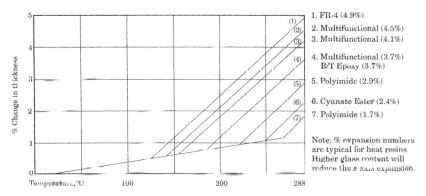

1. FR-4 (4.9%)
2. Multifunctional (4.5%)
3. Multifunctional (4.1%)
4. Multifunctional (3.7%) B/T Epoxy (3.7%)
5. Polyimide (2.9%)
6. Cyanate Ester (2.4%)
7. Polyimide (1.7%)

Note: % expansion numbers are typical for heat resins. Higher glass content will reduce the z-axis expansion.

FIGURE 13.12   Typical z-axis expansion via thermal mechanical analysis. (*Courtesy of Nelco International Corp.*)

- *Relative dielectric constant e_r*   This characteristic measures the effect that a dielectric has on the capacitance between a trace and the surrounding structures. This capacitance affects impedance as well as the velocity at which signals travel along a signal line. (See Figs. 13.3 and 13.4) Higher e_r produces lower impedance, higher capacitance, and lower signal velocity.

- *Loss tangent, tan (f), or dissipation factor*—A measure of the tendency of an insulating material to absorb some of the ac energy from electromagnetic fields passing through it. Low values are important for RF applications, but relatively unimportant for logic applications.

- *Electrical strength or dielectric breakdown voltage DBV*   The voltage per unit thickness of an insulator at which an arc may develop *through* the insulator.

- *Water absorption factor WA*—The amount of water an insulating material may absorb when subjected to high relative humidity, expressed as a percent of total weight. Absorbed water increases relative dielectric constant as well as reduces DBV.

13.5.1   Reinforcement Materials

The principal reinforcement for PCB substrate materials is cloth woven from glass fibers. A variation of this glass is cloth made from quartz fibers. The resulting material has a slightly lower dielectric constant than ordinary glass, but at a substantial cost premium and a more difficult drilling cycle. Kevlar is an alternate woven reinforcement that results in a lower-weight material system with a lower dielectric constant, also at a higher cost and higher difficulty in processing.

TABLE 13.3 Properties of Some Common PCB Materials Systems*

	T_g	e_r	tan (f)	DBV, V/mil	WA, %
Std. FR4 epoxy	125C	4.1	0.02	1100	.14
Multifunctional epoxy	145C	4.1	0.022	1050	.13
Tetrafunctional epoxy	150C	4.1	0.022	1050	.13
BT/epoxy	185C	4.1	0.013	1350	.20
Cyanate ester	245C	3.8	0.005	800	.70
Polyimide	285C	4.1	0.015	1200	.43
Teflon	N.A.	2.2	0.0002	450	0.01

* All with E-glass reinforcement, except Teflon.

The original reinforcement material for PCBs was paper or cardboard in some form. Paper impregnated with a resin system is still in use in consumer applications where lowest possible cost is necessary and where performance is not an issue.

13.5.2 Polyimide Resin Systems

Polyimide resin-based laminates are the workhorse of electronics that must withstand high temperatures in operation or in assembly or repair. Common applications include down-the-hole well-drilling equipment, avionics, missiles, supercomputers, and PCBs with very high layer count. The principle advantage of polyimide is its ability to withstand high temperatures. It has approximately the same dielectric constant as epoxy resin systems. It is more difficult to work with in fabrication, is more costly than FR4 systems, and absorbs more moisture.

13.5.3 Epoxy-Based Resin Systems

Epoxy resin-based laminates are the workhorse of virtually all consumer and commercial electronic products. There are several variations of this pervasive laminate family, each developed to answer a special need. Among these are standard FR4, multifunctional epoxy, difunctional epoxy, tetrafunctional epoxy, and BT or bismaleimide triazine blends. Each of these was developed to answer the need for a resin with a successively higher T_g or glass transition temperature. Multifunctional epoxy is the most commonly used form.

13.5.4 Cyanate Ester-Based Resin Systems

This resin system is a recent entrant into the high-performance resin system category. It is said to offer processing characteristics superior to those of the FR4 blends while offering a higher T_g.

13.5.5 Ceramics

A wide variety of ceramic or alumina substrate materials have been developed for use in hybrids and multichip modules. These materials are the subject of specialized manufacturing processes beyond the scope of this handbook. The reader needing information on this group of materials is advised to contact a major manufacturer of ceramic materials.

13.5.6 Exotic Laminates

Kevlar, Kapton, Teflon, and RO 2800 are materials developed for specialty applications. The first two, in the form of thin films, are commonly used as substrates for flexible circuits. The latter two are the principal dielectrics for microwave and RF circuits. All of these materials can be used with or without reinforcements.

13.5.7 Embedded Components Materials

Specialized materials have been developed to allow construction of passive components such as resistors and capacitors into the PCB structure itself. Most of these materials are patented and available from a very small supplier base.

13.5.7.1 Embedded Resistors.   These are formed by plating a very thin film of nickel or other metal onto a copper foil layer and laminating this foil, plated side in, to an FR4 or other substrate material. To form a resistor, a window is opened in the copper foil, exposing the underlying nickel resistor layer. A resistor of the appropriate value is formed in the resistance material layer. Contact is made with the resulting resistor by etching connecting pads in the copper foil layer and drilling holes through these pads and plating the holes. (See Fig. 13.13.)

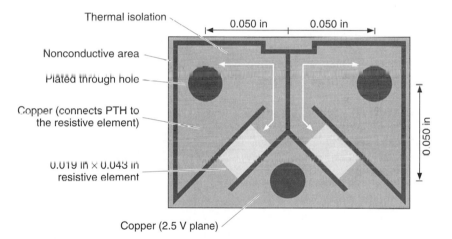

FIGURE 13.13   The two light diagonal shapes in the figure above are terminating resistors for ECL transmission lines. They are formed directly in the VTT powerplane by etching the copper away from the underlying nickel resistive layer. One end of each resistor is connected to the device terminal using a via and the other end is connected directly to the -2.5V plane. (*Courtesy of Ohmega Industries.*)

Resistive material is available in 25- and 100-Ω/square values. The principal application of embedded resistor technology is as terminating resistors for ECL transmission lines and as resistors on flexible circuits in products such as cameras and portable tape and CD players. Practical resistor values range from about 10 to 1000 Ω.

13.5.7.2 Embedded Capacitance.   This is formed by placing two copper planes close to each other using very thin dielectrics (1.5 to 2.0 mils). The principal application is in the creation of very high quality, high-frequency capacitance between two power planes. This does indeed result in high-quality capacitance, but usually at a high cost resulting from the need to add a pair of extra planes to a PCB in order to create the capacitance (see Fig. 13.14).

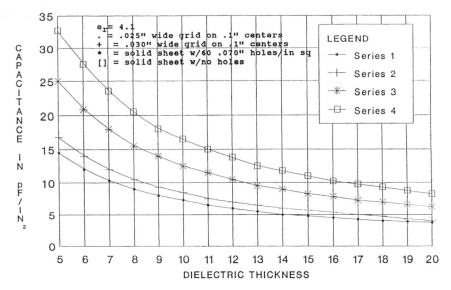

FIGURE 13.14 Capacitance per unit area vs. dielectric thickness.

13.6 FABRICATION METHODS

A wide variety of fabrication methods has been developed to meet the needs of the electronics industry. The following descriptions are quick summaries of each method intended to acquaint the reader with their advantages and disadvantages and likely applications. Detailed treatments are presented elsewhere in this book.

13.6.1 Punch Forming

Punch forming is used in the manufacture of very low cost, single-sided PCBs such as are used in many consumer electronic products. The process involves printing and etching the conductor patterns on one side of a laminate substrate, usually paper-reinforced epoxy. All holes are punched in a single stroke by a die containing a pin and opening for each hole. The PCB outline is formed in a second die that "blanks" it from the panel in which it is processed.

Often, several PCBs will be contained in a single panel sized to travel through the assembly process. After a PCB is punched out of the panel, it is forced back into the vacant hole and is held in place by interlocking fibers along the edges of the PCB. After assembly and testing is complete, each PCB is pressed from the panel. This is known as *crackerboarding* and is aimed at reducing overall manufacturing costs.

13.6.2 Roll Forming

Roll forming is a process used to manufacture flexible circuits in very large quantities. This is the lowest-cost method for the manufacture of flexible circuits. However, it involves substantial tooling, so it is applicable only for very high volume products. Examples of PCBs manufactured using this method are printer head connections, disk drive head connections, and the circuits used in cameras and camcorders. The PCBs can be single- or double-sided.

Roll forming resembles newspaper printing in that the process starts with a large roll of copper-clad laminate that is fed through a long process line containing stations which perform each operation on a continuous basis, starting with printing the conductor pattern, etching it, forming the holes, testing, and blanking from the roll itself. The process can include lamination of a cover insulator layer over the conductors as well.

13.6.3   Lamination

Lamination is the process by which PCBs of more than two layers are formed. The process begins by etching the conductor patterns of the inner layers onto thin pieces of laminate called *details*. These details are then separated by partially cured laminate called *prepreg* and stacked in a *book* with layers of prepreg on the top and bottom and foil sheets on the outside. This stack is placed into a press capable of heating the combination to a temperature that causes the prepreg resin to reach the liquid state. The liquefied resin flows into the voids in the copper patterns to create a *solid* panel upon cooldown. Once cooled, the panel is sent through the drilling and plating operations much like a two-sided PCB.

Note that some materials, such as polyimide, do not have a prepreg form to act as the glue during lamination. In these cases, a special glue sheet must be used during lamination to fasten the individual layers together.

13.6.4   Subtractive Plating

Subtractive plating is a method of forming traces and other conductive patterns on a PCB by first covering a sheet of laminate with a continuous sheet of copper foil. A layer of etch resist is applied such that it covers the copper pattern that is desired. The panel with protective coating is passed through an etcher that removes (subtracts) the unwanted copper, leaving behind the desired patterns. This is the dominant, almost only, method in common usage in the printed circuit industry today.

13.6.5   Additive Plating

As the name implies, this method of forming conductor patterns involves beginning with a bare substrate and plating on the conductor patterns. There are two methods for doing this: electroless plating on areas sensitized to accept electroless copper and electroplating by first applying a very thin coating of electroless copper over the entire surface to act as a conductive path, followed by electroplating to full thickness.

Additive plating is seen as a method for reducing the amount of chemicals required to manufacture PCBs, and this is true. However, the process does not yield copper sufficiently robust to withstand the handling of normal assembly and rework. As a result, it is not commonly available in production.

13.6.6   Discrete Wire

Discrete wire is a method for forming the wiring layers by rolling round wire into a soft insulating material coated onto the outsides of power plane cores. This method is often referred to as *multiwire*. It is available from only a very small number of manufacturers and offers few advantages over conventional multilayer processing. It is described more fully in Sec. 13.2.3.

13.7 *CHOOSING A PACKAGE TYPE AND FABRICATION VENDOR*

A key part of arriving at a successful design is choosing PCB materials, component-mounting techniques, and fabrication methods that meet the performance needs of the product being designed while achieving the lowest possible costs. Among the decisions that are part of this process are deciding whether to package a product on one large PCB or several smaller PCBs, whether to spread components out to hold the layer count down or increase layers, move components closer together, and design a smaller PCB, whether to package some components in a multichip module that is then mounted on the PCB, as well as other issues.

Part of this decision-making process is arriving at an overall package choice that can be manufactured by the mainstream fabricators and assemblers. Failure to do this will result in excessively high prices and long lead times stemming from the lack of a competitive supplier base from which to choose. At the extreme, where some of the more exotic materials systems are used, there may be as few as one supplier to turn to. In markets where there is substantial price pressure, such as with disk drives and PCs, it imperative that the design choices be made such that the PCBs can be manufactured at offshore fabricators. Not doing this will place the product at a competitive disadvantage.

13.7.1 Trading Off Number of Layers Against Area

Cost of the bare PCB is often a significant contributor to the overall cost of an assembly. As the number of layers in a PCB increases, the cost increases. A standard practice is to spread components out to make room for the connecting wiring as a way to avoid adding additional wiring layers. As might be expected, there is a point at which PCB size grows to where a smaller PCB with more layers yields a more economical solution. Determining where this breakpoint is requires some knowledge about the PCB fabrication process.

Table 13.4 shows typical costs of four-, six-, and eight-layer 18 in by 24 in standard process panels built at offshore manufacturers. This table can be used to calculate the relative cost of PCBs as layer count is increased to reduce area. While the absolute costs in the table are based on Spring 1995 pricing for Pacific Rim fabricators, the percentage relationships between the costs of panels of various layer counts are a good indicator of relative costs for deciding when to increase layer count and reduce area.

13.7.1.1 *Background Information.* Multilayer PCBs of six or more layers are normally built on standard 18 in by 24 in panels using pin lamination. Many four-layer PCBs are built offshore using a process called mass lamination with panels sizes of 36 in by 48 in (four times a "standard" panel). The pricing of individual PCBs is based on how many PCBs fit on a stan-

TABLE 13.4   PCB Panel Process Cost vs. Layer Count
Price per panel, $ U.S., Spring 1995

Number of layers	Panels per mo. 100	Panels per mo. 250	Panels per mo. 1000	Panels per mo. 5000
4 mass lam*	$260	$250	$240	$231
4 pin lam†	84	80	77	74
6 pin lam†	113	108	104	100
8 pin lam†	147	140	135	130

To determine the number of PCBs that will fit onto a panel, allow 0.125 in between PCBs and allow 0.75-in margin on all four sides. Net areas are 34.5 in by 46.5 in and 16.5 in by 22.5 in.
For gold plating on connector tips, add up to $2 per PCB.
* 36 in × 48 in panel.
† 18 in × 24 in panel.

dard panel. Therefore, designers need to choose finished PCB sizes with this in mind (assuming the size is negotiable).

The pricing matrix in Table 13.4 is in price per panel based on the following:

- Solder mask over bare copper, silk screen one side only
- Standard multifunctional FR4 laminate
- 1-oz copper inner layers, ½-oz outer layer foil plated up to 1.5 oz nominal
- Thickness accuracy: +10% overall, ±1.5 mil any dielectric layer
- No controlled impedance requirement or testing
- 1-mil minimum copper plating in holes
- No vias smaller than 13 mils
- Traces and spaces: 7 mil, 7 mil
- Standard delivery: (no acceleration premium)
- Fabrication site: **Pacific Rim Fabricators**
- No gold plating
- Tested to CAD netlist

Since PCB fabrication is based on standard panels, the cost of each PCB is affected by how much of each panel is used to form the PCBs built on it and the amount of the panel that is scrapped. Clearly, the more of each panel that is usable, the lower the average cost of each PCB will be. As one chooses the size and form factor of each PCB, attention should be paid to how many will fit onto a standard panel in order to minimize the scrap material created.

13.7.2 One PCB vs. Multiple PCBs

One way to keep individual PCB layer counts down is to divide the circuitry among several smaller, simpler PCBs. There is a hidden cost associated with doing this. The hidden cost is spread across several organizations, ranging from the design activity, through manufacturing, and into the sales and service organization. The costs are those associated with handling multiple assemblies, such as managing multiple designs and their documentation, managing the procurement, inventory, testing and stocking of multiple assemblies, and the cost of interconnecting the multiple assemblies. In almost all cases, these costs exceed any savings that might be realized by the creation of multiple assemblies.

CHAPTER 14
THE PCB DESIGN PROCESS

Lee W. Ritchey
3Com Corp., Santa Clara, California

14.1 OBJECTIVE OF THE PCB DESIGN PROCESS

The objective of the PCB design process is to engineer a PCB, including all of its active circuits, that functions properly over all the normal variation in component values, component speeds, materials tolerances, temperature ranges, power supply voltage ranges, and manufacturing tolerances and to produce all of the documentation and data needed to fabricate, assemble, test, and troubleshoot the bare PCB and the PCB assembly. Doing less than this in any area exposes the manufacturer and user of the PCB assembly to excessive yield losses, excessively high manufacturing costs, and unstable performance.

Achieving the objective involves carefully designing a process that matches the end product, selecting design tools with controls and analytical utilities, and selecting a materials system and components that match.

14.2 DESIGN PROCESSES

Figure 14.1 is a flowchart of the major steps in a complete PCB design process, beginning with specification of the desired end product and continuing through to archiving or storing away the design database in a form that permits subsequent design modifications or regeneration of documentation as necessary to support ongoing production. This process takes advantage of all the computer-based tools that have been developed to assure a "right the first time" design. The basic process is the same for either analog or digital PCBs. The differences in the design process for the two classes of PCBs center around the differences in complexity of these two types of circuits, as mentioned in Chap. 13.

14.2.1 The System Specification

The design team begins a new design by creating a *system specification*. This is a list of the functions the design is to perform, the conditions under which it must operate, its cost targets, development schedule, development costs, repair protocols, technologies to be used, weight and size, and other requirements as are appropriate. A rough definition of each of these variables is necessary at the start to permit proper choices of materials, tools, and instrumentation. For example, a project may involve the design of a portable computer that must weigh less

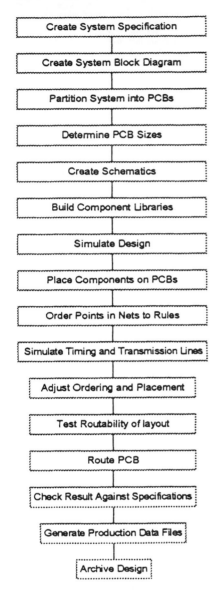

Create System Specification

Create System Block Diagram

Partition System into PCBs

Determine PCB Sizes

Create Schematics

Build Component Libraries

Simulate Design

Place Components on PCBs

Order Points in Nets to Rules

Simulate Timing and Transmission Lines

Adjust Ordering and Placement

Test Routability of layout

Route PCB

Check Result Against Specifications

Generate Production Data Files

Archive Design

FIGURE 14.1   PCB design process steps.

than 5 lb, fit into a briefcase, operate on batteries for two hours, have a mean time between failures (MTBF) of 200,000 hours or more, cost less than $2000, have 4 Mbytes of memory, 240 Mbytes or more of mass storage, and be MS-DOS compatible. This specification serves as the starting point for a new design.

14.2.2   System Block Diagram

Once the system specification has been completed, a block diagram of the major functions is created, showing how the system is to be partitioned and how the functions link or relate to each other. Figure 14.2 is an example of this partitioning.

14.2.3   Partitioning System into PCBs

Once the major functions are known and the technologies that will be used to implement them are determined, the circuitry is divided into PCB assemblies, grouping functions that must work together onto a single PCB. Usually this partitioning is done where data buses link functions together. Often these buses will be contained on a backplane into which a group of daughter boards are plugged. In the case of a PC (personal computer), partitioning often results in a mother board and several smaller plug-in modules such as memory, display driver, disk controller, and PC card (PCMCIA) interface.

14.2.4   Determining PCB Size

As soon as the amount of circuitry and the technology that each PCB must contain is known, the area and size of each PCB may be estimated. Often, the PCB size is fixed in advance by the end use. For example, a system based on VME or multibus technology will have to use the PCB sizes defined by the standard. In this case, system partitioning and component packaging technology will be dictated by what will fit onto these standard PCB sizes.

The finished cost of a PCB often turns on the number of layers and the quantity that will fit onto the standard manufacturing panel sizes. (For most PCB fabricators, this size is 18 in by 24 in with a usable area of 16.5 in by 22.5 in.) Sizing PCBs to utilize all or most of the panel area results in the most cost-effective bare PCB. (See Table 13.4.)

14.2.5   Creating the Schematic

Once the system function, partitioning, and technologies have been determined, the schematic or detailed connections between components can take place. Schematics and block diagrams are normally created on CAE (computer-aided engineering) systems. These systems allow the designer to draw the schematic on a CRT screen or terminal. The data needed by all of the following steps in the design process is generated by the CAE system from this schematic.

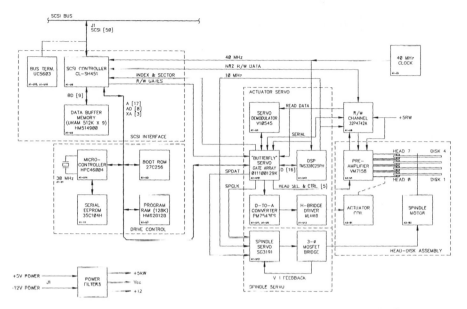

FIGURE 14.2   This is a block diagram of a digital device (in this case a disk drive) that has been segmented to its assembly levels. The dashed lines represent initial partitioning of the entire product to printed circuit assemblies and also shows the expected interface (connector) requirements.

14.2.6   Building Component Libraries

The tools used in the PCB design process must be supplied a variety of information about each part in order to complete each step. This information is entered into a library or set of libraries, one entry per component. Among the pieces of data needed are:

- Type of package that houses the component, e.g., through-hole, QFP, DIP
- Size of component, lead spacing, lead size, pin-numbering pattern
- Function each pin performs, e.g., output, input, power pin
- Electrical characteristics of each pin, e.g., capacitance, output impedance

14.2.7   Simulating Design

To be sure a design will perform its intended function over the intended range of conditions, some form of design verification must be done. These conditions may include component value accuracies, range of component speeds, operating and storage temperature ranges, shock and vibration conditions, humidity ranges, and power supply voltage range. Historically, this has been done by building breadboards and prototypes and subjecting them to rigorous testing. As systems and their operation software have grown more complex, this technique has proved to be inadequate. To solve this problem, simulators have been developed that allow a computer to simulate a function without having to build it. These simulators make it possible to perform tests far quicker, more rigorously, and more completely than any breadboard or prototype could ever be expected to.

Defects discovered by a simulator can be corrected in the simulation model with ease and the tests rerun before any commitment to hardware is made.

14.2.8 Placing Components on PCBs

Upon successful completion of the logical and gross timing simulation process, the actual physical layout can begin. It begins by placing the components of the design on the surface of the PCB in patterns that group logical functions together. Once this is done, the groups of components are located on the PCB surfaces such that functions that interact are adjacent, components that generate heat are cooled properly, components that interface to the outside world are near connectors, and so on. This placement operation can be done manually by the designer using graphics-based tools or automatically by the PCB CAD system.

14.2.9 Sequencing Nets to High-Speed Rules

Most logic families have sufficiently fast rise and fall times and short enough propogation delays to be subject to high-speed problems such as coupling and reflections. To ensure that these high-speed effects do not result in malfunctions, it is necessary to arrange the connections between the loads, terminations, and drivers to control these high-speed phenomena. This arranging of nodes or points in a net is referred to as *sequencing* or *scheduling*. Once the components have been placed on the surface of the PCB, the spatial arrangement of all the nodes on each net is known. At this point, it is possible to determine how to connect the driver to the loads and terminators to form proper transmission lines, ensuring that improper stubs are not created and that the terminator is at the end of the net.

14.2.10 Simulating Timing and Transmission Lines Effects

Upon completion of component placement and sequencing the nodes in each net, it is possible to estimate the length and characteristics of each net. This is possible because the x-y location of each point in a net is known, the order of connections is known, and the fact that the actual wiring must be done in either the x or y direction is also known. This length information can be used to model the high-speed switching characteristics of each net and predict the presence of excessive noise and reflections as well as to estimate the length of time required for signal to travel the length of each line, all before actually routing or building the PCB.

This simulation step makes it possible to detect potential malfunctioning signals prior to routing and take steps to fix the problem while the time invested in the design is still modest.

14.2.11 Adjusting Sequencing and Placement

If the simulations done in sec. 14.2.10 reveal excessive time delays or reflection problems, the placement may need to be adjusted to move critical parts closer together or add terminations to nets with excessive reflections. By doing this simulation and adjustment, a design can be assured of meeting the "right the first time" goal so important to high-performance designs.

14.2.12 Testing Routability of Placement

At this point, enough analysis has taken place to know that the design will function correctly if routed. However, it may not route in the number of signal layers required by the cost goal. Most CAD systems have tools, such as rats-nest analyzers, that help the designer determine if the design will fit into the allowed signal layers. If it will not, the routability analyzer may give clues on how to revise the component placement to achieve a successful route. Once the placement has been adjusted, the timing and transmission line simulation steps must be repeated to ensure that the set of goals has been met.

14.2.13 Routing PCB

This step involves fitting all the connections into the signal layers in the form of copper traces, following spacing and length rules. It usually involves a combination of hand routing special signals and automatically routing the rest.

14.2.14 Checking Routed Results

After all the connections have been routed into the signal layers, the actual shape and length of each wire is known, as is the layer(s) on which the wires have been routed and which nets are neighbors. This physical data can be loaded into the timing and transmission line analyzers to do a final check that all design goals have been met. Any violations that are detected can be repaired by hand rerouting as necessary. Once this set of checks has been completed and any adjustments made, the final routed result is checked against the schematic netlist to ensure that there are no discrepancies. A final check is performed on the Gerber data to ensure that the line width and spacing rules have been complied with and that there is no solder mask of silk screen on any pad, as well as that traces and other features that must be protected from solder are covered by solder mask.

14.2.15 Generating Manufacturing Files

This step involves generating the photoplotting files, pick-and-place files, bare and loaded board test files, drawings, and bills of material needed to do the actual manufacturing. Typical lists of these files are shown in Tables 14.1 and 14.2.

14.2.16 Archiving Design

Once all the manufacturing data has been created, the design database and all of the manufacturing data files are stored on a magnetic tape or other storage media for future use to incorporate changes and for backup in the event that the files and drawings created for manufacturing are lost or destroyed.

TABLE 14.1   A Typical Collection of Design Files Sent to a PCB Fabricator

File name	File contents
BBBBpCCC.arc*	Arc file of Gerber files containing:
applist.p	List of photoplot apertures for artwork
ly1 thru lyx.ger	Gerber photoplot data for x PCB layers
topmsk.ger	Gerber photoplot data for top solder mask
botmsk.ger	Gerber photoplot data for bottom solder mask
topslk.ger	Gerber photoplot data for top silk screen
botslk.ger	Gerber photoplot data for bottom silk screen
pc_356.out	IPC 356 data for blank PCB netlist testing
name0.rep	Drill allocation report for plated holes
name0.prf	Excellon drill file for all plated holes
name1.rcp	Drill allocation report for all nonplated holes
XX.XX.fab	Fabrication drawing in HPGL format, sheet XX of XX

* BBBBpCCC is the part number of the PCB.

TABLE 14.2 A Typical Collection of Design Files Sent to a PCB Assembler

File name	File contents
BBBBaCCC.arc*	Arc file of all assembly data containing:
applist.a	Aperture list for plotting paste mask
readme.asy	Readme file describing assembly
tpstmsk.ger	Gerber photoplot data for top paste mask
bpstmsk.ger	Gerber photoplot data for bottom paste mask
BBBB-CCC.dbg	Mfg. output, data format info.
BBBB-CCC.dip	Mfg. output, x-y loc. dip components
BBBB-CCC.log	Mfg. output, component log.
BBBB-CCC.man	Mfg. output, x-y loc. manual insert components
BBBB-CCC.smt	Mfg. output, x-y loc. top smt components
BBBB-CCC.smb	Mfg. output, x-y loc. bottom smt components
BBBB-CCC.unp	Mfg. output, parts not mounted
BBBB-CCC.vcd	Mfg. output
XX_XX.asy	Assembly drawing in HPGL format, sheet XX of XX
XX_XX.fab	Fabrication drawing in HPGL format, sheet XX of XX

* BBBBaCCC is the part number of the assembly.

14.3 DESIGN TOOLS

From the definition of the objective of the PCB design process it can be seen that the process extends from concept all the way through fabrication to assembly and test. Computer-based tools have evolved to automate or improve the speed and accuracy of every step in the process. These tools can be divided into three major groupings based on where they are used:

- CAE (computer-aided engineering) tools
- CAD (computer-aided design) tools
- CAM (computer-aided manufacturing) tools

It is apparent from the names of the tools that they are used in circuit design, physical layout of the PCB, and manufacture of the bare PCB and the PCB assembly.

14.3.1 CAE Tools

CAE tools is the name generally used to refer to the computer-based tools and systems that are employed in the stages of design before the physical layout step or to analyze and evaluate the electrical performance of the final physical layout. These include:

14.3.1.1 Schematic Capture Systems.   As the name implies, these tools are used by the design engineer to draw a schematic or circuit diagram. The simplest systems are graphical replacements for the classical drawing board, allowing the engineer to place logic and electronic symbols on the surface of the drawing and connect their terminals with lines. More advanced systems perform substantial error checking such as guarding against multiple uses of the same pin or net name. Failure to connect critical pins such as power pins can be done by using information contained in their component libraries about each part. In addition, these systems can generate netlists to be used by simulators and PCB routers and bills of material for use in the manufacture of the PCB assembly.

14.3.1.2 Synthesizers.   Synthesizers are specialized CAE tools that allow a designer to specify the logic functions that a design is expected to perform in the form of logical operations such as dual full adder, 16-bit-wide register, or other macro functions. The synthesizer will extract the equivalent logic circuit functions from a function library and connect them together as specified by the designer to arrive at a complete logic diagram. This synthesized circuit can then be used as part of a bigger design. Some advantages of synthesizers are that all functions of a given type will be implemented the same way and be error free and that time needed to compose a system schematic is reduced by eliminating the labor required to design repeating circuits.

14.3.1.3 Simulators.   Simulators are software tools that create computer-based models of a circuit and run them with input test patterns to verify that the circuit will do the intended function when implemented in hardware. Even when run on very large computers, simulators usually run at only a tiny fraction of the speed of the actual circuit. When circuits grow complex, as in a 32-bit microprocessor or digital signal processor, the time required to perform a complete simulation can be very long, sometimes so long that this method of circuit verification becomes impractical. Typical simulation speeds are 1 or 2 s for each machine cycle. A machine cycle may be as little as 2ns or 500 million/s for a system with a 500-MHz clock. This is a 500 million-to-1 slowdown! As circuits have grown more complex and simulation times have become excessive, engineers have been forced to build physical models of a proposed circuit and run actual code against the model as a method of ensuring that the design is accurate. Clearly, this adds both time and cost to the development cycle, both by adding the time needed to build the model and the time needed to locate design errors and fix them. The solution to this problem is circuit emulation.

14.3.1.4 Emulators.   Emulators or circuit emulators are collections of programmable logic elements, such as PLAs (programmable logic arrays), that can be configured to represent almost any kind of logic circuit. These emulators are commercially available as standard products from several EDA (electronic design automation) companies. The resulting hardware emulation of a circuit can be operated much faster than a software simulation, sometimes as fast as 1/100th of actual final operating speed. Due to this increased speed, the verification of a circuit can take place much faster. In some cases, the emulations are used as substitutes for the actual circuit to verify that the software created to run with the circuit is error free before any commitment is made to final hardware. This technique is used extensively in the design of complex ICs such as microprocessors and custom ASICs. In fact, the Intel Pentium™ microprocessor and its operating system was completely emulated and run successfully on a large hardware emulator prior to making the first silicon.

The use of emulation technology has eliminated the need to iterate or modify designs as errors are detected, saving very large sums of development money as well as development time. In some cases, it has made circuits practical that would not be without it. For example, most supercomputers and other advanced products are built with multilayer ceramic PCBs and MCMs. These packaging technologies do not lend themselves to modification with external wiring to correct errors. As a result, it is necessary to produce entirely new assemblies in order to correct design errors. The same is true for the integrated circuits in the system as well.

14.3.1.5 Circuit Analyzers.   Circuit analyzers are tools that examine circuits to ensure that they will perform properly over the range of timing variations in circuits and tolerances of components that are expected to be encountered in normal manufacturing. These analyzers do this by constructing mathematical models of each circuit and then varying the values of each component over their expected tolerance ranges. The behavior of the circuit is calculated by its model and the result compared to preestablished limits. Violations are flagged to alert the design engineer. Among the conditions checked for are freedom from excessive signal coupling from neighboring circuits and that transient behavior such as reflections, overshoot and undershoot, and ringing are within proper limits. This type of analysis is often referred to as worst-case tolerance and timing analysis. It can be run on a design before physical layout begins as well as after layout is completed.

Some examples of circuit analyzers are SPICE and PSPICE. SPICE and PSPICE build mathematical models of each circuit and then perform thousands of complex calculations to predict how the circuit will respond to input signals. Most suppliers of CAD systems also offer their own adaptations of these analysis tools.

14.3.1.6 Impedance Predicting Tools.   These tools are used to examine the cross section, trace sizes, and materials properties of a PCB to ensure that the resulting circuit impedance is within allowable limits or to interactively adjust these parameters to achieve a desired final impedance. This is an essential step in the design of the PCB itself. Most suppliers of CAD systems intended for high-speed design supply some form of on-line impedance analysis tools as part of their systems.

14.3.2 CAD Tools

CAD tools are used to turn the electrical circuit described by the schematic into a physical package or PCB. CAD tools are typically operated by PCB designer specialists who are skilled in the areas of PCB manufacture and assembly rather than by electrical engineers. These tools are fed netlists, component lists, wiring rules, and other layout information by schematic capture or CAE tools. In their simplest form, they allow the designer to create the pad patterns for the component leads and PCB shape and then manually connect the component leads with copper traces. The most sophisticated CAD tools can automatically determine the optimum location of each component on the PCB (autoplacement) and then automatically connect (autorouting) all leads while following high-speed layout rules. This is accomplished by providing the CAD tool with a table of rules specifying which components must be located in groups or near connectors as well as by specifying how much space must be maintained between neighboring traces, the maximum length allowed between points on a net, etc.

The outputs of CAD tools are the information files needed to fabricate, assemble, and test the PCB assembly. These are test netlists, photoplotting files, bills of materials, pick-and-place files, and assembly drawings. CAD tools are made up of circuit routers, placement tools, checking tools, and output file generation tools.

14.3.2.1 Placement Tools.   Placement tools are used to arrange the components on a PCB surface. Placement tools tend to be part of a complete CAD system rather than a module purchased separately. The inputs to a placement tool are:

- Component list or bill of material
- Netlist or manner in which the components connect to each other
- Shapes, sizes, and spacial arrangement of the component leads
- Shape of the PCB with areas into which components cannot be placed (keepouts)
- Instructions concerning fixed locations for components, such as connectors
- Electrical rules, such as maximum and minimum distance between points on a net
- Thermal rules, such as which parts must be kept apart or near sources of air flow

Placement tools range from completely manual to fully automatic. All have some form of graphical feedback to the designer that assesses the quality of the placement in terms of its ability to be routed or connected in the desired number of signal layers. Most have spacing rules that ensure that the components have enough room between them for successful assembly, rework, and testing.

14.3.2.2 Routers.   Routers are the part of a CAD system that makes the physical connections between the components as specified by the netlist. A router operates on the PCB netlist and placement after the placement step has been completed. Routers range from the completely manual, in which case the designer specifies where wires are to be located by using a graphical display and a mouse or light pen, to fully automatic, where a specialized software program takes the netlist, the placement, the spacing rules, and the wiring rules and makes all the decisions necessary to completely connect all components. The principal advantage of manual routing lies in the fact that the designer can tailor every connection to his liking. The principal disadvantage of manual routing is the fact that it is slow and time consuming, often taking several minutes to completely route and check a single net. Automatic or autorouters solve this speed problem. However, the ability to control the detailed shape of each net is limited by the ability of the autorouter to follow wiring rules. Some advanced autorouters are able to comply with very complex wiring rules. A significant problem with autorouters lies in the fact that they may not find ways to successfully route all of the wires. When this occurs, the designer must add more wiring space in the form of more layers or attempt to complete the routing manually. An important feature of a good autorouter is its manual routing option, as it substantially affects the ease with which this often necessary "finishing" operation is completed. Nearly all routers have a suite of checking tools that ensure that the final route matches the netlist and that all of the spacing rules have been followed.

Routers come in several forms and can be purchased as an integral part of a CAD system or as modules to add onto CAD systems. Some router types are:

Gridded Router. This type of router operates by placing wires on a predefined grid pattern. The routing surface is divided into a uniform grid that provides a proper gap between traces when wires are routed on every grid line. It is the first form of both manual and autorouter offered with CAD systems. The primary disadvantages of gridded routers are that it is difficult to manage more than one trace width without losing wiring density and it requires end points of nets to be on the routing grid in order to connect to them successfully. Offgrid component connections typically have to be made by hand and checked by hand.

Gridless Router. This form of router does not depend on a grid to locate wires on a surface. Instead, it places as many wires in a space as will fit and still maintain the spacing rules established by the design engineer to ensure proper electrical performance while optimizing manufacturability. Multiple trace widths on the same layer are handled easily by this type of router. Once the routing job is completed, the router divides up any unused space equally. The advantage of this technique lies in its ability to optimize manufacturability by keeping spaces as large as possible. The disadvantage of this type of router lies in the fact that it usually depends on a given wiring layer being all horizontal or all vertical—a real disadvantage in SMT applications where components do not need through-holes to connect them but a very powerful router for designs with a very regular array of high-pin-count parts, such as big CPUs and massively parallel processors. This router type is the workhorse of very high complexity digital designs where there is a great deal of regularity and a need to achieve predictable spacings and trace lengths for speed and performance reasons.

Shape-based Router. This type of router recognizes shapes already placed in a wiring surface and routes wires to avoid them. Spacing between wires and other objects, such as vias, used to change layers and component pads is maintained as the router places a wire in a space. This router is becoming the workhorse of SMT-based designs.

14.3.2.3 Checking Tools. These tools verify that the routed PCB complies with rules such as spacing between traces and trace and holes by comparing the actual spacings found in the finished artwork to rules provided by the designer. They also ensure that all nets are completely connected and are not connected to objects they should not be, such as other nets and mechanical features on the PCB, by comparing the routed results to data supplied by the CAD system. Some checking tools also check to ensure that transmission line rules are followed and that coupling from neighboring traces is within limits. Checking tools are usually an integral part of a CAD system.

14.3.2.4 Output File Generators. Once a PCB has been routed and all connectivity verified as accurate, the CAD system holds this information in a neutral form specific to the way in which its operating system is built. For this data to be useful in manufacturing, it must be converted into forms usable by other equipment such as photoplotters, testers, assembly equipment, and MRP systems. Output file generating routines do this conversion. Most CAD systems are equipped with a limited set of these when shipped. Additional generators or converters must be ordered as add-ons.

14.3.3 CAM Tools

CAM tools are CAD systems tailored to the needs of the fabrication process. The output of the PCB design process is a set of CAD files that describes each artwork layer of a PCB, the silk screen requirements, drilling requirements, and netlist information. This information must be modified before it can be used to build a PCB. For example, if a fabricator needs to build several copies of a PCB on a single panel, the fabricator will need to add specialized tooling patterns to the artwork and alter trace widths in order to compensate for etching. Initially, these operations were done manually with a high potential for error and significant labor costs. CAM stations or tools allow the fabricator to do all of these operations automatically and rapidly.

CAM stations can check artwork against spacing rules, breakout rules, and connectivity rules and make corrections if necessary. CAM stations can synthesize netlists (the manner in which points in a design are connected) from the Gerber data for use at bare PCB test, in those cases where no netlist was provided by the customer. In cases where a netlist was provided by the customer, the Gerber synthesized netlist can be compared to the CAD generated netlist as a final way of verifying that the artwork does, in fact, match the schematic—one more safeguard against data corruption anywhere in the translation processes.

14.4 SELECTING A SET OF DESIGN TOOLS

The task of selecting the proper set of CAE, CAD, and CAM tools for a project or company is often one of the most critical steps in determining project success. To reduce the problems associated with interfacing tools to each other, it is desirable to choose all of the tools from a single vendor. However, very few vendors are able to offer best-in-class tools for every phase of the design process. Further, it is very difficult for a single CAD/CAE system to handle the wide variety of design types that may be required across a very large process or across an entire company. For example, a CAD system with a powerful autorouter is of very little use to a design team laying out single-sided PCBs with many irregular parts such as transformers and power transistors. Conversely, a system tailored to do power supply layout well will fare very poorly on a CPU design with many high-speed buses. Knowing this, how does one select the "right" tool set?

14.4.1 Specification

The first step in the tool selection process is to characterize the types of design that the tool set will be required to handle. Next, the level of simulation and design rule checking required to ensure "right the first time" designs must be determined in order to select the right level of simulation and checking tools. For example, performing transmission line analysis on a stereo PCB may be an elegant step but it is a waste of resources. Choosing not to do this level of analysis on a high-speed disk drive PCB may result in a design that is unstable throughout its product life.

Key to success with any tool is the availability of designers qualified to operate it. This pool of designers needs to be assessed for each type of candidate system. If a system or tool is chosen that does not have a pool of qualified operators from which to draw, the result will likely be substantial delays while the necessary expertise is developed or designs that are substandard if the learning time is too short.

14.4.2 Supplier Survey

Once these facts are known, a survey of potential tool suppliers must be conducted. Basic elements include:

1. Determining how closely each tool candidate comes to meeting the need and its cost to acquire, set up, and maintain
2. Assessing the long-term viability of the supplier to ensure that the tool does not become an "orphan" should the supplier fail
3. Making a check with other users of the candidate tools to ensure that they perform as advertised

14.4.3 Benchmarking

Representative benchmark designs need to be made on each candidate system to assess how well they are done. Depending on the size of the potential sale and the size of the benchmark design, a vendor may do the benchmark free of charge. If not, one should be prepared to pay for this valuable step in the evaluation process. Only after completing all of these evaluation steps is it possible to make an informed selection. Doing less, such as relying on an outsider to recommend tools or taking a vendor's word for it, carries the risk of owning a tool that delays development or, worse, having to repeat the selection process in the middle of a project.

14.4.4 Multiple Tools

Knowing the wide variety of designs that may be encountered and the fact that virtually all CAD and CAE tools have been designed to be very good at some subset of design types, it is unrealistic to expect that a single set of tools from a single vendor will be able to deal with all problems equally well. A company with a wide range of design types, such as a company engaged in the design of computers and instrumentation, should expect to own more than one set of tools, each set optimized to its set of tasks. Trying to force one set of tools onto all types of problems is certain to result in overtooling simple designs and undertooling complex ones.

14.5 INTERFACING CAE, CAD, AND CAM TOOLS TO EACH OTHER

A major argument for buying all of the CAE, CAD, and CAM tools from a single vendor is to ensure that they all play together. In the past, this was a major concern because each vendor had proprietary data formats and there were no industry standard data formats. IPC, IEEE, and other trade associations have evolved standard forms of data interchange between systems. These have been adopted by suppliers, such that it is relatively easy to interface best-in-class tools from different vendors to each other.

14.6 INPUTS TO THE DESIGN PROCESS

14.6.1 Libraries

Each CAE and CAD tool uses a series of libraries that contain information describing each component that may be used in a design. These range from a simple description of the physical size of the pads and their relative positions to a full logical model that can be exercised in a simulator. Libraries do not usually come as part of a system. They must be purchased separately or developed one part at a time by the user. Libraries in mature systems can be quite large and represent a substantial investment in time to develop them. Unfortunately, libraries are usually unique to a given tool and cannot be transferred easily should a new tool be chosen.

14.6.1.1   Pad Shapes and Physical Features.   The most basic library used by a CAD system describes the physical characteristics of a part in a manner that allows the CAD system to create its mounting holes pattern and pads as well as its silk screen outline and solder mask pattern. This library entry will contain a *pad stack* that describes how large the component lead holes are and the size and shape of the pads that will appear in each type of PCB layer. For example, an outer layer pad will need to be large enough to ensure adequate annular ring, an antipad will be needed in a power plane to ensure that the plated-through-hole barrel does not touch the power plane, or a thermal pad will be necessary to make a connection to the plane in a manner that still allows reliable soldering. These library entries may contain the unique part numbers used by a company to build a bill of materials, in which case, the CAD system will be able to produce a bill of material in ready-to-use form.

Some physical feature libraries also contain information about the nature of a pin, such as whether it is an input, output, or power pin. This data is used by the checking programs to ensure that the points in a net are ordered properly for high-speed performance or to ensure that a net has the correct kinds of pins in it.

14.6.1.2   Functional Models.   CAE tools that simulate the operation of a PCB require a library of models that describe how each part operates logically. These are functional models. Functional models do not contain information about propagation delay or rise times needed to verify that timing rules are complied with. Functional models are often used to configure emulators.

14.6.1.3   Simulation Models.   Simulation models are extended versions of functional models. They contain all the functional information as well as detailed information about path delays through a part and rise and fall times. They are used to ensure that worst-case timing conditions result in a properly operating design.

14.6.2   PCB Characteristics

One of the sets of data required by the physical layout system is a description of the PCB or its physical characteristics. This includes its size, number and kinds of layers, thicknesses of insulating layers, copper thicknesses, and areas that are not available for parts or traces.

14.6.3   Spacing and Width Rules

To ensure compliance with manufacturing and transmission line rules, the trace widths and trace spacings for each layer must be entered into the CAD system. This is typically done in tabular form.

14.6.4   Netlists

Netlists describe to the CAD system how the pins of each device connect to each other. Systems that manage routing or layout to high-speed design rules will require netlists that contain instructions on how to handle each net, such as what impedance to use, what spacing to preserve with respect to neighbors, and whether terminations or special ordering is needed.

14.6.5   Parts Lists

Parts lists tell the CAD system what type of library entry to use for each part in the design.

CHAPTER 15
ELECTRICAL AND MECHANICAL DESIGN PARAMETERS

Ralph J. Hersey, Jr.
*Lawrence Livermore National Laboratory, Livermore,
California*

16.1 PRINTED CIRCUIT DESIGN REQUIREMENTS

The electrical characteristics of printed board and multichip electrical connection substrates have become a critical functional product definition, and design requirement, for many electrical and electronic products. Until the late 1980s, most printed board designs were printed wiring designs, in that, with the exception of power and ground distribution, component placement and the arrangement of conductive and nonconductive patterns were not critical for functional electrical requirements. This was particularly true for most digital applications. However, since the late 1980s, electrical signal integrity has become a more serious design consideration in order to meet both functional performance and regularatory compliance requirements.

15.2 INTRODUCTION TO ELECTRICAL SIGNAL INTEGRITY

Electrical signal integrity is a combination of frequency and voltage/current, depending on the application. In low-level analog, very small leakage voltages or currents, thermal instabilities, and electromagnetic couplings can exceed acceptable limits of signal distortion. In a similar manner, most digital components can erroneously switch by application of less than 1 V of combined dc and ac signals.

15.2.1 Drivers for Electrical Signal Integrity

Analog and digital signals are subject to many issues that can cause signal distortion and degrade signal integrity. Some of these concern the transmission of the signal, and some the return paths.

15.2.1.1 Signal Integrity Terms for Both Analog and Digital Signals. Table 15.1 provides a listing of many of the issues that affect the signal transmission.

TABLE 15.1   Representative List of Issues for Both Analog and Digital Signals

Rise time	Thermal offset voltage
Fall time	Thermal offset current
Skew	Low-level amplifier
Jitter	High-impedance amplifier
High skew rate	Charge amplifier
Intermodulation distortion	Integrating amplifier
Harmonic distortion	Wideband amplifier
Phase distortion	Video amplifier
Crossover distortion	Precision amplifier

15.2.1.2   Return Paths.   All electrical signals have a signal conductor and a signal return path. Very frequently, the signal conductor is shown on the schematic and the return conductor is neither shown nor mentioned on the schematic/logic drawing. This can also present a problem with PB CAD (computer-aided design) tools. Some CAD tools are "dumb" in that they will automatically route a transmission line as one of the signal conductors but will not route the necessary signal ground on one of the adjacent conductive pattern layers.

15.2.2   Analog Electrical Signal Integrity

The design of some analog printed circuits is a critical balance of all of the known parameters and characteristics of the complete design-through-use product development, manufacturing, assembly, test, and use processes. Analog designs cover all or portions of the complete electromagnetic spectrum, from dc all the way up into the GHz range of frequencies. Active and passive electrical/electronic components and materials have various levels of sensitivity to operating environments and conditions, such as temperature, thermal shock, vibration, voltage, current, electromagnetic fields, and light. In particular, the signal input terminals, voltage connections, and especially analog signal grounds are critical to analog signal integrity.

15.2.2.1   Sensitive Circuitry Isolation.   One of the key methods to improve analog signal integrity is to isolate or separate the more critical or sensitive portions of the design. Sensitive circuitry may be susceptible to one or more external forces, such as electromagnetic, voltage, and grounding systems, mechanical shock/vibration, and thermal. Sometimes the more sensitive circuitry is repackaged into a separate function module that provides its own isolation and separation from the offending condition. Isolation and separation can be provided by physical separation, electromagnetic and thermal barriers, improved ground practices and design, power source filtering, signal isolators, shock and vibration dampeners, and elevated or lowered temperature controlled environment.

15.2.2.2   Thermal Electromotive Force.   Below a few millivolts, thermal electromotive forces (EMFs) can have a significant impact on low-level analog signal integrity. Thermal EMFs of a nonsymmetrical sequence of various metal junctions (conductors) or symmetrical sequences of various metal junctions operating at different temperatures will generate and inject undesirable voltage (or induce unwanted currents) into the electrical signal path. This thermocouple effect is desirable in the case of temperature measurements. However, in the case of other low-level measurements it is an undesirable characteristic. Therefore, the requirement for low-signal-level PCDs is to ensure that all components and electrical interconnection networks and corresponding electrical terminations (such as soldered, welded, wire-bonded, or conductive adhesive) are symmetrical and isothermal. Electrical components, such as thin/thick-film resistors of different values (resistance), may have resistor elements manufactured from different formulations or compositions of materials and will have a designed-in thermal EMF error due to component selection.

15.2.3 Digital Electrical Signal Integrity

Each digital logic family of integrated circuits has manufacturer-specified electrical operating parameters and signal transfer characteristics, many of which have become industry standards due to multiple sourcing (manufacturing) of the family of components. The electrical signal integrity requirements for digital ICs are primarily the high and low electrical (voltage/current) requirements for the output, input, clock, set, reset, clear, and other signal names; the signal rise/fall times, clock frequency(s) and setup/hold times; and the voltage and ground connections as are necessary for the control and operation of the IC.

The input, output, and electrical signal transfer parameters and characteristics for digital ICs vary from logic or microprocessor family to family. Signal ICs are a large matrix of components, consisting of the semiconductor substrate materials, such as silicon, silicon/germanium, and gallium arsenide, that make up the various types of transistors. As shown in Table 15.2, the large number of digital IC families available create a complex matrix of design issues and requirements.

TABLE 15.2   Typical Digital Logic Rise and Fall Times

Logic family	Typical rise/fall time, ns	Logic family	Typical rise/fall time, ns
STD TTL	5	H	6
L	6.5	HCT	8
S	3.5	AC	3
ALS	1.9	ACT	
FTTL	1.2	10K ECL	0.7
BiCMOS	0.7	100ECL	0.5
		GaAs	0.3

The electrical signal integrity of the rise and fall times of electrical signals is a major driver and concern for high-speed and high-frequency printed circuits. Table 15.2 lists some of the typical rise and fall times of some of the popular digital logic families.

15.3 INTRODUCTION TO ELECTROMAGNETIC COMPATIBILITY

Electromagnetic compatibility (EMC) is a serious design requirement for both functional performance to design and regulatory compliance requirements. EMC encompasses the control and reduction of electromagnetic fields (EMF), electromagnetic interference (EMI), and radio frequency interference (RFI) and covers the whole electromagnetic frequency spectrum from dc to 20 GHz. Worldwide, the electronics industry has had to pay increasing attention to EMC to comply with both national and international standards and regulations. EMC involves major design considerations that ensure proper function within the electronic component, assembly, or system in order to:

1. Limit the emission (radiative or conductive) from one electronic component assembly, or system, to another.
2. Reduce the susceptibility of an electronic component, assembly, or system to external sources of EMF, EMI, or RFI.

There are three keys to EMC:

1. Design the product so that it produces less stray electromagnetic energy.
2. Design the product so that it is less susceptible to stray electromagnetic energy.
3. Design the product to prevent stray electromagnetic energy from entering or leaving the product.

15.4 NOISE BUDGET

Good design requires up-front determination of a *noise budget,* which should be included in the product definition requirements. The noise budget is the summation of all of the dc and ac voltages (or currents) that form a boundary within which the component, assembly, or system is designed to function.

$$e_{noise} = e_{dc} + e_{ac} \tag{15.1}$$

where e_{dc} = dc electrical noise
  e_{ac} = ac electrical noise

The *dc noise budget* consists of the voltage settings (preset) of the power supplies, the operating tolerance of the power supply, and the series dc voltage drops of the voltage distribution system.

The *ac noise budget* consists of the effectiveness of the local bypass capacitor, the amount of decoupling between the load, the bulk decoupling capacitor, and the power distributions system, the local voltage drops in the component's voltage/ground conductors, and the component's input voltage tolerance.

As mentioned in Sec. 15.5.3, many of the operating electrical, mechanical, thermal, and environmental parameters and conditions can have a major influence on the noise budget. With a limited focus on digital designs, additional noises that may need to be considered are EMC radiated and conducted emissions from other electromagnetic equipment and thermally generated voltages (thermocouple effect) due to electrical connections with differing layers of metals operating at different temperatures. The following is a list of most of the electrical voltages that should be considered for a noise margin analysis:

Switching noise*	Changes in supply voltage
Cross talk*	Changes in junction temperature
Impedance mismatch*	Changes in die ground voltage (IR)
Component wire bond (IR)[†]	Component lead (IR)

15.5 DESIGNING FOR SIGNAL INTEGRITY AND ELECTROMAGNETIC COMPATIBILITY

15.5.1 High Speed and High Frequency

Higher operating speeds and frequencies have had a dramatic effect on electronic packaging technology as a whole, but in particular on PBs and PBAs. Higher operating speeds have

---

* For digital circuits these three may amount to about 50 to 60 percent of the noise budget.
[†] (IR) voltage drop (current X resistance).

forced an evolutionary change from traditional PWD methods and practices, which were suitable for lower operating speeds, into the realm of serious PCD. Many of the technical design practices for most digital designs have had to adopt high-frequency analog techniques for design, synthesis, and analysis. In addition, component packaging densities have increased, functional electronic packaging densities have increased, more digital components are of the CMOS family design, and CMOS power dissipation increases with operating frequency, the end result being that thermal management has become more of a concern to PBD and other electronic packaging design personnel.

15.5.2 Leakage Currents and Voltages

Input guarding for stray leakage currents and voltages can be an important design consideration, in particular for some analog but also for some digital designs (especially CMOS). Many analog design requirements are based on thermal, pressure, force, strain, and other sensor technologies that have very low levels and ranges of electrical output voltages or currents. In addition, many of the requirements are for measurement accuracy and precision of a few percent or less. In combination, these requirements can present a challenge to electronic and PBD personnel. Very few sensors are robust in terms of their electrical output parameters; most of them have what is termed in the analog world "low-level" signal output for voltage and/or current. Therefore, many of these sensors require signal conditioning and amplifier circuits to improve the integrity of the electrical signal(s). Many of these signal conditioning and amplifier circuits have high-input-resistance characteristics which cause them to be more susceptible to erroneous signals. As a result, "guarding," as described later, becomes necessary to control stray voltages and currents.

Small, undesired, and unintended leakage currents and voltages can have a significant impact on the electrical signal integrity of some analog, as well as digital, applications. Leakage currents of a few nA or less, and leakage voltages of a few μV or less, can affect the designed functional performance of an electronic assembly. It is nearly impossible, from a practical point-of-view for a high-input-resistance circuit to distinguish between a desired and undesired electrical signal. Thus, extreme care must be used in the design, manufacturing, and assembly processes of high-input-resistance products. Therefore, suitable PBD concepts must be included in the design to reduce and control the leakage currents and voltages. The following list identifies some of the common causes for leakage currents and voltages in PBAs.

- Insufficient (surface or volume) insulation resistance of base laminate
- Environmental contamination, fingerprints and skin oils, human breath, residual manufacturing and processing chemicals, improperly cured materials, solder fluxes, and surface moisture, such as humidity.
- Surface and subsurface contamination, such as can be found:

  On or in the assembled component

  Between conformal coatings and the surfaces they are protecting

  On, in, or under the solder resist

  Between conductive patterns on or in the electrical interconnection substrate

15.5.2.1 Design Concepts to Control Leakage Currents and Voltages.   The primary concept for input guarding is to limit and control undesirable leakage currents and voltages or to prevent their formation in the first place. In theory, the principal is very simple: There are no leakage currents or voltages if there are no differences in electrical potential. In practice this is difficult to achieve and may not be a viable solution. However, by minimizing (the goal being to eliminate) the differences of potential between the critical electrical interconnection networks and components (leads and/or bodies) and all other materials, the formation of undesired leakage currents and voltages can be controlled (minimum effect) or eliminated (maximum effect).

- Form a *Faraday cage* around the critical conductive patterns and components using a combination of conductive patterns (frequently called *input guarding* and *guard rings*) and shielding enclosures.
- Keep all unguarded voltages out of the Faraday cage or protected area.
- Electrically connect the Faraday cage to a low-impedance voltage source that follows the critical (protected) voltage.

15.5.2.2 Guard Rings—A Design Method to Control Leakage Currents and Voltages.
The control of electrical leakage to the more critical signal terminals/leads can be optimized through the selection of components, various levels of implementing input guarding into the design, and the selection of the materials for the electrical interconnection substrate.

Some components provide unconnected, unused, or guard terminals/leads adjacent to the input terminals. Care must be exercised for the balance or trim terminals/leads, as in most cases these terminals are connected (internally) in the component directly into the input differential amplifier circuitry of the component; thus, any undesirable leakage currents or voltages to these terminals/leads may result in undesired operation. Some linear operational amplifiers (op-amps) and other linear components are more suitable for input guarding than others; some have two (or more) unused terminals/leads that are used to improve electrical isolation between the protected terminals/leads in the component itself, as well as the component land pattern and electrical interconnection substrate.

The simplest method for providing input guarding to control input leakage currents and voltages is through the use of guard rings of conductive patterns *on all layers of PB conductive patterns* that surround the terminals/leads and associated circuitry. The guard ring is attached to a low-impedance voltage source that best follows the input signal or, as recommended by some analog IC manufacturers in their application notes, to the metal case of the component. As a result, the input terminals of high-input-resistance, low-bias-current, low-offset-voltage Op-amps can be guarded from stray electrical leakage currents and voltages.

15.5.3   Voltage and Ground Distribution Concepts

There are a few main concepts for the distribution of voltage(s) and ground in PBs, their assemblies, and other electronic assemblies. In general, most serious PCDs use one or more ground planes for the common electrical connection(s), for the source(s) of electrical power, and for the reference or return electrical signal path. The keys to good voltage and ground distribution systems are:

- Providing a low-impedance voltage and ground distribution system
- Meeting functional performance product definition design requirements
- Optimizing EMC

Depending on the design, the ground system may also be used for the grounding conductor interconnection(s) for the electrical safety and similar compliance requirements. For electrical signal integrity considerations, it is generally desirable to have separate but parallel electrical interconnection networks for the grounded (signal and power) and grounding (electrical safety) conductors. Like ground, voltage distribution for serious PCDs generally consists of one or more voltage planes (or portions thereof), although for some designs routed conductive patterns or buses may be a functionally acceptable option. A bused voltage and ground system may be acceptable for some designs, but they are generally limited to the PCD with lower operating frequencies and slower rise and fall times. The voltage and ground distribution and the location and type of bypass capacitors can have a significant impact on EMC and electrical signal integrity.

15.5.3.1 *Grounding Concepts.*

Electrical grounding is one of the most important concepts and probably the least understood aspect of electrical signal integrity and EMC. All electrical conductors, including ground, form very subtle, but active, electrical interconnection network that can significantly compromise the product definition's requirements for electrical signal integrity and EMC. In particular, the grounding system is critical to ensure compliance to functional performance and regulatory requirements. Grounding (and voltage distribution) concepts are a matrix of requirements, concepts, concerns, considerations, and practices. In general, there is no universal solution suitable for all applications. Grounding is considered by some to be an art, which can be supported in that some grounding systems are completely unstructured and the reasons that some systems work, while others do not, are not clear. As a result, there has been an ongoing search to find a set of rules that can be used for the design of grounding systems and, unfortunately, many of the rules are conflicting. For example, a modular modem PCMCIA electronic assembly may have a suitable grounding system for normal telecom line operation. Yet, this may be totally inadequate if a 100-1kA electrical current, coupled through the PCMCIA assembly into the personal computer's electrical grounding system, is induced in the telecom line due to a nearby lightning strike or fallen power line. Similarly, a suitable electrical safety grounding system for power line frequencies may not be suitable near a high-power, high-frequency radio, television, or telecommunications transmitter. The following are some of the major concerns and considerations that are involved in good grounding practices:

- Integration of analog and digital signal converters, especially with more than 12 bits of resolution—ground loops and noise
- High-speed and high-frequency operation—ground pull-up and EMC
- High-speed and high-frequency bus line drivers and receivers—major ground pull-up and EMC concerns
- Low-signal level analog sensors (transducers)
- Length of the conductors in the voltage/grounding system as a considerable portion of the electrical wavelength of one of the frequencies of the signal range of EMC
- When designing a grounding system, consideration of developing a grounding map that identifies all grounding requirements and voltage/current/frequency requirements

15.5.3.2 *Grounding Systems.*

The following is an introduction to several grounding and voltage distribution systems and their electrical characteristics.

Single-Point and Point-Source Grounding Concepts. Single-point grounding is a method whereby the grounding electrical interconnection network is connected to ground at a single point at either the source or load end of the electrical interconnection network. The voltage drops between the various grounding nodes is a function of the interconnection network impedances, operating frequency, and current. Point-source or *star* grounding systems have a single-point grounding location for all electrical loads. The point-source grounding point is connected to another grounding point using a low-impedance bus or grounding conductor.

Multiple-Point Grounding Concepts. A multiple-point grounding system may be in the form of a loop or tree-like structure. In a loop grounding system the voltage drops around the loop may vary, depending on the electrical characteristics of each of the loads attached to the grounding loop. In a tree, the grounding system has good voltage regulation and allows leads to be independently attached or removed from the tree without a significant impact on the remaining loads.

Ground Planes. Ground planes are the grounding system of choice for most serious PCD requirements. Ground planes can improve the electrical signal integrity of the grounding system and EMC, provided all critical conductors are buried (lands only) on outer layer(s) of the printed board assembly.

Separation of Grounds. The identification and separation of natural groupings of grounds into similar requirements increases the assurance of conformance to product definition requirements. Some of the natural groupings are as follows:

- Electrical safety ground
- Power supply ground
- Low-level analog ground
- High-level analog ground
- Digital ground
- I/O ground
- Pulsed power/energy ground

When defining the printed circuit design requirements for a particular assembly, results of an analysis of the grounding elements necessary for proper functional performance must be included.

15.5.3.3 Bypass Capacitors.   The selection, location, number, and value of the bypass capacitor(s) for PCDs can affect functional circuit performance. The purpose of the bypass capacitor is to provide the necessary electrical energy to minimize the effect of a component's normal transient switching and load currents during functional operation. The selection, location, and placement of the bypass capacitor can have a significant impact on EMC and a lesser impact on functional performance. One of the keys to EMC management is to prevent or minimize the generation and subsequent radiation of electromagnetic fields in the first place.

15.5.3.4 Voltage and Ground Buses.   As operating frequencies and speeds increase, voltage and ground bus distribution systems may function as lump constant shock-excited oscillators. The frequency of oscillation is dependent on the series inductances of the voltage and ground bus system and shunt capacitors. One of the worst cases is when the voltage and ground buses are placed as railroad rails with the bypass capacitors, and digital integrated circuits are alternated in position like railroad ties bridging between the rails.

15.5.3.5 Voltage and Ground Planes.   Voltage and ground planes can be an effective means to provide a relatively low resistance and impedance[1] to distribute voltage and ground within a PBD. However, maximum effectiveness—solid metal, with no holes or cutouts for plated through-holes or other necessary features—is an unobtainable condition. Therefore, voltage and ground planes are a compromise in design and requirements due to the necessary holes in the planes for electrical interconnections and component mounting.

Voltage and Ground Plane Resistance.   The sheet resistance of solid copper voltage and ground planes is relatively low for most copper foil thicknesses. For 35-μm-thick copper foil, the dc (solid) sheet resistance is less than 1 mΩ/sq. The dc resistance of copper foil is shown in Eq. (15.2). The (solid) sheet resistance for selected copper foils is shown in Table 15.3. Due to the almost infinite number of possible variations in the size, placement, and shape of perforated mesh planes, the following data are presented for informational and comparative purposes only.

$$R_{\mathrm{DC}} = \frac{17.2}{t_{\mu\mathrm{m}}} \ \mathrm{m}\Omega/\mathrm{sq} \qquad (15.2)$$

t_{mm} is in μm

TABLE 15.3 Solid Area Copper Foil Sheet
Resistance

Copper thickness, μm	Sheet resistance, mΩ/sq
5	3.44
9	1.911
12	1.433
17	1.012
26	0.662
35	0.491
70	0.246

Resistance models for periodic mesh grid planes[2] is of interest for *uniform* grids, but is of limited use for grid planes that are not periodic and for ac electrical network analysis.

Voltage and Ground Plane Impedance.   The impedance of perforated mesh voltage and ground planes is difficult to perform due to the number of variations in the size, placement, and shape of perforated mesh planes.

15.6 MECHANICAL DESIGN REQUIREMENTS

The design intent of PBs and their assemblies is primarily to mount and lend mechanical support to components as well as to provide all of the necessary electrical interconnections. However, PBAs should not be used as a (major) structural member.

There are three general forms of PBAs:

1. *Functional module*—A plug-in and a mechanically mounted PBA. The functional module is in the form of a component, whereby leads or other types of electrical terminals provide both electrical interconnections and mechanical mounting of the module to the next higher level of electronic packaging.

2. *Plug-in module*—This provides all of the necessary electrical interconnections at one or more edge-board connectors. A plug-in module typically plugs into a mother board, or sometimes electrical cables are used. The plug-in module is mechanically supported on one edge by the edge-board connector and on one or more edges by card guides, rails, or a mounting frame.

3. *Mechanically mounted PBA*—This is mounted and/or supported in a mechanical assembly or housing with a number of mechanical fasteners around its periphery (and internally to the PBA for additional support, if required). The most common mechanically mounted PBA is the mother board, such as is frequently used in personal computers. Mechanically mounted PBAs are one of the members of an electrical/electronic assembly and are physically mounted in the assembly using one or more mechanical fasteners, such as screws, clips, and standoffs. A common example of a mechanically mounted PBA is a modular power supply.

All forms of PBAs have many common requirements necessary to meet their product definition requirements, although there will be significant variations in the product definition requirements for PBAs due to their specific form and application requirements. For example, the requirement for PBA flatness due to bow and twist may be different for a plug-in than for a mechanically mounted PBA. The requirements for the number and location of mounting fasteners for a mechanically mounted PBA will be different for a relatively thick MLB with low-mass components than for a simpler PB with high-mass components.

The following list of factors, based primarily on Ginsberg,[3] must be considered and evaluated in the design of a PB and its assembly:

- Configuration of the PB, size, and form factor(s)
- Need for mechanical attachment, mountings, and component types
- Compatibility with EMC and other environmental
- PBA mounting (horizontal or vertical) as a consequence of other factors such as dust and environment
- Environmental factors requiring special attention, such as thermal management, shock and vibration, humidity, salt spray, dust, altitude, and radiation
- Degree of support
- Retention and fastening
- Ease of removal

15.6.1 General Mechanical Design Requirements

The general mechanical design requirements for PBs and PBAs include the methods of dimensioning, mounting, guiding during insertion and removal of plug-in components or assemblies, retention, and extraction. Frequently, the PBA mounting method is predetermined as a design requirement to an established compatibility with existing hardware. In other cases, the printed board designer has a choice in determining which PB mounting method is more suitable after considering such design factors as the following:

- PB size and shape based on form, fit, and function requirements
- Input/output terminations and locations
- Area and volume restrictions
- Accessibility requirements
- Ease of repair/maintenance
- Modularity requirements
- Type of mounting hardware
- Thermal management
- EMC
- Type of circuit in its relationship to other circuits

15.6.1.1 Dimensioning and Tolerancing. The dimensioning and tolerancing system for PBs and PBAs must ensure that the product is appropriately defined for all of the product form, fit, and function requirements for the complete product life cycle, from definition through manufacturing to end use. Dimensioning and tolerancing are critical at least for the design, manufacturing, assembly, inspection, test, and acceptance phases.

Regardless of the dimension and tolerance standards that are used to establish and document the product definition's mechanical design and acceptance requirements,[4] there should be at least two (primary) data reference features in every PBD (see Fig. 15.1). The purpose is to ensure the integrity of a PB's and PBA's datum reference throughout the production and acceptance process. In general, it should be a nonfunctional hole (in the case of PBs) or a surface feature that is used as the primary datum reference for final dimensional measurements and acceptance of the product. The datum reference should not be a machined edge that is formed in the last phases of the manufacturing, fabrication, or assembly process in a secondary machining operation.

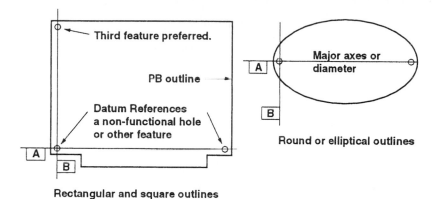

Rectangular and square outlines

FIGURE 15.1   All printed boards and their assemblies should have their datum references included into the design to ensure the integrity of the mechanical datum references throughout a product's production and acceptance cycles.

15.6.1.2   *Mechanically Mounted Printed Board Assemblies (PBAs).*   PBAs should be mounted to ensure their mechanical (and sometimes electrical grounding) integrity throughout their product life cycle. The following are some of the generally accepted requirements and practices for mechanically mounted PBAs:

- PBAs should be supported within 25 mm of the edge of the PBA on at least three sides.

- As good practice, fabricated printed boards (PBs) having a thickness of about 0.7 to 1.6 mm should be mechanically supported on 100-mm or lower intervals, and PBs thicker than 2.3 mm on less than 1.3-mm intervals.

- Fasteners should not be located on less than the PBs thickness or the fastener's head diameter (whichever is lower) from the edge of the PB.

15.6.1.3   *Guides for Printed Board Assemblies.*   A major advantage of using plug-in printed board assemblies rather than other electronic packaging techniques is the suitability of PBAs for use with mechanical PBA card guides for ease of maintenance, changing configuration, and up-grading function or performance. There are many PBA guide hardware systems that are available either commercially, as industry standard, or proprietarily. The PBA may be predetermined as a design requirement or may be developed based on the size and shape of the PBA, *the degree of dimensional accuracy needed to ensure proper mating alignment with the mating connector system,* and the desired degree of sophistication. Some PBA guide systems contain a built-in locking system that provides mechanical retention and thermal management (conductive).

   Caution: Some PBA card guide systems have become somewhat of industry standards and can be obtained or assembled to fit most PBAs. However, not all of the industry-standard-like PBA guide systems are compatible or interchangeable for retaining or extracting PBAs.

15.6.1.4   *Retaining Printed Board Assemblies.*   Quite often, shock, vibration, and normal handling requirements necessitate that the PBA be retained in the equipment by mechanical devices. Some PBA retaining systems are attached as hardware to the PB during assembly; other retaining systems are built into the PBA mounting hardware frequently called a *cage.* The selection of a proper PBA retaining system is important, since the retaining devices may reduce the amount of PB area available for component mounting and interconnections, and can add significantly to the cost of the electronic equipment.

15.6.1.5 Extracting Printed Board Assemblies. A number of unique principles have been developed and applied to solve the various problems of extracting PBAs from their plug-in enclosures. The result has been a proliferation of proprietary and a few industry-standard-like extraction systems. The most common industry-standard extractor is injection-molded plastic hardware that is free to partially rotate when attached to the PBA with a pressed-in pin. Many of these PBA extraction tools use a minimum of PB space, thereby maximizing available PB area for components and conductor routing. They also protect both the PBA and the associated mating connector(s) from damage during the extraction process.

The following should be considered when selecting among the many different types of PBA extraction tools:

- The area of the PBA available for attachment
- The extractor's effect on the PBA-to-PBA mounting pitch
- The need for special provisions in the PBD, such as mounting holes, mounting clearance holes, and notches
- The size of the extractor, especially if the extractor is to be stored in the equipment with which it is used
- The need for an extraction device that is permanently attached to the PBA, usually by riveting
- The need for specially designed considerations, such as load-bearing flanges, in the PBA mounting chassis or cage hardware
- The suitability of the extractor to be used with a variety of PB sizes, shapes, and thicknesses
- The cost of using the extractor, both in piece price and added design costs
- The degree of access required inside the equipment to engage and use the extraction tool

15.6.2 Shock and Vibration

Shock, vibration, flexing, and bowing can be functional performance and reliability concerns for PBAs—more so for larger PBAs. For many PBAs, the worst-case exposure to shock and vibration occurs in nonfunctional or operational usage, during shipping and other forms of transportation from one location to another, or possibly in functional use when the functionally operating product containing PBAs is inadvertently dropped on the floor. Other PBAs are designed to withstand specified levels of shock and vibration in transportation and in use. The design requirements for shock and vibration vary, depending on each family of general requirements. For example, there are nonoperating shock and vibration withstand requirements for vehicular, train, ship, and air transportation for domestic and international shipment for Level 1, 2, or 3 products, which include various procedures and requirements for packaging. The shock and vibration functional design requirements are many and varied, and are very dependent on the application. Some sources of shock and vibration are very obvious, while others are very subtle. The levels and duration of shock and vibration vary significantly in each application: An electronic sensor mounted on a vehicle's axle is different from the radio mounted on the dashboard. There are the differences among ground-based, rack-mounted, industrial control equipment, aircraft, aerospace, and munitions applications. Some vibrations are subtle, low-level continuous, and are frequently caused by electric or gas motor-driven rotating machinery and equipment. Continuous low-level vibrations can induce *mechanical fatigue* in some electrical/electronic equipment.

15.6.2.1 Shock. *Mechanical shock* can be defined as a pulse, step, or transient vibration, wherein the excitation is nonperiodic.[5] Shock is a suddenly applied force or increment of force, by a sudden change in the direction or magnitude of a velocity vector. With few exceptions, shocks are not easily transmitted to electronic equipment by relatively light mounting frames and structures. Most shocks to electronic equipment in the consumer, commercial, and industrial markets are due to dropping during handling or transportation, the exceptions being electronic sensors or equipment mounted to heavier mounting frames, such as vehicle axles and punch-press-like equipment, or military equipment subjected to air dropping or to explosive forces, such as munitions. Most shock impact forces result in a transient type of dampened vibration which is influenced by the natural frequencies of the mounting frame. Generally, shock either results in instantaneous failure or functions as a stress concentrator by reducing the effective strength of the connection or lead for subsequent failure due to additional shock(s) and vibration.

15.6.2.2   Vibration.   *Vibration* is a term that describes oscillation in a mechanical system, and is defined by the frequency (or frequencies) of oscillation and amplitude. PBAs that are subjected to extended periods of vibration will often suffer from fatigue failure, which can occur in the form of broken wires or component leads, fractured solder joints, cracked conductive patterns, or broken contacts on electrical connectors. The frequency(s) of vibration, resonances, and amplitude(s) all influence the rate to failure.

Flexing and bowing in PBAs is the result of induced shock and vibration into the PBA. Different PBA mounting methods have differing susceptibilities to shock and vibration. In general, most small PBA functional modules are manufactured as components and are frequently encapsulated with a polymeric potting material into a solid mass, and they therefore have minimal shock and vibration requirements within the module. The plug-in PBA is restrained on one edge by the edge-board connector(s), and to some extent along the two sides of the PBA by mechanical guides. This leaves only one edge of the PBA free to flex from shock and vibration or to bow from residual manufacturing or assembly stress in the PBA. However, a handle along the free edge of the PBA or a restraining bar can be located across the center of the free edge of the PBA and mechanically attached for support at the ends of the restraining bar to the PBA mounting hardware (card cage). Generally, the mating edge-board connector has a molded-plastic body that provides mechanical support to the mated edge-board connector and has sufficient compliance in the electrical contacts to maintain good electrical connections within the connector's performance specifications. Mechanically mounted PBAs can be more of a shock and vibration concern for three main reasons:

1. The PBA can be very large, sometimes as large as 600 mm^2. More frequently, the maximum width is 430 mm (the width of a standard electronic chassis) and less than 600 mm long. Although most are less than 300 mm^2, this still creates a large area and can be a problem if unsupported.

2. The PBA is misused as a mechanical support structure for high-mass components, such as magnetic components (iron-cored transformers and inductors), power supplies, and large (in physical size) function modules.

3. The PBA is not included in the mechanical design definition.

15.6.2.3   Major Shock and Vibration Concerns.   These include the following:

- Flexing between PBAs may cause shorts between adjacent PBAs or to the enclosure.

- The fundamental mode is the primary mode of concern because it has the large displacements that cause fatigue damage to solder joints, component leads, and connector contacts.

- Continuous flexing of a PBA will fracture component leads and, more important, surface-mounted component solder joints, due to mechanical fatigue failure. (Mechanically induced flexing or vibration in assembled PBs is used under controlled conditions to induce failures in solder joints for quality and reliability studies.)

- Movement of a PBA within its mechanical guides will be amplified due to shock and vibration resonance or harmonic resonance.

- Fatigue life modeling for PBA mounted components has become significantly more complex due to in-use simultaneous application of both vibration and thermal cycles. Vibration strains and thermal strains should be superposed for more representative modeling.

15.6.2.4 *Types of Edge Mounting.*   The problems, methods of analysis, and minimization of the effects of shock and vibration on PBAs are the same as in other engineering applications, and similar solutions can be used. PBAs are designed and manufactured in a wide range of shapes and sizes, with rectangular being the most common because of the shape of most electrical/electronic equipment, especially for plug-in PBAs. Though PBAs are a multi-degree-of-freedom system, the fundamental mode is of primary importance because it has the large displacements that are the primary cause of fatigue failure in solder joints, component leads, wires, and connector contacts.[6] Most vibrational fatigue damage occurs at the fundamental or natural frequency because displacement is highest and stress is maximized. Edge or boundary conditions are terms that are used to define the method of attachment of the PBA (or, more generically, a panel) to its mounting frame. The term *free edge* is used to define those edges that are not restrained and are free to move and/or rotate along the edge of the PBA out of their normal mounting plane. The terms *supported edge* or *simple support* used to define an edge that is restrained in out-of-plane movements but is allowed rotational movement around the PBA's edge. The terms *fixed edge* or *clamped edge* are used to define an edge that is restrained in both out-of-plane and rotational movements. Illustrations of the definitions for fixed edge, supported edge, and free edge, and their applications to plug-in mounted PBAs, are shown in Fig. 15.2.

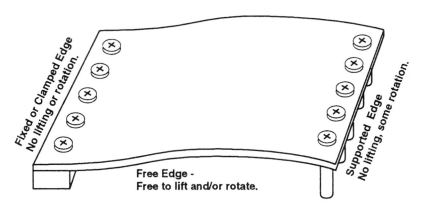

FIGURE 15.2   The method of mounting PBAs contributes to shock and vibration concerns, but can be reduced by good design practices.

15.6.2.5 *Board Deflection.*   The amount of strain on a PBA component is a function of the maximum deflection of the PBA when subjected to shock and vibration. Those components mounted closest to the center of the assembly are subjected to the greatest strain, as illustrated in Fig. 15.3.

A set of empirical maximum deflection (δ) calculations has been developed by Steinberg, and his latest equation[7] has more parameters than previous equations, reflecting the sophistication and requirements for modern PBDs (units adjusted from inches to mm).

$$\delta = \frac{k\,0.00022\,B}{ct\sqrt{L}} \tag{15.3}$$

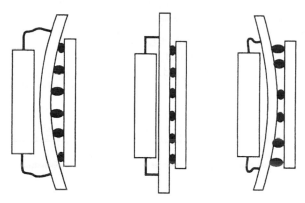

FIGURE 15.3 The PB bends in a PBA during shock and vibration, with the most severe stresses applied to the components mounted centermost on the assembly.

where k = units conversion coefficient; for inches $k = 1$, for mm, $k = \sqrt{25.4}$

B = length of PB edge parallel to component located at center of board (worst case), mm

L = length of component, mm

t = thickness of PB, mm

$\check{c}$ = 1.0 for standard DIP

= 1.26 for DIP with side brazed leads

= 1.0 for PGA with four rows of pins (one row extending along the perimeter of each edge)

= 2.25 for CLCC

Analysis of the maximum deflection calculation formula reveals (as expected) that components with some compliance built into their component mounting and electrical terminations (such as the DIP and PGA) can be subjected to about twice the vibrational deflection as an SMT CLCC, provided component size, PB size, and PB thickness are equivalent. The latest equation for maximum deflection calculations is rated for 10 million stress reversals when subjected to harmonic (sinusoidal) vibration, and 20 million stress reversals when subjected to random vibration. It must be understood that this equation is a first approximation for predicting solder joint life. There are many factors that must be included for a more rigorous analysis and prediction. A more thorough discussion is found in Barker.[6]

15.6.2.6 Natural (Fundamental) Resonance of Printed Board Assemblies. The mechanical mounting of PBAs and their components is a key design consideration in the ability of the PBA to withstand shock and vibration. The overall size of the PBA is not a major factor, provided a suitable mechanical support structure is included as a part of the PBA's product definition requirements. There is a large matrix of ways to mount panels (PBAs) using various combinations of free edges, supported edges, fixed edges, and point supports and by calculating fundamental resonance. The following four examples compare the fundamental natural resonances of the same rectangularly shaped PBA using different edge-mounting techniques. Additional formulas for calculating other natural resonances are found in Barker,[6] Steinberg,[7] and others.

In the following examples—demonstrations of the sensitivity of PBAs to their methods of mounting—the same PBA is used for direct comparative purposes (see Fig. 15.4). The following are the design requirements and material parameters that were used for the calculations:

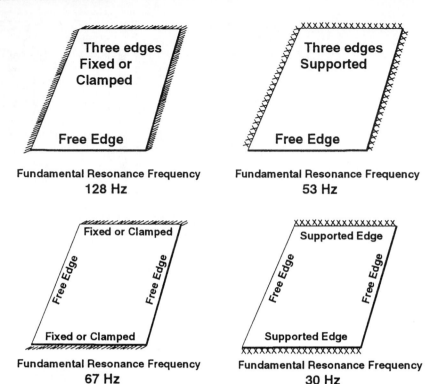

FIGURE 15.4 The same PBA can have a wide range of natural resonances depending on the mounting method, which is why shock and vibration concerns are a critical design consideration.

E Modulus of elasticity (type GF epoxy-glass)	1.378×10^4 MPa
μ Poisson's ratio	0.12 (dimensionless)
a Longer dimension of the PBA	200 mm
b Narrower dimension of the PBA	150 mm
t Thickness of the PBA	1.6 mm
W Weight of the PBA including components	0.25 kg

$$D = \frac{Et^3}{12(1 - \mu^2)} \qquad (15.4)$$

where D is the plate bending stiffness.

$$M = \frac{W}{g} \qquad (15.5)$$

$$f_n = C_0 P_0 \sqrt{\frac{Dab}{M}} \ \text{Hz} \qquad (15.6)$$

where f_n is the natural resonant frequency.

15.6.3   Methods of Reinforcement and Snubbers

Printed board assemblies are stiffened using one or more methods to raise the natural resonant frequency sufficiently above the shock and vibration threat. The most obvious is to change the method of retaining or mounting the PBA in the next higher level of assembly. Frequently, though, this may not be an acceptable option due to the resources and schedule changes that may be necessary for extensive redesign. However, some simpler modification in the design may meet requirements. Sometimes changing the plug-in PBA guide from a loose supportive guide to a tighter spring-loaded or clamping-type guide may be sufficient. Other methods are to add ribs or stiffeners, additional single-point mounting locations, or *snubbers* across the surfaces of the PBAs.

REFERENCES

1. Donald R. J. White and Michel Mardiguian, "EMI Control, Methodology and Procedures," *Interference Control Technologies, emf-emi Control,* 4th ed., Gainsville, Va., pp. 5.5–5.6.

2. Ruey Beei Wu, "Resistance Modeling of Periodically Perforated Mesh Planes in Multilayer Packaging Structures," *IEEE Transactions on Components, Hybrids, and Manufacturing Technology,* vol. 12, no. 3, September 1989, pp. 365 372.

3. G. L. Ginsberg, "Engineering Packaging Interconnection System," Chap. 4 in Clyde F. Coombs, Jr. (ed), *Printed Circuits Handbook* 3d ed., McGraw Hill, New York, 1988 pp. 4 17.

4. ANSI-YI4.5 "Dimensioning and Tolerancing," American National Standards Institute, New York, (date of current issue).

5. Cyril M. Harris and Charles E. Crede (eds.), *Shock and Vibration Handbook,* vol. 1, McGraw-Hill, New York, 1961, p. 1–2.

6. Donald B. Barker, Chap. 9, in *Handbook of Electronic Packaging Design,* Michael Pecht (ed.), Marcel Dekker, Inc., New York, 1991, p. 550.

7. Dave S. Steinberg, *Vibration Analysis for Electronic Equipment,* 2d ed., John Wiley, New York, 1988.

CHAPTER 16
CONTROLLED IMPEDANCE

Ken Taylor
Polar Instruments, Garenne Park, Guernsey, UK

16.1 INTRODUCTION

The function of a wire or track is to transfer signal power from one device to another. Theory shows that maximum signal power is transferred when circuit impedances are matched. Thus, when a signal-generating device, the connecting cable or printed wire, and the receiving device all have the same impedance, the maximum signal power transfer is achieved. This principle holds true at all frequencies, but in practice is of greater importance at higher frequencies.

This chapter describes:

- The basics of impedance
- The effect of impedance mismatches
- The key design elements that determine impedance in printed circuit boards (PCBs)
- A set of sample board constructions for the control of impedance
- Mathematical tools for determining impedance

16.1.1 Definitions

In any discussion of impedance it is important to have a clear definition of the terms used. Figure 16.1 provides a summary of the terms used in this chapter, along with some descriptive information to put them into context. These terms are used in the figures of alternative board constructions, as well as the equations that appear in the various sections.

16.1.2 The Digital Signal (Pulse)

One of the key drivers for the need for matching and controlling impedance has been the use of digital technology in all aspects of electronic equipment. The second major driver is the need to go higher in the frequency spectrum to support the needs for ever more bandwidth in the telecommunications equipment field.

The need for clear digital pulses, however, has the largest influence in volume of PCBs using controlled-impedance circuits. Since digital equipment operates on the presence or absence of a pulse at a given time, it is critical for the leading and trailing edges of the pulse to

Definitions

Dielectric: an electrical insulator

Permittivity: the ability of a dielectric to store or hold electrons

Relative permittivity: the ratio of a capacitor made from an insulator between the plates compared to one built with a vacuum

$$E = \text{permittivity}$$
$$0 = \text{zero (a vacuum)}$$
$$E_0 = \text{permittivity of a vacuum}$$
$$r = \text{relative}$$
$$E_r = \text{permittivity relative to a vacuum}$$

Various E_r: vacuum = 1.000, air = 1.0007, Teflon = 2.2, polystyrene = 2.5, Mylar = 3.0, epoxy = 3.2, paper, paraffin = 4.0, glass = 6.0, tantalum oxide = 25, ceramic = 10, water = 75, ceramic = 100–10,000

Impedance: the combined inductive reactance and capacitive reactance

FIGURE 16.1 Definitions of terms used in calculating impedance and designing controlled-impedance structures for printed circuit boards.

be as close to vertical as possible (see Fig. 16.2 for a generalized pulse shape and description). Since the slope of each edge is determined by the range of frequencies used to construct that pulse, a steep slope requires that the circuit be capable of transmitting very high frequencies. This makes almost all digital circuits candidates for controlled-impedance designs, and impacts not only the design but the fabrication of PCBs.

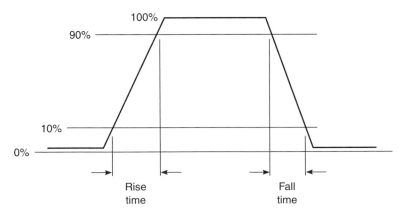

FIGURE 16.2 General shape of a digital signal (pulse) showing the rise and fall times between 10 and 90 percent of zero to maximum amplitude.

16.1.3 Controlled-Impedance Needs

High clock speeds of digital circuits require controlled impedance because of the fast rise and fall times of pulses (see the preceding text). Low clock speeds used to mean low rise or fall times, but semiconductor component manufacturers have done what is called "die shrinks" to older, slow-logic devices in order to get more die per wafer. This increases the efficiency of the use of the wafer. To match the slow clock requirements, these new dies of old chips have built-in delays, but with very fast risetimes, resulting in the requirement for matched termination of the circuit.

16.2   BASIC IMPEDANCE MISMATCH EFFECTS

When there is a mismatch of impedance at any place in the transmission line, maximum signal transfer does not occur and some signal power is reflected from the mismatch.

16.2.1   Signal Reflection Process

Consider a sending device connected to a receiver by means of a PCB track or a cable. If the sender impedance is not the same as the track impedance, and the track impedance is not the same as the receiver impedance, there will result a complex series of partial power reflections, as follows:

- At the connection from the sending device to the track, where the first impedance mismatch occurs, there will occur a partial reflection of signal power back to the sender. Only a portion of the available signal power will transfer successfully into the track.

- When that partial signal reaches the receiver, a second partial reflection will take place, resulting in some of the available signal power being reflected back along the track toward the sender and only the remaining fraction of power (the true signal) successfully entering the receiver. But that is not the end of the story.

- The partial reflected signal, which is now traveling back along the track from the receiver, will encounter the track-sender mismatch. As a result, some of the reflected signal power will now reenter the sending device and some of it will be reflected back along the track toward the receiver for a second time (some finite time after the sender first transmitted the signal).

- The initial result will be twofold:

  1. The first incident signal will be attenuated in power, possibly to such an extent that it will not successfully drive the receiving device.

  2. A smaller "ghost" signal will arrive some time later at the receiver, to be superimposed on the true signal, which was due to arrive at that later time.

- Depending on the extent of the mismatches, significant signal power may be reflected back and forth along the track more than once, but in any case the result will be some degree of corruption of received signal, whether analog or digital in nature.

16.2.2   Effect of Signal Frequency

If a circuit behaves as just described when driven by low frequencies, it is very possible that the ghost signal will arrive at the receiver while the true signal is still substantially present. Thus this state of affairs, while not ideal, presents less of a problem at the receiver because the incident true signal and the ghost signal (which is an image of the true signal) are largely time coincident and thus will superimpose. Corruption at the receiver will be minimal. However, the receiver's problem will increase as signal frequency increases, because at a sufficiently high frequency of signal it can be seen that, for a constant length of signal path, the round trip time for the reflected ghost signal will become a substantial part of a signal period. In this case the ghost signal will arrive at the receiver when the true signal has ended, and will appear as an interference signal corrupting a (later) time-coincident true signal.

High frequency and *low frequency* are relative terms that take into account the length of the signal path or track, and a mismatched track can be more easily tolerated in tracks that are short relative to the signal wavelength, all other things being equal. Throughout this discussion it has been assumed that the impedance of the track is constant along its length. If the track impedance should change for whatever reason (for example, where two tracks connect through a plug connector, or where a track branches or goes through a via), more reflections will occur at that change and will complicate events even further. At either high or low frequencies, however, it must be recognized that a signal path mismatch will return signal power to the sending device, and this input of power to the sender may cause problems of a different nature.

16.2.3   Effect of Components on Circuit Consistency

Any track impedance mismatch can be extremely difficult to analyze once a PCB is loaded with components. Components have a range of tolerances, so that one batch of components may tolerate an impedance mismatch while another batch might not. Moreover, a component's characteristics may change with temperature, so that the problems may come and go. Thus, if changing a component appears to cure a problem, the components may become the suspects instead of the track. Component selection becomes the solution, and building costs are driven up, while actually the real fault—track impedance mismatch—goes undetected! For all of these various reasons, a PCB designer will specify track impedance and tolerance, and should work with the PCB manufacturer to ensure that the PCB meets the specifications.

As recently as 1997, high-speed devices requiring controlled-impedance PCBs were considered exotic and comprised less than 20 percent of PCBs manufactured at that time. In 2001, around 80 percent of all multilayer PCBs carried high-speed devices and the designs required controlled-impedance tracks. These included boards for all types of popular technologies including low-cost devices routinely found around the home and office. This trend will continue, so that virtually all PCBs will include at least some controlled-impedance tracks. Controlled impedance has become a normal element in board design.

16.3   IMPEDANCE

For electrical devices, the controlled or characteristic impedance of a pair of conductors is the ratio of voltage to current of the propagating pulse or step voltage. For maximum power transfer from one device to another, as noted earlier, a required criterion is that the impedance of the two devices should be the same. This situation is often called a *matched impedance.* For maximum power propagation along a path, and for the pulse to remain unchanged as it propagates, the impedance must be the same at every point along the length of the path. This is often called a controlled impedance, and of course the controlled-impedance track is normally matched to the impedance of the driver and source devices.

16.3.1   Factors Determining Impedance

A number of factors taken together determine the impedance of a conducting track.

- The shapes and dimensions of the tracks and insulator
- The electrical characteristics of the insulator
- The position of the insulator
- The proximity and shapes of any nearby conductors

Since all of these combine to determine the charge distribution on the track, they will also determine the shapes, strengths, and interaction of electrical and magnetic fields, which, in turn, determine the inductance (L) and capacitance (C) of the track.

16.3.2   Characteristic Impedance

The characteristic impedance (Z_0) of the track is related to L and C according to the formula:

$$Z_0 = \sqrt{(L/C)} \tag{16.1}$$

It follows, therefore, that in the design and manufacturing process of a controlled-impedance track, all of these factors must be carefully specified and implemented.

16.4   TRANSMISSION LINES

16.4.1   Single-Ended Transmission Line

The theoretical circuit in Fig. 16.3 is an example of a single-ended transmission line. Controlled-impedance PCBs are usually produced using microstrip or stripline transmission lines. The single-ended transmission line is the commonest way to connect two devices. In this case a single conductor connects the output of one device (the signal source) to the input of another device (the receiver). The reference (ground) plane provides the return signal path. This is an example of an unbalanced line. The signal and return lines differ in geometry, because the cross-sectional profile of the signal conductor is different from the cross-sectional profile of the return ground plane conductor. The characteristic impedance of the line is generally symbolized as Z_0.

16.4.2   Differential Transmission Line

The differential configuration shown in Fig. 16.4 is used to provide better signal immunity from outside electrical fields. Differential connections are often used to minimize interference to the signal from outside sources. The signal lines are driven as a pair, with one line transmitting a signal waveform of the opposite polarity to the other. The signal is recovered at the output by subtracting the signal in one line from the signal in the other. If the lines are

FIGURE 16.3   Theoretical circuit, showing an example of a single ended transmission line.

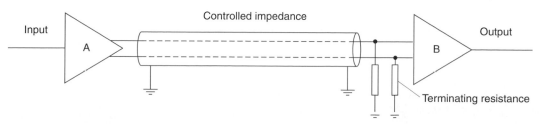

FIGURE 16.4   Differential configuration used to provide better signal immunity to outside electrical fields.

close together, external fields will introduce equal interference signals into each of the lines and the noise generated in them both will tend to cancel at the output.

However, if the lines are spaced widely apart, they will not share the same noise environment and thus will receive dissimilar interference signals that will not null at the receiver. Therefore it is important that the lines be relatively close together, or *closely coupled*. A true differential system requires lines that are also balanced. That is to say, they share exactly the same physical and electrical characteristics. In practice, there must be some detail dissimilarities between the two that result in a degree of imbalance. How much imbalance can be tolerated will depend on the system and its environment. When differential lines are discussed, balanced lines are assumed.

There are several ways to define the impedances associated with a differential pair, the most frequently of interest being the differential impedance:

- Z_{0o} (odd-mode impedance) is the impedance of one track when the other track is driven in an equal and opposite sense (driven differentially).

- Z_{dif} (differential impedance) is the impedance measured between the two tracks when they are driven in equal and opposite sense (driven differentially.) Z_{diff} is usually the impedance characteristic of interest, and is equal to $2 * Z_{0o}$.

- Z_{0e} (even-mode impedance) is the impedance of one of the two tracks when both are driven equally and in the same sense. Z_{0e} is always greater than Z_{0o}, since the capacitance between the tracks reduces the impedance of each of them when they are driven differentially.

- Z_{com} (common-mode impedance) is the impedance of the track combination when they are connected together and driven in parallel. $Z_{0e} = 2 * Z_{com}$.

- Z_0 (Characteristic impedance of one track) is the impedance of the track when only that track is driven.

16.5   TRANSMISSION LINE IMPLEMENTATION IN A PCB

The practical implementation of single-ended and differential transmission lines in the form of printed circuit tracks takes many different forms. A few examples follow.

16.5.1   Dimensions

In the United States of America, dimensions for PCB tracks are usually quoted in units of 0.001 in, often called *mils* and sometimes called *thou*. In other countries working in inches, the same unit is normally called *thou*. Countries working in metric units normally base dimensional specification on micrometers, sometimes called *microns*. The unit is 0.000001 meter (1/1000 of a millimeter) and may be symbolized variously as u, μ, um, or μm. One mil is equal to approximately 25.4 μm.

In the examples shown, tracks are depicted as if etched while bonded to the dielectric material beneath them. This gives tracks a trapezoidal cross section with a base surface wider than the top surface. This results from the upper edge spending a longer time in contact with the etching solution than the lower edge, and is known as etch-back ($W_1 - W$.) For half ounce copper (variously taken to be 0.0006 or 0.0007 in thick) a typical etch-back would be 0.0006 in (0.6 thou). Etch-back is an important parameter in determining track cross-sectional shape. The shape, in turn, affects charge distribution over the track, and thus electrical capacitance and impedance.

Dimensions variously labeled *T, W, W$_1$, H, H$_1$, S,* etc., are therefore important in determining the finished track impedance, where:

$$T = \text{thickness of the copper track}$$

$$W, W_1, W_2, \text{etc.} = \text{various widths at crests and bases of tracks}$$

H, H_1 = dielectric thickness at various locations

S = space between differential paths

D = distance between coplanar ground and signal tracks in coplanar structures

16.5.2   Factors of Influence*

In realizing a particular PCB construction of a controlled impedance in both design and fabrication, it is important to consider the elements that influence the actual circuit impedance and the accuracy with which the construction can be made. Table 16.1 shows these factors of influence, along with the amount of influence they exert. Table 16.2 shows the order of influence of these elements. The notes refer to Table 16.3, which shows improvements required to achieve ≤5 Ω.

TABLE 16.1   Factors of Influence

Factors	Influence
Dielectric thickness	51%
Dielectric constant	22%
Trace width	18.5%
Cu thickness	5.5%
S/M thickness	3%

The percentage is summarized from the results of our test lots.
    This percentage is also close to the calculation of the theoretical equation.
    S/M: solder mask.

TABLE 16.2   Order of Influence

Factors				Note
Dielectric thickness	Material	Core	Thickness uniformity	1
		Prepreg	Resin curtain	2
	Process	Lamination	Board thickness	3
Dielectric constant	Material	Core		4
		Prepreg		
Trace width	Process	Exposure/etch		5
Cu thickness	Process	Plating	Plating + foil	6
S/M thickness	Material	Ink viscosity	On top of trace	7
	Process	Printing		

TABLE 16.3   Improvements

Note	Items	For ±5 Ω (normal product)
1	Core thickness	±0.5 mils
2	Resin contain	50 ± 3%
3	Board thickness after lamination	±3 mils
4	Dielectric	4.4 ± 0.3
5	Trace width	±0.8 mils
6	Cu thickness	±0.5 mils
7	S/M thickness	±0.3 mils

---

*Additional material contributed by Happy T. Holden, Westwood Associates, Loveland, Colorado.

16.5.3 Single-Ended Examples

In these single-ended examples, it is generally true that the following will reduce impedance if all other factors remain unchanged: track is closer to the reference plane; track is wider; track is thicker; dielectric constant is lower; there are two reference planes (one above, one below the track).

16.5.3.1 Surface Microstrip (see Fig. 16.5).   The surface microstrip is usually plated, as are other surface tracks. The original track foil of 0.0006 (or 0.0007) in is usually plated up to three times its original thickness, to a finished thickness of between 0.0018 and 0.0021 in. The finished thickness will help determine the track impedance.

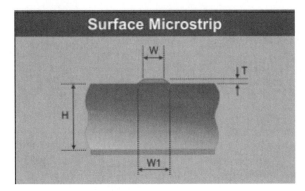

FIGURE 16.5   Surface microstrip.

16.5.3.2 Coated Microstrip (see Fig. 16.6).   Surface tracks are usually coated over with materials that assist in the production process, such as solder resist. The dielectric constant of the coating will significantly contribute to the finished impedance of the track, and the effect of the coating may not be ignored in the calculation. The profile of the coating may be of uniform thickness, as shown in Fig. 16.6, or may have a plane surface where the thickness varies and is thinner over the tracks than elsewhere. This profile also helps determine the finished impedance of the track.

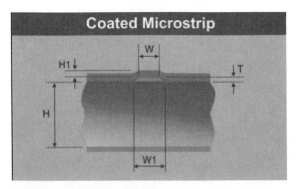

FIGURE 16.6   Coated microstrip.

16.5.3.3 *Embedded Microstrip (see Fig. 16.7).* The embedded microstrip may run orthogonally to another track, probably above it and on the surface of the dielectric. The orthogonality will minimize cross talk between the tracks in the absence of a ground plane between them. Since the tracks cross at right angles, a reasonable approximation ignores the area of overlap in calculating impedance. See the next example later for further illustration.

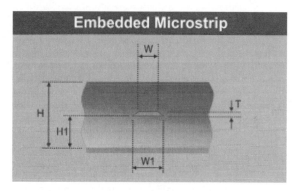

FIGURE 16.7 Embedded microstrip.

16.5.3.4 *Offset Stripline (see Fig. 16.8).* Stripline tracks are sandwiched between two reference planes. The stripline may be equidistant between the planes, or may be offset to permit another track orthogonal to it. In the latter case this would probably be to permit a second track to be positioned at a distance H_1 from the upper reference plane. The resulting structure would be referred to as a dual stripline—two signal conductors sandwiched between the two reference planes on adjacent layers. These two signal layers will be routed orthogonally to minimize interlayer cross talk. The structure is then behaving as two independent offset striplines. The dual-stripline structure is illustrated in Fig. 16.9.

Notice that when measuring the impedance of the tracks it is essential that the two reference planes be short-circuited together for AC signals.*

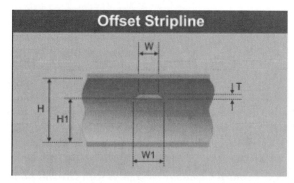

FIGURE 16.8 Offset stripline.

---

* For more information on this subject, see "Controlled Impedance PCBs—An Introduction to the Design and Manufacturing of Controlled Impedance PCBs," Polar Instruments Lit 145.

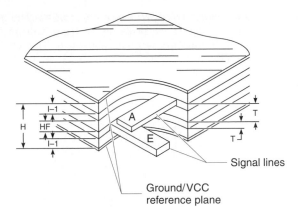

FIGURE 16.9 Dual stripline.

16.5.3.5   Surface Coplanar Wavguide (see Fig. 16.10).   The surface coplanar waveguide has a single controlled-impedance track with reference planes (or very wide strips) on each side. A variation on this structure is to have a ground plane on only one side of the track.

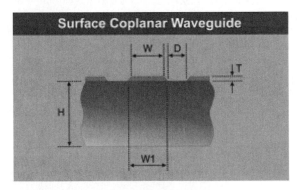

FIGURE 16.10 Surface coplanar waveguide.

16.5.3.6   Coplanar Coated Strips (see Fig. 16.11).   In this coated coplanar configuration there is a single controlled-impedance track with two ground tracks of a specified width (W_2 and W_3) on either side. All the tracks are coated. The ground strips are normally relatively wide, and restrictions may apply to the calculation; for example, greater than 10 times total trace separation ($10 \times \{W + 2S\}$), and 10 times the dielectric laminate thickness.

16.5.3.7   Coated Coplanar Waveguide with Ground Plane (see Fig. 16.12).   The controlled-impedance track has reference planes (or very wide tracks) on each side, and a continuous reference plane on the other side of the dielectric laminate. The signal track and coplanar reference planes are coated.

16.5.3.8   Coplanar Waveguide (see Fig. 16.13).   This configuration is similar to that in Fig. 16.12 except that it has a reference plane both above and below the coplanar configuration.

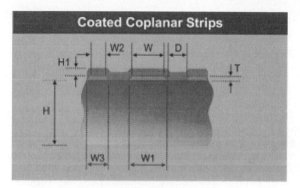

FIGURE 16.11   Coplanar coated strip.

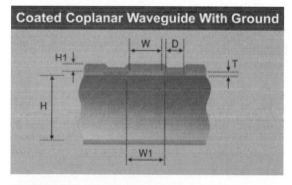

FIGURE 16.12   Coated coplanar waveguide with ground plane.

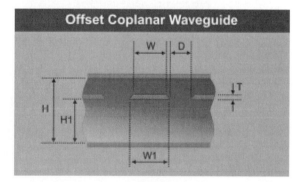

FIGURE 16.13   Coplanar waveguide.

16.5.4   Differential Examples

In these differential examples, it is generally true that the following will reduce differential impedance if all other factors remain unchanged:

- Track is closer to reference plane
- Tracks are closer to each other

- Tracks are wider
- Tracks are thicker
- Dielectric constant decreases
- There are two reference planes (one above, one below the tracks)

As explained earlier, the signal tracks should be physically similar and S should not be large relative to the other dimensions, e.g., $W < S < 3W$, $2S < H$. As the tracks are spaced further apart (for example, to increase differential impedance), and S begins to approach H, the tendency is for the tracks to become decoupled from each other. That is to say, they begin to behave as two separate tracks, each existing in its own environment, and they begin to lose their common mode properties.

To maintain a balanced differential pair, the tracks should have similar physical dimensions and follow the same route under the same conditions through a PCB. Most equations and commercially available impedance calculators assume well-balanced lines.

16.5.4.1   Edge Coupled Surface Microstrip (see Fig. 16.14).   Increasing S increases differential impedance. In this configuration, S should be less than $3 \times W_1$ for good coupling between the differential pair. It is not unusual to see one of these tracks "serpentined" so as to match track lengths, or routed away from its partner to avoid an obstacle, etc. This is bad practice that results in unplanned (and usually unknown) changes in differential impedance, unbalanced lines, and a loss of common mode signal rejection in the system. Recall that there are practical limits to S that help maintain close coupling and common mode properties. Beyond a certain point, increasing S does not increase differential impedance because the lines have achieved decoupling from each other, and are now coupled only to the reference plane. By becoming decoupled, the lines have lost their common mode properties, thus failing in a primary objective of differential systems.

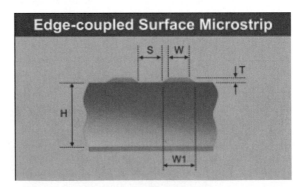

FIGURE 16.14   Edge coupled surface microstrip.

16.5.4.2   Edge Coupled Embedded Microstrip (see Fig. 16.15).   Similar to the surface model, the same considerations hold.

16.5.4.3   Edge Coupled Symmetrical Stripline (see Fig. 16.16).   In this model, the center of the tracks is halfway between two reference planes. For the same dimensions as the microstrip model in Fig. 16.15, the differential impedance will be lower due to coupling with the extra reference plane.

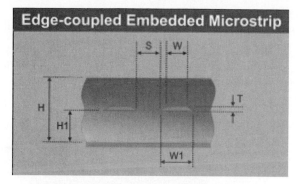

FIGURE 16.15   Edge coupled embedded microstrip.

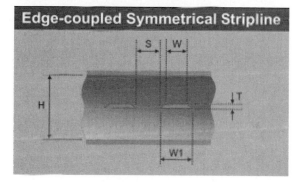

FIGURE 16.16   Edge coupled symetrical stipline

16.5.4.4   *Edge Coupled Offset Stripline (see Fig. 16.17).*   Tracks may be offset between the reference plane so as to permit another layer of signal tracks (not shown) orthogonal to these, as for the single-ended offset stripline structure earlier. The second layer would normally be spaced a distance H_1 from the upper reference plane, the minimal coupling between orthogonal layers would be disregarded. When considering the tracks in this structure, it is important to take account of the orientation of the trapezoidal cross section, as inverting it significantly affects the differential impedance. It should also be noted that making $H = 2 * H_1$ does not produce a symmetrical arrangement as the dimension T must also be taken into account. Thus for symmetry, $H = 2 * H_1 + T$.

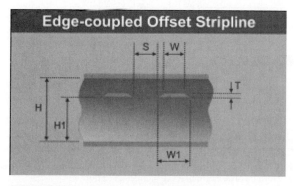

FIGURE 16.17   Edge coupled offset stipline.

16.5.4.5 Broadside Coupled Stripline (see Fig. 16.18). The broadside coupled pair must be not only closely matched in cross-sectional profile, but also symmetrically spaced between the reference planes in order to maintain close balance. This structure is normally fabricated on double-sided core material to assist in close registration of the broadside coupled differential pair, although in practice misregistration is far less significant than track profile deviation of the same order of magnitude. (See Fig. 16.19.)

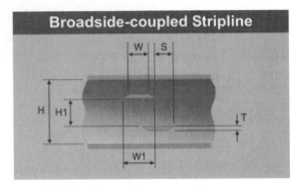

FIGURE 16.18 Broadside coupled stripline.

A track misregistration of 100 percent of track width changes the differential impedance by only 1 percent. A change of 3 percent in track width will produce the same error.

16.5.4.6 Differential Offset Coplanar Waveguide (see Fig. 16.20). Offset between two reference planes is a differential pair coplanar with a reference plane. An orthagonal track or tracks may exist at a distance H_1 from the upper reference plane.

Impedance change with misregistration of differential pair.

H	H1	W	W1	S	T	Er	Zo	(S/W)%	(ΔZo/Zo)%
31.4	10	4.85	5.35	0	0.7	4.3	100.00	0%	0.00%
31.4	10	4.85	5.35	0.5	0.7	4.3	100.04	10%	0.04%
31.4	10	4.85	5.35	1	0.7	4.3	100.10	21%	0.10%
31.4	10	4.85	5.35	1.5	0.7	4.3	100.18	31%	0.18%
31.4	10	4.85	5.35	2	0.7	4.3	100.28	41%	0.28%
31.4	10	4.85	5.35	2.5	0.7	4.3	100.40	52%	0.40%
31.4	10	4.85	5.35	3	0.7	4.3	100.54	62%	0.54%
31.4	10	4.85	5.35	3.5	0.7	4.3	100.70	72%	0.70%
31.4	10	4.85	5.35	4	0.7	4.3	100.87	82%	0.87%

Impedance change with deviation from nominal trace width.

H	H1	W	W1	S	T	Er	Zo	(ΔW/W)%	(ΔZo/Zo)%
31.4	10	5.25	5.75	0	0.7	4.3	96.66	8%	-3.34%
31.4	10	5.15	5.65	0	0.7	4.3	97.47	6%	-2.53%
31.4	10	5.05	5.55	0	0.7	4.3	98.30	4%	-1.70%
31.4	10	4.95	5.45	0	0.7	4.3	99.14	2%	-0.86%
31.4	10	4.85	5.35	0	0.7	4.3	100.00	0%	0.00%
31.4	10	4.75	5.25	0	0.7	4.3	100.87	-2%	0.87%
31.4	10	4.65	5.15	0	0.7	4.3	101.76	-4%	1.77%
31.4	10	4.55	5.05	0	0.7	4.3	102.67	-6%	2.68%
31.4	10	4.45	4.95	0	0.7	4.3	103.60	-8%	3.60%

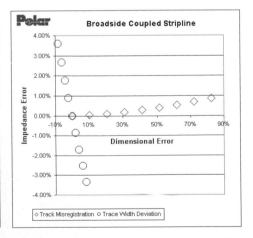

FIGURE 16.19 Comparison of the effect of dimensional error and trace misregistration on impedance error. This shows that track width has a bigger impact than misregistration.

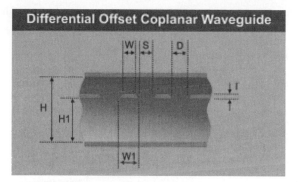

FIGURE 16.20   Differential offset coplanar waveguide.

16.6   CALCULATION OF PCB TRACK IMPEDANCE*

The use of high-speed circuits requires PCB tracks to be designed with controlled (characteristic, odd-mode, or differential) impedances. Wadell[1] is one of the most comprehensive sources of equations for evaluating these impedances. This source includes many configurations including striplines, surface microstrips, and their coplanar variants.

IPC publication IPC-2141[2] is another source of equations but has a smaller range of configurations, similar to those presented in IPC-D-317A. However, for some configurations there are differences between the equations given in these publications. The authors believe that it is now opportune to examine the origin of the equations and to update the method of calculation for use with modern personal computers.

16.6.1   Microstrip Example

As an example, consider the surface microstrip shown in Fig. 16.21. IPC-2141[2] gives the characteristic impedance as

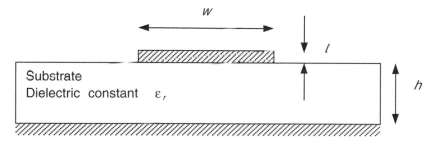

FIGURE 16.21   Surface microstrip structure.

* The material presented in Sec. 16.6 is adapted from a paper prepared for Polar Instruments Ltd., Guernsey, UK, by Andrew J. Burkhardt, Christopher S. Gregg, and J. Alan Staniforth. The references to the authors' opinions are theirs at the time of this writing.

$$Z_0 = \frac{87.0}{(\varepsilon_r + 1.41)^{1/2}} \ln\left[\frac{5.98\,h}{0.8w + t}\right] \tag{16.1}$$

Wadell[1] gives

$$Z_0 = \frac{\eta_0}{2.0\sqrt{2.0}\pi(\varepsilon_r + 1)^{1/2}} \ln\left[1.0 + \frac{4.0h}{w'}(A + B)^{1/2}\right] \tag{16.2}$$

where

$$A = \frac{14.0 + 8.0/\varepsilon_r}{11.0} \times \frac{4.0h}{w'} \tag{16.3a}$$

$$B = \left(A^2 + \frac{1.0 + 1.0/\varepsilon_r}{2.0} \times \pi^{1/2}\right)^{1/2} \tag{16.3b}$$

with

$$w' = w + \Delta w' \tag{16.3c}$$

The parameter w' is the equivalent width of a track of zero thickness due to a track of rectangular profile, width w, and thickness t. Wadell[1] gives an additional equation to determine the incremental value $\Delta w'$. The parameter η_0 in Eq. (16.2) is the impedance of free space (or vacuum), $376.7\,\Omega$ ($\approx 120\pi$). The quoted accuracy is 2 percent for any value of ε_r and w.

Table 16.4 shows the results of applying Eqs. (16.1) and (16.2) to a popular surface microstrip constructed from 1-oz copper track on $\frac{1}{32}$-in substrate.

TABLE 16.4 Comparison of the Results of Two Equations on the Accuracy of Characteristic Impedance Calculation

Width w (μm)	Numerical method, Z_0 (Ω)	Eq. (16.1) Z_0 (Ω)	Eq. (16.1) % error	Eq. (16.2) Z_0 (Ω)	Eq. (16.2) % error
3300	30.09	21.08	−29.94	29.89	−0.66
1500	50.63	49.46	−2.31	50.50	−0.26
450	89.63	91.79	+2.41	89.89	+0.29

$$t = 35\ \mu\text{m}, \qquad h = 794\ \mu\text{m}, \qquad \varepsilon_r = 4.2$$

The calculation of the error assumes the numerical method is accurate (see Sec. 16.6.4).

Table 16.4 shows that Eq. (16.2) is well within the quoted accuracy. The accuracy of Eq. (16.1) varies widely, but this equation has the advantage of simplicity and is useful in illustrating the general changes to the value of Z_0 as the width w and thickness t are varied. The example demonstrated by Table 16.4 highlights the general problem with published equations: complicated equations are usually more accurate. Ranges over which the equations are accurate are also usually restricted to a limited range of parameters (e.g., w/h, t/h and ε_r).

Equation (16.2) is complicated but with patience can be evaluated using a programmable calculator or computer spreadsheet. However, the complications increase greatly when two coupled tracks are used to give a differential impedance. For coupled surface microstrip, Wadell[1] gives seven pages of equations to evaluate the impedance. It is now a major exercise to evaluate the impedance using a calculator or spreadsheet.

16.6.2 Algebraic Equations

16.6.2.1 *Single Track.*

For the stripline of Fig. 16.22 with a symmetrically centered track of zero thickness, Cohn[3] has shown that the exact value of the characteristic impedance is

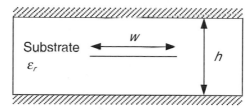

FIGURE 16.22   Stripline: centered track.

$$Z_0 = \frac{\eta_0}{4.0\sqrt{E_r}} \frac{K(k)}{K(k')} \tag{16.4}$$

where

$$k = \mathrm{sech}\left(\frac{\pi w}{2.0h}\right) \tag{16.5a}$$

and

$$k' = \tanh\left(\frac{\pi w}{2.0h}\right) \tag{16.5b}$$

K is the complete elliptic function of the first kind.[4] An equation for the evaluation of the ratio of the elliptic functions, accurate to 10^{-12}, has been given by Hilberg,[5] and also quoted by Wadell.[1]

When the thickness is not zero, corrections have to be made that are approximate.[1] These corrections are obtained from theoretical approximations or curve-fitting the results of numerical calculations based on the fundamental electromagnetic field equations.

When the track is offset from the center, the published equations become more complicated and the range of validity for a given accuracy is reduced. Attempts have also been made to include the effects of differential etching on the track, resulting in a track cross section which is trapezoidal.[1]

There is no closed-form equation like Eq. (16.4) for surface or embedded microstrip of any track thickness. Thus any equation used to calculate the impedance is approximate and is demonstrated in Table 16.4.

16.6.2.2 *Coupled Coplanar Tracks.*

Figure 16.23 shows two coupled coplanar centered stripline tracks. All the impedance equations for coupled configurations refer to both even-

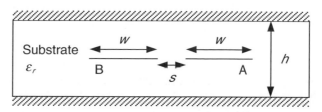

FIGURE 16.23   Stripline: coplanar coupled centered tracks.

mode impedance (Z_{0e}) and odd-mode impedance (Z_{0o}). These impedances are measured between the tracks and the ground plane. Z_{0e} occurs when tracks A and B are both at +V relative to the ground plane, and Z_{0o} occurs when track A is at +V and track B is at −V. When a differential signal is applied between A and B, then a voltage exists between the tracks similar to the odd-mode configuration. The impedance presented to this signal is then the differential impedance

$$Z_{\text{diff}} = 2 \times Z_{0o} \qquad (16.6)$$

All published equations[1] give Z_{0o}. The differential impedance must then be obtained using Eq. (16.6). For the zero thickness configuration of Fig. 6.23, Cohn[3] gives the exact expression.

$$Z_{0o} = \frac{\eta_0}{4.0\sqrt{\varepsilon_r}} \frac{K(k_0)}{K(k_0')} \qquad (16.7)$$

where

$$k_0 = (1 - k_0'^2)^{1/2} \qquad (16.8a)$$

and

$$k_0' = \tanh\left[\frac{\pi w}{2.0h}\right] \coth\left[\frac{\pi(w + s)}{2.0h}\right] \qquad (16.8b)$$

As before, K is the elliptic function of the first kind. There are no closed-form equations for coplanar coupled tracks.

16.6.2.3 *Effect of Track Thickness.*   When the track thickness is not zero, approximations must be made to obtain algebraic equations similar to Eqs. (16.4) and (16.7). Alternatively, equations based on curve-fitting of extensive numerical calculations are used. However, as the thickness increases, the impedance decreases, as can be noted from Eq. (16.1).

16.6.3 Numerical Principles

For pulses on a uniform transmission system,[1,6]

$$Z_0 \text{ (or } Z_{0o}) = \sqrt{\frac{L}{C}} \qquad (16.9)$$

where L is the inductance and C is the capacitance per unit length of line. For a stripline, where the electric (and magnetic) fields are in a uniform substrate, the dielectric constant ε_r (Eq. [16.9]) becomes

$$Z_0 = \frac{\sqrt{\varepsilon_r}}{cC} \qquad (16.10)$$

where c is the velocity of light in vacuum (or free space). The velocity of pulse travel along the transmission path is

$$v = \frac{c}{\sqrt{\varepsilon_r}} \qquad (16.11)$$

For a microstrip, the electric (and magnetic) fields are in air and the substrate. It can be shown that

$$Z_0 = \frac{1}{c\sqrt{CC_{\text{air}}}} \qquad (16.12)$$

where C_{air} is the capacitance of the same track configuration without substrate. The effective dielectric constant is

$$\varepsilon_{eff} = \frac{C}{C_{air}} \tag{16.13}$$

To find the impedance, the capacitance must be calculated. This can be done by applying a voltage V to the tracks and calculating the total charge per unit length Q, from which

$$C = \frac{Q}{V} \tag{16.14}$$

However, the surface charge on a track is not uniform. In fact it is very high at track corners. Therefore the total charge is difficult to calculate. From electrostatic theory, it is known that a charge produces a voltage at a distance r from the charge. Then a distribution of charge ρ (coulomb/unit width of track) gives a voltage

$$V = \int G \rho \delta l \tag{16.15}$$

where the integral is taken over the perimeter of the track cross section, δl is a small length, and G is the voltage due to a unit charge. It is also known as Green's function. The value of G depends on the configuration (or environment). For instance, a point charge in a two-dimensional dielectric space without conductors gives

$$V = -\frac{\rho \ln (r)}{2\pi\varepsilon_0\varepsilon_r} \tag{16.16a}$$

so that

$$G = -\frac{\ln (r)}{2\pi\varepsilon_0\varepsilon_r} \tag{16.16b}$$

In Eq. (16.15), the voltage V is known, and G is known for the particular configuration of tracks and substrate, but the charge ρ is unknown. Thus Eq. (16.15) is an integral equation that can be solved numerically by the method of moments (MoM).[7]

To proceed using MoM, the cross section perimeter of the track is divided into short lengths with a node at each end. Charges are assigned to each node. The voltage at each node is calculated from all the nodal charges and the estimated charge variation between nodes. This leads to a set of simultaneous equations represented by the matrix equation

$$\mathbf{A} \rho = \mathbf{V} \tag{16.17}$$

where ρ is a vector of nodal charges and $\mathbf{V}$ is a vector of nodal voltages. $\mathbf{A}$ is a square matrix whose elements are calculated from integrals involving Green's function. The size of the matrixes depends on the number of nodes. Equation (16.17) can be solved for the nodal charges ρ for given nodal voltages $\mathbf{V}$. The elements of $\mathbf{V}$ are usually +1 or –1 depending on the configuration. The total charge Q can be obtained by a suitable summation of the nodal charges.

This general approach has been used by most authors to evaluate the various impedances. Most of the calculations were published 15 to 20 years ago, when the principal calculator was a mainframe computer. Hence the need for equations that could be used with the pocket calculators available at that time.

The present authors have revisited the basic numerical approach and have developed software[8] that readily calculates the controlled impedances using a desktop PC. The software runs quickly on a modern PC and has been extended to include the calculation of configurations not well represented in the literature, including:

- Offset coupled stripline
- Broadside coupled stripline
- Embedded coupled microstrip

Thick tracks that have a trapezoidal cross section to allow for differential etching of the track are normally to be expected.

16.6.4 Numerical Results

This section describes in more detail some of the numerical techniques and compares the results with the exact Eqs. (16.4) and (16.7). In all cases Green's function for the configurations was obtained using charge images in the ground planes. There are an infinite number of these images. In the case of stripline, the sum of images converges with the result given by Sadiku.[9] Silvester[10,11] developed the image method for surface microstrips, which has now been extended by the authors for embedded microstrips. In all cases the sum of images converges, but the result has to be obtained numerically.

The distribution of charge over an element between nodes is assumed to be linear. A numerical singularity occurs when the charge node j coincides with the voltage node i. Sadiku[9] indicated how this can be resolved. The evaluation of the elements A_{ij} consists of both numerical and analytic integration in the same manner as that used in boundary element techniques.[12,13]

To avoid numerical inaccuracies at corners where there is a large concentration of charge, the length of an element at a corner is made very small. The other elements and nodes are then distributed by the method described by Kobayashi.[14] This means that wide strips require more nodes than narrow strips when the same small element is used. The results presented were performed on a PC with an Intel Pentium Pro running at 233 MHz using a compiled C program.

16.6.4.1 Single-Track Stripline. Figure 16.24 shows the variation of impedance with track width for the stripline of Fig. 16.22. Figure 16.25 shows the percent error of the numerical calculation compared with the exact values given by Eq. (16.4). Two curves are shown for different small elements at the corner (i.e., ends of the track). The figure shows that good accuracy can be obtained over nearly four decades of the width-height ratio. The computer processing time was less than 0.5 s for any of these values.

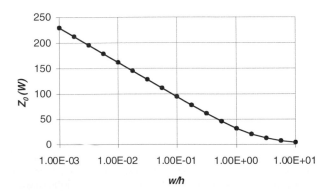

FIGURE 16.24 Impedance for different relative widths (substrate dielectric constant: 4.2).

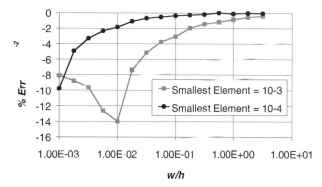

FIGURE 16.25 Error of numerical calculation compared with exact values given by Eq. (16.4) (substrate dielectric constant: 4.2).

16.6.4.2 *Coupled Coplanar Stripline.*

Figure 16.26 shows the variation of the odd-mode impedance for the stripline shown in Fig. 16.23. Figure 16.27 shows the percent error of the numerical calculation compared with the exact values given by Eq. (16.7) using 10^{-3} as the smallest element. The maximum processing time was less than 0.5 s. The maximum error can be reduced by decreasing the smallest element. For a maximum error of 6.0×10^{-2} percent, a processing time of 5.1 s is required.

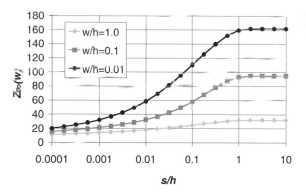

FIGURE 16.26 Odd-mode impedance for different separations and widths.

The results presented in Fig. 16.27 offer a very stringent test for the numerical method because of the sharp corners separated by seconds. In the odd-mode configuration this effect is enhanced even more because the tracks are of opposite polarity. This numerical validation is considered to be better than the results given by Bogatin et al.[15] for a pair of "round" tracks (i.e., a parallel wire transmission line) using finite-element software. In this latter case there are no singularities at the corners. Li and Fujii[16] state that the boundary element method (to which MoM is related) is more accurate for striplines and microstrips than the finite-element method.

16.6.4.3 *Surface Microstrip.*

As previously mentioned, there are no closed-form algebraic equations that are exact. But the discussion in the previous sections shows that the software

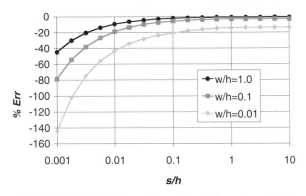

FIGURE 16.27 Variations of the odd-mode impedance for the stripline shown in Fig. 16.22.

can be made accurate, especially for practical purposes. Table 16.4 shows calculations for the configuration of Fig. 16.21. Because Green's function involves a summation, and two capacitances C and C_{air} are required, processing times are now longer than those for striplines. The longest time was less than 4.5 s for a width of 3300 μm. For coupled surface microstrips, two thick tracks of 3300 μm require a processing time of 5.1 s. The separation does not affect the time.

16.6.5 Practical Results

In order to verify the practical performance of the field-solving boundary element method, the authors commissioned production of a set of samples. During a 6-month period in 1998, over 1500 different printed circuit board tracks were manufactured. This sample consisted of both stripline and microstrip differential structures in surface and embedded configurations. Two types of coupled structures were included: edge coupled and broadside coupled. The track dimensions ranged from 75 to 1000 μm in width, with differential separations of one to four trackwidths using base copper weights of ½, 1, and 2 oz. The resulting differential impedances ranged from 80 to 200 Ω. Test samples were produced by three independent UK printed circuit board manufacturers* and the differential impedances were electrically measured by TDR at Polar Instruments using a CITS500s Controlled Impedance Test System. After electrical measurement, the samples were returned to the manufacturers for microsection analysis to determine the actual physical mechanical dimensions. The calculated impedance was predicted from the mechanical microsection data and a derived value of relative permitivity, ε_r, of the FR-4 material. Results[†] were analyzed, and comparisons of the electrically measured and the theoretically calculated results are presented in Figs. 16.28 and 16.29.

16.7 RESULTS DISCUSSION

Accuracy of the electrical measurements is estimated at 1 to 2 percent. This depends upon the impedance value and the quality of the interconnection between the test equipment and the

---

* The authors wish to acknowledge the assistance of Kemitron Technologies plc, Stevenage Circuits Ltd., and Zlin Electronics Ltd.
[†] Surface microstrip results were yet to be completed at the submission date for this chapter.

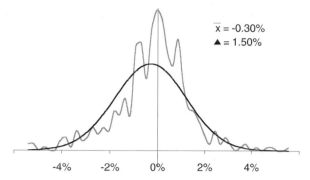

FIGURE 16.28   Distribution of differences between predicted and measured values for striplines.

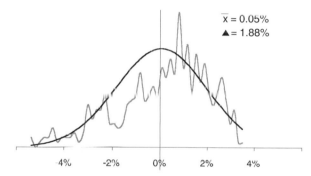

FIGURE 16.29   Distribution of differences between predicted and measured values for embedded microstrips.

test sample. Test samples were designed to be electrically balanced, but the manufacturing process will obviously not produce perfectly balanced tracks.

Microsection dimensions have an estimated accuracy of 1 percent; however, the model assumes symmetry and this will introduce a further small averaging error estimated at 1 percent. The total uncertainty in the experimental results is therefore estimated at 3 to 4 percent. Figures 16.28 and 16.29 show mean deviations of less than 0.5 percent with standard deviations of less than 2 percent.

These practical results clearly show that the differences between the measured electrical results and the numerically calculated results are well within the estimated uncertainty of the measurement method.

16.8   USE OF PERSONAL COMPUTERS FOR CALCULATIONS

This chapter shows that the early methods for calculating controlled impedance can now be used on desktop PCs. The accuracy is as good as, if not better than, that of the published algebraic equations. The processing times are less than 10 s, which is acceptable in most cases. Furthermore, the number of configurations can be extended and trade cross-sectional profiles can be readily incorporated.

REFERENCES

1. Brian C. Wadell, *Transmission Line Design Handbook,* Artech House, Norwood, MA, 1991.
2. IPC-2141, Controlled Impedance Circuit Boards and High-Speed Logic Design, April 1996.
3. Seymour B. Cohn, "Characteristic Impedance of the Shielded-Strip Transmission Line," *IRE Transactions MTT-2,* July 1954, pp. 52–57.
4. Milton Abramowitz and Irene A. Stegun, *Handbook of Mathematical Functions,* Dover, New York, 1965.
5. Wolfgang Hilberg, "From Approximations to Exact Relations for Characteristic Impedances," *IEEE Transactions MTT-17,* no. 5, May 1969, pp. 259–265.
6. Bryan Hart, *Digital Signal Transmission,* Chapman and Hall, 1988.
7. Roger F. Harrington, *Field Computation by Moment Methods,* MacMillan, New York, 1968.
8. CITS25, Differential Controlled Impedance Calculator, Polar Instruments Ltd., (www.polar.co.uk), 1998.
9. Matthew N. O. Sadiku, *Numerical Techniques in Electromagnetics,* CRC Press, Boca Raton, FL, 1992.
10. P. P. Silvester, "Microwave Properties of Microstrip Transmission Lines," *IEEE Proceedings,* vol. 115, no. 1, January 1969, pp. 43–48.
11. P. P. Silvester and R. L. Ferrari, *Finite Element for Electrical Engineers,* Cambridge University Press, Cambridge, UK, 1983.
12. C. A. Brebbia, *The Boundary Element Method for Engineers,* Pentech Press, 1980.
13. Federico Paris and Jose Canas, *Boundary Element Method: Fundamentals and Applications,* Oxford University Press, Oxford, UK, 1997.
14. Masanori Kobayashi, "Analysis of the Microstrip and the Electro-Optic Light Modulator," *IEEE Transactions MTT-26,* no. 2, February 1979, pp. 119–127.
15. Eric Bogatin, Mike Justice, Todd DeRego, and Steve Zimmer, "Field Solvers and PCB Stack-up Analysis: Comparing Measurements and Modelling," *IPC Printed Circuit Expo 1998,* paper 505–503.
16. Keren Li, and Yoichi Fujii, "Indirect Boundary Element Method Applied to Generalised Microstrip Analysis with Applications to Side-Proximity Effect in MMICs," *IEEE Transactions MTT-40,* no. 2, February 1992, pp. 237–244.

CHAPTER 17
MULTILAYER DESIGN ISSUES

Robert R. Holmes, Ph.D.
AT&T, Richmond, Virginia

17.1 RELIABILITY ISSUES

A multilayer board is a complex structure and it can have a wide variety of defects. Defects that actually prevent the board from working are called *functional defects*. Nonfunctional defects do not prevent the board from working, but they present a potential reliability threat. Functional defects, such as opens and shorts, are relatively easy to detect and are a clear cause for rejection. Some nonfunctional defects are very likely to cause future failures and must be avoided. For example, ionic contamination, although difficult to find, will almost certainly cause a future electrochemical activity and must never be tolerated. Other nonfunctional defects are cosmetic, affecting appearance but not reliability. Specifications vary widely on acceptance criteria for nonfunctional defects. Some specifications require only functionality. The IPC Class 1 specification for consumer products approaches this limit. Other specifications, such as the IPC Class 3 specification for high-reliability products, prohibit virtually all flaws.

Occasionally a fabricator will discover an entire production lot of MLBs with a nonfunctional defect, and the designer will be asked to accept boards on a one-time nonconforming basis. Often, there is little risk in agreeing to this request. If the nature of the flaw is understood and there is no serious reliability consideration, they can be accepted. However, caution is always required in giving approval, because even purely cosmetic flaws can be a sign of poor workmanship and may be a symptom of more serious problems. The goal of this chapter is to help the designer understand the nature of many common flaws so that an informed decision is possible.

Some defects are caused by poor design choices, while others are caused by poor processing. This section will review some common defects, describing each in enough detail to distinguish it from the others. The discussion will review potential reliability concerns, causes of the defect, and recommended corrective action. For the purposes of the discussion, defects are grouped into five broad categories:

1. *Substrate flaws:* Defects affecting MLB substrate integrity
2. *Surface damage:* Defects located on surface of copper features
3. *Mechanical problems:* Thickness and warp
4. *Internal misregistration:* Alignment of internal features to drilled holes
5. *Interconnection issues:* PTH and circuit flaws

17.1.1 Substrate Flaws

In a perfect laminate, the dielectric is uniform and free of flaws. In a real laminate, this is often not the case, and an entire lexicon has been developed to describe substrate flaws. Among the terms used are *blisters, delaminations, voids, silver streaks, tennis racket, weave exposure, weave texture, measles, foreign material,* and *pink ring.* Some specifications allow a minimum level of some of these flaws. Others allow none.

17.1.1.1 Blisters and Delaminations. Blisters and delaminations arise when there is an adhesive or cohesive failure in the dielectric. This failure can occur at a copper-to-dielectric interface, at a B- to C-stage interface, or within a dielectric layer. If the failure occurs during lamination, it is called a *blister,* otherwise it is a *delamination.* They appear as whitish areas often associated with a slight raised area on the board surface. In some cases, they are deep inside the board and can be detected only in microsection. Size varies from barely visible to several inches in diameter.

The internal void associated with a blister or delamination is very undesirable and is prohibited by most specifications. The void can grow with time. If it is cut by a drilled hole, it can trap process chemicals. If the board operates in a humid environment, it can become saturated with moisture due to diffusion. In any case, the internal surface of the void is an excellent site for electrochemical activity leading to internal shorts.

Blisters can be caused by poor design. For example, if a high-layer-count board has a region on several layers with no copper, the lamination pressure in this region may be inadequate to assure B-stage adhesion. The solution is to include dummy copper areas to equalize pressure. Another bad situation occurs when low-density signal layers have a solid copper border. The solution here is to replace the solid border with some type of stripe or dot pattern.

The most common cause of delamination is moisture. A fabricated board can absorb more than 1 percent water (by weight). At levels above 0.5 percent, boards can delaminate during soldering. The solution to moisture blisters is to store boards in low humidity or to bake before soldering. If neither solution is practical, there are materials such as the epoxy-PPO blends that have very low moisture absorption.

Another cause of blisters and delamination is poor processing. Out-of-tolerance B-stage or an out-of-control press can easily give poor adhesion. Trapped debris, such as pieces of release material, will cause delamination or blisters. Contaminated innerlayers will blister during lamination or delaminate during soldering. Failures of the B-stage to copper adhesion can be caused by a poor adhesion treatment. In the case of black oxide or silane treatments, an out-of-control bath may have been used. In the case of a reduced oxide treatment, there may be an excessive hold time between treatment and lamination (a 48-h maximum is recommended). In the case of double treated copper layers, the treated surface may be contaminated with photoresist debris.

17.1.1.2 Voids. The term *void* describes a variety of conditions. These include isolated voids that look like blisters, as well as clouds of small voids that give the laminate a hazy appearance. The primary distinction between a blister and an isolated void is the way they form. A blister is an adhesion failure, whereas a void is a failure of the resin to completely fill a space.

Like blisters, voids are a reliability risk due to the potential for moisture saturation and electrochemical activity. Unlike blisters, voids are not due to an adhesion failure, and they are unlikely to propagate. Under many specifications, voids are a cause for rejection, but if they are in an area with no circuitry or plated holes, they can be safely tolerated. Care must be exercised to assure that voids never span internal circuits.

Voids result from trapped air or volatiles and are found where the local lamination pressure is low. One such place is a panel edge. During B-stage flow, the panel perimeter is a boundary where the hydrostatic pressure in the liquid B-stage is zero. Near the perimeter, the pressure is too low to dissolve air and other volatiles, leading to voids. Generally the outer 0.5 in or so of any panel is subject to severe voiding. Insufficient B-stage flow arising from defective B-stage or an improper press cycle can greatly increase the size of the voided area. Using high lamination pressure or vacuum-assisted lamination reduces the width of the voided region, but does not eliminate edge voids altogether.

One cause of voids is a design feature requiring B-stage fill, coupled with B-stage that has poor flow characteristics. This condition is aggravated if the feature requiring fill is repeated on every layer. In this case, local pressure is low, reducing resin fill. A prevacuum may be of some help in preventing feature-related voids, but the real solution is to select a B-stage and press cycle combination that produces flow and fills all features before the cure is too far advanced.

17.1.1.3 Silver Streaks and Tennis Racket Appearance.
Occasionally an MLB shows streaks in the dielectric that follow individual glass bundles and are easily confused with voids. When they run preferentially in one direction, they are called *silver streaks.* Less frequently, they run in both directions, producing a tennis racket–like appearance.

SEM micrographs of silver streaks generally fail to find voids. The source of the streaks appears to be refraction or scattering of light at the interface between glass fibers and resin. The apparent cause of this optical discontinuity is poor wetting of the glass by the resin. Since there is no physical separation between the resin and the glass, silver streaks are not a serious reliability threat. Nevertheless, silver streaks and voids are easily confused, and, in serious cases, silver streaks are a cosmetic problem. Some specifications treat them like voids and prohibit them. Others allow a limited amount of silver streaking. One compromise is to allow them as long as they do not cross isolated conductors. In any case, silver streaks are evidence of marginal materials and should be avoided when possible.

Silver streaks often occur with one B-stage lot and then vanish with another. They affect glass running in one direction, but not both. Although most common near the panel edges, they often run many inches into the panel itself. Silver streaks can be reduced, but not eliminated, by adjusting the press cycle to increase resin flow. These observations suggest that the root cause of silver streaks is the glass yarn, probably an inconsistent application of the silane treatment used to enhance adhesion. Although the silver streak effect can be minimized by lamination process changes, the best solution is to work with the material vendor to obtain material not subject to silver streaks.

17.1.1.4 Weave Exposure and Weave Texture.
In a good laminate, glass is well encapsulated by resin and the surface has a uniform resin layer. In some laminates, the glass cloth is very close to the surface and an obvious weave texture is apparent. In the worst-case condition, glass yarn is actually exposed.

On innerlayers, exposed glass bundles are encapsulated during lamination. If the innerlayer process includes adequate cleaning, exposed glass is not a cause for concern. Process cleanliness should be verified by surface insulation resistance and extract conductivity tests. On finished MLBs, exposed glass provides a path for flux penetration and is a cause for rejection. Although weave texture does not expose glass and is generally accepted, it is a cosmetic problem that reduces assembly margin, and it is a sign of poor MLB lamination conditions. Corrective action should be taken when weave texture is observed.

Weave exposure and weave texture are signs that the surface dielectric has too little resin. This can occur if a very low resin content B-stage is used for the cap layer. A more likely cause is that the press cycle caused too much resin flow. In either case, the solution is to optimize the press cycle and materials to produce the correct surface conditions.

17.1.1.5 Measles.
Measles are tiny white spots that appear on or near the board surface. They correlate with the glass weave, occurring where horizontal and vertical yarns cross. They often appear after a mechanical or thermal shock. Measles generally do not propagate unless the board is subjected to additional thermal or mechanical abuse.

Books could be written about measles. There is great controversy about their cause and the risk they bring to an MLB. Although there is no real evidence for a reliability risk from measles, they are an undesirable cosmetic condition and most specifications limit them. One common limit is to prohibit overlap of adjacent measles. Another is to allow overlap so long as the measles do not bridge between two conductors. Some conservative specifications try to prohibit measles altogether.

Measles appear to be evidence of stress fractures at the resin–glass interface. Carefully prepared microsections often find evidence of this damage. Measles are most common with heavy glass yarns and with resin-poor surfaces. The best solution is to select a B-stage resin and press cycle combination that produces a resin-rich surface. If this approach is inadequate, then the outer-layer B-stage glass should be limited to the fine yarns such as type 106 or 1080.

17.1.1.6 Foreign Material.   There are many types of foreign material. The most common is organic. Laminates and B-stage are occasionally found with hairs or bugs trapped in the resin. MLB lamination can trap pieces of the plastic bag used to ship B-stage, or fragments from gloves and clean room garments. Sometimes, foreign material is visible through the board surface and can be detected by a careful visual inspection, but often it is hidden in the board and impossible to detect. The only real protection against foreign material is well-controlled clean rooms. Organic foreign material is a symptom of poor workmanship and a cause for rejection, but it is not always a serious reliability concern. Well-encapsulated organic material that does not lead to blisters or delamination is generally relatively harmless.

Another type of foreign material is metal slivers. Metal slivers are a much more serious problem than organic material because they often produce very small clearances, resulting in a serious reliability risk. The copper foil used for outerlayers is one source of metal slivers. Foil is cut to size and punched for tooling before it is used. If these mechanical operations are not carefully done, slivers are created. All copper foil used in MLB fabrication must be cleaned and inspected to ensure that it is completely free of copper slivers. Another source of slivers is the etched circuit lines themselves. When one innerlayer slides over another, tiny copper slivers are torn off the edges of the circuit lines. These slivers are a few tenths of a mil in diameter, and up to an inch long. The best way to avoid this source of slivers is to never stack etched layers without a protective separator.

17.1.1.7 Pink Ring.   Pink ring is a cosmetic defect in which a white or pink area is seen on an internal copper feature surrounding a plated hole. This appearance is caused by the presence of clean copper, free of the dark copper-oxide coating used as an adhesion promoter. Pink ring can extend 40 mil or more beyond the perimeter of the hole, but it never extends beyond the edges of the internal copper feature. In a worst case, pink rings from adjacent holes overlap. They are never seen on the vendor-treated side of a copper layer, and they can be hidden by large surface lands. To be detected in a finished board, an oxide-treated copper innerlayer must be visible. A microsection through a pink ring area often shows no evidence of a defect, but in some cases a small separation is visible between the copper layer and the adjacent B-stage.

There is considerable controversy on the reliability issues associated with pink ring, and most specifications tolerate limited amounts. Thermal shock studies show no evidence that the pink ring can propagate. The pink ring never spans isolated conductors, so there is no possibility of electrochemical activity. The crack is generally very narrow, and there is no evidence of significant chemical absorption. On the other hand, sometimes a large crack occurs and the resulting gap inhibits PTH plating, leading to severe neck-down in the PTH barrel. This creates a risk of a future PTH failure. Pink ring represents poor process conditions. There may be little risk in accepting a limited amount of pink ring, but the board fabricator should be held responsible for implementing root-cause solutions.

Pink ring occurs when a weak oxide bond is combined with aggressive drilling, resulting in a failure of the B-stage to copper bond during drilling. If the drill also heats the epoxy above T_g, it will pull epoxy upward during drill withdrawal, opening a small crack. During hole cleaning and electroless plating, acids dissolve the black oxide, leaving a pink ring.

One fix for pink ring is to bake the board after drilling. This allows the deformed epoxy around the hole to relax and seal the crack. Baking is not a root-cause solution for pink ring;

it is a repair operation that introduces a new process step. A better solution is to replace the black oxide surface treatment with a reduced oxide treatment. This increases adhesion and provides a surface resistant to acid attack, potentially hiding the problem. The best solution is to use less aggressive drilling conditions combined with an optimized reduced-oxide treatment. Less aggressive drilling includes fewer hits on a drill before it is changed and drilling fewer panels in a stack. Drilling is also improved if the designer omits nonfunctional lands. These lands reduce drill life and increase the heat that is generated by the drill.

17.1.2 Copper Surface Damage

A second class of MLB defects is dents and epoxy spots involving copper surfaces. For this discussion, dents include scratches and other forms of mechanical damage. Dents cause lifted photoresist, leading to shorts and clearance problems. Large dents in a conductor compromise conductor integrity. Dents in gold fingers are particularly serious because of the possibility of contact failures. Epoxy spots affect circuit yield by inhibiting plating and etching, leading to defective circuits. To avoid these problems, the MLB fabricator must strive to eliminate large dents and reduce smaller dents and epoxy spots to an absolute minimum.

17.1.2.1 Causes of Surface Damage. Whenever an MLB is handled, there is a risk of damage to the copper surface. The risk of damage during machine loading and unloading is generally recognized, and most fabricators carefully train operators on the need to handle panels with care. Well-designed automation will limit damage during handling steps.

Galling is a less well known source of surface defects. Whenever two clean metal surfaces come together, it is possible to form a small cold weld. Galling is the damage that occurs when the surfaces are separated, fracturing the cold weld. This is a well-known problem in relay contacts, but it also occurs between copper surfaces on MLB panels and between steel separators. In both cases, the solution is to keep the surfaces separated. The best way to do this is to use release sheets or slotted racks.

Poor cleanliness during lamination is another major source of surface defects. The separator sheets, which serve as a mold surface during lamination, must be clean. Foreign material trapped between the separator sheet and the MLB surface copper will cause dents. Epoxy dust will cause epoxy spots. Most fabricators make an effort to establish clean room conditions during stackup. However, their best efforts are frustrated by epoxy debris. Some solutions for epoxy debris are discussed in following sections.

17.1.2.2 Preventative Measures for Surface Damage. Epoxy dust that falls on internal layers is not a problem, because it melts into the B-stage during lamination, but dust falling on a surface copper sheet or on a separator leads to dents and epoxy spots. It is difficult to avoid this dust, because the stack-up operation uses dusty B-stage sheets. Among the techniques used to control epoxy dust are the following:

- Laminar flow air over the stack up area
- Ion sources in the incoming air stream to prevent static electricity
- Two operators at stack-up, one for B-stage and the other for Cu and separators
- Storage of B-stage on a low shelf below separators and copper
- Positive air flow over the B-stage storage shelf

Cured chunks of epoxy often find their way into the stack-up room, causing dents. One source is the resin that flows onto the separator sheet and into the tooling holes during lamination. If separator sheets are not completely cleaned after each use, cured epoxy will be carried into stack-up. One solution to this problem is to use a mold release on the separator. This inhibits epoxy sticking and eases cleaning. Lamination tooling pins and reusable compliant materials are another source of cured epoxy. Cured epoxy is shaved off dirty pins when separator sheets are added to the stack. The best practice is careful cleaning of anything that goes into the stack-up area and the use of mold release on pins and separators.

One way to minimize dents and epoxy spots is to preseal the copper foil to the separator sheet in a clean environment. A commercial product known as CAC™ does this. Figure 17.1 shows a cross section of a CAC sheet.

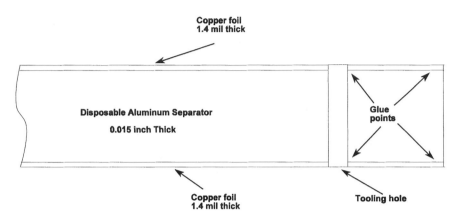

FIGURE 17.1   Diagram showing configuration of a CAC disposable separator.

The core of the CAC sandwich is a sheet of aluminum that serves as a disposable separator sheet. Copper foil is glued to each side of the aluminum, and tooling holes are punched through all three. As indicated, the glue joint connecting the copper foil to the aluminum is restricted to a small ring around each tooling hole and around the perimeter of the plate. The glue effectively seals out all foreign materials. In a standard stack-up, the operator completes an MLB by placing a sheet of copper foil, a steel separator, and another sheet of foil to start the next MLB. In a CAC stack-up, the separator and both sheets of foil are placed as a single part. This reduces stack-up labor and eliminates the possibility of anything getting trapped between foil and separator.

17.1.3   Mechanical Problems

A good MLB must be flat, satisfy a nominal thickness requirement, and be free of waviness and other types of thickness nonuniformity. These conditions are all difficult to achieve, and most specifications allow some deviation.

17.1.3.1   Warp.   The curvature of an MLB is called *warp*. This includes cylindrical curvature along one axis, as well as saddle shapes and spherical shapes in two axes. Warp is measured by supporting the board at three points and measuring the maximum deviation from the plane defined by these three points. The warp is the magnitude of this deviation divided by the greatest dimension of the board. For example, a board that measures 8 in by 16 in and has a 0.160-inch mil deviation from planarity is described as having a 1 percent warp (0.160 ÷ 16). Figure 17.2 shows an example of a board with a simple cylindrical warp and how warp is defined in this case.

Warped boards are a serious problem during assembly, and as device pitches decrease, the need for flat boards has grown. Older specifications allowed as much as 1.5 percent warp. Today most limit warp to 1.0 percent and some restrict it to as little as 0.5 percent.

Warp = D/L x 100%

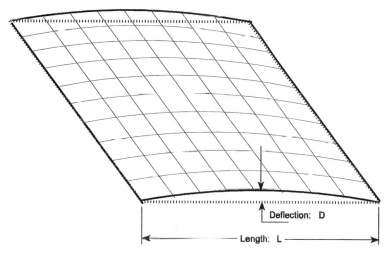

Deflection: D

Length: L

FIGURE 17.2   Example of stack-up asymmetry leading to warp.

The most common cause of warp is uneven cooling. This may occur during lamination, during board fabrication operations, or during assembly. The root cause is a large thermal gradient across the board as it cools through T_g. For example, when a board perimeter cools faster than its interior, the perimeter becomes rigid while the center is still rubbery. The perimeter then acts like a rigid band, producing out-of-plane distortions. This type of warp is called *extrinsic* to emphasize the fact that it is induced into an otherwise flat board. Extrinsic warp can be eliminated by heating the board above T_g and then cooling it slowly and uniformly.

Intrinsic warp is caused by a lack of symmetry in board design or materials. It does not disappear when the board is heated and cooled in a stress-free state. The only way to reduce intrinsic warp is to heat and cool the board under pressure. This "repair" is not advised, since the warp will generally return in subsequent thermal cycles.

For best results, MLB designs must be symmetrical with respect to a center plane. Asymmetric stack-ups often have intrinsic warp because differences in the coefficient of thermal expansion (CTE) cause internal stresses when the MLB is cooled from lamination. These stresses lead to warp. For best results, layer thicknesses and power planes placement must be symmetric. If this is not possible, a flat board can still be obtained if attention is given to the CTE values of the materials on the opposite side of the center plane.

Figure 17.3 shows two designs, each containing four innerlayers. The design on the left is asymmetrical. It has two 14-mil cores followed by two 5-mil cores. The arrangement of planes is also asymmetrical. The top four planes are S-P-S-P, while the bottom four are S-S-S-S. This MLB will show a large intrinsic warp. The design on the right stacks the same layers symmetrically. In this design the two 5-mil cores are on the outside and the two 14-mil cores are in the center. The planes are arranged symmetrically (SSPSSPSS). This structure will produce a flat MLB.

Material problems can cause random intrinsic warp. MLB materials are made by impregnating resin in a woven glass cloth. Warp can be caused by misalignment in the glass weave or when resin impregnation is heavier on one side of the cloth than the other. The weaving process is not symmetric and the two in-plane directions, called the *machine* and the *fill* directions, have different mechanical properties. If a sheet of glass is inadvertently rotated 90° it will cause severe warp. Some fabricators consciously cross-ply the B-stage material, but they maintain symmetry with respect to the center plane of the board.

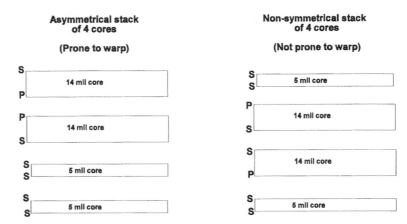

FIGURE 17.3   Diagram showing how warp is measured in an MLB.

17.1.3.2 Incorrect Thickness.   The nominal thickness of an MLB is an important parameter for many designs. An incorrect thickness is not a reliability problem, but it may affect system performance. The thickness of individual layers affects cross talk and impedance. Overall board thickness affects the proper function of connectors and latches. In some systems, excessively thick boards cause mechanical interferences. As a practical matter ±20 percent is a standard thickness tolerance. Tighter tolerances are possible, but they introduce design and manufacturing complexity.

The first step to thickness control is proper design. The B- and C-stage materials used to make an MLB come in discrete thicknesses. Often there is a conflict between a desired board thickness and the materials available. In this case, the design must be examined. Generally, a minor change in design parameters will allow the desired thickness to be achieved. As a rule of thumb, for a ±20 percent thickness tolerance, the design nominal must be within 5 percent of the target board thickness.

Material and process variations have a major impact on board thickness. During MLB lamination, resin flows out of the MLB. The final board thickness is determined by the initial resin amount, the glass thickness, and the resin flow. The first two parameters are controlled by the material vendor and are specified in terms of the treated weight of the B-stage. Resin flow is determined by lamination press parameters (temperature and pressure) and by B-stage rheology. The lamination process is generally well controlled and most flow variation comes from rheology variations.

B-stage rheology can be directly measured in a viscometer or indirectly determined by parameters such as gel time, resin flow, and scaled flow. Gel time and resin flow are not well correlated to press performance and are not recommended as material control parameters. The scaled flow test is a laboratory simulation of the lamination process. It uses an isothermal lab press operated at a pressure designed to give the same flow as the production press. Flow observed in this test is well correlated to the flow occurring during lamination. Many B-stage specifications include scale flow test limits.

17.1.3.3 Nonuniform Thickness.   Closely related to incorrect thickness is nonuniform thickness. This, too, causes mechanical problems with connectors, latches, card guides, etc. It is less of an issue for performance because the performance of a circuit is relatively insensitive to local thickness variations.

The root cause of nonuniform thickness is poor pressure distribution. MLBs are laminated in stacks of up to 12 panels. Separators are inserted between each board to isolate them from each other. However, even 0.062-in-thick steel is not completely effective, and a thinner separator, especially the aluminum used in CAC, is relatively ineffective. If a stack is all the same code, features can print through the entire stack. For the internal boards in a stack, this gives a slight dent on one side and a corresponding bump on the opposite. The resulting undulations of the surface are barely noticeable and there is little thickness variation. However, the top and bottom board in a stack are pressed against a rigid plate that prevents deflection. The compliant material used at the top and bottom of the lamination stack is designed to equalize pressure, but it is rarely completely effective and significant pressure variations occur over the area of a typical panel. Areas corresponding to a high copper density experience high pressure, resulting in thin spots in the panel. Areas corresponding to low copper density experience low pressure and are thick. This effect is particularly bad in areas adjacent to solid copper features such as panel borders or stiffeners. In this case, the stiffener region is thin and the adjacent board area is thick. This phenomenon is often described as a *wavy panel*.

Wavy panel problems can be reduced by using stiffer separators, thicker compliant materials, and lower lamination pressures. Unfortunately, these solutions increase costs and lower productivity. Another solution is to mix codes in a stack. This prevents the buildup of areas of high and low pressure. The problem with mixed codes is the difficulty in separating mixed lots. The best solution is to avoid sharp variations in copper density in the design. Noncircuit areas of a panel should be filled with a broken copper pattern such as dots or stripes. One approach uses a 50 percent density of 100-mil-wide dots of diamonds. Another approach uses a pattern of rays radiating from the panel center. The rays are about 1 in wide at the panel border and shrink to about 100 mil in the center. These rays fill any region not occupied by a board. In both cases, features on adjacent layers are staggered so that they do not fall on top of each other. Designs using these fillers have relatively uniform copper density and rarely give wavy panels.

17.1.4 Internal Misregistration

One of the greatest challenges in MLB manufacture is to obtain adequate innerlayer registration. All internal features must be registered accurately to each other, and they must all be accurately registered to the drilled holes. Hole-to-innerlayer misregistration creates two potential reliability problems: failure of the hole to line connection and shorts between holes and isolated conductors.

17.1.4.1 Breakout.   When a misaligned hole breaks out of its associated land, the area of connection becomes less than 360°. In an extreme case, where a misaligned hole cuts into a connecting circuit line, the connection area is reduced to the width of the line. When breakout occurs, poor drilling can cause land tearout. Many MLB users believe that these conditions represent a significant reliability hazard and prohibit hole breakout. Some even require a minimum annular ring as a margin of safety. Other users believe that tearout can be prevented by proper drilling, and allow breakout if the design includes *keyholes* to extend the land in the direction of connecting lines. Still others believe that a connection equal to the line width is adequate and allow unlimited breakout. The correct answer may depend on the PTH process used. In a well-controlled process with good drilling, hole cleaning, and plating, the metallurgical bond between an internal line and a barrel is as good as the line itself. There is no real reason to require a greater copper cross section at the barrel than elsewhere in the line. Indeed, technologies such as Multiwire™ that depend on a landless connection between a plated hole and an embedded wire are believed to be quite reliable. The only real reason for prohibiting breakout is to compensate for a poor PTH process. The real interconnection risk in a PTH comes from smear, and even a 360° connection is risky when covered with smear.

If breakout is allowed, a minimum clearance must be specified between a drilled hole wall and isolated internal features. A significant amount of substrate damage surrounds each drilled hole, and this region is affected by processing chemicals. If an isolated feature is too close to a hole wall there is a potential for shorts to develop when bias is applied. The minimum allowed clearance varies between specifications, but it generally falls in the range of 2 to 5 mil.

17.1.4.2 Region of Influence of a PTH.   If the process capability for the registration of a drilled hole to an internal feature is given by a value D, then for a drilled hole of diameter H there is an imaginary circle of diameter $H + 2D$ centered at the nominal center of the hole that defines all possible locations for hole wall. This is called the *region of influence* of the drilled hole. Figure 17.4 shows the relationship between the region of influence and the nominal location of the hole.

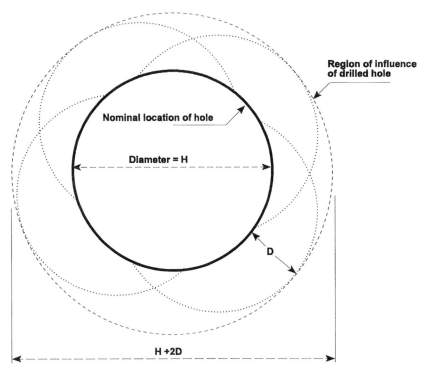

FIGURE 17.4   Region of influence of drilled hole with location uncertainty of D.

If breakout is prohibited, the size of all internal lands must be bigger than the region of influence of the hole. Figure 17.5 shows the relationship between minimum land size and the region of influence in the case where breakout is not allowed. As indicated by Fig. 17.5, the requirement of no breakout leads to rather large internal lands with a corresponding impact on internal routing density.

When breakout is allowed, the minimum land size can be reduced and the routing density increased. Unfortunately, the gain is relatively small because a minimum clearance is still needed between hole walls and isolated internal features. Figure 17.6 shows the situation for a hole of diameter H with a region of influence of diameter $H + 2D$ and a minimum hole to line spacing of C.

In this case, the designer must provide a spacing of at least $D + C$ between the nominal hole wall location and all isolated internal features. For example, a 13-mil hole may have a location

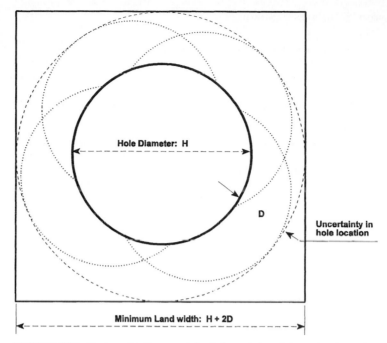

FIGURE 17.5 Region of influence of the hole and the minimum land size (no break out).

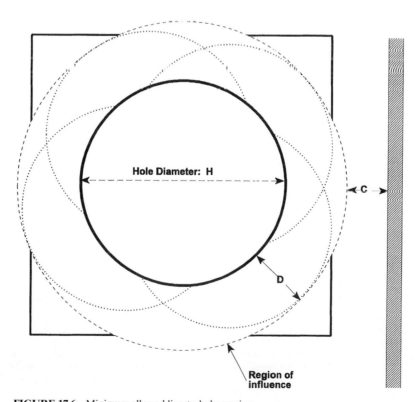

FIGURE 17.6 Minimum allowed line-to-hole spacing.

17.11

uncertainty of 7 mils and require a minimum 5-mil isolation. The result is that each 13-mil diameter hole actually excludes routing from a circle 37 mil in diameter ($13 + 2 \times 7 + 2 \times 5$). Replacing this hole with an 8-mil hole reduces the excluded region to 32 mil. The same gain can be achieved by reducing the location uncertainty from 7 to 4.5 mil. This emphasizes the importance of improving internal layer registration capability.

17.1.4.3   Mathematics of Registration.   The registration between drilled holes and the features on a layer can be described in terms of the superposition of two patterns. One pattern, the drilled holes, is considered to be accurate. The other pattern, representing the features on the layers, is treated as shifted and distorted.

Shifts can be represented as a combination of a rigid movement in *x*, a rigid movement in *y*, and a rotation. Figure 17.7 shows schematically how each of these affects the alignment between two patterns. The solid lines can be viewed as representing a hole pattern and the dotted lines a printed pattern. In the general case, all three components of rigid shift are combined.

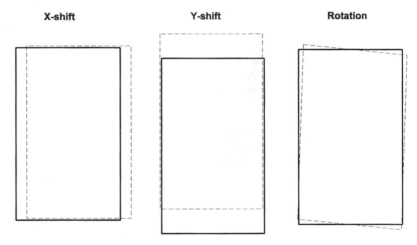

FIGURE 17.7   Example of rigid shift between two patterns.

Shifts can be caused by tooling errors at print, postetch punch (if used), stack-up, or drill. If only one layer is shifted, the problem is either print or artwork. If both sides of an innerlayer show the same shift, a problem probably occurred at postetch punch or at stack-up. If all layers are shifted with respect to the drill pattern, the most likely problem is a drill offset. Generally, careful diagnostics will determine the root cause.

Distortions observed in an MLB are generally linear. This means that size of the distortion is proportional to the *x* or *y* distance from the panel center. Any linear distortion can be represented as a combination of the following four deformations:

x growth	*x* deformation proportional to *x* distance from panel center
y growth	*y* deformation proportional to *y* distance from panel center
x shear	*x* deformation proportional to *y* distance from panel center
y shear	*y* deformation proportional to *x* distance from panel center

In reality, *x* shear, *y* shear, and rigid rotation are not independent, and it is often convenient to ignore rotation and capture it, as a combination of *x* shear and *y* shear.

Figure 17.8 shows schematically how each of the four linear distortions affects the alignment between two patterns. The solid lines can be viewed as representing a perfect pattern and the dotted lines a distorted pattern. In the general case, all four components of rigid shift are combined.

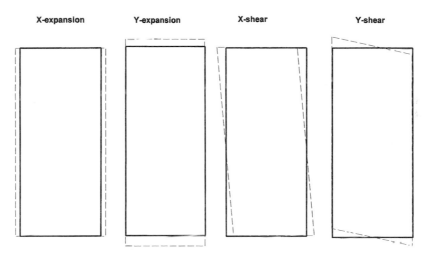

FIGURE 17.8   Example of the four linear pattern deformations.

One source of expansion is the laminate material. The innerlayer C-stage substrate shrinks when copper is etched. The etched panel shrinks further when its internal stresses are relaxed by heating above T_g. Another source of expansion is artwork which grows when heated or exposed to humidity. Careful environmental control is needed to avoid problems in this area. A third source of expansion is the lamination process. Although all three sources of expansion are important, the largest is the pattern shrinkage that occurs in lamination. Most fabricators compensate for this shrinkage by printing a slightly oversized image.

Shear deformation is generally small for symmetrical tooling. One source of shear deformation is a misalignment between the x and y axis on the drill machine. This will appear as a reproducible combination of x shear and y shear.

17.1.4.4 Lamination Effects on Registration.   With the exception of the lamination process, most of the contributors to misregistration can be minimized by paying attention to process details such as tooling accuracy, environmental control, and setups. Lamination effects are more complicated and account for most of the internal layer misregistration seen in a typical MLB process.

CTE Difference (C-Stage vs. B-Stage).   The CTE difference between B-stage and C-stage causes C-stage shrinkage during lamination. B-stage has more resin and a higher CTE than C-stage, so during the heat-up portion of the lamination cycle the B-stage expands more than the C-stage. After bonding, the two materials contract together with the result that the C-stage ends up in compression and the innerlayer pattern shrinks.

Resin Effects.   Resin cure and flow effects also contribute to misregistration. Epoxy shrinks slightly when it cures, causing some innerlayer shrinkage. The viscous drag associated with resin flow has a tendency to stretch the innerlayer, and if the flow is not perfectly centered, viscous drag will produce large lateral forces on an innerlayer, leading to innerlayer shifts.

CTE Difference (MLB vs. Tooling Plate). Tooling pins are used to ensure layer-to-layer alignment. In most processes, the pins are locked to steel caul plates or to steel separator plates. If the CTE of the steel does not exactly match the CTE of the MLB, the pins will induce stresses in the MLB. Some fabricators try to combat this problem by using an overdetermined tooling system that locks the MLB to the steel at many points. Other fabricators use a pin-in-slot approach that allows some relative movement between the steel and the MLB.

Lamination Tooling Effects. Some tooling systems use 20 or more pins. The hope is that by locking the MLB to steel, it will resist the lateral forces that tend to move the layers, thereby minimizing random shifts. It is also argued that the MLB will be constrained to expand and contract at the same rate as steel, giving a predictable lamination shrinkage. In reality, the forces associated with CTE mismatches are large, and even with 20 pins, innerlayer tooling holes tear slightly, resulting in unpredictable results. This approach also locks strains into the MLB that relax later, leading to unpredictable postlamination movement of tooling holes and internal features.

The four-slot system allows for thermal expansion mismatches during the heat-up portion of the lamination cycle. Unfortunately, resin flow fills the slots and prevents independent movement of the MLB during the cool-down cycle. One major weakness of the four-slot system is that it provides little resistance to lateral forces caused by uneven resin flow. Both overdetermined and four-slot systems have strong supporters. However, it is faster for an operator to stack on four pins, and the industry trend seems to be moving in that direction.

Insufficient Retained Copper. Another area of disagreement among fabricators is the benefit of retained copper. One school of thought says that copper will stabilize a layer and give a smaller, more predictable shrinkage factor. Adherents to this theory avoid signal–signal layers and use heavy copper borders around all board images. Other fabricators report little advantage from added copper and avoid solid copper borders because of the connection between substrate defects and abrupt changes in copper density. The best compromise is to replace solid copper borders with a dot or stripe pattern. This provides adequate stability for lamination without causing blisters or voids.

17.1.4.5 Registration Tests.
Most innerlayer features are hidden in a finished MLB, so special coupons are needed to measure internal registration. The linear model previously discussed predicts a maximum misregistration at one corner of a board, so a good test must examine all four corners. Some specifications allow sample testing on a few boards per lot, while others require registration testing on all boards. Sample testing identifies systematic errors such as incorrect compensation or tooling errors; but it misses board-specific problems such as layer shifts or shrinkage variations.

Microsection, x ray, and electrical test are all used to measure registration. The microsection method requires a coupon with a pair of holes and internal lands at every layer. The lands are sized so that the hole falls outside of the land when misregistration exceeds its allowed limit. To fully test a board, eight microsections are required, one in the *x* direction and one in the *y* direction for all four corners. If any one microsection shows breakout, the board is rejected. Figure 17.9 shows the appearance of a test hole with unacceptable breakout. This method has the advantage of providing specific feedback on the nature of the registration failure, but it is slow and costly.

X ray is a popular way to measure registration. Several varieties of x-ray coupon are used. One approach is to use a copper feature with a clearance hole on every layer. A test hole is drilled through the feature. If any layer is improperly registered, the test hole will hit the copper feature. A collimated x-ray beam can be used to determine if any layer contacts the hole. An alternative coupon has a solid copper land on each layer. The x ray is then used to look for evidence of breakout on any layer. The major drawback of x-ray methods is poor resolution. This is a particular problem for high-layer-count boards where it is very difficult to resolve a single layer.

An electrically readable registration test coupon is a fast and unambiguous way to find misregistration. Figure 17.10 shows one configuration for an electrically readable registration

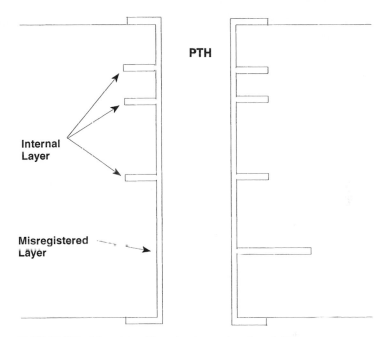

FIGURE 17.9   Misregistered layer in microsection of test hole.

Copper feature repeated on each layer

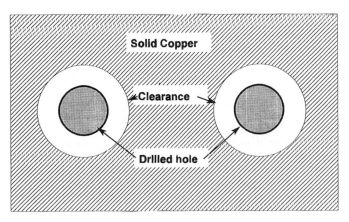

FIGURE 17.10   Electrically readable registration coupon (pattern printed on each innerlayer).

coupon. Each layer of this coupon has a copper feature with a clearance for two drilled holes. The size of the clearance is selected so that the nominal clearance between the edge of the feature and the hole wall is equal to the registration tolerance for that layer. On a properly registered layer, both holes pass through a clearance and are electrically isolated. However, if any layer is misregistered, the two holes short. This coupon is small and can be placed within a board image or on the MLB panel close to a board. The coupon is easy to test manually with an inexpensive continuity tester, or it can be tested automatically as part of final board electrical test.

17.1.5   Connection Flaws

Connection flaws are defined as flaws in the circuit or the PTHs. These include improper circuit width, nicks, reduced clearances, ionic contamination, and PTH flaws such as barrel cracks, epoxy smear, and innerlayer separation. None of these prevent the board from actually working, but if bad enough, all will cause future failure. Most specifications have limits on these flaws, and the limits are generally conservative. If specification limits are met, there is little reliability concern. The problem is that these defects are difficult to detect, and they often find their way into finished boards, leading to field failures. The answer is not to tighten specifications; rather it is to insist on root-cause solutions that will prevent the defects.

17.1.5.1   Improper Circuit Width.   Improper circuit width is caused by etching, plating, or artwork problems. On controlled-impedance boards, improper circuit width will cause impedance errors that compromise functionality. Wide circuits cause narrow clearances and the risk of electromigration failures. Narrow circuits limit current-carrying ability. If a narrow circuit is designed to carry a high current, perhaps for a short time during a fault condition, the circuit will function until the fault condition occurs. It then overheats, damaging the MLB.

Circuit width problems can be detected by impedance measurements and by dc resistance measurements. Most causes of improper width affect an entire lot, so a sample test on a few test circuits is a good screen for the lot. The root-cause solution is good statistical process control (SPC) during manufacture. Problems are generally caused by processes operating outside normal limits. A well-controlled MLB process will deliver 10 percent control on line width at a very high confidence level, and should be able to achieve 5 percent control with limited screening.

17.1.5.2   Nicks.   A nick in a circuit is defined as a local region where the circuit width is reduced. Nicks are caused by handling damage, artwork flaws, foreign material, and substrate dents. In practice, the length of a nick can vary from a few mil up to 100 mil, but is rarely any longer. It is not unusual to see nicks that reduce the circuit width to a few mil or less.

A nick is an alarming cosmetic feature, but unless it reduces circuit width to almost zero it is not a cause for failure. Since nicks are short they have no effect on high-frequency operation until the frequency approaches the 100-GHz range, which is well above the operating range of standard MLB materials. The high thermal conductivity of copper makes it a poor fuse, and even very deep nicks will carry high currents. The dc resistance added to a circuit by a nick is very small. As an example, a 1-oz copper circuit that is 10 mil wide by 3 in long has a dc resistance of about 300 mΩ. If that circuit has a nick that reduces its width to 1 mil over a distance of 20 mil (an extreme nick), the circuit resistance is increased by 20 mΩ (less than extending the circuit length by 0.2 in or reducing average width by 0.7 mil).

The major risk associated with nicks is the potential for circuit opens during mechanical handling or thermal cycling. This risk is minimized if the circuit is encapsulated by solder mask or, in the case of innerlayers, by cured epoxy. Nevertheless, nicks are a concern and most specifications limit them.

Automatic optical inspection (AOI), which is used on innerlayers, is a good screen for nicks, but it is rarely used on outerlayers. Most fabricators depend on human inspectors to find outerlayer nicks. Nicks occur at random, are very small, and are isolated. Human inspectors have difficulty finding them. The best defense against outerlayer nicks is a high-yield outerlayer process. Nicks and opens are caused by the same conditions, so any process with few opens also has few nicks. In fact, given their relatively low risk and the ineffectiveness of visual inspection, it may be cost effective to waive inspection for nicks in any process with a low frequency of opens.

17.1.5.3 *Reduced Clearances.*   Although clearance violations can be caused by wide circuits, they are more likely to be caused by unetched copper protrusion from a circuit line. Like nicks, protrusions are generally small and are caused by poor handling, damaged artwork, foreign material, and substrate dents. They can easily reduce clearances to a few mil or less.

The risk associated with narrow clearances is that electrochemical activity may produce dendrites bridging the gap. For clean boards operating at the voltages used in most electronic circuits, there appears to be little risk of this failure. The proof of this is that high-density boards perform reliably at a design *nominal* of 3 mil or less, and thin boards perform reliably with a single 2.8-mil-thick sheet of B-stage separating layers. In spite of this experience base, some specifications set a minimum line-to-line spacing of 5 mil or greater and require two sheets of B-stage for layer-to-layer spacings below 5 mil. This is very conservative.

For standard MLBs, the inspection situation for protrusions and clearance flaws is similar to that for nicks. AOI is an excellent screen for innerlayers, and the best defense against outerlayer defects is a high-yield process. A process with few shorts will also have few clearance faults, and it may be cost effective to waive inspection for clearances in a process with a low frequency of opens.

One situation in which a narrow space is very undesirable is in a circuit operating at high voltage. In circuits that must withstand high voltage for short periods, such as an occasional fault condition, dielectric breakdown across small clearances can lead to severe dielectric damage. Fortunately, the dielectric strength of epoxies and solder mask exceeds 1000 V/mil, and dielectric breakdown between completely encapsulated circuits is rarely an issue. Breakdown occurs at much lower voltages when a narrow clearance is combined with a dielectric void or a solder-mask skip. Voids and solder-mask skips are also a problem for circuits that must operate continuously at high voltage. In this case, the void is an excellent site for dendritic grown. In other words, *narrow clearances coupled with voids or solder-mask skips are a reliability problem for MLBs that must operate at high voltage.* The best approach for high-voltage applications is to test all high-voltage nets with an applied voltage of 750 to 1000 V. This will produce breakdown at potential failure sites.

17.1.5.4 *Ionic Contamination.*   Many MLB process steps include exposure to strong ionic chemicals. Any residue from these processes that contaminates innerlayers or finished boards is a serious reliability threat. Ambient moisture will produce ionization that will support leakage currents leading to dendritic growth. To prevent this failure, boards must be clean. The most effective way to ensure clean boards is good process controls coupled with regular process monitoring.

Two types of tests are used to monitor ionic contamination. One measures *insulation resistance* (IR) and the other uses extraction methods to measure ionic material. There are several variations of the IR test, all using a test coupon containing parallel lines. The coupon is exposed to a high humidity and the leakage current between the lines is measured. High leakage current is taken as a sign of contamination. The IR test is sensitive to small islands of contamination and it detects contaminants that have low solubility. On the other hand, it is slow to perform and requires a special coupon.

In the extraction method, the board is soaked in a mixture of alcohol and water to dissolve any ionic contaminates. The average density of ionic material on the board can be calculated from the change in conductivity of the bath and the area of the board. The extraction method is quick and can be performed on an actual board. However, it can miss insoluble salts, and it is not sensitive to small isolated spots of contamination.

For best results, both IR and extraction monitors should be used. A good strategy is to do extract conductivity tests once every shift at the following process points:

- Innerlayer prior to laminate
- MLB panel prior to solder mask
- Finished MLB

IR test coupons should be run through the entire process including soldering operations at least once a week.

17.1.5.5 PTH Flaws. The plated hole is the heart of an MLB. A modern MLB may contain more than 10,000 holes, and if one hole fails, the board fails. One type of PTH failure is the loss of barrel integrity. Incomplete plating can cause pinholes and ring voids. Thin or brittle plating leads to stress cracking during soldering or system operation. A second class of failure is related to connection integrity. Epoxy smear or poor copper adhesion can lead to marginal joints that separate in a thermal shock. All of these conditions raise serious reliability concerns.

With thousands of holes per board there is no way to inspect quality into plated holes. Even if a coupon is tested on every board, the sampling rate is low on a per-hole basis, and in a poor process there is no guarantee that boards with passing coupons are significantly better than boards with failing coupons. The only way to ensure reliable plated holes is to establish a robust process that produces no failures.

One way to maintain a robust process is to establish SPC limits for all critical variables at drill, electroless plate, and electrolytic plate. The process should never be operated if any variable falls outside limits.

Another safeguard is frequent PTH testing with immediate corrective action when any evidence of process drift is seen. As a minimum, metallurgical cross sections would be taken from the plated product hourly. If any evidence is seen of poor drilling, poor plating, voids, barrel cracks, epoxy smear, or interface cracks (postseparation), immediate corrective action must be taken. In addition, regular product samples should be tested for the same conditions. A good process should be able to pass these tests with few if any failures.

17.2 ELECTRICAL PERFORMANCE

MLB conductor layers can be described as either signal (S) or power/ground (P/G). As the names imply, S layers interconnect signals and P/G planes distribute power and ground. In addition, P/G planes provide impedance reference, shield cross talk, and control electromagnetic interference (EMI). These functions add design complexity and place demands on the dielectric properties of materials.

At low speed, the signal transit time is much less than its rise time, or in the case of analog signals, the transit time is much less than the period of the signal. In this limit, the signal line is an equipotential surface. The most significant signal loss is a small voltage drop due to the ohmic resistance of the trace. Other than isolating circuits, the properties of the insulator are unimportant.

At high speeds, the circuit can no longer be viewed as operating at one voltage, and the electric fields in the insulator become important. The conductor becomes a transmission line and the PCB a waveguide. Reflections occur at impedance discontinuities, and signal loss occurs through skin effect and dielectric loss. Signals couple through cross talk and radiate to the environment through EMI effects.

As a rule of thumb, low speed limits apply until the time of travel of a signal approaches half the signal rise time of a digital signal or half the period of an analog signal. When the travel time approaches these limits, voltage gradients along the signal path become important and the time for signal transmission adds significant delay time to the circuit.

The velocity of the signal is the speed of light divided by the square root of the dielectric constant. For epoxy MLBs, transmission speed is approximately 6 in/ns. This means that in a 6-in-long circuit, transmission line effects become important at a rise time of 2 ns, and for analog signals, at a frequency of 500 MHz. In low dielectric materials, transmission speeds increase to 8.5 in/ns and transmission line effects become important in a 6-in circuit at 1.4 n/s rise time and 708 MHz analog frequencies—not a big change.

17.2.1   Controlled Impedance (CI)

Characteristic impedance Z_0 is an important transmission line parameter. If the impedance of different parts of a circuit are mismatched, high-speed signals reflect at the discontinuity, much as light reflects at the surface of a pool of water. The result is a loss of signal strength and reflected pulses that cause timing problems and false triggers. For CI applications, the Z_0 of critical circuits must be within 10 percent of nominal for reliable performance.

The Z_0 of a circuit is equal to the square root of L/C, where L is the circuit self-inductance and C is the capacitance to ground. The parameters L and C are determined by the circuit geometry and the properties of the dielectric. Figure 17.11 shows three possible transmission line configurations. When the signal line is on the board surface and there is one buried P/G plane, the configuration is called *surface microstrip*. A *buried microstrip* configuration is identical to the surface microstrip except the signal line is covered by dielectric. In the *stripline* configuration, one or two signal layers are sandwiched between two P/G planes. For all three configurations, impedance is increased by increasing the dielectric spacing or decreasing line width. Analytical expressions for Z_0 have been published for these configurations.*

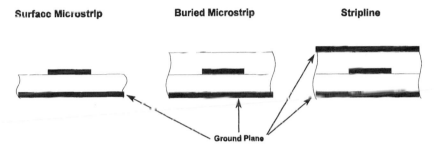

Surface Microstrip   **Buried Microstrip**   **Stripline**

Ground Plane

FIGURE 17.11   Comparisons of three transmission line configurations.

For standard epoxy, the dielectric thickness and the line width needed to achieve 50 Ω are approximately equal. For material systems with a lower dielectric constant, 50 Ω corresponds to line widths that slightly exceed the dielectric thickness. As a rule of thumb, the CI tolerance will be met if the percent tolerance on both line width and dielectric thickness are as good as the percent tolerance on CI. For example, in a design with 10-mil lines and a 10 percent CI tolerance, the required tolerances on both line width and dielectric thickness are approximately 1 mil. For 5-mil lines, the tolerance shrinks to 0.5 mil.

Tight tolerances present a severe challenge in manufacturing CI boards. Imaging processes must be carefully controlled to hold line width tolerances of 10 percent on fine lines. Some fabricators inspect innerlayers to ensure that line width tolerances are achieved. It is likely that successful manufacture of fine-line CI products will require new specifications on raw material to ensure tighter control of properties such as C-stage thickness, B-stage flow properties, and resin content. Dielectric constant variation is generally small, but care must be taken to account for changes in resin content that might be associated with material substitutions.

17.2.2   Signal Attenuation at High Frequency

At very high frequency, dielectric loss and skin effect become important. For analog circuits, these effects attenuate the signal strength. For a digital signal, they attenuate the high-frequency components of a pulse, resulting in an increase in the rise time.

---

* *Electronics Material Handbook,* vol. 1, Packaging ASM International, Materials Park, Ohio, 1989, pp. 601–603.

17.2.2.1 Dielectric Loss.   Dielectrics contain dipole molecules that are capable of absorbing energy from a high-frequency signal. This is how a microwave oven heats organic materials. In a transmission line, the amount of energy absorbed is proportional to the length of the line, the frequency of the signal, and a property of the dielectric called *tan δ*. For standard MLB materials, tan δ is 0.02. This gives serious signal losses at frequencies above 1 GHz. For circuits operating at 1 GHz or higher, a material like PTFE (tan δ = 0.001) is preferred.

17.2.2.2 Skin Effect.   The skin effect is the result of self-inductive effects that force high-frequency currents to the surface of a conductor. As the frequency of the signal increases, the thickness of the conducting "skin" decreases and the dc resistance increases. The skin effect resistance is proportional to the square root of the frequency and is independent of dielectric constant.* As an example, at a frequency of 0.1 GHz, the skin depth is 0.25 mil and the dc resistance of a 5-mil-wide, 1-mil-thick copper conductor doubles from 0.13 to 0.26 Ω/in. At a frequency of 10 GHz, the skin depth drops to 25 μin and the effective resistance increases to 2.6 Ω/in. This means that in the GHz frequency range, power losses due to the skin effect are important and wide traces are required.

17.2.3 Signal Coupling at High Frequencies

High-frequency electrical signals generate electric fields that radiate energy. When energy is coupled to a nearby circuit, it is called *cross talk*. When the energy is radiated to the environment, it is called *electromagnetic interference* (EMI). Both effects have a serous implication for high-speed MLBs.

17.2.3.1 Cross Talk.   When two electrical circuits are sufficiently close to each other, a signal in one induces a spurious signal in the other. This effect, called cross talk, is a serious problem for the high-speed, high-density circuits. Cross talk is measured by a coupling coefficient that gives the magnitude of the signal induced in the "quiet" or nondriven line as a function of the magnitude of the signal applied to the driven line. Coupling increases with signal speed and with proximity between nets.

Cross talk can occur between circuits on different layers (interlayer) or between circuits on the same layer (intralayer). Interlayer coupling is high for parallel, overlapping paths, but it is completely screened by intervening P/G planes Most designs are routed on layer pairs, using orthogonal routing rules, with circuits on one layer going in the *x* direction and circuits on the other going in the *y* direction. Layer pairs are isolated from other layer pairs by a P/G plane. The result is that there are no parallel circuits on adjacent layers and very little interlayer cross talk.

The quantitative details of the intralayer coupling coefficient are complex and are given elsewhere.† Qualitatively, intralayer cross talk is minimized by keeping a wide space between parallel lines and by avoiding long runs of parallel circuits. Unfortunately, these solutions reduce interconnection density, requiring a tradeoff between cross talk and interconnection density. Intralayer coupling is also reduced by nearby P/G layers. Coupling is not directly affected by the dielectric constant, but for any specified impedance, a lower dielectric constant allows a reduced line-to-layer spacing, giving lower cross talk for the same lines and spaces. This option is very important for high-speed designs and is one of the reasons for using low-dielectric-constant materials.

In summary, orthogonal routing on adjacent layer pairs prevents excessive interlayer cross talk. Intralayer cross talk is minimized by selecting the maximum line-to-line spacing consistent with routing needs, and by using low-dielectric-constant materials. Design audits must be used to minimize the length of parallel runs on the same layers.

---

* *Electronics Material Handbook*, vol. 1, Packaging ASM International, Materials Park, Ohio, 1989, p. 603.
† *Electronics Material Handbook*, vol. 1, Packaging ASM International, Materials Park, Ohio, 1989, pp. 35–41.

17.2.3.2 Electromagnetic Interference (EMI). A high-speed circuit radiates electromagnetic energy to the environment and can produce unacceptable interference in nearby electronics. Often, entire systems must be shielded to minimize EMI. Unfortunately, shielding is not 100 percent effective, and it is best to reduce EMI at its source. Outer ground planes effectively screen emissions coming from internal signal layers. One way to reduce EMI is to make the outermost innerlayer a P/G layer, and to restrict outerlayer routing to short fanout patterns.

EMI from RF circuits will also cause interference in circuits on the same board. One solution is to isolate RF circuits on a separate, well-shielded board. This works for some applications but is not always practical. Another solution is to isolate RF circuits in a well-shielded part of the board. If the board is large enough, this can be accomplished by dedicating one part of the board to RF and the rest to non-RF circuits. The RF portion is then totally enclosed by a metal shield. For applications like cellular phones, this approach makes the board too large. In this case, RF circuits can be restricted to one side of the board and non-RF to the other. This works well if the RF region is completely screened by a surface enclosure and a buried ground plane, and *no PTHs connect the two regions*. The restriction on PTHs leads to the use of blind holes for these applications.

CHAPTER 18
PLANNING FOR DESIGN, FABRICATION, AND ASSEMBLY

Happy T. Holden
Westwood Associates, Loveland, Colorado

18.1 INTRODUCTION

The purpose of this chapter is to provide information, concepts, and processes that lead to a thoughtfully and competitively designed printed circuit, ensuring that all pertinent design and layout variables have been considered.

Advances in interconnection technologies have occurred in response to the evolution of component packages, electronic technology, and increasingly complex functions. Therefore, it comes as no surprise that various forms of printed wiring remain the most popular and cost-effective method of interconnection. The response has been improvements in manufacturing, assembly, and testing technologies. These increased capabilities have made selection of technologies, design rules, and features so complex that a new function has developed to allow for the prediction and selection of design parameters and performance-vs.-manufacturing costs. This is the planning for design, fabrication, and assembly. This activity has also been called *design for manufacturing and assembly,* or, sometimes, *predictive engineering.* It is essentially the selection of design features and options that promote cost-competitive manufacturing, assembly, and test practices. Later in this chapter a process is offered to define producibility unique to each design or manufacturing process. Section 18.3 describes a process for defining producibility unique to each design.

18.1.1 Design Planning and Predicting Cost

The need for cost reduction in order to remain competitive is a principal responsibility of product planning. On the average, 75 percent of recurring manufacturing costs are determined by design drawing and specifications.[1] This was one of the conclusions found by an extensive study, conducted by General Electric, on how competitive products were developed. Manufacturing typically determines production setup, material management, and process management costs (Fig. 18.1), which are a minor part of the overall product cost.

Time to market along with competitive prices can determine a product's ultimate success. The first of a new class of electronic products in the market has the advantages. By planning the PWB layout and taking into consideration aspects and costs of PWB fabrication and assembly, the entire process of design and prototyping can be done with minimum redesign (or respins).

Costs Mainly Determined by Design Drawing and Specifications		Costs Mainly Determined by MANUFACTURING		
		Production Setup Cost	Material Management Cost	Process Management Cost
• Material cost • Purchased part cost • Process cost (Including assembly, adjustment and inspection	75%	13%	6%	6%
		(excluding advertising, sales, administration and design costs)		

FIGURE 18.1 Design determines the majority of the cost of a product.

18.1.2 Design Planning and Manufacturing Planning

Electronics is one of the largest enterprises globally. It is common for design to be done in one hemisphere and manufacturing in the other. It is also common for manufacturing to be done in a number of different places simultaneously. An integrated approach must be adopted when the intention is to rationalize fabrication and assembly as part of the entire production system and not as individual entities, as shown in Fig. 18.2. This dispersed manufacturing must be taken into consideration during the design planning and layout process. No finished product is ever better than the original design or the materials it is made from.

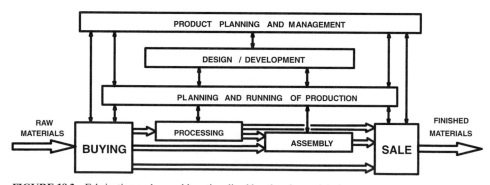

FIGURE 18.2 Fabrication and assembly rationalized by planning and design.

18.2 GENERAL CONSIDERATIONS

The central focus of the planning process is the trade-offs between the loss and gain in performance for layout, fabrication, assembly, and testing vs. the costs in these domains.

18.2.1 Planning Elements

Therefore, some major considerations are dealt with in the following sections:

1. New product design process (Secs. 18.3.1 and 18.3.2)
2. The role of metrics (Secs. 18.3.3 and 18.3.4)

3. Layout trade-off planning (Sec. 18.4)
4. PWB fabrication trade-off planning (Sec. 18.5)
5. Assembly trade-off planning (Sec. 18.6)
6. Tools for manufacturing audits (Sec. 18.7)

18.2.2 Planning Concepts

Planning for design, fabrication, and assembly (PDFA) is a methodology that addresses all those factors that can impact production and customer satisfaction. Early in the design process, the central idea of PDFA is to make design decisions to optimize particular domains, such as producibility, assemblability, and testability, as well as fit to a product family, etc., in manufacturing. Planning takes place continuously in the electronic design environment (Fig. 18.3). The data and specifications flow in one direction, from product concept to manufacturing. During the design process, 60 percent of the manufacturing costs are determined in the first stages of design when only 35 percent of the design engineering costs have been expended. The typical response is shown in Fig. 18.4.[1]

18.2.3 Producibility

Producibility is now regarded as an intrinsic characteristic of a modern design. As with the concept of quality in manufacturing, producibility must be built in, not inspected in. Producibility must be designed in; it cannot be a checkpoint in the design process or be inspected in by tooling.

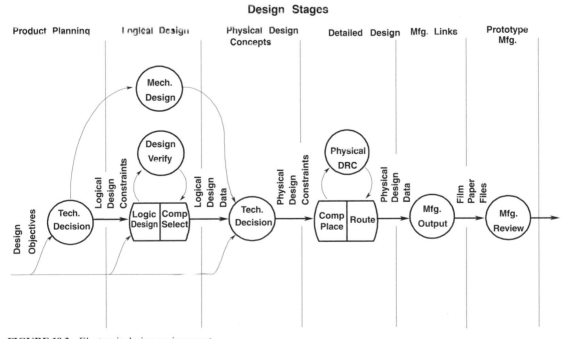

FIGURE 18.3   Electronic design environment.

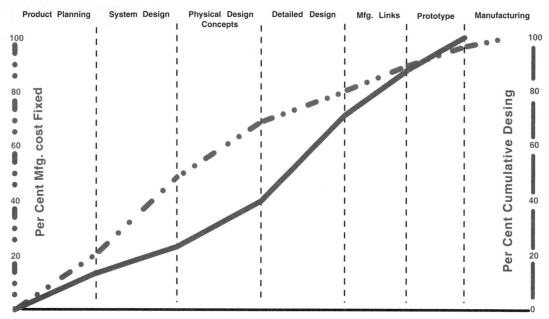

FIGURE 18.4   Design cost accumulation vs. intrinsic manufacturing costs.

18.3   NEW PRODUCT DESIGN

The keys to superior producibility in new product design can be found in the expanded design process. One of those keys is the role of metrics or data-based analysis of planning trade-offs.

18.3.1   Expanded Design Process

The new expanded design process that incorporates planning, trade-offs, and manufacturing audits is shown in Fig. 18.5. The process is made up of 12 separate functions (See Table 18.1) that incorporate the planning and trade-offs sections in this book.

This differs from the more conventional design process (as seen in Fig. 18.3) by the inclusion of four important functions:

1. The formal technology trade-off analysis during specifications
2. Detailed trade-off selection of features for layout, fabrication, and assembly
3. Design advice during component placement and routing
4. Manufacturing audits to review the finished layout for producibility, time to market, and competitiveness

18.3.2   Product Definition

The initial new product design stage is specification and product definition. This key step takes ideas, user requirements, opportunities, and technologies and formulates the executable

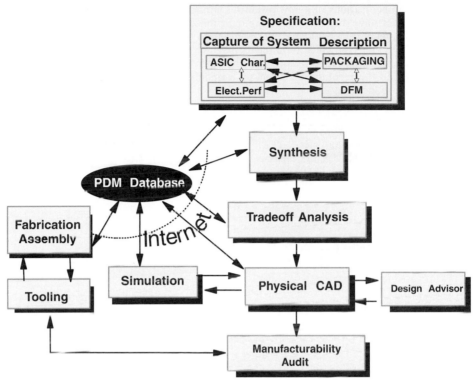

FIGURE 18.5   The expanded electronics design process.

TABLE 18.1   Separate Functions that Incorporate the Planning and Trade-Offs Sections in This Book

Specification	User-supplied constraints and ideas formulate executable specifications.
Capture of system description	Technology trade-off analysis: balance of loss and gain in various domains' performance vs. cost.
Synthesis	Generation of a netlist from the executable specifications.
Trade-off	Selection of layout, fabrication, and assembly features vs. cost.
Physical CAD	Conversion of netlists to system and module layouts.
Simulation	Detailed analysis of design structures to support all other design (CAD) activities.
Design advisor	Continuous display of design rated by performance rules.
Manufacturability audit	Check of design to manufacturing design rules and capabilities.
Tooling	Conversion of module layouts to panel layout.
Manufacturing	Conversion of module layouts to physical products (fabrication or assembly).
PDM database	Enterprise-wide database containing all product information (product data management [PDM]) including design files, libraries, manufacturing information and revisions, etc.
Internet	Multiteam designs via access over the Internet.

specifications of a new product. During this operation, the ability of technologists who may not have any manufacturing experience to predict what will happen in manufacturing can affect both time to market and ultimate product costs. Figure 18.6 shows the technology trade-off analysis, which requires the balance of loss and gain in various domains performance vs. costs. Size and partitioning for ICs and ASICs must be balanced with overall packaging costs and the resultant electrical performance. All of these factors affect the manufacturing and product cost.

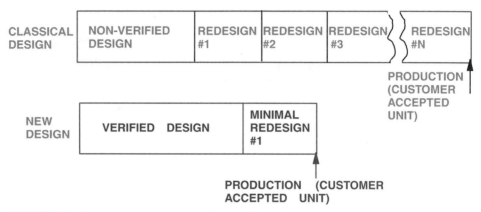

FIGURE 18.6   Design incorporating trade-offs vs. traditional design.

Another definition of this process could be called a "verified design."[2] A verified design is defined as one that was predicted from models or measures that have been correlated to past designs. This is compared to the traditional approach, which is a "nonverified design," or trial and error. This is diagrammed in Fig. 18.7. The advantage of the "verified design" can be significant reduction in redesigns in order to achieve the original product objectives.

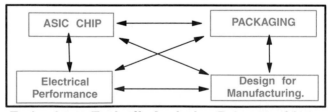

FIGURE 18.7   Specifications determine product partitioning and producibility.

18.3.3   Metrics for Predicting and Planning Producibility

Metrics are data and statistically backed measures, such as wiring demand W_d (Sec. 18.4.2). These measures can be density, connectivity, or, in this context, producibility. These measures are the basis for predicting and planning. When used in the design process, there are three categories of measures applied to a product. Only the metrics can be shared by all in the design team. The nonmetrics provide little assistance in design.

Metrics

- *Metrics:* Both the product and the process are measured by physical data using statistical process control (SPC) and total quality management (TQM) techniques (predictive engineering process).
- *Figure of merit:* Both the product and the process are scored by linear equations developed by consensus expert opinion (expert opinion process).

Nonmetrics

- *Opinion:* Opinion, albeit from an expert, is applied after or concurrent with design (manufacturing engineering inspection process).
- *No opinions:* No attempt to inspect or improve the design is done during the specification, partitioning, or design stage (over-the-wall process).

Metrics also establish a common language that links manufacturing to design. The producibility scores form a nonopinionated basis that allows a team approach that results in a quality, cost-competitive product (Fig. 18.8).

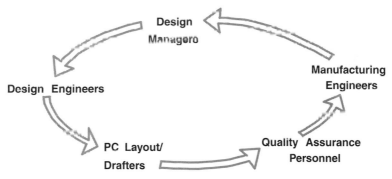

PDFA metrics are a means to an end:

- The means: To discuss alternatives, changes and improvements
- To reach the end: A quality, cost-competitive product

FIGURE 18.8   The benefits of metrics as a common design language.

The strategy in applying these measures is shown in Fig. 18.9. The analysis process is unique to every individual and company, but certain conditions have to be met and considered if the product is going to be successful. If the score meets producibility requirements, then select this approach; if not, then evaluate other opportunities and repeat the process. In the rest of this chapter, measures and metrics are introduced that provide insight for layout, fabrication, and assembly planning.

18.3.4   Nonmetrics

It is always preferable to have metrics when discussing producibility. However, if metrics are not available, then the opinions of experts are better than nothing (no opinions). The problem

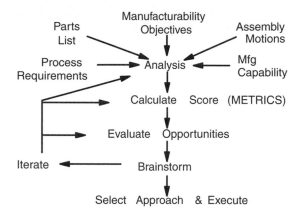

FIGURE 18.9 The process of using measures and metrics to obtain a producible product.

with opinions is that they are difficult to defend and explain and that, when used in conjunction with producibility, many times they vary with each person. Sometimes, the opinion process is implemented with good intentions by taking experienced production experts and having them review a new design. This is the expert opinion process, and, although it is sometimes successful, it is difficult to replicate and many times results in building barriers between manufacturing and design. That is why the figure of merit process is so popular. For a small amount of work by experts, it produces a scoring procedure that can be used and understood by all.

18.3.5 Figure of Merit (FOM) Metric

Metrics are the preferred measures for design planning, but their availability for predicting producibility is often limited. Metrics also can take many months to develop and the amount of experimentation may make them costly. The figure of merit measure is much more cost effective and quicker to develop. The figure of merit is the result of one or two days' work by a group of design and manufacturing experts. The process is an eight-step procedure, as shown in the flowchart seen in Fig. 18.10.

1. Define or identify new measure to be developed.
2. Determine why the measure was selected—ensure relevance of the measure to communication.
3. Survey customers—identify customer's Measure to be communicated.
4. Identify needs/expectations and collect data.
5. Brainstorm contributing factors and variables.
6. Determine the major contributors and normalize scores, using multivote, paired ranking, or ranking voting-pareto; verify data if available. These are the coefficients C_x of the equation.
7. Construct figure of merit factor weightings FW_x—fill in FOM table values for 1, 25, 50, 75, and 100.
8. Construct linear equation model (coefficient score times FOM weightings).

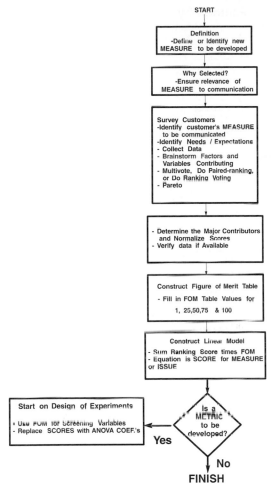

FIGURE 18.10 Process to develop a figure of merit as a
substitute for metrics.

18.3.6 Figure of Merit Linear Equation

This FOM procedure uses classical TQM techniques to brainstorm, rank, and formulate an
equation that will score producibility, assemblability, or any other measure that can be used in
design planning. The two factors used in the producibility score are (1) the coefficient C_x and
(2) the factor weighting FW_x.

18.3.6.1 Coefficient C_x.   The coefficients in the producibility score are the result of brainstorming all the possible contributors to producibility that can affect the product, as seen in
Fig. 18.11(a). These are grouped into common ideas or factors by techniques like clouds of
affinity or Kay-Jay techniques, as seen in Fig. 18.11(b). These factors are ranked by voting or
other pareto techniques such as paired ranking as seen in Fig. 18.11(c). Whatever the method
of voting, the values are normalized by dividing by the smallest nonzero value. The resulting

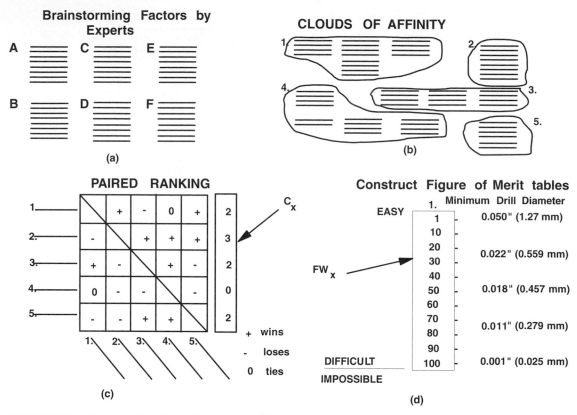

FIGURE 18.11 Elements of the figure of merit process. (*a*) Brainstorming factors; (*b*) grouping similar ideas; (*c*) paired ranking of factors; (*d*) assigning values to factors.

voting scores form the coefficients C_x. Those factors with no votes are zeros and drop out of consideration.

18.3.6.2 Factor Weightings FW$_x$. Each factor that emerges from the ranking process is calibrated by assigning values from 1 to 100, as seen in Fig. 18.11(*d*). A value of 1 stands for "easy to manufacture" and 100 for "impossible today, but merely very difficult in a few years."

The resulting scoring equation looks like the following linear equation and is used like this:

$$\text{Score} = (C_1)\,(FW_1) + (C_2)\,(FW_2) + (C_3)\,(FW_3) + (C_n)\,(FW_n) + \qquad (18.1)$$

where C_x = coefficients based on ranking
FW_x = factor weightings of assigned values (1 to 100)

For example, we assume the producibility of a bare PWB may be scored with the preceding equation if the following factors are established by the FOM process:

1. Size of the substrate ($C_1 = 1.5$)

2. Number of drilled holes ($C_2 = 3.0$)

3. Minimum trace width ($C_3 = 4.0$)

Where the proposed PWB design has:

1. Size of the substrate ($FW_1 = 36$)
2. Number of drilled holes ($FW_2 = 18$)
3. Minimum trace width ($FW_3 = 31$)

The producibility Score equals:

$$232 = 54 + 54 + 124 = 1.5 \times 36 + 3.0 \times 18 + 4.0 \times 31$$

The Score can be used if it is calibrated based on prior history of this kind of product. Prior products are scored with the FOM linear equation (Eq. 18.1). Those products that went into production smoothly and presented minimum problems determine the minimum Score that should be sought after. The Scores from products that were a problem, had to be reengineered, or presented delays in introduction determine the Score that the design should exceed if problems are to be avoided and producibility is to be obtained.

18.4 LAYOUT TRADE-OFF PLANNING

Predicting density and selecting design rules is one of the primary planning activities for layout. The actual layout of a PWB is covered in Chaps. 13 and 14. The selection of design rules not only affects circuit routing but profoundly affects fabrication, assembly, and testing.

18.4.1 Balancing the Density Equation

What with the need for more parts on an assembly, or the trend to make things smaller (portable) or for faster speeds, the design process is a challenging one. The process is one of balancing the density equation with considerations for certain boundary conditions such as electrical and thermal performance. Unfortunately, many designers do not realize that there is a mathematical process to determine the routing rules of a printed circuit. Here is a brief explanation. The density equation, as seen in Fig. 18.12, has two parts: the left side, which is the component wiring demand, and the right side, which is the substrate wiring capability.

$$\text{component PWB wiring demand} < \text{PWB design rules and construction wiring capabilities} \qquad (18.2)$$

where PWB wiring demand = total connection length required to connect all the parts in a circuit
PWB wiring capability = substrate wiring length available to connect all the components

Four conditions can exist between wiring demand and substrate capability.

1. *Wiring demand > substrate capability:* If the substrate capacity is not equal to the demand, the design can never be finished. There is not enough room for either traces or vias. To correct this, either the substrate has to be bigger or components have to be removed.
2. *Wiring demand = substrate capability:* While this is the optimum condition, there is no room for variability and to complete the design will take an unacceptable amount of time.
3. *Wiring demand < substrate capability:* This is the condition to aim for. There should be enough extra capacity to complete the design on time and with only a minimum of overspecification and costs.

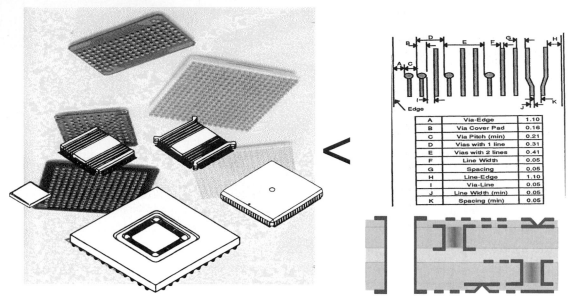

A	Via-Edge	1.10
B	Via Cover Pad	0.16
C	Via Pitch (min)	0.21
D	Vias with 1 line	0.31
E	Vias with 2 lines	0.41
F	Line Width	0.05
G	Spacing	0.05
H	Line-Edge	1.10
I	Via-Line	0.05
J	Line Width (min)	0.05
K	Spacing (min)	0.05

FIGURE 18.12   Balancing the density equation to achieve an optimal layout.

4. *Wiring demand* $\ll$ *substrate capability:* This is the condition that usually prevails. With PC layout, the schedule is tight and timing is all-important. Many choose tighter traces or extra layers to help shorten the layout time. The impact of this is to increase the manufacturing costs 15 to 50 percent higher than is necessary. This is sometimes called the sandbag approach. It is unfortunate, since the models above would help to create a more planned environment.

18.4.2 Wiring Demand W_d

Wiring demand is the total connection length (in inches or centimeters) required to connect all the parts in a circuit. When the design specifies an assembly size (in square inches), then the wiring density in in/in^2 or cm/cm^2 is created. Models used early in the design planning process can estimate the wiring demand. Three cases can control the maximum wiring demand:

1. The wiring required to break out from a component like a flip-chip or chip-scale package

2. The wiring created by two or more components tightly linked (e.g., a CPU and cache or a DSP and its I/O control)

3. The wiring demanded by all integrated circuits and discretes collectively

There are models available to calculate the component wiring demand for all three cases (see Sec. 18.4.5). Since it is not always easy to know which case controls a particular design, it is usually best to calculate all three cases to see which one is the most demanding and thus controls the layout.

Wiring demand is defined as:

$$W_d = W_c \times \varepsilon \text{ (in cm/cm}^2 \text{ or in/in}^2)\tag{18.3}[3]$$

where W_d = wiring demand
W_c = wiring capacity
ε = PWB layout efficiency (determined in Sec. 18.4.4)

18.4.3 Wiring Capacity W_c

Substrate wiring capacity is the wiring length available to connect all the components. It is determined by two factors:

1. *Design rules:* the traces, spaces and via lands, keepouts, etc that make up the surface of the substrate and its innerlayers.
2. *Structure:* The number of signal layers and the combination of through and buried vias that permit interconnection between layers and the complex blind, stacked, and variable-depth vias available in HDI technologies.

These two factors determine the maximum wiring available on the substrate. The maximum wiring times the layout efficiency is what is available to meet the wiring demand. The data are straightforward except for layout efficiency. Layout efficiency expresses what percentage of wiring capacity can be used in the design. The equation for wiring capacity for each signal layer follows. The total substrate capacity is the sum of all the signal layers, defined as:

$$W_c - T \times L/G \text{ (in cm/cm}^2 \text{ or in/in}^2) \qquad (18.4)^3$$

where T = number of traces per wiring channel or distance between two via pads
L = number of signal layers
G = wiring channel width or length between centers of via pads in T above

18.4.4 Layout Efficiency

Layout efficiency is the percentage of capacity from design rules and structure that a designer can deliver on the board.[3] Layout efficiency is the ratio of the actual wiring density it takes to wire up a schematic vs. the maximum wiring density or W_d divided by W_c. Typically, for ease of calculations, layout efficiencies are assumed to be 50 percent. Table 18.2 provides a more detailed selection of efficiencies.

TABLE 18.2 Typical Layout Efficiencies

Design scenario	Conditions	Efficiency*
Through-hole, rigid	Gridded CAD	15–25%
Surface mount/mixed	With/without back-side passives, gridless CAD	30–45%
Surface mount/mixed	With back-side actives, gridded CAD	35–55%
Surface mount only	With/without back-side passives, gridless CAD	up to 60%
Surface mount/mixed	1-sided blind vias, gridless CAD	up to 65%
Surface mount/mixed	2-sided blind vias, gridless CAD	up to 70%
Built-up technologies[3]	2-sided micro-blind vias, gridless CAD	up to 80%

  * Determined from analysis of printed circuit designs. Actual wiring capacity from CAD system divided by maximum wiring capacity, Eqs. (18.3) and (18.4).

18.4.5 Selecting Design Rules

To calculate a potential set of design rules and signal layers, first the wiring demand W_d should be calculated. Wiring models help accomplish this.

18.4.5.1 Wiring Demand Models.   Seven wiring models are reported in the literature, but only three are commonly used. The three wiring models are:

1. Coors, Anderson, and Seward statistical wiring length[4]
2. Toshiba technology map[5]
3. HP design density index[6]

The other four wiring models are:

1. Equivalent ICs per square inch[7]
2. Rent's rule[8]
3. Section crossing[9]
4. Geometric analysis[10]

Coors, Anderson, and Seward Statistical Wiring Length.[4]   This wiring demand model is based on a stochastic model of wiring involving all terminals; the probable wire length is calculated based on the distance of a second terminal and the spatial geometry of all other terminals. This is the most recently determined wiring model and represents the most practical approximation of surface mounting technology. Equation (18.5) presents the mathematical model that results.

$$d = D * N_i/A \text{ (in/in}^2) \tag{18.5}$$

where D = average interconnection distance (in)
$\quad\quad N_i$ = total number of interconnections
$\quad\quad A$ = routing area (in^2)
$\quad\quad D = E(x) * G$
$\quad E(x)$ = expectation of occurrence
$\quad\quad G$ = pad placement grid (in)

$$E(x) = \frac{1}{a} \left| \frac{((S-T)(S\,a-2))\,e{\wedge}aS + S(2-(S-T)a)\,e{\wedge}a(S\_T) - 2T}{(S-T)\,e{\wedge}aS - S\,e{\wedge}a(S-T) + T} \right| \tag{18.6}$$

where $S = M + N$
$\quad T = (M{\wedge}2 + N{\wedge}2){\wedge}0.5$
$\quad a = \ln \alpha$
$\quad \alpha$ = empirically derived constant = 0.94
$\quad M$ = board width of grid point = (width/G) + 1
$\quad N$ = board length of grid point = (length/G) + 1
$\quad N_i = 2 * N_t/3$

Toshiba's Technology Map.[5]   The packaging technology map[5] is a simple technique to predict a PWB, chip-on-board, or MCM-L wiring demand and its assembly complexity. By plotting components per square inch (or components per square centimeters) against average leads per component on a log-log graph (Fig. 18.13) the wiring demand W_d in in/in^2 (or cm/cm^2) and assembly complexity in leads per square inch (or leads per square centimeter) can be calculated.

The equations for these two metrics are as follows.

$$\text{Wiring demand } W_d = 3.5 \times (\text{comp})^{0.5} \times (\text{leads}) \tag{18.7}$$

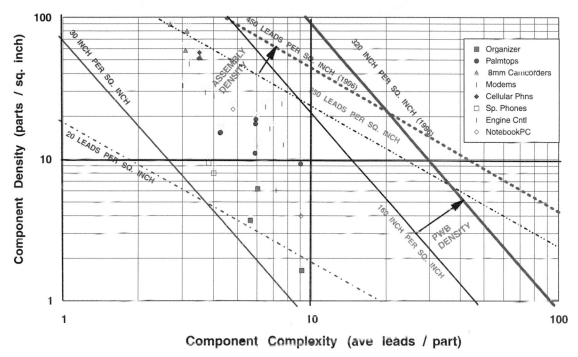

FIGURE 18.13 Packaging technology map. *(Source:* CircuiTree Magazine, *September 1994.)*

where comp = components per board area in cm^2 (in^2)
       leads = average leads per component

$$\text{assembly complexity} = (\text{comp}) \times (\text{leads}) \tag{18.8}$$

Using these two equations, Fig. 18.14 shows lines of constant wiring demand (in cm/cm^2 or in/in^2) and assembly complexity (in leads per square cm or square inch) that can be plotted on this chart (Fig. 18.13).

H-P's Design Density Index.[6]   Another metric is called the design density index (DDI). It is a collelation of the actual design rules for a PWB compared to the DDI. Equation (18.9) gives DDI, and a typical calibration chart looks like Fig. 18.15.

$$\text{DDI} = 13.6 \times (\text{EIC/board area}) ^\wedge 1.53 \tag{18.9}$$

where equivalent integrated circuits (EIC) = total component leads/16
       Board area = top surface area of printed circuit board (in^2)

This chart gives a good visual record of how efficient a company has been in PWB layout. As various PC boards are charted, their DDIs form a distribution. This distribution is a form of layout efficiency (ε) because more EICs are connected at the bottom of the distribution than at the top of the distribution.

Other Layout Metrics

EQUIVALENT IC DENSITY[7].   Equivalent integrated circuits (EICs) per unit area has been a traditional measure of density since the introduction of CAD systems in the early 1970s. A simple measure of the number of electrical connections required per unit area of the board, it

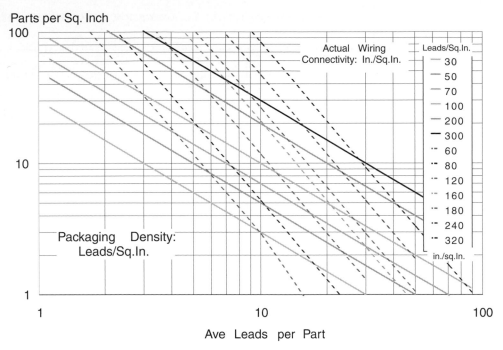

FIGURE 18.14 Wiring and assembly density.

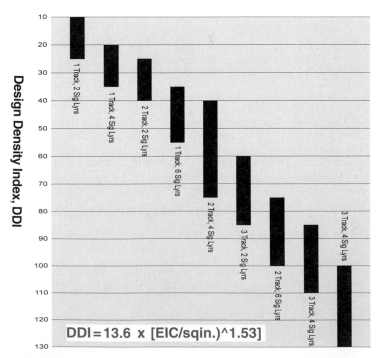

$$DDI = 13.6 \times [EIC/sqin.]^{1.53}$$

FIGURE 18.15 Design rules and layer count for various design density indexes.

remains in use with surface mount and is usually referred to as EIC density. An EIC is the total number of leads of the components divided by 14 or 16, the old number of pins on a DIP. Many people also use 20 as a divisor. Equation (18.10) defines EIC density mathematically in ICs per square centimeter (EICs per square inch).

$$\text{EIC density} = \text{connections}/16/\text{board area} \tag{18.10}$$

where   connections = total component leads
        Board area = top surface board area (in cm^2 or in^2)

RENT'S RULE[8].   Rent's rule is based on an estimation of the average interconnection length for a given design. The total wiring length required to connect a set of chips is predicted according to Eq. (18.11).

$$d = (f/f + 1)\, R(m)\, N_{io} F_b \text{ (in/in}^2) \tag{18.11}$$

where $R(m)$ = average interconnection distance (in)
        N_{io} = total number of interconnections
        A = routing area (in^2)
        N_c = expectation of occurrence
        f = average fan-out of ICs

$$R(m) = \frac{2}{9}\left[\frac{N_c{}^\wedge\beta - 0.5 - 1}{4^\wedge\beta - 0.5 - 1} - \frac{1 - N_c{}^\wedge\beta - 1.5}{1 - 4^\wedge\beta - 1.5}\right]\frac{1 - 4^\wedge\beta - 1}{1 - N_c{}^\wedge\beta - 1}$$

where $F_b = A/R(m)\,1.4$
        $N_{io} = N_t + N_i$
        β = empirically derived constant = 0.6
        N_c = number of ICs
        N_t = number of terminal leads
        N_i = number of terminal off board

SECTION CROSSING[9].   Developed by Sutherland and Oestreicher,[9] this technique estimates the number of connections that must pass through various board cross sections. The number of wires that cross the section yields the probability of crossing multiplied by the total number of nets in the system:

$$F_i I_c L = N_{io}/n\,(1 - P_1 n - P_2 n) \tag{18.12}$$

where $P_1 = P_2 = 1/2$
        F_i = actual wiring efficiency
        I_c = interconnect capacity
        L = board length
        N_{io} = total number of I/Os to be routed together
        n = average number of pins per net (i.e. the fan-out + 1)
        P_1 = the probability that the wire starts on side 1 and connects only to pins on side 1
        P_2 = the probability that the wire starts on side 2 and connects only to pins on side 2

GEOMETRIC ANALYSIS[10].   A geometrically based wiring estimation first developed by Moresco.[10] It is based on an extended neighbor counting method similar to that of Seraphim.[3] In Moresco's approach the wiring requirement is computed by assuming that a fraction of the nets (A) are nearest-neighbor routed and the rest of the nets ($1 - A$) are globally routed.

$$\text{Total wiring length} = A\,\frac{N_{\text{chip}} N_{ioc} F_p}{2} + (1 - A)(N_{\text{chip}} - 1)N_{ioc}\,F_p + \frac{N_{ioe} F_p - N_{\text{chip}}}{v} \tag{18.13}$$

where $A = 1$ is maximum parallelism, $A = 0$ is the minimum
N_{ioc} = number of signal and control I/Os per chip
N_{ioe} = number of signals and I/Os leaving the module
$v = 2$ for edge connector and 4 for an area array connector on the bottom of the module
A = fraction of nets that are nearest-neighbor routed $(0 < A < 1)$
N_{chip} = number of chips in the module
F_p = wiring-limited chip footprint dimension

18.4.6   Typical Example of Wiring Demand Calculation

As an example, take a typical consumer electronic board with these characteristics:

- *Design:* consumer PCB—all through-hole components
- *Number of components:* 86
- *Number of leads:* 1540
- *Size:* 19.6×19.5 in = 57 in^2

$$\text{EIC/in}^2 = 110/57 \text{ in}^2 = 1.93, \text{ using Eq. (18.10)}$$

$$\text{DDI} = 13.6 \times (1.93)^{1.53}$$

$$= 319.7\text{—Fig. 18.15 advises two tracks on two signal layers, using Eq. (18.9)}$$

18.5   PWB FABRICATION TRADE-OFF PLANNING

The metrics for PWB and COB fabrication deal with trade-offs between the performance objectives and the PWB price. Calculating prices requires the PCB characteristics and manufacturing yield. Manufacturing yield requires producibility estimates. Three items are required to predict PCB prices:

1. Fabrication complexity matrix
2. Prediction of producibility and first-pass yield
3. Relative price as a function of a price index

18.5.1   Fabrication Complexity Matrix

The fabrication complexity matrix is supplied by a PWB fabricator. It relates the various design choices on a PWB to design points. These points are based on the actual prices a fabricator will charge for these features. They are calculated by dividing up the actual costs and dividing the values by the smallest nonzero amount. Typical factors fabricators can use to price a printed wiring board are:

- Size of the board and number that fit on a panel
- Number of layers
- Construction material
- Trace and space widths
- Total number of holes
- Smallest hole diameter

TABLE 18.3 Example of a Fabrication Complexity Matrix

Factors	Points	Highest	Points	High middle	Points	Low middle	Points	Lowest
Number of layers	12	8	8	6	4	4	1	2
Trace width (mils)	8	4	5	5–6	3	7–8	1	10
Number of holes	10	5000–8000	5	3000–5000	3	1000–3000	1	>1000
Minimum hole diameter (mils)	8	8	5	12	1	16	0	<20

- Solder mask and component legends
- Final metallization or finish
- Gold-plated edge connectors
- Factors specific to design, etc.

A typical fabrication complexity matrix would look like Table 18.3. This is not a complete matrix, but does show the design factors and the design points assigned to each.

18.5.2 Predicting Producibility

The simple truth about printed circuit boards, multichip modules, and hybrid circuits is that the design factors like those listed previously can have a cumulative effect on manufacturing yield. These factors all affect producibility. Specifications can be selected that individually may not adversely affect yields but cumulatively can significantly reduce yields. A simple algorithm is available[11] that collects these factors into a single metric, in this case called the complexity index (CI). It is given in Eq. (18.14).

$$\text{complexity index} = (A)(H)(T)(L)(T_o) \tag{18.14}$$

where constants A = area of board

H = number of holes in this board
L = number of layers on the board
T = minimum trace width on this board
T_o = minimum tolerance, absolute number, for this board

and where area, number of holes, minimum trace width, number of layers, and minimum tolerance (absolute number) are the factors of the board being designed.

18.5.2.1 First-Pass Yield.
The first-pass yield equation is derived from the Wiebel probability failure equations.[11] This equation is a more general form of the equation typically used to predict ASIC yields by defect density and is as follows:

$$\text{FPY}\% = \frac{100}{\exp[(\log CI/A)^\wedge B]} \tag{18.15}$$

where FPY = first-pass yield
CI = complexity index

and A and B are constants.

To determine the constants A and B in Eq. (18.15), a fabricator will need to characterize the manufacturing process. This is done by selecting a number of printed circuits currently being produced that have various complexity indexes—hopefully some low, some medium, and some high. The first-pass yields (at electrical test without repair) of these printed circuits for several production runs are recorded. Any statistical software[12] program that has a model-based regression analysis can now determine A and B from the model, Eq. (18.16):

$$FPY = f[x] = 100 \div EXP\{LOG(complexity \_ PARM[1])^\wedge PARM[2]\} \qquad (18.16)$$

where   PARM[1] = A
        PARM[2] = B

The first-pass yield will follow the examples in Fig. 18.16. Constant A determines the slope of the inflection of the yield curve and constant B determines the x-axis point of the inflection.

Alternatively, any spreadsheet can be used to determine constants A and B. The [REGR] function in a spreadsheet like Excel™ or Lotus 1-2-3™ is used. The [REGR] function is defined as: (=LINEST(known_y's,known_x's,TRUE,TRUE). To use this function, first put the FPY function into the form $y = Ax + B$. This is done by creating two columns: (1) the complexity index, which we will call X1, and (2) yield. A third column is created for {log[log(X1)]}. A fourth column is created for {log[ln(−yield/100)]}. Provide the regression function with column 4 as 'known_ys' and column 3 as 'known_xs'. The regression function will return 10 values; FIT (slope and intercept), sig-M (slope and intercept), r^2, sig-B(slope and intercept), F,df (slope and intercept), and reg sum sq (slope and intercept). The constant B is equal to the FIT (slope) and the constant A is $10^\wedge[-FIT(int.)/FIT(slope)]$.

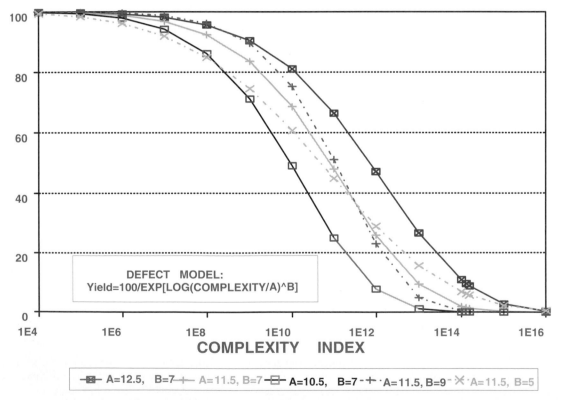

FIGURE 18.16   Estimated first-pass yield as a function of PWB design complexity.

18.5.3 Example of a Complete PWB Complexity Matrix

Here is an example of how one company approached this planning process as part of its PWB design for manufacturing program.[13,14]

18.5.3.1 PWB Fabrication Complexity Matrix (FCM). The PWB fabrication complexity matrix that this company developed is shown in Table 18.4. This FCM is based on a per panel basis of 18 × 24 in and does not use 'per board'. Also, volume is assumed to be in a preset amount.

18.5.3.2 PWB Complexity. Figure 18.16 shows this company's first-pass yield. The curve with $A = 11.5$ and $B = 9.0$ was current for 6 mo. Price index is the total points from the complexity matrix divided by the first-pass yield.

18.5.3.3 Relative Costs. The price index data for this company are shown in Fig. 18.17. The price index (PI) can vary from 150, which corresponds to a 70 percent price reduction, to 1000, which corresponds to a 275 percent increase in price.

18.5.3.4 PWB Fabrication Example. Continuing with the consumer electronics board example from Sec. 18.4.5, Tables 18.5, 18.6, and 18.7 follow an example from one company of the initial design characteristics of the printed circuit. The resultant total design points in Tables 18.5 and 18.6 show the calculation of the complexity index, estimated first-pass yield, and resultant price index and price adjustment. Table 18.7 shows the final board

The wiring demand indicates that an 0.007-in trace and an 0.008-in spacing (two tracks) for an 0.100-in grid (channel) is more density than is required (see Sec. 18.4.5). One-track wiring on two signal layers could achieve the required wiring density, or an 0.012-in trace with an 0.013-in spacing. The size requires a panel layout as shown in Fig. 18.18(a) of six per panel. A reduction of 14 in² in the board size would allow eight boards on an 18 × 24-in panel, as seen in Fig. 18.18(b).

A simple gate array ASIC was employed to reduce the small-scale ICs until 14 in² of space was freed up. The final optimized PCB had the fabrication factors seen in Table 18.6. The boards were eight up on a panel (Fig. 18.18(b) and had a higher producibility due to an 18 percent reduction in complexity and an overall price reduction of 24.7 percent.

TABLE 18.4 Example of One Company's Fabrication Complexity Matrix

Fabrication factors	Points	Highest	Points	High middle	Points	Low middle	Points	Lowest
Construction material	147	Polyimide	88	Cyanate ester	49	FR-4	40	CEM III
Number of layers	196	8 layers	137	6 layers	89	4 layers	36	2-sided
Number of holes/panel	270	<20,001	180	10,001–20,000	90	3,001–10,000	27	>3000
Minimum trace/spacing	25	>4 mils	10	4–5 mils	6	6–8 mils	1	<8 mils
Gold tabs	48	3 sides	32	2 sides	16	1 side	0	None
Annular ring	30	>2 mils	21	2–4 mils	8	4–6 mils	1	<6 mils
Solder mask	25	2 S dry film	17	2 S LPI	7	1 S LPI	5	screened
Metallization	75	Reflowed tin/lead	69	Selective solder coat	46	Electroless Ni/Au	29	SMOBC/ organic coat
Minimum hole diameter	166	=>8 mils	84	9–12 mils	69	13–20 mils	5	<20 mils
Controlled impedance tolerance	105	±5%	62	±10%	30	±20%	0	None

Points are per panel. For board, divide by number per panel.

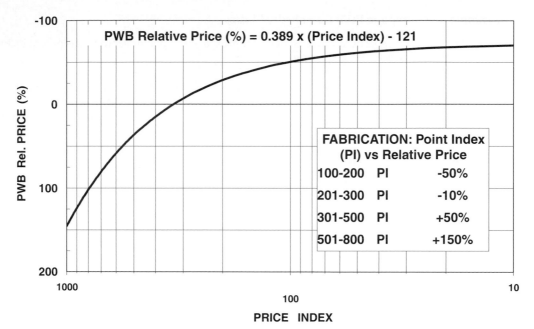

FIGURE 18.17 PWB relative price as a function of the price index.

TABLE 18.5 One Company's Fabrication Complexity Matrix

Fabrication factors	PWB design #1	Points	PWB design #2	Points
Construction material	FR-4	49	FR-4	49
Number of layers	6	137	6	89
Number of holes	$1,655 \times 6 = 9,930$	90	3,001–10,000	90
Minimum trace width	0.008 in	6	6–8 mils	1
Gold tabs	0	0	0	0
Annular ring	0.010 in	1	<6 mils	1
Solder mask	2-sided LPI	17	2-sided LPI	17
Metalization	SMOBC/SSC	69	SMOBC/SSC	69
Minimum hole diameter	0.025 in	5	<20 mils	5
Controlled impedance	0	0	0	0
Total points		374		321

TABLE 18.6 One Company's Initial Consumer Board Characteristics

Size	58.9 in^2
Number of layers	6
Number of holes	1,655
Minimum trace width	0.008 in
Tolerance	±0.003 in
Complexity index	68.06
First-pass yield	90.7%
Price index	412.3
Relative price	+14.5%

$A = 35, H = 1000, L = 2, T = 0.01, T_o = 0.01$

TABLE 18.7 One Company's Final Consumer Board Characteristics

Size	43.6 in^2
Number of layers	4
Number of holes	1,156
Minimum trace width	0.012 in
Tolerance	±0.003 in
Complexity index	55.8 (24.0)
First-pass yield	96.7%
Price index	331.9
Relative price	−10.2%

$A = 35, H = 1000, L = 2, T = 0.01, T_o = 0.01$

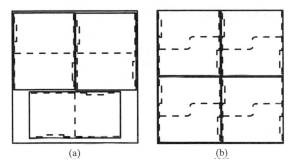

(a)   (b)

FIGURE 18.18   Fabrication panel layout of the consumer electronics board. (*a*) The original design; (*b*) the final layout.

18.6   ASSEMBLY TRADE-OFF PLANNING

The metrics of assembly trade-offs relate factors of process, component selection, and testing to assembly prices. Yields and rework are factored into the points of the assembly report card. The point total provides an estimation of the relative prices of assembly and testing.

18.6.1   Assembly Complexity Matrix

The assembly complexity matrix is a matrix supplied by the PCB assembler. It relates various process assembly and testing choices that the assembler provides along with various component sizes, orientations, complexities, and known qualities to the costs of providing for these design choices by design points allocation. Typical factors that affect assembly costs are:

- One-pass or two-pass IR reflow
- Wave solder process
- Manual or automatic parts placement
- Oddly shaped parts
- Part quality level
- Connector placement
- Test coverage
- Test diagnosability
- Assembly stress testing
- Repair equipment compatibility

By collecting all the costs associated with assembly, testing, and repair and then normalizing these costs with the smallest nonzero value, a matrix such as that shown in Table 18.8 can be produced.

TABLE 18.8   Example of an Assembly Report Card

Factors	Points	Highest	Points	Middle	Points	Lowest
Solder process	35	1-pass IR	20	2-pass IR	0	IR and wave solder
Placement	8	100% auto	5	90%–99% auto	0	>90% auto
Digital test coverage	9	<98%	3	90%–98%	0	>90%
Manual attachment	8	100% auto	25	Brackets	0	Post-solder assembly

18.6.2 Example of an Assembly Complexity Matrix

An example of the assembly complexity matrix is the assembly report card created by IBM-Austin.[15] This complexity matrix has 10 factors that range in point value from 0 to 35. The total points can affect the prices from a 30% discount to a 30% penalty. Table 18.9 illustrates this assembly complexity matrix trade-offs with design points.

TABLE 18.9 Assembly Complexity Matrix Defined as an Assembly Report Card

Assembly Factors	Points	A	Points	B	Points	C	Points	D
Assembly process	35	2-pass IR	25	IR/wave B/S passives	20	IR/wave ≤ 5 B/S actives	0	IR/wave > 5 B/S actives

- 2-pass IR is lowest-cost process due to low defect/repair level.
- Maximize SMT content, some PTH OK.
- Back-side SMT attachment with a wave solder process requires the use of adhesive.
- Back-side actives have a high defect level when the wave solder process is used.

Stress tests	15	0 hours	12	≤3 hours in situ	6	≤6 hours in situ and ≤3 hours static	0	<6 hours

- Stress test is a high-cost process step due to cost of chambers, fixtures, and process time.
- Stress test can be eliminated by using robust components, design, and process.

Parts SPQL	10	No high hitters	7	<2 high hitters	4	<4 high hitters	0	>5 high hitters

- Parts SPQL is the quality level of the parts used in the process (shipped product quality level).
- High hitters are parts known to have a high defect rate.

ICT digital test coverage	9	>98% coverage	6	>95% coverage	3	>90% coverage	0	<90% coverage

- The key to high test coverage is the selection of major components that have built-in testability and a PWB design with 1 test point per net.

Diagnosability	9	≤10 min	6	≤20 min	3	≤30 min	0	>40 min

- Effectiveness of diagnostic tools provided by the card designer influences diagnostic time.

Placement/ insertion	6	100% auto	4	≥95% auto	2	≥90% auto	0	<90% auto

- Elimination of manual component placement and insertion reduces process time, defects, and process cost.

Manual attachment	6	100% auto	4	Simple bracket	2	Difficult bracket	0	Post-solder assembly

- Simple bracket = <1 min assembly time
- Difficult bracket = >1 min assembly time
- Post-solder assembly = manual solder operation

Connector selection	4	Auto assembly and retention and keyed	2	Manual and retention and keyed	1	Manual and retention	0	Manual

- Manually inserted connectors that are keyed and have retention reduce defects.
- Prepackaged SMT connectors that can be automatically placed are considered equivalent to keyed with retention.

Handling damage exposure	3	No handling list violations	2	<3 handling list violations	1	<5 handling list violations	0	>6 handling list violations

- Single inline packages (SIPs) > 0.5 in high
- Memory SIMM connectors with plastic latches
- Unshrouded headers over 0.5 in high

Repair	3	100% auto	2	≥90% auto	1	<90% auto	0	<90% auto and difficult

- Auto repair refers to the use of semiautomated repair tools for the removal of large components and connectors.
- A difficult repair is one that takes >10 min.

18.6.2.1 SMT Assembly Report Card.   The IBM assembly report card's[15] 10 assembly trade-off factors are shown in Table 18.9. The report card was the work of many assembly engineers with the help of the accounting department. The report card was introduced in the early 1990s. Within 2 years, through the use of the report card, assembly point scores were averaging around 75 instead of the earlier 50, with its Gaussian distribution. Remember, higher scores indicate more producibility. This effect can be seen in Fig. 18.19. The scores are no longer normally distributed.

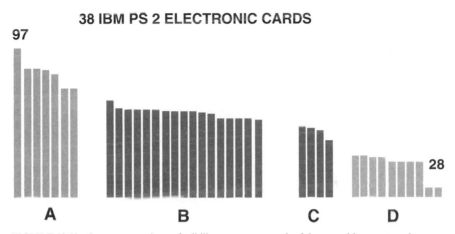

FIGURE 18.19   Improvement in producibility scores as a result of the assembly report card.

18.7   TOOLS FOR AUDITS

Chapters 13 and 14 go into the details of designing and laying out a printed circuit board. Part of that process is to execute the planning factors developed in this chapter. The other part is checking the design to see if the planned goals are achieved. A tool that will assist in executing that plan and checking the finished layout is manufacturability audit software.[16,17]

18.7.1   Manufacturability Audits Performed by the Designer

These tools provide checklists, audit rules, and design rules to ensure a correct design file and rapid introduction to fabrication and through holes/SMT assembly.[18]

18.7.1.1 Design Rule Checks (DRCs).   The design rules that these audit tools use are more detailed than those typically provided with CAD programs. Design rules can be stored by project, function, or application.

18.7.1.2 Machine, Component, and Producibility Checks.   The features these tools provide for producibility checks and audits are:

- Fabrication manufacturability and artwork audit
- Metal sliver clearance, shorts, solder shorts, copper islands, and unattached metal checks
- Checks for unconnected nets and subnets, loose ends, and nonfunctional vias
- Audits that compare drilling with manufacturability requirements
- Annular ring, pad stack, plane clearance, resist slivers, and silkscreen clipping
- Solder paste, solder mask coverage, mask-to-via, and other SMT checks
- Producibility enhancements such as teardrop additions

Figure 18.20 shows a typical analysis screen from a manufacturability audit tool and Fig. 18.21 provides a sample check sheet for a manufacturability audit.

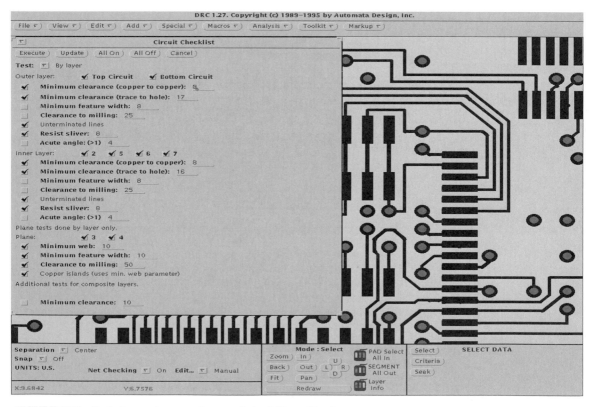

FIGURE 18.20   Typical computer screen for manufacturability audit software.

FEATURES	CHECKLIST
CAD netlist compare	
Annular ring error	
Pads tack checklist	
Plane clearance error	
Manufacturability analysis, DRC	
Thermal leg count violation	
Circuit Checklist	
Unterminated lines	
Resist slivers	
Copper islands	
Solder mask checklist	
Solder short violation	
Mask coverage	
Mask to via check	
Solder paste check	
Silkscreen clipping	
Teardrop pad addition	
Part-to-art clearance/automation	
Quality parts	
Hole audit/lead diameter	
Part density	
Height clearance	
Allowable machine span	
Part spacing	
Drill optimization	
Automatic solder mask generator	
Bare board test points	
In-circuit test point analysis	
In-circuit test checklist	
Boundary scan audit	
Test point management	
Design profile	
Registration generator	
Keep-out audit	

FIGURE 18.21 A checklist for items to review to assure PWB producibility.

REFERENCES

1. General Electric, "Review of DFM Principles," Internal DFM Conference Paper, Charlottesville, VA, 1982.

2. Robert Hawiszczak, "Integrating Design for Producibility into a CAE Design Environment," NEPCON EAST, June 1989, pp. 3–14.

3. D.P. Seraphim, R.C. Lasky, and C.Y. Li, *Principles of Electronic Packaging,* McGraw-Hill, New York, 1989, pp. 39–52.

4. G. Coors, P. Anderson, and L. Seward, "A Statistical Approach to Wiring Requirements," *Proceedings of IEPS,* 1990, pp. 774–783.

5. H. Ohdaira, K. Yoshida, and K. Sassoka, "New Polymeric Multilayer and Packaging," *Proceedings of Printed Circuit World Conference V,* Glasgow, Scotland, reprinted in *Circuit World,* vol. 17, no. 12, January 1991.

6. H. Holden, "Design Density Index," HP DFM Worksheet, Hewlett Packard, April 1991.

7. IPC-D-275 Task Group, "ANSI/IPC-D-275 Design Standard for Rigid Printed Boards and Rigid Printed Board Assemblies," *IPC,* September 1991, pp. 50–52.

8. W. Donath, "Placement and Average Interconnection Lengths of Computer Logic," *IEEE Transactions on Circuits and Systems,* no. 4, 1979, pp. 272–277.

9. Sutherland and D. Oestreicher, "How Big Should a Printed Circuit Board Be?" *IEEE Transactions on Computers,* vol. c-22, no. 5., May 1973, pp. 537–542.

10. L. Moresco, "Electronic System Packaging: The Search for Manufacturing the Optimum in a Sea of Constraints," *IEEE Transactions on Components, Hybrids and Manufacturing Technology,* vol. 13, 1990, pp. 494–508.

11. Holden, H.T., "PWB Complexity Factor: CI," *IPC Technical Review,* March 1986, p. 19.

12. STATGRAPHICS, ver. 2.6, by Statistical Graphics Corp. 2115 East Jefferson St., Rockville, MD 20852, 301-984-5123.

13. G. Boothroyd and P. Dewhurst, "Design for Assembly," Dept. of Mechanical Engineering, University of Massachusetts, Amherst, MA, 1983.

14. Daetz, Douglas, "The Effect of Product Design on Product Quality and Product Cost," *Quality Progress,* June 1987, pp. 63–619.

15. H. Hume, R. Komm, and T. Garrison, IBM, "Design Report Card: A Method for Measuring Design for Manufacturability," Surface Mount International Conference, September 1992, pp 986–991.

16. John Berrie and Andy Slade, "Knowledge Based Design Analysis for EMC," Zuken-Redac Systems Ltd., Tewkesbury, Gloucestershire, UK, 1995.

17. Barry O'Sullivan, "The Design Advisor: Capturing Design Practices and RFI/EMI Concerns," *The Board Authority,* vol. 2, no. 4, December 2000, pp. 47–52.

18. Robynne Hanus, "HDI DFM and Physical Design Verification," *The Board Authority,* vol. 2, no. 4, December 2000, pp. 42–46.

CHAPTER 19

MANUFACTURING INFORMATION DOCUMENTATION AND TRANSFER

Clyde Parrish
PC World, Division of Circuit World Corp., Toronto, Canada

19.1 INTRODUCTION

The manufacturing of printed circuit boards (PCBs) begins with the soft-tooling process. This process is the transformation of customer computer-aided design (CAD) data into the necessary tools required for manufacturing the bare printed circuit board. The typical tools required for manufacturing printed circuit boards include artwork for photoprinting of inner conductive layers, outer conductive layers, and solder mask patterns. Artwork is also created for screen printing patterns for nomenclature and via-plugging layers. Additional tools required include drill and routing numerical controlled (NC) programs, electrical testing netlists and fixtures, and CAD reference soft-tools. During the tooling process, the bill of materials (BOM) and process routing are also defined.

During the tooling process, the customer part numbers are analyzed to determine the compatibility of the design features with the manufacturing process capabilities. Additionally, attempts to optimize the manufacturing of the product at the lowest cost is a primary goal. However, the majority of the costs have been defined before the design is transmitted to the manufacturing site by the PCB designer. An early investment in time by the PCB design team and the manufacturing tooling team can result in the most significant savings in overall product cost.

This chapter describes the PCB tooling process, as defined in Fig. 19.1, including the transfer of information, design reviews, optimization of materials, definition of BOM and routings, tool creation, and additional processes that are required.

19.2 INFORMATION TRANSFER

The tooling process begins with the receipt of information from the customer. This information historically has been transferred via mail or overnight delivery. Increasingly over the past 10 years the communication between design site and manufacturer has been via electronic means to reduce lead times. Unfortunately, although the time required to send information to the manufacturer has decreased to minutes or hours from day(s), a significant issue with the provision of packages to manufacturers is the completeness of the information provided.

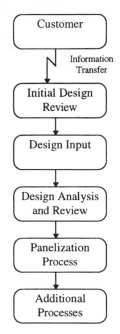

FIGURE 19.1   The soft-tooling process.

The following sections define the basic information required to transmit and methods for communication.

19.2.1   Information Required

Timely tooling of a part number depends on having the correct information. All features required to exist on the PCB must be defined to the manufacturer. The information is defined via design data, drawings, and textual information. The information and common data formats required to permit the tooling of the PCB include the following:

- Part number information

  Information: This information defines the part number to be built, including revision number, releases, dates, etc.

  Format: This information is typically provided in the part drawing or may be provided as an additional text file.

- Part drawing

  Information: This drawing may contain specific design requirements such as material requirements, controlled impedance requirements, solder mask type, nomenclature color, and dimensional tolerances.

  Format: Common formats for drawings are HP-GL and PostScript.

- Drill drawing

  Information: Although the drill data is provided via data files, this information typically contains only the location of holes and the tool number. The tool number is referenced against the drill drawings to determine the required sizes, plating status, size tolerances, and the total count for verification.

  Format: Common formats for drawings are HP-GL and PostScript.

- Subpanel drawing

  Information: Many assembly operations require boards to be provided in a subpanel form (many parts on one shippable unit). The drawings will define the orientation and position of each part, the subpanel dimensions, tooling hole information, special markings, and specific manufacturing processes and tolerances.

  Format: Common formats for drawings are HP-GL and PostScript.

- Artwork data

  Information: This data consists of files for each circuitry, coating (e.g., solder mask), or marking (i.e., nomenclature) layer.

  Format: The data required is usually RS-274, commonly called *Gerber data.* Gerber data is provided as a standard output from most PCB CAD systems.

- Aperture list files

  Information: The definitions of the shapes used for drawing are required for each layer provided of artwork data. Special shapes like thermal pads should specifically define its method of construction.

  Format: This information is usually provided as a text file, although the information may also be defined in the beginning of the individual artwork files, including constructions of complex apertures.

- Drill data

  Information: This information may consist of a single or multiple file(s), and defines the location and tool number used for each hole in the PCB. The files required should define all plated, unplated (can be combined with plated if fully defined), buried via, and blind via layers.

  Format: Common data files provided are Excellon format.

- Drill tool files

  Information: This information describes the size, plating status, layer-from and layer-to (in the case of buried and blind vias), and file names. This information is referenced against the drill drawing.

  Format: This information is usually provided as a text file, although the information may also be defined in the beginning of the individual drill files.

- Special requirements information

  Information: The drawing or a file should describe any special requirements not defined in other information. It is important for the PCB designer not to assume that requirements are understood, but to refer to specifications or clearly define the requirements.

  Format: This information is typically provided in the part drawing or may be provided as an additional text or drawing file.

- Netlist data

  Information: Netlist data defines the connectivity of the circuitry.

  Format: This information can be provided from the CAD systems in various formats, or it can be extracted from the drill and artwork data. Contact the PCB manufacturer for compatible formats, if the netlist data is to be provided directly. The Institute for Interconnecting and Packaging Electronic Circuits (IPC) has defined a neutral format, IPC-356, which provides all the information necessary for netlist and electrical test fixture creation.

  The IPC has defined an alternative neutral format for most of the previously defined data which provides simpler processing at the manufacturer, IPC-350. This format can be generated by most PCB CAD systems and processed by most PCB CAM/tooling systems. The PCB customer should review the compatibility of this format with the PCB manufacturer prior to sending.

19.2.2 Modem Transmission

Currently, the most popular means for transmitting data from design site to manufacturing site is via personal computer modem transmission. Two popular methods for communications are:

1. Dedicated connection to waiting compatible software
2. Bulletin board–type service

Method 1 is the simplest means to initiate communication. Almost any communication package can provide the capability, for example, X-Modem or Windows terminal. Method 2 provides a more robust environment similar in concept to CompuServe. Each user has an individual account logon and password.

19.2.3 Internet Transmission

Several methods are available for connecting to the Internet, depending upon the volume of data, the interactiveness required, and the budget. Typically, a company will utilize a UNIX-based machine for accessing the Internet. UNIX-based computers are used to provide multi-user/multitasking capabilities (i.e., two or more people may be performing a task at the same time). Personal computers can also provide access depending on the functionality and interactiveness desired. The following describes some of the alternatives available:

- *User-Account on Internet Provider's Computer.* This is the least expensive method for gaining access to the Internet. The account is accessed via dial-up modem connection, running a terminal emulation program. Many Internet providers offer this service. Monthly account fees are available for a minimal charge for each account, depending on disk storage requirements. Individual accounts may be maintained for each customer to provide security between customers. Transmission of data from the provider's machine may be provided via popular transfer protocols (e.g., X-Modem), dependent on your Internet provider. The communication speed is limited to modem speed, typically 14,400 baud.

- *Interactive Dial-Up.* This method provides direct communication to customers, through prior contact and the internal initiation of the connection. The communications between the internal computer and the Internet provider computer is typically via Serial Line Internet Protocol (SLIP) or Point-to-Point Protocol (PPP) software. These software packages provide the continuation of network communication from internal systems to the Internet provider. This service can cost less than $100 per month.

- *Dedicated Dial-Up.* Dedicated dial-up lines are similar to the interactive dial-up, except that a dedicated phone number is provided. The connection to the Internet can be used either for interactive or for continuous-connection dial-up service. This service will cost slightly more than the interactive dial-up service, due to the dedicated line costs at the Internet provider.

- *Leased Lines.* Leased lines provide continuous, high-speed communications between a company and its Internet provider. They are also the most expensive, ranging from several hundred to several thousand dollars per month.

Security of the machines connected to the Internet is of prime importance. Any direct connections (i.e., not the user account just described) to the Internet are usually handled by dedicated machines acting as a firewall to prevent unauthorized entry into a firm's network, as outlined in Fig. 19.2.

There are numerous books available about the Internet and how to connect to it. There are also numerous consulting firms available to assist in the process, which is nontrivial, to prepare a secure connection.

19.3 INITIAL DESIGN REVIEW

The purpose of the initial design review is to determine the potential fit of the product to the manufacturing facility, determine general cost information, and prepare for tooling. Proper upfront analysis of a product prior to manufacturing or tooling, results in a reduction in waste of time and materials.

It is the responsibility of the manufacturing site to determine the fit of a given product to its capabilities. PCB manufacturing sites should monitor and maintain a list of manufacturing capabilities and a technology roadmap of where the facility is developing additional process capabilities. This list of capabilities will define the acceptability of a product for manufacturing, or whether the product is a research and development project.

19.3.1 Design Review

Reviewing the incoming package for design requirements (e.g., line width and spacing) versus the PCB manufacturing capabilities will define the capability for manufacturing, and provide

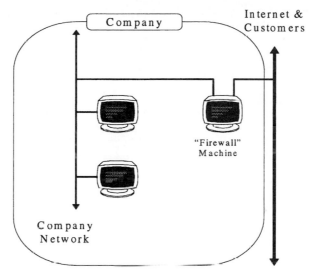

FIGURE 19.2   Typical Internet connection.

a prediction of the resulting yield. Among the design characteristics that should be reviewed against process capabilities are the following:

Item	Issue
Maximum number of layers	Higher-layer-count products require greater control of processes and tooling tolerances.
Board thickness	Some process or handling equipment may have limitations on board thickness (either too thin or too thick).
Minimum feature width	Finer line widths require better control of artwork and process tolerances. In addition, the relationship of line width to copper weight is significant in providing a well-defined line. Fine lines with lower copper weights are easier to manufacture than fine lines with higher copper weights.
Minimum feature spacing	Finer feature spacing requires better control of artwork and process tolerances. In addition, the relationship of feature spacing to copper weight is significant. Fine spaces with lower copper weights are easier to manufacture than fine spaces with higher copper weights.
Minimum finished hole size	Smaller holes require higher manufacturing process capabilities and imply (due to design characteristics) finer tolerances on tooling and internal registration systems.

Maximum aspect ratio	Small holes and thick boards result in difficulty during the plating processes and can result in defective products that may or may not pass electrical testing. Plating high-aspect-ratio holes requires chemistry and process parameter enhancements.
PCB dimensional tolerances	Fine PCB profile or cutout tolerances may result in punching/blanking requirements versus routing, or changes in the routing parameters or programming.
Feature-to-feature tolerances	Location of features on the PCB to other features may require alternative materials or process changes to reduce the tolerances.
Hole size tolerances	Consistency in the plating process versus selected drilling hole size and plating densities have significant impact on the capability to control hole tolerances. Adjustments to the PCB design, drill hole size selection, or process parameters may be required to produce an acceptable product with tighter tolerances.

19.3.2 Material Requirements

Determination of the bill of materials is required during the initial analysis of the design. The determination of the BOM and other material-processing requirements will define the manufacturing facility's capability to produce and its material cost structure. In addition, the definition of the material requirements will be the basis for the generation of the process traveler requirements.

The primary materials requiring definition are those included in the BOM, including laminates, prepregs, copper foil, solder mask, and gold. The materials may be explicitly defined by the PCB customer (e.g., usage of a specific solder mask), or may be implied in the drawings or specifications provided with the tooling package.

Several factors impact the selection of the raw materials, including the following:

- Customer-defined physical constraints, for example, the definition of the physical dimensions between conductive layers
- Customer specification of electrical properties, for example, the definition of the impedance requirements on certain layers
- Manufacturing process capabilities related to lamination thicknesses and tolerances
- Specification of material dielectric requirements, for example, the usage of FR-4 or polyimide
- Specifications of physical operating parameters, for example, the minimum requirements of the glass transition temperature

The determination of the laminates, prepregs and copper foils are based upon the following:

- Standard constructions for the PCB manufacturer of a defined PCB layer count, final thickness, copper weight, and dielectric spacing.
- Custom constructions based on defined physical constraints (e.g., minimum dielectric spacing). These custom constructions are defined through knowledge of the lamination pressing

thickness of materials vs. copper circuitry densities, and the availability of materials from suppliers.

- Custom constructions based on defined electrical property constraints. These custom constructions are typically defined via equations or software models provided with certain product parameters.

The determination of solder mask is based on the customer specifications and drawings. Once the acceptability of the solder masks is defined, the selection of the acceptable masks by the PCB manufacturer is based on either preferred process (due to volume or cost) or the design characteristics' interaction with the solder mask. These design characteristics include the following:

- *Tenting of vias*—A dry film solder mask may be preferred over liquid photoimageable solder mask with a secondary via plugging process.
- *Platable area densities/higher external copper weights*—Thin solder masks may not be able to ensure coverage of high plating.
- *Secondary processes*—Post-solder-mask processes may chemically or mechanically alter the appearance of certain solder masks.

The determination of gold requirements are based on the thickness and area of the gold. These factors can be used to calculate the requirement of gold per PCB product.

19.3.3 Process Requirements

The selection of the proper product routing (or traveler) is critical to the upfront analysis of the product acceptability to manufacturing. Considering a typical multilayer PCB product, the product routings can be broken into two parts: the innerlayer pieces and the outerlayer piece.

The product routings of the innerlayer pieces are fairly standard, and are typically as follows:

Step	Typical process requirements
Layer cleaning	Cleaning of the laminate surface via either mechanical or chemical cleaning processes
Imaging	Coating of the laminate with photoresist material and the exposure of the photoresist with artwork defining the innerlayer pattern
Develop-etch-strip (DES)	Developing of the photoresist, etching of the exposed copper, and stripping of the remaining photoresist
Innerlayer inspection	Inspection of the innerlayer piece to the PCB design intent
Oxide	Coating of the innerlayer piece with an oxide layer prior to lamination

One of the few decision points in innerlayer manufacturing is whether the product will require inspection. This decision can be based on the manufacturing facility's process capabilities and the design of the specific innerlayer piece. For example, if the process capability for innerlayer manufacturing of designs at 0.008-in lines and spaces is 100 percent yield, and the product has been designed at or above 0.008-in lines and spaces, then the product may not require inspection. Typically, the design package will identify the design technologies (e.g., line width and spacing), however, these should be confirmed during the design analysis and review stage.

The product routings of the outerlayer pieces define the finished product appearance and, as a result, are more complex. Assuming a pattern plating process, the typical outerlayer process routings for a SMOBC/HASL (solder mask over bare copper/hot-air solder level) product is as follows:

Step	Typical process requirements
Lamination	Lamination of innerlayer pieces with prepreg and copper foils to create the outerlayer piece
Drilling	Addition of the holes providing pathways for electrical conductivity between outer layer and innerlayer pieces, and other PCB design purposes
Electroless copper plating	Deposition of copper on the surface and in the holes of the product, providing the conductivity necessary for electroplating
Imaging	Coating of the panel with photoresist material, the exposure of the photoresist with artwork defining the outerlayer pattern, and the developing of the outerlayer pattern
Electroplate copper	The plating of the final circuitry of the product
Tin plating	Sacrificial plating over the final circuitry of the product
Strip-etch-strip (SES)	Stripping of the remaining photoresist, etching of the exposed copper, and stripping of the sacrificial plating
Solder mask	Coating of the bare copper product with either dry film or liquid photoimageable solder mask, exposing of the panel with the PCB customer-supplied artwork pattern, developing of the pattern (exposing the sites requiring solder), and curing of the solder mask
Nomenclature	Screening of the nomenclature onto the panel and curing
HASL	Coating the exposed copper sites with solder, leveling to the customer's requirements of thickness
Depanelization	NC routing of the PCB products from the manufacturing panel
Electrical testing	Electrical testing of the product for conformance to the PCB design
Inspection	Verification of product conformance to specifications prior to shipment to the PCB customer

After the electroplate copper operation, the process may change to meet the surface finish requirements. Some of the alternative steps affecting the outerlayer product routing include the following:

Item	Typical process requirements
Selective gold plating	Before the SES process, the panel is coated with additional resist and the selective sites are exposed. These exposed sites are then plated with nickel and gold. The panel will then proceed through the SES process. The HASL step would be omitted.
NPTH in planes without annular ring	A secondary drilling operation will be required to drill the holes, if the hole is to be located through the copper or the hole/feature size is beyond the process capabilities of tenting. This step may occur at the depanelization step.
Scoring	Additional tooling holes may be required to register the panel to the scoring machine blades. Programming of the scoring machine will be required to set the locations and depth of the cuts. This step may occur prior to NC routing of the product.
Countersink holes	The countersinking process is normally performed prior to depanelization.
Gold finger plating	Plating of nickel and gold over the finger area on connectors. This process typically requires shearing/NC routing separation of the panel to place the fingers at the edge of the remaining pieces, taping the product to expose only the fingers, stripping of the solder, plating nickel, and plating gold, and removal of the tape. These operations occur prior to the depanelization process. Finger plated products usually require chamfering of the fingers.
Via plugging	Plugging of vias may occur prior to solder mask or after the HASL operations.
Chamfer	After depanelization of the products from the manufacturing panel, the PCB products are processed through a machine to add an angle to the edge of the part or fingers.
Organic coating	Products requiring organic coating would omit the HASL step and be coated prior to inspection.

19.3.4 Panelization

The selection of the manufacturing panel is one of the most important steps in achieving product profitability. Several factors affect the selection of panel sizes to produce a specific product. These include:

- Material utilization
- Process-specific constraints
- Process limitations

All of these factors will impact the selection of the manufacturing panel size and the profitability of the product.

19.3.4.1 Material Utilization. The material manufacturing costs, which correspond to 30 to 40 percent of total costs, are directly related to the square inches of material processed. The material costs can include the following: laminate, prepreg, copper foil, solder mask, photoresists, drill bits, chemicals, etc. Generally, these materials are consumed relative to the panel area manufactured.

The appropriate manufacturing panel size (refer to Fig. 19.3) should be selected such that the shippable product consumes the highest percent of the manufactured panel as possible, thus reducing waste material and product costs.

During the design phase of the PCB, the board profile is defined. The impact of designing the profile, which results in poor manufacturing panel utilization, significantly impacts the cost of the product. During the definition of the PCB board profile, the PCB designer should consult with his manufacturing site to provide DFM feedback. Odd-shaped PCB profiles can still result in good material utilization, through nesting of the PCB on the panel.

General limitations in the selection of the appropriate panel sizes include the following: minimum spacing between products for depanelization processes (typically 0.100 in) and minimum panel border-to-product spacing to permit tooling and registration systems (typically 0.600 in).

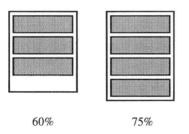

60%   75%

FIGURE 19.3 Selection of the best panel to increase the material utilization of the shippable product.

19.3.4.2 Product Process-Specific Constraints. In the selection of manufacturing panel sizes, there may be processing constraints which restrict the usage of specific panels or limit the method of placing the shippable product on the panel. For example, the manufacturing of products with gold tab plating may require additional spacing between products and restrictions on the rotation and nesting of products. These constraints may force the usage of panel sizes below the optimal material utilization.

19.3.4.3 Process Limitations. Process limitations may require the usage of nonoptimal panel sizes. For example, due to a product's registration requirements, additional tooling may be required, thus reducing the available area for shippable product and reducing the material utilization. Another example is the limitations of some processes to permit larger panel sizes due to the physical processing constraints of the equipment.

19.3.5 Initial Design Analysis

After the initial design review, the basics of the design are known and a decision on whether to manufacture the design has to be made, and if so, the cost must be determined. Once the cost has been determined, the product price can be provided to the PCB customer.

The material and process requirements with the manufacturing panel selection define the basics of the product costs. These factors, combined with the prediction of yield based on the design review, will allow the product to be costed.

19.4   DESIGN INPUT

The input of the design into the CAM system is primarily performed by the loading of Gerber data into the PCB CAM system, after defining the apertures and shapes to be used. Alternatively, most PCB CAM systems accept the IPC-350 format.

The information loaded into the PCB CAM systems includes all artwork layers (e.g., circuitry layers and solder mask layers) and the drill files. Although some PCB CAM systems can accept NC routing files (i.e., board profiling), these files are normally not part of the PCB design system's capabilities to create, and therefore are not provided to the PCB manufacturer.

Prior to the loading of the PCB design files, the aperture codes within the design file need to be related to physical shapes within the PCB CAM system. These shapes are usually round, square, and rectangular, but may also include complex shapes (e.g., thermal relief pads for innerlayer planes). The complex shapes may need to be created on the CAM system from information provided by the PCB designer. The complete definition of these complex shapes by the PCB designer is critical to the success of the resulting design. Incomplete descriptions could result in nonfunctional designs.

During the loading process of the design files, it is important to review any log files and screen messages for missing aperture definitions or damaged design files.

After loading the files, the design should be reviewed for any problems with the interpretation of the data or apertures while aligning the data for further processing.

The PCB CAM operator must ensure that the apertures match those defined by the PCB designer and that the design input process has proceeded without failure, since subtle errors could result in nonfunctional designs that may not be determined within the manufacturing site but only at the PCB customer site.

19.5   DESIGN ANALYSIS AND REVIEW

The design analysis and review step is a refinement of the initial design review performed earlier by focusing on the actual design data. The design analysis and review consists of the following steps:

- Design rule checking
- Manufacturability analysis
- Single image edits
- DFM (design for manufacturability) enhancements

These steps provide the final checking of the PCB design requirements against the capabilities of the PCB manufacturer and preparation of the design for manufacturing.

19.5.1   Design Rule Checking

Design systems lay out circuits to defined rules; however, these systems may fail to adhere to these rules, because of either system failure or manual intervention. The purpose of reviewing the data package, in addition to reviewing the documented design constraints, is to confirm that the product can be produced at the manufacturing site and to the expected manufacturing yields. The analysis of the provided data describing the circuitry of the PCB to the manufacturing facility's production capabilities is critical to the success of the product.

The following information defines typical design rule checking that is performed, the reasons for checking, and the method of checking. Also, see Fig. 19.4.

DRC item	Purpose of check	Method of checking
Drill layer duplicate coordinates	Duplicate drill holes consume manufacturing capacity and may result in broken drill bits.	Most PCB CAM systems have an explicit check for this problem, provided a given radius of tolerance.
Minimum drill hole-to-hole spacing	Drill holes too close may result in broken material between the holes.	Most PCB CAM systems have an explicit feature spacing check.
NPTH-to-board edge minimum spacing	Drill holes too close to the board edge may result in broken material at the board edge.	Merge routing profile onto a copy of the NPTH layer and execute a spacing check of the pad to trace (the route layer would appear as a trace).
Missing drill holes	Missing holes may result in nonfunctional designs or products that cannot be used due to missing mounting or tooling holes.	Pad registration checking against a circuit reference layer, with verification of pads without PTH holes.
Extra drill holes	Extra drilled holes could result in non-functional designs from cutting of traces or shorts (if the holes become plated).	Pad registration checking against a circuit reference layer, with verification of PTH holes without pads.
Board edge-to-board edge minimum spacing	Routing cuts leaving minimal material may result in fractured or broken material between routing.	Feature spacing checking of the routing layer as a design layer, without considering electrical connectivity.
Copper-to-board edge minimum spacing	Tolerances of the routing operation may result in copper exposed at the edges of the product.	Feature spacing checking of the routing layer as a design layer against the circuitry layer.
Minimum annular ring	Most customer specifications define the minimum acceptable annular ring on PCB products. Annular ring is dependent on the design and the registration tolerances of the manufacturing processes.	Most PCB CAM systems have an explicit annular ring check.
Minimum pad-to-pad spacing	In addition to feature spacing and potential shorts, pad spacing affects the capability to electrically test products.	Most PCB CAM systems have an explicit feature-to-feature spacing check.
Minimum pad-to-track spacing	Feature spacing below manufacturing capabilities could result in shorts and poor product yields.	Most PCB CAM systems have an explicit feature-to-feature spacing check.
Minimum track-to-track spacing	Feature spacing below manufacturing capabilities could result in shorts and poor product yields.	Most PCB CAM systems have an explicit feature-to-feature spacing check.
Copper-to-NPTH minimum spacing	NPTH locations too close may prevent proper tenting of holes and require secondary drilling operations. NPTH too close may result in damaged features (e.g., cut traces).	Feature minimum spacing against a copy of the design layer merged with the NPTH layers.
Minimum line width	Line widths below manufacturing capabilities could result in poor product yields.	Some PCB CAM systems have an explicit check for this problem, others may require review of the apertures used and highlighting of the apertures for visual inspection.
Track termination without pad	Although this may be design intent, missing pads may be the result of poor design information or loading failures. These problems can result in nonfunctional designs.	Most PCB CAM systems have an explicit check for this problem.
Pad stack alignment	Misaligned pad stacks may result in unpredictable annular ring results, incorrect compensations of registration, and product scrap.	Most PCB CAM systems have an explicit pad registration check.

DRC item	Purpose of check	Method of checking
Minimum solder mask pad clearance	Solder mask pad clearances below manufacturing capabilities could result in solder mask on the pads, and poor product yields.	Check solder mask layer (as a circuit layer) against the circuitry layer using an annular ring check.
Minimum solder mask edge-to-feature spacing	Solder mask edge to feature spacing below manufacturing capabilities could result in exposed features and poor product yields.	Feature minimum spacing against a copy of the design layer merged with the solder mask layer.
Minimum solder mask annular ring for NPTH	NPTH solder mask clearances may need to be larger to prevent ghosting of the solder mask from light diffraction through the product.	Check solder mask layer (as a circuit layer) against the NPTH layer, verify for no matches of the NPTH to solder mask layer.
Solder mask to board edge minimum clearance	PCB edge clearances may need to be larger to prevent ghosting of the solder mask from light diffraction through the product.	Feature minimum spacing against a negative copy of the solder mask layer merged with the routing layer.
Solder mask minimum web	Solder mask webs below manufacturing capabilities could result in solder mask breakdown and poor product yields. This problem potentially results in the PCB assembler with solder bridging defects.	Use minimum feature-to-feature spacing on the solder mask layer.
Plane-to-board edge minimum clearance	Spacing below the tolerances of the routing operation may result in copper exposed at the edges of the product.	Feature spacing checking of the routing layer as a design layer against the plane.
Minimum plane layer annular ring	Annular ring below manufacturing registration and tolerance capabilities could result in open connections, and poor product yields.	Most PCB CAM systems have an explicit annular ring check.
Minimum plane layer clearance	Plane clearances below manufacturing registration and tolerance capabilities could result in shorts and poor product yields.	Most PCB CAM systems have an explicit annular ring check.
Plane-to-plane isolation	Plane layer isolation typically results from either incorrect designs or interpretation of aperture lists; the result is a nonfunctional product.	Most PCB CAM systems have an explicit layer-to-layer isolation check.
Nomenclature-to-NPTH minimum spacing	Nomenclature spacing below manufacturing capabilities may result in nomenclature in the hole from misregistration or bleeding of the nomenclature ink.	Feature minimum spacing against a copy of the nomenclature layer merged with the NPTH layers.
Nomenclature-to-solder mask minimum spacing	Nomenclature spacing below manufacturing capabilities may result in nomenclature on product features from misregistration or bleeding of the nomenclature ink.	Feature minimum spacing against a copy of the nomenclature layer merged with each solder mask layer.
Nomenclature-to-feature minimum spacing	Nomenclature spacing below manufacturing capabilities may result in nomenclature on product features from misregistration, bleeding of the nomenclature ink or skipping of the screen over features resulting in illegible markings.	Feature minimum spacing against a copy of the design layer merged with the nomenclature layer.
Nomenclature minimum feature sizes	Nomenclature sizes below manufacturing capabilities could result in illegible nomenclature and poor product yields.	Some PCB CAM systems have an explicit check for minimum line width, others may require review of the apertures used and highlighting of the apertures for visual inspection.

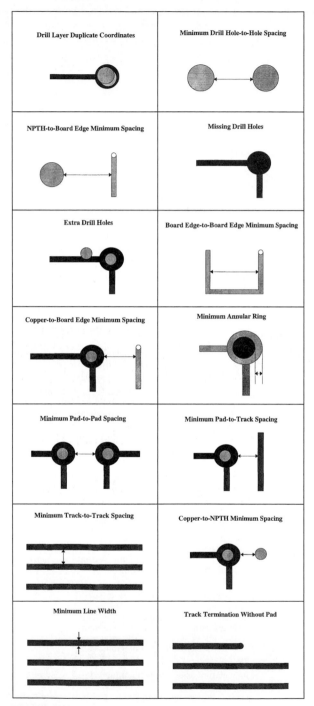

FIGURE 19.4 Examples of design rule check elements.

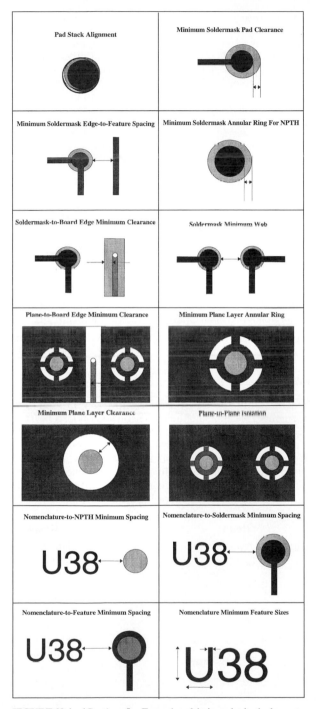

FIGURE 19.4   (*Continued*)   Examples of design rule check elements.

19.5.2 Manufacturability Review

The design rule checking results are reviewed against the PCB manufacturing capability matrix, with dispositions being made to the acceptability of violations. The PCB customer may be contacted regarding design violations for corrected designs to be retransmitted or to permit design changes to be performed by the PCB manufacturer.

Additionally, the factors reviewed during the initial design review are revisited to confirm the results based upon design actuals. If differences are noted, product yields and/or cost predictions may need to be altered.

19.5.3 Single Image Edits

In general, all designs have information which must be removed before the design can be manufactured. Most of these features are used as references during the PCB layout phase to assist the designer in understanding the available real estate. However, these references must be removed or modified to manufacture the board to design intent. Among these typical items are the following:

Item	Impact if not removed/modified
NPTH pads on outerlayers	Plated holes
Routing crop marks	Copper at board edge
Direct contact plane markings of drilled holes	Open circuits
Clip nomenclature with solder mask data	Nomenclature on pads/in holes

19.5.4 DFM Enhancements

The manufacturability of the PCB can be improved through design enhancements. Many of the enhancements are capable of being performed on CAD systems only (or through specialized post-CAD software); other enhancements can be performed at the manufacturing site.

Ideally, all design enhancements would be performed at the design site to maintain the consistency of products from multiple manufacturers; however, some of the DFM enhancements can impact the CAD system's capability to provide further design changes. Specifically, the optimization of track-to-feature spacing could result in blocked routing channels, preventing an autorouter from performing. The following are typical DFM enhancements (see Fig. 19.5a–e):

Item	Result from enhancement
Copper balancing	The balancing of copper on the surface of the design will improve the distribution of plating (in the pattern-plating process), preventing areas of overplating due to the isolation of features. This improvement is noticed in the ability to control the plated hole diameters to tolerances and the reduction of plating heights which impact solder mask coverage. (See Fig. 19.5a.)
Teardropping	As annular rings decrease due to tighter design specifications, the capability to have defects due to registration/annular ring failures increases. The addition of teardropping increases the pad-to-trace junction size, improving the reliability of

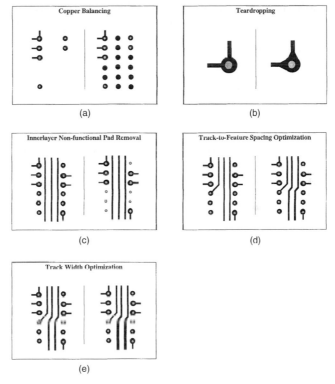

FIGURE 19.5 Typical DFM enhancements.

the connection and improving product yields. (See Fig. 19.5*b.*)

Removal of nonfunctional pads on innerlayers   Innerlayer shorts result from several factors, among them the proximity of circuitry to other circuitry. The occurrence of shorts can be correlated to the running length of the circuitry versus the running length at some minimum spacing between circuitry (at the process capability of the operation). Reducing the total distance at minimum spacing will reduce the probability of shorts, resulting in improved yields. (See Fig. 19.5*c.*)

Optimization of track-to-feature spacing   As defined in the nonfunctional pads discussion, the occurrence of shorts can be correlated to the running length of the circuitry versus the running length at some minimum spacing between circuitry (at the process capability of the operation). Reducing the total distance at minimum spacing will reduce the probability of shorts, resulting in improved yields. (See Fig. 19.5*d.*)

Track width optimization	The occurrence of opens can be correlated to the running length of the circuitry versus the running length at minimum track width (at the process capability of the operation). Reducing the total distance at minimum track width will reduce the probability of opens, resulting in improved yields. (See Fig. 19.5e.)

19.6 PANELIZATION PROCESS

The panelization process results in the placement of the PCB single images in the locations defined in the panelization definition step during the initial design review. In addition to the placement of the PCB single images, the following features may be added to the panel for manufacturing.

Item	Purpose
Customer coupons	Provided to the customers for their inspection of the PCB product. These coupons are typically treated as separate PCB single images.
Internal coupons	During the manufacturing process, destructive testing may be required to confirm product quality. These coupons are typically treated as separate PCB single images.
Tooling holes	Holes used for registration of manufacturing panels to artwork, drilling, or routing operations are added to the panel borders.
Outerlayer thieving patterns	Plating thieving patterns are sometimes needed to balance the platable area across the panel.
Innerlayer venting patterns	The addition of copper to innerlayers outside the PCB single image provides for more consistency in the lamination process.
Textual markings	Operator aides may be provided outside the PCB single image to improve manufacturing quality.

19.7 ADDITIONAL PROCESSES

An important step in all the processes already defined is the necessity for proper information management. Archival of PCB customer-supplied information and the files generated from the tooling process are critical for disaster recovery. Archiving systems exist in the market that also provide for centralized data management and distribution of the information to the various departments within an organization.

In addition to the basic tooling steps already defined, most PCB manufacturing sites require the following information to be created:

- CAD reference files
- Electrical test netlist and fixture creation files

The CAD reference files, if required, are generally created within the PCB manufacturing site. These files prepare *automated optical inspection* (AOI) systems with data that confirms that manufactured pieces match, within tolerances, the PCB design.

Electrical test netlist files may be either created internally or sent to an outside contractor to prepare the files and the resulting fixture. These files use as input either the Gerber or IPC data provided by the PCB customer.

CHAPTER 20
ELECTRONIC CONTRACT MANUFACTURING SUPPLIER SELECTION AND MANAGEMENT

Steven M. Allen
AVEX Electronics Inc., Fremont, California

20.1 INTRODUCTION

Intensified competition, short product life cycles, and budgetary constraints have caused electronics companies to create a boom to an emerging *electronic contract manufacturing* (ECM) industry. Electronic companies competing for a share of the world market rely a great deal on strategy, talent, and capital. With research and development costs to support up front, companies are hard pressed into dedicating a portion of their precious capital to justify a manufacturing facility along with the overhead that it requires.

Many *original equipment manufacturers* (OEMs) have adopted a strategy of focusing on core competencies in the areas of research and design, to ultimately develop patents that can be built by subcontractors whose goal is to manage capital utilization, inventory, and leading-edge manufacturing processes.

20.1.1 Development of the Electronic Manufacturing Services Industry

In the United States alone, the electronics industry will quadruple its outsourcing requirement from $7 billion to over $27 billion by 2000. EMSI sales for 1994 surpassed $10 billion, a 24 percent growth rate over 1993. Fueling a major portion of this growth is the advent of the OEM either selling off entire divisions to existing electronic manufacturing service providers, or subcontractors*, or actually jumping into the ECM business as a separate entity.

Lack of resources may make subcontracting an attractive choice; however, should an OEM fear the loss of control through outsourcing, risks may be minimized by putting together a thorough search for a competent and interested partner in the EMSI. The task of locating the "perfect" subcontractor may take a while, but by being prepared and making a thorough

---

* "Electronic Manufacturing Services Industry" is the general term established by the IPC to refer to all activities supporting the manufacture and assembly of electronic devices of original equipment manufacturers. For this chapter, however, we will use the terms *subcontractor* or *EMC* to denote *electronic manufacturing contractor* to describe an organization which provides either bare boards or loaded boards to higher-level users.

search, realistic expectations can be made by both parties to minimize risks and set the foundation for a long-term strategic relationship. In this chapter we will examine ways to evaluate a worthy ECM as well as to manage a relationship that is the lifeblood of the OEM.

20.2 BUSINESS PLAN

To understand what services an OEM requires to produce a quality product through the services of a contractor, a self-evaluation must first take place. The flowchart in Fig. 20.1 lists five basic stages in the life of a printed circuit board assembly (PCBA). By considering the major steps involved, a list can be put together to better understand which subcontractor is best suited to meet the needs of the OEM.

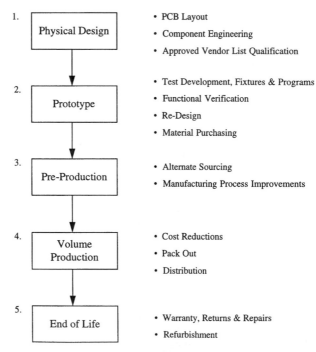

FIGURE 20.1 Customer requirements—five stages of product life.

20.2.1 Customer Requirements—Five Stages of Product Life

1. *Physical design:* Will the company need assistance in designing the PCBA or other engineering support? Is there a skilled PCBA design engineer available at the OEM? An experienced subcontractor will, at a minimum, have process/manufacturing engineers that can provide guidance in the areas of preferred location of tooling holes, fiducial markings, and test vias according to the requirements of the manufacturing equipment. But a skilled PCBA-designer will be necessary to ensure that the board can be manufactured in the most efficient way possible. As a *bill of materials* (BOM) list is developed, who will develop an *approved vendor list* (AVL)? A component engineer must be available to design-in vendors according to the board specifications. This is a job for which most subcontractors have limited resources. If a component engineer is not in-house (at the OEM), it may be necessary to render the services of a distributor to provide help.

2. *Prototype:*   In the prototype stage, quick turns in manufacturing are necessary to confirm the functionality of the product. Are there skilled technicians available with the proper tools and work environment to produce prototypes in-house? Are there test engineers available that can set the test parameters and develop fixtures and programs. Is *in-circuit test* (ICT) or *functional test* (FVT) a skill that will be kept in-house while outsourcing the production? Most subcontractors prefer to test what they build in order to provide timely feedback to the manufacturing process with regard to failures. On the other hand, some OEMs prefer control over test to keep a closer watch over quality levels.

3. *Preproduction:*   While prototype builds are meant to verify the functionality of a component, preproduction is intended to verify manufacturability. Quantities for a prototype build can range from 1 to 20 pieces and can be built by hand, while preproduction will require a *design for manufacturability* (DFM) verification on an automated line. If preproduction should remain in-house, skilled process engineers must be well versed in tracking yield information to avoid a second learning curve when moving to a subcontractor.

4. *Production:*   At the higher volumes, can the OEM handle the ramp to production? Will the OEM share manufacturing with a subcontractor? Are there additional services/items required to ship the final product such as pack-out and/or software disk duplication?

5. *End of life:*   Who will support the product at the end of its life as the quantities become minimal? Are there technicians in-house that can service returns, warranty support, refurbishment, or upgrades?

20.2.2 Supplier Requirements

Before visiting an ECM, keep in mind that they too have requirements of their prospects/customers. By understanding the needs of the ECM, time and effort can be saved by both parties before a substantial amount is invested. ECMs look at their customers as investments, and by understanding that their resources are limited, the OEM will be well advised to present itself as a worthy investment.

The largest risk any ECM will make is to purchase material to fulfill the turnkey requirements of the OEM. If a turnkey relationship is desired, ECMs will be interested in the company financials, how well an OEM is backed, and how their risk in exposure to inventory can be minimized. ECMs cannot survive if their customers become failures. Most ECMs survive on thin margins which leave little room for risk. Be objective in describing a business plan to an ECM, using an approach similar to the process of obtaining a loan from a bank, proving creditworthiness or reasons why the product and company are worth investing in. Be sure to include the product/company competitive advantage and marketplace. Partnerships require information to flow freely from both sides before determining whether or not to engage. With no written plan in place, information can easily be lost, forgotten, or made inconsistent. If the OEM product will experience an extended ramp to volume production, it will be necessary to get a buy-in from the ECM that it supports in the preproduction stage is critical. Another reason for the OEM to convince the ECM is to generate an unconditional interest in times of peak performance. It's easy for a subcontractor to fall into a pattern of emphasizing service to the largest customers in times of peak production. By convincing the ECM that the OEM has a promising future, the ECM is less likely to neglect its responsibilities in times of peak performance.

20.2.3 Presenting the Corporate Goals and Objectives

When approaching an ECM as a possible partner, being prepared to describe corporate goals and objectives will ensure that the candidate can best evaluate if the partnership makes sense from their point of view. It is also to the benefit of the OEM to best describe the corporate outlook in order to generate interest beyond the current requirements to emphasize both current and future opportunity. While some ECMs may desire any project, others may be selective for partners that minimize risk.

The OEM should set goals that are realistic in terms of volumes. If initial runs will be prototypes, be realistic about expecting a large shop to continually produce quick turns of small kits that can consume precious equipment time. Vendors look at customers as profit centers: If there is no volume shipping in the short term, then in the absence of revenue, it must be believed that volume production will definitely occur in the long term. This is not to say that some shops do not enjoy prototype (small-volume) manufacturing; however, it is more than likely that an ECM that specializes in small volumes will not be able to handle a quick ramp to higher volumes. Contractors with greater capacity to handle high-volume manufacturing can help avoid having to transfer a product when quantities exceed the capacity of a smaller shop. Being up-front and stating the requirements of the OEM from prototype to production with a time schedule will avoid the mistake of partnering with an ECM with short-term thinking.

20.2.4 Business Plan

- *Past:*   How has the company or division evolved? Who are the founders? What has been the history of financials? What is the growth and profitability record?
- *Present:*   What stage is the product in? What is changing; why the ECM search? What are the short-term goals? What are the main features, advantages, and benefits of the product? What is the competitive advantage that makes the product a winner? What are the requirements to build the product—both equipment and personnel? Describe the organization managing the product.
- *Future:*   What forecast or goals does the company have and how does it plan on getting there? What is the anticipated market share? Describe the corporate line on outsourcing strategy. How does the product ramp up and what plans are in place for end-of-life support? What is the next-generation product?

Depending on the thoroughness of the OEM presentation, it will then be the task of the ECM to present to the OEM why a mutually beneficial partnership is viable and what services can be provided to fill the OEM needs. The better the OEM requirements can be described, the more customized an ECM's presentation can address specific needs.

20.3 SUPPLIER CAPABILITIES

20.3.1 Overall Capabilities

ECMs come in many shapes and sizes, some specializing in a few services and others attempting to offer the one-stop shop. Two main classes of ECM are those that are strictly consignment shops and those that are turnkey. For consignment ECMs there is also the option to use a third-party distributor to provide material.

20.3.1.1 Turnkey.   The ability to plan, purchase, receive, and inspect material as specified by the OEM-approved vendor list requirements.

20.3.1.2 Kitting.   This involves the same function as turnkey but using a third-party distributor to provide the service. While smaller ECMs lack the credit line to purchase three or more months of a customer's inventory, or they may not want the risks associated with turnkey manufacturing, many will use distributors as a third party to provide a kitting service. Although using a distributor (kitting service) adds another supplier to manage, this can provide an OEM the flexibility of keeping its buying power and allocation priority closer to the manufacturer of the components. Since distributors keep on hand many common components, they often have the ability to turn a kit, at a premium, with shorter lead times than an ECM.

20.3.1.3 Consignment Labor. Provided the OEM has elected to purchase material on its own or use a distributor to provide a kitting service, a consignment shop will be a viable option. Consignment shops are generally smaller-run organizations that can provide a more personalized level of service. On the other hand, some smaller consignment shops do not have the resources or talent of larger turnkey ECMs.

20.3.2 Services Offered

20.3.2.1 PCB Layout. Many ECMs can provide a service to take OEM schematics and lay out a PCBA with DFM and DFT in mind. In addition, fiducial markings and tooling-hole requirements may be specified in order to place the assembly on a particular type of manufacturing equipment.

20.3.2.2 Design for Manufacturability (DFM) and Testability (DFT). In the rush to get a product to market, engineers and designers do not always adhere to volume manufacturing guidelines. To prepare the product for volume assembly, it may be necessary to make adjustments in lead spacing or surface-mount device land patterns. These changes may be intended to improve manufacturability and assure compatibility with automated systems. If the ECM was not involved in the layout of the PCBA, provide them with adequate time to evaluate DFM. Improvements made by experienced contractors can save a company thousands of dollars over the duration of the product. DFT is a service that should be integrated into the DFM study. A thorough DFM study will include the placement of test vias should ICT be required.

20.3.2.3 Testing Capability. ECMs with experience in any of the three primary areas of test (ICT, FVT, and environmental stress screening, or ESS), can provide recommendations through experience with other current and past assemblies. ICT, the most commonly outsourced method of test, allows an ECM to verify for opens and shorts before shipping to the customer. Many ECMs will either develop ICT programs and fixtures or use a preferred third party to design the fixtures for a specific test base. An experienced test engineer will be able to recommend test point placement to maximize coverage and minimize test time. FVT requirements are generally developed by the OEM and then performed either at the site of the ECM or at the OEM. ESS testing depends largely on the application of the product and is dictated by the OEM's end use of the product. Some ECMs which offer ESS can provide assistance in developing fixtures to fit the needs of the OEM.

20.3.2.4 Material Procurement. One of the clearest differentiators between the large and relatively small ECMs is the ability or desire to take on OEMs that require turnkey services, i.e., the ability to plan and purchase the raw material before assembly and delivery. While larger ECMs generally have more experience, capital equipment, talent, and customers, they may require a minimum volume of business from the OEM to make the partnership worth their time. Considering that the material cost in a PCBA may represent 80 to 90 percent of the total cost, with assembly labor, test, and packaging accounting for the remaining 10 to 20 percent, many ECMs consider turnkey manufacturing to be more profitable despite the inherent risks. Contractors' ability to provide turnkey services allows a single order to be placed for products that may have several hundred line items to order. While manufacturing lead times may take one to two weeks, the purchasing and receiving of material may take up to 20 or more weeks for hundreds of parts; thus, working in a turnkey relationship becomes a major factor in developing a strong relationship.

20.3.2.5 Program and Fixture Support.   Aside from test, other nonrecurring fixture costs include the development of a program to run automated manufacturing equipment and fixtures that enable the manufacturer to assemble the product during various stages of the process. Some fixture requirements may include:

Stencil	Solder paste application
Vacuum plate	To hold PCB in place during solder paste application
Surface-mount fixture	To hold PCB in place during component placement
Wave fixtures	To hold PCB if wave solder is required for pin-through-hole components

A contractor should be able to develop the requirements for fixtures, including quantities, according to the throughput required by the OEM. Most ECMs will use a third-party source to have fixtures developed and modified when changes occur.

20.3.3 Support Services

Aside from the core services, which include purchasing and assembly, there are many supporting roles that play a key role in ensuring that quality, communication, and delivery are not overlooked.

20.3.3.1 Quality.   Undoubtedly, all suppliers claim that their products are built with quality in mind. They may even boast of high first-pass yields at ICT, and share with the customer data that has been collected on a statistical level. While it's a great tribute that some ECMs can build dozens of different assemblies for a dozen different customers, it's not enough to just take their word for it. Quality does not happen by chance, and if a working model is not in place in every area and strictly adhered to, then there can be no assurance that a continuous improvement plan is working. If work is taking place at the ECM manufacturing site, quality measures should be evident at several checkpoints along the process, from solder paste application to pack-out. While a quality engineer should be able to clearly explain the quality process, each operator on the line should be able to explain what his or her contribution is to a quality product. In order to buy off on a quality system it must be understood how the ECM takes critical yield information and feeds it back to the process. If a closed-loop corrective action procedure is not in place to trace failures back to the source, then how is information compiled and constructively fed back to the source?

20.3.3.2 Information Systems.   When it comes time to transfer information, many tools are available to ensure that data is being received accurately and in timely fashion. Schedule and engineering changes, delivery schedules, shortage reports, and quality information are all forms of data that will be passed back and forth from the OEM to the ECM. Modern information systems that can be used by both parties can efficiently streamline the communication process, saving time and money. Although there is no substitute for meeting face to face with a supplier, leaving the office can be prohibitive at times. Information tools include *phone mail* to avoid playing phone tag or leaving lengthy messages with an operator; *pagers,* for when immediate accessibility of an individual is required; *facsimiles* (fax) to transfer written documentation; *electronic mail,* which is probably the most efficient means of communicating when several members need to be notified at the same time; and *EDI,* which allows users to transmit infinite amounts of data through a modem. Information systems can play an integral role not just in establishing efficient means of communication between the OEM and the ECM, but they also help individuals communicating in their own environments with other team members as well as other suppliers they have to manage. Even if an OEM is not able to utilize fully all the information tools offered initially by the ECM, the investment has been made to make communication more timely and accurate over the term of the relationship.

20.3.3.3 Packaging/Shipping.   Before a finished good can be delivered, packaging specification must be met to ensure that no damage takes place during delivery. Most ECMs can assist in recommending bulk packaging requirements through a preferred third party. Proper packaging can prevent both *electrostatic discharge* (ESD) and physical damage. Bulk packaging may include an ESD bag and a carton with partitions, or a box that has been ESD-protected without a bag. For finished goods that require retail packaging, another assembly process must be added, which requires additional equipment and operators if shrinkwrap is necessary. If neither the OEM nor the ECM has the capability to provide finished packaging, a third party must be identified to take on the task.

For close-proximity shipping, it is useful to find out if the ECM is able to deliver finished goods to the OEM dock. Other options include third-party delivery or picking up product at the ECM dock. If the product is to be packaged for retail, then consider having product delivered directly to distribution or the end user.

20.3.4 Technical Capability

20.3.4.1 Processes Offered/Equipment List.   Depending on the level of difficulty to manufacture a specific PCBA, certain process skills are necessary along with the appropriate equipment. If a PCBA has fine-pitch devices on both the top and bottom sides, then a proven process should be in place to ensure the OEM of the ECM's competence. Understand the ECM's experience relative to the technology at hand. Based on their customers' requirements, some ECMs can be way out in front of the learning curve. For technologies such as chip-on-board (COB), tape-automated bonding (TAB), and multichip module (MCM), special processes and equipment will be necessary. If the preferred supplier does not have the processes or equipment to place relatively sophisticated processes, then it must be determined if the OEM wants to invest in the learning of the process as well as the equipment with the ECM. A review of the equipment list can not only help identify if a current process is in place, but whether future technologies are being invested in, to stay ahead of the OEM requirements.

20.3.4.2 Capacity.   Understanding the factors of capacity flexibility will provide guidance for avoiding growing pains. Most ECMs will run two to three shifts per day, five days per week, with the weekends available at a premium. Identify what the current capacity is for machine utilization, and then learn what measures the ECM can take to expand the current capacity. Is there room for additional manufacturing equipment? Are there sufficient cash or credit lines to support the cost of additional equipment? Are there other facilities in other locations that can provide the same service without altering the process?

20.3.4.3 Engineers.   Engineers come with many different backgrounds with various levels of experience. Process, manufacturing, quality, and test engineering must be well represented in each area of the ECM to ensure that each process is well defined by qualified personnel to build a quality product. Look for examples from the ECM that exhibit experience which is relevant to the technology required by the OEM. What qualifications does the engineer require before being empowered as the lead engineer of a given OEM product?

20.4 *SUPPLIER QUALIFICATION*

In an effort to select the best ECM candidate, a qualification list should be developed to include services required and desired (see Fig. 20.2). Candidates can be scored based on how well they perform in the areas most desired, with less emphasis placed on areas that may add value at a later time.

By putting together a list of desired capabilities (see Fig. 20.2, Supplier Qualification Survey) an OEM can better understand the services as they are demonstrated by potential suppliers in the areas of importance to better choose a successful long-term strategic partner. One method of measuring a potential candidate might be to assign a score after each question/category on a scale of 1 to 4: 1=complete, 2=adequate, 3=incomplete, 4=inadequate. For preliminary questionnaires, to be filled out and returned by the proposed supplier, a simple yes or no is adequate to provide background information on the levels of service that are available. Categories may include technology, quality, responsiveness, delivery, cost, and business strategy. After the results have been tallied, suppliers can be ranked based on their respective scores.

20.5 *TIME-TO-MARKET ELEMENTS*

By understanding the process that must take place before a product is ready for delivery, an OEM can better commit to a realistic shipment date. The Gantt chart in Fig. 20.3 walks through many of the milestones that must be covered before a shipment can be made.

20.5.1 Specific Elements: Descriptions and Time Requirements

Quote.   A labor quotation, depending on the complexity, should take no longer than a week to complete. Turnkey quotation will add an additional 1 to 3 weeks, depending on the total number of line items of material to quote.

Manufacturing purchase agreement.   Before the order is placed, terms and conditions should be put in writing that specify payment terms, schedule flexibility, material liability, warranty, etc. Most ECMs will have a boilerplate agreement that can be modified per each OEM's requirements. A copy of the quote will become an attachment to the MPA.

Purchase order.   The official purchase order (PO) binds an agreement for the ECM to purchase material to meet a specified quantity to be delivered on a specified date at a specified price. An ECM may require 2 to 10 business days to verify acceptance of the OEM required delivery dates.

Material requirement planning.   Once the BOMs have been loaded onto the ECM's material requirement planning (MRP) system, the purchasing activity can take place. First-time MRP execution for a new assembly number may take two weeks to perform to compensate for the initial setup of the specific requirements. When recurring orders are placed, one week will suffice.

Purchase.   Through a quotation, it is advisable to identify all long-lead items—those that extend beyond 12 to 14 weeks. To avoid having to pay expedite charges or missing a target delivery date, buys should be placed, by the OEM in advance, for all long-lead items that may jeopardize a desirable delivery date. Purchase orders can be taken over by the ECM, or initial buys of long-lead items can be consigned to the ECM.

Receiving/inspection.   Receiving and inspection (RI) will add a week onto the lead time of components. If an ECM is handling the purchasing function, it is also their responsibility to verify that the components meet the specifications of the AVL. For a consignment mode, the ECM's responsibility is only to verify the count of the parts as supplied by the OEM or distributor.

FIGURE 20.2   Supplier Qualification Survey.

Technology

Item   1-2-3-4

1. Equipment in place is capable of meeting product need.   _____
2. Process is released and documented.   _____
3. Available technical support personnel are qualified to understand needs and provide required
   ongoing production/sustaining engineering: _____   Number of engineers   _____
                                                              Number of technicians   _____
4. A process development activity is in place and has defined goals.
   % for cost reduction   _____
   % for new processes   _____
   % for other _____   (describe) _____   _____
5. Prototypes can be run on line that typifies production conditions.   _____
          1=Complete   2=Adequate   3=Incomplete   4=Inadequate

Quality Plans and Methods

Item   1-2-3-4

1. Quality control manual is available and can be shown to reflect actual shop practices.   _____
2. Process flowchart and operation documents are available.   _____
3. Process characterization is documented.   _____
      Process Capability Index derivation process clear   _____
      Other process capability measures (define)   _____   _____
4. Documented process for translating customer requirements.   _____
5. Process changes are under engineering control and communicated to the customer
   prior to implementation.   _____
6. Process documents are current.   _____
7. Process control plan is based on SPC principles.   _____
8. Process control points are determined with customer.   _____
9. SPC is used routinely in operations (control charts visible).   _____
10. Operators and technicians are trained in SPC methods.   _____
11. Control chart information is used to make adjustments to process
    (examples were shown).   _____
12. Defective material is identified, isolated, and recorded.   _____
13. Preventive maintenance program is documented and records show that it is followed.   _____
14. Tool calibration and traceability process is documented and shown to be in operation.   _____
          1=Complete   2=Adequate   3=Incomplete   4=Inadequate

Responsiveness

Item   1-2-3-4

1. Process exists for accepting and implementing engineering and schedule changes.   _____
2. Responsibility is defined for support of all customer issues.   _____
3. Problems are communicated to customer prior to detection by the customer.   _____
4. Process exists to provide price, schedule, and technical information in a time
   acceptable to the customer.   _____
5. Capacity exists to respond to short-term surges in demand.   _____
6. Prototype and small runs can be run inside regular lead time.   _____
          1=Complete   2=Adequate   3=Incomplete   4=Inadequate

Delivery

Item   1-2-3-4

1. Process exists that allow changes in delivery schedules.   _____
2. Demonstrated ability to deliver within an on-time window.   _____
          1=Complete   2=Adequate   3=Incomplete   4=Inadequate

FIGURE 20.2 Supplier Qualification Survey. *(Continued)*

Cost

Item	1-2-3-4
1. Total cost of ownership, including part price, inspection cost, warranty cost, inventory cost, and customer support cost is competitive.	_____ _____
2. Cost improvement processes are in place that are coordinated with the customer.	_____
3. Quotation process ensures that costs reflect level of technology required by the customer.	_____

1=Complete   2=Adequate   3=Incomplete   4=Inadequate

Business Strategy

Item	1-2-3-4
1. Business strategies are consistent with customer's long-term needs.	_____
2. Business is financially sound.	_____
3. A process for disaster recovery is documented and evidence of its use is available.	_____

1=Complete   2=Adequate   3=Incomplete   4=Inadequate

Decision Criteria: The result of the survey process will be used in conjunction with the price quotation process (see Sec. 20.7.2) to make a final decision on whether a potential supplier should be further considered and what additional information is required prior to actually starting a business relationship.

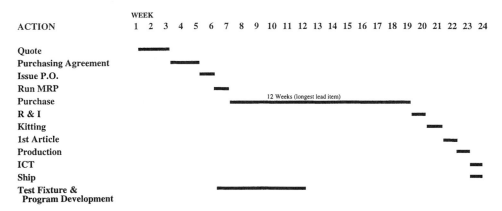

FIGURE 20.3 Typical timeline for starting a turnkey manufacturing relationship (assumes that long lead items beyond 12 weeks are ordered in advance).

Kitting. Parts are assembled in kits and segregated by assembly number. Time will vary depending on the number of parts per assembly.

First article. For first-time runs of new assemblies, an additional week may be necessary to verify the manufacturing process as well as the functionality of the product.

Production. Lead times can vary from 1 to 10 days, depending on the complexity.

In-circuit test. ICT can add an additional day or two, depending on complexity and first-pass yields. Failed units will have to be debugged and put through a rework process.

Ship. Shipping can range from same day or overnight to one week, depending on proximity and method of transportation.

Test fixtures and program development. Can take eight weeks or more, but should be accomplished within the materials lead time.

20.6   QUALITY SYSTEM

To understand how a particular product can be built to meet or exceed the expectation set by the OEM, ECMs must have in place a quality system that will aid the manufacturer information that can continuously improve the process. An effective quality system will not only ensure that a product's specification meets the needs of the end user, but can greatly contribute to the economic performance of the product.

20.6.1   Total Quality Management

Quality procedures and policies must be clearly written and understood by all employees who affect the outcome of a product's manufacturing cycle. In order to meet and/or exceed the customer's quality and reliability expectations, various quality policies must be in place and constantly managed by a quality management team or *total quality management* (TQM). This process will drive quality improvements based on a set of business and quality measurements that are reviewed at the highest level of the organization and then delivered to all employees through communication and/or action item meetings. Effective quality/continuous improvement will involve all personnel through meetings that solicit input from employees for improvements and problem resolutions.

20.6.2   ISO 9000

The International Standards Organization (ISO) has developed a list of procedures that enable corporations to adhere to guidelines that outline a method of demonstrating the adequacy of the quality system and its capability to achieve and service conformity with the specified requirements. ISO 9002 certification focuses on the manufacturing compliance per the specification as provided by another source while ISO 9001 certification focuses on compliance with regard to the design and the manufacturability of an assembly.

20.7   REQUEST FOR QUOTATION

20.7.1   Purpose

Aside from understanding what costs are associated with building a particular product, the quotation can serve as formal response from the ECM as to how efficiently it proposes to build the product, with any or all assumptions clearly documented. While some cost drivers can be assumed and built into the quote, others cannot be understood until several months' worth of production has gone by. Through a manufacturing agreement, terms should be discussed to identify and understand what costs are associated with the various cost drivers previously listed. Expectations should be put in writing to help avoid misunderstandings that will undoubtedly surface over the course of a relationship.

20.7.2   Contents and Sample

At a minimum, the information requested through a quotation should include assembly labor cost, test labor cost, material cost, nonrecurring expenditures, and all assumptions that may affect these categories. Figure 20.4 illustrates a quotation model.

FIGURE 20.4 Sample Quote.

Date   Bid # Revision

To: Customer Name
         Corporation

From: ECM

Re: Quote for Assembly #, Revision #, Assembly Name

Assembly # quantity/year	50 prototype	6,000	9,000	12,000
Assembly labor	$ 0.00	$ 0.00	$ 0.00	
Test labor	0.00	$ 0.00	$ 0.00	
Material cost	0.00	$ 0.00	$ 0.00	
Unit cost	$ 0.00	$ 0.00	$ 0.00	

Nonrecurring expenditures:

      Manufacturing startup: $0.00 (includes all setup costs)

      Router plate: $0.00

      PCB tooling: (if needed) $0.00

      ICT test fixture: $0.00

      ICT test program: $0.00

Notes and assumptions:

      Payment terms; build location; period of performance: 4/95–3/96; technology used; document baseline—(*date*).

      Manufacturing assumptions: PCB will be panelized (*number of boards per panel*); quote assumes kit size (*minimum kit size*);

      Test assumptions: (*time standards, assumed first-pass yields*)

      Material assumptions: All components are quoted per customer AVL; minimum piece buys (*list part numbers and quantities*); list all long-lead items beyond 12 weeks; list alternate vendors that may cost less.

To accurately quote an OEM's manufacturing requirements, the following documentation is required:

Documentation needed	*Purpose*
Bill of material (BOM)	To quote materials and identify reference designators
Approved vendor list (AVL)	To quote material as specified by the customer
PCB fabrication drawings	To quote board fabrication costs
Assembly drawings	To quote assembly labor
Schematics	To quote test times, fixtures, and programs
	Other items of interest:
Gerber files, aperture list	To have PCB fab produced
Sample assemblies	To assist in DFM study and assembly labor quote

Quantities: After establishing an estimated annual forecast, the volumes required by the OEM to be quoted should represent both up-side and down-side possibilities, as well as prototype and preproduction quantities (see Fig. 20.4). Quantities may be quoted on an annual or monthly basis. Material prices are usually quoted on an annual usage basis, while assembly labor is quoted on a per-run basis in order to understand the amount of setups on a manufacturing line as well as material turns on inventory.

20.7.3 Responses Required

Unit cost: Total cost of final product including all services and materials as specified by the customer and/or specified by the supplier.

Assembly labor: The cost to manufacture a product, including the overhead and profit-related expenses, as specified by the customer and/or specified by the supplier.

Material cost: The cost of all materials, per unit, including overhead, profit, freight, and attrition. Depending on material value and location of vendor, freight and attrition may average 2 percent.

Test costs: The cost to test, per unit, as specified by the customer. Time standards, throughput, and first pass yields, including estimated debug time per unit, should be understood up-front in order to verify time and yield estimates should they be grossly over- or underestimated.

Nonrecurring expenditures: The cost to develop all fixtures and programs proprietary to the customer's specific needs (See quotation model in Fig. 20.4).

In order to understand cost drivers, other information to gather that can be useful through the quote stage may include a costed BOM (by line item) and manufacturing processes and time standards. Suppliers may consider this information confidential, but depending on the relationship and/or the ability to leverage the supplier, an OEM may have access to virtually an unlimited amount of information.

20.8 MANAGING THE RELATIONSHIP

Because the vendor selection process can take a considerable amount of time and effort, managing the relationship effectively can prevent having to experience a learning curve in the manufacturing process as well as the relationship process. A failed partnership can mean disaster to the OEM—i.e., no product shipping.

20.8.1 Getting Started

A kick-off meeting should be assembled to introduce a program/product management team (see Fig. 20.5). In that meeting team members can explain their value-add or responsibilities to the program's success.

20.8.1.1 Program Management Team.   The *program manager* representing the ECM can be used as a focal point for all strategic information. Issuing purchase orders, engineering changes, and schedule changes should all be channeled through one individual to keep communication consistent. Surrounding team members should be looked on as support members who work in a tactical mode. Depending on the particulars of an ECM, the *manufacturing engineer* may be designated to develop the *manufacturing process instructions* (MPI). A *planner* will be responsible for assuring for the timing between manufacturing capacity and material availability. The *quality engineer's* responsibility may include collecting yield data and providing the information to the manufacturing engineer, who may need to fine-tune the process. The *test engineer* will provide timely yield information to the manufacturing engineer to point out areas of concern about failures. The *sales manager* may be responsible for processing quotations and proposals and/or

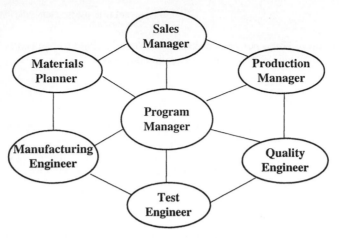

FIGURE 20.5 Example program management team.

Customer Satisfaction Survey

FROM: TO:
CUSTOMER: _____   PROGRAM MGR: _____
RESPONDENT: _____   FAX #: _____
FAX: _____   DATE: _____

4: Excellent at meeting and exceeding set requirements.
3: Generally meeting requirments most of the time.
2: Satisfactorily meeting requirements - could use improvement.
1: Generally missing requirements - needs development.
0: Unacceptable

Category	4	3	2	1	0
Quality					
Delivery					
Communication					
Service					
Overall					

COMMENTS:

SIGNED: _____ DATE: _____

FIGURE 20.6 Customer satisfaction survey.

monitoring the overall satisfaction level of the customer. While maintaining strategic information through the program manager, communication should be encouraged for each team member from all team members from both the ECM and the OEM. Meetings should be set on a regular basis either in person or by phone. The program manager, keeping an updated list of action items, can be in charge of seeing that all team members meet their obligations.

20.8.2   Evaluating Performance: Customer Satisfaction Survey (CSS)

By using a *customer satisfaction survey* (see Fig. 20.6) an OEM can quantify on a regular basis the level of service that is being provided by its supplier. The ECM can in turn use the CSS to provide guidance into the critical service areas as determined by their customer. ECMs can use the CSS report as a tool to provide corrective action plans as the customer identifies problems that have not been corrected through the day-to-day relationship. Although action items may be worked on and discussed on a regular basis, it's helpful for a supplier to understand exactly where they stand in the eyes of the customer. Depending on how frequent and heavy interaction is between a supplier and customer in relationship, a scorecard or CSS can be submitted weekly, monthly, or perhaps even quarterly.

Categories to score on a regular basis may include quality, delivery, communication, and service. A quality measurement may be based on first pass yields at the site of the OEM at functional test. A goal for first-pass yield may be put in place based on a mutual agreement between the OEM and the ECM. As the ECM gains experience from manufacturing a particular OEM requirement, quality yields may be expected to increase. Scoring an ECM's ability to deliver product may be based on the requirements as stated on the PO, with consideration given to account for material that may have developed unusually long lead times due to shortages. Communication can be more of a subjective category when critiquing the performance of an ECM.

Poor marks in communication may mean anything from a lack of understanding between parties to a lack of response on a timely basis. Parameters can be set as to what kind of response time is required from an ECM for most recurring requests. Service can be another subjective category that lends the ECM an idea of how well the OEM feels it is being taken care of. When any category score depicts above or below what is expected in the relationship, it is imperative that the OEM describe in detail what it is experiencing. Specific comments will allow the ECM to respond to criticism in a constructive manner through the use of a corrective action plan. The CSS can also be used for appraisal, which can help motivate team members who deserve recognition. By using a CSS effectively, OEMs and ECMs can better define their objectives to work towards the same end result.

P · A · R · T · 4

HIGH-DENSITY INTERCONNECT

CHAPTER 21
INTRODUCTION TO HIGH-DENSITY INTERCONNECTION TECHNOLOGY

Happy T. Holden
Westwood Associates, Loveland, Colorado

21.1 HIGH-DENSITY INTERCONNECTS (HDIs) DEFINED

The use of more complex components with very high I/O counts has pushed board fabricators to reexamine techniques for creating smaller vias, and many new or redeveloped processes have appeared on the market. These processes include revised methods of creating holes, such as laser drilling, micropunching, and mass etching; new methods for additively creating dielectric with via holes using photosensitive dielectric materials; and new methods for metallizing the vias, such as conductive adhesives and solid-post vias. All of these methods share some common traits. They all allow the designer to significantly increase routing density through the use of vias in SMT pads, to reduce the size and weight of product, and to improve the electrical performance of the system. These types of boards are generically called high-density interconnects (HDIs). An HDI board typically will have, as an average, over 120 to 130 electrical connections per square inch (20 connections per square centimeter) on both sides of the board. The IPC defines any via, blind, buried or through, of 150 μm (0.006 in) or less in size, as a microvia. HDI boards require microvias.

21.1.1 HDI Characterization

This generation of printed boards is characterized by microblind, buried, and through vias made by techniques other than mechanical drilling. In order to turn blind vias into buried vias, these process techniques are repeated and the layers are built up, hence the name build-up or sequential build-up (SBU) circuits.

This type of printed circuits actually came into being in 1980, when researchers started investigating ways to reduce the size of vias. The actual first innovator is not known, but surely projects using LaserVia™ by Larry Burgess of MicroPak Laboratories, or photodielectric vias produced by Dr. Charles Bauer at Tektronix,[1] or plasma-etched vias by Dr. Walter Schmidt at Contraves rank as some of the first. Laser-drilled vias were used in mainframe computer multilayers in the late 1970s. These were not as small as the laser-drilled vias of today and were produced only in FR-4 with great difficulty and cost.

The first production build-up or sequential printed boards were manufactured in 1984 with the Hewlett-Packard laser-drilled Finstrate computer boards,[2] then in 1990 in Japan with Surface Laminar Circuits™ by IBM[2] and in Switzerland with DYCOstrate™ by Dyconex.[3] Figure 21.1 shows one of the first Hewlett-Packard Finstrate boards and one of the first IBM SLC™ boards.[2]

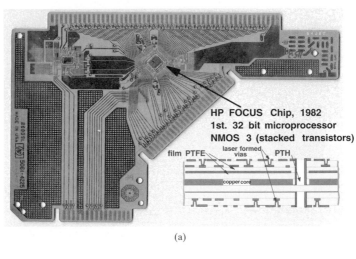

(a)

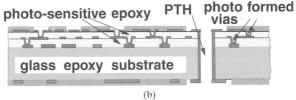

(b)

FIGURE 21.1 The first microvia PC boards in general production. (*a*) Hewlett-Packard's Finstrate was put into production in 1982. It was a copper-core, build-up technology that had direct wire-bonded ICs. Vias were mechanically drilled through-holes and laser-drilled 5-mil blind vias in six layers of PTFE dielectric. (*b*) The first photodielectric microvia board was produced in volume in Japan by IBM. This is the SLC technology with two build-up layers on one side of four conventional layers.[2]

21.1.2 Advantages and Benefits

Four main factors drive printed boards to require higher wiring densities:

1. More components can be placed on both sides of the printed circuit.

2. Components are closer together.

3. The size and pitch of components are smaller while the number of I/Os is increasing.

4. Smaller geometries allow faster transmission of signals and reduce signal crossing delay.

At the same time, enhanced performance is required for faster signal risetimes, reduced parasitics, reduction of RFI/EMI, fewer layers, and improved high-temperature performance and reliability. HDIs provide all of these advantages and more.

21.1.3 Comparison of HDIs and Traditional Printed Wiring Boards

The interactions between printed circuit boards, components, and assemblies are best seen using the packaging technology map (Fig. 21.2). Components are characterized by the average I/Os per part, assemblies by the components per square centimeter and I/Os per square centimeter, and the printed circuit by its wiring density, centimeters per square centimeters. Figure 21.2 shows the approximate crossover between traditional printed circuits and next-generation circuits with microvias.[4]

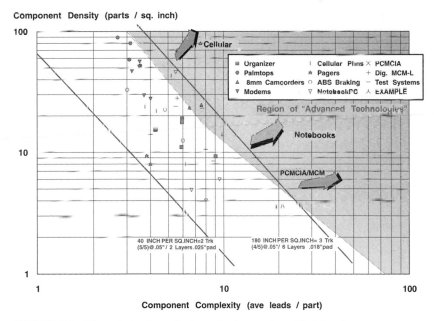

FIGURE 21.2 Microelectronics packaging technology roadmap. Each icon is a separate assembly. The diagonal lines are PC board wiring density. The shaded area is the crossover from conventional through-holes to HDI microvias.[4]

21.1.4 Design/Cost/Performance Trade-offs

The HDI structure is cost effective for the higher-density assemblies shown throughout this chapter. The relative price of HDIs vs. relative density is depicted in Fig. 21.3. Cost parity (for similar wiring densities) is achieved with a four-layer HDI microvia at approximately an eight-layer, through-hole printed-circuit multilayer. Wiring capacities and densities greater than an eight-layer multilayer can be achieved at a lower price with a properly designed HDI substrate. At very high densities there are no through-hole multilayers that can meet the demands for wiring capacity and density, while HDIs can easily meet the requirements.

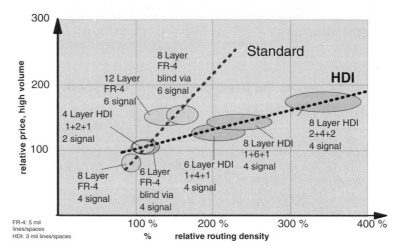

FIGURE 21.3   Price and density comparison between conventional and HDI printed circuits. The 1 + X + 1 represents the number of HDI layers on each side of the core multilayer of X layers.

21.1.5   Specifications and Standards

The IPC standards that cover HDI design rules, materials, and specifications are as follows.

- *IPC-2315:* educates users in microvia formation, selection of wiring density, selection of design rules, interconnecting structures, and material characterization. It is intended to provide guidelines for use in the design of printed circuit boards utilizing microvia technologies.

- *IPC-4104:* identifies materials used for high-density interconnection structures. A series of specifications (slash sheets) are defined for specific available materials. Each sheet outlines engineering and performance data for materials used to fabricate high-density interconnecting structures. These materials include dielectric insulator, conductor, and dielectric-conductor combinations. The slash sheets are provided with letters and numbers for identification purposes. For example, if a user wishes to order from a vendor and reference specification sheet number 1, the number 1 is substituted for the S in the designation example (IPC-4104/1). To start the ordering process, one can use the slash sheets in the IPC-4104 document in combination with relevant IPC documents for each material sets (e.g., IPC-CF-148, IPC-MF-150, IPC-4101, IPC-4102, IPC-4103, etc.).

- *IPC-6016:* contains the general specifications for high-density substrates not already covered by other IPC documents.

21.2   HDI STRUCTURES

The IPC HDI Structure Subcommittee responsible for defining performance requirements utilized the following methodology to specify HDI products. Since a microvia can be any shape, including straight wall, positive or negative taper, or cup, determining the methods used for producing microvias can be segmented into three methodologies noted in Table 21.1.

TABLE 21.1 Ten Process Methods Utilized
to Produce Microvias

Hole formation	Microvia process*
Mechanically drilled	A
Dry-etched (plasma)	A
Mechanically punched	A
Abrasive blast	A
Laser-drilled	A
Post-pierced	B
Photo-formed	A
Insulation displacement	B, C
Wet-etched	A
Conductive bonding sheets	C

* A: create hole, then make conductive; B: create conductive via, then add dielectric; C: create conductive via and dielectric simultaneously.

All of these technologies provide approximately the same high-density design rules. These design rules endow build-up technologies with four to eight times the wiring density of conventional technologies with all drilled through-hole vias. The 10 via profiles are depicted in Fig. 21.4.

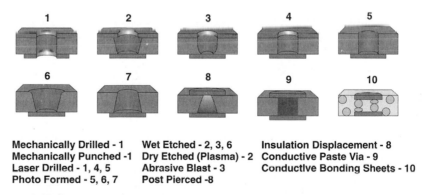

Mechanically Drilled - 1   Wet Etched - 2, 3, 6   Insulation Displacement - 8
Mechanically Punched -1   Dry Etched (Plasma) - 2   Conductive Paste Via - 9
Laser Drilled - 1, 4, 5   Abrasive Blast - 3   Conductive Bonding Sheets - 10
Photo Formed - 5, 6, 7   Post Pierced -8

FIGURE 21.4 Representative microvia profiles and the processes that produce them.

The structure of the high-density interconnection is by type (see Table 21.2). However, some of these construction types are based on which microvia material is to be utilized. Thus, the following definition applies to all high-density interconnect substrates (HDISs).

The core, defined as [C], can be identified as an A-, B-, or C-type core. Thus, [CA] is a core with internal vias only; redistribution makes contact to the surface. [CB] is a core with internal and external (through-microvia) structures. High-density interconnecting structures make contact into the innerlayers of the core. [CC] is a passive core with no interconnections.

TABLE 21.2   IPC-2315 Definitions of the Various High-Density Interconnect Substrates (HDISs)

Type I	1[C]0 or 1[C]1, with through-vias from surface to surface
Type II	1[C]0 or 1[C]1, with through-vias buried in the core and from surface to surface
Type III	≥2[C]≥0, 2 or more HDI layers added to through-via in core or from surface to surface
Type IV	≥2[P]≥0 where P is a passive "substrate with no electrical connecting functions"
Type V	Coreless construction using layer pairs
Type VI	Alternate construction using conductive pastes

[C], printed circuit core; [P], passive substrate core; 0, 1, and 2, number of build-up layers on each side of the core [C] or [P].

21.2.1   Construction Types

Type I to Type VI constructions currently describe all known HDI build-up structures, but, as the technology evolves, new ones are surely likely to be created. The notation used is:

$$x[C]x$$

where x = is the number of build-up layers on that side of the core and [C] denotes a standard laminated core of materials, with or without vias of n layers.

21.2.1.1   Type I Constructions (1[C]0 or 1[C]1).   This construction describes an HDIS in which both plated microvias and plated through-holes are used for interconnection. Type 1 constructions describe the fabrication of a single microvia layer on either one side (1[C]0) or both sides (1[C]1) of an underlined printed circuit substrate core. The printed circuit substrate core is typically manufactured using conventional printed circuit techniques. The substrate may be rigid or flexible and can have as few as one circuit layer or be as complex as a prefabricated multilayer printed circuit with buried vias. A single layer of dielectric material is then placed on top of the core substrate. Microvias are formed in the dielectric connecting layer 1 to layer 2 and layer n to layer $n - 1$. Through-holes are then drilled connecting layer 1 to layer n. The microvias and through-holes are then metallized or filled with conductive material. Layer 1 and layer n are circuitized and fabrication is completed.

21.2.1.2   Type II Constructions (1[C]0 or 1[C]1).   Type II has the same HDIS layers as Type 1. The difference is the core, [C]. Type II allows through-vias to be placed in the core before the HDI layers are applied. The processes are the same except for the core through-vias being filled before the HDI layers are applied.

21.2.1.3   Type III Constructions. (≥2[C]≥0).   This construction describes an HDIS in which both plated microvias and plated through-holes are used for interconnection. Type III constructions describe the fabrication of two microvia layers on either one side (2[C]0) or both sides (2[C]2) of an undrilled or drilled printed circuit substrate core. The printed circuit substrate core is typically manufactured using printed circuit techniques. The substrate may be rigid or flexible and have as few as one circuit layer or be as complex as a prefabricated multilayer printed circuit with buried vias. A single layer of dielectric material is then placed on top of the core substrate. Microvias are formed in the dielectric connecting layer 2 to layer 3 and layer $n - 1$ to layer $n - 2$. This first microvia layer is either metallized or filled with conductive material and then circuitized. A second layer of dielectric material is then placed on top of this circuitized layer and microvias are formed connecting layer 1 to layer 2 and layer n to layer $n - 1$. Through-holes are then drilled connecting layer 1 to layer n. The microvias and through-holes are then metallized or filled with conductive material. Layer 1 and layer n are circuitized and fabrication is completed.

21.2.1.4   Type IV Constructions. (1[P]0 or 1[P]1 or >2[P]>0).   This construction describes an HDIS in which the build-up layers are used over an existing drilled and plated passive substrate. The printed circuit or metal core substrate is typically manufactured using conventional printed circuit techniques. This substrate may be rigid or flexible.

21.2.1.5 Type V Constructions. The Type V construction describes an HDIS where there is no core. Both plated and conductive paste layer pairs are interconnected through a co-lamination process. The multilayer is created with an even number of layers (two-sided flex or rigid layer pairs) laminated together at the same time the interconnections between the even and odd layer are made. This is not a build-up process; it is essentially a single lamination parallel process.

Layer pairs are prepared using conventional processes of etching, plating, and drilling, etc., or by conductive paste processes. The layer pairs are laminated together using B-stage resin systems or some other form of dielectric adhesive into which conductive paste vias have been placed.

21.2.1.6 Type VI Constructions. Type VI describe constructions of HDISs in which the electrical interconnections and wiring can be formed simultaneously. Another variety forms the electrical interconnections and the mechanical structure simultaneously. The layers may be formed sequentially or co-laminated. The conductive interconnect may be formed by means other than electroplating, such as anisotropic films/pastes, conductive pastes, dielectric piercing posts, etc.

21.2.2 Design Rules and Categories

The designer should be aware that not all fabricators have equal capabilities in the areas of fine-pitch imaging, etching, layer-to-layer alignment, via formation, and plating. For this reason the HDIS design guide categorizes design rules into two categories, the preferred producibility range and the reduced producibility range. For simplicity of design, this handbook further divides the design rules into four category ranges—A, B, C, and D—with A being the easiest to produce and D being the most difficult. Selection of more stringent design standards limits the number of fabricators capable of producing such a board. Circuits produced with design rules in the A category will be easier to produce and will have higher yields and therefore can be fabricated at lower cost. To keep costs at a minimum, the design rules most appropriate for the application should be selected.

21.2.2.1 Category A. This specification allows conventional HDIS processes to be used with relaxed tolerances. It should have the highest yield and lowest cost. It is estimated that 95 percent of HDI fabricators can meet these design rules.

21.2.2.2 Category B. This is the conventional HDIS process. It is estimated that 60 percent of HDI suppliers can meet these design requirements under production conditions.

21.2.2.3 Category C. Top-level fabrication shops, representing 20 percent of HDI fabricators, can meet these design rules. Panel sizes are often reduced to increase yield, which increases final cost. Production volumes are presently limited, with special attention required during the production process.

21.2.2.4 Category D. These rules require smaller panels and more exotic fabrication techniques. They are generally required only in electronic packages, chip-on-board (COB), flip-chip interposers, or MCM applications. At present, yields are lower and costs are high. It is estimated that less than 1 percent of all fabricators can achieve these design rules in limited production or prototype volumes.

Symbol	Feature	Category A, μm (mil)	Category B, μm (mil)	Category C, μm (mil)	Category D, μm (mil)
a	Microvia hole diameter at target land (as formed, no plating)	125–200 (5–8)	100–200 (4–8)	75–250 (3–10)	35 (1.5) minimum 350 (14) maximum
b	Microvia diameter at capture land (as formed, no plating)	350/14	300/12	250/10	150/6
b-1	Via in SMT pad width	300/12	250/10	250/10	130/5
c	Landing pad diameter	350/14	300/12	250/10	130/5
d-1	Conductor/landing pad spacing on rigid innerlayer	125/5	100/4	75/3	50/2
d	Pad-to-pad spacing on innerlayer	125/5	75/3	75/3	75/3
e-1	Conductor line width on landing layer	125/5	100/4	75/3	75/3
e	Conductor width on inner-layers	125/5	75/3	75/3	75/3
f	Minimum finished hole size—fhs (plate through-hole)	250/10	200/8	150/6	50/2
g, o	Surface via pad (fhs + annular ring × 2)* minimum through hole pad diameter	fhs + 350/14 600/24	fhs + 300/12 500/20	fhs + 250/10 400/16	fhs + 100/4 150/6
h	Trace spacing on rigid outerlayer	125/5	100/4	87/3.5	50/2
i	Trace width on rigid outerlayer	125/5	100/4	87/3.5	50/2
k	fhs drill hole plane clearance on innerlayer (fhs + annular ring × 2)*	fhs + 700/128	fhs + 66/24	fhs + 500/20	fhs + 400/16
l	Surface conductor to unplated hole	250/10	200/8	200/8	200/8
n	Minimum unplated hole diameter	350/14	300/12	300/12	300/12
m-1	Minimum HDI dielectric thickness	75/3	62.5/2.5	50/2	25/1
ar	Minimum aspect ratio (m-1)/a	<1.0	1.0	1.0	1.3 (minimum 50 μm)
z	Minimum board thickness	725/29	600/24	500/20	200/16
m	Minimum core thickness	100/4	62.5/2.5	50/2	50/2
p	Minimum prepreg thickness	100/4	62.5/2.5	50/2	50/2
	Minimum plated thickness	25/1	25/1	30/1.2	30/1.2

FIGURE 21.5   Example of HDI design rules. Illustrated is IPC Type II HDI structure with design rules for categories A, B, C, and D. *(After IPC-2315, "Design Guidelines for HDI and Microvias."[5])*

Typical HDI design rules are given in Fig. 21.5. This bridges the two design categories from IPC-2315, "Design Guideline for HDI and Microvias."[5] This is a Type II HDI structure.

21.3 DESIGN OF HDI BOARDS

HDIs and microvias place a new burden of printed circuit design. The various IPC types are significant changes in the multilayer stack-up from a conventional board. Additionally, however, microvias can be implemented in many different ways and with different design rules. This is not an exhaustive list of design issues.

21.3.1 Design Tools

Printed circuit design has become one of the most important functions in the electronic product realization process.[6] The demands on printed circuit design keep mounting. Among the reasons are:

- To regain time for product schedules
- To lower costs for fabrication and assembly
- To enable new area array components: CSP, microBGA, and flip-chip
- To decrease time to market
- To improve electrical performance for high-frequency and EMI reduction

To bring the process back under control requires a methodology that involves planning the printed circuit layout process with predictive wiring density models. The other benefits include reducing printed circuit fabrication and assembly costs.

21.3.1.1 *Required Features of HDI CAD Toolsets and Autorouters.* Following is a partial list of features CAD tools will require (some of which have been discussed) to design HDI boards.

- Optimization of mixed vias by router
- Autorouting cost budget for mass via generation
- Buried/blind via control on layer
- Via pad stack control (landless)
- Staggered via control
- Pad within pad
- Blind via push/shove during manual routing
- Any-angle routing
- Manufacturing process rules at all design phases
- Buried components

21.3.1.2 *Autorouters.* Autorouters have been a part of printed circuit design systems for many years. An autorouter automatically places vias and traces on the printed circuit based on the schematic and part geometries. Elaborate configuration menus drive the appropriate placement of these features. A special autorouter is required for HDI structures, because many HDI processes employ mass via generation as utilized for surface laminar circuit (SLC) technologies. These processes produce all the vias simultaneously and at any desired diameter. Since the cost of vias is rather insignificant, the autorouter should have the capability of achieving a near-zero via cost. When this is done, a different but more optimized design results. Figure 21.6 shows the autorouter setup menu for the Zuken-Redac CAD System VISULA 8.0.

21.3.2 Trade-off Analysis

After a product has been partitioned, the circuit designed, and the components selected, the physical design must be planned with an eye toward the lowest manufacturing cost while meeting all performance and operating boundary conditions. This is especially true for HDI

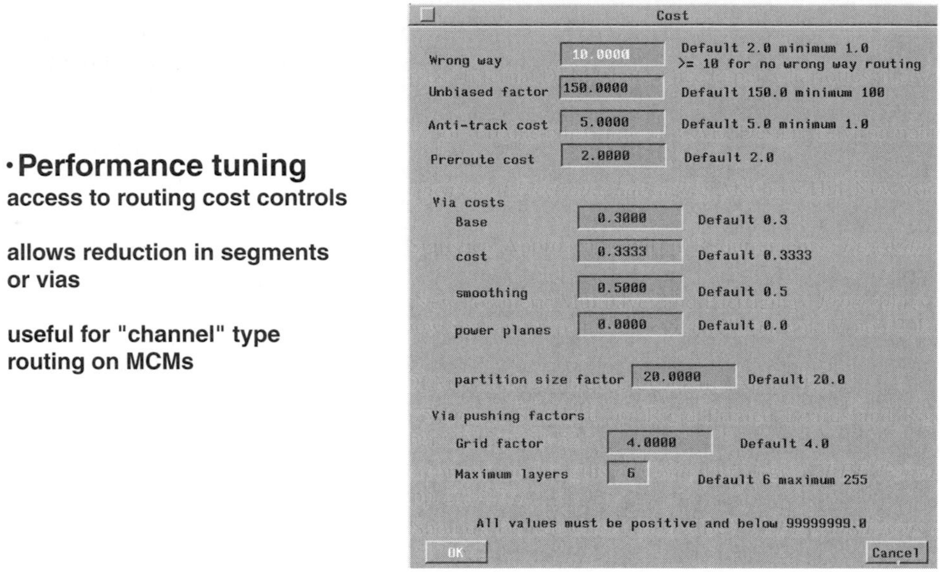

·Performance tuning
access to routing cost controls

allows reduction in segments
or vias

useful for "channel" type
routing on MCMs

Cost		
Wrong way	10.0000	Default 2.0 minimum 1.0 >= 10 for no wrong way routing
Unbiased factor	150.0000	Default 150.0 minimum 100
Anti-track cost	5.0000	Default 5.0 minimum 1.0
Preroute cost	2.0000	Default 2.0
Via costs Base	0.3000	Default 0.3
cost	0.3333	Default 0.3333
smoothing	0.5000	Default 0.5
power planes	0.0000	Default 0.0
partition size factor	20.0000	Default 20.0
Via pushing factors Grid factor	4.0000	Default 4.0
Maximum layers	6	Default 6 maximum 255

All values must be positive and below 99999999.0

OK   Cancel

FIGURE 21.6 Example of autorouter menu for an advanced CAD system. Tuning an autorouter can adjust the way it will design HDI boards.

designs. Conventional through-hole printed circuit design has not changed too much over the years. Finer geometries, more layers, and surface mounting have been added, but the design process remains essentially unchanged. Microvias or HDIs, however, have necessitated many changes and require new design rules and layer structures. Experience and history will not help here. The unfortunate truth about printed circuit layout is that there is an almost infinite number of combinations of layer structures and design rules that can satisfy the schematic and bill of materials. With all these choices, especially the new ones that HDIs offer, a trade-off tool is required to find the best set of design rules and features that allows the rapid design of a printed circuit and enables the design to be producible and meet all the performance expectations while providing the lowest total manufacturing costs. When used early in the design process, before the actual physical design of the printed circuit, the tools require predictive models that can anticipate cost and performance. Information and processes for planning an HDI design can be found in Chap. 18.[7-13]

21.4 MATERIALS

21.4.1 HDI Material Requirements

The materials used for HDI structures are relatively different from those used to manufacture standard printed circuit boards. There is a similarity in dielectric properties; however, the thin HDI materials are coordinated with the method to produce the microvias or small plated through-holes. Some materials are laminated to a core structure; others are deposited. If the method of microvia fabrication uses photosensitive techniques, then the dielectric contains a photosensitive polymer. There are several major techniques for producing microvias—mechanical drilling, laser drilling, plasma etching, chemical etching, etc.—each of which favors a different choice of material, although some can use a variety of thin HDI materials.

The fabrication of HDI structures requires that these thin materials be applied to a core. The core may be passive, such as a sheet of aluminum, or may provide an active part of the circuit, like a multilayer printed board. In addition, HDI redistribution layers are used for single-chip or multichip component assembly, as well as completed printed board assembly structures. The electrical and physical property requirements of an HDI product can dictate that the materials be slightly different. In some of the HDI structures the core is manufactured first, and then the HDI layers are added to one or both sides depending on the design's functional requirements, as in sequential build-up (SBU) structures.[14]

The IPC-4104 specification defines material requirements for HDI applications to meet user expectations necessary for electronic components, where die I/Os are redistributed from peripheral to array patterns, or for a completely functional electronic assembly.

21.4.2 Copper-Clad Dielectrics

Copper-clad dielectrics for HDIs can be reinforced, as in FR-4, or unreinforced, such as coated copper foil. The dielectrics can be epoxy, as in FR-4, or polyimide, cyanate ester, BT, PPE, or PTFE. Reinforcements are typically glass cloth, but there are a variety of glasses, as well as aramid paper and exotic fibers like quartz or carbon fiber. Figure 21.7 shows many of the resins and reinforcements for HDI materials.

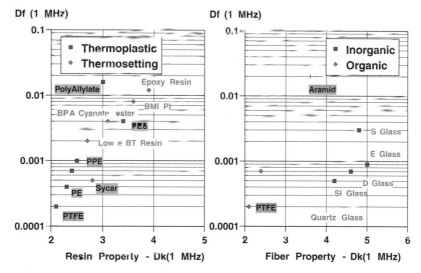

FIGURE 21.7   Electrical properties of typical printed circuit dielectrics, showing dielectric constants and dissipation factors (loss tangent) of dielectric reinforcements and bulk resins used in printed circuit laminates and films.

21.4.2.1 Coated Copper Foils

Resin-Coated Foils.   Resin-coated foils are the most common materials used for many build-up multilayer microvia applications. Many product variations are available, and fit well within the existing multilayer manufacturing infrastructure. Epoxy-based coated foils are the most common and have performance properties similar to those of FR-4 but have no glass reinforcement. Peel strengths, thermal performance, and electrical properties are excellent. A variety of other resin systems are also being developed and are starting to be used for coated foil build-up applications. Resin-coated foils come in two general types with either one or two resin layers. One-pass coated foils have a single B-stage layer designed to flow, fill, and provide thickness control at the same time. Two-pass coated foils have a C-stage resin layer adjacent to the foil and a B-stage layer for flow and fill. The fully cured C-stage layer acts as a "stop" during the lamination process, typically enabling better thickness control. Resin-coated foils are available in a variety of thicknesses, yielding finished dielectric layers between 1.0 mil (25 μm) and 3.0 mils (76 μm) thick. The copper foils most commonly used are ½ oz (18 μm) and ⅜ oz (13.34 μm), but there is substantial interest in thinner copper foils for improved laser efficiency and better fine-line circuitry definition.

Other Resins.   Since build-up technology is still in its relative infancy, but is evolving rapidly, many different and diverse approaches are being investigated relative to the resins used and variations on the via formation process. Low-D_k/D_f resins such as polyphenylene ether (PPE) are being used to address signal speed and integrity demands in resin-coated foil build-up structures. Another approach is to use additively platable resin and use the copper as a sacrificial carrier, eliminating the need to laser through copper. An additively platable resin has the characteristic of high surface adhesion to electroless copper or direct metallization. The sacrificial foil makes it compatible with normal multilayer lamination processes, saving the need for expensive coating machinery while also protecting the thin dielectric from harm until it is laminated to the printed circuit. This also gives excellent surface topography for good peel strength. The properties of a resin-coated foil (Allied Signal RCC® ViaFoil) are given in Table 21.3.

TABLE 21.3   Properties of a Typical Resin-Coated Foil (RCC®)

	Units	Via foil	Conditioning
Dielectric constant at 1 MHz	—	3.4	C-24/23/50
Dissipation factor at 1 MHz	—	0.0205	C-24/23/50
Electrical strength	V/mil	1776	D-48/50
Insulation resistance	MΩ	2.65×10^5	C-96/35/90
Surface resistivity	MΩ	6.60×10^8	E-24/125
	MΩ	4.71×10^8	C-96/35/90
Volume resistivity	MΩ-cm	7.17×10^7	E-24/125

21.4.2.2 Nonwoven, Non-Glass-Reinforced Laminate.

21.4.2.2   Nonwoven, Non-Glass-Reinforced Laminate.   From a mechanical standpoint, materials may be grouped as reinforced and nonreinforced laminates and prepregs (Fig. 21.8). Reinforced materials generally exhibit better dimensional stability and lower coefficients of thermal expansion (CTEs) than nonreinforced materials, which often have a lower dielectric constant (D_k) and may be photoimageable.

A variety of reinforced and nonreinforced materials are enabling further miniaturization in the high-reliability market segment, and also in consumer electronics. Some examples are PTFE-like laminate and aramid materials. Complete material performance information is provided in IPC-4104, sheet numbers 5 and 23. See also Chaps. 5–10, "Base Materials."

Aramid-reinforced prepregs and laminates have proven their functionality for a number of years in high-reliability applications and, more recently, in consumer electronics. Aramid-reinforced material is a cost-effective alternative when laser hole formation is used to form single- and multiple-layer interconnects, and is compatible with FR-4 materials used in the core. Due to the performance benefits of nonwoven aramid reinforcement, several laminators have adopted this technology, and laminates and prepregs are now widely available with multiple resin systems.[15]

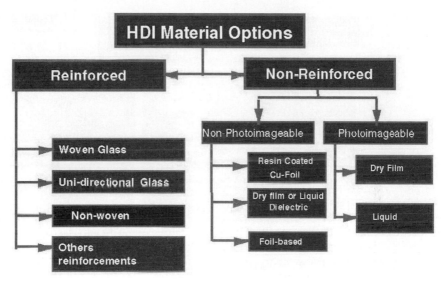

FIGURE 21.8   HDI laminate material options based on reinforcement. *(Courtesy of CircuiTree.)*

Tailorable CTE.   THERMOUNT® is a 100 percent nonwoven aramid-reinforced material that comes in prepreg (resin-coated but not fully cured yet) and laminate (prepreg fully cured and rigid with copper foil on the surface). It provides a low coefficient of thermal expansion (CTE) in HDI printed circuits. Depending on the resin and copper content of the laminate and the prepreg, the CTE of the printed circuit can be tailored between 10 and 16 ppm/°C. This allows the designer the capability of closely matching the CTE of the printed circuit to the component. Reliability can be designed in by printed circuit designers, as they know which component packages are used and the CTEs and life expectancy requirements of the electronic equipment. The ability to tailor in-plane CTE has made nonwoven aramid-reinforced printed circuits a favorite material option for avionics, satellites, and telecom applications, where lower weight, long life expectancy, and high reliability are required. In cellular phone applications, where chip-scale packages (CSPs) are now commonly employed, low-CTE, nonwoven aramid reinforcement extends solder joint life as much as three times over FR-4 and resin-coated foil (RCF). After more than 1000 thermal cycles (−40 to +125°C), nonwoven aramid-reinforced epoxy resin does not crack, as is common with various types of resin-coated foils.[16]

Processing.   Aramid-reinforced laminate and prepreg allow faster microvia hole formation and, at the same time, maintain the performance characteristic of a smooth surface for fine-line conductor imaging. The ablation speed of nonwoven laminates and prepregs is close to that achieved when using nonreinforced materials like resin-coated foil, dry film, or liquid dielectrics. However, the aramid-reinforced laminate allows the fabrication of double-sided etched innerlayers, which are then pressed to a multilayer package in a single pressing cycle. Laser drilling of stepped microvia holes is accomplished in one subsequent operation. This allows the designer to interconnect up to four layers on each side of the inner core without sequential processing—a substantial productivity advantage for the printed circuit fabricator that results in the lowest possible manufacturing cost, as seen in Fig. 21.9.

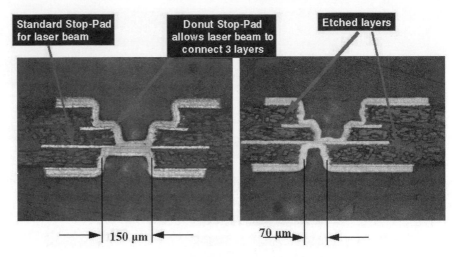

FIGURE 21.9   Vertical cross sections depicting the use of aramid laminate and laser drilling technology to connect two or three conductive layers. *(Courtesy of CircuiTree.)*

21.4.3   Unclad Nonreinforced Dielectric

21.4.3.1 Photoimageable Dielectrics.   Material chemistry options for this group include epoxies, epoxy blends, polynorbornenes, polyimides, etc. They can be applied as liquid or dry film or negative or positive imaging, and are solvent or aqueous developable. To improve the dielectric's adhesion to copper, most dielectric suppliers require a copper pretreatment with a black oxide or conversion coating (oxide replacement) process.

A unique advantage of photoimageable dielectrics is that the time required to make small or large vias is the same. With the growing need to make embedded reactives (passives), large rectangular areas need to be opened up to place these devices. Currently, this is only economical with photoimaging or plasma etching, because all geometries go in simultaneously, while to laser-ablate such large openings would require too much time of the laser. Some commercial photoimageable dielectrics available are given in Table 21.4. Figure 21.10 shows a typical photoimage-formed microvia.

Most of these materials are easy and fast to laser-drill because they are nonreinforced. The pure epoxy resin systems and optimized formulations used in many of these photoimageable dielectric materials make them extremely compliant and thus highly reliable.[17–21]

TABLE 21.4   Partial List of Commercially Available Photoimageable Dielectric Materials and Their Suppliers

Ciba	Probelec™ 81/7081 liquid dielectric
Dupont	ViaLux™ 81 photodielectric dry film
Enthone-OMI	Envision® PDD-9015 photodefinable liquid dielectric
MacDermid	MACu Via-C photodefinable liquid dielectric
Shipley	MultiPosit 9500 CC liquid dielectric
Morton	DynaVia2000™ photoimagable dielectric dry film

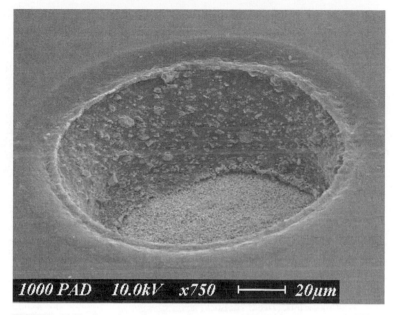

FIGURE 21.10 Scanning electron microscope photograph of a typical 130 μm microvia formed from unclad, nonreinforced photodielectric (Dupont ViaLux 81 PDDF, 75 μm thick). *(Courtesy of CircuiTree.)*

21.4.3.2 *Nonphotoimageable, Nonreinforced Dielectric.* This group can be laser-drilled, plasma-etched, and/or mechanically treated to form microvias. A partial list of commercial products and suppliers is given in Table 21.5.

TABLE 21.5 Partial List of Commercially Available Nonphotoimageable, Nonreinforced Dielectric Materials and Their Suppliers

Osada Ajinonomoto	ABF dry film
Tamura	TBR-25A-3 thermoset ink
Taiyo	HBI-200BC thermal cure ink
MacDermid	MACuVia-L liquid dielectric
Enthone-OMI	Envision® liquid dielectric
3M	Electronic bonding film
B. F. Goodrich	Polynorborene liquid dielectric

REFERENCES

1. Patent 4,566,186, "Multilayer Interconnect Circuitry Using PhotoImageable Dielectric," Charles E. Bauer and William A. Bold, Tektronix, Inc., January 28, 1985.

2. Y. Tsukada and S. Tsuchida, "Surface Laminar Circuit, A Low Cost High Density Printed-Circuit Board," *Proceedings of the Surface Mount International Conference & Exposition,* San Jose, CA, September 1992.

3. W. Schmidt, "A Revolutionary Answer to Today's and Future Interconnect Challenges," *Proceedings of the 6th PC World Conference,* San Francisco, CA, May 1993.

4. H. Holden, "Segmentation of Assemblies: A Way to Predict Printed Circuit Characteristics," *Proceedings of IPC T/MRC,* New Orleans, LA, December 6, 1994.

5. H. Holden, "Design Guidelines for HDI and Microvias," IPC-2315, IPC, Chicago, IL, 1998, p. 55.

6. H. Holden, "The Challenge: To Plan Successful Products When Packaging Is So Complicated," *Future Circuits,* iss. 1, vol. 2, 1997, pp. 106–109.

7. D. P. Seraphim, R. C. Lasky, and C. Y. Li, *Principles of Electronic Packaging,* McGraw-Hill, New York, 1989, pp. 39–52.

8. W. R. Heller, C. G. His, and W. F. Mikhail, "Wirability—Designing Wiring Space for Chips and Chip Packages," *IEEE Design Test,* August 1984, pp. 43–51.

9. G. Coors, P. Anderson, and L. Seward, "A Statistical Approach to Wiring Requirements," *Proceedings of the IEPS,* 1990, pp. 774–783.

10. I. E. Sutherland and D. Oestreicher, "How Big Should a Printed-Circuit Board Be?" *IEEE Transactions on Computers,* vol. C-22, no. 5, May 1973, p. 5323.542.

11. H. B. Bakoglu, *Circuits, Interconnections and Packaging for VLSI,* Addison-Wesley, Reading, MA, 1990.

12. R. J. Hannemann, "Introduction: The Physical Architecture of Electronic Systems," in *Physical Architecture of VLSI Systems,* R. Hannemann, A. D. Kraus, and M. Pecht (eds.), Wiley, New York, 1994, pp. 1–21.

13. L. Moresco, "Electronic System Packaging: The Search for Manufacturing the Optimum in a Sea of Constraints," *IEEE Transactions on Components, Hybrids and Manufacturing Technology,* vol. 13, 1990, pp. 494–508.

14. D. Maliniak, "Future Packaging Depends Heavily on Materials." *Electronic Design,* January 1992, pp. 83–97.

15. D. Powell and M. Weinhold, "Laser Ablation of Microvia Holes in Woven Aramid-Reinforced PWBs," *Chip Scale Review,* September 1997, pp 38–45.

16. D. Poulin, J. Reid, and T. A. Znotins, "Materials Processing with Excimer Lasers," *ICALEO Paper,* November 1987.

17. U.S. Patent 5,262,280, P. D. Knudsen et al., November 16, 1993.

18. U.S. Patent 4,902,610, C. R. Shipley, February 20, 1990.

19. U.S. Patent 5,246,817, C. R. Shipley, September 21, 1993.

20. E. Sweetman, "Characteristics and Performance of PHP-92: AT&T's Triazine-Based Dielectric for POLYHIC MCMs," *The International Journal of Microcircuits and Electronic Packaging,* vol. 15, no. 4, 1992, pp. 195–204.

21. Y. Tsukada, S. Tsuchida, and Y. Mashimoto. "Surface Laminar Circuit Packaging," *Proceedings of the 42nd ECTC Conference,* 1992, pp. 22–27.

CHAPTER 22
HIGH-DENSITY INTERCONNECT–BUILD-UP TECHNOLOGIES

Happy T. Holden
Westwood Associates, Loveland, Colorado

22.1 INTRODUCTION TO HIGH-DENSITY INTERCONNECT SUBSTRATES

High-density interconnect substrates (HDISs) are the newest organic printed circuit boards and integrated circuit die carriers. They achieve their density by using smaller geometries, and, more significantly, by using much smaller via structures. There is now also the option of using conductive pastes to form layer pairs and to attach substrates together in multilayers. This is the organic equivalent of cofired ceramic substrate technologies. The increasing number of new high-performance materials lend these organic structures to a growing number of high-frequency applications, high-I/O flip-chip carriers, high-reliability transportation and military uses, and an ever increasing number of portable products.

22.2 BUILD-UP TECHNOLOGIES

This section discusses processes that employ nondrilling via hole formation techniques. Through-via drilling is possible below 0.20 mm (0.008 in), but cost and practicality discourage this. Below 0.20 mm (0.008 in), laser and other via formation processes are more cost effective. Each of the five major via hole formation processes used for printed circuits are discussed in the following sections. They are as follows:

1. Mechanical drilling
2. Photosensitive dielectrics
3. Laser drilling
4. Plasma etching
5. Insulation displacement

22.2.1 Mechanical Drilling

Mechanical drilling has traditionally been the most widely practiced method worldwide for hole creation, but newer techniques have emerged because an increasing number of designs require microvias below 0.20 mm (0.008 in) in diameter. The use and popularity of blind and buried vias has accelerated that trend. Mechanical drilling is anisotropic, that is, the vertical walls are straight up and down. Many of the nondrilling processes are isotropic, that is, the walls of the via recede laterally as they go deeper or have sloped walls so the entry opening is larger than the exit opening. A larger entry opening facilitates metallization, but the isotropic receding wall can prevent metallization and is difficult to plate up. Thin plating on these hole walls is a definite reliability problem. Figure 22.1 shows these two hole wall profiles.

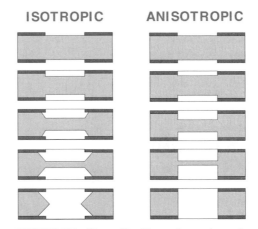

FIGURE 22.1 Via profile of isotropic vs. anisotropic drilling. Mechanical drilling is anisotropic, while chemical or plasma etching of vias is isotropic.

22.2.2 Photosensitive Dielectrics

Photosensitive dielectrics are materials that, when exposed to ultraviolet light, polymerize and cross-link to reach a state where they will not dissolve when sprayed with solvents or mild aqueous alkaline solutions. If the via holes are not exposed due to photomasks, then when processed they are dissolved away by the developer.

The earliest widely practiced technology utilizing photoimageable dielectrics (PIDs) is Surface Laminar Circuits (SLC®), developed by IBM-Yasu (Japan) in the early 1990s.[1] This is essentially a permanent photoresist that becomes the dielectric layer. The photosensitive epoxy is imaged to create the via just as a solder mask clearance is imaged to clear out an area around an SMT pad. The advantage of this type of process includes the ability to produce all vias in a batch process (mass via formation). The fact that the coating technology is based primarily on existing solder mask processing equipment, avoids the necessity of large capital equipment investment for implementation. Both dry-film and liquid-resist techniques are in use, so even if a fabricator does not possess a curtain-coater, flood-coater, or other method for dealing with liquids, this technology may be an option.

All PIDs have certain characteristics in general. Tables 22.1 and 22.2 list the typical characteristics, benefits, and processing factors in using PIDs. The equipment used in coating, curing, exposing, and developing PIDs is the same as for printed circuit fabrication and PBGA fabrication. A variety of coating methods are used, among them curtain coating, roller coat-

TABLE 22.1 Four Typical Photodielectrics (3 Epoxy and 1 Polyimide) and the Processing Parameters for Coating, Exposure, Developing, Desmearing, and Metallization

	Product A	Product B	Product C	Product D
Product	L-PID, Negative	L-PID, Negative	L-PID, Negative	DF-PID, Negative
Material	Epoxy	Epoxy	Polyimide	Epoxy
Preclean	Chemical clean	Pumice jet	Br. oxide	Chemical clean
Apply PID	Curtain coat 150 400 cps	Curtain coat 200–600 cps	Extrusion coat 12,000–25,000 cps	Vac. lam. 60 s @ 65°C
Thickness	50 µm	50 µm	37 µm	63 µm
Drying	15 h @ 90°C	6 h @ 25/40°C 3 h @ 140°C	5 h @ 25°C 15 h @ 125°C	N/A
Exposure	800–1,200 mJ/cm^2	800–1,600 mJ/cm^2	2,000–3,000 mJ/cm^2	700–1,200 mJ/cm^2
Heat bump	15 h @ 90°C	12 h @ 125°C	N.R.	20 h @ 85°C
Development	Aqueous proprietary 75 min @ 35°C	Organic GBL 60 min @ 30°C	Organic proprietary 150 min @ 30°C	Aqueous proprietary 60 min @ 35°C
Final cure	UV: 1.0 J/cm^2 + 60h @ 145°C	60 h @ 150°C	120 h @ 175°C	UV: 2 J/cm^2 + 60h @ 150°C
Roughen				
Swell	4 h @ 65°C	4 h @ 75°C	2 h @ 65°C	5 h @ 60°C
Etch	4 h @ 80°C	8 h @ 75°C	1 h @ 75°C	10 h @ 75°C
Neutralize	6 h @ 50°C	6 h @ 50°C	3 h @ 45°C	3 h @ 25°C
Metallize: electroless Cu	0.3–0.5 µm	0.7–1.0 µm	0.3–0.5 µm	0.4 µm
Bake	20 h @ 90°C	60 h @ 150°C	60 h @ 150°C	15 h @ 90°C
Electroplate				

TABLE 22.2 Electrical and Mechanical Properties of the PIDs in Table 22.1

	Product A	Product B	Product C	Product D
Product	L-PID, Negative	L-PID, Negative	L-PID, Negative	DF-PID, Negative
Material	Epoxy	Epoxy	Polyimide	Epoxy
Insulation resistance		1–10 E + 13 Ω		8 E + 13 Ω
MIR		1–4 E + 9 Ω		1.5 E + 11 Ω
D_k				
1 MHz			2.8	3.4
10 MHz	3.2			4.1
1 GHz		3.4–4.0		4.2
Loss factor				
1 MHz	<0.01	0.02	0.004	0.007
1 GHz		0.015		0.01
Breakdown voltage (between layers)		>2000		
E-migration	Pass	Pass		Pass
T_g	140–180°C	135°C	300°C	170°C
Peel strength	<9 N/cm >5 lb/in	14 N/cm 8 lb/in		
TCE (ppm/°C)	60–70	60–70		60–70
Tensile modulus		3000–3500 N/mm^2		4.0 E + 5 psi

ing, standard spray and electrostatic spray application, meniscus coating, and, in one case, spinning liquid PIDs. Dry-film PIDs are applied with a vacuum laminator. Curing employs conveyorized convection ovens, and exposing and developing use standard equipment. Via filling is required over the core holes and for each build-up layer. Most PIDs are employed with only one build-up layer and with via diameters of 100 μm or larger.

22.2.3 Laser Drilling

Laser drilling is one of the oldest microvia generation techniques.[2] The wavelengths for laser energy are in the infrared and ultraviolet region. Figure 22.2 shows the five major wavelengths used in current laser drills. The absorption curves of many organic dielectrics (epoxy, polyimide), matte copper, and glass fibers are shown in Fig. 22.3. Laser drilling requires programming the beam fluence size and energy. High-fluence beams can cut metal and glass, while low-fluence beams cleanly remove organics, but leave metals undamaged. A beam spot size as small as approximately 20 μm (<1 mil) is used for high-fluence beams and about 100 μm (4 mils) to 350 μm (14 mils) for low-fluence beams.[3,4]

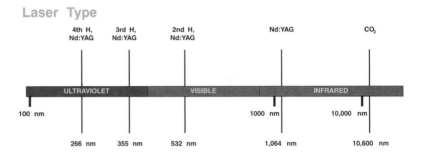

FIGURE 22.2 Primary wavelengths of commercially available laser drills for PWB drilling. The CO_2 and 4th harmonic of the YAG laser are the principal wavelengths used for laser drills.

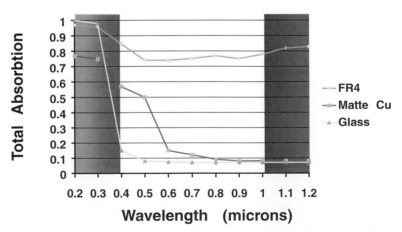

FIGURE 22.3 The advantage of ultraviolet wavelengths for laser drills is that copper and glass, as well as the dielectric, absorb these wavelengths. At lower wavelengths (>1 μm), CO_2 lasers are reflected by the copper and glass of laminates.

The laser is the most common method of production of microvias to be plated or filled with a conductive paste. Lasers are capable of ablating dielectric material and then stopping when intercepting the copper circuitry, so they are ideally suited for creation of depth-controlled blind vias.

Higher-power lasers (i.e., ultraviolet [UV]) can remove glass and copper and therefore can be used with conventional laminates, but are typically slower when going through copper and glass fibers.

Most laser processes utilize either CO_2 or UV lasers because those are the most readily available and economical lasers. When using a CO_2 laser to produce vias in epoxy laminates, the copper must be removed above the area to be ablated. The CO_2 laser is primarily used for laminates not supported by glass. This includes unsupported laminates such as flexible polyimide and resin-coated copper (RCC®) foil and laminates reinforced with alternative materials such as aramid fibers.

22.2.4 Plasma Etching

Plasma via etching evolved from the traditional process of plasma desmearing of throughholes. Different gas, magnetron, and equipment fixturing are employed by current plasma via-etching equipment. The plasma is generated in a partial vacuum filled with a mixture of oxygen, nitrogen, and chlorofluoro (CF_4) gases. Microwave magnetrons create the plasma field and special low-frequency kilowave units help provide rapid etching of organics. Figure 22.4 depicts the action of a plasma on organic materials and the schematic of a typical high volume plasma etcher with its six rotating printed circuit panels.

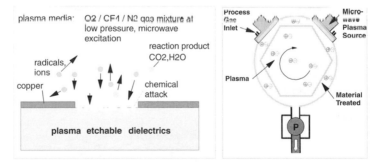

FIGURE 22.4 Plasma etching of a dielectric requires a metal mask and is done at radio frequency. The microwave plasma is generated in evacuated vessels where various gases have been introduced and excited to create the radical ions that make up the plasma.

22.2.5 Insulation Displacement

Physical displacement via generation processes consist of punching to displace the insulator material to create vias that are either screened or deposited with conductive pastes. The nature of the process creates pointed via metals that displace the glass fibers during lamination (thus the term *insulation displacement HDIs*), thus eliminating the need for drilling or other methods of forming the vias.

22.3   *PHOTODEFINED VIA TECHNOLOGIES*

22.3.1   Photoimageable Dielectric Technology

SLC technology was the first of the photoimageable dielectric (PID) microvia build-up technologies to reach the interconnect market in high volume. Development of SLC technology began during the late 1980s at IBM (Yasu, Japan), and the first product was introduced in 1990.[5] The first flip-chip direct chip attach (FC/DCA) product employing SLC technology was shipped from Yasu in 1992, and in 1995 IBM's Endicott, New York, location shipped the first printed wiring boards that utilized SLC technology.[6]

The most suitable photosensitive dielectric was the photoimageable solder masks, due to their proven compatibility with printed circuit assembly processes, ability to withstand exposure to service environments, and possession of the necessary via-imaging characteristics. SLC technology was originally developed and implemented with a liquid photoimageable solder mask applied by curtain coating. In 1995, it was also qualified with a dry-film photoimageable dielectric (PID) with virtually identical electrical properties. The dry-film version did not require the extensive surface grinding processes that the liquid PID process required.

PID technologies were originally conceived as an alternative to multilayer lamination as a way to make multilayer printed wiring boards (PWBs). PID technology builds up each surface signal layer on top of the previous layers in a sequential fashion. The base under all of the outer build-up layers is a conventional double-sided or multilayer circuit board containing voltage, ground planes, and even signal layers. Mechanically drilled and plated holes are used only in the core layers and to make connections to the back side of the laminate base, and to accommodate the mounting of pinned components. Greater wiring density is required for the use of plated through-holes (PTHs), which extend only through the base printed circuit structure and not through the build-up layers, especially when two or more build-up layers are used. Figure 22.5 shows a partial cross section with three SLC technology build-up layers (PIDs) on one side of a multilayer-base printed circuit board.

22.3.1.1   PID Applications.   PID technology has been used to manufacture high-density printed circuits and IC package substrates for computer and communications equipment and consumer electronics.[7]

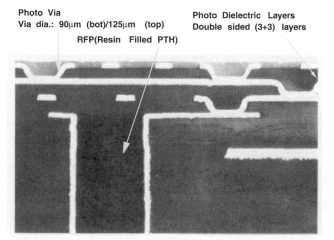

Photo Via
Via dia.: 90μm (bot)/125μm (top)

RFP(Resin Filled PTH)

Photo Dielectric Layers
Double sided (3+3) layers

FIGURE 22.5   Cross section of an IBM SLC printed wiring board with buried vias in the FR-4 core and three build-up layers sequentially applied to the core.

22.3.1.2 PID Technology Process.   PID technology is based on the use of photoimageable polymeric systems to form blind microvias in dielectric material between layers of circuitry. The use of PIDs allows all microvias on a panel to be formed simultaneously, with no incremental per-via cost. Its use is particularly advantageous in applications having high densities of vias (e.g., more than 50,000 on an 18×24-in panel).

Liquid vs. Dry PIDs.   A typical flowchart of the PID technology fabrication process, including options for multiple build-up layers and multilayer core, is given in Fig. 22.6. At this level in the process, PID technology is the same whether a liquid or dry-film PID is used. Figure 22.7 presents comparative flowcharts that list the PID processing steps, highlighting the process differences between liquid and dry-film PIDs.

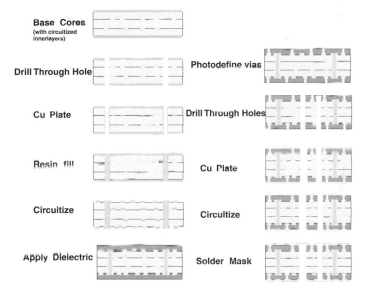

FIGURE 22.6   Typical flowchart of the PID technology fabrication process, including options for multiple build-up layers and multilayer core.

Liquid and dry-film PIDs do provide slightly different via wall profiles. The liquid PID has tapered via walls, while the dry-film PID has essentially vertical via walls. The tapered via walls provide good plating coverage on the via walls and base. Vertical walls allow for a smaller via top opening and correspondingly smaller capture land for a given via bottom diameter.[8, 9]

Tools.   The use of a liquid PID requires two unique pieces of equipment: a coater of some variety—either curtain, slot, or roller coater or screen printer (with associated drying ovens)— and a leveling tool (surface sander). A leveling tool is required to planarize the surface of the cured liquid PID to accommodate fine-line photolithography on the surface. Liquid PID provides a conformal coating over underlying circuit features, producing nonuniform planarity. The leveling operation also removes a lip of exposed PID that overhangs the via openings in these processes. Many liquid PIDs have a self-leveling characteristic and do not require leveling.

The use of a dry-film PID requires only one unique piece of equipment: a vacuum laminator. Vacuum laminators are common at printed circuit fabrication shops; in addition, the cap-

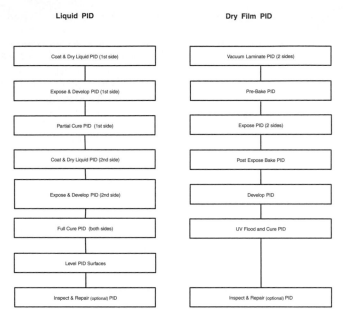

FIGURE 22.7   Comparative flowcharts listing the PID processing steps, highlighting the process differences between liquid and dry-film PIDs.

ital to obtain one is significantly less than for a curtain or slot coater. The dry-film PID has excellent planarization and does not require a leveling process due to the film's low solvent content, low shrinkage, and vacuum lamination process.[10,11]

22.3.1.3   Design Rules.   PID technology design rules are given in Table 22.3. Larger printed circuits can be designed with more relaxed rules, while the most aggressive rules are limited to small printed circuits, such as chip carriers, to achieve optimized yields and minimize cost.[12,13]

TABLE 22.3   Current and Future Design Rules for Fabrication of SLC Technology Using a PID

Feature	Standard	Feature size (μm)		
		High density	Prototype	Future
Photovia diameter (min)	125	100	70	50
Photovia land diameter, top/bottom (min)	300	200	110	80/70
Line width (min)	75	50	30	20
Space (min)	125	75	40	30
Line thickness (min)	18	12	7	5
Dielectric thickness (min)	40	30	30	20
Stacked photovias	No	No	No	Yes

22.3.2   IPN Polymer Build-up Structure System

IPN polymer is the proprietary photodielectric (PID) developed by Ibiden (Japan) and called IBSS. It is a liquid system applied by roller coater to a rigid core, typically BT. It is specifically formulated for flip-chip packages, has a high T_g, and is flexible. The metallization is fully additive copper, also proprietary to Ibiden. It is capable of two or more build-up layers per side. The higher-temperature PID and additive circuitry provides higher densities required for the fine pitch of flip-chip packages. The manufacturing process is simpler because of the fully additive circuitry and is shown in Fig. 22.8.

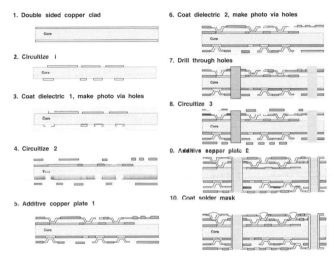

FIGURE 22.8   IPN polymer build-up structure system manufacturing process.

22.3.2.1   Design Rules.   The current IBSS design rules are among the most advanced in the world and are used for advanced die packaging, typically flip-chips.

22.3.3   Carrier-Formed Circuits

Meiko Circuits (Japan) implemented the PID process using stainless steel carriers. These provide the dimensional stability for fine geometries and microvias. The carriers also provide the pressure plates for lamination of multilayers. The surface circuits are flush with the dielectric and very suited to RF designs.

The process starts by taking a photoresist and coating it onto a stainless steel panel. The surface pattern is exposed and developed in the photoresist. The panel is successively plated with gold, then nickel, and finally copper. The resist is stripped, the photodielectric (PID) is applied over the entire panel, and via holes are developed in the dielectric. Once metallization is complete, the process can be repeated until the circuitry is complete or can be laminated to FR-4 materials that serve as stiffeners. The process can be repeated for multiple build-up layers, but one layer is reported. The advantages of this process are that surface geometries are not determined by etching or full additive metallization; the vias are located beneath the surface lands; and the circuits are all flush with the dielectric, permitting the elimination of sol-

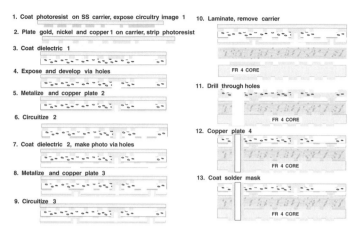

FIGURE 22.9 Carrier-formed circuit manufacturing process.

der masks. However, this is a more expensive process that involves carriers. The manufacturing process flow to fabricate flush circuits by the carrier process is shown in Fig. 22.9.[14,15]

22.4 LASER-GENERATED VIAS

Lasers have been producing vias in printed circuits for many years. The earliest documented use of lasers to form vias was in the fabrication of Hewlett-Packard's Finstrate, in production from 1984 to 1987, and Siemens' Microwiring, produced from 1987 to 1990.

Of the various processes for generating microvias, current estimates are that lasers produce 86 percent, of which 31 percent are produced by UV lasers, 50 percent by CO_2 lasers, and 5 percent by eximer lasers. Current production processes utilizing lasers are discussed in the following sections.

22.4.1 Laser-Formed Blind/Through-Vias

The use of FR-4 epoxy printed circuits has grown very rapidly because of new portable products such as cellular phones and camcorders. With the emergence of fine-pitch area array components, larger industrial, military, communications, and computer applications utilize laser-drilled blind and buried vias. A large variety of materials are compatible with lasers, including those listed in Sec. 21.4; nonglass laminates, as well as conventional glass-based laminates and ceramic-filled and PTFE laminates, can also be laser-drilled. Only UV lasers can make through-holes in many of these materials, but if the material is sufficiently thin and not glass-reinforced, then CO_2 can often be successfully utilized as well.

22.4.1.1 Structure. The structure for laser-produced blind vias in printed circuits is shown in Fig. 22.10. The core material is typically standard FR-4 epoxy fiberglass with a simple two-sided structure or complex multilayer printed circuit. A laser-produced blind via is the simplest structure to use when a few microvias are required to connect up a new high-density ball grid array or chip-sized package while not changing the design or structure of a board.

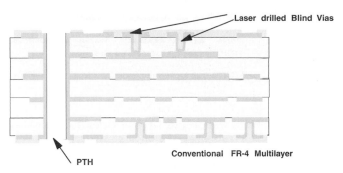

FIGURE 22.10   Structure for laser-produced blind vias.

22.4.1.2   *Manufacturing Process.* There are two major manufacturing processes for the laser production of blind vias. The first uses resin-coated copper foil or glass-reinforced laminates. A UV laser can pattern the via holes in the copper cladding directly, but a CO_2 laser requires the copper foil to be patterned with a photoresist and etching process identical to the one used to define the masks in plasma etching. The second process employs a liquid resin coated onto a laminated stack or finished board. After curing, the resin is lased to produce blind vias. A conventional additive or semiadditive metallization is used to provide a copper surface.[16–30]

22.4.2   Laser-Drilled Flex (ViaThin)™

Sheldahl has developed an advanced manufacturing process called ViaThin™ in which a bare polyimide film is drilled by a laser that uses holographic phase masks to produce all the holes simultaneously. This uses commercially available eximer lasers and phase masks. Because the metallization process is vacuum semiadditive, very fine geometries can be fabricated for tape ball grid array packages.

The expensive phase mask tooling has the advantage of accuracy and speed. Although these are small (no larger than 100×100 mm), they are large enough for integrated circuit packaging. Drilling in bare thin polyimide is also conducive to speed, and, when vacuum metallized, the metal covers the inside of the holes as well. A seed layer of 2000 Å is applied first, followed by 2000 Å of copper. Additional copper is electroplated up to a thickness of 5 μm. The flex material can be imaged and further plated to 15 μm.

22.4.2.1   *Structure.* The two-sided layer-pair structure that Sheldahl manufactures has either a 25- or 50-μm polyimide. Figure 22.11 shows a view of this structure. The material is adhesiveless due to the surface preparation process and the 5- to 15-μm-thick copper vacuum metallization. A photosensitive polyimide solder mask is employed to protect the copper and there are various protective coatings available to apply on the copper.[31,32]

22.4.2.2   *Application Examples.* The laser-drilled flex material called layer pairs is a major component of tape BGAs. A novel use of layer pairs is where 3 layer pairs are laminated to a 12-layer multilayer. This provides for local density and metal finish for a workstation CPU socket and cache, but is not required for the rest of the computer.

A roll of bare polyimide film is first tooled by aluminum masks that cover the film, and a CO_2 laser is used to accurately place sprocket holes in the film. The bare polyimide film is then drilled with an eximer laser that can use holographic phase masks to produce all the holes simultaneously. This produces 20-μm holes that are very accurately positioned. The film is

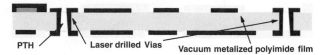

FIGURE 22.11   Example of laser-drilled flexible PWB.

vacuum-metallized with a seed layer of chrome or tungsten and then copper and is then plated with up to 5 µm of copper on large stainless steel drums. If this is a sufficient copper thickness, it is imaged and etched; if not sufficient it is plated with up to 15 µm of copper and etched. Protective coatings are applied and the copper is coated with electroplated gold or immersion coatings. The finished substrate is a layer pair. Layer pairs are sandwiched into multilayer structures by screening a conductive metal paste into a laser or punched polyimide adhesive. The multilayer structure is then sintered and the paste forms a metallurgical bond with the copper lands of the layer pairs. Four layer pairs or eight copper build-up layers can be connected in this manner.[33-35]

22.4.2.3 Design Rules.   Tape-based HDI design rules are listed in Table 22.4. These are typical of laser-drilled, tape-based processes where vias are drilled before metallization.

TABLE 22.4   Tape-Base HDI Design Rules

Design rules	Feature size (µm)	
	High density	Future
Outerlayer linewidth	50	25
Outerlayer line-line spacing	50	25
Minimum via diameter	25–75	18
Minimum via land diameter	125	70
Minimum via land pitch	508*	380
Electroless Ni/Au plating	Yes	Yes
Minimum polymide thickness	50	25
Imageable product area (mm)	150×150	300×300

* Currently using 0.508-mm grid, with plans for 0.38-mm grid in the future.

22.4.3 High-Density Interconnects

The HDI process developed by General Electric (now Lockheed-Martin) is very similar to IC processes. This is referred to as a "chips-first" process because assembly is done before the substrate is completed and ICs are directly bonded to the substrate. It does not use flip-chip or wire bonding. All ICs are bonded to a polyimide film layer pair flex circuitry and cured. The assembly is turned over and a laser drills down through the flex circuitry to make blind and through-vias, including opening up the bonding pads on the IC die. Gold, an under-bump metallurgy, or IBM's C4 bumps processes are not required. The panel is metallized via a tungsten sputtering process. A circuit pattern can be applied, plated, and then etched to complete the circuit. There are advantages in making multichip modules and in using standard die. The materials and the bonding technology have proven to be very reliable and suitable for military products. A number of build-up layers and very fine geometries are possible. The disadvantage is that the largest size panel that can be sputtered is typically 300 mm, and this equipment is very expensive.[36,37]

22.5 CHEMICAL/METALLURGICAL BONDED VIA IN PWB TECHNOLOGIES

Metallic pastes (conductive inks) describe a single-layer dielectric with microvias formed by photoimaging, laser, or insulation displacement. A conductive paste is used to fill the microvias and act as the conductive path between layers. Surface metallization may be accomplished either by laminating copper foil onto the dielectric surface or by chemical deposition.[38]

22.5.1 Paste-Bonded Solid-Via Laminate (ALIVH)

The any-layer interstitial via hole (ALIVH) process has been developed by Matsushita Components (Japan). This novel process eliminates additive metallization and plating, but defines all features by subtractive etching of the copper foil. The build-up process is not sequential; rather, it uses layer pairs and aramid-epoxy prepreg with copper-paste vias that can be laminated at one time into a three-dimensional structure. Six to 10 layers have been laminated in this way.[39]

22.5.1.1 Structure. The structure for the laser-produced blind-via printed circuit called ALIVH is shown in Fig. 22.12. The core material is an aramid-epoxy laminate. The man-made aramid filaments are ideal for cutting with a CO_2 or UV laser. If Dupont Kevlar™ filaments are added, the resulting material exhibits a very low x-y CTE. This is useful for mounting ceramic packages and for the direct attachment of flip-chip integrated circuits. The structure can be as simple as a two-sided or as complex as a multilayer printed circuit. The vias are a copper-epoxy paste that connects the top and bottom copper foil. If used as a prepreg layer without copper, it connects the various ALIVH layer pairs into a multilayer. This process is not a sequential build-up, but rather a parallel build-up process.

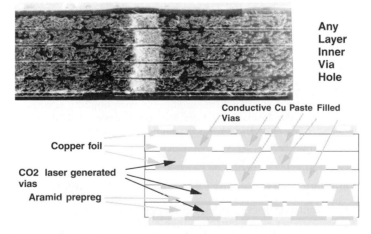

FIGURE 22.12 The sequential bonded solid-via structure for the laser-produced blind-via printed wiring board called ALIVH. *(Courtesy PrismarK Partners LLC.)*

22.5.1.2 Application Example. Panasonic has manufactured millions of phones for many markets using the ALIVH technology. The cellular phone has all its functions, including the keypad, integrated into one card. This type of structure is up to 20 percent lighter than normal FR-4 epoxy fiberglass multilayers.

22.5.1.3 Manufacturing Process. The base material is an epoxy-aramid B-stage prepreg. Cutting the holes by laser proceeds very rapidly. The material is printed with a conductive paste consisting of copper and epoxy to fill the holes. Copper foil is applied and the structure is laminated to attach the foil and cure the prepreg and conductive pastes that serve as vias. The sheet of material is imaged and etched to provide the various circuits. Registration is less critical because the vias are now beneath the surface lands. Several of these two-sided layer-pairs can be produced, inspected, and tested. The two-sided structures can then have additional single layers of B-stage/conductive paste layers with foil laminated to one or both sides. Alternatively the B-stage/conductive paste layer can be used to attach any number of layer pairs in one parallel lamination. The outside is imaged, etched, and completed as a normal printed circuit. The process can be seen in Fig. 22.13.

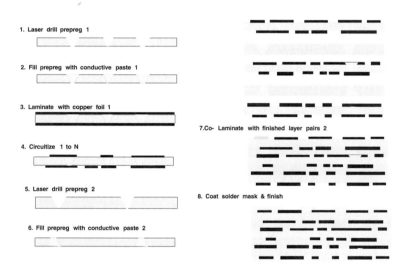

FIGURE 22.13 The ALIVH fabrication process.

An advanced manufacturing process produces a product known as ALIVH-FB. This has a fine-line, tight via structure suitable for wire-bonding and flip-chip substrates. As shown in Fig. 22.14, the ALIVH structure has added a laser-drilled 12.5-μm-thick polyimide film. Two of these fine-pitch layer pairs are added to the thicker aramid-ALIVH structure that serves as the center core.

22.5.2 Co-lamination with Conductive Paste/Adhesive Structures

Co-lamination involves the technology of taking layer pair printed circuits and laminating them together with a custom adhesive layer that provides the connections between the layer pairs. It is similar to the ALIVH process. The custom adhesive layer has small via openings in

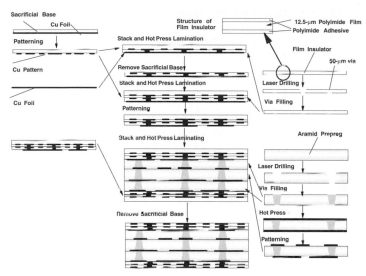

FIGURE 22.14   The ALIVH-F and FB production process. *(Source: CircuiTree, May 2000.)*

it produced by laser or other means and filled with a conductive paste that provides the connections between the layer pairs. One such technique is called ViaPly®.

ViaPly conductive paste is different from ALIVH in that it is a copper-tin organometallic matrix that sinters into a solid metallurgical via. It can now accommodate layer pairs from any other HDIS process (like Sheldahl's) and create a multilayer. Various materials can be mixed if a rigid core or heat spreader is required. Litronics (Allied Signal Substrates) has used this technology in addition to Sheldahl. Allied Signal currently has an exclusive arrangement with Ormet.

22.5.2.1 Structure.   A three–layer pair ViaPly structure is shown in Fig. 22.15. Each two-sided flex layer pair is laminated with a high-temperature bonding sheet and the conductive ink is sintered to form a solid metallurgical bond.

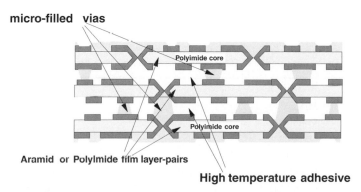

FIGURE 22.15   Three-layer pair ViaPly structure.

22.5.2.2   *Manufacturing Process.* The co-lamination structure is made up of polyimide layer pairs. Different materials can be mixed if a rigid core or heat spreader is required. The conductive paste is a transient liquid-phase sintering (TLPS) ink of copper-tin. A finished circuit consists of lasered-via polyimide layer pairs and TLPS solid-metal vias that connect the layer pairs.

22.5.2.3   *Unique Materials.* There are various commercially available conductive inks or conductive pastes available today. Table 22.5 is an incomplete list of several, noting the trademark name, manufacturer, basic composition, and application method of via fabrication.

22.5.3   Microfilled Via Technology (MfVia®)

The MfVia technology is used in the production of integrated-circuit plastic packages. Its singular HDIS focus has been on PBGAs. Two build-up layers is the current capability, but more are anticipated in the future. The process employs a PID (of the company's own formulation) coated on a carrier, typically copper or aluminum. The process is now a subtractive one and electroless or electrolytic plating is not employed.

The microfilled via structure is a photodielectric with silver-epoxy paste as vias. The PID is exposed and developed to create microvias that are then filled with a silver conductive paste, and copper foil is placed over the PID-paste layer. This is laminated to cure the PID and bond the copper foil to the dielectric. The circuitry is created by etching copper foil. As such, the structure has its vias under the surface lands. The structure is similar to the ALIVH technology and is shown in Fig. 22.16.

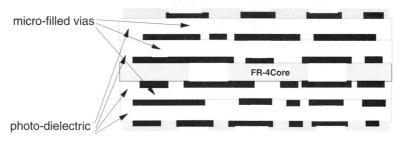

FIGURE 22.16   Microfilled via structure is a photodielectric with silver-epoxy paste as vias.

TABLE 22.5   The Seven Conductor Pastes to Cure or Sinter at Low Temperature (<220°C)

Manufacturer	Trade name	Via process	Via paste
Pronlinx	MfVia™	Photodielectric	Silver paste
CTS	ViaPly™	Laser or punch	Cu/Sn organometallic
Toranaga	Ormet™	Photodielectric	Cu/Sn organometallic
Matsushita	ALIVH™	Laser	Copper/epoxy
Toshiba	Bbit™	Insulation displacement	Silver/epoxy
Parelec	PARMOD™	Drill, laser, or photodielectric	Metalloorganic decomposition, Cu or silver
Namics	Unimec™	Punch, drill	Silver, palladium, copper pastes

22.5.4 Transient Liquid-Phase Sintering (TLPS) Conductive Circuits

A pure conductive paste circuitry exists in the use of TLPS pastes. Each layer of circuitry (traces and spaces) is built up on a base structure along with built-up via structures. This can theoretically be done with any of the conductive pastes in Table 22.5, but has been demonstrated only with the TLPS pastes.

22.5.4.1 Structure.   The sequentially built-up conductive paste structure is built on a solid base of either metal or an insulator. This is the IPC Type VI structure. A photodielectric (PID) is employed to define the circuit paths filled with the conductive paste. Alternate layers of conductive traces (in the x-y plane) and conductive vias (in the z axis) are applied to the photodielectric, cured, and sintered. A cross section of the structure is shown in Fig. 22.17. This shows three polyimide layer pairs sintered with TLPS pastes.

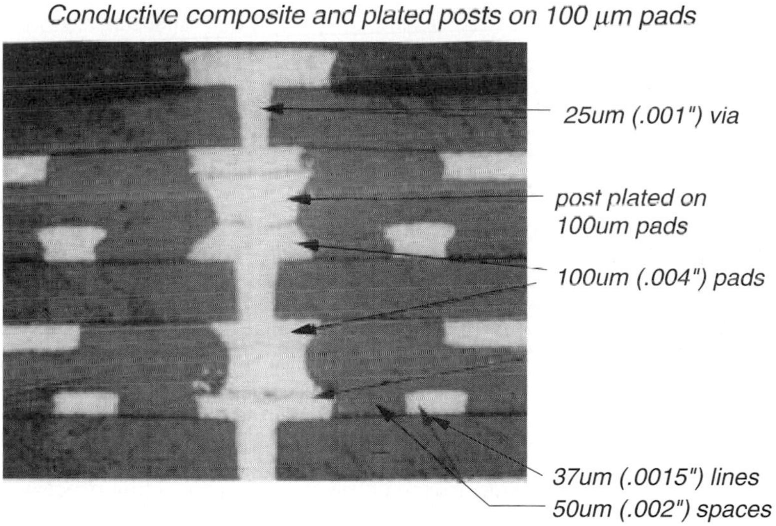

Conductive composite and plated posts on 100 µm pads

25um (.001") via

post plated on
100um pads

100um (.004") pads

37um (.0015") lines
50um (.002") spaces

FIGURE 22.17   Cross section of transient liquid-phase sintering (TLPS). *(Courtesy of Ormet Technology.)*

22.5.4.2 Manufacturing Process and Material.   The simple sequential steps in this built-up technology comprise just the application of a photodielectric to a rigid base. The photodielectric is imaged and developed with the pattern of circuitry. The TLPS paste is then flood-printed into these trace grooves with a thick film or SMT solder paste printed. The paste is dried for 30 min at 85°C. Another layer of photodielectric is applied over the existing layer. This layer is exposed for the via structure and again the paste is flood-printed into the via openings and dried. The alternate layers of circuitry and vias are applied until the circuit is completed. The conductive pastes have to be sintered in a condensing vapor of fluorocarbon at 215°C for 2 min. The structure is then postcured by baking for 40 min at 175°C. The process is detailed in Table 22.6.

TABLE 22.6   Properties and Curing Process for the Ormet Type of TLPS Conductive Pastes

Ormet® 2005 series ink	Specification and processing parameters
Electrical conductivity	Bulk 4.0×10^{-3} Ω-cm Sheet resistance 10.0×10^{-3} Ω/square
Adhesion (tensile pull) on various materials	
FR-4 ($T_g = 125°C$)	1300 psi (minimum)
Copper	2921 psi (average)
Printability	with 230 stainless steel wire mesh and emulsion thickness of 7.5 µm Sintered thickness 28–38 µm 200-µm traces on 400-µm pitch
Cure cycle	30 min drying at 85°C 2 min vapor cure at 215°C 40 min postcure at 175°C

22.6   WET/DRY-ETCHED VIAS

Contraves (Switzerland) developed the plasma etching process for microvias in the late 1980s. Evolving from the traditional plasma desmearing process, plasma etching of vias was jointly developed by Dyconex (successor to Contraves) and Technic Plasma, GmbH (Germany). Dyconex trademarked and patented their via-generation process as DYCOstrate.™ Hewlett-Packard licensed the technology in 1993 and developed it for mass, low-cost production, referring to it as plasma-etched redistribution layers (PERL).

22.6.1   DYCOstrate™

DYCOstrate and SLC are the oldest microvia production processes and have been used the longest. The DYCOstrate process was first employed for high-reliability military, aerospace, medical, and IC packaging starting in 1991. Since that time Dyconex has produced hundreds of different printed boards that use the process both in polyimide film and with the new nonreinforced laminates called resin-coated copper foils. These structures Dyconex calls DYCOstrate-C.[40]

22.6.1.1   Structure.   The structure of DYCOstrate (Fig. 22.18) can utilize a core material of polyimide film with plasma-etched holes, epoxy fiberglass that is drilled, or other plasma-

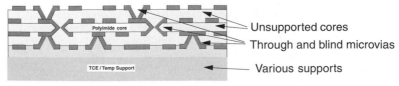

FIGURE 22.18   The structure of DYCOstrate.

etchable materials such as liquid crystal polymers or resin-coated copper foils. Multiple layers can be built up to increase density with buried and blind vias, but two-sided DYCOstrate is also used in products because of the very high density that 0.075-mm-diameter through-holes allow.

22.6.1.2 Application Examples.   Numerous examples have been produced by Dyconex. This product is unique because there are tiny blind holes around the perimeter that serve to lock in the plastic molding of the ear cavity after the unit is assembled and tested. Test of HDI is difficult under any circumstances. This flex printed circuit shows the assembly test points external to the circuit. Once assembled and tested, it is excised from its carrier and folded so as to be small enough to fit in the ear. DYCOstrate is used in satellites, avionics, military munitions, transportation (rail and commercial aircraft), automobiles, and high-temperature sensors.

22.6.1.3 Manufacturing Process.   The manufacturing process for DYCOstrate substrates utilizes commonly practiced printed circuit board techniques. Only the via generation process is different. In production of plasma-etched via holes, two or three process steps replace conventional mechanical drilling, deburring, and desmearing steps: (1) the location and geometry of the via holes are photographically defined. (2) Openings are etched in the copper foil that serve as the resist mask. (3) For blind vias, the thickness of the copper foil is reduced to eliminate the copper overhang.

Vertical cross sections of a plated through-hole and a blind via formed by plasma etching in a polyimide film are shown in Fig. 22.19. The plasma etching process is basically an isotropic process, as indicated by the undercut seen on the through-hole, which, when considering the actual dimensions, is too small to cause any plating problems.

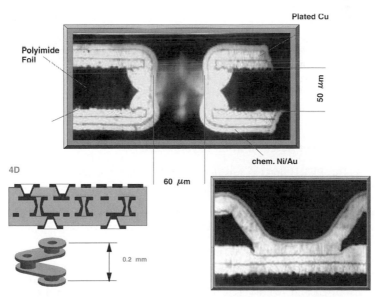

FIGURE 22.19   Vertical cross sections of a plated through-hole and a blind plasma-etched via in a polyimide film. *(©Copyright Dyconex Ltd., Zurich, Switzerland.)*

When the etching depth is increased, as in the case of a blind via, the resulting undercut is generally too big to allow reliable plating. The problem is overcome through copper reduction: etching the copper foil, eliminating the copper overhang, and providing a thinner copper foil for fine-line resolution.

The manufacturing process for DYCOstrate substrates with a standard four-layer build-up starts with a prefabricated double-sided DYCOstrate foil. Two single-sided copper-clad foils are bonded to the center core foil using standard lamination techniques. The resulting four-layer structure is then processed and structured analogously to the two-layer foil, producing blind vias instead of through-holes.

22.6.1.4 Design Rules.   The design rules for DYCOstrate are listed in Table 22.7. Refinements in magnetrons for IC manufacturing provide the opportunity for finer plasma-etched vias. Additionally, Dyconex has demonstrated landless vias to be simple and reliable, offering much higher densities at lower cost and simplifying fabrication registration.

TABLE 22.7   Design Rules for DYCOstrate

Design rules	Feature size (μm)		
	High density	Prototype	Future
Plasma diameter (min)	100	70	50
Plasma land diameter top/bottom (min)	200	110	80/70
Linewidth (min)	50	30	20
Space (min)	75	40	30
Line thickness (min)	12	7	5
Dielectric thickness (min)	50	35	25

22.6.1.5 Plasma Micromilling.   In addition to drilling holes, plasma etching can also be used to sculpture the substrate surface or to fabricate slots, grooves, stepped windows, etc. Even angled vias or tubular systems can be formed. This is useful because cavities can be formed for wire-bonding or recessed mounting of components. A simple definition of the periphery outline by artwork can provide very accurate final fabrication, or windows in the final etching step can define areas that flex.

22.6.2 Plasma-Etched Redistribution Layers (PERL)

The PERL process was developed by Hewlett-Packard to produce micro-blind and buried vias using the DYCOstrate plasma via formation process. The materials used are FR-4 epoxy resin-coated copper foils to replace epoxy-glass prepregs in normal multilayer production. These are copper foils coated with a B-stage epoxy or a dual C-stage/B-stage epoxy film. The Parlex/Allied Signal product has the name resin-coated Copper (RCC) or ViaFoil.[41]

22.6.2.1 Structure.   The DYCOstrate PERL structures are illustrated in Fig. 22.20. The core material can be standard FR-4 multilayer innerlayers or a through-hole plated two-sided or multilayer board. Multiple layers can be built up to increase density with the resulting buried and blind vias. Current production ranges from 4 to 12 layers with various buried board and buried via constructions. Like DYCOstrate, many materials are plasma-etchable and the resin-coated copper foils come in many thicknesses and resin types such as BT, cyanate ester, and PPE.[42–47]

22.6.2.2 Application Examples.   PERL HDI substrates have been used in a large number of applications. A multichip module converted from ceramic to PERL is shown in Fig. 22.21.

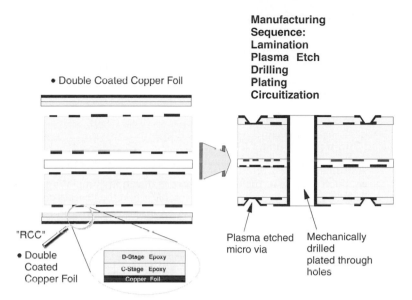

Manufacturing Sequence:
Lamination
Plasma Etch
Drilling
Plating
Circuitization

• Double Coated Copper Foil

"RCC"

• Double Coated Copper Foil

| D-Stage Epoxy |
| C-Stage Epoxy |
| Copper Foil |

Plasma etched micro via

Mechanically drilled plated through holes

FIGURE 22.20   DYCOstrate PERL structures.

Building Local Area Network

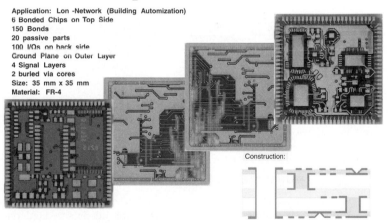

Application: Lon -Network (Building Automization)
6 Bonded Chips on Top Side
150 Bonds
20 passive parts
100 I/Os on back side
Ground Plane on Outer Layer
4 Signal Layers
2 buried via cores
Size: 35 mm x 35 mm
Material: FR-4

Construction:

FIGURE 22.21   A multichip module converted from ceramic to PERL.

The top side has seven wire-bonded dies and the back side is used for conventional surface mounting. The entire module has leads attached and is assembled as a conventional surface-mounted component. The module is used for local area networking in Europe and is mounted outside office buildings.

22.6.2.3   *Manufacturing Process.*   The manufacturing process for DYCOstrate PERL substrates is identical to Dyconex's process (Sec. 22.6.1.3).

22.7 INSULATION DISPLACEMENT TECHNOLOGY

Insulation displacement derives its name from the unique way that the via is formed in the insulator. The via metal actually displaces the dielectric and forms an attachment to the metal on the opposite side of the dielectric. Buried bump interconnect technology is the first example of such a technology.

22.7.1 Buried Bump Interconnect (BbiT)

A Toshiba process referred to as buried bump interconnection technology (BbiT) utilizes a conductive paste to replace additive metallization and plating. The process employs conventional FR-4 epoxy-glass prepregs. Notable is that the process does not require conventional or microvia drilling equipment. The via is formed by displacing the glass cloth and resin during lamination. Currently, the design rules and features focus on consumer products, but future anticipated uses are simple PBGAs.

22.7.1.1 Structure. The BbiT structure appears similar to other conductive paste via printed circuits, and, as seen in Fig. 22.22, the structure is very similar to ALIVH. The difference is that the BbiT structure uses standard FR-4 materials and a novel way to produce vias.

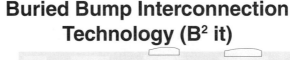

Buried Bump Interconnection
Technology (B² it)

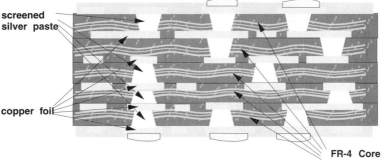

screened silver paste

copper foil

FR-4 Core

FIGURE 22.22 Buried bump interconnection technology.

22.7.1.2 Manufacturing Process. A silver ink is screened on copper foil in locations where vias are required. The copper foil is then used to manufacture standard FR-4 laminate. The silver paste thumbtack-like obstruction forces itself through the glass cloth during lamination to connect itself to the copper foil on the other side of the laminate and cure the prepreg. The sheet of material is imaged and etched to provide the various circuits. Registration is less critical because the vias are now beneath the surface lands. Several of these two-sided layer pairs can be produced, inspected, and tested. The two-sided structures can then have additional single layers of B-stage/conductive paste layers with foil laminated to one or both sides or the B-stage/conductive paste layer can be used to attach a number of layer pairs in one parallel lamination. The outside is imaged, etched, and completed as a normal printed circuit. The process is now capable of making PBGAs for wire-bonding and flip-chips.

22.8   *IMPRINTED CIRCUITS*

A technology from Dimensional Circuits is called imprinted circuits (Fig. 22.23). It is created using techniques for making CDs and laser discs. It involves the use of a master tool suitable for creating electroformed working tools that are used to mold circuits. The traces and lands all recede into the substrate. This is an ideal topology for fine-pitch parts and flip-chip devices. The advantage of the process is the elimination of laser drilling, conventional drilling, photo-resists, imaging and registration, and solder masks.

Imprint Patterned Circuits
Based on stamper technology for
LaserDiscs/Compact Discs
A CD has hundreds of millions of pits/grooves
(the LaserDiscs have billions)

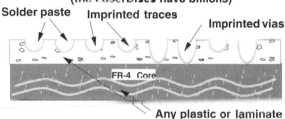

FIGURE 22.23   Imprinted circuits.

REFERENCES

1. Y. Tsukada, S. Tsuchida, and Y. Mashimoto, "Surface Laminar Circuit Packaging," *Proceedings of the 42nd ECTC Conference,* 1992, pp. 22–27.

2. A. Kestenbaum et al., "Laser Drilling of Microvias in Epoxy Glass Printed Circuit Boards," *IEEE Transactions,* vol. 13, no. 4, December 1990, pp. 1055–1062.

3. J. Morrison, "A Large Format Modified TEA CO_2 Laser Based Process for Cost-Effective Small Via Generation," *ISHM,* April 1994.

4. K. Arai, "High Reliability Laser Blind Hole Technology," *Hitachi Seiko Journal,* January 1997.

5. X. F. Bregman, "Multi Chip Module for Advanced Workstation Applications," *Proceedings of the 7th IMC,* Yokohama, Japan, 1992, pp. 30–36.

6. H. B. Bakoglu, G. F. Grohoski, and R. K. Iontoye, "The IBM RISC System/6000 Processor: Hardware Overview," *IBM Journal of Research and Development,* vol. 34, no. 1, 1992, pp. 12–22.

7. T. Kimura, T. Imura, K. Saitoh, and Y. Kohno, "Fabrication and Characterization of Silicon Carrier Substrate for Silicon on Silicon Multichip Modules," *Proceedings of the 1st International Conference on Multichip Modules,* Denver, CO, 1992, pp. 23–27.

8. C. Boyko, F. Bucek, V. Marovich, and D. Mayo, "Film Redistribution Layer Technology," *Proceedings of the 2nd International MCM Conference,* 1993, pp. 196–199.

9. O. Powell, T. L. Heleine, R. H. Magnuson, V. Markovich, and A. C. Bhatt, "Early Experience with Film Redistribution Layer Technology," *Proceedings of the 1994 IEPS Conference,* 1994, pp. 649–658.

10. R. Carpenter and I. Memis, "SLC: An Organic Packaging Solution for the Year 2000," *NEPCON West Conference,* 1996.

11. R. Tummala, E. Rymazewski, and A. Klopfenstein, *Microelectronics Packaging Handbook, Subsystem Packaging, Part III,* Chapman & Hall, 1997, pp. 265–268.

12. Y. Tsukada, S. Tsuchida, and Y. Mashimoto, "Surface Laminar Circuits and Flip Chip Attach Packaging," *Proceedings of The 4th International Microelectronics Conference,* Yokohama, 1992, pp. 20–27.

13. K. C. Norris and A. H. Landzberg, "Reliability of Controlled Collapse Interconnections," *IBM Journal Research and Development,* vol. 13, no. 3, May 1969, pp. 266–271.

14. N. Kawachi, T. Wada, and T. Miki, "Manufacturing Printed Wiring Board by Ultra High Speed Electroforming," *P.C.W.C.-5 U.K.,* no. B4 2, 1990.

15. T. Yoshino: "The Process of Printed Wiring Board Manufacturing, Using Photoelectroforming Process," *Surface Mount Technology,* Autumn 1990, p. 186.

16. H. Tourne, "Laser Via Technologies for High Density MCM-L Fabrication," *95 MCM Conference,* 1995, pp. 71–76.

17. D. A. Belforte, *Industrial Laser,* vol. 10, no. 8, 1995, p. 1.

18. R. Srinivasan, "Ablation of Polymers and Biological Tissue by Ultraviolet Lasers," *Science,* vol. 234, 1986, p. 559.

19. T. G. Tessier and G. Chandler, "Compatibility of Common MCM-D Dielectric with Scanning Laser Ablation Via Generation Processes," *IEEE Transactions on CHMT,* vol. 16, 1993, p. 39.

20. R. S. Patel et al., "Laser Via Ablation Technology for MCMs Thin Film Packaging—Past, Present, and Future at IBM Microelectronics," *ISHM,* 1994, pp. 31–41.

21. J. M. Morrison et al., "Small Via Generation," *Advanced Packaging,* November/December 1994, pp. 26–29.

22. T. G. Tessier and G. Chandler, "Compatibility of Common MCM-D Dielectric with Scanning Laser Ablation Via Generation Processes," *IEEE 1993,* vol. 16, no. 1, February 1993, pp. 39–45.

23. L. Burgess and P. Madden, "Blind Vias in SMD Pads," *Printed Circuit Fabrication,* vol. 21, no. 1, January 1998, p. 224.29.

24. T. G. Tessier, J. M. Morrison, and B. Gu, "A Large Format Modified TEA CO_2 Laser Based Process for Cost Effective Small Via Generation," *ISHM Paper,* April 1994.

25. J. Tourne, "Laser Via Technologies for High Density MCM-L Fabrication," *ISHM Paper,* April 1995.

26. M. Owen, "New Laser Technology for Drilling Through- and Blind Vias in Copper Clad Reinforced Circuit Boards," *CircuiTree,* February 1997.

27. R. Srinivasan, "Ablation of Polymers and Biological Tissue by Ultraviolet Lasers," *Science,* vol. 234, October 1986, pp. 559–565.

28. C. A. Pico et al., "Micromachining of Polyimides Using Ultraviolet Solid State Lasers," *FLEX-CONTM'95,* 1995, pp. 152–154.

29. M. Owen, "New Laser Technology for Drilling Through- and Blind Vias in Copper Clad Reinforced Circuit Boards," *IPC Exposition,* San Diego, CA, May 1995.

30. A. Cable and M. Owen, "High Density Packaging: Micro-Vias with the UV Laser," *Surface Mount Technology,* vol. 9, no. 7, 1995, p. 31.

31. L. Lemke, "Recent Developments in the Production and Application of Advanced Laminate Based MCMs and Chip Carriers," *95 MCM Conference,* 1995, pp. 23–24.

32. L. Lemke, "Advanced Laminate Based MCM Consortium," *Proceedings of the International Electronics Packaging Conference,* Atlanta, GA, 1994.

33. S. Roberts, "The Role of Flexible Materials in MCM Substrates," *IPC National Conference on MCM-L Proceedings,* Minneapolis, MN, 1994.

34. G. Gengel, "A Process for the Manufacture of Cost Competitive MCM Substrates," *Proceedings of the 1994 International Conference on Multichip Modules,* Denver, CO, 1994, pp. 182–187.

35. G. Gengel, "Quick Turn Manufacturing of Interconnect Systems Using Predrilled and Plated Via Arrays," *Proceedings of the First International Conference on Flex Circuits, Flexcon TM 94,* 1994.

36. R. A. Fillion, "Status and Update on the GE HDI Multichip Module Technology," *Proceedings of the Dupont Symposium on High Density Interconnect,* May 1990, p. 52.

37. R. A. Fillion, W. Daum, E. J. Wildi, and E. B. Kaminsky, "Multichip Modules—Chips First vs. Chips Last Analysis," *Proceedings of the ISHM International Symposium on Microelectronics,* San Francisco, CA, October 1992, p. 391.

38. Happy T. Holden, "Using Nonconventional PCB Build-Up Technology to Reduce the Cost and Design Complexity for MCM-Ls," *ISHM Advancing Microelectronics,* April 1995.

39. H. Holden, "Novel Approaches to Blind and Buried Vias Using Conductive Pastes," *Proceedings of the Chip Scale International Conference,* Santa Clara, CA, May 1998.

40. M. Buchwald, and P. Stimpfig, "DYCOstrate," *Expertise, the Newsletter for Zuken-Redac Customers,* Winter 1994/1995, pp. 9–10.

41. H. Holden, "Optimizing MCM-L Designs: The Focus on Metrics," *Proceedings of the 32nd ISHM-NORDIC Conference,* Helsinki, Finland, September, 1994, pp. 123–129.

42. H. Holden, "How to Use High-Density Interconnect Structures with Fine-Pitch Area Array Components," *Chip Scale Review,* September 1997, pp 54–57.

43. H. Holden, "Where Are Modern PCB Fabricators Today?" *Future Circuits,* iss. 1, vol. 1, 1997, pp. 15–20.

44. H. Holden, "Micro-Via PCBs: The Next Generation of Substrates and Packages," *Future Circuits,* iss 1, vol. 1, 1997, pp. 71–75.

45. H. Holden, "Microvias Build-up PWBs: The Next Generation Substrates," *Proceedings of the SMTA 4th Emerging Technology Symposium,* October 20–23, 1997, Bloomington, MN, pp. 1–9.

46. H. Holden, "Design and Fabrication Issues In Use of Flip Chip and High Density Area Array Packages, *Proc. of the SMTA 4th Emerging Technology Symposium,* October 20–23, 1997, Bloomington, MN, pp. 774–783.

47. H. Holden, "Microvia: The Next Generation of Substrates and Packages," *IEEE MICRO Magazine,* July August 1998, pp. 1–8.

CHAPTER 23
MICROVIA HOLE TECHNOLOGIES

Dr. Hayao Nakahara
N. T. Information Ltd., Huntington, New York

23.1 INTRODUCTION

Since the introduction of surface laminar circuit (SLC) technology (see Figs. 23.1 and 23.2) in 1991[1-3] (SLC is a trademark of IBM), many variations of making high-density interconnect (HDI) wiring boards have been developed and implemented for mass production. However, if one technology is to be picked as a winner judged in terms of volume produced, laser drilling technology is the one. Other methods are still used by a number of printed wiring board (PWB) manufacturers, but in a much smaller scale.

The purpose of this chapter is to examine a variety of microvia hole formation technologies. However, a greater emphasis will be placed upon the laser drilling process (*laser via* hereinafter) since it is the most popular process today and it seems its popularity will grow in the future. It must be understood that via hole formation is just one element of fabricating HDI wiring boards. Fabrication of HDI wiring boards with microvia holes involves many new processes not common to conventional board fabrication. Therefore, additional emphasis will be placed upon these new fabrication processes that are common to other microvia technologies.

23.2 DEFINITIONS

Printed wiring boards (PWBs) with microvia hole structures are called by different names: HDI, SBU (sequential build-up), BUM (build-up multilayer), and so on. However, HDI covers a broader range of high-density wiring boards such as extremely high-layer-count multilayer boards (MLBs) without microvia holes. MLBs with microvia holes are not necessarily built sequentially nor have build-up structures. These definitions are not appropriate for the discussions in this chapter, and therefore we shall address MLBs with microvia holes simply as *microvia hole boards* (all microvia hole boards are essentially multilayer boards).

Some trade and academic organizations define the microvia hole to be a hole of a certain diameter or less. For example, IPC defines a microvia hole as a hole with a diameter equal to or less than 150 μm. However, when a surface blind via hole (SBV) is formed between layer 1 (L1) and layer 3 (L3), the diameter of such a hole is typically 250 μm in order to facilitate reliable plating, but the hole is still considered a microvia hole. Since all microvia holes are essentially SBVs and are normally small in diameter in order to increase circuit density, it seems more appropriate to define the microvia hole as an SBV without limiting its diameter. As long as a hole has SBV structure, it is defined as a microvia hole throughout this chapter. When

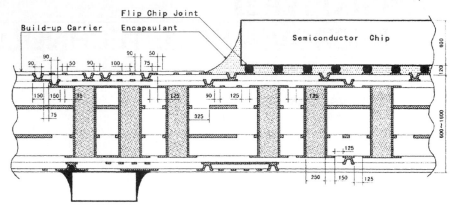

FIGURE 23.1 Cross-sectional view of microvia hole board made by SLC process. (*IBM Yasu.*)

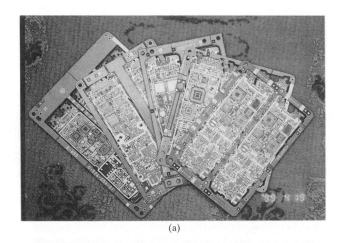

(a)

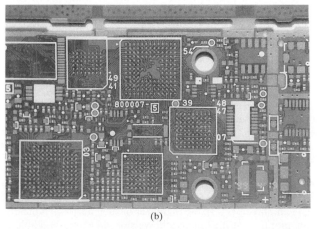

(b)

FIGURE 23.2 Microvia hole boards for cellular phones: (*a*) assortment of cell phone boards; (*b*) close-up photo of microvia hole board for cell phone. All ball grid array (BGA) pads have microvia holes right on the pads (hole on pad or HOP) and hundreds of holes on other pads and power and ground traces.

another wiring layer is built over the existing microvia holes, these holes become *buried via holes* (BVHs).

23.3 DIELECTRIC MATERIALS AND COATING METHODS

There are four basic methods of microvia hole formation being practiced today: laser drilling (laser via), photo formation (photovia), plasma etching (plasma via), and paste technologies (paste via). Each of these technologies has variations. There are two basic variations to paste via technologies. These are treated later in this chapter since microvia hole formation of these two are quite different from the first three.

Figure 23.3 shows compatibility of laser via, photovia, and plasma via methods with four basic surface dielectric structures on which microvia holes are to be formed. While laser via methods can cope with all four dielectric structures, photovia and plasma via methods are applicable to only one structure, respectively, as shown in the figure. This is one reason why laser via is more widely used today.

23.4 PHOTOVIA MATERIALS

Dielectric materials for the photovia process must be photosensitive. This limits the choice of fillers that can enhance peel strength, and the photosensitive nature makes the materials more expensive. After photoexposure of the hole pattern, microvia holes are formed by a developing process. There are two types of developer: solvent based and alkaline based. From the viewpoint of physical characteristics such as moisture absorption, peel strength, and so on, the solvent type is superior to the alkaline. Unfortunately, solvent-type developer is not environmentally friendly and is not preferred by the PWB industry in general. Alkaline-developable type, however, has generally inferior characteristics.

Photosensitive dielectric materials come in liquid or dry-film form. Dry-film dielectric is vacuum laminated to the base core. Liquid dielectric is coated to the base core by either the open-screen coating method (single side at a time or double sides simultaneously), spray coating method, roller coating method (normally double sides simultaneously), or slot coating method (single side at a time). There is no decisively superior method. The principle of slot coating is illustrated in Fig. 23.4.

There are two important factors in coating liquid dielectric. One is even thickness throughout the entire panel and coplanar surface. Even thickness is important to achieve good expo-

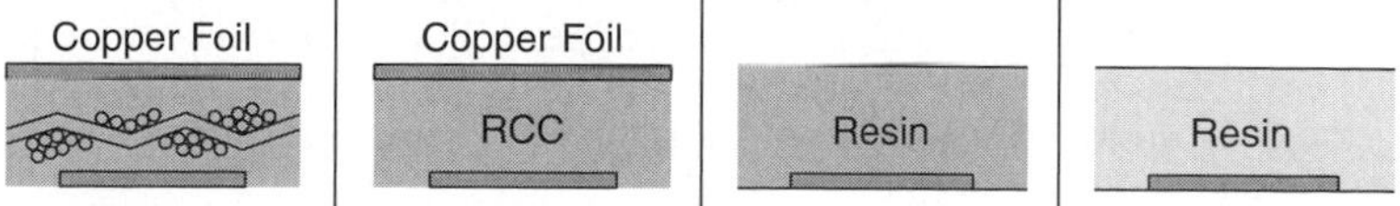

	Standard Configuration	RCC	Thermally Curable Resin	Photoimageable Resin
	Copper Foil	Copper Foil	Resin	Resin
Photo Via	X	X	X	O
Laser Via, CO^2	O*	O*	O	O
Laser Via, Yag	O	O	O	O
Plasma Via	X	O*	X	X

FIGURE 23.3 Compatibility of via hole formation methods with four basic dielectric layer structures.

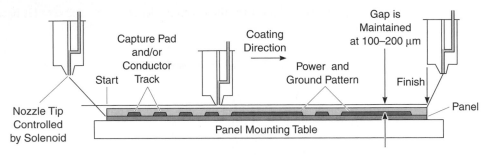

FIGURE 23.4 Principle of slot coating. (*Shipley Far East.*)

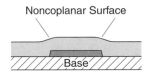

FIGURE 23.5 Coating conformal to inner layer circuit pattern.

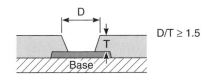

FIGURE 23.6 Desirable aspect ratio for reliable plating.

sure and developing. Liquid material tends to follow the surface curvature as shown in Fig. 23.5, which can create problems later in the process, particularly when conductor lines must be formed along the curvature. A brushing operation is used to scrape off the protruding portion of resin over the capture pad to make the surface coplanar when it is excessive.

When dielectric thickness of more than 60 µm is required, it is difficult to achieve in one coating. Normally, about 40 µm is coated first, the resin is tack-free cured, and then an additional 20 µm is coated. In the case of photovia processing, this curing must be controlled precisely. Otherwise, holes cannot be developed evenly. Experienced practitioners of photovia processing indicate that the limitation of a hole diameter that can be developed in the tapered shape desirable for subsequent copper plating is somewhere around the aspect ratio of 1:1.5 (ratio of dielectric thickness to hole diameter at dielectric surface). See Fig. 23.6.

The other important factor is cleanliness. In photovia processing, dust must be avoided at all cost because it is a main cause of defects. The resin coating room must have a cleanliness of at most Class 10,000. Manufacturers of semiconductor package substrates using the photovia process usually have Class 1,000 clean rooms and the cleanliness is maintained at less than Class 1,000.

23.5 LASER VIA MATERIALS

Dielectric materials for the laservia process come in resin form, either liquid or dry film, and in the form of resin-coated copper foil, RCC (RCC is a trademark of Isola AG). Other dielectrics are the traditional glass-reinforced prepreg and nonwoven aramid prepreg. Base aramid material is offered by Teijin of Japan and DuPont (DuPont's aramid base is called Thermount).

Liquid dielectric resin is coated more or less the same way as in photovia. However, there is a fundamental difference between photovia and laser via dielectric. In laser via processing, the resin is fully cured before laser drilling (microvia formation). This is a big advantage over photovia materials since the resin movement after hole formation is much more stable than in the case of photovia processing in which resin is fully cured after holes are formed. This resin (hole) movement makes pattern imaging registration difficult for photovia processing.

The RCC and prepregs (together with copper foil) are laminated much the same way as in normal multilayer board lamination processing. In recent years, a new lamination process and equipment were introduced using copper foil as a cap layer foil as well as a heating element (see Fig. 23.7). The press is called Adara (a trade name of Cedal s.r.l.). In laminating RCC, however, copper foil is used just as a means to conduct electric current for heating since copper foil of RCC when bent at 180 degrees can break, and flow of electricity will be disrupted. Copper thickness comes in different dimensions, typically 9, 12, 17.5, and 35 μm. RCC with much thinner copper foil (3 to 5 μm) is also available (see Fig. 23.8), although the price of ultrathin copper RCC is much more expensive. Carrier copper foil of 35 or 70 μm is coated with special electric-current-conducting release resin, facilitating easy removal after lamination, and is electroplated with thin copper foil. Subsequently, this laminate structure is coated with dielectric resin.

More than 80 percent of microvia boards used for cellular phones are made with RCC today, but this may change for laser via processing because of tighter registration requirements. This is explained in detail later in the chapter.

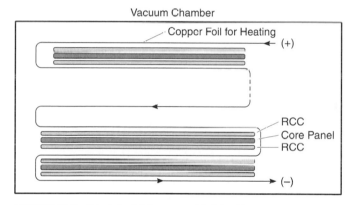

FIGURE 23.7   Principle of Adara press. (*Cedal s.r.l.*)

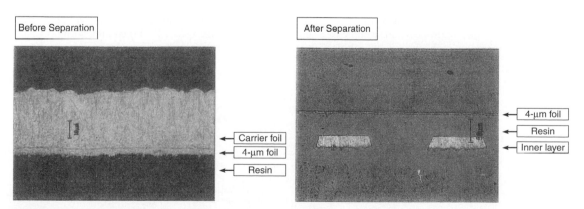

FIGURE 23.8   Ultrathin copper RCC. (*Mitsui Metal Smelting Co. Ltd.*)

23.6 *PLASMA VIA MATERIALS*

Plasma via processing requires a copper surface. A different approach, however, has been attempted in which a thin nickel layer is deposited by a spattering process after liquid resin is coated on the base core and the window is opened prior to plasma etching.[4] Plasma via processing is used to fabricate rigid PWBs with RCC and flexible PWBs, normally with copper foil coated with polyimide film.

23.7 *PASTE VIA MATERIALS*

There are two paste via technologies: ALIVH (any-layer interstitial via hole, ALIVH is a trademark of Matsushita Electric Industry)[5] and B²it (buried bump interconnection technology, B²it is a trademark of Toshiba).[6]

The ALIVH process uses nonwoven aramid prepreg as its dielectric material and copper paste for hole filling. B²it uses glass-reinforced prepreg for rigid wiring boards, liquid crystal resin for flexible wiring boards, and silver paste for holes.

23.8 *MANUFACTURING PROCESSES*

The manufacturing process for each microvia technology begins with a base core, which may be a simple double-sided board carrying power and ground planes or a multilayer board carrying some signal pattern in addition to power and ground planes. The core usually has plated through-holes (PTHs). These PTHs become BVHs. Such a core is often called an *active core.*

In fabricating a base core with a pattern, the core panel is usually panel plated and the pattern is made with a dry-film tenting process. Some makers seem to prefer pattern plating, however. The choice is up to the fabricator's familiarity with these processes. After the pattern is formed, dielectric material is laminated over the core (in the case of prepreg, with copper foil) and the holes can be filled with resins, depending on the plated hole diameter and the thickness of the core. It is generally agreed that when the diameter of plated holes is equal to or less than 0.3 mm *and* the core thickness is equal to 0.6 mm or less, these holes can be filled effectively by the lamination process (although the resin thickness of 80 µm is preferred in the case of RCC). See Fig. 23.9.

When the diameter/thickness conditions are not met, it is necessary to fill the holes by a separate process. This is done by a screening process from one side of the panel with polyester screen with an oversized hole pattern. After hole filling and curing the resin completely,

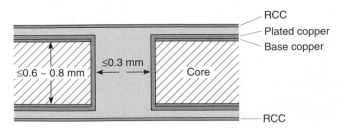

FIGURE 23.9 Conditions under which the through-hole can be filled with resin by RCC lamination.

excess resin is removed by a belt sander (#600 to 800) or ceramic brush. Figure 23.10 illustrates the process. This is a tricky but necessary operation for microvia board makers, particularly for photovia processing.

Hole filling is costly but yields a few advantages over simultaneous hole filling by lamination of RCC or prepreg. The edge of the plated holes is well protected; therefore, forming finer annular rings on the base core by tenting is easier. There is no concavity problem over the hole that can cause difficulties in later processes.

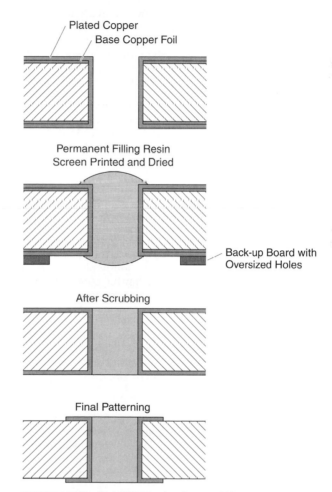

FIGURE 23.10   Hole filling and surface scrubbing step.

23.8.1 Photovia Process

Prior to dielectric material coating by any of the methods described previously, the copper surface of the base core must be treated by an adhesion promotion process to ensure good adhesion of dielectric material to the copper surface. Today, very few manufacturers use oxide treatment for this purpose. The most popular adhesion promotion treatment is a special etching process offered by many suppliers of chemicals. This step is common to all microvia processes.

Dielectric resin is semicured after coating to eliminate tackiness, and then the hole pattern is exposed by photoexposure processing. The usual photo-developing process creates microvia holes and the dielectric is fully cured, typically at 160°C for about one hour. Then, the panel goes through a permanganate etching process to remove any residual resin at the bottom of the hole and simultaneously create microporous surfaces that act as an anchor and ensure desirable peel strength after copper plating.

The level of peel strength is controversial. Minimum peel strength required for chip package substrates is about 600 g/cm², but motherboard users, particularly cell phone makers, demand a minimum of 1.0 kg/m² or more in order for cell phone handsets to withstand drop tests. Laser via materials usually yield stronger peel strength because of fillers that can be added to dielectric resin. These fillers when etched generate a superior microporous surface structure needed for strong peel strength.

After permanganate etching, the panel is catalyzed and metallized in an electroless copper bath and panel-plated galvanically to desired thickness. Some photovia process practitioners roughen the resin surface mechanically by brushing or liquid horning prior to catalyzing. Then, the conductor pattern is formed by dry-film tenting and etching. Some manufacturers prefer to use the pattern plating method for this purpose. Very few microvia board manufacturers use direct metallization methods for metallizing holes prior to galvanic plating. Several Japanese manufacturers use electroless copper all the way to the desired thickness in panel plating and use a positive electrodeposited (ED) system to achieve fine lines and very small annular rings.

One important step in microvia hole board fabrication when resin is the dielectric choice, whether the process is photovia or laser via, is the removal of residual catalysts (normally palladium) entrapped in a microporous surface that can cause migration. This step is normally a trade secret.

Photovia processing is now used primarily to fabricate semiconductor chip package substrates because a large number of holes can be formed in one photoexposure and development step. However, as mentioned previously, photovia processing suffers more from material shrinkage than laser via processing after full cure and hole locations tend to move randomly, which makes subsequent registration for patterning difficult. Because of this problem, photovia users limit the size of the panel to be much smaller than the usual panel size prevalent in motherboard fabrication to about 400 mm × 400 mm. Small hole formation is also difficult with the photovia process. As a result, even makers of package substrates are now resorting more to laser via processes as the laser drilling speed is being improved. Photovia process users engaged in mass production are found only in Japan today. A standard photovia process sequence is described in Fig. 23.11(a).

23.8.2 Plasma Via Process

Plasma via processing was developed by a Swiss PWB maker, Dyconex. Products made with the plasma via process are called DYCOstrate, which is a trademark of Dyconex Corporation. There are many variations to the plasma via process, one of which is illustrated in Fig. 23.11(b). Today, it is mainly used to fabricate sophisticated flex and flex-rigid wiring boards in small quantities.

First, an opening or window is made through copper foil by a normal etching process. When plasma etching is applied through this window, the shape of holes tends to be like a bowl as shown in Fig. 23.11(b), which is not suitable for reliable plating (though new plasma etching equipment claims to have solved this problem).[7] Another problem is related to how the microvia hole is formed. The copper edge of the window hangs out over the hole, which results in poor reliability after panel plating. Therefore, to ensure reliable plated holes, a secondary etching is necessary to remove this copper overhang. One good thing results from this secondary etching, however. Since surface copper is thinned, formation of finer conductors is made easier. Nevertheless, by the time the panel is ready for plating for subsequent conduc-

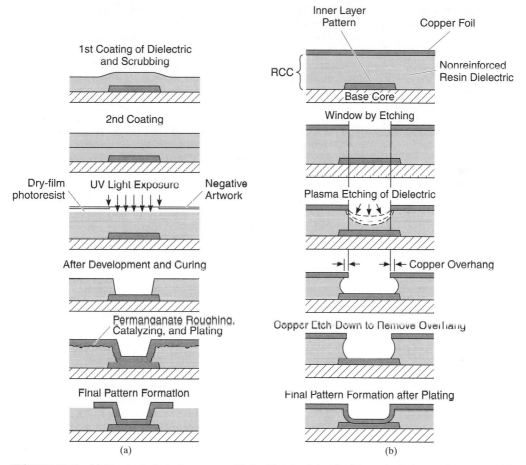

FIGURE 23.11 (*a*) Standard photovia process with liquid resin dielectric; (*b*) standard plasma via process with RCC.

tor pattern formation, it takes several times longer than other processes in a mass-production environment.

Plasma via processing is effective for forming through-holes on flexible materials since holes are formed by plasma etching from both sides of the film and the bowl effect is minimized. However, holes here are not the normal microvia type as defined previously in terms of SBV.

23.8.3 Laser Via Process

Laser via processing is by far the most popular microvia hole formation process. However, there are many variations. For the purpose of drilling microvia holes, there are four laser systems: UV/Yag laser, CO_2 laser, Yag/CO_2, and CO_2/CO_2 combinations. Then, there are three dielectric materials: RCC, resin only (dry film or liquid resin), and reinforced prepreg. Therefore, the number of ways to make microvia holes by laser systems is the permutation of four laser systems and these three dielectric materials.

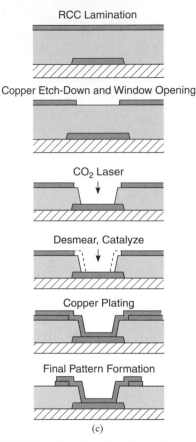

RCC Lamination

Copper Etch-Down and Window Opening

CO_2 Laser

Desmear, Catalyze

Copper Plating

Final Pattern Formation

(c)

FIGURE 23.11 (*c*) semiconformal laser via process.

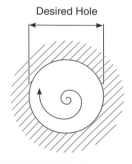

Desired Hole

FIGURE 23.12 Trepanning.

There are several factors to observe in laser via processing: position accuracy of lased holes (microvia holes), uneven diameters of holes, dimensional change of the panel after curing dielectric, dimensional change of the panel due to temperature and humidity variations, alignment accuracy of the photoexposure machine, unstable nature of negative artwork, and so on. These should be carefully monitored and are important for all microvia hole processes.

23.8.3.1 UV/Yag Laser. A Yag laser can penetrate copper; therefore, it is not necessary to pre-etch the window when the choice of dielectric is RCC or a copper foil/prepreg combination. Positioning accuracy in relation to the capture pad is good, but there are weaknesses associated with the Yag laser. The speed of drilling (ablation) is slower than the CO_2 laser, particularly when a hole diameter is equal to or larger than 125 μm because the laser beam is very small. Therefore, it is necessary to do trepanning (see Fig. 23.12), which results in longer drilling time. Another weakness is the Yag laser's sensitivity to the thickness of dielectric material. If the thickness of the RCC's resin is not even, the Yag laser beam whose power is adjusted to remove the thicker part of the resin may damage the capture pad. If adjusted to the thinner part, it may not reach the capture pad.

Although it can penetrate copper and glass-reinforced prepreg, the situation is worse in this material than with RCC. The laser power must be adjusted to penetrate through the crossover part of woven glass. Then, when the laser beam hits the part of prepreg where there is now a glass bundle (opening), the beam can really damage the capture pad. In order to avoid this problem, the laser power must be reduced and longer duration of laser ablation must be applied, which slows drilling time considerably. Depending on the construction of the dielectric layer through which the Yag laser beam must penetrate, the drilling speed must be adjusted to about 3 to 50 holes per second when glass-reinforced dielectric is used.

When the dielectric is RCC, the speed of the currently available UV/Yag laser is somewhere around 90 to 100 holes per second. When the dielectric is resin without copper foil, a CO_2 laser is the best since a CO_2 laser can offer much faster drilling speed (a single-head CO_2 laser can drill as many as 25,000 holes per minute including overhead such as loading panel onto the drilling table and panel registration). The best feature of the UV/Yag laser drilling machine is its ability to drill a very small hole (down to 20 to 30 μm while the smallest hole a CO_2 laser can drill is 50 μm in a mass-production environment) and its superior position accuracy through RCC.

For higher-speed drilling, even Yag laser users are now etching down copper foil to 6 to 9 μm, the process more frequently seen in CO_2 drilling.

23.8.3.2 CO_2 and Twin CO_2. A CO_2 laser beam cannot penetrate copper foil unless the foil is very thin, less than 5 μm, and the surface is treated to be dark to absorb the CO_2 laser beam (this process is called CO_2 laser direct drilling and will be treated later).

When the surface is resin only, a single-headed CO_2 laser machine can drill holes at the rate of 20,000 to 25,000 per minute depending on hole density and distribution. The drilling speed continues to increase. The denser the hole, the faster the drilling speed because time lost by table motion is minimized. A dual-head CO_2 machine can boost drilling productivity by about 70 percent compared to a single-head machine.

Manufacturers of dielectric resins mix fillers in the resins to enhance peel strength. Some fillers can cause slowdown of drilling speed by a CO_2 laser beam (beam absorption is weaker in this case). For example, when the required dielectric thickness is 60 μm, resin is coated twice, which is a frequent practice to produce a thick coating. The first coating is made with resin without filler (40 μm), and the second coating (20 μm) is resin with filler. Since the laser beam goes through a thin layer of filled resin, the speed degradation is minimized.

The most desirable shape of drilled microvia holes is tapered, which is accomplished by pulse drilling. Normally, two to three pulses are used. When three pulses are applied on the same spot continuously, the capture pad may be overheated and excessive epoxy smear may result. Therefore, when three pulses are to be applied, the entire panel is drilled first by two pulses and then the third pulse is once again applied to all holes. This way, the quality of the holes is ensured although it takes longer to drill.

When the choice of dielectric is RCC, the opening or window must be formed first on copper foil as in the case of plasma via. Prior to window formation, the surface copper is thinned, usually down to 6 to 7 μm, by an etching process. This process is sometimes called *half-etching*. Half-etching yields a few benefits. Window formation is made easier, and formation of fine-line conductors is also made easier. H_2O_2/H_2SO_4 etchant is often used for this purpose for even thickness. As mentioned previously, this half-etching procedure is also becoming common for Yag laser users.

There are two variations to CO_2 laser drilling through a window opening. One is called *conformal* and the other *semiconformal*. See Fig. 23.11(*c*).

In the conformal method, a laser beam with a diameter larger than the window opening is used. Straight ablation of dielectric results in a bowl-shaped hole as in the case of plasma via. Therefore, improved systems use laser beam diameters slightly larger than the window opening, and beams are pulsed to minimize the bowl effect. In the semiconformal method, a laser beam with a diameter slightly smaller than the window opening is used. This results in a slight step when plated afterward as shown in Fig. 23.11(*c*). In order to minimize this step, half-etching is important in semiconformal drilling.

Some manufacturers using CO_2 drilling etch down copper completely after RCC lamination. The dielectric surface below copper is already microporous, which generates good adhesion except prior to the RCC stage. Therefore, some special treatment is applied to the resin surface, such as electrolessly plated nickel layer, prior to copper plating, but such treatment is kept secret.

The RCC approach suffers from misregistration of window opening to capture pad when the surface and capture pad diameter becomes less than 250 μm, which causes serious defects (see Fig. 23.13). Laser via process practitioners have several solutions to this problem. They can use liquid resin or resort to a Yag or Yag/CO_2 laser. Since a Yag laser can break through copper foil, the beam can be directed to the capture pad position accurately. Another choice is to use laser direct imaging (LDI) for window etching if the speed of the CO_2 laser is to be enjoyed.

Another solution is to use CO_2 laser direct drilling. The advantage of using RCC is the pre-established peel strength. Manufacturers go through delicate steps to establish the appropriate peel strength when using liquid resin. The degree of resin cure, permanganate etching, control, and regeneration of the permanganate solution affect the resultant peel strength as well as cost. In this laser direct drilling method, copper foil is etched down to about 5 μm and its surface goes through oxide treatments such as black oxide or other methods, typically alternative to oxide treatment, which is normally an etching process. The copper thickness is further reduced by 1 μm, and the surface color becomes dark, allowing the CO_2 beam to penetrate.

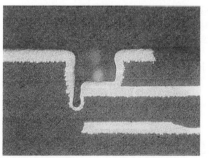

FIGURE 23.13 Misregistration of laser hole to capture pad. (*Ibiden.*)

Alternatively, ultrathin copper RCC may be used. Carrier foil, usually copper foil of 70-μm thickness (2 oz), is coated with a very thin layer (10 to 20 μm) of current conductive release film, and a thin foil of 3 to 5 μm is plated (see Fig. 23.8). After such RCC is laminated to the core, the carrier film is peeled off, and the surface condition is the same as etched-down copper. Drilling speed in this direct CO_2 laser drilling method is degraded by 30 to 40 percent compared to the case of straight resin drilling, but the registration to the capture pad is ensured.

23.8.3.3   Yag/CO_2.   This combination is particularly useful for a copper/prepreg structure. The Yag laser offers good position accuracy and cuts through copper, but it tends to leave glass fibers sticking out in the hole after prepreg drilling. And as mentioned before, it is sensitive to dielectric thickness variation. The CO_2 laser beam cuts through glass fiber more cleanly than the Yag. In this combination laser drilling, the power of the Yag laser beam is adjusted to cut through copper and remove a portion of glass fiber and leaves the finishing to the CO_2 laser. The speed of drilling is governed by the speed of the Yag laser and is therefore slower than a pure CO_2 laser. However, this is the preferred laser drilling machine for fabricating infrastructures types of PWBs, such as ones used for servers, network routers, and base stations that typically have copper/prepreg surface structures. The size of the panels used to fabricate such boards is usually large; therefore, the positioning accuracy of the Yag laser is preferred. Microvia hole boards used by the automotive industry also have this copper/prepreg structure.

The speed of drilling by a Yag/CO_2 machine is much slower than a CO_2 laser. However, the number of microvia holes to be drilled for infrastructure boards is normally one magnitude less than that seen in cell phone applications in which the number of microvia holes can reach as many as 650,000 to 700,000 per m^2. Furthermore, the price of infrastructure boards is much more than mass-produced cellular phone motherboards. Therefore, slower speed of drilling, which means higher drilling cost, does not affect the total cost of manufacturing.

Residual glass fibers in the hole that can affect reliability may be removed chemically or mechanically. Mechanical removal is preferred. Blasting with aluminum oxide (about 20-μm-diameter particles) is commonly used to remove glass fibers and smears as well. An excimer laser is also used to remove smears in the hole by a sweeping motion through the copper opening. In both cases, it takes about 30 seconds to desmear one side of the panel.

23.9   MULTIPLE LAYERS OF MICROVIA HOLES

For interconnecting IC substrates having hundreds of I/O terminals, a single layer of microvia holes may not be sufficient to satisfy interconnection requirements. Double, triple, or even quadruple microvia hole layers may be necessary whether the board is a motherboard or a chip package substrate. When microvia holes are made only between adjacent layers, all three microvia processes—photovia, plasma via, and laser via—can be used. However, when a design requires microvia holes that must connect beyond adjacent layers, say, L1 and L3 (skip-via), laser via processing is the only practical way. See Fig. 23.14.

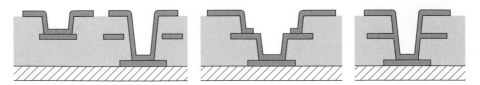

FIGURE 23.14   Microvia connections beyond adjacent layers.

23.10   *OTHER MICROVIA TECHNOLOGIES*

In this section, only those other microvia hole technologies that are used for mass production are discussed.

23.10.1   ALIVH

ALIVH uses nonwoven aramid prepreg as dielectric material.[5] The prepreg sheet is perforated by a CO_2 laser beam, and the resultant holes are filled with conductive copper paste by screening. This prepreg sheet with holes filled is laminated with copper foil on both sides and completely cured. Then, the pattern is formed by a conventional pattern etching process, and exposed copper is treated by an adhesion promotion process.

The core is now laminated with a similar prepreg processed in parallel, and the same process is repeated until the desired multilayer structure is completed. As can be noted, ALIVH is a totally sequential build-up process. There is no copper plating. See Fig. 23.15.

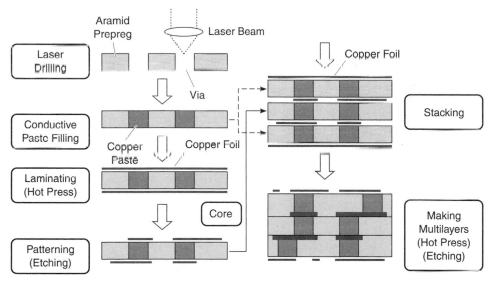

FIGURE 23.15   Manufacturing steps of ALIVH process.

The advantage of ALIVH boards is that because of the way they are constructed, conductor routing is always achieved in minimum distance and skip-via (L1-L3 connection) can be made effectively. The time required for circuit design is very short, which is vital in this fast-moving market. The disadvantage is the limitation on via diameter and that the size of the finished wiring boards cannot be larger than 200 to 250 mm. Also, since it is totally sequential, a practical limitation in layer count is probably about 10. If the prepreg thickness is 100 μm, holes of diameter smaller than this are not easy to implement because of difficulty of screen filling the hole with copper paste. Nevertheless, ALIVH offers vast applications, particularly for small, high-density portable products.

23.10.2   B²it

The B²it process uses silver paste for the microvia hole structure.[6] Silver paste is screened onto the adhesive side of copper foil through a perforated metal stencil. In order to build the triangular cones to a sufficient height ideal for vias, multiple screening passes are necessary.

When the silver paste cones are formed, glass prepreg and copper foil are laminated upon the core and the composition is fully cured thermally, as in the case of ALIVH. The conductor pattern is also formed by an etching process. The process is repeated to complete a desired multilayer structure. The glass prepreg acts as a stiffner as well as a dielectric.

All layers can be built by B²it processing. Some B²it products are made by a combination of a rigid base core fabricated as an ordinary multilayer board and B²it layers as illustrated in Fig. 23.16. The limitation of B²it processing is the minimum hole diameter and the size of the finished board, much the same as for ALIVH. Its main applications are for portable products.

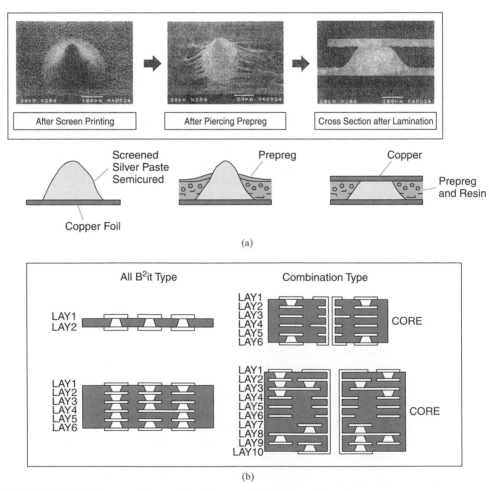

FIGURE 23.16   B²it process: (*a*) manufacturing steps; (*b*) product configuration.

23.10.3 Other Microvia Processes

There are many other microvia processes proposed, but they are not practiced in mass-production environments. Some of these processes are briefly introduced.

Hitavia (a trademark of Hitachi Chemical Co., Ltd.)[8] laminates mechanically predrilled RCC. Several RCC sheets can be drilled simultaneously. Then, RCC is laminated to the base core. The resin is formulated to be low flow so that when perforated RCC is laminated, resin flow onto the capture pad is minimum. However, because holes are mechanically drilled and resin still flows onto the capture pad, the practical limitation for the hole diameter is about 200 μm. Most applications of microvia boards require holes of diameter less than 150 μm; therefore, Hitavia has limited applications.

The sand blasting method (SBV® by Tokyo Ohka)[9] laminates thermosetting epoxy dielectric in dry film on the base core. Then, a photosensitive dry film is laminated on top of the dielectric, the hole pattern is exposed, and holes are created by a usual developing process. This dry film is elastic so that when the subsequent sand blasting is applied, it does not crack. Sand blasting removes a portion of the dielectric layer corresponding to the desired hole location. Here again, the minimum diameter of holes that can be formed is about 200 μm and not suitable for high-density microvia formation. However, the idea of sand blasting is utilized as a means to remove glass fibers that stick out in the hole after laser drilling.

23.11 TECHNOLOGY DRIVERS

The amount of microvia hole boards produced has been doubling every year. The largest application in terms of panel area is cellular phone handsets. More than 90 percent of all microvia hole boards are produced by laser via processes. As the capability of laser drilling machines continues to improve, this percentage will shift even more toward laser via.

REFERENCES

1. Yutaka Tsukada et al., "Surface Laminar Circuit and Flip Chip Attach Packaging," *Proc. 7th IMC,* 1992.

2. Yutaka Tsukada, *Introduction to Build-Up Printed Wiring Board* (in Japanese), Nikkan Kogyo Shinbun, Tokyo, 1998.

3. Happy Holden, "Special Construction Printed Wiring Boards," in *Printed Circuit Handbook,* 4th ed., Clyde F. Coombs Jr. (ed.), McGraw Hill, New York, 1995, chap. 4.

4. Akio Takahashi, "Thin Film Laminated Multilayer Wiring Substrate," *JIPC Proceeding,* vol. 11, no. 7, November 1996, pp. 481–484.

5. Kazuoki Shiraishi, "Any Layer IVH Multilayer Printed Wiring Board," *JIPC Proceeding,* vol. 11, no. 7, November 1996, pp. 485–486.

6. Yoshitaka Fukuoka, "New High Density Printed Wiring Board Technology Named B²it," *JIPC Proceeding,* vol. 11, no. 7, November 1996, pp. 475–478.

7. Tim Apol, "Directional Plasma Etching—Straight Sidewalls, No Undercut," *PC Fabrication,* vol. 20, no. 12, December 1997, pp. 38–40.

8. Koichi Tsuyama et al., "New Multi-Layer Boards Incorporating IVH: HITAVIA," *Hitachi Chemical Technical Report,* no. 24 (1995-1), pp. 17–20.

9. Tokyo Ohka company brochure.

FABRICATION PROCESSES

CHAPTER 24
DRILLING PROCESSES

Hans Vandervelde
LCOA (Laminating Company of America),
Garden Grove, California

24.1 INTRODUCTION

The purpose of through-hole drilling printed circuit boards is twofold: (1) to produce an opening through the board that will permit a subsequent process to form an electrical connection between top, bottom, and internal conductor pathways, and (2) to permit through-the-board component mounting with structural integrity and precision of location.

The quality of a hole drilled through a printed circuit board is measured by its ability to interface with the following processes: plating, soldering, and forming a highly reliable, nondegrading electrical and mechanical connection.

As with any process, the elements of the drilling process are:

- Materials
- Machines
- Methods
- Workers

When workers are properly trained and educated so that they possess a sound understanding of the other three elements, it is possible to drill holes meeting the aforementioned requirements with high productivity, consistency, and yield.

The goals of this chapter are as follows:

- Understanding the drilling process thoroughly
- Recognizing what might go wrong
- Locating where problems could occur
- Detecting whether problems do occur
- Finding root causes of problems
- Correcting undesirable conditions
- Striving for zero discrepancies
- Making improvement a team effort

24.2 *MATERIALS*

Materials that affect the drilling process are as follows and as shown in the fishbone diagram of Fig. 24.1.

- Laminate material
- Drill bits
- Drill bit rings
- Entry material
- Backup material
- Tooling pins

24.2.1 Laminate Material

A typical circuit board laminate panel consists of three basic components:

- Supporting fibers
- Resin
- Copper layers

The laminate substrate material is constructed of supporting fibers (most commonly a glass fiber weave) and a resin (most commonly an epoxy compound). Finished board thickness may range from 0.010 to 0.300 in or thicker, with the most common panel thickness ranging around $\frac{1}{16}$ (0.0625 in).

24.2.1.1 Supporting Fibers.   Generally, the larger the glass fibers in the weave, the lower the cost of the base material. However, larger fibers are less desirable from a drilling point of view because they are more likely to cause the drill bit to deflect, resulting in decreased hole

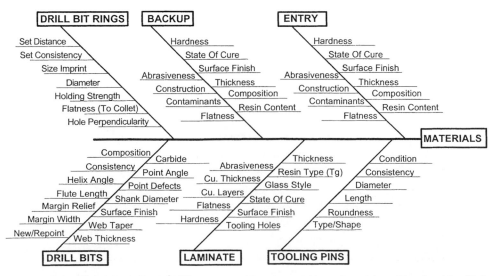

FIGURE 24.1   Materials used in the drilling process, with major variables and special considerations identified.

registration accuracy. In addition, larger fibers adversely affect drilled hole wall quality because they create larger (drilling) voids, defined as tear-out of supporting fibers. In simple terms: the larger the glass fibers, the larger the chunks torn out during drilling and thus the greater the hole wall roughness.

24.2.1.2 Resin. The most common resin system is FR-4 epoxy. However, general drilling considerations (see Fig. 24.1) are the same for all resin systems. The glass transition (T_g) rating of a material is defined as the temperature at which fully cured resin starts to soften. The T_g of FR-4 is typically around 130°C or higher. Many materials have resin systems that include additives or that are different from FR-4 resins (e.g., polyimide) in order to raise the T_g. The reason higher T_g is desired is because it implies a more stable material with lowered z-axis expansion rates. Although higher-T_g material may reduce the extent of hole wall resin smear, the compromise of a higher-T_g product is that the material is more brittle and more abrasive to the drill, resulting in increased tool wear and hole wall roughness. This increased abrasiveness to the drill bit may be offset by reducing the surface speed to reduce the spindle speed, resulting in less frictional heat being generated. Lowering drilling heat reduces drill wear.

24.2.1.3 Copper Layers. Outer and inner copper layers may be of various thicknesses, the most common being what is referred to as ½-oz and 1-oz copper. One-ounce copper equates to a thickness of approximately 1.4 mils (0.0014 in). The more copper layers within a laminate, the more balanced the panel from a drilling point of view, meaning reduced occurrences or extents of drilled hole defects such as voids (torn-out fiber bundles). However, in order to compensate for a greater number of copper layers, the chip load (advance per revolution) that determines the infeed rate usually needs to be adjusted to control nail-heading. Increasing the chip load reduces the amount of nail-heading. In addition, more copper layers wear the drill bit at a faster rate and may require lowering the maximum hit count per drill bit.

24.2.2 Drill Bits

24.2.2.1 Materials. Drill bits are made of tungsten carbide because its wear resistance (and relatively low cost) make it the most ideal material for cutting the very abrasive circuit board laminate materials. The compromise of this very hard (carbide) material is that it is also brittle and subject to damage in the form of chips if not handled carefully and correctly.

24.2.2.2 Handling and Inspection. When handling drill bits, do not permit the bits to come in contact with one another and be careful not to touch them to the sides of the tool pods. Modern drilling machines utilize drill cassettes to reduce time per load by eliminating manual tool changes. These cassettes may accommodate 120 or more drill bits and reduce drill bit handling damage. If drill bits are measured to verify diameter using a contact-type measuring device (such as a contact micrometer), take the measurement away from the point to prevent chipping of the cutting corners. Following diameter measurement, inspect the drill bit for damage using a microscope. After use, again use care when removing drill bits from the tool pods or cassettes and remember, if they are intended for repointing, to use the same careful handling practices as when they were new.

24.2.2.3 Geometric Attributes. The geometry of a drill bit very much affects the way it behaves during drilling. (See Fig. 24.2 for attribute nomenclatures.) The land is the area remaining after fluting. In order to reduce the amount of land that creates friction with the hole wall (thus generating heat), drill bits are margin relieved. The amount of land remaining in contact with the hole wall during drilling is referred to as the margin. The wider the margin, the greater the friction area and the higher the drilling temperature, resulting in higher extents of heat-related hole quality defects such as resin smear and plowing (defined as furrows in the resin).

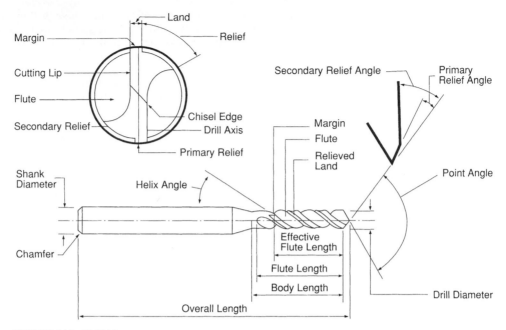

FIGURE 24.2 Drill bit geometry.

The result of increasing the land mass and the web is a smaller flute area. Less flute space implies reduced amounts of available area to remove drilling debris, which again raises the drilling temperature. It is important to understand that some drill designs that are meant to increase strength (particularly in smaller-diameter drill bits, with the intent to reduce breakage) may include what are referred to as partial margin reliefs (see Fig. 24.3). What this means is that when the drill bit is viewed from the point, a relieved margin is seen. However, viewing

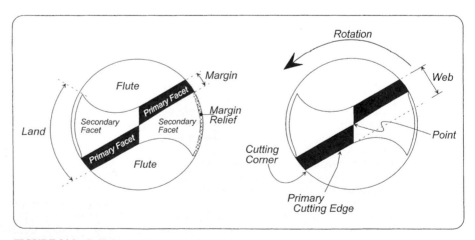

FIGURE 24.3 Drill tip attribute nomenclatures.

this type of design drill bit from the side reveals that the relief extends only part of the way, typically about one-fifth of the total flute length The major drawback of this particular design is that it increases drilling temperatures (documented to be as much as 25 percent or more), resulting in increased heat-related drilled hole defects (such as smear) and drill breakage due to packed margins.

24.2.2.4   *Flute Length.* Minimum flute length must equal total drilled depth (total laminate thickness + entry thickness + backup penetration depth) plus at least 0.050 in of unused drill flute remaining above the stack at the bottom of the drill stroke to allow debris to be removed by the vacuum system. If debris cannot be removed from the drill flutes during and between drill strokes, the results are extensive hole quality defects and drill breakage.

24.2.2.5   *New Drill Bits.* It is not necessary to inspect incoming orders of drill bits 100 percent. A procedure proved to be effective is to inspect drill bits according to an acceptable quality level (AQL) method (such as MIL-STD-105). This type of sampling plan allows the user to determine the percentage of drill bits that are inspected to judge an entire lot of drill bits to a predetermined quality level. Using this method, finding a defined quantity of drill bits that do not meet the specification is cause to return all received drill bits to the vendor; if the selected quantity of inspected drill bits meets specification, the entire lot of drill bits is accepted. Inspection criteria might include point geometry defects (refer to Fig. 24.4), damage (chips), diameter (drill and shank), and flute length, as well as correct ring set distance and size imprint (matching both the actual diameter and that labeled on the box).

24.2.2.6   *Repointed Drill Bits.* Drill bits are typically repointed for reasons of economics. The cost of repointing a tool may be as low as 15 percent of the cost of a new tool. The number of times a drill is repointed varies anywhere from 1 to as many as 10 times or more and typically depends on the drill diameter. The smaller the drill, the fewer times it is normally repointed. The reason is that smaller-diameter holes are more critical and require better hole quality.

There are two methods used to repoint tools.

1. The first is to specify a certain number of times the tool is to be repointed before being discarded. This number typically varies between one to three for smaller-diameter tools depending on hole quality experienced after repointing.

2. The second is to specify a minimum overall length of the tool at which it is discarded. Minimum overall length is determined by calculation based on minimum remaining flute length required to drill the required total drilling depth. This method does not allow determination of how many times a drill is actually repointed because stock removal during each regrind may vary from 0.002 to 0.005 in or more.

Repointed drill bits cannot be expected to perform as well as new drill bits because only the points are refurbished to a quality that may be as good as a new drill bit while the rest of the drill bit, including the critical margin, is not. The condition of the margin is very important because it is the part of the drill bit that finishes the hole wall. A rough margin results in a rough hole wall surface. When inspecting repointed drill bits, examine the sides of the drills for margin damage and/or fused or packed drilling debris remaining from previous use. These drill bits contaminate holes from the very first one drilled and may cause run-out due to an imbalanced condition of the drill bit resulting from the buildup. Drill bits must be repointed to the same point geometry specifications as those that apply to new drill bits. Point inspection criteria, therefore, is the same for repointed drill bits as for new drill bits.

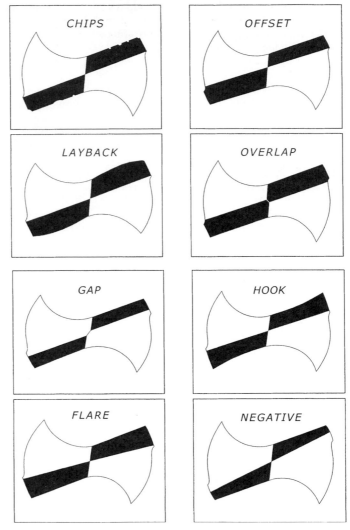

FIGURE 24.4 Drill point defect illustrations. (*Courtesy of LCOA Technical Center.*)

24.2.3 Drill Bit Rings

Drill bit rings are set to a common distance from the point to the back of the ring, thereby allowing a controlled drilling depth. The quality of these rings is as critical as the consistency of drill bit attributes because they can equally affect drill bit performance. Rings that fit loosely on the drill shank have been known to move during tool changes, resulting in insufficient drilling depth. Rings that fit too tightly may crack. Rings that have protruding material ("flash") on the inside diameter may cause improper seating of the drill bit in the collet (of the spindle) or may prevent the drill bit from being properly inserted into the tool pod or cassette, resulting in destructive tool change problems.

Rings have specific color codes assigned to each size and are commonly imprinted with size, diameter, and design or series number. Rings may be either machined or mold-injected. While machined rings are superior because of their consistency and quality, their cost is prohibitive for many drill bit manufacturers. Some of the drawbacks of mold-injected rings that must be monitored are inside diameter, affecting how well the ring fits on the drill shank, and remaining flash.

24.2.4   Entry Material

24.2.4.1   Purpose.   The purpose of the entry material is fivefold:

- Centers the drill
- Prevents drill breakage
- Prevents copper burrs
- Avoids contamination of the hole or drill bit
- Prevents pressure foot marks

24.2.4.2   Types.   There are many different available types (constructions) of materials used as entry material for printed circuit board drilling, although few are specifically designed and engineered for this purpose. Engineered products are designed to improve hole registration accuracy and reduce drill breakage. The performance qualifications of the most popular materials are discussed in the next section.

Commonly available entry materials, listed in order of performance quality with respect to the five characteristics listed in the preceding section, are:

- Aluminum-clad cellulose core composite
- Solid aluminum (various alloys and thicknesses)
- Solid or melamine-clad phenolic
- Aluminum-clad phenolic

24.2.4.3   Performance.   The right entry material will improve drilled hole registration and lower the risk of drill bit breakage by minimizing drill deflection upon contact with the stack. In order for the entry material to function properly, it must be flat and free of pits, dents, and scratches. Warped or twisted material will result in increased extents of entry burrs and drill bit breakage. Surface imperfections and materials that are too hard contribute to drill deflection, resulting in decreased hole registration accuracy and breakage of small-diameter drills.

Phenolic materials or phenolic composites (i.e., aluminum-clad phenolic) often warp and under most drilling conditions contaminate the hole wall, which results in problems with adherence of the plating because desmearing chemicals are not designed to remove phenolic resin. Solid aluminum materials of the correct composition and hardness that are not of an excessive thickness, yet are not too thin, may work satisfactorily with larger-diameter drill bits. However, drilling with solid aluminum materials (0.008 in and thicker) may increase the risk of breakage of smaller-diameter drills. Aluminum-clad cellulose core materials provide a hard surface to prevent burrs yet minimize drill deflection and breakage associated with solid aluminum.

24.2.5   Backup Material

24.2.5.1   Purpose.   The purpose of backup material is defined by the following criteria. An ideal backup material will:

- Provide a safe medium for drill stroke termination
- Prevent copper burrs
- Not contaminate the hole or drill bit
- Minimize drilling temperatures
- Improve hole quality

24.2.5.2 Types. Numerous materials are available that are sold as backup material. Few of the materials used as backup materials are actually specifically engineered for circuit board drilling. Many of the popular backup products are composites with a variety of surface coatings or skins bonded to several different core materials. Available backup products include the following:

- Epoxy-paper-clad, wood-core composite utilizing a bonding agent with lubricating properties
- Aluminum-clad, wood-core composite
- Epoxy-paper-clad, wood-core composite
- Melamine-clad, wood-core composite
- Urethane-clad, wood-core composite
- Solid phenolic
- Aluminum-clad phenolic composite
- Plain wood
- Hardboard

24.2.5.3 Performance. Desired qualities in a backup material are minimal thickness variations, flatness (no bow, warp, or twist), no abrasives or contaminants, a smooth surface, low cutting energy (minimizing drilling temperatures), and a surface hardness that supports the laminate copper surface (to prevent burrs) yet does not cause damage or extensive wear to the drill bit.

Backup materials with lubricating properties have been proven to significantly reduce drilling temperatures by as much as 50 percent or more, often resulting in temperatures below the T_g of the laminate product being drilled. This advantage greatly reduces hole wall defects such as roughness, smear, and nail-heading and often allows increased drill stack heights and/or greatly increased drill bit maximum hit counts. The importance of these benefits is significant reduction in drilling cost per hole and improved productivity and yield.

Remember that drilled backup debris exits the stacks by passing through the holes in the laminate material and that therefore contamination (from the backup material) is of great concern. Materials containing phenolic, or composed of solid phenolic, are not suitable for circuit board drilling. Phenolic materials or phenolic composites (i.e., aluminum-clad phenolic) often warp and under most drilling conditions contaminate the hole wall, which results in problems with adherence of the plating because desmearing chemicals are not designed to remove phenolic resin. Hardboard types of materials cannot be maintained to thickness variation tolerances acceptable for circuit board drilling and are a source for a great variety of contaminants (e.g., oils crystallized on the surface for hardening purposes).

24.2.6 Tooling Pins

Seldom is any due attention given to the tooling pins. They come in many shapes and sizes and their cost, with respect to how much they add to the cost of fabricating a circuit board, is insignificant. Yet, quite often, tooling pins are found to be damaged or deformed (e.g., "mushroomed" from being hammered into the stack) or do not fit snugly. Tooling pins that do not hold the stack tightly in place or that allow the stack to move create a large variety of problems from burrs and other hole defects to poor registration or drill bit breakage. These unnecessary problems may be prevented by simply replacing tooling pins when they start to show signs of wear or damage. Use tooling pins that are hardened to minimize wear and deformation, and (ideally) $\frac{3}{16}$ in in diameter. Pins that are less than $\frac{3}{16}$ in in diameter (i.e., $\frac{1}{8}$ in) do not hold the stack firmly in place during drilling and may result in stack movement.

24.3 MACHINES

Machine variables that affect the drilling process are as follows and as shown in the fishbone diagram of Fig. 24.5.

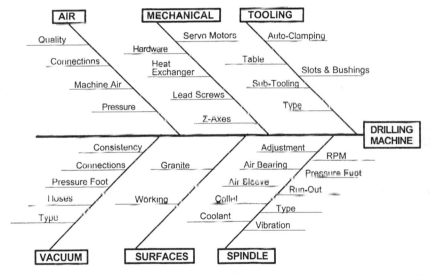

FIGURE 24.5   Drilling machine elements, with major issues and special considerations identified.

24.3.1 Air

24.3.1.1 Quality.   Maintaining high levels of air quality is important. Air supplied to the machine and the spindle assemblies needs to be clean and dry. This is accomplished by filters that remove dirt and moisture from the air. In areas where the relative humidity is high, or when using air-bearing-type spindles, an inline air drying and filtering system may be necessary to control air moisture levels and dirt. Moist air causes corrosion of air surfaces (such as the spindle sleeve and other spindle components operating on air) and reduces the life of these components, resulting in increased repair costs. Dirty air affects operation of components relying on air by clogging the channels that supply the air (e.g., to the spindle sleeves and table air shoes). In addition, dirt serves as an abrasive that causes faster wear of machine components, resulting in reduced life cycles between repairs, again increasing downtime and repair costs. Cleaning the machine air filters and purging air compressors and lines on a routine basis may minimize problems resulting from dirt or moisture.

24.3.1.2   Connections and Pressure.   Check connections, gauges, hoses, and switches for leaks and wear. Air hoses connected to the pressure foot pistons often crack near the piston connector; by simply bending the hoses in these areas it can be determined whether they are leaking. Other locations that commonly have poor connections and air leaks are the collet air connector on top of the spindle and the air manifold on the side of the spindle sleeve housing that supplies air to the collet and the spindle sleeve. Sufficient air pressure to both the collet and the spindle sleeve is critical and must be maintained adequately for proper operation. The tool table is designed to ride on a bearing of air supplied through the table air shoes. Verify proper operation of the air shoes by slightly rotating them back and forth; the shoes should move freely. If the shoes do not move with ease or do not move at all, the most likely reason is that the air channels are blocked due to trapped drilling debris, meaning the shoes need to be cleaned or replaced. Check the pressure foot air gauges for the correct air pressure; equal pressure must be supplied to each of the spindle stations. Insufficient or unequal air pressure results in burrs and increases the possibility of drill bit breakage.

24.3.2   Vacuum

Effective vacuum is absolutely essential because it greatly affects the resulting drilled hole quality. Heat (drilling temperature) seen by the hole walls depends on how effectively hot drilling chips are removed. Excessive temperature due to inefficient removal of drilling debris causes heat-related hole defects such as smearing and plowing, increases the opportunity for plugged holes, and is a major contributor to drill wear. Check hoses for proper connections, wear, and restrictions and examine the inside of the pressure foot for holes worn inside the connector attaching the vacuum hose.

24.3.3   Tooling

24.3.3.1   Bushings and Slots.   Maintain tight tolerances on tool table bushings and slots and replace bushings that are sunken below the tool table surfaces or are worn. Bushings and slots that do not hold tooling pins snugly and allow stack movement during the drill stroke cause burrs, registration problems, and drill breakage.

24.3.3.2   Subtooling.   Avoid subtooling plates that overlap across individual tool table stations and fasten the plates securely to the base plates with no separation or gaps. Ensure the subtooling is not warped, does not vary significantly in thickness, and has no surface protrusions (such as broken drill bits) to allow stacks to lay flat and to minimize variation in drill bit penetration depth into the backup material. If your subtooling is other than metal, watch for potential shrinkage or expansion problems indicated by difficulties in fitting the pinned stacks to the existing pinning holes in the plates. If this problem occurs, the results are the same as those caused by bushings and slots that allow stack movement.

24.3.4   Spindles

Spindle assemblies are one of the single most important components of the drilling machine. Proper operation is essential and requires maintenance (e.g., collet and collet seat cleaning) and verification on a regular basis.

24.3.4.1   Collet Maintenance.   For mechanical (ball) bearing spindles, cleaning of the collet and the collet seat (inside the spindle) needs to be performed a minimum of once per shift; dirty collets increase drill run-out. When processing certain laminate materials that create greater amounts of drilling dust, and depending on your vacuum system efficiency, a higher cleaning frequency may be necessary to maintain acceptable levels of drill run-out. It is recommended that after cleaning the collets be returned to the same spindle from which they were removed because collets tend to adjust to the respective collet seats. For air-bearing spindles, most manufacturers recommend cleaning the collets only when the run-out measures excessive.

24.3.4.2   Run-Out Measurement.   Drill/collet concentricity, or total indicated run-out (TIR), is a measure that indicates how true the assembly rotates. It can be determined while the spindle is running at various speeds (rpm), which is referred to as a dynamic form of measurement. Another method is performed while the spindle is not running and is called a static measurement. Static TIR is determined by rotating a ⅛-in (0.1250-in)-diameter pin installed into the collet by hand while reading movement on a dial indicator placed against the pin at a distance of approximately 0.800 in (simulating the distance of the drill point) from the collet nose (refer to Fig. 24.6). Of course, the pin used to measure TIR must itself be concentric.

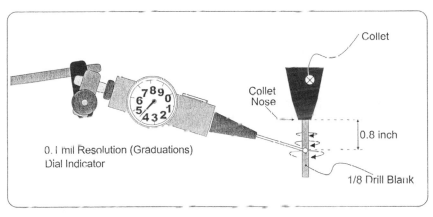

FIGURE 24.6   TIR measurement (static).

Maximum acceptable TIR for drills greater than 0.0200 in in diameter is 0.5 mils (0.0005 in). When using drills 0.0200 in or less in diameter, the maximum TIR needs to be maintained within 0.2 mils (0.0002 in) to prevent drill breakage. If excessive TIR is noted, it is wise to replace the pin and again measure the run-out. Drill bit blanks make ideal pins for measuring TIR and may be acquired from drill bit suppliers. Excessive spindle run-out results in drill breakage, causes burrs and other drilled hole defects, and adversely affects hole registration accuracy. Excessive TIR may often be corrected by simply cleaning the collet and the collet seat or by replacing a worn collet. Measure the run-out of spindles of each machine at a minimum frequency of once per week.

24.3.4.3   Spindle Speed.   Maintaining correctly adjusted spindle speed is important because spindles running at speeds other than those displayed on the monitor imply that drill bits are rotating at surface speeds other than those that are desired. When actual rpm is higher than the displayed rpm, it implies higher surface speeds (refer to Sec. 24.4), and vice versa. Higher surface speeds cause greater frictional drill heat, resulting in faster drill wear and greater extents of heat-related hole defects such as smearing and plowing. Noncontact-type tachometers able to measure up to 150,000 rpm or more are available for around $300 (a lot less than the cost of rebuilding just one spindle) and allow measurement of actual spindle rpm in a matter of minutes. Verifying spindle speeds once every 6 months is usually sufficient.

24.3.4.4   Pressure Foot.   The pressure foot insert lead distance to the point of the drill bit is set at approximately 0.050 in to give the pressure foot sufficient time to hold the stack flat before the drill bit contacts and enters the stack. If the lead distance is significantly less than the specified distance or if the drill bit point protrudes from the pressure foot, drill bits will break. To ensure the pressure foot assemblies function properly, check pressure foot inserts for wear or damage daily and verify that the pistons and guide rods are not bent and move smoothly.

24.3.4.5   Adjustment.   When adjusting the physical z-axis height of the spindle, for instance to accommodate a thicker subtooling plate, care must be taken to also adjust the pressure foot so its height remains identical relative to the spindle. The pressure foot adjustment is independent from the spindle and, if not performed correctly, may result in a gap between the bottom of the spindle casing and the top of the pressure foot window. This gap is noticed only while the spindle is engaged and will draw most of the vacuum air, effectively cutting off the vacuum from the pressure foot insert, resulting in a great variety of hole defects and drill bit breakage. If a gap is present, it is usually accompanied by great amounts of drilling debris spewed across the stacks and drilling machine.

24.3.5   Mechanical Factors

24.3.5.1   Heat (Coolant) Exchanger.   Drill motor temperatures are reduced by exchanging a fluid through the spindles. The fluid is processed through a heat exchanger that works much like a car radiator. As with a car, it is important to maintain the recommended coolant mixture, proper fill level, and flow rate indicated by the flowmeter. Check the operation of the fan and minimize coolant flow restriction (due to algae buildup) by cleaning filters regularly. Algae growth is promoted through exposure to ultraviolet light and may be controlled by using black hoses instead of clear ones in addition to adding growth inhibitors to the coolant mixture.

24.3.5.2   Hardware.   Although simple to perform, checking mechanical connections for loose or worn parts (i.e., causing travel slop of pressure foot assembly movement) is often neglected. It may be accomplished by observing the machine while it is running or checking by hand while the machine is idle.

24.3.5.3   Lead Screws and Servos.   The environment and cleanliness of the drill room determine how often lead screws need to be cleaned and lubricated to ensure smooth operation and minimize wear. Most machine manufacturers recommend performing this type of maintenance every 6 months. When lubricating the lead screws, only a light coat of the appropriate grease needs to be applied. Excessive amounts of grease on the lead screws defeat the purpose, as this causes more dirt to be attracted, resulting in faster wear. Constant searching for programmed x-y locations indicates problems with the servo motors or lead screws. If this occurs, the displayed x or y (actual) location displayed on the controller will change continuously while the machine is stopped.

24.3.5.4   z-Axis Stroke.   Because machines may have as many as six mechanical connections in the z axis, travel slop during the drill stroke or retraction may occur, causing chip loads to vary during the stroke, which results in burrs and other hole defects and drill breakage. Check connections by hand for slop while the machine is idle and observe and listen for improper operation while the machine is running.

24.3.6   Surfaces

A clean drilling room decreases repair costs and the possibility of improper operation of the drilling machines. Wipe surfaces clean using a lint-free cloth or use a vacuum to remove debris.

Never use compressed air to clean the drilling machine because debris will be blown into areas that need to be kept clean (e.g., lead screws).

24.3.6.1 Granite. Granite provides a stable platform that absorbs unwanted vibration. It also offers a smooth and level surface to support the movement of the table. Keep granite surfaces clean to prevent tool table air shoes from becoming clogged and to minimize collection of dirt on the lead screws.

24.3.6.2 Working Surfaces. Maintain clean working surfaces, such as the tool table, to minimize the possibility of dirt particles getting trapped between the stacked laminate, entry, and backup panels, causing stack separation that may result in burrs and possible drill bit breakage.

24.4 METHODS

Drilling parameter variables that affect the drilling process are as follows and as shown in the fishbone diagram of Fig. 24.7.

24.4.1 Surface Speed and Spindle Speed

As holes are getting smaller, higher spindle speeds are required to achieve the desired surface speed that determines throughput. However, higher surface speeds result in higher drilling temperatures that may increase heat-related hole defects such as smearing and plowing.

24.4.1.1 Definition. Surface speed is a measure of how much distance is covered by the drill's diameter while it is rotated by the spindle and is expressed in surface feet per minute (sfm). It is used to calculate spindle speed (rpm) for a given drill diameter. The formula to calculate spindle speed using the desired surface speed is shown in Eq. (24.1).

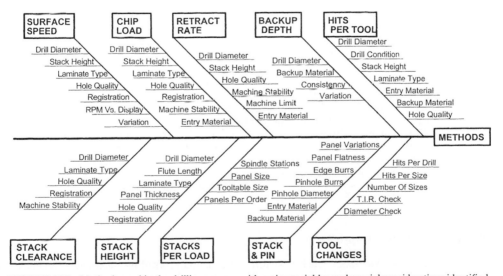

FIGURE 24.7 Methods used in the drilling process, with major variables and special considerations identified.

$$\text{spindle speed (rpm)} = \frac{\text{sfm} \times 12}{\pi \times \text{diam.}} \tag{24.1}$$

where diam. = drill diameter (in)
π = 3.1415

24.4.1.2 Effects. The higher the surface speed, the higher the spindle speed, and subsequently the higher the frictional heat that is generated, translating into greater extents of heat-related hole defects and drill wear. Conversely, lower surface speeds imply lower spindle speeds and less frictional heat. When more abrasive materials (e.g., materials with higher T_g such as multifunctional FR-4, polyimide, or cyanate ester) are processed or drilled stack height is increased, drilling temperatures increase. To offset the resulting increase in temperature under such conditions or when excessive extents of heat-related hole defects are apparent, decrease the surface speed to lower the spindle speed.

24.4.2 Chip Load and Infeed Rate

24.4.2.1 Definition. Chip load is defined as advance per revolution and is usually expressed in mils (1 mil equals $\frac{1}{1000}$ of an inch [0.001 in]). It implies the distance the drill travels during the drill stroke per each full revolution of the drill bit. Chip load is used to calculate the infeed rate in inches per minute (ipm).

$$\text{infeed rate (ipm)} = \text{chip load (mils/revolution)} \times \text{rpm} \tag{24.2}$$

24.4.2.2 Effects. Higher chip loads result in greater throughput. However, chip load and infeed rate affect hole registration accuracy, drill breakage, burrs, and mechanical types of drilled hole defects such as voids (tear-out of the supporting fibers) and nail-heading. Faster infeed rates translate into higher (top laminate) entry burrs, lower extents of nail-heading defects and (bottom laminate) exit burrs, increased occurrences of drill bit breakage, and higher extents of drilling voids. Lower infeed rates result in exactly the opposite but lower throughput.

24.4.3 Retract Rate

24.4.3.1 Definition. Retract rate is the speed at which the drill bit exits the stack following the drill stroke and is expressed in inches per minute (ipm). Machine default (maximum) settings vary between manufacturers and may range anywhere from 500 to 1000 ipm.

24.4.3.2 Effects. Higher retract rates imply lower processing times per load. While the maximum retract rate setting may be fine for larger-diameter drill bits, when using drill bits in the diameter range of 0.0250 in (size #72) to 0.0135 in (size #80), retract rates may have to be reduced to 500 ipm or lower to prevent drill breakage. When drilling with sizes smaller than 0.0135 in in diameter, retract rates may have to be reduced even further. The maximum retract rate that may be used with any given drill diameter without causing drill breakage greatly depends on the stability and vacuum system efficiency of the drilling machine as well as the drilled stack height, laminate construction and thickness, type of entry material, use of proper stacking and pinning procedures, and the design characteristics of the drill bit such as flute length, web thickness, and web taper.

24.4.4 Backup Penetration Depth

24.4.4.1 Definition. Backup depth is the distance a drill bit penetrates the backup material at the bottom of the drill stroke. The minimum backup penetration depth setting varies depending on drill diameter and is determined by calculating the drill bit point length (see Fig. 24.8) and adding approximately 0.010 in. As a rule of thumb, backup penetration depth may be set to a distance equal to the drill diameter or 0.040 in, whichever is less.

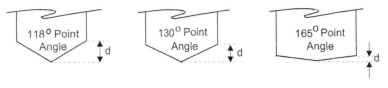

d(118) = tan 31 x Radius = ~ .600 x Radius or ~ .300 x Diameter
d(130) = tan 25 x Radius = ~ .466 x Radius or ~ .233 x Diameter
d(165) = tan 7.5 x Radius = ~ .132 x Radius or ~ .065 x Diameter

FIGURE 24.8   Point length calculation.

24.4.4.2 Effects.   Excessive backup penetration depth increases drill wear and the occurrence of breakage of small-diameter drill bits, adversely affects hole quality, and increases process time per load. Insufficient backup penetration depth results in incompletely drilled holes. This implies that thickness variations of the backup material are very important, meaning that minimal variations are a much desired characteristic of the backup and need to be considered when choosing a material suitable for your application.

24.4.5 Hits Per Tool

24.4.5.1 Definition.   The maximum hits per tool specified for any given drill size implies the number of drill strokes a drill bit is used for until its expected effective life is expired. Maximum hit count per tool is product and process specific and is affected by laminate material construction, panel thickness, drilled stack height, surface speed, and the type of entry and backup material used. Therefore no specific number of hits per tool can be arbitrarily specified.

24.4.5.2 Effects.   Excessive drill wear caused by excessive maximum hit count increases drilled hole defects and may prevent proper repointing. Conservative maximum hit counts greatly impact drilling cost per hole and increase time per load because of increased numbers of tool changes.

24.4.6 Stack Clearance Height

24.4.6.1 Definition.   Stack clearance height is the distance between the point of the drill and the surface of the stack at the top of the drill stroke. Maintain a minimum stack clearance distance of ⅛ in (0.125 in), which implies a space between the bottom of the pressure foot and the top of the stack of 0.075 in, assuming the pressure foot lead distance to the point of the drill is correctly set at 0.050 in. Stack clearance may be adjusted for each load by simply sliding a 0.075-in shim between the pressure foot and the stack and adjusting the upper limit ("UP#") until the pressure foot touches the shim.

24.4.6.2 Effects.   The less the stack clearance distance between the drill point and the top of the stack, the shorter the drill stroke and therefore the shorter the processing time per load. Increasing the stack clearance distance allows more time between drill strokes and gives the tool table more time to settle, which may improve hole registration accuracy and prevent small-diameter drill bit breakage. In addition, the greater the time between drill strokes, the more likely drilling debris will be removed from the drill flutes and, consequently, the lower the drilling temperatures, which results in reduced occurrences of drill breakage and lower extents of drilled hole quality defects.

24.4.7   Drilled Stack Height

Material construction (panel thickness, number of copper layers, laminate type, etc.), drill bit diameter, and flute length, as well as hole quality and registration accuracy requirements, all are factors that need to be considered when deciding on appropriate drilled stack heights. A greater number of panels in the drilled stack means higher drilling temperatures, accelerated drill wear, and greater drill deflection, affecting the resulting hole quality and registration accuracy. When using smaller-diameter drill bits, stack heights need to be reduced to prevent drill breakage and to accommodate the shorter flute lengths. As a rule of thumb, the maximum total drilled depth (number of panels × panel thickness + entry thickness + backup penetration depth) that can safely be handled by the drill bit without breakage is approximately 17 times its diameter.

24.4.8   Stacking and Pinning

24.4.8.1   Building the Stack.   Inspect all laminate panels and the entry and backup materials for surface damage and remove burrs from the panel edges as well as from the pinning holes. Even though the registration tooling holes on the laminate material may not be used to pin the stacked panels together, it is important to remove any resin buildup remaining around these holes after lamination (a common occurrence). Burrs and raised surface areas do not permit the panels to lay flat, causing gaps resulting in drilled hole registration problems, burrs, hole quality defects, and drill breakage. Reject entry and backup materials with excessive nicks, scratches, and other surface defects as well as those that are warped or twisted.

24.4.8.2   Pinning Procedures.   Wipe the surfaces of all laminate panels and the backup material using a lint-free cloth to remove any debris before stacking the panels (allowing intimate contact between the stacked panels). Verify that the pinning holes and pin insertion are perpendicular to the stack and avoid using pins that are damaged or deformed.

24.4.8.3   Installation.   Before placing the pinned stacks onto the drilling machine tool table, inspect the surface for burrs or broken drill bits protruding from the table that may prevent the stacks from lying flat. Do not continue if the pin bushings of the tool table are sunken or worn to the point that they do not hold the stack firmly in place. Loose or sunken tooling pin bushings cause movement of the stack during drilling and result in a variety of hole defects, registration problems, and drill breakage, yet are simple to replace at a minimum cost. After stacks are put in place, again wipe the surface of the top laminate material as well as the entry material. Place the entry material on top of the stack and tape it in place. The entry material size should be such that it clears the pins and does not extend beyond the stack edges. Pinning the entry material to the stack is not recommended because it tends to constrict the movement of the material, causing separation between it and the stack and resulting in entry burrs and possible drill breakage.

24.5   HOLE QUALITY

24.5.1   Definition of Terms

The terms in Tables 24.1 and 24.2 are commonly used to describe drilled hole defects observed on copper and substrate surfaces. It is important to be able to identify these defects specifi-

TABLE 24.1 Copper Defects

Defect	Definition	Type
Burr	Ridge left on external surface	Mechanical
Debris	Drilling residues	Mechanical
Delamination	Separation of the copper from the substrate	Mechanical/heat-related
Nail-heading	Burr on internal copper layer	Mechanical/heat-related
Smearing	Thermomechanically bonded resin deposit	Heat-related

TABLE 24.2 Substrate Defects

Defect	Definition	Type
Debris pack	Drilling residues packed into voids	Mechanical
Delamination	Separation of the substrate layers	Mechanical/heat-related
Loose fibers	Unsupported fibers in the hole wall	Mechanical
Plowing	Furrows in the resin	Heat-related
Smear	Thermomechanically bonded resin deposit	Heat-related
Voids	Cavities due to torn-out supporting fibers	Mechanical

cally rather than in general terms. Using a general term such as *roughness* may imply voids or plowing. While voids are a mechanical type of defect, plowing is a heat-related type of defect. Therefore, excessive voids would lead one to examine the chip load (infeed rate) used; plowing would lead one to look at surface speed (spindle speed).

24.5.2 Examples of Drilled Hole Defects

Examples of typical drilled hole wall defects are shown in Figs. 24.9 and 24.10.

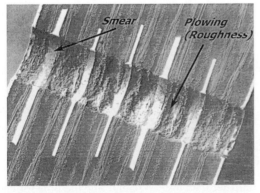

FIGURE 24.9 Cross section of drilled hole showing examples of smearing and plowing.

FIGURE 24.10 Cross section of drilled hole showing examples of nail-heading.

24.6 *POSTDRILLING INSPECTION*

A wealth of information is available by simply examining the materials from the drilled stack and the drill bits. For instance, inspecting the drill bits will allow one to determine if wear occurs at consistent rates (for drills of same diameter) or will reveal whether hit count maximums are excessive and the type of drilled hole wall defects to expect. Bonded debris and/or extensive wear to the drill corners imply high drilling temperatures (resulting in greater extents of defects such as smearing and plowing) or materials that are not fully cured and point to a problem with the materials (laminate, entry, or backup) or may suggest an excessive surface speed.

Extensive primary cutting edge wear indicates abrasive materials and may require lowering stack heights, reducing hit counts, or replacing entry or backup materials.

Burrs on the surfaces within the stack mean that there is a problem with the way panels are assembled and pinned or may be the result of warped panels. Entry or exit burrs on the outer laminates should cause one to question the entry and backup materials or the infeed rate.

The point of a postdrilling inspection process is that if one takes the time on a regular basis, to examine materials and drill bits after drilling, many drilling problems would be solved before they get out of hand.

24.7 *DRILLING COST PER HOLE*

Material and processing costs, as well as the resulting total drilling cost, may be determined by using an analysis matrix such as a cost model specifically designed for this purpose and generated with the use of a computer spreadsheet program. The advantage of using a spreadsheet is that it allows changes to be made in, for instance, specific material prices and processing times or parameters, and allows instantaneous viewing of the resulting effects on the total drilling costs, the cost per panel, and the average cost per hole. Knowing the cost per hole is important because it allows comparing different jobs or processing situations. Following is a step-by-step description of how to construct a drilling cost analysis matrix such as the one shown in Fig. 24.11.

24.7.1 Machine Time

Table A in the drilling cost analysis matrix is used to calculate the total time that is required to complete the job. First, the different drill sizes (a) and their respective total drilled holes per panel (b) as well as the total number of panels (c) to be drilled are determined and entered in the spreadsheet; this allows the spreadsheet to calculate the total number of holes for each size to complete the job (d).

Second, using the appropriate drilled stack height (e), the total number of drilled stacks (g) and the total number of drilled hits per drill size (f) can be calculated. The total number of drilled hits (drill strokes) is the total number of drilled holes per panel (b) divided by the number of panels per drilled stack (e).

Third, the number of total drilled stacks (g) is divided by the number of stations per machine (stacks per load [h]) to calculate the number of machine loads (i).

Fourth, the total drill time per load required per drill size (j) is entered to calculate the total machine time for each drill size (k).

Fifth, the total times of each of the drill sizes are simply added up to arrive at the total time required to finish the job.

An option is to enter the total drill time per load instead of entering the time for each of the drill sizes and multiplying the total drill time per load by the number of machine loads to determine total machine time.

TABLE A Machine Time

a	b	c	d	e	f	g	h	i	j	k
Drill bit size	Number of holes per panel	Number of drilled panels	Total number of drilled holes	Panels per drilled stack	Total number of drilled hits	Total number of drilled stacks	Stacks per machine load	Number of machine loads	Drill time per load (h)	Total machine time
0.0135	7,000	120	840,000	2	420,000	60	4	15.0	1.12	16.80
0.0160	5,000	120	600,000	2	300,000	60	4	15.0	0.80	12.00
0.0225	3,125	120	375,000	2	187,500	60	4	15.0	0.48	7.20
0.0350	1,500	120	180,000	2	90,000	60	4	15.0	0.19	2.85
0.0520	800	120	96,000	2	48,000	60	4	15.0	0.10	1.50
0.0700	250	120	30,000	2	15,000	60	4	15.0	0.04	0.60
	17,675		2,121,000						2.73	40.95

TABLE B Drill Bits

l	m	n	o	p	q	r	s	t	u	v
Drill bit size	Cost per new tool	Cost per repoint	Number of repoints per tool	Total cost per drill bit life	Number of uses per drill bit life	Average cost per drill bit use	Total number of drilled hits	Number of hits per drill bit use	Total number of drill bit uses	Total drill bit cost
0.0135	$1.25	$0.25	2	$1.75	3	$0.58	420,000	1,000	420.0	$245.00
0.0160	$1.20	$0.25	2	$1.70	3	$0.57	300,000	1,000	300.0	$170.00
0.0225	$1.15	$0.25	3	$1.90	4	$0.48	187,500	1,250	150.0	$71.25
0.0350	$1.15	$0.25	3	$1.90	4	$0.48	90,000	1,500	60.0	$28.50
0.0520	$1.15	$0.25	4	$2.15	5	$0.43	48,000	2,000	24.0	$10.32
0.0700	$1.30	$0.25	4	$2.30	5	$0.46	15,000	2,500	6.0	$2.76
										$527.83

TABLE C Entry and Backup Material

	w	x	y	z	aa
Material type	Cost per ft^2	Stack size ft^2	Cost per stack*	Total number of drilled stacks	Total material cost
Entry material	$0.53	3.00	$1.59	60	$95.40
Backup material	$0.65	3.00	$0.98	60	$58.50
					$153.90

* Backup material = cost/2 (uses per panel)

TABLE D Labor and Burden

	ab	ac
	Cost per hour	Cost total
Labor	$15.00	$614.25
Burden	$10.00	$409.50
		$1,023.75

TABLE E Total Costs

Cost variable	Cost per panel	Total cost	% of total
Drill bits	$4.40	$527.83	30.9%
Entry material	$0.80	$95.40	5.6%
Backup material	$0.49	$58.50	3.4%
Labor	$5.12	$614.25	36.0%
Burden	$3.41	$409.50	24.0%
Total drilling cost	$14.21	$1,705.48	100.0%

TABLE F Drilling Costs

Drilling cost per panel	$14.21
Average cost per 1000 holes	$0.804

FIGURE 24.11 Drilling cost analysis matrix. (*Courtesy of LCOA Technical Center.*)

24.7.2 Drill Bits

The cost of the drill bits needed to complete the job may be determined after the average cost per drill bit use has been calculated. To find the average cost per drill bit use, the typical number of repoints for the particular size (o) is multiplied by the cost of each repointing (n). The resulting cost, added to the new drill bit price (m), is the cost per drill bit life (p). By dividing the cost per life by the number of uses (q) per life (the number of times the bit is repointed + 1), you arrive at the average cost per drill bit use (r). Next, by dividing the total number of hits per drill size (s) by the number of maximum hits per drill bit use (t), the number of required drill bit uses (u) for each size may be determined. Then calculate the total cost per drill bit size (v) by multiplying the number of uses (u) by the average cost per use (r). The sum of the total costs of each of the drill bit sizes (v) brings you to the total cost of drill bits needed for the job. This cost, of course, is true only with the assumption of no drill bit breakage.

24.7.3 Entry and Backup Materials

Entry and backup material cost per stack (y) is determined by multiplying the stack size (square foot per panel [x]) by the cost per square foot. Remember to divide the backup cost by 2 since each backup panel may be used twice. Total material cost (aa) is calculated by multiplying cost per stack (y) by total number of drilled stacks (z).

24.7.4 Burden and Labor

Using typical burden and labor rates per hour (ab), these values are multiplied by the number of hours to complete the job (determined in Table A) in order to calculate the total burden and labor costs (ac).

24.7.5 Total Drilling Cost and Cost per Hole

After entering the required data in Tables A through D of Fig. 24.11, the total drilling cost and the cost distribution (see Table E) as well as the drilling cost per panel and the cost per hole (see Table F) can be calculated. Because the cost per hole typically ranges around $\frac{1}{10}$ of a cent, a more accurate and easier way to comprehend this value is by showing the average cost per 1000 holes, as is done in the cost model.

CHAPTER 25
HIGH-DENSITY INTERCONNECT DRILLING

Todd A. Barnett and Warren Wong
Excellon Automation Co., Torrence, California

25.1 INTRODUCTION

As circuit density continues to increase, along with the demand for higher and higher accuracy in hole location, the drilling machines and the environment in which the machines operate must be tightly controlled in order to achieve success in the drilling operation. High-density interconnect (HDI) has been defined as referring to holes with diameters of ≤0.006 in. HDI holes generally are expected to be made by nonmechanical means, such as lasers, plasma, or photoimaging. However, the bulk of holes continue to be created by mechanical means, and as the hole size approaches, or in some cases goes below, 0.006 in., special challenges face the mechanical drilling process. This chapter addresses these issues.

In this chapter, holes created by traditional mechanical drilling methods are referred to as *drilled,* and the process is referred to as *drilling*. Holes created by laser, plasma, or photoimaging are not really drilled, even though the term is often used to describe them. Where there is a reference to nonmechanical via hole creation processes, these are referred to by their specific type.

25.2 FACTORS AFFECTING HIGH-DENSITY DRILLING

PCB technology requires drilled holes as small as 0.002 in (0.05 mm) with extremely high accuracy, particularly when drilling dense hole patterns. The processes and machines used to drill these holes now constitute a highly developed science.

As the drilling process approaches these HDI dimensions, numerous factors become increasingly critical, such as:

- Hole location
- Predrilling process issues
- Drill room temperature/relative humidity
- Vacuum
- Drill bit condition

- Dynamic spindle run-out
- Backup and entry material (type and thickness)
- Maximum spindle speed
- Depth control

25.2.1 Positioning/Hole Location

The positioning system for conventional drilling machine consists of x, y, and z axes. The x- and y-axes refer to the position of the circuit board under the drill spindle, and the z axis to the plunging and retracting motions that allow the drill spindle to feed the drill bit into the substrate material. Machine accuracy is defined by the machine's ability to position the table under the spindles to the desired programmed x and y values. All process variables, such as tool design, stack height, feeds and speeds, and tooling methods, must be eliminated in order to evaluate machine accuracy.

The stability of the machine is critical when microdrilling dense drill patterns. When the drilling system is designed, stability is one of the most important factors taken into consideration. The foundation or base of the machine is usually constructed of granite or similar material. Granite is typically used because of its mass and low coefficient of expansion to temperature fluctuation. The upper structure of the machine is designed with a high natural frequency, which allows the structure to stabilize quickly, reducing the time the machine must pause before making a rapid movement to the next hole location.

Vibration caused by forces such as punch presses, drilling machines operating in close proximity, or any other type of equipment capable of causing severe vibration can have a detrimental effect on HDI drilling, such as drill bit breakage and hole mislocation. Each drilling machine should be isolated from these forces as thoroughly as possible. The ideal solution is to have each drill machine placed on its own isolated pad made of reinforced concrete. This setup serves to isolate the drill machine from a large percentage of the vibration transmitted through the shop floor.

25.2.2 Predrilling Issues

Control of the imaging and artwork generation process environments is extremely important as it relates to accurate mechanical hole placement. Artwork stability is predicted by temperature and humidity controls put in place at artwork generation, storage, and dry-film imaging. Any inconsistencies in the innerlayer pad location can be identified by measuring target-to-target values on plotted artwork and comparing them to the etched cores. Heat rise per minute in lamination measured from gel flow to gel set should match the material supplier's recommendations. Addressing any and all registration issues prior to mechanical drilling will assist in improving drilling registration.

25.2.3 Drill Room Temperature and Relative Humidity

All materials have coefficients that, because of the precision required, are critically important to the operation of drilling machines. Temperature has a profound impact on the accuracy of the drilling machines. With changes in ambient temperature, the materials used to build the machines may expand or contract, changing the position of the drilled hole in the printed circuit board.

Keeping a consistent room temperature is essential for maintaining accurate positioning of the machine. A stable temperature throughout the PCB process assists in decreasing the amount of potential growth and contraction of the substrate materials. The drill room should be kept at a constant temperature and humidity, typically $72 \pm 2°F$ ($22 \pm 1.1°C$) and 45 to 60 percent humidity.

25.2.4 Vacuum

Sufficient vacuum flow is imperative when drilling high-aspect-ratio holes in tightly drilled grids. The vacuum system serves two main purposes:

- To extract debris from the drilling surface
- To keep the tool cool by removing the debris, thus decreasing the amount of friction in the hole wall

Insufficient vacuum flow leads to hole wall quality problems such as gouging, nail-heading, and excessive smearing. Table 25.1 provides a hole quality troubleshooting matrix that addresses these problems. The minimum industry standard for vacuum strength is approximately 20 in of water measured at the pressure foot.

There are numerous applications such as high-aspect-ratio drilling that require an increased amount of vacuum pressure to remove debris sufficiently, and the equipment manufacturer can provide specific guidelines for special situations. Increasing the z-stroke distance between the pressure foot insert and the top of the drilled stack also assists in cooling the drill bit.

TABLE 25.1 Hole Quality Troubleshooting Matrix

Problem	Common causes	Parameter solutions	Process solutions	Possible equipment issues
Entry burrs	Excessive infeed rates	Lower infeed Decrease chip load Increase SFM	Change entry material type and/or thickness	Pressure foot lead set incorrectly; pressure foot insert damaged
Exit burrs	Warped panels Excessive infeed rate/SFM rate	Increase pressure foot, pressure value	Correct warpage issue in lamination; clean area between panel and backer	Verify pressure foot pressure is correct (45 psi)
Nailheading	Chipped/damaged tools Insufficient flute volume Poor drilling parameters	Modify feeds and speeds Decrease SFM	Remove defective drill bits Redefine minimum flute volume values	Machine not running at specified feeds and speeds
Rough holes	Chipped drill bits Poor drill concentricity Poorly cured laminate	Modify feeds and speeds	Inspect used tools for excessive wear patterns Replace worn drill bits Check laminate cure	Machine not running at specified feeds and speeds
Drill bit breakage	Poor drill concentricity Manufacturing defects Excessive drill depth Poor parameters	Check feeds and speeds Correct excessive runout Check z-axis offset values Clean and adjust collets	Check depth into backup Reset pecking depth values Replace drill bits Check program for double hits	Slop in lead screw Machine configured incorrectly Excessive spindle runout Poor machine maintenance
Excessive smear	Undercured laminate Excessive SFM Worn drill bits	Increase infeed rate Decrease SFM	Verify correct laminate cure	Machine not running at specified feeds and speeds

25.2.5 Drill Bit Condition

The condition of the used drill bit is a key indicator of the control and capability of the drilling process and provides an indication of the condition of the drilled hole wall. There is a wealth of process information that can be extracted from the used drill bit. Examining drills before the repointing process can be a warning indicator of drilling problems in a real-time mode soon after the process has been affected.

Drill machine operators, and drill bit resharpeners, should regularly evaluate the condition of used drill bits and report the findings to the process engineers. The areas of greatest concern when evaluating the condition of used drill bits are:

- Excessive margin wear
- Large amounts of burned substrate material attached to the relief areas
- Chips in the primary cutting edges of the tools

When tracked over time, this information can help alert the engineering team of any sudden changes in the drilling process. Although the used drill bit condition information is subjective, with properly trained personnel it can serve as a key indicator of the level of hole wall quality. (Chapter 24 contains a general discussion on drill bits and maintaining their points.)

25.2.6 Dynamic Spindle Run-out

The spindle houses the collet, which holds the drill bit that produces the hole in the printed circuit board. Modern spindles are capable of running at a wide range of rpm (15,000 to 170,000) with a power rating near 1 hp. Due to the accuracy required when drilling with small drill bits, the drill bit must run true in the spindle. Excessive spindle run-out causes the drill bit to rotate eccentrically, which leads to inaccurate hole location, drill bit breakage, and hole wall quality problems.

The spindle run-out must not exceed 0.002 in (0.005 mm) total indicated run-out (TIR) in a static situation. This trueness is achieved by using electrically powered spindles that rotate by the means of air bearings rather than mechanical ball bearings.

Other factors, such as poor collet condition and adjustment, can also lead to excessive run-out. Frequent collet cleaning and adjustment are necessary to ensure accurate mechanical hole location. Some modern drill machines are equipped with a tool metrology gauge (TMG). This feature measures the run-out for each drill bit during the tool change process, and, if the run-out value measures over the predefined range, the tool is rejected. TMGs can help increase productivity and quality by identifying potential problems before they occur.

25.2.7 Spindle Speed

As via holes get smaller and smaller, maintaining an adequate cutting speed is essential, particularly with chip loads exceeding 0.001 in per revolution. Increased spindle speed also allows for higher productivity because it corresponds to maintaining consistent chip load values while increasing z-axis feed rates. Spindle speed settings control the rotation of the drill bit during the drilling cycle. Measured in revolutions per minute, speed impacts hole wall quality as well as the condition of the drill bit's cutting edge. Excessive spindle speeds can result in premature wear on the cutting edges of the drill bit and may clog the margin relief area of the tool with burnt substrate material. Spindle speeds are dependent on the process and the application. The maximum spindle speed capability is a factor when microdrilling, particularly when the desired surface foot per minute (sfm) values are above 300. Note the maximum achievable sfm for the microdrilling diameter range when utilizing the 125,000-rpm parameters for FR-4 material listed in Table 25.2. When the maximum allowable spindle speed is

TABLE 25.2 Spindle Parameters for FR-4 Materials (Spindle Speed 125,000 rpm)

Tool size (in)	Infeed rate (in/min)	Chip load	sfm
0.0071	54	0.43	232
0.0080	66	0.52	261
0.0100	90	0.72	327
0.0120	106	0.84	393
0.0145	131	1.04	474

increased to 170,000 rpm, the 300-sfm value can be achieved when microdrilling, as shown in Table 25.3. As the table shows, the upper spindle speed range accommodates microdiameter drill bits.

TABLE 25.3 Spindle Parameters for FR-4 Materials at Various Spindle Speeds

Tool size (in)	Infeed rate (in/min)	Spindle speed (rpm)	Chip load	sfm
0.0071	68	170,000	0.04	315
0.0080	94	170,000	0.55	355
0.0100	128	170,000	0.75	444
0.0120	140	140,000	1.00	439
0.0145	152	131,000	1.16	497

There is an additional drilling parameter subset of spindle speed and feed rate values that is important to achieving the ideal drilling conditions.

25.2.7.1 *Chip Load.* Chip load is only uliited by dividing the feed rate by the spindle speed. The term identifies the amount of penetration the drill bit completes per revolution. As commonly used, the chip load accurately describes the proper cutting action needed to successfully drill certain applications.

25.2.7.2 *Cutting Speed.* The cutting speed in sfm is described as the distance the drill bit's outermost cutting edge travels in 1 min. Cutting speed is frequently used in the metal cutting industry, and the meaning is the same in the PCB drilling process. Particular substrate materials drill and fabricate favorably at certain cutting speeds. Cutting speed should remain consistent throughout the drill range (except when limited by spindle speed capability). For general guidelines, see Table 25.4.

TABLE 25.4 General Guidelines for Cutting Speeds

Cutting speed (sfm)	Material type
200	Plastics, acrylics, soft materials
300	Teflon, Kapton
400	Polyimide, cyanate ester, BT epoxy, High-T_g laminates
500	FR-4, tetrafunctional, multifunctional multilayers
600	FR-4, difunctional rigid, double-sided

25.2.8 Retraction Rate

The z-axis return stroke can be programmed as well. This rate, known as *retraction,* should be set to the value that will minimize the time the drill bit spends inside the drilled hole. The z-axis stability and drill bit diameter often influence the optimum retract values. An unstable or worn z axis can lead to poor depth control, poor hole wall quality, and high levels of drill bit breakage. Another advantage of high retraction rates is the speed of the drill stroke. Increasing production in this case through faster drill stroking is a high priority for the PCB manufacturer. Machine technology allows the user to set retraction rates upward to 1,400 in/min (35,560 mm/min).

25.3 DEPTH-CONTROLLED DRILLING METHODS

There are three common methods of depth control used in the drilling process:

1. Manual through-hole drilling
2. Depth-controlled drilling
3. Controlled-penetration drilling

25.3.1 Manual Through-Hole Drilling

Manual through-hole drilling utilizes a down limit value to determine the bottom of the z-axis stroke. This method does not utilize the top of the stack or the top of the backing material as a reference point. Careful consideration should be given when a down limit is set. The operator needs to ensure that the drill bit is penetrating all the way through the stack of laminated panels. Shallow penetration or incomplete holes can result in drilling scrap. Excessively deep penetration leads to increased drill bit cutting edge wear, excess debris to be removed, higher drilling temperatures, and drill bit breakage. The down limit should be set at a value that enables the drill bit to clear the point and also avoids excessive depth into the backing material.

25.3.2 Depth-Controlled Drilling

Depth-controlled drilling utilizes the top of the drilled stack as its reference point. Before this depth-control method can be utilized mechanically, the machine needs to know what the distance is from the pressure foot insert to the tip of the drill bit.

25.3.2.1 Tool Metrology Gauge Check.   This is achieved by performing a tool metrology gauge (TMG) check, using standard measuring devices. Once this distance is established, any negative (−) values entered into the depth box will cause the z axis to stop at the desired depth measured from the top of the stack.

25.3.2.2 Electronic Field Sensor—Depth-Controlled Drilling.   A sensor technology has been developed for higher depth control accuracy, which facilitates blind via applications. The system, which is integral to the pressure foot, incorporates the developments in electrical field sensor technology. Each sensor has its signals processed by its own dedicated microprocessor. This provides parallel processing for each sensor, permitting immediate and accurate tool analysis. The basis of operation is actual detection of tool contact to the board surface or top of the stack. This system permits the drill operator to specify controlled-depth drilling within ± 0.0002 in. Because the sensor detects tool contact with the surface, the accuracy is not affected by debris on the surface, variations in panel thickness, or adjacent burrs. The sensors can also detect broken tools ranging from 0.002 to 0.250 in.

Controlled-depth microvia drilling is currently being accomplished using this method. The drill bits are also undergoing modifications specifically for microvia drilling. Engineers continue to experiment with the flute design and the carbide in an effort to improve accuracy and increase tool life, which will essentially decrease the cost per hole.

25.3.3 Controlled-Penetration Drilling

Controlled-penetration drilling is utilized for drilling applications where the depth is referenced from the top of the backing material. This mode of depth control is not affected by stack height. The specified depth value or z-axis offset is treated as the desired amount of penetration relative to the top of the backing material. In order for the machine to be run in mode 3 depth control, a process called *mapping* must be performed prior to drilling any panels. During the mapping cycle, the machine touches down on the surface of the backing in 2-in increments. At each touchdown point, the thickness value of the backing is recorded. Once the mapping procedure is complete, the z-axis zero is established. Any negative values entered into the offset box represent the drill depth into the backing material. Controlling the penetration of the drill bit into the backing material assists in decreasing heat generation, improving hole wall quality by limiting the amount of debris, and decreasing drill bit breakage due to excessive depth.

25.4 HIGH-ASPECT-RATIO DRILLING

As circuits have gotten denser, not only have the traces gotten smaller, but the number of layers in a board has increased. The result is smaller holes penetrating thicker boards. To accommodate this increase in hole aspect ratio, two processes have been developed:

1. Peck drilling
2. Pulse drilling

25.4.1 Peck Drilling

Peck drilling is accomplished by dividing the total z-stroke into separate increments rather than completing the drill stroke in one action. (See the example shown in Fig. 25.1.)

There are several advantages to peck drilling and some disadvantages.

25.4.1.1 *Advantages.* The advantages of peck drilling are as follows:

- Decreased drill bit breakage
- Lower aspect ratio values
- Improved positional accuracy
- Decreased bottom-panel burring

Effective Aspect Ratio Reduction. The primary accomplishment of peck drilling is the decrease in effective aspect ratio value. Where aspect ratio is defined as the thickness of the board divided by the hole diameter, effective aspect ratio is defined as the total z-stroke divided by the smallest drill bit diameter. For example, if the total z-stroke equals 0.209 in (the total thickness of the board) and the smallest diameter is 0.025 in, the aspect ratio and the effective aspect ratio would have the same value of 8.36. By taking a total z-stroke of 0.209 in

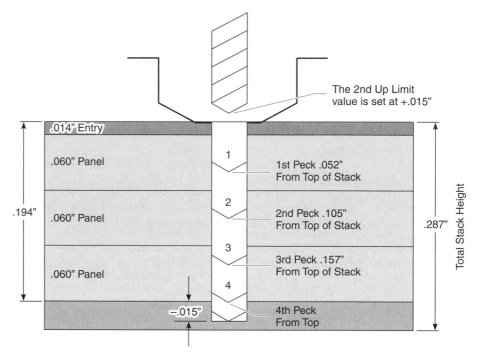

FIGURE 25.1 An example of through-hole peck drilling, showing a stack of three boards with entry and exit material creating a total stack height (or z-axis stroke) of 0.287 in. The effective aspect ratio is reduced when four pecks of the drill are made to complete the hole.

and dividing it into four pecks, the aspect ratio is decreased from 8.36 to 2.08. In general terms, an aspect ratio scale of 1 through 15 is commonly used, 1 being conservative and 15 being on the aggressive side of the scale.

Decision to Use Peck Drilling. The decision as to when to use peck drilling is very subjective. Drill bit breakage and bottom-panel burring are ordinarily the main factors considered when electing to peck drill.

Decreased Drill Bit Breakage. Decreased drill bit breakage is the primary benefit of peck drilling. Reducing the z-stroke means the drill bit has less debris to extract per revolution, decreasing the chances of clogged flutes, which can cause the drill bit to seize in the hole and break. Setting pecking parameters is a very subjective science; at best it is an approximation. Hole wall quality analysis should be completed before implementing any pecking procedures.

Hole Location Improvement. Another benefit of peck drilling is an improvement in hole location. Once a drill bit is deflected, it will continue at that deflection angle as it drills through the stack. Essentially, hole location is worse at the exit point than at the entrance point, particularly when drilling through thick panels with small-diameter tools (high-aspect-ratio drilling). Peck drilling can assist in achieving better hole location by lessening the effects of deflection by dividing up the total drill stroke, therefore decreasing the effective aspect ratio value.

Bottom-Panel Burring. Peck drilling can assist in decreasing the amount of bottom-panel burring by lessening the drilling temperatures resulting from drilling thick panels with small drill bits. Plowing, voids, and debris pack will be reduced by peck drilling due to the reduction in the amount of debris the tool must extract per z-axis stroke.

Drill Bit Cooling. When peck drilling, each individual stroke increment returns back to the upper limit value. For the example in Fig. 25.1, we used a value of 0.015 in above the stack and a total of four pecks. Increasing the distance above the stack (sometimes referred to as the second upper limit value) assists in cooling the drill bit before completing the next z-stroke. By cooling the drill bit, nail-heading values and smearing can be reduced substantially.

25.4.1.2 Disadvantages.   Peck drilling has some disadvantages. For example, it can cause:

- Increased nail-heading
- Increased smearing
- Hole wall roughness
- Increased cycle times

Hole Wall Quality   Peck drilling causes an increase in the overall cycle time due to the increased number of z-strokes necessary to complete one drilled hole. Nail-heading values tend to be worse for peck drilling due to the certainty that some copper pads will make contact with the drill bit several times as opposed to once when drilling using a single drill stroke. Hole wall smearing can also be a problem when peck drilling. The increase in heat caused by drilling the same hole several times can cause the substrate material to become heated to its melting point, which can lead to deposits of resin along the hole wall. A similar effect on hole wall roughness can also be seen.

Reducing the Effect of Disadvantages.   There are some methods of reducing the negative effects on the hole wall caused by peck drilling:

- Increasing the feed rate decreases the amount of time the drill bit spends in the hole.
- Limiting the number of pecks to a minimum also decreases heat generation and decreases overall drilling cycle time.
- Designing the tool has a major effect on hole wall quality when peck drilling. In order to decrease heat generation in the hole wall, the amount of friction caused by the drill bit making contact with the hole wall must be decreased. This can be accomplished by using an undercut tool design, sometimes referred to as a headed drill bit. The undercut tool design decreases the amount of contact by stepping down the diameter of the tool, which creates a larger clearance area, consequently decreasing the amount of flute making contact with the hole wall. The undercut tool design reduces the amount of nail-heading and smearing caused by peck drilling.

25.4.2 Pulse Drilling

Pulse drilling is designed primarily as a method to reduce problems related to "bird nesting," where excess debris builds up and remains attached to the flutes and shanks of the drill bits after the drill bit is retracted from the drilled hole. This condition also occurs in specialized drilling processes such as high-aspect-ratio drilling, drilling test fixture materials, and drilling abrasive materials used in high-temperature applications.

Pulse drilling acts as a chip-breaker by driving the z axis into the material using a short series of steps or pulses. Unlike peck drilling, where the tool is retracted completely out of the drilled hole, in pulse drilling the z axis pauses for milliseconds between each drill stroke while remaining in the drilled hole. This action allows the drill bit to extract debris up the flute before continuing the next stroke. Due to the rapid rate at which debris accumulates, the duration of each pulse must be kept to a minimum, resulting in a decreased infeed rate. To reduce this effect, pulse drilling uses a depth sensor on the drilling machine to detect the top of the stack. Once the top of the stack is detected, the spindle begins the pulse drilling process. Pulse drilling increases the amount of heat generated in the hole wall. Pulse drilling should only be used when hole wall quality is not the most significant aspect of the drilling operation.

CHAPTER 26
IMAGING

Brian F. Conaghan, Ph.D.
Parelec Inc., Rocky Hill, New Jersey

26.1 INTRODUCTION

Imaging is the process that patterns the metal conductor to form the circuit. This process involves a multistep integration of imaging materials, imaging equipment, and processing conditions with the metallization process to reproduce the master pattern on a substrate. Large features (200 μm and greater) can be very economically formed by screen printing. Feature sizes smaller than 200 μm, however, are formed using a photolithographic process, which is the main discussion of this chapter. As circuit densities have increased over the years, the imaging process has continually evolved to enable commercial production of finer features. This need for high-density interconnect (HDI) is driving the industry with feature sizes of 25 μm or lower. Imaging equipment and materials have been developed to meet this challenge.

To achieve a reproducible high-yield/low-cost imaging process for a given feature size, various factors must be carefully balanced. This chapter outlines the chemistry and equipment options for the photolithographic imaging process, highlighting trade-offs that must be considered so that process engineers, designers, and procurers of printed wiring boards (PWBs) and HDI structures have an overview of the process considerations that can enable a manufacturable product. Screen printing will not be discussed. (For information on the screen printing process, Chap. 11 of the third edition of this handbook is recommended.[1])

The photolithographic process sequence for imaging is given in Fig. 26.1. Basically the process involves applying a light-sensitive polymeric material (photoresist) onto the substrate of interest, exposing this photoresist to light with the desired pattern, and developing the exposed pattern. This developed pattern is used for either subtractive (etching) or additive (plating) metal pattern transfer. Details of these metallization processes are given in subsequent chapters. After metallization, the photoresist is stripped from the surface and the panel is ready for further processing.

Clean Surface

⇓

Apply Photoresist

⇓

Expose Photoresist

⇓

Develop Photoresist Image

⇓

Pattern Transfer Image
(plating or etching)

⇓

Strip Photoresist

FIGURE 26.1 The photolithographic process sequence.

26.2 *PHOTOSENSITIVE MATERIALS*

The photosensitive polymeric systems used as photoresists in the interconnect industry are either liquids or dry films formed from a liquid solution. Dry-film photoresists are the industry standard, but liquid photoresists have become more widely used, particularly for fine-feature circuits. Both types of photoresist can be used for a variety of processing requirements.

26.2.1 Positive- and Negative-Acting Systems

Photoresists may function in a photographic sense in either a positive or negative tone. The difference, illustrated in Fig. 26.2, is a result of the specific chemical reactions that occur on exposure to light. For the common negative-acting systems, exposure initiates cross-linking between one or more of the components in the polymeric matrix, reducing the photoresist's solubility in the developing solution. The exposed regions remain after development of the image. For positive-acting photoresists, the chemistry is often the novalac resin-based materials utilized by the semiconductor industry. Upon exposure, an acid-catalyzed reaction occurs that increases the solubility of the photoresist in the developer solution. In this case, it is these exposed regions that are removed in the developer solution.

Photographic tone can have an effect on product yield caused by contamination between the phototool with the master image and the photoresist. For a negative-acting photoresist,

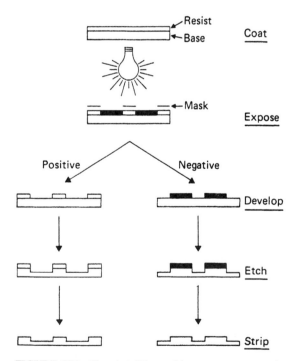

FIGURE 26.2 The photolithographic process sequence for positive and negative tone photoresists. (*Reprinted by permission of C. G. Willson,* Introduction to Microlithography, Theory, Materials and Processing, *ACS Symposium Series 219, Washington, D.C., 1983, p. 89.*)

contamination blocks the light and prevents cross-linking, leading to a mousebite (partial reduction in linewidth) or an open circuit in the conductor after etching. For a positive-acting photoresist, contamination blocks the light and prevents the acid-catalyzed solubilization, leaving excess metal and a possible short after etching. Since circuit patterns often have a greater area of spaces than lines, there is a lower probability of a contaminant causing an actual defect with the positive-acting photoresist. However, since many other factors impact yield and contamination must be minimized in any photolithographic process, negative-acting photoresists are the most widely used.

26.2.2 Decision Factors

The large variety of commercially available photoresists makes it a challenge to determine which one is likely to perform best for a given application. Key technical and economic factors that must be considered are summarized in Fig. 26.3. The primary consideration is the end use of the pattern.

Key Considerations in Photoresist Selection
Metallization Process
- Etching
  - Acid
  - Alkaline
- Plating
  - Copper
  - Tin
  - Tin/Lead
  - Gold
  - Nickel
  - Full Build Electroless Copper

Panel Design
- Substrate Topography
- Required Copper Thickness
- Number and Size of Holes
- Minimum Feature Sizes

Economic Factors
- Compatibility with Existing Application and Exposure Equipment
- New Capital Equipment Investment Required
- Productivity
- Process Latitude for High Yield
- Material Cost

FIGURE 26.3   Criteria for selection of photoresist materials.

- The photoresist must be chemically compatible with the subsequent pattern transfer steps (i.e., can withstand the chemical environment such as the solution pH) for it to function as an accurate mask for pattern transfer.
- The nature of the substrate itself in terms of topography and surface features must also be considered. Is the photoresist required to cover or tent plated through-holes or tooling features, and must it conform to existing circuitry? This consideration will dictate if dry-film or liquid photoresists are appropriate and, in some instances, the required tone.

- The required product feature dimensions are equally important. They are usually specified in terms of linewidth and spacing with a given allowable variation. Since these product specifications are for the final conductor features, the impact of photoresist processing on the critical dimensions must be established. Each photoresist has an inherent contrast (i.e., the rate of change of solubility on exposure to light). This contrast, combined with the resist thickness, phototool, light source used for exposure, and developing conditions, determines the finest feature that can be imaged. For photoresists used in the PWB industry, the minimum linewidth achievable in production is usually 10 to 25 μm greater than the photoresist thickness. Novel process sequences have demonstrated it is possible to overcome this traditional barrier.[2]

- The adhesion of the photoresist is another key attribute. The chosen photoresist must have good adhesion to the underlying substrate during pattern transfer but must also be able to release cleanly from the substrate during stripping.

26.3 *DRY-FILM RESISTS*

Dry-film photoresists are generally used for pattern formation prior to both plating and etching. These films have a three-layer structure, with the photosensitive material sandwiched between a polyester coversheet and a polyethylene separator sheet (see Fig. 26.4). The photoresist thickness is typically 25 to 50 μm, but thinner and thicker films are used for special applications. The photoresist is applied to the PWB substrate using lamination: the polyethylene separator is removed, and the photoresist flows under the heat and pressure from the laminating rolls to adhere to the substrate. The polyester coversheet performs several key functions. It protects the photoresist during handling, prevents sticking to the phototool during contact printing, and acts as an oxygen barrier during exposure. The optical properties of this polyester coversheet are important, especially for producing features less than 100 μm. Dry-film manufacturers have increased the optical clarity and reduced the thickness of the polyester sheet to achieve the required image quality. Thin dry films of 25 μm thickness and less generally use this high-quality coversheet.

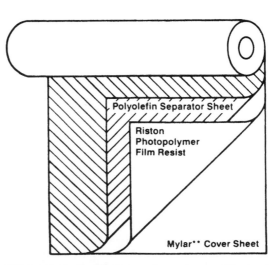

FIGURE 26.4   Dry-film photoresist components. (*Reprinted with permission of DuPont Electronics, DuPont Technical Literature.*)

The chemical and mechanical properties of dry-film photoresists have been tailored to withstand various plating solutions and acid or ammoniacal-etching solutions. They can be grouped in terms of the processing chemistry used to develop the image:

- Aqueous-developable, the industry standard
- Semiaqueous-developable, used in special applications
- Solvent-developable, the original system (see Table 26.1)

Since solvent development used chlorinated hydrocarbons, use of the solvent-developable photoresists has generally been abandoned.

26.3.1 Chemical Composition Overview

Dry-film photoresists vary in their exact chemical compositions, but all contain the following components:[3]

- Polymer backbone
- Photoactive compound
- Monomer
- Dye
- Additives

The polymer backbone sets the fundamental solubility and chemical resistance of the system. On exposure, the photoactive compound absorbs light of the appropriate wavelength and subsequently reacts with the monomer to alter the solubility of the film in exposed areas. The dye changes color on exposure to provide a visible latent image of the master pattern in the film, referred to as the printout image. Although the printout image decreases the efficiency of the desired photoreaction by consuming a portion of the exposed photoactive compound, it has proven to be a valuable manufacturing aid and is used in almost all dry-film photoresists. Additives include adhesion promoters, flexibilizers, and other compounds that can improve a desirable property. All the ingredients function together to provide the process latitude needed for the photoresist's specific range of applications.

26.3.2 Aqueous-Processable Dry Films

The majority of materials used are based on an acrylic polymer with various forms of photoreaction initiation. The initiator's absorption is designed to coincide with the major emission wavelength of the mercury arc lamps in the ultraviolet region of the spectrum at 365 nm. Many are based on the use of Michler's ketone and its subsequent sensitization of a triplet

TABLE 26.1 Summary for Dry-Film Photoresist Types and Chemistries

Photoresist type	Develop solution	Strip solution	Application
Aqueous-processable	Sodium/potassium carbonate	Sodium/potassium hydroxide	Acidic solution: etching, plating Limited alkaline solutions: alkaline etch
Semiaqueous-processable	Sodium tetraborate/Butyl cellusolve	Ethanolamine or Butyl carbitol/Butyl cellusolve	Alkaline solutions: etching, precious metal plating
Solvent-processable	(Methyl chloroform) Alternate to CFCs	(Methylene chloride) Alternate to CFCs	Strong alkaline: full-build electroless Cu plating

state promoted radical chain polymerization. This also explains the material's sensitivity to oxygen, an efficient triplet quencher. These materials are also very efficient at converting light energy into a chemical reaction due to the radical chain reactions; absorption of one photon of light results in many cross-links in the polymer matrix. Usual exposure doses measured as the irradiance from 330 to 405 nm to functionally cross-link these materials vary from 25 to 90 mJ/cm^2 depending on the exact chemistry used and the thickness of the materials, which compares favorably to positive-acting materials used for optical lithography in semiconductor IC production where doses of 200 to 500 mJ/cm^2 are common.

The images in these materials are developed in alkaline solutions of 1 percent or less by weight of sodium or potassium carbonate. The complete removal of the photoresist after pattern transfer is accomplished in 1M or greater solutions of sodium or potassium hydroxide at elevated temperatures, often with an antitarnishing additive to limit oxidation of the copper. Thus, in general, these materials exhibit excellent stability in acidic solution, but they do have varying stability to more strongly acid and alkaline solutions. In fact, there are three subclasses of materials: for acid etching of copper, for acid plating of copper, and for alkaline (ammoniacal usual) etching of copper. Increased acid stability is required for use in acid copper plating in which the pHs of the solutions often are below 1. For ammoniacal copper etching at pH 8 to 9, increased alkaline stability is required. This latter capability illustrates the flexibility in tailoring these materials to their end use since these materials are imaged using similar alkaline chemicals: develop at pH 10.3 and strip at pH 13. Often these alkaline-stable materials are developed at high temperatures.

In the past, several photoresists were formulated for exposure at a visible wavelength, 450 nm, using specialized exposure equipment—either visible lasers or magnified projection printing. More recently, laser direct imaging (LDI) based on the Ar$^+$ laser output clustered around 360 nm has been commercialized. Conventional photoresists can be exposed on LDI equipment, but higher photospeed has been key to the productivity and economics of LDI. Photoresists requiring only 10 mJ/cm^2 are now commercially available for the major applications. In general, these high-speed photoresist processes are similar to conventional ones, but improvements include photospeed, sensitivity to yellow safelights, and postlamination hold.[4]

Novel aqueous-developable photoresists have been formulated to process in dilute acidic media. A dry film based on electrophoretic depositable (ED) materials was shown to have excellent stability in strong alkaline solutions, potentially useful for etching polyimides or for full-build electroless copper plating.[5]

26.3.3 Semiaqueous- and Solvent-Developable Dry Films

Semiaqueous-developable photoresists are used when increased resistance to chemical attack is required, such as the highly alkaline solutions used in polyimide etching or gold and precious metal plating applications. Their composition is similar to the aqueous-developable photoresists, but the polymer backbone is less alkaline soluble. This increased chemical stability provides accurate image transfer by maintaining the integrity of the surface and sidewalls and minimizing underplating. Semiaqueous-developer chemistry is slightly alkaline (sodium tetraborate) and contains an organic assist, typically butyl cellusolve. Process control for this mixture is more complicated than that used for aqueous development.

Solvent-developable dry-film photoresists were the first to be commercially produced. Since they are the only materials that can withstand the harsh conditions in some chemical processing sequences, they still have some interest. They are based on methyl methacrylate polymers and a Michler's ketone initiation system similar to that used for aqueous-developable photoresists. Although they can have very high contrast, they are prone to scumming, leaving residue in the spaces between exposed photoresist features after development. Slight etching of the underlying substrate is often required to have a clean image-transfer step, especially for pattern-plating applications. Traditionally these materials were developed in trichloroethane and stripped in methylene chloride. However, as the hazards associated with chlorinated hydrocarbons were recognized, use of these solvents has been practically eliminated. This has forced the use of alternative develop and strip chemistries for solvent development.[6]

26.4 *LIQUID PHOTORESISTS*

Liquid photoresists come in a variety of chemistries.[7] Many are simply coatable forms of the dry-film materials. They have been used primarily for patterning innerlayers and are applied to the substrate by roller, spray, curtain, or electrostatic coating techniques. These methods are discussed later in the chapter (with equipment). Applied thickness is typically 6 to 15 μm, giving superior resolution compared to dry-film resists that are generally thicker. Since resolution is approximated by the thickness, they have the potential to resolve features less than 25 μm. Conformation of liquid photoresist to the copper substrate is generally very good, and it is possible to achieve high yield at 50 μm features and less.[8] However, cleanliness of the substrate and the coating and drying operation is critical. Liquid photoresists are dried to allow for hard contact with the phototool during exposure, maximizing resolution. If soft or off contact is used, resolution and yield decrease, especially for features 100 μm or less. For the majority of liquid photoresists, the developing and stripping conditions are similar to those for aqueous-developable dry films: mild alkaline sodium carbonate and strong alkaline sodium hydroxide. Exceptions are positive-acting novolac-based photoresists, thermal photoresists,[9] and negative-acting acid-developable photoresists.

26.4.1 Negative-Acting Liquid Photoresists

The chemistry of the typical materials is similar to that of the aqueous-developable dry films. They depend on a radical chain reaction to promote cross-linking in the acrylic polymer. Their sensitivities differ depending on the initiator system chosen. Materials are available with minimal solvent content, and a few are diluted with water.[10, 11] This reduces the emissions from coating and drying that must be contained and treated.

Acid-developable photoresists mentioned in the dry-film aqueous section can also be formulated as electrophoretic-depositable liquids. These materials have high contrast, good resolution, and chemical stability in strongly alkaline solutions since they are developed in aqueous acid solutions.

26.4.2 Positive-Acting Liquid Photoresists

The other alternative is a positive tone novolac resin-based system that finds its application most usually in the semiconductor industry for imaging integrated circuits. It has excellent resolution and chemical stability in acid and a large number of basic solutions (precious metal plating). The chemistry involves the conversion of a base-insoluble diazonaphthalquinone to the base-soluble indene acid. Thus, this photo-driven conversion alters the solubility of the base material in alkaline solutions, allowing for image formation (see Fig. 26.5).

26.5 *ELECTROPHORETIC DEPOSITABLE PHOTORESISTS*

Electrophoretic depositable (ED) photoresists[12, 13] were introduced in the late 1980s with expectations of broad adoption by the interconnect industry. The early materials were modifications of those used for coating automobiles and appliances. These fast-coating processes pro-

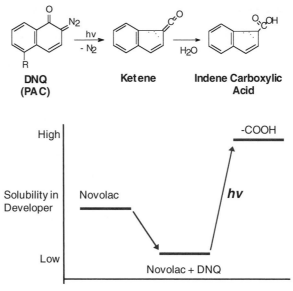

FIGURE 26.5 Chemistry mechanism of positive-acting liquid photoresists.

vide a defect-free film of high durability on conductive substrates and are well suited to large-volume production. Anodically and cathodically deposited materials that are both positive and negative acting have been introduced but have achieved limited market penetration in the interconnect industry. Advances in dry-film photoresists and liquid photoresists reduced the interest in implementing a new photoresist process. (A detailed discussion of electrophoretic depositable photoresists is included in Sec. 17.1.6 of the fourth edition of this handbook.)

26.6   RESIST PROCESSING

Imaging is a sequential process, and the various steps are very interdependent. Equipment and processing conditions play a large role in the image quality, yield, and productivity that can be achieved. The choice among the numerous options for each step depends on the type of substrate to be imaged, the feature sizes to be produced, the existing equipment, the type of photoresist, and the productivity and economics of the production line. Alternatives for each process step are described and compared.

26.6.1   Cleanliness Considerations

Cleanliness is an issue throughout the entire photolithography process, including yellow room contamination level (cleanliness class), solution filtration, equipment cleanliness class, and product handling. Its importance increases as feature size decreases. The level of cleanliness needed to give acceptable yield for the finest features must be established and maintained. Sticky rollers, either as free-standing equipment or as handheld items, are one type of equipment used to remove contamination from flat surfaces, either panels or equipment.

26.6.2 Surface Preparation

Good adhesion between the photoresist and the substrate is essential for the remainder of the process to be successful. All laminates arrive at the imaging area dirty, and the exact process sequence for surface preparation is chosen based on the nature of the contamination. Epoxy dust generated during trimming and processing of the laminate is removed by mechanical cleaning. The copper surface is either foil or foil with electroless or electroplated copper. Foil typically has an antitarnishing agent of chromium and zinc that must be removed for reproducible imaging results. During cleaning, the texture or surface roughness of the copper is altered, promoting mechanical adhesion between the copper and the photoresist. The extent of this texturing is measured by the usual parameters for surface roughness, the height and spacing of the topography measured by profilometry (see Fig. 26.6).

R_a Arithmetic mean of all departures from the mean line

R_{max} Maximum peak to valley height within sampling area

S_m Mean spacing between peak measured at the mean line

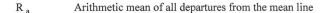
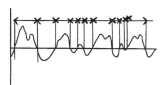

FIGURE 26.6   Surface roughness parameters.

Chemical cleanliness can be determined in a variety of ways. Wetting is often tested using a *water break test* or more analytically in terms of contact angle, with a low value being desirable. In addition, analytical techniques (Auger and x-ray photoelectron spectroscopies) can be used to evaluate the chemical composition of the surface (see Table 26.2).[14]

26.6.2.1 Mechanical Cleaning.   Most mechanical cleaning equipment physically abrades the copper surface using a pumice or aluminum oxide slurry either directly (scrubbing or vapor blasting) or impregnated into brushes. Scrubbing provides an even final surface texture, while brushing creates grooves (see Fig. 26.7). The copper thickness must be sufficient

TABLE 26.2 Copper Surface Chemical Composition as Measured by X-Ray Photoelectron Spectroscopy (XPS)

Sample	% Cu	% O	% C	% N	% Zn	% Cr
Initial		46	24		16	12
Preclean 1	12	19	68			
Preclean 2	6	10	70	14		

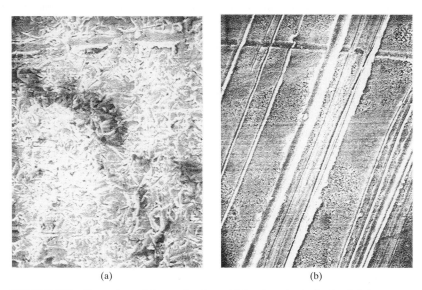

(a) (b)

FIGURE 26.7 Comparison of the pumice cleaned (*a*) and mechanically abraded copper surfaces (*b*), 1200×. (*Reprinted with permission of D. P. Seraphim, R. C. Lasky, and C-Y Li,* Principles of Electronic Packaging, *McGraw-Hill, 1989, p. 383.*)

to provide the desired final conductor height after cleaning. Mechanical cleaning is more difficult for very thin flexible panels.

The texture of the mechanically cleaned surface affects the adhesion between the photoresist and the copper. The resist should conform to the topography, and thus, the extent and type of texture must be tailored to the type of photoresist. If the gouges are too deep, a dry film applied with traditional lamination will have difficulty conforming to the surface and create defects: near or full opens after etching and underplating or shorts after pattern plating. Thus, the cleaning process, photoresist type, and application method must be matched to image at high yields.

Conveyorized equipment is used that passes the panels through the brushes or slurry and then to rinse chambers before exiting through a dryer section. If rinsing is insufficient, residual pumice can remain on the copper surface. Maintaining proper functioning of the mechanical parts that contact the product is essential since brushes and slurry deteriorate with use. Equipment designs that isolate the abrasive from the majority of the tool are more durable.

26.6.2.2 Chemical Cleaning. Chemical cleaning includes a variety of solutions since the solubility of contaminants can be quite different (see Table 26.3). Grease and fingerprints require a soap solution or solvent to dissolve them. Antitarnish treatment removal and copper roughening is done with mild etchants for the copper, such as ammonium persulfate, sodium

TABLE 26.3 Cleaning Solutions Used to Remove Various Contaminants

Cleaning technique	Process chemistry	Contaminant removed
Abrasion	Pumice, Al oxide	All
Plasma	CF_4/O_2, O_2/H_2O	Organic
Chemical solutions (aqueous etchants)	H_2SO_4, HCl, etc.	Inorganic
Chemical solutions (nonaqueous)	Alcohols, etc.	Organic
Thermal	N_2	H_2O

Source: Adapted from *Introduction to Microlithography,* 1st ed., L. F. Thompson and M. J. Bowden, *American Chemical Society Symposium Series 219,* Washington, D.C., 1983, p. 184.

persulfate, or peroxide/sulfuric acid. Depending on the oxidation level and the amount of antitarnish treatment, there is often an induction time for copper removal. Solutions used can be common chemicals or proprietary mixtures in which surfactants and other additives have been included. Alternatively, the oxide and antitarnish treatment can be removed initially, followed by a surface roughening. In this case, the oxide removal is accomplished with mild acidic cleaning solutions such as sulfuric acid.

The selection of the cleaning solution also depends on the copper thickness. Thin "seed" layers used for electrolytic pattern plating can tolerate very little etching, and extremely dilute solutions or dry methods are used. Thus, chemical adhesion can be more important than mechanical. A review of adhesion considerations is found in Ref. 15. Selection of the proper process chemicals and sequence depends on the overall conductor formation process and the photoresist to be used.

Chemical cleaning sequences are often contained in conveyorized spray equipment with rinsing between each step. Batch processing in a tank system is also used, with a hoist for basket movement between solutions. Uniformity of the etching of panels within a basket must be measured, but is generally quite good and reproducible as long as the immersion time is short and the bath's chemical composition is consistent.

26.6.2.3 Electrolytic Cleaning.   A unique inline conveyorized tool built by Atotech uses electrolytic cleaning.[16,17] The antitarnish is removed electrolytically along with oils and fingerprints in a uniform process with very little copper removal. The surface texture is then altered by microetching and passivated prior to dry-film photoresist lamination or other coatings.

26.6.3 Photoresist Application

The technique used to apply the photoresist to the substrate depends on the type of photoresist selected. The various techniques used with dry-film and liquid photoresists are discussed. For any of these techniques, the cleanliness of the operation—both the equipment and the room—and the handling of the product are critically important, especially for high density. It is also advisable to minimize handling at this stage by using automated inline equipment for the preclean step and automatic loading and unloading.

26.6.3.1 Dry-Film Hot Roll Lamination.   In this process, both temperature and pressure are used to laminate the dry-film photoresist to the panels. After stripping off the polyethylene separator sheet, the photoresist is contacted to the substrate in a nip between two heated rolls. Heat reduces the viscosity of the photoresist and pressure causes it to flow, conforming to the irregularities of the copper surface.

The weave of the laminate affects the topography of the substrate and how much flow is required to achieve good conformation, an essential condition for high yield. Dry-film photoresists are tailored to have rheology that maximizes flow during lamination while preventing flow in the roll prior to lamination, which could cause edge fusion. Panels are often heated with heated rolls and infrared heaters prior to lamination to improve photoresist conformation. The key process parameters affecting conformation are laminator speed, laminating roll temperature, and rigidity of the rolls with respect to the core material and rubber durometer. For aqueous-developable materials, wet lamination has proven to be effective in improving conformation and yield for innerlayers. By using a water-alcohol mixture, wet lamination can also be applied to tent and etch inner via holes to minimize the land diameter.[18]

Both automatic and manual tools are available for film lamination (see Fig. 26.8). With manual tools, the panel is placed in the nip and pulled through by the rotation of the rollers. The continuous film of photoresist must be trimmed for each panel on at least two sides. This technique is labor intensive and extremely dirty due to the chips of resist generated by trimming. However, it is useful for very thin materials and nonstandard-size panels in small lot sizes. With automatic equipment, the basics are the same but the panel enters from an automatic loader or a conveyorized preclean tool directly into and through the dry-film laminator. This is a relatively clean operation with little handling of the panels since the photoresist is automatically trimmed to the appropriate size and placed within the board edge. Equipment is available to convey even very thin panels. Panels are often heated, wetted (wet lamination), or cleaned before the actual lamination.

26.6.3.2   Dry-Film Vacuum Lamination.

This method is similar to hot roll lamination but the temperature and pressure used to adhere the photoresist to the substrate is applied by a heated vacuum platen instead of laminating rolls. Vacuum lamination is useful for products with tall or closely spaced features that make it difficult for the photoresist to conform. The vacuum removes the air that would be trapped and pulls the photoresist into tight spaces, providing conformal coverage. This equipment is also used for material that will deform nonuniformly under the pressure of laminating rolls, such as thin unreinforced polyimides. With vacuum lamination, the pressure is evenly distributed over the panel, improving dimensional stability control.

26.6.3.3   Liquid Coating.

A great variety of methods are used to apply liquid photoresist to panels. Roller, spray, electrostatic, and electrophoretic coating provide double-sided application and are well suited to high-volume manufacturing. Curtain coating is single sided, requiring either two passes of the panel through the equipment with slight drying between coatings or an inline equipment layout with two coating and drying units. This causes a difference in the solvent content of the two sides. Screen coating equipment is available for either single-side or double-side coating. In all instances, the cleanliness of the surroundings, equipment, and solutions is essential for high-yield processing. Most equipment is designed to minimize material consumption by collecting and filtering excess liquid so it can be recycled to the feed tank.

26.6.3.4   Roller Coating.

This method places liquid photoresist on the board by transferring the liquid from one set of rollers to another (see Fig. 26.9). The exact number, physical configuration, and surface of the rollers vary, as well as how the material is metered. Often, gravure roller coating is used. Grooves or crosshatches in the rollers that contact the panel determine the amount of liquid deposited on the substrate and are set with a measured overlap or interference so that the board is squeezed between the rollers.[19] Photoresist viscosity also influences the coating thickness applied. Improvements in panel handling and in confining the solution to the center of the panel allows for simultaneous two-sided "postage stamp" coverage similar to that obtained with dry-film photoresists, leaving the tooling and location holes clear of material. In commercial systems, the coater is conveyorized, with panels entering a clean oven immediately after coating. The throughput of these systems can be as high as 240 panels per hour.

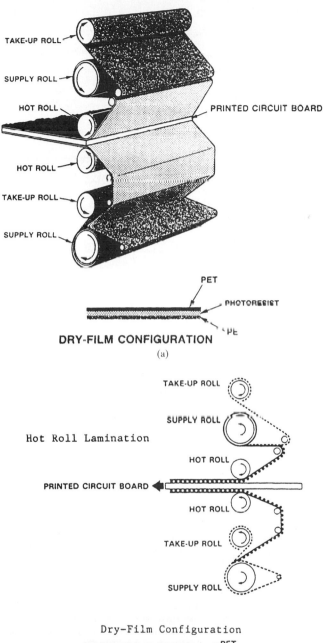

FIGURE 26.8   Configuration of a hot roll laminator: (*a*) three-dimensional schematic; (*b*) two-dimensional schematic. (*Reprinted with permission of E. S. W. Kong,* Polymers for High Technology, *ACS Symposium Series 346, Washington, D.C., 1987, p. 280.*)

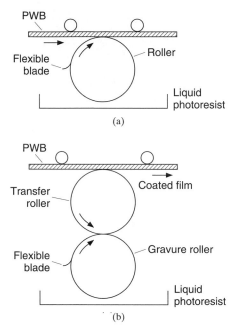

FIGURE 26.9   Configuration of a roll coater: (*a*) direct coating; (*b*) indirect coating.

26.6.3.5   Spray Coating.   Although very thin and uniform coatings are possible with this method, its use in resist imaging has been limited due to concerns about waste caused by overspray. The spray head traverses a larger area than the substrate to be coated, and the amount of liquid photoresist that becomes overspray can be significant. Exact characteristics of the coating in terms of thickness and topographical coverage depend on spray-head configuration, nozzle backpressure, droplet size, and conveyor speed.[20]

26.6.3.6   Electrostatic Coating.   This method is similar to spray coating except that the rotation spray head is charged and the panel is grounded. Since the atomized photoresist is attracted to the substrate, there is less material waste than with spray coating.[21] Electrostatic coating is used for solder mask since it provides good coverage of conductor sidewalls. However, its use with photoresist is limited to innerlayers. It is unsuitable for outerlayers because it deposits a thicker coating on the rim of a plated through-hole while leaving its interior uncoated.

26.6.3.7   Electrophoretic Coating.   This method requires materials specially formulated for anodic or cathodic deposition. The equipment schematic in Fig. 26.10 illustrates the process sequence: preclean, coat, permeate rinse, rinse, and dry.[22] Panels are placed in the photoresist solution wet or are often sprayed with solution to ensure wetting, especially if the parts have blind or through-vias. Voltage is applied, and within 20 seconds to 3 minutes an insulating film forms. Rinsing removes loosely bound material that is returned to the plating cell after ultrafiltration to separate the photoresist and the counterion. This aids in maintaining the ionic balance. After a final water rinse, the panel is dried to remove water and consolidate the film. The film is now tack free and can be imaged.

The chemical composition of the photoresist determines whether the equipment components are made from PVC, polypropylene, or polyethylene. Electrodes are made of stainless

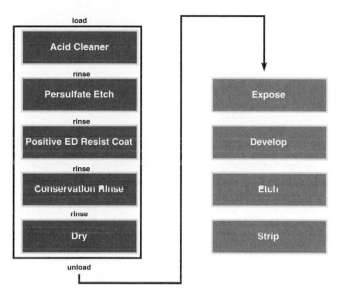

FIGURE 26.10   Configuration of an electrophoretic photoresist coater. (*Reprinted with permission of J. Dubrava et al., "Development of a Novel Positive-Working Electrodeposited Photo Resist Process for the Production of High Density PWB Outerlayers," Presentation at the IPC Conference, Spring 1995, San Diego, CA.*)

steel and encased in an ion-selective membrane that is flushed to prevent the buildup of counterion migrating there during deposition. The permeate and ultrafiltration unit is important for long-term stability of the coating solution. The coating unit is often enclosed in a clean environment to reduce the amount of particles found in solution or resting on the panel as the process proceeds.

26.6.3.8   Curtain Coating.   This method is commonly used to coat liquid solder-mask materials. The panel is conveyed at high speed through a vertically falling curtain of the liquid photoresist. The coating width can be adjusted so that the edges are not covered. Temperature and viscosity are controlled for reproducible coatings, and coating thickness is determined by the conveyor speed and the solution flow rate.[23] Material in the curtain is returned to the sump and reused, giving very efficient material utilization. This method is one-sided, and the first-side photoresist coating must be partially dried before coating the reverse side. Due to the extra handling required to coat one side at a time and the inability to convey very thin panels through the curtain at high velocity, this method has seen only limited use with photoresist.

26.6.3.9   Screen Coating.   Another common method for coating solder mask with either a pattern or flood coverage is screen coating. The screen mesh size and the liquid solution viscosity set the wet coating thickness. The screen is placed above the panel to be coated, and the material is forced through the openings in the screen onto the panel surface where it forms a film. Equipment is available for either single-side or double-side coating. With single-sided screen printers, the coating must be partially dried before the reverse side is coated. This causes a difference in the solvent content between the two sides that would result in nonuniformity for very precise lithography. In addition, the sequential nature of the coating allows debris and contaminants to be embedded into the first coating. Double-sided screen printers eliminate a drying step and minimize panel handling. The advantages of screen coating are that the equipment is relatively inexpensive to purchase and operate, "postage stamp" (small embedded areas) coating is possible, and waste is minimized.

26.6.4 Expose

Expose is the actual imaging step, reproducing the master pattern in the photoresist. The relief image is created after subsequent development. The expose process elements are phototool generation, registration of the phototool to the panel, and exposure through the phototool by the light source. Light-source alternatives include contact printing, either collimated or uncollimated; proximity printing; projection printing; and laser direct imaging. The noncontact methods separate the phototool from the substrate, reducing yield losses caused by contamination between the phototool and photoresist. With laser direct imaging, there is no phototool and the design data file of the master pattern drives the laser beam to expose the photoresist.

Phototools are generated on either film or glass substrates, depending on the feature sizes that are being patterned and the durability required. The panel is either mechanically aligned, with pins holding the phototool with respect to the product, or optically aligned, with alignment features dictating the movement of the phototool and product. The alignment requirements are interrelated to those achieved at composite lamination and drilling. The registration scheme is contained within the exposure equipment.

26.6.4.1 Conventional Imaging

Artwork Generation. Polyester and glass are the substrates used for phototools. They differ in optical properties, dimensional stability, and durability. Silver halide on polyester is used to create first-generation phototools on a laser plotter. This image is often contact printed onto diazo on polyester to make the copies used in production. Optical absorbance of the different phototool materials versus wavelength is plotted in Fig. 26.11. Diazo phototools are the least expensive, and glass is the most expensive. Although the less-expensive materials have poorer transmission, this property is generally not limiting, except for applications requiring high intensity. Table 26.4 illustrates the impact on exposure time. Edge definition is a major difference among the options, as shown in Fig. 26.12. The photos illustrate that chromium on glass gives the sharpest edge definition. Edge definition is also affected by the pixel size of the

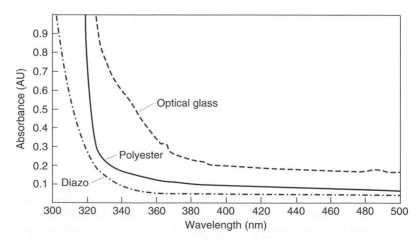

FIGURE 26.11 Optical absorbance of common phototool materials in the spectral region of PWB photoresist exposure.

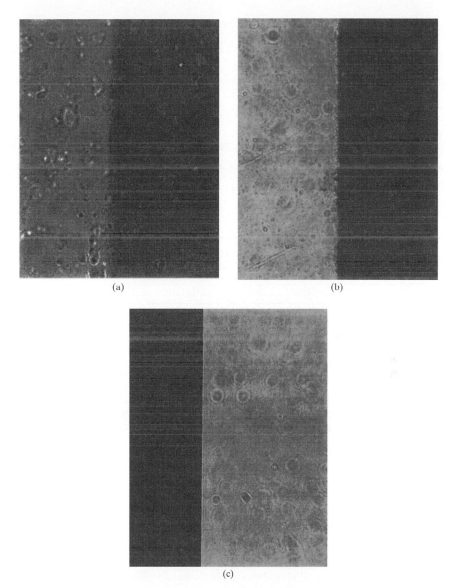

FIGURE 26.12   Optical micrograph of the image edge of common phototool materials, 1600×: (*a*) diazo on polyester; (*b*) silver on polyester; (*c*) Cr on glass.

phototool plotter and the spot's addressability (see the section on LDI). For applications that use a low-contrast photoresist and poor light collimation, these differences are not measurable on the printed photoresist image or in the patterned conductor. However, for high-density imaging, edge definition is important, as is the dimensional stability of the phototool. Film dimensions change more rapidly with temperature than glass and are also affected by humidity. Careful control of both temperature and humidity is required throughout the life of a phototool (see Table 26.5).

TABLE 26.4 Comparison of the Optical Transmission of Various Phototool Materials

Property	Diazo	Polyester	Glass
Absorbance at 365 nm	0.271	0.107	0.047
Transmission at 365 nm	54%	78%	90%
% increase in exposure time vs. glass	40	13	0

TABLE 26.5 Dimensional Stability of Phototool Substrate Materials with Respect to Temperature and Relative Humidity

Substrate	Expansion coefficient, ppm/°C	Expansion coefficient, %RH/°C
Soda lime glass	9.2	0
Low-expansion glass	3.7	0
Pyrex	3.2	0
Quartz	0.5	0
Polyester	18	9

Source: Kodak Technical Literature (ACCUMAX 2000) for polyester film; *Tables of Physical and Chemical Constants*, Longman, London, 1973, p. 254; and *Formulas, Facts and Constants*, H. J. Fischbeck and K. H. Fischbeck, Springer-Verlag, Berlin, 1987, for glass data.

The durability of the phototool is also important, especially for use in a manual contact printer. Damage depends on phototool handling, the exposure tool, the cleanliness level of the operation, and the surface of the product being contacted. Commonly, glass artwork is usable for 100 to 400 contacts with repair, while film artwork is usable for 20 to 50. For large numbers of the same part, number cost per contact may actually be lower with glass artwork, while for prototype parts film artwork is ideal. However, the majority of production panels are exposed with film artwork. High-density applications such as chip-scale packaging and multichip modules using laminate (MCM-L) may require glass artwork. The requirements of the product determine the type of phototool that should be used.

Registration. The image is placed on the panel with respect to a point of reference so that the image is well aligned to previous and future features. In multilayer construction, the inner-layer images are aligned front to back for successful lamination and drilling of the PTHs. The outerlayer image is then aligned to the drilled holes. The same considerations are relevant for single-layer boards. Thus, accurate alignment is clearly part of the imaging process and also depends on knowledge and control of the dimensional stability of the product. The image must be scaled to match the dimension of the board at the exposure step; for multilayer structures with several photolithographic steps, a series of dimensional measurements are required prior to product manufacture.

Mechanical registration entails the use of fixed pins or other mechanical devices to hold the phototool and the panel in place during exposure. The actual configuration of the pins and their shape varies. Both two- and three-point systems are used. The three-point systems well define the center location since they are located at the extremes of the area to be patterned. The shapes of the pins are either round or elongated with one flat side. In the latter case, the flat side defines the edge, while round pins center within the hole in the panel and phototool. Film artwork alignment holes are punched with respect to the product pattern by plotting alignment targets as part of the product pattern and optically aligning and punching a slot or hole. The edges of the punched feature wear, and registration will deteriorate with use. For glass artwork, the glass is drilled at the alignment locations and a bushing is placed at the center of the hole. The pin is then inserted through the bushings in top and bottom artwork and through the panel. With use, the bushings do move and must be reset for maximum reproducibility.

Optical alignment systems operate either manually or automatically. In both instances, the phototool is plotted with alignment targets—usually an opaque dot that is smaller than a drilled or punched hole in the panel. With backlighting, the dot (phototool) is aligned to the center of the hole (product). Often, three locations at the edges of the panel are used. In manual systems, micrometers are used to move either the part or artwork. In automatic equipment, a vision system calculates the necessary movement, and motors then move either the phototool or the product into alignment. Since there is no wear on the phototool, when it is optically registered the accuracy is maintained with usage. The absolute accuracy achieved with optical methods is superior to mechanical registration, with the best accuracy and reproducibility obtained with automatic optical systems. The extra expense for automated equipment is warranted if it is needed to meet stringent product requirements.

Exposure Control and Measurement.   The role of exposure is to chemically change the photoresist and its solubility in the developer solution. The appropriate energy dose is determined experimentally by measuring the combination of dose and development that is needed to produce features with straight sidewalls. The photoresist is coated on an optically clear substrate and exposed from the backside. Contrast curves, plots of the log of the exposure dose versus the thickness of the film remaining, are used to identify the functional cure point, for example the dose that gives a thickness loss less than 10 percent (see Fig. 26.13). Exposure in the region of stability ensures that the base of the material has reacted. Step wedges, film strips with a series of neutral density filters, are also used to determine appropriate exposure doses. The dose is varied to obtain photoresist residue on the area with the manufacturer's recommended step value. This technique depends on appropriate and consistent developer conditions. Step wedges are often used to control the expose or expose-and-develop processes.

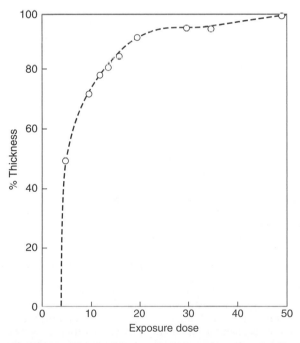

FIGURE 26.13 Contrast curve for a negative-acting dry-film photoresist, percent film thickness remaining versus exposure dose (mJ/cm²).

The energy incident on the photoresist is the product of the lamp intensity and the time of exposure:

$$\text{Energy} = \text{intensity} \times \text{time} \qquad (26.1)$$

Thus, the exposure dose can either be measured directly by the use of an integrating radiometer or indirectly by measuring exposure time and light intensity with a radiometer. The radiometer response must be matched to the spectral output of the light used so that representative measurements are made (see Fig. 26.14). Most exposure equipment provides the option of using either direct energy or time measurement. With a stable light source, both approaches are equally reproducible.

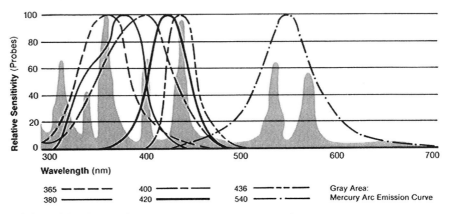

FIGURE 26.14   Output of Hg/Xe light source and radiometers. (*Reprinted with permission of Optical Associates, Inc., Milpitas, CA, Technical Literature.*)

Contact Exposure Tools.   This equipment places the phototool and the panel in direct contact and draws a vacuum between the pieces for hard contact. A range of light wavelengths is used, often from a mercury or mercury/xenon light source, which is stable and has a high-intensity output at ultraviolet and visible (UV/Vis) wavelengths (see Fig. 26.14). The light is distributed over the area to be exposed by placing the board at a distance from the source, either noncollimated or collimated by the use of optical elements, as shown in Fig. 26.15. Collimation refers to the angle of incidence of the light onto the photoresist, and it is critically important for fine-line images. Defining fine spaces is the challenge for both additive and subtractive conductor formation, and exposure under the opaque areas cannot be tolerated. The lamp intensity for noncollimated sources is greater, and therefore, less time is required to expose the panel. Process throughput is noticeably enhanced for photoresists requiring a large dose.

In addition to collimation, good contact between the phototool and the panel is the most important factor to control in fine-line formation. Any gap will allow exposure under the opaque, resulting in poor linewidth control and reduced resolution. The sensitivity to off-contact exposure varies among photoresists, and newer resists are formulated with increased off-contact latitude.[24] Since the lamp's spectral output degrades over time, the lamp must be routinely replaced to maintain a stable photoprocess. The exact frequency for replacement depends on the line and space definition required, but the lamp life is typically 1000 hours. Failure to replace a lamp can result in it exploding and damaging the optical elements in the exposure unit, which is far more costly than routine maintenance.

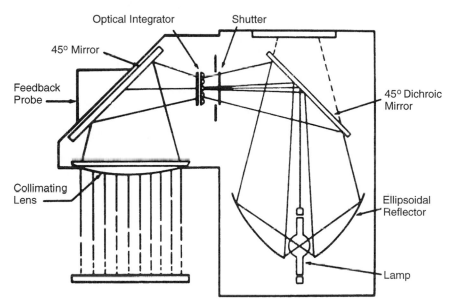

FIGURE 26.15 Configuration of lamp and optics for a collimated contact printer. A noncollimated source is identical except for the collimation lens in the lower left. (*Reprinted with permission of Optical Associates, Inc., Milpitas, CA, Technical Literature.*)

Noncontact Exposure Equipment. Semiconductor imaging demonstrated the yield limitations in using contact-exposure equipment. Noncontact methods had to be developed to achieve acceptable conductor yield as critical dimensions decreased to 0.18 μm and below. The PWB industry is following this evolution, driven by the increased demand for products with fine features of 50 μm and less. Semiconductor tool concepts cannot be used directly due to important differences between IC and PWB products, namely the physical size of the substrate and features to be imaged and the flatness of the substrate. The key to noncontact methods is the use of appropriate optics for clear image placement in the photoresist (Fig. 26.16), even if the image transfer function has degraded. For PWB applications, this includes the depth of focus to allow for panel warpage and the resolving power, given the contrast of the resist. The following approaches address these concerns and are viable alternatives to contact printing.

Proximity Printing. This is the oldest method for off-contact printing and requires no modification to the equipment optics. The phototool is held out of contact from 125 to 500 μm. The image in the photoresist suffers since the distance between the phototool and the part allows exposure under the opaque area and imaging of debris on the phototool. Nevertheless, resolution as fine as 75 μm has been reported for proximity printing with thin coatings of liquid resist.[25]

Projection Printing. Three approaches are possible for this technique: scanning, stitching, and magnified projection printing. Scanning and stitching evolved from semiconductor and MCM-D thin-film technology, where smaller, flatter substrates are used. They require movement of the phototool and/or the substrate during exposure. Magnified projection imaging is unique to the PWB application and has no moving optical components. A schematic of all three systems is given in Fig. 26.17.

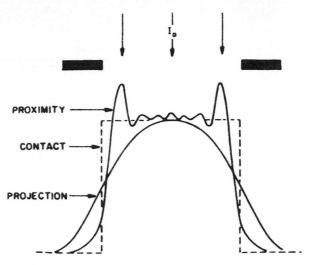

FIGURE 26.16 Image transfer for contact, proximity, and projection printing.

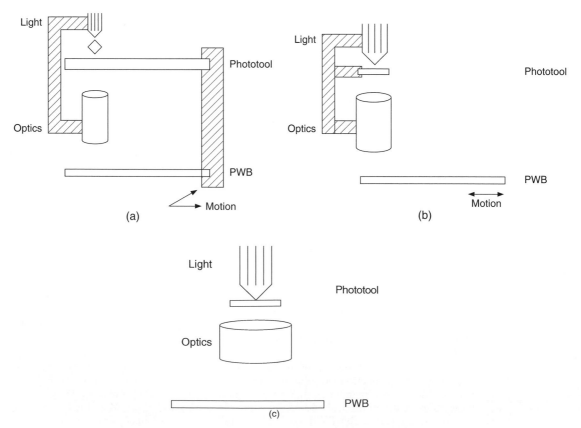

FIGURE 26.17 Configuration for projection imaging equipment: (*a*) scanning, (*b*) stitching, and (*c*) magnified.

- The scanning printer requires the synchronized movement of the part and the phototool during exposure. Precise alignment of the two pieces must be maintained during the motion. Periodic adjustment for flatness is usually required since this equipment has a small depth of focus. For typical PWB formats, a large phototool is needed, requiring expensive glass phototools for their flatness, rigidity, and sharp line definition. This exposure technique has demonstrated high resolution over a large, relatively flat area using a high-contrast resist.[26] Few PWB applications exist, but the technique is useful for flat panel display manufacture.

- Stitcher projection printers are step-and-repeat tools that use a repeating image or "stitch" together several images. The artwork generation and management in the equipment is complicated, which has hampered its widespread use for large panels. The usual active area size as defined by the optics is approximately 6-in × 6-in. Thus, for a standard 18-in × 24-in panel, 12 artwork changes are required to expose one side. These systems have very high resolving capability and large depth of focus. Again, glass artwork is used, but these smaller plates are easier to generate and use than in a scanning projection printer.

- The magnified projection imaging concept was first demonstrated in the late 1980s with the SeriFLASH exposure tool, utilizing 436-nm illumination and a 5-in × 5-in liquid crystal display as the phototool. The image was magnified 6 times to expose an 18-in × 24-in panel, but demonstrated photoresist definition was limited to 125 μm since the liquid crystal features were one-sixth of the final conductor width. The equipment was subsequently modified for 365-nm exposure with a glass phototool to improve resolution. Fifty-μm line and 63-μm space resolution has been demonstrated.[27] The exposure time required is similar to that of contact exposure equipment. With optimized optics, this may become a high-resolution tool that utilizes conventional photoresist and phototools.

26.6.4.2   Laser Direct Imaging.   The LDI method of exposure provides the ultimate in flexibility for customized pattern compensation factors and for engineering changes to the product design since it does not require a phototool. Systems introduced in the 1980s used an Ar$^+$ laser with visible optics, requiring specialized photoresists that were sensitive to visible light. More recent Ar$^+$ systems have relied on the UV wavelengths, making it compatible with conventional photoresists. However, high-speed photoresists are needed to achieve the productivity for high-volume production. Concerns about the lifetime (3000 to 5000 h) and power consumption (60 to 80 kW) of gas lasers has led to development of solid-state laser systems at various wavelengths.[28,29,30] Systems have also been introduced that use no photoresist at all, laser ablating either tin for use as an etch resist or PVD copper on polyimide for use as a plating base.[31,32] The various LDI systems are summarized in Table 26.6.

In addition to laser type and wavelength, specific equipment designs differ in optics, either single- or multiple-beam operation; platen design; single- or double-side exposure; pixel shape;

TABLE 26.6   Laser Direct Imaging Systems

Laser	Wavelength	Photoresist material	LDI suppliers
Gas (argon ion)	488 nm	Special visible-sensitive resist	Pentax
	351–364 nm	Conventional or high-speed UV resist	ETEC Systems, Orbotech
	333–364 nm	Conventional or high-speed UV resist	Automa-Tech
Solid state	810–900 nm	Positive-working thermal resist	Creo
	visible	Special visible-sensitive resist	Dainippon Screen
	UV	Conventional or high-speed UV resist	Barco
Nd:YAG	1064 nm	None	Siemens
Excimer	248 nm	None	LPKF

and spot size and resolution. Multiple-beam operation increases exposure productivity, a key requirement for economic feasibility in large-volume manufacturing. Target production rates for the various systems range from 60 to 180 panels per hour (18-in × 24-in with 50 μm features). Much of the experience with installed equipment has been in small-lot, quick-turn production of high-density boards, where savings are achieved due to artwork elimination and reduced labor used in setup. For volume production of medium- to high-technology boards, several companies have demonstrated a quick return on investment is possible due to yield savings and the ability to produce designs that had not been technically or economically possible with conventional imaging.[33] Although the UV systems can expose conventional photoresist, use of high-speed photoresist is needed to maximize tool throughput.

Most LDI systems use a flatbed platen design and can image both innerlayers and outerlayers. Systems with an external drum architecture, similar in design to laser photoplotters, are limited to innerlayers because the panel must have sufficient flexibility to conform to the drum (Fig. 26.18). The pixel shape can be either square or gaussian. Square pixels are made up of many point sources of light and can theoretically achieve perfect stitching. However, in real systems there is intensity variation due to mechanical vibration and edge rounding. Gaussian

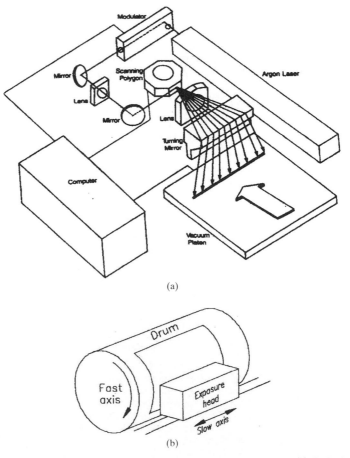

(a)

(b)

FIGURE 26.18 Schematics of laser direct imaging equipment: (*a*) flatbed platen; (*b*) external drum.

pixels use a single point source and may have less intensity variation. Due to the diffraction-limited nature of light, greater depth of focus can be achieved.[34] Figure 26.19 illustrates that imaging with a smaller laser spot size gives steeper aerial image sidewalls, increasing process latitude.[35] In addition to spot size, addressability of the beam is important in determining the system resolution, as shown in Fig. 26.20.

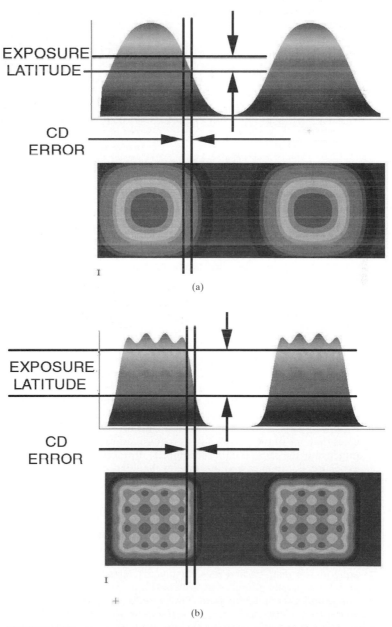

FIGURE 26.19 Computer simulation of aerial image for a 25-μm feature: (*a*) four 12.5-μm laser spots on 6.25-μm pixel spacing; (*b*) four 6.25-μm laser spots on 6.25-μm pixel spacing.[35]

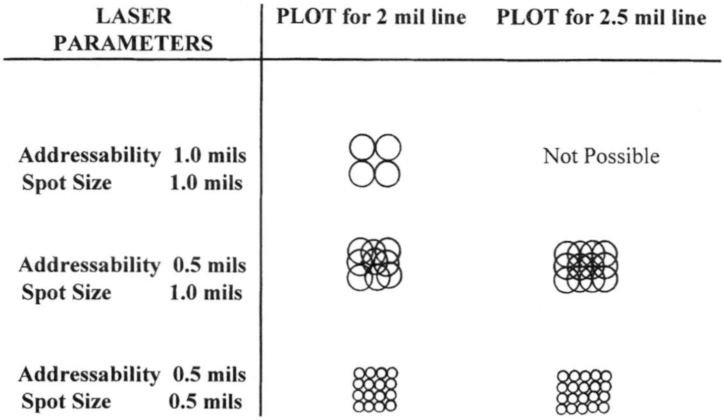

LASER PARAMETERS	PLOT for 2 mil line	PLOT for 2.5 mil line
Addressability 1.0 mils Spot Size 1.0 mils		Not Possible
Addressability 0.5 mils Spot Size 1.0 mils		
Addressability 0.5 mils Spot Size 0.5 mils		

FIGURE 26.20   Final image dimension as governed by spot size and addressability of a laser plotter or laser direct exposure tool.

Many factors must be considered in selecting an LDI system. Both the capital cost and operating cost vary among the various systems. However, the critical factor is that the equipment combined with the photoresist material is capable of achieving both the resolution needed for the product mix and the productivity needed for the production volume. High-speed photoresists are available that are compatible with the various metallization processes.

26.6.5 Develop

In this process step, the solubility difference between the exposed and unexposed areas of the photoresist creates a relief image of the master pattern. The panel is immersed in an appropriate solvent, and the process conditions are adjusted to control the clearing time for dissolving the unexposed areas for negative photoresists or exposed areas for positive ones. Total dwell time is set to approximately double the time to clear, commonly called a 50 percent breakpoint. Solution concentration, temperature, and agitation are key variables. The resulting photoresist images should be distinct, with vertical sidewalls. Failure to achieve this indicates that one of the previous steps requires adjustment. For images larger than the phototool dimension, the cause is either incomplete development, overexposure, or poor contact during exposure. For images that are smaller than expected, either the exposure dose is too low or development is too aggressive. Distorted images can be caused by problems with preclean, application, or exposure.

The common equipment for developing is spray conveyorized, either horizontally or vertically. Additives are used in the developer solution to prevent foaming. The solution is filtered to remove resist particles and either replenished with fresh solution to maintain a consistent dissolved resist content and solution concentration or operated continuously for a certain amount of product and then replaced. Waste-developer solution is treated (aqueous and semi-aqueous) or distilled and reused (solvent). Rinsing is important in stopping the dissolution and, for aqueous photoresists, water with a high-mineral content often improves the resist image and the conductor yield. Tank systems can also be used with photoresists that have a wide-process latitude. Ultrasonic agitation is often used to aid in the dissolution.

There are additional steps that improve the resist removal in the line channels and the conductor formation yield. Plasma treatment has been used effectively to improve product yield, especially with respect to shorts in a print-and-etch process. In addition, for some aqueous-developable dry films, a heat treatment after exposure has improved the space definition, and spaces equal to or smaller than the resist height have been resolved. Thus, these process steps ensure that tight resolution requirements can be met.

26.6.6   Strip

After the pattern transfer has been completed, the photoresist is removed from the substrate using equipment similar to that used for development. The stripping solution swells and dissolves the photoresist, stripping it either in sheets or as small particles. The equipment design must effectively remove and separate the skins. Often, brushes and ultrasonic agitation are added to aid in the resist removal. As with developing, filtration is important to keep the spray nozzles clean and keep fresh solution reaching the panel. For stripper chemistries that oxidize the copper, an antitarnishing agent is often added either to the stripping solution or as part of the rinsing.

26.7   *DESIGN FOR MANUFACTURING*

As design features are continually reduced to produce higher-density interconnects, tighter control of every step in the conductor formation process is required to achieve high yield. The maximum possible yield with an etching or plating process depends on the conductor dimensions such as the conductor pitch in terms of line and space, the conductor thickness, and the size and shape of the capture pads around plated through-holes and vias.

26.7.1   Process Sequence: Etching vs. Plating Considerations

For a specific circuit design, there is often a question of the appropriate process sequence. Although the capabilities of imaging and the pattern transfer step dictate the overall limitations, the decision depends on the unique capabilities of the manufacturing line to be used. Some general considerations can clarify the true issues for most production situations.

Photoresist patterning has different constraints for etching and plating. For etching, a thin photoresist is desirable to maximize the etchants' attack in the channel developed in the photoresist. Resolution does not limit the process, since very small spaces can be resolved in either liquid or thin dry-film photoresists. The etching process itself is key to conductor formation. Etching the copper from the channel becomes increasingly difficult as feature size is reduced, particularly for thicker metal layers. Therefore, the key criteria defining the capabilities of the print-and-etch process is the minimum space cleared, constrained by the final conductor height and the etchant chemistry and equipment.

For pattern plating, the photoresist thickness must be greater than or equal to the final thickness of the conductor, and it must be possible to develop a photoresist channel equal to the final conductor width. As conductor thickness increases, it becomes more challenging to produce channels with a higher aspect ratio. Since photoresist resolution is on the order of its thickness, it would be difficult to resolve smaller than a 38-μm photoresist channel for 25-μm-thick plating, irrespective of the final conductor spacing. Thus, the challenge for pattern plating is to resolve and develop fine channels in thicker photoresist materials and then ensure that the plating solution wets the bottom of the narrow photoresist channels.

Thus, in choosing the conductor formation process, the thickness of the conductor and the linewidth and spacing are the key parameters. A generalized relationship between them is found in Fig. 26.21. It is important that this type of plot is known for the production area before the process sequence is determined.

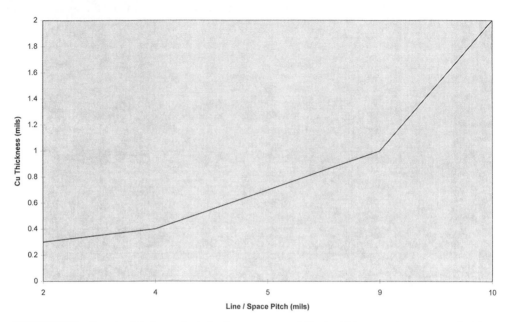

FIGURE 26.21   Sample plot of manufacturing line process capability to aid in processing sequence decisions. Feature dimensions above the line would be processed with pattern plating; those below the line, with print and etch.

26.7.2   Line and Space Division for a Fixed Pitch

It is common for product designs to have a fixed pitch, whether it refers to the I/Os for direct-chip or packaging attachment or to the spacing between PTHs. This space is often divided equally between conductors and spacing. For etching or pattern-plating processes, avoiding their respective resolution limitations can increase the pattern transfer yield.

In etching, the etchant undercuts the photoresist and an etch factor is used to obtain straighter sidewalls and improved linewidth control. If lines and spaces are equally allocated in the circuit design, then the photoresist must resolve a smaller space than a line. This is difficult for fine features, except for very thin photoresists. For fine pitch, having a larger space than line will give higher yield. For pattern plating, the spacing in the photoresist is the limiting factor. After pattern transfer, this will become the line. In this instance, equal line and space is more acceptable, but based on the photoresist concerns, a wider final line than space is preferred. At the same time, it is desirable to increase the spacing for reduced incidence of line shorting, which translates to a wider "line" in the photoresist. The former is usually more important. Therefore, for both processes there is preferably balance between linewidth and spacing.

26.7.3   PTH Capture Pad Size and Shape for Optimum Line Formation

Just as the conductor line and spacing can be optimized, the PTH capture pad shape and size can also be altered to increase the yield. The absolute dimension of the feature is dictated by the placement accuracy of the drilling process and the overall dimensional stability of the product. The feature is sized to ensure that the PTH and conductor are connected. There are varying specifications as to the extent of capture that is required.

Depending on the direction and magnitude of the dimensional stability and the drill wander and accuracy, the shape required to capture the PTH can be changed. This would reduce the size of this feature in at least one direction. In consequence, the spacing between the line and pad increases. Since this location is a change to the nominal line-to-line spacing, the narrowing of the channel results in shorting between the features in both a print-and-etch process and additive processes. In the former case, it is more difficult to clear the space, and in the latter, the narrower resist width is often underplated. Thus, when possible, an elongated pad will benefit the final conductor yield.

REFERENCES

1. L. R. Wallig, "Image Transfer," in *Printed Circuits Handbook,* 3d ed., Clyde Coombs (ed.), McGraw-Hill, New York, 1988, pp. 11.2–11.3.

2. R. Stoll, "High Definition Imaging," *PC Fab,* vol. 17, no. 6, June 1994, p. 31.

3. E. Hayes, "An Overview of Dry Film Imaging Chemistry," *PC Fab,* May 1988, p. 74.

4. M. R. McKeever, "Laser Direct Imaging—Trends for the Next Generation of High Performance Resists for Electronic Packaging," *Proceedings, S12-1, 2000 IPC Expo and Technical Conference,* San Diego, CA.

5. J. H. Choi, "Chemistry and Photoresist for Electroless Deposition," *Presentation at the IPC Spring 1992 Conference,* Bal Harbor, FL.

6. U.S. Patent 5,268,260, Nov. 7, 1993.

7. T. C. Sutter, "Liquid Photoresist Systems—An Overview," *The Board Authority* supplement to *CircuiTree,* vol. 1, no. 3, October 1999, p. 22.

8. J. Gangei, "Pushing the Envelope in Innerlayer Primary Imaging," *The Board Authority* supplement to *CircuiTree,* vol. 1, no. 3, October 1999, p. 30.

9. I. Taff and H. Benron, "Liquid Photoresist for Thermal Laser Direct Imaging," *The Board Authority* supplement to *CircuiTree,* vol. 1, no. 3, October 1999, p. 66.

10. R. J. Almond, M. E. Goewey, and B. F. Jobson, "Lowering Innerlayer Fabrication Cost with Liquid Resists," *IPC Spring 1995,* San Diego, CA.

11. M. Gurian and N. Ivory, "Performance Requirements of Primary Liquid Resists," *Electronics Packaging & Production,* March 1995, p. 49.

12. J. Murray, "ED Processes Revisited," *PC Fab,* May 1992.

13. H. Nakahara, "Electrodeposition of Primary Photoresists," *Electronic Packaging & Production,* February 1992, p. 66.

14. K. H. Dietz, "Surface Preparation for Primary Imaging," *Presentation and Paper at IPC Spring 1992 Conference,* Bal Harbor, FL, p. TP-1025.

15. W. Moreau, *Semiconductor Lithography,* Plenum Press, New York, 1989, pp. 651–664.

16. S. Crum, "Surface Preparation Process Improvements," *Electronic Packaging & Production,* July 1993, p. 24.

17. Atotech technical information.

18. H. Yamada, "Wet Lamination Technology and New Application For Image Transfer," *Poster Presentation at Electronic Circuits World Convention 8,* September 1999, Tokyo, Japan.

19. R. Patel and H. Benkreira, "Gravure Roll Coating of Newtonian Liquids," *Chem Eng Sci,* vol. 46, no. 3, 1991, p. 751.

20. C. Young, "The In Line Spray Conformal Coat Process," *Proceedings of the Technical Program, NEPCON West '89,* Cahners, Des Plaines, IL, vol. 2, p. 1676.

21. D. Marks and T-C. Lee, "Electrostatic Applications of Resists and Solder Masks," *PC Fab,* September 1991, p. 100.

22. J. Dubrava, D. Pai, J. Rychwalski, and J. Steper, "Development of Novel Positive-Working Electrode-posited Photoresist Process for the Production of High Density Outerlayers," *Presentation at the IPC Spring 1994 Conference,* Boston, MA.

23. J. Nguyen, "Curtain Coating in Soldermask Application," *Proceedings of the Technical Program, NEPCON West '89,* Cahners, Des Plaines, IL, vol. 1, p. 869.

24. S. Cox, "Wide Exposure Latitude Photoresists and Fine Line PWB Manufacturing," *The Board Authority* supplement to *CircuiTree,* vol. 1, no. 3, October 1999, p. 34.

25. W. R. Grace & Co., ACCUTRACE® technical information.

26. H. G. Muller, Y. Yuan, and R. E. Sheets, "Large Area Fine Line Patterning by Scanning Projection Lithography," *IEEE Transactions on Components, Packaging and Manufacturing Technology Part B: Advanced Packaging,* vol. 18, no. 1, 1995, p. 33.

27. N. B. Feilchenfeld, P. J. Baron, R. K. Kovacs, D. T. W. Au, and R. D. Rust, "Further Progress with Magnified Image Projection Printing for Fine Conductor Formation," *Presentation at the IPC Fall 1995 Conference,* Providence, RI.

28. K. H. Dietz, "Alternatives to Contact Printing," *CircuiTree,* May 1999, p. 120.

29. C. Vaucher, "Solid or Gas?" *CircuiTree,* January 2001, pp. 96–97.

30. D. Stone, "Use of a Solid State Laser to Expose High Speed LDI Resists in PCB Fabrication," *Presentation at EPC 2000 Conference,* Maastricht, Belgium, October 4, 2000.

31. Siemens news release at IPC Expo 2000, San Diego, CA.

32. J. Kickelhain, "New Excimer Laser Technology—Ultra Fine Lines (15 μm) without Etching," *Poster Presentation at Electronic Circuits World Convention 8,* Tokyo, Japan, September 1999.

33. S. Waxler and S. Spinzi, "Direct Imaging Implementation: Real World Case Studies," *Presentation at IPC Printed Circuit Expo 2000,* San Diego, CA.

34. M. Kesler, "Direct Write for HDI Substrates," *Presentation at HDI Expo,* Phoenix, AZ, September 2000.

35. J. M. Tamkin, "The Impact of HDI on Fine-Line Lithography," *The Board Authority* supplement to *CircuiTree,* October 1999, vol. 1, no. 3, p. 59.

CHAPTER 27
MULTILAYER MATERIALS AND PROCESSING

C. D. (Don) DuPriest
Lockheed Martin Missiles and Fire Control, Dallas, Texas

27.1 INTRODUCTION

The structure of multilayer printed wiring board (PWB) has evolved immensely. While the building block processes of the PWB industry are still prevalent, the sequences and material utilization employed continue to emerge into new technologies. The driver for this is the "density revolution," which is caused by changes in electronic component technology. Since the circuit board was born as a means to an end, that is, to provide electrical connectivity between components, it has had to evolve to meet new demands as the components evolve. Nowhere is that more evident than in the materials and processes used to make multilayer PWBs. It is that need for interconnectivity that brings the challenge.

One of the components most responsible for the enabling of advanced technologies is the ball grid array (BGA). The on-board density of the area array component family creates a unique challenge to the circuit routing of the multilayer configuration. This new configuration, known as high-density interconnect (HDI), drives the new structure of PWB. High-density designs, driven by the need for smaller, faster, cheaper products in all fields, have common issues to deal with, e.g., high operating signal speeds (1 GHz and higher), thermal management, and mechanical fatigue, all of which affect multilayer construction. In addition, life cycle management is critical in determining the correct multilayer board technology (MLT) to employ. Product requirements place serious demand on all aspects of the MLT process, e.g., the materials employed, the design techniques used, and the process methods implemented.

This chapter discusses the multilayer materials, construction options, and processes of the typical PWB configuration. The discussion provides the necessary background information to build a foundation for the advanced MLT designs. Section 27.2 provides discussion of the critical properties and various resin systems available. Material performance specifications introduced highlight the separation of standard materials from those employed in high-density applications. Section 27.3 provides insight into the options available to meet specific performance criteria and density requirements. Section 27.4 documents the key manufacturing steps that are employed in basic multilayer printed wiring board (ML-PWB) construction and the introduction to some of the methods of advanced MLT build-ups.

27.2 MULTILAYER PRINTED WIRING BOARD MATERIALS

A basic understanding of material performance is necessary for both the designer and fabricator of the PWB (for a detailed discussion of the various materials used in PWB manufacturing, see Chaps. 5 to 10). Understanding the materials available is one of the primary elements in designing for manufacturability. It is necessary to match the product end use performance requirement as well as the environmental exposure the board experiences in the fabrication and circuit card assembly (CCA) processes. It is common for boards to experience as many as five thermal excursions in CCA. To facilitate the many choices of laminates and their associated properties, industry standards groups such as the IPC have defined minimum performance specifications. Several specifications are available to facilitate the selection process. Some of the most commonly used material specifications are those that deal with laminate/prepreg and copper foil. IPC-4101, "Specification for Base Materials for Rigid and Multilayer Printed Boards," and IPC-4652, "Metal Foil for Printed Wiring Applications," are the primary specification for clad laminates, prepregs, and foils. Another specification, IPC-4104, "Specification for High Density Interconnect (HDI) and Microvia Materials," deals with many of the new materials for HDI, e.g., the epoxy coated microfoils, as discussed in this chapter. Tables 27.1 and 27.2 list some of the materials available by slash sheet, the specification method by which properties are itemized. Table 27.1 shows three key elements defined in slash sheet IPC-4101, while Table 27.2 show these elements defined in slash sheet IPC-4104.

TABLE 27.1 Slash Sheet IPC-4101 Series

Slash sheet IPC-4101	T_g (°C)	Resin system	Reinforcement
4101/21	110	Epoxy, flame resistant	Woven E-glass fabric
4101/24	150–200	Majority epoxy modified or unmodified, flame resistant	Woven E-glass fabric
4101/40	≥200	Polyimide	Woven E-glass fabric
4101/41	≥250	Polyimide	Woven E-glass fabric
4101/42	200–250	Polyimide	Woven E-glass fabric
4101/53	≥220	Polyimide	Nonwoven aramid
4101/55	135–190	Modified epoxy	Nonwoven aramid
4101/70	≥230	Cyanate ester	Woven S-2 glass fabric
4101/71	≥230	Cyanate ester	Woven E-glass fabric

TABLE 27.2 Slash Sheet IPC-4104 Series

Slash sheet IPC-4104	T_g (°C)	Resin system	Reinforcement
4104/12	>140	Epoxy	N/A
4104/19	240	Polyphenylene ether	N/A
4104/20	180	Epoxy blend	N/A
4104/21	150	Epoxy blend	N/A
4104/22	120	Epoxy liquid, epoxy coated	N/A
4104/23	240	Epoxy	Nonwoven aramid

27.2.1 Critical Properties

In addition to providing circuit interconnection, an ML-PWB provides the electrical and mechanical platform for the system. This means that the electrical and thermal properties of the ML-PWB material are very important for the proper functioning of the system. Among the properties of importance are dielectric constant D_k, dielectric loss (tan δ), glass transition temperature T_g, degradation temperature T_x, and moisture absorption. The following sections discuss the importance of these and other properties of an ML-PWB substrate.

27.2.1.1 Dielectric Constant D_k. Impedance and transmission speed are both affected by D_k. Since impedance must be matched throughout a high-speed circuit, this means that the impedance of other circuit elements determines the required impedance of a circuit line. The designer can use standard equations to select values of linewidth, dielectric thickness, and dielectric constant that will achieve the desired impedance (often 50 Ω). However, factors such as interconnection density and ML-PWB imaging capability generally determine the linewidth. This means that the trade-off available to the designer is dielectric thickness vs. dielectric constant, with thinner dielectrics requiring lower D_k values. Since thinner layers have lower cross talk and give thinner, more reliable ML-PWBs, there is a serious need for low-D_k materials.

Transmission speed is important because the signal transit time affects device timing and determines at what circuit length transmission line effects become important. The transmission speed of an electromagnetic wave in a dielectric medium is the speed of light divided by the square root of D_k. Air has a D_k of 1.0, and electromagnetic waves travel at the speed of light (12 in/ns). At 1 GHz, standard FR-4 ML-PWB materials have a D_k of 4.4 and the transmission speed is reduced to 6 in/ns. With low-D_k materials, transmission speeds reach 8 in/ns. Although this is a small improvement, the higher transmission speed provided by a low-D_k material may be important in some high-speed applications.

27.2.1.2 Dielectric Loss (tan δ). The energy absorbed by the dielectric media is measured by tan δ. Attenuation is proportional to tan δ and signal frequency. For standard FR-4 ML-PWB materials, tan δ is 0.02, which translates into serious losses at frequencies above 1 GHz. For circuits operating at 1 GHz or higher, a material such as PTFE with a tan δ of 0.001 is preferred.

27.2.1.3 Moisture Absorption. Moisture is the enemy of an ML-PWB. Absorbed water raises D_k, expands the board, and causes thermal defects such as substrate blisters and barrel cracking during soldering. Increases in D_k are generally small, but the resulting impedance drift can be a problem in a circuit with a critical impedance requirement. In-plane expansion can approach 1 mil/in and cause registration problems during stenciling and device placement. Thermal defects during assembly and rework soldering can cause serious yield problems. The severity of these problems depends on storage condition prior to soldering. If storage times are short or if the humidity is low, moisture will not be a serious issue. However, if the ML-PWB is subjected to high humidity for several months or more, special care must be taken. This can include the use of moisture-resistant materials, baking prior to assembly, or storage with a desiccant.

27.2.1.4 Thermal Stability. The important thermal properties of a laminate are the glass transition temperature T_g and the degradation temperature. These properties measure how well the material performs in the assembly soldering and in high-temperature use conditions. The T_g of a resin is the temperature at which the resin reversibly changes from a glassy state to a rubbery state. This loss of modulus creates an effective limit on the operating temperature of the system. T_g also affects the thermal fatigue life of the plated holes in the ML-PWB. Higher values of T_g translate into greater thermal cycle life. The degradation temperature is a measure of the temperature where the epoxy begins to degrade irreversibly. This is generally much higher than T_g. One measure of degradation is the time at 260°C (called the T_{260} time) when expansion due to delamination is detected by a thermal mechanical analysis (TMA). For common materials the T_{260} varies from a few minutes to hours. Another measure is the degradation temperature T_x where a 50 percent weight loss is seen in a TMA scan operating at 2°C/min. Generally the T_{260} time and T_x are correlated with each other, but not correlated to T_g.

27.2.1.5   Coefficient of Thermal Expansion (CTE).   The CTE for standard FR-4 (14 to 20 ppm/°C) is higher than those for ceramic and silicon. The resulting thermal expansion mismatch between an ML-PWB and assembled devices can lead to solder joint fatigue failures when the system undergoes multiple heat cycles during power-up and power-down. Packages with compliant leads can accommodate the CTE mismatch, which allows the system to use standard ML-PWB material systems.

In cases where the device package (leadless area array with ball grids) is incompatible with the CTE of a standard material, a low-expansion substrate must be used. This can be achieved in several ways. In the past, leadless SMT focused on replacing the woven glass in standard FR-4 with woven quartz or aramid fibers. Although this reduced the expansion of the substrate, both materials were expensive and difficult to process. Another past approach was to laminate a low-expansion metal such as Invar into the ML-PWB. The Invar plane could double as a heat sink or a power/ground plane. This approach is in limited use and is difficult to process, requiring a backfill operation to insulate the hole wall. A better solution for CTE match is to use a nonwoven aramid mat material that is resin impregnated. Nonwoven aramid comes in either modified epoxy or polyimide types. This material solves the processing problems and is compatible with laser drilling.

27.2.2   Standard Properties of the Epoxy Systems

Materials of the GF epoxy family are in common use. This is a highly cross-linked brominated epoxy resin reinforced with woven glass cloth. The bromine is reacted with the epoxy matrix and is used to provide fire retardancy. Most GF materials satisfy the UL classification of V-0 for fire retardancy. There is some environmental push to remove bromine from laminate production. A clear substitute has not been named at the time of this writing. GF epoxy materials are sold by many suppliers, and have become essentially a commodity material.

Two resin systems are used to make GF laminate: difunctional and tetrafunctional. These systems are distinguished by the nature of the epoxy cross-linking. In a difunctional system, the epoxide molecule has two cross-linking sites, and the cured epoxy contains long linear molecular chains. Pure difunctional laminates have excellent physical properties, and for many years were the mainstay of the industry. They have a T_g of 120°C, which is adequate for most use environments, but is low for some applications.

The epoxide molecule of a tetrafunctional epoxy has more than two cross-linking sites. This allows a high cross-link density and a high T_g. A pure tetrafunctional system is expensive and difficult to work with. To meet the need for a T_g above 120°C, laminators blend difunctional and tetrafunctional resins, producing a mixture referred to as multifunctional. Around 1985, some laminators began to sell a multifunctional epoxy blend with a T_g between 130 and 145°C. This blend was called tetrafunctional, even though it actually contained both di- and tetrafunctional epoxies. This blend is available at little or no price premium over a difunctional laminate. Laminates with a higher fraction of tetrafunctional resin are available. These systems, which are called multifunctional, have T_g values in excess of 170°C and are sold at price premium of approximately 10 percent over difunctional systems. Multifunctional blends often have lower moisture absorption and higher thermal degradation temperature than difunctional systems. However, in some multifunctional systems these properties are not improved. Care must be taken in selecting a multifunctional system to ensure that all properties of importance are enhanced.

27.2.3   Materials with Enhanced Thermal Properties

Modern ML-PWBs often operate at elevated temperatures due to the heat output from semiconductor devices. When the system is turned off, the board sees a large thermal cycle. As boards become thicker and holes smaller, these thermal cycles result in an increasing threat to the reliability of plated holes. For example, plated holes have been shown to fail when subjected to multiple thermal cycles up to temperatures near the T_g of the base material. These cycles can easily occur when a high-power device is turned on and off. One solution to this problem is to use materials with a higher T_g.

ML-PWBs often have surface-mount devices on both sides of the board, and they receive three or more solder operations during the assembly of connectors and devices. In addition, because of the value of a completed assembly, the board must be able to withstand additional soldering operations needed for occasional removal and replacement of defective devices. Boards made with difunctional GF epoxy can suffer from lifted lands, cracked PTH barrels, or substrate blisters during these multiple soldering operations. The solution is to use materials with low moisture absorption and high thermal degradation temperature.

As indicated earlier, multifunctional epoxy is emerging as an inexpensive way to enhance thermal performance. With a T_g in excess of 170°C, multifunctional epoxy boards have excellent PTH fatigue life when subjected to thermal cycles up to 150°C. However, as noted, some multifunctional epoxies offer little or no improvement in moisture absorption or thermal degradation. There are three alternate resin systems commonly available for applications requiring thermal properties superior to those available from multifunctional epoxies: polyimide (PI), cyanate ester blends, and polyphenylene oxide (PPO) blends.

27.2.3.1 Polyimide.

Probably materials in the GI polyimide family provide the best thermal stability. This family of materials has a T_g in excess of 250°C, a high thermal degradation temperature, and a CTE less than that of epoxy. Polyimide resins can be coated on glass fabric to produce an ML-PWB substrate that processes like epoxy GF. With a high T_g and high thermal degradation temperature, polyimide ML PWBs provide high temperature reliability. For systems that must operate at temperatures above 200°C, polyimide is a good choice. The combination of a high T_g and a relatively low CTE results in excellent fatigue life for plated holes. This makes polyimide a good candidate for very thick boards and for applications where the system must survive multiple thermal cycles over a wide temperature range.

Polyimides have several disadvantages. Early versions used a solvent called MDA, which is widely viewed as carcinogenic. Although there was no evidence of MDA release during ML-PWB fabrication or use, many fabricators refused to process MDA-containing polyimide. Fortunately material suppliers have been able to formulate MDA-free versions of polyimide, and most polyimide on the market today is rated as MDA free. A second disadvantage of polyimide is fast moisture uptake. This leads the problems discussed in Sec. 27.2.1.3. Generally, polyimide users combat moisture by including multiple bakes in the board fabrication and assembly processes. The third disadvantage of polyimide is cost. The cost of a polyimide laminate is up to two to three times that of GF. In addition, ML-PWB fabrication requires a high-temperature lamination.

Polyimide material has been used in low volume since the early 1980s. Broad acceptance was always limited by cost and fabrication issues. This material family should only be specified for systems that must operate with high reliability in an extreme environment, such as a high ambient temperature or extremes in thermal cycling. This includes some military applications and potentially a few consumer applications such as under-hood automotive electronics. For most commercial applications, lower-cost materials—such as conventional epoxy blended with CE or PPO resin—may be used.

27.2.3.2 Cyanate Ester Blends.

The second most stable resin system is the triazine or cyanate ester family. In its pure form, the cyanate ester resin is brittle and is difficult to drill without cracking. Low peel strengths are typically associated with its use. In addition, cyanate ester is an expensive resin system. As a result, cyanate ester resin is often blended with epoxy and a small amount of polyimide. This blend, called BT after its two ingredients, bismaleimide and triazine, can be coated on conventional glass cloth to produce a laminate.

BT laminates have a T_g of 180°C and a high degradation temperature. For most high-temperature applications, they are a direct substitute for polyimide. They have the added advantage that the moisture sensitivity and processibility of BT are much closer to those of conventional epoxy than those of polyimide. Other than a high-temperature postlamination bake required for a full cure, the BT process is the same as an epoxy process. The major drawback of the BT laminate is cost. Although BT is much less expensive than polyimide, it is still up to 1.5 times the cost of epoxy-based GF. The result is that BT is a popular replacement for polyimide, but its use is limited to specialty applications. The trend for most high-temperature commercial applications is either a multifunctional epoxy or an epoxy-PPO blend.

27.2.3.3 Polyphenylene Oxide (PPO)-Epoxy Blends. One of the cost-effective high-temperature laminates in use is a material based on a blend of PPO and epoxy resin. This material has a T_g of 180°C and a T_{260} of 1 h or more. When coated on glass, the PPO blend processes like epoxy. The only exception is the need for a high-temperature bake to achieve full cure. The major advantages of this laminate family are very low moisture absorption and a cost that is only 20 to 50 percent higher than that of conventional epoxy. The disadvantage of PPO is that it has a broad T_g with softening beginning well below 150°C. This has an adverse effect on the ultimate fatigue life of high-aspect-ratio holes.

PPO blends are relatively new, but they have the potential to capture much of the demand for high-temperature materials, including some that is now being met by multifunctional epoxy systems. They are an excellent candidate for any application requiring high operating temperatures or moisture resistance. Care should be taken in using this material in applications where the system must operate over a very large number of thermal cycles that include temperatures in excess of 130°C.

27.2.4 Materials with Enhanced Electrical Properties

The resin and the reinforcement used determine the D_k and tan δ of a composite material. Standard GF epoxy, made up of epoxy resin and woven glass, has a D_k in the range of 4.4 and a tan δ of 0.02. This may be reduced somewhat by replacing some or all of the epoxy resin with PPO, cyanate ester, or PTFE. Further reductions require replacing the glass reinforcement.

27.2.4.1 Cyanate Ester Blends. Both the tan δ and the D_k of the cyanate ester resin system are much lower than those of epoxy. The best results are achieved with pure cyanate ester. However, as mentioned earlier, this resin is brittle and difficult to drill. The BT blend provides an excellent compromise. BT has a D_k of 2.94 and a tan δ of 0.01. This improvement is useful in some applications; however, it is inadequate for many high-performance applications and the approximately 100 percent cost penalty for BT has prevented its broad acceptance.

27.2.4.2 PPO Blends. As with cyanate ester, both the tan δ and D_k of the PPO resin are lower than those of epoxy. For processibility and cost reasons, a blend of approximately 50 percent PPO and 50 percent epoxy is generally used. This gives electrical performance similar to that of BT, but with a cost penalty of only 20 to 50 percent above standard GF epoxy. It is likely that PPO will find wide use in those applications such as high-performance workstations that need a small improvement in electrical performance. Unfortunately, the characteristics of a PPO-epoxy blend are often inadequate for very high-speed applications such as supercomputers and wireless (RF) applications.

27.2.4.3 PTFE-Based Laminates. This resin system is commonly known as Teflon.* Of all resin systems in common use, the best electrical performance is provided by PTFE. The D_k of PTFE is close to 2.0 and the tan δ is less than 0.001. This is a significant improvement over all other materials. In addition to excellent electrical properties, PTFE has excellent thermal properties. It is a thermal plastic that can operate at temperatures above 300°C without softening, oxidation, or other forms of degradation. It is naturally fire retardant, so it does not need bromine addition to achieve the UL rating of V-0. It has very low moisture absorption.

---

* Teflon is a registered trademark of E. I. DuPont de Nemours & Co.

For ML-PWB applications, PTFE has three serious disadvantages:

1. *Poor processibility:* Because PTFE is a high-temperature thermoplastic, it is generally laminated in a high-temperature lamination cycle not achievable in conventional lamination presses. This can be avoided by using a low-temperature adhesive, but this can lead to serious compromises with electrical and thermal properties of the laminate. The second major difficulty in processing PTFE is the hydrophobic nature of the material. This makes it difficult to clean and metallize the holes. Generally plating must be immediately preceded by a special fluoride etch (TETRA-ETCH*) that activates the hole wall and enhances wetting. As an alternative, some PTFE systems can be activated in plasma etch; however, this does not work with all PTFE materials.

2. *High CTE:* PTFE maintains full strength up to and above solder temperature. When this strength is combined with a high CTE, PTH reliability is a serious concern. This is a particular problem with glass-reinforced PTFE, where the x-y constraint of the glass magnifies the out-of-plane expansion. A second problem attributable to high CTE is excessive warpage whenever Teflon® is combined with GF in an unbalanced hybrid structure.

3. *Cost:* PTFE-based laminates are often 100 times more expensive than epoxy-based GF. This makes it difficult to justify the use of PTFE if any alternative exists.

At least three types of laminate systems are available with PTFE resin:

1. *PTFE resin coated on a conventional glass mat:* Although the construction of this laminate is most like standard GF, it is not recommended for ML-PWB applications. The large CTE of PTFE coupled with the in-plane strength of the glass reinforcement leads to a very high out-of-plane expansion that can easily fracture a plated hole during soldering. In addition, the glass cloth raises the D_k of the composite. This material is commonly used in applications such as antennas or RF microstrip cables where the circuit is placed on one side of the substrate and a ground plane on the other. Plated holes are not needed, and a very low D_k is not critical.

2. *Expanded PTFE film impregnated with cyanate ester or epoxy resin:* This material has no glass cloth reinforcement. The resulting increase in in-plane CTE reduces the out-of-plane expansion, allowing the reliable use of plated holes. Although the non-PTFE resin impregnating the PTFE film increases D_k and tan δ over those of a pure PTFE layer, this composite is at least as good as glass-reinforced PTFE. The conventional resin surface of this material means that it can be laminated using conventional methods and materials, allowing the fabrication of ML-PWBs containing a mixture of PTFE and epoxy layers. The resulting hybrid board can contain any symmetrical mix of epoxy and PTFE layers. The high in-plane CTE of the PTFE layer leads to severe warp in any nonsymmetrical stack-up. Hybrid boards are less expensive than pure PTFE boards because they are laminated in a conventional press and they minimize the use of costly PTFE layers. However, the high cost of PTFE and the need for a fluoride etch still makes these boards expensive. The major application for this material is in supercomputer designs that have high-speed circuits requiring enhanced electrical properties mixed with others that operate on FR-4. This approach does not work in applications where a mixture of RF signals and digital logic requires a nonsymmetrical design with low-loss material on one side and standard material on the other.

3. *PTFE mixed with a low-D_k ceramic:* A third approach, in which up to 60 percent ceramic (by weight) is combined with PTFE resin, results in a very interesting material. This material has a low CTE, low tan δ, and low D_k, and, by a proper choice of the ceramic, it is possible to have an in-plane CTE that matches that of GF. This allows the fabrication of an unbalanced hybrid structure without excessive warp. The high ceramic loading typical of this material minimizes PTFE use and reduces costs. A commercial version of this material is available with a cost in the range of four to five times that of standard GF, rather than the

---

* TETRA-ETCH is a registered trademark of W. L. Gore & Associates.

100 times typical of other PTFE options. The only serious drawback of this material is the use of a special process such as plasma etch to ensure hole wall wetting. Success has been reported with using H_2 or He mixed with O_2 to activate the hole for plating.

27.2.5   Materials Summary

Material properties affect PWB operation in many important ways. The preceding discussion is focused on thermal and electrical properties. The use of small, high-aspect-ratio holes, thick boards, and high-power devices drives the need for materials with improved thermal properties. Similarly, the use of high-speed devices, including the RF circuits in wireless applications, drives the need for improved electrical performance. Materials exist with improved properties, but the ideal material that combines low cost with improvements in both thermal and electrical properties does not exist. Therefore the designer must consider the needs of each design before selecting a cost-effective material.

The need for improved thermal properties includes three needs: High T_g, high thermal decomposition temperature, and low moisture absorption. Polyimide meets the first two needs at the expense of increased moisture absorption and high cost. BT satisfies all three needs, but cost remains an issue. The most economical solutions are the multifunctional epoxies and the epoxy-PPO blends. Multifunctional epoxies are relatively inexpensive, provide a significant improvement in T_g, and, in some cases, provide small improvements in thermal degradation temperature and moisture absorption. The PPO blends provide significant improvement in all three areas at a slightly higher cost. If moisture absorption is an issue, the PPO blend may be the best choice; otherwise the multifunctional epoxies are likely to be the lowest-cost option.

Both BT and PPO-epoxy boards provide small improvements in electrical properties. If the real need is to reduce D_k so that the board thickness can be reduced, these materials may be useful. However, if dielectric loss is a serious problem, the only real solution is to use PTFE. Unfortunately, PTFE is very expensive and difficult to process, so it should be specified only where absolutely necessary. For example, RF circuits will nearly always require PTFE. The ceramic-PTFE blends used in a hybrid construction may be the most cost-effective solution for these applications.

In addition to the issues discussed previously, there are other problems that become critical in high-density designs, and many new materials are emerging. Desired material characteristics for high-density designs include the ability to assure low warp to facilitate solder stenciling. Also, materials need to be extremely flat across the device footprint. High-density material also needs to have very good dimensional stability. A need for registration tolerance often limits interconnect density and, when building with high-density features, a material with significantly better dimensional stability is desirable.

27.3   MULTILAYER CONSTRUCTION TYPES

The construction of the rigid multilayer PWB (ML-PWB) can take on many variations. In order to help categorize the various constructions, the IPC has developed industry PWB design specifications, defining them by class and type. Grouping the ML-PWB into categories facilitates the ability to communicate in a common format among designers and fabricators.

27.3.1   IPC Classifications

27.3.1.1   IPC-2221—Generic Standard on Printed Board Design

- *Class 1, general electronic products:* Includes consumer products, some computers and computer peripherals, and general military hardware suitable for applications where cosmetic imperfections are not important and the major requirement is function of the completed printed board or printed board assembly.

- *Class 2, dedicated service electronic products:* Includes communications equipment, sophisticated business machines, and instruments and military equipment where high performance and extended life are required and for which uninterrupted service is desired but is not critical. Certain cosmetic imperfections are allowed.

- *Class 3, high-reliability electronic products:* Includes the equipment for commercial and military products where continued performance or performance on demand is critical. Equipment downtime cannot be tolerated, and equipment must function when required for life support items or critical weapons systems. Printed boards and printed board assemblies in this class are suitable for applications where high levels of assurance are required and service is essential.

The feature size of the ML-PWB product further delineates the technology differences. The most predominant feature responsible for the division in types is the microvia. The following discussions are limited to methods and materials employed to fabricate some of the commonly produced ML-PWB types as well as some of the more advanced types.

27.3.1.2 IPC-2222—Rigid Organic Printed Board Structure Design.   This covers products with conventional feature sizes.

- *Type 3:* Multilayer board without blind or buried vias (see Fig. 27.1)
- *Type 4:* Multilayer board with blind and/or buried vias (see Fig. 27.2)

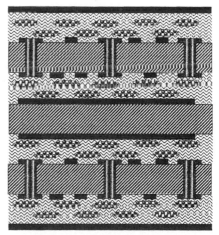

FIGURE 27.1 Type 3 (eight-layer ML-PWB) multilayer board without blind or buried vias.

FIGURE 27.2 Type 4 (eight-layer-B/V ML-PWB) multilayer board with buried vias.

27.3.1.3 IPC-2226—Design Standard for High Density Interconnect (HDI) Printed Boards.   This covers products with high-density feature sizes.

- *Type I:* 1[C]0 or 1[C]1—has through-vias connecting the outerlayers (see Fig. 27.3).
- *Type II:* 1[C]0 or 1[C]1—has buried vias in the core and may have through-vias connecting the outerlayers (see Fig. 27.4).
- *Type III:* >2[C]>0—may have buried vias in the core and may have through-vias connecting the outerlayers (see Fig. 27.5).

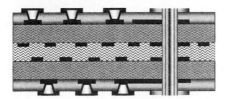

FIGURE 27.3 Type I (six-layer HDI ML-PWB) high-density multilayer board with blind vias from top and bottom layers and through-vias connecting the outerlayers.

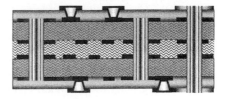

FIGURE 27.4 Type II (six-layer HDI ML-PWB) high-density multilayer board with vias as well as buried vias in the core and through-vias connecting the outerlayers.

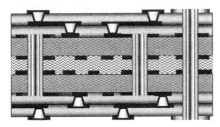

FIGURE 27.5 Type III (eight-layer HDI ML-PWB) high-density multilayer board with blind vias as well as buried vias in the core and through-vias connecting the outerlayers.

27.3.2 Basic Type 3 ML-PWB Stack-ups

The Type 3 construction can be said to be the most basic of PWB multilayer technologies. An ML-PWB is fabricated by the bonding (laminating) of clad details consisting of imaged laminates (typically double-sided). The bonding media, known as prepreg, is a B-staged (partially cured) reinforced resin. The imaged details consist of C-staged (fully cured) laminate. These material components are arranged by layering according to the design documentation. This layering method in fabrication, known as the stack-up, follows the layer numbering order of the design. The stack-up formation method is often loosely defined in the design documentation; therefore a good understanding of the lamination options is necessary. The lamination options described in this discussion refer to methods used to form the outerlayers and to form layer pairs. The material resin system should be predefined on the design documentation. Refer to IPC-2221/2222 for minimum suggested design documentation requirements.

The basic ML-PWB stack-up can be constructed with two options to produce the outerlayers and to form layer pairs. A third option that employs single-sided "capped" clads is rarely used and is not discussed. Often, in the case of designs having an odd number of layers, a combination of these methods is employed. The following three-layer stack-up methods are discussed:

- *Foil outerlayer construction (foil outer)*: A stack-up in which the outerlayers are formed by using a sheet of pure copper foil.
- *Clad outerlayer construction (clad outer)*: A stack-up in which an imaged clad detail provides the copper for the outerlayer.
- *Odd-layer construction:* A stack-up example of how to balance an odd-layer-count construction.

27.3.2.1 Foil Outer Stack-up. A board produced with copper foil outers is fabricated from one or more patterned innerlayer details and two copper sheets. The copper sheets form the outerlayers of the fabricated ML-PWB. This stack-up is the least expensive way to fabricate an

ML-PWB and is by far the most popular design option. Figure 27.6 shows a typical stack-up for a foil outer board consisting of eight layers.

The stack-up shown contains three imaged clad details bonded with two sheets of prepreg at each opening and having two sheets of copper foil on the outside. When possible, the higher-resin-content prepreg ply should face the signal layer side. This is especially important when the signal is a heavyweight copper (2 oz or more). The copper layers, numbered from L1 to L8, begin with the top foil layer (typically called the primary side). In the design pictured, the layers L2, L4, L5, and L7 represent signal layers. Layers L3 and L6 represent power/ground layers. When the final imaged patterned is produced, it can provide another signal layer pair. These now can be used for a "pads-only" pattern to support vias and component holes, or for the device footprints of the electrical component and their associated fan-out patterns.

Some of the advantages of this bonding method are:

- *Lower raw material cost.* The costs of loose sheet copper foil and prepreg sheets are more economical than those of clad laminate.

- *Lower consumable material cost.* The reduction in imaging resist and chemistry results in a cost savings.

- *Lower touch labor cost.* A reduction in material handling and process time in imaging and prelamination processing results in less labor.

27.3.2.2 Clad Outer Stack-up.   Figure 27.7 depicts the same eight-layer design as Fig. 27.6, except the outerlayers are formed with a clad laminate. This stack-up arrangement requires four

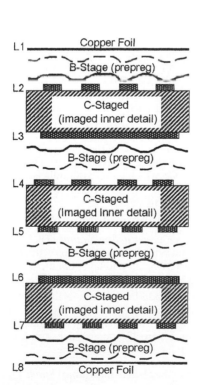

FIGURE 27.6 Foil outer stack-up (eight-layer ML-PWB), a typical stack-up for a foil board.

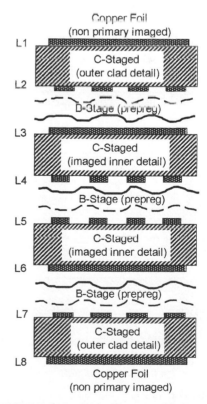

FIGURE 27.7 Clad outer stack-up (eight-layer ML-PWB) multilayer board design with the outerlayers formed with a clad laminate.

clad details, as compared to only three for the foil-stacked board shown in Fig. 27.6. This makes the clad outer board more expensive than the more commonly used foil outer construction. However, the clad outer design has one less B-stage opening and no copper foil. In addition, the clad outer boards are patterned on only one side prior to lamination. These factors partially offset the higher cost of this design. In Fig. 27.7, notice how the layer pairs have changed position placement. Layers L3 and L6 are now paired with a signal layer, which may be advantageous for controlling the dielectric thickness. See Sec. 27.3.2.4 for a discussion on controlled impedance.

Some of the advantages of using the clad outer bonding method include:

- *Improved surface topography.* When surface flatness is critical, the clad outer construction provides a smoother surface. When a signal layer of heavyweight copper is the first layer down, it is possible for weave texture imprinting to occur. This topography condition, sometimes referred to as "telegraphing," can sometimes imprint the circuit pattern when using a foil buildup.

- *Improved handling.* Depending on the layer pairing, sometimes additional copper can be retained, as in the case of the example, where a signal-to-signal imaged detail can be avoided. This is mostly the case when the layer pair details are being produced on very thin core laminate.

- *Improved dielectric thickness control.* Sometimes it is necessary to control the thickness at a specific layer pair tightly. This could be due to high voltage considerations or to aid in the precision spacing of a signal plane to a ground plane where controlled impedance is necessary. This is possible because the C-staged (fully cured) clad laminate typically has an improved thickness tolerance.

27.3.2.3 Odd-Layer Stack-up.

Sometimes, due to signal routing or the need to have a greater dielectric space, an odd-layer circuit is formed. Similar in characteristics to the clad stack-up, the odd-layer construction technique employs a nonimaged single-sided clad laminate. Material can be procured with a single side, but a warning is associated with its use due to a lack of bondability. When this material is manufactured at the laminator supplier with only copper on one side, a release sheet is used against the noncopper side. The release sheet produces a slick, smooth surface. This causes the noncopper surface to require an aggressive surface preparation to promote bondability. One alternative to gain the same benefit of the single-sided clad is to etch the copper off one side of a double-sided clad. The etched side becomes the bond surface, which has greater adhesion due to the copper tooth imprint left behind. See Fig. 27.8 for an example of a balanced odd-layer construction.

Some of the uses of the single-sided odd-layer stack-up are:

- *Odd-layer construction option.* The single-sided construction technique allows an opportunity to supply a layer for circuit designs having an odd layer count.

- *Odd-layer core balance.* One of the problems associated with a build-up of odd-layer circuits is the ability to maintain a balanced construction of the cores. Depending on the overall design thickness and layer count, this can become a major concern to reduce warping. Figure 27.8 depicts an example of an odd-layer stack-up with three equal core thicknesses.

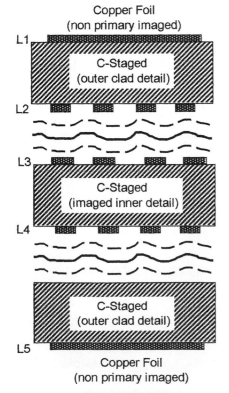

FIGURE 27.8 Odd-layer clad stack-up (five-layer ML-PWB), an example of a balanced odd-layer construction.

27.3.2.4 Controlled-Impedance Stack-up.

The number of ML-PWBs containing controlled impedance signals has increased significantly due to the growing number of applications in the telecommunications industry. The signal lines in an ML-PWB have a characteristic impedance value expressed in ohms. Active devices, connectors, and transmission lines also have characteristic impedance. In a high-speed circuit environment (typically starting at 2 to 3 MHz and above), if the signal energy is reflected rather than transmitted, the signal impedance mismatch is known to have discontinuity. This means that, to avoid signal losses due to reflections, all impedance characteristics must be matched. The physical relationship between a signal line and the surrounding power/ground layers determines the impedance of that signal line.

The controlled-impedance stack-up requires a specific manufacturing discipline that must be adhered to in order to maintain the integrity of the designed circuit. An ML-PWB design requiring a controlled-impedance fabrication is strongly recommended for a design for manufacturing (DFM) discussion between the designer and manufacturer in order to avoid producibility traps and induced error. Several resources are available to assist in developing the proper manufacturing parameters to ensure the product meets the specified tolerance. IPC-2141, "Controlled Impedance Circuit Boards and High Speed Logic Design," and IPC-317, "Design Guidelines for Electronic Packaging Utilizing High-Speed Techniques," are two of the industry standard documents that contain formulas and examples of various impedance configurations. Typical producible process tolerances for common impedance values are as follows:

- 50-Ω traces typically have a tolerance around 5 percent.
- 75-Ω traces typically have a tolerance around 15 percent.
- 100 Ω traces typically have a tolerance around 20 percent.

When tolerances are required to be tighter than this, it places more significance on the stack-up and etch of the circuit. The design requirement should be clearly specified in the ML-PWB documentation. The preferred way to specify an impedance requirement is in a drawing note rather than in applying dimensional boundaries. Firm dimensional boundaries do not allow the manufacturer freedom to meet the end requirement based on process knowledge. It is recommended that during the design process software modeling be performed and then confirmed with the manufacture before fabrication. Several excellent simulation software tools are available commercially. Some also exist as freeware and shareware. However, it is best to confirm the models' predicted values by actual process verification. The most versatile method of verification is with a test coupon. The test coupon should accurately represent the construction configuration of the stack-up, except it requires a few special prerequisites in order to yield reliable test data. Figure 27.9 shows a typical impedance test coupon stack-up. Note that the signal ref-

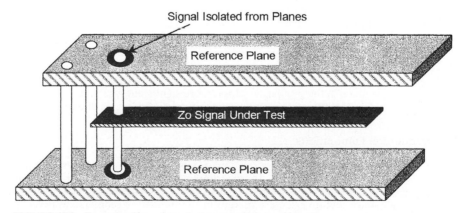

FIGURE 27.9   Example of impedance test coupon design and layout.

Symmetrical Stripline

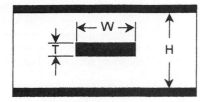

Offset Stripline

Edge-Coupled Symmetrical Stripline

FIGURE 27.10 Examples of common impedance stacks.

erence planes of the coupon are shorted together. Also, the simulated signal line is isolated and open-ended; the signal lines must be between 6 and 12 in long. The footprint of the coupon via holes that feed through to the surface should be verified to match to the spacing of the time domain reflectometry (TDR) instrument probe. This will reduce the risk of rendering the specimen not testable or necessitating the use of expensive adjustable probes. The IPC test method IPC-TM-650 2.5.5.7 contains suggested spacing, but it is best to confirm this with the manufacturer's preference. Examples of some common impedance types are shown in Fig. 27.10.

The factors most influential on characteristic impedance are:

- *Dielectric separation H:* The separation of the signal line between the reference planes has a significant influence on characteristic impedance. The variability of the dielectric layering must be reduced to minimize the effect on tolerance. This is where stack-up selection becomes critical when determining if the signal-to-plane opening will be made with a clad laminate or within the bondable area of the B-staged (prepreg) resin. This is where glass weave selection becomes important. The glass style and subsequent resin content will have different effects on the nominal thickness obtained.

- *Conductor width W:* The finished width of the conductor can produce variance, which is likely to occur from lot to lot. Therefore, process control measures are necessary when producing controlled impedance circuits. The density of neighboring circuits will have an effect on the final etch. Often it is advantageous to modify the artwork linewidth for predicted variance at phototool generation.

- *Dielectric constant:* Choosing a laminate resin system with a consistent dielectric constant has an influence on characteristic impedance. The influence of dielectric constant becomes most critical when the ML-PWB design is a high-layer-count design, i.e., backplanes. The lower the D_k value of the resin system, the thinner the overall board can be. Typically, the manufacturer has little opportunity to affect the dielectric constant because the material type is specified by the design. Here it is important that the designer/manufacturer know the D_k value range of the laminate supplier's resin system. Caution: do not use the D_k value of the neat resin, but that of the composite laminate, which will vary somewhat with glass style.

- *Conductor thickness T:* The conductor thickness or copper weight can also affect the final impedance value. Here, as in conductor width, manufacturing process variations can have an inverse effect on the precision of the impedance value. Some modern software simulation tools, such as Polar Instrument, Inc., allow values for conductor profile (area) to be included in the simulation. This is more significant when lines are of heavy copper. It is best to avoid routing of impedance lines on plated subcores due to the added variability.

27.3.3 Sequential Laminations

When a design includes buried and/or blind vias, it typically requires a set of sequential lamination and plating cycles. These technologies are defined in the industry design standard IPC-2221/2222 and are known as Type 4 ML-PWBs. These build types, when employing industry standard feature sizes, are mature in the industry. Complementing technologies, employing use of sequential processes for ML-PWB designs containing microvias ≤0.15 mm, are considered as build-up technologies. The terminology of build-up technologies encompasses many design stack-up variations. They can take on many forms and employ a multitude of new methods. The build-up technologies, defined in IPC-2315 and IPC-2221/2226, include new categories named Type I, II, III, IV, V, and VI. Many of the new advanced build-up technologies engage use of many new materials, which can be found in IPC-4104. This includes new materials for layer forming, dielectric insulation, and interconnectivity. These include photoimageable and nonphotoimageable materials (liquid, paste, or dry-film nonreinforced dielectric), adhesive-coated dielectrics (reinforced and nonreinforced), and conductive foils and paste (coated and noncoated, photoimageable). For a detailed discussion of these processes, see Chaps. 21, 22, and 23. This discussion will limit itself to addressing processes for standard technology Type 4 and advanced technology build-up Type I, II, and III, which can be manufactured with conventional processes.

27.3.4 The Buried Via Stack-up

To avoid hopelessly complex routing, each signal net is generally routed using only one pair of layers with what is called *Manhattan geometry*. This means that diagonal routing is avoided and all signal lines run in a horizontal or a vertical direction. To avoid blockage and side-to-side cross talk problems, horizontal lines are run on one layer and vertical lines on the other. This means that, in addition to a via at each end of the net (I/O via), most nets need one or two additional vias (routing vias) to change direction from horizontal to vertical. Figure 27.11 shows an example of Manhattan geometry.

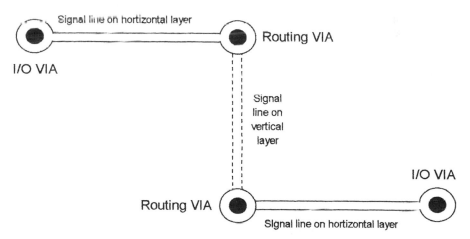

FIGURE 27.11   An example of a net routing using Manhattan geometry.

The net shown in Fig. 27.11 has two I/O vias and two routing vias. The I/O vias connect the signal lines to the board surface, where the net connects to an input or output point of an active circuit. The routing vias are used to change directions from horizontal to vertical. With through-hole vias, the routing vias pass through all layers, consuming valuable routing space. A high-layer-count board with many signal layer pairs can run out of via sites. In this case, additional layers will not improve routing completion and we say the board is via-starved.

One solution to a via-starved board is to use buried vias. The buried via connects two adjacent signal layers and provides routing vias without affecting routing on other layers. Figure 27.12 shows an eight-layer board with two buried via layers.

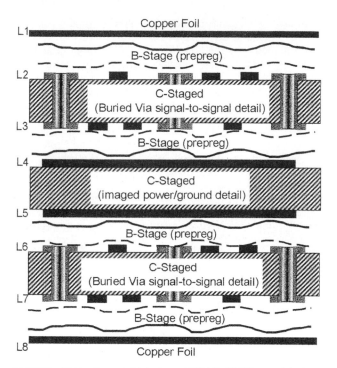

FIGURE 27.12 Type 4 (eight-layer B/V ML-PWB), an eight-layer board with two buried via layers.

Buried vias do not pass through the board, so they do not congest the routing on other layers. In addition, the same via site can be used simultaneously on different layer pairs. Since buried vias are drilled and plated in the thin laminate prior to lamination, they can be very small and positioned very accurately, saving additional routing space. In some applications, buried vias are placed where needed with no reference to a predetermined grid. This gridless routing approach gives very high CAD autorout completion.

To benefit from buried vias, signal layer pairs must be routed on opposite sides of the same C-stage component. In Fig. 27.12, L4 and L5 are the only buried via candidates. In Fig. 27.7, no signal layers can benefit from buried vias. The design shown in Fig. 27.12 is able to use buried vias on two layer pairs because a power/ground layer pair separates them. In non–buried via designs it is best to use one power/ground plane between each signal layer pair. This gives cross talk isolation and impedance reference. A buried via design uses a second redundant layer to force the next signal layer pair onto the same imaged detail. In other words, for a high-layer-count design to use buried vias on all signal layer pairs, power/ground plane pairs must be inserted between each signal layer pair. This increases the number of layers in the board and increases cost. Another disadvantage of buried vias is the cost associated with the extra drilling and plating operations.

27.3.5 The Blind Via Stack-up

Another via option is the blind via, which connects the surface layer to one or more internal layers but does not go through the board to the opposite outside layer. Figure 27.13 shows an example of an eight-layer ML-PWB with both blind and buried vias. The design shown in Fig.

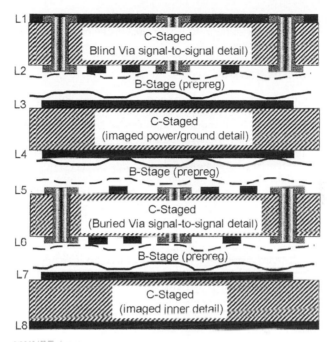

FIGURE 27.13 Type 4 (eight-layer blind and buried via ML-PWB). The design contains buried vias connecting L5 and L6. Blind vias connect layers L1 and L2.

27.13 contains buried vias connecting L5 and L6. Blind vias connect layers L1 and L2. The buried vias can be fabricated in layer details as described earlier or can be fabricated by blind drilling.

Blind vias may become very important in very dense double-sided surface-mount designs that have interference between I/O vias from opposite sides on the board. If this is particularly troublesome, a via-in-pad (VIP) approach or dog-bone escape pattern is used. The VIP approach places a blind hole directly in the device I/O pad. The dog-bone approach places a blind hole in an adjacent pad next to the solderable land.

A second application for buried vias is to ensure complete side-to-side electrical isolation. This is particularly important in wireless designs where the RF circuits must be shielded from other circuits. Through-vias allow RF electric fields to escape from a shielded region. A blind via eliminates this problem and allows RF functions to be combined with logic and control functions on the same board.

The ultimate use of blind vias is to effectively convert a dense double-sided surface-mount design to a pair of less dense single-sided surface-mount designs. To see how this is possible, visualize an ML-PWB with fine-pitch SMDs on both sides as two separate boards with some level of side-to-side interconnection. If the board is manufactured as two separate subassemblies, the I/O connections on one side will not interfere with the I/O connections on the opposite side. For example, consider a 16-layer board fabricated as two 8-layer subassemblies. The subassemblies are laminated, drilled, plated, and patterned like a standard ML-PWB, with the exception that the sides that will become the board exterior are blanket-metallized. After patterning, the two subassemblies are laminated together and then processed as a standard ML-PWB. This process is another form of sequential lamination. The only through-holes required in this design are the relatively few that provide side-to-side interconnection. Since each via site is used twice, a 100-mil grid may replace a 50-mil grid, providing a significant increase in innerlayer routing resources. This type of ML-PWB structure is very costly and has mostly been replaced with the routing features of the high-density design.

27.3.6 The High-Density Stack-up

The buried via stack-up design came about for a solution to gain signal routability when CAD routing solutions were pressed to the limit. Early on, Type 4 designs were considered specialty products and utilized standard-feature-size characteristics. With the advent of more sophisticated CAD routing tools, it became possible to autorout with greater efficiency, sometimes avoiding the necessity and cost of buried vias. Soon high-I/O, full-matrix components created new interconnectivity demands. Today buried via sections defined as cores in IPC-2221/2226 are used for the central starting point of manufacture and then build-up layers are applied to complete the design as an HDI structure. As noted earlier, the common discriminator that separates the high-density design from a conventional Type 4 is mainly the feature size. For example, the HDI Type II as shown in Fig. 27.4 uses a conventionally processed ML-PWB Type 3 board as the core (L2 to L5) and the PTHs become buried after the high-density features of L1 and L6 are added. Microvias are used to form a connection to L2 and L5; then through-holes are formed to tie L1 through L6. The microvias and holes are now metallized in one cycle. The build-up layers in the example use thin dielectrics to provide a close proximity to the adjacent layer. This is required to accommodate the microvia connection and maintain producibility for metallization.

Figures 27.14 to 27.16 show a comparison of the HDI build-ups for Types I, II, and III. In each case complexity is added by means of building add-on layers sequentially. The build-up

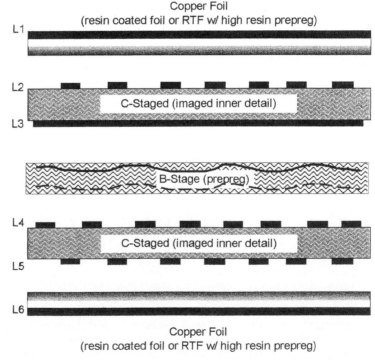

Copper Foil
(resin coated foil or RTF w/ high resin prepreg)

L1

L2

C-Staged (imaged inner detail)

L3

B-Stage (prepreg)

L4

C-Staged (imaged inner detail)

L5

L6

Copper Foil
(resin coated foil or RTF w/ high resin prepreg)

FIGURE 27.14   Type I (six-layer HDI stack-up) shows one example of how to stack up an ML-PWB with HDI features.

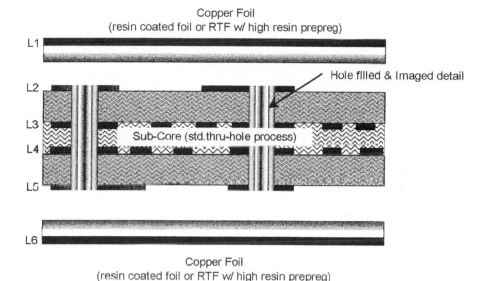

FIGURE 27.15 Type II six-layer stack-up having at its core a four-layer board with through-holes, which later become buried.

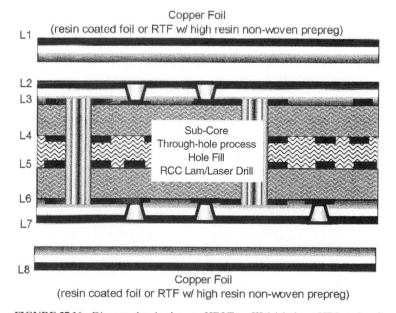

FIGURE 27.16 Diagram showing how an HDI Type III (eight-layer HDI stack-up) might be produced employing conventional lamination techniques.

layers are typically for signal routing and contain low copper weight. Dielectric-coated (non-reinforced) microfoils (as thin as 9 to 12 μm) are employed to provide a low-profile thickness after copper plating. This is needed in order to facilitate the image of fine linewidths typically associated with HDI. These styles of circuits offer some economy through use of a variety of dielectrics. The use and style of construction type must be matched with the product's expected application environment and operating life. Often the need for CTE match of area array components facilitates the application of a nonwoven aramid layer at the surface. The resin system selection should be based on the expected CTE mismatch concerns, such as when a ceramic device is surface-mounted to the assembly. Thermal cycling of power on and off can cause earlier failure when a large mismatch is present.

Manufacturing capability becomes critical when targeting HDI features. The IPC has taken the responsibility of developing industry standards to assist in benchmarking and quantifying capability limits. Manufacturing features such as imaging, etching, hole formation, plating, and lamination registration are all strained and must be optimized. Other processes that are equally challenged are the end product continuity testing to access the high-density features.

27.3.6.1   HDI Type I Stack-up.   Figure 27.14 shows one example of how to stack up an ML-PWB with HDI features. The economies of this design are obvious because it only requires one lamination cycle. The density push in this level of HDI focuses on imaging and microvia routing. By routing at a high circuit density, the surface area of the board is better utilized. The outerlayers are typically formed with low-profile dielectric. When utilizing conventional lamination, this is usually a resin-coated foil or a microfoil bonded with a high-flow nonwoven aramid prepreg. Special nonwoven series of aramid fiber laminates have been developed to yield an equivalent dielectric thickness of 1.9 mils.

Special handling is required for the microfoils, which are usually 9 to 12 μm thick. To facilitate in handling, a sacrificial carrier foil is sometimes cohered to the microfoil externally for added stiffness. After lamination, one option is to process the microvias through laser drilling and conventional through-hole drilling. Sometimes a clad outer construction may be best, depending on the laser technology chosen. For example, IR (CO_2) lasering using an etched mask (small etched openings the size of the via) in the outer copper can be performed at primary print. This helps reduce possible misregistration error. One metallization cycle is required to electrically tie in all vias. This is a cost savings compared to sequential cycles.

27.3.6.2   HDI Type II Stack-up.   Density demands, when not met by the single lamination approach of Type I, must then address the use of sequential processing when employing standard ML-PWB techniques. The Type II stack-up shown in Fig. 27.15 has at its core a four-layer board with through-holes, which later become buried. The lamination of the HDI outerlayers (with thin 2-mil dielectric) often does not have enough flow to fill the PTHs of the core structure and remain thin enough to facilitate metallization of the blind holes. Depending on the core thickness, the reliability requirement, and the end product specifications, these holes are filled. The hole-filling process uses specific methods based on which commercially available material is used. Often the hole-filling material is an epoxy-based material with fillers that closely match the expansion characteristics of the base material. The filler can be conductive or nonconductive depending on the end use application. Some methods after filling require a sanding operation to produce a coplanar surface. Once the outerlayer of the structure is formed, it can be processed to form the high-density features utilizing both through-vias and microvias.

27.3.6.3   HDI Type III Stack-up.   The use of a Type III structure may be employed when routing densities are greatly pressed. At this point of complexity and above, alternative HDI approaches should be investigated because of the extra cost associated with performing repetitive cycles with conventional ML-PWB processes. Alternatives to lamination provide build-up of the dielectric and conductive layering through other means. These should be discussed with the manufacturer prior to selection. It is important to remember that the manufacturing method related to fabrication can have a great impact on the design rules chosen at CAD layout. Figure 27.16 represents how an HDI Type III might be produced employing conventional lamination techniques. Here the core substrate is a four-layer ML-PWB detail similar to the Type II starting construction discussed earlier. Processing for the first pair of build-up layers follows the processing analogy of the Type II structure. Here, as in other foil-laminated buildups, the starting copper thickness should be kept to a minimum. Plating uniformity becomes critical when attempting to image/etch high-density features. After L2 and

L7 have been defined, the final copper layer pair (L1 and L8) is laminated. At this final buildup, since no through-holes are present, the flow of the resin-coated foil or prepreg should be sufficient to fill the microvias and flush the circuits. Once the additional layer is laminated, another level of microvias may be produced.

27.4   ML-PWB PROCESSING AND FLOWS

27.4.1   Process Flow Charts

Attempting to visualize the process flow of the manufacture of multilayer printed wiring products can be overwhelming. One way to help picture the multiple paths a board travels is through the use of flowcharts. The flowchart in Fig. 27.17(*a*) is a typical process flow for the beginning innerlayer process identified as 1 through 4. Figure 27.17(*b*) is a typical finish board flow process after lamination and drill. Three additional diagrams are provided in Fig.

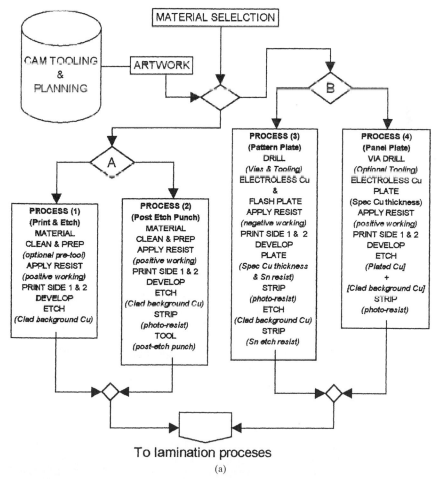

(a)

FIGURE 27.17   (*a*) A typical process flow for the beginning innerlayer processes identified as 1 through 4.

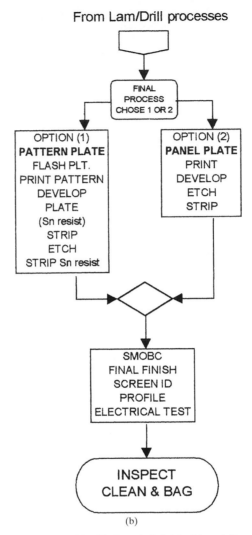

FIGURE 27.17 (*b*) A typical finished board flow process after lamination and drill.

27.18(*a*) through (*c*) that represent possible sequential flows for HDI products. They are identified as processes 5, 6, and 7, referring to how the different HDI types might flow. It should be noted that alternative flows and methods are possible in HDI when employing alternative processes such as conductive hole fill. This discussion is limited to mostly conventional processing. Major aspects of these processes are discussed in detail in the following sections.

27.4.2 Innerlayer Materials

The multilayer process begins with the accumulation of the innerlayer clad dielectric laminate.

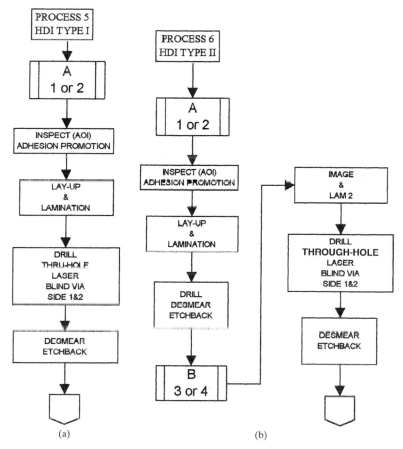

FIGURE 27.18   Diagrams representing possible sequential flows for HDI products. They are identified as processes 5, 6, and 7, referring to how the different HDI types mentioned in the text might flow.

27.4.2.1   *Documentation and Specifications.*

The ML-PWB design documentation should specify the specific material system to be used in the fabrication. Typically materials are identified by IPC-4101 callout or slash sheet designation, e.g., IPC-4101/24. Copper weights as well will be specified within the drawing documentation. The specification IPC-4562, "Metal Foil for Printed Wiring Applications," provides reference to the various weights of copper foils along with an applications use guide. The common terminology refers to foils less than oz/ft^2 as a metric thickness of micrometers. The most commonly used foil is STD-Type E electrodeposited (IPC-4562/1) with a weight of 1 oz/ft^2 and a nominal thickness of 1.35 mils (35.5 μm).

27.4.2.2   *Copper Foil.*

High-current applications use heavier (thicker) copper of 2 oz or above. With copper weights above 3 oz, processing difficulty is increased. High-density circuit designs that have low voltage and are mainly concerned with passing signal may use thinner-copper-weight foils, e.g., 18 μm or less. When sequentially processing clads as buried via pairs, thin starting foil is necessary to promote line definition and image integrity.

Copper foil is fabricated in an electroplating process on a rotating drum that produces a coarse-grained columnar copper. The economy of the foil process always yields copper at the minimum thickness tolerance. Due to the speed of the foil process, often a course grain struc-

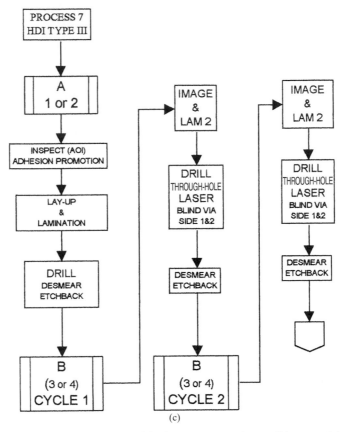

FIGURE 27.18 *(Continued)*   (*c*) Diagrams representing possible sequential flows for HDI products. They are identified as processes 5, 6, and 7, referring to how the different HDI types mentioned in the text might flow.

ture can yield poor elongation. Elongation is a property of major importance in reducing trace fracturing. Standard copper foils typically fail at elongations around 3 percent. IPC-4562/3 high-temperature elongation (HTE) provides a very slight improvement in elongation. High-ductility electrodeposited foil, called HD-Type E (IPC-4562/2), is specified to withstand an elongation of 10 percent minimum. Foil vendors also sell special fine-grain-structure, annealed, or wrought foils. The foils with higher elongation are advertised to have superior etch performance for fine-line densities.

Standard foils have a rough surface called *tooth* on one side (drum side out) and a smooth or shiny surface on the opposite side (drum side). The rough side, treated with an adhesion promoter, is laminated against the C-stage dielectric to ensure good adhesion. Since the shiny side of the copper has poor adhesion characteristics, the ML-PWB fabricator must include adhesion promotion steps prior to resist lamination and prior to final board lamination. Double-treated copper foil has an adhesion promoter applied to both sides and is attractive in high-volume applications. Although double-treated foil requires no further adhesion-promoting treatments, it has several process disadvantages to overcome:

- Cleanliness and material handling sensitivity is more critical.
- The cost often offsets the savings from the eliminated processes.

- It is somewhat fragile, so it is difficult to rework.
- Complete resist development is difficult, leading to a high incidence of shorts.
- It is not compatible with the plating processes used to make buried and blind vias.

Another foil type offered by laminate suppliers, reverse-treated foil (RTF), offers an advantage for producing fine lines. The RTF copper has adhesion promoter applied to both sides and is classified by 4562 as code R (reverse-treated bond enhancement [cathode side] stain-proofing on both sides). This approach provides advantages to imaging fine lines. When the copper tooth is reversed, line definition is improved by allowing the etch chemistry process to stop at the surface of the laminate.

27.4.3   Innerlayer Process

An innerlayer detail is essentially a thin double-sided printed circuit. The standard innerlayer process contains no plated holes because it is produced by using a print-and-etch process. Blind, buried via layers and laminated cores contain holes that must be plated in either a pattern-plate or a panel-plate process. Fiure 27.17(*a*) shows a typical flowchart for four inner-layer process options. Processes 1 and 2 support standard innerlayers and processes 3 and 4 can be used for buried-via innerlayers or core subassemblies. All four processes start with a bare copper-clad laminate and end with a patterned double-sided circuit. The patterned circuit must be inspected and treated to enhance adhesion prior to further ML-PWB lamination. All four of these sequences work equally well with any of today's materials systems.

27.4.3.1   *Process Sequence Flows 1 and 2.*   Process sequence flows 1 and 2 are variations of how to produce a standard innerlayer with no plated holes by print-and-etch or post-etch-punch processes.

Process 1: Print and Etch.   The laminate layers receive an application of photoresist. After photoresist application, the panel tooling holes are punched with the predefined pattern to match the lamination tooling. This is followed by the image printing (exposure) process, where the artwork pattern is aligned by the punched tooling holes. After print exposure, the inner panel is ready for development, which clears the nonimage area for etching. The artwork is a negative image and produces a retained pattern of "what light sees stays." The resist that stays behind protects the copper, which becomes the circuit pattern. The etch process then removes the unprotected background copper. The stripping process then removes the resist material and the layer is ready for adhesion promotion.

Process 2: Post-Etch-Punch.   The difference in this sequence is the order in which the laminate material is punched (tooled). The innerlayer detail is resist-coated and placed in the exposure cabinet. The artwork layers have already been prealigned as the top and bottom layer pattern pair within the exposure frame. This is defined as a pinless registration system. The process then is followed by etching and stripping. The final step is to punch the tooling holes, which are aligned optically to etched targets. The advantage of this process is improved innerlayer registration because it compensates for panel in-plane dimensional movement that occurs after stresses are relieved after etching.

27.4.3.2   *Process Sequence Flows 3 and 4.*   Process sequence flows 3 and 4, as shown in Fig. 27.17(*a*), are plating processes that may be used for blind and buried-via layer pairs or lami-nated cores. The flows represent two distinctly different plating methods referred to as pat-tern plate and panel plate. Note that in these processes, the printed image ultimately is registered to tooling holes as well. Depending upon the tooling system, often the tooling holes are drilled at the same time as the via holes. This assures optimum alignment between drilled holes and printed features due to dimensional movement of the materials.

Process 3: Pattern Plate.   The distinguishing factor of this process is that final build-up plating is performed after imaging. Extra processing steps are required in this process to pro-tect the primary metallization. Typically, a deposit of tin metal acts as etch resist, then is stripped after etching. Note that this process is suitable for defining small circuit patterns. The copper plating is selectively built up on the circuit pattern. The etch process only need remove the base clad copper to define the circuit. Photoresist thickness must be matched with respect to the image resolution of the circuit.

Process 4: Panel Plate.   The diagram of process 4 represents the flow for a panel plate substrate. Here the plating deposit covers the entire surface area of the substrate (panel). Photoresist is applied postplating and provides the only protection to the defined circuit and the plated via holes. The via hole protection is referred to as tenting. Photoresist materials chosen must have characteristics with tenting properties to avoid resist breakout and voiding of the plated hole. Note that this method places a burden on plating uniformity across the entire surface of the panel due to high current densities at the perimeter of the panel. Imaging of fine features is difficult because the base clad copper and plated copper are etched together. Additionally, the panel plating of thin-core laminates can induce dimensional shrinkage of the substrate. Process parameters should be optimized through experimentation to applying a low-stress deposit.

27.4.4   ML-PWB Tooling

The tooling system employed for ML-PWB fabrication is one of the most critical aspects of the process and requires much forethought. The investment of tooling is significant and not easily changed. Planning is required to determine the degree of flexibility required in a tooling system. Typically, a minimum number of panel sizes are chosen in order to standardize the process and reduce cost. The two most common panel sizes are 12×18 in and 18×24 in. These two are prominent in industry because they yield the best material utilization from raw stock of the laminate supplier. Another panel size, 24×36 in, is popular in high-volume fabrication.

The ML-PWB tooling system can be broken down into three aspects: the front-end tooling, the method of generating tooling holes, and the investment tooling plates.

27.4.4.1   *Front-End Tooling.*   The first aspect of the ML-PWB tooling is commonly called front-end tooling or simply computer automated manufacturing (CAM) tooling. The front end is where manufacturing personnel using a CAM software package generate all phototools and associated computer numerical control (CNC) electronic files. The computer CAM station mirrors the tooling method chosen for the shop and the job work order. The CAM software is capable of overlaying the tooling pattern on each circuit layer to produce the master pattern alignment. The master artwork pattern is then photoplotted to reproduce the artwork film tools. Several important manufacturing functions fall within the bounds of the front-end tooling responsibility.

27.4.4.2   *Design Rule Check (DRC).*   Design rule check (DRC) is where the electronic design data are analyzed against a specific set of fabrication rules, a virtual design for manufacturing (DFM). The capabilities and attributes of these rules are based on sound, industry-proven values that match the technology targeted for construction. The default values embedded into the CAM are input by the manufacture as technology files. These files should reflect the capability of the fabricator's equipment and processes in light of how they are utilized against the ML-PWB type being produced.

27.4.4.3   *Panelization.*   To gain economies in the manufacturing process, the individual board design is stepped and repeated within the artwork tooling. *Panelization* is the term used to describe this method. Here the elemental goal is to yield as many parts on one panel as possible. The CAM software allows manipulation of the one-up design where it is placed or nested within the outline of the chosen panel dimension.

27.4.4.4   *Artwork Generation (Photoplotting).*   Once the design has been formatted into the desired panelization, it is ready to be made into a working phototool. The CAM data are output to a server that converts the data into a language recognizable by the plotting equipment. The plotter exposes the circuit pattern on the film tool. Typically, the film tool is polyester based with a silver emulsion and is exposed by means of a laser or fiberoptic light source. The resolution of the light source directly affects the ability to image the technology type. Note that when the technology type approaches the feature size of HDI, the use of an alternative approach to phototooling may be employed. This is known as direct imaging, and, as the name implies, the image is directly exposed onto the production innerlayer, thereby eliminating the use of a phototool.

27.4.4.5 CNC Drilling Routing Tools.   The machine communication method commonly employed in the automation of the drilling and routing process is output from the CAM tooling as CNC programs. The front-end tooling software facilitates the organization and optimization of the CNC routine. The routine must match the flow of the technology process steps being produced. When sequential processing is required, the drilling routing data must be broken out into separate routines. When the technology type involves HDI features that require laser drilling, specific tooling accommodations have to be considered to match the discipline of the laser equipment, for example, the use of special imaged fiducial targets or a special dot mask aligned to the drill locations.

27.4.4.6 Automatic Optical Inspection (AOI).   Another aspect of the tooling set is the data configuration to support the automated optical inspection (AOI) process. AOI is considered part of the detailed inspection process. When the AOI equipment supports what is known as CAD reference, the original design data (which were reproduced in the phototool) are now output as a configured file to compare the imaged circuit layer to the original designed circuit (see Chap. 50 for a complete discussion of inspection).

27.4.4.7 Electrical Testing.   The final aspect of the tooling set is preparation of the electrical design data to support an electrical testing routine. The purpose of the electrical testing within the ML-PWB production process is to verify the integrity of circuit continuity. Electrical testers can take on many forms, but are generally divided into two major categories, the "bed-of-nails" and the "flying probe." The electrical data required for the test must be extracted from the circuit pattern unless a separate net list is provided that contains the connectivity. The standard data format for electrical testing is IPC-356. Tooling for a bed-of-nails tester consists of outputting data to manufacture a fixture to accommodate the pins, which make contact with the circuit nets. When a flying probe tester is employed, the net data are fed into the machine, which uses its own software to configure the routine to probe continuity and perform isolation. The throughput of a flying probe unit is less than that of a bed-of-nails tester, but the flying probe does not have the expense of the fixturing. The chosen method of test tooling requires an examination of the economics based on the volume and technology of the ML-PWB. Typically, when the technology type involves HDI features, the only alternative is to use a flying probe due to the limitation of the grid spacing in a bed-of-nails tester. High-density test requirements continue to drive the flying probe technology, with machine probe head counts increasing to 16 or more to achieve higher speed. (See Chaps. 51 and 52 for a complete discussion of electrical testing.)

27.4.5 Tooling Hole Formation

The shape, size, and location of tooling holes are dependent in part on the investment tooling system predetermined by manufacturing engineering for a given manufacturing facility. The tooling holes range in size and shape from 0.125 to 0.250 in in diameter to a slotted hole of 0.187×0.250 in. It is generally believed that copper should be retained around the tooling hole/slot to reduce dimensional slip. Three common methods are used to produce the tooling holes in innerlayers:

- Prepunch
- Post-etch punch
- Drilled holes

27.4.5.1 Prepunch.   A fixed punch and die set is used to generate tooling holes/slots prior to imaging/etching. To avoid having to cut away the photoresist material over the tooling hole, the punching is sometimes performed after resist lamination. In this case, cleanliness has to be significantly refined. If a cut sheet resist laminator is used, it can be adjusted so the resist falls inside the boundary of the tooling locations. The punching occurs in a hydraulic punch press, where alignment is ensured by means of an edge stop on two axes of the panel to square the alignment. The die set penetrates the laminate in one stroke, punching all holes at one time while holding the material flat. Material is punched one piece at a time, yet high throughput can be achieved by this method. Punching of multiple laminates at once is not recommended because the quality and size of the tooling hole/slot is affected. Additionally, thicker laminates (greater than 0.032 in) require the punch and die set to have a slightly larger tolerance to accommodate the spring-back of the hole within the composite material. Overall, the pre-punch method is very accurate and repeatable. Materials with a high degree of movement (low dimensional stability) are at a significant disadvantage in this method due to the drift of the tooling location after innerlayer processing.

27.4.5.2 Post-etch Punch.   An optically aligned punch is used to generate holes/slots after imaging/etching. The fully processed innerlayers are edge-aligned into the punch; a pair of video cameras locates on a diagonal pair of printed fiducials; a servo-system aligns the panel, minimizing alignment errors due to panel distortion; and tooling holes are punched. This process punches an optimized set of tooling holes/slots in the layer just prior to lamination. Post-etch punch can be automated so it has high productivity, but it is slower than a simple punch and die. However, the productivity gain achieved by printing without tooling may off-set any productivity loss from using post-etch punch.

27.4.5.3 Drilled Tooling.   When processing follows the process sequence 3 and 4 described earlier, tooling locations may be drilled. This allows the tooling locations to be formed at the same time as the via drill cycle. In order to minimize distortion, which affects final registration accuracy, it is recommended that the drill program be scaled. Scaling factors must be characterized through experimentation. Scaling helps minimize the dimensional effect that processing has on the true positions of the holes. Alternatively, extra sets of tooling holes may be produced just for the imaging process, then followed by a post-etch punch operation to optimize the lamination registration.

27.4.6 Tooling System

There are numerous rationales on innerlayer tooling schemes, and this has resulted in a wide assortment of tooling systems being used across fabricators. Over the years, many tooling systems have evolved, ranging from full-perimeter distribution of holes to as few as four holes located on the midaxis. The tooling system comprises a substantial investment and therefore merits considerable engineering conceptual judgment. Many shops become trapped with poorly conceived tooling schemes that are difficult to migrate from due to the cost to convert earlier designs. Therefore, careful analysis should be given to the product mix and technology types to be manufactured prior to choosing a scheme.

The primary purpose of the tooling system in ML-PWB manufacturing is to facilitate layer-to-layer alignment during the lamination cycle while maintaining a positional reference for subsequent processing. The tooling hole arrangement is mirrored within the front-end tooling routines mentioned earlier. The master alignment of each part keys off the tooling scheme. The

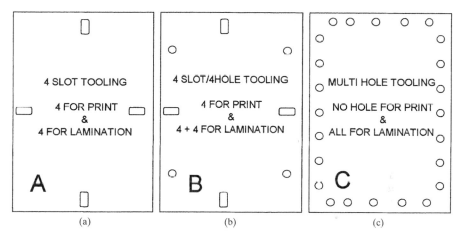

FIGURE 27.19   Three different tooling schemes. (*a*) Four-slot tooling. (*b*) Four-slot/four-hole. (*c*) A tooling system with a full perimeter of holes placed.

primary tooling locations may be used at all process steps or sometimes used to introduce secondary reference holes. Figure 27.19 shows three different tooling schemes. Figure 27.19(*a*) shows a system known as four-slot tooling. This is an excellent system, because the use of four slots allows for growth and shrinkage of the layers during processing. The four slots may be used for printing, inspecting, and lamination, or, if post-etch punch is used, they are used for inspection and lamination only. Figure 27.19(*b*) shows a system known as four-slot/four-hole. This is common in a prepunched system where the slots are used for image printing and the extra holes are engaged at lamination. Figure 27.19(*c*) shows a tooling system with a full perimeter of holes placed. This style would be considered overdetermined. It has a disadvantage at lamination because imaged panels typically have to be stretched over the alignment pins.

27.4.7 Imaging

27.4.7.1 Surface Preparation.
The first step in the imaging operation is surface preparation to enhance photoresist adhesion. Double-treated foils (DTFs) and foils laminated with the tooth side up (RTFs) require minimum surface preparation. Generally, a process through a tacky roller machine to remove dust and foreign material is adequate. For standard-thickness layers, a more aggressive treatment is needed. Common options include a conveyorized chemical cleaning and/or an abrasive cleaning. Automated equipment is available to perform this function with pumice or aluminum oxide slurry. Scrubbing the layer by hand is acceptable. However, acidic 4F grit pumice is required for single-step cleaning. Standard copper foils are coated with a chromate conversion to eliminate formation of heavy copper oxidation during storage. This coating, and any oxides, must be removed to provide a microroughening of the surface. When using any of the mechanical cleaning methods, care must be taken to avoid mechanical deformation (stretching) of the thin laminates. Stresses induced will be relieved after etching of the image, causing positional movement.

27.4.7.2 Photoresist.
Photoresist is supplied in both dry-film and liquid forms. Dry films are a popular choice because of the simplicity of application. For buried via innerlayers, dry films are preferred because liquids are difficult to use with through-holes. The major weakness of dry-film resists is sensitivity to surface flaws and a tendency to lift on poorly prepared surfaces.

27.4.7.3   Dry Film.   Resist film is typically 1.5 to 2.0 mils thick. Both thinner and thicker resist is available as a specialty product. Dry-film resist is applied with a hot roll laminator. To aid in production of fine lines and spaces, the use of a wet nip applicator aids in filling surface imperfections. Precision flatness and uniform temperature are equally important as well. Popular resists for high-density applications include films formulated with high-speed sensitivity for laser.

27.4.7.4   Liquid Photoresist.   Liquid resists work well in a print-and-etch process. They have excellent adhesion and a tolerance for surface flaws and are relatively low in cost. The disadvantage of a liquid photoresist is the need to produce a perfect coating. Foreign material, skips, thin spots, and dewetting all cause serious image problems. The use of a wet coating can also cause problems due to resist contamination of the transport system.

Roller Coating.   The least expensive and most popular way to coat liquid resists is with a pair of roller coaters consisting of pinch rollers in which one or both rollers is used for coating. The coating roller has a closely spaced, precisely cut spiral grove. Liquid is metered onto this roller and then transferred to the panel. A good roller coater produces an extremely uniform coating. By carefully selecting coater parameters, it is possible to achieve a coating thickness control of 0.1 mils. Some roller coaters use the top roller as a coating roller, allowing the panel to be transported on a clean conveyor system. In these systems, the resist can be dried or the panel can be printed wet. Other roller coaters use both the top and bottom rollers for coating. These roller coaters require a handling system that holds the panel by its edge until the resist is dry. The coating roller is the weak link in a roller coating. Alignment problems result in nonuniform coatings. Flaws produce repeating defects in the coating. Worn or improperly cut grooves result in a low-quality coating.

Curtain Coating and Electrodeposition.   Curtain coating and electrodeposition can also be used to apply liquids. The curtain coater operates by pumping a waterfall of liquid through a narrow slot. By careful control of the slot width, pumping pressure, and viscosity, a well-controlled curtain of liquid is created. When a panel moves through the curtain, a thin coating is applied to one side. Curtain coaters produce a good-quality coating, but not all photoresists have the proper viscosity for curtain coating. With electrodeposition, a polymer is deposited on a biased metal surface submerged in a liquid medium by a process that is analogous to plating. It gives a well-controlled, high-quality coating. Both curtain coating and electrodeposition are expensive and are not widely used for innerlayer processing.

27.4.7.5   Photo Print Exposure.   The third step in imaging is photo printing. The three types of photo printing machines are flood, collimation, and direct image. Most flood printers consist of high-UV (5000 W) light sources housed within large reflectors to distribute the light uniformly across the imaged surface. Hard contact by means of a vacuum is required between the film and the innerlayer. The silver emulsion side of the polyester film tool makes direct contact with the resist. This allows good image reproduction without collimated light. The drawbacks of contact printing are defects associated with poor contact, poor productivity due to the time required to establish hard contact, and the potential for artwork damage. The alternative, collimation, an off-contact printing method, holds the film a small distance above the innerlayer surface and prints with collimated light. The collimated machine uses a high-UV lamp to reflect intense light by means of mirrors straight down on the innerlayer. This method permits a high degree of automation. Off-contact printing is absolutely necessary for printing wet layers. The disadvantage of off-contact printing is a loss of resolution due to incomplete collimation and a great sensitivity to dust and scratches on the artwork. The other style of printing is referred to as direct imaging and has become a leader in the production of high-density line features. In this process, the electronic data are input into the direct image machine instead of being used to produce a phototool. The direct image, typically output through a laser, is used to expose the image by means of a series of on/off pulses manipulated within an x-y axis. The quickness of this process is combinations of resist speed (sensitivity) and the ability to reposition. This method is gaining in popularity; however, throughput and cost continue to be issues in its wider use.

27.4.8 Develop, Etch, and Strip

27.4.8.1 Developing. All modern innerlayer photoresist materials are aqueous soluble. This means they develop in a mild caustic solution. Note that resists that use solvent developing are no longer in use due to significant health and environmental concerns. The developing process is easily automated by means of conveying the part through a sprayed caustic solution. The developing solution removes the nonimaged resist, which was not polymerized by the light source. Therefore, in effect, "what light sees stays." The quality of the developed image is often quantified by means of a density step chart that was previously exposed. Comparing the step held on both sides of the imaged panel allows process control to be determined.

27.4.8.2 Etching. Etching chemistry is determined by the process flow sequence as outlined in Fig. 27.18. Etching can be classified as print and etch when using processes 1, 2, or 4 and using a cured photoresist as an etch resist. Etching is classified as print-plate-etch when using process 3 employing a sacrificial metal layer as an etch resist.

The preferred etch chemistry of the print and etch process is cupric chloride. Cupric chloride is easier to control than the ammoniacal etchant used with finished boards, and it can be easily regenerated. Etching occurs by spray erosion with the highly loaded (~25 oz/gal) copper-soluble solution. A high degree of automation is possible in the print and etch process. Since there is no plating step, it is possible to feed printed innerlayer panels directly into a conveyorized inline machine that develops the resist, etches the circuits, and strips the resists. These automated machine lines are called develop, etch, strip (DES) and improve productivity by reducing the defects caused by handling.

The print-plate-etch process follows the same general methods as used in the finished board steps (see Chap. 29). The use of a metal etch resist requires ammoniacal or acid etch. Resist developing is separated from etching by a plating step, so a standalone resist developer is required.

27.4.9 Inspection

Once innerlayer etching is complete, the innerlayer must be inspected. The innerlayer inspection criteria follow industry-adopted performance limits (IPC-600 and the IPC-6000 series). These limits must be confirmed prior to lamination into an expensive multilayer to prevent decreasing yields. Flaws are categorized to reflect processing defects related to laminate materials and circuit definition. Linewidth and spacing are of primary importance on dense signal layers. In the case of controlled-impedance layers, it may be necessary to gauge line widths. Power and ground layers are less likely to have defects and are sometimes excluded. Most modern circuitry precludes manual visual inspection due to the inability to accurately detect flaws and the associated intense labor. Automated optical inspection (AOI) is the industry standard for inspecting innerlayers.

AOI has the advantage of being able to detect circuit nicks and protrusions. False errors can be a problem and time-consuming to confirm. AOI technology using CAD reference (the comparison of the circuit image to the design data) has a lower escape rate. Cleanliness and copper oxidation are factors that can influence poor performance in most AOI technologies. Two common methods exist for imaging the circuit. In one, the innerlayer is scanned with a tiny laser spot that causes the substrate to fluoresce. The circuit is seen as a dark image against the light emitted from the substrate. The other method illuminates the circuit with a bright light so that its image appears bright against the dark background of the substrate. Both methods have strengths and weaknesses. The fluorescent method is blind to surface flaws and substrate flaws can confuse it. The top-light method is oversensitive to the surface appearance of the copper. The AOI inspection of high-density features presents new challenges to the technology related to minimizing false errors. Modern machines employ a combination of image acquisition techniques while performing virtual reasoning against a background scan of design rules that are compared to the design image data.

27.4.10 Adhesion Promotion

Epoxy does not adhere well to untreated copper surfaces. This means that some type of treatment must be applied to the innerlayer before lamination. One option is to use double-treated copper, discussed earlier. Double-treated copper has a rough surface with a treatment supplied by the material vendor. Many ML-PWB fabricators report excellent results with double-treated copper. Others report problems with contamination and difficulty with rework. The alternative to using pretreated copper foil is to use a chemical treatment after etching. Two basic chemistries are used for this process: copper oxide and silane.

27.4.10.1 Copper Oxide. There are many variations of the copper oxide treatment, but they all work by producing a rough surface topology that enhances adhesion. Depending on the exact nature of the treatment, the color varies from a light brown to a velvet-like black appearance. The differences among the treatments include the density of the oxide and the ratio of cupric oxide to cuprous oxide. To maximize bond strength, oxide formulations that form tall vertical crystalline structures should be avoided. Low-profile, self-limiting formulations generally provide the most consistent adhesion.

Most copper oxide treatments are applied by dipping the innerlayer in a hot (85 to 95°C) caustic bath for 1 min or more. The oxide process generally uses a batch process that requires vertical racking of the layers. Rack design is critical to prevent problems with innerlayer damage and contamination. By minimizing the contact points, this concern can be eliminated. Racks require coating to insulate conductive contact to reduce electromotive effect in the oxide chemistry, which will polarize conductors and prevent oxide from being applied on random isolated traces. Since the oxide treatment is done after inspection and immediately prior to lamination, it can be the source of quality problems if improperly done.

The copper oxide surface treatment is very soluble in acids such as those found in the electroless copper line. If the epoxy-copper interface is fractured during drilling, the dark oxide will dissolve during the plating process, leaving a distinct pink ring around holes. This is not a serious reliability threat; it is generally considered a cosmetic problem that often causes false concern. One way to minimize pink ring problems is to use an oxide reduction step after oxide formation. This step converts the copper-oxide crystals back to copper metal, preserving their topology. This reduced oxide surface has limited shelf life, so lamination should follow this process within 48 h. Reduced peel strength should be expected as well.

Copper oxide treatments are not recommended for material systems that require processing temperatures significantly above 180°C. This includes Teflon and some forms of polyimide film materials. Oxide reduction begins to occur spontaneously above this temperature, reducing adhesion. The use of a thin brown oxide, or a reduced oxide, will reduce the risk of this failure mode. A second problem with polyimide film is that it is soluble in strong caustics, so if the treatment time in the oxidization process is excessive, significant substrate damage can occur. When processing circuits with microfoils having high-density features, processing times should be reduced because the oxide chemistries can reduce copper thickness.

27.4.10.2 Silane-Based Adhesion Promotion. The silane-based process is an alternative to the copper-oxide process. Silane can be used to bond epoxy to other materials. One end of the silane molecule bonds to the epoxy. If the other end of the molecule is modified to bond to the secondary material, the silane can serve as a bridge, greatly enhancing adhesion. It is commonly used in this mode to enhance the adhesion of epoxy to glass. It can be also used to enhance the adhesion of epoxy to copper.

This silane process attaches a thin silane layer to the copper surface. In lamination, the active epoxy molecules bond to the silane. This causes it to be a binding layer holding the epoxy to the copper. The objective is to achieve a stable coupling between the silane and the copper. This can be done by precoating the copper with a metal such as tin that reacts with the silane. When properly applied, the silane treatment is very stable and is resistant to chemical attack and delamination. The silane process has the advantage that it can be conveyorized in an inline process system. The major weakness of the process is that silane layers can absorb water and fail in some environments. The proper choice of silane chemistry minimizes the risk.

27.4.11 Drilling

The point at which the drilling process occurs is determined by the ML-PWB sequence chosen.

27.4.11.1 Standard Drilling of Innerlayers.

Standard buried via innerlayers are drilled and plated prior to lamination using the same procedures as a finished ML-PWB. Buried via innerlayers are typically thin, and it is easy to drill very small holes. Whereas an 8-mil hole is difficult to drill and plate in a 62-mil-thick ML-PWB, it is relatively easy in a 5- or 10-mil-thick innerlayer. Some manufacturers report success with holes as small as 4 mils. However, very small bits are expensive and easy to break. Special handling is required to load the bits and the drill machine must be vibration free with very low run-out. For most ML-PWB shops, 8 mils is a practical lower limit for mechanically drilled buried and blind vias. At 8 mils it is generally possible to drill thin layers in stacks up to 100 mils thick, greatly increasing drill productivity.

The major challenge in buried via processing is handling. Care must be taken in mechanical operations such as deburring to avoid mechanical damage or distortion. Often a frame is used to stiffen the layer during plating. Blind vias may be fabricated like buried vias and drilled prior to lamination, or controlled-depth drilling may be used after lamination.

Controlled-depth drilling has the advantage that standard innerlayer processing is used, including foil stacking. Controlled-depth drilling has several limitations:

* Blind holes cannot be stack-drilled, severely limiting drill productivity.
* It is difficult to plate a blind hole whose depth exceeds its diameter, limiting the maximum hole depth.
* Drill depth tolerances make it easy to under- or overdrill a blind hole, producing a reliability risk.

All of these limitations are avoided when the blind vias are drilled through a two-sided board used for the outerlayer and next innerlayer in a stack-up. These holes are drilled prior to lamination just as a thin two-sided plated through-board. In this case, the process sequence for a blind via layer is identical to the sequence for a buried via layer. Layers can be stack-drilled. There is no aspect ratio limitation and the need for controlled-depth drilling is completely eliminated. To be able to drill blind vias prior to lamination, a clad outer stack-up is required. One side of the outer component layer is patterned prior to lamination, while the other is patterned after lamination. This means that when the blind via is metallized, the unpatterned side is blanket-metallized. This is true for either pattern plate or panel plate. This outside layer is metallized again when the holes in the finished ML-PWB are metallized. The result can be very thick plating on the exterior of a blind via board. To minimize this problem, the blind via innerlayer should be plated with the minimum possible current density on the blanket-metallized side.

27.4.11.2 Drilling Related to Laminated Subcores.

The drilling process for laminated subcores usually follows the rules for finished board drilling and is design dependent. When designs follow this methodology, they typically push the limits of plating aspect ratio. Density constraint typically drives this requirement due to routing availability. In this case additional density can be gained through the use of laser blind vias.

27.4.11.3 Laser Drilling.

Laser-drilled blind vias can be placed on both sides of the subcore and have very accurate controlled-depth ability. Productivity of the laser is very high. When employing laser blind drilling, it is necessary to use thin insulating dielectrics to produce a thin stack height to the sublayer. Depending on the type of laser employed, it is sometimes necessary to image fiducial targets the next layer down for alignment. This is the case when employing an UV or Nd:YAG laser, which can penetrate copper. Other laser drilling alignment targets are required when employing a carbon dioxide (CO_2) laser, which penetrates the dielectric. Very high drill rates can be achieved on a scanning CO_2 laser. Here a mask pattern is etched in the copper to allow exposure of the laser beam to the dielectric. The laser removes the dielectric, stopping at the copper pad below. Multiple via stacks can be accomplished with this method by repeating the mask opening at different layer depths. Aspect ratio should always be kept in mind. Most success has been reported when the mask diameter is kept greater than 0.003 in. For best accuracy, a minimum of three etched targets is required and should be referenced to the CNC drill data.

27.5 LAMINATION PROCESS

The lamination process is an essential step in the fabrication of an ML-PWB. The lamination process is also one of the longest cycle time operations. Therefore, when the process method calls for repeat cycles, or what is called sequential laminations, it can be a significant cost driver. The lamination process involves two distinct yet linked operations: layup and laminating.

27.5.1 Layup and Materials

Layup occurs in a clean and controlled room environment. The level of environmental control depends on the circuit feature technology being fabricated. The layup process for standard multilayer processing is relatively forgiving. However, when working with material that is hygroscopic, additional measures are required. The innerlayer details require a bake cycle (typically 120°C for 1 h minimum) to remove moisture. A reduced bake cycle duration is indicated when certain oxide reduction chemistries are employed, as stated previously. Once layers are readied for layup, they should be processed as soon as possible. Should extended hold times be necessary, storage in a nitrogen-purged dry box is recommended.

The layup operation is often referred to as "building up a book." The operation follows a guide, often referred to as a stack-up sheet. The guide sheet, which depicts the engineering design of the ML-PWB, is highly recommended to minimize error. The written and illustrated guide to the book build-up process is made a part of the job planning/traveler. Due to the complexity of some products, the operator follows this guide in a recipe-type manner to perform the systematic construction. The layup may consist of some four or more PWB material components or other subassemblies. The layup operation produces a large assembly when complete that consists of the tooling plates, consumable press materials, and the ML-PWB detail. Refer to Figure 27.20 for illustrations of the following.

27.5.1.1 Tooling Plates. The outermost item in the stack is termed a caul plate or carrier plate. These are thick, oversized metal plates, generally made of ⅜-in-thick steel. The 4130 alloy of steel is often chosen due to its precision machining capability for placement of the holing holes. Sometimes a hardened aluminum alloy is chosen, but is not highly recommended due to its high in-plane expansion. The purpose of the caul plate is to provide a stable base to transport the ML-PWB stack.

27.5.1.2 Consumable Press Materials. The purpose of consumable press materials, placed between the tooling plate and the top layer, is to ensure that pressure is applied uniformly and to provide thermal lagging to correct heat rise. Among the materials used for this purpose are multiple sheets of Kraft paper, silicone rubber pads, expanded mat paper, and composite board. Kraft paper and composite board have the advantage of low cost, but pro-

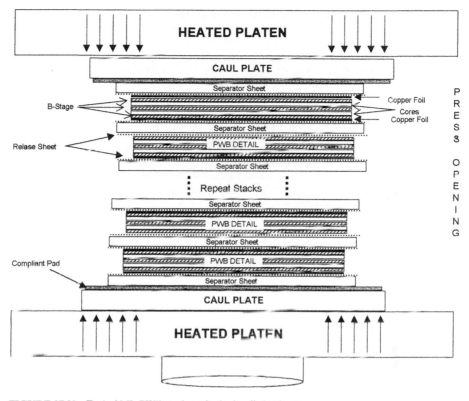

FIGURE 27.20 Typical ML-PWB stack-up for hydraulic lamination.

duce an odor that some press operators find objectionable. Silicone rubber pads have the advantage of being reusable, but are limited by the number of heat cycles they can withstand. Silicone rubber pads, when close to the end of their life, go though reversion and leach out silicone oil, which can become a source of contamination. They also have limited success in controlling edge voids. Excellent results are reported with expanded paper mat products that are commercially available. Mat paper press pad materials come in different thicknesses. Some are even produced with a sandwich layer of release material cohered. Other products exist that flow to act as stop-off for laminating predrilled blind vias. Some form of release material is required against the outer foil surface of the ML-PWB. The release material acts as a non-stick slip sheet to keep the copper surface smooth, minimize the plate cleanup, and catch the resin run-out. Alternatively, oversized foil can be used to catch the resin run-out.

27.5.2 Lamination Stack-up

The lamination process is key to building a reliable ML-PWB. In the lamination process, the board is subjected to heat and pressure that melts the B stage (bonding sheets) and causes it to flow. This encapsulates the circuits and fills any buried vias. The B stage then cures, establishing a good mechanical bond to the inner detail layers. A variety of materials can be utilized in standard laminating press cycles.

A typical ML-PWB stack-up is shown in Fig. 27.20. For productivity reasons, multiple stack heights are repeated within each opening of the press. The various components and their placement within the book are shown.

A standard lamination uses tooling pins that go from caul plate to caul plate, passing through each board and all of the separator sheets. Since the CTE of stainless steel roughly matches the in-plane CTE of a multilayer board, a tight fit to the pin is possible. Aluminum has a much higher CTE, so if aluminum is used, a loose fit to the tooling pin is required. Some fabricators use as few as four tooling pins. Others may use 20 or more. In addition to easy stacking, the four-slotted pin system minimizes problems arising from material growth and shrinkage in an overdetermined tooling system. On the other hand, users of systems with a large number of pins believe that firmly anchoring the ML-PWB to stainless steel obtains better dimensional stability.

The multilayer boards in the stack are isolated from each other by metal separator sheets. The separator sheets provide a mold surface for the laminated ML-PWB. It is extremely important that the separator sheets be clean and free of debris. Both aluminum and steel separator sheets are used. Separator sheets should be cleaned regularly. The surface finish of the plates becomes very critical when laminating against microfoils that will be printed with fine lines. The most common type of steel is one of the 400 series stainless steels that have a very durable surface. Hardened 300 series steels are also occasionally used. Separator sheet thicknesses range from as thin as 0.015 in to as thick as 0.062 in. The thicker plates are more rigid and resist the tendency of internal layer features to print through from one ML-PWB to another. Thin aluminum plates have the advantage of being disposable, eliminating the need for cleaning.

27.5.3 Lamination Process Methods

27.5.3.1 Standard Hydraulic Lamination.   A standard hydraulic press generally has a top or bottom ram and several floating platens that create multiple press openings. Typical presses have four to eight openings. The output of the press is based on the stack height in each opening. Low-layer-count designs can yield as many as 96 panels per press cycle based on a stack height of 12 by 8 openings. High-layer-count and thick laminated boards have the stack height reduced to ensure uniform heat rate from the outside to innermost stack. Steam, hot oil, or electrical resistance heaters may heat hydraulic-style presses. The steam and hot oil presses have the advantage of a fast heating rate, but the temperature of the heating fluid limits their maximum temperature. For steam, this is generally below the lamination temperature needed for high-temperature materials. If a steam press must be used, polyimide, PPO, and cyanate esters can be oven-baked to complete their cure. However, the thermal plastic adhesive layers used with Teflon are not compatible with a steam press.

27.5.3.2 Vacuum-Assisted Hydraulic Lamination.   Many hydraulic presses use vacuum to eliminate volatiles. Users of vacuum presses report a reduction of edge voiding and the ability to laminate at a lower pressure. Almost all hydraulic presses sold since 1990 are equipped with a vacuum chamber. In a typical process, the ML-PWB caul plate is loaded onto a carrier sheet where a spring-loaded rail holds it off the press platen and limits the heat transfer to the ML-PWB stack. Next the vacuum chamber is closed and a vacuum is drawn and held. A typical vacuum cycle may be from 15 to 60 min, depending on the nature of the materials being laminated. This gives the vacuum time to pull air, moisture, and other volatiles out of the ML-PWB stack. At the completion of this prevacuum process, the press is closed, compressing the spring-loaded rails and establishing good thermal contact with the press platens.

27.5.3.3 Autoclave.   The autoclave is a popular tool in the composites industry and has found limited use by board fabricators. The autoclave is a sealed cylindrical chamber that subjects the ML-PWB stack to a high-pressure heated gas. The ML-PWB stack is sealed in a vacuum bag, and the hydrostatic pressure from the gas produces the force necessary for lamination. In principle, the autoclave is an excellent machine for producing a void-free lamination. Since the pressure is hydrostatic, it eliminates the problem of low pressure at the panel edge. This results in void-free panel borders and increases the usable area on a panel. In practice, an autoclave requires a long prevacuum cycle and has a slow heat-up rate. It is also vulnerable to problems with vacuum bag failures. The result is that autoclaves have not found broad acceptance.

27.5.4 Critical Lamination Variables

Since hydraulic presses are by far the most common type of press, the following discussion focuses on that process. The quality of the ML-PWB lamination is affected by both the pressure ramp and the temperature ramp. There are almost as many unique cycles as there are fabricators. However, in general the lamination cycle can be divided into four regions:

- B-stage melt
- B stage flow
- B-stage cure
- Cooldown

Figure 27.21 shows a typical lamination cycle with these critical variables identified.

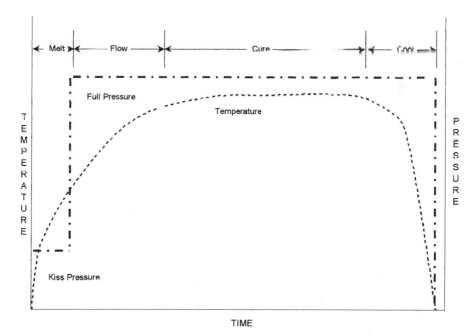

FIGURE 27.21   Typical temperature and pressure cycle for hydraulic lamination.

27.5.4.1 B-Stage Melt.   During the melt cycle, the B stage is solid and the pressure should be low. Excessive pressure during this portion of the cycle will damage the glass cloth and exaggerate problems with image print-through. Most press cycles start with a low pressure known as a kiss pressure. The kiss pressure is high enough to assure good thermal contact without damaging the ML-PWB. The length of the kiss cycle depends on the heating rates and the cure kinetics of the B stage. In a hot press cycle, the ML-PWB stack is loaded into a preheated press, and the heating rate approaches 20°C/min. In such a cycle, the kiss cycle should be limited to a few minutes, or the B stage will begin to cure before flow is complete. At the opposite extreme, the press is loaded cold, and the heating rate is determined by the heating ramp of the press. For a slow heating rate of 5°C/min, a kiss cycle of 15 min is appropriate.

27.5.4.2 B-Stage Flow. The second portion of the cycle begins when the B stage liquefies, but before its viscosity begins to rise due to curing. In this region of the cycle, the liquid B stage flows and encapsulates the circuitry. As long as the internal features are surrounded by liquid B stage, print-through is not an issue. The key to good results is to select a pressure that will allow reproducible resin flow but that does not squeeze out all available resin before the flow is stopped by the progressing cure. Once again the exact pressure depends on the temperature cycle, the B-stage viscosity characteristics, and the B-stage cure kinetics. For a fast-curing B stage and a fast temperature ramp, pressures as high as 600 psi may be needed to ensure complete circuit encapsulation. On the other hand, with a slow ramp and a long B-stage working time, high pressures give excessive flow and best results are obtained at 200 psi. Improper melt staging can result in footballing, a term used when the center of the panel is significantly thicker than the outside. To determine proper melt, it is recommended to install a thermocouple in the center stack edge to chart the actual heat rise through the flow.

27.5.4.3 B-Stage Cure. In the third part of the cycle, flow has stopped and the resin cure is proceeding. The temperature is held at its maximum value to minimize the time to obtain full cure. For a standard epoxy system this is generally 170 to 180°C for 60 to 90 min. For high-temperature materials such as polyimide and Teflon, the cure temperature can be significantly higher.

27.5.4.4 Cooldown. The last part of the cycle is the cooldown cycle. In Fig. 27.21 it is suggested that the pressure is released after some cooling has occurred, but before the stack reaches room temperature. In many modern systems, the ML-PWB stack is transferred hot to a low-pressure cooling press. It is important to control the cooling rate to minimize warpage. It is generally desirable to cool through T_g in a stress-free state without any significant thermal gradients present. A properly designed cooling press will meet these conditions.

27.5.5 Critical B-Stage Variables

During a typical lamination cycle, the B stage undergoes several significant changes. At the beginning of the cycle, the B stage is a solid with a low cross-link density and a melt temperature near 90°C. As the temperature rises, the B stage melts and becomes a high-viscosity liquid. As the press heats further, the viscosity of the liquid drops. When the B stage begins to cure, viscosity reaches a minimum and begins to rise. The region around the viscosity minimum is called the region of maximum flow. The wider this region and the lower the minimum viscosity, the more flow occurs. Figure 27.22 shows a schematic viscosity curve for a typical cure cycle.

In a high-flow B stage, the initial cure level is low. This results in a longer time at temperature before the B-stage viscosity rises due to cure. This is often described as a long gel time. In terms of Fig. 27.22, a high-flow B stage has a low minimum viscosity and a wide region of maximum flow. A low-flow B stage has a higher degree of initial cure and may include flow restrictors to increase the minimum viscosity. High-flow B stages are useful in presses with a high heating rate where the resin may begin to cure before flow is complete. They give excessive flow if used in a press cycle with a very slow heating rate.

27.5.6 Buried and Blind Via Considerations

Buried vias are generally relatively small when drilled as a clad pair, so the amount of resin that it takes to fill them is not significant. Therefore, the use of buried vias has minimal impact

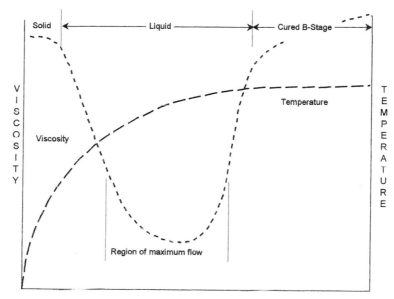

FIGURE 27.22   Typical viscosity curve for B stage during lamination.

on the lamination cycle when they are part of a thin clad. However, when buried vias are drilled as part of a buried laminated core, they may increase in diameter due to plating aspect ratio. There is some evidence that suggests that in some severe environments a partially filled via is detrimental to life cycle. This may indicate a separate filling operation. Blind vias create a special problem when laminated as predrilled details within the clad. They provide a direct path for the melted resin to escape to the board surface. For best results, a low-flow B stage minimizes the resin flow onto the board surface. In this case, normal deburring and hole cleaning processes remove the resin. Use of a special compliant release material that flows under heat will help minimize this problem.

27.6   LAMINATION PROCESS CONTROL AND TROUBLESHOOTING

A good lamination cycle produces a flat ML PWB that is free of voids and has a fully cured substrate. All layers must be well registered and the ML-PWB must be free of warp. Each of these requirements puts special demands on the lamination cycle.

To assist in setting control measures, the following process indicators (see Table 27.3) should be monitored with an SPC method.

TABLE 27.3   Process Variables and Limits

Process indicator	Specification limit	Gauge type
Thickness control	Per engineering drawing or internal spec requirement	Over-arm micrometer
Heat rise	°/min through melt per resin system requirement	Chart recorder
Cure	T_g of resin system	TMA
Postcure	Dwell time/temp per resin system requirement	Chart recorder

- *Poor thickness control:* One cause of poor flatness is print-through or "telegraphing." If a group of boards are laminated in a common press opening and all have several layers with a clearance in the copper pattern, these clearances will telegraph to the board surface. One cure for this problem is to use thick separator sheets. Another is to use a kiss cycle that delays the application of pressure until the B stage has melted. A third solution is to avoid stacking identical part numbers in the same press opening.

- *Voids and moisture:* Substrate voids are a serious problem in many lamination processes. One source of these problems is moisture. B stage is very hygroscopic and must be stored in a low-humidity environment to avoid serious void problems. C-stage components also have a tendency to absorb water, and many fabricators use a bake to dry layers prior to lamination. However, for a fast innerlayer line with a good dryer in the oxide line, an innerlayer bake is not necessary. Voids generally cluster in the low-pressure regions near the edge of the panel. This effect is minimized by the use of vacuum lamination. Increasing the lamination pressure can also reduce voids. However, the use of high pressures with high-flow materials can result in excessive flow, which leads to other substrate flaws such as resin starvation.

- *Blisters and delamination:* Blisters and delamination are also caused by trapped volatiles that collect in the low-pressure regions associated with print-through. If a board has wide copper borders on every layer, blisters are often found in the lower-pressure circuit areas adjacent to the borders. The best solution to this problem is to avoid areas of heavy copper adjacent to low-density circuit areas by replacing solid borders with a dot or stripe pattern. Also, the glass style and resin content of the prepreg should be matched to the weight of copper thickness adjacent to the prepreg.

- *Undercure:* The requirement of full cure is relatively easy to obtain if proper cure time and temperature are used. One measure of the cure level is T_g. Periodic measurements of T_g are an excellent check for material and process consistency. Another way to check cure is to make two consecutive measurements of T_g using a thermal cycle that goes to 180°C. If the epoxy is partially cured, it will cure during the first measurement and a higher T_g will be detected on the second measurement. A shift in T_g of more than 5°C is an indication of poor cure. This measurement is typically performed on a thermomechanical analyzer (TMA). Figure 27.23 shows an example of an epoxy TMA run; note that the second TMA run has a delta of 3°C.

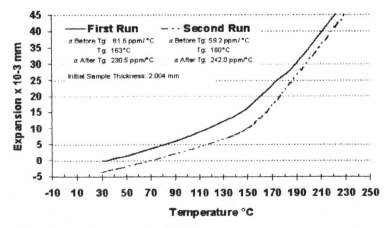

FIGURE 27.23   TMA analysis of GFG epoxy. *(Courtesy of Microtek Labs, Anaheim, CA.)*

- *Postlamination bake:* Many fabricators bake ML-PWB at 150°C for up to 4 h after lamination. One purpose of this baking is to assure a complete cure. Although a bake will advance the cure, it is unnecessary if a proper lamination cycle is used. As discussed earlier, this can be verified by T_g measurements. Additional baking beyond full cure degrades the material and reduces T_g. A second purpose of the bake is to reduce the warpage often seen in the outer boards of a stack-up. Although a postlamination bake will flatten the boards, it is more of a repair than a root solution. If the panels are cured in the press in a way that ensures they go through T_g in an isothermal stress-free state, warpage should not occur. The warpage seen in the outer boards of a stack is generally a symptom of nonuniform cooling. The third purpose of a bake is to relax internal stress and improve registration. Internal stresses are a symptom of an overdetermined tooling system. If such a system is used, a bake may improve registration. If the more popular four-pin system is used, a bake is unnecessary. Always confirm the need of a postcure with the recommendations of the material supplier.

27.7 LAMINATION OVERVIEW

The standard hydraulic vacuum press cycle is flexible and has very high output. Through the use of multiple openings and stacked lamination, high productivity is obtainable. Hydraulic lamination works effectively with all modern materials systems of today.

ML-PWB fabricators use many different press cycles, and there are B-stage formulations available for each. The most significant difference among press cycles is in the heating rate. At one extreme the ML-PWBs are loaded into a cold press. This gives a very slow heating rate, and a B stage with a low flow is needed to avoid excessive flow. This material works well at low pressures, minimizing print-through and innerlayer distortion. A vacuum cycle is recommended to minimize voiding in low-pressure regions near the panel edge. The other extreme is a hot-loaded press with a very fast temperature rise. This cycle needs a high-flow material and a high pressure to complete the resin flow cycle before the onset of cure. Although a vacuum is less important with this cycle, it will minimize edge voids.

Some fabricators use bakes both before and after lamination. The bake before lamination is designed to ensure moisture removal in layers. It is mostly needed if innerlayers are stored at high humidity prior to lamination. Other fabricators use a bake after lamination to complete the cure, reduce warp, and relieve stress. Although a postbake will achieve these goals, it is usually unnecessary in a controlled lamination process. In the case of high-temperature materials such as polyimide, cyanate ester, and PPO, a postbake is a useful way to achieve a full cure in a process where the maximum press temperature is limited.

27.8 ML-PWB SUMMARY

The printed wiring board manufacturing industry is experiencing the highest rate of change ever. Not too many years ago, the only difference in ML-PWB manufacturing was related to what brand of chemistry was being run. Today's global market has introduced innovative change to meet the technology demands of a robust electronics industry. The manufacturing flow no longer resembles a standard methodology. The discussion in this chapter is a review of tried-and-true manufacturing processes for insertion of some of the more recent packaging innovations. A successful implementation of any of the high-density constructs discussed should always be complemented with sound statistical process verification. Sound benchmarking standardization is one tool for establishing manufacturing capability. Additionally, design standardization is required to clearly communicate technologies based on application. This will provide the needed stability within the rapidly changing environment.

CHAPTER 28
PREPARING BOARDS FOR PLATING

Edward F. Duffek, Ph.D.
Adion Engineering Company, Cupertino, California

28.1 INTRODUCTION

A major part of manufacturing printed circuit boards involves wet process chemistry. The plating aspects of wet chemistry include deposition of metals by electroless (metallization) and electrolytic (electroplating) processes. Topics to be described here are multilayer processing, electroless copper, direct metallization, electroplating of copper and resist metals, nickel and gold for edge connector (tips), tin-lead fusing, and alternative coatings. Specific operating conditions, process controls, and problems in each area will be reviewed in detail. The effects of plating on image transfer, strip and etching are also described in this chapter. See printed circuit plating flowchart in Fig. 28.1.

Two driving forces have had major influence on plating practices: the precise technical requirements of electronic products and the demands of environmental and safety compliance. Recent technical achievements in plating are evident in the capability to produce complex, high-resolution multilayer boards. These boards show narrow lines (3 to 6 mil), small holes (12 mil), surface-mount density, and high reliability. In plating, such precision has been made possible by the use of improved automatic, computer-controlled plating machines, instrumental techniques for analysis of organic and metallic additives, and the availability of controllable chemical processes. Mil-spec-quality boards are produced when the procedures given here are closely followed.

28.2 PROCESS DECISIONS

Process and equipment needs dictate the physical aspect of the facility and the character of the process, and vice versa. Some important items to consider are the following.

28.2.1 Facility Considerations

1. *Multilayer and two-sided product mix:* Need for lamination presses and inner layer processes.
2. *Circuit complexity:* Need for dry film, photoimageable resist, and clean room.

3. *Level of reliability (application of product):*   Need for extra controls and testing.

4. *Volume output:*   Need for equipment sizing and building space.

5. *Use of automatic versus manual line:*   Need for productivity, consistency, and workforce.

6. *Wastewater treatment system:*   Need for water and process control capability.

7. *Environmental and personnel safety; compliance to laws.*

8. *Costs.*

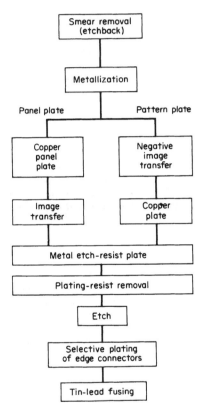

FIGURE 28.1   Printed wiring board plating flowchart.

28.2.2 Process Considerations

1. *Material:*   The principal printed board material discussed will be NEMA grade FR-4 or G-10, i.e., epoxy-fiberglass clad with ½-, 1-, or 2-oz copper. Other materials will be briefly mentioned because they can significantly alter plating and related processes.

2. *Standard:*   Plated-through-hole (PTH) is the current standard of the industry. The following purposes, objectives, and requirements apply to both multilayer and two-sided boards.

   a.   Purposes
      Increased circuit density
      Double-sided circuitry
   b.   Objectives
      Side-to-side electrical connection
      Ease of component attachment
      High reliability
   c.   Requirements
      Complete coverage
      Even thickness
      Hole-to-surface ratio
      0.001 in minimum
      No cracks
      No voids, nodules, inclusions
      No pullaway
      No epoxy smear
      Minor resin recession
      Optimum metallurgical structure
      M/L compatibility

3. *Image transfer.*   Photoimageable, dry film or screening of plating resists will depend on board complexity, volume, and labor skills.

4. *Electroless copper.*   The type chosen will depend on the method of image transfer as well as on the need for panel plating. These processes are readily automated. Currently, 95 percent of printed circuit board manufacturers worldwide rely on the electroless copper method for hole metallization.

5. *Direct metallization technology (DMT).*   Some of the remaining printed circuit board manufacturers have eliminated the electroless copper step and converted to DMT. Developed in the 1980s, DMT methods produce a conductive surface on the nonconductive through-hole surfaces. Electroless copper baths contain formaldehyde and chelators. In addition, the baths use large volumes of water and are difficult to control and waste treat.[1] DMT claims include increased productivity, ease of control, and lower hazardous material involvement. Because of these characteristics, DMT will probably become the principal

method in the next 5 to 10 years.[2] At this time, acceptance of the DMT process has been delayed by the high cost of conveyorized equipment and chemicals.

DMT primary technologies include:
  a. Palladium
  b. Carbon-graphite
  c. Conductive polymer
  d. Other methods

See Chap. 30 for further information.

  6. *Electroplating processes.*   Deposit requirements are as follows:

  Electrical conductivity

  Mechanical strength

  Ductility and elongation

  Solderability

  Tarnish and corrosion resistance

  Etchant resistance

  Compliance to Mil-specification

  Details emphasizing operation, control and mil-spec plating practices are given elsewhere. Metal plated structures of completed PC boards are as follows:

  Copper/tin-lead alloy

  Copper/tin (SMOBC)

  Copper/tin-nickel (nickel)/tin-lead

  Copper/nickel/tin

  Nickel/silver

  7. *Strip, etching, tin-lead fusing.*   Methods required by these steps are determined by the preceding processes and by the need for automation.

28.3   PROCESS FEEDWATER

28.3.1   Water Supply

Printed circuit board fabrication and electronic processes require process feedwater with low levels of impurities. Large volumes of raw water must be readily available, either of suitable quality, or else treatable at reasonable cost. New facilities must consider water at an early stage of the site selection and planning process. Zero discharge, although a desirable goal, is very costly and difficult to achieve.

28.3.2   Water Quality

Highly variable mineral content causes board rejects and equipment downtime, as well as reduced bath life, burdened waste treatment, and difficult rinse water recovery. Many water supplies contain high levels of dissolved ionic minerals and possible colloidals that can cause rejects in board production. Some of these impurities are calcium, silica, magnesium, iron, and chloride. Typical problems caused by these impurities are copper oxidation, residues in the PTH, copper-copper peelers, staining, roughness, and ionic contamination. Problems in the equipment include water- and spray-line clogging, corrosion, and breakdown. The best plating practices suggest good-quality water for high yields. The need for water low in total dissolved solids (TDS), calcium hardness, and conductivity is well known. Good water eliminates the concern that the water supply may be responsible for rejects. Although water quality is not well defined for plating and PC board manufacturing, for general usage, some guidelines can be assigned as follows. Where high-purity water is required, see Sec. 28.3.3.

Typical quantities are:

Total dissolved solids (TDS)	4 to 20 ppm
Conductivity	8 to 30 μS/cm
Specific resistance	0.12 to 0.03 MΩ
Carbonate hardness (CaCO$_3$)	3 to 15 ppm

Somewhat higher values are acceptable for less-critical processes and rinses.

28.3.3 Water Purification

Two processes widely used for water purification are reverse osmosis and ion exchange. In the reverse osmosis technique, raw water under pressure (1.4 to 4.2 MPa or 200 to 600 lb/in^2) is forced through a semipermeable membrane. The membrane has a controlled porosity which allows rejection of dissolved salts, organic matter, and particulate matter, while allowing the passage of water through the membrane. When pure water and a saline solution are on opposite sides of a semipermeable membrane, pure water diffuses through the membrane and dilutes the saline water on the other side (osmosis). The effective driving force of the saline solution is called its *osmotic pressure.* In contrast, if pressure is exerted on the saline solution, the osmosis process can be reversed. This is called the *reverse osmosis process* (RO), and involves applying pressure to the saline solution in excess of its osmotic pressure. Fresh water permeates the membrane and collects on the opposite side, where it is drawn off as product. Reverse osmosis removes 90 to 98 percent of dissolved minerals and 100 percent of organics with molecular weights over 200, as shown in Table 28.1.

A small quantity of dissolved substances also facilitates deionized (DI) water production, wastewater treatment, and process rinse water recovery, since it makes recycling less costly and more feasible. An RO system will result in lower costs for DI water preparation and for process water recycling. The setup for recycling requires additional equipment for polymer addition, filtration, and activated carbon treatment.

TABLE 28.1 Purified Water Supply Values
Typical in/out RO values

TDS, ppm	SiO$_2$, ppm	Conductivity, μS	Hardness-CaCO$_3$, ppm
170/4	30/1	130/8	24/1
240/7	45/2	200/14	35/2
300/10	60/2	250/20	45/3

Deionized (DI) water purification is used when high-purity water is required, for example, in bath makeups, rinses before plating steps, and final rinses necessary to maintain low ionic residues on board surfaces. Mil-spec PC boards must pass the MIL-P-28809 test for ionic cleanliness. This is done by final rinsing in DI water. Deionized water is made by the ion exchange technique. This involves passing water containing dissolved ionics through a bed of solid organic resins. These convert the ionic water contents to H$^+$ and OH$^-$. Deionized water systems are more practical when using feedwater low in ionic and organic content.

Other requirements are:

pH	6.5–8.0
Total organic carbon	2.0 ppm
Turbidity	1.0 NTU
Chloride	2.0 ppm

28.4 MULTILAYER PTH PREPROCESSING

Two-sided printed circuit boards are usually processed by first drilling and deburring, followed by the through-hole metallization. Multilayers require treatment involving resin smear removal and etchback prior to the electroless copper or DM

28.4.1 Smear Removal

Drill smear refers to the epoxy resin that coats the innerlayer copper surface in the drilled hole and is due to heating during the drilling operation. Control of this smear is difficult due to variability in dielectric materials, inconsistency in curing stage, and poor drill quality. This smear must be removed before metallization to get full electrical continuity from the innerlayer copper to the PTH. The innerlayer connections will be flush with the drilled hole after smear removal. See Figs. 28.2 to 28.4.

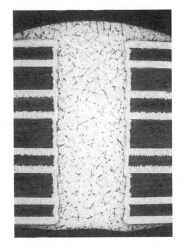

28.4.2 Etchback

This term refers to the continuation of the smear removal process to expose 0.5 mil on the top and bottom surface of the innerlayer copper. Physically, the innerlayer copper will now protrude from the drilled hole three-point connection for copper bonding, which is required on some mil-spec boards (see Fig. 28.5.)

28.4.3 Smear Removal/Etchback Methods

The four methods commonly used utilize hole-wall epoxy or dielectric oxidation, neutralization reduction, and glass etching.

28.4.3.1 Sulfuric Acid.   This process has been used extensively for many years because of its ease of operation and reliability of results. A major disadvantage is a lack of control, which leads to hole-wall pullaway and rough holes. Operator safety is crucial since concentrated sulfuric must be used.

28.4.3.2 Chromic Acid.   This method provides more control and longer bath life. However, problems with copper voids due to Cr^{+6} poisoning, wastewater pollution, and contamination of plating processes must be considered. Etchback is possible by double processing.

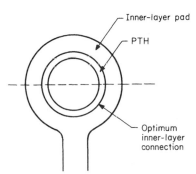

Inner-layer pad

PTH

Optimum inner-layer connection

FIGURE 28.2 PTH vertical and horizontal cross sections illustrating optimum innerlayer connection and smear removal.

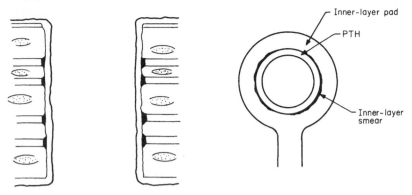

FIGURE 28.3 PTH vertical and horizontal cross sections illustrating innerlayer smear.

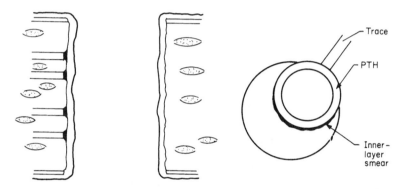

FIGURE 28.4 PTH vertical and horizontal cross sections illustrating innerlayer smear and misregistration.

28.4.3.3 Permanganate. This method is rapidly gaining acceptance due to improved copper adhesion to the hole-wall (less pullaway), smoother PTHs, and better control.[3] Problems result from sludge by-product formation, dark copper holes, and the operators' lack of experience. Permanganate is also used as a second step in conjunction with the other methods to enhance hole quality.[4]

28.4.3.4 Plasma. This is a dry-chemical method in which boards are exposed to oxygen and fluorocarbon gases. Persistent problems with the process are nonuniform treatment of holes—that is, higher etch rates on the edges—and the high cost of equipment. This process has few steps and eliminates the use of large quantities of chemicals. Controls must be provided to prevent air pollution.

28.4.4 Process Outline: Smear Removal and Etchback

The four common methods for smear removal and etchback are given in the following table. Combinations of these methods are also in use because of added reliability to both process and product.

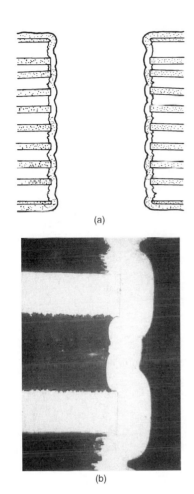

(a)

(b)

FIGURE 28.5   PTH vertical cross section illustrating optimum innerlayer connection and etchback.

28.4.4.1 Smear Removal and Etchback Processes*

Sulfuric Acid	**Chromic Acid**
Rack panels	Rack panels
Sulfuric acid, 96% 20-s dwell, 15-s drain, room temperature	Chromic acid 3 min, 140°F
Neutralizer 8 min, 125°F	Reducer 3 min, room temperature
Ammonium bifluoride 3 min, room temperature	Ammonium bifluoride 3 min, room temperature
Unrack panels	Unrack panels
High-pressure hole cleaning	High-pressure hole cleaning
Release to electroless copper/DM	Release to electroless copper/DM

Permanganate	**Plasma Etch**
Solvent conditioner 90°F, 5 min	Plasma Oxygen, CF_4, 30 min
Alkaline permanganate 170°F, 10 min	Glass etch (optional)
Neutralizer-120°F, 5 min	High-pressure hole cleaning
Glass etch 4 min, room temp	Release to electroless copper/DM
High-pressure hole cleaning (optional)	
Release to electroless copper/DM	

Polyimide and polyimide-acrylic systems are processed in chromic acid or plasma. Teflon® and R T Duroid®† materials are treated (before operations) in sodium-naphthalene mixtures to yield void-free, high-bond-strength copper in the PTH.

28.5 ELECTROLESS COPPER[5–10]

28.5.1 Purpose

This second series of chemical steps (after smear removal) is used to make panel side-to-side and innerlayer connections by metallizing with copper. The process steps needed include racking, cleaning, copper microetching, hole and surface catalyzing with palladium, and electroless copper. Typical steps are as follows:

---

* Water rinses after each step are not shown.
† Teflon is a registered trademark of E. I. Du Pont de Nemours & Co. R T Duroid is a registered trademark of Rogers Corporation, Chandler, Ariz.

1. *Cleaner-conditioner.* Alkaline cleaning is used to remove soils and condition holes.

2. *Microetch.* This slow acid etching is used for removal of copper surface pretreatments, oxidation, and presentation of uniformly active copper. Persulfates and sulfuric acid–hydrogen peroxide solutions are commonly used.

3. *Sulfuric acid.* Used for removal of persulfate residues.

4. *Predip.* Used to maintain balance of the next step.

5. *Catalyst (activator).* Neutral or acid solutions of palladium and tin are used to deposit a thin layer of surface active palladium in the holes and on the surface.

6. *Accelerator (postactivator).* Used for the removal of colloidal tin on board surfaces and holes.

7. *Electroless copper.* Alkaline chelated copper reducing solution that deposits thin copper in the holes (20 to 100 µin) and surfaces.

8. *Antitarnish.* This is a neutral solution that prevents oxidation of active copper surfaces by forming a copper conversion coating.

28.5.2 Mechanism

Equations 28.1 and 28.2 illustrate the process.

$$Pd^{+2} + Sn^{+2} \rightarrow Pd + Sn^{+4} \tag{28.1}$$

$$CuSO_4 + 2\ HCHO + 4\ NaOH \xrightarrow{Pd} Cu + 2\ HCO_2Na + H_2 + 2\ H_2O + Na_2SO_4 \tag{28.2}$$

28.5.3 Electroless Copper Processes

Selection from several types available depends on the type of image transfer desired. Operation and control of three bath types and the function of constituents are given in Tables 28.2 and 28.3.

TABLE 28.2 Electroless Copper Processes
Operation and control

	Low deposition	Medium deposition	Heavy deposition
Copper	3 g/l	2.8 g/l	2.0 g/l
HCHO	6–9 g/l	3.5 g/l	3.3 g/l
NaOH	6–9 g/l	10–11 g/l	8 g/l
Temperature	65°–85°F	115° ± 5°F	115° ± 5°F
Air agitation	Mild	Mild/moderate	Moderate
Filtration	Periodic	Continuous	Continuous
Tank design	Static	Overflow, separate sump	Overflow, separate sump
Heater	Teflon	Teflon	Teflon
Panel loading	0.25–1.5 ft²/gal	0.1–2.0 ft²/gal	0.1–2.0 ft²/gal
Replenish mode	Manual	Manual or continuous	Automatic
Idle time, control	70–85%	Turn off heat	Turn off heat
Deposition time	20 min	20 min	20–30 min
Thickness	20 µin	40–60 µin	60–100 µin

TABLE 28.3 Electroless Copper
Function of constituents

	Constituent	Function
Copper salt	$CuSO_4 \cdot 5 H_2O$	Supplies copper
Reducing agent	HCHO	$Cu^{+2} + 2e \rightarrow Cu^0$
Complexer	EDTA, tartrates, Rochelle salts	Holds Cu^{+2} in solution at high pH; controls rate
pH controller	NaOH	Controls pH (rate) 11.5–12.5 optimum for HCHO reduction
Additives	NaCN, metals, S, N, CN organics	Stabilize, brighten, speed rate, strengthen

The problems encountered with this system are as follows:

1. *Uncoppered holes.*   This may appear as dark, hazy, or voided copper in the holes. To correct this, check the operation of the smear remover, cleaner, catalyst, accelerator, and the items listed in Table 28.2. Voids may also be due to copper plate precleaner attack.

2. *Hole wall pullaway.*   This refers to copper pulling off the PTH, and is observed either as large blisters or by cross sectioning. This may be due to spent sulfuric acid smear removal, or to items given above. Pullaway is controlled by maintaining copper plating smooth and over 1-mil thickness. This may also be observed on solder mask on bare copper (SMOBC) boards due to shock of hot solder immersion.

3. *Bath decomposition.*   This is rapid plating out of the copper. Common causes are bath imbalance, overloading, overheating, lack of use, tank wall initiation, or contamination.

4. *Electroless copper to copper-clad bond failure.*   Review initial step in process and items in 1.

5. *Staining.*   Copper oxidation is due to moisture or contamination on the copper surface. Corrective action involves dipping boards in antitarnish or hard rinsing in DI water.

28.5.4 Process Outline

This outline presents the typical steps in an electroless copper process:

1. Rack
2. Clean and condition
3. Water rinse
4. Surface copper etch (microtech)
5. Water rinse
6. Sulfuric acid (optional)
7. Water rinse
8. Preactivator
9. Activator (catalyst)
10. Water rinse
11. Postactivator (accelerator)
12. Water rinse
13. Electroless copper

14. Rinse

15. Sulfuric acid or antitarnish

16. Rinse

17. Scrub (optional)

18. Rinse

19. Copper flash plate (optional)

20. Dry

21. Release to image transfer

REFERENCES

1. M. Carano, *Proceedings 16th AESF/EPA Pollution Prevention & Control Conference,* 1995, p. 179.
2. K. Nargi-Toth, *Printed Circuit Fabrication,* vol. 15, no. 34, September 1992.
3. C. A. Deckert, E. C. Couble, and W. F. Bonetti, "Improved Post-Desmear Process for Multilayer Boards," *IPC Technical Review,* January 1985, pp. 12–19.
4. G. Batchelder, R. Letize, and Frank Durso, *Advances in Multilayer Hole Processing,* MacDermid Company.
5. F. E. Stone, *Electroless Plating—Fundamentals and Applications,* Chap. 13, G. Mallory and J. B. Hajdu (eds.), American Electroplaters and Surface Finishers Society, Inc., Fla. 1990.
6. C. A. Deckert, "Electroless Copper Plating," *ASM Handbook,* vol. 5, 1994, pp. 311–322.
7. J. Murray, "Plating, Part 1: Electroless Copper," *Circuits Manufacturing,* vol. 25, no. 2, February 1985, pp. 116–124.
8. F. Polakovic, "Contaminants and Their Effect on the Electroless Copper Process," *IPC Technical Review,* October 1984, pp. 12–16.
9. K. F. Blurton, "High Quality Copper Deposited from Electroless Copper Baths," *Plating and Surface Finishing,* vol. 73, no. 1, 1986, pp. 52–55.
10. C. Lea, "The Importance of High Quality Electroless Copper Deposition in the Production of Plated-Through Hole PCBs," *Circuit World,* vol. 12, no. 2, 1986, pp. 16–21.

CHAPTER 29
ELECTROPLATING

George Milad
SHIPLEY RONAL, Inc., Marlborough, Massachusetts

29.1 INTRODUCTION

The reality in printed circuit evolution is that the parts are continuing to increase in degree of difficulty and complexity. A new genre of high-density interconnect (HDI) boards is making the transition from leading edge to mainstream. These boards are characterized by combining a series of complexity features that include buried and blind vias, high-aspect ratio plating, small-hole plating (as low as 6-mil-diameter holes), and fine lines and spaces side by side with ground plane areas of different sizes.

The industry still has a need to produce simpler product, single-sided, and double-sided boards; these are still in demand. Lower-layer-count (4 to 6), multilayer boards fall in this category also.

This chapter will focus on all aspects of electroplating. Emphasis will be on acid copper plating as the main process for providing interconnection. Tin, tin/lead, nickel, and gold electroplating will also be covered as important plated materials. In addition, the latest innovations in technology for meeting the challenges of product complexity, such as pulse plating and horizontal conveyorized plating, are discussed in detail.

29.2 ELECTROPLATING BASICS

Electroplating is the production of adherent deposition of conductive surfaces by the passage of electric current through a conductive metal-bearing solution. The rate of plating depends on current and time and is expressed by Faraday's law:

$$W = (ItA)/(nF) \tag{29.1}$$

where W = metal, g
I = current, A
t = time, s
A = atomic weight of the metal
n = number of electrons involved in metal ion reduction
F = Faraday's constant (96485 C/mol)

Plating occurs at the cathode (the negative electrode). Accordingly, deposition thickness is determined by time and by the current that is impressed on the surface being plated; for example, by using the preceding formula, one can readily calculate the weight of the deposited metal. The weight can then be converted to a specific thickness over a known area. The rate of deposition of the most common metals is shown in Table 29.1.

TABLE 29.1 Rate of Deposition of the Most Common Metals

Metal	Grams deposited per amp.hr	Amp. hr. per sq.ft. to deposit 0.001 in	Amp. hr. per sq.dm. to deposit 25 μm
Copper	1.186	17.8	1.88
Tin	2.214	7.8	0.82
Lead	3.865	6.9	0.73
Nickel	1.095	19.0	2.00
Gold	7.348	6.2	0.65

To plate 1.0 mil of copper, one would need 17.8 amperes per square foot (ASF) or 1.88 amperes per decimeter (ASD) for 1 h (60 min). Properties of electrodeposits are shown in Table 29.2.

TABLE 29.2 Properties of Electrodeposits

Property	Cu	Ni	Au	SnPb	Sn
Melting point, °F	1980	2600	1945	361	450
Hardness, VHN	150	250	150	12	4
CTE X 10^{-6}/°F	9.4	8.0	8.2	12.2	12.8
Conductivity, %IACS	101	25	73	11.9	15.6
Electrical resistivity, μΩ/cm	1.67	6.8	2.19	14.5	11.1
Thermal conductivity, CGS°C	0.97	0.25	0.71	0.12	0.15

29.3 HORIZONTAL ELECTROPLATING

As the degree of part complexity continues to increase, new innovative plating solutions are being put into place. *Periodic pulse reverse* (PPR) is an example of this, and another major development is the advent of *horizontal acid copper plating*.

Here, the panel is transported in a horizontal mode through the plating equipment with stationary anodes below and above it. The panel continues to plate as it travels, interfacing with one anode after the other. The panel comes off the end of the line plated.

The plating thickness is a function of time and current density. The dwell time is determined by the length of the plating module and the conveyor speed. Plating at 20 ASF for 60 min in horizontal equipment running at 1 m/min would require a plating module that is 60 m long. This length is not practical. Current density has to be increased to 40, 60, 80, and higher ASF. At 80 ASF, the equipment length needs only to be 15 m.

New chemical additives were formulated to accommodate high-current-density plating. However, this current density offered new challenges: at high-current densities, most plating solutions lose throwing power. Another problem that was encountered early on was that the anode film particulates would land on the top side of the panel and eventually create nodulation. This did not occur on the bottom side.

Two major developments helped make horizontal processing popular:

1. Pulse plating, coupled with solution impingement, resolved the throwing power problem of plating at high current densities.

2. Nodulation was resolved by taking the copper dissolution out of the module and replacing the standard anode with an inert or insoluble anode. Copper dissolution or oxidation was achieved by different means, such as the use of rectification, the use of ozone, or the dissolution of copper oxide.

29.3.1 Advantages of Horizontal Processing

The most dramatic advantage of horizontal plating is the uniformity from panel to panel. All panels go through the equipment in an identical fashion. They all see the entire bank of anodes, and they are in a single plating solution. All panels coming off a horizontal line are alike. If, for some reason, one edge of the panel has higher plating, then all panels will have the same edge equally high. In vertical plating, different panels are plated in different parts of the cell, in different positions in the rack, and even in different tanks or baths. Although vertically plated panels may all be within specifications, there will always be inherent variability that the shop has to contend with in downstream processing beyond the plating line.

Horizontal plating machines can be easily automated. Loaders and unloaders are usually integrated in the system. Automation of chemical analysis and dosing is incorporated with minimum cost increase. The equipment is easily linked to pre- and postprocesses. Horizontal processing lends itself well to continuous production flow and reduced cycle time.

Anode uniformity is a major advantage in horizontal plating. The size and geometry of an insoluble inert anode do not change. Anode variability and maintenance are eliminated. The latter is a very labor-intensive and hazardous activity associated with vertical plating.

Horizontal equipment is completely enclosed. This minimizes the operator and shop personnel's exposure to acid and chemical fumes and creates a better and safer working environment.

Horizontal plating gives superior surface thickness distribution. Variation of less than 10 percent across a panel-plated surface is normal. This is due to the anode's close proximity to the cathode; usually, the anode-to-cathode spacing is between 8 and 10 mm, or 0.3 and 0.4 in.

In addition, surface-to-hole thickness variation is minimized. Excellent throwing power (80 to 100 percent) is achievable in blind vias (aspect ratios greater than 1:1 [i.e., with depth exceeding hole diameter]) and in through-holes (with aspect ratios of 10:1). This is the result of solution dynamics coupled with periodic pulse reverse rectification. This feature makes horizontal pulse plating an enabling technology for HDI applications. Figures 29.1 and 29.2 show the difference in throwing power of DC vs. PPR plating in a 1:1 via hole. Figure 29.3 shows an example of horizontal PPR plating of a via with an aspect ratio greater than 1, i.e., the depth is greater than the diameter of the via.

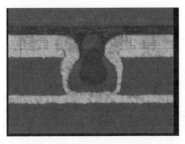

Horizontal DC	
Via Diameter	90 μm
Via Depth	72 μm
Surface Thickness	19 μm
Base Thickness	4 μm
Throwing Power	25%
Current Density	7 ASD

FIGURE 29.1 Horizontal PTH/DC-Plated Resin-Coated Copper™ microvia.

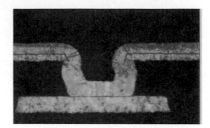

Horizontal Pulse	
Via Diameter	90 μm
Via Depth	72 μm
Surface Thickness	17 μm
Base Thickness	15 μm
Throwing Power	88%
Current Density	7 ASD

FIGURE 29.2   Horizontal PTH/Pulse-Plated Resin-Coated Copper™ microvia.

Copper Gleam PPR-H	
Via Diameter	120 μm
Via Depth	160 μm
Surface Thickness	17 μm
Base Thickness	17 μm
Throwing Power	100%
Current Density	8 ASD

FIGURE 29.3   Horizontal PTH/pulse-plated microvia.

Horizontal equipment is capable of handling thin material (as thin as 10 mils). In the vertical mode, thin parts are a challenge to rack and to transport. Core plating requires the use of a metal frame to add rigidity to the part so that it can be handled and agitated in the bath during the plating cycle. This adds labor and creates handling defects.

29.3.2 Drawbacks of Horizontal Processing

To date, most horizontal equipment is designed around panel plating. Pattern plating has its advantages and is being worked on by a series of equipment suppliers and could be available in the near future.

The equipment requires a substantial investment and is a major deviation from standard vertical processing. There is both a learning curve and engineering know-how that will be needed to run this equipment smoothly. The equipment is integrated, and either it all works or it all doesn't. It requires a large parts inventory and trained equipment maintenance crew to ensure uninterrupted operation.

29.4 COPPER ELECTROPLATING GENERAL ISSUES

Copper is the most commonly used metal in the structure of a printed circuit board. Copper has high electrical conductivity, strength, ductility, and low cost. It is readily plated from simple solutions, and it is readily etched. MIL-STD-275 states that electrodeposited copper shall be in accordance with MIL-C-104550, and shall have a minimum purity of 99.5 percent as determined by ASTM-E-53, and shall be 0.001 in (1.0 mil). Requirements for good soldering also indicate the need for 1.0 mil of copper and smooth holes. Copper plating is generally regarded as the slow step in manufacturing PC boards.

Copper electrolytic plating of printed circuit boards is a key process step in circuit manufacturing. After all, it forms the traces that carry the signal. The physical properties of the deposit, specifically tensile strength and elongation, have to be controlled so the product can withstand the temperature excursions of assembly and of use without cracking. A successful copper-plating system must be able to provide a copper deposit with tensile strength between 40 and 50 kpsi and percent elongation between 10 and 25 percent. As demands for speed and impedance control continue to increase, the thickness uniformity of the plated copper becomes extremely critical.

In the late 1990s, significant developments occurred in HDI boards, usually for portable products. HDI has traditionally focused on the high-end telecommunications, imaging, and computing markets. Handheld electronics products present unique challenges in terms of size, weight, functionality, and cost. This poses a series of challenges to the board manufacturer in every processing area. Acid copper electroplating is quickly becoming the limiting factor in most board shops today. High–aspect ratio plating, 3-mil lines and spaces, laser-drilled microvias, as well as buried vias, are forcing the industry to look for innovative approaches to the traditional methods of plating.

29.4.1 Pattern Plating vs. Panel Plating

Panel plating does not involve any patterning or imaging before plating and hence has a uniform surface geometry; with good plating practice, less than 10 percent thickness variation across the panel and from panel to panel is achievable. The challenge here is to get good throwing power so as not to overplate the surface to achieve 1.0 mil in the hole. An overplated surface limits the line width and spaces that can be circuitized by subsequent etching.

Pattern plating, as the name implies, occurs after pattern or image transfer and, accordingly, is the real challenge for the plating process. Here, the surface geometry is not uniform and may have extremes of isolated traces and pads, as well as ground plane areas. This gives rise to great disparity in primary current distribution, which can result in two to four times thicker copper in the isolated areas, compared with the ground plane areas.

29.4.2 Thickness Distribution

A formidable challenge facing acid copper plating of printed circuit boards is thickness distribution on the surface of the panel (whether pattern or panel plated) and thickness distribution in the barrel of the hole or the via, namely throwing power. If the throwing power is 100 percent, then plating 1 mil of copper in the hole would result in plating 1 mil on the surface. At 50 percent throwing power, to achieve 1.0 mil in the hole would result in 2.0 mils on the surface, which would limit the etching capabilities in the panel plate and increase the thickness variation in pattern plate.

HDI compounds the plating challenges with microvias, small holes, and fine lines. Different approaches are being practiced to resolve the following four plating issues:

1. Low-current-density plating
2. Chemically mediated process
3. Electrically mediated process
4. Horizontal conveyorized systems

29.4.3 Additives in Acid Copper Plating

Additives used for bright acid copper plating are divided into three categories: (1) the carrier or suppressor, (2) the additive or brightener, and (3) a leveling component. Each component has a specific role in controlling the quality of the deposit. Polarization curves that show the ensuing current as the voltage is increased are used to study the effects of the various components on the plating system and how they affect the deposit.

Figure 29.4 shows a polarization curve where the current is plotted against the voltage. The data can be obtained from a potentiostat setup, shown in Fig. 29.5. Here, we have three electrodes in the plating solution. The first is a *counter electrode* and is usually made of copper or platinum; the second is the *working electrode* and is a rotating platinum disk; and the third is the *reference electrode,* like a double-junction calomel. Voltage is applied between the working and the reference electrodes, and the ensuing current is measured between the working and the counter electrodes. This is the principle upon which *cyclic voltametric stripping* (CVS) is based. CVS equipment is commonly used to assay the additives in the plating solution. A closer look at the polarization curve shows four distinct areas:

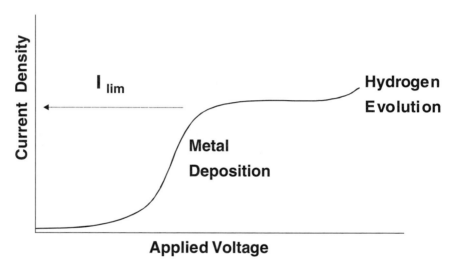

FIGURE 29.4   Acid copper: cathodic polarization curve.

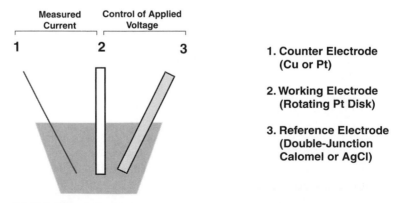

FIGURE 29.5   Acid copper: polarization study—three-electrode cell.

1. *Low-current area.* Here, there is a minimum increase in current as the voltage is increased.
2. *Metal deposition area.* This is an area where there is a significant increase of current with applied voltage.
3. *Limiting-current-density area.* This area is marked by a plateau, where there is no increase of current as the applied voltage continues to rise.
4. *"Hydrogen evolution" part of the curve.* This is the last area.

The area of most interest to plating, of course, is the area of metal deposition. The slope of this part of the polarization curve can be altered (suppressed or accelerated) by the addition of organic additives to the plating bath.

29.4.4 Carrier/Suppressor

Carriers are large-molecular-weight poly-oxyalkyl-type compounds. Figure 29.6 shows the effect of the carrier on the polarization curve. The addition of a carrier alone did not alter the polarization curve; however, when this is done in the presence of 10 ppm of the chloride ion, a marked suppression is observed initially but is not sustained as voltage continues to increase. At 60 ppm of chloride, the suppression is sustained; the result is that it now takes more voltage to produce the desired current.

The *suppression* is a result of the effect of the carrier on the diffusion layer (also referred to as the *Helmholtz double layer*). The carrier is adsorbed to the surface of the cathode; this results in increasing the thickness of the diffusion layer. The result is better organization. This gives rise to a deposit with a tighter grain structure (see Fig. 29.7). The carrier-modified diffusion layer also improves plating distribution and throwing power, without burning the deposit.

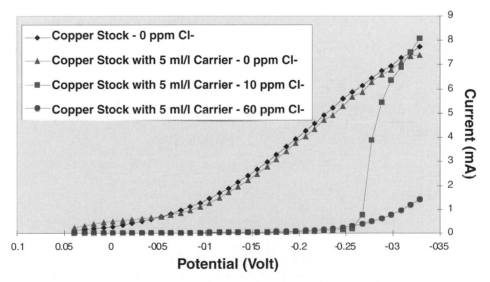

FIGURE 29.6 Effect of chloride concentration in a copper stock solution with carrier.

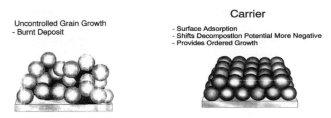

FIGURE 29.7 The effect of a carrier on the grain structure of the deposited copper.

29.4.5 Additive/Brightener

These are typically small-molecular-weight disulfide compounds, with the general structure R-S-S-R. The R groups are organic functional groups and vary from one brightener system to the other. Figure 29.8 shows the effect of a brightener on a suppressed polarization curve. Basically, the brightener increases the current and reduces the suppression. The brightener plays a key role in determining the physical properties of the deposit. It is a grain refiner and, as such, it has a direct enhancing effect on the physical properties of the deposit, such as tensile strength and percent elongation.

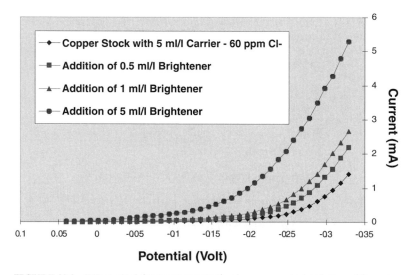

FIGURE 29.8 Effect of brightener concentration in a copper stock solution with carrier.

29.4.6 Levelers

Levelers are used to offset localized high-current-density areas that result from high points or sharp edges in the plating surface. An example would be the dog-bone effect, occasionally seen at the knee of the hole.

Levelers are a class of compounds, typically aromatic, which adsorb preferentially at high points in the plating topography due to the increased mass transport possible at these sites. The adsorption of the leveler at these sights creates localized suppression, allowing plating in the lower-current-density areas to catch up with high areas—hence, the leveling effect.

29.4.7 Low-Current-Density Plating

One way to achieve good distribution is the use of a low-current acid copper-plating system, which is designed specifically to give a deposit with the required physical properties at reduced current density. Some shops are presently plating HDI boards at 6 to 8 ASF for a period of 4 to 6 h. This will give excellent plating distribution, due to the increased suppression at these low plating voltages.

Low-current-density plating baths or electrolytes are characterized by lower copper concentration and increased sulfuric acid concentrations (refer to Table 29.3). The additive package of carrier and brightener is specifically formulated to operate at these low current densities.

These types of products are in use throughout the industry and are tolerated as long as the numbers of HDI boards are a small percentage of the overall production. This practice, however, dramatically reduces the output of the plating line and increases the cost of the part. As the percentage of HDI boards increased, the need for a different answer became apparent.

29.4.8 Chemically Mediated Process

Here, the carrier is engineered for maximum chemical suppression, even at a high current density. This is demonstrated by the shift in the Tafel slope (see Fig. 29.9). These baths are capable of good throwing power and surface distribution while plating at relatively high current densities. "Super-throw" baths are capable of 85 percent throw in a 5:1 aspect ratio board while plating at 30 and at 35 ASF. This kind of bath contrasts with the low-current bath and can improve productivity significantly, while still operating under conventional direct-current (DC) plating. The system fits readily into existing in-house equipment and in-house know-how. Figure 29.10 shows an example of a blind via that is 5.0 mil in diameter and 3.0 mil deep, plated with this type of bath at 20 ASF for 90 min.

TABLE 29.3 Operation and Control

Operating variable	Low current plating	Conventional density plating	High-speed CD* plating	plating
Copper	1–2 oz/gal		2–3 oz/gal	7–8 oz/gal
Copper sulfate	4–8 oz/gal		8–12 oz/gal	28–32 oz/gal
Sulfuric acid	25–32 oz/gal		22–28 oz/gal	5–8 oz/gal
Chloride	40–80 ppm		40–80 ppm	40–80 ppm
Additives	As required		As required	As required
Temperature	75–85°F		70–85°F	75–100°F
Cathode CD	5–15 ASF		20–40 ASF	40–150 ASF
Anodes				
Type	Bars or baskets			
Composition	Phosphorized 0.04–0.06% P			
Bags	Closed-napped polypropylene			
Hooks	Titanium or Monel			
Length	2 in shorter than rack			
Anode CD	25–50% of cathode CD			
Properties				
Composition	99.8% (99.5% min, ASTM-E-53)			
Elongation	10–25% (6% min, ASTM-E-8 or ASTM-E-345)			
Tensile strength	40–50 kpsi (36 kpsi min, ASTM-E-8 or ASTM-E-345)			

* CD, current density.

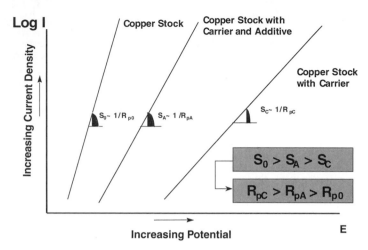

FIGURE 29.9 Tafel plots: direct current.

29.4.9 Pulse Plating (Electrically Mediated Process)

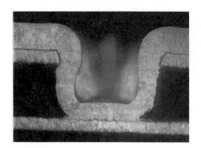

FIGURE 29.10 Example of a blind via that is 5.0 mils in diameter and 3.0 mils deep, plated with chemically mediated bath.

In this plating system, the bath is engineered to respond to a periodically pulsed reverse current. The required Tafel slope shift is accomplished through rectification (see Fig. 29.11). The rectifier produces a pulsed wave with a forward cathodic current that is perturbed by short anodic pulses. The forward current at $1X$ (e.g., 30 ASF) is maintained for 10 ms and the reversed at $3X$ (e.g., 90 ASF) for 0.5 ms, for example. The duty cycle may vary (e.g., 20 ms forward with 1.0 ms reverse). There is also room to optimize the forward-to-reverse current ratio; 1:3 is only one example. The shape of the wave is very important here, also. A square wave with minimum rise time gives the best results (see Fig. 29.12).

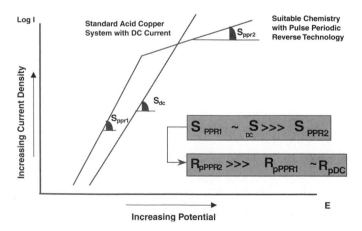

FIGURE 29.11 The pulse periodic reverse effect.

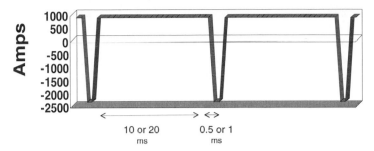

FIGURE 29.12 Periodic pulse reverse rectifier output wave.

Chemical suppliers have designed specific additive packages that ensure maximum response to the periodic pulse reverse wave. During the reverse cycle, the additive is preferentially desorbed from the high-current-density areas. This results in less plating acceleration due to the additive, and more suppression due to the carrier. Since the low-current-density areas of the panel will also receive a lower reverse pulse, the acceleration is reduced to a lesser degree than in the high-current-density areas. This leads to greatly improved plating thickness distribution. Often, the difference in plating thickness between isolated surface features and ground planes will be no greater than 0.5 mil. This effect is so powerful that throwing power greater than 100 percent is commonplace.

Magnification 5x

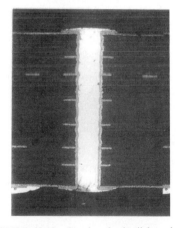

FIGURE 29.13 One-hundred-mil board with 13-mil-diameter hole plated with Copper Gleam PPR™ for 60 minutes. Surface-to-hole ratio is 0.75.

Figure 29.13 shows the plating in a hole that is 0.013 in in diameter in a 0.100 in thick board (8:1 aspect ratio). The board was plated using Copper Gleam PPR (a proprietary additive) under the following conditions:

Forward current density, 30 ASF

Forward to reverse (F:R) CD ratio, 1:2.67

Duty cycle, 20.1 ms

Plating time, 60 min

Result

Average thickness on the surface, 1.1 mils

Average thickness in the hole, 1.4 mils

Reverse pulse plating gives dramatic improvements in copper thickness distribution beyond the capabilities of the chemically mediated process. It is clearly the way of the future as the HDI boards become more and more complex. However, it involves capital investment for the pulse rectifier, which may be 5 to 6 times the cost of an equivalent DC rectifier. It also involves a learning curve on how to match the new parameters (F:R ratio, duty cycle, ASF, and the shape of the wave) to meet the specifications of the part number being plated.

29.4.10 Key Factors for Uniform Plating

To have day-to-day control and achieve ductile, strong deposits and uniform copper thickness, the following controls are required:

1. Maintain equipment following best practices, such as uniform air agitation or "e-ductors" (also known as "ser-ductors") in the tank, equal anode/cathode distances, rectifier connection on both ends of tank, and low resistance between rack and cathode bar.

2. Maintain narrow-range control of all chemical constituents, including organic additives and contaminants.

3. Conduct batch carbon treatment as needed.

4. Control temperature at 70° to 85°F, or as specified by supplier.

5. Eliminate contaminants in the tank from preplate cleaners, microetchants, and impure chemicals.

The plating thickness distribution of your plating bath can be readily evaluated. Use non-copper-clad or bare epoxy laminate. First, plate 80 to 100 μin of electroless copper, then rack the panels and plate at the specified ASF to plate 80 to 100 μin of plated copper (usually 1/10 of the plating cycle time). Place the rack in a microetch solution until 60 to 80 percent of the copper is etched away. Remove and examine. The remaining copper is where the high current density is. This is a useful tool in optimizing bath geometry, especially anode placement and panel placement on the rack. It is also useful in designing thieving and shielding for the best thickness distribution.

The plated-copper thickness distribution for a specific pattern can also be studied the same way. After plating the electroless copper on non-copper-clad bare laminate, image the pattern. Plate for a limited time to get approximately 80 to 100 μin, or 0.2 to 0.3 μm. Strip the resist off the panel and place in microetch until 80 percent of all copper is removed. The remaining copper is where the high current density is.

29.5 ACID COPPER SULFATE SOLUTIONS AND OPERATION

The preferred industrial process uses an acid copper sulfate solution containing copper sulfate, sulfuric acid, chloride ion, and organic additives.

29.5.1 Solution Makeup by Current Density

Using the proper additives, the resultant copper is fine-grained with tensile strengths of 50,000 lb/in^2 (345 MPa), a minimum of 10 percent elongation, and 1.2 surface-to-hole thickness ratio (see Table 29.3).

29.5.2 Operation and Control

    I. *Agitation.* Air (vigorous) from oil-free source, at 70° to 80°F. Or e-ductors (airless high-volume, low-pressure circulation).

   II. *Filtration.* Continuous through 3- to 10-μm filter to control solution clarity and deposit smoothness.

  III. *Carbon treatment.* New baths do not require activated carbon purification. Circulation through a carbon-packed filter tube is recommended to control organic contamination; for design and number, consult with supplier. The need for batch carbon treatment is indicated by corner cracking after reflow; dull, pink deposits; and haze, haloing, or comet trails around the PTH. Carbon-treat about every 1500 (Ah) per gal.

  IV. *Procedure for batch carbon treatment.*
      A. Pump to storage tank.
      B. Clean out plating tank.
          1. Rinse and clean tank.
          2. Leach with 10 percent H$_2$SO$_4$.
          3. Adjust agitators.
          4. Clean anodes.

C. Heat solution to 120°F.

D. Add 1 or 2 qt of hydrogen peroxide (35 percent per 100 gal of solution). Dilute with 2 pt water, using low-stabilized peroxide.

E. Air-agitate or mix for 1 h.

F. Maintain heat at 120° or 140°F.

G. Add 3 or 5 lb powdered or granular carbon per 100 gal of solution. For a specific carbon source, seek recommendation from your supplier. Mix for 1 to 2 h.

H. Pump back to plating tank promptly or within 4 h.

I. Analyze and adjust.

J. Dummy plate at 10 A/ft^2 for 6 h. Panels should be matte and dull. Replenish with additive.

K. Follow supplier instructions for electrolyzing and start-up.

V. *Contaminants.*   In general, acid copper tolerates both organic and metallic contaminants. Organic residues may come from cleaners, resists, and certain sulfur compounds. Dye systems usually are more resistant than dye-free systems with respect to certain cleaner constituents. Metals should be kept at these maximums: chromium, 25 ppm; tin, 300 ppm; antimony, 25 ppm. Nickel, lead, and arsenic may also cause roughness, etc.

29.5.3 Process Controls

29.5.3.1 *Bath Composition*

- Copper sulfate is the source of metal. Low copper will cause deposit burning; high copper will cause roughness and decreased hole-to-surface thickness ratios or throwing power.

- Sulfuric acid increases the solution conductivity, allowing the use of high currents at low voltages. However, an excess of sulfuric acid lowers the plating rate, whereas low acid reduces hole-to-surface ratio (throwing power).

- Chloride ion (Cl$^-$) should be controlled at 60 to 80 ppm. Below 30 ppm, deposits will be dull, striated, coarse, and step-plated. Above 120 ppm, deposits will be coarse-grained and dull. The anodes will become polarized, causing plating to stop. Excess chloride is reduced by bath dilution or by dragout.

- Additive components analysis and control are critical for consistent product quality. The primary analytical tool is *cyclic voltametric stripping* (CVS), and the Hull cell continues to be a useful complementary tool. Excessive or insufficient additive will cause deposit burning and corner cracking. This condition can be judged by metallographic cross-sectioning and etching. Optimum quality plated metal is fine grained and equiaxed (nondirectional) and shows no laminations or columnar patterns.

- Concerning water quality, the use of DI water and contamination-free materials such as low chloride and iron will give added control and improved deposit quality.

29.5.3.2 *Temperature.*   Optimum throwing power and surface-to-hole ratios are obtained by operating at room temperature (i.e., 70 to 80°F). Lower temperatures cause brittleness, burning, and thin plating. Higher temperatures cause haze in low-current-density areas and reduced throwing power. Cooling coils may be necessary during a hot summer or under heavy operation.

29.5.3.3 *Deposition Rate.*   A thickness of 0.001 in (1.0 mil) of copper deposits in 54 min at 20 A/ft^2, in 21 min at 50 A/ft^2.

29.5.3.4 *Hull Cell.*   Operation at 2 A will show the presence of organic contamination, chloride concentration, and overall bath condition. However, an optimum Hull cell panel is

only a small indication that the bath is in good operating condition, since test results are not always related to production problems. More reliable results are obtained by adjusting the bath before Hull cell testing. See Sec. 29.12.4 for procedures on Hull cell.

29.5.4 Cross-Sectioning Results—Troubleshooting

Sectioning with etching provides information on the plated copper that explains PTH quality in terms of processing factors. Besides showing the overall quality, cross-sectioning gives information on thickness and on possible problems, such as drilling, cracking, blowholes, and multilayer smear. Copper deposits with columnar or laminar patterns indicate inferior copper properties. Cross sections of optimum copper deposits are fine-grained and equiaxed (structureless) upon etching.

29.5.5 Inferior Copper Deposits

These may be caused by any of the following:

- Either low or excess additives
- Chloride out of range (i.e., too high or too low)
- Organic, metal, or sulfur (thiourea) contamination
- Excess DC rectifier ripple (greater than 10 percent)
- Low copper content with bath out of balance
- Roughness in drilling, voids in electroless steps, or other problems introduced in earlier processing

29.5.5.1 *Cracking and Ductility.*   Resistance to cracking is tested by the following:

Solder reflow: Or wave soldering and cross-sectioning.

Elongation: Two-mil copper foil should exceed 10 percent elongation. Acid copper elongation should range between 10 and 25 percent. Frequent testing gives more meaningful results.

Float solder test: This test includes prebaking and flux, using a 5- to 20-s float in a solder pot (60:40) at 550°F, followed by cross-sectioning for evaluation.

Copper foil bulge test: This test measures tensile strength by puncturing copper at high pressure.

DC ripple: High values of rectifier ripple (8 to 12 percent) may cause inferior copper deposits and poor distribution of thickness.

29.5.5.2 *Visual Appearance.*   Plated, copper has a semibright appearance at all current densities. Unevenness, hazy or dull deposits, cracking, haloing around the PTH, and low-current-density areas indicate organic contamination. Carbon treatment is required if these conditions persist. Burned, dull deposits at high current densities indicate low additive content, contamination, solution imbalance, or low bath temperatures. Dull, coarse deposits at low current densities mean that the chloride ion is not in balance.

When chloride is high or bath temperature too low, anodes may become heavily coated and polarized (the current drops). Decreased throwing power (surface-to-hole ratio), reduced bath conductivity, or poor-quality plating may also indicate contamination and are corrected by:

TABLE 29.4 Printed Wiring Board Copper-Plating Defects

Defect in copper process	Cause
Corner cracking	Excess additive, organic contamination in solution
Nodules	Particulate matter in solution; also drilling, deburring residues
Thickness distribution	See Sec. 29.5.2
Dullness	Off-balance solution, organic contamination
Uneven thickness in PTH	Organic sulfur (thiourea) contamination
Pitting	Additive malfunction, defective electroless, or preplate preparation
Columnar deposits	Low additive, rectifier malfunction
Step plating, whiskers	Excess or defective additive

Defect, overall manufacturing process	Cause
Voids	Malfunction of electroless copper steps, also preplate cleaner etching
Inner-layer smear	Drilling or malfunction in smear removal
Roughness	Drilling or drilling residues
Hole-wall pullaway	Malfunction of smear removal or electroless copper steps
Copper-copper peeling	Surface residues from electroless and/or image transfer
Soldering blowholes	Drilling roughness, voids, and thin plating

- Maintaining solution balance and chloride content at 60 to 80 ppm
- Circulating solution through filter continuously, passing through a carbon canister periodically, or by batch carbon treatment
- Analyzing organic additives by CVS or Hull cell
- Checking metal contaminations every 3 months
- Controlling temperature between 70 and 85°F
- Checking anodes daily and replacing bags and filters (rinsed in hot water) every 3 to 4 weeks

29.5.5.3 Problems.   Table 29.4 lists problems that appear after copper plating. Two groups are listed, with the first group readily correlated to the copper-plating process. Thin, rough copper plating in the PTH may also be exhibited by outgassing and blowholes during wave soldering.[30, 42-45] Figures 29.14 through 29.23 illustrate some of these effects.

29.6 SOLDER (TIN-LEAD) ELECTROPLATING

Solder plate (60 percent tin, 40 percent lead) is widely used as a finish plate on printed circuit boards. This process features excellent etch resistance to alkaline ammonia, good solderability after storage, and good corrosion resistance. Tin-lead plating is used for several types of boards, including tin-lead/copper, tin-lead/tin-nickel/copper, SMOBC, and surface-mount (SM). Fusing is required on all tin-lead-plated surfaces. Thickness minimums are not specified.

The preferred composition contains a minimum of 55 percent and a maximum of 70 percent tin. This alloy is near the tin-lead eutectic, which fuses at a temperature lower than the melting point of either tin or lead and, thus, makes it easy to reflow (fuse) and solder. (The composition of the eutectic is 60 percent tin, 37 percent lead, with a melting point of 361°F.) Fusing processes include infrared (IR), hot oil, vapor phase, and hot-air leveling for SMOBC.

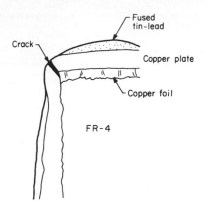

FIGURE 29.14 PTH vertical cross section illustrating corner cracking

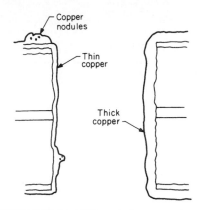

FIGURE 29.15 PTH vertical cross section illustrating uneven, thick/thin copper plating. Nodules due to particle contamination are also shown.

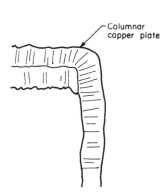

FIGURE 29.16 PTH vertical cross section illustrating columnar copper deposit structure.

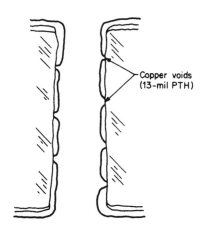

FIGURE 29.17 PTH vertical cross section illustrating copper voids.

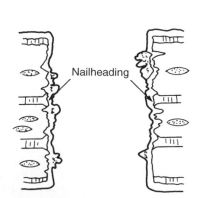

FIGURE 29.18 PTH vertical cross section illustrating rough, nodular copper plating due to drilling roughness. Nail-heading is also shown.

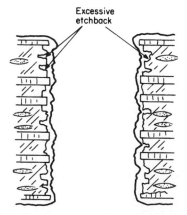

FIGURE 29.19 PTII vertical cross section illustrating hole roughness due to excessive etchback.

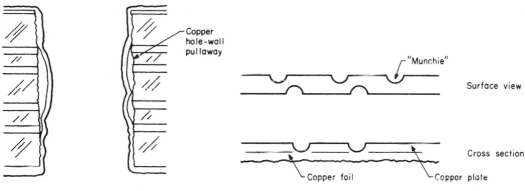

FIGURE 29.20 PTH vertical cross section illustrating copper hole wall pullaway.

FIGURE 29.21 Trace surface view and vertical cross section illustrating "munchies" ("mouse bites") and pitting.

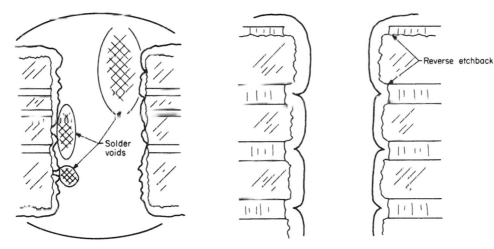

FIGURE 29.22 PTH vertical cross section illustrating wave-soldering blowholes and thin, rough copper plating.

FIGURE 29.23 PTH vertical cross section illustrating reverse etchback.

Plating solutions currently available include the high-concentration fluoboric acid–peptone system as well as low-fluoboric, nonpeptone, and a nonfluoboric organic aryl sulfonic acid process. These processes are formulated to have high throwing power and give uniform alloy composition.[1] The sulfonic acid process has the advantage of using ball-shaped tin-lead anodes. Table 29.5 gives details of operation and control of two high-throw tin-lead (solder) baths.

29.6.1 Agitation

The solution is circulated by a filter pump, without allowing air to be introduced.

TABLE 29.5 Tin-Lead Fluoborate: Operation and Control

	High HBF$_4$/peptone	Low HBF$_4$/proprietary
Operating conditions		
Lead	1.07–1.88 oz/gal	1.4–3.0 oz/gal
Stannous tin (SN^{+2})	1.61–2.68 oz/gal	3.0–4.5 oz/gal
Free fluoboric acid	47–67 oz/gal	15–25 oz/gal
Boric acid	Hang bag in tank	Same
Additive	Use as needed by Hull cell and Ah usage.	Use as needed by Hull cell and Ah usage.
Temperature	60–80°F	70–85°F
Cathode current density	15–18 A/ft^2	10–30 A/ft^2
Agitation	Solution circulation	Mechanical and solution circulation
Anodes		
Type	Bar	
Composition	60% tin-40% lead	
Purity	Federal Specification QQ-S-571[30]	
Bags	Polypro	
Hooks	Monel	
Length	Rack length minus 2 in	
Current density	10–20 A/ft^2	

29.6.2 Filtration

A 3- to 10-μm polypro filter is needed to control solution cloudiness and deposit roughness.

29.6.3 Carbon Treatment

Solution is batch-treated at room temperature every 4 to 12 months. If clear and colorless, additive is added based on Hull cell and analysis. Do not use Supercarb or hydrogen peroxide.

29.6.4 Contaminations

Organic: Comes from peptone or additive breakdown and plating resists. Periodic carbon treatment is needed.

Metallic: Copper is the most serious of these contaminations. It causes dark deposits at low current densities (in the plated through-hole) and may coat the anodes. The maximum levels of metallic contaminants allowable are copper, 15 ppm; iron, 400 ppm; and nickel, 100 ppm.

Nonmetallic: The maximum levels allowed are chloride, 2 ppm; sulfate, 2 ppm.

Dummy: Copper is removed by dummy plating at 3 to 5 A/ft^2 several hours each week. (Dummy plating is plating at a low current density, using corrugated metal sheets or scrap panels.) Other metals such as iron and nickel in high concentrations may contribute to dewetting and cannot be easily removed. Ensure that the copper is covered by a tin-lead coating before reducing the current density for dummying.

29.6.5 Solution Controls

Stannous and lead fluoborates are the source of metal. Their concentrations and ratio must be strictly maintained, as they will directly affect alloy composition. Fluoboric acid increases the conductivity and throwing power of the solutions. Boric acid prevents the formation of lead fluoride. Additives promote smooth, fine-grained, tree-free deposits. Excess peptone (3 to 4 times too much) may cause pinholes (volcanoes) in deposit when reflowed. Testing by Hull cell and periodic carbon treatments is indicated. The peptone add rate is about 1 to 2 qt per week for a 400-gal tank. Only DI water and contamination-free chemicals should be used; for example, <10 ppm iron-free and <100 ppm sulfate-free fluoboric acid. A clear solution is maintained by constant filtration.

29.6.6 Deposition Rate

A layer of 0.5-mil tin-lead is deposited in 15 to 17 min at 15 to 17 A/ft^2. The best practice is to plate at 10 to 25 A/ft^2 (⅔ copper current). Higher currents lead to coarse deposits and more tin in the alloy. Excessive current causes treeing and sludge formation.

29.6.7 Deposit Composition

The composition should be 60 percent tin to 40 percent lead. The composition should be confirmed with a deposit assay. Alloy composition is determined by the ratio of tin and lead in the solution; thus

$$Sn \ (oz/gal) + [total \ Sn + Pb \ (oz/gal)] \times 100 = \%Sn \qquad (29.2)$$

29.6.8 Hull Cell

This test shows overall plating quality and the need for peptone, additives, or carbon treatment, as well as the presence of dissolved copper in the solution.

29.6.9 Visual Observation of Plating

Plated, solder has a uniform matte finish. The deposit should be smooth to the touch. A coarse, crystalline deposit usually indicates the need for additives or peptone or a current density that is too high. Peeling from copper is generally related to the procedure. The fluoboric acid predip should not be used as a holding tank (for more than 5 min) prior to solder plating. Acid strength and cleanliness are also important in this case. Load the tank with some residual current (5 to 10 percent). Rough, dark, thin, or smudged deposits may be due to organic contamination and may require carbon treatment. Dark deposits, especially in low-current-density areas and in the plated through-hole, are due to copper contamination or to thin plating. A Hull cell test will confirm deposit, and a dummy plate will remove copper.

29.6.10 Corrective Actions

- Keep bath composition in balance
- Use contaminant-free chemicals.
- Dummy-plate once a week at 3 to 5 A/ft^2.
- Circulate the solution through a filter continuously.
- Maintain additives by Hull cell and by analysis.
- Analyze for copper, iron, and nickel at least once a month.
- Carbon-treat on schedule.

29.7 TIN ELECTROPLATING

Tin is used extensively for plating electronic components and PC boards due to its solderability, corrosion resistance, and metal etch-resist properties. The current MIL-STD-275 does not include tin plating, although earlier versions stated a required thickness of 0.0003 in. Specifications covering tin plating are MIL-T-10727 and MIL-P-38510, which say that tin must be fused on component leads.

29.7.1 Acid Tin Sulfate

This is the most widely used system. Among the many processes available, some produce bright deposits for appearance and corrosion resistance; others give matte deposits, which can be fused as well as soldered after long-term heating. Tin sulfate baths are somewhat difficult to control, especially after prolonged use.[2, 3] Operation and control are given in Table 29.6.

TABLE 29.6 Acid Tin Sulfate: Operation and Control

Operating conditions:	
Tin	2 oz/gal
Sulfuric acid	10–12% by volume
Carrier, additives	Replenish by Ah usage and spectrophotometry
Temperature	60–65°F for bright, 65–85°F for matte
Cathode current density	10–30
Current efficiency	100%
Plating rates	0.3 mil @ 25 A/ft^2
Anodes:	
Type	Bars
Composition	Pure tin
Bags	Polypropylene
Hooks	Monel or titanium
Length	Rack length minus 2 in
Current density	5–20 A/ft^2

29.7.1.1 Process Controls

- *Agitation.* Solution is circulated by a filter pump without allowing air to be introduced. Cathode rod agitation is also useful for a wider range of plating current densities.
- *Filtration.* A 3- to 10-μm polypropylene filter is needed to control excess cloudiness and sludge formation.
- *Temperature.* The preferred temperature for deposit luster and visual appearance is 60 to 65°F. Use cooling coils. Baths can operate up to 85°F, but may result in smoky, hazy deposits.
- *Carbon treatment.* Light carbon filtration at room temperature removes organic contaminations. New baths are also made up if problems continue with deposit quality, solderability, thickness control, and cost-effectiveness.
- *Contaminations.* These include three groups of contaminants.

    1. *Organics.* These come from additives and breakdown of resist.

2. *Metallic.*   The effects of metallic contaminants on the plated deposits and the maximum levels allowable are as follows:

Copper	Darkness	5 to 10 ppm
Cadmium	Dullness	50 ppm
Zinc	Dullness	50 ppm
Nickel	Streaks	50 ppm
Iron	Dullness	50 to 120 ppm
Chromium	Dullness	5 ppm

3. *Nonmetallic.*   The maximum level allowable is 75 ppm of chloride.

- *Anodes.*   Remove when bath is idle to maintain tin content.

29.7.1.2   Solution Controls

- *Bath constituents.*   Stannous sulfate and sulfuric acid are maintained by analysis, additives by spectrophotometry, Ah usage, Hull cell, and the percentage of sulfuric acid additions. Electronic-grade chemicals must be used to control metallic contaminations of cadmium, zinc, iron, etc.
- *Control additive.*   Low levels of additives must be maintained.
- *Hull cell.*   This test is useful for control of additive levels and plating quality. See Sec. 29.12.4 for the procedure.
- *Rinsing after plating.*   Adequate rinsing after plating is important to control white or black spots on tin surfaces.
- *Visual observation.*   As plated, tin has a uniform lustrous finish. The deposit should be smooth to the touch.

29.7.2   Problems with Tin Electroplating

- *Dull deposits.*   These are due to an out-of-balance condition of the main solution constituents; i.e., low acid (<10 percent), high tin (>3 oz/gal), improper additive levels, contamination by metals or chlorides, or high temperatures (>65°F).
- *Peeling tin.*   This comes off due to low acid (<10 percent) or organic contamination.
- *Slivers.*   These are caused by overetching. A review of etching practice is indicated, as well as the use of ½-oz copper foil.
- *Pitting.*   If substrate is not the cause, check precleaning, solution balance and contaminants, current efficiency, and current densities. High current densities may cause pitting.
- *Strip and etching residues.*   Tin is attacked by strong alkaline solutions. To control residues and spotting, use mild room-temperature stripping and alkaline-ammonia etching.
- *Poor solderability.*   This may be caused by excess additives or contaminations in the bath, poor rinsing, bath age, or excessive thicknesses (>0.3 mil).

29.8   NICKEL ELECTROPLATING

Nickel plating is used as an undercoat for precious and nonprecious metals. For surfaces such as contacts or tips that normally receive heavy wear, the uses of nickel under a gold or rhodium plate will greatly increase wear resistance. When used as a barrier layer, nickel is effective in preventing diffusion between copper and other plated metals. Nickel-gold combinations are frequently used as metal etch resists. Nickel alone will function as an etch resist against the ammoniacal etchants.[4] MIL-STD-275 calls for a low-stress nickel with a minimum thickness of 0.0002 in. Low-stress nickel deposits are generally obtained using nickel sulfamate baths in conjunction with wetting (antipit) agents. Additives are also used to reduce stress and to improve surface appearance.

29.8.1 Nickel Sulfamate

Nickel sulfamate is commonly used both as an undercoat for through-hole plating and on tips. Conditions given in Table 29.7 are applicable for through-hole and full-board plating.

TABLE 29.7   Nickel Sulfamate
Operation and control

Operating conditions:	
pH	3.5–4.5 (3.8)
Nickel	10–12 oz/gal
As nickel sulfamate	43 oz/gal
Nickel chloride	4 oz/gal
Boric acid	4–6 oz/gal
Additives	As required
Antipit	As required
Temperature	$130 \pm 5°F$
Cathode current density	20–40 ASF
Anodes:	
Type	Bars or chunks
Composition	Nickel
Purity	Rolled depolarized, cast, or electrolytic; SD chips in titanium basket
Hooks, baskets	Titanium
Bags	Polypro, Dynel, or cotton
Length	Rack length minus 2 in

29.8.1.1   Process Controls

- *pH.*   With normal operation, pH increases. Lower pH with sulfamic acid (not sulfuric acid). A decrease in pH signals a problem; anodes should be checked.
- *Temperature.*   The preferred temperature is 125°F. Low temperature causes stress and high-CD burning. Higher temperatures increase softness of nickel deposit.
- *Agitation.*   Solution circulation between panels is done by filter pump and/or cathode rod agitation.
- *Filtration.*   This is done continuously through 5- to 10-μm polypro filter, which is changed weekly.
- *Deposition rate.*   For 0.5 mil, plate 25 to 30 min at 25 A/ft². For 0.2 to 0.3 mil, plate 15 min at 25 A/ft².
- *Contaminations.*   These include three groups of contaminants.
  1. *Metals.*   Maximum allowable by atomic absorption: iron, 250 ppm; copper, 10 ppm; chromium, 20 ppm; aluminum, 60 ppm; lead, 3 ppm; zinc, 10 ppm; tin, 10 ppm; calcium, 0 ppm. These metal contaminants lower the deposition rates and cause nonuniform plat-

ing. Cooper and lead cause dark, brittle deposits at low CD. Dummy at 3 to 5 A/ft^2. Iron, tin, lead, and calcium cause deposit roughness and stress.

  2. *Organics.*   These cause pitting, brittleness, step plating, and lesser ductility.

  3. *Sulfates.*   These cause solution breakdown and should not be added.

- *Carbon treatment.*   Circulate through carbon canister for 4 h (more than one canister may be required), or batch-treat to remove organics. The basic six steps to batch-carbon-treat are as follows:

  1. Heat to 140°F.

  2. Transfer to treatment tank. Do not adjust pH.

  3. Add 3 to 5 lb carbon per 100 gal of solution. Premix outdoors.

  4. Stir 4 h at temperature.

  5. Let settle 1 to 2 h.

  6. Filter back to cleaned tank.

29.8.1.2 Common Problems.   Pitting, stress (cracking), and burned deposits are common problems.

- *Pitting.*   This is caused by low antipit, poor agitation or circulation, boric acid imbalance, or the presence of organics. In testing for antipit, the solution should hold bubble for 5 s in a 3-in wire ring and for less than 5 s in a 5-in ring. In the case of severe pitting, cool to room temperature, add 1 pt/200 gal hydrogen peroxide (35%), bubble air, and reheat to 140°F. Carbon-treat as above. Adjust pH before restarting.

- *Stress.*   This refers to the cause of deposit cracking. While low nickel chloride causes poor anode corrosion (rapid decrease in nickel content), high chloride causes excess stress. To prevent this, maintain pH, current density, and boric acid for control below 10 kpsi.

- *Burns.*   A low level of nickel sulfamate causes high-current-density "burns."

- *Hull cell.*   Plate at 2 A, 10 min with gentle agitation. This test is useful for bath condition and contaminants.

- *Visual observation.*   The plated metal has a matte, dull finish. Burned deposits are caused by low temperature, high current density, bath imbalance, and poor agitation. Rough deposits are due to poor filtration, pH out of spec, contamination, or high current density. Pitting is due to low antipit, poor agitation, bath imbalance, or organic contamination. Low plating rates are due to low pH, low current density, or impurities. Gassing at the cathode is a sign of low plating rates.

29.8.2 Nickel Sulfate

This is typically plated with an automatic edge connector (tip) plating machine. Table 29.8 gives the operating conditions that apply to these systems. For additional instruction, follow the details in Sec. 29.8.1.

Nickel anodes are preferred in tip machines because the pH and the metal content remain stable. The pH will decrease rapidly when insoluble anodes are used. The pH should be maintained at 1.5 or higher with additions of nickel carbonate. Stress values are higher than in sulfamate baths, with values of about 20 kpsi.

29.9 GOLD ELECTROPLATING

Early printed circuit board technology used gold extensively. In addition to being an excellent resist for etching, gold has good electrical conductivity, tarnish resistance, and solderability

TABLE 29.8 Nickel Sulfate: Operation and Control

Operating conditions:	
pH	1.5–4.5
Nickel	15–17 oz/gal
Nickel chloride	2–4 oz/gal (with soluble anodes)
Boric acid	5–7 oz/gal
Stress reducer	As required
Antipit	As required
Temperature	$130 \pm 5°F$
Cathode current density	100–600 A/ft^2
Current efficiency	65%
Anodes:	
Composition	Nickel or platinized titanium

after storage. Gold can produce contact surfaces with low electrical resistance. In spite of its continued advantages, the high cost of gold has restricted its major application to edge connectors (tips) and selected areas, with occasional plating on pads, holes, and traces (body gold). Both hard-alloy and soft, pure gold are currently used. Plating solutions are acid (pH 3.5 to 5.0) and neutral (pH 6 to 8.5). Both automatic plating machines for edge connectors and manual lines are in use.

29.9.1 Acid Hard Gold

To a large extent, acid golds are used for compliance to MIL-STD-275, which states that gold shall be in accordance with MIL-G-45204, Type II, Class 1. The minimum thickness shall be 0.000050 in (50 μin); the maximum shall be 0.000100 in. Nonmilitary applications require 25 to 50 μin. A low-stress nickel shall be used between gold overplating and copper. Type II hard gold is not suited for wire bonding. These systems use potassium gold cyanide in an organic acid electrolyte. Deposit hardness and wear resistance are made possible by adding complexes of cobalt, nickel, or iron to the bath makeup. Automatic plating machines are being used increasingly because of the enhanced thickness (distribution) control, efficient gold usage, productivity, and quality. A comparison of automatic versus manual plating methods for edge connectors is given in Table 29.9.

29.9.1.1 Process Controls

- *Analysis.* Gold content, pH, and density should be maintained at optimum values. Operation at very low gold content causes early bath breakdowns with loss of properties, less current efficiency, and less cost savings. The pH is raised by using potassium hydroxide and is lowered with acid salts. Solution conductivity is controlled by density, which is adjusted with conductivity salts. Hull cell is not recommended for this purpose.

- *Anodes.* Platinized titanium should be replaced when operating voltages are excessive or when thick coatings develop on the anode surfaces.

- *Recovery.* Plating solutions should be replaced when contaminated, when plating rates decrease, or after about 10 total gold content turnovers.

29.9.1.2 Problems. Difficulties can be controlled by proper gold bath and equipment maintenance. Some typical situations follow.

- *Discolored deposit.* This may be due to low brightener; to metal contaminants such as lead, low pH, low density, or organic contaminations; or to leaking from tape. Gold plate is stripped to evaluate nickel. For these problems, first try to lower the pH, raise the den-

TABLE 29.9 Acid Gold-Cobalt Alloy
Operation and control

	Manual	Automatic
Gold content, troy oz/gal	0.9–1.1	1–3
pH	4.2–4.6	4.5–5.0
Cobalt content	800–1000 ppm	800–1200 ppm
Temperature	90–110°F	100–125°F
Solution density	8–15 Be	12–18 Be
Replenishment per troy oz gold	8 Ah	6.5 Ah
Current efficiency	50%	60%
Agitation	5 gal/min	50 gal/min
Anode-to-cathode distance	2–3 in	¼ in
Anodes, composition	Platinized titanium	Platinized titanium
Cathode current density	1–10 A/ft²	50–100 A/ft²
Thickness	40 ± 10 μin	40 ± 2 μin
Deposition rate for 40 μin	3–6 min	0.3–0.6 min
Deposition composition	99.8% gold, 0.2% cobalt	99.8% gold, 0.2% cobalt
Hardness	150 Knoop	150 Knoop

Gold solution contaminants			
Metal	Maximum ppm	Metal	Maximum ppm
Lead	10	Iron	100
Silver	5	Tin	300
Chromium	5	Nickel	300–3000
Copper	50		

Organics: Tape residues, mold growth, and cyanide breakdowns.

sity, or increase the brightener before replacing the bath or using decreased current densities.

• *Gold peeling from nickel.*   This is generally due to inadequate solder stripping, leaky tape, or poor activation. Methods to increase nickel activation include increasing acid strength after nickel plating and using a fluoride activator, cathodic acid, or gold strike. A gold strike that is compatible with gold plate and has low metal content and low pH is available. Its main purpose is to maintain adhesion to the nickel substrate.

• *Wide thickness range.*   To narrow thickness range, improve solution movement between panels; clean, adjust, or replace anodes; and adjust or replace solution.

• *Low deposition rate.*   This is characterized by excessive gassing and low efficiencies. To correct this condition, adjust solution parameters and check for contaminants such as chromium.

• *Pitting.*   Strip gold and evaluate nickel and copper by cross-sectioning.

• *Resist breakdown.*   Low current efficiencies cause this condition. Use solvent soluble screened or dry film for best results.

29.9.2 Pure 24-Karat Gold

High-purity 99.99 percent gold processes are used for boards designed for semiconductor chip (die) attachment, wire bonding, and plating solder (leaded) glass devices, for their solderability and weldability. These qualities comply with Types I and III of MIL-G-45204. The processes are neutral (pH 6 to 8.5) or acid (pH 3 to 6). Pulse plating is frequently used. Table 29.10 gives typical conditions for a neutral bath.

TABLE 29.10   Neutral Pure Gold: Operation and Control

Gold content	0.9–1.5 troy oz/gal
pH	6.0–7.0
Temperature	150°F
Agitation	Vigorous
Solution density	12–15 Be
Replenishment	4 Ah/troy oz
Current efficient	90–95%
Cathode current density	1–10 A/ft^2
Deposition rate for 100 μ-in, @5 A/ft^2	8 min
Deposition composition	99.99% + gold
Hardness	60–90 Knoop

29.9.3   Alkaline, Noncyanide Gold

Various processes for alloy and pure gold deposits are available. Solutions are based on sulfite-gold complexes and arsenic additives and operate at a pH of 8.5 to 10.0. A decision to use this process is based primarily on the need for uniformity (leveling), hardness (180 Knoop), purity, reflectivity, and ductility. PC board use is limited to body plating, since wear characteristics of the sulfite-gold are not suitable for edge connector applications. The microelectronics industry uses these processes for reasons of safety and gold purity. Semiconductor chip attachment, wire bonding, and gold plating on semiconductors are possible applications and are enhanced by using pulse plating, without metallic additives.

29.9.4   Gold Plate Tests

Several routine in-process and final quality control tests are performed on gold.

- *Thickness.*   Techniques are based on beta-ray backscattering and x-ray fluorescence. Thickness and area sensitivity are as low as 1 μin with 5-mil pads.
- *Adhesion.*   Standard testing involves a tape pull test.
- *Porosity.*   Tests involve nitric acid vapor and electrographics.
- *Purity.*   Lead is a common impurity that must be controlled to <0.1 percent. Other tests included discoloration by heating, electrical contact, and wear resistance.

29.10   PLATINUM METALS

Interest in these systems usually soars in a climate of high gold prices, even though plating results are not always as dependable as those obtained when using gold.

29.10.1   Rhodium

Deposits from the sulfate or phosphate are hard (900 to 1000 Knoop), highly reflective, extremely corrosion-resistant, and highly conductive (resistivity is 4.51 μΩ/cm). Rhodium plate is used where a low-resistance, long-wear, oxide-free contact is required. In addition, rhodium as a deposit on nickel for edge connectors has been replaced by gold. This is due to difficulty in bath control problems with organic and metallic contaminants, and cost. Table 29.11 shows details of rhodium plating.

TABLE 29.11   Rhodium Sulfate: Operation and Control

Rhodium	4–10g/L
Sulfuric acid	25–35 mL/L
Temperature	110–130°F
Agitation	Cathode rod
Anodes	Platinized titanium
Anode-to-cathode ratio	2:1
Cathode current density	10–30 A/ft^2
Plating rate	At 20 A/ft^2, 10 μin will deposit in 1.4 min, based on 70% cathode current efficiency

29.10.2 Palladium and Palladium-Nickel Alloys

Deposits of 100 percent palladium, 80 percent palladium–20 percent nickel, and 50 percent palladium–50 percent nickel find use as suitable deposits for edge connectors. Deposits are hard (200 to 300 Knoop), ductile, and corrosion-resistant. A palladium-nickel undercoat for gold shows good wear and electrical properties.

29.10.3 Ruthenium

Deposits of ruthenium are similar to rhodium but are plated with easier control, greater bath stability, and high current efficiency at lower cost. Deposits are usually stressed.

29.11 SILVER ELECTROPLATING

Silver is not widely used in the printed circuit industry, although it finds applications in optical devices and switch contacts. Thicknesses of 0.0001 to 0.0002 in (0.1 to 0.2 mil) in conjunction with a thin overlay of precious metal are specified.

Silver plating should not be used when boards are to meet military specifications. The reason for this is that, under certain conditions or electrical potential and humidity, silver will migrate along the surface of the deposit and through the body of insulation to produce low-resistance leakage paths. Tarnishing of silver-gold in moist sulfide atmospheres also produces electrical problems on contact surfaces due to diffusion of the silver to the surface.

Another reason for the lack of acceptance of silver is that silver is plated from an alkaline cyanide bath, which is highly toxic. Bright plating solutions that produce deposits with improved tarnish are usually related to *black anodes* and are due chiefly to solution imbalance, impurities in anodes, or solution roughness and pitting. Most metals to be plated, particularly the less noble metals, require a silver strike prior to silver plating to ensure deposit adhesion.

29.12 LABORATORY PROCESS CONTROL

29.12.1 Conventional Wet Chemical Analysis

The traditional wet chemical methods for metals and nonmetal plating solution constituents are available from suppliers and in the literature.[5, 6] These methods also make use of pH meters, ion electrodes, spectrophotometers, and atomic adsorption. The composition of liquid concentrates for plating solutions is shown in Table 29.12.

TABLE 29.12   Composition of Liquid Concentrates

Chemical	Formula	Weight, lb/gal	Percent	Metal, oz/gal
Acids				
Sulfuric	H_2SO_4	15.0	96	—
Hydrochloric	HCl	9.8	36	—
Fluoboric	HBF_4	11.2	49	—
Alkaline				
Sodium hydroxide	NaOH	12.8	50	—
Ammonium hydroxide	NH_4OH	7.5	28	—
Metals				
Copper sulfate	$CuSO_4 \cdot 5H_2O$	9.7	27	9
Copper fluoborate	$Cu(BF_4)_2$	12.9	46	25.4
Stannous fluoborate	$Sn(BF_4)_2$	13.3	51	44.3
Lead fluoborate	$Pb(BF_4)_2$	14.4	51	65.0
Nickel sulfate	$NiSO_4 \cdot 6H_2O$	11.0	44	17.8
Nickel sulfamate	$Ni(NH_2SO_3)_2$	12.9	50	24
Nickel sulfamate	$Ni(NH_2SO_3)_2$	12.3	43	20
Nickel chloride	$NiCl_2 \cdot 5H_2O$	11.2	54	23.7

29.12.2   Advanced Instrumental Techniques

New techniques have been developed for the control of organic additives in copper plating. Continued development is in progress in the area of measurement of such additives in nickel, gold, and tin solutions. Methods used include liquid chromatography, ultraviolet/visible (UV/VIS) spectrophotometry, cyclic voltametric stripping (CVS), ion chromatography, UV-persulfate oxidation, and polarography. These techniques can detect contaminations in various processes, and they show the need for an effectiveness of carbon treatment. Table 29.13 lists and references these techniques, which are having a major influence on plating process capabilities. Literature references list 136 entries on these techniques.[7]

TABLE 29.13   Advanced Instrumental Analysis Techniques

Technique	Constituent
Cyclic voltammetry stripping	Organics and inorganics
Liquid chromatography with UV/VIS	Organics and inorganics
Ion chromatography	Ionic species
Polarography	Organics and inorganics
Ion selective electrode	Ionic metals, nonmetals
Atomic absorption (AA)	Metals, nonmetals
UV oxidation	Total carbon

29.12.3   Metallographic Cross-Sectioning[8,9]

A method for cross-sectioning PC boards is as follows:

1. *Bulk cutting.*   Removal of a manageable-sized piece of board or assembly by shearing or abrasive cutting.
2. *Precision cutting.*   Low-speed sawing with a diamond wafer blade to produce vertical sections about $1 \times \frac{1}{2}$ in, and horizontal sectioning cut next to PTH pads.
3. *Mounting.*   Vertical and horizontal encapsulation of sections in epoxy resin.

4. *Fine grinding.* Hand grinding using 240-, 320-, 400-, and 600-grit silicon carbide papers. Rinse sample between grits.

5. *Polishing.* Diamond polishing (6 μm) on nylon cloth and alumina polishing (0.3 μm) on nap cloth on a rotating wheel. To polish the sample, place it on the rotating-wheel polisher and move it slowly in the opposite direction. Polish for 4 min, if using a 6-μm diamond on nylon, and 1 min on 0.3-μm alumina on nap cloth. Clean and dry between polishing compounds.

6. *Etching.* Apply a cotton swab for 2 to 5 s, soaked in a solution of equal parts of ammonium hydroxide and 3 percent hydrogen peroxide. Rinse in water and dry carefully.

7. *Documenting.* Observe and photograph the sample with a microscope at 30 to 1500× magnification.

29.12.4 Hull Cell

Although the advanced techniques discussed previously provide precise control of plating solutions, Hull cell testing is still widely used in the industry. Its advantages are low cost, simplicity of operation, and its actual correlation with plating production. Its main disadvantage is that defects in copper plating frequently are not shown by this method. For example, Hull cell testing will not help in detecting dull plating, roughness, or pitting. The procedure starts with brass panel preparation, in the following order:

1. Remove the plastic film.
2. Treat with cathodic alkaline cleaner.
3. Soak in 10 percent sulfuric acid.
4. Rinse.

Repeat these steps until the panel is water-break free. Proceed with the Hull cell as follows.

1. Rinse with test solution.
2. Fill to mark.
3. Adjust temperature and agitation.
4. Attach panel to negative terminal.
5. Plate.

Agitation should be similar to tank operation—that is, vigorous air bubbling for copper, gentle stirring for tin and tin-lead, and none for tin-nickel. Plate copper and nickel at 2 A and other metals at 1 A. The effects of bath adjustment, carbon treatment, dummy plating, etc., are readily translated from the Hull cell to actual tank operations. See previous sections on metal plating for Hull cell results, and consult supplier for test equipment.

REFERENCES

1. B. F. Rothschild, "Solder Plating of Printed Wiring Systems," *Proceedings of the Printed Circuit Plating Symposium,* California Circuits Association, November 5–6, 1969, pp. 10–21, and November 12–13, 1968, pp. 61–65.

2. G. F. Jacky, "Soldering Experience with Electroplated Bright Acid Tin and Copper," *Circuit Technology Today,* California Circuits Association, October 1974, pp. 49–74.

3. P. E. Davis and E. F. Duffek, "The Proper Use of Tin and Tin Alloys in Electronics," *Electronic Packaging and Production,* vol. 15, no. 7, 1975.

4. R. G. Kilbury, "Producing Buried Via Multilayers: Two Approaches," *Circuits Manufacturing,* vol. 25, no. 4, 1985, pp. 30–49.

5. D. G. Foulke, *Electroplaters' Process Control Handbook,* rev. ed., R. E. Krieger Publishing Company, Huntington, NY, 1975.

6. K. E. Langford and J. E. Parker, *Analysis of Electroplating and Related Solutions,* 4th ed., R. Draper Ltd., Teddington, England, 1971.

7. See Ref. 64, Chap. 7, "Additives."

8. J. A. Nelson, "Basic Steps for Cross Sectioning," *Insulation/Circuits,* 1977.

9. P. Wellner and J. Nelson, *High Volume Cross Sectional Evaluation of Printed Circuit Boards,* IPCWC-4B-1, Evanston, IL, 1981.

CHAPTER 30
DIRECT PLATING

Dr. Hayao Nakahara
N.T. Information Ltd., Huntington, New York

30.1 DIRECT METALLIZATION TECHNOLOGY*

For 40 years the PTH process of choice has been palladium followed by electroless copper, but there have been at least 12 DMT processes challenging that established process, with several hundred installations, a significant percentage of the total number of PCB shops. The basic idea for the palladium systems dates back to the Radovsky patent of 1963,[1] which claimed a method of using an electrically nonconductive film of palladium in a semicolloidal form to directly metallize through holes in printed circuit boards. Radovsky's invention was never commercialized. The basic idea for the carbon/graphite systems dates back to the very early days of eyelet boards when Photocircuits was experimenting with graphite, silver, and other media to turn their single-sided PCBs into reliable double-sided boards.

30.1.1 Direct Metallization Technologies Overview

From the many media and technical variations, some common elements have evolved.

30.1.1.1 Direct Metallization Technologies Common Elements. There are two elements common to all DMT:

1. Holes must be conditioned more specifically and thoroughly than for electroless copper.
2. Conductive media must be removed from the copper foil in a majority of the techniques (an exception is DMS-E).

It is understood that additional desmear steps are necessary or advisable when processing multilayer PCBs. Common elements to all horizontal conveyorized DMT systems are:

- Throughput is typically 6 to 15 min for a panel, with the next panel following 1 in behind
- Tremendous economies in rinse water use
- Lower consumption of chemicals

---

* All proprietary names are trademarked by their respective owners.

- Fewer steps than in its vertical mode
- Panel-to-panel uniformity

Many DMT processes work better than electroless copper on substrates with "difficult" resin systems such as PTFE, cyanate ester, or polyimide.

30.1.1.2 Direct Metallization Categories. DMT falls into four broad categories:

1. Palladium-based systems
2. Carbon or graphite systems
3. Conductive polymer systems
4. Other methods

30.1.2 Palladium-based Systems

30.1.2.1 Palladium/Tin Activator with Flash Electoplating. EE-1,[2] the first commercialized direct metallization technique was invented in 1982 at Photocircuits and PCK. It uses a palladium/tin activator followed by mandatory flash electroplating. The flash-plating bath contains a polyoxyethylene compound to inhibit the deposition of copper on the foil surface without inhibiting deposition on palladium sites on nonconductive surfaces (holes, edges, substrate). Deposition occurs by propagation from the copper foil and grows epitaxially along the activated surface of the hole. Coverage takes about 5 to 6 min. This flash is pattern- or panel-plated subsequently to full thickness in any electroplating bath. Microetching is incorporated in the accelerator to remove palladium sites and nail heads from innerlayers. A special cleaner/conditioner is used. This process has never been adapted to horizontal equipment. (See Fig. 30.1.)

30.1.2.2 Palladium/Tin Activator with Vanillin. DPS[3] was invented in Japan in the late '80s. This method uses a palladium/tin activator with vanillin, followed by pattern or panel electroplating. It employs a special cleaner/conditioner and a carbonate accelerator. The three key solutions—cleaner/conditioner, activator, and accelerator—all operate at elevated temperatures. DPS has recently been adapted to horizontal equipment, but works well in the vertical mode, both manual and automatic. Following the last step, the *Setter,* DPS yields a stable, grayish conductive palladium film in the holes. It is believed that the cleaner/conditioner slightly solubilizes the activator, attracting it to the nonconductive surface, and that the vanillin lines up the palladium molecules, and directs them towards the work, hence giving lower electrical resistance and better adhesion to nonconductive surfaces. It is also claimed that there is little palladium/tin left on the copper foil, so it is easily soft-etched away in the normal electroplating preplate cycle. (See Fig. 30.2.) DPS was the first DMT process to suggest an Ω meter as the standard quality assurance (QA) tool.

30.1.2.3 Converting Palladium to Palladium Sulfide. Crimson,[4] invented by Shipley, employs a conversion step after the activator where palladium is changed to palladium sulfide, which is claimed to be more conductive for subsequent electrolytic copper plating. The *enhancer* stabilizes the conductive film so that it is chemically resistant to imaging steps. The *stabilizer* neutralizes residues from the enhancer, thereby preventing contamination of subsequent steps. The *microetch* selectively removes activator from copper surfaces to achieve optimum copper-to-copper bond and reliable dry film adhesion. The process works best in conveyorized horizontal equipment and can be followed by pattern or panel electroplating. (See Fig. 30.3.)

30.1.2.4 Process Variations. ABC,[5] invented in Israel by Holtzman et al., is similar to EE-1. It has been adapted to conveyorized horizontal equipment, but must be followed by a flash electroplating in a proprietary bath.

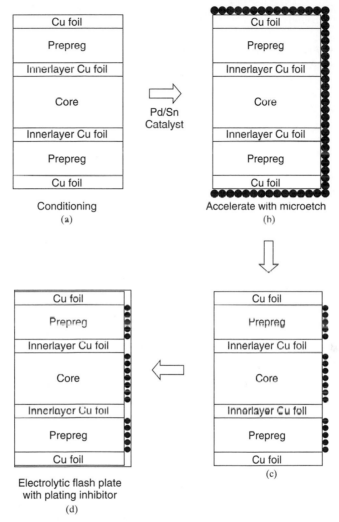

FIGURE 30.1 Principle of EE-1 process: (*a*) conditioning—making glass and epoxy surfaces receptive of Pd catalyst sites; (*b*) palladium catalyst sites; (*c*) acceleration removes excess tin and microetch removes catalysts from the surface of Cu foil; (*d*) flush Cu plate coppers enter surface in drilling hole walls.

Conductron, from LeaRonal, is similar to DPS with the addition of a special cleaner/conditioner and a glass-etch step. It has been adapted to conveyorized horizontal equipment and can be followed by pattern or panel electroplating.

Envision DPS, from Enthone-OMI, and Connect, from M & T (now Atotech), are fairly similar to each other, and to DPS, though each has a specific cleaner/conditioner and modified accelerator. No adaptation has been made to horizontal processing. Both processes can be followed by pattern or panel electroplating.

Neopact,[6] from Atotech, uses a tin-free palladium activator in colloidal form. The subsequent postdip removes the protective organic polymer from the palladium, leaving it exposed and with increased conductivity. It has been adapted to conveyorized horizontal equipment, works well in vertical, and can be followed by pattern or panel electroplating. (See Fig. 30.4.)

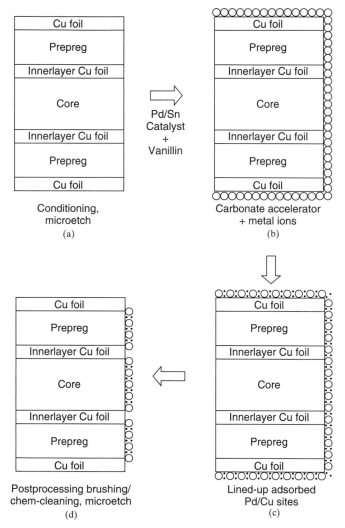

FIGURE 30.2 Principle of the DPS process: (*a*) conditioning for receptive surface; (*b*) catalysts adhering to the surface; (*c*) stronger adhesion; (*d*) ready to plate.

30.1.3 Carbon/Graphite Systems

30.1.3.1 Carbon Suspensions. Black Hole,[7] the second direct metallization technique, was patented by Dr. Carl Minten in 1988 and pioneered by Olin Hunt, who sold their technology to MacDermid in 1991. MacDermid improved the process considerably and called it Black Hole II. Instead of palladium activator, Black Hole II uses carbon suspensions as its conductive medium. Polyelectrolyte conditioned nonconductive surfaces absorb carbon sites, and they "line up" after heating. To ensure sufficient conductivity, the carbon treatment is performed twice. Residues of carbon sites must be removed from the copper foil surface by a

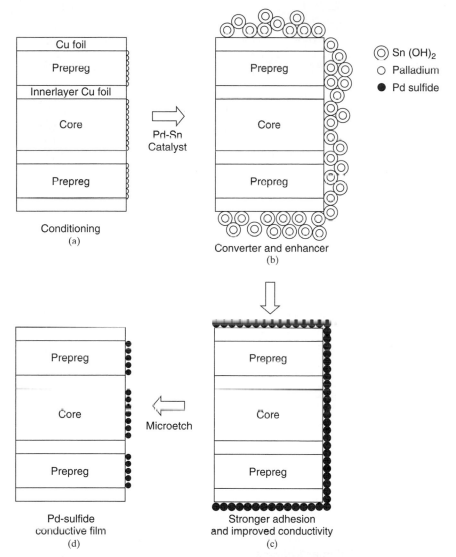

FIGURE 30.3 Principle of Crimson process: (*a*) conditioning; (*b*) colloidal Pd sites adhered to the surface; (*c*) Pd-sulfide (stronger adhesion of Pd catalysts); (*d*) microetch to remove Pd from Cu surface for better Cu-Cu adhesion.

microcleaning step. Black Hole II has been well adapted to conveyorized horizontal equipment and can be followed by pattern or panel electroplating. (See Fig. 30.5.)

30.1.3.2 Graphite. Shadow[8] is from Electrochemicals (division of LaPorte Industries, UK) and uses graphite as its conductive medium. The process sequence of Shadow is very simple and involves fewer steps than most DMTs. Electrochemicals and one of their fabricators, Eidschun Engineering, made the breakthrough in inexpensive, compact, conveyorized horizontal equipment, and the Shadow process is well adapted to this mode. It can be followed by pattern or panel electroplating.

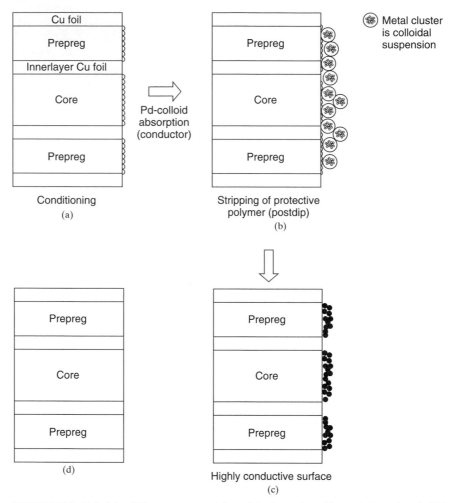

FIGURE 30.4 Principle of Neopact process: (*a*) conditioning surface; (*b*) metal cluster in colloidal suspension adhered to the hole wall; (*c*) stripped Pd adhering to the hole wall; (*d*) Cu reduced.

30.1.4 Conductive Polymer Systems

30.1.4.1 DMS-E. DMS-E[9] from Blasberg is a second-generation DMS-2 process with which they pioneered in this field. DMS-1 was similar to EE-1. After microetch and conditioning, a potassium permanganate solution forms a manganese dioxide coating in the holes which acts as an oxidizing agent during subsequent synthesis reaction. In the catalyzing step, an EDT* monomer bath wets the manganese dioxide surfaces especially well. During the sulfuric acid fixation step, a spontaneous oxidative polarization takes place, forming a black conductive poly-EDT film on the nonconductive areas of the PCB. This technique is very suitable for use in conveyorized, horizontal equipment, since the oxidative conditioning step is very hot (80 to 90°C), and there are solvents involved. It can be followed by pattern or panel electroplating. (See Fig. 30.6.)

———————————————
* EDT = 3,4 Ethylendioxythiophene.

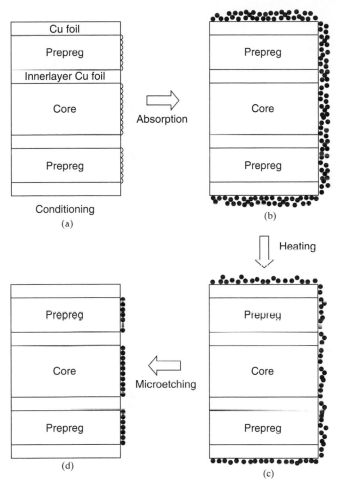

FIGURE 30.5 Principle of carbon/graphite process: (*a*) conditioning; (*b*) carbon particles adhered to the surface; (*c*) dense more conductive carbon conducting film; (*d*) removal of conduction film from Cu surface resulting in stronger Cu-Cu interface.

30.1.4.2 Compact CP. Compact CP[10] was developed by Atotech in 1987 and is essentially similar to DMS-E, except that it combines the catalyzing and fixation steps, it uses an acid permanganate, and the conductive film is a polypyrrole. The technique is very suitable for use in conveyorized, horizontal equipment. It can be followed by pattern or panel electroplating.

30.1.5 Other Methods

There are several other novel ways to metallize holes in PCBs, such as Phoenix and EBP from MacDermid, Schlötoposit from Schlötter. There are conductive ink and laser scribing/filling techniques; there are sputtering and sequential plating, to name a few, but they do not fall within the scope of this chapter.

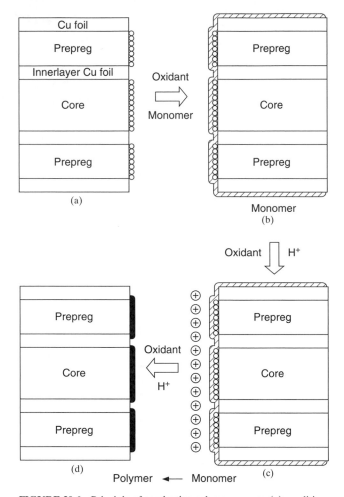

FIGURE 30.6   Principle of conductive polymer process: (*a*) conditioning; (*b*) monomer created on the surface; (*c*) oxidant reduces monomer to polymer yielding highly conductive surface; (*d*) further improved conductive surface.

30.1.6   Comparative Steps of DMT Process

30.1.6.1   DMT.1:   Palladium-based Systems.   All vertical modes only, for comparison purposes, are shown in Table 30.1.

30.1.6.2   DMT.2:   Carbon/Graphite Systems—Conveyorized Horizontal.   Conveyorized horizontal systems are shown in Table 30.2.

30.1.6.3   DMT.3:   Conductive Polymer Systems.   Conveyorized, horizontal systems with optional electrolytic flash are shown in Table 30.3.

TABLE 30.1 DMT.1
Palladium-based systems—all vertical mode only

EE-1	DPS	Crimson	ABC	Conductron	Envision DPS	Neopact
Cleaner conditioner	Cleaner conditioner	Cleaner conditioner	Cleaner conditioner etch	Glass conditioner	Conditioner	Etch cleaner
Rinse	Rinse	Rinse	Rinse	Rinse	Rinse	Rinse
Microetch	Microetch	Predip	Activator	Cleaner conditioner	Carrier	Conditioner
Rinse	Rinse	Activator	Rinse	Rinse	Activator	Rinse
Predip	Predip	Rinse	Salt remover	Microetch	Rinse	Predip
Activator	Activator	Accelerator	Rinse	Rinse	Generator	Conductor
Rinse	Rinse	Rinse	Dry	Predip	Rinse	Rinse
Accelerator	Accelerator	Enhancer	Microetch	Activator	Stabilizer	Postdip
Rinse	Rinse	Rinse	Rinse	Rinse	Rinse	Rinse
EE 1 electrolytic flash	Setter	Microetch	Reactivator	Accelerator	Microetch	Dry
Rinse	Rinse	Rinse	Rinse	Rinse	Rinse	
Dry	Dry	Dry	ABC electrolytic flash	Acid Dip	Dry	
			Rinse	Rinse		
			Dry	Dry		

TABLE 30.2 DMT.2
Carbon/graphite systems—conveyorized horizontal

Black Hole II	Shadow
Cleaner	Cleaner conditioner
Rinse	Rinse
Black Hole I	Shadow bath
Dry	Heated dry
Rinse	Inspection chamber
Condition	Microetch
Rinse	Rinse
Black Hole II	Antitarnish
Heated dry	Dry
Microclean	
Rinse	
Antitarnish	
Rinse	
Dry	

TABLE 30.3   DMT.3
*Conductive polymer systems—conveyorized,
horizontal, shown with optional electrolytic flash*

DMS-E	Compact CP
Microetch	Microetch
Rinse	Rinse
Conditioner	Cleaner/conditioner
Rinse	Rinse
Oxidative conditioning	Permanganate
Rinse	Rinse
Catalyzing	Polyconductor
Fixation	Rinse
Rinse	Soft-etch
Acid dip	Rinse
Electrolytic copper	Acid dip
Rinse	Electrolytic copper
Dry	Rinse
	Dry

30.1.7   Horizontal Process Equipment for DMT

Although many DMTs fit easily into existing electroless copper lines, whether manual or automatic, and perform well in a vertical mode, DMT is becoming inextricably linked to horizontal processing in a conveyorized machine. Like so many inventions, necessity was its mother. Certain DMT processes are marginal in the vertical/basket mode, and handling each panel individually was the only solution. The advent of compact and inexpensive horizontal DMT equipment has been a catalyst. Although Atotech pioneered horizontal electroplating equipment with their Uniplate system, until the development of horizontal DMT machines it was not feasible to engineer a horizontal PTH machine. Now, however, for those who want only flash plating, or even full electroplating, the entire process—both PTH and galvanic—may be horizontal, conveyorized, and automated with the advantages of reduced chemical consumption; radically reduced use of rinse water; panel-to-panel uniformity, reliability, and quality; reduction in operating personnel; reduction of handling; a fully enclosed operating environment; and JIT delivery. Stackers/accumulators further minimize handling and optimize efficiency.

30.1.8   DMT Process Issues

Anything brand new experiences some "teething problems" at the beginning, and DMT is no different. But most of the processes referred to in this chapter are in at least their second generation and some are already in their third, so there are no serious contraindications to the use of DMT. However, there are a few caveats:

1. Esoteric base material does not run well with all DMT.
2. Not all DMT performs as well in the vertical mode as in the horizontal.
3. Certain DMT is better suited to the rigors of multilayer production.
4. Some perform better than others with very small holes.
5. Some processes are cleaner than others.
6. Certain DMT is more sensitive than electroless copper to organic contamination in electrolytic copper plating.

7. Rinsing is important, and some DMT needs special rinsing.

8. Rework is simple in most DMT, but it can be abused, i.e., it changes the cost structure.

9. Quality assurance tools for DMT (how do you know that holes will plate void-free?) are being developed, but at the moment, the only absolutely certain method is to flash electroplate, because, unlike electroless copper, there is not much to see in the hole after DMT.

10. Some DMT analyses and operating controls are rudimentary.

30.1.9 DMT Process Summary

Strong ecological and health reasons will drive the use of DMT for the manufacture of PTH boards. Since the cost of PTH metallization constitutes only 2 to 3 percent, at most, of the total process cost of a PCB, the direct saving resulting from DMT is very small. However, the fringe benefits of this technology are rather significant: water conservation, no noxious chemicals, minimal panel handling, reduced waste treatment, lower labor costs, fewer rejects, etc.

REFERENCES

1. Radovsky, U.S. Patent 3,099,608, 1963.

2. Morrissey, et al. (Amp/Akzo), U.S. Patent 4,683,036, July 1987.

3. Okabayashi (STS), U.S. Patent 4,933,010, June 1990.

4. Gulla, et al. (Shipley), U.S. Patent 4,810,333, March 1989.

5. Holtzman, et al. (APT), U.S. Patent 4,891,069, January 1990.

6. Stamp, et al. (Atotech), PCT WO 93/17153, September 1993.

7. Minten, et al. (MacDermid), U.S. Patent 4,724,005, February 1988.

8. Thorn, et al. (Electrochemicals), U.S. Patent 5,389,270, February 1995.

9. Blasberg, Europatent 0489759.

10. Bressel, et al., U.S. Patent 5,183,552, 1993.

CHAPTER 31
PWB MANUFACTURE USING FULLY ELECTROLESS COPPER

Dr. Hayao Nakahara
N.T. Information Ltd., Huntington, New York

31.1 FULLY ELECTROLESS PLATING

Fully electroless plating has been recognized as a viable technology for some time. It is especially useful for the formation of fineline conductors and considered excellent for plating small, high-aspect-ratio holes because of its high throwing power when compared with that of galvanic plating (see Fig. 31.1). However, its use was limited to the manufacture of double-sided and simple multilayer printed wiring boards (PWBs) for some time after the commercial introduction of the additive process called CC-4 began at Photocircuits Corporation in 1964. This was due to some relatively poor physical properties of electrolessly deposited copper, such as elongation of 2 to 4 percent, compared to 10 to 15 percent achieved by galvanically deposited copper.

The view on electroless plating technology began to change in the mid-1970s when IBM decided to utilize the technology for the fabrication of multilayer boards (MLBs) to package its then top-of-the-line mainframe computers.[1] IBM and other PWB makers using fully electroless copper plating have continuously improved the properties of electrolessly deposited copper since the early 1980s. IBM has continued to use the technology for the fabrication of more advanced MLBs for mainframe and supercomputers.[2,3] Stimulated by IBM's work, and because of technical necessity, NEC Corporation and Hitachi Ltd. of Japan also applied electroless plating technology for the fabrication of MLBs for their mainframe and supercomputers.[4,5]

Today, electrolessly deposited copper is considered as reliable as galvanically deposited copper, and fully electroless plating technology is finding its way to many applications.[6–20] In this chapter, we will discuss various methods of PWB fabrication by means of fully electroless copper-plating technology.

PWB fabrication technology using electroless plating is often referred to as *additive technology*. Therefore, throughout this chapter, the words *electroless* and *additive* will be used interchangeably. Electroless copper plating for through-hole metallization, which deposits a thin film of copper on the wall of plated-through-holes (PTHs), typically from 0.3 to 3 μm thick, is a technology different from the one under consideration and will not be discussed in this chapter (see Chap. 29, "Electroplating").

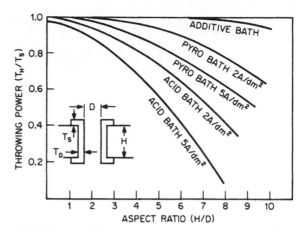

FIGURE 31.1 Ability of electroless bath to plate hole of high aspect ratio without reducing throwing power.

31.2 THE ADDITIVE PROCESS AND ITS VARIATIONS

There have been many variations to additive processes,[20,21] but there are three basic additive processes commercially practiced, as illustrated in Fig. 31.2:

1. Pattern-plating additive
2. Panel-plate additive
3. Partly additive methods

31.3 PATTERN-PLATING ADDITIVE

Pattern-plating additive methods can be classified further into three different approaches, as described in the following sections, depending on the base substrates used.

31.3.1 Catalytic Laminate with CC-4®*[22]

CC-4 stands for *copper complexer number 4,* EDTA, the fourth complexing agent successfully tried by Photocircuits in the early 1960s for full-build electroless copper-plating solution. Over time, the term CC-4 has been used frequently as an adjective, such as in "CC-4 process" or "CC-4 bath."

31.3.1.1 Process Steps. The CC-4 process starts with catalytic laminates coated with catalytic adhesive. The process sequence is as follows:

1. Catalytic base laminate coated with catalytic adhesive (both sides)
2. Hole formation

---

* CC-4 is a registered trademark of AMP-AKZO Corporation.

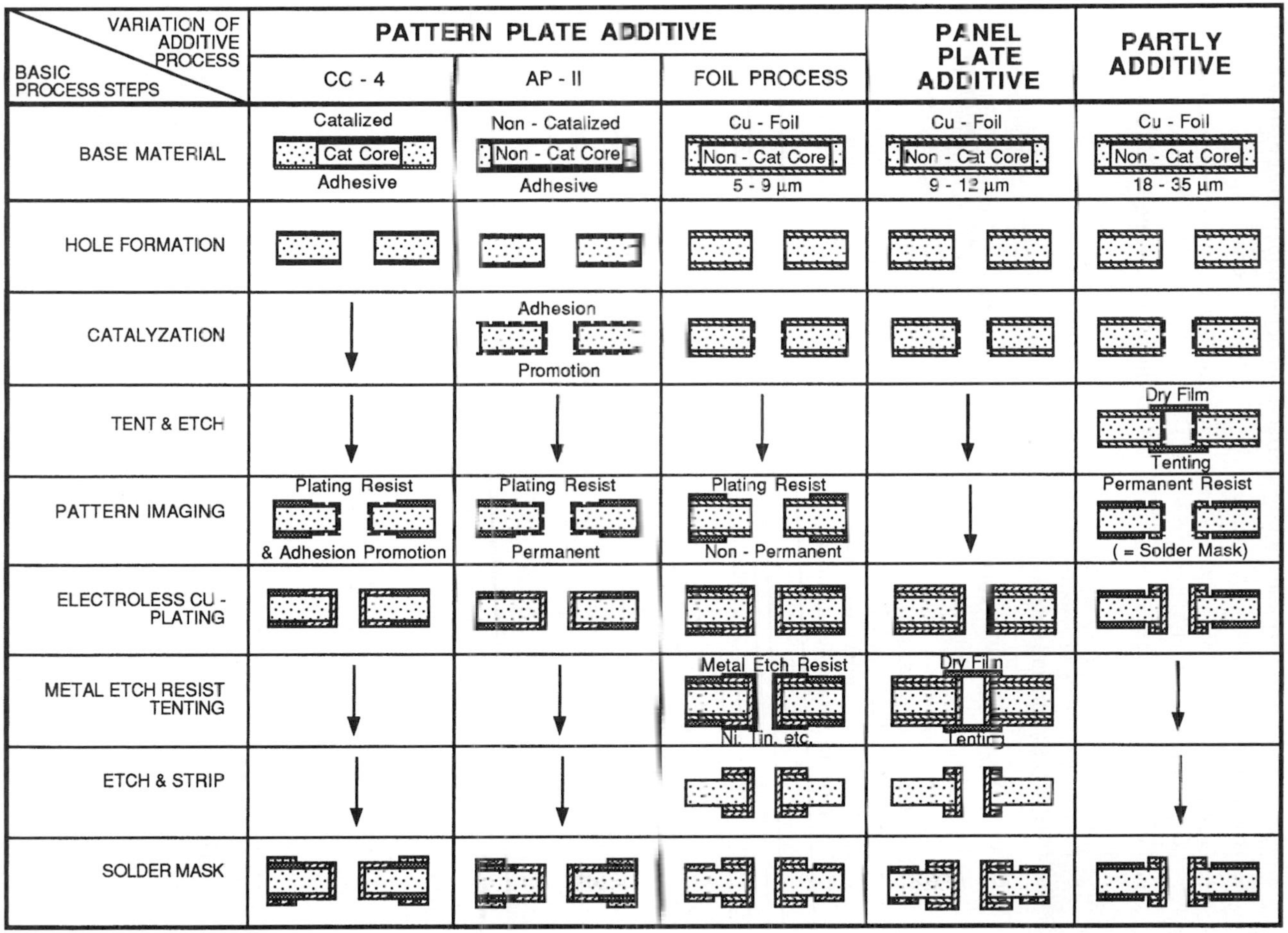

FIGURE 31.2 Variation of additive process.

3. Mechanical abrasion of the adhesive surfaces for better adhesion of plating resist

4. Application of plating resists (screening ink or dry film resist)

5. Formation of microporous structure of exposed adhesive surfaces by chemical etching in acid solution (CrO_3/H_2SO_4 or CrO_3/HBF)

6. Fully electroless copper deposition on the conductor tracks and hole walls

7. Panel baking

8. Application of solder mask and legends

9. Final fabrication and test

31.3.1.2 Resist Issues.   When the conductor width is 8 mil (0.2 mm) or wider, image transfer (step 4) for a majority of CC-4 boards is still done by screen printing with thermally curable or UV-curable ink, since the cost of imaging by screening method is less than one-third that of dry film imaging. The plating resist is permanent; that is, it stays permanently as an integral part of the board. UV-curable ink is preferred because of its shorter curing time that tends to minimize lateral ink flow while being cured. Lateral ink flow narrows the conductor width since the conductor is formed in the trench between plating resists. See Fig. 31.3.

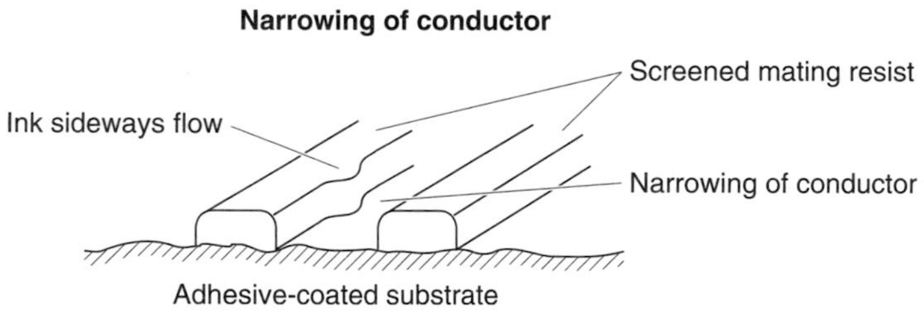

Narrowing of conductor

FIGURE 31.3   Narrowing of conductor due to lateral flow of plating resist after application.

31.3.1.3 Typical Plating Conditions.   Although the practical limitation of screen printing conductor pattern is 8 mil (0.2 mm), in a mass production environment, 6-mil (0.15 mm) conductor patterns have been successfully screened, but such a pattern is confined to a limited area of the panel. Therefore, a conductor pattern which is less than 6 to 7 mil is normally formed by dry film resist. E.I. du Pont de Nemours & Co. and Hitachi Chemical Co., Ltd., are two major suppliers of permanent dry film plating resists which withstand the hostile plating bath environment. Typically, additive bath temperatures range from 68 to 80°C and pH from 11.8 to 12.3, and panels must stay in the bath for more than 10 h with plating speed of typically 2.0 to 2.5 μm/h.

Since the use of solvent such as 1-1 trichloroethane is being banned, semiaqueous-developable permanent dry film resist has been developed by Hitachi Chemical to withstand the plating condition as previously described and is used extensively in Japan.

Since paper-based laminates are still the dominant materials in Japan (although CEM-3 and FR-4 laminates have been used increasingly in recent years), some allowance for this material in the plating solution is made. Therefore, plating baths for paper-based laminates are usually maintained at the lower end of the operating temperature (68 to 72°C) in order not to delaminate them.

31.3.1.4 Permanent Plating Resist.   The use of a permanent plating resist creates a flush surface, where the copper and resist are at the same height from the base laminate. This offers two distinct benefits. In the board fabrication process, after the conductors are formed, the chance of their destruction due to scratch is minimized. (See Fig. 31.4.) In addition, the flush surface makes the application of solder mask easier and reduces the consumption of solder resist by as much as 30 percent when compared to that incurred in a conventional subtractive, etched foil process. In the board assembly process, a permanent plating resist acts as the solder resist and tends to minimize solder bridges during the soldering operation, particularly for fine-pitch devices.

FIGURE 31.4 Flush surface resulting from additive plating with permanent plating resist. *(Photo courtesy of Ibiden Co. Ltd.)*

31.3.2 Noncatalytic Laminate with AP-II[22]

AP-II is an abbreviation of Additive Process II developed by Hitachi Ltd. This process also makes use of permanent plating resist. However, unlike the CC-4 process, it starts with noncatalytic laminate coated with noncatalytic adhesive. Its process sequence is as follows.

31.3.2.1 Process Steps

1. Noncatalytic laminate coated with noncatalytic adhesive
2. Hole formation
3. Mechanical abrasion of adhesive surfaces followed by chemical roughening (adhesion promotion)
4. Catalyzation of the surfaces and hole walls
5. Application of plating resists

The rest of the process is essentially the same as the CC-4 process (steps 6 through 9).

31.3.2.2 Advantages/Disadvantages.   This catalyzing version of the pattern-plating additive process has one advantage over the CC-4 process, that is, the use of noncatalytic base laminates which are cheaper than catalyzed laminates. On the other hand, this process has a disadvantage also in that pattern repair before plating is not possible because resist which

protrudes sideways and makes the conductors narrower cannot be removed since the Pd catalyst underneath the protruding part of the resist tends to be removed as well if repair is attempted, resulting in plating voids. However, in a really fineline conductor case such as 4 mil (0.1 mm) or below, repair attempts will not be made; hence, this disadvantage does not usually impede this catalyzing version of the pattern-plating additive process.

31.3.2.3 AP-II Variation AAP/10. A variation of the AP-II process called the AAP/10 process has been developed and practiced by Ibiden Co., Ltd., of Japan to fabricate high-layer-count MLBs with fineline conductors down to 3 to 4 mil.[23,24] The AAP/10 process is essentially the same as AP-II except that the adhesive used for the AAP/10 process has better insulation characteristics, suitable for higher performance.

31.3.3 Foil Process

Although an adhesive-coated surface provides adequate insulation characteristics for most applications, some boards, such as ones used for mainframe computers, require higher insulation resistance than adhesives can render. When IBM decided to use fully electroless plating to form fineline conductors on innerlayers of high-aspect-ratio MLBs in the mid1970s, it took an approach different from the adhesive system.[1]

31.3.3.1 Process Steps. The process developed at IBM is as follows:

1. Copper-clad laminate (5-μm copper foil).
2. Drill holes.
3. Treat copper surface with benzotriazol.
4. Catalyze the panel.
5. Dry film laminate for conductor pattern (reverse pattern).
6. Do fully electroless copper deposition on the conductor track.
7. Metal overplate as etch resist (tin-lead, tin, nickel, etc.).
8. Strip dry film and etch copper.
9. Strip metal etch resist.
10. Do oxide treatment for subsequent lamination.

31.3.3.2 Process Issues. This foil process requires strong adhesion of dry film plating resist on copper surface. Benzotriazol treatment of the copper surface improves adhesion of dry film. However, the most crucial step is cleaning the copper surface (virgin copper) to obtain an optimum pH for best reaction between copper and benzotriazol.

Hitachi Ltd. found that the use of compounds having amino and sulfide groups is effective in suppressing delamination between copper surface and plating resist.[21] It uses dry film resist which contains a trace of benzotriazol rather than treating the copper surface with the chemical. Baking the panel after exposure/development of dry film at about 140°C for 1 h is found to improve the adhesion between copper surface and dry film during plating operation and at the same time prevents this chemical from breaching out into the plating bath, which could poison the bath.

Table 31.1 shows representative high-tech MLBs fabricated by means of additive technology. Innerlayer patterns of all MLBs listed in the table are made by the foil process. The outerlayer patterns are formed by the panel-plate additive process, which is the subject to be discussed in the following section.

TABLE 31.1 Large-Scale MLBs

Computer	IBM ES-9000	NEC ACOS-3900	Hitachi M-880
Materials	Brominated epoxy glass	Polyimide glass	Maleimide styryl
Dielectric C	NA	4.6	3.6
Size, mm	$600 \times 700 \times 7.4$	$446 \times 477 \times 8.0$	$534 \times 730 \times 7.1$
No. layers	22	42	46
Cond. width, μm	81	70	70
Cond. height, μm	43	30	65
PTH dia., mm	0.46	0.6/0.3	0.56/0.3
Aspect ratio	16/1	27/1	23/1
No. of PTHs	42,000	17,000	100,000
No. of IVHs	8,000	11,000	NA
Innerlayer cond. formation	Additive pattern 5-μm copper	Additive panel 12-μm copper	Additive pattern 12-μm copper
Outerlayer formation	Additive panel	Additive panel	Additive panel
Impedance, Ω	80 + 10, −9	60 ± 6	NA

Source: N.T. Information Ltd.

31.4 PANEL-PLATE ADDITIVE

Maintaining uniform thickness on the panel surface in a galvanic plating operation is difficult. As the aspect ratio becomes higher, maintaining uniform thickness on the hole wall is also a problem.

Electroless plating offers two distinctively superior features in this respect:

1. Its excellent throwing power for high-aspect-ratio holes (Fig. 31.5)

2. Its ability to deposit copper film of even thickness over the entire circuit panel.

FIGURE 31.5 Excellent hole coverage is achieved by fully additive plating. This shows the cross section of a hole with a 27:1 aspect ratio. *(Photo courtesy of Hitachi Ltd.)*

These features of fully electroless plating are ideal for the fabrication of high-tech MLBs, as described in Table 31.1. All signal conductors of the MLBs are formed on innerlayers and the outerlayer surfaces provide pads only for through-holes and some bonding pads for decoupling capacitors. The process steps of the panel-plate additive method are essentially the same as those of the galvanic panel-plate method:

31.4.1 Process Steps

1. Copper clad laminate (thin copper foil is preferred).

2. Perform hole formation.

3. Catalyze holes.

4. Perform fully electroless copper deposition (panel plate).

5. Perform dry film tent-and-etch.

31.4.2 Process Issues

Tenting with dry film becomes difficult when annular rings are very small or in some cases when there are no annular rings at all. In such cases, a positive electrodeposition system may be applied to overcome this difficulty.

It should be cautioned, however, that there is a limitation to this panel-plate additive approach, even though the thickness of deposited copper film is much more even across the panel. If very fine conductors are to be formed by this method, the conductor cross section becomes intolerably distorted from the ideal rectangular shape as the ratio of the width to the height of the conductor approaches unity, as shown in Fig. 31.6.

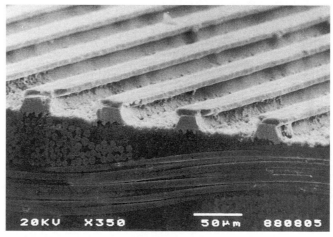

FIGURE 31.6 Formation of fineline conductors by etching process results in undesirable, nonrectangular conductor cross section. *(Photo courtesy of Ibiden Co., Ltd.)*

31.5 PARTLY ADDITIVE

As in the case of the panel-plate additive method, the partly additive process starts with copper-clad laminate.[11–14] The essence of this process is to minimize the problem encountered in etching fineline conductors through thick copper which results from through-hole plating (in the conventional etched foil process, through-hole is plated first), by forming conductors first, prior to through-hole plating.

31.5.1 Process Steps

1. Copper-clad laminate.
2. Perform hole formation.
3. Catalyze hole walls and remove catalyst sites from the surfaces by brushing.
4. Perform image transfer by screen print or dry film tenting.
5. Form conductor pattern by etching and stripping the resist.
6. Apply plating resist over the entire surface, but leaving pads and holes uncovered.
7. Perform electroless copper deposition onto the pads and hole walls.

31.5.2 Process Issues

By proper selection of catalysts and etching solutions, the catalyst on the hole walls survives during the etching operation, posing no problem in the subsequent plating step.

In step 6, plating resist can be applied by screen printing when the circuit density is not so high. When the density becomes higher, dry film or liquid photoimageable resist can be used. Liquid resist can be applied by open screen coating, roller coating, or curtain coating.

In some cases, step 6 can be bypassed when sideways line growth is not a problem (Fig. 31.3). In such a case, the plating speed is reduced by 20 to 30 percent, from the normal speed of 2.0 to 2.5 μm, in order to avoid copper deposition on unwanted areas (bare laminate areas). Conductors must be made narrower than the intended final width at the design stage to compensate for the side growth.

31.5.3 Fine-Pitch Components

As the lead pitch of multipin devices becomes narrower than 20 mil (0.5 mm), bridging between adjacent pads during plating can occur even when plating resist is placed between them. As a result, the practitioners of the partly additive process are abandoning this process and converting to the panel-plate additive method based on the experiences they gained from practicing partly additive technology.

31.6 CHEMISTRY OF ELECTROLESS PLATING

Chemical reactions which take place in electroless plating have been treated well in the literature.[4,21] However, they are discussed again here because some of the due comments are relevant to the chemistry of electroless plating.

31.6.1 Electroless Plating Solution Chemical Reactions

Main ingredients of most electroless plating bath formulations consist of

NaOH   pH adjustment
HCHO   Reducing agent (formaldehyde)
$CuSO_4$ Source of copper iron
EDTA   Cheleting agent

31.6.1.1 Copper Deposition.   The chemical reaction of copper deposition may be represented by

$$Cu(EDTA)^{2-} + 2HCHO + 4OH^- \rightarrow$$
$$Cu^\circ + H_2 + 2H_2O + 2CHOO^- + EDTA^{4-}$$

(31.1)

Under the alkaline pH condition, the cupric ions would normally combine with OH^- to produce cupric hydroxide $[Cu(OH)_2]$, a useless precipitate. When a cheleting agent such as EDTA is added, it prevents cupric hydroxide formation by maintaining the Cu^{2+} in solution. Once the deposition of metallic copper starts through catalytic sites, the reaction in Eq. (31.1) continues because of the autocratic nature of the reaction.

31.6.1.2 Cannizzaro Reaction and Formaldehyde Concentration.   While this main reaction, represented by Eq. (31.1), takes place, other undesirable side reactions proceed, also in competition with the reaction. One of the major difficulties is the lowering of formaldehyde concentration in the solution due to its disproportionate consumption in alkaline solution. This is known as the *Cannizzaro reaction,* which may be characterized by

$$2HCHO + OH^- \quad CH_3OH + HCOO^- \tag{31.2}$$

This reaction continues independently. Fortunately, methanol (CH_3OH) one of the by-products of the Cannizzaro reaction, tends to shift the equilibrium of the reaction in Eq. (31.2) to the left and thus prevent the decrease of formaldehyde concentration by an unproductive reaction. By controlling the plating conditions properly, the wasteful consumption of form-aldehyde by the Cannizzaro reaction can be retained within 10 percent of the consumption in the reaction in Eq. (31.1).

31.6.1.3 Fehlings-Type Reactions.   In addition to the Cannizzaro reaction, the following side reaction also competes for formaldehyde:

$$2Cu^{2+} + HCHO + 5OH^- \quad Cu_2O + HCOO^- + 3H_2O \tag{31.3}$$

31.6.1.4 Other Formaldehyde-Related Reactions.   Under conditions which would favor the Fehlings-type reaction [Eq. (31.3)], spontaneous decomposition of uncatalyzed plating solutions produces precipitation of finely divided copper with attended vigorous evolution of hydrogen gas. The finely divided copper produced is due to the disproportionate quantity of Cu_2O under alkaline conditions:

$$Cu_2O + H_2O \quad Cu^\circ + Cu^{2+} + 2OH^- \tag{31.4}$$

Furthermore, formaldehyde may act as a reducing agent for the cuprous oxide to produce metallic copper:

$$Cu_2O + 2HCHO + 2OH^- \quad 2Cu^\circ + H_2 + 2HCOO^- + H_2O \tag{31.5}$$

31.6.1.5 Filtering.   The Cu° nuclei thus produced according to Eqs. (31.4) and (31.5) are not relegated to deposition on substrate but are produced randomly throughout the solution and become the catalytic sites for further undesirable copper deposition. Continuous filtration with a filter of pore size 20 μm or smaller can improve extraneous copper due to these catalytic sites.[25] The use of continuous filtering also improves the ductility of copper by eliminating codeposition of impurities. In most mass production facilities of PWBs by additive technology, two or three stages of filtering with filters of pore sizes down to 5 μm are common.

31.6.2 Use of Stabilizers

Since the reaction represented in Eq. (31.3) indicates the greatest degree of instability of the plating bath, various measures are taken to counteract this reaction. Alkaline cyanide has been a popular stabilizer used for the CC-4 bath, since cyanide forms strong complexes with Cu^+ ($Cu^+ + 2CN^-$ $Cu(CN)_2^{2-}$), but relatively unstable complexes of Cu^{2+}. However, since cyanide also reacts with HCHO, it is difficult to control. 2,2'-dypiridil also chelates Cu^+ and does not react with HCHO. Hence, it is a more favored stabilizer, particularly for those baths used in Japan. On the other hand, a dypiridil bath deposit begins to have less ductility after several plating cycles and weekly bath makeup may be necessary.

Aeration of the plating bath is known to stabilize the solution, and vigorous aeration is commonly used to operate modern full-build electroless baths.

31.6.3 Surfactant

Reduction in surface tension assures that the plating solution is in intimate contact with the catalyst nuclei and permits thorough surface solution interaction. For this reason, a wetting agent or surfactant is included in bath formulations. Polyethelene oxide, polyethelene glycol (PEG), etc., are popular surfactants. Proper usage of surfactants also prevents the accumulation of gaseous material (air and hydrogen bubbles) on the surface of the board, particularly around the entrances of holes, which would also tend to lower the surface-solution interaction and thus seriously impair proper metal deposition.

31.6.4 Reliability of Deposited Copper and Inorganic Compounds

Generally, copper film with high elongation and reasonable tensile strength is preferred for through-hole reliability, although the relation between through-hole reliability and ductility and tensile strength is still not understood clearly.

The use of inorganic compounds containing vanadium or germanium as a part of chemical ingredients in plating baths enhances the quality of deposited copper.[13] Most full-build electroless baths used today contain either V_2O_5 or G_2O_5 to improve the physical properties of deposited copper.

31.6.4.1 Typical Copper Properties. Typical physical properties of reliable electrolessly plated copper are (as plated):

| Elongation | 8 to 10 percent |
| Tensile strength | 35 to 45 kg/mm^2 |

After annealing the copper, elongation increases to 12 to 18 percent and tensile strength decreases slightly. These properties are equivalent to the ones resulting from acid copper. Since all panels are baked at about 140 to 160°C for 30 to 60 min during the course of the production process after plating, deposited copper gets annealed, thus assuring high through-hole reliability.

31.6.5 Removal of Impurities

Use of continuous filtering has been mentioned to remove some of the impurities. In addition there are other methods.

31.6.5.1 Electrodialysis. Electrodialysis[26,27] is effective to remove formate ions, sulfate ions, and carbonate ions which are by-products of reactions defined in Eqs. (31.1), (31.2), (31.3), and (31.5).

31.6.5.2 Overflow Method. Another very effective way to remove these unwanted by-products is the *overflow* method. Since essential components of the plating solution are consumed continuously, they must be replenished to maintain desirable plating conditions. In so doing, a considerable amount of the solution overflows because of the addition of water. The overflow also contains the reaction by-products as well as chelated copper ions.

After a certain amount of overflow solution is accumulated, sodium hydroxide and formaldehyde are added to the overflow solution and heated to promote spontaneous decomposition of the plating solution to precipitate copper metal. Addition of copper powder accelerates the precipitation. This so-called *chemical bomb-out* recovery method can be operated in bulk or continuously. The remaining solution can be treated with sulfuric acid to recover over 95 percent of the EDTA in the solution. Usually, the purity of EDTA thus recovered is higher than that of purchased EDTA.

Ninety-eight percent of the solution is recovered as permeate and is used in such processes as rinsing. The concentrate, the other two percent, is high in salt content (Na_2SO_4 and HCOONa) and may be evaporated to produce a small amount of solid waste or discharged into the ocean, if the plant is located near the ocean and locally approved.[6] This chemical bomb-out method can achieve two objectives at once: removal of unwanted impurities and recovery of chemicals. With this system, it is easy to maintain Baume at the desired 9°.

31.6.6 Environmental Issues of Formaldehyde

Formaldehyde is considered to be a carcinogen. If exposed to formaldehyde fumes of certain concentration for a long time, it can cause allergy conditions and perhaps cancer, although cancer due to long-time exposure to formaldehyde in the PWB industry has never been reported.

With a good exhaust system and scrubbing through water and peroxide solution, formaldehyde can be made quite harmless. It is a matter of economy. EDTA can be treated as previously mentioned.

Reducing agents alternative to formaldehyde such as glyoxylic acid[27] and hypophosphite (M_3PO_2)[28] have been suggested and tried, but they have not been able to replace formaldehyde as an effective reducing agent at this stage.

31.7 FULLY ELECTROLESS PLATING ISSUES

31.7.1 Efficiency of Electroless and Galvanic Plating

PWB manufacturers which are not familiar with electroless plating often have the misconception that electroless plating takes "too long." On the contrary, electroless plating is faster than galvanic plating, and its volumetric efficiency is twice as much as galvanic plating.

Electroless plating baths operate at the speed of 2.0 to 5.0 μm/h, 2.5 μm being the medium speed. Because of its high throwing power, a deposition thickness of 25 μm on the panel surface also assures 25 μm in the hole. That is, 10 h are needed to plate this thickness. The loading factor in electroless plating can be maintained at 4 dm^2/liter. A 2500-gal (10,000-l) tank can plate about 400 m^2 (4000 ft^2) of panels in one shot in 10 h. That is, 400 ft^2/h, or approximately 130 (18-in × 24-in) panels/h from a 10-m^3 tank. This is equivalent to twice the speed of galvanic plating from the same volume of plating solution. When small, high-aspect-ratio holes are involved, this figure becomes more favorable to electroless plating because the current density of galvanic plating must be reduced to 10 to 12 A/ft^2, which requires more than 2 h of plating for 25 μm in the hole.

Figure 31.7 shows the plating racks filled with panels waiting to be placed in electroless plating tanks. In large-volume operations, the panel-racking operation is fully automated.

31.7.2 Photochemical Imaging Systems

Various photochemical imaging systems have been developed and tried to be put into practice.[6] However, no photochemical imaging systems are used for real production today. They are all unstable and the lateral growth of conductors poses a problem for the fineline applications for which they are intended.

31.7.3 Molded Circuit Application

A handful of three-dimensional molded circuit manufacturers (or, the preferred identification, *molded interconnection device* or MID) have continued their quest for market with this

FIGURE 31.7   Plating racks filled with panels to be plated. *(Photo courtesy of Adiboard, Brazi.)*

product and have met with some success. (See Fig. 31.8.) There are several alternative methods to manufacture MIDs, of which the *two-shot* molding method can produce the most sophisticated three-dimensional interconnection devices.

A part which is to be metallized (circuitized) is molded first. Then, a portion of the first mold which is not desired for metallization is surrounded by the second molding, exposing

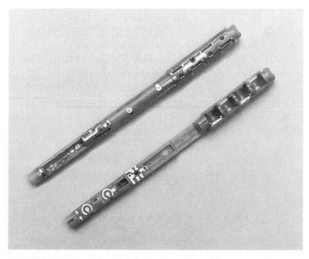

FIGURE 31.8   An example of a three-dimensional molded circuit application, in this case a light pen for computer input. *(Photo courtesy of Mitsui Pathtek.)*

only the portion desired to be metallized. This two-shot molding method can be classified further into two alternative methods, which are similar to CC-4 and AP-II.

One is to use catalyzed engineering plastic for the first molding and noncatalyzed plastic for the second molding. Adhesion promotion consists of a swell-and-etch step followed by chromic acid etch. The other alternative is to mold the first part with noncatalyzed plastic materials and expose the molded part to an adhesion promotion process and catalyze. The second part is then molded, as in the first case.

These parts are placed in an electroless plating bath for circuitization. A high rate of growth is expected for MIDs, with the automotive market being the most promising application area.

REFERENCES

1. W. A. Alpaugh and J. M. McCreary, IPC-TP-108, Fall Meeting, 1976.
2. R. R. Tummala and S. Ahmed, "Overview of Packaging for the IBM Enterprise System 9000 Based on the Glass-Ceramic Copper/Thin Film Thermal Conduction Module," *IEEE Trans. on Comp., Hybrids and Mfg. Tech.,* vol. 14, no. 4, Dec. 1991, pp. 426–431.
3. L. E. Thomas, et al., "Second-Level Packaging for the High End ES/9000 Processors," *IBM J. of Research & Development.*
4. H. Murano and T. Watari, "Packaging Technology for the NEC SX-3 Supercomputers," *IEEE Trans. on Comp., Hybrids and Mfg. Tech.,* vol. 15, no. 4, Aug. 1992, pp. 411–417.
5. A. Takahashi, et al., "High Density Multilayer Printed Circuit Board for HITAC M-880," *op. cit.,* pp. 418–425.
6. Clyde F. Coombs, Jr. (ed.), *Printed Circuits Handbook, 3d ed.,* McGraw-Hill, New York, 1986, Chap. 13.
7. U.S. Patent 2,938,805, 1960.
8. U.S. Patent 3,628,999, Dec. 1971.
9. U.S. Patent 3,799,802, Mar. 1974.
10. U.S. Patent 3,799,816, Mar. 1974.
11. U.S. Patent 3,600,330, Aug. 1971.
12. Japanese Patent 43-16929, July 1968.
13. Japanese Patent 58-6319, Feb. 1983.
14. H. Steffen, "Additive Process," Ruwel Werke GmbH, private communication.
15. K. Minten and J. Toth, "PWB Interconnect Strategy Using a Full Build Electroless Plating System: Part I," *PC World Conference V,* Paper No. b-52, Glasgow, Scotland, June 1990.
16. K. Minten, J. Seigo, and J. Cisson, "Part 2: Etch Characteristics," *Circuit World,* vol. 18, no. 4, Aug 1992, pp. 5–12.
17. K. Minten and J. Cisson, "Part 3: Characterization of a Semi-Additive Naked Palladium Catalyst," *Circuit World,* vol. 19, no. 2, Jan. 1993, pp. 4–13.
18. K. Minten, K. Kitchens, and J. Cisson, "Part 4: The Future of FBE Process," private communication.
19. N. Ohtake, *Electronics Technology* (Japanese), 27(7), June 1985, pp. 55–59.
20. N. Ohtake, "New Printed Wiring Boards by a Partly Additive Process," *PC World Conference IV,* Tokyo, Paper No. 44, June 1987.
21. S. Imabayashi, et al., "Partly-Additive Process for Manufacturing High-Density Printed Wiring Boards," *IEEE Trans.* 0569-5503/92/0000-1053, 1992.
22. H. Akaboshi, "A New Fully Additive Fabrication Process for Printed Wiring Boards," *IEEE Trans. on Comp., Hybrids and Mfg. Tech.,* vol. CHMT-9, no. 2, June 1986.
23. R. Enomoto, et al., "Advanced Full-Additive Printed Wiring Boards Using Heat Resistant Adhesive," *PC World Conf V,* Paper No. B6-1, Glasgow, Scotland, June 1990.

24. K. Ikai, "Full Additive Boards," *Electronic Materials* (Japanese), Oct. 1993, pp. 58–63.

25. H. Honma, K. Hagihara, and K. Kobayashi, "Factors Affecting Properties of Electrolessly Deposited Copper," *Circuit Technology* (Japanese), vol. 9, no. 1, 1994, pp. 31–36.

26. T. Murata, "Chemistry of Electroless Plating," *Circuit Technology* (Japanese), vol. 8, no. 5, 1993, pp. 357–362.

27. J. Darken, "Electroless Copper—An Alternative to Formaldehyde," *PC World Conf V,* Glasgow, Scotland, Paper B/6/2, June 1990.

28. J. P. Pilato and Joe D'Ambrisi, "Process for Metallization by Copper Deposition Without the Use of Formaldehyde," MacDermid paper, private communication.

29. R. E. Horn, *Plating and Surface Finishing,* Oct. 1981, pp. 50 52.

30. A. Iwaki, et al., *Photographic Science & Engineering,* vol. 20, no. 6, 1976, p. 246.

CHAPTER 32
SURFACE FINISHES

George Milad
Shipley Ronal, Inc., Marlborough, Massachusetts

32.1 INTRODUCTION

Surface finish is about connectivity. It is at the surface where a connection from the board to a device is to occur. Initially soldering and insertion were the preferred methods for connecting components to the board and the board to the system. The dominant method of assembly was wave soldering. Hot-air solder leveling (HASL) was the finish of choice for component holes on the board. Electrolytic nickel/gold was used to plate the tabs for edge connectors.

The evolution from through-hole components to surface-mount components offered great opportunities to reduce the size and weight of the final electronic product. Surface-mount pads required screen pastes and reflow assembly. At its inception, surface mounting still fell within the capabilities of the HASL finish; however, as the need for further miniaturization continued, the surface-mount pad continued to shrink and screening paste on the pads became challenging. The HASL finish was not flat or coplanar enough for successful screening of paste on fine-pitch pads.

32.1.1 Surface Finishes—Alternatives to HASL

New designs required innovative solutions; ball grid arrays, wire-bonding pads, press fitting, and contact switches were all outside the traditional realm of HASL and electrolytic nickel/gold. In addition, environmental concerns are being focused on the elimination of lead. This will change HASL from a process that uses a standard solder to a new lead-free process.

A series of surface finishes emerged to fill these needs. They include:

- Organic solderability preservative (OSPs)
- Electroless nickel/immersion gold (ENIG)
- Electroless nickel/electroless palladium/immersion gold (ENEPIG)
- Immersion silver
- Immersion tin

The following aspects of each of these finishes are described in this chapter:

- Chemical and metallurgical principles
- Manufacturing process

- Applications
- Limitations

Other elements are described as appropriate for the understanding of the finish in the context of an electronic assembly.

32.1.2 Surface Finish Capability Summary

Each one of these surface finishes offers connectivity solutions in some areas (see summary in Table 32.1). However, only ENEPIG is capable of meeting all the different assembly requirements. It is often referred to as the universal finish.

TABLE 32.1 Surface Finish Capability Summary

Surface finish	Coplanarity	Solderability	Au wire bonding	Al wire bonding	Contact surface
HASL	No	Yes	No	No	No
OSPs	Yes	Yes	No	No	No
ENIG	Yes	Yes	Yes	No	Yes
Ni/Pd/Au	Yes	Yes	Yes	Yes	Yes
Pd	Yes	Yes	Yes	No	Yes
Ag	Yes	Yes	Yes	Yes	No
Sn	Yes	Yes	No	No	No

32.2 ORGANIC SOLDERABILITY PRESERVATIVE (OSP)

32.2.1 Principles

Organic solderability preservatives (OSP), as the name implies, is an organic coating that preserves the copper surface from oxidation until it is soldered. The two most widely used preservatives are both nitrogen-bearing organic compounds. Benzotriazoles constitute one class and imidazoles form the other. Both of these organic chemicals have the ability to complex with the exposed copper surface. In that respect they are copper specific and do not adsorb to the laminate or the solder mask.

Benzotriazoles form a monomolecular layer and protect the copper until it is exposed to a single thermal excursion at assembly. The coating readily volatilizes under reflow thermal conditions. Imidazoles form a thicker coating and survive multiple thermal excursions at assembly.

32.2.2 Process

Process steps for OSP are detailed in Table 32.2.

- *Cleaner:* the purpose of this step is to clean the copper surface in preparation for processing. The cleaner removes oxides and most organic and inorganic residues and ensures that the copper surface will be uniformly microetched.

TABLE 32.2 OSP Basic Process

Process step	Temp (°F)	Temp (°C)	Time (min)*
Cleaner	95–140	35–60	4–6
Microetch	75–95	25–35	2–4
Conditioner	85–95	30–35	1–3
OSP	120–140	50–60	1–2

* For conveyorized equipment, the dwell time must be reduced. Consult with equipment and chemical supplier.

- *Microetch:* the purpose of this step is to microetch the copper and to expose a fresh copper surface that will complex the OSP uniformly. An appropriate etchant (for example, sodium persulfate, peroxide/sulfuric acid, etc.) may be used here.
- *Conditioner:* this is an optional step depending on vendor requirements or recommendations.
- *OSP:* the OSP solution is usually run at a temperature of 120 to 140°F (50 to 60°C) and a short dwell time of 1 to 2 min per vendor recommendation. The process is self-limiting.
- *Rinsing:* rinsing is required after each chemical process step to remove all chemical residuals from the previous process step. This cleaning may be achieved using a single or dual rinse. Excessive dwell time in a rinse bath may cause oxidation or tarnishing of the product and is undesirable. Vendor specifications for temperature, dwell time, agitation, and turnover must be followed.

The short dwell times required for application of OSP to the copper are very conducive to the use of horizontal conveyorized equipment.

32.2.3 Applications

- *Product:* a thin layer of organic compound coats the copper surface. The coating can be as thin as 100 Å for benzotriazoles and as thick as 4000 Å for imidazole-type preservatives. The coating is transparent and not easily discernible, making inspection difficult.
- *Assembly:* at assembly the organic coating is readily dissolved into the screened paste or into the acidic flux, in either case leaving a clean active copper surface to solder to. The solder then forms a copper/tin intermetallic joint.

32.2.4 Limitations

- *Inspection:* due to the fact that the coating is transparent and colorless, it is difficult to inspect.
- *Electrical testing:* the organic coating is nonconductive. Benzotriazole, being a very thin coating, does not interfere with electrical testing. Some imidazoles, on the other hand, are thick enough to interfere with electrical testing. Most shops that use these thicker coatings do their electrical testing prior to OSP applications.
- *Assembly:* imidazoles may require a more aggressive flux after the first and second thermal excursions. Assemblers that use OSP are very familiar with the fluxing requirements for this finish.

32.3   *ELECTROLESS NICKEL/IMMERSION GOLD (ENIG)*

32.3.1   Principles

In this finish a layer of electroless nickel with a thickness of 120 to 240 µin (3 to 6 µm) is deposited on the copper surface. This is then followed by a thin coating (2 to 4 µin) of immersion gold. The nickel is a diffusion barrier to copper and is the surface to which the soldering occurs. The function of the immersion gold is to protect the nickel from oxidation or passivation during storage.

32.3.2   Process

Process steps for ENIG are detailed in Table 32.3.

TABLE 32.3   Electroless Nickel/Immersion Gold (ENIG)

Process step	Temp (°F)	Temp (°C)	Time (min)
Cleaner	95–140	35–60	4–6
Microetch	75–95	25–35	2–4
Catalyst	RT	RT	1–3
Electroless nickel	180–190	82–88	18–25
Immersion gold	180–190	82–88	6–12

- *Cleaner:* the purpose of this step is to clean the copper surface in preparation for processing. The cleaner removes oxides and most organic and inorganic residues and ensures that the copper surface will be uniformly microetched for catalyst uptake.

- *Microetch:* the purpose of this step is to microetch the copper and to expose a fresh copper surface that will catalyze uniformly. An appropriate etchant (for example, sodium persulfate, peroxide/sulfuric acid, etc.) may be used here.

- *Catalyst:* the purpose of this process step is to deposit an immersion coating of a catalyst on the copper surface. The catalyst lowers the activation energy and allows the nickel to deposit on the copper surface. Examples of catalysts are palladium and ruthenium.

- *Electroless nickel:* the purpose of this bath is to deposit the required thickness of electroless nickel on the catalyzed copper surface. The term *electroless* implies the presence of a reducing agent (sodium hypophosphite is the most common) to sustain the continuity of nickel deposition. The deposit is a nickel with 6 to 11 percent phosphorus. The thickness of the nickel should be adequate to create a diffusion barrier to copper migration and to serve as the solderable surface or as the contacting surface, depending on the intended use. The nickel bath has a relatively high deposition rate and its active chemical components must be replenished and maintained in balance on a continuous basis. Electroless nickel baths run at a high temperature and a long dwell time to achieve the required deposit thickness. One must ensure that material compatibility issues are understood and allowed for, such as choosing a compatible solder mask for the ENIG chemistry.

- *Immersion gold:* in this step, a thin, continuous layer of immersion gold is deposited. Immersion deposits are displacement reactions—gold displaces the nickel on the surface. They do not require a reducing agent and are self-limiting as the nickel surface is coated with the immersion gold. The gold protects the nickel from oxidation or passivation. This bath also runs at relatively high temperatures and dwell times.

- *Rinsing:* rinsing is required after each chemical process step. The purpose is to remove all chemical residuals from the previous process step. This cleaning may be achieved in a single or dual rinse. Excessive dwell time in a rinse bath may cause oxidation or tarnishing of the product and is undesirable. Vendor specifications for temperature, dwell time, agitation, and turnover rate must be followed.

In some instances predip acid baths may be required to minimize drag-in of water to the plating bath. Postdip acid baths may also be used to activate the surface as needed. The long dwell times needed for this process make horizontal conveyorization impractical.

32.3.3 Applications

- ENIG gives a flat coplanar surface. It is solderable and wire-bondable and is ideal as a switch contacting surface.
- ENIG has excellent solder wetability characteristics. The gold readily dissolves in the molten solder, leaving a fresh nickel surface to form the solder joint. The amount of gold that dissolves in the solder is insignificant and will not cause solder joint embrittlement. The nickel forms a tin/nickel intermetalic joint.
- ENIG is compatible with both aluminum and gold wire bonding. However, the gold wire bonding is marginal and not recommended for this finish. Gold wire bonding requires more gold thickness than ENIG is capable of producing. Aluminum wire bonding works well on this surface because the aluminum wire will eventually form a bond with the underlying nickel.
- ENIG is an ideal surface for soft pad contacts. These contact surfaces are needed for applications such as telephones and pagers and any device that requires multiple on/off switching. The hardness and thickness of the nickel lends itself well to this type of application.

32.3.4 Limitations

- The process is fairly complex and requires good process control.
- The electroless nickel bath operates at 180 to 190°F (82 to 88°C) and the dwell time usually exceeds 15 min. Material compatibility must be taken into account here. Continuous replenishment is needed as nickel is plated out. Good process control is of particular importance here to achieve the desired thickness and morphology of the nickel deposit.
- The gold bath also operates at similarly high temperatures. An 8- to 10-min dwell time is adequate to deposit the required immersion gold thickness. Excessive dwell times and operating the bath out of vendor's specifications are known to cause corrosion of the underlying nickel. This corrosion, if excessive, can interfere with the functionality of the nickel surface.

32.4 *ELECTROLESS NICKEL/ELECTROLESS PALLADIUM/IMMERSION GOLD (ENEPIG)*

32.4.1 Principles

In this finish an electroless nickel layer of 120 to 240 μin (3 to 6 μm) is deposited on the copper surface. This is then coated with an electroless palladium layer of a thickness of 4 to 20 μin (0.1 to 0.5 μm) and finally topped with immersion gold at 1 to 4 μin (0.02 to 0.1 μm). The electroless palladium layer prevents any probability of corrosion that may be caused by the immersion reaction and creates an ideal gold wire bonding surface. The gold layer caps the palladium and ensures that its catalytic activity is contained.

32.4.2 Process

Process steps for ENEPIG are detailed in Table 32.4. The process is very similar to the ENIG process except for the addition of the palladium catalyst and the electroless palladium step.

TABLE 32.4   Electroless Nickel/Electroless Palladium/Immersion Gold (ENEPIG)*

Process step	Temp (°F)	Temp (°C)	Time (min)
Cleaner	95–140	35–60	4–6
Microetch	75–95	25–35	2–4
Catalyst	RT	RT	1–3
Electroless nickel	180–190	82–88	18–25
Catalyst	RT	RT	1–3
Electroless palladium	120–140	50–60	8–20
Immersion gold	180–190	82–88	6–12

- *Cleaner:* the purpose of this step is to clean the copper surface in preparation for processing. The cleaner removes oxides and most organic and inorganic residues and ensures that the copper surface will be uniformly microetched for catalyst uptake.

- *Microetch:* the purpose of this step is to microetch the copper and to expose a fresh copper surface that will catalyze uniformly. An appropriate etchant (for example, sodium persulfate, peroxide/sulfuric acid, etc.) may be used here.

- *Catalyst:* the purpose of this process step is to deposit an immersion coating of a catalyst on the copper surface. The catalyst lowers the activation energy and allows the nickel to deposit on the copper surface. Examples of catalysts are palladium and ruthenium.

- *Electroless nickel:* the purpose of this bath is to deposit the required thickness of electroless nickel on the catalyzed copper surface. The term *electroless* implies the presence of a reducing agent (sodium hypophosphite is the most common) to sustain the continuity of nickel deposition. The deposit is a nickel with 6 to 11 percent phosphorus. The thickness of the nickel should be adequate to create a diffusion barrier to copper migration and to serve as the solderable surface or as the contacting surface, depending on the intended use. The nickel bath has a relatively high deposition rate and its active chemical components must be replenished and maintained in balance on a continuous basis.

  Electroless nickel baths run at a high temperature and a long dwell time to achieve the required deposit thickness. One must ensure that material compatibility issues are understood and allowed for, such as choosing a compatible solder mask for the ENEPIG chemistry.

- *Catalyst:* the catalyst is an immersion palladium that is used to activate the nickel surface to facilitate the deposition of electroless palladium. It is similar to the catalyst used to activate the copper surface before electroless nickel deposition.

- *Electroless palladium:* the bath includes a reducing agent and thus is an electroless and not an immersion bath. Hypophosphite and bisulfite are reducing agents commonly used for this application. The hypophosphite gives a deposit with 1 to 2 percent phosphorous; the bisulfite gives a phosphorus-free deposit. The bath operates at 120 to 140°F (50 to 60°C). Dwell time varies from 8 to 20 min, depending on desired thickness.

- *Immersion gold:* in this step a thin, continuous layer of immersion gold is deposited. Immersion deposits are displacement reactions—gold displaces the palladium on the surface. They do not require a reducing agent and are self-limiting as the palladium surface is coated with the immersion gold. This bath also runs at relatively high temperatures and dwell times.

- *Rinsing:* rinsing is required after each chemical process step to remove all chemical residuals from the previous process step. This cleaning may be achieved using a single or dual rinse. Excessive dwell time in a rinse bath may cause oxidation or tarnishing of the product and is undesirable. Vendor specifications for temperature, dwell time, agitation, and turnover rate must be followed.

32.4.3 Applications

- ENEPIG gives a flat coplanar surface. ENEPIG is the universal surface finish. It is capable of functioning as the ENIG finish. In addition, the electroless palladium at this thickness makes an ideal surface for gold wire bonding.
- During soldering, the palladium and the gold both eventually dissolve in the solder and the joint forms a nickel/tin intermetallic. During wire-bonding, the aluminum and the gold wires bond to the palladium surface. Palladium is a hard surface and is suitable for contact switching.

32.4.4 Limitations

- The primary limitation for this finish is the additional cost of palladium. Palladium prices reached $1000.00 an ounce in the year 2000, compared to $280.00 for an ounce of gold in the same time frame. Most of the world's palladium resources are in the former Soviet Union and, to a lesser extent, Canada. Another limitation is the length of the process and the need for good process control as with the ENIG process.

32.5 IMMERSION SILVER

32.5.1 Principles

Immersion silver deposits provide a thin (5 to 15 μin or 0.1 to 0.4 μm), dense silver deposit incorporating an organic. The organic seals the surface and allows for extended shelf life. Silver offers a flat, extremely solderable surface that may be applied with high productivity in conveyorized equipment. The surface is also bondable for both aluminum and gold wire.

32.5.2 Process

Process steps for immersion silver are detailed in Table 32.5.

TABLE 32.5 Process Steps for Immersion Silver

Process step	Temp (°F)	Temp (°C)	Time (min)*
Cleaner	95–140	35–60	4–6
Microetch	75–95	25–35	2–4
Predip	RT	RT	0.5–1
Immersion silver	95–115	35–45	1–2

* For conveyorized equipment, the dwell time must be reduced. Consult with equipment and chemical supplier.

- *Cleaner:* the purpose of this step is to clean the copper surface in preparation for processing. The cleaner removes oxides and most organic and inorganic residues and ensures that the copper surface will be uniformly microetched.

- *Microetch:* the purpose of this step is to microetch the copper and to expose a fresh copper surface. This allows the immersion reaction to proceed uniformly. An appropriate etchant (for example, sodium persulfate, peroxide/sulfuric acid, etc.) may be used here.

- *Predip:* the primary function of the predip bath is to prevent drag-in of contaminants into the immersion silver bath.

- *Immersion silver:* the bath deposits a thin coating of silver by an immersion reaction in which the silver displaces the copper on the surface of the pads. The deposit also incorporates organic additives and organic surface-active agents. The purpose of the additives is to ensure uniformity of the deposit. The surface-active agents protect the silver from humidity during storage.

- *Rinsing:* rinsing is required after each chemical process step to remove all chemical residuals from the previous process step. This cleaning may be achieved using a single or dual rinse. Excessive dwell time in a rinse bath may cause oxidation or tarnishing of the product and is undesirable. Vendor specifications for temperature, dwell time, agitation, and turnover rate must be followed.

32.5.3 Applications

- Immersion silver is an ideal surface for soldering. During assembly the silver dissolves readily into the solder, allowing the formation of a copper/tin intermetallic solder joint, similar to HASL and OSPs. It offers the coplanarity that HASL lacks and it is also lead free. Unlike OSPs, it offers ease of inspection with no compromise in performance after the third thermal excursion. Immersion silver lends itself well to electrical testing in the board shop.

- Silver also serves as a platform for bonding with aluminum and with gold wire.

- At this time the application as a contact surface remains to be demonstrated.

32.5.4 Limitations

- The concern with silver has always been silver migration in electronic environments. This is due to the property of silver to form water-soluble salts when exposed to moisture and electrical bias. The incorporation of organics into the immersion silver minimizes this phenomenon.

- The immersion silver as a surface completely dissolves into the solder during the assembly process.

- After wire-bonding, the exposed silver is encapsulated and thus is isolated from the environment.

32.6   *IMMERSION TIN*

32.6.1 Principles

Immersion tin is a very logical replacement for HASL for two reasons; first, it is flat and coplanar, and second, it is lead free. However, tin readily forms a copper/tin intermetallic that is not solderable.

Immersion tin became viable as a surface finish when two problems, namely grain size and copper/tin intermetallics, were overcome. The deposit was engineered to be very fine-grained and nonporous. A thick deposit of 40 μin or 1.0 μm was feasible, thus ensuring a copper-free tin surface. A new class of immersion tin processes also referred to as *white tin* do exactly that. They are based on methyl-sulfonic acid, tin methylsulfonate, and thiourea.

32.6.2 Process

Process steps for immersion tin are detailed in Table 32.6.

TABLE 32.6   Immersion Tin Process

Process step	Temp (°F)	Temp (°C)	Time (min)*
Cleaner	95–140	35–60	4–6
Microetch	75–95	25–35	2–4
Predip	75–90	25–30	1–2
Immersion tin	140–160	60–70	6–12

* For conveyorized equipment, the dwell time must be reduced. Consult with equipment and chemical supplier.

- *Cleaner:* the purpose of this step is to clean the copper surface in preparation for processing. The cleaner removes oxides and most organic and inorganic residues and ensures that the copper surface will be uniformly microetched.
- *Microetch:* the purpose of this step is to microetch the copper and to expose a fresh copper surface to allow the immersion reaction to proceed uniformly. An appropriate etchant (for example, sodium persulfate, peroxide/sulfuric acid, etc.) may be used here.
- *Predip:* the primary function of the predip bath is to activate the copper surface and to prevent drag-in of contaminants into the immersion tin bath.
- *Immersion tin:* although this bath is an immersion bath, it continues to build up in thickness over time. Most immersion baths are self-limiting. This occurs because the copper forms copper/tin intermetallics that sustain the immersion reaction. When all is said and done, a minimum of 10 to 15 μin of pure tin is left on the surface, which is the solderable surface.
- *Rinsing:* rinsing is required after each chemical process step. The purpose is to remove all chemical residuals from the previous process step. This cleaning may be achieved in a single or dual rinse. Excessive dwell time in a rinse bath may cause oxidation or tarnishing of the product and is undesirable. Vendor specifications for temperature, dwell time, agitation, bath loading, and bath chemical control must be followed.

32.6.3 Applications

- Immersion tin is a highly solderable surface and forms the standard copper/tin intermetallic solder joint. Tin provides a dense uniform coating with superior hole wall lubricity. This characteristic makes it the choice for backplane panels that are assembled by pin insertion.

32.6.4 Limitations

- The bath makeup entails the use of thiourea, which is banned in certain geographical locations for environmental reasons. During processing, in the board shop, the primary by-product in the bath is copper thiourea. Waste treatment allowance must be made for the containment of the thiourea and its copper salt by-product.

- The shelf life of the surface is, to some extent, limited (less than a year). This is due to the progression of the copper/tin intermetallic until it reaches the surface and renders the product nonsolderable. This could be accelerated under excessive temperature and humidity conditions.

32.7 SURFACE FINISH ATTRIBUTE AND COST COMPARISON

A series of surface finish alternatives to HASL are now available to board designers, manufacturers, and assemblers. Each has it strengths and its limitations. Table 32.7 summarizes the primary attributes of each finish and compares the cost in terms of cost of HASL.

TABLE 32.7 Summary of Primary Attributes and Relative Cost of Surface Finishes Discussed in This Chapter

Surface finish	Relative cost	Solder joint	Attribute
HASL	1.0	Cu/Sn	Number one choice; losing ground to the new flatter surfaces
OSPs	1.2	Cu/Sn	For soldering only; concerns with third thermal excursion
ENIG	4.0	Ni/Sn	Excellent contact surface; solderable; requires attention to process control
ENEPIG	5.0	Ni/Sn	The universal finish; a costly and complex process
Immersion Ag	1.1	Cu/Sn	Low-cost alternative that is also bondable for both Al and Au wire
Immersion Sn	1.1	Cu/Sn	Low-cost alternative with pin insertion applications

CHAPTER 33
ETCHING PROCESS AND TECHNOLOGIES

Marshall I. Gurian, Ph.D.
Marshall Gurian Consulting, Tempe, Arizona

33.1 INTRODUCTION

One of the major steps in the chemical processing of subtractive printed boards is etching, or removal of copper, to achieve the desired circuit patterns. Etching is also used for surface preparation with minimal metal removal (microetching) during innerlayer oxide coating and electroless or electrolytic plating. Technical, economic, and environmental needs for practical process control have brought about major improvements in etching techniques. Batch-type operations, with their variable etching rates and long downtimes, have been replaced completely with continuous, constant-etch-rate processes. In addition, the need for continuous processing has led to extensive automation along with complete, integrated systems.

The most common etching systems are based on alkaline ammonia and cupric chloride. Other systems include peroxide–sulfuric acid, persulfates, and ferric chloride. Process steps include resist stripping, precleaning, etching, neutralizing, water rinsing, and drying. This chapter describes the technology for etching high-quality, fine-line (0.003 to 0.005 in) circuits in high volume at a practical cost, as well as continuous processing, constant-etch rates, and control at high dissolved copper capacities. Increasingly, the production of uniform and constant feature geometries calls for precision and statistically robust control of the circuitization processes and materials.

There remains the ever increasing need to balance the process selection according to the factors of cost, environmental and regulatory compliance, stable factory productivity, low worker intervention requirements, compatibility with board design, and construction innovations. It is significant that environmental concern has eliminated the use of chromic–sulfuric acid and ammonium persulfate etchants and chlorinated solvents. Limitations on chlorine gas and volatile organic emissions loom as a significant factor in determining future process choices, and pressure to eliminate brominated compounds and lead content may have consequences for processing mandated by material developments. More than ever, the simplification of process selection and vendor support appear to drive process selection in the industry.

Typical procedures are given for etching organic (i.e., dry film) and metal-resist boards, and for innerlayers. Strippers and procedures for resist removal are described based on resist selection, cost, and pollution problems. The properties of available etchants are also described in terms of resist compatibility, control methods, ease of control, and equipment maintenance. Other considerations include chemical and etchant effects on dielectric laminates, etching of thin-clad copper and semiadditive boards, solder mask on bare copper (SMOBC), equipment selection techniques, production capabilities, quality attained, and facilities.

33.2 GENERAL ETCHING CONSIDERATIONS AND PROCEDURES

Good etching results depend on proper image formation in both organic innerlayer print-and-etch and plated-metal etch resists. Etch personnel must be familiar with screened, photosensitive, and plated resists commonly used. The etching of printed boards must begin with suitable cleaning, inspection, and pre-etch steps to ensure acceptable products. Increasingly, metal foil or plating uniformity in structure thickness and freedom from coatings and defects must be maintained to achieve uniform small-feature definition. Plated boards also require careful and complete resist removal. The steps after etching are important because they remove surface contamination and yield sound surfaces. This discussion considers the various types of resists and outlines typical procedures used to etch printed boards using organic and plated resist patterns.

33.2.1 Screened Resists

Screen printing is a common method for producing standard copper-printed circuitry on metal-clad dielectric and other substrates. The etch-resist material is printed with a positive pattern (circuitry only) for copper etch-only boards or with a negative image (field only) when plated through-holes and metal resist are present.

The type of resist material used must meet the requirements for proper image transfer demanded by the printer and for stripping chemistry compatibility. From the metal etcher's point of view, the material needs to provide good adhesion and etch-solution resistance; be free of pinholes, oil, or resin bleed-out; and be readily removable without damage to substrate or circuitry.

Typical problems are excessive undercutting, slivers, unetched areas, and innerlayer shorts in multilayer boards. In addition, conductor line lifting may occur when the copper-to-laminate peel strength (or surface contamination) is below specification.

33.2.2 Hole Plugging

Plugged-hole, copper-only boards use alkaline-soluble screen resists in a unique manner. The technique, called *hole plugging,* makes the SMOBC board possible. This technique can be used with screened images and to augment *tent-and-etch* processing for small annular ring structures.

33.2.3 UV-Cured Screen Resists

Ultraviolet-cured solventless systems are available for print-and-etch and plating applications. These products are resistant to commonly used acidic plating and etching solutions. Stripping must be evaluated carefully.

33.2.4 Photoresists

Dry-film and liquid photoresist materials are capable of yielding the fine-line (0.003 to 0.005 in) circuits needed for production of surface-mount circuit boards. Like screened resists, photosensitive resists can be used to print either negative or positive patterns on the metal-clad laminate. Although dry-film and liquid materials differ in both physical and chemical properties, they are considered together for our purpose.

In general, both positive- and negative-acting resists offer better protection in acidic rather than alkaline solutions; however, negative-acting types are more tolerant of alkaline solutions. Negative resists, once exposed and developed, are no longer light sensitive and can be processed and stored in normal white light. The positive resists remain light sensitive even after developing and must therefore be protected from white light. Liquid photoresists, although less durable, are capable of finer line definition and resolution.

Positive-acting resists are subject to the same problems as are negative-acting resists, although they may be easier to remove cleanly, after exposure, from areas to be etched or plated. Problems related to plating photoresist-coated boards are reviewed in Chap. 29.

33.2.5 Plated Etch Resists

At present, the most extensive use of metal-plated resists is found in the production of double-sided and multilayer plated-through-hole circuit boards. The most commonly used resists are tin (matte and bright surface). Solder plate (60 percent Sn, 40 percent Pb) continues to be used on some reflow-treated solderable coating situations, but it is not favored for a sacrificial resist that is to be removed for SMOBC processing. Nickel, tin/nickel, and gold are occasionally used. Silver is used to some extent for light-emitting and liquid crystal applications. Details concerning the deposition of these metals are given in Chap. 29. The use of these resists is described in the following paragraphs.

33.2.5.1 Tin Plating. Thin tin deposits (0.0002 in) are used to make SMOBC boards where the tin etch resist is stripped after etching. Alkaline ammonia etchants are usually favored. Other etchants, such as sulfuric acid–hydrogen peroxide and ammonium persulfate–phosphoric acid, have been especially formulated for bright tin. Tin plating (directly over barrier layers of nickel or tin/nickel) has been used because of its optimum solderability. Cupric and ferric chloride etchants attack tin and are not used.

33.2.5.2 Solder Plate. Tin/lead solder (0.0003 to 0.001 in thick) is the plated etch resist for "fusing" into a solderable finish. The 60Sn/40Pb alloy offers good etchant resistance with few problems but must be treated with a brightening solution to retain solderability. Increased reliability may be achieved by the use of solder plate over tin/nickel.[1] Thin solder deposits (0.0002 in) can be used for the SMOBC process but have no benefit over tin plating for this purpose. The most suitable etchants are alkaline ammonia and sulfuric acid–hydrogen peroxide. Ferric and cupric chloride acid etchants cannot be used because of solder plate attack. Postetch neutralization rinses are needed, especially with alkaline systems, to rinse away etchant residues and to maintain optimum surface properties.

33.2.5.3 Tin/Nickel and Nickel. Tin/nickel alloy (65 percent Sn, 35 percent Ni) and nickel plate, used as is or overplated with gold, solder, or tin, are also capable metal resists for etching copper in alkaline ammonia, sulfuric acid–hydrogen peroxide, and persulfates.

33.2.5.4 Gold. With an underplate of nickel or tin/nickel, gold provides excellent resistance to all the common copper etchants. Some etchants have a slight dissolving effect on gold.

33.2.5.5 Precious Metals and Alloys. Rhodium has been described as being a suitable resist for edge connectors on boards; however, the plating process is difficult to control. When plated over nickel, rhodium tends to be thin and porous and to lift during etching. Because of varying surface properties, 18-karat alloy gold and nickel/palladium must be evaluated carefully when used as substitutes for pure gold systems.

33.2.5.6 Silver. Although silver is not used on most printed boards (MIL-STD-275 states that it shall not be used), it has found some applications for camera, light-emitting, and liquid crystal devices. Copper etching using silver as resist can be done with alkaline ammonia solutions. Silver loss is about 0.0001 in/mm.

33.2.5.7 Etching Procedures for Plated-Metal Resists. Etching of the metal-resist-plated boards begins with removal of the plating resist using commercial strippers. The stripping process must be sufficient to remove resist trapped under plating edges. Gold, solder, and tin resists must be handled carefully because they scratch very easily. Tin/nickel alloy and nickel plate, however, are very hard and resistant to abrasion.

The procedure after etching includes thorough water rinsing and acid neutralizing to ensure removal of etchant residue on the board surface and under the traces. Alkaline etchants are followed by treatment with proprietary ammonium chloride acidic solutions, ferric and cupric chloride with solutions of hydrochloric or oxalic acid, and ammonium persulfate with sulfuric acid. Alkaline cleaning is used for tin/lead boards etched in chromic–sulfuric acid. Etchant residue not removed before drying or reflow results in lowered electrical resistance of the dielectric substrate and in poor electrical contact and soldering on the conductive surfaces.[2] Care must be taken to minimize undercutting and overplating to eliminate slivers of plating breaking off and bridging circuitry.

A problem common to etching printed boards is not having the entire area etch clean at the same time. This occurs when etch action is more rapid at the edges of printed areas than in a broad expanse of copper. If very fine patterns and lines are required, the result can be loss of the pattern due to undercutting, especially when the board is left in the etcher until all the field copper has etched away. Modern etchers with reduced pooling of etchant have reduced this problem.

Fine-line boards in high-volume production may require special fine-line etchants, high-resolution photoresists, thin-clad laminates, controlled plating distribution, and thin base foil processing. Care must be taken to balance etching and thickness factors. In some cases, fine-line etchant additives have actually made cleanout more difficult in spaces of 0.003 in and narrower.

33.3 RESIST REMOVAL

The method used for resist stripping must be carefully evaluated when a resist is selected. The effect on board materials, cost and production requirements, and compliance with safety and pollution standards must be taken into account. Solvent-based stripping solutions are under significant environmental pressure. Chlorinated solvents and cyclic compounds (toluene, xylenes, etc.) have been banned, and many glycol ethers are restricted. In spray applications, care must be taken to capture or eliminate VOC emissions. For these reasons, aqueous or primarily aqueous stripping systems are necessarily widely used.

33.3.1 Screen Resist Removal

Alkali-soluble resist inks are generally preferred. Stripping in the case of thermal and UV-curable resists is accomplished in 2 percent sodium hydroxide or in proprietary solutions. The resist is loosened and rinsed off with a water spray. Adequate safety precautions must be taken, since caustics are harmful.

Conveyorized resist-stripping and etching machines use high-pressure pumping systems that spray hot alkaline solutions on both sides of the boards. Certain laminate materials such as the polyimides may be attached by alkaline strippers. Measling, staining, or other degradation is noted when strippers attack epoxy or other substrates. Control of concentration, temperature, and dwell time usually can prevent serious problems.

Screened vinyl-based resists are removed by a dissolving action in solutions of chlorinated, petroleum, or glycol ether solvents. Methylene chloride and toluene (both currently not usable) were used extensively in cold stripper formulations. The most common strippers are acidic formulations that contain copper brighteners and swelling, dissolution, and water-rinsing agents. The usual procedure for static tank stripping involves soaking the coated boards in at least two tanks of stripper. Excessive time in strippers is to be avoided because of attack on the butter (top epoxy) coat, especially on print-and-etch or single-sided boards. Water is a contaminant in most cold strippers.

33.3.2 Photoresist Removal

33.3.2.1 Dry Film. Dry-film resists have been formulated for ease of removal in aqueous-alkaline solutions. Strippers are available for both static tank and conveyorized systems, although conveyorized systems are preferred because the spray action aids in separating the resist from the copper surfaces and washing it off the panel. Prompt processing after etching avoids resist adhesion "lock-in," which can make removal difficult. Aqueous-alkaline stripping results in partially undissolved residues of softened resist films. These residues are captured in a filter system and disposed of in accordance with waste-disposal requirements. Design and availability of specific filters to match the character of the "skins" can be a significant factor in the selection of the proper resist. Environmental caution must be taken to properly evaluate and dispose of resist residues. Metal content (particularly lead) may lead to classification of these residues as toxic waste.

33.3.2.2 Negative-Acting Liquid Photoresist Removal. Negative-acting, liquid-applied photoresist can be readily removed from printed boards that have not been baked excessively. Baking is critical to removal because it relates to the degree of polymerization. Since over-baking is also damaging to the insulating substrates, processes should stress minimal baking—only enough to withstand the operations involved. The negative-acting resists are removed by using solvents and commercial strippers. In this case, the resist may not dissolve; instead, it softens and swells, breaking the adhesive bond to the substrate. Once that has taken place over the entire coated area, a water spray is used to flush away the film.

33.3.2.3 Positive-Acting Photoresist Removal. Positive-acting photoresists are removed in commercial organic and inorganic strippers if baking has not been excessive. Removal by exposure to UV light and subsequent dipping in sodium hydroxide, TSP, or other strong alkaline solutions is also effective. Overbaking makes removal difficult. Machine stripping is done in a solution of $0.5N$ sodium hydroxide, nonionic surfactants, and defoamers.

33.3.3 Tin and Tin/Lead Resist Removal

In SMOBC processing, the metal-plated resist is removed to present a flat, clean copper surface for solder mask definition. Tin/lead alloys can be stripped in oxidizing fluoride solutions such as fluoboric acid and hydrogen peroxide or ammonium bifluoride with hydrogen peroxide or nitric acid. (Caution: machine construction must be made compatible with fluorides by elimination of titanium and glass components.) Commercial formulations are available to be used inline after the etch machine rinses. Accumulations of spent solution or filtered lead-fluoride deposits must be treated as hazardous waste and have been accepted by solution vendors for treatment and disposal costs. Modern applications usually use lead-free tin plating resists, which can be fluoride containing as previously discussed, or compounds of ferric chloride followed by nitric acid. In either case, feed-and-bleed solution management with filtration and periodic chamber cleanout has been successful.

33.4 ETCHING SOLUTIONS

This section is a survey of the technology and chemistry of the copper etching systems in common use. Selection of practically available etchants has been limited by economic, operational, and (environmental) regulatory concerns. Fabricators have been forced into practical trade-off decisions to suit situations. Two etchants in particular, chromic–sulfuric acid and ammonium persulfate, are no longer practical considerations due to environmental pressures. Other formulations and choices have been modified to suit these pressures.

There are two basic etchant needs to be met. The first is traditional foil etching for print and etch, plate/tent and etch, and pattern plate and etch. Virtually all processes in the United States and Europe use constant-rate systems for alkaline ammonia or cupric chloride etchants for this purpose. The second need is developing technology for specific precision very-fine-line etching—including foil thinning and thin metallization clearout for HDI constructions and fine features. (See Sect 33.7 for additional discussion and mention of additional chemistries that may be useful for these applications.)

Continuous constant-rate systems with process automation represent current practice in production etching. These systems feature feed and bleed of replenishment chemicals under control of process instrumentation that monitors and responds to real-time changes in properties of the working solution. The "bleed" stream of the working composition etchant is usually returned to suppliers for copper reclamation and chemical recycle. The resulting degree of constant etching rate allows for stable and repeatable performance to achieve practical manufacturing processes. However, as demands for higher precision increase, further sophistication in controls will be necessary.

Key factors for selecting an etching chemistry include:

- Board design requirements
- High yield
- Compatibility with resist
- Etch rate (speed)
- Equipment required for process control of etch rate, regeneration, and replenishment
- Ease of equipment maintenance
- By-products disposal and pollution control
- Operator and environmental protection

The preceding factors serve to evaluate copper etchants to be used. Introduction, chemistry, properties, and problems are given in this section, along with suggestions for selection and control.

33.4.1 Alkaline Ammonia

Alkaline etching with ammonium hydroxide complexing is increasingly used because of its continuous operation, compatibility with most metallic and organic resists, high capacity for dissolved copper, and fast etch rates. Continuous (open-loop) spray machine chemical control systems are universally used. This operation provides constant etch rates, high work output, ease of control and replenishment, and improved pollution control. However, rinsing after etching is critical, and the ammonium ion introduced to the rinses presents a waste-treatment problem. On-site closed-loop regeneration with chemical recycling is commercially available but not routinely practiced because of facility requirements, capital cost, fluctuating economics depending on copper commodity pricing, and worker requirements. The general economical and environmentally appropriate operating strategy is to recycle the by-product etchant products under contract to a supplier who reclaims or reconstitutes the copper contained and regenerates the ammoniacal constituents into a reformulated replenisher solution for return to fabricators.

33.4.1.1 Chemistry. The main chemical constituents function as follows:

1. Ammonium hydroxide (NH_4OH) acts as a complexing agent and holds copper in solution.
2. Ammonium chloride (NH_4Cl) increases etch rate, copper-holding capacity, and solution stability.
3. Copper ion (Cu^{2+}) is an oxidizing agent that reacts with and dissolves metallic copper.
4. Ammonium bicarbonate (NH_4HCO_3) is a buffer and as such retains clean solder holes and surface.
5. Ammonium phosphate [$(NH_4)_3PO_4$] retains clean solder and plated through-holes.
6. Ammonium nitrate (NH_4NO_3) increases etch rate and retains clean solder.
7. Additional additives are included in most formulations to enhance speed and/or sidewall protection. Thiourea or its derivatives are often used, although a newer thiourea-free formulation with improved undercut protection is available.
8. Continuous operations consist of single-solution makeup buffered to a pH of 7.5 to 9.5.

Alkaline etching solutions dissolve exposed field copper on printed boards by a chemical process of oxidation, solubilizing, and complexing. Ammonium hydroxide and ammonium salts combine with copper ions to form cupric ammonium complex ions [$Cu(NH_3)_4^{2+}$], which hold the etched and dissolved copper in solution at 18 to 30 oz/gal.

Typical oxidation reactions for closed-loop systems are shown by the reaction of cupric ion on copper, and air (O_2) oxidation of the cuprous complex ion:

$$Cu + Cu(NH_3)_4^{2+} \rightarrow 2Cu(NH_3)_2^+ \tag{33.1}$$

$$2Cu(NH_3)_2^+ + 2NH_4^+ + 2NH_3 + \tfrac{1}{2}O_2 \rightarrow 2Cu(NH_3)_4^{2+} + H_2O \tag{33.2}$$

There is strong evidence that the etching rate is dependent on the diffusion of $Cu(NH_3)_2^+$ from the copper surface (Eq. 33.1) into the bulk of the active solution, where oxidation per Eq. 33.2 occurs.[3] Etching can be continued with the formation of $Cu(NH_3)_4^{2+}$ oxidizer from air during spray etching and as long as the copper-holding capacity supported by Cl^- ions is not exceeded.

33.4.1.2 Properties and Control. Early versions of alkaline etchants were batch operated. They had a low copper capacity, and the etch rates dropped off rapidly as copper content increased.[4,5,6] It was found to be necessary to add controlled amounts of dissolved oxidizing agents to speed up the rate and increase copper capacity at a constant temperature. Batch operation is no longer supported by commercial suppliers.

Etching solutions are operated at 120 to 130°F and are well suited to spray etching. Efficient exhaust systems are required because ammonia fumes are released during operation.[7] Etching machines must have a slight negative pressure and moderate exhausting to retain the ammonia necessary for holding dissolved copper in solution. Care must be taken that sufficient fresh air to supply needed O_2 is introduced to balance the extraction. Currently available solutions offer constant etching of 1 oz (35 μm) copper in 1 min or less, with a dissolved copper content of 18 to 24 oz/gal.

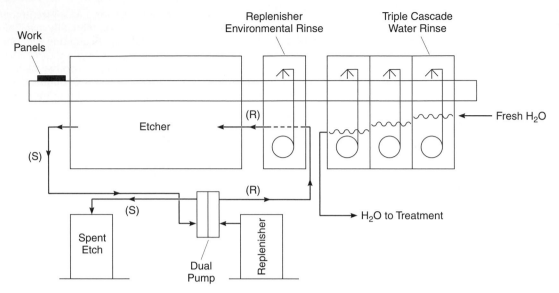

FIGURE 33.1 Automated flow-through alkaline etching system.

33.4.1.3 Continuous Systems. A practical method of maintaining a constant etch rate with minimal pollution uses automatic feeding controlled by specific gravity or density.[6] The process, which is generally referred to as *bleed and feed,* is illustrated in Fig. 33.1. As the printed boards are etched, copper is dissolved, and the density of the etching solution increases. The density of the etchant in the etcher sump is sensed to determine the amount of copper in solution. When the density sensor records an upper limit, a switch activates a pump that automatically feeds replenishing solution to the etcher and simultaneously removes etchant until a lower density is reached. An elegant by-product of this technique is the introduction of the replenisher into the first rinse stage following the etcher. The rinse captures the copper from the panel and reintroduces it to the etcher, where it is eventually eliminated to the by-product tank with the rest of the copper. When used with aqueous processing photoresists, it has been found that low pH (7.9 to 8.1) aids etch reliability. However, maintaining this pH by adjustment of exhaust extraction has been difficult. Measurement of pH and controlled addition of anhydrous ammonia into the etchant has improved consistency. It has been established that control of free ammonia as well as ammonium chloride and oxygen levels are needed to stabilize etching rates. Direct injection of oxygen has also been attempted to stabilize the etching rates between low and high copper etching demand process utilization periods.

Typical operating conditions are as follows:*

Temperature	120 to 130°F (49 to 50°C)
pH	8.0 to 8.8
Specific gravity at 120°F (49°C)	1.207 to 1.227
Baumé, Be°	25 to 27
Copper concentration, oz/gal	20 to 22
Etch rate, 0.001 in/mm	1.4 to 2.0
Chloride level	4.9 to 5.7 mol/L

* PA Hunt Chemical Corporation, West Paterson, NJ.

A study of the etching rate vs. dissolved copper content shows the following effects:

0 to 11 oz/gal	Long etching times
11 to 16 oz/gal	Lower etch times but solution control difficult
18 to 22 oz/gal	Etch rates high and solution stable
22 to 30 oz/gal	Solution unstable and tends toward sludging

All work must be thoroughly rinsed immediately upon leaving the etching chamber. Techniques such as replenisher rinsing and multiple cascade water rinsing assist in thorough rinsing with controlled effluent and limited water use.[8] Do not allow the boards to dry before rinsing. Etched circuitry with plated tin/lead solder resist also requires an acid application of solder brightener for reflow of the coating. Thiourea-containing brighteners have largely been eliminated due to use of sacrificial plating resist (SMOBC) and environmental pressures. Rinsing must be sufficient to remove etchant from under circuit edges and to completely clean the circuit surfaces and plated through-holes.[8] Multiple-stage cascading water rinsing and air knife drying result in clean, stain-free surfaces (see Fig. 33.1). Modern processes often follow the etching rinse with metal-resist stripping and rinsing chambers before exiting the machine conveyor.

33.4.1.4 *Closed-Loop Regeneration.*   True regeneration requires the following:

1. Removal of portions of the spent etching solution from the etcher sump under controlled conditions, according to the amount of copper added to the etch bath.

2. Chemical restoration of spent etchant (i.e., removal of excess by-products and adjustment of solution parameters for reuse).

3. Replenishing of etchant in the etching machine, balancing the actual production use demands. Constant etching conditions are achieved when regeneration is continuous. Regeneration by these methods is expensive and is thus limited to large printed-circuit facilities. The principal methods of regeneration are crystallization, liquid–liquid extraction, and electrolytic recovery.

   - Crystallization reduces the copper level in the etchant by chilling and filtering precipitated salts. This is followed by refortification and adjustment of operating conditions.
   - Liquid–liquid extraction[9,10] is gaining acceptance because of its continuous and generally safe nature. The process involves mixing spent etchant with an organic solvent (i.e., hydroxy oximes) capable of extracting copper. The organic layer containing copper is subsequently mixed with aqueous sulfuric acid, which extracts copper to form copper sulfate. The copper-free etchant is restored, and the copper sulfate is available for acid copper electrowinning. Closed-loop regeneration systems reduce chemical costs, sewer contamination, and production downtime, but require floor space, resources, and technical attention. Economics are directly affected by copper market prices.
   - Electrolytic recovery of copper directly and through membrane cells from an ammonia-complexed copper sulfate etchant yields possible benefits such as reduced waste shipments and cost savings.[11,12]
   - These processes are usually employed by the large-scale vendor/recycling plants that offer a spent removal and replenisher replacement service as part of the product contract. This practice allows for economies of scale and environmental responsibility to be removed from the fabrication plant site.

33.4.1.5 *Special Problems Encountered During Etching*[13]

1. *Low etch rate with low pH, <8.0.* This is caused by excessive ventilation, heating, downtime, and spraying when the solution is hot, under conditions of adequate replenishment or low ammonia. The pH must be raised with anhydrous ammonia. Automatic replenishment equipment must be checked.

2. *Low etch rate with high pH, >8.8.* This is caused by high copper content, water in etchant, or underventilation.

3. *Low etch rate with optimum pH.* This is due to error of copper thickness, oxygen starvation in the etcher, or contamination of the etchant.

4. *Solder attack.* This is caused by excess chloride in the etchant, improper tin/lead deposits, or insufficient phosphate content.

5. *Residue on solder plate, holes, and traces.* These may be caused by etchant solution imbalance or by spent finisher.

6. *Under- or overetching.* This may be due to improper pH or equipment adjustments.

7. *Sludging of etch chamber with low pH, <8.0.* See special problem 1. Sludge will be gritty and dark blue. This may be corrected by the addition of anhydrous ammonia.

8. *Sludging of etch chamber with high pH, >8.8.* See special problem 2. Sludge will be light blue and fluffy. This may be due to the copper concentration exceeding the capacity of the chloride concentration. It may be corrected by adding ammonium chloride. It may also be due to water added to the etchant.

9. *Presence of ammonia fumes.* The cause of this is leaks in the etcher. Ventilate for operator safety.

10. *Pollution.* This results when water coming from the etchers contains dissolved copper. If so, it must be chemically treated and separated from ammonia-bearing rinses. Thin-clad copper laminates present a further problem because their faster movement through the etcher increases the transport of etchant into the rinses. Evaluate isolation roller placement and cleanliness, internal etcher overspray, and proper chemical and water rinsing.

33.4.2 Cupric Chloride

Cupric chloride is the second mainstream etchant of choice. (See Table 33.1 for characteristic cupric chloride solutions.) It is more compatible with photopolymer resists and is the etchant of choice for precision etching of small features. Because the cuprous ion buildup renders the etchant impractically slow, this etchant must always be used as a regenerative system with continuous oxidizer management and control. These systems have been the subject of many innovations to achieve continuous regeneration, lower costs, and a constant, predictable etch rate. Steady-state etching with acidic cupric chloride permits high throughput, material recovery, and reduced pollution. Regeneration in this case is somewhat complex but is readily maintained. Dissolved copper capacity is high compared with that of batch operation. Cupric chloride solutions are used mainly for fine-line multilayer inner details, print-and-etch boards, and panel-plate/tent-and-etch boards.[14] In addition to photoresists, screened inks, gold, and tin-nickel have been used. Solder and tin resists are not compatible with cupric chloride etchant.

TABLE 33.1 Characteristic Cupric Chloride Etching Solutions

Component	Solution 1[15]	Solution 2[16]	Solution 3[17]	Solution 4[18]
$CuCl_2 \cdot 2H_2O$	1.42 lb	2.2 *M*	2.2 *M*	0.5–2.5 *M*
HCl (20° Be)	0.6 gal	30 mL/gal	0.5 *N*	0.2–0.6 *M*
NaCl		4 *M*	3 *M*	
NH$_4$Cl				2.4–0.5 *M*
H$_2$O	*	*	*	*

* Add water to make up 1 gal.

33.4.2.1   Chemistry.   The etching reaction is as follows:

$$Cu + CuCl_2 \rightarrow Cu_2Cl_2 \tag{33.3}$$

Chloride ions, when added in excess as hydrochloric acid, sodium chloride, or ammonium chloride, act to solubilize the relatively insoluble cuprous chloride and thereby maintain stable etch rates. The complex ion formation can be shown as

$$CuCl + 2Cl^- \rightarrow CuCl_3^- \tag{33.4}$$

The formation and diffusion of this chloro-complex form of the Cu(I) ion determine the etching rate for cupric chloride. Thus, the rate is affected by chloride ion concentration as well as the typical diffusion concentration and film thickness variables.

The etchant is regenerated by reoxidizing the cuprous chloride to cupric chloride by several chemical choices as illustrated by the following equations:

Air

$$2Cu_2Cl_2 + 4HCl + O_2 \rightarrow 4CuCl_2 + 2H_2O \tag{33.5}$$

This method of regeneration is not used because the oxygen reaction rate in acids is slow and the solubility of oxygen in hot solution is low (4 to 8 ppm). Spray etching induces air oxidation, but is not sufficiently reactive to support practical etching rates. An additional option of using ozone can accelerate the rate of this process, but ozone concentrations in an oxygen stream are difficult to maintain at above 3 percent, thus limiting utility.

Direct Chlorination

$$Cu_2Cl_2 + Cl_2 \rightarrow 2CuCl_2 \tag{33.6}$$

As the traditional regeneration choice, the chlorine reaction is rapid, positive, and easily controlled. Liquefied chlorine gas has been generally available, and standardized commercial addition systems have been available. Recently, pressure on chlorine distribution and use due to safety and environmental regulations have made availability and implementation more expensive and difficult. It is significant that the reaction simplicity indicates that all etched copper is converted directly to cupric chloride (combination of Eqs. 33.3 and 33.6). Thus, there is no effect on water or acid balances from the etching reaction itself.

Hydrogen Peroxide

$$Cu_2Cl_2 + 2H_2O_2 + 2HCl \rightarrow 2CuCl_2 + 2H_2O \tag{33.7}$$

This process is more complicated because both hydrogen peroxide and hydrochloric acid add water in the reaction process. Since hydrogen peroxide is unsafe to store and handle in concentrations above 35 percent, there is considerable water added with both the oxidizer and the hydrochloric acid. The result limits the copper-holding capacity of the etchant. The addition is reasonably and simply controlled as a direct ratio of HCl, and peroxide is added to the bath to maintain the copper oxidation. This formulation has been favored in Europe and to some extent in Asia.

Sodium Chlorate

$$3Cu_2Cl_2 + NaClO_3 + 6HCl \rightarrow 6CuCl_2 + NaCl + 3H_2O \tag{33.8}$$

Sodium chlorate has become one of the most widely used formulations. The oxidizer solutions can be mixed from powder and are also commercially supplied as 45 percent solutions. The water balance must also be carefully controlled with this oxidizing plan, and the amount of sodium chloride in the working bath can be boosted by addition with the oxidizer. This can be a rate enhancer by supplying chloride ions needed by Eq. (33.6), and since the chloride ion is separately added from the hydrochloric acid, the free acid can be maintained at very low levels (<0.1 normal).

Electrolytic

(Cathode) $$Cu^+ + e^- \rightarrow Cu \qquad (33.9)$$

(Anode) $$Cu^+ \rightarrow Cu^{2+} + e^- \qquad (33.10)$$

Electrolytic systems of direct and membrane cell design have been employed. However, this technology is not a major factor in current manufacturing practice.

33.4.2.2 Properties and Control. Early cupric chloride formations had slow etch rates and low copper capacity and were limited to batch operation.[15–21] Regenerated continuous operation using modified formulations has brought useful improvements. Etch rates as high as 55 s for 1 oz copper are obtained from cupric chloride–sodium chloride–HCl systems operated at 130°F with conventional spray-etching equipment. Copper capacities are maintained at 20 oz/gal and above. However, more recently, higher copper contents and adjusted chemistries at 125°F typically etch at 75 to 90 s for 1 oz copper.

33.4.2.3 Continuous Etching and Regeneration. Systems in use include chlorination, sodium chlorate, hydrogen peroxide, and electrolysis.

Chlorination. Direct chlorination has been the preferred technique for regeneration of cupric etchant because of its historically low cost, high rate, efficiency in recovery of copper, and pollution control. The cupric chloride–sodium chloride system (Table 33.1, no. 3) is suitable. Figure 33.2 shows a generalized process.[17] Chlorine, hydrochloric acid, and sodium chloride solutions are automatically fed into the system as required. Sensing devices include oxidation-reduction instruments (Cu oxidation state), density (Cu concentration), level sen-

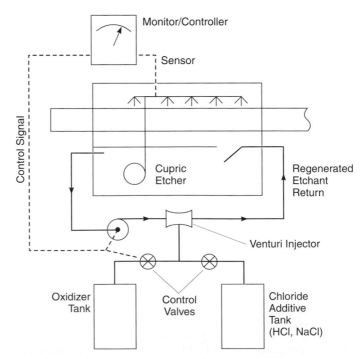

FIGURE 33.2 Cupric chloride chlorination regeneration system.

sors, and thermostats. Chlorination is reliable and controllable. Other factors are safety and solution control. Optical colorimetric sensors can be used to detect Cu(I) levels. However, these cells are subject to fogging by organic contamination and crystalline buildup with the hazardous defect of continuing to add chemicals past the proper control points and liberating chlorine into the work area.

1. *Safety.*   Use of chlorine gas requires adequate ventilation, tank and cylinder storage, leak-detection equipment, emergency protocols, personal protective equipment, operator training, and fire department approval and inspection of installations.

2. *Solution control.*   An increase in pH will cause the copper colorimeter to give erroneous readings caused by turbidity in the solution. Organic deposits can also foul controls. Excess NaCl at 18 to 20 oz/gal copper causes coprecipitation of salts when the solution is cooled. Solution filtration and etch tank cleanliness must be maintained. Replacement of electrodes, instruments, and metering devices should be scheduled regularly.

Chlorate Regeneration.   This method, currently widely supported, uses sodium chlorate, sodium chloride, and hydrochloric acid and is an alternate method similar to chlorination. A control sequence similar to the one shown in Fig. 33.2 may be used. Sodium chlorate is available in commercial solutions of several strengths with some degree of formulated agents (usually sodium chloride) added to the mixture. These solutions are added with hydrochloric acid either as a coordinated addition or in a special control logic sequence that allows very low free acid. This control system uses a multiple sensor color or turbidimetric optical system. The addition of one component (e.g., oxidizer) to a flowing sample cell occurs when the primary cell senses are triggered. A second cell allows this addition so long as there is a change from the first cell input. When there is no longer a change response and the first cell is still in "triggered" condition, the second component (e.g., HCl) is added. The logic continues to flip-flop until the first cell is satisfied. The etch sump conditions seem to have a partially delayed response to the chemical addition reaction such that oxidizing seems to occur over a longer period. Therefore, proper addition and mixing in the sump must be considered for stable control operation.

1. *Safety.*   Sodium chlorate is a potent oxidizing agent that can support combustion. It is necessary to ensure that solution spills are completely cleaned up and that the rags or other materials are not allowed to dry out. Excess addition of reagents above amounts to support the actual copper etched may form chlorine gas that will be liberated into the etch machine and environment. Proper instrumentation and training for this eventuality must be implemented.

2. *Solution control.*   If there is insufficient free acid content in the etch sump to react with the oxidizer and copper (Eq. 33.8), a dangerous imbalance can occur. First, the instruments and controls may not respond properly, and the etch reaction may slow down. Second, any subsequent addition of acid or mixing with acid-containing wastes can liberate chlorine gas in an uncontrolled and dangerous manner. Proper chemical balance is another issue. Water is added in all of the chemical streams and is produced in the chemical reaction. Therefore, any excess chemical addition can dilute the chemistry and change the copper-holding ability of the etchant.

Hydrogen Peroxide Regeneration.   Hydrogen peroxide uses a two-part chemistry much like sodium chlorate. ORP instrumentation is identical in the control of adding proportional amounts of oxidizer and hydrochloric acid. Peroxide tends to decompose on sitting so that the spills are not as dangerous. Older application of colorimetric cells for the control of dosing has largely been eliminated.

1. *Safety.*   Hydrogen peroxide solutions can become unstable and disassociate into oxygen and water with the liberation of heat. This can occur with explosive force in a confined vessel such as a drum stored in sunlight or heat. It can also occur in plumbing with closed valving. The decomposition is catalyzed by many metals, including Cu, Ni, and Fe. If traces of metal or metal salts are allowed to contaminate piping runs, the piping may also rupture explosively. Facilities should be constructed with pressure relief capabilities.

Electrolytic Regeneration. The electrochemical reversal of the etching of copper shown in Eq. 33.9 and 33.10 is claimed to be effective and economical. Descriptions of this system are given in the literature.[15] On a large scale, electrolytic regeneration requires a high investment in equipment and materials, as well as high power consumption. The chemical formulation is relatively high in acid content and low in copper, a condition not optimized for etching effectiveness or plating efficiency.

The etchant is a solution of cupric chloride and hydrochloric acid (Table 33.1, no. 1). Etchant flows continuously between spray-etching machines and a plating tank. In the plating machine, two processes take place simultaneously: copper is plated at the cathode, and regeneration of the spent etchant occurs at the anodes. Copper recovery may not return copper value and may be expensive, inconvenient, and cause difficulty with recycling.

33.4.2.4 *Problems with Cupric Chloride Systems*

1. *Slow etch rate.* Depending on chemical formulation, cupric chloride is inherently slower than ammonia and must be properly evaluated before production expectations are established. Check to ensure no resist residues or chromate coatings remain on the panel surface before etching. A slower than usual rate is frequently due to low temperatures, insufficient sump mixing, or poor solution chemical control. If temperature and agitation are under control, slow etching with a dark green solution may result from a low cupric ion content, indicating insufficient oxidizer. Acid must also be added to clarify cloudy solutions. (Before acid addition, make sure excess oxidizer is not present.) In regenerative systems, the source of oxidation may be depleted. It is advisable to analyze specific gravity, free acid, and total chloride on a regular schedule and maintain control charts to ensure that bath controllers are consistent and operating properly.

2. *Sludging.* This occurs if acid is low or if water dilution occurs.

3. *Etcher goop.* This is a common floating accumulation of photoresist compounds leached into and precipitated from etch baths. Accumulations appear to increase with higher acid formulations. Formation can be limited by proper exposure of photoresists and continuous recirculation of the etch bath through carbon filtration. Periodic, complete drain-down of the etching machinery and thorough mechanical and chemical equipment cleaning with a sulfamic acid–type commercial cleaner can minimize problems.

4. *Yellow or white residues on copper surface.* Yellow residue is usually cuprous hydroxide. It is water insoluble and is left when boards are etched and alkali-cleaned. A white precipitate will probably be cuprous chloride, which can remain after etching in solutions that are low in chloride ion and acid. To eliminate both conditions, the solution in which the board is rinsed just before final water-spray rinsing should be 5 percent by volume hydrochloric acid.

5. *Waste disposal.* Spent or by-product etchant is usually sent off-site for copper reclamation. There is usually a fee for this service depending on copper content and distance to the reclamation facility. The solutions must be free of unreacted oxidizer (see previous chlorate system discussion). Etchant can contain traces of zinc, chromium, and arsenic from the foil treatments.

33.4.3 Sulfuric Acid–Hydrogen

Sulfuric-peroxide systems are used extensively for copper surface preparation by microetching the surface to provide texture and active surfaces. These formulations have been used in the past for primary circuitization etching but are no longer favored because of slow etch rate and needs for accompanying control of exotherm and copper crystallization. However, the increasing need for fine-line etching has renewed use possibilities for this and other etchants (see Sec. 33.7.4). In essence, these call for etching foils thinner than ½ oz and the control precision offered by slower etchants. They are compatible with metal resists and many organics.

33.4.3.1   *Chemistry.*   Typical constituents of both immersion and spray etchants and their functions are as follows:

1. Hydrogen peroxide is an oxidizing agent that reacts with and dissolves metallic copper
2. Sulfuric acid makes copper soluble and holds copper sulfate in solution.
3. Copper sulfate helps to stabilize etch and recovery rates.
4. Molybdenum ion is an oxidizing agent and rate exaltant.[22]
5. Aryl sulfonic acids are peroxide stabilizers.[23]
6. Thiosulfates are rate exaltants and chloride ion controllers that permit lower peroxide content.[24]
7. Phosphoric acid retains clean solder traces and plated through-holes.[25,26]

The etching reaction is as follows:

$$Cu + H_2O_2 + H_2SO_4 \rightarrow CuSO_4 + 2H_2O \tag{33.11}$$

33.4.3.2   *Properties and Control.*   The earlier technical problems of slow etch rates, peroxide decomposition, and foaming in spray systems have been solved, but critical concerns still remain. Among these are process overheating, etchant composition balance with by-product recovery, etchant contamination, and the dangers of handling concentrated peroxide solutions. The specific problem of decomposition of peroxide during idle hours has caused disastrous equipment meltdowns, and thus requires thermal management while equipment is otherwise not in operation.

33.4.3.3   *Closed-Loop Systems.*   Production facilities require continuous recirculation of the etchant through the etching tank or machine and the copper sulfate recovery operation. Etchant replenishment is controlled by chemical analysis and by additions of concentrates.

Copper sulfate recovery is based on lowering the solubility of $CuSO_4 \cdot 5H_2O$ by decreasing the etchant temperature to 50 to 70°F. Crystallizer equipment and conveyorized discharge has been available for this process.

33.4.3.4   *Problems Encountered with Peroxide Systems*

1. *Reduced etch rates.* This problem can be caused by operating conditions, solution imbalance, or chloride contamination.
2. *Under- and overetching.* A review of etching conditions, solution control, and the resist stripping process may show deviations from normal. In the case of immersion etching, the solution and panel agitation may need to be increased. When spray-etching, a check of nozzles and line clogging is indicated.
3. *Temperature changes.* Recirculating water rates and thermostats need to be examined regularly. Overheating may be due to high copper content, contamination, or rapid peroxide decomposition.
4. *Copper sulfate recovery stoppage.* Examine solution balance, heat exchanger, and other recovery equipment. Exit conveyor jamming difficulties from thermal excursions and varying crystal size and liquid content cause major shop floor problems.

33.4.4   Persulfates

Ammonium, sodium, and potassium persulfates modified by certain catalysts have been adopted for the etching of copper in PC manufacturing. Continuous regenerative systems and a batch system using ammonium persulfate are no longer used. Wide use is made of persulfates as a microetch for innerlayer oxide coating and copper electroless and plating processes. Persulfate solutions allow all common types of resists on boards including solder, tin, tin/nickel, screened inks, and photosensitive films. Formulations of ammonium persulfate catalyzed with mercuric chloride are no longer used, primarily because of environmental issues, costs, and improved alternatives. In general, persulfate etchants are unstable and will exhibit decomposition, lower etch rates vs. copper content, and lower useful copper capacity.

33.4.4.1   Chemistry.   Ammonium, potassium, and persulfates are stable salts of persulfuric acid ($H_2S_2O_8$). When these salts are dissolved in water, the persulfate ion ($S_2O_8^{2-}$) is formed. It is the most powerful oxidant of the commonly used peroxy compounds. During copper etching, persulfate oxidizes metallic copper to cupric ion as shown:

$$Cu + Na_2S_2O_8 \rightarrow CuSO_4 + Na_2SO_4 \tag{33.12}$$

Persulfate solutions hydrolyze to form peroxy monosulfate ion (HSO_4^{1-}) and, subsequently, hydrogen peroxide and oxygen. This hydrolysis is acid catalyzed and accounts for the instability of acidic persulfate etching solutions.

Ammonium persulfate solution, normally made up at 20 percent, is acidic. Hydrolysis reactions and etchant use cause a reduction of the pH from 4 to 2. The persulfate concentration is lowered, and hydrated cupric ammonium sulfate [$CuSO_4 \cdot (NH_4)_2SO_4 \cdot 6H_2O$] is formed. This precipitate may interfere with etching.

Solid persulfate compounds are stable and do not deteriorate if stored dry in closed containers. Solution makeup composition includes various catalysts, including organic matter and transition metals (Fe, Cr, Cu, Pb, Ag, etc.). Materials for storage must be chosen carefully. Persulfates should not be mixed with reducing agents or oxidizable organics.

The useful capacity of the etchant is about 7 oz/gal copper at 100 to 130°F. Above 5 oz/gal of copper, it is necessary to keep the solutions at 130°F to prevent salt crystallization. The etch rate of a solution containing 7 oz/gal of dissolved copper is 0.00027 in/min at 118°F.

33.4.4.2   Batch Operation.   Sodium persulfate is preferred because it has minimal disposal problems and somewhat higher copper capacity and etch rates. A composition of 3 lb/gal sodium persulfate with 15 ppm of $HgCl_2$, 1 gal of proprietary additive, and 57 ml/gal of H_3PO_4 has been successfully used for batch-type spray etching.[27] Etch rates vary throughout bath life and range from 0.0018 to 0.0006 in/min for copper content of 0 to 7 oz/gal. Prepared solutions must be aged for 16 to 72 h before etching when proprietary additives are used.

33.4.4.3   Problems with Persulfates

1. *Low etch rates.* Since the solution may decompose, it will be necessary to replace the bath. If solution is new, add more catalyst and check for iron contamination.

2. *Salt crystallization.* Salts crystallize on the board and cause streaks, damage the solder plate, and plug the spray nozzles or filters. When copper content is high, blue salts may precipitate.

3. *White films on solder surface.* This may occur normally or when the lead content in the solder plate is too high.

4. *Black film on solder.* This condition can result when the solder alloy is high in tin. If solder reflow or component soldering is to follow, activate by tin immersion or with solder brighteners. Adjust phosphoric acid content in etchant and solder-plating conditions.

5. *Spontaneous decomposition of etch solution.* This breakdown is due to contaminated, overheated, or idle solutions. Ammonium persulfate etchants are unstable, especially at higher temperatures. At about 150°F, the solution decomposes quickly. Use it soon after mixing.

6. *Disposal.* The exhausted etchant consists mainly of ammonium or sodium and copper sulfate with a pH of about 2. Two methods for disposal are:
  • Electrolytic disposition of the copper on the surface of passivated 300 series stainless steel is one disposal method. The spent etchant is acidified with sulfuric acid prior to electrolysis. Once the copper has been removed, the remaining solution can be diluted, neutralized, checked, and discarded. The copper can be removed from the cathode. Spent sodium persulfate can be treated with caustic soda.
  • Addition of aluminum or iron machine turnings to a slightly acidified solution is another practical but possibly more difficult means of removing the dissolved copper. The reaction, especially in the presence of chloride ions, will be violent, and considerable heat will be given off if the solutions are not diluted.

33.4.5 Ferric Chloride

Ferric chloride solutions are used as etchants for copper, copper alloys, Ni/Fe alloys, and steel in PC applications, electronics, photoengraving arts, and metal finishing. Current use of ferric chloride etchant in printed wiring fabrication is extremely limited in the United States because of costly disposal of the copper-containing etchant, and the much better commercial support for ammoniacal and cupric chloride etchants. There is still considerable use for alloy etching and photochemical machining applications.

Ferric chloride can be used with screen ink, photoresist, and gold patterns, but it cannot be used with tin or tin/lead resists. However, ferric chloride is an attractive spray etchant because of its ease of use, holding capacity for copper, and ability to be used on an infrequent batch application basis.

The composition of the etchant is mainly ferric chloride in water, with concentrations ranging from 28 to 42 percent by weight. Free acid is present because of the hydrolysis reaction and the need to maintain an acid environment. The natural acidity is usually supplemented by additional amounts of HCl (up to 5 percent) to prevent formation of insoluble precipitates of ferric hydroxide. Commercial formulations for copper alloy etching are usually 36 Be, or approximately 4.0 lb/gal FeCl₃, and may contain antifoam and wetting agents. Customary acid operating HCl content is 1.5 to 2.0 percent.

The effects of ferric chloride concentration, dissolved copper content, temperature, and agitation on the rate and quality of etching have been reported in the literature.[28,29]

33.4.6 Chromic-Sulfuric Acids

These etchants for solder- and tin-plated boards were preferred for many years. More recently, their use has been completely eliminated due to Cr(VI) listing as a critical environmental hazard. Other problems with chromic-sulfuric etchants are difficulty in regeneration, inconsistent etch rate, the low limit of dissolved copper (4 to 6 oz/gal), and dangerous degradation of PVC and polypropylene equipment. Chromic acid etchant is suitable for use with solder, tin/nickel, gold, screened vinyl lacquer, and dry or liquid film photoresists. Chromic-sulfuric mixtures etch copper slowly, and additives are needed to increase the etch rate. For example, sodium sulfate and iodine have been used for rate increase. Alkaline etchants have become so well controlled that there is no justification for the risks and costs of chromic acid formulations.

The following listing demonstrates some of the operational difficulties with chromic-sulfuric etchants.

1. *Solder attack.*   The protective value of solder depends on the formation of insoluble compounds on the surface. Solder is attacked if the sulfate content of the bath becomes very low or contains chloride or nitrates. The solder plate composition can also cause etchant attack. When the lead content becomes low, the sulfate film protection is lowered, and protection is lost.

2. *Slow or no etching of copper.*   This can be caused by low chromic content, low temperature, insufficient acid, or high copper content. The solution is maintained as close as possible to 30 Baumé (pH about 0.1, temperature 80 to 90°F) and should be discarded when copper metal content exceeds 5.5 oz/gal.

3. *Staining of board materials.*   The surfaces of dielectric substrates such as paper-based phenolics are attacked by chromic acid etchants. Removal is difficult, and the boards are generally rejected.

4. *Disposal.*   Spent chromic acid etchants present a serious disposal problem. Disposal must comply with pollution standards and approved practice.

5. *Safety hazards.*   Chromic acid is an extremely strong oxidizing agent. It will attack clothing, rubber, plastics, and many metals. Safety measures require adequate ventilation to keep fumes out of room air, synthetic rubber gloves, aprons, face and eye shields, and storage away from combustible materials. Dermatitis and nasal membrane damage are possible dangers.[7]

33.4.7 Nitric Acid

Etchant systems based on nitric acid have not found extensive application in PC manufacture. Copper etching is very exothermic, which may lead to violent runaway reactions. Problems with this system include solution control, attack on resists and substrates, and toxic gas fuming. However, nitric acid has certain advantages. These include rapid etching, high dissolved copper capacity, high solubility of nonsludging products, availability, and low cost.

33.4.7.1 Chemistry.   Reaction in strong acids is shown by the following equation:

$$3Cu + 2NO_3^- + 4H+ \rightarrow 3Cu^{2+} + 2NO_2 + 2H_2O \tag{33.13}$$

Recent work shows that process improvements are possible.[30,31] Controlled etching has been attained in solutions containing 30 percent copper nitrate, water-soluble polymers, and surfactants. Dry-film resists work well in this etchant. An important finding was that straight wall trace edges were achieved using nitric acid. This could result in higher yields and density of fine-line boards. The major difficulty in obtaining the sidewall results is the requirement for specific crystollagraphic grain structure to the copper foil that has been difficult to reproduce in production quantities.

33.5   *OTHER MATERIALS FOR BOARD CONSTRUCTION*

Printed board laminates are usually composed of copper foils bonded to organic dielectric materials, ceramics, or other metals.

1. *Organic dielectrics.*   These are thermosetting or thermoplastic resins usually combined with a selected reinforcing filler. Thermoset-reinforced materials used for rigid and flexible boards provide overall stability, chemical resistance, and good dielectric properties. Thermoplastic materials are also used for flexible circuit applications. A factor in material selection is the effect of process solutions, etchants, and solvents on the material. In addition, the adhesives used in laminating metal to substrate can be softened, loosened, and attacked by some solutions. Flat-surface planarity is an aid to precision etching. This can be improved by selection of laminate with fine glass woven structures that do not express a waviness to the free copper foil surface.

2. *Thin-clad copper.*   Etched printed boards with epoxy laminate of ¼ oz (9 μm) copper or less show minimal lateral etching, thus enabling higher fidelity to pattern-plated traces. The problem with very thin copper foils is pinhole density and fragility in the laminate manufacturing process. One method of avoiding these problems is to start with a more conventional ½-oz (18-µm) foil and reduce its thickness by etching to a thinner state (3 to 9 µm).[32] This technique depends on a uniform starting foil and a very well-controlled etching process. Although the particulars of the published process were not disclosed, cupric chloride and sodium persulfate formulations have been used for this process.

3. *Reverse foil laminate.*   This construction is receiving attention for improving precision and control of etched lines. In order to achieve this foil structure, the *tooth* or crystalline roughness must also be controlled in the foil. The tooth side is usually placed on the dielectric interface of the laminate to increase peel strength and present a flat surface for processing. However, one variation is to place the *drum* or flat side of the laminate toward the dielectric and leave the tooth side upward. This forms a flat interface so that the etching stops at this surface, instead of the conventional extra etching to remove the teeth that are buried in dielectric.

4. *Semiadditive copper.*   A copper thickness of 0.000050 to 0.000200 in with subsequent copper and resist metal plating shows no overhang or sliver formation. These layers must be of sufficient thickness and uniformity to withstand the cleaning and preparation steps for plating the required conductors.

33.6   METALS OTHER THAN COPPER

33.6.1   Aluminum

Aluminum-clad flexible circuits find use in microwave stripline[33] and radiation-resistance applications. Aluminum and its alloys have good electrical conductivity, are lightweight, and can be plated, soldered, brazed, chem-milled, and anodized with good results. Laminate dielectrics include PPO,[33] polyimide,[34] epoxy-glass, and polyester.

Precleaning for resist application includes nonetch alkaline soak, water rinsing for 5 to 10 s in chromic-sulfuric acid, rinsing, and drying. Preferred etchants include ferric chloride (12 to 18° Baumé), sodium hydroxide (5 to 10 percent), inhibited hydrochloric acid, phosphoric acid mixtures, solutions of HCl and HF, and ferric chloride–hydrochloric acid mixture.

Screen-printed vinyl resists and dry-film photoresist are the most durable for deep etching or chem-milling. A dip in a 10 percent nitric or chromic acid solution will remove residues from the surface or edges of conductor lines that may be left on some alloys. Dilute chromic acid has also been used for this purpose. Spray rinse thoroughly with deionized water after etching.

33.6.2   Nickel and Nickel-Based Alloys

Nickel is increasingly used as a metal cladding, electroplated deposit, or electroformed structure for printed wiring because of its welding properties. Nichrome- and nickel-based magnetic alloys are other examples of materials requiring special etching techniques.

The methods previously described are adaptable to image transfer and etching of nickel-based materials. Etching uses ferric chloride (42° Baumé) at about 100°F. Other etchants include solutions made from one part nitric acid, one part hydrochloric acid, and three parts water, or one part nitric acid, four parts hydrochloric acid, and one part water.

33.6.3   Stainless Steel

Alloys of stainless steel are used for resistive elements or for materials with high tensile strength. Etching of the common 300 to 400 series can be done with the following solutions:

1. Ferric chloride (38 to 42° Baumé) with 3 percent HCl (optional).

2. One part HCl (37 percent) by volume, one part nitric acid (70 percent) by volume, one to three parts water by volume. Etch rate is about 0.003 in/min at 175°F, useful for high 300 to 400 series alloys.

3. Ferric chloride + nitric acid solutions.

4. One hundred parts HCl (37 percent) by weight, 6.5 parts nitric acid by weight, 100 parts water by weight.

33.6.4   Silver

Silver, the least expensive precious metal, has excellent properties, including superior electrical and thermal conductivity, ductility, visible-light reflectivity, high melting point, and adequate chemical resistance. As such, it is widely used throughout the electronics industry. Flexible circuit structures with silver are used in electronic cameras and LED products.

Standard image-transfer methods are suitable. Pre-etch cleaning should include a dip in dilute nitric acid. Mixtures of nitric and sulfuric acids are effective etchants. With silver on brass or copper substrates, a mixture of 1 part nitric acid (70 percent) and 19 parts sulfuric acid (96 percent) will dissolve the silver without attacking the substrate. The solutions should be changed frequently to prevent water absorption and the formation of immersion silver on the copper.

Etching can be done with a solution containing 40 g chromic acid, 20 mL sulfuric acid (96 percent), and 2000 mL water.[35] This is followed by a rinse in 25 percent ammonium hydroxide. Thin films of silver are etched in 55 percent (by weight) ferric nitrate in water or ethylene glycol. Solutions of alkaline cyanide and hydrogen peroxide will also dissolve silver. Use extreme caution. Electrolytic etching is also possible with 15 percent nitric acid at 2 V and stainless-steel cathode.

33.7   BASICS OF ETCHED LINE FORMATION

The basic process of chemical etching to form features has been studied and modeled extensively. However, the best scientific knowledge can only proceed to an understanding of factors influencing the process. The practical execution of the process to manufacture useful circuits contains elements of experience in the best practice of several processes and choices in the selection of materials and process variations to achieve technically sound, cost-effective, and manufacturable results. In order to make fine-line circuit products, it is necessary to understand the process fundamentals so that the limitations may be understood and then overcome. The intent of this section is to briefly review the underlying science and discuss factors and practices for precision etching.

33.7.1   The Image

33.7.1.1   Phototools.   It seems obvious that artwork is the first defining step of the image. If the phototool lacks definition, integrity, or dimensional stability, there is no possibility of improving the image in subsequent processing. The edge definition and contrast are particularly important. Images of multiple pulses or *spots* must be overlapped so that there is minimum waviness of edges. As the feature to be resolved becomes smaller, the image must have better and cleaner integrity. It is desired that the artwork be at least an order of magnitude (factor of 10) better than the final tolerance required for the final product. Therefore, if the line tolerance is ±0.0001 in, then the artwork should be ±0.00001 in.

33.7.1.2 Image Integrity.   The effectiveness of the image translated from the artwork/ exposure process can be best achieved by optimization through all of the processing. Surface preparation, resist application, exposure, handling, development, and cleanliness must be maintained sufficiently to produce images with proper integrity. Image integrity may be gauged periodically with a resource such as the Conductor Analysis Technologies etch evaluation protocol. Defects such as shorts and opens have been determined to be related to resist integrity, and persistent linewidth repeating patterns may be image related. General concepts indicate that the thinner the resist, the shorter the light path; and the less optical interface layers between the light source and the copper surface, the better fidelity of the image to the phototool. However, constant improvement in film, exposure, and phototool products makes fine-image options dependent on proper optimization and constancy of technique and cleanliness as well as choice of technology.

33.7.2 Basics of Processing

33.7.2.1 Diffusion—The Controlling Mechanism.   In Sec. 33.4, the chemical reactions of the etching process are discussed. In order to understand the formation of the actual shapes of the foil cross section, the influences of diffusion and fluid flow must be understood.[32] As a result of the etching reaction at the reaction point on the copper surface, a complex ion concentration is built up and active etching chemicals are depleted. In order to complete the reaction, this complex ion must move through a static *boundary layer* into a place where there is fresh etchant supplied to complete the reaction and carry away the etched product. (See Fig. 33.3.) This boundary-layer shape is dependent on the specific shape of the resist and etching wall, the fluid flow over the surface, and the critical fluid flow in the channels formed by the resist and the etched side wall.

33.7.2.2 Fluid Flow Contribution.   Examination of Fig. 33.3 shows that the contour of the boundary layer and its thickness vary with the shape and narrowness of the channel. If the surface is flat with no resist image, the boundary layer is thin and only dependent on the fluid properties and velocity across the surface. On the other hand, with resist images forming channels with depth, and the etched copper forming further channels, it is easy to see how the flow of fluid in the channel can be much different than the surface flow. In order to make matters even more confusing, the circuit traces and resist images form actual channels in the surface.

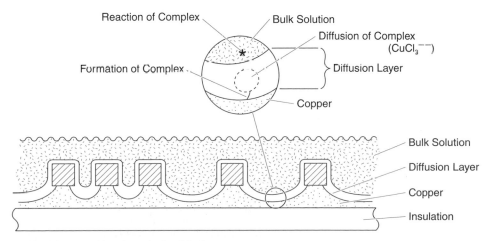

FIGURE 33.3   Etching dynamics for diffusion.

Much like a river bottom, the fluid flow in these individual channels is influenced by but different from the overall surface flow. In general, the narrower and deeper the channel, the slower the flow; and, critically, the diffusion boundary layer is more significant and etchant slows. In general terms, this is the reason that the middle section of a group of parallel, closely spaced lines etches slower than the outside lines. Similarly, this is the reason why a sharp 90° angle has a slower-etching section in the inside edge of the bend.

The concept of spray etching adds another dimension to the fluid dynamics. The surface of the panel is never really free of a liquid film. The top side of a panel has the etch liquid held up between delivery by the sprays and removal by flow off the edges. This flow is influenced by the number, flow, and pattern of the nozzles, the interaction of the flow between the sprays, the motion of the panels, and the effects of rollers and other contacting and shadowing objects. Even the bottom side retains fluid by surface tension and is influenced by the amount of liquid moving under the direction of the nozzles. The end result is complicated flow patterns that are the result of testing and optimization by the machine manufacturers.

33.7.3 Trace Shape Development

33.7.3.1 Etched Trace Shape Description.   During etching, the gap between resist areas is removed evenly at first, and then progressively cup-shaped until the center area is broken through. After this, the area of exposed laminate opens progressively and the etching works on the side walls, etching in and under the resist as the side wall is relieved to expose more insulation and isolate the desired feature. (It has been noted that very little undercutting is apparent until the center is opened.) Figure 33.4 illustrates the attack with standard definitions for R (resist width), B (base of trace), T (top of trace), and t (foil thickness). There are two descriptive measures of the etching process—undercutting and etch factor—which are defined in Eqs. (33.14) and (33.15). These definitions are not universal, so it is necessary to clarify the definitions to correctly interpret discussions using these factors. Undercutting defines the average overhang of resist after top width reduction, and etch factor defines the average side wall inward taper of the etched feature per unit thickness. Occasionally, the etch factor is inverted or different measured values are used.

Undercut: $$U = (R - T)/2 \qquad\qquad (33.14)$$

Etch factor: $$F = 2t/(B - T) \qquad\qquad (33.15)$$

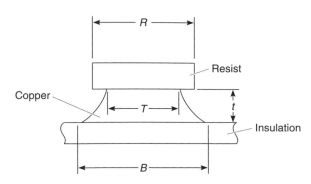

FIGURE 33.4   Trace cross-section measurement.

It is significant that these two units of measure are useful in describing the development of the shape of an etched part. It is desirable that U be minimized and F maximized for best feature definition.

33.7.3.2   Undercut and Etch Factor Development.   The development of trace shapes has been studied by observing etched traces at progressive times.[36,37] This allows one more parameter to be developed, the extent of etch, which is the ratio of R/B. This factor equals 1.0 when the bottom of the trace equals the resist width, one measure of ideal etching. Note that if $R < B$, this implies that the trace is underetched, and $R > B$ indicates overetching. The traces were etched in cupric chloride using 3.0-mil line and space patterns of 1.0-mil resist on 1-oz (1.4-mil) copper foil. The progression of etch values is found in Table 33.2 and Fig. 33.5.

TABLE 33.2   Etch Progression

Etch time, S	Undercut, U mil	Etch factor, F	Extent of etch, R/B
90	0.05	0.90	0.5
110	0.30	1.75	0.75
125	0.45	2.33	1.1
140	0.525	2.67	1.0
165	0.75	3.11	1.25

There is an important conclusion to be drawn from these data. In order to compare or characterize an etch process (either to to other, etchant to etchant, condition to condition, etc.), it is not sufficient to only use the undercut or etch factor alone. The data must be evaluated at a given R/B point for the same foil, resist, artwork, and panel size. If these factors are not held constant, the statement that "process A gives less undercut" has little meaning—additional etching gives more undercutting and straighter side walls.

33.7.3.3   Sensitivity Analysis.   One of the extensions of the previous analysis is the ability to evaluate the sensitivity of the process. This is a study of the change in the result (R/B) due to the additional etching (relative time). Relative time is the ratio of the actual etching time to the nominal etch time taken to reach R/B of 1.0. In Table 33.2, it took 140 s to reach $R/B = 1$, but only 25 s (18 percent) more etching to achieve 25 percent overetching ($R/B = 1.25$). This sensitivity indicates that a small amount of overetch time could cause overetched 3-mil circuits. In this example, the etching process is slow enough to be controlled with the adjustment of the conveyor speed in the etcher. However, if the etchant were twice as fast (alkaline, for example) with the same sensitivity, it would be difficult to accurately "dial in" the proper result. Further evaluation of the cumulative variation from several variable properties on the overall product capability statistics can be found in the literature.[38]

33.7.3.4   Implication for Improved Processes.   As a practical process application, it is often desirable to improve the etch factor to get a more square shape of the etched profile. One method for accomplishing this is to design the phototool for wider lines than desired, and then overetch to the desired endpoint. This practice increases F at the expense of also increasing U. However, as spaces get smaller, there is no space available for additional resist linewidth, because the fluid activity is restricted so that the etchant is stagnant between the traces. Therefore, resist width near the final line specification and precise control of the etch process are required to obtain desired shape profiles within the $R/B = 1$ degree of etching. As feature specifications tighten, the process to do this may need to be altered. One patented approach is to finish the etch with a specific slower-speed fine-tune etching chemistry to be able to stop the etching process at precisely the correct point.[39]

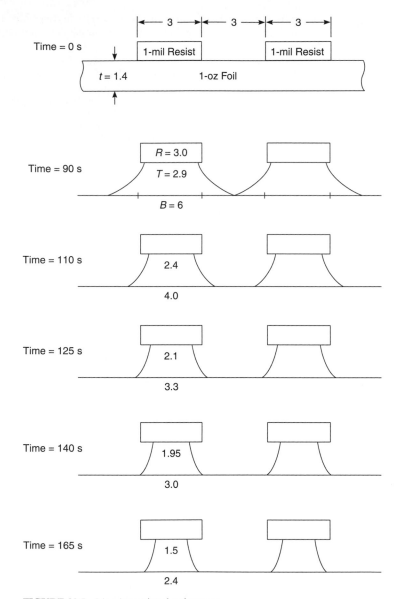

FIGURE 33.5 Line formation development.

33.7.4 Fine-Line Formation Etching Requirements

33.7.4.1 Fine-Line Definition. The term *fine line* is a relative one because the current state of processing by conventional available technology is constantly improving. Therefore, fine lines can be considered the next extension of capability that requires precision higher than that afforded by available technology. This is understood by means of statistics and standard gauging tools. The measure C_p (simple process capability) evaluates the variability of output results ($6s$, where s is the standard deviation of the linewidth distribution) to the specification limits of variation.[40] The data for the performance can be obtained by using a standardized test vehicle (IPC-9251) or a commercial testing program by CAT (see Chap. 36). In general, fine lines can then be defined as the linewidth point at which yield performance begins to drop below the run of production. Note that it is up to the customer and shop to determine what product line variation (usually a percent of line dimension) is allowed. It is apparent that this definition depends on the available technology and operation of the individual shop.

33.7.4.2 *Limitations—Practical Rule of Thumb.*

It is useful to have a practical limit to express an understanding of where the technologies may limit performance. In the case of etched linewidth, it has been expressed that the total of the resist thickness plus the etched foil thickness would limit the gap between the etched features. Therefore, for 1.2-mil dry-film resist over 1-oz copper foil, the total thickness is 2.6 mils. This could be used as a practical limit for both the trace and gap. Further limits can be determined by the undercut and etch factor experienced for the same type of etched features. Therefore, using the $R/B = 1$ data from Table 33.2 ($U = 0.525$ mil), a 2.6-mil resist line would be 2.6 mils at the trace bottom and would have a 1.5-mil etched top reduction, leaving only a 1.1-mil top surface. It must then be determined if these dimensions (with allowance for variations) are sufficient for the design functionality.

33.7.4.3 *Thinner Is Better.*

In general, it follows that thinner foil and thinner resist allow for smaller features. However, if thin foil (9 μm, ¼ oz) is used, the traces may be plated to achieve better current capability. Thin resist may be used, but the image may be susceptible to handling damage or damage in the process equipment. However, for coated resists of 0.4 mil and ¼-oz foil (0.35 mil), the result could indicate a limit of 0.75 mil (19 μm). In fact, there has been a process using 3- and 5-μm foil to produce 30-μm line- and spacewidth patterns with 14 μm additional plated copper (total trace height of 17 and 19 μm). This method uses both panel and pattern plating.[41] The original foil was etched from thicker commercial foil laminate to the working thickness.

33.7.4.4 *Changing the Microchannel Flow.*

As previously explained, the flow in the microchannels surrounding the features is critical in achieving uniform and accurate etching results. There is a unique process development using fibers to affect localized fluid mixing in these channels.[42] This approach requires specific patented equipment and processing methods.[43] Reported results down to 50-μm (2-mil) lines and spaces in 1-oz (34-μm) copper have been produced with 1.4-mil (37-μm)-thick resist. This is a very unique and important result because conventional foil and resist technology can be used to make features significantly smaller than conventional capability through improved fluid application mechanics.

33.7.4.5 *Beyond Fine? HDI Impacts.*

High density interconnect (HDI) technology involves several technology changes to form the circuitry. The highlight, as the name implies, is small via structures made in a layer-by-layer methodology with a built-up thin insulator and thin foil structure. To achieve the interconnection, thin and precisely controlled traces are required. The technologies, such as the two previous items, are becoming available to attack the needs of the products. As layer counts increase, the accumulating yield implications are significant, so that each process must produce minimum defects. For etching, one of the challenges is to make the resist conform to the surface—which forces the structural materials and processes to engage planarity of the end stage process as an issue. Etching capability requires improving the precision, stability, and capability of processing. The flowchart in Fig. 33.6 graphically illustrates the effects required for etching concerns and approaches to address these.[44] It is significant that there are no new concerns added to previous discussions.

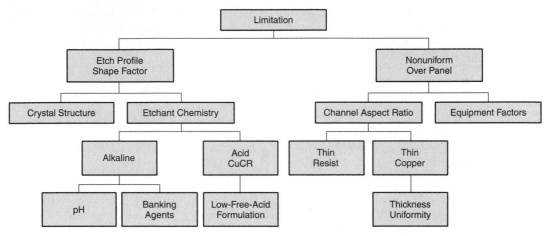

FIGURE 33.6 Flowchart for study of resolution of HDI etched product limitations. Profile and uniformity limitations may be studied for process improvement by progressing down the chart for contributing elements.

33.7.4.6 Stability and Control.   It is important to measure and understand sources of variation in order to eliminate them. The gauging tools mentioned previously are very useful for collecting the data for analysis. One important feature is the positional and orientational variation in line geometry. Various conclusions can be taken from feature variation by location across the panel and orientation relative to the direction of motion. Repeating the gauging technique on a regular basis can allow observation of changing conditions within the process (such as plugged nozzles and other defects).

Another type of variation can be obtained by examining a series of gauge panels progressively processed with specific time-lag variations. If etching chemical controls are not adequate, the etch rate may vary in the sequence and therefore cause line variation with time. Usually, regeneration equipment can be adjusted or modified to minimize these variations, but the value of such instrument upgrades can be determined easily by data recorded about the type and extent of defects incurred.

33.8 EQUIPMENT AND TECHNIQUES

Etching equipment for circuit line formation has developed many variations on spray processing equipment. Often, as members of a coordinated automatic immersion plating process line, there are immersion etching tank varieties for microetching and surface preparation. However, for individual panels, conveyorized spray processing is clearly established because of efficient part handling and processing effectiveness.

33.8.1 Spray Equipment Basics

33.8.1.1 Construction.   Etching equipment must be made from materials for long life, good dimensional stability, and resistance to chemicals and temperatures (130°F and higher). The basic construction includes rigid plastics (PVC, polypropylene, CPVC, and PVDF), metals (titanium, Hastalloy C), elastomers (vinyl, Viton A, Kel F, ethylene-propylene), specialized composites (epoxy-glass, graphite, and carbon-filled polymers, and other specialized formulations), and glass or clear and translucent polymers for visibility. The challenge is to select proper grades, compositions, and properties for mechanical strength, functionality, and durability. Equipment manufacturers perform these tasks in their design process.

33.8.1.2   Controls.   Controls and electronics are significant because the increasing robustness and precision can have a large impact on cost. Therefore, the compromise on cost-effectiveness/technology trade-off often is made with a market-price primary goal. Therefore, cost-benefit analysis must be made by purchasers to evaluate available upgrades in instrumentation. The key controls are temperature, pressure, conveyor speed, and safety interlocks. Temperature is especially important because the etching reactions are usually exothermic and heat must be removed by cooling water coils to maintain stable rates. Conductivity, pH, density, ORP, and other process chemical control instruments may be added to monitor and control the processes. These instruments may be integrated into a PLC or computerized control system or may be individual control-panel stand-alone instrumentations. The wiring and the control cabinet itself must be suitably fume and hazard protected for water and chemical splashes as well as corrosive fumes. It is desired to have critical condition displays available to operators in the machine environment as well as some means of monitoring and recording process information.

33.8.1.3   Spray Strategies.   Individual equipment designers have implemented very different types and mechanisms of spray attack on the panels. It is important to realize that the spray system design must be carefully integrated with the conveyor and mechanical design to minimize the effect of rods, wheels, panel control devices, and other mechanisms to achieve both uniformity and geometry of the features produced. It must be realized that the effect of the sprays is not merely the attack by droplets on the surface, but that the sprays are the prime drivers of the direction and effectiveness of the microchannel fluid flow on the panels. Therefore, separate controls and gauges must be provided for top and bottom spray arrays. For maximum precision, four quadrant controls are often beneficial. The mechanical spray arrays can be described as a variation of the following types.

Fixed Spray Array.   This is a system of spray tubes containing multiple nozzles connected transversely across the conveyor to a header pipe on one or both sides. The spray nozzles are placed into the tubes in a pattern designed to optimize the effect. Care must be taken to install and orient the nozzles properly for designed effect. Mechanical intervention to selectively activate (or deactivate) portions of the design to effect the fluid dynamics of particular patterns should be included.

Oscillating Sprays.   This system uses spray tubes that are connected generally to be aligned with the conveyor direction (often at a slight angle). The ends of the spray tubes are made with rotating seals so that the tubes can be moved back and forth by an oscillating crank or gearing mechanism. The nozzles move the liquid in a pattern generally across the conveyor travel. Individual spray tubes may be throttled and/or selectively activated.

Reciprocation.   These nozzles are usually placed in an array similar to the fixed system, and the entire array is moved by a cranking mechanism in a direction across the conveyor. Maximum precision offerings include various patterns of nozzles that can be separately valved and selectively activated by computer controls.

Spray Nozzles.   Spray nozzles are available in several types, sizes, and materials. For etching it has been established that PVDF nozzles are robust and consistent. The nozzle patterns are flat fan, full cone (round and squared), and open cone designs. Generally, flat fan nozzles have large openings and high impact for a given flow rate, while round cones have lower impact and orifice restrictions to ensure the pattern is filled. One of the principal failings of nozzles is simply clogging. The best array design can become dangerously nonuniform simply by plugging a small number of sprays. There is a commercially available nozzle that has a self-clearing fan design. Many are installed. For some applications, filtration between the pump and sprays has been installed to prevent plugging.

Selection.   All of these combinations of nozzles, arrays, and mechanisms offer benefits and drawbacks that are often counterintuitive to the end user. It is necessary, therefore, to test and evaluate the performance of the spray systems for the work type and working environment. Testing of performance must be evaluated in a carefully controlled trial or by comparison with a standardized gauging tool.

33.8.2 Spray Equipment Options

33.8.2.1 Vertical Panel Orientation

Batch Operation. Some of the earliest equipment is still useful for prototype manufacture. This equipment holds a single panel on a workpiece that is generally rotated about an axis and sometimes oscillated as well. The sprays are addressed from one or both sides of the work and may be fixed or oscillating. This type of equipment usually is limited to small panels. The liquid all drains off the panel and falls into a collection sump. The panels must be removed after etching is complete and placed into a separate rinse vessel. (Some equipment has been designed to divert the drain and spray water onto the same work orientation.) This equipment is not capable of high capacity but has been able to etch high-quality work.

Conveyorized Operation. There are several versions of large machines that accept vertically oriented workpieces, convey them on a series of rollers from beneath the panel, and stabilize the work with rollers or wires. There are some machines that suspend the panels from the top surface. Vertically oriented nozzle arrays, either stationary or moving, spray the panel surfaces. These machines have been more successful in developing than etching because the depth of the image to be developed is formed by the exposure rather than completely depending on the etching fluid dynamics. All the liquid sprayed on the panels is removed by running downward off the bottom panel edge; therefore, the nozzles must be specifically designed to offset the effects of higher amounts and speeds of downflowing etchant progressively toward the panel's lower edge. It is obvious that very thin workpieces can have difficulty being conveyed through this design.

33.8.2.2 Horizontal Panel Orientation

General Operation. The most widely used configuration of etching equipment handles the panels horizontally on a roller table support. The panels placed on the rollers move constantly through arrays of sprays above and below the work. One of the principal design challenges is the fluid flow control of the liquid pool that builds up on (and runs off) the top surface of the panel. The various designs of the spray system must allow equally good and equally fast etching beneath the panel as on top. In a well-conceived design, the simple adjustment of spray pressures can achieve this balance.

Conveyance. The conveyor design is particularly critical. Rollers must turn at constant speed to ensure that the surfaces are not rubbed and damaged. Chains and gear designs have been successfully used. The potential problem areas are the entrance and exit squeegee rollers and the changes from one machine drive section to another. The roller wheels themselves need to be made of compliant materials that have flat area contact with the panels (rounded contact areas allow points that can cause resist damage, especially if the rollers harden with age). The wheels must be spaced far enough apart to allow the bottom spray to hit the panel with minimum shadowing, but close together enough to convey thin work without sagging or deforming.

Thin panels (0.0015- to 0.003-in insulation core) are difficult to convey because the leading edges must feed consistently from one roller to the next in spite of the downward force of the fluid. In addition, the deflection of the panels between rollers due to fluid loading can cause lack of uniformity in etching. Often, special mechanical guides such as clips or wires are used to overcome these difficulties. Problems occur when the guides stretch or are misplaced during production. If one panel jams, then the others behind it continue to jam and pile up, resulting in a major problem when discovered. The clearing of the jam often requires removal of rollers that are usually enmeshed in the mass of displaced guides and work. Not only are damaged panels costly, the downtime to renew the equipment and reestablish production causes significant production delay. One particular patented design features a combination of rollers and spray force that assures robust transport while simultaneously improving planarity for spray action.[45]

Truly flexible circuitry requires even more enhancements for reliable transport. Frames and screens have been used in the past, but these fixtures require extra handling and cause drag-over and cleaning problems. For sheet-cut panels, the best solution is a leading edge support attached to the panel by tape or mechanical clips. Usually these panels are attached and removed manually, but automatic systems could be provided. Flexible circuits can also be manufactured in continuous rolls that can be threaded through multiple processes. In addition to the previous difficulties with planarity and drag-over, the precise synchronization of roller speed throughout all the processes is mandatory. Therefore, machine and processing speeds must be carefully planned at original manufacture to allow proper dwell settings and latitudes. Rinsing and drag-over present problems.

33.8.3 Rinsing

Rinsing is a process technology that needs particular attention because of the multiple needs of uncontaminated panels, conservation of rinse waters, and environmental discharge limitations. It is no longer sufficient to flood the panel with large amounts of water that are then discharged to sewers. The capability of technical analysis of this process has been published.[46]

33.8.3.1 Cascade Rinsing.   The principal tool to effect the required change is the concept of multiple-stage cascade rinsing. This process includes the concept that a robust flow recirculated on a panel can reduce the concentration to a level in equilibrium with a much lower circulated water stream. The lower stream passes through a series of chambers of increasing concentration, while the work passes through a series of chambers of increasing cleanliness (in reverse order from the water). The end result is a small use of water capturing a relatively high level of contamination. The contamination level is many times the discharge limit for direct environmental discharge. However, the fact that the stream is relatively small and highly concentrated allows for reasonable treatment and recovery of by-products by ion exchange, direct electrowinning, or membrane cell concentration. In certain circumstances, the waste stream may be comingled with a main process stream for recycling (see Fig. 33.1).

33.8.3.2 Drag-over Minimization.   All of the contamination that eventually ends up in the water stream is that introduced by drag-over from the process itself (or from a recirculated environmental rinse). Therefore, the first step in successful rinsing is the reduction or prevention of materials from the panel. The conventional means of attempting this is by a set of pinch or squeegee rollers at the output of the chemical module. These rollers may be increased in effectiveness by softness or by pressure. Often, rollers made of PVA foam have been employed effectively. However, these rollers must be kept clean and moist to be effective. Other methods include improved baffling, ensuring capture of overspray, and low-velocity air blow-off. Proper management of drag-over on conveyorized equipment has reduced residues by a factor of 10 or better. Thin panels cause difficulties because the compromise made between conveying reliability devices and planarity and contact of squeegee rollers and surface tension reduces effectiveness of retention. Thick panels (over 0.125 in) also cause problems because squeegee rollers and baffles must lift to clear the panels, allowing gaps for solution escape. It is therefore a subject for engineering study and process management trade-off decisions that must be considered during the specification process.

REFERENCES

1. E. Armstrong and E. F. Duffek, *Electronic Packaging and Production,* vol. 14, no. 10, October 1974, pp. 125–130.

2. W. Chaikin, C. E. McClelland, J. Janney, and S. Landsman, *Ind. Eng. Chem.,* vol. 51, 1959, pp. 305–308.

3. Jieh-Hwa Shyu, "Electrochemical Studies of Etching Mechanisms in Ammoniacal Etchants," IPC TP-751, October 1988.

4. U.S. Patent 3,466,208, J. Slominski, 1969.

5. U.S. Patent 3,231,503, E. Laue, 1966.

6. U.S. Patent 3,705,061, E. King, 1972.

7. I. Sax, *Dangerous Properties of Industrial Materials,* rev. ed., Reinhold Publishing Corp., New York, 1957, p. 464.

8. Marshall I. Gurian, "Rinsing as a Process Technology," Paper T14, *Proceedings of Printed Circuit World Convention VI,* San Francisco, May 1993.

9. U.S. Patent 3,440,036, W. Spinney, 1966.

10. *Solvent Extraction Technology,* Center for Professional Advancement, Somerville, NJ, 1975.

11. Galvano Organo, *Printed Circuit Fabrication,* vol. 16, no. 1, January 1993, pp. 42–47.

12. Atotech USA, Inc., State College, PA.

13. K. Murski and P. M. Wible, "Problem-Solving Processes for Resist Developing, Stripping, and Etching," *Insulation/Circuits,* February 1981.

14. C. Swartzell, *Printed Circuit Fabrication,* vol. 5, no. 1, January 1982, pp. 42–47, 65.

15. G. Parikh, E. C. Gayer, and W. Willard, *Western Electric Engineer,* vol. XVI, no. 2, April 1972, pp. 2–8; *Metal Finishing,* March 1972, pp. 42, 43.

16. L. Missel and F. D. Murphy, *Metal Finishing,* December 1969, pp. 47–52, 58.

17. F. Gorman, "Regenerative Cupric Chloride Copper Etchant," *Proceedings of the California Circuits Association Meeting,* 1973; *Electronic Packaging and Production,* January 1974, pp. 43–46.

18. U.S. Patent 3,306,792, W. Thurmal, 1963.

19. L. H. Sharpe and P. D. Garn, *Ind. Eng. Chem.,* vol. 51, 1959, pp. 293–298.

20. J. O. E. Clark, *Marconi Rev.,* vol. 24, no. 142, 1961, pp. 134–152.

21. O. D. Black and L. H. Cutler, *Ind. Eng. Chem.,* vol. 50, 1958, pp. 1539–1540.

22. U.S. Patent 4,130,454, B. Dutkewych, C. Gaputis, and M. Gulla, 1978.

23. U.S. Patent 3,801,512, C. Solenberger, 1974.

24. U.S. Patent 4,130,455, L. Elias and M. F. Good, 1978.

25. U.S. Patent 3,476,624, J. Hogya and W. J. Tillis, 1969.

26. A. Luke, *Printed Circuit Fabrication,* vol. 8, no. 10, October 1985, pp. 63–76.

27. U.S. Patent 2,978,301, P. A. Margulies and J. E. Kressbach, 1961.

28. E. B. Saubestre, *Ind. Eng. Chem.,* vol. 51, 1959, pp. 288–290.

29. W. F. Nekervis, *The Use of Ferric Chloride in the Etching of Copper,* Dow Chemical Co., Midland, MI, 1962.

30. U.S. Patent 4,482,425, J. F. Battey, 1984.

31. U.S. Patent 4,497,687, N. J. Nelson, 1985.

32. Don Ball, "The Surface Mechanics of Fine-Line Etching," *Printed Circuit Fabrication,* vol. 21, no. 11, 1998.

33. F. T. Mansur and R. G. Autiello, *Insulation,* March 1968, pp. 58–61.

34. H. R. Johnson and J. W. Dini, *Insulation,* August 1975, p. 31.

35. P. F. Kury, *J. Electrochem. Soc.,* vol. 103, 1956, p. 257.

36. Marshall Gurian, "Process Effects Analysis for Fine Line Production," IPC Paper WCIV-28, *Printed Circuit World Convention IV,* Tokyo, 1987.

37. Marshall Gurian, "Reliable Fine Line Wet Processing," *Printed Circuit Fabrication,* vol. 10, no. 12, 1987.

38. Marshall Gurian, "Fine Line Processing: The '90's Are Here!" *Printed Circuit Fabrication,* vol. 13, no. 5, 1990.

39. U.S. Patent 5,904,863, Michael Hatfield and Marshall Gurian.

40. Michael Brassard and Diane Ritter, "The Memory Jogger II," Goal/QPC, Methuen, MA.

41. Yasuo Tanaka, Hireyuki Urabe, and Morio Gaku, "Three Micron Copper Foil Clad Laminate for 30/30 Micron Line/Space Circuit," *CircuiTree,* vol. 10, no. 11, 1997.

42. Igor Kadija and James Russel, "New Wet Processing for HDI's," *CircuiTree,* vol. 12, no. 5, 1999.

43. U.S. Patents 5,024,735, 5,114,558, and 5,167,747, Igor Kadija.

44. Karl Dietz, "Process and Material Adaptations for HDI Requirements," *CircuiTree,* vol. 13, no. 12, 2000.

45. U.S. Patent 4,607,590, Don Pender.

46. Marshall Gurian, "Rinsing as a Process Technology," Paper T14, *Proceedings of the Printed Circuit World Convention VI,* 1993; summarized in *Printed Circuit Fabrication,* vol. 20, no. 7, 1997.

CHAPTER 34*
SOLDER RESIST MATERIAL AND PROCESSES

Dr. Hayo Nakahara
N.T. Information, Ltd., Huntington, New York

34.1 INTRODUCTION AND DEFINITION

IPC T 50† b defines a solder resist (mask) as "a coating material used to mask or to protect selected areas of a printed wiring board (PWB) from the action of an etchant, solder, or plating." A somewhat more useful working definition for a solder resist is as follows: a coating which masks off a printed wiring board surface and prevents those areas from accepting any solder during reflow or wave soldering processing (see Fig. 34.1).

34.2 FUNCTIONS OF A SOLDER RESIST

The prime function of a solder resist is to restrict the molten solder pickup or flow in those areas of the PWB, holes, pads, and conductor lines that are not covered by the solder resist. PWB designers, however, often expect more functionality out of the solder resist than just a means to restrict the solder pickup. Table 34.1 lists the functions of a solder resist.

34.3 DESIGN CONSIDERATIONS FOR SOLDER RESISTS

34.3.1 Design Goals

The design goals for the selection and application of a solder resist should be carefully considered. As with all design goals, one should try to achieve maximum design flexibility, reliability, and functionality at a cost consistent with the required level of system performance.

---

* Sections 34.1 to 34.9 have been adapted from Lyle R. Wallig, "Solder Resist," in Coombs (ed.), *Printed Circuits Handbook,* 3d ed., McGraw-Hill, New York, 1988, Chap. 16. Sections 34.10 to 34.12 were prepared specifically for this edition by Dr. Nakahara.
† Institute for Interconnecting and Packaging Electronic Circuits (IPC), Lincolnwood, Illinois.

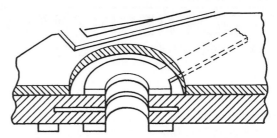

FIGURE 34.1 Important factors of solder resist on a printed wiring board: (*a*) mask should be away from plated-through-hole and its associated land or pad; (*b*) trace should be covered; (*c*) laminated area should be completely covered; (*d*) adjacent conductors should not be exposed.

TABLE 34.1 Functions of a Solder Resist

Reduce solder bridging and electrical shorts.
Reduce the volume of solder pickup to obtain cost and weight savings.
Reduce solder pot contamination (copper and gold).
Protect PWB circuitry from handling damage, i.e., dirt, fingerprints, etc.
Provide an environmental barrier.
Fill space between conductor lines and pads with material of known dielectric characteristics.
Provide an electromigration barrier for dendritic growth.
Provide an insulation or dielectric barrier between electrical components and conductor lines or via
    interconnections when components are mounted directly on top of the conductor lines.

34.3.2 Design Factors

The system's performance and reliability requirements are the keys in determining the selection process for a solder resist. Critical life-support systems will require different materials and standards than a less critical system such as a VCR.

Table 34.2 outlines some of the factors to consider in the design process when selecting a solder resist.

It is not very likely that a single solder resist material or application technique will satisfy all the design considerations that are viewed as necessary. It should also be noted that not all the design factors listed in Table 34.2 carry the same weight or value, so the designer needs to prioritize those design factors, analyze the necessary tradeoffs, and then specify the solder resist material and process that gives the best balance of properties or characteristics.

34.4 ANSI/IPC-SM-840 SPECIFICATION

In an effort to aid PWB and systems designers who wanted to know what performance properties or characteristics they would receive when they asked for a solder resist (mask), an IPC industry group developed the first solder resist specification in 1979. This specification is IPC-SM-840, entitled *Qualification and Performance of Permanent Polymer Coating (Solder Mask) for Printed Wiring Boards*.

The IPC-SM-840 specification calls out three classes of performance which the designer can specify:

TABLE 34.2 Design Factors

Criticality of system's performance and reliability
Physical size of PWB
Metallization on PWB, i.e., SnPb, copper, etc.
Line and space (density) of the PWB
Average height of conductor line (amount and uniformity of the metallization)
Size and number of drilled plated-through-holes (PTHs)
Annual ring tolerance for PTHs
Placement of components on one or both sides of PWB
Need to have components mounted directly on top of conductor line
Need to tent via holes in order to keep molten solder out of selected holes
Need to prevent flow of solder up via holes, which may have components sitting on top of them
Likelihood of field repair or replacement of components
Need for solder resist to be thick enough to contain the volume of solder needed to make good
   solder connections
Choice of specifications and performance class that will give the solder resist properties that are
   necessary to achieve the design goals

Class 1: Consumer—noncritical industrial and consumer control devices and entertainment electronics

Class 2: General industrial-computers, telecommunication equipment, business machines, instruments, and certain noncritical military applications

Class 3: High reliability—equipment where continued performance is critical; military electronic equipment

In addition to calling out the performance classes, the specification assigns responsibility for the quality of the solder resist to the materials supplier, the PWB board fabricator, and, finally, the PWB user.

Responsibilities per IPC-SM-840

Role	Responsibility
Materials supplier	The solder material and the appropriate data on the chemical, electrical, mechanical, environmental, and biological testing. Data are gathered on the standard IPC-B-25 test PWB.
PWB fabricator	The entire PWB fabrication and production process. This includes the application and curing of the solder resist material. The fabricator is also responsible for determining the end use application and the conformance of the PWB and the solder resist to the specification class call-out on the PWB fabrication print.
PWB user	Monitoring of the acceptability and functionality of the completed PWBs.

34.5 SOLDER RESIST SELECTION

The solder resists available are broadly divided into two categories, i.e., permanent solder resists and temporary resists. The breakdown of the solder resist types is shown in Fig. 34.2.

The permanent solder resist materials are classified by the means used to image the solder resist, i.e., screen printing or photoprint. In addition, the screen-printed resists are fur-

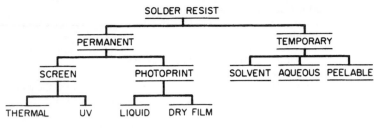

FIGURE 34.2 Solder resist selection tree.

ther classified by the curing technique, i.e., thermal or ultraviolet (UV) curing. The photo-print solder resists are distinguished from each other by whether they are in the form of liquid or dry film. The temporary resist materials are classified by chemistry or means of development.

34.5.1 Temporary Resists

A distinction is made between permanent and temporary solder resists. The temporary resists are usually applied to a selected or limited area of a PWB to protect certain holes or features such as connector fingers from accepting solder. The temporary resist keeps solder out of the selected holes and thus allows for certain process- or temperature-sensitive components to be added manually at a later time.

The temporary solder resists usually consist of a latex rubber material or any of a variety of adhesive tapes. These materials can be applied by an automatic or manual dispenser. Some of the temporary mask materials dissolve in the solvents or cleaning processes that are used to clean off the soldering flux residues. This is really a benefit, since it eliminates the need for a separate manual removal and/or cleaning step for the temporary resist.

34.5.2 Permanent Resists

Permanent solder resists are not removed and thus become an integral part of the PWB.

The demand for permanent solder resist coatings on PWBs has greatly increased as the trend toward surface mounting and higher circuit density has increased. When the conductor line density was low, there was little concern about solder bridging, but as the density increased, the number and complexity of the components increased. At the same time, the soldering defects, such as line and component shorts, greatly increased. Inspection, testing, and rework costs accelerated as effort went into locating and repairing the offending solder defects. The additional cost of the solder resist on one or both sides of the PWB was viewed as a cost-effective means to offset the higher inspection, testing, and rework costs. The addition of a solder resist also had the added value for the designer of providing an environmental barrier on the PWB. This feature was important and dictated that the materials considered for a permanent solder resist should have similar physical, thermal, electrical, and environmental performance properties that are in the laminate material. See Table 34.3 for a comparison of properties for permanent solder resist types.

34.5.3 Selection Factors

General considerations are as follows:

- Reliability and performance data on the solder resist material

TABLE 34.3   Permanent Solder Resist Selection Guide

Feature	Screen print		Dry film		Liquid photoresist
	Thermal	UV	Aqueous	Solvent	
Soldering performance	1	1	1	1	1
Ease of application	1	1	2	2	2
Operator skill level	2	2	2	2	2
Turnaround time	2–3	2–3	2	2	2
Inspectability	2–3	2–3	3	3	3
Feature resolution	3	3	1	1	1
Adhesion to SnPb	1	3	1–2	1–2	1
Adhesion to laminate	1	1–2	1	1	1
Thickness over conduct or lines	3 4	3 4	1 2	1 2	2
Bleed or residues on pads	3–4	3	1	1	1
Tenting or plugging of selected holes	4	4	1	1	4
Handling of large panel size with good accuracy	3	3	1	1	1
Meeting of IPC-SM-840 Class 3 specification	3–4	3–4	1–2	1	1
Two-sided application	4	4	1	1	3–4
Capital equipment cost	4	4	1–2	1–2	1

Key: 1 = good or high; 2 = moderate; 3 = fair; 4 = poor or low

- Cost effectiveness
- Past experience
- Vendor reliability and technical support
- Number of panels required
- Appearance and cosmetics of solder resist
- Number of sources for application and supply

Material considerations are as follows:

- IPC-SM-840 class designation call-out
- Cost and availability of materials
- Lot-to-lot consistency record
- Setup and cleanup times
- Working time and shelf life of solder resist
- Safety concern for release of noxious or toxic fumes during processing or curing steps
- Degree of workability

Process considerations are as follows:

- Operator skill level requirement
- Need for special applications or curing equipment
- Cleaning requirements for PWB before and after application
- Size of PWB
- Need for solder resist on one or both sides of PWB

- Number of panels to be processed
- Turnaround time required
- Machinability
- Need to tent selected holes
- Touchup or rework limitations
- Inspectability and conformance to specification

Performance considerations are as follows:

- Testing to IPC-SM-840 specification requirements
- Adhesion after soldering and cleaning
- Bleed-out of resist onto pads or PTHs
- Solvent resistance to flux and flux cleaners
- Ionic contamination levels
- Integrity of resist after soldering and thermal cycling

34.6 *SOLDER MASK OVER BARE COPPER (SMOBC)*

A major solder resist application technology is called SMOBC. This name stands for the application of *solder mask over bare copper.* A problem for conventional copper-tin-lead electroplated PWBs is the flow of tin-lead solder under the solder resist during the wave or vapor phase or infrared soldering. This flow of molten metal underneath the resist can prevent the resist from adhering to metal or laminate. If the resist fractures because of this hydraulic force, the surface integrity is lost and the effectiveness of the resist as an environmental or dielectric barrier can be severely impaired. In fact, such breaks in the resist can actually trap moisture, dirt, and soldering flux and serve as a conduit to direct liquids down to the resist-laminate interface. This solder resist situation could lead to serious reliability and/or performance concerns.

The SMOBC process addresses the tin-lead flow problem by eliminating the use of tin-lead electroplating on the conductor lines under the solder resist. An all-copper PTH printed wiring board is often produced by a tent-and-etch process. This process is one in which the PWB is drilled and plated with electroless copper, which is immediately followed by copper panel plating with the full-thickness copper required for the PTHs. A dry film resist process is often used with a negative phototool (clear conductor lines and pads) to polymerize the resist in only those clear areas of the phototool. This polymerized resist will now protect the lines and PTHs during a copper-etching process which will remove all the unwanted background copper. The photoresist is then stripped off, and the solder resist material is applied and processed through curing. Tin-lead is next added to the open component pads and PTHs by the hot-air leveling process.

An alternative process uses conventional procedures to create a pattern-plated board. After etching, however, the metal etch resist is removed chemically, leaving the underlying copper bare. Subsequent process steps are the same as SMOBC.

The primary function of tin-lead in the PTHs and on the component pads is to improve solderability and appearance. It is important to demonstrate the solderability of the holes and pads on the SMOBC panel. This is accomplished by a hot-air leveling process which places a thin coating of molten tin-lead on only those copper areas of the PTH that have not been covered by the solder resist. This hot-air leveling process proves the ability of the copper surface to be soldered and also improves the appearance and solderability after longer-term storage of the PTHs and pad surfaces.

Since there is no flowable metal under the solder resist during the hot-air leveling step or later during component soldering, the resist maintains its adhesion and integrity.

The lower metallization height of the conductor lines allows the use of a thinner dry film resist and also makes the liquid and screen-printing application somewhat easier.

One variation on the basic SMOBC process is to make the PWB by the conventional pattern-plate copper and tin-lead process followed by etching of the background copper. Then another photoresist step is used to tent the holes and pads so that the tin-lead can be selectively stripped from the conductor lines. This would be followed by infrared or oil reflow, cleaning, and application of solder resist. A second process variation strips off the tin-lead plating completely and is followed by cleaning, solder resist application, and hot-air leveling. There are still other PWB fabricators who do not like either of these processes and are opting to use a nonflowable, copper-etchant resistant metal like tin-nickel under the solder resist. The major shortcoming to tin-nickel is that it is considered more difficult to solder with low-activity soldering fluxes.

34.7   CLEANING AND PWB PREPARATION PRIOR TO SOLDER RESIST APPLICATION

Optimum solder resist performance and effectiveness can be obtained only if the PWB surfaces are properly prepared prior to the application of the resist.

Surface preparation usually consists of a mechanical brush scrubbing for the non–tin-lead PWBs followed by an oven-drying step. The tin-lead PWBs should not be scrubbed and require less aggressive cleaning procedures. The cleaning options prior to solder resist application are shown in Table 34.4.

TABLE 34.4   Preparation for Solder Resist

Operation	Copper	Tin-lead	Other
	\multicolumn{3}{	}{Panel metallization}
Mechanical brush	Yes	No	Yes
Pumice	Yes	No	Yes
Solvent degrease	Yes	Yes	Yes
Chemical cleaning	Yes	Yes	Yes
Oven drying	Yes	Yes	Yes

34.7.1   Surface Preparation

Dry-film and liquid-photopolymer resists applied to meet the IPC-SM-840 Class 3 requirements are particularly sensitive to the cleaning processes and baking step used to remove volatiles prior to the application of the resist.

Tin-lead-coated PWBs should not be mechanically brush-scrubbed because of the smearing potential of such a malleable metal as tin-lead. Scrubbing can cause a thin smear of the metal to be wiped across the substrate, leaving a potentially conductive path or, at a minimum, a decrease in the insulation resistance between the conductor lines. Pumice scrubbing is also unacceptable for tin-lead circuitry, since the pumice particles may become embedded in the soft metal, which can lead to poor soldering performance.

Solvent degreasing with Freon* or 1,1,1-trichloroethane cleaners is necessary with tin-lead PWBs in order to remove the light process oils from the solder reflow step, dirt, and finger-

---

* Registered trademark of E. I. du Pont de Nemours & Company.

prints that are usually found on the PWB at the solder resist step. Solvent degreasing will not remove metal oxides or contaminates that are not soluble in the degreasing media.

The last step prior to the application of a solder resist to the PWB should be an oven-drying step in which absorbed surface moisture and low boiling volatiles are removed. This drying step should immediately precede the resist application step in order to minimize the reabsorption of moisture.

The most stringent performance specifications will in turn require the most stringent cleaning procedures prior to the resist application. Not all resists and performance specifications require the same degree of cleaning. In certain cases, where performance requirements are less stringent, some screen-printed resists may be used successfully without any particular cleaning or oven-drying steps.

34.8   SOLDER RESIST APPLICATIONS

Permanent solder resists may be applied to the PWB by any of several techniques or pieces of equipment. Screen printing of liquid solder resists (ink) is the most common; with regard to photoprint solder resists, the dry film solder resists are applied to one or both sides of the PWB by a special vacuum laminator, and the liquid-photoprint solder resists are applied by curtain coating, roll coating, or blank-screen-printing techniques.

34.8.1   Screen Printing

Screen printing is typically carried out in manual or semiautomatic screen-printing machines using polyester or stainless steel mesh for the screen material. If solder resist is required on two sides, the first side is coated and cured or partially cured and then recycled to apply the solder resist on the second side using the screen pattern for that side, and then the entire PWB is fully cured.

34.8.2   Liquid Photoprint

For some liquid-photoprint solder resists, a screen-printing technique is used to apply the resist in a controlled manner to the surface of the PWB. The screen has no image and serves only to control the thickness and waste of the liquid solder resist. There is no registration of the screen, since there is no image. The actual solder resist image will be obtained by exposing the coated PWB with ultraviolet light energy and the appropriate phototool image. The unexposed solder resist areas defined by the phototool are washed away during the development step.

Some liquid-photoprint solder resists require a highly mechanized process (roller or curtain coating) and therefore, because of the equipment costs and the setup, cleanup, and changeover costs, are best suited to high-volume production.

The liquid solder resists do not tent holes as effectively as the dry film solder resist materials.

34.8.3   Dry Film

Dry-film solder resists are best applied using the vacuum laminators that have been designed for that purpose. The equipment removes the air from a chamber in which the PWB has been placed. The solder resist film is held out of contact from the PWB surface until atmospheric pressure is used to force the film onto one or both sides of the PWB.

The roll laminators that apply the dry film resists for plating or etching are usually found to be unacceptable for solder resist application. The roll laminators were designed to apply a resist to a smooth, flat surface such as copper foil, and not to a three-dimensional surface such as an etched and plated PWB. Air is usually trapped adjacent to the conductor lines as the lamination roll crosses over a conductor line running parallel to the lamination roll. Entrapped air next to the conductor lines can cause wicking of liquids, which in turn causes reliability and/or performance concerns.

The dry film resist thickness, as supplied, is usually 0.003 or 0.004 in and will meet the requirement of the Class 3 specification for a 0.001-in minimum thickness of solder resist on top of the conductor lines. The resist thickness chosen depends on the expected thickness of the copper circuitry to be covered, allowing for filling of the spaces between circuit traces with resist.

34.9 CURING

Once the solder mask has been applied to the PWB, it must be cured according to the manufacturer's recommendations. Typically, curing processes are thermal curing by oven baking or infrared heating, UV curing, or a combination of the two processes. The general objective of the curing process is to remove any volatiles (if present) and to chemically cross link and/or polymerize the solder resist. This curing toughens the resist to help ensure that it will maintain its integrity during the chemical, thermal, electrical, and physical exposure the PWB will see during its service life.

Undercuring, or an out-of-control curing process, is usually the prime cause for solder resist failure. The second leading cause for failure is inadequate cleaning prior to solder resist application.

Special Note—Inspection before Curing.   All PWBs should be carefully inspected for defects prior to curing. Once a solder resist has been cured, it is usually impossible or impractical to strip the resist for rework without seriously damaging the PWB.

34.10 LIQUID PHOTOIMAGEABLE SOLDER RESIST (LPISR)

With the advent of surface-mounting technology (SMT) in the early 1980s, the requirements for tighter registration of solder mask to circuit features have become ever more demanding. As lead pitch of SMT components became finer as illustrated in Table 34.5, conventional thermal and UV curable screen resists could no longer satisfy the requirements to deposit material completely and consistently between the board features, such as adjacent traces or traces and pads. Although dry film solder resist (DFSR) is able to satisfy many of these tighter tolerance requirements, it is expensive and sometimes it has difficulty in coverage of spacings between tightly formed fineline conductors at their base area, leaving small air pockets which tend to erupt during the soldering operation. Therefore, the use of DFSR has declined as SMT has proliferated and liquid photoimageable solder resist (LPISR) has gained acceptance. The usage of DFSR seems to be confined to some special cases when requirements such as hole tenting, small lot size, and thicker mask (3 mil or more) are present.

When Ciba-Geigy introduced the Probimer 52 liquid photoimageable solder resist (LPISR) system in 1978, SMT was not yet in place and most of the printed wiring board (PWB) manufacturers were reluctant to adopt it because the cost of the Probimer system was expensive compared to what they were then using. However, some PWB manufacturers catering to the telecommunication industry started to adopt it because of its excellent corrosion

TABLE 34.5   Standard Design Rules for Surface-Mounted Device (SMD) Pad
(Unit: mm)

Kind of SMD	Number of pins	Pad			Solder mask		Pattern wiring diagram
		Pitch A	Width B	Space C	Clearance D	Width E	
SOP (PLCC)	8 ~ 28	1.27	0.5 ~ 0.6	0.77 ~ 0.67	0.1 ~ 0.15	0.37 ~ 0.57	
			0.6 ~ 0.7	0.67 ~ 0.57	0.1 ~ 0.15	0.27 ~ 0.47	
QFP	64	1.0	0.6	0.4	0.135	0.2	
	80	0.8	0.5	0.3	0.085	0.13	
	100	0.65	0.35	0.3	0.085	0.13	
	48	0.5	0.3	0.2	0.05	0.1	
	224	0.4	0.22	0.18	0.05	0.08	
	300	0.3	0.15	0.15	0.04	0.07	

Source: NEC Corporation.

resistance; then, as SMT started to gain momentum in the mid1980s, other solder resist ink makers saw the value of LPISR and followed Ciba-Geigy into the marketplace. Today, there are a great number of LPISR manufacturers, offering material and process alternatives, and nearly all SMT boards are coated with LPISR.

This section discusses the scope of these LPISR technologies, particularly from the viewpoint of coating methods.

34.10.1   LPISR Makers and Products

Table 34.6 shows a list of major LPISR makers and their representative product line-ups. The reader should be aware, however, that there are many variations to the listed products and they are continuously changing; therefore, this list is presented as a set of examples, rather than a current exhaustive set of alternatives.

After Probimer 52 became successful in the marketplace, many LPISR products were developed. Initially, most of LPISRs were the solvent-developable type (1,1,1-trichloro-ethane) to match the performance of Probimer 52 in corrosion and electrical characteristics. Because of the environmental concerns, however, the use of trichloroethane is discouraged. As a result, nearly all LPISRs marketed today are the aqueous-developable type, with the exception of Probimer 52, 61, and 65, which are solvent-developable (a mixture of nonchlori-nated, biodegradable solvents).

At some time, there were nearly 40 LPISR makers worldwide, but only a dozen seem to remain in the business as listed in Table 34.6. Improvements have been made continuously on these products, as mentioned previously. All products in the table satisfy IPC-840B, Class 3, MIL-P-55110D, and Bellcore requirements. Some makers list only a few products in their gen-

TABLE 34.6   Major LPISR Makers and Their Product Line-up

Maker	Trade name	Products	Coating method	Comments
Ciba-Geigy Ltd	PROBIMER	Probimer 52,61,65	Curtain	Nonchlorinated solvent-developable
		Probimer 71,77	Curtain	Aqueous-developable
		Probimer 74	Screen	Aqueous-developable
Coates Circuit Products	IMAGECURE	AQ-XV500	Screen	Aqueous-developable
		AQ-XV501	Curtain	Aqueous-developable
Dexster	HYSOL	SR-8100 01	Screen	Aqueous-developable
Electra Polymers & Chemicals	CARAPACE	EMP 110	Screen	Aqueous-developable, also curtain and spray-coatable
Enthone-OMI	ENPLATE	DSR-3241 A-G	Screen	US Version of DSR-2200, green
		DSR-3241 A-U	Screen	*ibid.,* clear
		DSR-3300	Screen	*ibid.,* matte finish
Lackwerke Peters	ELPEMER	GL-2461 SM, SM-G	Curtain	"G" designates green
		SD-2461 SM, SM-G	Screen	"G" designates green
		SD-2431 SM	Screen	Red transparent
		SD-2451/2461	Screen	Blue-transparent
		ES-2461 SM	Spray	Electrostatic or conventional spray
Morton Electronic Material	EPIC	EPIC 200	Screen	Aqueous-developable
		EPIC CC-100	Curtain	Aqueous-developable
		LSF 60	Screen	Former Hoechst products
		LSF 60 MATT	Screen	*ibid.,* matte finish
Taiyo Ink Mfg	PHOTO FINER	PSR-4000 H,Z	Screen	Aqueous-developable
		PSR-4000 CC	Curtain	Aqueous-developable
		PSR-4000 SP	Spray	Airless spray
Tamura Kaken	FINEDEL	DSR-2200 (C,F,G)	Screen	Clear, fineline, and green
		DSR-2200 K	Curtain	
		DSR-2200 SP-AL	Spray	Airless
Tokyo Ohka (Lea Ronal)	OPSR	OPSR-5600	Spray	Electrostatic
Toyo Ink	SOLDEREX	SOLDEREX	Curtain	
		D14,D43,D44	Screen	Fast exposure
		SC 40	Spray	
W.R. Grace	ACCUMASK	CM 2001	Spray	Curtain and spray-coatable versions available

eral catalogs, but they offer extensive variations. All of them offer screen-coatable, curtain-coatable and spray-coatable versions.

34.10.2   Coating Methods

It is a standard practice for users of LPISR to select a coating method and the associated equipment, and then select the LPISR, rather than the other way around. Characteristic differences among various LPISRs being small, the selection of a coating method for the requirements of the individual user's goal dictates the success of the LPISR operation.

Once a PWB manufacturer selects a coating method and related equipment, it tends to be locked up with the method, preventing the selection of other options because of the heavy capital investment involved. Therefore, the initial selection of a coating method is very important. For this reason, the focus in this chapter is on the coating method rather than the resist ink itself.

34.10.3   Panel Preparation

Before proceeding with the topic of coating method, a discussion of panel preparation is appropriate because it is the starting point of all processes.

When cleaning copper circuitry for solder mask application, it is important to remove all intermetallic compounds, oxides, and organic and ionic contaminants. A typical cleaning process may consist of the following steps:

- Acid spray rinse (5% hydrochloric acid, for example)
- Water spray rinse
- Mechanical abrasion (jet scrub, pumice scrub, brush scrub, etc.)
- High-pressure water and deionized water rinse
- Dry and optional bake at about 160 to 180°F for 30 min

Mechanical abrasion ensures better adhesion of solder resist to copper and helps resist ink to flow into spacings more naturally by removing sharp conductor edges which sometimes block the smooth flow of resist ink.

34.10.4   Screen Coating

Open screen coating is the simplest entry-level method for most PWB manufacturers, although screen coating can be very sophisticated when the process is to be automated.

34.10.4.1   Single-Sided Screen Coating.   On a worldwide basis, the most popular coating method by far is single-sided open screen coating because it is relatively easy to do successfully and has a low entry cost. By its nature, however, screen coating tends to remove resist ink at the conductor edge that makes the first contact with the squeegee and leave spacings between densely spaced conductors uncovered, whether it is a single-sided or simultaneous double-sided coating to be explained in the next subsection. (Figure 34.3 shows this weakness.) To overcome this weakness, the users of the screen-coating method normally screen the panel twice or even three times to ensure sufficient coverage at conductor edges and spacings, particularly at their bottom area. Users of PWBs usually demand that the thickness of the solder resist be about 0.6 mil (or 0.15 mm) at the edges of the conductors.

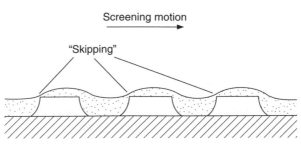

FIGURE 34.3   "Skipping" problem associated with screen coating of LPISR.

Multiple Screening.   When the panel is screened more than once, holes tend to be filled with ink at their entrance, which can cause problems later in the process. Earlier, when LPISR was first introduced, PWB manufacturers used to blot the screen with a piece of paper after two or three panels were screened to eliminate excessive ink buildup around holes. Alternatively, the screen or the table (panel) was moved slightly on the second screening to offset the positions of holes in relation to the previous locations of holes on the screen. These practices made the LPISR process very slow and expensive due to the poor productivity. Modern screening machines for LPISR coating are equipped with various features which minimize ink falling into holes. These include elements such as scrapers, which scrape off excess ink buildup at hole locations at the bottom side of the screen while the squeegee returns to the starting position.

Tent-and-Etch Process Application.   Some high-volume manufacturers make plated-through-hole (PTH) boards with a panel-plate/tent-and-etch process. This tends to yield a flusher surface than the pattern-plating process, using a patterned screen, which totally prevents ink from going into the holes. In such cases, the speed of screening can be as fast as 10 s per panel or 360 panels per hour. Screening the panel twice or three times naturally slows down the process time to 20 to 25 s per panel, resulting in throughput of only 120 to 150 panels per hour at most.

High-Volume Screening.   In high-volume applications, two screening stations are connected in series. After the first side is coated, the panel is semicured for tack-free operation through a convection oven, and the other side is coated and tack-free cured again before exposure.

Screened Ink Curing.   Most resist makers recommend the first-side tack-free cure at about 160°F (70°C) for 15 to 25 min and the second side at about the same temperature for 25 to 30 min.

Some manufacturers screen coat one side, tack-free cure, expose, develop, and repeat the same process for the other side of the panel, and then give a final cure to the panel.

Exposure Alternatives.   Japanese PWB manufacturers prefer exposure of a single side at a time for better registration, while the rest of the world seems to prefer double-side simultaneous exposure. Energy required for exposure ranges from as low as 200 to upward of 1000 mJ, averaging 450 mJ. High-pressure mercury lamps of 5-, 7-, and 8-kW power are popularly used for photoexposure. Depending on the products and lamp intensity, exposure time ranges from 8 to 60 s.

Development of Exposed Material.   Development is done in aqueous solution of 0.8 to 1.2% Na_2CO_3 at a temperature in the vicinity of 85°F (30°C) for all the products except Probimer 52, 61, and 65 which are developed in a nonchlorinated solvent.

Postcure Bake.   Postcure bake is done at about 300°F (150°C) for 30 to 60 min in a convection oven. LPISR makers also recommend IR bump at about 340°F (170°C) for 5 min. Some high-volume producers using fully automated line screen legends right after drying the panel, following the final rinse and before the final cure, and combine the final LPISR cure with the legend ink cure in one shot.

Dealing with Warped Material.   The panel is usually warped after the conductor pattern is formed. Some screening machines are equipped with a clamping mechanism to hold down the warped panel on the screening table.

Screen Coating Advantages/Disadvantages.   The advantage of screen coating over other methods is its ability to block the four peripheral edges of the panel from being coated, thus minimizing the ink waste. However, unlike other methods, the screen must be changed for each different part number and, therefore, setup tends to be time consuming.

34.10.4.2 *Simultaneous Double-Sided Screen Coating.*

This method was conceived by a Japanese PWB manufacturer, Satosen, and the hardware implementation was made by another Japanese company, Toshin Kogyo. Depending on the models, double-sided screening equipment can coat between 120 and 180 panels per hour. These screening machines are equipped with back-side scrapers to prevent resist ink from falling into holes.

There is another maker of double-sided screening equipment: Circuit Automation. The equipment made by this company is essentially the same as the Japanese machine. However, the Circuit Automation machine, sold under the trade name of DP series, avoids ink getting into holes by moving the screen slightly and changing the skew angle of the squeegee on the second screening.

Double-Sided Screen Coating Advantages/Disadvantages.   One major advantage of the double-sided simultaneous screening method is that the panel receives tack-free curing only once, and therefore the degree of cure on both sides of the panel is equal while single-sided screening makes the first-side cure more than the second side, and, in some extreme cases, the colors of the two sides may become different.

Double-sided screen coating has one weakness in that it is difficult to screen thin panels. To overcome this problem, the makers of such equipment provide a special frame to mount thin panels, which gives tension to the panel and makes screening possible. However, the throughput by the thin-panel version is somewhat reduced with such a scheme due to the more-intricate panel-mounting scheme.

Basic Process.   The conditions for tack-free curing, exposure, development, and final bake in the case of double-sided screening are essentially the same as for single-sided screening. Over 300 double-sided screen coaters have been installed worldwide.

34.10.5 Curtain Coating

Ciba-Geigy introduced the first curtain coating system, Probimer 52, in 1978. Coates Circuit Products of the United Kingdom followed. Years later, Maas of Germany started to offer curtain-coating equipment, but no resist ink. Figure 34.4 shows a typical fully automated configuration of a curtain-coater line.

The speed of the belt conveyor under the curtain of resist ink determines the thickness of the coating. One pass is sufficient to secure the required coating thickness. The quality of coverage by curtain coating is excellent. Curtain coating can accommodate mixed panel sizes of different thicknesses without any setup change, but it can coat only one side of the panel at a

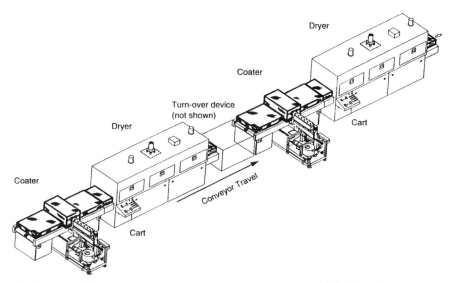

FIGURE 34.4   Typical fully automated configuration of a curtain-coater line.

time. Because of the nature of coating, the entire panel surface gets coated. To provide room for tooling holes and test coupons, 15 to 25 percent of the panel area is used as trimming, and resist ink coated on this trimming area is wasted in curtain coating. That is, "useful" ink utilization in curtain coating is usually between 75 and 85 percent. However, the newer curtain coaters have the provision to block two edges of the panel from being coated, thus improving resist ink utilization.

In the early days of curtain coaters, it was difficult to coat thin panels since the leading edge of a thin panel tends to droop down at the end of the fast conveyor belt under the curtain. However, this problem is overcome by a flipping mechanism provided at the end of the conveyor belt.

To avoid the "lap-around" effect of resist ink at the leading edge of the panel, the panel is usually fed into the curtain at a slightly skewed angle. Curtain coating can process typically 180 panels per hour; a high-productivity model can handle in excess of 300 panels per hour.

The process after coating is more or less the same as in screen coating. Newer types of tack-free curing ovens are made much shorter than the original one and the entire length of a fully automated line with two curtain coaters in series is no more than 60 to 70 ft.

34.10.6 Spray Coating

Spray coating is practiced in three different modes:

1. Electrostatic spray
2. Airless spray
3. Air spray (very small amount of air is mixed with ink)

There are two variations to spray coating:

1. Vertical spray
2. Horizontal spray

34.10.6.1 Vertical Spray Systems. When it comes to the arrangement of spray guns (atomizers), there are also a few variations. In one variation, a single gun sways sideways back and forth, perpendicular to the direction of panel travel. In a second single-gun system, the gun is stationary. In other coaters, two stationary guns are arranged in staggered position. Each one of these gun arrangements has its strength and weaknesses.

In electrostatic spray systems, effective grounding is essential for good results. Also, it is important to keep a distance of about half an inch between adjacent panels. If the adjacent panels get closer than this clearance, sparks may be induced at the edges of these panels, leaving uncoated spots.

In such a system, the typical coating speed is about 340 panels per hour.

34.10.6.2 Horizontal Spray Systems. All horizontal systems coat one side at a time. In most such systems, the panel is tack-free cured before being coated on the second side. In some systems, however, the panel is carried on V belt. After the first side is coated, the panel is flipped and the other side gets coated. In this case, tack-free curing is done only once, but it is difficult to process thin panels in such a system. Thin panels are usually spray coated vertically.

34.10.6.3 Overspray. Unlike screen and curtain coating, spray coating creates an overspray. There are various ways to treat oversprayed ink. In one system, a roll of paper, 800 to 1000 yards long, is used to absorb oversprayed ink. When the entire length of paper is used up, it is removed and treated for waste disposal. In other systems, a container tank is provided underneath the carrying belt to collect oversprayed ink. When the tank gets filled to a certain level (about once every two to three months), chemicals are added to coagulate the ink, and the coagulated ink is carted away for waste disposal.

Curtain-coating and spray-coating methods can process panels of different sizes and shapes without special setup, which is an advantage in dealing with small lot sizes. On the other hand, when panels of mixed sizes are passed through a spray coater, ink utilization can be very poor—as low as 40 percent. Some resist ink manufacturers claim oversprayed ink can be used again by adding solvent, but in reality, reutilization of oversprayed ink is not done because such solvents are not a part of ink formulation and are not compatible with the main ingredients.

Spray coating gives the best conformal coverage on the peaks and valleys of the panel surface. Because of this, however, spray coating can create skips in subsequent legend screening when valleys are too low, as illustrated in Fig. 34.5. After years of struggle, ink makers have now corrected this "skipping" problem by formulating spray inks to fill the valleys high.

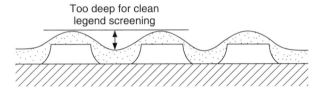

FIGURE 34.5 Spray coating can create good conformity with circuits, which, however, can cause skipping problems in legend screening.

34.11 TENTING HOLES

All coating methods described so far have one problem in common. None of the methods can effectively and reliably tent holes. Closely placed feed-through holes can cause bridging at the time of soldering. Flux can be entrapped in small holes. Therefore, some users of PWBs demand that the holes be tented or filled with solder mask.

Dry film solder mask can provide tenting, but it is usually expensive and often fails to fill the narrow valleys between tightly spaced conductors. To overcome this difficulty, Du Pont came up with a solution with the VALU system, in which the panel is first coated thinly with LPISR and then dry film solder mask is laminated on top of it.

When hole tenting is required in conjunction with screen, curtain, or spray coating, holes are filled with epoxy-based ink by screening after the panel is coated with LPISR. Some PWB manufacturers fill the holes first before coating with LPISR. Such processes add extra cost, but there seems to be no better alternative way to accomplish hole tenting.

34.12 ELECTROLESS NICKEL/GOLD PLATING
ISSUES FOR SOLDER RESISTS

Hot-air leveling (HAL) is the most popular surface finish for SMT. As the density of SMT boards becomes higher, first-pass assembly yields tend to be poorer and repair is very costly and error prone. In recent years, electroless nickel/gold finish has been gaining popularity as an alternative finish to HAL. MLBs for cellular telephone applications and a large portion of PCMCIA cards (now called PC cards) are finished with electroless Ni/Au plating, which provides excellent protection against oxidation before the soldering operation. Boards with Ni/Au finish can withstand a few soldering cycles without oxidation. Initially, some LPISRs could not withstand the electroless Ni/Au plating operation. Improvements have been made on most LPISRs available in the market to accommodate electroless Ni/Au plating. If not suitable for electroless Ni/Au plating, LPISR makers offer versions of their resist inks that can satisfy the plating requirement.

CHAPTER 35
MACHINING AND ROUTING

James Cadile
NOVA Drilling Service Inc., Santa Clara, California

Leland E. Tull
Contouring Technology, Mountain View, California

Charles G. Henningsen
Insulectro, Mountain View, California

35.1 INTRODUCTION

Laminate machining consists of the mechanical processes by which circuit boards are prepared for the vital chemical processes of image transfer, plating, and etching. Such processes as cutting to size, drilling holes, and shaping have major effects on the final quality of the printed board. This chapter will discuss the basic mechanical processes which are essential to producing the finished board.

35.2 PUNCHING HOLES (PIERCING)

35.2.1 Design of the Die

It is possible to pierce holes down to one-half the thickness of XXXPC and FR-2 laminates and one-third that of FR-3 (Fig. 35.1). Many die designers lose sight of the fact that the force required to withdraw piercing punches is of the same magnitude as that required to push the punches through the material. For that reason, the question of how much stripper-spring pressure to design into a die is answered by most toolmakers: "as much as possible." When space on the dies cannot accommodate enough mechanical springs to do the job, a hydraulic mechanism can be used. Springs should be so located that the part is stripped evenly. If the board is ejected from the die unevenly, cracks around holes are almost certain to occur. Best-quality holes are produced when the stripper compresses the board an instant before the perforators start to penetrate. If the stripper pressure can be made to approach the compressive strength of the material, less force will be required and the holes will be cleaner.

If excessive breakage of small punches occurs, determine whether the punch breaks on the perforating stroke or on withdrawal. If the retainer lock is breaking, the cause is almost certain to be withdrawal strain. The remedy is to grind a small taper on the punch, no more than

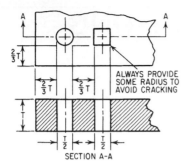

FIGURE 35.1 Illustration of the proper sizing and locating of pierced holes with respect to one another and to the edge of paper laminates. Minimum dimensions are given as multiples of laminate thickness t.

1½ in and to a distance no greater than the thickness of the material being punched. If the grinding is kept within those limits, it will have no measurable effect on hole quality or size. The other two causes of punch breakage are poor alignment, which is easily detected by close examination of the tool, and poor design, which usually means that the punch is too small to do the job required.

35.2.2 Shrinkage of Paper-Base Laminates

When paper-base laminates are to be punched, it must be remembered that the materials are resilient and that their tendency to spring back will result in a hole slightly smaller than the punch which produced the hole. The difference in size will depend on the thickness of the material. Table 35.1 shows the amount by which the punch should exceed the print size in order to make the holes within tolerance. The values listed should not be used for the design of tools for glass-epoxy laminates, the shrinkage of which is only about one-third that of paper-base materials.

TABLE 35.1 Shrinkage in Punched Hole Diameters, Paper-Base Laminates

Material thickness	Material at room temp.	Material at 90°F or above
¹⁄₆₄	0.001	0.002
¹⁄₃₂	0.002	0.003
³⁄₆₄	0.003	0.005
¹⁄₁₆	0.004	0.007
³⁄₃₂	0.006	0.010
⅛	0.010	0.013

35.2.3 Tolerance of Punched Holes

If precise hole size tolerance is required, the clearance between punch and die should be very close; the die hole should be only 0.002 to 0.004 in larger than the punch for paper-base materials (Fig. 35.2 and Table 35.2). Glass-base laminates generally require about one-half that tolerance. Dies have, however, been constructed with as much as 0.010 all-around clearance between punch and die. They are for use where inspection standards permit rough-quality holes.

A die with sloppy clearances is less expensive than one built for precision work, and wide clearance between punch and die causes correspondingly more break and less shear than a tight die will cause. The result is a hole with a slight funnel shape that makes insertion of components easier. Always pierce with the copper side up. Do not use piercing on designs with circuitry on both sides of the board, because lifting of pads would probably occur.

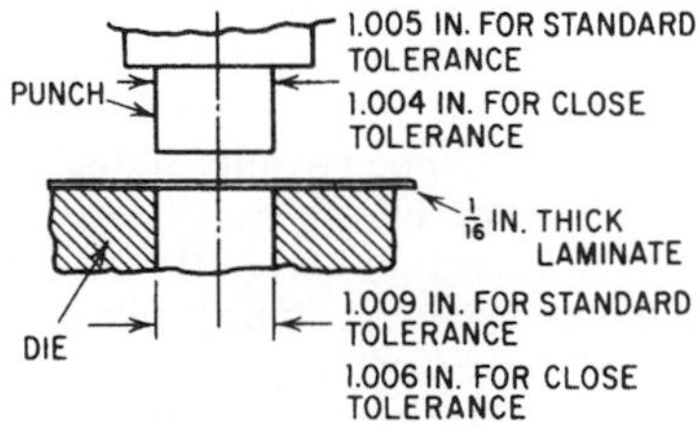

PUNCH — 1.005 IN. FOR STANDARD TOLERANCE
1.004 IN. FOR CLOSE TOLERANCE
¹⁄₁₆ IN. THICK LAMINATE
DIE — 1.009 IN. FOR STANDARD TOLERANCE
1.006 IN. FOR CLOSE TOLERANCE

FIGURE 35.2 Example of proper tolerance of a punch and die.

TABLE 35.2 Tolerances for Punching or Blanking Paper-Base Laminate

Material thickness	Base material	Tolerance on hole size, in	Tolerances, in, on distance between holes and slots, 90°F				Tolerances for blanked parts, overall dimension, in
			Up to 2 in	2 to 3 in	3 to 4 in	4 to 5 in	
To and including ⅟₁₆ in	Paper	0.0015	0.003	0.004	0.005	0.006	0.003
Over ⅟₁₆ in to and including ³⁄₃₂ in	Paper	0.003	0.005	0.006	0.007	0.008	0.005
Over ³⁄₃₂ in to and including ⅛ in	Paper	0.005	0.006	0.007	0.008	0.009	0.008

35.2.4 Hole Location and Size

Designs having holes whose distance from the edge of the board or from other holes approaches the thickness of the material are apt to be troublesome. Such designs should be avoided; but when distances between holes must be small, build the best die possible. Use tight clearance between punch and die and punch and stripper, and have the stripper apply plenty of pressure to the work before the punch starts to enter. If the distance between holes is too small, cracks between holes may result even with the best of tools. If cracks between holes prove troublesome, plan the process so that the piercing is done before any copper is etched away. The reinforcing effect of the copper foil will help eliminate cracks. Most glass-epoxy laminates may be pierced, but the finish on the inside of the holes is sometimes not suitable for through-hole plating.

35.2.5 Warming Paper-Base Material

The process of punching paper-base laminates will often be much more trouble-free if the parts are warmed to 90 or 100°F. That is true even of the so-called cold-punch or PC grades. Do not overheat the material to the point at which it crumbles and the residue is not ejected as a discrete slug. Overheated material will often plug the holes in the die and cause rejects. Opening the taper on the takeaway holes will reduce plugging, but the most direct approach is to pierce at a lower temperature. Glass-epoxy is never heated for piercing or blanking.

35.2.6 Press Size

The size of the press is determined by the amount of work the press must do on each stroke. The supplier of copper-clad sheets can specify a value for the shear strength of the material being used. Typically, the value will be about 12,000 lb/in² for paper-base laminate and 20,000 lb/in² for glass-epoxy laminate. The total circumference of the parts being punched out multiplied by the thickness of the sheet gives the area being sheared by the die. If all dimensions are in inches, the value will be in square inches. For example, a die piercing 50 round holes, each 0.100 in in diameter, in 0.062-in-thick laminate will be shearing, in square inches:

$$50 \times 0.100 \text{ in} \times 3.1416 \times 0.062 \text{ in} = 0.974 \text{ in}^2$$

If the paper-base laminate has 12,000-lb/in² shear strength, 11,688 lb of pressure, or about 6 tons, is required just to drive the punches through the laminate. Bear in mind that, if a spring-loaded stripper is used, the press will also have to overcome the spring pressure, which ought to be at least as great as the shear strength. Therefore, a 12-ton press would be the minimum which could be considered. A 15- or 20-ton press would be considerably safer.

35.3 BLANKING, SHEARING, AND CUTTING OF COPPER-CLAD LAMINATES

35.3.1 Blanking Paper-Base Laminates

When parts are designed to have shapes other than rectangular and the volume is great enough to justify the expense of building a die, the parts are frequently punched from sheets by using a blanking die. A blanking operation is well adapted to paper-base materials and is sometimes used on glass-base ones.

In the design of a blanking die for paper-base laminates, the resilience, or yield, of the material previously discussed under Piercing applies. The blanked part will be slightly larger than the die which produced it, and dies are therefore made just a little under print size depending on the material thickness. Sometimes a combination pierce and blank die is used. The die pierces holes and also blanks out the finished part.

When the configuration is very complex, the designer may recommend a multiple-stage die: The strip of material progresses from one stage to the next with each stroke of the die. Usually in the first one or two stages, holes are pierced, and in the final stage, the completed part is blanked out.

The quality of a part produced from paper-base laminates by shearing, piercing, or blanking can be improved by performing the operation on material which has been warmed. Caution should be exercised in heating over 100°F because the coefficient of thermal expansion may be high enough to cause the part to shrink out of tolerance on cooling. Paper-base laminates are particularly anisotropic with respect to thermal expansion; that is, they expand differently in the x and y dimensions. The manufacturer's data on coefficient of expansion should be consulted before a die for close-tolerance parts is designed. Keep in mind that the precision of the manufacturer's data is probably no better than ±25 percent.

35.3.2 Blanking Glass-Base Laminates

Odd shapes that cannot be feasibly produced by shearing or sawing are either blanked or routed. Glass blanking is always done at room temperature. Assuming a close fit between punch and die, the part will be about 0.001 in larger than the die which produced it. The tools are always so constructed that a part is removed from the die as it is made. It cannot be pushed out by a following part, as is often true when the material has a paper base. If material thicker than 0.062 in is blanked, the parts may have a rough edge.

The life of a punch, pierce, or blank die should be evaluated with reference to the various copper-clad materials that may be used. One way to evaluate die wear caused by various materials is to weigh the perforators, or punches, very accurately, punch 5000 pieces, and then reweigh the punches. Approximately 5000 hits are necessary for evaluation, because the initial break-in period of the die will show a higher rate of wear. Also, of course, the quality of the holes at the beginning and end of each test must be evaluated. Greatly enlarged microphotos of the perforator can be used for visual evaluation of changes in the die.

35.3.3 Shearing

When copper-clad laminates are to be sheared, the shear should be set with only 0.001 to 0.002 in clearance between the square-ground blades (Fig. 35.3). The thicker the material to be cut, the greater the rake or scissor angle between the top and bottom shear blade. The converse also is true: The thinner the material, the smaller the rake angle and the closer the blades. Hence, as in many metal shears, the rake angle and the blade gap are fixed; the cutoff piece can be twisted or curled. Paper-base material can also exhibit feathered cracks along the edge that are due to too wide a gap or too high a shear angle. That can be minimized by supporting both piece and cutoff piece during the shear operation and decreasing the rake angle. Epoxy-glass laminate, because of its flexural strength, does not usually crack, but the material can be deformed if the blade clearance is too great or the shear angle is too large. As in blanking, the quality of a part produced from paper-base laminates by shearing can be improved by warming the material before performing the operation.

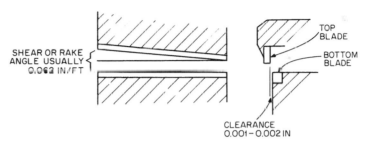

FIGURE 35.3   Typical adjustable shear blades for copper-clad laminates.

35.3.4 Sawing Paper-Base Laminates

Paper-base laminates are much harder on sawing tools than are the hardest woods, and therefore a few special precautions are necessary for good saw life. Sawing paper-base laminates is best accomplished with a circular saw with 10 to 12 teeth per inch of diameter at 7500 or 10,000 ft/min. Hollow-ground saws give a smoother cut; and because of the abrasive nature of laminated materials, carbide teeth are an excellent investment. (See Fig. 35.4 for tooth shape.)

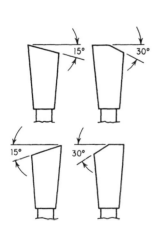

When a saw does not last long enough between sharpenings, use the following checklist. (These steps could have a cumulative effect and change saw life by a factor of 4 to 5.)

1. Check the bearings for tightness. There should be no perceptible play in them.
2. Check the blade for runout. As much as 0.005 in can be significant.
3. When carbide teeth are used, inspect them with a magnifying glass to make sure a diamond tool no coarser than 180 grit was used in sharpening them.
4. If the saw has a thin blade, use a stiffening collar to reduce vibration.
5. Use heavy pulleys with more than one V belt. Rotating parts of the system should have sufficient momentum to carry the sawtooth through the work smoothly and without variation of speed.
6. Check the alignment of the arbor and the motor mounting.

FIGURE 35.4   Commonly used sawtooth designs for paper and cloth laminates. At left, two successive teeth on a 15° alternate-bevel saw. At right, two successive teeth on a 30-ft alternate-corner-relieved (AC-30) saw.

All these steps are intended to reduce or eliminate vibration, which is the greatest enemy of the saw blade. If vibration is noticed, find the source and correct it.

35.3.5 Sawing Glass-Base Laminates

When glass-base laminates are to be sawn, carbide-tooth circular saws can be used; but unless the volume of work is quite low, the added investment required for diamond-steel-bonded saws will be paid for in future savings. The manufacturer's recommendation of saw speed should be followed; usually it will be for a speed in the neighborhood of 15,000 ft/min at the periphery of the saw blade. When economics dictate the use of carbide-tooth circular saws for cutting glass, use the instruction previously given for paper-base laminates (see Fig. 35.4 for tooth shape) and remember that each caution regarding runout, vibration, and alignment becomes more important when glass-reinforced laminates are sawn.

35.4 ROUTING

Modern circuit board fabricators rely principally on routing to perform profiling operations. The high cost and extended lead times for blanking dies, combined with the problem of design inflexibility of hard tooling, limit the punching operation generally to very high volumes or designs specific to die applications. Shearing or sawing are limited to rectangular shapes and generally are not accurate enough for most board applications.

In the modern circuit board fabrication industry, rapid response to customer lead times and economies of universal process application are well served by routing, especially multiple-spindle *computer numerically controlled* (CNC) routing

Routing consists of two similar, yet vastly different fabrication processes:

1. CNC multiple-spindle routing
2. Manual pin routing

The similarities consist of the use of high-speed spindles, utilizing carbide cutting tools, and generating high cutting rates.

35.4.1 Pin Routing

Pin routing is a manual routing process utilizing a template which has been machined of aluminum, FR-4 laminate, or a fiber-reinforced phenolic. The template is made to the finished board dimensions and has tooling pins installed to register to the board's tooling holes. The package (which can have up to four pieces in a stack) is routed by tracking the template against a pilot pin protruding from the router table. The pin height is less than the template thickness. Usually, the machine pilot pin is the same diameter as the router bit and can be offset adjusted to give the operator flexibility in optimizing dimensions. Work should be fed against the rotation of the cutter to prevent the cutter from grabbing.

Pin routing can be an economical process when a small generation of boards is profiled, or if the shapes required are relatively simple. For pin routing to be effective, generally a very skilled operator is required to fabricate the template and to route the boards. Outside machine shops can build aluminum route fixtures for each customer application; however, lead times and costs per order must be considered. Pin routing is usually used by small shops not able to invest in the CNC equipment and its associated support, or as a specialty process, off-line from CNC routing.

In the best pin-routing operations, the volumes produced cannot be compared with multiple-spindle routing.

35.4.2 CNC Routing Applications

The applications of NC routing extend well beyond merely cutting board profiles. The ability to produce boards in multiple-image modules reduces handling, not only in the board shop, but in every subsequent operation from packing, component assembly, wave solder, and test. This is of special value when dealing with a postage-stamp-size part or wire-bondable gold surfaces. Where handling must be minimized, the module acts as a pallet throughout these operations. In addition, unusual or irregular shapes, small or large, can be palletized to simplify handling and conveyance. In Fig. 35.5, examples of tab-routed, or multiple-image, modules are shown.

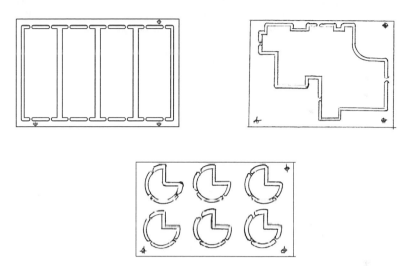

FIGURE 35.5   Examples of tab-routed, or multiple-image, modules.

In Fig. 35.6, each part is shown attached to the frame with breakable or removable tabs and can include features permitting tab removal below the board edge of the image periphery. Note on Fig. 35.6 the use of a score line to ease tab removal. Scoring, alone or in concert with routing (as discussed in Sec. 35.5) can play a large role in board and pallet separation.

Beyond multiple-image palletization, as shown in Fig. 35.7, CNC routing can provide a variety of board requirements including:

1. Internal cutouts
2. Slots
3. Counterbores
4. Board edge conditioning for plating

There are many benefits available through CNC routing beyond those of efficient part profiling. A little planning before the release to production can improve manufacturability as well as provide many no-cost benefits to assembly, soldering, etc.

35.4.3 Computer Numerical Controlled (CNC) Operation

CNC router equipment has the ability to process high volumes of circuit boards very accurately and economically, yet it is coupled with features to enable quick program and setup. This coupling enables the same processing used for high volume to be utilized for prototypes and short-lead-time production.

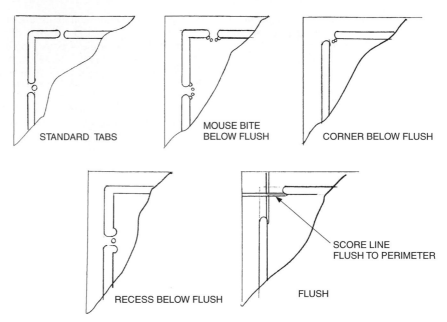

STANDARD TABS

MOUSE BITE
BELOW FLUSH

CORNER BELOW FLUSH

RECESS BELOW FLUSH

FLUSH

SCORE LINE
FLUSH TO PERIMETER

FIGURE 35.6 Each part is shown attached to the frame with breakable or removable tabs. A variety of breakable tabs is shown. Also shown is a score line for ease of tab removal.

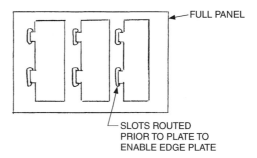

FULL PANEL

SLOTS ROUTED
PRIOR TO PLATE TO
ENABLE EDGE PLATE

FIGURE 35.7 Multiple-image palletization.

With circuit board data files so universally available, the part programming time has dropped to a few minutes, as opposed to the hours it once took, while setups remain at about 15 to 20 minutes, plus cutter labyrinth and first article routing.

Router operation consists of multiple spindles (2 to 5), capable of operating from 6,000 to 36,000 r/min, or more. The router path (x-y table movement and spindle plunge and retract) is determined by program. This permits any number of paths and any location.

The preferred method of registration of the panel to the machine table is to use the full panel and the tooling holes previously drilled. Tooling holes internal to the part provide for the highest accuracy, although manufacturing panel tooling may be used if considered earlier in the process.

35.4.4 Cutter Offset

Since the cutter must follow a path described by its centerline, it must be offset from the desired board edge by an amount equal to its effective radius. This is the basic cutter radius, and it will vary with the cutter tooth form. Most newer-generation equipment will automatically adjust for the cutter radius. However, either manually or automatically, this is a basic element of routing planning. Since the cutters deflect during the routing operation, it is necessary to determine the amount of deflection to be added to the basic cutter radius before expending large amounts on programming parts (Fig. 35.8). Cutter compensation values can be varied.

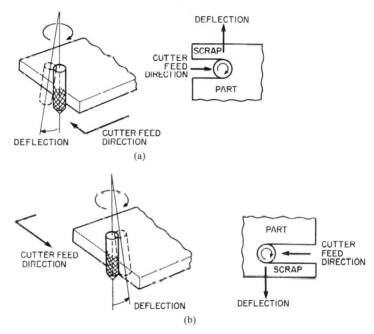

FIGURE 35.8 Effect of cutter deflection on part size and geometry. (*a*) Clockwise cutting (recommended for outside cuts) deflects cutter away from part. That leaves outside dimensions large on first pass unless compensated for in programming. (*b*) Counterclockwise cutting (recommended for inside cuts and pockets) deflects cutter into scrap. Therefore, inside dimensions of holes or cutouts will measure small unless compensated for in programming.

Variables which affect deflection are thickness, type of material, direction of cut, feed rate, and spindle speed. To reduce those variables, the manufacturer should:

1. Standardize on cutter bit manufacturer, selected diameters, tooth form, and end cut.
2. Fix spindle speed (24,000 r/min recommended for epoxy-glass laminates).
3. Rout in clockwise direction on outside cuts, counterclockwise on inside cuts.
4. Standardize on single or double pass.
5. Fix feed rates for given materials. (Note that higher rates will increase part size and slower feed rates will decrease part size.)
6. Develop documented process controls after experimenting with varied parameters.

35.4.5   Direction of Cut

A counterclockwise direction of feed (climb-out) will leave outside corners with slight projections and inside corners with small radii. A clockwise direction of feed (rake cut) will give outside corners a slight radius, and perhaps give inside corners a slight indentation. These irregularities may be minimized by reducing the feed rate or cutting the part twice.

35.4.6   Cutter Speed and Feed Rate

The variables affecting cutter speed are usually limited to the type of laminate being cut and the linear feed rate of the cutter. A cutter rotation of 24,000 r/min and feed rates up to 150 in/in may be used effectively on most laminates, although cutter feed direction may require a lower feed rate. Teflon-glass and similar materials, the laminate binder of which flows at relatively low temperatures, require slower spindle speeds (12,000 r/in) and high feed rates (200 in/in) to minimize heat generation. The graph in Fig. 35.9 shows recommended feed rates and cutter offsets for most standard laminates at various stack heights. The cutter used is a standard ⅛-in-diameter burr type.

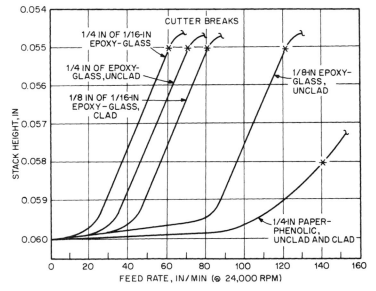

FIGURE 35.9   Recommended feed rate, using ⅛-in-diameter burr cutter at 24,000 r/min, for varying stack heights of specific thickness of material.

35.4.7   Cutter Bits

Because of the precise control of table movement in NC routing, cutter bits are not subjected to shock encountered in pin routing and stylus routing, and therefore small-diameter cutters may be used successfully. However, the fabricator would do well to standardize on ⅛-in-diameter cutters because they are suitable for most production work and are readily available from a number of manufacturers in a variety of types. The resulting 0.062-in radius on all inside corners is usually acceptable if the board designer is aware of it.

Cutter tooth form is more important in NC than in other routing. Because of the faster feed rates possible, it is important that a cutter have an open tooth form that will release the chip easily and prevent packing. Many standard diamond burrs available on the market will load with chips and fail rapidly. The carbide cutting bit will normally cut in excess of 15,000 linear inches of epoxy-glass laminate before erosion of the teeth renders the cutter ineffective or too small.

If extremely smooth edges are required, a fluted cutter may be used. Single- or two-flute cutters with straight flutes should be used when cutting into the foil if minimum burring is desired. It should be noted that such cutters will be more fragile than a standard serrated cutter, and feed speeds should be adjusted accordingly. When a slightly larger burr can be tolerated, two- and three-flute left-hand spiral cutter bits should be used because of their greater strength. The left-hand spiral will force the work piece down rather than lift it, assuming a right-hand turning spindle.

35.4.8   Tooling

To simplify tooling and expedite loading and unloading operations, effective hold-down and chop-removal systems should be provided as part of the machine design. Various methods may then be devised to mount the boards to the machine table while properly registering them to facilitate routing the outline. Some machine designs will have shuttle tables available so that loading and unloading may be accomplished while the machine is cutting. Others will utilize quick-change secondary tooling pallets or subplots that allow rapid exchange of bench-loaded pallets with only a few seconds between boards.

35.4.8.1   Tooling Plates   Tooling plates utilize bushings and a slot on the centerline of the active pattern under each spindle. They are doweled to the machine table (Fig. 35.10). The plates may be made by normal machine shop practice, or the router may be used to register and drill its own tooling plate. Mounting pins in the tooling plate should be light slip fit.

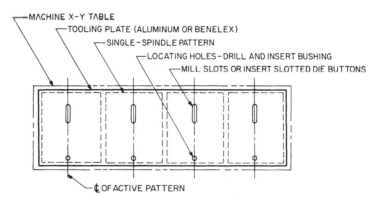

FIGURE 35.10   Typical tooling of numerically controlled routing.

35.4.8.2   Subplates.   Subplates should be made of Benelax, linen phenolic, or other similar material. Subplates should have the pattern to be routed cut into their surfaces. The patterns act as vacuum paths and aid in chip removal. Part-holding pins should be an interface fit in subplates and snug to loose fit in the part, depending on cutting technique used (Fig. 35.11).

It is recommended that the programmer generate the tooling and hold-down pinholes in addition to the routing program. That will provide absolute registration between the tooling holes and the routing program.

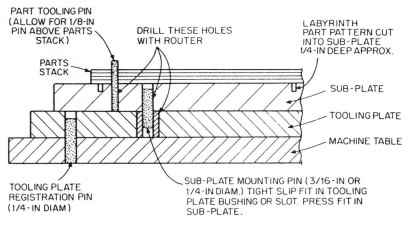

FIGURE 35.11   Tooling schematic for numerically controlled routing.

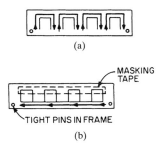

FIGURE 35.12   No-internal-pin method. Step 1: Cut three sides (*a*). Step 2: Apply masking tape (*b*). Step 3: Cut parts away.

35.4.9 Cutting and Holding Techniques

Since the precision required for cutting board outlines, as well as the placement of tooling holes for registering boards, will vary, a number of different cutting and holding methods may be used. Three basic methods are illustrated here. Experimentation will determine which method or combination of methods is most applicable to a particular job. With all methods, the minimum dimension for board separation with a 0.125-in cutter is 0.150 in.

1. *No-internal-pin method:*   If no internal tooling pins are used, the procedure of Fig. 35.12 may be employed, but it is normally used only when no other method is possible. Characteristics of this method are as follows:

Accuracy:	±0.005 in
Speed:	Slow (best used with many small parts on a panel)
Load:	One panel high for each station

2. *Single-pin method:*   The single-pin method is illustrated in Fig. 35.13. Characteristics of this method are as follows:

Accuracy:	±0.005 in
Speed:	Fast (quick load and unload)
Load:	Multiple stacks

3. *Double-pin method:*   In the two-pin method, there is a double pass of cutter offset; see Fig. 35.14. Make two complete passes around each board, the first pass at a recommended feed rate and the second at 200 in/min. Remove scrap after the first pass. Characteristics of this method are as follows:

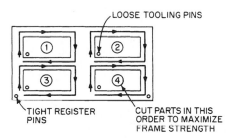

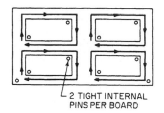

FIGURE 35.13 Single-pin method.   **FIGURE 35.14** Double-pin method.

Accuracy: ±0.002 in

Speed: Fast (highest-accuracy system—loads and unloads slower than single-pin method due to tight pins)

Load: Multiple stacks

35.5 SCORING

Scoring is a circuit board fabrication method used to make long, straight cuts quickly, and therefore is often used to create rectangular profile board shapes. More commonly, however, it is used in concert with CNC routing for complex shapes, enabling each tool to be used to its unique advantage. When used with routing, scoring has a much wider application and can provide simple breakaway for complex profiles (Fig. 35.15).

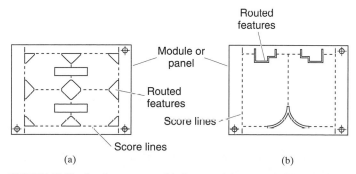

FIGURE 35.15 Scoring processes: (*a*) shows module or panel with typical scoring lines and routed corners; (*b*) shows scoring lines and routed complex features.

35.5.1 Scoring Application

Scoring is accomplished by machining a shallow, precise, V-groove into the top and bottom surfaces of the laminate, generally with the use of CNC equipment. The two most significant elements of the score line are as follows:

1. The positional accuracy from the reference feature (usually the registration hole)

2. The depth of the score, which determines the web thickness

The final edges of a scored circuit board are yielded by breaking the panel, or border, at the score line (Fig. 35.16). The angle of the cutting tool is reflected in the V-groove geometry, and limiting this angle to 30 to 90° will minimize the score line intrusion into traces near the edges of the circuit board. The score line exposes the laminate glass fibers and resin. Measurements from these surfaces will vary greatly, even though the score line is precisely machined (Fig. 35.13). These irregular surfaces will be noticed as dimensional growth and should be considered in design or planning when designated as a scored edge.

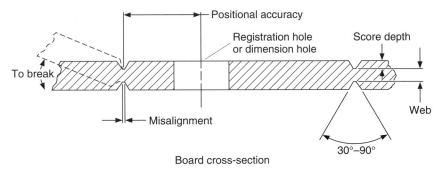

FIGURE 35.16 Cross section of board, showing finished V-groove and break.

The dimensional accuracy of the final board is determined by the degree of precision with which the following are performed:

1. Misalignment within ±0.003 in of the score line from the desired location
2. Web thickness within ±0.006 in of the designated dimension

Typically, a nominal web thickness is 0.020 in for 0.060-in-thick FR-4 boards and 0.014 in for 0.030-in-thick FR-4 laminate. For CEM-1 or CEM-3 materials, 0.040 in and 0.024 in nominal web thickness apply, respectively. These web thickness values enable sufficient module strength to avoid accidental or premature score separation, while providing simple breaking efforts without excessive edge roughness or growth.

35.5.2 Operation

Two major types of panel scoring systems are available:

1. Dedicated CNC scoring machines utilizing high-speed carbide or diamond-embedded cutter blades, operating as a pair, one on each side of the board. This generates the V-groove on each side simultaneously

2. Drills or driller/routers equipped with scoring software and spade type carbide bits, generating score lines on one side of the panel at a time (Fig. 35.14).

35.5.2.1 Dedicated Scoring Equipment. The dedicated CNC scoring machines are high-production, precision computer-driven machines. With an exception or two, they utilize blade-type cutting tools of all-carbide or with carbide inserts, as well as diamond-embedded varieties, and are designed to self-center the panel.

The panel feed rate is high due to the blade's ability to operate at high-surface-feet cutting rates. Scoring both sides simultaneously with one pass contributes greatly to elevated production processing. This equipment utilizes pin or edge registration, with positioning of score lines and steps by programmed instructions. The vertical adjustment of the cutter blades permits variation in V-groove depth and on many models *jump scoring* is available. The ability to do jump scoring, or score/no-score segments along a simple line at desired points is programmable.

35.5.2.2   Multiple-Role Machines.   Scoring with CNC drillers or driller/router machines equipped with scoring software produces score lines only on one side of the panel per machine cycle, although each spindle can be used. The panel and program data must be flipped to score the second side. The panel registration method is similar to that of routing, using existing tooling holes to pin the panel to the machined tooling plates. The tooling plates must be machined flat to assure uniformity of score depth. Brush-type spindle pressure foot inserts should be used to apply downward pressure during score line cutting. Spade-type carbide tools of various angles and configurations are used in the spindles. Typically, multiple passes (two or three) may be required to produce a clean, uniform score line.

CHAPTER 36
PROCESS CAPABILITY AND CONTROL

Timothy A. Estes
*Conductor Analysis Technologies, Inc., Albuquerque,
New Mexico*

Ronald J. Rhodes
*Conductor Analysis Technologies, Inc., Green Brook,
New Jersey*

36.1 INTRODUCTION

Pressures to reduce package sizes and increase electrical performance have mandated technological developments in the design and manufacture of printed circuit boards. Development efforts have been focused primarily on smaller sizes of features such as conductors, spaces, via hole diameters, and via land diameters. As their sizes become smaller, features are inherently more difficult to manufacture, and a greater number of vias and a greater length of conductors and spaces fit within a printed circuit board manufacturing format. Therefore, smaller features must be fabricated at even lower defect densities than larger features to achieve equivalent yields.

The increased defect densities associated with smaller features, and the demands for greater electrical performance, typically increase the price of the circuit boards. When manufacturing yields are lower than expected, delivery schedules are impacted, delaying product introduction or even missing market windows completely. Additionally, quality issues with printed circuits can result in a variety of problems during assembly, cause poor circuit performance, and ultimately result in product failure once delivered to the end user.

Purchasers of printed circuit boards must manage the complexities of technology, quantity, delivery, and price for boards they are responsible for procuring, while keeping in mind the capability of each of their suppliers and the quality of the boards the suppliers produce.

36.1.1 Circuit Density

The technology used to manufacture a printed circuit board is often determined early in the design process. The circuit designer works within the constraints of overall size, thickness, weight, electrical performance, and thermal demands, but may have discretion on parameters such as layer count, feature sizes, and material properties to achieve the design objectives.

Interconnect density is increased with narrower lines and spaces and smaller-diameter via holes and via lands. The major layer interconnect technology choices for multilayer board fabrication are through vias, blind vias, and buried vias. In many instances, all three technologies are incorporated into the finished board. High-density interconnect structures employ blind microvias to interconnect one or more layers without impacting routing on the remaining layers. The ultimate reduction in via land diameter is illustrated in Fig. 36.1, which shows a landless via.

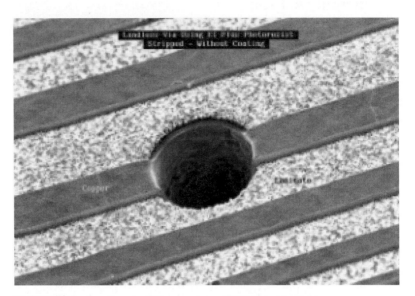

FIGURE 36.1   Landless via. *(Photo courtesy of PPG Industries, Inc.)*

The interconnect technology and feature sizes selected by the designer impact the manufacturability, quality, performance, reliability, and cost of the final product.

36.1.2 Supplier Capability and Quality

Quantitative statistical measures that distinguish one supplier or process from another can assist procurement staff in decisions regarding supplier capability and quality. Capability implies the ability to successfully form features such as conductors, spaces, and vias. Given that the features were formed successfully, quality asserts the degree to which they conform to specifications.

By utilizing measures of capability and quality, purchasers of printed circuits can minimize the potential for shipment delays caused by lower-than-expected manufacturing yields while ensuring the highest possible product quality.

36.1.3 Design for Manufacturability

Capability and quality data collected from suppliers can be used to optimize designs for manufacturability. By minimizing or eliminating features that are difficult to manufacture, suppliers are able to manufacture designs at higher yields, lower costs, improved quality, and minimal risk of shipment delays. When designers follow predefined design rules based on supplier capability and quality, the purchaser and the suppliers find themselves in a win-win situation.

36.2 MEASURING CAPABILITY AND QUALITY

The approach to measurement of capability and quality relies on three key elements:

- Specialized test patterns, referred to as *process capability panels,* designed to reproduce the features present in printed circuit boards
- Methods of testing the completed capability panels to extract raw capability and quality data
- Data analysis techniques to generate relevant statistics

The printed circuit board manufacturing process can be represented by a transfer function, as seen in Fig. 36.2(*a*). The input to the transfer function is the circuit board design data, typically Gerber data that contains design features, and the output is the actual printed circuit boards. As with most transfer functions, the output does not exactly equal the input but is "blurred" by the transfer function—the printed circuit manufacturing process. When applied to process capability panels, as seen in Fig. 36.2(*b*), the input contains a range of known design features, and the output is the capability and quality data that are collected from the process capability panels.

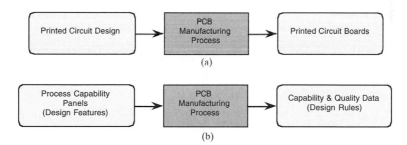

FIGURE 36.2   (*a*) PCB manufacturing transfer function. (*b*) Capability transfer function.

The printed circuit board manufacturing process is blurred by defects and variations in the design features caused by process limitations, processing conditions, and process nonuniformities.

36.2.1 Process Capability Test Panels

The test patterns used to collect data from printed circuit board manufacturing process should be designed to be conceptually as close as possible to the product they are intended to model. This includes considerations such as conductor and space sizes, via hole and land sizes, registration and impedance requirements, number of layers, stack-up, board thickness, and materials.

A set of test patterns that have been specially designed to collect detailed process capability and quality data on a range of feature sizes is shown in Fig. 36.3. When distributed over the

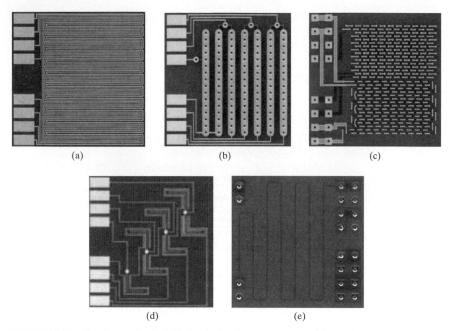

FIGURE 36.3 Conductor Analysis Technologies, Inc. test patterns. (*a*) Conductor/space module; (*b*) via registration module; (*c*) via module; (*d*) soldermask registration module; (*e*) impedance module.

area of the manufacturing panel format and manufactured in sufficient numbers, they provide the basis for relevant statistics on the capability and quality of the process that was used in their manufacture. Table 36.1 details the information that that can be obtained from each of these patterns.

TABLE 36.1 Test Pattern Statistical Attributes

Test pattern	Defect type	Capability information	Quality information
Conductor/space	Opens and shorts	Conductor and space defect density	Conductor width and height uniformity
Via registration	Breakout	Via probability of breakout	
Via	Opens	Via defect density	Via net resistance coefficient of variation
Soldermask registration	Misregistration	Clearance yield	
Impedance			Impedance uniformity

36.2.2 Testing Methods

Many testing technologies are available for collecting data from process capability panels. These include electrical, optical, x-ray, cross section, and other specialized techniques. Regardless of the method employed, the data must be collected in a timely manner, and procedures must be established to ensure the accuracy and precision of the data.

36.3   DATA ANALYSIS TECHNIQUES

36.3.1   Defect Density

The calculation of defect density from capability data normalizes the probability of having defects, and can be used to predict product yields. Defect density may be calculated for conductors, spaces, and vias. Equation 36.1 is used to calculate conductor defect density from process capability panel capability data.

$$\lambda_C = \frac{-\ln\left\{\dfrac{Y}{100}\right\}}{l}$$
(36.1)

where Y = conductor yield from test pattern (defects/millions of inches)
l = length of individual conductors (inches)
λ_C = conductor defect density

Similar equations are used to calculate defect density for spaces and vias.

36.3.2   Predicted Yields

The fraction yield on product due to opens in conductors is calculated by:

$$Y_{fo} = e^{-\lambda_C L_C}$$
(36.2)

where λ_C = opens defect density determined from process capability panels
L_C = total conductor length on product

Similarly, the fraction yield on product due to shorts between conductors is calculated by:

$$Y_{fs} = e^{-\lambda_S L_S}$$
(36.3)

where λ_S = shorts defect density determined from process capability panels
L_S = total space length on product

The product of the fraction yield due to opens and the fraction yield due to shorts provides an estimate of yield on product due to opens and shorts:

$$Y_p = 100 \, Y_{fo} Y_{fs}$$
(36.4)

36.3.3   Capability Potential Index

Control indices provide a numerical result that is indicative of the quality of the process used to manufacture conductors. The capability potential index (C_p) (Eq. 36.5) is the ratio of the difference between the upper specification limit (USL) and lower specification limit (LSL) to six times the standard deviation (σ). Larger capability potential indices indicate that the manufacturing process has greater potential to provide features that fall within the specification limits.

$$C_p = \frac{\text{USL} - \text{LSL}}{6\sigma}$$
(36.5)

36.3.4   Capability Performance Index

The capability performance index (C_{pk}) relates the mean and standard deviation to the specification limits by the relationship in Eq. 36.6. The capability performance index is always less than or equal to the capability potential index. If the mean (μ) is centered between the lower and upper specification limits, then the capability performance index equals the capability potential index; otherwise, it is less than the capability potential index. Larger capability performance indices indicate a greater probability of the data falling within the specification limits.

$$C_{pk} = \min \left\{ \frac{\text{USL} - \mu}{3\sigma}, \quad \frac{\mu - \text{LSL}}{3\sigma} \right\} \tag{36.6}$$

36.3.5 Coefficient of Variation

The coefficient of variation (CoV) is defined as 100 multiplied by the standard deviation, divided by the mean (Eq. 36.7). This expression normalizes the spread in the data to the mean and is sometimes called the relative standard deviation. For a given mean, smaller standard deviations characterize improved performance so that the smaller the CoV, the better.

$$\text{CoV} = 100 \left\{ \frac{\sigma}{\mu} \right\} \tag{36.7}$$

36.3.6 Probability of Breakout

Probability of breakout is a measure of the chance of a via hole being misregistered from its land by greater than the annular ring of the via. The probability of breakout is calculated for each designed clearance by Eq. 36.8.

$$P_{bi} = 100 \left\{ \frac{N_{fi}}{N_{ti}} \right\} \tag{36.8}$$

where N_{fi} = the number of failures at clearance i
N_{ti} = the total number of opportunities at clearance i

36.4 CONDUCTOR AND SPACE CAPABILITY AND QUALITY

The ability to successfully manufacture conductors and spaces, as measured by defect density, provides a convenient means of estimating expected yield on product and illustrates the impact of defects on yield. Figure 36.4 shows predicted product yield due to opens plotted vs. feature length for 2-, 3-, 4-, and 5-mil conductors formed on outerlayers. The curves show that, for a fixed defect density, yield falls off with increased conductor length on product. At 156 defects per million inches of 4-mil conductor, predicted product yields (due to opens) for 100, 1,000, and 10,000 in of conductor are 98.5, 85.6, and 21.0 percent, respectively. When defect levels increase to 598 defects per million inches, as shown for the 3-mil conductor in Fig. 36.4, predicted yields drop to 94.2, 55.0, and 0.3 percent for 100, 1,000, and 10,000 in of conductor, respectively.

Figure 36.5 shows predicted product yield due to shorts between conductors vs. space length on product. Defect densities in this example are higher for spaces than for conductors.

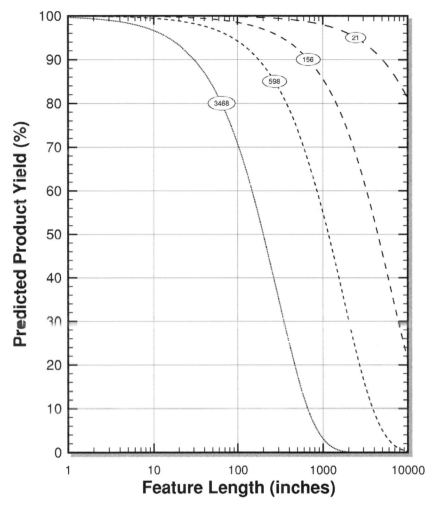

FIGURE 36.4   Predicted product yield vs. conductor length. The curves correspond to defect densities for 2- (3468), 3- (598), 4- (156), and 5-mil (21) outerlayer conductors.

A defect density of 644 defects per million inches of spaces was recorded for the 4-mil space. Once again, predicted yield falls off with increased space length on product for a fixed defect density.

   To estimate the combined effects of opens and shorts on product yield, the fraction yield due to opens is multiplied by the fraction yield due to shorts and converted back to percent by multiplying by 100. As an example, if there were 1000 in of 4-mil conductors and 500 in of 4-mil spaces on product, the predicted yield would be $0.856 * 0.725 * 100 = 62.0$ percent.

   Given that conductors were successfully fabricated, their quality may be characterized by examining the accuracy and precision with which they were formed. Figure 36.6 displays the distributions of calculated conductor width plotted vs. target conductor width for 2-, 3-, 4-, and 5-mil outerlayer conductors. The distributions are portrayed by notched box plots with the dotted lines showing the upper and lower specification limits (± 20 percent in this example).

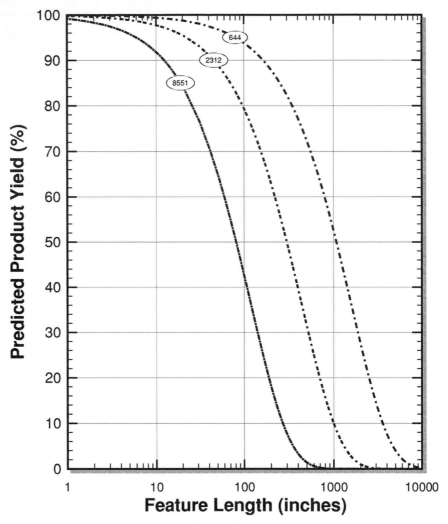

FIGURE 36.5 Predicted product yield vs. space length. The curves correspond to defect densities for 2- (8551), 3- (2312), and 4-mil (644) outerlayer spaces.

When the conductors are formed accurately, the notched box plots are centered between the upper and lower specification limits. When the conductors are formed precisely, the distribution of widths indicated by the notched box plot is tightly clustered.

The shaded bars in Fig. 36.6 show the control indices for each of the four conductors. In all cases, the capability potential index was much greater than the capability performance index, because the conductors were not formed with high accuracy. The median widths were approximately 0.4 mils narrower than the target conductor widths.

The quality of conductors is also affected by the accuracy and precision of their height. Figure 36.7 shows calculated conductor height plotted vs. target conductor height for the outerlayer conductors. While the height was more accurately controlled than the width (the notched box plots are more closely centered between the upper and lower specification limits), the pre-

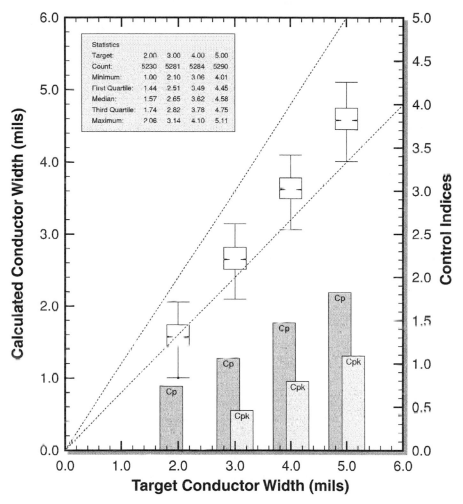

FIGURE 36.6 Conductor width vs. target conductor width with control indices. The dashed lines correspond to ±20 percent specification limits.

cision of the height (indicated by the extent of the notched box plot and the lower value of capability potential index) was much worse than that of the widths.

36.5 VIA CAPABILITY AND QUALITY

The capability of forming and interconnecting vias, and the quality and reliability of the interconnection, depend upon many steps, including the registration, drilling, cleaning, metallization, and patterning processes. Misregistration of the via with respect to the land(s) can lead to marginal interconnections that exhibit increased resistance, and perhaps to reliability problems.

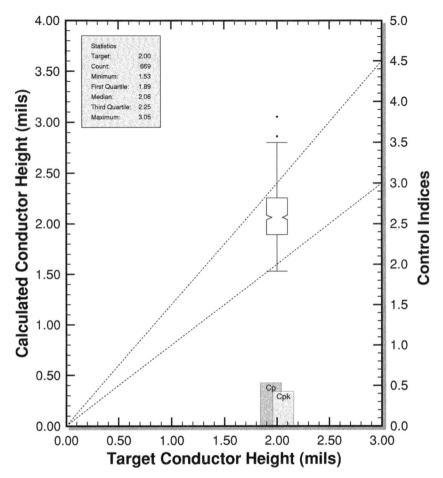

FIGURE 36.7 Conductor height vs. target conductor height with control indices. The dashed lines correspond to ±20 percent specification limits.

Registration capability of blind microvias is shown in Fig. 36.8, which displays probability of breakout plotted vs. annular ring. The smallest clearance (–1 mil), included to verify that the proper hole size was drilled, is intentionally designed to fail. The probability of breakout decreases from 98 percent for the 2-mil designed clearance to zero for the 6- and 7-mil clearances. If breakout is not allowed, then designs fabricated by the supplier in this example would require annular rings of 6 mils or larger for the microvias. By designing to 5-mil annular rings, the data estimate that 5 percent of the microvias will exhibit breakout.

The capability of forming microvias is summarized in Fig. 36.9, which displays predicted product yield due to opens in vias plotted vs. the number of vias. Similar to predicted product yield for conductors and spaces, the estimated yield for a given defect density drops off with an increased number of vias in the design. For example, the 5-mil microvias recorded a defect density of 106 defects per million vias. At this defect level, the estimated yield drops from 89.9 percent to 34.7 percent when the number of vias is increased from 1,000 to 10,000.

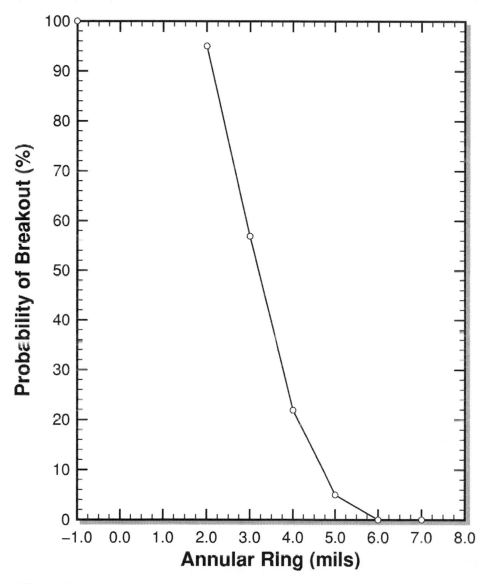

FIGURE 36.8 Probability of breakout vs. annular ring for microvias.

The quality of vias may be investigated by examining the distribution of resistance measurements made from daisy chain patterns. Figure 36.10 shows the distributions for 4-, 5-, 6-, and 7-mil microvias. The results indicate that 6- and 7-mil-diameter microvias were fabricated with high quality, but the 5-mil and especially the 4-mil-diameter microvias exhibited very poor quality, indicated by the broad distributions and numerous outside values.

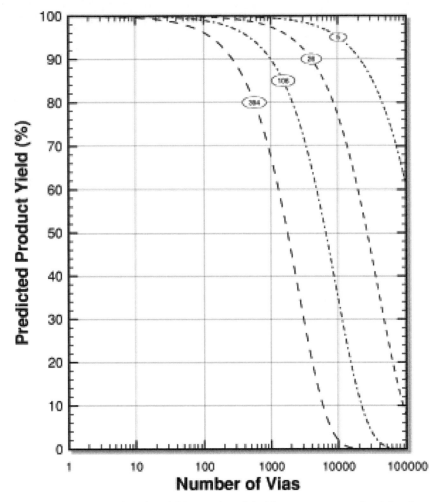

FIGURE 36.9   Predicted product yield vs. number of vias. The curves correspond to defect densities for 4- (394), 5- (106), 6- (26), and 7-mil (5) microvias.

36.6   *SOLDERMASK REGISTRATION CAPABILITY*

The soldermask prevents shorting between pads during the assembly process and protects surface features from mechanical and environmental damage. Registration of the soldermask becomes increasingly important as surface mount and ball grid array pitch sizes decrease. Therefore, the soldermask must be applied and patterned to stringent tolerances.

Results from soldermask registration modules are shown in Fig. 36.11. Clearance yield is plotted vs. soldermask clearance. The clearances ranged from 1 to 3.5 mils in 0.5-mil increments to provide a range of clearances that establish local registration. The clearance yield at 3.5 mils is nearly 100 percent in this example. The yield falls off gradually to 2 mils, and then drops more significantly at 1.5 mils and 1 mil. The data from this process suggest that a clearance greater than 3.5 mils is necessary to ensure clearance between the soldermask and copper pads.

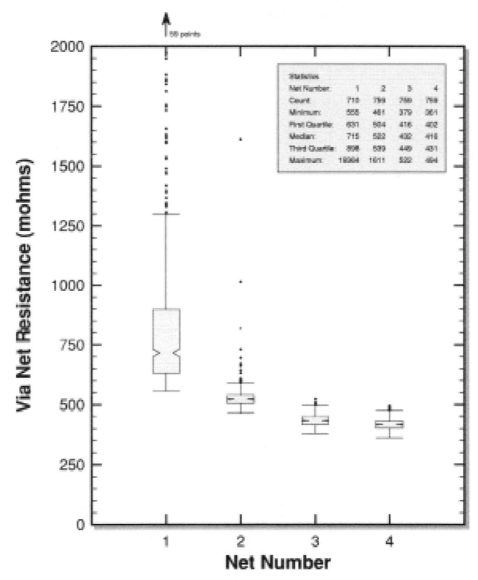

FIGURE 36.10   Via net resistance vs. net number. The distributions correspond to 4-, 5-, 6-, and 7-mil microvias, respectively.

36.7 *CONTROLLED-IMPEDANCE CAPABILITY*

Many applications require controlled-impedance lines on critical signals to ensure signal integrity. The impedance of a net is primarily affected by the dielectric constant, the width and height of the conductor, and the separation between the conductor and the ground plane(s). Once the dielectric materials are selected for the design, process variations that affect the conductor width and height, and distance to planes, impact the impedance.

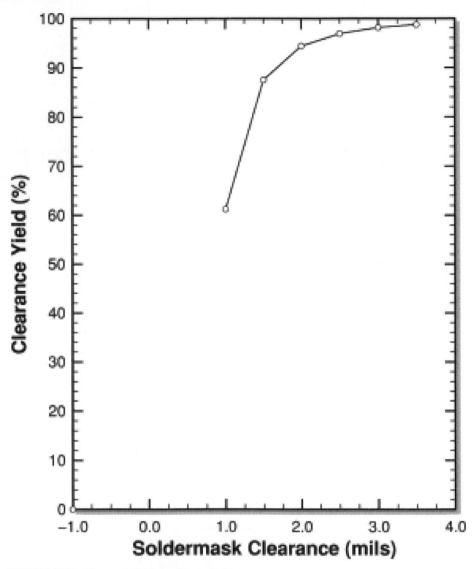

FIGURE 36.11   Clearance yield vs. soldermask clearance.

Processes that are in control have the best opportunity to achieve controlled-impedance requirements.

Data collected from embedded microstrip modules show the distribution of calculated
impedance as a function of target impedance of 51 Ω in Fig. 36.12. The data are tightly controlled and nearly centered within the specification limits, registering a capability potential
index of 2.2 and capability performance index of 1.9.

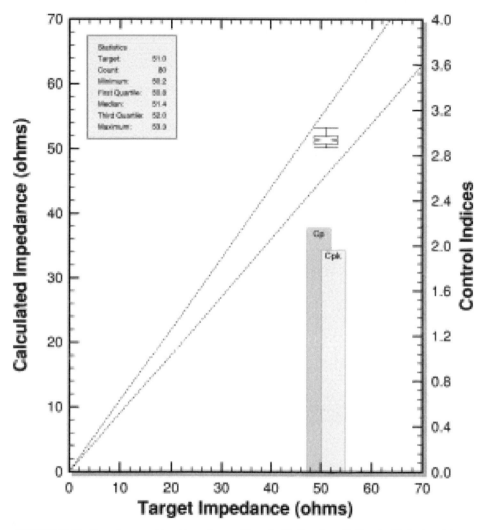

FIGURE 36.12 Impedance vs. target impedance with control indices for a 51-Ω embedded microstrip.

36.8   STANDARDIZATION

In order to establish industry standardization for the design of process capability panels, the IPC has formed a working subcommittee (D-36). The charter of this subcommittee is defined as establishing and maintaining a family of process capability panel designs for use by PCB suppliers, original equipment manufacturers (OEMs), and contract electronic manufacturers (CEMs).

CHAPTER 37
BARE BOARD TEST OBJECTIVES AND DEFINITIONS

David J. Wilkie
Everett Charles Technologies, Pomona, California

37.1 INTRODUCTION

Advances in packaging technology resulting in finer board geometry, including the various forms of high-density interconnection (HDI), have combined with increasing data rates to put significant pressure on the electrical test area. Fixture construction is more expensive and requires improved process control. Advanced test methods, such as RF impedance testing, are more often required. Global price competition demands reduced costs. Meanwhile, OEMs demand that board manufacturers accept increased liability for defective product—thereby demanding improved fault coverage. This chapter is devoted to the why, what, where, when, and how of current electrical test methods useful in meeting these requirements.

37.2 THE IMPACT OF HDI

At the present time test engineers are confronting significant changes in test requirements that derive from changes in the product itself. Notable among these changes are those driven by the growth of various high-density interconnect (HDI) technologies. Examples of common HDI applications are direct chip attach (DCA), high-density ball grid array (BGA), and variants of the above often referred to as chip-scale packaging (CSP). In addition, higher-density I/O connectors are commonly applied. In addition to the changes in physical geometry, HDI also implies increasingly high data rates and clock speeds. Advanced means are being applied that verify not only the ability of the product to interconnect electronic components, but to do so in a manner that guarantees signal fidelity.

An analogy is possible comparing the impact of HDI on board technology with the impact of the sound barrier on aerodynamic flight. This analogy is readily applied to final electrical testing. It was popularly imagined that the sound barrier was impassible, or could be passed only by Buck Rogers in his silver rocket—that beyond the speed of sound was no place for an airplane to go. As it turned out, passing the sound barrier *was* difficult, and required re-thinking many aerodynamic principles. But the resulting vehicles are still recognizable as aircraft. They have wings, they burn fuel. They meet all of the objectives of earlier craft, and do so in a superior manner. The test processes that are resulting from the impacts of HDI upon final electrical testing are similarly recognizable evolutions of previous methods, employing familiar base technologies, but in new combinations and with added features that actually improve test coverage rather than diminishing it.

Proposals have circulated for radical new approaches to electrical testing. These have included the use of electron beams, laser-stimulated photoelectric effects, and gas plasma techniques in configurations similar to existing flying probers. In each case, the board is scanned without use of a custom test fixture. To date, all of these methods involve compromise of fault coverage, or add little or no fine-pitch performance beyond conventional methods. Relaxation of test criteria is not a relief available to most users. HDI product types demand increased fine-pitch capability. As a result, none of these approaches has seen widespread adoption.

In place of radical breakthroughs, the market has seen a steady improvement in the availability of software and hardware tools that provide for the effective combination of the best features of conventional test systems, as well as significant extension to the capability of test systems and their related fixtures. Also new are test systems dedicated to specialty markets, in particular the small-format laminated chip carrier product type.

37.3 WHY TEST?

Why pay for testing? Why do we need to test? The answer has several components. The primary assumption is that not all boards produced are good. If we could look at the yields through the history of the industry in printed wiring boards using similar technology, we would see exponential reductions in the percentage of bad boards produced. Process improvements and a wide variety of quality improvement programs continue to make a difference. In spite of these improvements, the need for electrical testing remains.

37.3.1 The Rule of 10s

One reason to test is to block the addition of further value to defective product. Consider the printed wiring board as a component of a completed assembly. A commonly accepted relative measure of the cost of faults in a completed electronic assembly can be expressed by the rule of 10s (see Fig. 37.1). The idea is that the earlier a fault is caught, the less it will cost. An example might be an open in a bare board that is not found at bare board test. The faulty bare board is now loaded with the components, soldered, and tested. If the fault is found at the loaded board level, repair or scrapping of the assembled board is much more expensive than repair or scrapping at the bare board level. There are situations where the assembler of the board charges back to the manufacturer of the bare board some portion of the cost of the scrapped components, assembly labor, and/or production cost. This is increasingly true with the high-cost and high-capability ICs present in so many designs. A fault that passes a system-level test and makes it to the end user entails even higher cost. Some of these costs are very tangible in the form of field service labor, downtime, parts, and labor. Other costs are counted in lost business and reputation.

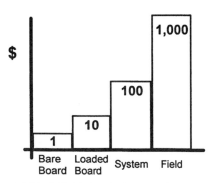

FIGURE 37.1 Rule of 10s: cost of a found fault.

37.3.2 Satisfying Customer Requirements

Where the rule of 10s is a logical big-picture approach to the necessity of testing, requirement from a customer is quite the opposite. The board user has already determined that electrical testing is appropriate and necessary and has made it a requirement as part of supplying the board. It is the obligation and responsibility of the bare board manufacturer to meet this requirement of purchase contracts. In practice we find such customer requirements stated in various ways.

37.3.2.1 *Generic or 100 Percent Board Test.* Unfortunately, this requirement may be nothing more than the words *100 percent electrical test* appearing somewhere in a purchase order or a larger package of documentation. Not only is such a specification relatively meaningless, but the issuer of such a requirement is led to believe that the boards will be tested to the most stringent requirements—that it will not be possible for a single bad board to be delivered. Nothing could be further from the truth. Such a specification fails to define what a 100 percent test consists of. The purchaser of the bare board will certainly be dissatisfied when he or she receives a bad board. Although the specific fault on the bare board in question may have a perfectly viable explanation (might not be detected electrically, or may be undetectable by the specific methods employed), the 100 percent test specification implies that no explanation is acceptable. In many cases, application of this general sort of specification is combined with a customer-imposed limitation on test cost, which in turn requires cost-driven limits on the thoroughness of electrical testing performed by the board manufacturer.

37.3.2.2 *Written Test Specification.* It is the responsibility of both the bare board manufacturer and the bare board purchaser to agree upon a test specification that is specific in terms of test thresholds and method of testing, while keeping in mind the appropriateness to the application for the bare board design. Some limits on test thresholds and test methodology may be due to equipment and capabilities of the manufacturer. It is the responsibility of the bare board manufacturer to educate the end user as to the implications of test specifications in thoroughness and cost of testing. A matrix of test options, fault detection capabilities, escape risks, and associated costs should be presented to the end user. It is the responsibility of the end user to work with the bare board manufacturer to develop specifications appropriate to the bare board being produced and to accept and plan for any testing compromises imposed by cost or specification limits. Areas that should be covered in the specification are:

- Data-driven testing, source of data, format, and integrity
- Test thresholds: maximum allowable for continuity verification
- Test thresholds: minimum allowable threshold and applied voltage for isolation verification
- Test fixture methods: single fixture with simultaneous access of all test points vs. split-net fixtures vs. flying probe
- Test point optimization
- Board marking indicating electrical test satisfaction
- Resolution of discrepancies in process (contacts and procedures)
- Requirements for special test methods (TDR, embedded passives, etc.)
- Resolution of escapes

Such agreements make for good communication, clear understanding, happy customers, good reputation, and good business.

37.3.2.3 *Using Available Standards.* Industry standards or guideline documents such as IPC-ET-652, "Guidelines and Requirements for Electrical Testing of Unpopulated Printed Boards," are useful in developing agreed-upon written specifications for particular product. Standards documents must be applied in consideration of the target application. An excessively rigorous specification would result in unacceptable test cost with little quality improvement. Arbitrarily high isolation voltages or continuity test currents can actually lead to product damage. Conversely, relaxing isolation resistance specifications on boards containing sensitive amplifiers would be inappropriate. These are but a few examples. In ET-652, the IPC has made efforts to distinguish between general classes of board applications, suggesting specific electrical test requirements that attempt to recognize different levels of criticality in testing. Final determination of the suitability of a given set of requirements must be made in cooperation with technical representatives of the bare board purchaser.

37.3.3 Electrical Testing as a Process Monitor

A third benefit of testing is the improvement of processes and the resultant reduction in costs. To raise yields, reduce scrap, or generally improve quality, there must be a system of measurements able to express results of changes. One of the best areas to collect data is electrical testing. This usually requires an integration of testing and repair data. While the test system can identify a bad board and the location of a fault, a repair operator can further classify the fault to associate it with a specific process (imaging, plating, etching, solder masking, etc.). These data can be quantified and analyzed in a variety of ways, and corrective action may then be directed to one or more of the following levels.

37.3.3.1 Faults Specific to the Operation.   Fault types may be detected that are not typical of the industry or competition. It can be difficult to assess this type of situation accurately because data commonly available from the customer, alternative vendor, or competitor are likely to be narrow or biased. It may be better to analyze fault types specific to an operation in terms of cost or profitability. Are certain product types profitable to manufacture? What would be the cost to make them profitable?

37.3.3.2 Faults Specific to the Process.   An analysis of fault data may lead to a specific process that regularly induces faults in a wide variety of part numbers, or in a variety of part numbers sharing key characteristics. This may point to requirements for new settings, procedures, materials, training, personnel, equipment, etc.

37.3.3.3 Faults Specific to a Part Number.   Fault data on a specific part number are useful in correcting and optimizing processes for future runs of the same part number, and often for similar future part numbers. It may be possible to feed back electrical test data in real time using software systems, depending upon lot size, factory flow, and the type of test and repair analysis equipment available for use. In this situation, results from electrical testing may help drive some adjustments or quick changes in the process that will immediately improve the yield and reduce scrapping, repair, and retesting costs.

37.3.4 Quality System Improvement

Quality control analysis of fault data is only a piece of what needs to be in place already, that is, a quality control system or process. The value of electrical test (or other process) data for process monitoring is only as great as the quality control framework making use of it. It is possible for electrical test results to become a vehicle for improvements to a board manufacturer's quality system.

37.4 CIRCUIT BOARD FAULTS

For electrical test purposes, faults may be defined as test system measurement results other than those programmed to be representative of a good board. The faults detected may or may not impact the functionality of the circuit board, though in most cases they will. Some guidelines may prove useful when more specific guidance from the original specifier or designer is lacking. Obviously, gross shorts and opens in a circuit are likely to cause problems. Will an electrical test system find all the faults? No. The definition of "all the faults" is too subjective. The electrical test system will not detect all faults related to aesthetics, annular rings, layer-to-layer registration, etc., unless they present an effect measurable by the test system. Further, the electrical measurement employed, type of test fixture, test program generation method, and end use requirements vary too widely to broadly state that electrical testing will find all the faults.

For purposes of the following discussion, it is worthwhile to clarify the distinction between a defect and a fault. A fault is a test system designation for an item that does not meet the expected criteria. A defect refers specifically to the board and a defect in its design, fabrication, appearance, etc. Not all defects can be detected by the test system.

37.4.1 Fault Types

Table 37.1 presents common fault types. For those fault types that are commonly detected, it is valuable to distinguish types of tests from types of faults. It may be preferable to refer to isolation testing or continuity testing rather than to shorts or opens testing, as the latter meaning is often confused. (Does opens testing refer to isolation testing [the testing of opens], or is it referring to continuity testing [making sure that no opens exist]?) Shorts and opens are results, not test types. Test methods themselves are discussed in detail elsewhere in this chapter.

TABLE 37.1 Test and Fault Types

Test type	Fault type identified by test
Continuity	Open
Isolation	Short and/or leakage
TDR or network analysis	RF impedance fault
Hi-pot	Voltage breakdown

37.4.1.1 Shorts. Shorts, hard shorts, or short circuits are defined here as erroneous (undesired and unexpected) low-resistance connections between two or more networks or isolated points, typically exhibiting a fairly low electrical resistance value. Shorts are reported as failures of the isolation test of the product. Shorts are produced in a variety of ways, including exposure problems, underetching, contaminated phototools, poor alignment of layers, defective raw material, and improper solder leveling.

37.4.1.2 Opens. Opens represent an absence of expected circuit continuity, or in other words, a missing connection. This divides a circuit network such that the network is split or divided into two or more pieces. Opens are reported as failures of the continuity testing of the product. Opens are produced in a variety of ways, including overetching, underplating, contaminated phototools, contaminated raw material, layer registration errors, and mechanical damage. A common problem during electrical testing is "false open" errors, typically the result of localized contamination on the product or test probe that prevents proper connection to the test system.

37.4.1.3 Leakage. A leak or "leaking network" is essentially a type of short. Leaks are also referred to as high-resistance shorts, and differ from hard shorts in that they exhibit a higher resistance value. The precise division between the two types of error reports varies according to the type of equipment used, and some equipment does not attempt to distinguish them in fault reports. As in the case of hard shorts, leakage is a failure of the isolation testing of the product.

Common leakage causes are moisture, chemicals, or debris. Contamination can occur during innerlayer fabrication, lamination, plating, solder masking, or at any stage due to handling. Chemical contaminants are often deposits of metal salts left as artifacts of the chemical processes used to manufacture the product. With sensitive test methods, even fingerprints can result in detectable leakage between networks. Such contaminants are often spread over an area of the board such that several networks become interconnected. Consideration of the potential for multiple network involvement is useful in selecting the particular isolation test algorithm or test method, as these methods differ in their sensitivity to this situation. Isolation test methods are discussed elsewhere in the chapter. Some circuits are not sensitive to the high-resistance loads immediately presented by such contaminants. But it is important to note that, in the presence of time, electric fields, and moisture, it is possible for the resistance of a high-impedance short to decrease greatly. Contamination sites may facilitate the growth of metallic crystals that reach out as thin metal threads between networks, forming hard shorts. Thus an area of the product exhibiting unusual leakage may, at some future time, exhibit hard shorts between networks. This amplifies the need for effective high-impedance testing of high-reliability product types as a means of preventing latent field failures. Note that the nature of board materials is such that they can absorb moisture relatively easily, and they do so over time. Thus even a fairly "dry" contamination providing a very weak electrical path may eventually result in sufficient metal migration during product operation to cause a serious field failure.

37.4.1.4 *RF Impedance Fault.*   Many circuits produced today are required to operate at very wide bandwidths. Examples include fast microprocessors, fast general digital circuits, radio frequency (RF) amplifiers in wireless devices, etc. Just as we must use a proper type of cable to connect a television antenna to a TV receiver, it is important for specific RF characteristics to be maintained in interconnections between components of fast electronic circuits on printed wiring boards. One parameter commonly specified and measured is the RF transmission line impedance of the signal traces. This parameter is strongly affected by the materials used in fabricating the board, the trace thickness and width, and the spacing from ground planes and adjacent signals. A common method of measuring RF impedance is to employ time domain reflectometry (TDR).

The TDR measurement provides a statement of RF impedance as a function of distance along the trace. (Distance and time are related here, as the electric signal flows through the board at velocities approaching the speed of light.) TDR testing is often performed on a test coupon attached to the product during manufacture and subsequently disconnected. TDR testing on actual product traces is also done on selected traces, but is complicated by the need for a trace length of several inches, uninterrupted by branches or other constructions. Common values for RF impedance on circuit boards range from the low tens of ohms to several hundred ohms.

RF impedance should not be confused with ordinary DC resistance and cannot be measured with common ohmmeters, even though the same unit of measure—the ohm—is used. RF parameters of interconnections can also be characterized in the frequency domain using instruments referred to as network analyzers, but this method is not common in bare board testing. Requirements for RF impedance testing are more commonly applied as signal frequencies exceed 100 MHz.

37.4.1.5 *Hi-Pot Faults.*   Hi-pot or high–potential voltage breakdown testing is often confused with isolation testing. The tests are very similar, and, to the extent that high voltage is employed in an isolation test, they might achieve a similar result. But isolation testing of bare boards commonly occurs with 250 V or less applied between networks, and hi-pot testing is often performed at values from 500 V to several kilovolts. Hi-pot testing attempts to verify the strength of the insulating material between networks by subjecting it to so high a voltage that a catastrophic or avalanche-type voltage breakdown will occur if the insulator is subpar. In contrast, isolation testing on bare boards attempts to detect the small current flowing through contamination (or, for that matter, a hard short) before voltage breakdown occurs. Of course, if the insulator is very weak, then avalanche failures can occur at almost any voltage. Hi-pot testers tend to be benchtop devices with a pair of test leads and no switch matrix, and hi-pot test requirements usually specify that the voltage be applied for a sustained period of time.

Hi-pot testing is valuable for inspection of very thin insulating core material, before circuits are etched. This can serve to detect z-axis faults or contamination in the material before value is added in subsequent processes. It can be impractical to perform hi-pot tests between all conductors of finished fine-pitch boards, as the atmospheric environment (air) between conductors will break down before high voltage is reached. The slow speed at very-high-voltage hi-pot testing, and the costs of suitable fixtures and electronics, present problems. For final product inspection, it is arguable that a very-high-impedance isolation test provides the superior solution.

CHAPTER 38
BARE BOARD TEST METHODS

David J. Wilkie
Everett Charles Technologies, Pomona, California

38.1 INTRODUCTION

Although the main bare board testing technology is electrical, it is important to consider that nonelectrical methods are also important in the acceptance or rejection of bare printed wiring boards. This chapter therefore includes detailed descriptions of both electrical and nonelectrical testing methods.

38.2 NONELECTRICAL TESTING METHODS

There are two nonelectrical acceptance/rejection methods, both based on inspection processes:

1. Visual inspection
2. Automatic optical inspection

38.2.1 Visual Inspection

Visual inspection is a very manual approach in that it makes use of people, good lighting, some type of training defining what is acceptable and what is not, and good operator judgment. Usually a comparison to a known good product or the artwork is made. If the operator has seen the board often, he or she becomes more skilled at finding faults and looking for faults in likely locations. As product complexity has increased, we find that many modern products are not suited to this method. Many innerlayer defects are completely undetectable, and even the external layer complexity is visually overwhelming. Visual inspection often remains appropriate for detecting cosmetic defects, such as poor solder masking or physical damage. Such defects generally fall outside the realm of electrical testing as they are not detected by electrical means.

38.2.2 Automatic Optical Inspection

There are computer-based visual inspection methods, referred to as automatic optical inspection (AOI). AOI equipment compares the board or its innerlayers to expected data and/or design rules that have been programmed into the controlling computer. These can be generally accepted parameters or design-rule-based parameters, or windows of acceptable dimensions for each specific feature on the board. As with manual visual inspection, faults found with this method can imply that there may be an impact on the board's functionality, but the board's functionality and interconnect are not directly tested. Distinctions between the aesthetics of the board's features and its fitness for use are difficult to differentiate, and may result in false failures. Rather than being used in final testing, AOI is typically used for inspection of innerlayers prior to lamination, with the goal of increasing final yield by weeding out the majority of defective layers prior to the addition of further value. As such, AOI can achieve significant financial benefit. AOI can detect some defect types not readily detected by electrical means, particularly "mouse bites" (brief narrowing of conductor cross section). But AOI is not effective in examining internal layers after lamination and is not generally accepted as a quality assurance substitute for final electrical testing.

38.3   BASIC ELECTRICAL TESTING METHODS

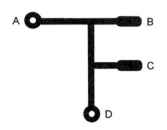

FIGURE 38.1   Sample network.

Electrical testing is the final test method frequently used to determine whether a board should be shipped. Electrical testing emulates the intended function of the board conductor and insulator patterns by passing currents through conductors and applying voltages across insulators. Such direct electrical measurement requires that the board come into physical contact with a measurement system. Two test types are almost universally performed: continuity and isolation testing. Some other tests may be applied selectively, depending upon the product and customer requirements. Test order is usually such that the continuity test is performed first. This verifies that each network is intact within itself, and that contact is established between any test fixture and the product. The isolation test can then be performed using only a single test point per network. Some test methods attempt indirect inference of continuity and isolation without making direct DC measurements. These methods are commonly employed in flying probe systems.

38.3.1   DC Continuity Test Method

Continuity testing checks for the expected continuous path within each electrical network. This is done in a series of point-to-point measurements within each network. The resistance found in each measurement is compared to the selected continuity resistance threshold. If the measured value is higher than this threshold, then a fault report is generated. For complex networks, multiple measurements are required in order to ensure that all extremities of the network are interconnected. For example, the network with test points labeled A through D, shown in Fig. 38.1 could be tested in the sequence illustrated in Fig. 38.2. For a network with four test points, the minimum number of tests would be three to determine if all points are connected. If a board contains a total of N isolated networks, and contains a total of X test points, we may calculate the number of continuity measurements C as $C = X - N$. In assigning test points, the software system often is programmed to delete those that are unnecessary. This is referred to as test point opti-

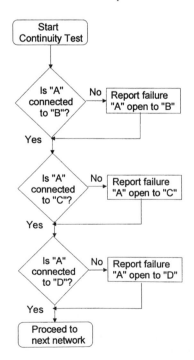

FIGURE 38.2   Continuity test algorithm.

mization. In Fig. 38.1, a test point located along the path between D and the branch to C would not be useful. Various optimization rules may be applied, but should be applied with care to ensure that adequate test coverage remains in place.

38.3.1.1   Continuity Test: Two-Wire vs. Four-Wire Switching.   Switch matrixes are constructed using either two-wire or four-wire circuits. A diagram comparing a single continuity measurement being performed on both two-wire and four-wire matrixes is shown in Fig. 38.3. Grid test systems use solid-state switches to connect appropriate test points to an internal measurement system (which is effectively an ohmmeter). Current I is driven from the upper test point through the product network and returned to the measurement system through the lower test point. The resultant voltage V across the network is measured. The continuity resistance R is then determined using the relationship $R = V/I$.

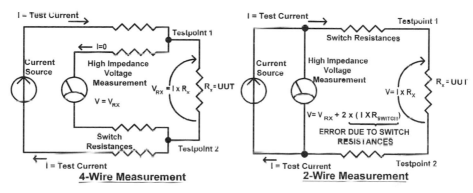

FIGURE 38.3   Continuity measurement circuits.

38.3.1.2   Two-Wire Switch Matrix Construction.   The simplest and most affordable construction of the switch matrix is to have a pair of switches for each test point, which can connect the test point to either the high side or low side of the ohmmeter. This technique is illustrated on the right side of Fig. 38.3. However, as the solid-state switches exhibit a certain amount of on-resistance, they contribute error to the measurement. The resistance of the switches is added to the resistance we measure, increasing the likelihood of failing the test. These errors can be reduced in various ways.

- Use switches with low on-state resistance (and thus a small amount of variation in on-resistance). This means using physically larger transistor dies. Such devices are not expensive, but are harder to integrate into integrated circuits. Thus, most two-wire designs rely upon discrete output transistors as switch elements.
- Use software to subtract the estimated typical on-resistance. As switch resistance effects vary from device to device, and change with temperature, some error remains.
- Make multiple measurements for each continuity to be inspected, and mathematically subtract most of the switch resistance. However, this slows the measurement process somewhat by requiring extra measurements.
- Attempt to learn the resistance of each switch. Some systems require the operator to install a shorting plate over all test points, and the system then learns the sum of the switch resistances plus contact resistances to this plate. Note that this technique can actually add error if a high contact resistance value is learned for a particular test point, perhaps due to dirt on a probe. The excessive resistance value learned is subtracted from all subsequent measurements made with that probe, masking some high product resistances. The net result is a slightly increased possibility of passing a board that should be failed.

38.3.1.3 Four-Wire Switch Matrix Construction. The left side of Fig. 38.3 represents a Kelvin or four-wire switch matrix construction. Notice that each test point will require four switches, so that any test point can be connected to the high-side current drive, high-side voltage sensor, low-side voltage sensor, or low-side current return via separate paths. Because the voltage-sensing portion of the measurement system has an extremely high input resistance, almost zero current flows through the extra pair of switches that connect the test points to this circuit. The result is that there is zero voltage drop across these sense switches, and the measurement unit sees exactly the voltage across the unknown load. Because the measurement system knows exactly the current flowing through the load, and exactly the voltage across it, it can calculate the load resistance accurately. The dominant error term in this situation is generally the contact resistances in the fixture and fixture-product interface. Also, the smaller transistors leak less current in the off state, thereby permitting isolation testing at higher thresholds with smaller errors. The continuity accuracy benefit of this technique is only realized if high-quality fixtures with low contact resistance are employed.

38.3.1.4 Continuity Threshold. The continuity resistance threshold parameter is usually specified in the range from a few ohms to 1000 Ω. Several standards useful in suggesting continuity thresholds are summarized in Table 38.1. As discussed earlier in the chapter, these must be applied with judgment. Generally, a lower continuity threshold provides a more stringent test of the board. Networks with resistances of 5, 10, or 25 Ω, although rare when using copper traces of moderate length, can significantly impact the functionality of precision measurement instruments or high-speed computer products. At the same time it should be noted that there are practical and economic considerations in determining how low the continuity test threshold should be set. Part of the limitation comes from the test system's measurement and switch matrix capabilities and, to a greater part, from the type of test fixture used. At the time of this publication, a 10-Ω continuity resistance test threshold is a common lower limit for production testing with good-quality systems and fixtures.

TABLE 38.1 Examples of Continuity Test Parameter Standards

	IPC-ET-652	MIL-55110D (obsolete)
Maximum continuity Resistance test threshold	Class 1: general electronic, 50 Ω Class 2: dedicated service, 20 Ω Class 3: high reliability, <20 Ω	10 Ω

38.3.1.5 Continuity Test Current. Continuity test current is not addressed in IPC-ET-652 or in most other publications. Use of high current has been proposed as a means of burning out weak traces or mouse bites. But such currents may also damage good traces. If this occurs after the test system has already determined that there is a good connection, the result is a board that once tested good but is now bad. It is preferred that the continuity test not be invasive or destructive. Typical test currents today are in the range of 5 to 50 mA.

38.3.1.6 Continuity Test False Opens. A common problem with continuity testing is a high incidence of false failures due to fixture and product contamination, poor product registration, or fixture damage. Dramatic improvement is often possible with the addition of product and fixture cleaning methods. Separation of the board testing environment from such dust-producing processes as drilling/routing can be invaluable in increasing throughput. High-voltage pulses are sometimes used to overcome thin-film contaminants or oxides coating the surface of the board and preventing good contact with the test probe. As no current is initially flowing through the oxide, some test systems offer a feature delivering a high-voltage pulse of strictly limited current and duration, and therefore limited total energy. While brief, the energy level is higher than for a normal continuity test and a small risk of damage at defect sites remains. Adding test time for automatic or manual retesting of the product yields additional boards, but this common approach becomes expensive when large numbers of false opens occur. It may be advisable to correct the root cause.

38.3.2   DC Isolation Test Method

Isolation testing verifies the presence of adequate electrical isolation between networks that are not intended to be connected to one another. Typically a resistance measurement is made from a given network to another net (or group of nets). If the measured value exceeds the specified isolation resistance threshold while the specified voltage is applied, then the measurement is considered to have passed. Otherwise, a fault report is generated. So long as contact with the net has been ensured (during the continuity test), only a single test point per network is required to perform the isolation test. The actual number of isolation measurements required to test a given board can vary substantially with details of the algorithm employed, with subtle impacts upon fault coverage. These issues are discussed in a separate section.

38.3.2.1   Isolation Resistance and Voltage.   As isolation testing is a means of evaluating the ability of product insulation to withstand voltage and prevent current flow, it is common for the isolation test specification to include not only a statement of minimum resistance, but also an applied voltage. This is the voltage that the insulator must withstand while exhibiting at least the minimum isolation resistance. Given the relationship $R = E/I$, increasing the applied voltage is also a means of increasing the measurement current level up out of the noise floor internal to the test system. Thus, high-voltage-capable bare board test systems usually are able to verify isolation at higher resistance levels and do so at reasonable speed. Typical values for isolation resistance can range widely, from as little as 1 kΩ to as much as 1000 MΩ. Values of 2 to 10 MΩ are common, but higher values are useful in detecting trace contamination. Excessive humidity in the test area may preclude the use of very high thresholds and may affect accuracy at lower thresholds. Values below 50 to 55 percent relative humidity are desirable. Common isolation test standards (see Table 38.2) have not always kept pace with the capabilities of new equipment to test at elevated and sensitive thresholds. In any event, assignment of specific test thresholds would ideally be based upon a competent analysis of the intended application of the specific circuits on the product to be tested.

TABLE 38.2   Examples of Isolation Resistance Test Threshold Standards

	IPC-ET-652	MIL-55110D (obsolete)
Minimum isolation resistance test threshold	Class 1: general electronic, 500 Ω Class 2: dedicated service, >2 MΩ Class 3: high reliability, >2 MΩ	>2 MΩ

There is somewhat less emphasis on test voltage than was true previously, probably as a result of increasingly fine geometry in the product. Older or simpler test systems may be limited to 10 to 40 V, but most are able to apply 100 to 250 V during the isolation test. While the latter values are commonly used, excessive voltage is inappropriate for very fine-pitch substrates and may result in damage from arcing in normal environments (see Table 38.3). The insulating properties typical of modern board material suggest that it may be of little value in specifying elevated voltages while using low (relaxed) threshold resistances. Raise the resistance threshold first, then use enough voltage to get adequate speed and accuracy.

38.3.2.2   True Isolation Test Method.   Several different algorithms have been developed for sequencing the switching state during the isolation test, and the choice may affect test cov-

TABLE 38.3   Examples of Isolation Test Voltage Standards

	IPC-ET-652	MIL-55110D (obsolete)
Minimum isolation resistance test threshold	High enough to provide sufficient current for the measurement in question, but low enough to prevent arc-over	At least 40 V, or twice the maximum rated voltage of the board, whichever is greater

erage. In the most rigorous method each network is individually tested to determine the total parallel leakage resistance to all other networks on the product. This requires one measurement per network, as illustrated in Fig. 38.4 for a board with three networks. Each network is, in turn, given a chance to charge to an elevated voltage. All other networks are connected together and to 0 V at this same moment. Notice that only one test point is needed per network. If the network under test is shorted or leaking to any of the other networks, or to any combination of them, it will fail to charge adequately and the test will fail. The pattern continues until each network has been tested.

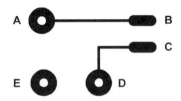

Test 1: Raise net AB positive while nets CD, E are Grounded.
Test 2: Raise net CD positive while nets AB, E are Grounded.
Test 3: Raise net E positive while nets CD, AB are Grounded.

FIGURE 38.4   True isolation test of three-network board.

Figure 38.5 illustrates the status of the measurement system during the three tests. Notice that test points A and C are used to access their respective networks, and that test points B and D are not needed for the isolation test. These test point switches remain open through all measurements.

A key feature of this method is the ability to answer the rigorous question: "How well isolated is this network from the rest of the board?" Consider the example illustrated in Fig. 38.6. Here the isolation threshold is 100 MΩ. Four leakage paths exit network A, one each to networks B, C, D, and E. Each path, measured separately, is well above the 100-MΩ pass/fail threshold. But electrically we can only guarantee that network A is isolated by the parallel combination of these four resistances, given by:

$$R_A = \frac{1}{(1/R_{AB}) + (1/R_{AC}) + (1/R_{AD}) + (1/R_{AE})} = 68 \text{ M}\Omega$$

Thus, the true isolation test method will correctly fail this measurement. A not unlikely real-world situation would be a smear of contamination that touched A, B, and C.

38.3.2.3 Log of (N) Isolation Test Method.   Table 38.4 illustrates the log of N isolation test method, and compares it to the true isolation method described earlier. The log method offers

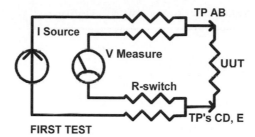

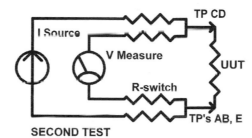

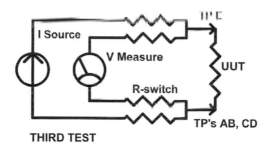

FIGURE 38.5 Measurement sequence for three isolations.

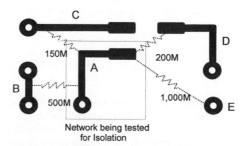

FIGURE 38.6 Parallel leakage detection.

TABLE 38.4 Isolation Methods Compared

Measurement number	Network name							
	A	B	C	D	E	F	G	H
Log method								
1	+	+	+	+	−	−	−	−
2	+	+	−	−	+	+	−	−
3	+	−	+	−	+	−	+	−
True method								
1	+	−	−	−	−	−	−	−
2	−	+	−	−	−	−	−	−
3	−	−	+	−	−	−	−	−
4	−	−	−	+	−	−	−	−
5	−	−	−	−	+	−	−	−
6	−	−	−	−	−	+	−	−
7	−	−	−	−	−	−	+	−
8	−	−	−	−	−	−	−	+

the powerful advantage that only a small number of measurements are required to test a complex board. That number is given by:

$$\log_2(N)$$

rounded up to the next whole number, where N is the number of networks.

Thus, for a board with eight networks, we need only three measurements. The gain is even greater for large boards. For a realistic board with 4000 networks, only 12 measurements would be needed.

This method involves a compromise of fault coverage for parallel leakage detection. Consider the example shown in Table 38.4. The top half of the table considers a very simple board with eight networks and illustrates the pattern of measurements required by the log method. Networks noted with a + in a given measurement are connected to the upper side of the measurement system. Networks with a − are connected to the lower side. Notice that in each measurement about half the networks are positive and half are negative.

A hard short between any two networks in Table 38.4 will be detected by either method. In at least one measurement, a + will be at one end of the short, and a − at the other. Both methods work well for hard shorts, and parallel resistance effects play no significant role. But consider the effect of two 15-MΩ leaks, one from network A to network B and the other from network A to network C. If the threshold is 10 MΩ, this fault will not be detected by the log method, but the true method will fail test measurement #1. The log method is ineffective because there is no one measurement that has the + end of the measurement system on network A while the − end is connected to networks B and C simultaneously. (Were the A-B resistor moved to A-D, then we would detect an error on measurement #2 of the log method.) The log method may fail to detect certain combinations of high parallel leakage. For product types where a low isolation resistance threshold is acceptable, this may not be a significant issue. Note that this risk may be reduced by substantially increasing the test threshold, such that a larger proportion of individual leaks are likely to trigger a fault report.

38.3.2.4 Isolation Failures: Distinguishing Shorts and Leaks.   Earlier discussion made a distinction between hard shorts and leakage failures, although both reports result from the isolation test. Consideration of either the true isolation test method or the $\log(N)$ method shows us that neither method immediately tells us the opposite end of the short or leak (refer again to Table 38.4). In the true isolation method we know that the network under test is leaking or shorted to some other network, but we don't know which one. Initially the log(N) method tells us slightly less: we know only that the short is between that half of the networks switched positive and the half switched negative (or ground). Upon fault detection either method must call a subroutine to systematically search all network pairs to localize the fault. Searching all networks in pairs for leakage can be slow as well as ambiguous. Therefore some test systems allow the user to disable searching for leakage.

In the case of distributed leaks, a clear answer may be impractical because searching pairs of networks never connects all of the leakage paths in parallel. The effort to localize a high-resistance leakage may report the entire problem, part of the problem, or none of the problem. The exact result depends upon the actual leakage resistances present, the pattern of networks involved, and the isolation resistance threshold. Note that a flying probe system seeking to "verify" defects may be unable to detect some or all leakage for the same reason. It may be advisable to return such boards to a grid (or other fixtured tester) for final pass/fail.

38.4 SPECIALIZED ELECTRICAL TESTING METHODS

Certain special testing methods have evolved to provide for detection of defects more subtle than simple shorts, opens, or leaks. Additional methods have evolved as suitable to specific types of test equipment, notably the flying prober. These are discussed in the following text.

38.4.1 Hi-Pot Testing

Hi-pot or high-potential testing is very similar to isolation testing, but is commonly distinguished by the magnitude of the applied voltage and, to some extent, by the expected behavior of the detected failure. In the context of bare printed circuit boards, hi-pot testing usually refers to voltages over 250 V, often 500 to 3000 V. The objective is to locate faults in the dielectric (insulating) layers of the board that may result in subsequent field failures at lower voltages. When the dielectric is subjected to a voltage substantially in excess of the expected working voltage, certain types of material defects can result in an "avalanche" mode of failure of the insulation. Ionization of the intervening material or atmosphere occurs, resulting in a sudden increase in current flow to some relatively large value. Visible arcing and/or burning may result at the failure site. This contrasts somewhat with the isolation test, where a very subtle leakage current flow is usually detected using a somewhat lower voltage. Most often, the isolation test is less destructive as the total energy delivered is less. However, depending upon product conditions, arcing can occur at many voltages and in such cases an ordinary isolation test is essentially the same as a hi-pot test.

It is often impractical to perform hi-pot tests on finished products, except at limited numbers of test points. Equipment limitations are one factor, it being expensive to construct test equipment and fixtures capable of routing such high-voltage signals to a large number of test points. However, the product itself is often a poor target for such tests when evaluated in finished form. Modern products are often constructed with relatively fine spaces between conductors. On the surface layers of the board the exposed component connection sites are usually too close together to withstand very high voltage stress without surface arcing. Such arcing is destructive to perfectly good product. The voltage at which such arcing will occur is a function of product geometry and atmospheric conditions.

In consideration of this, the most common practical application of hi-pot testing is to inspect raw material for defects before etching. For example, a thin FR-4 core, clad with copper on both sides, might be evaluated by placing a high voltage across the opposing sides. Any crack or other defect in the insulating material may ionize, resulting in a large current flow. Thus the defective material is rejected before substantial value is added in subsequent processing.

38.4.2   Embedded Component Tests

Methods of embedding certain electronic components *within* the board have been developed. The most common example is the embedded or "buried" resistor. Such resistors are constructed by embedding a layer of partially conductive material within the board. By selectively removing (or adding) material, the resistance value is adjusted. Accuracy ranging from a few percent to many tens of percentage points is realized. Typical resistance values range from a few ohms to thousands of ohms. The most common use of this technology is within high-speed digital circuit designs to replace large numbers of termination resistors with a resistance of $200\ \Omega$ and lower. Measuring these values accurately is challenging, requiring good fixture construction and cleanliness. Another difficulty with testing of buried resistors is obtaining usable expected value data for the resistors. At the board shop most resistors are specified as a shape and a type/thickness of resistance material, rather than as a specific resistance value and tolerance. It may be necessary to analyze the circuit pattern and compute the net resistance of series and parallel combinations, arriving at a measurable final resistance value for the tester.

Embedded capacitors may also be present. All boards exhibit capacitance between various traces, and especially between planes. The amount of capacitance is determined by the parallel surface area of the conductors, the thickness of the insulator between them, and the dielectric constant of the insulating material. Some product designs seek to maximize the desirable noise suppression benefits of capacitance between power and ground planes by making the intervening insulator thin. Some go further and use special insulating cores for these layers. This is the common form of buried capacitance. Most bare board test systems are not well suited to measuring capacitance values, particularly small ones. In some cases, no measurement is specified and it is only important that the test system tolerate the presence of the capacitance. When measurements are specified, it is usually true that few test points are involved. It may be expedient to use benchtop equipment.

38.4.3   Time Domain Reflectometry

Time domain reflectometry (TDR) is a measurement method often used to verify the RF impedance of a signal conductor on a circuit board. The RF impedance is important to the proper function of high-speed digital or radio frequency (RF) applications. Examples include computer products, cellular telephones, radios, etc. The RF impedance of a signal path should not be confused with the DC resistance, as verified during the continuity test. It is common for a trace with seriously errant RF impedance to demonstrate a very solid DC connection (i.e., low DC resistance). The RF impedance of the trace is most strongly affected by the trace's width, thickness, z-axis spacing from the ground plane, the location of adjacent traces, and the relative dielectric constant of the type of insulating core used to build the board. These parameters are usually rather constant within a particular panel, justifying use of coupons as a means of monitoring the delivered product. Standard bare board testing systems (other than flying probes) do not incorporate TDR capability, as the signal paths through the fixture will not pass the fast-risetime TDR signal. TDR is commonly conducted on a test bench manually.

TDR test systems inject a very fast-risetime voltage step into one end of the conductor. Discontinuities in the RF impedance level along the conductor result in reflected voltage waves being returned to the driving point, where they are collected by the same probe that injected the signal. The result is usually presented as a graph of RF impedance vs. distance from the point of injection. Because of reflections and disturbance occurring at the point of injection, a minimum trace length is needed to get a meaningful measurement. Extremely short networks are not ideal candidates for TDR measurement. A trace 2 in or more in length is generally required to obtain a useful reading. Longer traces may provide improved accuracy of the result. A branch-off to a second signal path along the measured length will disturb measurement badly; thus an undisturbed signal path provides the best measurement target. A typical TDR result graph is illustrated in Fig. 38.7. This illustrates a region of approximately $50\ \Omega$ RF impedance, rising toward infinity as the trace ends in an open circuit at the right, but falling between the upper and lower pass/fail bounds in the region of interest.

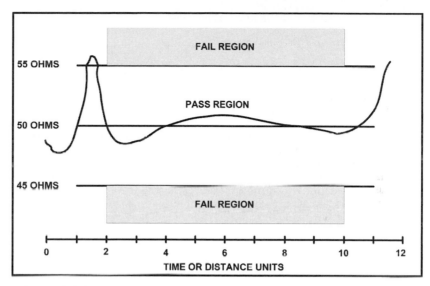

FIGURE 38.7   Typical TDR measurement.

38.4.4   Test Methods Unique to Flying Probe Systems

Because flying probe systems contact pairs (or other limited numbers) of points at any one time, they cannot directly perform precisely the same isolation measurements as can universal grid (and other fixtured) systems. DC isolation is performed between pairs of networks, with the number of measurements reduced by determining which networks are adjacent to the network being tested and therefore likely to cause isolation problems. Otherwise, flying probers are capable of ordinary DC continuity and isolation measurement, as discussed earlier. In addition, most flying probe vendors have developed alternative measurement methods that reduce the number of measurements and therefore reduce time lost to mechanical motion. These methods are discussed in the following text.

38.4.4.1   Indirect Measurement of Isolation or Continuity.   Different vendors have developed various implementations of indirect measurement, but in general these methods share a general assumption that a given product network will display a certain amount of capacitive or other electromagnetic coupling to neighboring planes or traces. The amount of coupling is affected by the geometry of the traces involved. If a trace is broken, the remnant will display reduced coupling. Similarly, if a trace is shorted to another trace, the amount of coupling will increase substantially. If the test system measures the amount of coupling from each trace to a ground plane (or other electrical environment), a degree of confidence can be obtained that all traces are intact. No direct measurement of isolation or continuity is necessarily made. Instead, the coupling signature is used to imply the correctness of the configuration. As no direct measurement of isolation or continuity is made, we refer to these methods as indirect measurements.

Typically these methods are highly reliable in detecting hard shorts and opens, but may be less effective in detecting distributed contamination or high-resistance connections of several megohms or more. Specific methods include constant-current capacitance measurement (charging or discharging variations), voltage-source measurement of RC time constant, AC capacitance measurement, and measurement of the electromagnetic coupling of an AC signal between adjacent networks.

The capacitive techniques using constant currents are generally based upon the approximation:

$$C = i \frac{\Delta t}{\Delta v}$$

where C is the network capacitance, i is the charge or discharge current, and Δv is the change in voltage that occurs during the measurement time Δt. Related, though more complex, behaviors occur for RC time constant and AC coupling measurements.

Using the preceding example, note that if two networks are shorted, the resulting capacitance is the sum of their individual capacitances. The larger capacitance causes a slower charge rate than is expected for either net individually. The system may note that these two networks are adjacent and displaying suspiciously similar charge time behavior. The system tags the networks as suspicious, and either fails them or verifies the presence of a short with a DC measurement. Networks containing an open will have reduced capacitance and will charge or discharge too quickly. These conditions are illustrated in Fig. 38.8.

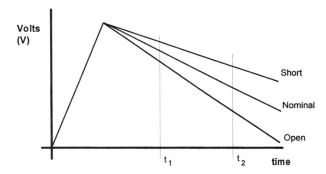

FIGURE 38.8   Indirect measurement: constant-current discharge.

To use these indirect methods, most systems require testing of a first article board using standard direct measurement methods for both continuity and isolation. Once the board is known to be good, the signal-coupling behavior is learned for each network on this board. The learned values are saved and compared to measured results from subsequent boards as described previously. Tremendous speed increases are obtained by eliminating repetitive probing of each network at multiple locations, particularly in the case of isolation testing. Some users combine a traditional DC continuity measurement with the indirect method as a substitute for isolation.

38.4.4.2 Adjacency Analysis: Isolation Testing on Flying Probe Systems. Adjacency analysis simplifies isolation testing on flying probers by reducing the number of measurements, even for direct measurement methods. In adjacency analysis a database is prepared listing each network and all traces found to be immediately adjacent to that network according to a set of geometric criteria. Such locations are assumed to represent the sole opportunities for shorts and leaks. Isolation testing is then performed only between adjacent pairs, this measurement being well suited to the nature of the flying prober. It is important that attention be paid to whether such analysis is three-dimensional. Networks may cross over a network horizontally, but on a different layer of the board. A hole or defect in the intervening insulating layer may allow a short or leakage to occur.

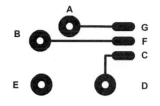

— = ADJACENCY DISTANCE
SPECIFICATION

FIGURE 38.9 Network adjacency.

In the example illustrated in Fig. 38.9, network A would probably be judged adjacent to B, but not to C. B would be considered adjacent to A and D. Assuming that the qualifying dimension was a bit smaller than the distance from E to either B or D, E might not be judged adjacent to either.

38.5 DATA AND FIXTURE PREPARATION

An essential element in any test operation is the means of generating test programs, and, in the case of fixtured tests, the design of an appropriate test fixture. There are several methods for deriving this program and fixture information. These methods vary from simple to complex and from escape-prone to very sound. The most common terms used in the language of bare board testing for the test program development are self-learning and netlist testing. Netlist *testing* is a misnomer that confuses CAD output with test inputs and implies that there is not much in between. In this discussion we will use the term *data-driven test program* to indicate the program to be loaded into the test system is derived from the board's design data (which may be provided in any of several different forms). On the other hand, a self-learned program is derived by placing an assumed good board on the test system and causing the test system's internal computer to automatically create a program that can be used to compare other boards to the detected pattern. It is worth noting that the description *known good* is often used to describe this first board, when in fact our knowledge of its quality is often imperfect.

Flying probe test systems generally require no fixturing, except sometimes a frame to hold thin boards or to load multiple small boards simultaneously. Flying probers still require test program generation and optimization. Instead of a fixture, the product program contains probe position information for planned test. Data preparation processes are therefore similar to those of fixtured tests, but only test program data are output.

Note that the extractions of program and fixture information are quite naturally linked. In fact, the final test program cannot be created until the fixture is defined. The fixture determines how the individual test points of the test system are connected to specific product locations. Knowledge of how the fixture has taken hold of the product is essential to determining which system test points to employ in performing specific measurements. Thus, the same software system generally outputs both the final test program and the fixture drilling/loading information.

38.5.1 Self-Learning

Self-learning has increasingly limited application and is subject to certain escape risks derived from assuming that the original circuit board is good. The self-learning method requires that a fixture be already available, as well as a known good board (preferably). A shorting plate is placed on the fixture in place of the product, and the test system uses this to identify which test points are employed in the test fixture. (All test points found shorted to the plate are considered active. This eliminates test points of no interest, shortening the test program and saving considerable execution time during both learning and testing operations. This set of active points is sometimes referred to as a mask.) The known good board is then substituted for the shorting plate, and the pattern of product interconnections is learned and saved as a test program. A few boards are tested, and, if results appear reasonable, the program is considered valid. The key drawback of the method is that self-learned test programs determine that all the boards are the same, not that they are good. Moreover, for economic fabrication of the test fixture it is necessary to process product data anyway; thus we may as well have output a data-driven test program. With virtually 100 percent of board designs being CAD-driven, there is little motivation to use self-learning today.

38.5.2 Data-Driven Testing

The preferred method of deriving program and/or fixture data is data-driven programming (DDP), sometimes referred to as netlist testing. The basic idea of data-driven test programming is to test the board using the same database used to specify its manufacture—in other words, the original design database. The fixture and DDP development process can be divided into two stages:

- Input/extraction, the preparation and processing of the various possible input sources into a format usable by the second stage
- Fixture data, test program, and repair file output, and subsequent application of the data

Factors such as data quality and completeness, the format of data available, and fixture design can significantly reduce or extend software processing and engineering time. Board technology, size, and complexity tend to increase time requirements. While software tools are not always the most immediately visible tools in the testing department, the completeness and automation of software tools can have a large impact upon whether you are able to set up and test boards quickly and efficiently vs. struggling to build fixtures quickly—while testing very few boards. Ideal fixture software readily accepts the product data, analyzes the data, recommends an optimal fixture design including all fabrication details, and outputs CNC files for fixture fabrication, material requirements, special procedures required, and a test program in the proper format for the selected test system(s).

38.5.2.1 Selecting Fixture Software.
A complication arises due to the mechanical incompatibility of many types of test systems. If the available equipment includes something of a grab bag of different equipment types, the test manager's task is made considerably more difficult in several ways. Aside from the obvious problems of manufacturing different fixture types, the manager is also faced with difficult load-leveling tasks. In such circumstances the test manager may be forced to select general-purpose fixture software packages. By virtue of offering many configuration options, these require additional operator training and present increased opportunity for error. If the testing equipment uses a shared fixture type it becomes possible to standardize the software process (and raw material). Optimized configurations may be locked into the software, minimizing error opportunities.

38.5.2.2 Test Data Extraction Methods for Fixtures and Programs.
The initial stages of processing include the acceptance and inspection of the input data to ensure that they have been presented in a valid format. These data are then scanned to determine which sites on the board deserve test points, and what measurements are required among these test points in order to ensure proper continuity and isolation of the product. There are a variety of input sources, summarized in subsequent sections.

A usual goal of the extraction process is to eliminate unnecessary test points. According to predefined rules, the software identifies those end points of each network segment where test pins can and should be placed. For example, a T-shaped network probably requires three test points, located at the extreme extensions of each arm of the T. This provides sufficient connectivity to the product to permit a complete continuity test. One point per network is the minimum needed to attempt isolation testing, and two points per network make the isolation test more robust by permitting the tester to ensure that these points are in proper contact with the product.

The following is a generic description of the software process, where CAD or photoplot information is available.

- *Data orientation.* The toughest side of the board in terms of test centers is usually oriented face down. It is often easier to manage a heavier, more complex fixture on the bottom side of the test system. This may require flipping or inverting the data.

- *Test site identification and optimization.* Depending upon customer preference, it is often true that some test points can be deleted from the program. Figure 38.10 shows a network with a midpoint B that does not need to have a test point on it. The test points at A and C are all that are necessary to determine whether this network is isolated and contains no opens. Test point optimization can reduce the number of points in a fixture by 10 to 50 percent, thereby reducing the cost of the fixture and the time needed to test the board. Although optimization provides some economic and minor throughput benefits, this must be offset by any impact upon the repair process and by any fault coverage risk. Midpoints can provide failure data that help the repair station more easily pinpoint the actual fault site. Where an operator is required, or when flaws appear in the algorithm, the deletion of test points can add to the risk of test escape. Optimization is sometimes not performed, speeding data preparation and reducing escape risk but adding cost to the fixture.

FIGURE 38.10   Optimizing a network.

- *Pin type assignment.* Once the test sites have been identified, test pins can be assigned to the features to be tested. Different pin types may be required for different features. For example, a through-hole might require a large head or larger-diameter pin than an SMT pad on 0.025-in centers. The software usually includes a table relating feature types to pin types.

- *Test system type identification.* Mechanical details of the fixture must suit the target test system. The fixture in a universal grid system requires assignment to a grid of uniform centers. A more detailed description of fixture and system types is included elsewhere in this chapter, but obviously the fixture must be built in the correct size and grid density to match the specific test system to be employed. Figure 38.11 shows three of the most common grid configurations available. Single density (100 points per square inch or 0.100-in grid) is the most common, with double-density systems being added as product technology demands. Of course, a few systems still use wired fixtures, and here the output will be a set of drill files, a wire list, and the test program. The balance of this discussion focuses on tilt pin translator–equipped grid systems.

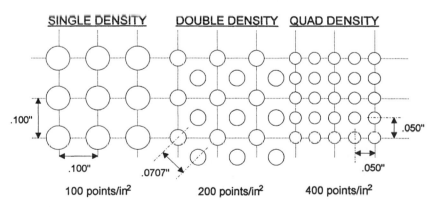

FIGURE 38.11   Single-, double-, and quad-density universal grid patterns.

- *Grid pin mapping.* The fixture software determines which system grid location will be mapped to each of the previously determined on-product test targets. Several factors are considered:

  - No two targets can be assigned to the same grid location.
  - No assignments can allow the pins to touch each other as they tilt through the fixture (cross pins).

  – Drilled holes in fixture plates cannot break out into adjacent holes (adequate spacing on all plates).
  – No assignment can require a pin to tilt beyond the predetermined deflection limit (determined by the fixture and pin design).

FIGURE 38.12   Test point staggering.

  – Test sites that are very close together may require that on-product targets be staggered such that the test pins are more widely separated. (See Fig. 38.12.)
  – Hole diameter and placement relative to the pin diameter and pin deflection must be optimized. A nearly vertical pin might be assigned a hole diameter just slightly greater than its shaft diameter. This clearance balances avoidance of friction against pointing accuracy during testing. Other pins of the same diameter that are highly tilted will require a slightly larger hole to avoid binding. The fixture adjusts drill sizes for each hole, depending on the amount the pin is deflected. This maintains pointing accuracy while eliminating pin binding. Hole locations may also be shifted slightly to compensate for the fact that the pin is pointed at an angle and that the point of contact will not be at the center of the hole. The result is a more reliable test fixture.

- *Test program output.* With all pins fully mapped to grid locations, the test program file can now be constructed and saved.

- *Tooling.* The last step in data preparation requires the marrying of the job being processed with the tooling. The tooling includes tooling to the test system and tooling for the specific fixture design utilized.

- *Fixture file outputs.* The completed CNC files for fixture drilling and fabrication can now be created and the test fixture prepared.

38.5.3   Data Formats

Modern circuit boards are designed using computer-aided design (CAD) systems. These systems output data files used by the board shop to generate phototools, drill files, routing patterns, and other tools used to fabricate the boards. While several standards have been developed for data outputs from this process that can provide useful data input to the final test process, adoption has not been universal. As a result, test engineers and software providers have developed means of hijacking and converting the photoplot data stream, and can use this to drive the test process with certain compromises and a degree of inconvenience. Both approaches are discussed in the following text.

38.5.3.1   Gerber Format Data Extraction.   The primary example of "hijacked" photoplot data arrive in Gerber photoplotter format. Even here a standard was eventually developed. RS-274X is a standard that affords a degree of consistency to the photoplotter files. This data stream is a series of drawing commands that direct the photoplotter in drawing the copper patterns for each layer of the board. The drawing is made with a selection of "apertures," which define different pen tip shapes and dimensions. Thus a fine line is drawn by choosing a small aperture, issuing a light-on command, and directing a move. These data are contained in a very large set of files, containing a picture of each layer of the board. Combined with a drill file and aperture definitions, the appropriate information for generation of fixture and test program files is present, but a considerable amount of processing is necessary to develop understanding of layer-to-layer connectivity and finally identify test targets.

Although the standardization of Gerber data has been improved with efforts such as RS-274X, it remains a data format originally optimized for the purpose of photoplotting and not for testing. CAD system developers and board designers have a great degree of freedom in how they design board features and in how these features are represented in Gerber. As a child will readily demonstrate, many different sequences can be used to draw even simple shapes. A rectangle might be created by "flashing" a square pattern, or might be created by an elaborate spiral using a much finer aperture, a zigzag pattern, or a sequence of overlapping flashes. This variety leads to occasional errors, because it is difficult for any Gerber extraction routine to be capable of handling all possible permutations. The software system must successfully interpret any of these examples (and many others) as defining the same intact square area of copper conductor pattern on the final product. Some methods result in huge file sizes.

Despite these problems, this method is far superior to self-learning, digitizing, or drill file extraction, and remains in common use in a variety of software packages. The data requirements to start from this format include:

- Photoplot files for each layer, solder mask layers, etc.
- Photoplot files for silk screen legend(s) (required for computer-aided repair)
- Aperture file
- Drill file
- Board outline

38.5.3.2 CAD/CAM Data Extraction Method.   Recognizing the problems inherent in photoplot extraction, system vendors have agreed to add support for test oriented output formats, in particular IPC-D-356 and 356A. These formats provide data in a much more readily digested format, and eliminate most ambiguities in processing the data. These formats include the following data:

- Signal ID, network name, and/or network number
- Reference designator or PIN number (e.g., U14, 12, R11) for related component on board
- *x-y* coordinate of pad center (minimum data set requirement if grouped as connected)
- Pad dimensions relative to center, and hole size (if any)
- Resistor or other component values (if appropriate and not usual)
- Board side (top or bottom)
- Mid-network flag suggesting that test point placement may not be required

Often this data set is converted to a standard such as IPC-D-356 from the CAD/CAM system's internal data formats by means of an intermediate converter. These software converters, although simple, are usually customized for each individual CAD system. A converter must be updated or modified with any output changes made by a CAD vendor due to a CAD system software update or new product introduction. The number of converters required by an independent PWB manufacturer could be quite high, as data will likely be received from many different CAD system types. Fortunately, many of the CAD vendors now include direct output of IPC-D-356 data or readily provide converters. Acceptance of these standards is now such that in many cases even Gerber input data are first converted into IPC-D356 format prior to final processing.

38.5.4 Outputs from Data Extraction

Once all the preparation steps and processes have been performed, there are several outputs generated by the fixture software.

- *Test file.* A data-driven test program is outputted in a format compatible with the test system type. This file informs the test system of the measurements expected and of pass/fail criteria. Some test system formats permit the data-driven test program to support a graphical representation of the fixture and/or board on the test system monitor, provided that sufficient data are included in the test program file. This graphical presentation is useful in fixture and/or program debugging.

- *Fixture fabrication files.* Probably the most significant output — the drill files, one for each plate or pass required by the fixture design—is needed in drilling to start building the fixture. (In the case of wired fixtures, a wiring list is also required.)

- *Repair/verification files.* Files can also be outputted that support the repair or verification function. These relate the graphical image of the board to the assigned test point locations in the program. Preparation often relies on inclusion of Gerber data in the input data stream to the software system. Extensions to IPC data formats are planned to better support the repair function without this recourse to photoplot data. Some test systems may carry trace image information in the test program file as well, for enhanced debugging support on the test system.

38.5.5 Setting Up a Fixture

With the fixture assembled and CAD data–derived program prepared, the next step is to set up the fixture on the test system and validate the fixture and program. Details vary with the fixture and system used, but in general the following steps are taken:

- Program data are loaded into the test machine from the disk or network.

- Test thresholds are set to the desired continuity and isolation values.

- Compression settings on the test system are adjusted to the proper values for the fixture type. In some cases, these values may already be included in the test program file.

- The new (or recalled to duty) fixture is compressed with a nonconductive material of similar thickness to the board being tested.

- An all-isolations version of the test program is executed to verify that the fixture does not contain any internal shorts.

- The fixture is then compressed again with a conductive plate in place of the product, intentionally shorting all fixture test points together. Often this plate is a simple piece of copper-clad G-10. Copper oxidation can prevent reliable contact, and users sometimes wrap the plate in fresh aluminum foil or use a flash–gold covered plate. In this step, an all-shorts version of the test program verifies that all expected test points are continuous through the fixture, up to the shorting plate. The intent is to verify that the number of test points is correct and that, under multiple compressions or closures, all points are present and remain in contact with the shorting plate.

- Assuming proper alignment and cleanliness, product can then be tested with the final test program. If certain errors immediately repeat on all boards, then it is reasonable to suspect a pin loading error or other error in the fixture, and investigation is warranted.

The basic techniques just described are generally applicable to all fixture types, whether tilt pin, wired-dedicated, or other specialized types.

Do not underestimate the value of removing dust and debris from the product and fixture, both during setup and occasionally during operation. Target sizes on modern products are not tolerant of debris, and false open circuit reports are an immediate result. Tacky roller-type cleaning systems may be helpful in periodically removing debris from test fixtures, system grids, and the products themselves. Follow the test equipment manufacturer's recommendation regarding the use of electrostatic discharge (ESD)-safe cleaning materials near the test system.

Adjustment of fixture compression is also important, especially on press-type systems where the amount of compression stroke is controllable. (In vacuum fixtures the dimensions of various fixture plates and components usually fix the amount of compression.) Undercompression usually leads to poor contact and false open results. Overcompression can cause excessive marking of the product by probe tips, probe damage, and fixture damage.

Overcompression is a very common problem, and, unfortunately, so is product damage in the form of excess probe marking. It seems intuitive to just press harder when you experience contact problems, and this is often the first step taken. But the actual change in force per spring probe is very small as it travels further, until the spring probe hits bottom—and at that point you begin damaging the product almost immediately. A typical spring probe in a grid system presents about 139 g of force when compressed 0.167 in (two-thirds of travel). At zero travel (uncompressed) the spring inside is already preloaded to about 55 g. The force increases at a spring rate of 503 g/in. Just before bottoming, the spring force peaks at about 180 g. Thus, there is only modest force gain possible before serious danger of bottoming the probe occurs. Consult the spring probe manufacturer for details, but most probes function well at about two-thirds of full travel. If contact problems remain at this level of compression, leave further compression adjustment as a last resort and look elsewhere. Generally, you should never cause the probes to hit bottom. (Some older vacuum fixture probes are a notable exception).

38.6 COMBINED TESTING METHODS

As product density and geometry become more challenging, the cost of fixture construction increases. In a large number of cases, the most challenging test sites occur at only a few areas of the product. It is possible to combine test techniques such that less expensive means are used to test the majority of product features, while more advanced and costly techniques are reserved for the most challenging test sites. Some examples follow.

38.6.1 Split Net Testing

This is perhaps the oldest and most primitive example of a sequential or combinational test. When the product complexity exceeds the capacity of a fixture design to solve the test requirement with a single fixture, it is possible to divide the test responsibility between multiple fixtures. The flip test method of dual-side access described earlier is one example of this. More typically, the test department has dual-side-access equipment, but the density of test sites exceeds the capability of the fixture and/or test system.

One way to split the responsibilities is to divide continuity and isolation between fixtures. For example, one fixture could be limited to two test points per network. These two would be checked for continuity with each other (to verify that there are no open pins in the fixture), and would then be used to perform a 100 percent isolation test of the product. A second fixture would complete the continuity test. In very difficult cases, multiple fixtures can be used.

Handling boards in this scenario is complicated. Boards that pass the first fixture must be stacked and prepared to run on the second fixture. But boards that fail the first fixture must be tagged with failures, and in many cases must be run through the repair process, retested, and combined with the boards being sent to fixture 2. Such failed boards must not be confused with any failed boards from fixture 2, which have already completed the first phase of testing. Unless there is a large continuous flow of boards with two different test systems/fixtures set up, it is also likely that the two fixtures will have to be installed/removed from the test system a number of times.

38.6.2 Manual Combination of Methods

It is also possible to divide the test burden by testing the majority of ordinary test sites with a test fixture, followed by a flying probe system to test strictly limited portions of the product.

Without special software, one has the same handling problems as in the split net case. But it is possible to automate the data handling between the two test systems using network resources, such that a single error tag results from both tests. All boards then flow through both test systems, and any rejects flow to repair. This flow is very similar to that of ordinary testing, resulting in less confusion. An interesting additional benefit of this combination is that the flying prober can not only test those sites that you wish to avoid fixturing, but can also retest any of the failures noted by the universal grid. This can eliminate the majority of false open reports caused by stuck pins and contamination, improving the total first-pass yield.

38.6.3 Integrated Sequential Testing

With increasing application of automated material handling to bare board test systems, it is possible to physically dock a universal grid test system to one or more flying probe systems, as illustrated in Fig. 38.13. The resulting unit is capable of a complete combinational test of complex products without intermediate handling of the product. While the flying probe systems may not completely keep pace with the grid units, the grid units can certainly improve the total productivity of the flying probe systems by decreasing the number of measurements required. The flying probers reduce false error reporting, improving the net throughput in the testing department.

FIGURE 38.13 Integrated sequential test system. *(Courtesy of atg Test Systems GmbH.)*.

CHAPTER 39
BARE BOARD TEST EQUIPMENT

David J. Wilkie
Everett Charles Technologies, Pomona, California

39.1 INTRODUCTION

The largest volumes of bare boards are tested on fixtured systems. However, for smaller volumes or special purposes, "flying probe" systems may be preferable. This chapter discusses them both to enable the user to make the most effective decision based on testing objectives and volumes.

39.2 SYSTEM ALTERNATIVES

39.2.1 Fixtured Systems

Each dedicated, or hard-wired, test system presents some sort of standardized interface pattern of test points. This may consist of a universal grid's continuous array of points (bed of nails), or may be a connector pattern of some type. In the most primitive systems, a simple cable connector may be presented. Each of these types of equipment uses one or more types of customized test fixtures to connect this standard electrical interface pattern to the unique contact pattern of a particular product to be tested.

Each fixtured system includes a measurement unit that can be connected to any of some thousands of test points by a solid-state switching matrix. A central computer controls the measurement unit and switching matrix. This computer also controls the press (or vacuum mechanism), which can compress the product against spring-loaded test points presented to the product by a customized test fixture.

To imagine the system's operation, first consider the measurement unit to be an ohmmeter. The computer commands the switch matrix to connect the red lead of the meter to, say, test point 17. It then connects the black lead of the meter to test point 1027. The measurement system can then be commanded to measure the resistance of the product under test between test points 17 and 1027. The test program dictates a long series of such measurements organized to completely test the product. Because the switching and measuring occur at the high speeds possible with modern electronics, a complex board can be tested in seconds.

39.2.2 Dedicated (Hard-Wired) Fixture Systems

Originally the most popular system type, dedicated systems are generally being displaced by systems employing less expensive and more accurate fixturing methods. The word *dedicated*

39.1

refers to the fact that it is generally impossible to salvage very much material from these fixtures for reuse in other fixtures. The cost of these fixtures derives from this fact, and from the high labor and material costs involved in the original construction.

39.2.2.1 Advantages of Dedicated Fixture Systems. A primary advantage of these systems is that, because each test point is routed within the fixture by a flexible wire, it is possible to position any test point at just about any location. If your product requires 8000 active test points, then you need no more points than this in the test system. 100 percent utilization of available test points is possible. This contrasts with the universal grid approach discussed later, wherein points can only be shifted a small distance from the original location on the grid. In that case, you must have enough electronic test points to cover the entire surface of the product at some constant density, and any given fixture will generally use only a portion of these points. Thus, the capital equipment cost of a dedicated system may be lower, but the ongoing fixture costs may be much higher. Dedicated systems are generally built with no more than about 10,000 test points, with smaller numbers common.

39.2.2.2 Construction of Dedicated Fixtures. The test fixture for such systems generally consists of a rigid box structure with one side facing the product to be tested and some other side or surface presented to the interface pattern of the test system. The side facing the product includes the probe plate. This plate of insulating material is drilled to mount an array of spring-loaded pins able to make electrical contact with the product. The reverse side of these pins is connected within the box by physical wires to the system interface connector on the opposite side of the fixture.

Fabrication of the fixture involves the usual data-processing steps, followed by spring probe type selection, drilling of the probe plate, installation of spring probes, drilling of the interface plate, and installation of system mating contacts or connectors. Many spring probe tip styles are available to suit different diameters of target pads and holes. (See Fig. 39.1). In most cases the

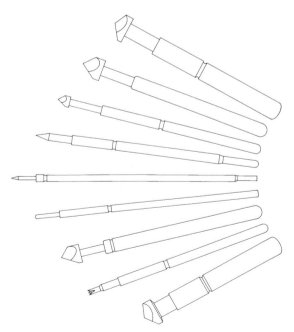

FIGURE 39.1 A variety of spring probes and receptacles. *(Courtesy of Everett Charles Technologies.)*

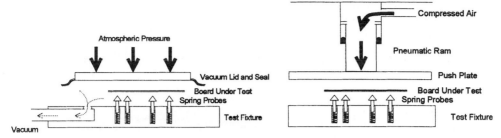

FIGURE 39.2 Vacuum vs. pneumatic compression.

probe plate actually carries a socket or receptacle into which a replaceable spring probe is inserted, simplifying service. The underside of the receptacle protrudes into the fixture box and carries a wire-wrap tail. Each of these receptacle tails must be wired to an individual system interface contact on the bottom of the fixture.

Compression of the product onto the fixture spring probes may be accomplished with electric drives, pneumatic air cylinders, or vacuum (see Fig. 39.2). Vacuum compression can result in the lowest system cost, but adds cost to the fixture and limits the total number of probes that can be compressed.

Many forms of fixture interfaces have been provided, with many appearing as small universal grid patterns. Some very simple test systems have been constructed with a large number of ribbon (or other) cable connectors presented as the only connection to the fixture. The matching wired fixture may be troublesome to connect (because of the large number of cables), but is left connected for days or weeks at a time to a single fixture type due to the quantities of boards being tested.

39.2.3 Flying Probe–Type Test Systems

Smaller volumes and specialty parts are tested on flying probe–type test systems. These consist of a small number of robotic probes with independent measurement abilities. These are moved among the various product test target locations, and a sequence of measurements is made. Such systems offer the powerful advantage that no fixture preparation is required (although data must still be processed to prepare the test program). The best of these systems are also able to provide extremely high probe placement accuracy, generally exceeding that of fixtured systems. The primary drawback of such systems is low throughput, as a result of the need to mechanically reposition the probes between measurements.

39.3 UNIVERSAL GRID SYSTEMS

The most flexible and widely used electrical test solution is the universal grid test system. (See Fig. 39.3). Most systems now include upper and lower grids, permitting simultaneous dual-side access for test of SMT-type products. Grids offer very high test speeds, and moderate the cost of test fixtures by permitting the reuse of many fixture components. Thus, today, the majority of product is tested on universal grid systems and fixtures.

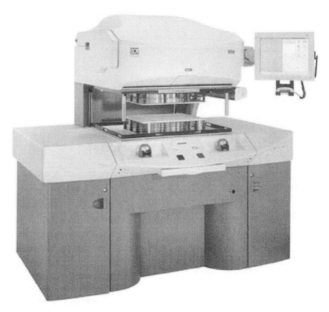

FIGURE 39.3 Universal grid test system. *(Courtesy of Everett Charles Technologies.)*

39.3.1 Universal Grid Test System Design

The universal grid test system presents a rectangular array of equally spaced test points. Generally this is chosen to be large enough to cover the test area of the largest product type to be tested (Fig. 39.4). It is common to speak of the density of test points presented. A single-density system presents points spaced at 0.100 in (10 per inch). Thus, there are 100 points per

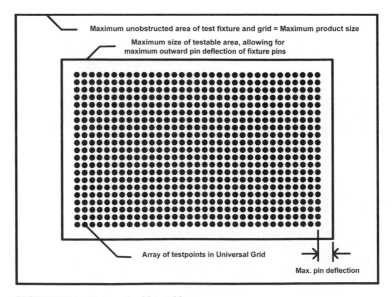

FIGURE 39.4 Universal grid testable area.

square inch. Similarly, a double-density system has test points spaced at 0.0707 in and provides 200 points per square inch, and quad-density spacing is 0.050 in for a density of 400 points per square inch. As grid cost is largely a function of the number of test points, larger sizes and/or higher-density configurations become expensive. With modern grid designs, the test system grid size can be upgraded in the field by addition of electronic modules. With older designs where wire is used between the grid and the electronics, upgrades may be less practical.

39.3.2 Exclusion Mask Fixtures for Universal Grid Systems

Occasionally, some applications involve product whose test point spacing exactly matches the grid pattern. This situation may permit use of a very simple fixture referred to as an exclusion mask. This is composed of a thin glass epoxy sheet, perhaps 0.030 in thick, drilled at those locations where test probes are desired to pass through from the grid to make contact with the product. The exclusion mask (see Fig. 39.5) prevents unused grid probes from contacting the product unnecessarily, preventing erratic contacts and marking of the product. In order to use such a mask, it is necessary that the grid be constructed with a pointed or chisel-tip probe. Unfortunately, few products exist that are so simple to test.

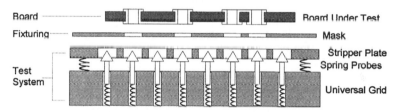

FIGURE 39.5   Exclusion mask fixture.

39.3.3 Pin Translator Fixtures for Universal Grid Systems

Most boards manufactured today are tested on universal grids through the use of pin translator fixtures. These fixtures are rapidly designed and produced using available software systems, and the most expensive material components (the contact pins and plate spacers) are generally reuseable. These fixtures are also referred to as *tilt pin fixtures* or *grid fixtures*. Grid fixtures and related software remain the fastest-evolving area in electrical testing. Particular fabrication details will not remain current for long. Therefore this discussion employs current practice only as means of illustrating key issues, such as density, registration, etc.

In a tilt pin fixture, a rigid pin serves the dual roles of providing an electrical contact path from a grid test point to the product and translating the grid test point location a small distance horizontally (in x and y coordinates) such that contact is made to the desired target on the product surface. Because the product targets are unlikely to be perfectly centered above grid test point locations, these pins are usually somewhat tilted from vertical. The amount of tilt or displacement varies according to the amount of x-y shift required. Several lightweight plates of plastic material support these pins. The individual hole locations in each plate are offset the amount required to achieve the desired pin tilt and spaced apart the required distances (see Fig. 39.6). These pins are variously referred to as *tilt pins, translator pins,* or *fixture pins.*

In order that grid test points may be translated to exactly the correct target location on the product, it is important that each translator pin fall into the correct sequence of holes while the fixture is being assembled. If sufficient intermediate guide plates are employed, then the geometry of drilled holes can be so arranged that a virtual tunnel is created for each translator pin. It becomes impossible for a pin to fall into an incorrect sequence of holes. A larger

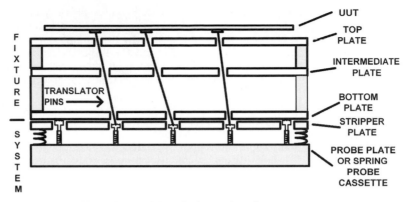

FIGURE 39.6   Single-side multiplate tilt pin translator fixture.

number of plates are required when the maximum degree of pin tilt is increased. Calculation of plate count, position, and drill hole locations is a key function of the data extraction and fixture software, whose process is described elsewhere in this chapter.

39.3.3.1   Test Pins for Pin Translator Fixtures.   There is a wide variety of pins available for use with different system types and for different applications (see Fig. 39.7). These vary in length, tip style, material, cost, and thickness. For a given angle of tilt, a longer pin is generally able to translate the test point a larger horizontal distance (larger pin displacement). Pins

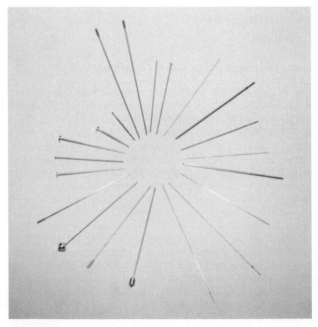

FIGURE 39.7   Examples of translator pins. *(Courtesy of Giese International.)*

with enlarged heads of various shapes are useful for probing large-diameter holes. Aggressive shapes such as chisels provide the pin with enhanced abilities for cutting through product contaminants, but may be unsuited to closely spaced test applications. At present, the vast majority of pins are headless music wire pins with lengths ranging from 2.5 to 3.75 in. When probing a large-diameter through-hole, common practice now is to probe the annular ring around the hole, rather than pay the penalty for an odd-headed pin type.

39.3.3.2 Pin Displacement in Grid Fixtures. Pin displacement capability is one of the more significant performance measurements of a pin-fixture design combination. A larger displacement ability provides greater freedom in allocating test points, resulting in improved ability to fixture complex products on a test system of given density. With an arbitrary fixed rule of thumb of 10° maximum tilt, we can see from Fig. 39.8 that the longer the pin, the further the pin can be displaced from its on-grid location to its on-product target location. The maximum pin displacement that can actually be obtained is a function not only of pin length, but also of fixture design and the resultant maximum tilt angle that can be tolerated in the design. The fixture software usually seeks to arrange pins such that the maximum tilt in the fixture is minimized. This minimizes the number of guide plates required, minimizes the number of different hole sizes required in the plates, and tends to reduce friction and binding. Nonetheless, a design that permits a maximum tilt capability ultimately provides the greatest flexibility in assigning test points and generally permits the solving of denser products on grids of a given density.

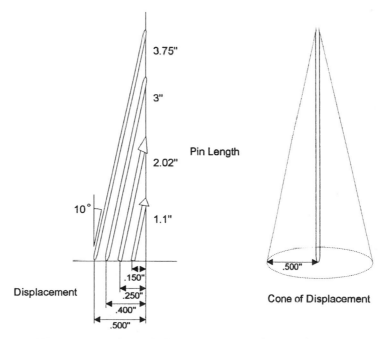

FIGURE 39.8 Approximate pin displacement as a function of pin length.

39.3.3.3 Pin Binding and Friction in Grid Fixtures. As noted, the grid fixture pins must lean some distance to connect the grid to the on-product target location. Thus, the relationship between the plate thickness, hole diameter, and angle of displacement are critical. If the plates capture the pin too tightly, binding and friction can result. A pin that is held down below other pins in the fixture by binding will fail to make contact with the product (see Fig. 39.9).

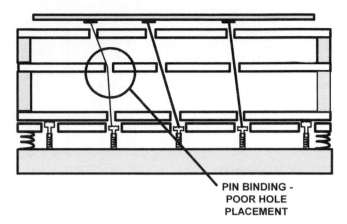

PIN BINDING -
POOR HOLE
PLACEMENT

FIGURE 39.9 Pin binding in multiplate tilt pin fixture.

The spring force of the spring probe in the universal grid is intended to supply the force that presses the tilt pin against the product. Every bit of friction in the tilt pin motion subtracts some of this force. If the friction is great, not only may the pin fail to contact, but if the pin is stuck high (toward the product surface), the product may press on it with undue force in shoving it back into the fixture—resulting in excess product marking. Binding can be caused within a single plate by hole diameters that are too small for the degree of pin tilt. Binding can result between plates if the plates are offset from one another, forcing the pin to bend as it snakes its way through the various plates. Binding can also result if the plate spacing is incorrect as compared to the design values.

39.3.3.4 *Calculations for Tilt Pin Fixtures.* Modern fixture construction software should perform all of the necessary calculations automatically. Yet it is valuable to understand the basic decision processes.

Minimum Hole Diameter Calculation. The relationships between pin displacement, pin diameter, plate thickness, and drilled hole diameters are trigonometric. Given:

HS_{min} = minimum hole size in a particular fixture plate

PT = fixture plate thickness

SD = pin shaft diameter as it passes through the plate

PL = overall pin length

PD = horizontal displacement distance of the pin tail relative to the pin head location

A = angle of pin tilt expressed as degrees from vertical. (i.e., a vertical pin has a tilt of 0°)

HT = minimum acceptable tolerance to ensure that the pin slides freely in the hole

We calculate:

$$A = \sin^{-1}(PD/PL)$$

and

$$HS_{min} = [SD/\cos(A)] + [PT \times \tan(A)] + HT$$

The first term in the equation derives from the effective increase in shaft diameter in the horizontal plane as the pin tilts (see Fig. 39.10). The second term derives from the horizontal dis-

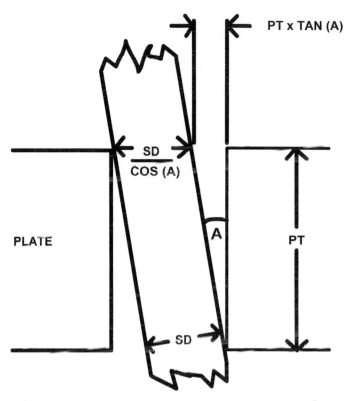

FIGURE 39.10 Illustration of hole diameter calculations for tilt pin fixture.

tance that the center line of the pin travels as it passes through the plate thickness. The third term, as mentioned earlier, is to provide a minimal clearance.

For angles of tilt below 15°, the value of the cosine is close to 1 (over 0.96), and for rough calculations the term may be omitted. (For a 20-mil-diameter pin, the contribution of pin tilt to effective shaft diameter is less than 0.7 mils). Similarly, for the same small angles, the calculation of the tangent can be simplified to $\tan(A) \approx PD/PL$. The error incurred increases as the plate thickness increases. For a 0.1-in-thick plate, the error at 15° will again amount to something less than a 1-mil hole diameter.

With these simplifications, for low tilt angles and thin plates, the formula may be reduced to:

$$HS_{min} \approx SD + [PT \times (PD/PL)] + HT$$

with an extra mil or two added to the clearance.

For the highest-accuracy applications, use the exact calculation.

Drill Bit Size Selection. Note that the minimum hole size calculated previously is often not the ideal drill size. The most common material used for grid fixtures is polycarbonate. Polycarbonate tends to have a finished hole size that is about 0.001 in smaller than the drill size for plate thicknesses typically employed in fixtures. Therefore, 0.001 in should be included in the tolerance figure *HT* to permit accurate drill bit selection.

Pin Lean and Pointing Accuracy. Pointing accuracy is defined as the pin's ability to hit an intended target, usually described as a radial dimension (the distance from the intended target to the actual contact location). Because the pin is typically not perfectly vertical, there are several possible sources of error, depending upon the type of pin employed. With the older headed pin types the head sat above the top plate of the fixture. With modern headless approaches, the pin tip is usually flush with the upper surface of the top plate while in contact with the product. As the pin tip projects less (nearly zero) above the top plate surface, its tip position in x-y coordinates is better controlled by the hole in the top plate. For this reason, most modern fixture designs operate with the probe tips arriving flush with the top plate surface while the product and fixture are compressed.

It is also preferable that the pins remain approximately flush when compression is released. If they project above the top plate surface, the tip moves out of position in x-y coordinates as illustrated in Fig. 39.11. When a product is placed on such a pin, the tip initially contacts an incorrect target location. During compression, the pin must drag across the product surface as the pin retracts into the flush position. Thus some spring mechanism, which supports the fixture and pins above the array of grid spring probes, is commonly employed. The fixture may rest on a spring-loaded stripper plate over the grid of spring pins, or the fixture may be designed with spring feet on its underside. In some designs the fixture kit itself is compressed.

FIGURE 39.11 Pointing accuracy.

Some fixture designs have used bent or curved test pins. In such cases the pin is bent by means of additional fixture plates such that the tip of the pin as it comes through the top plate is perpendicular to the test pad. This allows the top plate holes to remain on the product and mirror the board under test and thus eliminates concerns about the effects of pin contact at an angle. Common examples of such fixtures were built as spring boxes, such that the overall thickness of the fixture reduces during compression. This eliminates the need for spring feet or a stripper plate, but complicates friction and binding issues as the entire array of test pins must bend and flex during every compression. Contact reliability, product marking, and ease of pin loading all suffer.

Offset Error Compensation.   Figure 39.12 represents the problem associated with the deflection of a pin retained in the top plate and its ability to hit the center of the pad or target. As described earlier, the hole diameter must be sufficiently oversized to allow the pin to pass through the hole at its angle of deflection without binding. Vertical force on the pin from the product above tends to press the pin against one side of the hole, as illustrated. (This has the curious effect that tilted pins can display more consistent targeting performance than do perfectly vertical pins). As the pin tilts, the center line of the pin becomes substantially shifted from the center line of the hole (note the offset distance OS in the illustration). Finally, the conical tip associated with most modern pins has a certain finite radius—the pin is not infinitely sharp. Thus, in the illustration, the product will not contact the pin on its center line. As

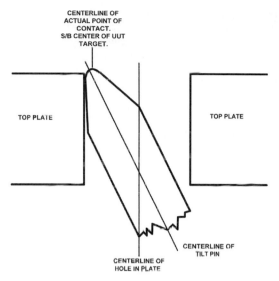

FIGURE 39.12   Pin tip offset error.

illustrated, contact will come at the highest point of the tip radius, slightly to the right of the center line. Thus, the theoretical contact point will be at neither the center line of the pin nor the center line of the hole. Probing targets only a few mils wide with probes two or three times as wide requires probes of high quality displaying consistent tip geometry, and software able to make careful calculation of the actual point of contact.

   Countersink Drilling of Top Plate.   The calculations and considerations just discussed illustrate that thicker plates and higher pin tilt present some interesting problems when combined. A highly tilted pin in a thick plate requires a substantially oversized hole if binding is to be prevented. Looking through the hole at a viewing angle equivalent to the tilt of the pin, the hole will appear to be oval, with the narrow axis considerably shorter than the drill bit diameter. Of course, the pin is thereby allowed somewhat more freedom of movement along the major axis of the oval. This bit of looseness can be a problem with very fine-pitch products.

   A better solution is to use a very thin plate for the upper fixture plate, backed up by a thicker plate drilled oversize. Alternately, a single plate can be countersunk. The effect is the same, as illustrated in Fig. 39.13.

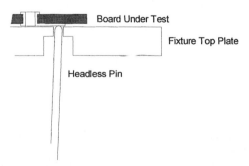

FIGURE 39.13   Countersink drilling of top plate to avoid binding or oversize holes.

39.3.3.5 Retaining Fixture Pins. It is necessary to retain the pins in a translator fixture such that they do not fall out of the fixture during ordinary use and handling. Older headed pin types were often used in single-sided fixtures. The oversized head protruded above the top fixture plate, and therefore the pin was unable to fall out the bottom of the fixture. (although they fell out all too easily if one accidentally inverted the fixture). Early top-side fixtures attempted to duplicate this method for the upper fixture by employing a different pin in the upper fixture half, using an oversized "tail" on the fixture. However the oversized geometry of the heads is inappropriate for fine-pitch testing, and therefore the use of such pins has fallen out of favor. The desire to use thin, headless music wire pins necessitated the invention of new means of pin retention (see Fig. 39.14). These have included:

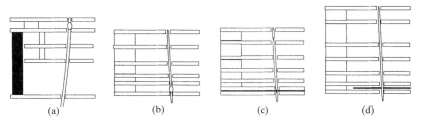

FIGURE 39.14   Alternative pin retention methods.

- Crimping a wide flat into the pin, and trapping the oversize flat between two plates
- Trapping the pin in a sheet of foam rubber such that friction retains the pins
- Passing the pin through a sheet of spandex cloth that grips the pin
- Passing the pin through a sheet of drilled latex rubber that grips the pin with flexible flaps

39.3.3.6 Examples of Real-World Fixture Designs.   While many combinations of fixture techniques have been employed, it is worthwhile to compare several popular examples. No substantial discussion of older headed rigid pin types is included here, as their use is largely disappearing. Most fixtures today are built with headless pins that compress flush into the fixture without bottoming on the spring probe below, such that the contact force is entirely controlled by the spring (excluding friction or binding effects).

Figure 39.14(*a*) represents a common bent or curved pin design using 3.75-in music wire pins. The upper and lower fixture plates are held apart by a spring mounting, and the entire fixture compresses during actuation. The pins hit the intended target perpendicularly, because they bend toward vertical as they reach the product area. But extra force is required, increasing pad marking. Hole diameter in the top plate can be tightened to improve pointing accuracy because all pins pass through perpendicularly, and a fairly thick top plate is possible. For a given pin length, the maximum possible horizontal displacement is more limited than in straight pin designs, because all displacement must occur in the lower portion of the pin length. The hanging middle plates do not contribute to torsional stability of the fixture design, increasing distortion under compression. This in turn can increase pin binding as the pins try to bend and slide through the compressing fixture envelope. The pins usually used have a flattened crimp trapped between the upper two plates, and must be loaded or serviced with the upper plate removed. Pin loading is therefore difficult. The increased diameter in the crimp region limits close spacing of the probe. Finally, the crimp cannot be employed on very fine-diameter pins, as they would weaken excessively.

Figure 39.14(*b*) illustrates a significant evolution from Fig. 39.14(*a*). Here the pins are not bent or curved. Shorter pins (3 in) are sometimes used, providing about the same horizontal translation as in Fig. 39.14(*a*). Deleting the bend significantly reduces pin binding. A retention crimp similar to the previous example is used, relocated toward the bottom of the fixture where the pins are more widely spaced. Pins are loaded from the bottom side with the lower plate removed. This design still requires disassembly of plates for pin servicing, but the bottom plate is somewhat easier to reinstall due to the possibility of larger hole sizes. Fixtures are usually rigid as opposed to spring-loaded, reducing friction effects during compression.

Variations of the fixtures illustrated in Figs. 39.14(*a*) and 39.14(*b*) are produced using a foam rubber pin retention method. In these versions the crimp on the pin is deleted, and a sheet of foam rubber material is sandwiched between the two lower plates. During loading, pins are forced through this foam, and friction holds them in place. This provides the important advantage that individual pins can be replaced without fixture disassembly. Unfortunately, high-density zones incorporating many pins stretch the foam rubber material. The stretched foam holds adjacent pins so tightly that a group of pins becomes reluctant to move except as an almost solid block. The resulting pin binding effect causes contact reliability and product marking problems. Also, the foam tends to deteriorate with time and/or heavy use. As densities have increased, this fixture has become difficult to use.

Figure 39.14(*c*) illustrates a retention technique employing a very thin sheet of drilled Mylar combined with a unique feature on the pin. The Mylar sheet is sandwiched between standard polycarbonate plates. The pins employed begin as straight-sided music wire pins 3-in long. A grinding operation reduces the pin diameter in a short region (about 0.2 in) of the pin length. The Mylar retaining sheet is drilled with a hole diameter that is slightly smaller than the original pin diameter. As the pin is inserted, it is forced through this undersized hole until the reduced-diameter zone reaches the Mylar sheet. At that point the pin can slide freely over the 0.2-in narrowed length. Only in this region are the pins able to move quite freely up and down. The height of the Mylar plate in the fixture is matched to the reduced-diameter portion of the pin, such that the pin is free to move a short distance up and down from its nominal position, without falling out of the fixture. Pins are generally specified with the grinding operation performed at locations near both ends of the pin, such that the pin is symmetrical and can be loaded without regard to orientation. This pin demonstrates low friction and good contact, but a few problems remain. The Mylar sheet is thin and fairly brittle. It is not uncommon for a particular hole to fracture when extracting/reinserting a pin, such that the damaged hole can no longer retain a pin. Replacing the Mylar sheet requires unloading all pins and removing one or more plates. Pins for this fixture are expensive, as the grinding operation is difficult. The need to machine the pin undersize limits the minimum pin diameter that is practical. As product pitch and density demands have increased further, these issues have become significant limiting factors.

Figure 39.14(*d*) represents techniques employed in more recently developed fixtures. These fixtures include nonoriented, nonfeatured pin designs, usually 3.75 in long. A key feature is retention via drilled latex rubber or undrilled spandex cloth. The drilled latex is not sandwiched but rather floats in a gap between two plates, amounting to 0.1 in or so. Drilling tends to cut or tear a slit in the rubber, rather than to cut a round hole. The result upon pin insertion is that the pin is gripped by a pair of tiny rubber flaps, allowing short motions of the pin to occur without actually rubbing the pin against the rubber. (The flaps are able to move short *z*-axis distances with the pin.) This provides essentially frictionless pin motion over the short range of travel needed to conform to surface irregularities on the product. Properly sized to match pin diameter, the presence of the holes/slits tends to relieve any accumulation of stress in the rubber sheet in high-density areas, providing good contact reliability under such conditions. The lack of motion against the rubber sheet also promotes long life of the fixture. The presence of the hole (or slit) provides stress relief in the rubber in high-density situations. Molded one-piece plate spacers speed assembly and disassembly, lower cost, and provide accurate *z*-axis position control of the plates. Use of the longer 3.75-in pin provides maximum pin displacement. The featureless music wire pin is economical and widely available. A countersunk top plate hole or two-ply top plate is generally used for high-performance applications. Notice that no disassembly is required to service a pin.

A variation of this fixture employs a stretchable fabric web in place of the latex, generally referred to as spandex. Under close examination this material appears as a very fine stretchable net. The stretchable net requires no drilling operation—pins are just pressed through during the pin loading operation. However, some stress accumulation is observed in very dense areas.

39.3.3.7 *Intermediate Guide Plates in Fixtures.* Early pin translator fixtures were constructed with only two plates, top and bottom. Modern fixtures add intermediate plates to speed fixture building, produce a more reliable fixture, and make the fixture determined. Determined fixtures ensure that a pin started into a fixture hole during pin loading is able to follow one and only one path through the fixture (without bending). This is required if data-driven test programs are to be used because the test program assumes a specific mapping of grid test points to product targets. The determined pin load must remain constant if the fixture is taken apart and later reloaded, otherwise the test program will become invalid.

Intermediate plates simplify and shorten the pin loading process because each pin must fall into the intended track. Assemblers do not have to pick between several possible holes or spend time checking for misloads and/or short circuits. Determined pin loads permit the fixture software to calculate the side loads imposed on fixture plates by the combination of tilted probes in each region of the fixture, adjusting the pin loading plan until these side loads are balanced. This prevents the fixture from collapsing sideways under compression. The intermediate plates also support finer-diameter pins, ensuring that they do not buckle under load.

As density increases, holes in the various plates become more closely spaced. At some point, for a given plate stackup, there is some risk that some pair of holes on the next plate will become close enough that a pin being loaded might fall into the wrong one. At that density, the fixture software must add an additional intermediate plate to the design. Thus, denser fixtures generally require more plates. Fixture software should analyze the requirements of each fixture and delete unneeded plates.

39.3.3.8 *Fixture Plate Spacing Techniques.* Errors in the vertical (z-axis) position of fixture plates can result in pin binding. As a tilted pin travels through the plate stackup, the hole position in each plate shifts slightly in the x and y directions, according to the amount and direction of tilt. If a plate were to be mounted either too high or too low in the plate stackup, the drilled hole would no longer be at the correct x-y position, forcing the pin to bend in order to pass through the hole. Pins would become difficult or impossible to load; binding and friction would occur during operation. Sensitivity to this problem increases as the amount of tilt increases.

Early multiplate fixtures used a plate spacing technique referred to as *post and doughnut,* which had mediocre control of z-axis plate position. A long screw passed through a stack consisting of the fixture plates and thick plastic washers. The washers were drilled in the center, and the plates were drilled with corresponding mounting holes. The screw passed through all of these in turn. The thickness of a given washer determined the space between two plates. The problem was that the errors in washer and plate thickness accumulated through the stack-up. At greater pin tilts, the result was binding. It turned out that inexpensive fixture plate material was not terribly precise as to thickness.

Alternative means of plate spacing have been devised in which a single molded (or machined) part is responsible for the entire fixture stack-up, and which eliminate intermediate plate thickness as an error contribution. One method is illustrated in Fig. 39.15. This molded spacer provides a series of flat surfaces upon which individual plates sit. In a top view, the shape of each surface is modified such that plates slip easily over the top of the spacer, but are locked into position when the spacer is rotated 90°. As long as

FIGURE 39.15 Molded spacer prevents pin binding.

the spacer is accurately molded, no cumulative plate spacing error results. For midfixture plate supports, a similar part of smaller diameter is used. Short screws at the top and bottom plate secure the support in place. Alternate methods use machined bars around the periphery of the fixture. Either method results in accurate fixtures, reduced parts count, and ease of assembly.

39.3.3.9 Headed Music Wire Pins. For very large through-holes, the preferred probing method is to probe the annular ring of the plated through-hole. If, however, this ring is insufficient, a headed pin may be required. Such pins are manufactured with a machined tip attached to a standard music wire pin. Various head shapes are available. These headed music wire pins work in the fixture design much the same as the headless pin, i.e., flush in the top plate (see Fig. 39.16), with the exception that the headed pin must be loaded from the top individually.

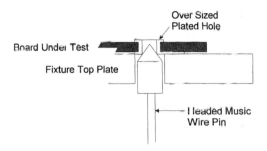

FIGURE 39.16   Headed pin in tilt pin fixture.

39.3.4 Dual-Side Testing Considerations

Modern SMT boards often have networks that terminate at component locations on both sides of the board. Continuity testing of such networks requires that the top and bottom terminations of such networks be probed simultaneously. There are three ways to test boards that require dual-side access:

- Flip test
- Clamshell upper fixture
- Dual-side universal grid fixture

These methods vary in terms of test coverage, fixture cost, labor cost, risk of test escape, and required infrastructure. Each method is discussed in the following text.

39.3.4.1 Flip Testing.   Flip testing is a last-resort method employed where up-to-date dual-side equipment is not available and compromising in test coverage is acceptable. The flip test method requires that either two fixtures, one with the top-side image and one with the bottom-side image, or one fixture with both images on it, be built. The board is then tested one side at a time. Although this does provide partial fault coverage, it does not test those plated holes that link the top-side half of a network to the half on the bottom of the board. This poses a significant risk that bad boards will escape the test process. The problem only becomes worse if these interconnecting vias are tented (have solder mask over them), preventing probing. In this case the traces that run from the tented via to the first probeable pad on the network are also untested at both sides of the board.

Flip testing is expensive. Not only are two fixtures built, but also many test points are duplicated between the fixtures. Much additional board handling and potential for confusion are involved. The cost and risk incurred would likely justify an equipment upgrade to dual-side test capability.

39.3.4.2   Clamshell Upper Fixture.   Clamshell testing is another compromise method employed where dual-side equipment is not available. This method provides top-side access through the use of a wired (dedicated) top-side fixture. The method is superior to flip testing in that it can provide complete simultaneous access and full test coverage. However, fixture costs are quite high, especially for top-side test points. In this scenario, a more or less normal pin translator grid fixture is built for the bottom side of the board. The board should be oriented such that the busy side of the board faces down, in order to minimize the costly top-side fixture (see Fig. 39.17). Additional test points (called transfer points) are added outside the product area on both the top and bottom fixtures, conducting test points from the lower fixture and grid to the upper fixture assembly. Normal wired fixture construction methods are used within the upper fixture to relocate these points to desired top-side product target positions. In order to use this method, it is necessary that the lower grid be larger than the product under test such that a number of transfer test points adequate to provide for the upper fixture can be placed outside the product area. Again, the cost of the extra contact points, spring probes, wiring, and other special material and labor processes associated with the top-side fixture may well justify an equipment upgrade.

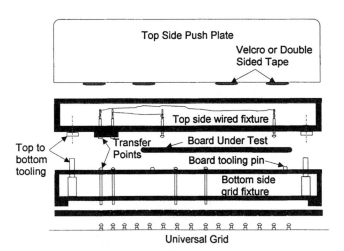

FIGURE 39.17   Transfer point clamshell fixture for dual-side access.

39.3.4.3   Dual-Side Access Universal Grid.   The third and most widely accepted method of providing for volume testing of dual-side product is the double-sided universal grid. Such systems provide electronics and grid probe fields on both the top and bottom sides. Upper and lower translator fixtures are built for dual-side products using the standard techniques discussed earlier, providing economical fixtures, reusable pins and hardware, full fault coverage, excellent throughput, reliable contact, and accurate targeting.

39.3.5 Press Units

The major mechanical component of the grid is a press unit. The three major compression methods used to compress the product onto the fixture are hydraulic, pneumatic, and electric. The total force requirement can be quite large. The spring probes employed in grids usually require a force in the range of 4 to 10 oz each. A large, complex board may have 20,000 active points per side. This amounts to 5,000 to 12,500 lb of force, depending upon the probes used. Hydraulic drive is probably the least used. Although it provides tremendous power, it is difficult to maintain and fluid leakage is a problem. Pneumatic and electric compression are popular and successful. Electric drives are easy to control, but become expensive and/or slow when larger forces are required (as in the case of larger grids). Pneumatic systems are fast and powerful, but require a good supply of clean, dry compressed air. With any method, accurate compression control is critical to reliable fixture performance.

As noted, universal grids present the test points as an array of spring-loaded contacts. A variety of contact tip shapes have been used, but most contacts used today offer waffle tips. Waffle tips are nominally flat, but with an H-shaped pattern of grooves cut into the surface in order to provide a somewhat aggressive contact surface. Such a surface is better able to cut through thin contaminants on any surface it contacts.

39.4 FLYING PROBE/MOVING PROBE TEST SYSTEMS

Moving probe, flying probe, and x-y prober are all names for test systems that make use of two or more test points that can be accurately positioned anywhere on the board surface by means of a computer-controlled motion system (see Fig. 39.18). Probe tips can be retracted in a z-axis direction away from the board surface, then moved in the x and y directions to a new board location and extended again to make contact. Figure 39.19 shows two probe tips contacting board. Dual-side systems generally provide a minimum of two independently moving probes per side for a minimum total of four heads.

FIGURE 39.18 Flying probe test system. *(Courtesy of atg Test Systems GmbH.)*

FIGURE 39.19   Two probe tips contacting board. *(Courtesy of Probe Inc.)*

39.4.1 Advantages of Flying Probe Systems

The major advantage of these systems is elimination test fixtures, making these systems ideal for small to moderate volume production. Advanced flying probe systems provide highly accurate probe placement (on the order of 0.001 in or better), contact the board with minimal force, and are very well suited to testing the finest pad sizes. While not subject to limitations due to test point density, these systems do slow down as additional test points are added.

DC continuity test measurements are accomplished by placing one probe tip at each end of the continuity to be verified and performing a measurement. The probes then retract, move to the next measurement site of interest, make contact, and make the next measurement. Similarly, it is possible to perform DC isolation testing of a network by placing one probe (probe A) on the network, while another (probe B) checks all other networks in sequence for the presence of a short or leak. Probe A can then be stepped to the next network to be tested and the process repeated. This continues until each network has been checked against all others. This requires a huge number of measurements. For a board with N networks, the total number of measurements M required would be:

$$M = (N^2 - N)/2$$

For example, a fairly ordinary board with 1,000 networks would require 499,500 measurements. Fortunately, several means of reducing the measurement burden have been developed.

The program preparation software can analyze the conductor pattern on the board, and, noting that short circuits should occur only between physically adjacent conductors, reduce the number of measurements required. This is termed *adjacency analysis*. As further time savings, an indirect measurement method may be used. These methods are discussed in Sec. 26.3.4.4. Some indirect methods require only a single probing action of each net, combining continuity and isolation testing with some extra probing to verify any suspect results. Theoretically, our example board with 1000 networks might have isolation verified with only 1000 probe placements.

39.4.2 Economics of Flying Probe Systems

The primary limitation of flying probe test systems is throughput. While systems with up to 16 test heads are commonly available, these systems still lose a substantial fraction of the operating time due to mechanical positioning of the heads. Yet the cost savings derived from elimination of direct fixture costs, fixture support infrastructure, fixture debugging, etc., are substantial. Accurate comparison of the costs associated with grid and prober methods requires inclusion of all the costs associated with each approach. Common omissions include:

- Labor overhead
- Floor space allocated to fixture storage, materials, and assembly
- Costs outside the test department (drilling?)
- Stranded investment in completed fixtures, pins, other fixture materials
- Stranded investment in drills and other equipment
- Lost throughput of test systems due to maintenance
- Lost throughput of fixtured systems during fixture setup, debugging, and maintenance
- Peak throughput demand

It is quite possible that, depending upon product mix, a testing department would be made more efficient with a larger investment in capital equipment (in the form of flying probe systems) and a smaller investment in fixtures. With smaller runs and/or smaller lots necessitating frequent fixture teardown/bring-up, this becomes more likely.

39.5 VERIFICATION AND REPAIR

The test system identifies faults on the board. Assuming that repaired boards are acceptable to the end user and that it makes economic sense to repair boards, a repair process normally follows testing. Because fixture problems, contamination problems, product registration problems, and other issues often result in false error reports, there is usually a verification process added between testing and repair. During verification a technician reads the fault data and makes confirming measurements to determine whether the reported faults are genuine.

If verification of the fault is valid, the board proceeds through the repair function. If the fault is not valid, the board may be retested or moved on, depending on customer requirements, in-house policy, etc. When deciding on a ship-from-verification policy, be mindful of the possibility of multiple defects masking one another (an open hiding a short) or appearing only under the pressure of the fixture. If possible, it may be preferable to perform a complete retest.

In some cases, simple tabletop meter equipment is used for verification, and verification may actually be done at the repair station by the repair technician. More commonly, some form of computer-assisted search-and-display tool is employed. The computer has information concerning the product and test program available, and searches the artwork pattern for areas likely to contain the fault being considered. The resulting risk area is displayed to the operator or superimposed as a visual projection onto the suspect board, simplifying the task of placing meter leads.

Automatic verification may be accomplished by a flying probe–type system. The flying probe performs a similar analysis, but does the retesting itself and displays a final result. Video camera systems may provide image capture of suspected fault sites, storing the image for subsequent recall at the repair station. Advanced systems may even suggest the repairability of particular fault sites, according to user-defined rules.

Figure 39.20 shows an example of a fully automatic verification system and CAR station. At the repair station, a display computer similar to that described for verification purposes is used to highlight the proposed repair site to the operator, who performs the appropriate cuts or welds.

FIGURE 39.20   Fully automatic verification system and CAR station. *(Courtesy of Everett Charles Technologies.)*

39.6   *TEST DEPARTMENT PLANNING AND MANAGEMENT*

In most cases the new manager inherits a selection of existing equipment, processes, and personnel. Nonetheless, over a period of time, the manager has the same opportunity to shape the test floor operation as does the unlikely individual starting from a clean sheet of paper. To the extent that the past suggests the future, the new manager may expect increased density, finer pitch, higher total test point counts, decreased tolerance for test escapes, shorter product delivery times, shorter product lifetimes, and larger numbers of smaller batches of boards for just-in-time delivery.

39.6.1 Equipment Selection

Most equipment being purchased for general bare board testing consists of universal grids and flying probers. The increased capabilities of flying probers and increased demand for small lot and prototype testing have increased the number of probers selected. This trend seems likely to continue, but the appropriate solution must be determined individually for each shop.

A first consideration in selecting equipment lies in recognition of the type of business you anticipate. For volume production, the universal grid is a likely participant. For prototype or small lot production, a quantity of flying probe systems may be a better alternative due to the elimination of fixture expense. Dedicated (wired) fixture–type equipment is rarely demanded due to fixture cost. The sole exception may be very simple systems that remain devoted to an extremely high volume run of a single product type for extended periods of time.

A further important concern is the level of technology you anticipate. Verify that it does not outrun the capabilities of the fixture or test system types you plan to use. Plan for change, as the equipment you purchase must last for several years. Also consider the level of support available from the equipment vendor, including ordinary maintenance service, emergency service, employee training support, and applications support.

In high-volume operations, substantial throughput and consistency of operation can be achieved by adding automatic handling systems. For prototypes, the ability to place a stack of boards on a flying probe system and walk away for an hour or so (rather than wait for a board to come out every three minutes) can result in huge labor savings. Note that some very large, very thin, or very small product types may not be well suited to automatic handling or may require that the product be run through the machine in simple carriers machined from G-10 or similar material.

An often overlooked but practical consideration is maintaining some degree of commonality in equipment and fixtures. It is not uncommon for fixture problems to lead a test engineer to change equipment types several times, resulting in a motley assortment of incompatible (but expensive) test systems but failing to resolve the fundamental underlying issue. With data-driven test processes and automated fault data management in the mainstream, the ability of equipment to share data, fixtures, and operating procedures will decrease cost, increase throughput, simplify training, and enhance opportunities for combinational test solutions. This is harder to achieve with a random mix of equipment. If this seems to state the obvious, try this experiment: visit some shops, stopping in drill rooms and in the electrical testing area. Do you see a difference in the assortment of equipment? Department planning should include not only new equipment, but also an exit strategy for old and inefficient equipment you wish to retire.

39.6.2 Fixtures: Build or Buy, and What Type?

If you plan to use fixtures, the fixture selection is probably more important than the equipment upon which you choose to run the fixtures. Certainly you must have equipment at least adequate to the fixture, and with sufficient test point density for the product. This is where a careful evaluation of product technology is crucial.

If the product includes even fairly simple SMT boards, then you should plan immediately for a dual-side capability. If the product includes fine-pitch SMT, dense BGA patterns, etc., then you must consider higher-density universal grids and modern tilt pin fixtures consistent with the test site density on these products.

If you plan to build fixtures in house, recognize that a complete manufacturing operation is being created. Measurement of complete material, labor, inventory, and overhead costs as well as operating efficiency is critical to effective management of this operation. Don't just measure how fast you build a fixture, but include measurement of the time it takes to get it running. Consider whether you would be better served by outsourcing fixture fabrication. Avoid manufacturing multiple styles of fixtures. Ensure that you have an inventory management process that does not run out of pins at midnight, but does not have half a million dollars tied up in obsolete pin inventory. Consider outsourcing at least the more difficult fixtures to specialists. You may save time, plus gain a partner who provides exposure to alternative techniques and materials.

39.6.3   Selecting Fixture Software

Today's tilt pin translator fixture is largely a software product. It should be possible for you to establish a flow of information from your CAM department to the front-end fixture software that automates the design of fixtures and the creation of test programs. (The same goal holds true for processing data to create flying probe test programs.) A premium tool will pay for itself many times over in reduced fixture scrap, reduced late deliveries, and elimination of test escapes. If you can select a tool that is also used by several fixture outsourcing vendors, you will have a backup supply of fixtures in the event of equipment problems, personnel problems, or workload peaks. If you have special and specific demands for the fixture design, verify that the proposed software package readily supports your requirements. And, as in the case of the test system, evaluate the software vendor's ability to sustain and support your operations.

CHAPTER 40
HDI BARE BOARD SPECIAL TESTING METHODS

David J. Wilkie
Everett Charles Technologies, Pomona, California

40.1   INTRODUCTION

Printed circuit technology is now being applied to high-density interconnect (HDI) applications that require consideration of new testing techniques. Notable examples include many types of advanced IC packaging introduced in recent years, such as area array devices, ball grid arrays (BGAs), laminate chip carriers (LCCs), and circuit boards employing direct chip attach (DCA) or chip scale packages (CSPs). In some of these cases semiconductor dies are being directly attached to a printed circuit surface, and in the case of CSP the package size is approximately the same as a raw die. Flip-chip die (or CSPs) may be attached to a circuit board by an array of solder balls. Other devices may attach with a peripheral array of wire-bond pad sites just outboard of the die area. For high-pin-count devices, this last example can produce the finest pitch as efforts are made to squeeze a large number of contacts onto the limited periphery of the die. The flip-chip or CSP approaches allow the designer to distribute the solder ball contacts over the full surface area of the die or package.

The general goal of electrical testing for HDI substrates remains about the same—detection of opens, shorts, and leakage. The measurement electronics are often the same, and are not further discussed here except for special cases. For some applications TDR or other RF tests are desired, but are complicated by the short signal runs on some of these substrates. Sensitivity to pad marking is increased as there is not much pad area to mark. Marking that penetrates the metallization layers on the board can affect the chemistry of the solder joint, adversely impacting reliability. The major challenge, then, is usually the test fixture, amplified by the need to minimize witness marks. New methods and modifications to old methods have been developed. Because of the high costs associated with HDI fixtures, some methods compromise test coverage such that the fixture can be radically simplified.

40.2   FINE-PITCH TILT PIN FIXTURES

Many HDI structures can be tested with advanced examples of tilt pin fixtures. Using such techniques as test pins that taper (see Fig. 40.1) to a very fine diameter at one end, fixtures have been built for devices with pitch down to about 0.010 in, with R&D examples at about 0.008 in. These fixtures contain additional plates to support the very thin pins (so they do not

FIGURE 40.1 Tapered pin for fine-geometry fixture.

buckle under pressure or short to adjacent pins). Those fixture plates very close to the product must be fairly thin, or the closely spaced holes will break out into one another. Such fixtures increase the demand for higher grid densities. These fixtures and pins are moderately more expensive to prepare than standard fixtures, but use similar technologies and processes. Excellent process control is important, as is excellent pin tip symmetry.

Areas of the product requiring large numbers of probes are subjected to considerable total force. If no equal and opposing force is applied on the opposite side of the product, bowing or "potato chipping" of the product will occur. Ultimately this can collapse the opposing test fixture, but long before this the probing accuracy of a fine-pitch fixture will be affected because the product no longer lies in a flat horizontal plane. To avoid this, the fixture software can add support within the opposing fixture, directly opposite the densely probed site. This can be accomplished with a spacer affixed between the top and bottom fixture plates or by using blind pins. Blind pins are ordinary test pins located in the opposing fixture at sites where the top (product side) plate of the probing fixture is not quite drilled all the way through. If a quantity of blind pins equal to the quantity of dense product site probes are employed, then the opposing forces are perfectly balanced in this region.

When dealing with fine-pitch fixtures, the problem of product/fixture registration must be considered. Even if the fixture is manufactured perfectly, the product may not align. Because the tooling features of the product (edge or tooling hole) are added in a process separate from the artwork production, the artwork and tooling features are usually misaligned. With HDI substrates, it doesn't take much for this error to move pads completely away from the assigned test pins. To overcome this, a variety of test systems are available with optical and/or electrical means of sensing misregistration. Movable plates in the fixture are controlled by servomotors, which reposition the product through motion of the tooling pins. As the product grows in size, or as the manufacturing process deteriorates, the registration errors may not be sufficiently constant over the product area to permit the use of motorized alignment. In this case, none of the probe-per-pad fixtured approaches may be effective.

40.3 BENDING BEAM FIXTURES

These fixtures are somewhat similar to tilt pin fixtures, except that they use extremely thin test pins manufactured from a special alloy. As in the tapered pin fixture, a significant number of thin supports are required near the product, where the pin tends to be held fairly straight to avoid conflict with other nearby pins. These pins are not held rigid, but are expected to buckle under force. This buckling displaces the pin sideways. The buckled portion may be located at some distance from the product plane, where the probes can be more widely separated. Once the pin has buckled, the force applied to move through a substantial travel is constant. This is very different from the case for a spring probe, where the force increases at a constant rate determined by the spring constant. This provides good immunity to witness mark damage. The basic technique has been in use in specialty situations for many years.

At the interface side, these pins may mate to spring probes, instead contacting a rigid contact surface. In some commercial examples the pin continues away from the product as a piece of wire routed to widely spaced contacts on the interface portion of the fixture. These in turn interface with traditional spring probes. Within the fixture, the probe wire is bonded in place at some distance from the product, defining the separation between the probe and wiring portions of its overall length. Replacing damaged probes can be challenging, and this technique is usually restricted to small areas and limited test point counts.

40.4   FLYING PROBE

The more precise examples of flying probe systems are very well suited to HDI testing. With on-head optical pattern recognition guiding the probe tips, they can contact extremely small targets—even in the presence of significant product registration problems. With bending beam probe tips of the sort used commonly in semiconductor testing, they can probe small features without marking or disturbing the surface. The principle drawback is, of course, test speed. These systems are ideal for prototype applications, where the cost of fixtures would be painful.

40.5   SCANNING CONTACT

Scanning contact testers (see Fig. 40.2) specialize in the testing of laminate chip carrier substrates. They take advantage of a unique characteristic of many chip carrier products. In these products, most or all signals pass from a contact on one side of the board to a contact on the other, and many signals stop nowhere else. On one side of the chip carrier are very fine pitch contacts for bonding to the die (either by flip-chip or wire-bond means). On the other side are more widely spaced ball sites (or wire lead sites) intended to mount the device to a larger main circuit board. It is fairly easy to fixture and contact these latter pads, but hard to contact the fine-pitch pads.

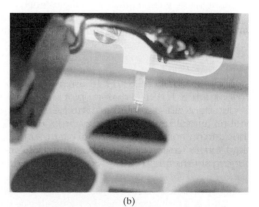

| (a) | (b) |

FIGURE 40.2   (a) Scanning contact system; (b) detail of wiper. *(Courtesy of Everett Charles Technologies.)*

An isolation test of the product can be performed by contacting only the larger pads and ensuring that the networks are isolated. The scanning tester therefore contacts the large-scale side of the product with a more or less traditional test fixture. To address continuity testing, the scanning tester adds a single robotically controlled wiper contact that scans over the fine-pitch surface of the fine-pitch contacts. As the wiper moves over the surface, the measurement system rapidly scans the fixture contacts on the other side. Direct DC measurement of continuity from the wiper to specific fixture targets is thereby achieved through the product conductor pattern. There are only two untestable situations of note. If a network appears that joins together two sites on the fine-pitch side and does not pass through to the fixtured side, then it is not possible to test the continuity of that network using the scanning technique. A secondary test using a pair of flying probes must be added. In the second case, consider large networks such as ground. The network appears on both sides of the board, but at many points. This can be tested as long as the wiper size and motion path can be selected such that no two ground points are wiped at the same instant. (If they were, the wiper would short them together and mask any open circuit to one or the other.) The net effect of these limitations is that the scanning tester is ideal for single-chip substrates and not very useful for multichip modules (MCMs).

Scanning testers are fairly fast, lying between flying probers and universal grids in terms of throughput. With the addition of automatic material handling, they are well suited to volume production. Because they are performing genuine DC electrical testing of continuity and isolation, there is no test compromise. These systems are also virtually fixtureless. This is because the chip carrier products are generally built to common spacing patterns on the coarse-pitch side, such as 1 mm for 1-mm BGAs, 0.8 mm for 0.8-mm BGAs, etc. Thus, one fixture may serve a broad family of products, and the fixture cost is amortized over a large number of products. For most products, the fixture already exists, and there is no delay or cost associated with obtaining it. The primary limitation is that the tester is generally limited to single-die chip carrier–type product. If a sufficient volume of applicable product is being run, this technique is moderate in cost.

40.6   COUPLED PLATE

In a situation where most of the target product is relatively easy to fixture, but a DCA or CSP site is difficult, it may be possible to use coupled plate–type testing. A variety of proprietary products are available, but the basic technique is fairly similar. The primary assumption is that each signal network arriving at the DCA/CSP site is accessible by a traditional probe at some other site on the board. These probe sites are used to perform a normal isolation test of the product. Continuity testing of most of the board is performed in the usual manner also.

To verify the continuity of the signals to the DCA/CSP site, a small metal plate or antenna is suspended just over the site, perhaps insulated from the site only by a thin dielectric. One at a time, the standard probe sites are used to inject some sort of AC signal or pulse into each network in turn. If the network is properly continuous to the DCA/CSP site, then a signal of appropriate amplitude is detected at the antenna. Substantial deviation from the proper amplitude means that something is wrong. The different methods vary in terms of the applied signal, antenna characteristics, etc.

This method eliminates the need for super-fine-pitch probing at the DCA/CSP site and also avoids marking of the product at this site. However, the test does not perform a true DC test of continuity and may miss high-resistance connections, which would be detected by a low-threshold DC continuity measurement. Signals that loop from one DCA/CSP pin to another and go nowhere else may be untestable as there is no outboard probe site at which to inject the test signal.

40.7 SHORTING PLATE

This method is employed in circumstances similar to the coupling plate method just described. In this case, however, the plate must be movable during the test. Generally a pencil-sized pneumatic actuator is mounted within the test fixture to accomplish this motion. The plate employed here is a small flat metal plate that is the size of the DCA/CSP site and covered with an electrically conductive rubber (Fig. 40.3). When pressed against the product, it shorts together all of the pads at the DCA/CSP site. Continuity testing is performed in this condition by using the outboard probe sites to confirm that all of these (otherwise isolated) networks are shorted together via the path to the DCA/CSP site. All other networks are tested normally for continuity. Then the shorting plate is removed and a normal isolation test is performed. Key advantages of this method are the accomplishment of true DC measurements for both isolation and continuity, and the use of standard bare board test systems.

In some applications there is concern about any trace chemicals that may be left behind by the conductive rubber, although outright marking is minimal. Cleanliness of the product is critical, as the rubber/plate must be replaced when excessive dirt is embedded. As with the coupled plate, certain signal topologies are untestable or difficult to test. Signals that loop from one DCA/CSP pin to another and go nowhere else are untestable, as there is no external probe site with which to verify either continuity (to the DCA/CSP shorting pad) or isolation. Any signal that connects two DCA/CSP pads on the same package may be partially untestable, even if the signal continues to an external probe site. In this case, it is possible to verify that the external signal arrives at the device, and it is possible to verify isolation of this segment. But there is no simple way to discern whether the two device pads are joined to each other. (They are already shorted together on the board and engaging the conductive rubber has no effect.) This latter case may sometimes be mitigated by segmenting the shorting pad, such that each of the target device pads is in a separate segment. Then, if the connection on the board is good, the halves will be joined through the board.

FIGURE 40.3 Pneumatically actuated shorting plate.

40.8 CONDUCTIVE RUBBER FIXTURES

Several designs for fixture systems employing conductive rubber as the basic probe element have been offered commercially. In some cases, the rubber is a specialized material in sheet form and is conductive only in the z axis (through the thin sheet vertically, but not sideways across its surface). The fixture is itself made from a circuit board, with slightly raised pads to tightly compress the rubber against desired product sites. The fixture board connects to the grid electronics at its reverse side. Other designs have included various types of locally deposited rubber dots, usually of conductive rubber that is not sensitive to orientation. Again, a rubber probe is formed. Problems with cost, complex manufacturing, complex repair, dirt sensitivity, and suitability to very small pad areas (which limit contact quality to the rubber) seem to have prevented widespread adoption.

40.9 OPTICAL INSPECTION

Optical inspection has been discussed elsewhere. It is generally applied early in the fabrication process as a yield improvement and data collection tool, not as a means of final product qualification. However, with improvements in resolution, the type of defect that may escape undetected becomes somewhat more limited, and it is argued as a possible means of final testing. With complex multilayer product, optical inspection will not be able to identify assembly/contamination-induced defects internal to the board in any case, and may still be limited in cases of contaminants or very fine-geometry shorts and opens on external layers. The equipment is somewhat slower than universal grid test systems, particularly when run at very high resolution. For such reasons, it is still not common practice to employ optical inspection (by itself) as a substitute for electrical testing. This may be a method that develops further in the future or that finds acceptance in special circumstances.

40.10   NONCONTACT TEST METHODS

Quite possibly the greatest daily irritant (and cost) in the operation of a typical test area today revolves around the cost of building test fixtures. Customers still hate to pay for them, and the creation of fixtures burdens the board shop with an entire manufacturing process that seems to add no value and that distracts from the main productive purpose of the factory. No fully successful noncontact method has yet emerged, yet some brief description of various techniques seems warranted.

40.10.1   Electron Beam Methods

When a test system contacts a product with an electromechanical contact, it uses this contact to inject or remove electrons. That is, current flows. The sort of electron beam common to the picture tube in a television might be used to do the same thing without contacting the product physically. Experimental systems have been built by several firms to date, but have shared some common problems. First and foremost, the amount of current delivered by the electron beam is so small that only a very poor continuity test is possible. Lab systems have been limited to continuity thresholds of 100,000 Ω or more, and this only very slowly. As most users wish to test at 10 to 100 Ω, this is a major compromise. (A smear of contaminant across an open circuit would appear as a perfectly good conductor.) Test speeds are affected by product capacitance, as it requires a longer time for the limited current to achieve significant voltage effect in a highly capacitive environment. Also, such systems must operate in an extremely high vacuum of laboratory grade. Expensive pumping systems are required. Costly air-lock systems with multiple stages and robotic product handling are probably required to keep any reasonable flow of material through the system. (Staging product in and out through a series of chambers avoids the time delay of constantly pumping down the main chamber when loading new product.) At the present time, flying probe systems seem to offer superior test coverage at more practical operating costs.

40.10.2   Photoelectric Methods

Subjecting a metal surface to an intense beam of light, such as that from a laser, can cause electrons to be ejected from the metal. As in the case of the electron beam, a very small current flow can be induced. Again a vacuum is required (so that the ejected electrons can be measured before they collide with air molecules), though the vacuum requirement is less stringent than that for the electron beam technique. The resistance at which continuity measurements can be made is quite limited, as in the case of the electron beam. (See discussion of electron beam method in Sec. 40.10.01.) Again, test speed and accuracy may interact in the presence of product capacitance. Work continues on this method, but no commercial equipment is available at this writing, and the method may not offer significant cost/performance advantages.

40.10.3 Gas Plasma Methods

Fluorescent light bulbs emit light because a gas is subjected to an electric field, which adds energy to the electrons orbiting the gas molecules until some break free. As they attempt to reattach themselves to the gas molecules, they eject their excess energy as photons (light). The plasma consists of a mixture of gas molecules, ionized gas molecules (missing electrons), and free electrons. In this state, the gas can conduct an electric current. If a jet of plasma is directed from a small nozzle to the surface of a circuit board, the effect is that of building a "gas probe." In broad terms, the gas probe can be used just as any other probe and test is completely possible. Generally a noble gas such as argon is used. The gas residue is nontoxic, only a tiny amount of gas is consumed, and substantial current can flow to the product. Several companies have developed experimental systems in the form of flying probe systems.

While offering the benefit of little or no product marking, it is difficult to achieve a gas probe that is as fine as the best mechanical probes. Adjacent jets tend to combine, producing a short circuit just as if two mechanical flying probe tips shorted while testing. Thus, the commercial system offered to date includes completely traditional mechanical probes as well, using these to probe closely adjacent (i.e., HDI) sites. As a flying probe mechanism is still used, there is little speed advantage to date. Eventually some benefit may be obtained due to the elimination of any wait time for z-axis travel of the prober head, though this is already the fastest motion axis of most flying probe systems. The cost of this technique is modestly higher than that of an ordinary flying prober.

40.11 COMBINATIONAL TEST METHODS

One technique that offers immediate practicality in resolving difficult testing situations for high-density product—with proven and practical technology—is generally described as combinational testing or sequential testing. As the name implies, this is testing in one or more stages, using a combination of test techniques. Combining techniques inevitably adds complication. However, modern software tools have greatly lessened the practical impacts of combining test methods. Where the majority of the board is testable at high speed by traditional fixture techniques, use of combinational techniques such as shorting plungers, coupling plates, and flying probe systems can provide extremely cost-effective results.

P · A · R · T · 6

ASSEMBLY

CHAPTER 41
ASSEMBLY PROCESSES

Richard Boulanger
Universal Instruments Corporation, Binghamton, New York

41.1 INTRODUCTION

41.1.1 The Density Revolution

Traditional market segments like consumer, communications, and computers are converging, and new applications are straddling two or more segments. An Internet device or a Sony Play-Station II could actually fit in all three market segments. Hence, some attributes that were linked to one particular segment (e.g., smaller and cheaper for the consumer, performance for computers, more bandwidth for communications) are rapidly becoming the prerequisite attributes for the new breed of consumer-communication-computer (C3) application.

These new applications create enormous challenges to the electronic "food chain," from semiconductors to packages to boards to electronics assembly to final assembly. Board fabrication especially is under a lot of strain to keep up with the increased density of electronic assemblies. Assembly equipment suppliers need to handle smaller and smaller components that are placed even closer together. As features shrink more and more to add more function and more components per square centimeter, the assembly equipment suppliers need to find novel ways of not only handling the components and the substrates but imaging them while constantly being pressured to increase throughputs and yields.

41.1.2 Printed Circuit Assembly

Printed circuit boards and automated electronics assembly have come far since the early 1960s. New applications keep pushing the edge for smaller areas, lighter products, faster speeds, and more bandwidth. A proliferation of packages and new materials has helped revolutionize new machine development. In the electronics market, characterized by ever-shorter product development, intense global competition, and rapidly evolving technology, understanding the assembly equipment and processes is necessary as a means to ensure survival. This chapter will cover major assembly types, accuracy definitions and considerations, network communications, and some key machine selection criteria.

Electronics assemblies can be grouped into three major categories:

1. Through-hole
2. Surface-mount
3. Mixed-technology (which can be any combination of through-hole, surface-mount, odd-form, or bare-die assembly)

Through-hole is still very much alive and new machines are still being introduced in the new millennium, but the total volumes continue to shrink. Surface mount has become the workhorse, and new developments include new machine types, several line configurations, and a dependency on network communications. Globalization is also driving the need for multilingual applications and user-friendly interfaces. Odd-form components are being automated more and more to keep up with line throughput and tact times. Bare-die, including flip-chip, assemblies are growing at a faster rate than the overall electronic assemblies. To complicate things further, several applications actually require mixed technologies, which puts a heavier burden on the equipment manufacturers.

41.2 INSERTION MOUNT TECHNOLOGY (THROUGH-HOLE)

41.2.1 Introduction

Through-hole assembly refers to the process of inserting electronic components through holes in a printed circuit board (PCB). The leads of the components are then soldered to the PCB to create an electronic circuit. This assembly method was first introduced in the late 1950s and automated in the early 1960s. The through-hole components and assembly process have traditionally been less expensive than the surface-mount components and assembly process, but they cannot achieve the same board density. The typical applications are in the cost-sensitive consumer market, which does not require a very high-density board. Examples include televisions, lighting ballasts, garage door openers, VCRs, and so forth.

41.2.2 Design Considerations

It is important to understand the manufacturing processes that will be employed when creating any design for manufacturing. It is not only important to understand the current manufacturing process, but also to be aware of the latest manufacturing advances in automated processes and design for those processes. Through-hole printed circuit assembly design is no exception to this concept.

The following elements are important design considerations to ensure that through-hole products can be assembled by machine:

- Tooling holes
- Registration holes
- Edge clearances
- Component lead-hole sizes
- Physical sizes

41.2.2.1 Tooling Holes.   The first consideration that must be given when designing for through-hole automation is tooling clearances. Unlike surface-mount components where the tooling sizes are in most cases as small as or smaller than the device that it is placing—the tooling for through-hole components is much bigger and usually handles the component about its perimeter. This means that special consideration to the tooling footprint must be given to ensure that the placement tooling has adequate clearance to adjacent, previously inserted, components as well as to tooling locators that hold the board to the assembly equipment. It is important to understand that axial components will have a different tooling footprint than radial components and that tooling footprints sometimes vary by the equipment manufacturer. Equipment specifications that outline tooling footprint sizes are readily available from each of the equipment manufacturers.

41.2.2.2 Registration Holes.   Registration holes to locate the printed circuit board to the component insertion equipment are mandatory when using a through-hole assembly process. These holes will be used to locate the printed circuit board to the component insertion equipment. They should also be used as the datum references to the component insertion holes to reduce variability in hole locations during the automated assembly process. It is important that standards be established as to the locations and diameters. Only two holes in the same plane are required and should be located at one of the following: 3.5, 4.0, 5.0, 6.35, or 7.62 mm from the front edge of the board. The distance from the left and right edges in the x axis is recommended to be 5.0 mm. The diameter can be as small as 3.17 mm or as large as 6.35 mm. It is important to note that standardized dimensioning for tooling holes should be established and maintained throughout all the through-hole designs. By establishing these standards, equipment setup time can be minimized.

41.2.2.3 Edge Clearances.   Edge clearances on both the front and rear edges of the printed circuit board need to be considered. These clearances need to be maintained for the conveyor belts that will be used to transport the printed circuit board through the various assembly processes. Minimum edge clearances should be no less than 3.0 mm. This means no component bodies or lead should violate this clearance zone.

41.2.2.4 Component Lead-Hole Sizes.   Component lead-hole sizes are another important consideration in the through-hole printed circuit design when designing for automation. Because of tolerance stack-up between hole location, equipment positioning accuracy, equipment insertion tooling accuracy, component lead variations, and board location accuracy, clearances around the lead and hole become important to minimize insertion errors. Insertion errors can occur if the lead contacts the edge of the hole during an insertion attempt. The nominal hole diameter should be 0.48 mm larger than the component lead diameter to achieve adequate clearance.

41.2.2.5 Physical Size.   The physical size of the end product usually dictates printed circuit board size and shape. Although this is usually the case, it is not always feasible to assemble a board using automation in the exact configuration in which it will be used, especially if it is very small or if it has some irregular shape. Boards or circuits smaller than 76×100 mm or configured with edges that are not straight can be difficult to assemble using automation. The most common practice to overcome such circumstances is to panelize the circuits. This means to design multiple circuits in a square or rectangular panel that will be cut out, sheared, or broken apart at some point after the assembly process is complete.

41.2.3 Assembly Process

The use of through-hole components and the through-hole assembly process can be the most cost-effective method to produce a printed circuit board. Production of through-hole assemblies requires less material costs, process engineering costs, and assembly costs, as compared with the assembly process of surface-mount technology. Component types include dual inline pins (DIPs), axial, radial, and odd-form components.

The automatic assembly of through-hole printed circuit boards almost always follows the same sequence of events. The automated insertion process and the automated soldering process dictate this sequence of events:

1. Dual inline assembly
2. Axial leaded assembly
3. Radial leaded assembly
4. Odd-form components

5. Soldering

6. Cleaning

41.2.3.1   Dual Inline Pin (DIP) Assembly.   The first autoinsertion process in a through-hole printed circuit board is the assembly of DIPs. The typical DIP (see Fig. 41.1) is manufactured in two common widths, 0.3 and 0.6 in. Their lengths will vary with the number of pins. A 0.3-in DIP with 6 pins has a length of approximately 0.3 in, and a 0.6-in DIP with 42 pins will have a length of 2.15 in.

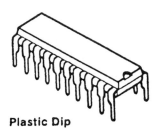

Plastic Dip

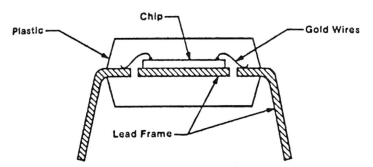

FIGURE 41.1   Dual in-line package (DIP).

DIPs are typically packaged in a plastic extruded tube, approximately 2 ft in length. The number of DIPs in each tube varies depending upon the length of the DIP. Each tube carries only one type, or part number, DIP. The automated DIP inserter carries these tubes vertically in magazines. A shuttle device selects the appropriate DIP and carries it to the center of the machine for automatic insertion. Typical insertion tooling for DIPs is designed to hold the DIPs by their leads during the insertion process. This insertion tooling is also designed to insert the various lengths of DIPs. The same tooling used to insert a 6-leaded DIP (0.3 in in length) is also used to insert a 28-leaded DIP (36.45 in in length).

41.2.3.2   Axial Leaded Components.   The second insertion process is the assembly of axial leaded components. To facilitate automation, axial components are taped by their lead ends. [See Fig. 41.2(*a*).] Most all-axial components are procured from their source in this manner. During the assembly process, the axial leaded components must be sequenced in the order of insertion. A separate machine that cuts the axial components from their tape, places them in a sequential order, and then retapes them accomplishes this. This sequenced reel of components

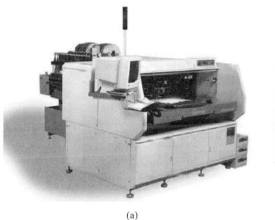

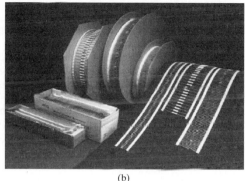

(a)   (b)

FIGURE 41.2 Axial leaded components in various assembly formats. (*a*) Automatic axial inserter. (*b*) Axial leaded components.

is then fed into an axial insertion machine for PCB assembly. The second method is where a single machine cuts the components from the tape and either sequences or shuttles the components in the proper order to the insertion tooling. Choosing which machines to employ in a given assembly process depends upon production volume requirements and the number of different assemblies that must be produced in a given time frame.

Axial components are inserted before radial-leaded components due to the relative size of the components. Typically, axial components are smaller in overall size than radial components. [See Fig. 41.2(*b*).] By inserting the axial components, prior to the radial components, greater population density can be achieved. The tooling footprint for axial components is small: approximately 5.2 mm^2 of tooling meets the printed circuit board at each end of the axial leads.

41.2.3.3 Radial Leaded Components.   After DIP and axial insertion, automatic insertion of radial components can then take place. Radial components can vary greatly in size, shape, height, and weight. (See Fig. 41.3.) Radial leaded components are also taped by their lead ends and are provided by the component manufacturer this way. Unlike the axial process, only one assembly method exists for the radial insertion process. Radial insertion machines remove the

FIGURE 41.3 Radial leaded components in various assembly formats.

components from the tape and insert them into the printed circuit board in the proper order. This is accomplished either by a shuttle system or a sequencing chain. Most insertion tooling used for radial components is designed to avoid previously inserted components.

41.2.3.4 Odd-Form Components. After the automated components are inserted (i.e., DIP, axial, and radial), odd-form component assembly then takes place. The term *odd-form* is often used to describe components that cannot be inserted by the previously mentioned processes. These may include axial, radial, or DIP components if the volume of them is too low to justify automated insertion but can also include connectors and such.

Manual Assembly. Most often, these components are installed manually using a variety of assembly-line methods. The most common method is the *slide line.* (See Fig. 41.4.) A slide line is a progressive assembly line whereby operators install a few components and then slide the PCB to the next operator via a conveyor belt. Other manual-assembly methods include continuous moving conveyors and offline workstations.

FIGURE 41.4 Manual slide line.

Automated Assembly. Automated assembly of odd-form components can be achieved in some cases. The greatest challenge in automating these components is packaging them for pickup by an automated machine. To automate any component, the location of the component must be known. In automated axial and radial insertion, the taping process accomplishes this. A governing body specifies the output of these taping processes: in this case, the Electronic Industry Association (EIA). However, most odd-form components are packaged just to protect them during the shipping process. For automated odd-form inserters to operate, they require the components to be packaged in matrix trays, tape and reel, or tubes. Most odd-form inserters are designed to accommodate all three packaging styles.

Odd-form inserters typically employ a robotic base: either scara, Cartesian, or overhead gantry. Using the robotic base, the odd-form inserter is then equipped with platforms to hold various component feeders. Optional equipment may be added, such as cameras to locate component leads, component gauging stations to straighten leads, component reject stations, and nozzle or gripper nozzle change station.

41.2.3.5 Soldering. After all through-hole components have been inserted into the PCB, it is then either transferred to an automated soldering process or, depending upon the technology, to surface-mount placement. The most commonly used soldering process for through-hole components is the *wave solder machine.* The word *wave* suggests how molten solder is pumped up through nozzles and forms a wave. Printed circuit boards are carried either in a pallet or by a finger conveyor; in either case, the bottom of the PCB is exposed. The PCBs are first passed over a fluxing station where a solder flux is applied to the solder pads and component leads on the bottom of the PCB. The PCB then continues over heating elements that preheat the solder joints in preparation for the actual soldering process. After the PCB has been brought to a predetermined temperature, it continues over the molten solder wave and all the solder joints are soldered in a single pass.

41.2.3.6 Cleaning.   The final process is cleaning off any excess flux if required. PCBs that do need cleaning are sent through a cleaning machine, either in a batch process, similar to a dishwasher, or through an inline machine where the PCBs are conveyed through each cleaning process.

41.2.4 Cells vs. Process Lines

The through-hole assembly process typically passes printed circuit boards from axial to radial to odd-form/hand assembly and then on to wave solder. Within this process are basically two schools of thought regarding how boards get through each of the assembly process steps. The first and oldest of the two is the cell concept. Typical applications include consumer products like televisions and VCRs. Some companies prefer the cell concept, and others prefer the process line concept.

41.2.4.1 Cells.   The cell concept configures each assembly process as an individual stand-alone machine, or as it is sometimes referred to, automation islands. In this concept, boards are routed in batches from one assembly process to another. The machines can be loaded and unloaded by hand and batches of boards manually carried to each of the insertion assembly machines. However, the most common cell configuration utilizes automatic board-handling equipment to process boards through each individual assembly process step. Boards are automatically shuttled out of magazines through the insertion processes to another magazine waiting on the output side of the machine. These magazines are then manually routed through the entire insertion assembly process. Each of these methods requires boards to be manually routed on to the next process step. Table 41.1 lists the advantages and disadvantages of the cell concept.

41.2.4.2 Lines.   The other concept for processing boards through each insertion process is the process line concept. In this concept, machines are connected together utilizing automatic board handling to shuttle each board automatically through the entire process. By connecting machines together in this fashion, multiple machines tied together act as one to shuttle boards

TABLE 41.1 Advantages and Disadvantages of the Cell Concept

Advantages	Disadvantages
Machine down does not cause an immediate stoppage to down-line processes	Work in process increases throughout the assembly process as product travels through the process in batches
Greater flexibility to create alternate workflow routing	Difficult to control or manage workflow routing in multiple-machine process
	Line-balancing discrepancies are not always apparent
	Excessive manual material handling required that have potential for handling damage

through the entire assembly process. Table 41.2 lists the advantages and disadvantages of the process line concept.

TABLE 41.2   Advantages and Disadvantages of the Process Line Concept

Advantages	Disadvantages
Work in process decreases throughout the assembly process due to process batches	One-machine-down situation immediately affects the output of the entire line
Simplified management of workflow routing in multiple-machine process	Less flexibility to create alternate workflow routing
Line-balancing discrepancies become more apparent	
Eliminates the potential for operator handling damage	
Reduced operator labor to run the line	

41.2.5   Insertion Machine Architectures

41.2.5.1   Axial.   There are two basic processes for axial automation in use:

1. For every high-volume, low-mix configuration, many electronic manufacturers prefer a two-step process. The first machine receives reels of a single-part-number axial component. These components are taped axially with the ends of their leads held by tape as per the Electronic Industry Association (EIA) specification number 296-E. As many as 220 single-part-number axial components can be loaded to this machine, known as a *sequencer*. The sequencer then processes the components in a predetermined order. The output is a reel of different components prearranged in a certain order. This prearranged reel is loaded onto an axial inserter machine. The components are then inserted onto the printed circuit board. Machines equipped with two insertion heads can populate two printed circuit boards simultaneously, resulting in insertion rates of up to 40,000 components per hour (CPH). Machines equipped with only one head can deliver up to 25,000 CPH.

2. For the insertion of axial components by one machine, this process combines the sequencing and inserting. The machine sequences the components first in a predetermined order and delivers them directly to the insertion head of the machine. This process can insert up to 25,000 CPH.

The level of automation for the insertion of axial leaded components has increased significantly since the early days. Because the demand for through-hole insertion remains strong, machine improvements continue to be introduced. Brushless servomotor technology and state-of-the-art motion controllers have replaced most pneumatic drive assemblies. These technology advances have allowed machine manufacturers to advertise insertion reliability rates of less than 150 parts per million defects.

41.2.5.2   Radial.   To meet the demands of high-volume assemblers, a single radial machine at speeds in excess of 20,000 CPH can insert components. For high-mix environments, components can be loaded onto integrated sequencers or feeder carriages, which provide random access to 100 or more unique part numbers. Given the requirement for high-quantity products, radial components can be electrically verified immediately prior to insertion, and fewer than 300 defects per million insertions is achievable in real production.

The roots of radial lead insertion date back into the early 1970s, when radial lead components began coming out of Japan with their leads taped at a fixed span of 5.0 mm. Soon thereafter, additional taping spans were adopted as standards in order to increase the range of components that could be presented on radial tape. These specifications are described in the EIA standard 468-B. Insertion machine developers have responded by providing the capability to reliably handle multiple lead spans in one machine.

The typical radial machine can handle components with 2.5- and 5.0-mm lead spans. However, due to the increased needs of the marketplace, radial machines have been developed to handle three differing lead spans. These triple-span machines operate at the same high level of performance as their dual-span counterpart. Triple-span machines can be configured to handle either 2.5/5.0/7.5-mm or 5.0/7.5/10.0-mm components. The advantage of these triple-span machines is improved flexibility. In cases where dual-span tooling is required, the triple-span machine can be configured with dual-span tooling so that the footprint of the insertion guide jaw is not limiting.

The radial taping format has proven to be an effective means of packaging and delivering a wide variety of component types for automatic assembly. Devices such as tact switches, potentiometers, resistor networks, LEDs, SIPs, connectors, and fuse holders used to be available only in bulk and had to be inserted manually. These and many other components that were once considered odd forms can now be inserted reliably by radial insertion machines because they are now packaged appropriately for automated insertion. This trend is expected to continue and cause machine suppliers to further expand the insertion capabilities of their machines.

41.3   SURFACE-MOUNT TECHNOLOGY

41.3.1   Assembly Process

For most surface-mount assembly boards, the bottom side contains mostly passive components, and the topside contains both passive and active components. Most often, as seen in Fig. 41.5, topside components are placed first, following the printing of solder paste onto the pads of the board; then, the board is moved through an oven, which causes the solder paste to reflow and form the solder joints. If any through-hole components are used, they are inserted next. The board is then flipped over, an adhesive is applied on the board for some of the component locations, these components are placed, and the board is moved through an oven (sometimes ultraviolet-light-curing adhesives are used) to cure the adhesive. The board is flipped once more and passes through a wave-soldering process, which secures the through-hole and bottom-side surface-mount components in place.

Increasingly, assemblers of pure surface-mount boards are moving to a two-sided reflow process that uses solder paste with different reflow temperatures on each side of the board. The higher-temperature solder is used first so those components do not reflow again and fall off when the board is flipped and the second side is assembled and reflowed. This process eliminates the need for different top- and bottom-side processes and eliminates the need for adhesives and wave-soldering equipment.

41.3.2   Component Placement Machines

41.3.2.1   Types.   The electronics industry has experienced explosive growth over the last two decades. Increasing demand from consumer, telecommunications, and PC markets has fueled this growth. During this time, the turret-style chip shooter and the gantry-style flexible fine-pitch (FFP) machines have both enjoyed worldwide industry dominance and acceptance. However, steep increases in production volume requirements and very aggressive board changeover demands have forced the industry to examine alternate high-end machine architectures such as high-speed gantry and massively parallel machines.

The basic turret-style concept of chip placement has remained common between all suppliers since the early 1980s. (See Figs. 41.6 and 41.7.) Multiple heads are equally positioned around a horizontally rotating turret. A moving feeder carriage positions tape feeders under

Top Side SM Assembly Process   Bottom Side SM Assembly Process

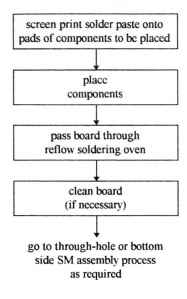

screen print solder paste onto
pads of components to be placed

↓

place
components

↓

pass board through
reflow soldering oven

↓

clean board
(if necessary)

↓

go to through-hole or bottom
side SM assembly process
as required

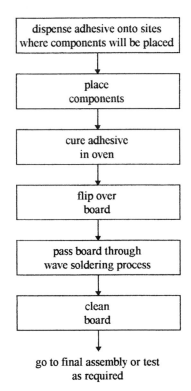

dispense adhesive onto sites
where components will be placed

↓

place
components

↓

cure adhesive
in oven

↓

flip over
board

↓

pass board through
wave soldering process

↓

clean
board

↓

go to final assembly or test
as required

FIGURE 41.5 Process flows typically followed in bottom-side/topside surface mount applications.

FIGURE 41.6 Typical turret used for component placement.

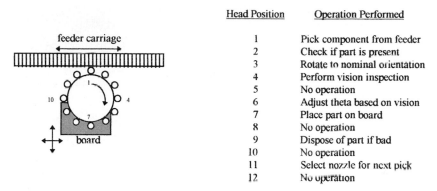

Head Position	Operation Performed
1	Pick component from feeder
2	Check if part is present
3	Rotate to nominal orientation
4	Perform vision inspection
5	No operation
6	Adjust theta based on vision
7	Place part on board
8	No operation
9	Dispose of part if bad
10	No operation
11	Select nozzle for next pick
12	No operation

FIGURE 41.7   Simplified illustration of how a turret-style chip shooter works.

a single pickup point. After the part is acquired, the turret rotates the component to a vision-processing station. At this station, a CCD camera visually acquires an image of the part. As the turret continues rotating, a moving table positions the PCB under the vision-corrected part placement location. With the PCB positioned under the single placement point, the part is lowered to the target placement location. After the component is placed, the heads are rotated back to the feeder pickup point and the cycle is repeated. The rotation of all heads on the turret results in the simultaneous pickup, inspection, and placement of individual components. (See Table 41.3.)

TABLE 41.3   Turret Architecture

	Common usage	Lends itself to
Part range	0402–20 mm	0201–32 mm
8 mm feeder capacity	160	300
Average speed range	20,000–25,000 CPH	25,000–40,000 CPH
Characteristics	Chip placement	Chips, small-area arrays
	Moving feeders and PCB	Moving feeders and PCB
	Tape-fed components	Tape and bulk

The traditional application of the gantry-style machine architecture (Fig. 41.8) has been in the area of flexible fine-pitch placement. (See Table 41.4.) The basic concept of gantry operation uses a fixed feeder and stationary board design. A multispindle placement head is mounted to either a single- or dual-gantry positioning system. The positioning system moves the placement head over the desired feeder locations where one or more components are picked. After the components are acquired, the head moves the components over a fixed upward-looking vision station. After the components are visually inspected, they are placed on the fixed PCB using the vision-corrected coordinates.

A variation of the gantry-style machine architecture is replacing the turret. Compared with the turret, these machines offer several options that help them to achieve higher speeds:

- Tape splicing for uninterrupted machine production
- Simpler bank-changing capability
- Stationary feeders and PCB
- Smaller footprint

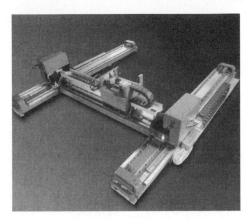

FIGURE 41.8   Gantry architecture.

TABLE 41.4   Flexible Fine-Pitch (FFP) Gantry Architecture

	Common usage	Lends itself to
Part range	Full SMT	Full SMT, flip-chip, odd-form
8 mm feeder capacity	70	130
Average speed range	1,000–6,000 CPH	5,000–15,000 CPH
Characteristics	Larger parts, higher accuracy	More odd-form third-party feeder
	Fixed feeders and PCB	integration
	Tape-, tube-, bulk-, and tray-fed parts	

The high-speed gantry uses a dual-beam positioning system. (See Table 41.5.) One inline or rotary placement head is mounted on each beam. After components are picked, part inspection usually occurs with a head-mounted camera while the gantries are in motion. This eliminates time normally associated with part inspection using a stationary upward-looking camera. After the vision inspection, components are placed on a fixed PCB. Significant machine speed is realized by having one placement head picking components, while the other placement head is placing components on the PCB. Overall chip placement is accomplished by linking together multiple high-speed gantry machines. This style of machine will continue to capture market share at the expense of the turret.

Another type of machine architecture impacting the turret dominance involves parallel processing to achieve very high-speed chip placement. (See Table 41.6.) The basic concept of operation relies on the cumulative operation of multiple placement modules. Each individual

TABLE 41.5   High-Speed Gantry Architecture

	Common usage	Lends itself to
Part range	0402–20 mm	0201
8 mm feeder capacity	80	150
Average speed range	14,000–18,000 CPH	15,000–21,000 CPH
Characteristics	Electric feeders	Smart feeders
	Turret alternative	Expanded part range
	Big-board capability	Increased feeder capacity

TABLE 41.6 Massively Parallel Architecture

	Common usage	Lends itself to
Part range	0402–10 mm	0201–25 mm
8 mm feeder capacity	96	192
Average speed range	50,000–85,000 CPH	60,000–100,000 CPH
Characteristics	Chip placement	Expanded part size
	Difficult to change over	Bank changing
	Restricted board size	Large-board capability
	Tape splicing	

placement module is capable of picking, inspecting, and placing components on a portion of the PCB. During the assembly process, boards are advanced along an indexing conveyor, progressively bringing the whole PCB area within reach of every placement module. The net effect of parallel module operation is high throughput. (See Fig. 41.9.)

41.3.3 Machine Vision Technology

As the surface-mount process has evolved into such areas as high-density printed circuit boards, area array devices, and fine-pitch packages, the use of mechanical techniques to ensure reliable, accurate placements has become obsolete. Modern printed circuit boards have achieved densities and sophistication that demand that some technique of artwork registration be used. In addition, with the development of packaging technologies such as flip-chip (with bump pitches as small as 0.1 mm), 0402 and 0201 package type discrete devices, and quad flat packs with lead pitches as small as 0.3 mm, extraordinary precision is required to maintain yields in the overall surface-mount process. In addition, beyond standard surface-mount packages, a myriad of odd-shaped devices exists, such as RF shields and connectors. Machine vision technology is typically employed to solve the metrology issues associated with such a breadth of application. Machine vision provides the speed, accuracy, reliability, and flexibility that is required in both high-volume and high-mix assembly challenges.

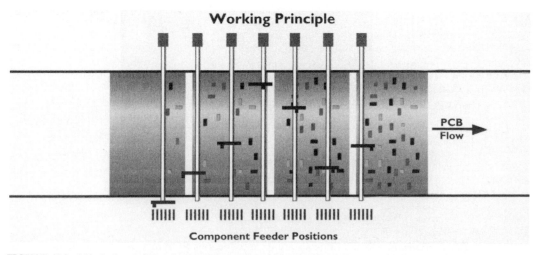

FIGURE 41.9 Massively parallel architecture showing working principle and component feeder positions.

41.3.4 Machine Vision Operating Overview

In the surface-mount process, machine vision is a collection of electronic cameras, optics, specialized computing hardware, and algorithms that together perform important metrology and analysis functions. The first step in machine vision is to acquire an image of a subject such as a surface-mount component or a PWB registration mark. Electronic cameras (typically CCD devices) are generally used for this purpose. That image is then transferred to a vision computer or vision system, which extracts the location of features such as fiducial marks or surface-mount component leads, and reports them to the host machine's controller. That information is then used to guide the placement process to ensure that accurate placements are maintained.

41.3.5 The Function of Machine Vision in the Surface-Mount Process

The typical use of machine vision in the surface-mount process is twofold:

1. The registration of a circuit board or substrate in a pick-and-place machine's workplace and location
2. The geometric verification of surface-mount devices prior to placement

41.3.5.1 Locating a Printed Circuit Board or Substrate in a Machine's Work Area. To populate a PCB or substrate accurately, a machine must have the ability to locate a printed circuit board in its workspace. The general technique for this is for a machine to first transfer a PCB into its work area. Next, the PCB is affixed in place by mechanical clamp, vacuum chuck, or similar technique. The machine then positions a camera over the PCB at the location where an artwork registration mark, or fiducial, is expected to be. Then, using machine vision, the fiducial is located on the PCB. (See Fig. 41.10.) This process is repeated for several fiducials on the PCB (two or three, typically). The measured locations of the fiducials are then used to modify the nominal placement coordinates such that they are aligned with the artwork on the PCB.

FIGURE 41.10 A typical fiducial on a printed wiring board.

Local Fiducials. When fiducials are used to locate a PCB in a workspace, an implicit assumption is that there is a known relationship between the artwork and the fiducial marks. In some cases, such as ceramic substrates, it is possible that enough local distortion exists in the

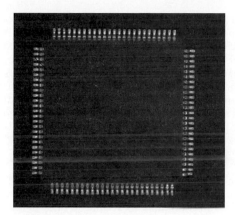

FIGURE 41.11 An image of quad flat pack (QFP) leads, as viewed by a vision system.

artwork that the known relationship assumption breaks down. In this case, the placement site must be measured directly either by the use of fiducials in close proximity or by measuring the placement artwork itself. By doing the direct measurement of the placement site, the local distortion can be compensated for and an accurate placement can be made. Note that local fiducials do not enhance overall placement equipment accuracy. Instead, they may enable a piece of equipment to successfully populate distorted substrates.

Component Vision. The goal of any surface-mount process is to maximize the yield of the solder process. One factor in this is ensuring maximum lead-to-pad coverage. From a machine vision perspective, this is achieved by using those critical features [e.g., leads on a quad flat pack, or solder bumps on a flip-chip (see Figs. 41.11 and 41.12)] of a surface-mount component that are intimately involved with the solder process as key metrology points. Using the locations of the critical features, the final placement location is calculated such that the maximum lead-to-pad coverage is achieved.

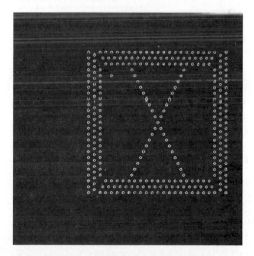

FIGURE 41.12 An image of flip-chip bumps as viewed by a vision system.

41.3.5.2 Component Inspections. Some surface-mount components, such as quad flat packs, are prone to damage due to worn die sets or improper handling. Surface-mount placement equipment can serve as a screening tool for these devices. Often, there are inspections for errors in lead pitch and overall component geometric integrity. In general, however, pick-and-place machines are not component-quality statistical tools. Instead, they provide pass-fail screening for surface-mount devices that are about to be placed.

41.3.5.3 Optical Considerations. Machine vision is not magic. Certain physical limitations do exist. For example, any given feature must have a minimum number of pixels in order to be recognized by a vision system. Very small features can be made to appear larger with higher-magnification optics. The downside to this is that the maximum size of a component that can be viewed is reduced proportionally (i.e., the field of view is reduced). The choice of optics is critical for maximum efficiency in the surface-mount process. For example, the same camera that processes a 32-mm QFP such as a 208-pin QFP in a single image generally does not have enough resolution to process a flip-chip that has bump features of 0.1 mm in diameter. For maximum efficiency, either the two components must be processed on separate machines, or, if multiple cameras are available, the appropriate combination of cameras must be chosen to cover the application space.

41.3.6 Line Architecture

The machines described earlier can be combined to constitute several types of line configurations that will be reviewed in this section. Different variables such as mix, volumes, footprint, ease of software interconnection, optimization tools, area, and cost of ownership will drive different configurations.

Assemblers will combine various individual machine architectures in an attempt to create the optimal manufacturing line that is tailored for their specific manufacturing environment. Many factors are used in determining the ultimate line, including line footprint, cost of ownership, throughput, changeover capability, equipment commonality, etc. Each specific customer will weigh these factors differently, depending on their own unique application requirements. However, there are three line configurations that have gained industry acceptance. Table 41.7 gives a summary comparison of the three types of line architecture.

TABLE 41.7 Comparison of Line Architectures

	High-speed turret line	Ultra-high-speed line	Modular line
Flexibility	Medium	Low	High
Throughput	Medium	High	Low
Moving feeders	Yes	No	No
Moving PCB	Yes	No	No
Board size range	Medium	Low	High
Line reliability	High	Medium	Low
Tape splicing	No	Yes	Yes
Ease of changeover	Medium	Poor	High
Feeder commonality	No	No	Yes
Installed base	First	Third	Second

41.3.6.1 High-Speed Turret Line.   This line type combines the two dominant machine architectures of turret-style chip placement and gantry-style fine-pitch placement. This category represents the largest installed base of manufacturing lines, having penetrated a majority of OEM and CEM facilities. As such, a tremendous infrastructure has grown as a result of this line acceptance. This infrastructure includes:

- Trained technicians, programmers, and maintenance personnel
- Large inventories of feeders and spare parts
- Common software to support programming and balancing programs for transfer to any production line in any geographical location
- Perception of reliable, familiar, and safe performance

These attributes will continue to promote the acceptance of this type of mature manufacturing line. However, two other growing line types will threaten this dominant position.

41.3.6.2 *Ultra-High-Speed Line (UHSL).*   This line combines one or more massively par-allel machines for chip placement with one or more gantry-style architectures for flexible fine-pitch placement. One of the most powerful features of the UHSL is a very high through-put in a relatively small footprint. It is very common to have line outputs reach 65,000 com-ponents per hour, or around double the output of a traditional high-speed turret line. The overall line length of the UHSL is also smaller than the turret line.

The UHSL line is common at high-volume automotive, telecommunications, and consumer-oriented accounts. These accounts have short-line pulse rates. Completed PCBs can exit the last machine in the UHSL line at a rate of less than 60 s. Due to the inflexibility of the massively par-allel machine, product PCB changeover is kept to a minimum.

41.3.6.3 *Modular Line.*   This line integrates multiple gantry-style architectures in an at-tempt to combine flexibility and high-speed placement. The modular line offers the highest level of flexibility in terms of PCB changeover, board size range, and component handling range. Compared with the high-speed turret line, changeover is simplified by the use of com-mon feeders across the machines in the modular line. Most turret lines require different feed-ers for the turret and flexible machines. The modular line also permits the handling of PCBs, which are larger than 20 × 18 in. This capability would allow a CEM to explore additional big-board business, which would not be possible in any of the other line types. Two different ap-proaches have emerged:

1. An approach that uses machines that are physically smaller versions of dedicated chip place-ment and flexible fine-pitch machines. In this scenario, each machine is only capable of han-dling a fixed component range. As such, constraints associated with dedicated equipment remain. For example, line flexibility suffers as components are forced to specific machines. Forcing parts on specific machines complicates line balancing and usually results in machine-specific bottlenecks.

2. An approach that uses machines, each of which is capable of placing the full part range. This approach has answered the flexibility limitation problem. It simplifies line-balancing concerns and eliminates any bottlenecks caused by dedicated machine types.

41.3.7 Dispensing

Adhesives have been used in electronic assembly processes for years. Examples are epoxies and silicones, which have been used for encapsulation and assembly operations. Adhesives are applied to surface-mount devices to secure them to the substrate prior to soldering.

Surface-mount devices (SMD), unlike through-hole components, are not inserted into the board; therefore, leads are not clinching to hold the components in place. SMDs are simply placed onto the surface of the printed circuit board and then soldered, by either a reflow-soldering or a wave-soldering process. In bottom-side applications, it is necessary to apply adhesive to the components in order to hold the component in place until the assembly can be sent through the wave-soldering operation.

The adhesive must hold the component in its desired orientation and placement, maintain this placement during final assembly and handling, and withstand the high temperature and fluxing concerns associated with the wave-soldering operation.

After the soldering operation has been completed, the adhesive's purpose is finished. If con-siderations are not taken concerning the selection of the surface-mount adhesive, it can become a possible problem area. Some epoxies, for example, might absorb moisture and cause problems for the correct operation of the circuit. This is why it is important to carefully consider the type of surface-mount adhesive that is correct for your particular application.

41.3.7.1 *Adhesives.*   In determining which adhesive is correct for your operation, a list of ideal adhesive characteristics should be made for reference during the adhesive evaluation.

The following list of characteristics includes properties of the adhesive itself, as well as its properties for dispensing, curing, and its environmental concerns.

Adhesives are categorized into the following four groups, each differing in their chemical composition with their own advantages and disadvantages:

1. *Thermosetting adhesives.* These materials cure by chemical reaction to form a cross-linked polymer. Thermosetting adhesives are strong adhesives and once cured, by heat or catalytic action, cannot be softened to establish a new adhesive bond. One example of this type of material is epoxy.

2. *Thermoplastic adhesives.* These materials do not change chemically when a bond is established. When reheated, these materials can be softened and reformed. Materials from this group are typically used in nonstructural application where only a small amount of stress is applied. One example of a thermoplastic material is nylon.

3. *Elastomeric adhesives.* These materials are a subset of thermoplastic adhesives and have a high degree of elasticity. One example of this type of material is rubber.

4. *Toughened alloy adhesives.* These materials are formulated by blending rubbers and resins to form a special type of thermosetting adhesive. They provide high structural strength with resistance to shock loading. One example of this type of material is epoxy-nylon.

In selecting the surface-mount adhesive, the first thing to be considered must be the application process. When the process has been chosen, a suitable machine should be selected to complete the process. A machine can be chosen that is a stand-alone dispenser or one that is integrated into a pick-and-place machine. When these aspects have been determined, a suitable adhesive can be chosen.

Thermosetting epoxies are typically best suited for use in electronics manufacturing processes. These epoxies exhibit excellent precure properties (green strength, fill properties), effective curing properties (no undesirable gas release or adhesive migration), ideal postcure properties (resistance to damage during handling), and no long-term damage to the devices or the printed circuit board. These adhesives are also compatible with all major dispensing methods.

41.3.7.2 Methods. There are five major methods for applying surface-mount adhesive to a printed circuit board. They each have their advantages and disadvantages.

1. Pin transfer
2. Screen or stencil printing
3. Time-pressure
4. Archimedes screw
5. Positive displacement pumps

Pin Transfer. Pin transfer (see Fig. 41.13) is the simplest mass-application method for high-volume automated assembly machines. In this process, a pin is dipped in a reservoir and picks up a droplet of adhesive. The pin is lowered above the surface of the printed circuit board, and the surface tension causes a portion of the adhesive to be deposited on the printed circuit board. It is very important that the pin does not touch the board because this will cause inconsistent dot sizes and shapes. This system requires the substrate to be relatively flat and free from distortion. The pin is also smaller at the end to help keep the major portion of the diameter of the pin off the board in order to keep the dots as consistent as possible. The nature of the pin array allows for adhesive to be applied to the board after through-hole parts have been placed. The adhesive used must have sufficient strength to hold the component when placed with the placement machine or head. One major problem is that the bath of adhesive absorbs water from the air. This requires the glue to have lower water absorption properties in normal atmospheric conditions.

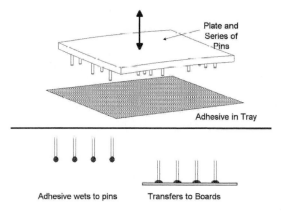

FIGURE 41.13   Pin adhesive transfer example.

Screen/Stencil Printing.   Similar to the pin transfer technique, screen printing or stencil printing (see Fig. 41.14) deposits all of the adhesive dots at the same time. This also makes it useful for mass-production applications. Stencil printing also requires the printed circuit board to be flat and distortion-free. Unlike the pin transfer method, stencil printing requires the board to be free of any obstructions, such as through-hole or surface-mount components.

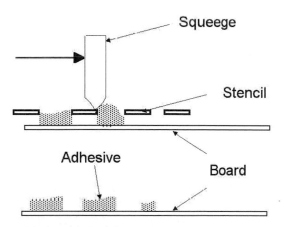

FIGURE 41.14   Stencil printing.

This limits screen printing to single-pass, unmixed-process assemblies. When screen-printing adhesive, a stencil with apertures for each component is placed over the printed circuit board. A squeegee passes across the stencil and forces the adhesive through the openings and onto the substrate. The adhesive dot size is determined by the size of the exposed area in the stencil, the thickness of the stencil, and the viscosity of the adhesive. One of the drawbacks of this technology is the extensive cleaning that is required to maintain this process. The stencil or screen must be kept clean in order to allow the adhesive dots to form properly. Cost is another problem associated with stencil printing. A new stencil is required for each product. Water absorption could also be a problem in this type of process.

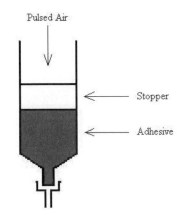

FIGURE 41.15   Time pressure dispensing.

Time-Pressure. A machine that utilizes time-pressure technology dispenses adhesive from an adhesive syringe by applying air pressure for a fixed length of time (see Fig. 41.15). The adhesive dot size is determined by the nozzle diameter and by the duration and magnitude of the applied pressure. Machines utilizing this technology typically incorporate multiple spindles so that multiple dot sizes can be dispensed without added setup. Another advantage is the flexibility that comes with being able to simply set up the machine to run a different board based on the programming. One problem with this technology is that the dot size dispensed decreases as the syringe empties because more pressure will be required to push the same amount of material out of the nozzle. This pump type is slow compared with the mass-application methods that are available, although the dot characteristics are much more controllable.

Archimedes Screw. This type of dispenser utilizes an Archimedes screw to push the adhesive out of a nozzle. (See Fig. 41.16.) The friction of the adhesive against the threads of the screw causes the material to move along the screw. These systems also typically utilize multiple spindles to dispense various dot sizes and configurations. As well as using multiple spindles, these machines are typically software controlled so that multiple dot sizes can be achieved with each spindle. The adhesive dot size dispensed by

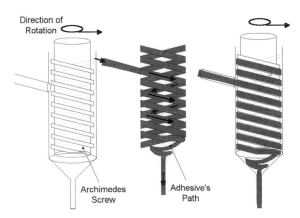

FIGURE 41.16   Archimedes screw dispenser.

these types of machines is determined by the amount of time and the number of encoder counts the screw turns, the pressure that is applied to the syringe, the type of screw utilized, and the diameter of the nozzle. One advantage of this technology is that multiple dot sizes can be achieved with each spindle. The fact that this is an open system is one disadvantage because low-viscosity materials can run out of the nozzle if too much pressure is applied or too little friction is generated between the material and the screw.

Positive Displacement Pumps. The concept behind this type of pump is that a piston seats at a specific distance inside the nozzle. (See Fig. 41.17.) The piston retracts, allowing adhesive to fill the hole in the nozzle. When filling is complete, the piston fires down into the hole in the nozzle, forcing an exact amount of material out the nozzle and onto the printed circuit board.

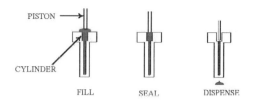

FIGURE 41.17   Positive pump displacement.

This technology is very consistent because the exact volume of material displaced is dependent upon the volume of the cylinder and the distance that the piston displaces within that cylinder.

41.3.7.3   Material Considerations

Dot Height.   The height of the adhesive dot is important because care needs to be taken to make sure that the adhesive will contact the bottom of the component when it is placed. Care must also be taken to be sure that there is not too much adhesive. Excessive adhesive could cause problems related to contamination of the component pads.

Contamination.   If even a very thin layer of adhesive is on the pads, it could cause problems with the soldering application. To avoid this problem, the dot size must be tightly controlled so that the dot will not spread onto the pads. Also, the method of application must be controlled so that extra material does not contaminate the pads.

Adhesive Curing.   The method that is used to cure the adhesive depends on the type of adhesive used. For automated electronics manufacturing, adhesive is typically cured in a reflow or a curing oven. The manufacturer's data sheet should be consulted for information concerning the recommended curing profile, including the maximum temperature and cycle time.

41.3.7.4   Traditional Bottom-Side Line

Process Description.   Typically, a traditional bottom-side line (see Fig. 41.18) includes a method of applying adhesive, a placement machine to place the bottom-side components, and an oven to cure the adhesive. This line will be followed by a wave solder machine, which will in turn be followed by an inspection station.

Process Considerations.   The first thing that must be considered when setting up any manufacturing line is the type of components and assemblies that are going to be used or built on it. This type of line can be used simply to apply glue to a printed circuit board, place components on the board, and then cure the glue in order to hold the parts onto the board prior and during wave soldering. In this type of application, the green strength of material determines whether components stay in place during placement operation on the chip shooter. This makes the choice of glue very important. After wave solder, using this type of line, parts may be missing due to missing adhesive dots or the postcure strength of the adhesive. Some of these parts may be knocked off of the board during manual assembly or handling, so care

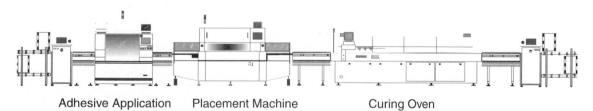

Adhesive Application   Placement Machine   Curing Oven

FIGURE 41.18   Traditional bottom-side surface-mount assembly line.

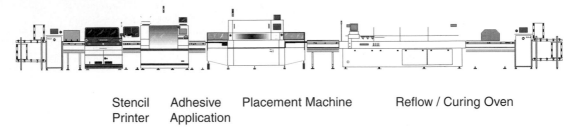

Stencil   Adhesive   Placement Machine   Reflow / Curing Oven
Printer   Application

FIGURE 41.19   Bottom-side line with solder paste application.

should be taken to control the forces to which these assemblies are subjected. This line is very basic in its functionality, but it can reliably build products when implemented correctly.

41.3.7.5 Bottom-Side Line with Solder Paste Application

Process Description.   A bottom-side line that includes solder paste application includes a system for applying the solder paste (stencil printer or high-speed dispenser), a method of applying adhesive (high-speed dispenser or pin transfer), a placement machine to place the bottom-side components (flexible placement or chip shooter), and an oven to cure the adhesive and reflow the solder paste. (See Fig. 41.19.) A wave solder machine and an inspection station will then follow this line. The inspection station, however, should see limited use because of the robustness of this process.

Process Considerations.   This type of manufacturing line is more flexible than the previously discussed line. For bottom-side applications, this configuration provides greater torque strength due to the adhesive being combined with solder paste. This will assist in reducing the number of missing-part defects present in the assembly. This type of line also helps to reduce problems that are related to the wave-soldering operation (insufficient solder). In this type of application, the dot height is important to consider because the dot must be tall enough to contact the component even above the solder paste deposit. Consideration must also be given to the design of the stencil used to print the solder paste and the design of the nozzle used for high-speed dispensing operations. Both of these points can turn into problems later if not considered properly. It is also very important, in this type of assembly line, to consider the accuracy of the machine because this parameter can be critical to the success of the manufacturing line.

41.4 ODD-FORM ASSEMBLY

41.4.1 Overview

In electronic PCB assembly, odd-form components are identified as those through-hole or surface-mount components whose height, weight, or shape traditionally would not allow them to be automatically placed using standard pick-and-place assembly machines. (See Fig. 41.20.)

While most of the processes in electronic printed circuit board assembly have enjoyed great advances in automation over the years, until recently, odd-form component placement has seen little movement toward automation. For the most part, odd-form assembly has remained a manual process for many reasons, including a lack of suitable component packaging, feeding, and placement equipment.

Several factors have combined, however, to lead electronic manufacturers to take a new look at automating odd-form component assembly. The key drivers behind this new push are manufacturers' increasing needs for higher production volumes, improved product quality, reduced time to market, lower assembly cost, and increased global competitiveness.

FIGURE 41.20   Examples of odd-form components.

41.4.2   Manual Assembly

Manual assembly involves operators working in sequence within a PCB assembly line. The operators pick components from bulk bins and their primary task is to repeatedly locate and position those components onto the PCB, while keeping pace with the line's pulse rate. (See Fig. 41.21.)

FIGURE 41.21   Manual printed circuit assembly line.

The unusual nature of the odd-form components means that the operators pay special attention to certain devices. For example, they will generally need to put extra effort into handling components with high pin counts so they don't damage the leads. Also, components that incorporate snap-in features (connectors) to retain them in the PCB require high insertion force, thus requiring the operator to manually exert this force or to use an alternate tool to press the component into the PCB.

Generally, manual assembly is located as the last assembly operation in a PCB assembly line and when the operators have completed their task the board proceeds to solder reflow.

41.4.3 Automation

Automated assembly involves the use of an odd-form assembly cell consisting of a placement machine, conveyor, placement head with pick-and-place tooling, and component feeders. The general sequence consists of a PCB being transported into the machine by means of a conveyor, the placement head picks the odd-form component from a feeder, and moves to a preprogrammed site on the PCB where it places the component. Upon completion of placing all the components, the PCB is transported out of the machine to the next assembly operation.

Once picked, the component may be inspected over a vision system for accurate X-Y offset and θ correction to increase placement accuracy and reliability. Automated equipment has the flexibility to place odd-form components, fine-pitch and ball grid array SMT components, and straddle-mount connectors—all in one machine, resulting in substantial floor space reduction. (See Fig. 41.22.)

The speed and accuracy at which automated odd-form component insertion is performed is the critical advantage. A machine provides you consistent, reliable insertion speed and positional accuracy, yielding favorable results in quality.

The advantages of automation include:

- Increased quality—less likely to have missing or misoriented or damaged components.
- High throughput through increased speed.
- Steady-state cycle time—no breaks.
- Floor space savings.

FIGURE 41.22 Close-up of an odd-form component placement machine showing the flexible servodriven gripper tooling used to handle a wide range of component types.

- For intrusive reflow processes, automation will not disturb the pattern of solder paste when inserting odd-form leaded devices (manual assemblers must find the holes buried beneath the paste).
- Mobility—allows for rapid deployment of manufacturing around the globe.
- Computer integrated manufacturing compatible—ability to be integrated with a host computer to gather required management data and download of new products when product changeover is required.
- Eliminates operator's repetitive strain-stress injuries.

The disadvantages of automation include:

- Significant start-up investment.
- Cost of packaging components for automation.
- Product changeover may be time-consuming and require additional investments in feeders.
- Odd-form feeder tooling is generally dedicated.
- Machine programming requires up-to-date programs.
- Equipment maintenance and utilities cost.
- Training new employees on equipment.

41.4.4 Components

Odd-form components consist of a wide variety of parts, including larger components such as slot 1 connectors, universal serial bus (USB) connectors, dual inline memory module (DIMM) connectors, headers, phone jacks, zero-insertion-force (ZIF) connectors, and speakers as found on computer motherboards, to miniature components such as surface-mount and ball-grid-array connectors, vibrators, shields, and miniature speakers found in palm-top and cellular phone products. (See Fig. 41.23.)

The primary odd-form component package has been through-hole components that were processed using wave solder technology. The challenge that today's designers and assemblers face to produce products that are faster, more complex, more reliable, and at a lower cost has resulted in a major shift for odd-component packages. This challenge has driven component manufacturers to change or release new odd-component package types such as:

- Replacing the standard through-hole odd components with surface-mount equivalents
- Changing the component body material of the through-hole components to a high-temperature material that can be processed using reflow ovens
- Introducing solderless through-hole connectors incorporating compliant pin technology originally developed for backplane applications
- Supplying straddle-mount connectors that mount to the side of the printed circuit board for miniature and high-speed communication products
- Adapting the latest ball grid array technology into connectors

The volumes of odd-form components are connectors. Connectors currently make up 75 to 80 percent of odd-form components, with the remaining percentage being transformers, relays, crystal oscillators, or electrolytic capacitors. The introduction of new high-temperature components has allowed manufacturers to implement odd-form components into their standard surface-mount assembly process and withstand the heat of reflow ovens.

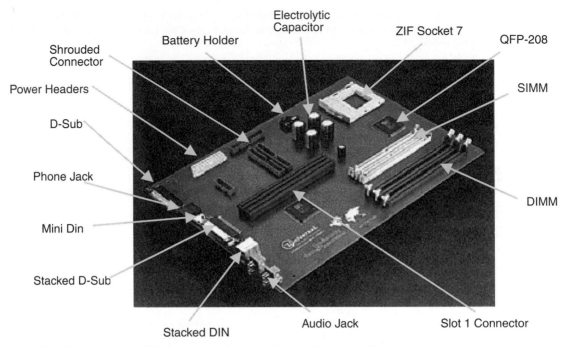

Shrouded Connector

Power Headers

D-Sub

Phone Jack

Mini Din

Stacked D-Sub

Battery Holder

Electrolytic Capacitor

ZIF Socket 7

QFP-208

SIMM

DIMM

Stacked DIN

Audio Jack

Slot 1 Connector

FIGURE 41.23 Examples of odd-form components that can be assembled by machine.

41.4.5 Component Packaging

A key factor to the viability of automating odd-form component assembly is the component package. The packaging formats used are tape and reel, radial and axial tape, extruded tube, tray, continuous strip, and bulk. Each of these formats has benefits and drawbacks, depending on the situation in which they are used.

41.4.5.1 Tape and Reel. Tape and reel provides for a higher volume of parts than either tubes or trays. (See Fig. 41.24.) Tapes are available in flat-carrier tape, as well as deep-pocket and semipocket tape. The Electronics Industry Association (EIA) standard for these tapes is

FIGURE 41.24 Tape packaging.

EIA-481. All three tape formats are available in standard configurations and can provide a high volume of online parts inventory. However, components used with flat-carrier, deep-pocket, or semipocket tape often require repackaging, which can add cost to the component. Deep-pocket tape is covered by adhesive tape as supplied for surface-mount assembly. In deep pockets, tall or heavy parts may change position or break the cover tape. Semipocket tapes have shallower pockets and cover tapes that loop over components. In the shallower pockets, the cover tape holds the parts snugly. The packaging of multiple parts per pitch is possible with semipocket tape. The advantage of packaging multiple components per pitch is that feeder space is conserved.

41.4.5.2   Radial and Axial Tape.   Radial and axial tape formats are covered by EIA standards. For radial, the standard is EIA-468-B; for axial, it is EIA-296-E. These packages work for a wide range of components and are generally produced at a low cost. Because of its configuration, radial tape should not be used for components that are tall and can easily be deformed on the tape. (See Fig. 41.25.)

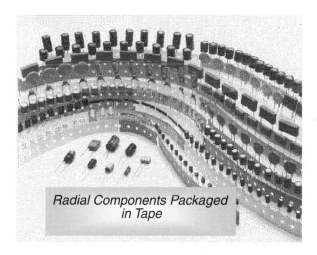

Radial Components Packaged in Tape

FIGURE 41.25   Radial components packaged in tape.

41.4.5.3   Extruded Tube.   Extruded tube is the most commonly used format and can be a low-cost solution. (See Fig. 41.26.) Components such as D-Sub connectors, phone jacks, transformers, relays, and shrouded headers are generally supplied in tube packaging. Recently, component manufacturers have begun to supply long connectors such as SIMMs, DIMMs, and other connectors with a long aspect ratio in edge-stacked tubes.

41.4.5.4   Tubes.   Tubes are a good package format for protecting the component and achieving high-quality automated production. Although tube packaging can be inexpensive to implement, one needs to verify that the tube used for shipping is acceptable for automation. Tubes can be constructed with a thin cross section that allows the tube to deform easily, resulting in jamming of the components within the tube. The thin cross section may eliminate the ability to stack full tubes of components in a feeder depending on the component weight. Heavy components packaged in thin cross-sectional tubes will deform adjacent tubes when stacked in an automated multitube feeder. Other potential areas of concern when selecting tube packaging is the need for a specially designed feeder track, the need to verify tube design will not allow shingling of

*Components Packaged in
Extruded Tubes*

FIGURE 41.26   Component tube packaging.

components, and the need to be aware of tube orientation when loading the feeder. Today's equipment manufacturers provide a number of tube feeders from low-cost to quick-change stackable, thus making tube packaging one of the most widely used formats.

41.4.5.5 Matrix Trays.   Matrix trays are an inexpensive means to supply components. (See Fig. 41.27.) One of the key issues with trays is that in most cases the components are not packaged for automation. Component leads are not always oriented in the insertion plane, and

FIGURE 41.27   Components packaged in matrix trays.

pocket registration of the trays is often not acceptable for reliable automated pickup and insertion. Other factors to consider when choosing trays include: trays lack rigidity required by many assembly machines; vacuum-form trays provide a limited online parts inventory; there are no standards for vacuum-formed trays for use with odd-form components; and tray size may take up a significant amount of feeder space in a placement machine.

41.4.5.6 Continuous Strip.   Continuous strip packaging is designed for high-volume automation applications. (See Fig. 41.28.) In most cases this packaging technique eliminates the need for a secondary means of component packaging. Generally, the component remains on the carrier strip used during the manufacturing process. Example components are continuous headers, battery clips, and motor brushes. This type of packaging provides high levels of online parts inventory. However, such strips are application-specific and may carry a higher cost to automate due to the need for a specially designed feed mechanism.

41.4.5.7 Bulk.   Bulk eliminates the expense of component packaging and reduces environmental issues. A significant number of odd-form components are shipped to assemblers in this format. However, bulk remains one of the most difficult packaging formats to automate with odd-form components. Generally, bulk components require a specially designed and manufactured bowl feeder, which results in higher cost and long lead times. The feeders usually take up considerable space and are not easily retooled for product changeover. Bulk packaging and bowl feeding are good alternatives for high-volume dedicated assembly cells.

41.4.6 Equipment

As advances have been made in odd-form component packaging formats, so too has odd-form assembly machinery advanced. (See Fig. 41.29). Unlike SCARA robotic systems designed to perform a single task and expensive to reconfigure for other uses, modern automatic odd-form assembly systems can be integrated into existing assembly lines, reconfigured to produce different PCBs, and can achieve accurate high speed placement at a competitive cost.

Odd-form automation systems utilize an overhead gantry-style positioning system with vision to precisely position each component for accurate placement. With the emphasis toward surface-mount odd-form components and pin-in-paste printed circuit board assembly, accuracy is an essential part of a reliable process. As companies eliminate wave soldering and move to pin-in-paste, the component must be placed accurately into the solder paste the first time so that the solder paste is not disturbed and solder joint integrity is not compromised.

Equipment manufacturers have made major advances such as enhanced vision systems, specially designed interchangeable grippers and nozzles, the ability to handle straddle-mount connectors in the same machine as standard surface-mount, and improved component feeding

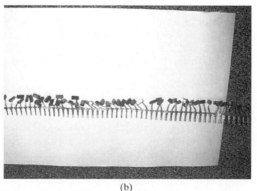

(a)   (b)

FIGURE 41.28   Motor brushes packaged in continuous strip. (*a*) Feeder for continuous strips. (*b*) Strip of motor brushes.

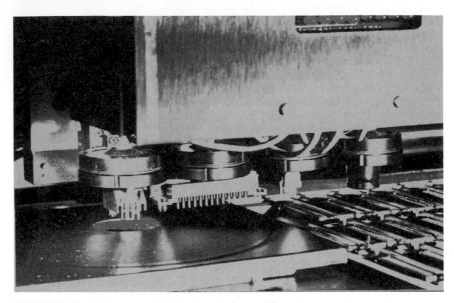

FIGURE 41.29   Odd components being inspected over vision system.

mechanisms. Automated assembly systems can be configured to work with a variety of odd-form component types, feeders, and package formats.

41.5 DIRECT CHIP ATTACH

41.5.1 Overview

By the late 1990s, approximately 6 to 7 percent of semiconductors, or 4 billion units, were assembled directly on printed circuit boards instead of individual packages such as small-outline integrated circuits (SOICs), quad flat packs, or ball grid arrays. The density requirements of wireless applications are driving this trend upward. To increase the number of functions in a cellular phone, a personal digital assistant, or laptop computer requires more components per square centimeter. On one hand, components are shrinking, and, on the other hand, more semiconductors are assembled directly on the printed circuit board of the final product or on higher-density modules that are then assembled in the final product. In either case, the integrated circuit is assembled directly on the board.

41.5.2 Types

Chip-on-board, chip-on-ceramic, or chip-on-flexible-circuit are typical semiconductor applications that are attached to a substrate by a thermally conductive adhesive, which is then cured. The die is wirebonded to the substrate and subsequently encapsulated and cured. This is the prime technology for smaller chips (<200 IOs). The extra costs of known good die (KGD) and the inability to rework after encapsulation are the major barriers for this technology. These barriers are then weighed against the extra density gains and substrate cost savings.

41.6   ACCURACY AND REPEATABILITY

41.6.1   Considerations

Several aspects of the equipment, the process, and the materials ultimately affect the accuracy and repeatability of the end product. From an equipment standpoint, the positioning system, the head architecture, the frame, the cameras, the vision engines, the encoders, and the overall tolerances of the nozzles are very important variables that have a high degree of interaction.

41.6.1.1   The Machine Structure.   The stability of the frame to withstand higher speeds, more weight, and higher acceleration factors has become a critical part of the design and manufacturing process. In addition to the traditional steel structures, there are some novel composite materials and structural cross-members that are being implemented.

The combination of camera resolution, vision engine, and on-the-fly scanning has put more pressure on optics to minimize the trade-off between field of view, resolution, and speed. New cameras with 1-million-pixel resolution, vision engines with parallel vs. serial processing, new focal schemes, and faster microprocessors are keeping pace with the needs of the industry.

The heads are getting larger and the nozzles are getting smaller. The dichotomy is handled by several clever approaches, including ceramic and composite materials, dual vision stations for rough and fine alignment, as well as new technology sensors for faster responses and smaller areas.

41.6.1.2   The Drives.   Belt-type drives are being replaced by leadscrew drives, which are being replaced by linear motors with finer and finer resolution encoders. The motors are growing larger to handle the larger heads, and new materials are being developed to reduce the power so that less heat will be generated to achieve repeatability in the sub-10-μm range.

41.6.2   Definitions

One critical feature of the assembly equipment is its placement accuracy—if components are placed in such a way that prevents the formation of an adequate solder joint, the product will not be functional. Due to the high costs that are associated with rework or scrapping defective product, it is wise to ensure that the accuracy of the machine is adequate for the products that will be assembled. The amount of inaccuracy that can be tolerated is a function of the size of the leads/terminations and the size of the pads where the terminations will take place. While it is ideal for the entire termination to be placed on pad, the IPC standards that pertain to the assembly process and are recognized by most players in the industry allow for 25 percent (or even 50 percent in many cases) of a lead's width to be off pad without jeopardizing the solder joint. Traditionally, x, y, and rotational (θ) errors made when placing components have been characterized separately, but now many companies are using equations to model the combined effects of these errors in order to estimate the percentage of a component's leads that will be on the pad, which is called the lead-to-pad (LTP) ratio. While each component has multiple leads, the LTP is computed for the lead that is most impacted by the particular combination of x, y, and θ errors, so each component only has one LTP value.

Furthermore, the development of an industry-wide standard (by key supplier and user companies) that will lay out a full complement of optimal metrics as well as associated test vehicles and verification methods is nearing completion (IPC-9850). This will give users additional tools in evaluating placement machine accuracy, which will only grow in importance as the miniaturization trends continue, which is leading to smaller parts (such as 0201 capacitors and resistors, as well as flip-chips) that require higher levels of accuracy. One of the metrics will be the process capability index (C_{pk}) for the machine, relative to its ability to place certain parts on pads with particular dimensional characteristics with no more than 25 percent of the leads being off pad.

The generic equation of C_{pk} is the minimum of $(USL - \mu)/3\sigma$ and $(\mu - LSL)/3\sigma$, where μ represents the process mean and σ represents the process standard deviation. Since there is no upper spec for LTP, C_{pk} reduces to $(\mu - LSL/3\sigma)$, where μ refers to the mean of the lead-to-pad percentages for a group of components, and σ refers to the standard deviation of these values.

The resulting value of C_{pk} designates the number of sets of 3 standard deviations that lie between the process mean (for LTP) and the spec limit, for instance, 75 percent LTP.

41.7 NETWORK COMMUNICATION

41.7.1 Overview

Several factors such as globalization, inventory turnover, yield costs, traceability, and manufacturing cycle time are driving the need to obtain and transmit data for products and machines faster and faster. Networks started slowly at first in the early 1990s, mostly limited to a few buildings within a company, but now several sophisticated networks link suppliers and customers to the manufacturing facilities worldwide.

The need to create, optimize, and download programs for new products; the requirement to collect machine uptime data for production, reliability, availability, and maintainability; the mandatory traceability of mission-critical automotive parts or medical products; and remote monitoring are all factors necessitating network communications.

To communicate, a machine has several options. Serial communication via an RS-232 port is relatively common, but too slow for most applications. Ethernet communication via transmission control protocol/Internet protocol (TCP/IP) is a preferred method. Regardless of the communication method, a machine supplier must decide if it is necessary to utilize real-time communication, or if file transfer and remote storage are acceptable. This decision should be based on what needs to be accomplished with the network connection. Most of the equipment suppliers in the electronic assembly industry offer standard networking via TCP/IP, and a "Semiconductor Equipment Communication Standard" (SECS)/"Generic Equipment Model" (GEM) host interface (for real-time communication) on their assembly machines.

Communication via TCP/IP requires an Ethernet network card and TCP/IP software. Some operating systems include TCP/IP, while, with others, it must be purchased separately. This software must be installed on the machine, along with network file system (NFS) software, if the NFS remote mount function is desired.

TCP/IP is the name that is given to a set of protocols developed to allow computers to share resources across a network. TCP and IP are actually only two of many protocols used to communicate across networks, but the name is taken from these two. TCP/IP is supported by a variety of networks, and is available for many operating systems, including Apple®, Microsoft® Windows®, Windows 95®, Windows NT®, IBM OS/2®, and various UNIX operating systems such as Solaris®, HP-UX, and SCO® UNIX. TCP/IP includes the protocol functions discussed in the next section.

41.7.2 Protocol Functions

41.7.2.1 *File Transfer.*   The file transfer protocol (FTP) provides the means to transfer files from one computer to another via a network. FTP requires that the user log into the remote computer to provide some level of security.

41.7.2.2 *Telnet.*   Telnet is a network terminal protocol that provides the means to log into a remote computer via a network. By selecting a computer to connect to, then logging into that remote computer from a local computer, anything the user types at the local keyboard is sent directly to the remote computer. Once the user is done with the remote computer, the session is ended.

41.7.2.3   *Mail.*   TCP/IP allows mail to be sent from one computer to another.

41.7.2.4   *Servers.*   The aforementioned functions are standard with any implementation of TCP/IP. Network design has changed to include servers—computers or applications that are simply designed to perform specific functions. The client-server architecture includes servers that provide specific services to a network, and clients, computers, or applications that use the services. These services utilize functions within the TCP/IP protocols. While there are many different types of servers, the most common are described in the following paragraphs.

Network file system (NFS).   Unlike FTP, where a file must be transferred from a remote to a local computer in order to access it, NFS allows one computer to remotely mount or access the hard disk on a remote computer. As far as the local computer is concerned, this remote disk is another partition on the local drive. Files do not need to be transferred, because the local computer is reading, writing, and storing files directly on the remote computer's hard disk. The advantages to this include:

- Large files can be stored on a remote drive, reducing the space used on a local drive.
- Many users can access the same file without having to make individual copies.
- File backups are easier, as they can be done by backing up one drive as opposed to many.

In the case of assembly machines, pattern programs can be generated, edited, and stored on a file server and the machine can download these files and run them as simply as it would if they were on the local disk.

Printer.   Printer servers allow a computer to send print requests to a printer attached to another computer.

Application Server.   Application servers store software applications in a central location and allow remote computers (clients) to execute these applications across a network. This eliminates the need to take up space by storing applications on local drives.

Name Server.   Name servers allow the management of names, passwords, and addresses of all of the computers on a network in a central location. A domain name system (DNS) is an example of this.

FTP and NFS are the two functions that are most useful for a networked assembly machine. The majority of users do not build pattern programs for a machine on the machine itself. These programs are designed in an engineering department and optimized prior to loading on a target machine. In the past, this program would be transferred to the target machine via *sneaker net,* that is, by copying the file to a floppy disk, then copying it from the floppy to the target machine's hard disk. FTP allows a file to be *put* by a remote computer onto a target machine, or to be gotten by the target machine from a remote computer. NFS allows the hard drive from the target machine to be *mounted* by a remote computer. Files created on a remote computer would actually be stored on the target machine's hard disk.

FTP also provides a method for gathering management data from a machine. Data that are stored in the management data file on an assembly machine include information regarding products, productivity, reliability, and other aspects of machine use. This file can be transferred via FTP to a remote computer where the data can be parsed out and manipulated to give results that may increase productivity. There are weaknesses with this approach, although the method is easy to implement. If the file becomes corrupt, for whatever reason, the user will lose data that may be critical to the operation or analysis of the machine. In addition, the file holds data thought to be relevant by the machine designer. There is a multitude of data created by the machine that are not available in this file.

In addition to the need for file transfer, there is a growing need within our industry to communicate *real time* to assembly machines. These requirements include program upload and download, data collection (as it occurs), alarm or exception data handling, and event logging.

Events are descriptions of functions in a machine. Every time a particular function is executed, an event is generated and recorded by an event ID number. For example, every time a placement machine completes a printed circuit board, it generates an event such as "board complete," tied to an event ID number. Events that are significant to a host are referred to as *collection events*. Collection events can include the complete list of machine events or a subset thereof.

In addition to events, machines produce data that are stored in variables categorized as *status variables*, *data variables*, and *equipment constant variables*. Variables are reported to a host as part of a report defined by the host and executed based on the occurrence of a particular event. For example, the host can define a report to be sent by the machine upon each occurrence of the event "board complete." The report might include such variables as the number of components picked, the number of components placed, and the number of components rejected, along with the name of the product and the board count for the product.

41.7.2.5 *SECS/GEM Host Interface.* Many assembly machines are capable of real-time communications via an SECS/GEM host interface. This interface is based on a standard referred to as the "Semiconductor Equipment Communication Standard" (SECS), published by Semiconductor Equipment and Materials International, Inc. (SEMI). The standard is subdivided into two parts: SECS—I and SECS—II. SECS was developed out of the need to communicate from a host computer to many types of equipment with a number of constraints. These include a variety of equipment that would use the protocol, the need to minimize the resource requirements of the equipment, and the variety of networks used within the industry.

The SECS/GEM interface is an implementation of the SECS protocol. The easiest way to understand the interface is to understand SECS—I and SECS—II.

The SECS—I standard describes a protocol that is used to exchange data between a host computer and semiconductor processing equipment or electronic assembly equipment. It specifies how data will be transferred between computer and machine. It is a layered protocol used for point-to-point communication and includes a physical layer, a block layer, a message layer, and a transaction layer. SECS—I communicates serially over RS-232.

SECS—I describes *how* data are transferred between a host computer and an assembly machine. Standards are outlined in SEMI E-4-936.

SECS—II describes *what* data are transferred between a host computer and an assembly machine. The purpose of this standard is to define messages at a level of detail that will allow vendors to develop host software for a variety of process or assembly equipment, without having esoteric equipment knowledge. Similarly, a machine designer can develop a host interface without having a specific host in mind. While the standard defines messages for the typical or common functions of a machine, it also describes how to define messages to support functions that are specific to a unique machine. SECS—II is fully compatible with the protocol described by SECS—I.

SECS—II messages are divided into streams and functions. *Streams* are categories of messages intended to support similar or related activities. *Functions* are specific messages for a specific activity within a stream. Standards are outlined in SEMI E-5-1296.

Functions are specific messages within a stream. For example, stream 1 deals with equipment status. Function 1 is an "Are You There?" request. When sent by a host, the machine responds with function 2, "Online Data," which includes the equipment model type and the software revision running on the machine. Equipment model type and software revision are two variables defined by the Data Item Dictionary (DID), a part of the SECS—II standard. Machines must produce variables that coincide with the elements defined in the DID or can produce variables defined by the machine supplier, as long as these variables follow the requirements as defined by the standard.

Among the other SEMI standards that apply to the electronic assembly industry, the "High-Speed SECS Message Service" (HSMS) and "Generic Equipment Model" (GEM) standards are important to understand. HSMS is intended as an alternative to SECS—I for applications where higher speed communication is needed or when a simple point-to-point topology is insufficient. HSMS simply defines a communication interface for use over a TCP/IP network.

It provides reliable data transmission, and the actual implementation is very much like SECS—I. Standards are outlined in SEMI E-37-95.

41.7.2.6   GEM.   By the late 1980s, many semiconductor companies and equipment suppliers were using SECS—I and SECS—II. While both standards greatly assisted the industry in becoming interconnected, there were a couple of major roadblocks.

One such roadblock was that the SECS—II standard specified many different options, and equipment suppliers used a variety of them in different combinations. In essence, there was little consistency among machine suppliers, and it became extremely difficult to network equipment to a common host. Also, machine operators had different communications interface requirements—different communication functions were needed depending on where the machine went. On top of that, the minimum requirements specified by the SECS—II standard were too limited to be of value in establishing a common communication capability. For these reasons, SEMI published the GEM standard in 1992 (SEMI E-30-1296).

> The scope of the GEM standard is limited to defining the behavior of semiconductor equipment as viewed through a communications link. The GEM standard defines which SECS—II messages should be used, in what situations, and what the resulting activity should be. The GEM standard does not attempt to define the behavior of the host computer in the communications link. The host computer may initiate any GEM message scenario at any time and the equipment shall respond as described in the GEM standard.

The GEM standard specifies the following:

* Behavior to be exhibited by manufacturing equipment in a SECS—II environment
* Description of information and control functions that are needed in a manufacturing environment
* Definition of the basic SECS—II communication capabilities of manufacturing equipment
* Single means of accomplishing an action when SECS—II provides multiple possible methods
* Standard message dialogues that are necessary to achieve useful communication capabilities

GEM includes a set of minimum requirements along with additional capabilities. Any equipment supplier who designates a machine *GEM-compliant* must build that set of minimum requirements into the machine. Beyond these minimum requirements, the machine may include any of the additional GEM capabilities. If 100 percent of that capability is implemented, the supplier may state that the particular GEM capability is implemented.

Using GEM.   The typical implementation of a GEM interface defines what (and when) data are going to be passed from the machine to the host. To begin, when the host is connected and powered up, it establishes communication with the machine. Once this is accomplished, the host disables all machine events from being sent. Using a predetermined list, the host informs the machine of which events to enable as collection events. The host also informs the machine of defined reports that list the variables, or data items, needed from the machine, and which event generates each report. The host interface subsequently monitors collection events, sending the appropriate report to the host when they occur.

The advantage of this type of data collection—as opposed to FTP—is that these data are generated as they occur and are collected with a date-and-time stamp by the host. The data can be stored in a file or database to be used in a variety of ways. As long as the connection is maintained between the host and the machine, data are collected. GEM includes a spooling capability that precludes the loss of data due to a host or connection failure.

In addition, GEM includes the Stream 7 Process Program Management function, which is used to manage and transfer process programs between the host and the machine.

GEM Host Interface.   The GEM host interface provided by a machine supplier can be viewed as only one-half of the whole. In addition to the driver on the machine, a driver needs to be developed to reside on the host along with an application that uses this driver to communicate with the assembly machines. There are various toolkits available to assist in the development of a host driver, as well as completed host applications that can be purchased if the user wants an off-the-shell solution for the host application.

41.8   KEY MACHINE SELECTION CRITERIA

41.8.1   Cost of Ownership

As the products become more and more complex, so, too, are the associated assembly machines and lines. The selection of the right machine or best line becomes more difficult because of all the variables involved. Throughput and labor rates are always very important, but other variables, such as occupancy rate, line length, flexibility for changeovers, and uptime, have become more and more critical. Many companies have devised their own cost-of-ownership model to take into consideration their own priorities as well as country rates since several large companies are truly becoming global players. There is no industry model yet. Sematech has devised an excellent tool for semiconductor manufacturers.

41.8.2   Utilization

Manufacturing efficiency and equipment utilization are significant cost factors for manufacturers in the electronic board assembly business. Unfortunately, participants in discussions about efficiency and utilization bring their own definitions based on personal experience and knowledge. The purpose of this document is to determine the most appropriate definition for equipment utilization, then derive a parsimonious formula for component placement equipment. The definition and resulting formulas are based on definitions from well-known industry groups as outlined in SEMI E-10, E-79, E-35, and Sematech 95032745, as well as the experiences of equipment manufacturers.

Equipment utilization can be defined as the fraction of theoretical production rate attained by a piece of equipment during a specified period of time. It may be expressed as follows:

$$\text{Equipment utilization} = \frac{\text{actual output}}{\text{operations time} \times \text{theoretical production rate}} \tag{41.1}$$

For placement machines, *actual output* rate can be based on either board throughput or component placements. However, because component placement speed is only one of the control factors involved with board throughput, board-handling technique and board-handling speed are also involved. The more appropriate definition is based on board throughput. *Operations time* can also be called *net working hours;* it encompasses all but nonscheduled time. See Fig. 41.30 for more details. The *theoretical production rate* is the maximum rate that might be obtained for an optimized placement pattern under ideal conditions. Further, the theoretical production rate must include board transfer and fiducial read times. The following terms are based on daily computation of equipment utilization.

Term	Units
EU = equipment utilization	Percent
AO = actual output	BPD = boards per day, or CPD = components per day
OT = operations time	HPD = hours per day

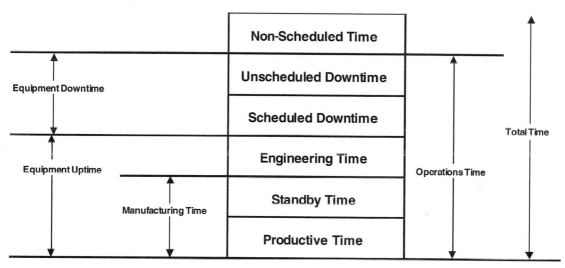

FIGURE 41.30   SEMI E-10-0699(E) Equipment States Stack Chart.

Term	Units
CT = cycle time	SPB = seconds per board
PT = productive time	H = hours per day
EqU = equipment uptime	H = hours per day
TR = theoretical production rate	BPH = boards per hour

All of the preceding standards make use of the *equipment states* concept, outlined in SEMI standard E-10-0699 (E). It is the foundation on which utilization-and-efficiency metrics is built. (See Fig. 41.30.)

According to the E-10 standard, equipment utilization is defined as: "The percent of time a piece of equipment is performing its intended function during a specified time period." Depending on the user, the specified time period may include nonscheduled time. This definition, however, is incomplete. It does not capture how well equipment is used during productive time. A piece of equipment might be used continuously during normal working hours but at a rate much lower than its potential capability (peak performance). The standards listed in Fig. 41.30 build upon the SEMI E-10 definition but include an efficiency component.

Equipment utilization is the product of:

1. The amount of time that a resource is in a condition capable of producing a product
2. The effectiveness of an organization in utilizing this resource during the time that it is available
3. The theoretical efficiency of the resource

The word *resource* is used in a generic sense and may refer to a production line or to a single piece of equipment. In equation form, the definition becomes:

$$\text{Equipment utilization} = \text{availability} \times \text{performance efficiency} \qquad (41.2)$$

Equipment utilization should not be confused with *equipment productivity,* which adds a *quality efficiency* component to the definition. Metrics used to track equipment productivity go by labels such as *overall equipment efficiency* (OEE) and *intrinsic equipment efficiency.* The equations are of the form:

$$\text{OEE} = \text{availability} \times \text{performance efficiency} \times \text{quality efficiency} \tag{41.3}$$

Availability is the fraction of time that the equipment is in a condition to perform its intended function during a predefined time period. This time period may be specified in a number of ways, but the two most basic are *total time* and *operations time.* Total time is calendar time and is typically set to 24 hours. Operations time can also be called *net working hours,* and it encompasses all but nonscheduled time. So availability becomes:

$$\text{Availability} = \frac{\text{equipment uptime (EqU)}}{\text{operations time (OT)}} \tag{41.4}$$

Performance efficiency is equipment productivity while the equipment is in a state that is capable of performing its intended function. It breaks down into two components as follows:

$$\text{Performance efficiency} = \text{operational efficiency} \times \text{rate efficiency} \tag{41.5}$$

Operational efficiency is the fraction of equipment uptime that the equipment is in a productive state:

$$\text{Operational efficiency} = \frac{\text{productive time (PT)}}{\text{equipment uptime (EqU)}} \tag{41.6}$$

Rate efficiency is production rate expressed as a fraction of its theoretical output capability:

$$\text{Rate efficiency} = \frac{(\text{actual output [AO]/productive time [PT]})}{\text{theoretical production rate (TR)}} \tag{41.7}$$

The *theoretical production rate* is the maximum rate that might be obtained for an optimized placement pattern under ideal conditions. For placement equipment the theoretical production rate is product-specific and must include board transfer time and fiducial read time. It may be calculated from optimization runs or derived from software simulation.

Substituting Eqs. 41.4 through 41.7 into Eq. 41.2, we have:

$$\text{EU} = \left(\frac{\text{EqU}}{\text{OT}}\right) \times \left(\frac{\text{PT}}{\text{EqU}}\right) \times \left(\frac{[\text{AO/PT}]}{\text{TR}}\right) = \left(\frac{\text{EqU}}{\text{EqU}}\right) \times \left(\frac{\text{PT}}{\text{PT}}\right) \times \left(\frac{[\text{AO/OT}]}{\text{TR}}\right) = \frac{\text{AO}}{\text{OT} \times \text{TR}}$$

or

$$\text{Equipment utilization} = \frac{\text{actual output}}{\text{operations time} \times \text{theoretical production rate}} \tag{41.8}$$

For placement machines, actual output shall be based on board throughput. To summarize:

$$\text{Basic formula:} \quad \text{EU} = \frac{\text{AO}}{\text{OT} \times \text{TR}} \tag{41.9}$$

$$\text{Calculation:} \quad \text{EU} = \frac{\text{BPD}}{\text{HPD} \times \text{BPH}} \tag{41.10}$$

41.8.2.1   Example Calculation

Machine type: Universal GSM2 (2-beam)
Board-handling scheme: SMEMA compatible
Transfer type: Left-to-right
Operating mode: High-speed, full-run
Optimization method: Manual

300 placements per board, 150 placements per beam
   Component type: 0402 resistors
   Feeder setup: 58 feeders
   Tact time: 19,300 CPH (this figure includes fiducial read and transfer time)
Operations time: 8 HPD
Actual production rate: 480 BPD

$$\text{Theoretical production rate:}\quad TR = \frac{19{,}300\ CPH}{300} = 64.33\ BPH$$

$$\text{Utilization:}\quad EU = \frac{BPD}{HPD \times BPH} = \frac{480}{8 \times 64.33} = 0.9326 = 93.26\%$$

P · A · R · T · 7

SOLDERING

CHAPTER 42*
DESIGN FOR SOLDERING AND SOLDERABILITY

42.1 INTRODUCTION

This chapter deals with the specification of the materials system for the printed circuit and the design parameters which must be considered before a final circuit is laid out. The soldering operation must be considered from the inception of the board layout in order to ensure satisfactory performance. The rules are simple and straightforward; if they are followed, the operation should run smoothly and efficiently. If they are disregarded, however, the result is invariably recurring problems with bridges, icicles, and imperfectly formed fillets that will need handling by a large number of touch-up operators.

42.2 DESIGN CONSIDERATIONS

During the layout of the board, several soldering parameters should be carefully considered. They are (1) the wire-to-hole ratio, (2) the size and shape of the terminal areas, (3) the number and direction of extended parallel circuit runs, (4) the population density of the solder joints.

42.2.1 Wire-to-Hole Ratio

The wire-to-hole ratio represents a compromise between the ideal situation for assembly (large hole and small-diameter lead) and the ideal situation for soldering (smaller lead-to-wire ratio). The minimum hole size can be established by the rule of thumb that it should be no less than the lead diameter plus 0.004 in. The maximum hole diameter should be no more than 2.5 times the lead diameter. Of course, if the board is a plated-through-hole or a multilayer circuit board, the hole-to-wire ratio should be lower than 2.5 to encourage the capillary action of the flux and solder during the soldering operations.

42.2.2 Size and Shape of the Land Area

The pad area around the solder joint is normally either circular to slightly elongated (teardrop). It should not be more than 3 times the diameter of the hole in the board. There is

---

* Adapted from Coombs, *Printed Circuits Handbook, 3d ed., McGraw-Hill, New York, 1988, Chap. 23,* by Hugh Cole.

sometimes a tendency, particularly on low-density boards, to leave large irregular land areas around the holes. That should be avoided! Excessively large land areas expose too much copper to the solder pot, cause excessive quantities of solder to be used in joint formation, and promote bridging and webbing.

If the leads are to be clinched during assembly, the land should be so oriented that the clinched lead will be in the center of the elongated pad. Both the pad and the component should be so oriented that the clinched lead is parallel to the direction of solder flow in the solder wave and not perpendicular to it.

42.2.3   Number and Direction of Extended Parallel Lines

The use of automated printed board layout programs and the trend toward high-density circuit packaging have resulted in a tendency to group large numbers of circuit paths together and run them parallel to one another for long distances. If those paths are oriented perpendicularly to the direction of flow in the solder wave (i.e., at right angles to the direction of the conveyor), then they can contribute to bridging and webbing. Every effort should be made to maximize the spacing between lines which must be oriented perpendicular to the direction of the conveyor.

42.2.4   Population Distribution

An excessive number of joints in one area promotes bridging, icicling, and webbing. It may also cause a heat-sinking effect and interfere with the formation of a good solder joint.

42.3   MATERIAL SYSTEMS

In the soldering process there are two surfaces which must be considered before a solder flux is selected. They are the lead surface and the pad surface. The average printed circuit assembler has little control over the material systems used in component leads, since the selection is usually made by the component manufacturer. Furthermore, most components are mass-produced and supplied on large reels. It is not economically justifiable to treat each lead according to an individual assembly shop's specific requirement. Therefore, in selecting the components, care should be taken that the leads are solderable, and an incoming inspection should be established to ensure lead solderability.

The board itself, however, is a different story. Since each board is custom-manufactured, the assembly or soldering engineer can exercise a great deal of control over the material systems used on the board. Again, it is important, in order to keep defects to a minimum, that the board be made of a solderable material and that the solderability of the board be checked as a part of incoming material inspection. The next section will deal with some typical material systems encountered during the soldering process.

42.3.1   Common Metallic Surfaces

42.3.1.1   Bare Copper.   Because of its low cost and ease of processing, one of the most common metallic surfaces encountered is bare copper. Chemically clean copper is the easiest material to solder; it can be soldered with even the mildest fluxes. But unless it is protected with a rosin-based protective coating, its solderability will rapidly degenerate because of oxides and tarnishes. As we shall see later in this chapter, however, the solderability of tarnished copper surfaces is easily restored with surface conditioners. If boards with bare copper surfaces are used, care should be taken to maintain solderability during handling and storage (storage time should be minimal), and the boards should not be stored in the presence of sulfur-containing material such as paper, cardboard, or newsprint. Sulfur produces a tenaciously adhering tarnish on copper which seriously impairs the solderability.

42.3.1.2 Gold.   Gold is encountered most commonly on component leads and plug-in finger surfaces. It is a highly solderable material, but it is extremely expensive and it rapidly dissolves in the molten solder. Because it affects the properties of the solder joint, causing the joint to become dull and grainy, it is usually avoided or eliminated by pretinning the lead before soldering. Various studies have shown that all gold on a gold-plated lead can be dissolved in a solder pot of eutectic tin-lead solder within 2 s (plate thickness of about 50 µin). Therefore, pretinning is economical as well as easy.

42.3.1.3 Kovar.   Many dual in-line packages (DIPs) and related integrated circuitry are supplied with Kovar leads. Kovar is a very difficult metal to solder because it doesn't wet well. For that reason component manufacturers and/or assembly shops prefer to pretin Kovar. The pretinning is normally accomplished only with organic acid fluxes and certain proprietary acid cleaners.

42.3.1.4 Silver.   Although silver was once very popular in the electronics industry, it is not used on terminal areas or component leads. The reason is the problem of silver migration, a phenomenon discovered in the late 1950s and extensively researched during the early 1960s. Silver should be avoided. If it must be used, it is an easily soldered material and should be treated similarly to the bare copper surface (i.e., avoid sulfur-bearing materials and minimize storage and handling).

42.3.1.5 Immersion Tin.   Immersion tin coatings are electrolessly deposited coatings of tin metal on bare copper surfaces. When tin is initially deposited, the coating is extremely solderable. It does, however, deteriorate rapidly, and it becomes more difficult to solder than bare copper. Originally, immersion tin coatings were used to protect the solderability of bare copper surfaces and thereby extend the shelf life of the board. Experience has shown, however, that fused tin-lead plate is far superior for the purpose.

42.3.1.6 Tin-Lead.   Tin-lead coatings are put on printed boards and component leads to preserve the solderability of the material. They can be applied by electroplating, hot dipping, or roller coating. The mechanics of the processes are discussed elsewhere in this chapter. A properly prepared tin-lead surface should exhibit excellent solderability and long shelf life (about nine months to one year). Tin-lead coatings can be soldered with most rosin-based fluxes, even the nonactivated types. However, optimum results are obtained with the activated rosin fluxes.

42.4 WETTING AND SOLDERABILITY

Soldering is defined as a metallurgical joining technique involving a molten filler metal which wets the surface of both metals to be joined and, upon solidification, forms the bond. From the definition it is apparent that the materials to be soldered do not become molten and therefore the bonding occurs at the interface of the two metals and is strongly dependent on the wettability or solderability of the base metal by the molten alloy. Although the base metal does not become molten, some alloying can take place if the base metal is soluble in the filler metal. The bond which is formed is strictly metallic in nature, and no chemical reaction which covalently or ionically bonds the metal to the surface occurs.

To understand the basic mechanism of soldering, it is necessary to understand the thermodynamics of wetting. Fortunately, however, in order to understand wetting it is not necessary to understand thermodynamics. Wettability or solderability of two materials is a measure of how well one material "likes" the other. The property can be easily visualized by using a water drop resting on the surface, such as the one shown in Fig. 42.1. When the water drop doesn't like the surface on which it rests, it pulls up into a ball and touches the surface, in the idealized case, at only one point. The angle between the drop and the surface at the point of contact is called the dihedral angle. If the drop likes the surface, it spreads out all over the surface and comes in intimate contact with it. Various degrees of wettability are therefore related to the ability of the drop to spread out or wet the surface. Figure 42.2 shows the relation between the dihedral angle and the various wetting states. Wettability or solderability is related to the surface energy of the material. Wetting is substantially improved if the surface is clean and active (i.e., if all dirt and grease are removed and no oxide layer exists on the metal surface).

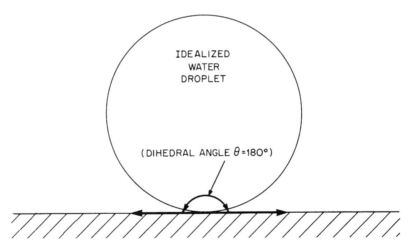

FIGURE 42.1   Complete nonwetting of an idealized surface.

Therefore, to form a solder bond efficiently, we must start with a material system which can be wet by the molten solder, and the cleanliness of the parts must be maintained.

42.5 SOLDERABILITY TESTING

Solderability testing is an important quality control procedure in the electronics industry. It is a simple procedure but problems can occur if the fundamental test principles are not understood thoroughly. *Solderability* is a measure of the ease (or difficulty) with which molten solder will wet the surfaces of the metals being joined. When molten solder leaves a continuous permanent film on the metal surface, it is said to wet the surface. Wetting is a surface phenomenon which depends on cleanliness. Fluxing facilitates wetting by cleaning the surface, and the degree of surface cleanliness depends on the activity of the flux.

However, there is often a limitation on the activity of the flux in the electronics industry. Most electronics soldering operations require relatively weak rosin-based fluxes to avoid the possibility of current leakage caused by flux residues remaining on the part. To enhance the

TOTAL NONWETTING (θ=180°)

PARTIAL WETTING (180°>θ>0)

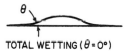

TOTAL WETTING (θ=0°)

FIGURE 42.2 Relation between dihedral angle θ and the degree of wetting.

solderability of some surfaces and eliminate the need for active fluxes, electroplating often is employed to deposit a solderable coating over a base metal that tarnishes easily or is difficult to solder.

42.5.1   Testing Procedures

Testing for solderability can be a simple procedure of inspecting production parts or of dipping an appropriately fluxed lead or portion of a printed board in a solder pot and observing the results. Good and bad wetting are then identified visually. The problem is to recognize borderline cases which simulate effective solderability but quickly deteriorate. To alleviate borderline solderability, the mildest possible flux should be used at the lowest soldering time-temperature relations which will give adequate results.

An effective solderability test involves the use of water-white rosin flux and a solder pot. The surfaces to be checked are fluxed and immersed for 3 to 4 s in the solder pot, which is maintained at approximately 500°F. The solder is then permitted to solidify, and the components are cleaned of flux residues prior to visual examination. The inspection is usually performed with either no or low magnification (5 to 10×). Most solderability tests will permit up to a 5 percent imperfection of the total surface, provided the entire imperfection is not concentrated in one area.

More elaborate solderability tests are described in governmental and industrial specifications. Testing of component leads is covered by Electronic Industries Association (EIA) Test Method RS17814, which is similar to the solderability method described in Military Specification 202, Method 208. The test incorporates a dip fixture which provides identical dip ratio and immersion times (Fig. 42.3).

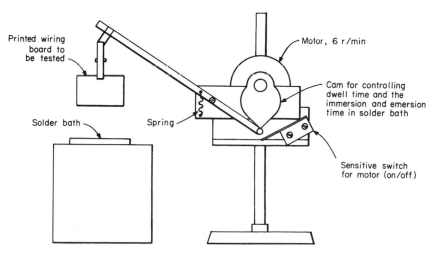

FIGURE 42.3   Solderability dip tester.

42.5.1.1 Dip Test.   For printed boards, the edge dip test, described in the EIA specification RS319 and by the IPC* in standard S801, often is employed. The edge of a printed board is dipped first in a mild flux and then in a solder pot for a predetermined time and temperature. After the flux residues are removed, the board is inspected visually for the quality of wetting. A similar test is employed to determine the solderability of solid wire leads, terminals, and conductive accessories of component parts normally joined by soft solder. Applicable test standards are EIA Standard RS178A, Solderability Test Standard, and MIL-STD-202C, Method 208 A, Solderability.

To perform the dip test, the operator places the item to be tested into a holding arm. The arm lowers the sample section into the solder pot. After the preset dwell time has elapsed, the arm automatically raises the sample. Then a visual determination of solderability is made. The dip test can also be performed as a manual operation, but that leaves too many variables to the discretion of the operator.

Since interpretation of results is based on a subjective judgment by the operator, it is essential that the operator be provided with examples of good, marginal, and poor solderability. It is imperative, too, that pot temperatures, cleaning and fluxing procedures, dwell times, and solder purity be controlled carefully to obtain meaningful results.

42.5.1.2 Globule Test.   This test is one that is prevalent throughout Europe and is mandatory for European suppliers whenever specifications dictate. It is described in the International Electrochemical Commission Publication 68-2 Test T, Solderability. The globule test provides a numerical designation for the solderability of wires and component leads. It measures the ability of the solder to wet the lead.

A lead wire, coated with a nonactivated rosin flux, is placed in a holding fixture. It is gripped and straightened and then lowered into a globule of molten solder. The volume and temperature of the globule are controlled. As the lead wire bisects the globule, the timer actuates and measures the time between the moment the wire contacts the solder and the moment the solder flows around and covers the lead. At this second point, the timer stops, and the elapsed time is registered on a read-out. Elapsed time is indicative of the solderability of the lead: the shorter the time, the greater the solderability.

This globule test method is completely automatic and is designed for continuous operation. Time is measured to $\frac{1}{100}$ s. Wires that are 0.008 to 0.062 in in diameter can be tested, and special heat regulators can hold solder temperature to ±2°F.

When wire plated with a soluble or fusible coating is tested, it is advisable to perform a dip test to supplement and verify the globule test findings. The reason is that, under certain conditions, the plated coating might be totally reflowed or dissolved during the soldering operation, which would give misleading results.

42.5.2 Standard Solderability Tests

Although many solderability tests are used throughout industry, the IPC has issued one described in its document IPC-S-804A that has been shown to give acceptable and repeatable results. If a printed board has a special surface treatment, a specifically designed test for that material may be required, but for copper, or solder surface boards this is an effective test.

Examples, and pictures, of standards of wetting, dewetting, and nonwetting are shown in Chapter 35 "Acceptability of Printed Boards," which can be used to evaluate the results of the IPC solderability test.

In addition to the preceding, Fig. 42.4 shows acceptable and nonacceptable examples of solder filling of plated-through-holes. This blow-hole formation is not a solderability issue, although it is often considered one. It is caused by moisture left from the plating process forming bubbles during the exposure to high temperature during soldering. This outgassing will usually leave voids in the hole after the solder solidifies. Often the reaction is to question the

---

* The Institute for Interconnecting and Packaging Electronic Circuits is referred to as IPC.

effectiveness of the solder process when, in fact, the drilling and/or plating processes most likely are the sources of the problem.

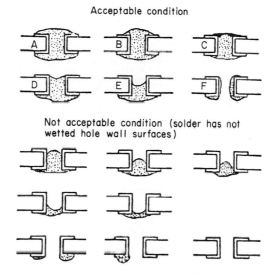

FIGURE 42.4 Effectiveness of solder wetting of plated-through-holes. *(From IPC-5-804A.)*

42.6 PLATED COATINGS FOR PRESERVING SOLDERABILITY

The three commonly used types of plated coatings are generally referred to as fusible, soluble, and nonfusible and/or nonsoluble. Fusible electrodeposited coatings provide corrosion protection to a surface that has been activated for soldering. Whether the solder bond of a soluble electrodeposited coating is to the base metal or to the deposit depends on soldering conditions and coating thickness. Nonfusible and/or nonsoluble electrodeposited coatings are used frequently as barrier layers in electronic applications to prevent diffusion of solder and base metal.

42.6.1 Fusible Coatings

Tin and tin-lead electrodeposited coatings are commonly used in electronic applications to preserve solderability because they are fusible and do not contaminate the solder pot or fillet. Contamination can adversely affect tensile, creep, and shear strengths at a solder connection. Also, contamination of a part can reduce the flow and spread of the solder on the part.

If the coating operation is not closely controlled, an electrodeposited coating may be applied over a partially contaminated surface. If that happens, dewetting will occur in the contaminated areas because the electrodeposited coating is fused during soldering. Therefore, adequate cleaning prior to plating is essential to obtain good solderability. Figure 42.5 shows a surface which has dewet after a reflow operation.

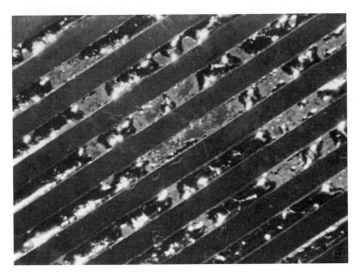

FIGURE 42.5   Tin-lead-plated surface which exhibits dewetting after the reflow operation. *(Alpha Metals, Inc.)*

Another point to remember is that plating thickness should be sufficient that porosity is virtually eliminated. Porosity and codeposited impurities will lower the protective value of the electrodeposited coating and will eventually cause poor solderability.

42.6.2 Soluble Coatings

Soluble coatings commonly used in electronic applications include gold, silver, cadmium, and copper. During soldering, these metal coatings are either completely or partially dissolved. The amount of dissolution depends on solubility of the coating metal, thickness of the deposit, and the soldering conditions. Silver and copper tend to tarnish, and if a mildly activated flux is called for, they should be protected with a thin rosin coating. Cadmium offers sacrificial corrosion protection which often necessitates the use of highly activated fluxes to promote effective soldering.

Soluble gold coatings provide excellent corrosion and chemical resistance. However, because of the high cost, fairly thin coatings are used. Moreover, care should be taken because gold coatings of under 50-μin thickness tend to be porous and, as a result, lower the protective value of the metal. Corrosion of the base metal or barrier plate via pores causes soldering problems because the gold usually dissolves completely during soldering and it is difficult to wet the corroded base metal. Also, solderability decreases with the amounts of alloying elements that are used to increase hardness and are often codeposited with the gold. Thicker gold coatings, on the other hand, may cause brittleness in a solder connection because of formation of gold-tin intermetallic compounds.

42.6.3 Nonfusible and Nonsoluble Coatings

Nickel and tin-nickel electrodeposited coatings are considered to be nonfusible and/or non-soluble because they provide an effective barrier to prevent alloying of solder to the base metal in electronic applications. They allow for effective soldering to such materials as aluminum and silicon. However, problems can occur from passivity, which is caused by code-posited impurities, or from some additional agents that are used in nickel plating to increase hardness and reduce internal stress. In such cases, nickel and tin-nickel electrodeposited coatings should be protected with tin or tin-lead to improve the shelf life of the soldered part.

Furthermore, nickel and tin-nickel have limited solubility in solder, and a flash coating is not an effective barrier. When electroless nickel is used as a barrier coating on aluminum, the required thickness will depend on soldering conditions. A thickness of 50 to 100 μin is sufficient for most operations.

42.6.4 Organic Coatings

Although the metallic coatings for preserving solderability are the most reliable and effective coatings, they are also the most expensive ones. For applications that do not require a long shelf life, considerable economies can be obtained by using an organic protective coating. There are several basic types of organic protective coatings: water-dip lacquers, rosin-based protective coating, and the benzotriazole-type coating. Organic protective coatings must be easily removable, and they must be compatible with the rosin-based fluxes normally used in the printed board industry.

Water-dip lacquers were once a very popular means of protecting the solderability of printed boards. However, they have fallen into disfavor because of their tendency to set up and polymerize with age. As they age, they become extremely difficult to remove. They also become insoluble in the flux solvent. If they are not properly removed after fluxing, they leave white residues which are not corrosive but are extremely unattractive.

Today, rosin-based protective coatings are much more prevalent than water-dip lacquers. They are applied by dip, spray, or roller coating, and, depending on the thickness, they will provide solderability protection for six weeks to four months. They are composed of the rosin solids material, and as such they are quite compatible with the rosin thinners and the normal cleaning method used to clean rosin fluxes. When those materials are used, it is important to ensure that adequate fluxing occurs and that the preheating time and temperature are sufficient to allow the protective coat to melt and be displaced by the flux. If the coat is not displaced, soldering will occur with a rosin nonactivated flux, no matter how active the solder flux really is.

A third alternative which some people have explored is the benzotriazole-type coating. Benzotriazole is an organic compound which is applied to the board surface during the final rinse operation of the plating line. It forms a thin nonporous film on the copper surface and prevents oxygen from reacting with the surface copper molecules. Benzotriazole films are very fragile, and they cannot be subjected to handling or scraping abuse.

42.7 TIN-LEAD FUSING

Fused coatings ordinarily are electrodeposited, low-melting metals or alloys that have been heated sufficiently above their melting points to become completely molten. In the molten state, alloying between the liquid and the basis metals is accelerated; on solidification, the deposit usually is dense and nonporous. The procedure is commonly known as *reflowing*, and it usually connotes tin or tin-lead electrodeposits. However, it can also apply to the remelting of hot-dipped coatings.

Fused coatings are employed to guarantee that the cleaning procedure prior to plating is adequate, to produce a slightly denser deposit with less porosity, and to improve the appearance of the coating. Reflowing leaves a bright deposit which has a definite sales appeal.

42.7.1   Thick Fused Coatings

Tin-lead electrodeposits that are reflowed are usually between 300 and 500 μin in thickness. That thickness provides adequate protection with a minimum of reflow problems. When printed boards are reflowed, the tin-lead deposit forms a meniscus on the conductor pad. For that reason the edges of the pad have a much thinner coating than the original plating, whereas the center is thicker. The average thickness is the same, but it is distributed differently. As the plating thickness increases over 500 μin, the surface forces are not always sufficient to hold the solder in the meniscus—especially on wide pads; the solder shifts upon solidification and the deposits appear uneven. On inspection, that may be mistaken for a dewetting condition, but it actually is a shifting of the molten solder before solidification.

When tin-lead with a thickness over 500 μin is reflowed, the boards must be in a horizontal position and withdrawn in a smooth manner. Otherwise, shifting of the molten solder is inevitable. Because of that problem, there is a practical limit on the coating thickness. In recent years, some military specifications have been calling for 0.001 to 0.0015 in (1000 to 1500 μin) in reflow tin-lead plating. The main objective is to obtain maximum corrosion protection. Deposits of that thickness have been reflowed, but not without processing difficulties.

Improvements in infrared reflow equipment have increased the use of thick fused coatings in recent years. Reflowed tin-lead electrodeposits offer several advantages. One advantage is that they provide a 100 percent quality control check immediately after etching. By examining the boards after reflow, problems in hole drilling, cleaning procedures, plating, and etching can be detected and corrective action can be taken immediately. The procedure has had a dramatic effect in improving printed board reliability.

Another benefit is that solder slivers are eliminated. During etching there is undercutting where the tin-lead coating overhangs the conductor pad. Under certain vibration conditions, the overhang can fall off and cause a short circuit. Fusing eliminates the condition and adds to board reliability.

42.7.2   Thin Fused Coatings

Thin fused coatings level the solder during fusing when they are applied by procedures such as roller coating, spin coating, and hydro-squeegeeing. The leveling is accomplished by means of a hot liquid ejected from spray nozzles. The techniques usually result in coatings under 50 μin, hardly enough for adequate corrosion protection. Also, thin coatings can mask poor solderability when tin-lead is plated over unsolderable copper. For those reasons, use of leveling techniques has diminished.

42.7.3   Problems in Reflowing Plated Coatings

Codeposited impurities, especially copper in tin-lead, can cause dewetting in a reflowed deposit. Variations in alloy composition raise the melting point and cause reflow problems. That is particularly true when organic contamination results in poor solution throwing power and consequent high lead deposit in the plated-through-holes. Heavy oxidation or tarnish films that result from chemical attack by etching solutions must be removed prior to reflowing. The films can act as insulating barriers and interfere with reflowing.

42.8   SOLDERABILITY AND THE PLATING OPERATION

Blowholes in plated-through-holes on printed boards normally are caused by solution entrapped in the hole, but they also can be caused by an excess of organics occluded in the electrodeposits. The heat involved in soldering causes moisture and entrapped chemicals in the laminate to build up in pressure and escape through voids or cracks in the plating. That results in blowholes. The problem can usually be alleviated by prebaking at 180 to 200°F.

In some cases blowholes can be caused by an excessive amount of occluded organics in the electrodeposit. The problem is more prevalent with bright plating deposits such as tin and tin-lead. When blowholes are caused by excessive organics occluded in the deposit, prebaking will not remedy the problem. The occluded material can be released only when the metal liquefies.

Figure 42.6 shows a cross section of a fused tin-lead surface that had an excessive amount of organic material. Note that the organic material tends to create voids in the plate which rise to the surface. The result is a grainy, pitted appearance after solidification.

FIGURE 42.6   Organic material from plating bath occluded in reflowed tin-lead deposit. (*Alpha Metals, Inc.*)

42.8.1   Effect of Organic Plating Additives

To solder successfully to electrodeposits that employ addition agents, it is extremely important that the additives be carefully controlled. Additives usually are essential to produce sound deposits of tin and tin-lead alloys. In electroplating, chemicals are added to the basic formulation of a plating solution to enhance the properties of the deposit. Some properties that can be improved by additives are throwing power, smoothness, hardness, leveling, brightness, and speed of deposition.

Normally, when foreign metals or organic materials are present in a plating solution, the properties of the deposit are adversely affected. In some rare cases, beneficial effects can be produced by codepositing small amounts of other metallic ions or occluding organic material. When that occurs, the materials are classified not as contaminates but as addition agents. When they produce a bright deposit, they are called brighteners. Strictly speaking, they are controlled impurities.

When organic additives are employed, some forms of the compounds are absorbed at the cathode surface during electrolysis. Often, the compound absorbed is a decomposition product of the original material. The amount that is absorbed is proportional to the nature and concentration of the compound and the time of electrolysis. Frequently, the organic decomposition products develop over a period of time and may affect the deposit adversely. Six months to a year may pass before critical concentrations are reached. If the breakdown products can be removed by an activated carbon treatment, the problem can be controlled. It always is a worthwhile practice to remove organic addition agents and their breakdown products by a carbon treatment at least two or three times a year. That assures continuous operation of the plating solution without unscheduled purification treatments during peak production.

42.8.2 Effect of the Plating Anode

When inorganic contaminants build up in a plating solution, increased concentrations of the additives are normally required to produce the desired effects. Poor-quality anodes constitute the major source of inorganic contaminants. In tin and tin-lead plating, at least a 99.9 percent purity anode is required. High-purity chemicals and efficient rinsing before plating are also essential to maintain a high-purity plating solution. Operation of tin-plating solutions at temperatures below 65°F can reduce the amount of organics occluded in the deposit.

When excess organics are occluded in a tin or tin-lead electrodeposit, bubbling is often observed during soldering. That often can be the cause of blowholes in soldering a printed board. However, it should be noted that the majority of blowholes are caused by entrapped plating or cleaning solutions. When postbaking does not alleviate the problem, occluded organics in the tin or tin-lead electrodeposits are a likely cause.

42.9 USE OF PRECLEANERS TO RESTORE SOLDERABILITY

In the electronics industry, a restriction is often placed on the activity of the flux that can be employed in soldering, because an assumption has been made that postsoldering cleaning may not always be 100 percent effective. If ionic residues are left on a printed board after cleaning, there is a possibility that voltage leaks could develop under high-humidity conditions. Because of that restriction, a situation in which effective soldering cannot be accomplished with the specified flux can arise. In that event, solderability must be restored or the parts must be scrapped.

42.9.1 Causes of Poor Solderability

It is important that the cause of poor solderability be understood. In some cases, oil, grease, or organic films may be responsible. A simple solvent or alkaline cleaning can remedy that situation. The most common cause of poor solderability, however, is heavy tarnish or oxidation on the surface of the metal being soldered. Precleaning in an acid cleaner usually will restore solderability. After acid cleaning, it is essential that the acid residues be thoroughly rinsed off. In critical applications, a neutralization step, followed by another rinse, is employed to ensure that all acid residues are removed. A quick, thorough drying is required after rinsing to prevent reoxidation.

It is also possible that the solderability problem may be a combination of the two situations. That would necessitate a solvent or alkaline cleaning to remove organic films and an acid cleaning to remove tarnish and oxidation. If the dual cleaning operation is impractical, then cleaning with an organic solvent containing acid should be considered. With that type of solution, effective cleaning can be performed in one operation followed by rinsing and thorough drying. That type of cleaning, which requires minimum space and equipment, is ideal when organic films on the surface are not extensive but do prevent 100 percent removal of oxides and tarnish.

A solvent containing acid cleaner is ideal for copper and brass, since it dissolves most organic films that could be on the surface and assures complete removal of tarnish and oxidation. Straight acid cleaners such as hydrochloric, sulfuric and fluoroboric acids, and sodium acid sulfate (sodium bisulfate) are completely effective only when organic films are removed in prior operations. In severe cases of copper oxidation, etching type cleaners such as ammonium persulfate sometimes are used. Although the solutions are very effective, it must be noted that they leave the metal in an active state so that it can easily reoxidize. Hence, it is a good practice to follow this procedure with a mild acid dip and thorough rinsing and drying.

42.9.2 Cleaning Tin-Lead Surfaces

Leads and printed boards are often coated with tin or tin-lead to preserve solderability. When that type of coating is applied by electroplating, it is extremely important that the deposit be applied on a solderable surface. Adequate cleaning prior to electroplating is essential. Tarnishing of the tin or tin-lead coating during etching or storage can detract from solderability. Acid cleaners for removing tarnish from tin or tin-lead usually contain thiourea, fluoboric acid, wetting, and complexing agents. If spray equipment having titanium heating coils or rollers is used, then an acid cleaner containing fluoboric acid cannot be used. However, there are available equivalent cleaners that do not contain fluoboric acid that will clean tin or tin lead effectively in a spray operation. It should be noted that if the tin or tin-lead is plated over an unsolderable surface, then the only recourse is to strip the tin or tin-lead and replate. Solutions containing glacial acetic acid and hydrogen peroxide often are employed for the purpose. Stripping and replating can be a costly operation, and economic considerations may rule out the procedure in some cases.

CHAPTER 43
SOLDER MATERIALS AND PROCESSES

Gary M. Freedman
*Compaq Computer Corp., Alpha High Performance Systems,
Marlboro, Massachusetts*

43.1 INTRODUCTION

The theme for this latest edition of *Printed Circuits Handbook* is the density revolution. When first thinking about this, a parochial view toward soldering and the printed wiring board itself was taken. It is clear that the complexity of printed circuit assemblies continues to increase. Component count is on the rise, component I/O numbers are multiplying, board size is increasing, layer count and overall board thickness is growing, and component-to-component spacing is decreasing. The density revolution is certainly upon us across all classes of products, be they consumer portables or high-end servers. Now, personal computer motherboards have taken on the look of yesterday's workstations in terms of complexity and functionality. These industry-wide changes impart an increasing level of sophistication and difficulty to the assembly of today's electronics products. But there is more to the revolution than just soldering alone. The onset of more dense assemblies was and is inevitable. One need only look at the enormous lag between the integrated circuit density revolution and the much more staid circuit assembly technology progression to predict how things will proceed.

Simultaneously other revolutionary changes are in progress in terms of electronics assembly. There has been an upsurge in outsourcing of electronic assembly. The move to electronics manufacturing services (EMSs, or contract assembly houses) has necessitated that printed circuit assembly (PCA) design must be robustly amenable to EMS conventions for assembly. Where once it was possible to cut corners in design and take chances in violating internal design guidelines, it is difficult to track these changes when boards are outsourced. If there is an assembly problem, it will be harder to determine the root cause when the work is farmed out compared with building within one's own plant. Contract manufacturers' guidelines may be more restrictive in terms of component pad shapes, component spacing, through-hole sizes, surface finishes, component types, etc., and there may be extra build fees assessed for any such variations.

At the same time some package sizes are increasing, others are decreasing. Surface-mountable area array packages (BGAs) are slowly transitioning from a 1.27-mm pin pitch to a 1.0-mm pitch. Some ceramic column grid array packages have close to 1700 solder joints to the circuit board. Chip-scale packages (CSPs), in which the overall package size is roughly no larger than 1.2 times the silicon die size, are starting to appear in consumer portable products and high-end computers alike. Pin pitches on these area array devices are typically between 0.8 and

0.5 mm. Soldering these can be tricky and very much dependent on the quality of the stenciled solder paste deposit. The small stencil aperture sizes requisite for these devices inhibit solder paste release due to rheologic properties of the paste and capillary forces between the paste and stencil aperture walls. So, the equipment, stencils, and every aspect of PCA is being challenged and revolutionized.

Although surface-mount reigns, there are still some through-hole components that are not amenable to conversion for surface mounting, so wave soldering is still in wide use. The difficulty of wave soldering has increased with the increased complexity of today's PCAs. Often selective soldering fixtures (wave pallets) are needed to mask bottom-side surface-mount components from contact with the solder wave. The pallet is also necessary as a thermal insulator to keep top-side surface-mount device solder joints from remelting during the wave soldering process.

The revolution has even spread to solder metallurgy. In the USA there have been attempts to limit the use of lead (Pb) in electronics assembly, but legislative action was defeated due to the impact on the electronics industry and consumer and the dearth of studies on lead's impact on the environment from electronic assemblies. But today, there is a strong drive in Europe for reduction and elimination of lead-bearing solders. The proposal is embodied in the Waste Elimination in Electrical and Electronics Initiative, generally referred to as the WEEE Initiative. WEEE was proposed and funded by the European Union, and is likely to result in EU legislation within the next 4 to 8 years. With that in mind, many European electronics manufacturers are scrambling toward lead-free soldering solutions. The move to lead-free solders will impact the entire electronics industry. Since nearly all the practical no-lead alloys have higher melting points or melting ranges, a reexamination of IC construction and packaging materials, board laminates, and molded parts such as connector bodies is under way. In some cases, current reflow ovens may not have the higher temperature capacity or tunnel length required for use with some of the alloys being investigated. Fluxes, board and lead surface finishes, and reliability models will all need updating in time for a transition to non-lead-based soldering.

Japan has jumped on the marketing bandwagon with Pb-free products. Some Japanese manufactured consumer products are touted as lead-free, although board finishes or component terminations may still contain lead (Pb). The Pb-free movement will take hold in the United States as the threat of European or Japanese import trade restrictions on Pb-bearing products increases.

There is a density revolution in progress that is both exciting for electronics designers and challenging for circuit assemblers. Never has so much circuit integration been offered on ICs. Instead of the amount of circuitry needed outside of the chip being limited, the opposite seems to be true owing to increased chip and product functionality. The result is the densest circuitry ever and a host of new component types and process technology choices. Circuits run hotter and heat sink materials and types are hard pressed to keep pace. Board sizes are on the increase as system performance increases. Even heat dissipation in some portable products has become a challenge.

To most, this density revolution is characterized by familiar consumer products that we touch every day in personal and business pursuits. Cell phones, pagers, CD players, microwave ovens, camcorders, electronic organizers, portable computers, and other personal electronic items continue to drastically influence our lifestyles. All of these and a variety of other consumer and business electronics goods continue to shrink in size and increase in functionality. The Internet has spawned several new opportunities such as links to cell phones and personal data assistants (PDAs, also known as palmtop devices or handheld computers) that have wireless links to the Internet. Where the computer in the home was once a rarity, now many homes have multiple personal computers, which in some cases are even networked to one another. Palm-type portables are now available with more on-board memory than was available in the best home computers just a few years ago. There are wristwatch televisions, filmless digital cameras, and breakthrough medical instrumentation, all part of this miniaturization and density revolution.

Supporting the burgeoning market for miniaturized electronic goods and services is a less visible set of products: high-end computers. These are all-pervasive in their scope and range, running our telephone exchanges, stock exchanges, banks, governments, schools, universities, police and fire departments, hospitals, businesses, and the Internet. Their performance is astounding. Processor chips now clock above 1 gHz and memory and disk farms are measured in terabytes of storage. While the computers were once dependent on a single central processing chip, now they may have up to 32 processors within a single computer cabinet capable of running more than one operating system simultaneously. These were key to recently deciphering the human genome, which on a home-type PC would have taken tens of years. This daunting performance necessitated the quieter side of the density revolution in electronics. It is the high-end computer infrastructure that has permitted the consumer goods design, interconnection, and circuit density miniaturization revolutions.

The package types and assembly technologies are very much enabling factors of the density revolution that is altering the electronic product landscape. Several of these elements are discussed subsequently.

Press-fit is a sleeper technology just now awakening. Around for years, this connector methodology is suddenly becoming very important to the density revolution. Particularly suited to thick, high-end circuit boards, it relies upon compliant pins forced into plated through-holes in the board. This type of connector is used on thick circuit cards and backplanes. It is a means of avoiding additional thermal steps—a boon for reliability. This type of connector can be used on thick boards where wave soldering is impractical. Furthermore, press-fit connectors can be applied to either or both sides of a circuit board as a final step after all soldering steps are complete.

Until recently, most high pin count ICs, such as processor chips, were in the form of pin grid arrays (PGAs), which are plated through-hole devices and generally wave soldered. More recently the PGA has given way to surface-mountable area array packages such as ball grid arrays (BGAs) and column grid arrays (CGAs). The next generation of high-pin-count, high-performance packaging is the land grid array (LGA). These are solderless interconnect packages that rely upon mechanical pressure to contact the device to the PWB. The LGA looks like a BGA without solder ball contacts on its underside. Instead, there are hard gold pads. The circuit board has a land pattern to match, composed of hard gold pads. An electrically conductive, pattern matched, somewhat compliant interposer is sandwiched between the PWB and the LGA and the system is bolted in place to complete the circuit by pressure contact. The flexibility of the interposer serves as a means of ensuring electrical contact even if the circuit board is somewhat wavy or locally warped. It adds to the system reliability by absorbing mechanical stresses imposed by differing coefficients of thermal expansion between board and LGA. Stiffener plates are applied on both sides of the board, one in contact with the LGA and one in contact with the bottom side of the PWB. When the bolts are applied, the pressure required for contact is made. LGAs seem to be gaining in popularity for high-end computers. There are several brands of interposers available; however, there is little in the way of published field reliability data.

The LGA offers a means by which complicated ICs can be attached and replaced without the impact of a soldering cycle. LGAs can be reused many times as long as their gold pads and the corresponding gold pads on the PWB remain intact. Care must be taken in assembly to ensure that the system will yield reliable contacts. It is important that all surfaces be free of dirt, flux, or other contaminants. The LGA must be torqued according to interposer vendor's specifications and the torque sequence must be even so as to avoid damage to the LGA package, the interposer, or the PWB surface.

In addition to the changing face of electronic circuitry, there are a myriad of details that one must pay attention to in order to be successful in PWB assembly. The most important part of manufacturing is board design and process preparation. This chapter delves into the basic techniques of soldering and reviews critical background and details to ensure process success. Even though many of the techniques contained in this chapter are older technologies, much is being learned about them. No process engineer can engage in process optimization without having a proper understanding of the basics. In terms of circuit assembly, one has to have a breadth that reaches into many disciplines: chemistry, physics, metallurgy, thermodynamics, mechanical engineering, reliability, and material science, to name a few. The more accumulated knowledge in the basics, the more success the engineer will have in the process. This chapter is written with the intent of providing a fundamental understanding of the main processes of circuit assembly. Heed the details. Measure thermal profiles carefully. Gauge the performance of each piece of equipment and process. Learn the basics of each process for best insurance of process success.

43.1.1 Designing for Process and Product

There are numerous considerations that are crucial for circuit assembly and solder joint success. The exposure environment must be understood:

1. What thermal conditions must the product endure?
2. What mechanical shock and vibration will the assembly experience?
3. How many power cycles are expected of the product?
4. To what airborne gases and humidity regimes will the soldered assembly be exposed?

These considerations and others have implications for materials, solder joint configuration, and assembly design rules. Unfortunately, design rules are quite specific to these considerations, coupled to the bill of materials for the soldered product. So the notion of listing specific design criteria for broad use is not a simple or practical matter. Instead, it is necessary for the solderer to understand the basic phenomena that influence the soldering process and the reliability of the finished assembly, which is the essence of this chapter. Additionally, there are certain key rules one could follow for incoming printed circuit board cleanliness, solder purity, etc. These are best described in certain standards documents, such as those published by the IPC.*

The old-fashioned idea and trap of offering up generalized rules for all soldering process applications will be strictly avoided. The materials and solder properties for a desktop computer are vastly different from those for the ignition system printed circuitry under the hood of a car. Similarly, mobile electronics have added the need for mechanical shock-resistant assemblies for calculators, pocket pagers, cell phones, laptop computers, etc. This, coupled with the reality of high-volume production for the mobile electronics market, has driven the industry to ever increasing soldering yields as well as better joint-to-joint and assembly-to-assembly reproducibility. The cost to repair, in some cases, may be steeper than the combined material and labor content of the as-assembled product. This further reinforces the need to solder properly the first time.

There are numerous rules of thumb that will become evident in discussions of metallurgies, fluxes, and processes that, if taken to heart, will result in best soldering yield and reliability. However, the only way to ensure that the assembly will endure its use and environment is to develop appropriate accelerated testing methods that can accurately predict product field reliability. This is discussed in Chap. 53.

There are also abundant pitfalls in soldering. Too much gold in a tin-lead solder will cause catastrophic embrittlement of the solder joint. Too high a process time or temperature can have the same effect on tin-lead solder even in the absence of gold. The residues left by some fluxes can be corrosive. Electrochemical migration of flux corrosion products is a common mode of circuit failure that is difficult to pinpoint by electrical testing. Dendritic filaments can form between two board features of differing electrical potential. Because these microscopic dendrites have limited current-carrying capability, they may act as microscopic fuse links

---

* Institute for Interconnecting and Packaging Electronic Circuits, 7380 N. Lincoln Ave., Lincolnwood, IL 60646-1705.

that blow only to grow back again. The electrical shorting effect can be transient, complicating board-level or system-level failure diagnosis.

Another soldering process implication on circuit testing is that of flux residues from "no-clean" soldering fluxes and solder pastes. These flux residues, formulated to be noncorrosive, are meant to remain on the printed circuit board for the life of the product. These residues are electrically insulative and typically hard, which inhibits probe penetration during electrical testing. Residue characteristics are mainly an attribute of the flux or solder paste chosen, but the soldering process can influence residue characteristics. So, some forethought and testing of materials prior to electrical assembly will minimize impact on new product start-up.

43.1.2 Common Metal-Joining Methods

The three most popular techniques for joining metals are:

- Soldering
- Brazing
- Welding

While a rigorous review of the three techniques will be avoided, a cursory comparison of these metal-joining methods is useful in setting the stage for an in-depth discussion of circuit board soldering (see Fig. 43.1).

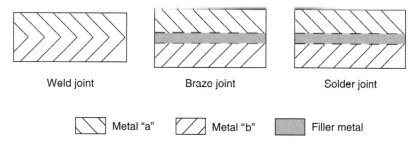

FIGURE 43.1   Welding involves heating of the metals to be joined to the point that they fuse into one another. Brazing and soldering involve the use of a filler material that is fluxed and combines metallurgically with the metals to be joined.

43.1.2.1 Soldering.   In order for a metallurgical bond to result from the soldering process, the solder material has to be compatible with basis materials or any coatings applied to the underlying metals. The basis materials are those bulk metals that provide the shape and stiffness of the metallurgical structure, be it printed circuit board solder land, plated through-hole, or component lead. The basis metal may be of single elemental composition, such as copper, or an alloyed metal such as Kovar. The basis metal may require a coated surface to permit, preserve, or enhance its solderability. Gold, tin, or solder over Kovar are common examples. Metal system alternatives are discussed in subsequent sections.

43.1.2.2 Brazing.   There are only subtle differences between soldering and brazing. The primary distinction is the temperature required for joint formation. Brazing is generally accomplished at relatively high-temperature regimes as compared to soldering. Most agree that when the filler metal has a melting temperature greater than about 500°C, it qualifies the process as brazing, although texts are not consistent on the cutoff temperature.

The filler metal is referred to as the *solder* or *brazing alloy,* depending on which process is being employed. Solders are generally thought of as soft, having a low melting point and high ductility, while the opposite is generally true of the brazing alloys. The bulk metal being joined experiences only superficial metallurgical changes where in contact with the solder or brazing alloy. The bulk metals being joined, with the exception of any filler metal (solder or brazing alloy), are never heated to their melting points, as is the case with welding.

Since circuit board assembly is done almost exclusively by means of soldering, this chapter is devoted to that traditional joining method. The focus is on the soft soldering process or soldering with relatively ductile, easily fused solders typified by the eutectic tin-lead system with a melting point at 183°C. Compatibility with the circuit board materials and ease of processing dominate the list of reasons soldering is chosen over brazing for circuit board assembly. This, along with discussions of alloy selection, is reviewed in subsequent sections.

43.1.2.3 Welding.   Welding involves heating one or more compatible metals to the point where they join one another. They may be of the same material, in which case they merely melt into one another. Welding of two dissimilar materials is commonly accomplished where one metal alloys with another. Technically there is no need for the addition of a filler material for the welding process, since the bulk specimens being joined are altered to the point that they alloy or *wet* to one another. For various reasons, filler metals can be added. Welding may be highly localized or may extend across entire mating metal surfaces. Welding is diametrically opposed to the soldering and brazing processes. In the latter two, a separate material—a filler metal—is necessarily sandwiched between the two metal surfaces to be joined. The filler must metallurgically combine, or *alloy,* with the surfaces of the parts to be joined. This fusible alloy, the filler, is the *cement* of the joint.

43.2   SOLDERING

When molten, the solder must wet to the materials in contact with it, forming a metallurgical bond by diffusion, dissolving, or alloying with the basis metal. To demonstrate this point, especially in the context of soldering for electronics, the most common example is the case of tin-lead solder joining the basis metal, copper. For this example, it is not necessary to discuss the exact alloy of tin-lead (Sn-Pb), but it is important to know how the materials involved alloy with the copper (Cu) basis. To develop an understanding of joint formation, one must examine the tenets of processes operant during soldering. Primary steps involved are as follows:

1. Intimate contact of the solder to materials being joined
2. Application of heat sufficient to melt the solder
3. Oxide removal from the basis and solder metallurgies
4. Solder wetting to basis metal(s) and intermetallic formation
5. Quenching of the solder liquidus

Each of these is summarized in the following text.

43.2.1   Application of Heat to Solder

Heating is the one step that has the most bearing on the entire solder process. It should not be trivialized by merely associating thermal rise with the onset of solder liquidus. Warming solder will cause it to soften. Heating to the melting point will allow solder to flow, which further enhances intimacy of the solder mass with the adjacent mating surfaces of parts to be joined. Note that at room temperature, eutectic tin-lead alloy (63 wt % Sn/37 wt % Pb) is already at 65 percent of the way to liquidus (183°C or 456 K) when referenced from absolute zero.

Thermal input is also responsible for imparting the necessary energy to activate the chemical processes that both aid and impede the soldering process. Most important, heating drives the metallurgical processes required for intermetallic formation. All of these concepts are covered in this chapter.

43.2.2 Oxidation Formation

Most materials, when in equilibrium with our oxygen-rich environment, possess an oxidation coating. Upon heating, surfaces of the solder as well as the metallic surfaces of the basis metal will further and more thoroughly oxidize in our normal air environment. If a silver-bearing solder alloy is heated in the presence of a sulfur-containing ambient (sulfur-tainted air), sulfuration will occur and that tarnish will also inhibit soldering. Generally, the higher the soldering temperature or the longer the process time, the more oxidation or tarnish will be a problem. Oxides and tarnishes provide a physical barrier preventing alloy formation between the solder and the metal(s) to be joined by means of the solder process. Of course, in the case of gold (Au) there is insufficient oxidation to degrade soldering, but unless oxidation is removed from the solder itself, solder alloying will not be possible. Note that in most soldering processes, the ambient can be altered to mitigate the detrimental effects of oxygen or other airborne contaminants. More on that later.

43.2.3 Chemical Preparation of Solder Surfaces

The most generally applied remedy for the effects of oxidation and tarnish during soldering is the application of a chemical agent, flux, which attacks and removes tarnish and oxidation and, further, protects metal surfaces from reoxidation during the joining process. The word *flux* comes from the Latin *fluxus*, which means flowing. Flux ensures that the solder, once molten, will flow over the surfaces to be bonded, unconstrained by oxide skins on the solder or the metals to be joined.

43.2.4 Solder Wetting/Intermetallic Formation

Intermetallic compound (IMC) formation is the key to soldering. It is a local alloying between the boundary of the solder liquidus and the surfaces of the basis metals with which it is in contact—the essence of soldering. There is much misinformation about IMCs. One thing is certain: without intermetallic formation, there is no solder joint formation. This is stressed because some erroneously claim lack of IMC in such processes as laser soldering. In the case of that technology, the IMC can be exceptionally thin immediately after soldering, but it is and must be present for soldering to occur. High temperature and intimate contact of the liquid solder enhance the rate and area of intermetallic compound formation.

43.2.5 Solder Quenching

Once below the melting point again, the solder cools, intermetallic compound formation rate decreases significantly, and the solder joint is formed. It is critical that the solder attain its solid state at process completion and prior to circuit board handling to preclude accidental movement of the components in the molten solder.

43.3 SOLDER FILLETS

The solder fillet is an overt manifestation of surface tension and wetting. Figure 43.2 shows fillet formation when solder wets to the solder land and component lead. Figure 43.3 shows cross sections of fillets for different solder joints.

In Fig. 43.2, the menisci are easily discernible as anchored webs of solder extending from the bonding pad to the perimeter of the component lead. Looking at the side view, Fig. 43.3(*a*), fillets are readily apparent at the heel and toe of the component lead. In Figs. 43.3(*b*) and 43.3(*c*), side fillets between the bonding pad and the component lead are also quite distinct. The fillet is a reasonable indicator of the degree of wetting and, therefore, of process goodness. The

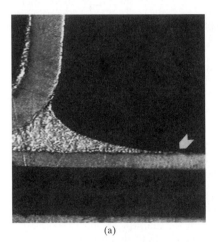

(a)   (b)

FIGURE 43.2   (*a*) Solder has wet well to the pad (see arrow), as characterized by a very shallow contact angle; (*b*) wetting has been interfered with, as characterized by the steep dihedral angle (see arrow).

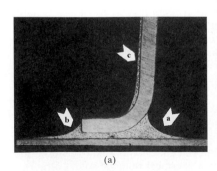

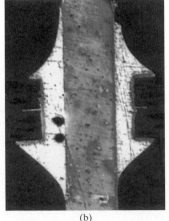

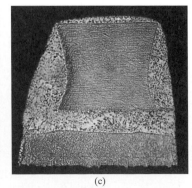

(a)   (b)   (c)

FIGURE 43.3   (*a*) Cross section of a surface-mount lead exhibiting heel and toe fillet formation, with solder wicking up the lead. (*b*) Cross section of a plated through-hole solder joint. Rich fillet formations extend from the barrel lands on both surfaces of the circuit board to the lead. (*c*) Transverse cross section of a surface-mount component lead to a circuit board pad, showing rich side fillets from pad to lead.

higher the fillet, the better the joint. This can be confusing. When comparing two boards that were not processed simultaneously, a higher fillet may just indicate a longer time above liquidus. If this is the case, embrittlement of the solder joint may also result from intermetallic compound buildup. Intermetallic compounds are brittle and degrade joint strength. Large solder fillets may also be indicative of larger solder volumes. Therefore, fillet examination alone is not sufficient to accurately assess soldering process performance. But, for a given time-temperature profile, solder alloy, species and conditions of metallurgical contacting surfaces, solder volume, flux, and soldering atmosphere, the solder fillets may provide a first-pass comparative assessment of the solderability of various parts of the bonded assembly. Too much solder wicking up the lead may cause solder shorting, particularly near the package body where the lead spacing may be finest. Excessive solder on the leads may reduce the flexibility of the lead and result in premature joint failure when thermally cycled through use in the field.

Note that up to this point in the discussion on the soldering process, the system has been treated as solid or liquid metal(s) interfacing with a gas (the soldering ambient). This simplistic view has been used to demonstrate the fundamentals of the process. However, the realities dictate a modified view. To keep the solder and surfaces to be joined bondable, they must be kept oxide free. Fluxes are therefore formulated to not only remove existing oxide on the pre-reflowed solid solder, but to stem oxidation of the cleaned metal surfaces during the high-temperature excursions of the soldering process. This is accomplished by using fluxes that have a high enough molecular weight or configuration to prevent them from burning off entirely at solder alloying temperatures. Therefore, the solder flow and surface tension are influenced by a liquid-liquid interface, so solder spreading must be examined in terms of this liquid solder–liquid flux system.

43.4 INTERMETALLIC COMPOUNDS AND METALLURGY

At the boundary between the solder and the metal to which it is joined, there is an intermediate alloy phase of some or all of the contacting metal constituents. This phase is characterized by stoichiometric association and metallic bonding. This high-alloy-concentration region generally has properties that are vastly different from those of either the solder or the contacted base metal. It may be, and generally is, of brittle behavior. It may be electrically conducting or even semiconducting. This crystalline material is called intermetallic compound (IMC). IMC is also known as an electron compound or Hume-Rothery compound. Most people think of solder as a soft, forgiving material of low melting point. It is ironic that the essence of soldering—the actual bond, or more correctly, the bond's anchor—is, in fact, a brittle, high-melting-point composition of generally poor electrical conductivity. Without intermetallic formation, though, there is no solder joint.

43.5 SOLDERING TECHNIQUES

There have been many soldering methods devised over the millennia. The most common are the soldering torch and the soldering iron. These are direct outgrowths of the most ancient methods, but they are inadequate for printed circuit board soldering. The latter is characteristic of a directed-energy bonding method where the heating is localized to a specific small piece of board real estate to accomplish soldering of one or a localized set of joints. Conversely, and hypothetically, the torch could be viewed as a means of mass soldering, encompassing a much larger area of heating than was possible by the iron. Thus, today's techniques are divided into categories according to area of heating or number of joints formed simultaneously.

The iron is too slow and clumsy for bonding the large number of fine-leaded components found on today's circuit cards. For the most part, a torch's flame is not controllable enough and is prone to damaging most circuit board substrate materials and component packages, but this technology has been reincarnated for fine-circuit soldering, though it is not a popular choice. As the boards and components have matured, so have the soldering assembly methods. Although the tools of soldering for electronic assembly have changed dramatically over the last five decades, the dichotomy of methods persists. There are other terms applied to differentiate the two, such as flow soldering vs. reflow soldering. These are explored later.

The topic of modern soldering methods can be divided into two distinct categories that encompass the following:

Mass soldering	*Directed energy methods*
Wave soldering	Hot-gas soldering
Oven reflow	Hot-bar soldering
Vapor phase reflow	Laser soldering
	Soldering iron
	Pinpoint torch

Each method has its place, purpose, and advantage. Techniques must be judged by their applicability, cost, and end result. Mass soldering techniques are by far the most common for high-volume manufacturing circuit board assembly. Some of the directed-energy soldering methods are now catching on in volume manufacturing and, as PCBs are miniaturized, suitability of the directed-energy methods may become increasingly important.

43.5.1   Mass Soldering Methods

The mass flow or reflow methods are suited for high-volume manufacturing. The entire board is heated and large numbers of components on the board are soldered at one time. The two most common of these methods are oven reflow soldering and wave soldering. A third technique, vapor phase reflow soldering, has dwindled in popularity due to environmental concerns regarding the use of chlorofluorocarbon-based solvents that were key to this process. These techniques are reviewed along with mention of some other soldering methods in this chapter.

Choice of soldering method will lie in the types of components and boards being soldered, the required throughput rate, and requisite solder joint properties. There are no clear-cut rules. These days, some plated through-hole components are being assembled along with surface-mount components in reflow ovens. In that method, sometimes referred to as *intrusive reflow,* a generous amount of solder paste is deposited over the circuit board's plated through-holes. The through-hole components (axially leaded, pin grid array, etc.) are inserted in the circuit board just before or after surface-mount component placement insertion. The paste-bearing board is then mass-reflowed in a reflow oven. Solder paste coalesces into liquid solder, which is drawn into the interstices between the through-hole barrel and the inserted component lead concurrent with surface-mount component joint formation.

If intrusive reflow (also known as pin-in-paste soldering) is to be used, check that components are compatible with mass reflow soldering. The high temperatures and long exposures associated with oven-type soldering may cause unsuited molded component bodies to melt or warp. Components internal to some devices may become disbonded and some electrolytic capacitors may leak or even explode as a result of an oven reflow cycle.

One of the major shortcomings of intrusive reflow is the fact that often it is not possible to apply enough solder paste to meet standard solder joint acceptability criteria. Requisite solder volume is dictated by component pin pitch (available printing space between component leads); stencil thickness, which is generally dictated by the smallest components on board; plated through-hole annulus pad size; plated through-hole volume; and associated lead diameter and length. As a rule, intrusive reflow is difficult but not impossible to apply to boards thicker than

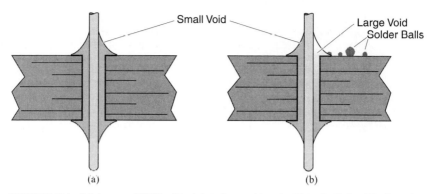

FIGURE 43.4 **FIGURE 43.4** (*a*) A normal PTH solder joint where void content is typically low. The intrusive soldering process (*b*) is more prone to void formation due to the rich organic content of the solder paste. Large voids and explosive solder ball formation can result.

0.062 in; however, solder foil preforms can be added to through-hole parts to augment solder volume.

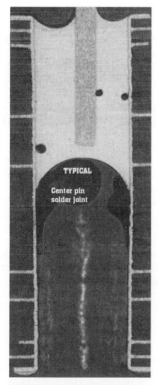

FIGURE 43.5 Nonstandard solder joint with solder retained by surface tension around pin. Very little solder has drained down the plated-through-hole barrel.

Since requisite solder volume for pin-in-paste reflow is a function of the ratio of PTH lead volume to component lead volume, it follows that reducing PTH barrel size is advantageous, especially in the case of thick PWBs. But as the volumetric annulus between lead and barrel is reduced, there is a tendency for excessive void formation.

Solder pastes are formulated for surface-mount applications. They consist of solder spheres in a creamy matrix of organic chemicals composed of soldering flux, thickening agents to give the paste the required consistency for the printing process, and materials to impart tack to hold SMT parts onto solder lands during processing. Voids are caused by vaporization of organics in the solder paste during the soldering process. As the temperature of the soldering process increases, the gases expand to the point that pocket pressure gently displaces the dense, high-surface-tension liquid solder. In this case, there is little in the way of entrapped flux volatiles and the resultant solder joint will meet acceptable standards upon solidification. Conversely, if the gas pockets expand too rapidly and pocket pressure is high enough, the gases may explosively eject solder from the plated through-hole barrel, resulting in loss of solder around the lead and unwanted solder balls on the board's surface (see Fig. 43.4). Voids detract from solder joint quality and reliability. Since some solder pastes are more prone to void formation, care must be taken to select an appropriate paste and soldering process time-temperature profile in order to minimize void formation.

In an unlikely twist, enlarging a plated through hole in a thicker board may also benefit the pin-in-paste technique. Figure 43.5 shows that if the volumetric ratio of the plated hole–to–component pin diameter is within the correct range and if the component pin is shorter than its associated through-hole, capillarity will retain the solder around the pin and minimize drainage of liquid solder down the through-hole barrel. Of course, the

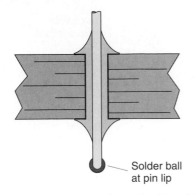

FIGURE 43.6 The effects of solder wicking down the lead and solder dome formation at the tip of the intrusively reflowed pin.

Solder ball
at pin lip

result is a nonstandard solder joint configuration that must be assessed for the intended application.

The advantage of intrusive reflow is that if there are but a few through-hole components and they are compatible with the reflow process, the wave soldering process can be eliminated—a cost savings to the finished product. There are some other characteristics of the pin-in-paste reflow process that differentiate it from the usual SMT process. These are:

- Solder joints may not look as good as those formed by the wave soldering process.
- Leads may have some domed solder on their tips (see Fig. 43.6).
- There is generally a thick deposit of flux around and between pins on the bottom side of the printed circuit assembly in the vicinity of the intrusively reflowed device. The thick flux residue, besides being unsightly, may be an impediment to electrical test station probe penetration.

43.6   OVEN REFLOW SOLDERING

In reflow soldering, primarily used for surface-mount components, solder paste is forced through a metal stencil by either a metallic or polymeric squeegee. The paste contains both solder flux for preparing the metal surfaces for solder attachment and sufficient solder for joint formation. Components are placed in the solder paste on the board and the populated PCB is inserted into an oven for reflow. As in the wave-soldering machine, the boards are transported through the oven by means of a conveyor. The oven gradually raises the circuit board and component temperatures and activates the flux in the solder paste, and, finally, enough heat is imparted to cause the solder to flow (meaning to liquefy, also known as *reflow*). When built and implemented properly, the oven reflow process results in a controlled and predictable heating and cooling cycle, allowing reflow without spattering of the flux and volatiles that compose the solder paste. Rapid heating of the paste is widely known to be a source of solder ball formation. Solder balls (isolated spheres of solder not necessarily connected to the solder mass of the joint) can be problematic for printed circuit board assemblies. They can induce electrical shorts, especially with finer-pitch components where solder ball diameters may be on the order of component pad spacing. Too rapid a heating cycle or direct infrared irradiation of certain fluxes may cause flux charring or "caramelizing," which results in unsightly, difficult-to-clean flux residues. It also lessens the efficacy of the fluxing process.

The circuit board itself may also fall victim to the reflow process if temperatures are not maintained properly or evenly. It is for this reason that much development has ensued, resulting in an excellent selection of reflow ovens, accessories, and board processing methodologies that are the mainstay of today's surface-mount manufacturing. Different approaches to reflow soldering are discussed here that should help the reader to choose the best features when picking a reflow oven or to identify or minimize problems in processing with new or existing reflow oven equipment.

43.6.1   Reflow Oven Subsystems

Even the simplest reflow ovens consist of several subsystems:

- Insulated tunnel
- Board conveyor

- Heater assemblies
- Cooling
- Venting

See Fig. 43.7 for details.

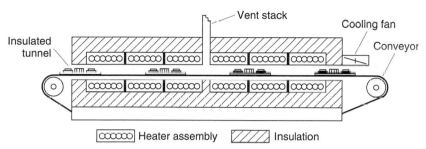

FIGURE 43.7 The five subassemblies of a reflow oven; heater assemblies with temperature controllers, insulated tunnel, conveyor, cooling fans, and venting systems.

Reflow ovens have reached a high level of sophistication, and there are many other items that enhance oven suitability for the manufacturing floor. Those items, beyond the aforementioned reflow oven subsystems, are niceties, accessories, and gimmicks offered by oven manufacturers and will not be discussed in this section; however, those subsystems previously listed are reviewed to allow readers an understanding of the basics of oven construction and operation as well as the most advantageous configuration for their applications.

43.6.1.1 Tunnel. The tunnel is a thermally insulated passage through the length of the oven that allows for a continuous reflow process. It serves to insulate the heaters and boards from the external (room) environment just as much as it is designed to maintain thermal conditions as prescribed by the process and demanded of the heaters. Boards are moved through the tunnel and past multiple heaters by an adjustable constant-speed conveyor. This permits controlled and gradual preheating, reflow, and post-reflow cooling of the circuit board. Each heater's thermal output is sensed via a thermocouple, which is used to close the loop to the heater controller. In larger, production-worthy systems, heaters are located both above and below the plane of the circuit board and are at least as wide as the conveyor. Thermal uniformity across an 18-in tunnel width can be better than $\pm 5°C$ on some top-of-the-line ovens. To a large extent, this is a function of the tunnel insulation, heater performance, and convective mixing of the heated air or gas.

Consideration of tunnel dimensions is critical for the application. Short tunnel ovens may not permit a profile adequate for attaining prescribed reflow temperature-vs.-time slopes for larger, thermally massive PCB assemblies. Tunnel height dimensions must also be adequate to accommodate the tallest components or component heat sinks.

43.6.1.2 Conveyors. There are two main conveyor systems used in today's reflow ovens:

- Pin chain conveyor
- Mesh belt conveyor

One is required and both are recommended for any reflow machine. The pin chain conveyor, also known as an *edge-hold conveyor,* looks like a bicycle chain with a pin protruding inward from certain evenly spaced links, as shown in Fig. 43.8. There is one such chain on each side of the reflow oven. The circuit board rests on these pins on the inner aspect of the chains and is

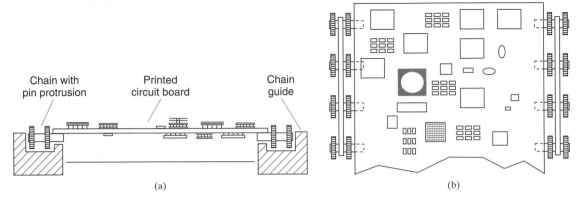

FIGURE 43.8   Cross-sectional view of circuit board being carried by pin chain conveyor. (*a*) The board is held on pin protrusions from the chain links; (*b*) plan view of pin chain conveyor.

transported through the oven during the soldering process. The two chains are driven from a common motor, appropriately geared to ensure that the circuit board is conveyed through the oven evenly and to preclude angling of the board and jamming of same in the oven. Conveyor speed is adjustable and is one of the determinants in the thermal (time-vs.-temperature) profile development. The chains are slung on rails that are parallel and run through the entire oven tunnel. Generally, one rail is fixed within the oven while the other one, parallel to the first, is adjustable with respect to the fixed rail and is movable across the width of the oven tunnel to accommodate various circuit board widths. The conveyors should have sufficient exposure area outside of the oven's tunnel to facilitate manual loading and unloading of the circuit boards.

Pin chain conveyors are best for circuit boards with components on both sides. Because the circuit board rests on its edges, and contact with the pins is minimal, there should be little thermal influence from the conveyor in a properly designed oven. Since the pin chain does not contact components on either side of the board, induced misregistration of to-be-reflowed or previously reflowed components on the opposite side of the board is minimized. This is sometimes a factor with the mesh conveyor system, either through direct contact with the mesh conveyor on side two or from rocking around on the uneven surface of a mesh belt system. When the pin chain conveyor is absent, double-sided boards can be run in pallets to keep reverse-side components from contacting the mesh belt. Any mass added to the reflow process will have an effect on the time-temperature (thermal) profile. The pallet will have such an effect, so its impact will have to be gauged and accounted for.

When reflowing thin, large circuit boards, pin chain conveyance can enhance board sagging, which will be quite noticeable after reflow. The sagging is due to thermal excursion above the glass transition temperature (T_g) of the circuit board epoxy during the reflow process. T_g is the temperature at which a partially crystalline polymer will change from a hard structure to a rubbery, viscous state, as in the case of FR-4 and other such laminate materials.

The T_g of most circuit board laminates is in the range of about 135 to about 170°C, well below the peak process temperatures in the reflow of tin-lead solder. Sagging will be pronounced due to the fact that the board was unsupported at its middle as it surpassed its T_g while riding on the edge pin conveyor. Copper traces within the board provide some support, but often not enough to counter this. Sagging alone or together with rail-to-rail nonparallelisms at process temperature can even lead to dropped boards. Many of today's oven manufacturers have suffered through numerous design changes to keep pin chain rails, the guides for the chains, from bowing or twisting when subjected to the differential heating incurred along the length of the tunnel. Most of the major vendors in the reflow oven market have conquered the rail twist problem, but hot testing with precisely toleranced test vehicles should be performed prior to final acceptance of an oven.

Board sagging can be mitigated, to some extent, by means of mechanical stiffeners affixed permanently or temporarily to the circuit board. It should be noted that stiffeners may affect the thermal mass of the board, making reflow more of a challenge. Care should be taken to ensure that the attachment of stiffeners does not interfere either mechanically or thermally with components that may be positioned close to the board edge. Design rules should preclude the use of reflow-challenging components or component fields too close to the circuit edges.

There are potential problems with some types of rail-based board conveyors. Some systems have been seen to be problematic in terms of apparent heat-sinking by the pin chain–rail combination. In some systems, the board is guided through slots on each rail and pushed along either by the conventional pin chain or by upward protruding pins or "fingers" on the chain. These edge-guided systems can have a dramatic impact on the thermal transfer at board edges overheating or heat-sinking, depending on the oven's performance and rail position. Some vendors offer rail heaters to counteract this effect. It is best to avoid this type of system because it adds further complexity to the oven and makes the job of process control all the more difficult.

This should not be a problem with the conventional pin chain conveyor because in actuality the pin chain is only point-contacting the circuit board. Thermal transfer through these points is very poor. Further, most of the newest reflow oven manufacturers have replaced carbon steel chains with stainless steel chains. Stainless steel has better wear characteristics and is a poorer thermal conductor than other commonly available chain materials.

Mesh belts used in reflow ovens are fabricated of coarsely woven stainless steel. The belt traverses the oven tunnel. Mesh belts offer exceptional versatility, as there is no need for adjustment for various board sizes. Mesh belt width is comparable to the full width of the oven's tunnel. And board dropping is mitigated completely.

The best situation, however, is a combination of conveyors with the pin chain running above a mesh belt. In this case, redundancy can save boards and maintenance time and perhaps can enhance personal safety. The mesh belt here serves as either an optional means of conveyance for boards populated on a single side or as a safety net to catch any boards that may fall off the pin chain conveyor. If the mesh belt were not used in conjunction with the pin chain, any dropped boards would "cook" on the heater assemblies, resulting in the release of noxious, irritating fumes. Additionally, the heater or heaters affected could become encrusted with the decomposed board laminate, which could impact their thermal performance. If a thermocouple, the sensing instrument for oven temperature control, is damaged or insulated by the laminate decomposition products, overheating of the zone could occur, causing oven damage and manufacturing downtime.

Reflow ovens with silicone rubber or other high-temperature polymer edge conveyors should be strictly avoided. Uneven tensioning or thermal differentials can result in stretching and one edge belt becoming longer than the other. Being two different lengths, they will necessarily travel at two different speeds, even though propelled from a common motor drive and shaft. This can lead to board canting on the conveyor and eventually conveyor jamming by the board, as shown in Fig. 43.9.

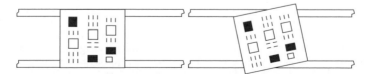

FIGURE 43.9 Plan view of board traveling on polymeric belt oven. If the belts are not traveling at the same speed, the boards will cant and jam.

43.6.1.3 Heaters. There are several heating schemes used in reflow ovens, the result of years of technological evolution. Focused infrared (IR) lamps have given way to secondary emission panel heaters and, finally, to forced hot air convective ovens.

IR Heater Types. Alternative IR heater types are shown in Fig. 43.10. Early ovens utilized focused and nonfocused IR lamps, mounted in the reflow oven tunnel. These bathe the solder paste–coated circuit boards and associated components with a broad spectrum of photonic energy heavily weighted to the IR end of the electromagnetic spectrum. The radiant energy absorbed by the circuit board materials causes the solder and the components on the PCB to heat eventually to the point required for sustained solder reflow and joint formation. As the board travels beyond the last reflow heaters at the exit end of the oven, board heat is lost to the environment or the boards are actively cooled via forced unheated air, permitting the molten solder to return to its solid state, completing the soldering cycle.

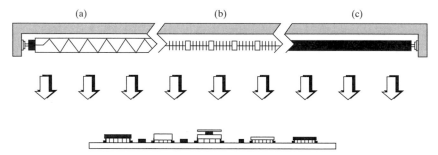

FIGURE 43.10   There are three prominent IR heater types: (*a*) lamp; (*b*) open resistance heater wire; (*c*) resistance rod (calrod-type heater).

Since the materials introduced into the reflow process (component body materials, lead/pad metallurgies, solders, printed circuit board materials, solder paste fluxes, adhesives) are constantly being reformulated by their manufacturers, the notion of a repeatable, predictable reflow process by direct radiation IR must be discarded. Hot spots due to localized IR absorption can result in overheating of some components and portions of the PCB, while other areas may not see enough heat to attain reflow. Worse yet, radiant IR heating has caused the loss of much circuit board product revenue through board charring and component body cracking, especially of plastic-packaged ICs. Nearly all major oven manufacturers have ceased building this type of reflow equipment.

IR Panel Heaters. The next evolutionary step in oven design came with secondary IR emitter panels lining the tunnel interior, as shown in Fig. 43.11. These are metal or ceramic platens heated either conductively via attached resistance heaters or by direct IR irradiation on the back side of the panel. During reflow, the circuit board is shielded from direct short-wavelength IR impingement from IR lamps or filament heaters. Instead, it is heated by black-body emission of the heated platen. This type of heating results in much longer-wavelength IR emissions, allowing slower heating rates and more even heating, which is a significant improvement over direct IR lamp oven reflow soldering methods.

A variant of this method is the convective IR oven. This technique also relies on radiant or panel IR heating, but the air in the oven is stirred by fans to enhance uniformity of heating. Today, however, techniques such as IR lamp, IR emitter, and combination convective units have been supplanted by a more favorable method: forced-air convection.

Other Heaters. There are two other heaters that are widely used in today's oven manufacture. The first is an enlarged version of a resistive cartridge heater embedded in a metal-finned or channeled thermal transfer jacket. The other is an open, coiled resistive wire, much as in

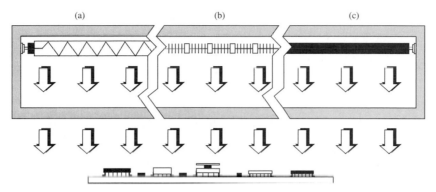

FIGURE 43.11 Secondary IR emitter panels rely on radiant IR sources such as a lamp (*a*), resistance wire (*b*), or solid resistance heated rod (*c*). In the case of this type of heating panel, the broad-spectrum IR radiation from these sources impinges on an absorbing platen, which itself becomes an emitter. The reemitted radiation is of a longer wavelength and is prone to heat the circuit board and component materials more evenly.

older (and some new) kitchen ovens. Both of these are used in connection with the forced convection reflow oven, which will be examined next. The latter heater, being of considerably lower thermal mass, is much more responsive than the cartridge-type heater. This is advantageous if the heater control circuits and thermometry methods are suitably matched to the responsiveness. If not, an overtemperature condition can result, which can cause the in-process circuit boards to overheat. Conversely, the embedded cartridge-type heater may be too slow to respond to certain oven loading conditions such as the first boards in a long queue, especially if the boards are thermally massive. The result is that those first into the oven may not be heated sufficiently, while those just upstream may be overheated because the heaters respond to the increased thermal mass. In this case, the long time constant associated with the thermally massive, embedded cartridge-type heaters may work against the process goals.

43.6.2 Forced-Air Convection Reflow Oven

This method relies on heated air recirculated at high velocity in the oven tunnel. The only direct irradiation of the PCBs is by the longest IR wavelengths. This radiation is incidental, resulting from blackbody radiation of heated oven components, and accounts for only a small fraction of the total board heating. Instead, boards are bathed in heated air that is circulated at high velocity. This method has superior thermal uniformity and controllability of the reflow process. It precludes overheating of the circuit board and/or components caused by preferential absorption, which was a common problem with the old lamp IR reflow ovens. Although the combination IR convective method has helped in this regard, it is no substitute for the hot-air convection technology that now dominates reflow oven design.

Top-of-the-line forced-air convection ovens have heating units mounted above and below the board conveyor. Air is heated by either passing the air stream over resistance heaters (such as the cartridge-type heater or the open resistance wire-type heater), through perforated resistance heated platens, or a combination of the two. Either these heaters are well shielded from the board or the air circulation velocity is high enough to mitigate the effects of any direct or indirect IR impingement on the circuit board. The high-velocity air streams do a remarkable job of evening out the temperature of the air impinging on the PCB. The result is uniform thermal control all through the entire reflow process, as shown in Fig. 43.12, and the tightest process control of any of the heating methods.

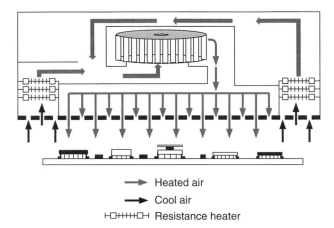

→ Heated air

→ Cool air

⊢O⊩⊩⊩⊩O⊣ Resistance heater

FIGURE 43.12 Cross-sectional view of a single convection cell in a reflow oven. Cool air or nitrogen is drawn in from the oven tunnel ambient and passed through resistance heater elements. It is then forced through a diffuser, which baffles the flow and evens its velocity and temperature distribution. In this type of reflow environment, the board is predominantly heated by the forced air. This is an even heating and also avoids direct impingement of short-wavelength IR radiation.

43.6.3 Cooling

Some ovens rely upon passive cooling to lower the temperature of the PCB assembly below the solder liquidus point. The board would merely traverse a region of the oven devoid of heaters. This is generally adequate for thin, low-thermal-mass boards. But the reality of to-day's multilayer, densely packed circuit boards necessitates some means of cooling the board so that upon oven exit, the solder joints are solid. This calls for an area of active cooling within the oven. Many silicon IC or passive component device manufacturers require a stringent thermal ramp rate that needs to be observed on heatup or cooldown of the device. The ramp that is generally recommended ranges between 2 and 4°C/s.

Active cooling comes in many forms. Forced-air cooling by fans is the most common. Fans can be deployed on top, on the bottom, or in combination for board cooling. Among the most exotic but the most efficient is the water-to-gas heat exchanger found in some ovens. This provides a stream of temperature-controlled cool air at the exit end of the oven for board cooldown and has distinct advantages in inert reflow systems. This method can take advantage of whatever process gas is in the oven, cooling it and redirecting it onto the circuit board. This method does not require additional nitrogen for cooling if running in an inert mode. Further, it precludes the need for cooling fans adjacent to the oven exit. This minimizes air turbulence at that end of the oven. Air turbulence can cause unwanted air entrainment into the oven's inerted environment. Water-to-gas cooling is efficient but it must be included in the reflow profile to ensure that thermal ramp rate is not exceeded.

Some ovens use polymeric "muffin" fans. These must be kept on at all times to prevent overheating of the fan blades and their motor's plastic fan support structure. Others use air rakes or even air amplifiers to direct a flow of cooling compressed air at the board.

The disadvantage of in-oven cooling is the condensation of flux volatiles and flux decomposition products. As deposits build up on the coolest surfaces within the oven, they change the thermal transfer characteristics of the cooling system. This alters the thermal profile of the oven over a long period of time. Flux condensates may also drip onto circuit boards as they pass by. Nitrogen atmosphere ovens running no-clean solder pastes face the biggest challenge in trying to control these condensates. Ambient air entrainment is necessarily low for precise control of the inert ambient within the oven, while nitrogen volume is minimized to keep board processing costs reasonable. The result is low gas exchange within the oven's tunnel and little dilution of condensable vapors. Condensates build up rapidly on the cooler surfaces and are difficult to remove.

There are many elaborate schemes devised by oven manufacturers for flux condensate management. Gas stream filtration in and around cooling zones, replaceable filters, cold traps, and cold fingers with self-cleaning burn-off cycles (kitchen oven style) have been incorporated into some of the newest ovens. Several ovens have quick change finned radiators that are swapped out, cleaned, and readied for the next change-out. None of the flux management systems has proven wholly effective. Aside from reduced nitrogen consumption, flux condensate management remains one of the last challenges in oven design, especially since nitrogen/no-clean reflow is becoming the de facto industry standard.

43.6.4 Venting

An often overlooked reflow oven subsystem is exhaust venting, as shown in Fig. 43.13. This is most important from an industrial hygienic point of view, but it can also have a profound effect on the soldering process itself. There are three good reasons to insist on exhaust venting of any reflow soldering oven. During the soldering process, minute quantities of lead-rich dust may accumulate in the oven atmosphere or coat the oven surfaces. The dangers of prolonged exposure to microquantities of lead or lead compounds are well understood and necessitate venting of any soldering process.

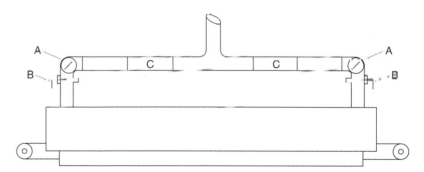

FIGURE 43.13   Oven with exhaust stacks at each end of the oven. Manometers (A) are mounted symmetrically on each branch of the line (C), with blast gates (B) mounted symmetrically in the branches.

43.6.4.1   Volatiles and Fumes.   During reflow, solder paste volatilizes and paste reaction products are given up. Since most solder pastes and fluxes are highly guarded proprietary formulations, the paste and flux vendors may neither accurately disclose the composition reagents nor report the decomposition products in the mandated Material Safety Data Sheet (MSDS). Additionally, if a PCB should fall off the pin chain conveyor or a component should fall through the mesh belt, it may land on a high-temperature surface—either the heater itself or, in the case of forced-air convection reflow machines, the perforated baffles above the heater assembly. Either board or plastic component will overheat, decompose, and release unpleasant or even dangerous fumes.

43.6.4.2   Venting and Performance.   Note that a high-velocity exhaust may have a profound influence on the performance of the oven. Too much exhaust will cause a significant end-to-end or end-to-middle flow through the oven, which can result in unwanted turbulence in a direction counter to the controlled convective flows established by the oven manufacturer. This will make temperature regulation much more difficult, decrease zone separation, decrease ease of profile establishment, and result in process variability. It is best to follow manufacturer exhaust requirements and recommendations. Further, one should install manometers or other exhaust pressure or exhaust velocity measuring analogues in the exhaust line(s) to monitor its setup and routine performance. This will ensure oven venting consistency for repeatable reflow processing and operator safety.

43.6.4.3 Oven Exhausting Schemes.   There are two types of reflow oven exhausting schemes in wide use today. These are passive and active exhaust systems.

A passive exhaust system comprises an outdoor vented exhaust duct without a powered exhaust fan. This method relies upon the chimney effect to vent oven gaseous or particulate effluent. Passive system performance is highly dependent on local weather conditions. Temperature inversions, rapid changes in barometric pressure, and wind vector changes will all contribute to a change in venting performance. The vagaries of this system make it undesirable for safe and repeatable oven operation.

Active exhaust denotes the use of some powered devices to vent the oven to the outdoors. There are three types in use today:

- Eductor
- Oven-mounted exhaust fans
- Oven-mounted exhaust fans with powered exhaust duct

The eductor relies on a stream of compressed air to run a venturi device that actively draws on the tunnel's atmosphere. In this mode, as in most other exhaust methods, make-up air is entrained through the tunnel entrance and exit openings. The eductor is a noisy and inefficient method of exhausting waste products. Due to small orifice openings of the venturi itself, its performance can be affected by flux or paste residues and should be checked routinely as part of a preventive maintenance program. The eductor effluent is fed into a fanned or unfanned exhaust duct.

Some reflow ovens rely strictly on the fanned exhaust duct. This is one step above the passive system, since exhaust performance is still dependent on the vagaries of prevailing weather conditions. Another of the active exhaust systems is the oven-powered exhaust fan. An integral hood at each end of the oven contains a small fan that withdraws a small, controlled amount of atmosphere from the tunnel entrance and exit. This in turn is blown into the main powered exhaust vent line and pushed out of the building. This system is the most efficient and reliable of the exhaust methods and minimizes meteorological interferences.

Active exhaust systems should be checked to ensure that impeller blades are not being fouled by flux or flux decomposition product residues. Again, the installation of exhaust stack monitors should help pinpoint such problems.

43.6.5 Reflow Oven Characteristics

Reflow ovens are multizoned; i.e., they have multiple heating zones top and bottom, each independently controlled and settable to a different temperature. Some reflow ovens have as many as 24 heaters (12 top and 12 bottom heater zones). Note that a zone here is the top and opposing bottom heater pair, as shown in Fig. 43.14.

Zone separation—the influence of one thermal zone upon another—is exceptionally important. The ability to separate the influence of one zone from another makes the job of profiling that much simpler and oven performance more predictable.

Temperature uniformity across the width of the tunnel is of primary importance. This will determine the quality and uniformity of the reflow and allows for controllable and consistent

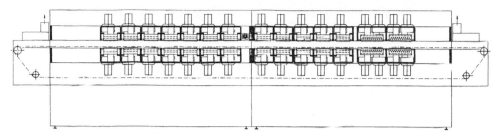

FIGURE 43.14 Forced hot air oven with 12 top and 12 bottom forced-air cells. Each cell is independently heated and controllable. The last two cells on the right are for active cooling of the circuit board as it exits. On this machine, cooling zones are aided by fans blowing over water-cooled radiators, and vent stacks are located at each end. *(Courtesy of Heller Industries.)*

soldering. Top-quality ovens have rail-to-rail thermal uniformity better than ±5°C, even for board widths on the order of 20 in.

Another important characteristic of an oven that requires consideration is the responsiveness of the oven's heaters and accompanying thermal controller circuitry. As a board or multiple boards pass through the oven, the oven's heat is absorbed by those cooler boards on their way to reflow. The oven has to respond to that net loss. The thermocouples built into the oven sense the thermal impact on the tunnel's ambient. The heater controllers try to compensate for the thermal deficit by increasing energy to the heaters. Just as some oven heaters are more responsive than others, so, too, is the circuitry. Today's ovens have reached a high degree of sophistication, including computer-controlled thermostatic circuitry. If matched properly to the job and tuned accordingly, the oven should be able to reflow a train of closely spaced circuit boards just as well as a single board in its tunnel. It is important not to space the boards too closely together. This can fully segregate the top from the bottom heaters and thoroughly alter oven performance. The spacing between the boards in a forced-air convection oven must be sufficient to obviate this top-to-bottom divisional isolation of the heater zones. The spacing will be dependent on line throughput requirements and thermal mass of the product design. Board separation may be as small as a few inches, but the effects of board loading and spacing need to be determined empirically and checked to detect any impact on the process temperature vs. time profile. Any assessment of new equipment ought to include load testing to ensure that the oven can handle anticipated process throughput and capacity. The oven's profile needs to be examined before, during, and after loading of the board train.

43.6.6   Reflow Profile

The reflow profile is of greatest importance in the oven soldering process. It is the relationship of temperature with respect to time required to bring a PCB assembly to solder liquidus and back to below the melting point of solder. This profile is dependent on many variables:

- Solder paste composition
- Circuit board composition
  - Number of layers
  - Board material
- Component types
- Component layout density

The profile in any soldering operation is largely dependent on the solder and flux in use. The solder will dictate the temperature regime, while the solder flux, or more commonly the flux contained in the solder paste, will impose its reflow requirements. Discussion once again will be restricted to reflow of eutectic or near-eutectic Sn/Pb solder, in this case Sn/Pb solder paste.

Paste manufacturers will generally provide a rough estimate of the profile required for each of their solder paste formulations. There is no universal profile recipe for reflow, but it should be consistent from batch to batch of a single paste formulation for a given solder assembly or product. Each profile will be determined by the requirements of the solder paste in use. How well the profile is maintained will be dependent on the materials to be reflowed and the process equipment.

There are four distinct regions in a reflow oven to accommodate the requirements of the soldering profile. These are:

1. Preheating/drying
2. Thermal soaking or activation
3. Reflow
4. Cooling

Refer to Fig. 43.15 for a depiction of a generalized profile.

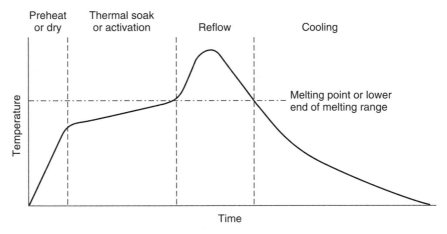

FIGURE 43.15 Generalized oven reflow time-temperature profile.

43.6.6.1 Preheating/Drying. The first region is the preheating/drying zone in which the paste is relieved of vehicle volatiles. The vehicle regulates the paste viscosity, making it suitable for stenciling or dispensing. If the volatiles are given up too rapidly, then liquids in the paste will boil and cause the paste to spatter, resulting in unattached solder ball formation, as shown in Fig. 43.16. If there is enough explosive boiling, it may adversely affect the local volume of the solder and result in solder opens. If the spattered solder balls are very large, or small but prolific, they may result in electrical shorts between two adjacent conductors.

Another consideration of presolder heating is oxidation. Naturally, the longer the board and devices to be soldered are held at elevated temperature, the greater the oxidation of the metals, including the solder, assuming that the reflow is conducted in an air ambient.

Ramp rate considerations are crucially important in this first step. The board sees its first large thermal differential in this step. The board is cold and is being ramped to about 150°C. There are restrictions, however, on the ramp rate aside from those imposed by the paste's

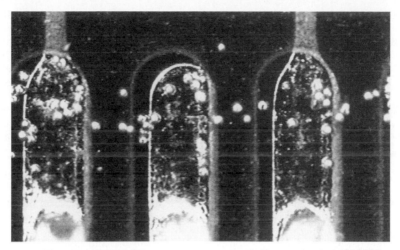

FIGURE 43.16 Solder balls may be loosely attached or they may be partially soldered to one another, with the potential for electric shorting.

propensity to boil and spatter. The electrical components and their packaging are heat sensitive and can be damaged if the heating rate is excessive. Device manufacturers generally restrict this ramp rate to between 2 and 4°C/s. A heating rate greater than that may cause component cracking, also known as *popcorning,* which is attributed to entrapped water that vaporizes internally to the package, stressing the packaging to its rupture point. This is particularly germane to plastic encapsulated devices.

Other package types, such as ceramic, may be subjected to thermal shock and differential expansion that can cause packaging failure. Excessive ramps could also be deleterious to the IC itself, either causing it to delaminate from the packaging material, crack, or damage the electrical connections to the die. It is wise to check with the device vendors for the ramp rate specification for each of the devices being reflowed. Take time to measure and analyze the chosen time-temperature profile programmed into the oven to ensure that the heating and cooling ramp rates correspond with those recommendations. Last, qualify the packages for reflow to see that they are meeting the vendor's claims.

43.6.6.2 *Thermal Soaking.*

The next step, thermal soaking, is a much slower, prolonged heating exposure. It is here that the flux is raised to a temperature sufficient to begin activation. Were the flux to activate very early in the process, it could be spent too soon. If its potency is lost, fluxed metal surfaces may begin to reoxidize in the critical moments before the onset of reflow. This inhibits solder wetting (alloying) at liquidus, precluding effective solder joint formation. This region of the reflow profile has to be long enough and hot enough to ensure that the flux has time and energy to do its job in terms of stripping the metal surfaces of the component lead, pad, and solder of oxides and surface coatings that would otherwise interfere with the solder wetting process. Overheating the circuit board would stimulate intermetallic formation. This is especially true of copper pads previously treated by the HASL process or plated and reflowed solder or tin coatings. These surface finishes have begun to alloy with the copper, which means that an intermetallic layer exists prior to the reflow soldering operation. During the reflow soldering process in which electrical devices are soldered to the circuit board, any intermetallic layer will be augmented by time and temperature, resulting in a yet thicker intermetallic compound buildup. This step also has to be at a low enough temperature and short enough time so as not to degrade the epoxy or other materials of the circuit board. The soak region is generally thought to begin at about 150°C and extend to the onset of liquidus. It is wise to follow solder paste vendor recommendations and trial reflow runs. Those values are only rough guidelines to demonstrate the various aspects of the reflow time-temperature profile.

43.6.6.3 Spike.   The reflow spike is characterized by a rapid rise in temperature to and slightly beyond the liquidus point, as shown in Fig. 43.15 (reflow segment). For eutectic or near-eutectic tin-lead solder, liquidus onset is 183°C. The reflow spike is generally chosen to be well above the melting point of the solder chosen, generally on the order of 25 to 50°C. This temperature overage is to ensure that all the elements to be soldered, including the board, component, and solder, are heated sufficiently for reflow. It is meant to overcome variations in oven performance, compositional variations in the solder itself, and disparity in thermal mass of the board layout/loading. This high temperature also ensures that the solder's viscosity will be low, which will help it wet out on pads and leads, which is the essence of the reflow soldering process.

During this step, the component leads should be awash in liquid solder. If conditions of molten solder viscosity and wetting forces on the leads and pads are adequate, surface tension effects will draw component leads into best registration with circuit board pads.

As mentioned previously, the slope—the rise and fall of the reflow spike with respect to time—is critical. If it is exceeded, package integrity can be jeopardized.

43.6.6.4 Cooling.   The fourth and final stage is cooling. It is here that the board's temperature is lowered beyond that of solder liquidus prior to exiting the oven. Once again, recommendations for device heating/cooling ramp rates should be heeded. Most electronic packaging and board materials are reluctant to shed heat quickly. If the board is thick, the laminate, a relatively poor thermal conductor, will remain hot. It must be cooled to below solder liquidus prior to exiting the oven to preclude dislodging of any soldered components. On the other hand, with the trends toward thinner circuit boards, slimmer components, and hot-fast profiles (to keep up with high-volume product demand), it is wise to ensure that thermal ramp requirements are considered for this step also.

As mentioned previously, the solder paste manufacturer will provide a recommended profile. This is a starting point only. The reflow engineer should optimize the profile based on actual product runs. Older textbooks and industrial references cite completeness of fillet formation and solder reflectivity as typical characteristics to examine, but these are subjective measures that may not be true indicators of joint quality. The best measure is tensile pull/peel testing.

If the reflow profile is too hot (i.e., if the solder is maintained at liquidus for too long a time or at too high a temperature), the ensuing solder joints will be brittle owing to intermetallic buildup. These solder joints will be unreliable and will crack if subjected to either mechanical shock or thermal cycling.

43.6.7 Success in Reflow

There are several factors for successful oven reflow:

1. Adequately suitable, maintained, and controlled reflow equipment
2. Good-quality solderable parts
3. Thermally balanced and process-in-mind board designs
4. Reliable solder (including solder paste) and proven thermal profile
5. Good thermometry techniques

43.6.7.1 Adequately Maintained and Controlled Reflow Equipment.   The oven should have small thermal differentials across the width of the tunnel and should be capable of adequate heating and cooling ramps as well as adequate zone separation characteristics.

43.6.7.2 Good-Quality Solderable Parts.   The best reflow equipment, profiles, and solder paste will not make up for inadequacies in part quality (lead form coplanity) or restore solderability. Extended storage should be avoided and parts should be kept cool and dry prior to use. Sufficient quality control methods should be established to ensure that the solder and components are solderable per vendor claims. Plastic molded parts such as PQFPs and PBGAs can absorb moisture from the atmosphere over time. When moisture-laden plastic packages are subjected to reflow soldering, the entrapped vapor may expand to the point where the package fractures, often damaging the silicon die and associated wires internal to the package. This phenomenon is commonly known as popcorning throughout the industry.

Ensure that plastic packaged components are stored in nitrogen-filled bags as supplied by the component vendor until time of use. Once the components are removed from the manufacturer's nitrogen-filled bag, follow IPC guidelines for useful shelf life in the unbaked state and bake-out requirements prior to soldering.

43.6.7.3 Thermally Balanced and Process-in-Mind Board Designs.   It is a must for reliable and adequate reflow to ensure, when possible, that thermally massive components are not relegated to one portion of the board. Also, it is desirable to distribute smaller components rather than to create component fields to permit uniform thermal balance across the board. Components should be spaced adequately to prevent shadowing of smaller devices on the board by larger, neighboring components. This is particularly important in IR ovens without the forced convection option. If the oven is characterized by poor uniformity across the tunnel width, the board should be oriented, if possible, to take advantage of the imbalance; i.e., the board edge with greatest thermal mass should be aligned to the oven edge with the highest recorded process temperatures. Board components should be designed far enough away from board edges to preclude any pin chain conveyor influence on the reflow or damage of bottom-side components.

43.6.7.4 Reliable Solder (Including Solder Paste) and Proven Thermal Profile.   The profile provided by the solder paste vendor is only a recommended starting point that needs to be optimized for the reflow conditions and process time requirements. Also, bear in mind that solder pastes with finer solder particles are more troublesome to flux owing to their greater total surface area and corresponding greater surface oxide volume per unit volume of solder.

It is important to limit exposure to high temperatures and particularly time above reflow temperature. It is in this regime that the intermetallic layer grows most vigorously for most of the common IMCs that accompany and distress the resulting solder joint. The thicker the IMC layer at the interface of a joint, the more likely the joint is to fail.

43.6.7.5 Good Thermometry Techniques.   Understanding the thermal impact of the oven on the board and the board on the oven is critical for a controlled and reproducible reflow process. Thermometry is the only practical method for validating these influences. Some of the most important process nodes benefiting from or requiring adequate thermometric methods are:

- Identifying oven impact to the board as it traverses the oven length
- Establishing heating/cooling ramps and ensuring that they are within the recommended bounds described by the IC and discrete device manufacturers
- Determining peak process temperature and duration
- Considering dependency on the heater types, quantities, cooling methods (whether active or passive), board and component materials, and conveyor speed
- Determining how the heat is distributed along the oven's width
- Determining if center or one edge of board heats adequately for reflow while the other edge or center may not be heated sufficiently
- Determining rate at which the oven reacts to the board's presence
- Mimicking the preferred solder paste profile, including paste manufacturer recommendations

43.6.8 Product Profile Board

All of the preceding and many more variables have dramatic impact on the soldering process. The only way to assess these is to accurately gauge the thermal energy distribution in the oven in the presence and absence of the board. Most process engineers trust their favorite profiling board to determine the health of their reflow oven or assess the performance of a new oven prior to purchase. The profiling board is simply a thermocouple-instrumented product board or test vehicle that is run through the oven. The thermocouples trace the temperature profile of the oven as seen by the lead or package as the board is conveyed through the oven for reflow.

However, reliance on a product board to determine oven performance may mask some inherent problems with the reflow oven. If the board is not thermally balanced in terms of componentry, it may not be able to detect a strong transverse (rail-to-rail) thermal differential. It is for this reason that a balanced oven diagnostic profiling device should be made and run to assess oven performance. That is not to say that the oven diagnostic board is a direct replacement for the product profiling board. The latter is to verify that the critical sectors of the board or, best yet, device leads are heating per set profile. Thermocouples for this board are deployed at the leading edge, trailing edge, center, and two sides, with the thermocouple beads ideally embedded within the solder joints of disparate component types. Additional thermocouples can be placed on unusually massive components as well as on adjacent smaller components to determine if the large component is thermally shadowing its miniature neighbor. The number of thermocouples deployed will be dependent on the package types, known oven characteristics, board thickness, board layout complexity, etc. The following sections describe good thermometry practices and the oven diagnostic board, as well as some of the thermocouple measuring innovations that are now a mainstay of large-scale mass reflow manufacturing operations.

43.6.9 Oven Diagnostic Board

The oven diagnostic board can be a very simple or a rather complex device. Its job is to probe the normal operating conditions of the oven and determine if all the heaters and circulating fans (if installed) are working to expectation. It is useful in assessing not only the oven's time-temperature profile parallel to the tunnel's long axis, but also perpendicular to it at the rail edges. The profiles recorded by this board will not necessarily be translatable to the establishment of a product time-temperature profile. The diagnostic board, whether a product board or a separate test board, is balanced in design. It is designed to preclude thermal imbalance, high thermal conductivity, and high thermal mass. These qualities may profoundly affect oven performance and diagnosis of its operating behavior. For this reason, materials are chosen that correspond to these properties. The ideal material will be thermally insulative. Attached to this are thermocouples mounted top-side and bottom in an evenly spaced array across the width of the material. They should be mounted close enough to the conveyor edges to determine how that region is influenced by such things as the conveyor rails or the lateral ends of the heater assemblies. Thermocouple beads should be identical in size and spacing from the insulating material. Furthermore, they should also be mounted slightly off the insulating material such that the heated oven environment can freely swirl around them. The result of measuring with such a device will be the operational condition of the oven, a true picture of the oven's thermal environment.

43.6.10 Printed Circuit Board Thermometry

Knowledge of good thermometry practices and thermocouple use is key to acquiring accurate board time-temperature profiles and oven thermal performance information.

43.6.10.1 Good Thermocouple Practices. Thermocouples are nothing more than two atomically disparate metals or alloys in contact with one another. The contact is generally accomplished in one of two ways: twisting the wires tightly together or welding the two. Owing to the thermoelectric effect, when one thermocouple is tied to another held at constant temperature or electronically compensated, the EMF generated by the electronic difference between the two results in measurable voltage that is temperature dependent.

Although thermocouples can be simple to make, as shown in Fig. 43.17, they can also work against the reflow engineer. As previously mentioned, it is only necessary for the two halves of the couple to touch one another to be useful for thermometry. If the two halves contact inadvertently, thermocouple output will be related to the temperature at that accidental juncture. This temperature may be vastly different from what is being experienced at the far end of the thermocouple pair where one is trying to measure temperature. A thermocouple wire pair should be welded to help preclude noise-induced measuring errors from intermittent contact. The two halves of the thermocouple pair should touch at only one point—that closest to the area desired for temperature measurement. Avoid the twisted pair.

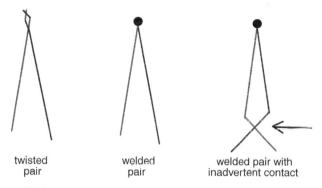

twisted   welded   welded pair with
pair   pair   inadvertent contact

FIGURE 43.17   Diagram of welded pair vs. twisted pair, and welded pair with inadvertent contact.

The welded pair has a significant advantage in that the resulting weld bead can be made small and uniform in shape. As a rule of thumb, the bead diameter can be made as small as about 1.5 times the diameter of the individual conductors that compose the thermocouple. The temperature will be sensed only at the weld bead. The twisted pair output may not be so unambiguous and is rarely as small. Regardless of whether you are making your own welded bead assemblies or buying them, always check them, with a magnifier or microscope if necessary, to ensure that there are no other contacts between the legs of the thermocouple wires other than the bead. During manufacture of a thermocouple assembly, it is usual practice to twist the thermocouple wires together prior to welding. Upon welding, the twisted material will fuse and melt back. It may be necessary to untwist the pair just behind the bead. Of course, this is risky because many of the thermocouple materials are brittle before welding and are even more so after. On the other hand, any twist below the bead will have the greatest influence on the measured temperature, regardless of how well the bead is formed or positioned.

The most common thermocouples used for mass reflow thermal assessment are of the K type. This is a Ni-Cr/Ni-Al (chromel/alumel) couple and has a thermal measurement range best suited to the most commonly used solder systems. There are two main devices in use for measuring the output of the thermocouple. The most primitive is the electronic (digital) thermometer with output tied to a chart recorder. The other (much more sophisticated) device is the reflow profiler or tracker, an electronic package that can not only output the time-temperature plot of multiple thermocouples but can also make certain assessments of the data, such as:

- Heating ramp rate
- Time below liquidus
- Time above liquidus
- Cooling ramp rate

Tracker systems are computer-based and can generate detailed reports and graphs of a reflow cycle.

43.6.10.2   Thermocouple Deployment.   The size and deployment of the thermocouple is important, especially when trying to gauge the temperature of very small solder joints. There are many opinions and methods for securing the thermocouple bead to the PCB. Among those that have been suggested are:

- Adhesives
- High-temperature solders
- Tape
- Pressure contact

Recall that the goal is to ensure that all lead/solder/pad combinations are being exposed to enough thermal input for sufficient time to effect robust solder joints. Therefore, thermocouples adequately positioned on a product profile board should be able to monitor this process, ensuring an adequate profile. But there are numerous interferences that can preclude this from happening. Many items on the preceding list are, in fact, not adequate for the job. By examination of those suggested thermocouple hold-down techniques and materials, it is evident from a fundamental chemical and physical perspective that many on that list could be problematic.

As in every other facet of engineering, one should keep in mind the end goal and the fundamental phenomena that are operant in the process—even in a process as simple as thermometry of a circuit board. In fact, the board temperature is only one piece of the equation. Most important are the solder joints of the most demanding and disparate components on the circuit board. Let us start with the assumption that understanding the thermal behavior of a potential solder joint during the reflow process by thermocouple thermometry requires that certain criteria be met:

1. The thermocouple bead must be exposed to the same thermal conditions as for the solder joint reflow process.
2. There can be but one junction between the two legs of the thermocouple and that junction is in contact with object(s) to be measured.
3. The measuring technique must not interfere with the outcome of the measurement.

From the first assumption, it should be obvious that the thermocouple bead must be in the vicinity of the lead/pad combination. Where better than between the device lead and the circuit board bonding pad? After all, that is where the solder joint will result. The second has been described previously but should always be checked before and after measurement. In keeping with the third assumption, there are many factors that can confound the solder joint temperature measurement process, regardless of whether the thermocouple bead is properly deployed between lead and pad. Among these are the size of the bead and the thermal mass of the thermocouple assembly. In general, use fine-gauge thermocouples such as 20 to 36 AWG. See Table 43.1 for examples of wire size and bead diameter.

Also strive for the shortest possible length of thermocouple wire pair that is practical for the application. This is discussed later. The finer gauges such as 30 AWG will allow insertion between lead and pad even at very fine lead/pad pitches. Mechanically flattening the bead

TABLE 43.1   AWG vs. Wire Diameter and
Anticipated Thermocouple Bead Diameter

AWG	Wire diameter (in)	Bead diameter (in)
20	0.032	0.048
24	0.020	0.030
30	0.010	0.015
36	0.005	0.0075

with a small hammer, vise, press, or smooth-jaw pliers will enhance its placement between lead and pad.

If an adhesive is to be used to immobilize the thermocouple bead, its characteristics must be taken into account. It must be able to survive the reflow cycle. Were it to release the thermocouple bead during the process, then measurement data would be useless. Most adhesives will not tolerate the >200°C regime of a typical reflow cycle. Some of the UV curables will. Most of the fast-setting epoxies and adhesives (methyl methacrylate super glues) decompose below that temperature, or at least soften to the point that they could not be trusted for bead hold-down. Furthermore, it is important that, if an adhesive is used, its thermal transfer properties must be understood. Some are characteristically insulative and inhibit exposure of the thermocouple to the true thermal environment. The best-case adhesive would be temperature compatible and filled with a fine, thermally conductive material such as some metals or ceramics. There are few of these, though, that are both commercially available and affordable.

43.6.11   Reflow Profiler or Tracker

In tracker or profiler systems, the thermocouples on the circuit board are attached to a battery-powered, thermally insulated electronics box, the tracker, which accompanies the circuit board through the reflow cycle. Once a selected threshold temperature has been reached, such as 30°C, the tracker begins recording the experienced thermal environment per unit time. It samples and records the circuit board's thermocouples at short, programmable time intervals (several times per second) for the entire trip through the oven's tunnel. When the tracker emerges from the exit end of the oven's tunnel, it is removed from its protective thermal barrier and connected to the host computer. The data stored in the electronic tracker can be conditioned, displayed, reported, and printed for analysis. The user defines the reporting format, including the type of data sought, thermocouple plots, and other relevant, run-related data. One system is even able to output its data telemetrically as it is being recorded in the oven. Some systems are capable of making predictive corrections to the profile based on current measured conditions and recorded oven settings. This can further enhance the ease of adjusting a complex multizoned reflow oven to meet the requirements of a particular job. Whether simple or complex models are selected, these systems are invaluable for accurate reflow profiling and process repeatability.

When an electronic tracker is used, its presence in the oven may have an impact on the resulting profile data. It may have an effect on the local aerodynamics within the oven. The tracker has thermal mass. Some trackers are covered with insulative blankets, while others may have a stainless steel cover with insulation beneath. In either case, one should determine empirically what influence the tracker has on the oven performance and measured reflow profile. To do this, use long thermocouple wires and, once the oven has warmed to a steady-state operating condition, run the profile tracker with the tracker a fixed distance behind the board.

Note that the tracker is usually placed to follow the board in reflow. Were the tracker to precede, it might cause the heaters to ramp up as the oven attempts to compensate for the thermal mass of the tracker. Although that is what happens during a reflow cycle as a cooler board enters a heated zone of the oven, the oven would attempt to compensate for the board's thermal mass. The added thermal mass or disruption of the air flows from the profile tracker would not be a factor in the normal reflow process.

Once the profile is recorded, shorten the distance between the tracker and the board and once again record the oven's performance profile and compare it to the first in this series. Keep repeating this until the distance between the tracker can be standardized, adding some length as a safety factor.

43.6.12 Atmospheres for Reflow

The most common reflow atmosphere is ambient air. It is either passively entrained into the reflow tunnel or is heated and blown into the tunnel as in forced-air convection ovens. Air, especially at high temperatures, will cause most metals to oxidize, so the soldering flux in the system has to work very hard to remove native oxidation on the lead, pad, and solder as well as to prevent further oxidation at all phases of the reflow cycle until all elements of the joint (pad, lead) are wetted by the solder. Moist air is more of an oxidizing agent than dry air. Another item that can be subject to oxidation is the flux itself. Oxidation imparts a variability to the soldering process that is difficult to control except with the strongest solder-fluxing formulations. These highly activated fluxes are not generally used for electronic manufacturing. They are linked to reliability problems of the final electronic assembly; the main problem is corrosion from flux residues that have not been properly removed from the circuit board.

Were one to control the level of oxygen in the reflow process, weaker formulations of solder fluxes could be used with better success. Less oxidation would occur during the process. The flux could work on the oxides native to the metallurgy and would not be required to tackle new oxidation as a result of the heating process.

In many classical soldering texts there is discussion of how reflowing in an inert environment, such as nitrogen (N_2), can have profound effects on wetting. These texts describe the effect as surface tension related. This is correct in a sense but seems to be widely misunderstood in the soldering community, because it is not that the surface tension of liquid solder varies much between an 80 percent N_2 atmosphere, as in air, and a pure (100 percent) N_2 ambient. Instead, the solder, when exposed to a mixture of N_2 and O_2 (~20 percent), rapidly develops a skin of oxide. This encapsulation has a profound effect on the underlying solder, preventing it from wetting and spreading unencumbered. It is not the same as water and ice having vastly different surface tension values. That is a one-material system. In the case of oxidized liquid solder, it is a two-material system. It is true that, even as the solder flux is working, there are rafts of predominantly tin-rich oxide afloat on the solder's surface. These can inhibit the solder from spreading and can even insulate it from wetting to adjacent metal surfaces.

In summary, the greatest effects of soldering in an inert or reduced oxygen atmosphere are as follows:

- The prevention of further oxidation of the metal component of the soldering system
- The elimination of oxidation of the flux itself
- Better fluxing action, resulting in cleaner parts
- More thorough fluxing and dissolution of metal oxide rafts on the surface of the solder
- Lower level of residues when using no-clean solder paste formulations
- Better compensation for misalignment due to higher wetting forces
- Reduction of solder ball formation
- Fewer voids in the solder joint
- Less board discoloration

This last item is perhaps of little value, as board discoloration is rarely a concern. It used to be of much more concern in the older IR ovens, where occasionally boards would char.

Unfortunately there is ambiguity in the reports as pertains to the reflow parameters, inerting levels, impact on manufacturing costs, and resultant soldering yields. There is, however, universal agreement that N_2 betters the soldering joint formation process. Given the number of area array devices in use today (BGA, CBGA, CCGA, CSP, etc.) and the difficulty and expense of their inspection and repair, the addition of nitrogen provides added insurance for best solder joint formation. It is a natural fit with a high-yielding no-clean soldering strategy.

Forced-air convection is, in a sense, the best and worst circumstance for reflow soldering. By far, it is the most uniform approach to heating a circuit board and has rapidly gained status as the preferable machine. But it is the very nature of this reflow method that detracts from the process. As a metallurgical system is heated in the presence air, it is more prone to oxidation. This method delivers a continuously replenished supply of air blown at the metal surfaces and the circuit board at high velocity. This ensures maximum oxidation of the soldering system elements. Use of a nitrogen atmosphere will counter this.

Forced convection systems are also notably hard to inert. In this type of oven, when specially configured for nitrogen convective recirculation, fresh air is not blown in. Instead, nitrogen is metered in, lowering the oxygen content of the oven's atmosphere. The inherent turbulence internal to the oven can enhance entrainment of room ambient air. Special precautions are taken to prevent this from happening. Most oven manufacturers have inert reflow oven models, each with their own special inerting improvements installed.

Oven mass reflow will continue to dominate surface-mount manufacturing for many years to come. The sophistication of today's machines appears to be sufficient for future needs.

43.7 WAVE SOLDERING

Once the predominant method for mass assembly of circuit boards, wave soldering has taken a back seat to oven-based reflow. The rise in popularity of surface-mount, along with a greater variety of surface-mountable packages, especially in the finer-pitch range (≤ 1.27 mm [0.05 in]), has displaced much of the work done by this method. Nonetheless, through-hole componentry persists and mixed-mount (surface-mount plus through hole) still may be the only alternative for some assemblies. It is unlikely that wave soldering will disappear from PWB manufacturing.

Wave soldering utilizes a reservoir of molten solder pumped and circulated to form a standing wave. The circuit board is prepared with devices for wave soldering in one of three ways:

1. Coarse-pitch surface-mount components, especially passive devices, are affixed to the bottom side of the PWB using surface-mount adhesive that is cured prior to wave soldering.

2. Solder-tailed components such as connectors, PGAs, or other through-hole devices are slipped into plated through-holes from the top side of the board.

3. Solder-tailed components such as axially leaded devices are inserted from the top side of the circuit board and the leads are clinched on the bottom side of the board.

The board is placed on a motorized, edge-hold conveyor where it is fluxed, preheated, and skimmed over the crest of the solder wave. Only the bottom of the circuit board is exposed to the molten solder (Fig. 43.18). The surface-mount devices, held by the adhesive, pick up solder on their metal contacts. The solder bridges from the contact to the metal pad on the bottom side of the PWB. The through-hole parts, clinched or unclinched, pick up solder. The molten solder is drawn by capillary action between the lead and the PTH barrel. If the barrel and lead are hot enough and well fluxed, the solder will fill the barrel and wick up to form fillets on the top side of the component lead. As the board continues past the wave, it cools and the solder solidifies, completing joint formation. Figure 43.19 shows a cross section of a through-hole pin that has been wave soldered.

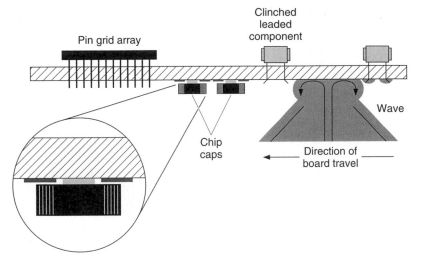

FIGURE 43.18   Diagram of combined SMT and through-hole board going over wave, with detail of glued chip capacitor and clinched leaded component.

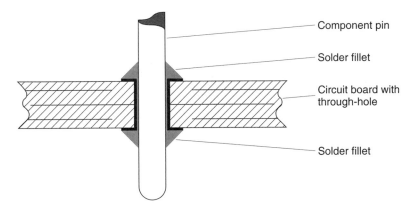

FIGURE 43.19   Cross section of through-hole solder joint.

43.7.1   Types of Wave-Soldering Systems

There are many types of wave-soldering systems, each with its unique advantage as claimed by the manufacturer. The soldering engineer has to assess these improvements as they relate to the type of assembly being soldered. The technology has matured significantly but the degree of equipment complexity attests to the complexity of the process. There are many process variables associated with this operation. If these are not understood or properly controlled, wave-soldering defects such as skip soldering (electrical opens) and bridges (electrical shorts) will occur as in any other soldering operation. Another increasingly important defect is the influence of side 2 wave soldering on side 1 surface-mounted components. It is possible to re-reflow those side 1 SMDs, inducing opens or solder-starved joints. This is most commonly associated with thin, densely populated, double-sided PWBs with fine-pitch surface-mount devices such as quad flat packs and BGAs that are soldered on the opposite side from the wave. Thermal conduction from the wave through vias and along electrical traces in and on the board provides enough heating to cause previously soldered parts to re-reflow. Solder can be drained away or wicked up side 1 surface-mount component leads or down traces and through-vias. This can result in an open or weakly soldered joint with an hourglass-shaped attachment of solder from the bottom of the lead to the bonding pad.

43.7.2   Flux Application for Wave Soldering

One of the key subprocesses of wave soldering is the application of flux. There have been many developments in this area. The most important criterion is the uniform application of a sufficient quantity and activity for the job at hand. Short preheat and liquidus times, characteristic of the wave soldering process, may not meet the time-temperature requirements of some fluxes. This is particularly the case with no-clean fluxes, as they are generally the weakest of fluxing agents and may not have adequate time during the wave process to work at removing the metal oxides.

The flux must be heated to an ample temperature to permit best reactive conditions without drying it out or denaturing it. For the sake of economy, the thinnest application practicable is necessary such that the board will have sufficient flux to remove innate oxides on surfaces to be soldered but will be limited in quantity to minimize the remaining residues, which may be difficult to clean. In the case of no-clean fluxes, thick applications of flux may leave unacceptably dense residues that may be difficult to penetrate when the boards are probed at the electrical testing step.

Flux quantity may also be of concern in the wave soldering process for yet another reason: fire hazard. Flux-laden boards are preheated going into the wave. If the flux application is too heavy, the flux may drip onto preheater elements. This may cause the flux to volatilize rapidly, combine with oxygen in the atmosphere, and provide the right conditions for flame initiation. Even if there is not direct exposure of the liquid flux to preheaters, if the quantity of volatile, flammable components is high enough in the vicinity of an ignition source, then an explosive condition may develop.

43.7.3   Flux Application Techniques

Foam and wave fluxing have given way to spray fluxing as the predominant flux application method. All three are discussed here.

43.7.3.1   Foam Fluxing.   Foam fluxing is done by flowing and aerating a stream of liquid flux through a porous metal nozzle, a fritted glass, or porous stone. The nozzle, also called a chimney, shapes the flow of the aerated flux stream. The board to be soldered is run over the flux foam and then heated to activate the flux before reaching the solder wave. As it moves past the solder wave, the solder wets to solderable metals and solidifies to complete the soldering process. Foam fluxing is particularly effective for soldering of plated through-hole assemblies. The foam, drawn into the plated through-hole barrels, thinly and uniformly fluxes the walls of the plated barrel, component leads in the barrel, and any surface-mounted discrete components that are in contact with the foamed flux.

43.7.3.2   Wave Fluxing.   Wave fluxers work much as the solder wave itself. The printed circuit board is moved over a standing wave of solder flux. The height of the wave and depth of the board penetration into the wave are adjusted to allow for proper flux application thickness. Capillary action, as in all the fluxing methods for through-hole components, draws the flux into the space between the component lead and the barrel. As with the foam-fluxing technique, it is difficult to control the amount of flux delivered. For the most part, flux applied for wave soldering is largely removed by the turbulence of the solder wave; however, it is possible to bake on the flux if the preheaters are set to too high a temperature. This may impede fluxing and the residue can even mask otherwise solderable contacts from the solder during initial contact with the wave.

43.7.3.3 *Spray Fluxing.* Spray fluxing has gotten to be an accurate technology. Precise amounts of a low-solids flux can be applied generally to a board or can be selectively delivered to small areas of the board to be soldered. One of the difficulties associated with this technique is the formulation and consistency of the flux. Very volatile flux solvents must be used to thin it sufficiently for spray application. Therefore, flux formulation and maintenance is much more critical. Fluxes meant for this method of deposit may be slightly more expensive than those for other flux deposition methods. Spray fluxing is also messy, due mostly to the airborne mist and condensed volatiles that deposit wherever the air or process gases carry them. Fire hazard is also maximized. Ultrasonic methods minimize flux volumes consumed in manufacturing.

43.7.3.4 *Fluxer Maintenance.* As fluxes are exposed to the atmosphere, they are vulnerable to evaporation of the volatile constituents. Of course, this is enhanced in systems such as wave or foam fluxers where the flux is open to the atmosphere or processing environment, has significant exposed flux surface area, and is constantly recirculated. This necessitates monitoring and maintenance of the flux. Although some automatic systems are now available, most require routine measurement and adjustment of the flux's specific gravity. A hydrometer is generally used for specific gravity determination. Flux thinner must be added to restore the specific gravity to compensate for evaporative losses. In addition, the volume of the flux in the system must be adjusted to the proper level. The fluxer must be maintained to prevent impact on process yield. The flux manufacturer will provide information as to the target specific gravity per flux and will also recommend an appropriate thinning formulation.

It is wise to periodically empty the flux reservoir, clean it thoroughly, and refill with a fresh charge of flux. Sometimes the flux will develop a polymeric residue that will change its surface tension characteristics or clog nozzles or dispensing pores in foaming systems. In addition, it may become contaminated with debris carried in by the PCB to be soldered. These, too, will impact assembly quality. Since the plated through-hole relies on a minute capillary to be filled first by flux and ultimately by solder from the solder wave process, any small particulates entrained in the flux—or the solder, for that matter—may be forced into the interstice between the through-hole component lead and the through-hole barrel, thus inhibiting solder filling of the barrel. Also, while the flux reservoir is empty, it is a good idea to inspect it to ensure that the materials of construction are holding up to the rigors of system operation and prolonged contact with fluxing agents that may be corrosive in the long term. An inspection of the materials of construction for the entire system should be made prior to committing to the purchase of a wave-soldering machine. Unfamiliar or untested materials should be avoided unless there is sufficient literature, test results, or experience to assure compatibility with fluxes that will be used in that system.

43.7.4 Preheating

As previously discussed as regards the fundamentals of soldering, heat is a significant component of the soldering process from three points of view:

1. The heat must be sufficient to melt the solder.
2. Materials (components and PCB) must be hot enough to permit alloying of the solder to the board.
3. The flux must reach a high enough temperature to allow its activation, reaction, and disruption of oxides and tarnishes on the metals of the materials to be soldered.

This last step, PCB/flux heating (part of the preheating stage), is exceptionally important in the wave-soldering process. A component's resident time in the solder wave is brief; in fact, its time at liquidus is 10 to 30 times shorter than for joint formation in a comparable oven reflow soldering cycle. By the time the circuit board and associated componentry hits the wave, it must be fluxed sufficiently to allow adequate solder joint formation. The preheater is responsible for this in the wave process. It brings metal surfaces up to a temperature at which the flux can start its work to remove the oxidation, which can perturb soldering. There must be enough flux left on the circuit board after preheating to protect the newly fluxed surfaces to permit soldering when the assembly reaches the solder wave. As for any step in circuit board soldering, there are several process choices, although the method used for preheating will be dictated by the equipment and available options. Often the type of equipment purchased will limit types of preheaters available. Preheating is also critical in the avoidance of component cracking due to thermal shock when components contact the molten solder wave. In the case of thermally massive printed circuit boards, if not sufficiently preheated, they may soak up so much heat from the wave that rapid, localized solidification at the wave's surface may occur. This can cause widespread solder shorts because it affects the surface tension–defined meniscus from wave to board.

Just as in reflow soldering systems, there are two systems in prevalent use. These are radiant preheaters (direct and indirect IR) and forced-air convective preheaters. Both are effective, although the advantages of the latter are significant in terms of uniform heating of the components and board materials. An in-depth discussion and comparison of these two heating methods can be found in Sec. 43.6.1.3. Additionally, some wave-soldering machines permit both top- and bottom-side preheaters. This can be advantageous for thermally massive boards. As in all cases of mass soldering, the use of profile boards is encouraged to ensure that critical areas of the board are maintained at a proper temperature for the duration of the process. Top-side SMT component solder joints must be kept well below melting point. At the same time, PTH barrels and PTH components must surpass liquidus to guarantee sufficient wetting and capillary rise of the molten solder to yield sound solder joints. With the complexity of today's mixed-mount boards, this is more of a challenge than ever.

43.7.5 The Wave as a Process

There are numerous solder wave styles being touted. A detailed discussion of each will not be presented here, but a basic overview of the process in relation to the wave will be presented. As previously mentioned, the molten solder is pumped to form a standing wave. This is accomplished by an impeller on the bottom or side of the solder reservoir. Once the solder is molten, the impeller motor is activated and the solder wells up between baffles and nozzle that reside within the solder reservoir, dictating overall wave dimensions. The nozzles and baffles are generally adjustable, as are impeller speed, molten solder temperature, board introduction angle, and board conveyor velocity. These, along with preheater settings, define the profile parameters or process variables that must be tamed to accomplished high-yield wave soldering.

Since the solder is a molten, turbulent liquid, it makes intimate contact with the underside of the circuit board. Therefore, thermal uniformity is generally not difficult to control as long as the wave contact area with the underside of the board is uniform. But hot solder is prone to rapid oxidation at the air-liquid solder interface. Although the wave is in constant motion, the solder is actually flowing beneath a stationary film. Thin and plastic, the skin is composed chiefly of tin oxide but also contains oxidized lead and other contaminants. These impurities that form or float to the surface in the reservoir are broadly encompassed by the term *dross*. The skin has a beneficial aspect in that it helps to limit oxidation of the constantly recirculating wave. When adjusted properly, the board meets the crest of the wave, disrupting this skin. In doing so, the fluxed components and board are immersed in the flowing, oxide-free molten solder. The solder will wet to these component leads, and solder joints will result on exiting the wave and cooling below the solder liquidus temperature.

Solder wave manufacturers have devised a number of options that may be helpful to the process. The gas knife—more commonly called the air knife—directs a high-velocity stream of heated air or nitrogen at the bottom side of the board directly after the board emerges from the wave. It can be effective in relieving solder bridges from tight interstitial pin fields or closely spaced passive device fields. When used with air, this feature will increase rate of drossing. Also, if not set properly, it could have an effect on the process. If the angle is wrong and the velocity is too high, the gas knife could disrupt the solder wave. If it is set too cool, the knife will have little effect or can even encourage solder bridging. Another accessory imparts high-amplitude sonic waves into the solder wave. This can help pump the solder into the plated through-holes and increase wetting, especially on thick printed wiring boards. Care must be taken as very high amplitudes may pump so much solder up the barrel that it may result in top-side solder bridging.

43.7.6 Dross

The more the system is used to solder boards, the faster the contamination and dross buildup. The dross is of concern from three points of view:

1. *Economic.* Dross is solder lost from the manufacturing process. In high-volume manufacturing, it can mean hundreds of dollars of lost solder per week per machine. The dross can be returned to some solder founders for recycling, however.

2. *Process.* Excessive dross on the surface of the solder can disrupt normal wave dynamics. Since tin oxidizes more easily than lead, the solder can become tin depleted over the long term. This is known as tin drift.

3. *Hygienic.* Airborne tin and lead oxide fumes are not healthy to breathe. Risks associated with lead oxide intake are well documented, but inhalation of any particulate should be considered potentially hazardous. Precautions should be taken, especially during system maintenance procedures, to preclude health-related problems. Donning a personal particulate mask and washable or disposable outer garment (contamination suit) is recommended. So too is proper hygienic venting of the work area, not only during maintenance procedures but for normal soldering operation.

Molten solder droplets can become entrapped in the dross as a result of wave turbulence and mixing of the oxides with the liquid solder. Once oxidized, they are unable to rejoin solder in the reservoir. This thickened layer can impact wave-soldering results, especially if rafts of dross are entrained in the wave. They interfere by blocking the solder wave from effectively contacting land and lead, and opens result. They can also change the dynamics at the wave-land-lead interface, discouraging adequate pull-back of the solder and encouraging solder-induced electrical shorts (bridges). Various schemes have been devised to tame the dross, such as co-mixing the pumped solder with a mineral oil, which floats to the surface, blanketing the solder from the atmosphere. Liquid reducing agents can be added to the solder as well as fluxes. In the long term, none of these has proven popular. Inerting is effective in reducing dross formation and ensuring best fluxing from no-clean or other weakly activated fluxes. Nitrogen as a cover gas has become the most effective method of dross management.

Iman et al.[1] have explored the use of formic acid in gaseous form as an additive to the nitrogen atmosphere for the soldering process. The nitrogen-formic atmosphere not only prevents dross formation, but also thins and removes the oxide on the parts to be soldered and on the surface of the wave itself. In that work, formic's action was supplemented with adipic acid, a weak, water-soluble organic acid serving as a liquid flux. The fundamentals of this process were demonstrated by Hartmann,[2,3] who had shown formic acid to be an effective gaseous fluxing agent for soldering.

43.7.7 Metal Contaminants

Metal contaminants can also have an effect on wave soldering. They stem from two sources: (1) contaminants from the bar solder used to charge the solder pot, and, to a greater degree, (2) from the soldering process itself. As the circulating wave washes over component leads and PCB through-hole lands and pads, there is a leaching of their materials into the wave. Even if leads and pads are solder plated or coated, there is the opportunity for adulteration of the solder in the reservoir, dependent on coating composition, thickness, and underlying basis metals. Copper, gold, silver, additional tin or lead, and intermetallic compound precipitates are all common contaminants derived from the slight dissolution of lead and pad or coatings thereon during soldering. Of course, their contamination level is very small on a per-board basis, but in high-volume manufacturing the quantities can rapidly rise. Even in low-volume applications, if the solder pot contents are not changed frequently enough, the solder and resulting joints can be of inferior quality. Solder pot contamination can have significant process impact, eventually altering the liquidus temperature or melting range of a solder. This may lead to an increase in shorts or opens. It may also result in brittle solder joints. The composition of the wave reservoir can be assayed by a testing service to determine its impurity content. A small sample of the solder can be scooped out to check its solidus and liquidus points (melting range), although this is a less accurate method for contamination assessment and will not be adequate for assaying some impurities such as intermetallic precipitates.

Excessively high temperatures in the solder reservoir and low speeds through the wave should be avoided to limit dissolution of lead, pad, and through-hole metals. Note, though, that the solder wave is maintained at fairly high temperatures, often above 250°C. Board exposure is short (2 to 8 s). In fact, the board generally experiences less thermal impact than in the oven reflow process, where the entire board is typically maintained at solder reflow temperature for 30 to 120 s. As in oven reflow, thermal shock in wave soldering can lead to component cracking or degradation problems. So the preheating rate should be tempered such that the maximum slope corresponds to that recommended by the component manufacturer, often in the range of 2 to 4°C/s.

Bar solder impurities such as aluminum, gold, cadmium, copper, and zinc can increase surface tension of the solder and make the process more prone to bridging. Bernier[4] reviews some of these contaminant effects and describes empirically derived impurity limits.

43.7.8 Design for Wave Soldering

In many plants, defect levels at the wave step are now higher than those for the oven mass reflow process. Although wave soldering has been around for a long time, it is still not very well understood, due mainly to the various machine configurations and number of process variables. There are numerous wave designs available from the various wave-solder machine manufacturers. There are wave machines that provide multiple smaller, turbulent wave(s), which are best for leadless components such as surface-mount chip resistors and capacitors. Smoother-flowing waves are recommended for leaded components and through-hole as well as coarse-pitch surface-mount devices. Wave dynamics are dictated by process values as well as the materials in contact with the wave. As the solder wets to the circuit board materials, solder wetting contact angle and solder viscosity will impose wave peel-off characteristics, as shown in Fig. 43.20. In extreme cases, the wave crest can be caused to collapse on itself or snap back and disrupt the process, promoting unwanted solder bridges (shorts).

Hot-air knives—jets of hot air directed at the wave-air-solder interface on the bottom side of the board—are sometimes deployed to help the solder drain from the components and separate from the board. This discourages solder bridge formation between adjacent leads/pads.

Pin fields in components such as pin grid arrays (PGAs) and connectors have grown large and pitches smaller. The introduction of the interstitial pin grid array (IPGA), also known as the staggered pin grid array, has caused a reexamination of wave dynamics. This device has

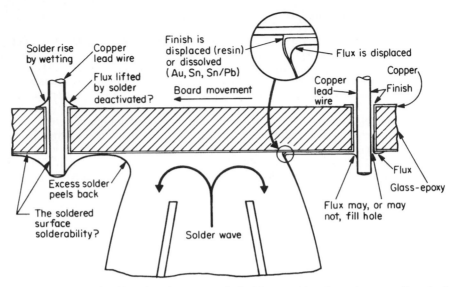

FIGURE 43.20 Printed board passing over a typical solder wave. Note the tendency to pull a web of solder from the wave as the board passes over. *(Source: Alpha Metals, Inc.)*

pins positioned interstitial to the normal field of a PGA, as shown in Fig. 43.21. As the wave moves through this frustrated pin field, areas of fluid flow stagnation may occur and opens or insufficient solder may result. In contrast, the IPGA may not drain solder well enough in other areas, causing solder bridge formation. There are numerous wave designs that are aimed at improving the soldering yields of these large packages, but the final choice is not yet clear.

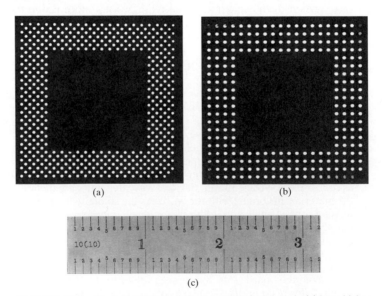

FIGURE 43.21 The interstitial pin grid array (IPGA) package in (*a*) has a higher pin density than the orthogonal PGA, which frustrates solder flow, making the array more prone to opens and bridges and requiring a higher degree of process understanding and control.

In designing circuit boards for the wave-soldering process, it is important to anticipate the flow of the wave over the board. It is wise to avoid placing tall components in front of short-component fields. Component spacing should be maximized. In both cases, the problem is the occlusion of solder from the wave. The taller component may shadow the flow of the wave, causing areas of flow eddying in the vicinity of shorter components behind it. Also, via placement should be as far away from the leads as is practical to avoid the aforementioned problems of re-reflow. This phenomenon has been known for a while but only recently reported.[5,6] Component spacing from the edge of the board will be dictated by wave flows and surrounding structure. There are no hard rules here, only hard lessons. Each machine is different, just as every wave has its own peculiarity and every PCB its own unique topography. System familiarity and experimentation with test boards are required to determine the efficacy of various circuit board design features and their impact on the wave process.

Kear's review of the thermal aspects of through-hole solder joint formation offers a glimpse of the physical phenomenon of the wave-soldering process.[7] A number of papers and patents have recently been written regarding this development.[8-11]

In some cases circuit boards are too large and flimsy to travel safely through the wave solder process, further softening as T_g is reached. In other cases, certain bottom-side areas of the board must be shielded either from the heat of the solder wave or from the solder wave itself. These situations warrant the use of a wave solder pallet or shield. The pallet is nothing more than a masking device into which the board is nested. There are cutouts that expose those areas in need of soldering to the solder wave (see Fig. 43.22, A–D).

BGA via dispersion patterns on the bottom side of the board provide excellent thermal conduction from the solder wave to the top-side solder joints. This can cause re-reflow of

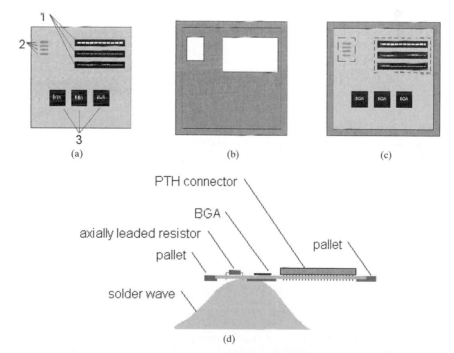

FIGURE 43.22 (*a*) Circuit board with PTH connectors (1), PTH axially leaded resistors (2), and surface-mount BGAs (3). (*b*) Pallet with cutouts for PTH components. It also shields BGA via pattern and other components and features on the bottom side of the circuit board. (*c*) Circuit board nested in pallet. Dashed lines indicate areas in the pallet cut out to expose bottom side of the board to the wave. (*d*) Masked board passing over solder wave showing open and shielded areas.

these solder joints. Area array devices such as BGAs do not offer the opportunity for convenient inspection for re-reflow. Repair is also difficult. For these reasons, shielding these areas from the wave is the safest way to ensure that SMT solder joints will stay cool enough through the wave solder process to remain intact.

It is important that pallet cutout clearance is large enough that it does not frustrate the wave and preclude good bottom-side coverage of the solder in critical areas requiring soldering. Also, when components requiring soldering are too close to the pallet material, the pallet itself may act as a thermal sink and the would-be joints may not get hot enough to effect good wetting.

Pallets are generally made from high-temperature epoxy-glass composites such as Delmat®. Although it is difficult to machine, there are shops that specialize in creating wave solder shields. Pallet materials can be ESD compatible, so check product needs before fabrication. Include features to lock down the PWB during wave soldering. If none are included, the PWB will float up when contacted by the wave. Include enough areas of support so that a thin, large PWB will not sag as it traverses the wave. Over-clamps can be built in to hold connectors and other nonclinched parts from buoying up when contacted by the wave.

Some other areas to consider when designing for wave soldering are the following:

- Complex PCAs with BGAs and fine-pitch SMT components on boards thicker than about 0.093 in may not be wave solder compatible. The heat required for wave soldering thick, high-density PCAs may result in re-reflow of the top side of SMT components or insufficient capillary rise within the PTH barrels.

- Carefully mask off gold fingers, plated through-holes used for press-fit components, and hand-solder or hand-assembly areas of the board. Gold fingers, once exposed to solder, are rendered useless. Use pallet, Kapton tape, or peelable spot solder masks to protect these areas from contact with the solder wave.

- Solder in press-fit plated through-holes is difficult to clean sufficiently for proper press-pin process. Use pallet, Kapton tape, or peelable spot solder masks to protect these areas from contact with the solder wave.

- Use appropriate heat reliefs around vias internal to the board. The IPC has recommendations regarding the size, shape, and application of these heat relief features.

43.8 VAPOR-PHASE REFLOW SOLDERING

Until very recently, vapor-phase reflow was popular but not in as widespread use as oven reflow soldering. Because of safety and environmental concerns and compliance with the Montreal Protocol for the reduction of ozone-depleting chemicals, this soldering technique has fallen out of favor. Due to this diminished status, vapor-phase reflow is covered only in abbreviated fashion.

43.8.1 Basic Process

As in any other reflow technique, the board must be supplied with sufficient solder for joint formation. Most commonly, solder paste is stenciled onto circuit board pads. Components are placed onto the solder paste in preparation for reflow. The board is conveyed into the reflow chamber where it is exposed to the vapor phase of a boiling liquid. This liquid is inert with respect to the solder and board. It is a dense synthetic of high boiling point (slightly higher than that of solder liquidus but not high enough to damage the circuit board or components). Chlorofluorocarbons, hydrochlorocarbons, and hydrochlorofluorocarbons, some of which cost several hundred dollars per pound, are used for this purpose. In batch processors, a CFC is used as a cover to retard evaporation of the very expensive reflow medium. When conditions are optimized, the hot vapors begin to condense on the cooler PWB, heating it. As the process progresses, sufficient energy for sustained solder reflow results. Solder wets to lead and pad and, as the board is removed from the hot vapor, the molten solder solidifies, bonding component leads to circuit board pads.

43.8.2 Machine Subsystems

The vapor reflow soldering machine is composed of three main subsystems: conveyor, reservoir/vessel, and heaters. Most interesting to note is the fact that the vessel has cooling coils surrounding it, placed well above the level of the liquid in the reservoir. These condense the vapors, returning the majority to the reservoir, as shown in Fig. 43.23.

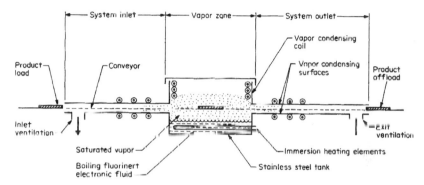

FIGURE 43.23 In-line single-vapor heating system schematic. (*Reprinted with permission from* Electronic Packaging and Production, *November 1982, p. 63, Fig. 1.*)

43.8.3 Advantages/Disadvantages

The process has some distinct advantages. It is exceptionally uniform in heating and is also very precise in temperature. Problems associated with varying topologies and areas of high thermal mass are not a problem for this method of reflow as is the case in wave soldering or oven reflow. Since the reflow is occurring in a relatively inert atmosphere, joint quality is generally excellent. Fluxes are required that have lower activation levels than those for processing a comparable assembly in air-ambient oven reflow. Although vapor phase reflow has the appearance of being a fast process, there are some hidden time factors that must be considered. First, the use of solder paste dictates a preheat just as in other reflow processes. If the paste is heated too rapidly, paste volatiles may boil, resulting in explosive solder ball formation. Also, preheating is required to preclude damage to components. As in other soldering methods, the maximum heating ramp rate has to be commensurate with component manufacturer recommendations. Plastic package popcorning is pronounced with this method of reflow.

CFC restrictions top the list of negative attributes that, for the most part, have driven this technology off the manufacturing floor. The liquids used in vapor-phase reflow are expensive. With their continued recycled use, there are toxic materials formed such as hydrofluoric acid and perfluorisobutylene. These must be neutralized and their by-products must be disposed of properly.

When massive, densely populated boards are introduced to the vapor reflow oven, another well-known problem can occur: vapor collapse. This is characterized by a condensation rate that outpaces that of vaporization. The result is that the internal atmosphere of the oven thins dramatically to the point that it cannot sustain adequate reflow. Vapor-phase machines that rely on immersion heaters are prone to this phenomenon. More recent machines include massive heating element housings that provide sufficient thermal inertia to preclude this problem.

Increased incidence of tombstoning, solder ball formation, and component displacement have been noted in this method of reflow. The condensing vapor transfers heat directly to the best thermal conductors, component leads, and pads.

Excessive solder wicking can occur, transporting solder from between the lead and pad up the lead where it is not needed. This can cause solder bridges proximal to the component body. Excess solder on the component lead decreases the lead's flexural compliance, detracting from its reliability. Furthermore, the lead-pad interface is now solder starved and an inferior solder joint has been formed. At worst, so much solder is depleted that an open can occur.

As previously noted, this soldering method does have some merit. The search is on by some vapor-phase reflow machine manufacturers for safer, ecologically sound replacements for the CFC-based synthetics that compose the reflow medium.

Additional information can be gleaned from older reference books and other authoritative publications, including previous editions of this book.[12-14]

43.9 LASER REFLOW SOLDERING

The main features of the laser reflow soldering technique are its versatility, wide process window, and future promise.

43.9.1 Laser Soldering Applications

Laser soldering can accommodate bonding of the finest or coarsest peripherally leaded surface-mount packages. The device population density on the board, the thickness of the printed circuit board, the PCB materials, and the presence or absence of package heat sinks do not preclude the use of the laser for soldering. At the same time, lead, pad, or solder metallurgical systems do not necessarily preclude laser use either. When laser soldering is implemented properly, high soldering yields are possible. Although laser soldering has much potential for manufacturing, very few manufacturers have realized its potential; its use is currently relegated mainly to low-volume, high-mix lines or niche applications. In order to use this technique successfully, it is necessary to use tooling to hold the component—or, more specifically, the component leads—down to circuit board pads.

Recently the laser has been introduced as a versatile rework tool. Laser systems are packaged expressly for this purpose and sold commercially for rework applications. Boards are heated to about 100°C within the system by means of a bottom-side resistance heater. A diffuse laser beam is then scanned around the package by a rapidly moving galvanometer. The impinging energy heats surface-mount leads or the entire BGA to the point of solder reflow. Once solder joints are molten, the package can be removed. This system has the advantage of such localized heating that only the package targeted for rework reaches reflow, while everything around it stays close to the system preheat temperature. There is little thermal spillover as seen in hot air repair where the hot, rapidly moving gas stream used for package heating and reflow can inadvertently melt solder joints on adjacent or back-side components.

There are several advantages that distinguish laser bonding from other soldering techniques. Its biggest edge is the photonic advantage: the use of photons to do the bulk of the process work—more specifically, heating. While there are several ways of harnessing the photon to accomplish soldering, none possesses the ease and elegance of laser-based manufacturing. Lasers offer a constancy in performance, predictability in behavior, and versatility in use that can be found in no other tool.

Photons (packets of electromagnetic energy) are massless and will not disrupt component lead position on a PCB bonding pad. Photons are focusable to small spot size and are easily moved around the circuit board. While it is true that other photon sources of a nonlaser variety can be used for soldering, the laser brings with it distinct advantages. It is constant in wavelength and therefore its effect on various materials is predictable—a necessity for ease of process setup and reproducibility. A laser's beam can be focused to smaller diameters more easily than other sources due to its monochromaticity. This quality permits its beam to be characterized easily as to spatial energy distribution and average power with off-the-shelf equipment. This facility to monitor is key to process control and consistent processing results.

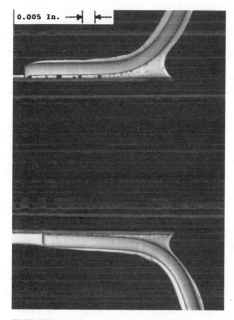

FIGURE 43.24 Laser-soldered, 0.025-in-pitch surface-mount components bonded back to back on a 0.062-in printed circuit board. It has been shown that heating of the second side during laser soldering has no effect on the first side at board thicknesses of 0.016 in or less.

There are numerous myths surrounding laser soldering as regards price, safety, materials compatibility, and throughput. This section will help to unravel these. But first, let us examine the essence of the laser and the methods associated with this form of bonding. Note that the plural *methods* was used. That is because laser-bonding techniques are numerous and diverse in practice as well as application. Only the most notable examples are studied here.

It should be noted that this is a highly specialized area and should not be attempted without the requisite knowledge of soldering, laser fundamentals, laser safety practices/regulations, and laser equipment. Although there seems to be much mystery associated with lasers and their application to the bonding space, the reader should find that the basic principles are quite simple.

Laser soldering requires no hot zones to profile or maintain, does not induce substrate warpage, and requires no substrate preheating to accomplish bonding of peripherally leaded components. There is no heated bonding head that can degrade or vary with usage. Experience has shown that package configuration is of little consequence and the presence of component heat sinks is immaterial as long as it does not shadow the leads from the laser's bonding energy. Because the laser's beam can be highly localized, current circuit board design wisdom can be successfully challenged. Components can be spaced exceptionally close to one another—closer than currently permitted for other techniques. In addition, large active components can be placed on both sides of the printed circuit board because laser bonding on side 2 will have no influence on previously soldered side 1 components, as shown in Fig. 43.24.

43.9.2 Lasers

The word *laser* is derived from the acronym for *light amplification via stimulated emission of radiation*. Although postulated by Einstein in 1917, the first practical demonstration of the laser took place in the late 1950s. Over the course of the 1960s and 1970s, lasers became ruggedized, capable of sustained use in manufacturing. Over the years they have proven to be an excellent production tool in many industries due to their versatility, simplicity, and high uptime. Although the PCB assembly industry has been slow to embrace lasers in manufacturing, there has been growing acceptance in terms of nonsoldering applications such as metrology and marking.

43.9.2.1 Laser Elements.   All lasers are comprised of three main subassemblies:

- Power supply
- Cavity or oscillator
- Beam delivery optics

Although there are many configurations and additional accessories that may enhance a laser's output, only the basics are covered here. Figure 43.25 illustrates the basic laser sections.

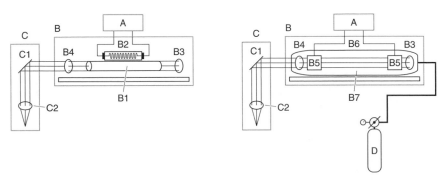

FIGURE 43.25 Diagrams of YAG laser and CO_2 laser.

43.9.2.2 Cavity/Oscillator. Lasing, which is a laser photon generation and amplification process, occurs in the cavity, also known as the oscillator. The cavity contains the lasing medium, a material that is both the origin and the amplifier of the laser beam. Photons, electrons, or other high-energy sources are used to raise certain atoms or ions in the lasing medium to a temporarily stable electronic transition, a metastable state. When the medium is returned to its stable, ground state, energy is released in the form of heat and fluoresced photons. These photons are, for the most part, reflected at the ends of the cavity and focused back through the lasing medium by the cavity optics.

There are always two types of reflectors that bound a laser's cavity. The rear or retroreflector focuses and returns nearly 100 percent of the emission wavelength photons back through the cavity, further stimulating the lasing medium to emit photons. The front reflector is a leaky mirror/lens, one that is reflective but somewhat transmissive also. It reflects most of the emission beam back into the oscillator for further stimulation of the lasing medium while allowing a small fraction of the laser beam to pass through it. The component of the beam that is transmitted by the front partial reflector is the working laser beam. It is monochromatic by virtue of the lasing medium, which limits photon emission to a set of discrete wavelengths, or laser lines, which are characteristic of the medium material, and by virtue of the optical coatings on the front reflector, which allow only one laser line through it.

43.9.2.3 Delivery Optics. Delivery optics placed in the path of the laser beam direct it to the work piece. In the case of laser soldering, the work piece is the circuit board or, more specifically, the leads, pads, or lead-pad combinations. The beam can be easily steered by means of wavelength-appropriate mirrors and can be focused to the required spot size via a final objective lens to accomplish soldering. The delivery optics can be held fixed and the circuit board moved beneath, or, conversely, the optics can be moved to direct the beam as needed. Either can be done with great precision sufficient for any PCB soldering task. Moving beam or moving board both have their own advantages. Moving the optics in any direction is very simple and will not interfere with any PCB or surface-mount device fixturing requirements. Also, moving the circuit board may cause misalignment of component leads and circuit board pads. Fixed optics are more stable and require fewer adjustments, although, if engineered properly, moving optic beam delivery systems can be exceptionally stable and should not need frequent adjustment.

Fiberoptics are in vogue for laser beam delivery, but it should be noted that, when dealing with high-energy-density laser beams as required for soldering, fibers are subject to damage if not precisely maintained. This maintenance requires cleaning and, most importantly, centering of an appropriately sized laser beam onto the face of the fiberoptic or into the fiberoptic coupling lens. Fiber diameter will dictate to some degree the ultimate minimum size of the laser beam. A good rule of thumb for delivered fiberoptic beam size is that the resultant minimum beam diameter will be on the order of the diameter of the fiber bundle. Fixed optic systems require much less maintenance and are less expensive than fiber delivery systems. Of course, fiber delivery does have the advantage of allowing for easier steering of a beam in a complex machine that may preclude the use of orthogonal mirror and lens beam steering.

43.9.2.4 *Beam Characteristics.*

There are many key differences between a laser beam and other light sources. (The term *light* is generally reserved for those wavelengths that are part of the visible spectrum and detectable by the human eye, but for the purposes of this discourse, *light* and *photonic emission* will be used interchangeably.)

Laser beams have several distinctive characteristics. They are generally set to emit monochromatic radiation, which is also coherent radiation. Coherence here relates to the synchronized propagation of photons; i.e., all waves of the emission radiation are in phase with one another. There are a number of other attributes that make the laser beam a useful manufacturing tool. These are discussed in the context of laser soldering also.

Although most lasing media are capable of emitting more than one wavelength, laser cavity optics are generally coated for an output emission that is restricted to a tightly distributed set of wavelengths—so tight that the beam is considered to be monochromatic. Coherence and monochromaticity are two important factors that permit fine focus of a beam. But that is not to say that a laser's emission has to be focused very finely to allow for soldering. This aspect of the soldering equation is dealt with later.

43.9.3 Criteria for Lasers for Soldering

43.9.3.1 *General Criteria.*

The choice of a laser for soldering is predicated on several criteria:

- Wavelength
- Required power
- Required beam diameter
- Reliability
- Price

43.9.3.2 *Optical Properties Criteria.*

The optical properties of the target material are of primary importance in the laser selection process. In the case of surface-mount soldering, the absorption, reflection, and transmission characteristics of the PWB laminate are important, as are the reflectivity and absorptivity of the metallics to be involved in the soldering process for leads, pads, and solder. The vast majority of circuit boards produced today are composed of fiber-reinforced epoxy resin, although there are some MCMs of thin film, multilayer polyimide, ceramic, and polyimide over ceramic, among others. Each has its own unique optical properties and characteristic laser damage threshold (LDT). LDT can be defined as the energy required to cause denaturation or damage of the target material. In the case of an organic-based laminate PWB, it would be the laser energy necessary to char the board or cause the bonding pads to lift from the surface. In a ceramic MCM, it is the energy required to scribe, drill, or crack the ceramic.

43.9.4 Laser Alternatives

As in other sections of this chapter, discussion is largely restricted to the bonding of eutectic, or near eutectic, tin-lead solder alloys on industry-standard substrate materials such as FR-4.

There are few practical choices for laser selection; only a few possess the characteristic energy and production-tested reliability necessary for PWB assembly. The most common are the neodymium drifted yttrium aluminum garnet (Nd:YAG) laser, an example of a solid-state laser; and the carbon dioxide laser, which has a gaseous lasing medium. These two types of laser are among the most common of the industrial machining lasers. Each has been on manufacturing floors in various industries for at least a quarter of a century. They are versatile in terms of applications and capable of the output required to weld, braze, solder, cut, and mark. While there are other lasers that could be used for soldering, the Nd:YAG and CO_2 types are the most commonly available with proven track records and are also the most commonly reported in terms of application to soldering.

43.9.5 Carbon Dioxide (CO_2) Lasers

In the case of the CO_2 laser, there are mirrors at both ends of the tube that contains the lasing gas—a mixture of carbon dioxide, nitrogen, and helium. Each of the components of the mixture helps in the lasing process. The CO_2 is the lasing medium per se, and electricity is discharged directly into this gaseous lasing medium. The discharge results in photons characteristic of the wavelength of the CO_2 gas, or, more specifically, of that of the carbon-to-oxygen bonding. These photons are reflected back through the laser cavity to further stimulate the production of photons. As this process continues, the beam intensifies and emissions from the cavity compose the working laser beam. The introduction of N_2 helps in the transfer of energy to excite the CO_2 molecules into their metastable state. When the metastable state decays to its stable ground state, energy is released in the form of photons and heat. The helium, with its high thermal conductivity, helps to transfer some of this thermal energy to the laser's cooled cavity walls. Were the cavity gases to become too hot, the CO_2 would dissociate and would be less effective as a lasing medium. The majority of the photons generated are reflected back through the lasing medium, which helps to stimulate further photon emissions. As previously indicated, one of the mirrors is a full reflector while the other is leaky, allowing a small fraction of the incident laser beam through it.

CO_2 lasers are known for their reliability and stability in manufacturing. Their emission is well into the IR spectrum at 10.6 μm (10,600 nm). It has limited use in PWB soldering since that wavelength is well absorbed by most organic materials (such as epoxy laminates) and is well reflected by most metallurgies. This is disadvantageous, since to get sufficient energy into a lead-pad combination, the CO_2 beam has to be kept to a small size to prevent it from spilling onto the FR-4. If the FR-4 were irradiated either directly or by errant reflection off a specular surface, it would char and the resultant carbon-rich residues might be electrically conductive enough to cause an electrical short circuit. Since the reflectivity of most metals is high at 10.6 μm, it is necessary to direct large amounts of energy at the solder target to start the heating process. As the temperature of a material increases, so does its optical absorption through a process known as free-carrier (free-electron) absorption. This is characterized as a runaway process. The hotter a solid metal or semiconductor is, the more absorptive it becomes. There are two other points that detract from the use of the CO_2 laser as the preferred tool for circuit board soldering. First, the minimum practical spot size is large—10 times that of an Nd:YAG laser. The theoretical diffraction-limited spot size is directly proportional to wavelength. Since a laser's wavelength is fixed, one can calculate the spot size with the following formula:

$$S = f\lambda \backslash D \tag{43.1}$$

where S = diffraction limited spot size
f = focal length of the lens
λ = wavelength of the laser
D = lens diameter

Thus, for a lens of 25 mm diameter and 100 mm focal length, in conjunction with CO_2 or YAG lasers, the spot sizes shown in Table 43.2 are theoretically possible. Imperfections in the lens or beam shape and other factors prevent practical achievement of these minimal spot sizes. Generally, the attainable focused beam diameter on the factory floor is about two to three times larger than the ideally calculated spot size. Note that for an Nd:YAG laser, the spot size is at least a factor of 10 smaller than that of the CO_2 laser, permitting a fine, high-energy-density spot for soldering.

TABLE 43.2   Theoretical Spot Size for Alternate Laser Type

Laser type	Emission wavelength (μm)	Theoretical spot size (μm)
CO_2	10.6 (10,600 nm)	42
Nd:YAG	1.064 (1064 nm)	4.2
Nd:YAG (frequency doubled)	0.532 (532 nm)	2.1

Lastly, the CO_2 laser's output is not compatible with fiberoptic delivery for all practical purposes. The wavelength is well absorbed by the most common fiberoptic material, fused silica. The use of fibers for CO_2 wavelength of 10.6 μm is a topic of intense research. Despite these drawbacks, the use of CO_2 lasers for soldering has been widely reported in the literature.[15]

43.9.6 YAG Lasers

The Nd:YAG laser, commonly referred to as the YAG laser, has a solid-state medium. It relies on lamp excitation of a cut and polished yttrium-aluminum-garnet crystal that has been doped with neodymium. An intense broad-spectrum lamp is used to stimulate the YAG crystal solid-state lasing medium. The photons released by the solid-state lasing medium are used to stimulate more lasing medium-derived photons. Otherwise, the same lasing principles apply to the operation of that laser as to the gaseous CO_2 laser. Nonlinear optical materials can be used in conjunction with the YAG laser to double the output frequency, halving the operating wavelength. This can be used to advantage when working with highly reflective materials such as gold and copper. As is the case with many metals, reflectivity is lower in visible and ultraviolet wavelengths than in the infrared. This, however, detracts from laser performance by decreasing available operating power, adding complexity to the system as well as increasing maintenance requirements.

Both YAG and CO_2 lasers can be operated in a variety of modes, each of which can be used to advantage in soldering. Continuous wave operation, also referred to as CW, is a constant emission analogous to the continuous output of a light bulb that is powered by a direct current source. Diametrically opposed is the pulsed laser output, which, furthering the light bulb analogy, is akin to a bulb operated by an alternating current or pulsating power supply—a strobe with intense bursts. The pulsing can be attained through a variety of methods, including switched power supply, capacitive discharge, or mechanical shutter, or by means of an optically manipulated shutter as by acoustooptic, electrooptic, or magnetooptic methods.

43.9.7 Laser-Soldering Fundamentals

There are relatively few variables associated with the laser in its application to soldering. This is one of the big advantages for laser processing. Beam wavelength, irradiation time, and beam power are important to the process, as are the properties of the materials being joined. Reflectivity, thermal conductivity, and laser damage threshold must be understood before soldering is attempted. The wavelength will be fixed by the laser of choice, as shown in Table 43.3.

TABLE 43.3 Emission Wavelength for Alternative Laser Type

Laser type	Emission wavelength (nm)
CO_2	10,600
Nd:YAG	1064
Nd:YAG (frequency doubled)	532

The shorter the wavelength, the smaller the theoretical spot size. This is important for single-point laser bonding of finest geometries. (More on that will follow.) Generally, the reflectivity of a metal is lower at shorter wavelengths. This means that a metal is more easily heated by an Nd:YAG laser as compared to a CO_2 laser. The converse is noticeably operant for many polymeric materials. Most are quite absorptive at longer wavelengths and prone to burning. Many polymeric materials are also absorptive at the UV end of the spectrum. Therefore, a carbon dioxide laser beam is more likely to impart damage to a circuit board than the beam of an Nd:YAG laser. Common circuit board laminates such as FR-4, G-10, and polyimide can be easily damaged by any laser beam if the energy is not regulated properly.

The reflectivity of metals varies widely with composition and surface condition. Every metal can be heated with a laser as long as the energy density of the beam is high enough and the dwell time of the beam sufficient to stimulate free-carrier absorption. This is also the case with laser-irradiated component leads, board pads, and solder during the laser bonding process. Measurements of the reflectivity of Sn/Pb solder show that a eutectic alloy can be as high as 74 percent at 10,600 nm versus 21 percent at 1060 nm. Therefore, in the case of the CO_2 laser at 10,600 nm on, say, an Sn/Pb-plated lead and solder-coated pad, much energy will have to be directed at the metals to start the absorption process, since only about 26 percent is being absorbed by the solder and converted to heat. This can be problematic owing to the specular reflectance of solder. The reflected or multiply reflected beam may impinge on adjacent components and damage package bodies or even the circuit board itself and cause it to char. That is why Nd:YAG is preferable for circuit board soldering. CO_2 should be reserved for bonding of exceptionally large components, such as board-mountable transformers and power supply tabs. Further discussion of laser soldering is restricted to the Nd:YAG.

As mentioned previously, beam diameter is critical to laser processing. If the beam is larger than the target—in this case, the lead-pad combination—then the energy input must be compatible with the optical properties of the circuit board substrate. There are laser-soldering techniques in which the beam is purposely large and does encroach on the circuit board laminate. Such techniques can be acceptable and successful. These techniques as well as others are discussed in a review of the most common laser-soldering methods.

43.9.8 Through-Lead vs. Through-Pad Bonding

Generally, the laser beam is directed at the component lead to accomplish soldering, but when the lead material is highly reflective, as would be the case with a gold-plated finish, heating is slow and irradiation times impractically long. An alternative has been demonstrated whereby the beam is directed at the circuit board bonding pad. If there is leeway in design, the bonding pad can be extended to aid in this method, dubbed *through-pad bonding.* In contrast to the usual beam impingement on the component lead or *through-lead bonding,* the beam is directed at the more absorptive solder on the circuit board land, increasing process efficiency and locally melting the solder in the vicinity of beam impingement. Since the molten solder is in intimate contact with the lead and pad, the heat from the bonding process is efficiently transferred. The process can result in soldering along the entire length of the component lead and circuit board pad with rich solder fillets evident if the process is conducted properly, as shown in Fig. 43.26.

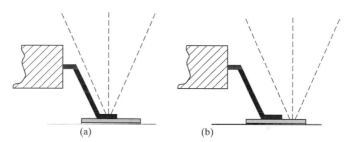

(a)   (b)

FIGURE 43.26 (*a*) For through-lead laser soldering, the laser's beam is directed at the foot of the component. (*b*) For through-pad laser soldering, the beam is directed at a pad extension; this heats the solder directly. This method can be used to solder leads coated with highly reflective materials such as gold.

43.9.8.1 *Single-Point Laser Soldering.*

This method requires a laser beam that is smaller in diameter than the length or width of the component lead or bonding pad. The beam is stepped to each lead/pad, constrained to the single lead or pad, and left on long enough to cause the solder to flow. The beam can be continuous wave (CW), pulsed, or multiply pulsed to accomplish soldering as long as there is enough radiation delivered to cause the solder to undergo the phase change required for soldering.

Owing to the small beam diameter, as required by this technique for fine-pitch components, the energy density can be exceptionally high. If the lead is not in good contact with the bonding pad and if the energy density is too high, then the laser's beam can damage the lead, perhaps cutting through it rather than soldering it. Overly intense irradiation can also cause the bonding pad to delaminate from the PCB.

In this variation of laser soldering or in other methods where very small spot size is required, one must take precautions to precisely control the beam diameter. A very small variation in beam diameter, either through change in laser setup parameters or in working distance, will have a dramatic effect on the focused spot energy density. As an example, take the case of a 10-W Nd:YAG laser beam focused to a spot of 0.1 mm (0.004 in) and 0.2 mm (0.008 in). A seemingly small change in spot size will result in a large change in power density.

$$P = \frac{p}{d} \qquad (43.2)$$

where P = power density
        p = average power (W)
        d = focused beam diameter

For this example, the power density would vary from 1273 W/mm^2 for the 0.1-mm beam to 318 W/mm^2 for the 0.2-mm beam diameter, a reduction factor of 4. So, with finer spot sizes, it is crucial to maintain strict process control, because a small change in beam diameter will translate to a large change in delivered power.

It should be noted here that in most cases a gaussian distribution of the beam's energy is assumed, and in the optics world it is customary to measure the beam at the $1/e^2$ point (13.5 percent of the peak height), so, in fact, the beam is impinging on a slightly larger area, but the most intense portion of the laser's beam is confined to the region bounded by the $1/e^2$ points in two dimensions, as shown in Fig. 43.27.

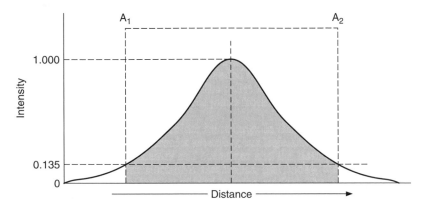

FIGURE 43.27 Laser output energy is conventionally measured from points on the distribution curve where the beam's intensity is $1/e^2$ (13.5 percent) of peak intensity, or between points A_1 and A_2 in this plot.

Single-point laser bonding has been applied to both inner lead bonding of a silicon die to TAB (tape automated bonding) leadframe, as well as for bonding TAB components to circuit boards. It has also been demonstrated for soldering conventionally rigid leaded surface-mount packages to either a circuit board or multichip module.

43.9.8.2 Continuous-Wave Scanning.   In sharp contrast to single-point laser soldering, this technique relies on a larger, intense moving beam to accomplish the heating. The beam may be many times larger than the narrowest lead or pad dimensions. The laser beam purposely spills over onto the substrate material, irradiating the lead, pad, or lead-pad combinations as well as the interpad board surfaces. Because larger beam diameters are used, longer-focal-length lenses can be employed. The longer-focal-length lens, along with the generous spot size, allows for much less stringent process control in terms of the final beam diameter and tolerances in terms of working distance of lens to circuit board surface. The laser spot is moved at a rate such that the beam exposure to the board is below the laser damage threshold limit. It heats the lead-pad combination, the solder flux, and the board sufficiently to cause the solder to change state.

A variation of this technique was reported by Raytheon for surface-mount soldering. It utilizes orthogonally mounted galvanometrically controlled scanners to move an Nd:YAG laser beam around the periphery of a surface-mount device sitting atop solder lands on a circuit board. The beam is driven at high velocity repetitively around the surface-mount device until all leads are heated to the point of solder liquidus. As the beam is turned off, it solidifies and solder joints result. As previously mentioned, this same galvanometer-driven technique is gaining popularity for rework of BGAs. A slightly defocused laser beam is aimed at the body of the BGA. The beam is rapidly moved around the package body at a rate below the laser damage threshold of the packaging material(s). With each pass the component heats up a little more. Eventually the BGA solder joints go molten and the package can be removed. A new BGA can be attached using this same technique. This method can be applied to any type of device rework, although component leads/pads would be targeted rather than the body as was the case for the BGA.

43.9.8.3 Multiple Beam.   One of the attractions of working with a laser is the fact that its beam can be split for multiple use within one station or even shared between two or more stations. The split can be accomplished by the use of bifurcated fibers or beam splitters. The beam of a single laser cavity can be duplexed to solder two sides of a surface-mount component simultaneously. It is entirely possible to share a common beam between two or more soldering stations either simultaneously or in a time-shared manner.

43.9.8.4   *Tooling.*   In all modes of laser bonding, the component leads must be in contact with solder lands on the PCB to accomplish joint formation. It is therefore necessary, in most cases, to use a specialized hold-down tool to ensure that leads do contact pads. This detracts from the ideal of noncontact bonding, but there are several methods that have been developed and reported. Much development work is continuing in this area. It is worthwhile to review a few of the reported methods.

Many investigators have used transparent hold-down media. Included are glass, quartz, and transparent high-temperature plastics. There are several significant problems associated with this type of approach, however, and it can result in an inconsistent soldering process. First, the materials are rigid, which is an impediment when trying to overcome lead-to-lead, pad-to-pad, or lead-to-pad coplanarity differences inherent in most circuit boards, components, and presoldered assemblies, as shown in Fig. 43.28.

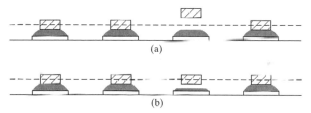

(a)

(b)

FIGURE 43.28   In order for laser soldering to be effective, it is necessary to have lead-to-lead, pad-to-pad, and lead-to-pad coplanarity: (*a*) noncoplanar leads on constant-volume solder pads; (*b*) coplanar lead array on noncoplanar PCB solder lands.

Second, during the soldering cycle, the glass can accumulate spattered flux and flux byproducts that may change the delivered laser beam intensity, adding variability to the process. Comb or pin arrays have been applied that match the lead configuration of a package.

Probably the most common method for rigid-leaded surface-mount packages is the body-push method, in which a force is applied to the component body, springing the leads just slightly. As the solder melts below the lead, the lead drops down to board level, where it is frozen in the cooling solder. It is necessary to control the amount of push such that the solder joint is not stressed by a lead in compression. When board pads and component leads are reasonably coplanar, then this method is adequate for high-yield, high-reliability soldering.

A compliant hold-down method has been reported that is inexpensive to fabricate and easy to implement. This consists of a silicone rubber foot and an aperture that allows the beam unobstructed access to the lead-pad combination. This has been proven effective for TAB as well as for rigid leaded surface-mount device bonding. It obviates the problems associated with transparent hold-down methods in terms of flux buildup. The compliant foot holds leads to pads and can accommodate lead-to-pad noncoplanarity differences.

43.9.8.5   *Bonding Rates.*   Lasers are capable of operating at very high speeds; even so, it is difficult to compete with mass reflow methods. It is anticipated that this will not be the case one day, especially as pitches decrease and mass reflow and ancillary processes such as placement and solder paste stenciling struggle to tackle finer and finer lead pitches. Bonding rates in laser soldering are very much dependent on the device pitch and outline. It can take as little as 10 μs to solder a lead, but the beam must be accurately translated to the next lead-pad combination. This translation can be orders of magnitude slower than the soldering itself. Practically speaking, for inner lead bonding, rates as high as 65 leads per second have been reported.* Outer-lead soldering rates vary with package type, pitch, and footprint, but per-lead rates vary between about 15 to 60 joints per second. With exceptionally fine-pitch/leaded, small-outline packages such as TAB, bonding rates can be considerably higher.

43.9.8.6 Flux for Laser Bonding. Requirements are much the same as for other reflow methods. Flux must be active enough to remove oxides from the lead and from the solder. Its optical characteristics in terms of reflection, absorption, and transmission must not interfere with the solder process. If too absorptive, it could caramelize or char and impart damage to the underlying circuit board.

In laser soldering, particularly in the scanned methods, the flux adds a path of heat transfer. The heated flux transfers its energy to the board, preheating the next joint to be soldered. In the absence of a liquid flux, there is more propensity for the board to char if the laser impingement is not stringently controlled.

During soldering, the laser beam's output is quick and intense. The high energy of the process results in exceptionally high processing temperatures for very short durations. The high processing temperature aids in flux activation, making even the mildest of no-clean fluxes quite an effective soldering aid.

There have been many reports of fluxless laser soldering. As used for inner-lead bonding to silicon die, these utilize a very high-energy, short-duration pulse. Under these conditions, when the mass of the lead and the bonding pad are small enough, the process is more akin to welding. The component lead is reported to melt and alloy with an underlying metal such as tin.

However, to accomplish conventional reflow soldering by means of a laser, or any other technique, it is necessary to employ a liquid solder flux or gaseous analogue pretreatment to permit effective bonding. Plasma cleaning and inert storage may be one such method. Ultrasonic assisted laser soldering can be effective but is slow, requiring precise placement of an ultrasonically agitated head and running the laser beam either through the ultrasonic tool or precisely adjacent to it. The ultrasonic agitation breaks up the oxide surrounding the molten solder and the lead, allowing joining of the liquid solder to the lead metallurgy.

This author has demonstrated the use of gaseous phase carboxylic acid fluxing in conjunction with laser bonding.[16] This precludes the use of conventional liquid fluxes and leaves no visible residues on the circuit board. No cleaning is required and the assembly reliability is not degraded from the use of a gaseous fluxing approach.

43.9.9 Laser Solder Joint Characteristics

There are very few differences between solder joints prepared by laser as compared to those formed by other methods. Laser soldering results in intermetallic formation per usual, but the layer is extremely thin if the laser-soldering cycle heating is kept to a minimum. The layer is much thinner than that found in solder joints manufactured by more conventional methods. Upon cooling, which is very local and very rapid, the laser-soldered joint typically possesses exceptionally fine solder grain growth, distinctive of this process. The fine grain growth leads to greater joint strength initially. This strength advantage, although significant in magnitude, tapers off with age. After about a year of storage at room temperature, the metal grains coarsen and typical nonlaser solder joint metallurgical properties dominate.

---

* See "Laser Based System for Tape Automated Bonding to Integrated Circuit," *Proceedings of IEEE/Electronic Components and Technology Conference,* May 1990, pp. 757–761. These rates are for TAB inner lead bonding of ultrafine leads in a small area.

Another characteristic of laser soldering is rich solder joint filleting. This is particularly true of joints prepared by the CW scanned method of laser bonding. The solder joints are generally full length and very strong compared to bonds made by conventional solder methods.

43.9.10 Solder Sources and Defects Associated with Laser Reflow

Solder requirements are the same as for any other process. There are no alloy composition requirements specific to laser soldering. It is possible, however, to use some of the more exotic materials such as 10:90 or 5:95 Sn/Pb with melting points of about 302 and 312°C, respectively.[17] These temperatures are generally considered too high for conventional reflow methods. The board will darken, burn, or warp in a reflow oven set high enough for the onset of high-lead-alloy liquidus. When single-point laser reflow is applied, the board quality and integrity are not compromised if parameters are chosen and adequately controlled.

The thickness of solder required on the board is a function of the required product reliability and the component pitch. As the pitches decrease to about 0.5 mm and below, hot-air solder-leveled pads may bear sufficient solder for adequate joint formation. Electroplated reflowed or even unreflowed solder platings will be sufficient for laser soldering. Some of the coatings, such as SiPad® and Super Solder®, have also been demonstrated as useful methods of solder deposition for laser reflow.

Laser soldering is not prone to defects if implemented correctly. Perhaps the most common characteristic defect unique to laser soldering is charring or burning of the circuit board. This can occur if too high an energy density is used and the laser damage threshold is exceeded. Charring or burning can also occur if the circuit board is grossly contaminated with grease or other organic contaminants. That is not to say that boards for laser soldering have extraordinary requirements for cleanliness. Requirements should be considered the same for this technique as for any other reflow process.

Another characteristic defect is the presence of solder balls when the laser's beam is focused onto a component lead embedded in solder paste or when the paste is irradiated directly. The use of raw solder paste is not recommended for laser soldering. Just as oven reflow of paste requires a gradually ramped drying of the solder paste, so the same is required of laser processing. If the paste is heated too rapidly, as would be the case with laser soldering, the solvents and vehicles in the paste would vaporize much too rapidly, causing explosive spattering of some of the paste mass. These volatile explosions are linked to solder ball formation. In general, laser soldering of plated or hot-air-leveled solder or other solid solder coatings will not result in appreciable solder ball formation.

Solder bridging is generally of little concern in laser soldering. In fact, if a solder bridge exists on the substrate prior to soldering, the laser-soldering process may relieve the bridging condition and cause the redistribution of the solder onto the component leads. This is particularly true with the scanned CW laser-bonding method.

The occurrence of solder opens will be proportional to the quality of the leadframe of the component, its solderability, and the effectiveness of the component hold-down method. Very high yields are possible, and, if the process is implemented correctly, few opens will occur. Although some may feel that this technique is too futuristic or too slow, it is ripe for commercial exploitation. Equipment costs can be modest, with the laser cavity available for about $30,000. Add about $5000 for each axis and funds for a computer or CNC to drive the axes. An enclosure is easy to fabricate out of sheet metal.

43.9.11 Laser Safety Issues

Lasers are categorized by their safety hazard potential. A full review of these will not be provided, but suffice it to say that Class 1 is an intrinsically safe laser, posing no intraoccular danger, while Class 4 lasers pose the greatest eye hazard. All lasers considered for soldering use are of the Class 4 variety. Because of this, they are generally embedded in appropriate cabinetry with laser-safe viewing ports or CCTV incorporated. Also, they are equipped with latches on all cabinet covers that are interlocked to the laser's safety shutter. This precludes direct ocular exposure to the laser's intense beam. When the Class 4 system is embedded in an interlocked cabinet as described, it is considered a Class 1 system. As such, it poses no hazard to the immediate area; laser-safe eyewear need not be worn except on occasion for servicing the laser. As mentioned previously, lasers, especially Nd:YAG, are known for their high uptime and lack of required expendables.

As the need for finer bonding continues, lasers will continue to be scrutinized, because they yield a method that exhibits much potential. Easy to automate, highly reproducible, extensible to bonding across the entire pitch spectrum, and without foreseeable limitations in pitch applicability, this technique is likely to persist and gain in acceptance.

There are numerous excellent supplemental texts available for additional detail and instruction in this technology. Those by Charschan, Hecht, and Ready are particularly useful.[18–21] Additional information on CO_2 lasers in the realm of soldering can be found in Refs. 22–26.

43.10 HOT-BAR SOLDERING

Specifically suited to surface-mount assembly of leaded packages, hot-bar soldering has been in use for several years. The technique relies on a resistance-heated element to push component leads into contact with solder and bonding pads, simultaneously reflowing the solder. Compression of the leads onto the circuit board lands is continued as the heat is ramped down. On cooling, the solder solidifies and the heating element is withdrawn from the newly formed solder joints. The heated element is commonly referred to as the *hot bar,* although the term *thermode* is also in widespread use.

43.10.1 Solder Application

The use of solder paste is discouraged because blade heat-up is fast and explosive solder ball formation, caused by rapidly volatilized paste constituents, will result. Also, the paste is likely to squeeze out from between the component lead and circuit board pad, which can cause the solder to bridge to adjacent joints. In fact, even with solid solder coatings on the circuit board, bridging can be problematic in hot-bar bonding. This is usually a function of the volume of solder on the pad, the quantity of flux and its degree of activity, and lead-pad pitch. As the solder is melted, it oozes out from between lead and pad in the rudimentary stages of solder wetting. This displaced solder may bulge laterally to the point that the solder masses of two or more adjacent pads may touch one another, forming a solder bridge. Once the bridge has formed, the forces associated with lead-pad wetting and capillarity may not be strong enough to overcome the capillary conditions established during bridge formation. If that is the case, the bridging defect(s) will persist, as shown in Fig. 43.29. Solid solder coatings such as hot-air-leveled pads or solder-plated boards are recommended for this bonding method.

43.10.2 Fluxes and Fluxing

Liquid flux is applied just prior to soldering. Fluxes chosen should be tested to resist caramelization, the development of polymerized decomposition products, or lacquers, which adhere to both circuit board and hot bar. Residue buildup on the bar can adversely inhibit hot-bar performance by diminishing thermal transfer. It can also become thick enough to keep the hot bar from squarely contacting component leads. Baked-on residues make flux cleaning more difficult and also detract from the visual appearance of the printed circuit assembly.

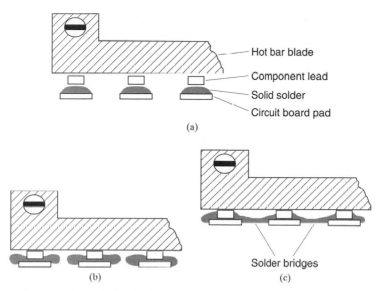

FIGURE 43.29 As the hot bar is forced into contact with component leads and circuit board pads (*a*), the solder may be displaced laterally (*b*), resulting in solder bridges (*c*). Careful control of the solder volume, applied pressure, and temperature during hot-bar bonding can preclude this.

43.10.3 The Soldering Operation

During the soldering operation, liquid solder is squeezed out from between the lead and pad unless a mechanical stop is employed to keep the bonding head from driving this far down. Of course, a thin film of solder remains that composes the solder joint. Solder-starved joints are notably weak, though, and it is preferable not to displace so much solder. One equipment vendor offers a system that automatically aligns the component with the bonding pads and pushes component leads into contact with them; on reflow, the bond head and component are retracted several hundredths of a millimeter. This results in a thick coupling of solder between lead and land for robust joint formation. This retraction step is also claimed to discourage solder bridge formation.

As in any other soldering process, it is important to maintain proper reflow time-temperature characteristics. Most hot-bar systems are configured with a fine-gauge thermocouple welded to the bar, an integral part of a closed-loop, temperature-controlled hot-bar heater system. Generally, hot-bar blades are relatively small in size; therefore, the thermal mass of the circuit board and components being bonded have profound influence on the hot bar's performance. Often they provide so much thermal conduction that accessory heaters must be utilized to augment the hot-bar heating process. This can be in the form of hot gas blowing on the top or bottom surface of the circuit board or of direct contact with a heated platen. This permits circuit board bonding in a reasonable time period. Care must be taken to avoid overheating the circuit board so as not to re-reflow previously soldered components or otherwise damage the circuit board. It may also hinder bonding of the next sites as global preheating may hasten oxidation and intermetallic compound formation on pads that are not yet soldered.

43.10.4 Construction

A hot-bar bonding head can be composed of one or several hot bars. They are designed to solder one side, two sides, or all four sides of a component simultaneously. Each hot bar of the assembly is configured to accommodate the maximum span of a lead set; i.e., the bar's length is manufactured to be slightly longer than the lead array to be soldered. This permits concurrent bonding of all leads on a package side. The bar length is also generous enough to facilitate bar-to-component alignment. In the case of some very long connectors or other large packages, the thermal uniformity of a single blade may not be adequate to reflow all leads simultaneously without overheating some of the joints. This is sometimes remedied by using a smaller hot bar and stepping it along the length of the lead set until all leads are bonded. Bar thickness is somewhat dependent on the lead form and should not interfere with the lead form. The bar should sit flatly on the foot of the lead, neither contacting the radius area of the lead form nor significantly overhanging the lead toe, as shown in Fig. 43.30.

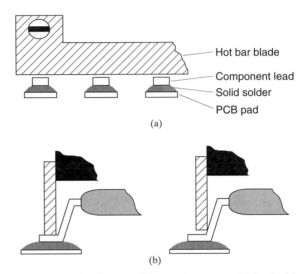

FIGURE 43.30 Diagram of hot bar in contact with leads: (*a*) transverse view; (*b*) lateral view. The hot bar must be positioned so as not to interfere with the leg of the formed lead. It should sit squarely on the foot of the lead.

43.10.5 Hot-Bar Design and Materials

The bar itself can be manufactured to nearly any dimensions, but there are limitations on its size. The longer the blade, the worse its longitudinal thermal uniformity. Variations in blade temperature may cause it to distort due to differential thermal expansion/contraction. (This is discussed later.) Thermal uniformity is critical for consistent lead-to-lead soldering and joint quality per component side. The tolerable variation in bar temperature is limited by the component and board to be soldered, the solder itself, and product reliability requirements. As in any other soldering process, the variation in process temperature must be understood and controlled to assure highest board assembly yields. The bonding head must also be able to shed heat rapidly to allow the solder to resolidify within a reasonable time.

There are numerous bar designs and materials of construction. Tungsten, titanium, and molybdenum are commonly chosen, not only for their electrical resistance and thermal conductivity but for their immunity to flux damage and solder wetting. Some ceramics are also used for blade construction.

The bar must be designed for uniform heating across its length. Additionally, it has to expand and contract uniformly throughout the soldering cycle. Some blades have been seen to develop a "frowning" or "smiling" profile in their z axis during heating, owing to differential thermal expansion or built-in stress in the metal, as shown in Fig. 43.31. For the same reasons, transverse warping of the blade is also commonplace.

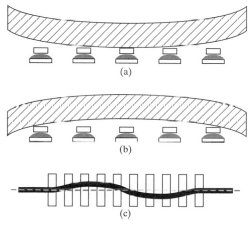

FIGURE 43.31   A hot bar, if not constructed properly, can develop a "smiling" profile (a) or "frowning" profile (b), or may warp laterally along its length (c).

The "smile" or "frown" will not provide full contact of the blade with all lead-pad combinations. This can result in solder opens if the curvature is sufficiently large. There have been many solutions proposed in the realm of structure, materials of construction, and electrical input to permit highest uniformity of blade heating and contact to the bonding pads. Waller et al. have patented a rigid molybdenum truss blade that has been found to be stable dimensionally and thermally uniform over long spans (in excess of 3 in).[27]

Other solutions include independently heated sectional bars and compliant ribbon blades. The former is a linear array composed of two or more hot-bar segments, each with its own heating control. This is more expensive to machine and maintain. It is also more difficult to use because its use includes an additional alignment of the bonding tool to ensure that interblade sections fall between and not on leads to be bonded.

Although the concept of a self-planarizing ribbon blade seems ideal for the task of accommodating variations in topography, these are also prone to distortion that can cause solder opens; this is especially true for long span ribbons.

Thermal uniformity along this blade is less controllable than that for the conventional rigid hot bar. When comparing the options for electrical input for these two types of blades, the reason becomes obvious. The ribbon can only be connected at its ends, while the rigid hot bar can have multiple taps allowing adjustment of current to each section to establish a thermally balanced output.

Generally the hot-bar blade is attached to a self-leveling, spring-loaded bonding head that allows the blade to planarize itself in relation to the underlying circuit board. Nonetheless, variations in board contour, such as a localized high or low spot, can render the concept of self-planarization useless and solder opens can result.

Rigid-member hot bars are excellent for accommodating differences in lead coplanarity because the blade forces lead to pad. Driving a heavily sprung lead to a pad can, however, build stress into the resultant solder joint because the solder is working to resist the spring force of the lead. This can detract from the joint quality and cause premature failure of the joint.

43.10.6 Maintenance and Diagnostic Methods

It is important to maintain the hot bar properly. Check it frequently for distortion. Ensure that it is orthogonal to the circuit board to be bonded. Scrub the hot bar on a ceramic flat to release any flux residue scale or polymerized coating that may develop on it. This may need to be done every few bonding cycles, depending on the flux and the criticality of the assembly.

There are several diagnostic tools that can be applied in assessing hot-bar conditions, ranging from the mundane to the exotic. The two most important characteristics that must be understood and monitored are thermal performance and planarity.

43.10.6.1 *Thermal Monitoring.*

One of the most common methods of assessing hot-bar performance is the use of a thermocouple-instrumented board. Although this is true for all bonding methods, making such a board for thermode use has its own requirements. Fine-gauge thermocouples are attached to leads of the component or the bonding pad of the circuit board to be soldered. Best positioned at both ends of the lead set and also near its center, this will help quantify the thermal uniformity of the bar during soldering. It is also useful to deploy thermocouples on adjacent components to ensure that the hot-bar soldering process is not jeopardizing the integrity of previously soldered joints.

Since it is important that the coplanarity of the board and the component are not compromised, it is wise to avoid placing the thermocouple bead(s) between lead and pad. The added height of the bead would prevent the bar from contacting adjacent component leads and could result in point contact heating that is not indicative of the bar's normal operation. Instead, place the bead in the pad extension area either just in front of the lead foot or to the rear of the lead's heel. Attaching the thermocouple can be done with a small dot of high-temperature solder or thermally conductive epoxy. The amount of attachment material added must be kept small to minimize change to the thermal mass of the would-be joint. The amount of attachment material also must be restricted so that it does not interfere with the component seating plane, because leads must sit on pads absent of obstructive, nonfusible materials. It should be noted that the coplanarity of solder on the PWB pads does not have to be exact. The hot bar will be reflowing this solder and it will self-planarize, as does any liquid. It will remain so into the solid phase.

Once thermocouples are correctly installed, the component site on the board can be fluxed and subjected to a bonding trial. Thermocouple output can be directed to a calibrated strip chart recorder to assess the time-temperature profile of the bonding cycle. Alternatively, a computerized thermal profile tracker can be used.

IR cameras have been successful in allowing visualization of hot-bar thermal performance. This can be an excellent but expensive approach to evaluating bar-heating uniformity and its impact on the bonding operation.

43.10.6.2 *Planarity.*

As stressed throughout this section, the orthogonality of the hot-bar blade or multiblade assembly with respect to the bonding surface is extremely important for the preclusion of solder opens. There are numerous measurement methods that can help diagnose this; unfortunately, though, they are meant for evaluation of the blade(s) at room temperature. As previously mentioned, a blade may distort temporarily or permanently during heating, but planarity measurements on hot blades are difficult and often not practical. Therefore, the majority of techniques encountered are performed at room temperature. Wilkins[28] suggests the use of a colorant, such as from a marking pen, applied to the cold bonding surface of a freshly cleaned thermode. Once the ink is dry, the blades are then scrubbed over a clean, flat ceramic plate. Low spots on the bar are indicated by the presence of colorant remaining on the bar after several circular swipes on the ceramic flat. Regrinding of the blade(s) to restore flatness is recommended and the procedure is repeated to ensure that the machining is effective.

Driving the cold blade assembly down onto carbon paper is another commonly used method. The carbon paper is sandwiched between a mechanical flat and a piece of thin white paper. The carbon leaves an impression on the paper. The boldness of imprint is a rough indicator of bar planarity. If the bar is nonplanar, its impression on the paper will run from dark on the low side of the blade to light on the high side. There are also pressure-sensitive papers that can be used in the same way. The problem with all of these diagnostic techniques is that they are used to assess a cold bar blade. When at temperature or during ramp-up, the bar blade can twist and warp, so the flatness at room temperature may not be an adequate indicator of blade condition during bonding.

Single-bar and two- or four-sided blade assembly planarity can also be evaluated using an array of ground, leveled, rigidly mounted pressure transducers. Bar pressure differential, an indicator of blade planarity, can be adjusted so as to be uniform from end to end and from blade to blade on the two-up or four-up hot-bar assembly. Of course, some hot-bar assemblies are self-leveling, but even these should be checked for planarity and force per blade to ensure best uniformity in bonding.

Hot-bar soldering is best suited for low-thermal-mass, single-sided surface-mount assemblies. Each component type requires its own hot-bar bonding head assembly. These can be expensive, costing several hundred dollars to well over $1000 per assembly. Price is dependent on the complexity, the materials of construction, the precision required, and the overall size of the head.

43.10.6.3 *Process-Induced Defects.*

Solder bridging is the most prevalent problem associated with this method of soldering. Recall that the solder can be squeezed out of the solder joint and that conditions may favor solder bridge formation. Solder opens result from lack of coplanarity between the hot bar and the plane of the circuit board surface. Lead misalignment during the bonding cycle is another defect that detracts from this method. As pressure is applied to component leads by the hot bar prior to the onset of solder liquidus, leads are sometimes forced to slide down the solder domes. This displacement causes misregistration of component leads to bonding pads, and may build stress into soldered joints. If the forces are great enough, it may also cause the whole package to move and misregister the entire lead set.

Because the heating is rapid, the thermode temperature is necessarily well above the solder liquidus temperature. If the time-temperature cycle is not carefully controlled, intermetallic compound formation can be a problem. This is especially true in this process, where the solder may be largely displaced from between lead and pad. Within the joint, the ratio of volume of intermetallic compound (hard and brittle) to the remaining solder (soft and compliant) may be large. If this is the case, solder joints will be less reliable and more susceptible to fracture failure.

All of these obstacles have prevented the widespread acceptance of hot-bar bonding in manufacturing. Hot-bar bonding is most useful for low-volume, fine-pitch surface-mount soldering and rework. If configured properly with a suction device to retain a surface-mount package, all four sides of a component can be reflowed simultaneously and the component removed. The addition of flux is helpful in the component removal process, since it serves as a heat spreader and also improves the contact area of the blade with the solder prior to reflow. After component removal, the pads are reconditioned (smoothed with a hand soldering iron, solder replenished, and fluxed). Braided copper wicking can be used to remove excess solder. A new component can be reinstalled via hot bar. Care must be taken once again to preclude intermetallic compound formation through the removal, dressing, and component resoldering process. Aside from the reliability risk imposed by a thickened intermetallic compound layer, the intermetallic layer is difficult to bond to and may require a more activated flux to complete the soldering.

43.11 HOT-GAS SOLDERING

In hot-gas soldering, pressurized gas is heated and directed at component leads, solder, circuit board lands, and soldering flux. The thermal energy is absorbed by these four items, and if temperature rise is sufficient, melting of the solder will occur. Upon cooling, solder joints are formed. This noncontact, directed-energy method is most suited to the bonding of surface-mount components. Although hot-gas soldering has been around for years and after numerous machine offerings in this vein, this technique is not a popular method for soldering despite its evolutionary improvements. In its most common incarnation, hot-gas soldering is used for reworking components, i.e., removal of previously soldered devices (through-hole or surface-mount) from a circuit board and replacement of same.

One of the disadvantages of this soldering method is that the thermal energy is not very well localized. Most machines typically emit a hot-gas jet too large to be isolated to reflow only the lead or leads of interest. The gas jet, once impinged upon the board and component leads, is deflected and its backwash can be problematic. It may cause unwanted reflow of previously formed joints, especially on closely spaced adjacent components. This problem is typically overcome by the use of baffles that are either applied to adjacent components or by a singular baffle that confines the gas jet to the component to be soldered.

As shown in Fig. 43.32, nozzles are available in a variety of forms. The simplest of nozzles is the single orifice. This can be translated around the entire periphery of a component. Some machines offer a double translatable nozzle assembly that can solder two opposing sides of a component simultaneously. A plenum nozzle, comprising multiple gas ports, can solder or unsolder all sides of a component simultaneously without the need for nozzle translation except in the vertical direction. This last option relies on dedicated tooling per component type. The cost for inventorying numerous individual tools can be expensive.

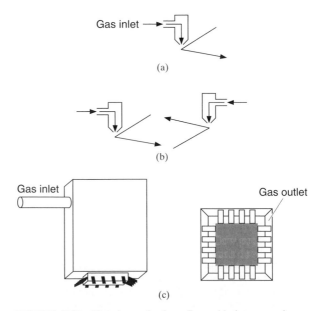

FIGURE 43.32   Hot air nozzles for reflow soldering or repair may be deployed singly (*a*) or multiply (*b*). In addition, a single plenum-type assembly (*c*) can be used to reflow all sides of a component simultaneously.

The jetting gas, which is heated, is forced through a small nozzle orifice or nozzle array. The gas can be any pressurized gas compatible with the system and circuit board materials. Air is commonly employed for its economy, but it is an oxidizing medium and may preclude the use of weakly activated fluxes. Nitrogen is recommended because it is nonoxidizing, inexpensive, and safe. The use of hydrogen-nitrogen mixtures, argon, and other gases is also common to aid this method of reflow.

Solder can be applied to the board as a paste, solid preform, or solder-coated pad. If anything but solder paste is used, a means of component hold-down must be applied to ensure contact of component lead with bonding pad. Gas pressures must not be excessive, as this could move an unconstrained component during the reflow cycle. Proper gas pressure, temperature, nozzle translation speed, and flux are required to effect joint formation. Otherwise, the same reflow considerations are required for this technique as for any other. Heating ramp rate, solder paste preheating, peak temperature, liquidus duration, etc., must be observed for successful joint formation and joint reliability considerations.

The use of thermocouples in the vicinity of reflow on anything but test boards is not practical, so sensing of the soldering temperature is not easily accomplished during a hot-gas soldering cycle on actual product. Therefore, a dummy board with thermocouple-instrumented component(s) is useful in determining the effects of the various operating parameters on the process performance and reflow profile. Additionally, thermocouples deployed in the solder joints of adjacent components are recommended to ensure that operating conditions chosen do not inadvertently re-reflow these neighboring devices. As in any soldering operation, care must be taken to prevent overheating of the component and circuit board during hot-gas soldering. Overheated circuit traces may delaminate, flux may char, and excessive intermetallic compound formation may result. All of these can impact product reliability.

43.12   ULTRASONIC SOLDERING

This method relies on a heated, ultrasonically vibrated soldering tip (see Fig. 43.33) that simultaneously melts and agitates the solder. The ultrasonic energy is transferred from the tip through the molten solder droplet beneath it and ultimately to the component lead and circuit board pad. This high-energy agitation of the solder droplet helps to cleanse bond-inhibiting materials from the solder and solder-metal interfaces. The integrity of metal oxides that encapsulate the solder, circuit board pad, and component lead is disrupted to the point that unoxidized underlying metals are exposed and wetted by the solder. This precludes the need for the addition of chemical fluxing agents.

Aluminum and other difficult-to-join metals can be soldered by this method. The viability of this technique has been well proven on a commercial scale in the manufacture of air con-

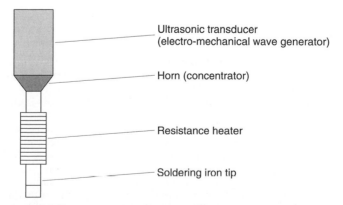

FIGURE 43.33   The ultrasonic soldering iron is composed of four main components: an ultrasonic transducer; a horn for concentrating and directing ultrasonic energy; a resistance heater; and the soldering tip, which emits both thermal and ultrasonic energy.

ditioner heat exchangers.[29,30] More important, this joining method has implications in the assembly of printed circuit boards.

Ultrasonic soldering has also been applied as a batch or continuous mass reflow process. In these instances, the molten solder is ultrasonically agitated while the assembly to be soldered is immersed in it. Similar arrangements have been made for ultrasonically vibrating the part while dipping it in a molten bath or wave of solder. These mass processes are more common for nonelectronic assembly.

Care must be taken in ultrasonic soldering to tune the tip amplitude and/or frequency to the mass of the system being soldered. Overagitation results in excessive cavitation in the liquid solder, causing it to splash. This generates solder balls, which could short finer-spaced leads or pads. Additionally, ultrasonic agitation increases the dissolution rate of any soluble metals into the solder at any given temperature, which may degrade solder joint strength.

This technique can be useful for the repair of opens or the installation of new or change-order components onto completed circuit assemblies. Because no flux is required, a previously cleaned board will stay clean through repair or upgrading operations. This technique is applicable to all peripherally leaded surface-mount components. It can also accommodate through-hole device soldering. Equipment availability is limited, with only a few manufacturers worldwide. Several past and recent publications provide a comprehensive review of this technology's applicability and attributes.[31–34]

43.13   REPAIR AND REWORK

As boards become more densely populated, the probability of rework increases. So does the complexity of rework steps. Deep preheating cycles, careful thermal profiling, and thermal masking of adjacent components become critical. Printed circuit assemblies (PCAs) today are often mix-mount (SMT and PTH). Additionally, they may contain press-fit components that may be affected by overheating.

Aside from hand repair with a soldering iron, there are two techniques that are the mainstays of repair: hot gas and solder fountain. The former has been discussed in Sec. 43.11, while the latter is a variant of wave soldering (Sec. 43.7). Hot-gas (also known as hot-air) soldering is used primarily for removal and replacement of SMT components, while the solder fountain has become the standard for PTH repair. Because repair of dense assemblies is difficult, some additional descriptions and tips for repair are covered below.

43.13.1   Hot Gas

This technique consists of directing a heated stream of air or other process gas such as nitrogen at the component to be removed. Once the solder joints are molten, the component can be pulled from the board. Next, the solder lands on the PWB can be prepared for component replacement. A new component is placed, and the hot gas is used to heat the package and the board to the point that solder joints are formed.

During the component removal step, solder surface tension capriciously allows some of the solder from the joint to be removed with the package lead. In other instances on the same package, the solder remains with the board. In the case of the BGA, this can be extreme because the ball may remain with the package or with the land. Ceramic column grid arrays (CCGAs), which have high-melting-point solder columns for interconnection, may leave some columns with the boards while the balance come off with the package. The same is true for some ceramic ball grid arrays that have noncollapsible, high-melting-point solder balls that may separate from the package during rework, remaining on some solder pads. Therefore, solder land preparation is a labor-intensive operation. But care at this step is critical for success in resoldering a fresh component onto the used lands.

To prepare the pads for soldering, excess solder, flux, and debris must be removed. This is typically accomplished with a hot-air solder vacuum iron or a hand iron with braided copper wick. The hot-air solder vacuum iron is a hand variant of the hot-gas soldering system with a built-in vacuum nozzle that can pick up molten solder. Any remaining CCGA columns or CBGA balls on the pads need to be tweezed off. Solder lands have to be smoothed with a hot soldering iron and some solder may need to be added to achieve a reasonably smooth pad and pad-to-pad uniformity. Care must be taken to eliminate any solidified solder peaks that will interfere with fresh component placement.

In the case of CCGAs, CBGAs, or other surface-mount components, solder is necessarily added to the PWB lands. This can be done with flux-cored solder wire and a hot iron, but it is difficult to achieve uniformity within a pad or from pad to pad. To ensure consistency in solder deposition, solder paste can be deposited in preparation for component replacement. This is accomplished with the use of a miniature solder paste stencil tailored for the component site. A miniature spatula or even a razor blade is used to squeegee the solder paste onto the site once the stencil apertures are visually aligned to the solder lands. Solder paste can also be applied with a dispensing machine if the solder lands are not too fine in geometry.

In the case of a typical BGA, fresh solder does not have to be added. Instead, the solder from the ball of the BGA is considered sufficient for joint formation upon repair. But solder paste can be applied to gain increased joint reliability. It can also yield better process control because the paste can compensate for differences in ball sizes or solder volumes on the prepared pads. Therefore, opens are less likely to occur, although more process precision is required to preclude solder bridging from paste squeeze-out.

Carefully thermocouple the area to be reworked as well as surrounding solder joints if possible. A dedicated populated board for this is most useful. Thermocouples can be embedded in solder joints and in the board. Time-temperature profiles can be established and transferred to the actual board being reworked.

There are several hot-gas nozzle styles. There are nozzles that direct the hot gas stream to the top of the package. Others direct the hot gases to the package periphery or even underneath the package (useful for BGAs). Some nozzles exhaust the gas stream from the top, others from the sides or from a single side. Package type, thermometry, and experience dictate nozzle effectiveness and choice for a particular application.

Most assemblies can be reworked multiple times. There is no firm rule, but PCAs can be reflowed twice (once for each side of SMT), can be wave soldered once, and can undergo at least two to three rework cycles. The key in all of these steps is to limit the amount of time at or near solder liquidus. Each time the PCA is soldered, the intermetallic compounds (IMCs), which form at the interface of the solder and the pad, thicken. The thicker the IMC, the more brittle the solder joint, so it is always prudent to limit thermal exposure of the assembly. During component rework and replacement, this rule is often overlooked. Too often, overly hot preheating cycles are used and dwells at liquidus are prolonged as compared to a normal mass reflow soldering cycle. Remember that when a component is removed and replaced, the IMC from previous soldering operations or other thermal excursions remain on the pad. When a new component is soldered to the same pads, the IMC is given further opportunity to grow. Because the IMC is brittle, this is where solder joint fatigue fracture is most likely to initiate.

Many of the hot-gas repair machines on today's market incorporate placement capability and a mechanism to remove the component from the board surface once all solder joints are molten. The machine does not necessarily sense that the solder joints have reached liquidus. Instead, the machine is programmed to lift the component at a certain point in a preprogrammed thermal profile or a gentle upward pressure is applied to the component throughout the thermal cycle. When the solder joints go liquid, the mechanism is free to retract the component from the board surface and the cooldown cycle begins. This is a much safer method than relying on pliers, gloved hands, or some other probe to remove a component, because if some of the solder joints have not reached liquidus, the leverage of the operation may be enough to rip SMT pads off the circuit board along with the reworked device. Generally, the lift mechanisms in the hot gas rework stations are gentle enough or are settable, precluding this type of irreparable damage to the PCA.

43.13.2 Solder Fountain

Plated through-hole components are generally repaired by solder fountain, which is akin to a wave-soldering system with a pumped standing wave of solder. The solder fountain is generally smaller than a wave-soldering machine and often lacks conveyors and preheaters. Dense assemblies can be batch-heated in a box oven to make the repair process more efficient and limit exposure of the board to the turbulence of the molten solder wave.

As mentioned earlier, limiting the soldered assembly's time near or above solder liquidus is key for solder joint reliability. But in using the solder fountain, there is another time-based consideration also. If the PCA is exposed to the turbulent molten solder of the fountain for too long, the PTH pads and barrels can be thinned or even dissolved. For this reason, precise process control, including strict exposure time limits and minimum solder pot temperature, must be imposed for this largely manual operation.

The PCA is generally masked by a glass-epoxy carrier (pallet) and the solder fountain is shaped by means of a nozzle that limits solder flow to the area of the board requiring repair. If the volume of the solder in the fountain reservoir is small and the thermal mass of the board is large, necessarily long dwell times in the molten solder may be required. This can lead to the pad and barrel thinning and dissolution discussed previously, so ensure that a machine is matched to the type of PCAs that it will be used to rework. Once the component solder joints go molten, the component can be pulled out with gloved hand or pliers, some flux can be added, and a new component can be added to complete the rework within one thermal excursion. Care must be taken to ensure that all solder joints have gone molten on the component being reworked. If not, PTH barrels can be ripped out of the board or barrel annuli can be damaged.

43.13.3 Laser

Laser is the latest innovation in rework and repair. The fundamentals of laser for initial soldering or for repair have already been discussed in a previous section. In the commercial incarnation of this technique, a laser beam is quickly scanned around component leads or, in the case of area array devices (BGA, CBGA, CCGA, CSP, etc.), the body itself is heated. Otherwise, there is little difference between this and alternate rework/repair techniques.

Since the energy of the laser beam is more tightly confined to the area it irradiates, the heating is more localized. This can be advantageous when trying to remove or replace a component on a densely populated, double-sided PCA. Otherwise, the same caveats must be kept in mind regarding IMC formation and pulling components before fully reflowed. Add to this list the need to carefully assess the assembly for laser damage threshold so as not to char components or PCAs.

43.13.4 Considerations for Repair

43.13.4.1 Chemical Considerations. Care must be taken when using fluxes for the rework and repair process. If an aqueous clean solder chemistry is used for building the PCA, then either no-clean or aqueous flux formulations can be used. If an aqueous chemistry is used for the repair, then the board will need to be subjected to another aqueous cleaning cycle. Be sure to check that all components at that stage are compatible with the aqueous cleaning process.

No-clean solder flux formulations can be used for rework and repair even if the PCA was manufactured with an aqueous clean chemistry, although the reverse is not recommended. Generally, when a no-clean board is subjected to an aqueous cleaning, the no-clean flux residues take on a white, gooey characteristic that is conducive to dendritic crystal growth and soft or hard shorting as a result. Saponified aqueous cleaning can be used, but must be tested for effectiveness in removing the polymerized flux residue from the no-clean process.

Another consideration with no-clean solder chemistries is that of pooled, unactivated flux residue. No-clean solder flux is generally harmless once it has gone through the thermal cycle associated with solder reflow. It is activated and denatured by the heat. If the flux is applied generously and the flux pool does not see the high temperatures associated with solder joint formation, the unheated flux can cause reliability problems due to dendritic corrosion in the long term. There are commercial flux solvents that can be used to dilute and wash away excess flux; otherwise, the assembly can be baked briefly to about 120°C to denature the flux. Check with the flux manufacturer for specific recommendations regarding flux residue management.

43.13.4.2 *Component Bake-out.*

As previously mentioned, components that have been sitting out may be subject to water absorption. Then, when the components are heated during rework, where the heating cycle is more localized and perhaps more extreme than the slow heating experienced during mass oven reflow soldering, moisture-induced popcorning may be a more serious problem. Follow IPC or component manufacturer recommendations for pre-soldering moisture bake-out.

43.13.4.3 *Gold Finger Protection.*

There is ample opportunity to ruin gold edge connector fingers on a circuit board during the repair operation. Care should be taken to keep solder and flux away from gold fingers during repair. Mechanical shields or polyimide tape (Kapton®) can be used to protect fingers from smeared solder paste, escaped solder debris, or flux. If solder comes into contact with gold, the gold will rapidly dissolve into the solder mass. The solder cannot be removed from the gold finger without scraping and replating, a costly process.

Keep solder away from gold fingers. Keep finger temperature as low as possible. Ensure that gold fingers are never handled, even with gloved hands. After rework, clean gold fingers with flux residue removal solvent and a brush or lint-free cloth. Clean each side separately. Avoid folding the lint-free cloth over both edges and swiping repeatedly; the cloth will wear against the glass/epoxy and shred, leaving lint on the board. A better alternative is to use a brush with solvent to clean the gold fingers and follow with a filtered air blow-off to dry.

REFERENCES

1. R. Iman et al., "Evaluation of a No-Clean Soldering Process Designed to Eliminate the Use of Ozone Depleting Chemicals," IWRP CRADA No. CR91-1026, issued by Sandia National Laboratories, 1991.

2. H. H. Hartmann, "Soft Soldering Under Cover Gas— A Contribution to Environmental Protection," *Elektr. Prod Und Prftechnik,* H.4, 1989, pp. 37–39.

3. II. J. Hartmann, "Nitrogen Atmosphere Soldering," *Circuits Assembly,* January 1991.

4. D. Bernier, "The Effects of Metallic Impurities on the Wetting Properties of Solder," *Proceedings of the First Printed Circuit World Convention,* Uxbridge, UK, 1978, 2.5/1.5, vol. 2.

5. C. Hallmark, K. Langston, and C. Tulkoff, "Double Reflow: Degrading Fine Pitch Joints in the Wave Soldering Process," *Technical Proceedings of NEPCON West,* 1994, pp. 695–705.

6. I. N. Sax, *Dangerous Properties of Industrial Materials,* 5th ed., Van Nostrand Reinhold, New York, 1979, pp. 766–769, 1032.

7. F. W. Kear, "The Dynamics of Joint Formation," *Circuits Assembly,* October 1992, pp. 38–41.

8. S. Adams, "New Applications in Electronic Soldering," *Technology Magazine,* October 1992.

9. R. Trovato, "Soldering Without Cleaning," *Circuits Magazine,* April 1990.

10. U.S. Patent 5,121,875, "Wave Soldering in a Protective Atmosphere Enclosure Over a Solder Pot," L. Hagerty et al., June 1992.

11. C. F. Coombs, Jr. (ed.), *Printed Circuits Handbook,* 3d ed., McGraw-Hill, New York, pp. 27.18–27.30.

12. Manko, *Solders and Soldering,* 2d ed., McGraw-Hill, New York, 1979.

13. R. J. K. Wassink, *Soldering in Electronics,* Electrochemical Publications, Ltd., Ayre, Scotland, 1984.

14. C. L. Hutchins, "Soldering Surface Mount Assemblies," *Electronic Packaging & Production,* Supplement, August 1992, pp. 47–53.

15. U.S. Patent No. 5,083,007, "Bonding Metal Electrical Members with a Frequency Doubled Pulsed Laser Beam," P. J. Spletter and R. R. Goruganthu, January 1992.

16. U.S. Patent No. 5,227,604, "Atmospheric Pressure Gaseous-Flux-Assisted LASER Reflow Soldering," G. Freedman, July 1993.

17. "Properties of Lead and Lead Alloys," Lead Industries Association, Inc., 292 Madison Ave., New York, NY 10017.

18. S. Charschan, *LASERs in Industry,* LASER Institute of America, Toledo, OH, 1972, p. 116.

19. E. F. Lish, "LASER Attachment of Surface Mounted Components to Printed Wiring Boards," *6th Annual Soldering Technology Seminar,* Naval Weapons Center, China Lakes, CA, February 1982.

20. J. Hecht, *Understanding LASERs,* Howard W. Sams & Co., Indianapolis, IN, 1988.

21. J. F. Ready, *Industrial Applications of LASERs,* Academic Press, New York, 1978.

22. F. Burns and C. Zyetz, "LASER Microsoldering," *Electronic Packaging & Production,* May 1981, pp. 109–120.

23. D. U. Chang, "Analytical Investigation of Thick Film Ignition Module Soldering by LASER," *Proceedings of ICALEO '85,* November 1985, pp. 27–38.

24. M. Hartmann et al., "Experimental Investigations in LASER Microsoldering," *SPIE,* vol. 1598, LASERs in Microelectronics Manufacturing, 1991, pp. 175–185.

25. S. Hernandez, "Wirebonding with CO_2 LASERs," *Surface Mount Technology,* March 1990, pp. 23–26.

26. E. Wright, "LASER vs. Vapor Phase Soldering," *Proceedings of the 30th National SAMPE Symposium,* March 1985, pp. 194–201.

27. D. Waller, L. Colella, and R. Pacheco, "Thermode Structure Having Elongated, Thermally Stable Blade," July 1993.

28. J. A. Wilkins, "Heat Transfer Control for Hot Bar Soldering," *Proceedings of Surface Mount International,* 1993, pp. 186–192.

29. R. Gunkel, "Solder Aluminum Joints Ultrasonically," *Welding Design and Fabrication,* vol. 52, no. 9, September 1979, pp. 90–92.

30. J. L. Schuster and R. J. Chilko, "Ultrasonic Soldering of Aluminum Heat Exchangers (Air Conditioning Coils)," *Welding Journal (USA),* vol. 54, no. 10, October 1975, pp. 711–717.

31. F. M. Hosking, D. R. Frear, P. T. Vianco, and D. M. Keicher, "SNL Initiatives in Electronic Fluxless Soldering," *Proceedings of the 1st International Congress on Environmentally Conscious Manufacturing,* Santa Fe, NM, September 18, 1991.

32. P. T. Vianco, J. A. Rejent, and F. M. Hosking, "Applications-Oriented Studies in Ultrasonic Soldering," *Proceedings of the American Welding Society Convention and Annual Meeting,* Cleveland, OH, 1993.

33. J. N. Antonevich, "Fundamentals of Ultrasonic Soldering," American Welding Society, *4th International Soldering Conference, Welding Journal, Research Supplement,* vol. 55, July 1976, pp. 200-s–207-s.

34. A. Shoh, "Industrial Applications of Ultrasound," *IEEE Transactions on Sonics and Ultrasonics,* vol. SU-22, March 1975, pp. 60–71.

CHAPTER 44
NO-CLEAN ASSEMBLY PROCESS

Paul W. Henderson
*Hewlett-Packard Electronic Assembly Development Center,
Palo Alto, California*

44.1   INTRODUCTION

Changes in environmental regulations governing the use of chlorine-containing solvents (particularly chloroflurocarbons or CFCs) forced the electronics industry to seek other methods of ensuring that printed circuit assemblies are free of harmful flux residues. Through 1992, the most common means of defluxing PCAs was to use some type of CFC defluxing process. When the laws changed, assemblers were faced with the choice of either converting to some other type of defluxing or changing their manufacturing processes in ways such that no defluxing of the finished assemblies would be required. The motivation behind the transition to no-clean processing continues to strengthen as environmental regulations for air, water, and waste discharge become progressively more stringent and opportunities to reduce costs are realized.

The transition to no-clean processing has been nearly as sweeping and traumatic as was the shift from through-hole to surface-mount assembly. Large electronics manufacturers spent tens of millions of dollars each on engineering, capital equipment, product requalification, and initial quality problems, all associated with the conversion to no-clean. The pain was also felt by printed circuit fabricators, flux, solder paste and solder mask suppliers, and equipment vendors who suddenly were called upon to meet new requirements imposed on them by their downstream customers. The lessons which were learned will allow others to avoid many of the pitfalls and expense which early adopters of the technology encountered.

This chapter will discuss some different approaches to no-clean processing, highlight some of the differences between no-clean and other processes which require cleaning, discuss the key success factors of no-clean implementation, and present a troubleshooting guide for no-clean. This chapter is intended to be of use to fabricators of printed circuit boards, assemblers, and those purchasing finished assemblies.

44.2   DEFINITION OF NO-CLEAN

The definition of a no-clean printed circuit assembly process is one in which the post-solder defluxing step has been eliminated. It does not mean, as the name implies, that cleaning is eliminated entirely. Wave-solder pallets, misprinted boards, and tools must still be cleaned.

Generally speaking, there are two main types of no-clean processes: low-residue processes and leave-on processes. There are some novel processes which deliver very low residue performance, but for various reasons (chiefly cost related), they have not become mainstream technologies.

44.2.1 Definition of a Low-Residue No-Clean Process

The low-residue category uses special fluxes, solder pastes, soldering atmospheres, and equipment all designed to minimize the amount of residue left on the board after assembly. The small amount of residue left from these materials can be difficult to detect without magnification. Low-residue materials interfere least with pin testing.

44.2.2 Definition of a Leave-On No-Clean Process

The leave-on category uses fluxes and solder pastes which are frequently low-activity versions of standard materials and have two to three times more residue than the low-residue materials. Leave-on materials do not require special soldering atmospheres or equipment but perform better in them. Leave-on materials have the advantage of being compatible with standard equipment and are usually more robust than the low-residue materials. The drawback of leave-on processes is that they produce flux residues that are readily detected by the naked eye and may interfere with pin probe testing. Because the activators are encapsulated in the residue, spot cleaning to enhance testability can be detrimental to reliability of the assembly, because those activators can be exposed. With leave-on materials it is best to clean thoroughly or not at all. Leave-on residues can also interfere with conformal coating.

44.3 CLEANING OR NO-CLEAN?

If the goal is to minimize cost on simple assemblies with no need for pin testing or concern for cosmetics, then a leave-on process is a good choice. Leave-on processes are attractive because they represent the least dramatic shift and are more tolerant to process variation. If there is no compelling reason to move to a low-residue process, the simpler and less expensive route is to use a leave-on process or some combination of low-residue and leave-on materials to achieve the optimal result. Unfortunately, committing to a leave-on process also means limiting the types of assemblies that can be built in the factory to ones which are compatible with a high-residue process.

Cleaning may be absolutely required in a very few circumstances. Choosing which type of process to implement depends mainly on the requirements of the assembly customer. In other cases, when the board is complex enough to require pin testing or cosmetics are an issue, a low-residue process may be the only choice that does not require cleaning. Assemblers frequently choose a low-residue process because they are forced to by the requirements of the product. To help select a strategy, consider the questions in Table 44.1 to identify the constraints which would direct the choice of one technology or another.

Replacing a mature, well-understood cleaning process can be expensive and requires a substantial effort by all parts of the supply chain. Most assemblers who have made the shift to no-clean have not done so out of environmental altruism alone, but have made the commitment to change only after a careful economic analysis of their operations indicated that no-clean would represent a lower overall cost of ownership.

Though it can be costly to implement, the shift to no-clean is frequently less expensive than aqueous or semiaqueous defluxing. The true cost of ownership can be evaluated only if the assemblers' cost models are detailed enough to provide a true understanding of the current and proposed processes. An accurate analysis must be total cost oriented and consider things that may not be reflected in the current cost model. Table 44.2 illustrates some examples of costs which may be eliminated and some new costs which are associated with changing to no-clean.

TABLE 44.1   Choosing Which Process to Use

Question:	If so, choose:
Will any residues have a negative effect on the electrical performance of the product during its functional life (e.g., corrosion or current leakage)?	For products with a long service life or products intended for a harsh service environment, cleaning or a more rigorous no-clean process qualification may be required.
Does the product operate at very high frequencies (e.g., over 50 MHz) or have very high impedance requirements ($>10^{12}$ $\Omega/\square$)?	A low-residue process or cleaning may be the only option to remove nonionic residues that can act as insulators, which can contribute to cross talk at high frequencies[1] and reduce the impedance between lines.
Does the product use guard traces to protect sensitive circuitry which could be rendered ineffective by residues?	If the guard traces are only on the wave-solder side of the board, a hybrid process using low-residue no-clean materials and a water or semiaqueous cleaning step would be effective for those assemblies. All other assemblies could be handled as no-clean.
Is the customer concerned about the cosmetic appearance of residues?	If so, a low-residue no-clean process may be preferable to a leave-on process.
Would residues have a negative effect on other downstream processes such as in-circuit test or conformal coating?	Cleaning or a low-residue no-clean process can be used with materials chosen, with particular emphasis on compatibility with test and coatings.
Is the assembly dense enough to require probing of solder pasted features (leads or pads)?	Cleaning or use of new types of test probes which rotate or use higher force have been shown to be effective with a low-residue process.

TABLE 44.2   Costs Eliminated and Added by No-Clean

Costs eliminated	Costs added
Floor space for defluxing and water purification equipment	New equipment for reflow and wave solder
Engineering and technical support for defluxing and water purification equipment	Process development
Less maintenance, material used, and waste generated from dross formation	Nitrogen facilities and material
Elimination of hand soldering of cleaner sensitive components	Better hand-soldering tools
Less nonvalue-added touchup	Operator training
	Product requalification

Other cost savings which are real but difficult to quantify are:

- Concerns about component compatibility with cleaning are eliminated.
- Water or solvent is entrapped in connectors or electrolytic capacitors.
- Flux deposits in connectors and switches are reduced.
- Contact poisoning by cleaning materials is eliminated.
- Many no-clean solder pastes are more robust than comparable water-clean pastes, resulting in fewer open solder joint defects and requiring less frequent replacement on the stencil.

No-clean processes are generally less expensive to operate than clean processes, given that the defluxing step and all of the associated operating, support, and occupancy costs are eliminated. However, the idea that all cleaning is eliminated from the process is not, strictly speaking, correct. There remains the need to wash some incoming PCBs or components if contaminated, remove solder balls from some assemblies, wash misprinted PCBs and stencils, and clean wave-solder fixtures. There are also several other new costs added by the switch to no-clean. For certain types of components, the stencil apertures are different for the no-clean process, requiring the replacement of stencils for existing designs. The stencil-washing operation may require a different chemistry (e.g., saponified water or semiaqueous), which may require different equipment. Some special provision must be made to deal with misprinted PCBs; previously, misprinted boards would simply be passed through the cleaning equipment. Facilities are required to support nitrogen storage or generation and have significant associated operating costs. Reflow is best accomplished in a nitrogen atmosphere, which requires that existing equipment be replaced or retrofitted. Wave solder performance is enhanced by nitrogen, which may require replacement, or at least a retrofit to existing equipment.

Whether a process is considered low-residue or leave-on is based on the amount of residual solids remaining in the flux after reflow, or wave, solder, as shown in Table 44.3.

TABLE 44.3   A Comparison of the Solids Content of Low-Residue and Leave-on Materials

Percent solids	Leave-on	Low-residue	Very low residue
Wave-solder flux	15–40%	1.5–4.0%	<1.5%
Solder paste	40–70%	20–30%	<20%

The role of the residues in the solder paste is very important. Rosin or resins serve to protect the solder particles in the paste from becoming oxidized in reflow, promote tack, activate the surfaces to be joined, and promote wetting. Other additives promote better rheology of the paste. All of the materials which do not volatilize in the reflow process are left behind on the assembly.

These categorizations are only meant to be representative. In practice, a wide variety of different combinations of materials may be used to achieve different results. For example, many assemblers will use a leave-on solder paste with a low-residue wave-solder flux to maximize the robustness of the SMT portion of the process and minimize the residues left on the wave-solder side of the assembly.

A few assemblers use a hybrid process where they use a leave-on solder paste and an organic acid wave-solder flux followed by DI water cleaning to compensate for components which are marginally solderable. The leave-on residues are unaffected by the DI water clean while the organic acid flux residues are removed. Other pure surface-mount products follow the same process with the exception of wave solder and cleaning.

44.4 IMPLEMENTING NO-CLEAN

Before beginning to implement no-clean, it is important to study the present cleaning equipment very carefully and log everything that goes into it. In addition to soldered assemblies, there are many other things that are routinely put through the cleaner: misprinted boards, parts in baskets, and reworked assemblies. There are many quality issues that can be hidden by cleaning: large numbers of misprinted boards, assemblies that are difficult to test, use of an aggressive flux to compensate for poor solderability, etc.

Nearly every assembly process step requires some adaptation to work successfully in the no-clean environment. Other parts of the organization not normally associated with production are also affected. Cost models will change to reflect differences in process. Procurement must get involved to control soldering materials coming into the factory. Materials engineering must determine which parts will be qualified and how; information systems and databases may change to include which parts are qualified and which are not. Inspection criteria and SPC procedures will change, requiring new training for operators and inspectors. Documentation must be updated to include new procedures to maintain ISO compliance.

44.4.1 Incoming Quality Assurance (IQA)

Procedures in IQA must be modified to include verification of solderability and cleanliness of incoming boards and components. If the assembler chooses not to conduct these tests in-house, the supplier's methods or other service must be evaluated. Additional resources may be required to work with component and PCB suppliers to bring their products into compliance.

44.4.2 Stencil Aperture Design and Stencil Finish

In order to prevent the formation of solder beads (small solder balls next to passive components on the solder-pasted side of the board), the stencil apertures must be modified to prevent the solder paste from being squeezed underneath the component during placement. An example of solder beading and several possible aperture modifications designed to eliminate them are shown in Fig. 44.1.

FIGURE 44.1   Stencil apertures used to eliminate solder beads.

The surface finish of the stencil can make a significant difference in the robustness of the stencil-printing process for pitches ≤0.65 mm. There are many materials and processes available to make stencils (plated Ni, molybdenum foils, laser etching, etc.), but the most important principle is that the walls of the aperture must be smooth in order to consistently release the paste. The least expensive way to accomplish this is to use a standard stencil material (e.g., alloy 42), with a subtractive etching process and electropolish the stencil afterward to smooth the surface. Electropolishing adds 10 to 20 percent to the cost of the stencil, but the result is well worthwhile.

44.4.3 Stencil Washing

The stencil-washing operation, which is a frequently overlooked part of the surface-mount process, is critical to the success of the process overall. If the stencil is inadequately cleaned, flux residues will adhere to the side walls of the apertures, making it difficult for the solder paste bricks to release from the stencil during printing. This problem causes insufficient solder joints and open solder joints on finer pitches, and is difficult to diagnose. It is extremely important to actively monitor the concentration of saponifier or the condition of the cleaning medium. If the saponifier concentration falls to an out-of-control low level, partially saponified flux residues will be left behind which are not visually obvious. These residues can be removed by applying alcohol liberally to the stencil and wiping dry with a clean, lint-free cloth. Scraping excess solder paste from the stencil with a polymer squeegee will aid in cleaning and help minimize the lead content of the effluent. Stencils should be washed as soon as possible after removing them from the stencil printer to prevent solder paste flux residues from drying in the stencil apertures.

44.4.4 Cleaning Misprinted PCBs

Cleaning misprinted boards requires some special consideration. Merely scraping the solder paste off and wiping the board with an alcohol wiper does not do an adequate job of removing the solder particles from between the lands and the solder mask. In some cases, the same equipment can be used for washing stencils and washing misprinted boards. With most solder pastes, misprinted boards should be washed within an hour of printing.

44.4.5 Stencil Printing

Low-residue solder paste rheology is more sensitive than most RMA pastes and anything that can be done to improve the consistency of stencil printing is worthwhile. The most important factor is to have a consistent process for setting up and operating the stencil printer which is followed by all of the operators.

Proper choice of squeegee material can have a significant impact on stencil-print process yield. Polymer squeegees have a tendency to deflect into stencil apertures in the direction perpendicular to squeegee travel and scoop out solder paste, leaving peaks of solder paste at the end of the printed bricks. This effect contributes to variability in the amount of solder paste deposited. This variability can be reduced dramatically by using metal squeegees which stay planar with the top surface of the stencil and do not scoop out the paste.[2]

Since low-residue solder pastes have more volatile components, the useful working time of the paste on the stencil is limited to 4 to 6 h. At the end of this time, the paste should be removed from the stencil and discarded, the stencil washed, and production resumed with fresh paste. The best indication that the paste is past its useful life is when insufficient paste defects suddenly begin to occur on 0.5-mm pitch patterns. Mixing fresh paste into the old paste on the stencil is not recommended; it only extends the usefulness of the paste for a few more prints. Humidity above 50 percent RH or below 10 percent RH will further reduce the life of the paste on the stencil. Making a plot of insufficients versus time will show at what point the material breaks down.

44.4.6 Pick and Place

The pick-and-place process itself is not significantly affected by no-clean, but queuing times before component placement can be limited with a low-residue process. Because the tack time is shorter with a low-residue process, there is less flexibility with printing a large number of boards and holding them in a buffer before pick and place. Similarly, tack will decrease after the component has been placed and before the board is reflowed. In practice, if a board has not been placed before 75 percent of the tack time has passed, it should be treated as a misprint, cleaned, and printed again. Humidity above 50 percent will reduce tack time significantly. See Table 44.4.

TABLE 44.4 Tack Times of Solder Pastes

Solder paste type	Average tack time, h
RMA/leave-on	>24
Low-residue no-clean	2–6
Water-clean	1–2

44.4.7 Nitrogen

Depending on the volume and purity of nitrogen required for the facility, nitrogen can be either delivered to the site in bulk liquid form or generated on-site through a variety of methods[3] (membrane separation, pressure swing absorption, etc.). Nitrogen promotes solder wetting by increasing the wetting force and decreasing the wetting time.[4] Since little oxygen is present, discoloration of laminates, light-colored plastic connectors, and flux residues is eliminated. The purity of nitrogen required is determined by the type of process. (See Table 44.5.)

TABLE 44.5 Nitrogen Purity Requirements

No-clean process type	Maximum allowable oxygen concentration
Leave-on	21% (air)
Low-residue	100–500 ppm
Very low residue	50–100 ppm

Nitrogen represents a significant operating cost ranging from $10 an hour per reflow oven or wave-solder machine in the United States to more than $80 in Singapore. For this reason alone, many assemblers in the Far East have chosen a leave-on process which can tolerate processing in air. However, in wave solder, it can be difficult to determine the difference in defect rates between soldering in a 5-ppm oxygen atmosphere and a 1000-ppm atmosphere. At 1000 ppm there is still a benefit of reduced dross formation, and on-site generation is a better alternative than not inerting the process.

44.4.8 Reflow

Many studies have indicated that reflow performance is significantly enhanced by the use of nitrogen.[5,6] Many assemblers use the transition to no-clean as an opportunity to upgrade their reflow equipment to fully convective ovens. A highly convective reflow environment enables better temperature control of the whole assembly with a smaller total temperature difference between the most and least dense parts of the assembly.

The penalty for this superior performance is more complicated, time consuming, and expensive maintenance of the equipment. The portion of the solder paste which volatilizes tends to redeposit on the cooler surfaces in the oven, requiring regular disassembly of the equipment for thorough cleaning, particularly if the oven uses a special cooling zone at the end. The lower the residue of the solder paste material, the more material there is to redeposit in the oven and exhaust. The volatile residues also tend to block the sampling ports of the oxygen analyzer, giving false readings (this can be readily diagnosed by checking the flow meter on the analyzer).

44.4.9 Wave Solder

Flux selection for wave solder depends primarily on the priority placed on different performance characteristics of the flux:

- Solder defect rate (skips, icicling, solder webbing, etc.)
- Visible residue level
- Compatibility with other process chemistries
- Compatibility with in-circuit test
- Surface insulation resistance
- Ionic contamination level
- Volatile organic compound (VOC) content
- Consistency of deposition

In many cases, one factor may dominate the decision-making process. For example, if a customer is very sensitive to any visible residues, residue level may be the driving factor. In locations where VOC emissions are strictly regulated, a VOC-free flux may be the only option. These fluxes use water instead of alcohol as a solvent and are not flammable. They do require additional preheating to evaporate the water. If any water remains on the bottom side of the board when it enters the solder wave, it will cause the molten solder to spatter, causing solder balls on the bottom surface of the board.

Flux can be applied with standard foam or wave fluxers, but most spray application methods are superior.[7] Foam and wave fluxing are difficult to control for most low-residue fluxes because of the high alcohol content (about 98 percent) and the rapid evaporation rate. Specific gravity monitoring is not possible for fluxes with a solids content of less than 5 percent. Instead, frequent titration is required to monitor the flux composition. The rapid evaporation of alcohol in the flux contributes significantly to VOC emissions. Foam fluxers require frequent cleaning, and contaminated flux must be disposed of as hazardous waste.

Spray fluxers may use a rotating mesh drum, an ultrasonic head, air spray, air-assisted airless spray, or other novel approaches. The most robust methods keep the flux in a sealed container so that the flux is never contaminated, composition does not vary, and it does not need to be controlled.[8] Spray fluxing offers more control over the quantity of flux deposited on the board, so that the minimum amount of flux needed may be applied. Spray flux methods can easily reduce flux and thinner usage by 70 to 90 percent compared to foam or wave fluxing. Table 44.6 illustrates some of the advantages and disadvantages of the different approaches to flux application.

The biggest difference in wave-solder equipment for no-clean is the addition of nitrogen inerting. Some benefits of inert wave soldering are 80 to 95 percent less dross formation, fewer solder skips, less icicling, solder webbing, and less makeup solder is required. Switching to inert wave soldering can often be justified on the basis of the reduced maintenance, downtime, and hazardous waste disposal associated with dross removal. Inert soldering will not, however, overcome poor design or solderability.[9] Dross formation is directly related to the amount of oxygen in the soldering atmosphere, as shown in Table 44.7.

Soldering equipment and retrofit options vary in the degree to which the equipment is inerted. The wave-solder machine may be fully inerted, have a short hood which covers a portion of the preheat and the solder pot, or only inerting the solder pot. Some of the advantages and disadvantages[10] of the different types of inerting are shown in Table 44.8.

Board design issues for no-clean are mostly generic to wave soldering. The biggest problems are solder skips on SOTs, ICs, Tantalums, and thick chip components. Component orientation is critical: The surfaces to be soldered must be perpendicular to the wave. Care must be exercised to prevent components from being shadowed by one another. Pad design is an important secondary effect.[11]

TABLE 44.6 Advantages and Disadvantages of Flux Application Methods

Flux application method	Advantages	Disadvantages
Foam or wave	Inexpensive equipment	Difficult to control flux composition and deposition quantity, high solvent evaporation loss, freqeuent cleaning required
Rotating drum spray	Simple, flux deposition quantity is more controllable than foam or wave	Same problems as foam and wave since the drum rotates in an open reservoir of flux
Ultrasonic spray	Good control of flux deposition quantity, excellent composition control	Sensitive to variations in exhaust, some variation from center to edge on large assemblies
Air spray	Good control of flux deposition quantity, good center-to-edge distribution control, excellent composition control, low maintenance	Significant overspray
Air-assisted airless spray	Good control of flux deposition quantity, good center-to-edge distribution control, excellent composition control, low maintenance	Significant overspray, more variation than airspray

TABLE 44.7 Dross Formation Rates as a Function of Oxygen Level

Oxygen level, ppm	Relative dross formation rate
5	1
50	2
500	4
1000	5.2
5000	9
10,000	10.8

TABLE 44.8 Advantages and Disadvantages of Inerting Methods

Wave solder inerting method	Advantages	Disadvantages
Fully inerted	Prevents oxidation of board and components in preheat, lowest oxygen consumption (600–1200 SCFH)	Dedicated equipment—very expensive
Short hood retrofit	Reduces oxidation of board and components in preheat	Moderate oxygen consumption (>1400 SCFH)
Pot-only retrofit	Simple	Highest oxygen consumption (up to 2400 SCFH), does not prevent oxidation of board and components

44.4.10 Hand Soldering

Hand soldering and rework are made significantly more difficult by the use of low-residue no-clean materials. Leave-on rework fluxes and cored solders are similar to those which require cleaning, except that they are less active. The relatively high solids content of leave-on materials performs several important functions in soldering: it wets the surface to be soldered, activating it and providing good heat transfer between the tip of the soldering tool and the workpiece, it protects the tinned surface of the tip, and it provides protection against oxidation during the heating cycle. Because the flux residues remain on the workpiece, longer heating cycles can be used to overcome marginal solderability and less attention needs to be paid to soldering temperatures and tip maintenance.[12]

Low-residue hand soldering has none of these advantages. The low residues activate more quickly and at lower temperatures, volatilize while soldering, and require good solderability of the workpiece. The residue does not protect the tinned surface of the soldering tool well and requires that the tips be maintained much more frequently. Soldering must be done at the lowest possible temperature and as quickly as possible to prevent oxidation buildup on the tip. In addition to the flux in the cored solder, many instances will require a liquid flux to be applied with a fine brush to enable both the tip and the workpiece to be adequately cleaned. After soldering, some of the residues may be further volatilized by passing a stream of hot nitrogen over the hand-soldered joints with a hot-air pencil; this also ensures that all of the flux has been fully activated. Even so, some assemblies may require spot cleaning to enable testing and enhance cosmetics.

The key to successful implementation of low-residue hand soldering is training, discipline, use of appropriate soldering tools and rigorous maintenance. Even highly experienced hand-soldering operators will need to significantly change the way they work and will not readily accept the new materials without a good understanding of how they perform. Operators will need to:

- Learn to solder at the lowest possible temperature.
- Turn off soldering irons when not in use.
- Tin tips before storing them.
- Use tips that maximize contact between the tip and the workpiece (usually short, blunt ones).

Periodic review of soldering practices will help prevent operators from reverting to old practices. Trying to economize by purchasing cheap hand-soldering tools is a very bad idea. The better tools work by controlling the heat delivery to the tip and, though more expensive initially, last much longer and produce better quality results if maintained properly with tip tinners and tinning blocks.[13]

44.4.11 In-Circuit Testing

Test is one area which requires significant adaptation to be compatible with no-clean. In some cases, test pins should be replaced with high-force or rotating pins. In many cases where users of no-clean have had problems with test, the root cause was related to fixture design—rigorous adherence to fixture design rules is a must.

44.5 RELIABILITY OF NO-CLEAN PRODUCTS

The biggest concern related to no-clean processing is whether the assemblies will be reliable or not. But before discussing reliability, it is important to define the chemical nature of the residues which are left on the assembly after processing. The chemical nature of the residues will determine what tests to perform and what kind of results to expect. The most easily tested components of any residue left on the assembly are the residues contributed from the solder paste vehicle, wave-solder flux, rework flux, and the interactions of all of them. (See Table 44.9.) If a no-clean material is not fully activated (e.g., when flooding an area with liquid flux in a hand-soldering operation), the unreacted activators left behind can cause corrosion. The interactions between materials are important because two materials which are benign when separate can sometimes cause current leakage, corrosion, or dendritic growth when combined and exposed to high temperature or humidity, or are biased.[14,15] It is less easy to test other materials that may be present in the residue: ionic contamination from the PCB or components, lead and tin metallic salts which form when their oxides are removed, and organic byproducts which may be evolved when the PCB and components are elevated to reflow temperatures.

TABLE 44.9   Chemical Nature of No-Clean Residues

Residue type	Low-residue process	Leave-on process
Solder paste	Rheological agents, tackifiers, rosin or resin	Rosin, activators
Wave-solder flux	Adipic acid, succinic acid, surfactants	Rosin, activators
Rework flux	Adipic acid, succinic acid	Rosin, activators

Three of the most common reliability tests performed to qualify no-clean processes are surface insulation resistance testing (SIR), ionic contamination testing, and highly accelerated stress testing (HAST). In addition, product-specific reliability qualification testing which is appropriate to the intended service environment is essential to detect any unanticipated effects on the finished assembly. There are other methods for detecting residues (e.g., ion chromatography, high-performance liquid chromatography), but most are not well suited to a production environment because they are expensive to set up and operate and require highly trained people to operate and interpret them.

44.5.1   Surface Insulation Resistance Testing for No-Clean

Measuring SIR is probably the most effective and widely used method of determining if residues on an assembly are safe. One of the most accepted protocols is given in IPC-TM-650 section 2.6.3. It is particularly important to evaluate the interactive effect of the various materials used in the process. In some instances, incompatibilities have been found between solder paste flux and wave-solder flux, rework fluxes, and certain types of solder masks.

This test is commonly done by applying the material(s) to be tested to an IPC B-25 comb pattern, processing the test board or coupon through the normal process steps, and measuring the leakage current in an elevated temperature and humidity-controlled environment. Test conditions which are commonly used are 85°C, 85 percent RH and 40°C, 95 percent RH. It is important to choose a set of conditions to use and apply them consistently—it is very difficult to meaningfully compare results obtained at different temperature and humidity conditions. If a comb pattern is designed onto a sparse area of a production PCB, SIR can be used to monitor process performance.

44.5.2   Ionic Contamination Testing for No-Clean

Testing for ionic contamination via a solvent extraction (e.g., Omegameter, Ionograph or similar instrument) has long been used to monitor the effectiveness of cleaning processes used on organic acid fluxes. The methods used with each of the instruments are similar in that they rely on dissolving and dissociating any remaining flux residues in an alcohol/water solution and monitor the conductivity of the solution to estimate the level of contamination. Unfortunately, many no-clean materials do not dissociate well in the alcohol/water solution and the instruments are not able to detect them efficiently, rendering the current ICT standards largely irrelevant. The method may be useful for monitoring incoming materials (to evaluate the degree and consistency of solder mask cure, for example) and the baseline performance of the no-clean process, but will not, in fact, ensure that an assembly is acceptably clean.

44.5.3 Highly Accelerated Stress Testing (HAST) for No-Clean

Evaluating the effect of residues can be done with HAST[16] testing. This involves applying the flux material(s) of interest to SOT-23 diodes on a test board, processing the board normally, and exposing the board to an 85°C, 85 percent RH environment with a reverse 20-V bias for 1000 h. This method is sensitive to ionic residues, particularly chlorides and bromides. Since some flux formulations advertised as no-clean actually contain some amount of chloride activators, this test is particularly useful in evaluating their long-term effects.

44.6 THE IMPACT OF NO-CLEAN ON PRINTED CIRCUIT BOARD FABRICATION

Because the assembly is not intended to be cleaned again after the printed circuit board is fabricated, the bare board must be clean before it enters the assembly process. While the current mil-spec[17] for ionic contamination is aimed at the cleanliness of finished assemblies, it does not specify cleanliness levels for the raw board. The specific cleanliness level required varies depending on the instrument used to perform the test. A useful rule of thumb is to aim for a consistent bare-board cleanliness level of less than half the allowed level for the finished assembly. For example, if the specification for the finished assembly is 20 $\mu g/in^2$, the bare board should have no more than 10 $\mu g/in^2$ prior to assembly. This allows a reasonable margin for the additional contamination that is added to the assembly via flux residues, incoming contamination on components, and contamination from handling. The most effective way to implement this is for the printed circuit fabricator to sample the boards just prior to packing and shipment operations.

No-clean fluxes are significantly less active than fluxes which are intended to be removed by a cleaning operation. Poor solderability of the printed circuit board cannot be compensated for in a no-clean assembly process by using a stronger flux. The use of nitrogen in wave solder and reflow cannot promote wetting on inherently unsolderable surfaces. Unfortunately, no reliable, inexpensive method to quantify solderability has replaced the widely accepted "dip-and-look" method.[18] This test involves dipping a sample of the PCB coated with a standard flux into a solder pot and looking for any signs of nonwetting or dewetting of the solder. While far short of ideal, the dip-and-look test is better than no test at all and should be performed routinely by the printed circuit fabricator.

In the wave-solder process, it is common for small solder balls to adhere to the bottom of the board as it exits the solder wave. This is a phenomenon which has always occurred but was never specifically addressed because the solder balls were nearly always removed during postsolder cleaning. There are a variety of causes for solder balls that occur during wave soldering.[19] Some of the most common are:

- Solder mask material
- Solder mask surface roughness (can vary some depending on material surface treatment)
- Solder mask cure profile

- Type of flux used in hot-air leveling
- Type of wave-solder flux

The most significant factor is the choice of solder mask material. Some reduction in the number of solder balls can be made by experimenting with different combinations of wave solder and HAL flux types and solder mask cure profiles, but these measures offer only incremental improvements. Though troublesome and expensive, qualifying a new solder mask may be the only way to truly eliminate solder balls.

Hot-air-leveled solder has traditionally been applied to protect and promote solderability of the printed circuit board metallization. However, both fabricators and assemblers alike have been plagued with problems associated with the difficulty of applying a complete, even, and consistent thickness of solder to the board. An overly thin solder coating results in the copper-tin intermetallic layer being exposed, which is much more difficult to solder (largely negating the intended effect of ensuring a more solderable surface on the board). An overly thick or inconsistent solder coating makes assembly difficult because the bumpy topography of the board can hold the stencil away from the board, resulting in excess solder paste deposition on the board and bridging during reflow. Also, a thick solder coating results in crowned pads which are difficult to reliably place parts on. When a placement machine applies pressure to the part, the leads slide off of the crown, resulting in part misalignment and bridging.

To avoid the problems associated with hot-air-leveled solder surfaces, many assemblers have switched to organically coated bare copper boards. The metallization is perfectly flat, solders well, and has a lower ionic contamination level than similar hot-air-leveled boards.[20] The organic coating process is also vastly simpler and less expensive to operate than hot-air leveling. Most of the concerns about the solderability, shelf life, and robustness of the organically coated surface (e.g., imidazole) have turned out to be unfounded.

44.7 TROUBLESHOOTING THE NO-CLEAN PROCESS

See Table 44.10.

TABLE 44.10 Troubleshooting the No-Clean Process

Problem	Probable causes	Remedy
Poor wetting of solder paste to PCB	• Oxygen level in reflow oven too high (low-residue process)	• Check oven for leaks, check exhaust balance
	• PCB surface has poor solderability	• Test PCB for solderability
Solder beads next to passive components	• Full-sized apertures on stencil	• Check stencil apertures, modify stencil design if not updated to new design rules
Open solder joints on fine-pitch components	• Insufficient solder paste (clogged stencil apertures)	• Check stencil to see if apertures are clogged, clean stencil, discard solder paste and apply new paste to the stencil; if problem persists, check saponifier concentration in the stencil washer and correct
	• Poor component coplanarity	• Measure 30 components and calculate mean and standard deviation of coplanarity, if Cpk is <1.33, contact supplier

TABLE 44.10   Troubleshooting the No-Clean Process   (*Continued*)

Problem	Probable causes	Remedy
Solder shorts on solder paste side	• Excessive solder paste	• Check PCB for excess HASL
	• Component misalignment	• Check part alignment after placement and before reflow
Fine solder particles around pads on the paste side	• Stencil-to-board misalignment	• Check print alignment to see if solder paste is being printed in between the pad and the solder mask, correct alignment
	• Poor cleaning of a misprinted board	• Check to see if misprinted boards are being completely cleaned
	• Insufficient wiping of the underside of the stencil	• Wipe the bottom of the stencil with a dry wiper
Open solder joints on wave-solder side	• Assembly not in intimate contact with the solder wave	• Check wave height, board position in the conveyor, pallet condition
	• Incomplete flux coverage	• Run flux coverage test, check board position in the conveyor
Solder shorts on wave-solder side	• Component orientation	• Check component orientation with respect to the solder wave, rotate 90°
Components misaligned	• Tack failure of solder paste (low-residue process)	• Check to see if time between stencil print and placement has been more than 3 h; if so, clean pasted boards and reprint them.
		• Check to see if paste has been on the stencil for more than 6 h; if so, clean the stencil and use new paste
	• Crowned surface on PCB metallization (excess HASL)	• Rework bare PCBs
	• Placement machine error	• Check placement accuracy and correct if necessary
Poor print definition of solder paste	• Wiping	• Wipe bottom of the stencil with a dry, lint-free cloth
	• Alignment	• Check alignment of stencil to board
	• Stencil cleanliness	• Clean stencil
	• Saponifier concentration	• Check saponifier concentration
	• Old paste	• Paste on stencil more than 6 h— clean stencil and replace paste
Grainy solder joints	• Reflow profile too cool	• Check reflow profile
	• Oxygen level in oven too high	• Check oxygen level
		If the oxygen level reads OK, check to be sure that gas is flowing through the analyzer
		If the oxygen level is too high, check for changes to the oven setup and changes in the exhaust system from varying demands

TABLE 44.10 Troubleshooting the No-Clean Process (*Continued*)

Problem	Probable causes	Remedy
Large solder beads on top side of board next to passive components	• Stencil aperture shape	• Redesign stencil using an aperture shape that prevents solder from being squeezed underneath the component
Small solder particles on top side of the board between pad and solder mask	• Stencil alignment	• Check stencil printer alignment
	• Pad size on board less than stencil aperture size, paste is printed between the pad and the solder mask and may not coalesce with the rest of the joint during reflow	• Check to see if board feature size is within spec
		• Redesign stencil with slightly smaller apertures

REFERENCES

1. Allan Beikmohamadi, "Post-Reflow, No-Clean Solder Paste Residue and Electrical Performance," *Circuits Assembly*, March 1994.
2. Mark Curtin and Andrew Davis, "The Micro-Mechanics of Fine-Pitch Printing," *Circuits Assembly*, Aug. 1992.
3. Kevin McKean, "An Atmosphere-Purity Selection Guideline," *Surface Mount Technology*, Nov. 1994.
4. Mark Nowotarski, "Oxygen's Effect on the Surface of Solder," *Nepcon East*, Boston, June 1994.
5. Frank de Klein, "Open vs. Closed Reflow Soldering," *Circuits Assembly*, April 1993.
6. Steve Christensen, "An Observation of Nitrogen Atmospheres in a Convective IR Reflow System," *Proceedings of PWB Assembly Section of NEPCON East*, 1990, p. 655.
7. Brad Stoops, "Precision Application of No-Clean Fluxes," *Circuits Assembly*, March 1994.
8. Santiago Rodriguez-Sallaberry, "Spray Fluxing vs. Foam Fluxing," *Circuits Assembly*, June 1992.
9. Edward K. Chang, Mark T. Kirschner, and Sean M. Adams, "A Study of Several No Clean Fluxes under Different Atmospheres," *Circuits Assembly*, March 1994.
10. W. H. Down, "Wave Soldering Issues," *Circuits Assembly*, Nov. 1994.
11. John Maxwell, "Wave Soldering," *Surface Mount Technology*, Aug. 1994.
12. Edwin Oh, "Extending Soldering Iron Tip Life," *Circuits Assembly*, June 1994.
13. Kathi Johnson, "Hand Soldering and No-Clean Fluxes," *Circuits Assembly*, June 1994.
14. John Savi, Allan MacCormack, and Curt Gold, "No-Clean Flux Incompatibilities," *Circuits Assembly*, June 1992.
15. Chantal Hemens-Davis and Roy Sunstrum, "No-Clean: Material Compatibility Issues," *Circuits Assembly*, March 1993.
16. JEDEC Std. 22-A110.
17. MIL-P-28809 or IPC TM650.
18. MIL-STD-202F, Method 208F, or ASTM 0577.
19. Theo Langer and Torsten Reckart, "A Solder Ball Study, Part 2," *Printed Circuit Fabrication*, Dec. 1992.
20. Jerry Murray, "Beyond Anti-Tarnish: An SMT Revolution," *Printed Circuit Fabrication*, Feb. 1993.

CHAPTER 45
LEAD-FREE SOLDERING

Gary M. Freedman

*Compaq Computer Corp. Alpha High Performance Systems,
Marlboro, Massachusetts*

45.1 INTRODUCTION

With sufficient mechanical proprieties for consumer, commercial, military, and automotive product reliability and relatively low liquidus temperature, tin/lead (Sn/Pb) solder has been the mainstay of electronics manufacturing from the outset. For over 100 years the materials database on Sn/Pb for electronic interconnection has grown slowly. Sn/Pb solder has helped to define products rugged enough to last in a child's hands or travel to the limits of our galaxy.

45.1.1 Tin/Lead Solder Alloys

The most prevalent alloy composition is at, or close to, the eutectic composition: 63wt%Sn/37wt%Pb (melting point 183°C). Some component devices (CBGAs, CCGAs, etc.) have contacts that are made of high-lead alloy (10Sn/90Pb or 5Sn/95Pb) with melting ranges in the vicinity of 300°C, but these are not meant to liquefy during board assembly, only to serve as a compliant contact for soldering and electrical interconnection. Component leads are commonly coated with 90Sn/10Pb or 95Sn/5Pb to make them solderable and whisker free. The most common board surface finish worldwide is hot-air solder leveling (HASL), which involves eutectic or near-eutectic Sn/Pb flowed onto exposed surface pads and traces on a finished PWB at the board shop prior to electronic assembly.

45.1.2 Candidate Metals for Solder Alloys

Of the 90 naturally occurring elements, only 11 can be combined with each other to form a soft solder usable for circuit board assembly. These are listed in Table 45.1. Of these, only copper, silver, zinc, tin, antimony, and bismuth in combinations are practical for a lead-free solder due to either resource constraints or raw cost. Silver is just marginal in this respect. Other metals on the periodic table are either too high in melting point, too rare, too reactive, or poisonous or do not alloy well with other materials and so are not useful for a low-melting-point (<230°C), fusible formulation. Some alloys, such as several of the indium and gallium alloys, are too low in melting temperature to be useful for electronic applications. While lead-free solders have been in use for many years (in jewelry making, plumbing, brazing, etc.), none have been studied sufficiently for electronics assembly, most possess too high a melting point to be useful, and none come with the well-documented physical properties database of tin-lead solder.

TABLE 45.1   Candidates for Solder Alloys

Element	Symbol	Melting point (°C)
Antimony	Sb	630.5
Bismuth	Bi	271.5
Copper	Cu	1084.5
Gallium	Ga	29.75
Gold	Au	1063
Indium	In	156.3
Lead	Pb	327.50
Palladium	Pd	1550
Silver	Ag	960.15
Tin	Sn	231.89
Zinc	Zn	419.6

45.2   THE IMPETUS TO ELIMINATE OR REDUCE LEAD

The incredible rise in output and rapid obsolescence of consumer and commercial electronic goods has prompted the European Union to consider environmental-based legislative action to ban the use of lead for electronics manufacturing. Entitled "Waste Electrical and Electronic Equipment" (WEEE), Europe's initiative outlines potential steps and a schedule for the manufacture and disposal of electronic equipment and the elimination of environmentally hazardous materials associated with the electronics industry. It is intended to address Europe's shrinking available landfill space along with a perceived environmental threat of certain landfilled materials including circuit boards. The initiative seeks to reduce or eliminate mercury, cadmium, heavy metals including lead (Pb), and a variety of other potentially toxic substances from electronic goods.

Although the electronics industry consumes only a small fraction (<10 percent) of all lead (storage batteries are by far the largest Pb-based application), there is serious movement under way to eliminate lead in electronics manufacturing in Europe. Japan is also working toward reduction or elimination for the same reasons and has showcased some consumer products that are touted as lead free. There is conflicting evidence of solder's toxicity and its ability to leach into the water table from landfills at a rate that warrants concern. It is well known, however, that lead is a toxic material to handle and process, so it is in everyone's best interest in the long term to look for alternative alloys in solder as well as batteries. The legislative timeline as outlined in the WEEE Initiative draft targets 2006* as the deadline for elimination of lead in most electronics assemblies.

There is no drop-in replacement for Sn/Pb solder. Each potential substitute alloy has its disadvantage, as noted in the following text. The materials database for anything but Sn/Pb solder is meager; therefore, critical factors for understanding reliability and field life are absent, as are practical experience, acceptability criteria, and workmanship standards.

45.3   THE ROLE OF LEAD IN THE SOLDER JOINT

In addition to lowering the melting point of pure tin (232°C), Pb retards Sn-Cu intermetallic formation by piling up at the intermetallic boundary, frustrating tin-copper intermixing. Sn-Cu is crucial for solder joint formation, but if the intermetallic layer is overly thick, the resultant solder joint will be brittle and subject to failure during thermal cycling and mechan-

---

* Data subject to change.

ical shock. Although Sn/Pb solder is known for its ability to wet well to a number of component lead and circuit board finishes, Pb actually inhibits wetting and keeps the solder localized to the targeted solder joint area. Excessive solder spreading can be detrimental in three ways: (1) if solder wicks away from the intended solder joint area, the resultant joint will be solder starved and weaker than intended. (2) If the solder is too mobile, it can wick up connector leads and into the connector, decreasing inner contact flexibility, decreasing contact gap, and changing connector contact physics, resulting in a less reliable interconnect. (3) If the solder wicks up too high on gull-wing component leads, it inhibits the flexibility of the component lead and makes it more susceptible to mechanical failure.

45.4   THE IMPACT OF GOING LEAD FREE

Most lead-free solders being investigated require much higher process temperatures than present tin/lead solders. Consider that eutectic Sn/Pb solder with a melting point of 183°C requires overheating by 20 to 40°C to ensure thorough thermal soak of the board and components and sufficient wetting and wicking for solder joint formation. Assume the no-lead "substitutes" require at least the same 25 percent differential as typically used for tin/lead. Since the bulk of the lead-free solders have melting points or melting ranges in the vicinity of 220°C, applying this differential results in peak temperatures of 240 to 260°C. Some soldering equipment on the manufacturing floor will not cope with the higher temperatures for many of the no-lead solders. Note that consumer products with low component density, low-thermal-mass PWBs, and small components may not require much above liquidus for thorough soldering. Large, dense, electronic assemblies with high-lead-count devices typically exhibit considerable thermal insulation and shadowing by the components. The shadowing is the physical resistance of reflow oven–forced convective air flows due to component height, type, and population density. Without good convective circulation, components are slow to heat. These situations necessitate longer thermal soaks and higher overall temperatures to compensate and result in good quality solder joints.

45.4.1   Impact on Wave Soldering

In traditional wave soldering, solder pot temperatures are typically 240 to 260°C with eutectic Sn/Pb. If the most commonly considered no-lead solders require the same 60 to 80°C differentials, then the requisite solder pot temperature will tend toward 280 or even 300°C. Higher solder pot temperature coupled with constant mixing motion results in a higher drossing rate (rate of solder oxide formation) for the solder in the wave and reservoir. With Sn/Pb solder, the dross, which manifests itself as rafts of oxides afloat on the surface of the solder wave, can entrap liquid solder and cause solder bridging (drossy shorts). It can also mask solder contact to the fluxed through-hole parts, resulting unsoldered joints (skips or opens).

Some of the no-lead solder constituents, such as copper, zinc, and bismuth, are notorious for oxidizing. Nitrogen inerting in wave-soldering equipment will be mandatory to ensure good soldering and to conserve solder composition by retarding oxidation.

45.4.2   Impact on Components

Even if the equipment and the circuit board can withstand higher process temperatures, there are other considerations. Generally, ICs, passive devices, and connectors are not rated to survive temperature excursions much above that of eutectic Sn/Pb solder and only for a brief thermal cycle. This has caused IC makers and component manufacturers to work to understand the impact of the higher process temperatures required for no-lead processing, but results are not conclusive at the time of this writing.

Plastic-encapsulated components are known to absorb water from the atmosphere. As these components are heated to reflow soldering temperature, the water expands and can cause component cracking. This phenomenon is popularly known as *popcorning* because the plastic package often bulges prior to cracking and remains deformed upon cooling. To counter this, plastic-packaged components are baked out for many hours prior to soldering. The higher-temperature regimes of most lead-free solders will exacerbate the issue of popcorning, requiring more attention to component drying cycles. Longer bake-out times will likely be required.

45.4.3 Supply Constraint Impact

Some of the solder materials being considered for lead-free alloys cannot be mined or refined in sufficient quantity to satisfy the needs of the electronics industry in the event of a worldwide shift to a single solder alloy. Indium (In) and silver (Ag) fall into this category. Supplies of bismuth (Bi), a by-product of lead refining, would be marginally sufficient. Indium, silver, and some other elements discussed in the context of lead-free solders are deemed too expensive for widespread use.

45.4.4 Impact of Oxidation

Elements such as bismuth, copper, and zinc oxidize readily. This makes fluxing all the more difficult. In addition, copper and zinc are characteristically poorly wetting. All three will impact the shelf life of solder pastes. The use of nitrogen inerting is highly recommended for most of the lead-free solders, especially solder alloys containing bismuth, copper, and zinc oxide.

45.4.5 Impact on Solder Flux

Solder fluxes used with traditional tin/lead alloys are not formulated for the high-temperature range required of some of the proposed lead-free alloys. Critical materials in the flux may boil off or decompose before thorough fluxing can take place. Solder fluxes for no-lead solders have to be tailored to the composition of the new alloys and their inherently higher soldering process temperatures. Also, because many of the lead-free alloys contain elements that tend to oxidize rapidly, fluxes will have to be more reactive. This may buck the industry trend toward environmentally preferred no-clean flux and necessitate a return to water-soluble (aqueous-clean) flux formulations.

45.4.6 Impact on Electrical Testing

Since the process temperature for most lead-free solders are higher than those for solders that contain lead, flux residues will be more thoroughly baked onto PWB surface metals. This will inhibit electrical test probe contact. Even with today's no-clean solder pastes, electrical probing seems to be at the limit. Often the residues that cover test points necessitate multiple seating cycles of the test probes to penetrate the flux residue.

45.4.7 Impact on PWB Laminate

The higher process temperatures of most lead-free solders will impact the most commonly used epoxy-glass PWB laminates. Board softening, which leads to sagging during the process, may be a problem. Laminates will be more prone to delamination also. High-temperature PWB laminates are available. These are currently expensive and limited in supply.

45.4.8 Impact on Other Materials and Processes

SMT glue, wave pallet material, and solder mask will have to be optimized for compatibility with higher process temperature. They will need to be checked for chemical resistance and interactions with new flux formulations at these elevated temperatures.

45.5 SURFACE FINISHES

45.5.1 Hot-Air Solder-Leveled Surfaces

Lead-free solders mark the demise of the Sn/Pb hot-air solder leveling (HASL) surface finish. There are efforts to find other metallurgies to extend the useful life of HASL equipment.[1]

45.5.2 OSP-Copper

OSP-copper surface-finished PWBs, which are coated with an organic coating meant to retain solderability through several reflow cycles, are probably not compatible with the generally higher process temperatures associated with lead-free solders. The organic coating will burn off early in the process, allowing the copper to oxidize. This will inhibit solder wetting to the copper board surface finish. There could be impacts on other surface finishes also.

45.6 ALLOY SYSTEMS

The properties of an alloy are dependent on its constituent metals and the ratio of the number of atoms of one constituent to the atoms of the other. Metals can be combined in many ratios to result in alloys of vastly different properties, always tailored to suit a particular application. Melting points can be adjusted, as can hardness, durability, etc. But there are always trade-offs. In trying to achieve a suitable melting point, other material properties may degrade. Such is the case in the search for lead-free alloy substitutes. To date, there are no direct replacements that approximate the properties of our well-understood standard of eutectic Sn/Pb (Sn63) solder.

Families of alloys created from the same metals are called alloy systems. Sn/Pb is a system available in any number of compositions (50Sn/50Pb, 60Sn/40Pb, etc.). Some solder alloys are composed of two elements. These are binary alloys such as Sn/Pb or Sn/Bi. Others are ternary systems (Sn/Ag/Cu), quaternary alloys composed of four elements (Sn/Ag/Bi/Cu), or pentanary alloys (Sn/Ag/Cu/In/Sb) composed of five. The pentanary alloys are least understood, and binary systems are best understood and easiest to formulate. Binaries are less likely to interact with elements within a soldering system to form undesirable alloys than is the case with higher-order systems. In some higher-order alloys, by varying one metal or another, multiple eutectic alloys can be achieved. Therefore formulation is critical, as is exposure to other soluble metals during the soldering process. Some of the small differences in compositions (0.5 wt%) as seen in ternary, quaternary, and pentanary systems, will be difficult for solder vendors to accurately control. Also, in wave soldering, where the solder pot contents are in constant contact with metals from component leads and circuit board surface finishes, materials will be leached into the solder pot and will affect solder composition over the long term. Minor constituents of a solder may be depleted more quickly from a solder pot. As it turns out, most of the higher-order (ternary, quaternary, pentanary) alloys are too costly for the large solder volume required within a wave solder pot.

45.7 CANDIDATE LEAD-FREE SOLDERS

At the time of this writing, there is neither consensus nor industry plan for wide adoption of a common replacement for eutectic or near-eutectic tin/lead solder. The lead-free alloy systems receiving the most attention are listed in Table 45.2.

TABLE 45.2 Candidate Lead-Free Solders

Alloy	Melting point or melting range (°C)*
93.6Sn/4.7Ag/1.7Cu	216
95.5Sn/3.9Ag/0.6Cu	217
96.5Sn/0.5Ag/3Cu†	225 → 296
96.2Sn/2.5Ag/0.8Cu/0.5Sb	216 → 218
99.3Sn/0.7Cu	227
42Sn/58Bi	138
43Sn/1Ag/56Bi	136.5
91.8Sn/3.4Ag/4.8Bi	211
94Sn/2Ag/4Bi	223 → 231
78Sn/6Zn/16Bi	134 → 196
96.5Sn/3.5Ag	221
95Sn/5Ag	221 → 240
91Sn/9Zn	199
93.3Sn/3.1Ag/3.1Bi/0.5Cu	209 → 212
92Sn/3.3Ag/3Bi/1.7In	210 → 214
95.5Sn/3.5Ag/1Zn	217

* Eutectic alloys have a distinct melting point. Noneutectic alloy melting is characterized by a melting range where the alloy is present as both solid phase(s) and liquid phase simultaneously (pasty range). At the upper temperature of the melting range, only the liquid phase exists.
† There are many noneutectic formulations composed of the same elements in varying ratios—too many to list. Each variation in composition has its own melting range. Those listed in this table are examples only to show approximate temperature range and composition.

Bismuth expands upon freezing, while tin contracts. A phenomenon called fillet lifting has been reported.[2,3] It is mostly associated with bismuth ternary alloys (see Fig. 45.1) such as Sn/Cu/Bi and Sn/Ag/Bi used in PTH wave soldering.

Bi-containing solders are not recommended for Sn/Pb surface finishes on component lead or solder land. Small amounts of Pb can result in low-temperature ternary eutectic (Sn/Bi/Pb alloy, melting point 96°C). In some applications, solder joints fall apart if the service temperature is high and the low-melting-point Sn/Bi/Pb alloy is formed.

Some alloys are too low in temperature for many applications. Despite its excellent tensile strength and thermal cycling endurance (better than those of Sn/Pb), Sn/Bi eutectic (melting point 138°C) would not be practical for use in most automotive applications or even for high-end computer assembly.

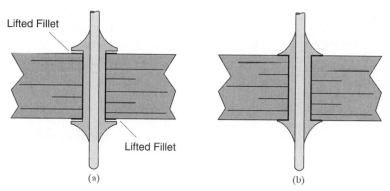

FIGURE 45.1   Expansion upon solidification in a bismuth-containing ternary solder alloy can cause solder fillets to fracture and lift from the plated through-hole annular pads (*a*). Compare this to the schematic in (*b*), which is indicative of a conventional tin/lead solder joint.

45.8 FINANCIAL IMPACT OF LEAD-FREE AND LEGAL CONSTRAINTS

During the last two decades, there was a threat of legislation against use of lead-based solders in the USA and other countries. This gave rise to research for solder replacement alloys. Several corporations and universities devised non-lead solders, some of which were patented. As is the case with any patented item or process, the rights to the invention reside with the inventor or sponsoring corporation and several of these alloys may not be used freely. In some cases only slight modifications in alloy composition separate commonly available solder alloys from patented solder compositions.

Cost is a major factor when considering solder substitution. The costs of most lead-free alloys range from two to four times those of Sn/Pb solder. Both tin and lead are abundant, easy to refine, and thus inexpensive to purchase. This is not the case for some of the constituents of lead-free solders being considered, such as indium (In), gallium (Ga), and silver (Ag). In wave soldering, a large volume (hundreds of pounds) of solder is required to fill the wave solder reservoir (solder pot). The cost and availability of some of the lead-free alloys make them impractical for the large volume of the solder pot. Compare Sn/Pb solder at roughly $1.50 per pound to Sn/0.7Cu at $2.30 and Sn/2Ag/0.8Cu/0.5Sb at $4.00 per pound. Granted, with lead being such a heavy element, price per pound is not the best metric. Price per unit volume would be a better metric; nonetheless, moving away from the Sn/Pb binary eutectic will mean significantly higher manufacturing materials costs. For this reason, it may not be practical to use the same lead-free solder alloy for wave soldering as for reflow. The NEMI* consortium has proposed using two different alloy systems; one for oven reflow (Sn/3.9Ag/0.6Cu, melting point 217°C) and one for wave soldering (Sn/0.7Cu, melting point 227°C or Sn/3.5Ag, melting point 221°C. This strategy complicates manufacturing and repair of mixed-mount soldering because there will be exposure of one solder alloy to the other, especially at repair. The effects of solder alloy intermixing on solder joint reliability will have to be thoroughly tested.

---

* National Electronics Manufacturing Initiative, Herndon, VA.

Some of the overt costs and impacts of moving to lead-free solders are mentioned in the preceding text. There are other potentially hidden costs. Soldering equipment may not have the bandwidth to handle the higher temperatures of some of the lead-free formulations. Reflow ovens may need to be constructed out of different materials to handle long-term effects of the higher temperatures. Fan bushings—polymeric seals to maintain inert atmosphere integrity—may not withstand much above the temperature regimes required for Sn/Pb processing. These same considerations apply to wave soldering. Hand-soldering irons will require higher temperatures and it is likely that more boards will have to be scrapped due to localized charring incurred while trying to attain reflow during the hand-soldering operation or at repair. Board materials, fluxes, etc. may require reformulating and that is likely to drive costs higher too.

45.9 CHARACTERISTICS OF LEAD-FREE SOLDERS

No-lead solders exhibit poorer wetting/spreading characteristics as compared to Sn/Pb alloys, but many offer an advantage in tensile strength and creep resistance. In terms of the soldering process, a whole new understanding of the various no-lead alloys is required. Propensity of solder void formation, solder paste shelf life, solder paste service life (while on the stencil), printability, effects on stencil and squeegee life, fatigue characteristics, alloy interactions with board and component surface finishes, corrosion resistance, mechanical shock resistance, and numerous other properties will have to be understood, especially for the dense, high-end electronic assemblies expected to last many years in the field. It may be less of an issue for consumer electronics that are considered obsolete within a couple of years; however, ruggedness, especially for portable electronics, is a concern. Many of the lower-melting-point solders contain precious elements such as In and Ga. These tend to be expensive alloys. While Sn/9Zn is inexpensive, it wets poorly, tends to corrode, and is prone to oxidation during soldering. Sn/0.7Cu also oxidizes rapidly in air and especially at reflow temperature.

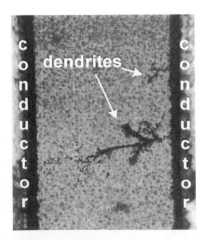

FIGURE 45.2 Dendritic growth between two finely spaced conductive traces (0.5-mm pitch) on FR-4 PWB.

Solder corrosion has long been known to be a failure mechanism in electronic assemblies. Ionic contaminants, especially in the presence of moisture, can lead to corrosion. Contaminants may come from PWB manufacturing, handling, and flux residues to name a few sources. The corrosion may cause metal species migration and corrosion crystallites on the surface of the PWB between two oppositely charged surface traces. These crystallites, called dendrites, are conductive (see Fig. 45.2). They can have enough current capacity to persist, causing an electrical short, or they may heat up to the point of fusion, melt, interrupt current flow, and return the assembly to normal operation. This can make diagnosis difficult. Finer-pitch, surface-mount geometries are particularly susceptible to this phenomenon. Abtew and Selvaduray published a chart of the EMFs of various metal couples present in some lead-free solder alloys.[4] The lower the EMF, the more corrosion resistant the alloy. Eutectic Sn/Pb, the basis of comparison, had by far the lowest value on the list at 0.010 V. The next nearest, Sn/51In, was 0.201 V or 20 times the voltage as compared to Sn/Pb solder. Other values reported were Sn/57Bi at 0.323 V, Sn/9Zn at 0.624 V, Sn/3.5Ag at 0.937 V, and Sn/80Au at 1.636 V. So tin-lead solder is significantly more resistant to corrosion and dendritic growth than the other seven alloys studied.

45.10   RECOMMENDATIONS

Were everyone's favorite alloy introduced in manufacturing, much would be learned in the short term, but to the detriment of some manufactured goods in terms of reliability and cost. More industry collaborative work needs to be completed. Within a few years there may be widespread consensus as to which alloys are viable in terms of equipment compatibility, component compatibility, solder joint reliability, and economic affordability. At this point there is no clear choice or direction in which the industry is moving as regards lead-free soldering alloys for volume manufacturing. There are several potential alloys mentioned in this chapter that could be made to work. Careful data collection, peer and industry-wide review, and analysis of lead-free alternatives will help shake out the preferred alloys in the next few years. By far, the best and most expeditious means of lead reduction would be the establishment of recycling programs for electronic assemblies and waste.

Many products manufactured today are meant for long field life. High-end computers, telecommunication base stations, and some consumer products such as refrigerators, etc., may be in service for many years. With the transition to lead free solders on the near horizon, it is prudent to start labeling PCAs as to the solder alloy used for assembly. This may preclude field repair with the wrong alloy in the future and subsequent reliability problems due to solder alloy incompatibility.

Check with equipment vendors to ensure that equipment in place or new equipment purchases will be compatible with lead-free soldering alloys in terms of processing temperatures and materials compatibility. The solder pot on a wave machine may degrade more rapidly when using lead-free alloys. Reflow oven heaters may not be capable of sustained operation at the higher temperatures required for no-Pb solder alloys. Keep in mind that the ultimate process temperature will depend not only on the soldering alloy of choice, but on the effectiveness of the machine heater scheme and the size, density, and complexity of the PCA.

Last, look for ways to limit the use of lead in today's circuit board assemblies. Moving away from the hot-air solder-leveled finish, reducing or eliminating PTH components requiring wave soldering, and increasing reliance on surface-mount and press-fit connectors will help in lead use abatement.

REFERENCES

1. Hot Air Level (HAL) users group (www.info@huggroup.org).
2. NCMS Lead-Free Solder Project Final Report, August 1997.
3. J. Bath, C. Handwerker, and E. Bradley, *Circuits Assembly,* May 2000, p. 31.
4. M. Abtew and G. Selvaduray, "Lead Free Solders for Surface Mount Applications," *Chip Scale Review,* September 1998, pp. 29–38.

CHAPTER 46
FLUXES AND CLEANING

Laura J. Turbini
*Centre for Microelectronics Assembly and Packaging,
University of Toronto, Canada*

46.1 INTRODUCTION

Computers, portable phones, pagers, and personal digital assistants (PDAs) have one thing in common. They each contain an electronic assembly composed of integrated circuits and discrete components soldered to a printed wiring board (PWB) to create an electrically functional circuit. In the past, printed circuit assemblies were made exclusively with through-hole components. Today's assemblies contain components that are both through-hole and surface-mount and some even contain unpackaged integrated circuits (ICs) attached directly to the PWB. Except in the case of direct chip attachment by wire bonding, the use of a soldering flux to ensure good joining is almost universally practiced. Thus, soldering flux choice has become an important consideration in the manufacturing of electronic assemblies. As electronic product designs have moved to more densely packed printed wiring boards with fine lines and spacing, double-sided surface-mount technology, fine-pitch components, and ball grid array and chip-scale packages, the role of the soldering flux has become a critical link in the chain of events required for a successful manufacturing process.

In the past, the soldering process was usually followed by a cleaning step to remove flux residues and to ensure the reliability of the assembly. This practice has been challenged by the elimination of a major class of cleaning materials that had been used for flux residue removal. The targeted cleaning agents include chlorofluorocarbons (CFCs) and methyl chloroform (1,1,1-trichloroethane), which have been shown to destroy stratospheric ozone.[1]

Twenty years ago, the predominant flux chemistry was rosin based. Today, great advances have been made in fluxes and a large variety of formulations with very different chemistries are being used. This chapter reviews the assembly, soldering, and cleaning processes; discusses the soldering flux in terms of its role in soldering; reviews the variety of flux chemistries available along with their advantages and disadvantages; and comments on the clean/no-clean options as they relate to flux choice. In addition, characterization techniques for soldering fluxes and pastes and electrochemical migration failure mechanisms are included.

46.2 ASSEMBLY PROCESS

The basic elements of an electronic assembly are the PWB, the components, and the interconnecting metallization. The majority of PWBs are made of epoxy-glass substrate with copper or solder-coated copper metallization. In many cases there is a solder mask that covers most of the circuit traces, leaving exposed only those patterns where soldering will occur during assembly. Boards designed for through-hole components usually contain copper-plated through-holes that may be solder coated; those designed for surface-mount components have metallized pads for interconnecting the surface-mount parts to form the assembly.

Through-hole components and bottom-side surface-mount discrete components are soldered to the board in a wave-soldering machine. The process flow for this is diagrammed in Fig. 46.1. The PWB is populated with bottom-side discrete components, which are placed on a small drop of adhesive. When the adhesive is cured, it holds the component to the board. Next the board is turned over and the through-hole components are inserted. The assembled board is then placed on a conveyor that moves the assembly over (1) the flux applicator, (2) the preheat station, and (3) the solder wave. The most common method of flux application is by foam created when compressed air is forced through a porous stone immersed in liquid flux. As the assembled board passes over the fluxer, the foam flux covers the bottom surface of the board and penetrates up through the plated through-holes. Spray fluxers that spray a fine mist of flux on the board are also used, particularly for applying low-solids fluxes. A third type of flux application involves a wave fluxer in which flux is pumped up through a nozzle producing a wave of liquid flux.

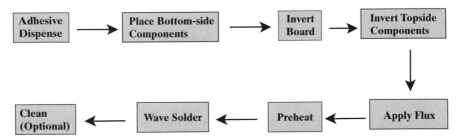

FIGURE 46.1 Traditional through-hole soldering involves the insertion of components followed by flux application, a preheating step, wave soldering, and a final optional cleaning step. Today, bottom-side discrete components are also included in the process. This requires that they be placed first before the board is inverted and top-side components are added.

The preheater station is responsible for volatilizing the flux solvent and heating the flux ingredients to activate them. When the assembly moves on to the solder wave, the metallization on the board and components, having been cleaned through the action of the soldering flux, is bathed in molten solder that pushes up through the plated holes and metallurgically bonds the components to the board. In most instances the solder alloy used in the solder wave is 63Sn/37Pb (melting point of 183°C), which is heated to as high as 260°C. Finally, the conveyor moves the assembly past the solder wave, allowing the molten solder to solidify upon cooling. In a high-volume manufacturing facility, the completed assembly may continue on a conveyor into an inline cleaner unless a no-clean soldering process is used.

Today many assemblies are composed exclusively of surface-mounted components—traditional chip carriers, quad flat packs, and discrete resistors and capacitors, as well as ball grid array packages, and direct chip attach for the thin, hand-held electronics. Figure 46.2 shows the steps required for this process. Solder paste is stenciled onto the metallized pads

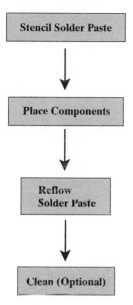

FIGURE 46.2 Surface mount soldering requires stencil printing of the solder paste followed by component placement, reflow soldering, and a final optional cleaning step.

of the PWB, the component leads are accurately placed into the solder paste, and the assembly is reflowed in a convection oven. As the conveyor carries the assembly through the oven, the first zone acts as a preheat zone where the solder paste becomes active in removing oxidation from the board and component metallization as well as the solder powder of the paste. The zones following the preheat zone allow the solder to melt, forming a bond between the metallization on the board and that on the component leads. The final zones allow cooling and solidification of the solder.

46.3   SOLDERING FLUX

Soldering[2] is defined as the process of joining metallic surfaces with solder without the melting of the basis metal. In order for this joining to take place, the metal surfaces must be clean of contamination and oxidation. This cleaning action is performed by the flux,[2] a chemically active compound that, when heated, removes minor surface oxidation, minimizes oxidation of the basis metal, and promotes the formation of an intermetallic layer between solder and basis metal. The soldering flux has several functions to perform. It must:

- React with or remove oxide and other contamination on the surface to be soldered
- Dissolve the metal salts formed during the reaction with the metal oxides
- Protect the surface from reoxidation before soldering occurs
- Provide a thermal blanket to spread the heat evenly during soldering
- Reduce the interfacial surface tension between the solder and the substrate in order to enhance wetting

To perform these functions, soldering flux formulations contain the following types of ingredients: (1) vehicle, (2) solvent, (3) activators, and (4) other additives.

46.3.1   Vehicle

The vehicle is a solid or a nonvolatile liquid that coats the surface to be soldered, dissolves the metal salts formed in the reaction of the activators with the surface metal oxides, and, ideally, provides a heat transfer medium between the solder and the components or PWB substrate. Rosin, resins, glycols, polyglycols, polyglycol surfactants, polyethers, and glycerine are among the major chemicals used. Rosin or resins are selected when more benign chemicals are required because their residues are less apt to cause reliability failures. Glycols, polyglycols, polyglycol surfactants, polyethers, and glycerine are used in water-soluble flux formulations because they provide excellent wetting of the board surface and dissolve the more active materials used in these formulations.

46.3.2   Solvent

The solvent serves to dissolve the vehicle, activators, and other additives. It evaporates during the preheating and soldering process. The solvent chosen will depend upon its ability to dissolve the flux constituents for a given formulation. Alcohols, glycols, glycol esters, and glycol ethers are common solvents used.

46.3.3 Activators

Activators are present in the flux formulation to enhance the removal of metal oxide from the surfaces to be soldered. They are reactive even at room temperature, but their activity is enhanced as the temperature is raised during the preheating step of the soldering process. Amine hydrochlorides; dicarboxylic acids such as adipic or succinic acids; and organic acids such as citric, malic, or abietic acids are among the common activators found in flux formulations. Halide- and amine-containing activators give excellent soldering yields but may cause reliability problems if not properly removed in a well-controlled cleaning step.

46.3.4 Other Additives

Soldering fluxes often contain small amounts of other ingredients, which serve a specialized function. For example, a surfactant may be added to enhance the wetting properties. This constituent can also assist in the foaming characteristics required for foam flux application. Other additives may be included to lower the interfacial surface tension between the molten solder and the PWB as it exits the solder wave, decreasing the chance of solder bridges formation. Solder paste formulations require the presence of additives to provide good viscosity or flow characteristics, low slump during the preheating step, and good tack characteristics for holding the component in place until reflow occurs. Finally, cored-wire flux used for hand soldering contains a plasticizer to harden the flux ingredients that are present in the core of the wire.

46.4 FLUX FORM VS. SOLDERING PROCESS

Although one usually thinks of soldering flux as a liquid, in fact the soldering flux can exist in several different forms. Liquid flux is commonly used for wave-soldering or hand-soldering applications. Paste flux, a thick, viscous flux, is used to hold components to the board prior to reflow of the already present solder. Direct attachment of a solder-bumped chip or ball grid array (BGA) packages to solder pads on a PWB is an example of its usefulness. Solder paste contains solder-paste flux, which is the solder paste without the solder particles. Solder paste is used in surface-mount attachment of components such as chip carriers, tape automated bonding (TAB)-leaded packages, and discrete resistors and capacitors. Finally, flux-cored solder wire is used for hand soldering and flux-cored solder preforms are used as the solder/flux source in some applications such as backplane connector pin reflow soldering.

46.5 ROSIN FLUX

Early flux formulations for electronics used rosin, which is a naturally occurring resin obtained from the sap of a pine tree. Its exact composition varies depending upon the part of the world where it originates as well as the time of the year. Rosin contains a mixture of resin acids, the two most common ones being abietic acid and pimeric acid (Fig. 46.3). Rosin has been a favorite material for soldering because it liquefies during the soldering process, dissolves the metal salts, and then solidifies when cooled, entrapping and for the most part immobilizing the contaminants. In addition, rosin has some innate fluxing activity because its molecular struc-

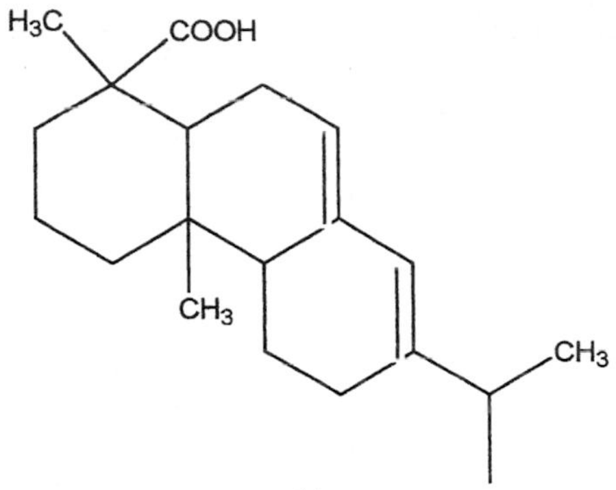

(a)

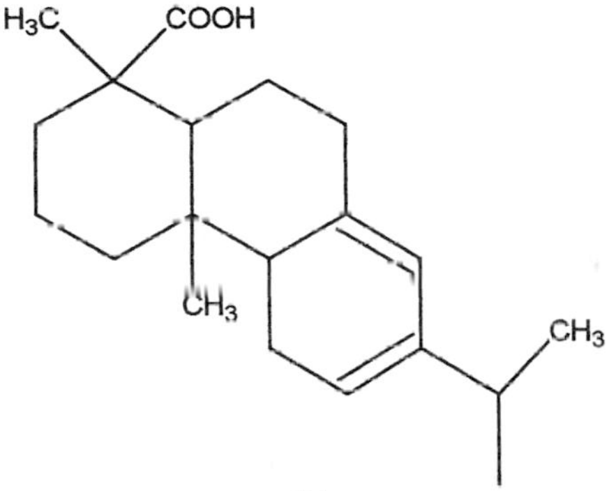

(b)

FIGURE 46.3 The two most common rosin iso-
mers are abietic acid (*a*) and pimaric acid (*b*).

ture contains a weak organic acid. Finally, rosin-based fluxes work well in hand-soldering and repair operations because rosin provides good heat transfer characteristics.

The activity of a rosin-based flux is determined by the activators and surfactants, which are part of the formulation. Some activators help to remove metal oxide but also leave residues that are essentially noncorrosive. Halide-containing activators leave residues that could be corrosive if too much is left on the board. A variety of specifications are available that help to define the effect of flux residues. One of these specifications is detailed later in this chapter.

Military specifications[3] in the past have required the use of rosin-based fluxes. These were defined as pure rosin (R), rosin mildly activated (RMA), rosin activated (RA), and rosin superactivated (RSA), based upon the level of halide activators they contained. Typically only R or RMA fluxes were approved for high-reliability military applications. In the 1970s tele-communication companies in the US and Europe predominantly used rosin fluxes for their wave-soldering requirements. These companies had their own internal set of testing methods for selecting noncorrosive rosin flux formulations. Using their selection criteria, they believed the fluxes to be sufficiently safe. For their applications they cleaned only the bottom side of the assembly, where removal of rosin residues was needed to ensure good electrical contact during bed-of-nails testing.

46.6 WATER-SOLUBLE FLUX

Water-soluble fluxes have also been called "organic acid" fluxes. This name is misleading because all fluxes used for electronic soldering contain organic ingredients and many contain organic acid activators. The term *organic acid flux* probably originated from the designation of water soluble fluxes as organic and those activated with organic acid activators as organic acid fluxes. Other activators for these fluxes include halide-containing salts and amines. Although the correct name for this category of fluxes is *water-soluble*, it should also be noted that the flux solvent is normally not water, but rather alcohols or glycols.

As the name implies, water-soluble fluxes are soluble in water and their soldering residues are also expected to be water-soluble. These fluxes are much more active than rosin fluxes, have a wider process window, and give a higher soldering yield with reduced defects. This means that the final assembly requires less touch-up or repair. On the down side, water-soluble fluxes contain corrosive residues that, if not properly removed, cause corrosion in the field and long-term reliability problems.

As indicated earlier, water-soluble fluxes usually contain glycols, polyglycols, polyglycol surfactants, polyethylene oxide, glycerine, or other water-soluble organic compounds as the primary vehicle. These provide good solderability for the activators, which are usually the more corrosive amines, and halide activators. With the onset of highly efficient cleaning equipment, this type of flux became popular for computer and telecommunication applications.

Zado[4] raised concern in the late 1970s that water-soluble fluxes affect the electrical characteristics of the epoxy-glass laminate by reducing the insulation resistance. This reduction was due to the dissolution of the polyglycols of the flux formulation into the epoxy substrate during the soldering process. Later work by Brous[5] indicated that some polyglycols were much more deleterious than others. In general, polyglycol-containing fluxes (and fusing fluids used by board manufacturers to fuse tin-lead plating) can cause an increase in moisture absorbance of the epoxy-glass substrate.

There are several considerations the user must include in determining whether to use water-soluble fluxes in a given application. One important factor is the operating environment. If the assembly will experience extremes of temperature while under power, it is possible that localized condensation can occur and dendrites will form, shorting out some of the circuit elements. Assemblies of this nature should be conformally coated. A second critical factor is the use of a board design and cleaning process, which ensures that corrosive residues are removed. A third consideration is the voltage gradient in the electrical design of the circuit. Within reduced lines and spacings, a failure mechanism called conductive anodic filament (CAF) formation[6] has been associated with high humidity and high voltage gradients. This is described in detail later in the chapter.

46.7 LOW-SOLIDS FLUX

Until the mid-1980s, liquid soldering fluxes were formulated in 25 to 35 percent (weight percent) solids or nonvolatile liquid. Then flux chemistries changed and new formulations that were lower in total solids content came on the scene. These fluxes are composed principally of weak organic acids, often with a small amount of resin or rosin. Early formulations had 5 to 8 percent solids, but today's low-solids fluxes are 1 to 2 percent solids composition.

Nomenclature of these fluxes moved from *low-solids flux* to *low-residue flux* to *no-clean flux*. The thinking was that the amount of residue left by these fluxes after soldering was so minimal that they did not need to be removed. This is only true if the residues are noncorrosive.

Flux application became a challenge. Guth[7] noted that when higher levels of these low-solids fluxes were applied to assemblies and soldered, dendritic growth could be observed later in field trials. The need to limit the amount of flux applied to the assembly led to the increased use of carefully controlled spray fluxers. These can ensure that flux is available in the through-holes of the PWB, but also limit the amount of flux that pushes up to the top side of the board during wave soldering. The corrosive properties of the residues from the partially heated flux on the top side of the board can be very different from those of the flux on the bottom side of the board exposed to the solder wave.

Another challenge of low-solids fluxes is the processing window. Unlike water-soluble fluxes, which give very low defect levels and have a wide processing window, the soldering process for low-solids fluxes must be carefully designed. To begin with, the recommended preheating temperatures are different from those for rosin flux and the preferred solder wave temperature is lower than that for rosin. Additionally, the solderability of incoming boards and components must be ensured. While water-soluble flux can cut through heavy metal oxide layers, the amount of fluxing ingredients in low-solids fluxes is not sufficient to accomplish the task.

Low-cost manufacturing can be achieved with low-residue fluxes if the cleaning step can be eliminated. This assumes that the incoming components and boards are clean, and that operators handling the boards are careful not to introduce contamination. This requires a flux whose residues are noncorrosive and one that will not impede or contaminate the electrical bed-of-nails test probes.

It should be noted here that a new category of low-solids fluxes was introduced in the early 1990s to meet the needs of localities where volatile organic compounds (VOCs) are regulated. These fluxes are marketed as VOC-free or low-VOC fluxes. The solvent in this case is 100 percent water or at least greater than 50 percent water. Use of these fluxes requires special care in the preheating step, where the water (solvent) must evaporate before the assembly reaches the solder wave. Failure to do this results in excessive solder ball formation.

46.8   CLEANING ISSUES

Flux removal is motivated by several factors, the importance of which will vary with the end use of the product. In general, the purpose of cleaning is (1) to remove corrosive residues from the soldering process or handling operations (some fluxes have high levels of halide activators that can cause corrosion if they are not removed); (2) to remove rosin, resin, or other insulating residues that can interfere with bed-of-nails electrical testing; (3) to remove rosin, resin, or other residues that attract dust, dirt, and other airborne contaminants; and (4) to ensure that the board is free of residues and contamination prior to conformal coating.

During the late 1970s there was a push to replace some of the chlorinated cleaning solvents used for rosin flux removal. The EPA had listed cleaning agents such as perchloroethylene and trichloroethylene as suspected carcinogens, and the allowed level of these in the operator breathing zone was severely curtailed. At the same time, chlorofluorocarbon azeotropes with methyl alcohol, ethyl alcohol, and methylene chloride found increasing application as replacements.

46.8.1   The Montreal Protocol

In September 1987 at the United Nations Environment Program Conference in Montreal, 24 countries signed the Montreal Protocol on Substances that Deplete the Ozone Layer. The protocol was the beginning of an international agreement on the reduction of chlorofluorocarbons (CFCs), which had been shown to have a deleterious effect on the stratospheric ozone layer that protects the earth from harmful ultraviolet radiation. The initial protocol targeted CFCs and halons (bromine-containing CFCs) and proposed a 50 percent reduction in production by 1998 based on 1986 levels. Revisions to the protocol led to a ban on the manufacturing of CFCs and other ozone-depleting chemicals by the end of 1995. This has been accomplished in the major countries around the world, although developing countries have been given a longer time along with technical assistance in implementing alternatives.

46.8.2   Solvent Replacements

Until the Montreal Protocol dictated the reduction and elimination of CFCs and methyl chloroform use, these chemicals were the major cleaning solvents for removing rosin residues from electronic assemblies. The elimination of their availability has forced manufacturers to rethink their flux and cleaning choices. Alternative cleaning agents were needed to remove flux residues left on the assembly after soldering. These alternatives must be capable of dissolving the particular chemical residues from a given flux formulation. Several detailed cleaning references are available.[8-12]

Solvent cleaning options for rosin fluxes have focused on semiaqueous cleaning and on aqueous detergent cleaning. Semiaqueous cleaning involves the use of an organic solvent capable of dissolving rosin or other residues. Terpenes such as *d*-limonene were the most widely used solvents, and other chemicals such as dibasic esters and nonionic surfactants were added to improve cleaning and rinsing. These materials provide good cleaning properties, but contamination of the solvent with as little as 1 percent water can reduce the cleaning efficiency. An alternate semiaqueous cleaning solvent is composed of high-molecular alcohols. The solvent is usually sprayed onto the assemblies in its pure form. This step is followed by a rinse step that emulsifies the semiaqueous solvent and dissolves the ionic residues. Innovations[13] have shown that an aqueous emulsified solution of the solvent can also provide good cleaning. Semiaqueous cleaning agents are primarily used for rosin- or resin-based fluxes because these cleaning materials require specialized capital equipment that is more costly than that needed for other clean chemicals.

Saponifying detergents have been used for removing rosin flux residues since the late 1970s. The detergent is diluted to a 4 to 10 percent aqueous solution, which removes rosin in the form of a rosin-soap along with the ionic contaminants in a water rinse step. A final blow-dry step using heated air ensures that the assembly is dry prior to electrical testing. In addition to rosin flux, detergent cleaning can also be used to remove water-soluble flux or low-solids flux.

The cleaning agent is based on an organic amine such as monoethanolamine (MEA), which reacts with rosin or resin to form a "soap" in the same manner that the detergent used to wash your greasy dishes reacts with the grease and helps it dissolve, or at least be emulsified and removed in the rinse water. Detergents for electronics are designed not to foam. The saponifying reaction can be simply written:

$$rosin + MEA = rosin\text{-}MEA\ soap + water$$

Once the rosin or resin is removed, the rinse water solubilizes the ionic residues that were on the surface of the board as well as those trapped within the solid rosin matrix.

A more recent type of aqueous saponification involves the use of buffered metal salt solutions such as NaOH/sodium carbonate/sodium bicarbonate. The moderately high pH of this solution (similar to MEA) allows the rosin to dissolve and react with the metal salt to form a rosin soap. If we represent rosin as RCOOH (since it contains a COOH group attached to its backbone) and the buffered NaOH as MOH, the following represents the reaction.

$$RCOOH + MOH = RCOOM + H_2O$$

The detergent also contains a wetting agent to improve the penetration under components and into tight spaces. Thus, a small amount of detergent (2 to 4 percent) is often added to the water when cleaning water-soluble fluxes to enhance wetting and to dissolve oily fingerprints and other processing residues that may not be water soluble.

The environmental concerns related to these cleaning agents differ. In the case of semi-aqueous systems, the solvent may have a relatively low flash point and commercial cleaning equipment must add specialized features to ensure safety. The EPA lists all organic cleaning agents as VOCs and their emissions are regulated. Closed-looped recycling systems exist for semiaqueous cleaning processes, and the spent solvent can be reused as an alternate fuel source. However, contamination of the solvent with temporary solder mask can severely limit separation and reuse.

In the case of detergents that are dissolved in water, their aqueous emissions are regulated. Detergents usually have a high pH (10 to 11) that exceeds the level acceptable for release to drain. Therefore, the spent solution and the rinse water must be neutralized prior to disposal. In addition detergents generally place a high biological oxygen demand (BOD) on the local municipality's waste treatment plant. In some parts of the United States, such as the Northeast, the BOD load associated with detergent cleaning exceeds the local EPA regulations. As a final environmental concern, heavy metals (Pb, Sn, Cu) can be dissolved in the detergent waste or in the rinse water from the semiaqueous process. These levels must be monitored and reported. Proper filtration units on the cleaner remove undissolved solid solder residue and help ensure that levels of any dissolved heavy metals are well below those allowed by regulations.

Aqueous cleaning can only be used to remove water-soluble residues. Thus, this cleaning method is limited to water-soluble fluxes and to low-solids fluxes whose residues are soluble in water. An advantage of aqueous cleaning is that closed-loop systems allow the removal of contaminants and the recycling of the water.

There are few solvent cleaning options that are not ozone depleting. Alcohol cleaning with 75/25 percent isopropanol/water has been used by some, but white residues may appear after cleaning due to the partial dissolution of baked-on rosin residues left from the soldering process. n-propyl bromide is another new cleaning solvent that can be used in specialized applications. Recently cyclohexanol has been promoted, but this requires expensive batch equipment to ensure a safe process. Table 46.1 shows a matrix of flux/cleaning compatibility.

TABLE 46.1   Soldering Flux/Paste Types Matched with Possible Cleaning Agents

Flux/paste type	Possible cleaning agents	Comments
1. Rosin/resin	Detergent cleaning Semiaqueous Cleaning Solvent cleaning	
2. Low-residue flux	No cleaning step Aqueous cleaning Detergent	If flux is rated L0 or L1 in ANSI-J-Std-004 If flux does not contain rosin or resin If flux contains rosin or resin
3. Water-soluble flux	Water Detergent cleaning	

46.9 FLUX CHARACTERIZATION TEST METHODS

Soldering fluxes for electronics have traditionally been characterized by their chemical composition, i.e., rosin, resin, or organic. In the past, military specifications have limited the manufacturer to rosin-based fluxes:

R   Pure rosin
RMA   Rosin mildly activated
RA   Rosin activated
RSA   Rosin super activated

The activation level of these has been defined in part by the use of water extract resistivity testing, where the resistivity is greater than 100,000 Ω for RMA and 50,000 to 100,000 Ω for RA. This characterization measurement developed as a means of controlling the amount of corrosive flux activators such as amine halides in fluxes used for military product.

In the early 1980s, the IPC developed flux characterization criteria based on the flux and flux residue activity. Thus, fluxes were classified as:

L   Low or no flux/flux residue activity

M   Moderate flux/flux residue activity

H   High flux/flux residue activity

These designators were determined by a series of tests that include the copper mirror test; a qualitative silver chromate paper test for chlorides and bromides; a qualitative spot test for fluorides, a quantitative test for halides (chloride, bromide, and fluoride); a corrosion test for flux residue activity; and a surface insulation resistance (SIR) test at accelerated temperature and humidity conditions. The latest industry standard, "Requirements for Soldering Fluxes" (J-STD-004A), updates the earlier IPC-SF-818 solder flux specification and includes some international elements from the International Standards Organization (ISO-9454). In addition to defining the flux categories L, M, and H, it notes the absence or presence of halides by a 0 or 1 that is added to the descriptor. To parallel the international standards, the basic chemical constituents further classify fluxes. Thus, fluxes may be listed as rosin (RO), resin (RE), organic (OR), or inorganic (IN). The test methods for this specification are contained in the test methods document (IPC-TM-650). A brief description of each of the test methods is included in the following text. The test results are used to characterize the fluxes as L, M, or H. Table 46.2 summarizes the test results required for flux classification.

46.9.1 Copper Mirror Test (TM 2.3.32)

In this test a drop of flux is placed on one end of a glass slide that is coated with 5000 Å of vapor-deposited copper. A drop of 25 percent water white (WW) rosin flux is placed on the other end. The fluxed slide is held at 25°C/50 percent RH for 24 h. Then the slide is rinsed with isopropyl alcohol. If there is no evidence of mirror breakthrough where the test flux was placed (no white showing through the slide when placed on white paper), this classifies the flux as L-type flux (provided that the flux passes all of the other tests for L-type fluxes). Breakthrough in less than 50 percent of the test area defines an M-type flux. Breakthrough in more than 50 percent of the test area classifies the flux as H type.

TABLE 46.2   Test Results Required for Flux Classification (ANSI-J-0004)

Flux type	Qualitative copper mirror	Qualitative halide (optional) Silver Chromate (Cl,Br)	Spot Test (F)	Quantitative halide (F,Cl,Br)	Qualitative corrosion test	Conditions for passing 100 MΩ SIR	Conditions for passing ECM requirement
L0	No evidence of mirror	Pass	Pass	0.0%	No evidence of	Both cleaned	Both cleaned
L1	breakthrough	Pass	Pass	<0.5%	corrosion	and uncleaned	and uncleaned
M0	Breakthrough in less	Pass	Pass	0.0%	Minor corrosion	Cleaned or	Cleaned or
M1	than 50% of test area	Fail	Fail	0.5 to 2.0%	acceptable	uncleaned	uncleaned
H0	Breakthrough in more	Pass	Pass	0.0%	Major corrosion	Cleaned	Cleaned
H1	than 50% of test area	Fail	Fail	>2.0%	acceptable		

46.9.2   Halide Content (TM 2.3.33)

The absence of chloride and bromide in a flux can be tested using the silver chromate paper test. When a drop of flux is applied to the test paper, the appearance of a yellow color indicates the presence of these halides. This test is subjective but can reach sensitivity levels down to about 0.07 percent. If the flux fails this simple test, a titration test with silver nitrate can be used to determine the exact percentage of halides from chloride and bromide, based on the solids content of the flux. A total percentage of halide (including the fluoride component of the flux) of less than 0.5 percent classifies the flux as L type; if the value is between 0.5 and 2.0 percent, the flux falls into the M-type category; and if the halide content is greater than 2 percent, the flux will be an H-type flux. Ion chromatography can also be used to determine the halide concentration quantitatively.

46.9.3   Fluoride Test (TM 2.3.35)

In the revised J STD 004 document, a test for fluoride has been added. The qualitative test is a spot test using zirconium alizarin purple lake as the developing agent. A change in color from purple to yellow indicates the presence of fluoride. The presence of oxalate can give a false positive result. For quantitative analysis for fluoride, a fluoride-specific electrode is used to identify and quantify the fluoride present. The concentration of fluoride from this test is added to the chloride/bromide concentration results to determine the total percentage of halides in the flux.

46.9.4   Qualitative Corrosion Test (TM 2.6.15)

This test uses a small copper sheet that has a well (indentation) created in it with a ball pinning hammer or other device. After the coupon is precleaned with sulfuric acid and ammonium persulfate to remove the metal oxide, a sample of flux solids and some solder wire is placed in the well. The coupon is then floated on a solder pot at $235 \pm 5°C$ ($455°F$) for 5 s after the solder has begun to reflow, creating flux residues. Then the coupon is cooled and placed in an oven set at $50°C$/65 percent RH for 10 days. The absence of corrosion products indicates an L-type flux. Partial corrosion at the edges defines an M-type flux. The presence of heavy corrosion places the flux in the H category.

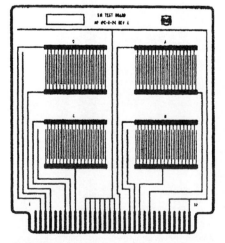

FIGURE 46.4 The IPC-B-24 test coupon for SIR testing contains four comb patterns with 0.4-mm lines and 0.5-mm spacing between conductors.

46.9.5   Surface Insulation Resistance Test (TM 2.6.3.3)

The substrate of a printed wiring board has certain electrical properties that include a bulk and a surface resistivity. The test measuring the resistance to current flow between two surface conductors is termed a surface insulation resistance (SIR) test. This is different from hole-to-hole resistance measurement, which defines the bulk insulation resistance.

Frequently, interdigitated comb patterns are used to measure SIR. Here the comb patterns are electrically powered at a bias voltage and periodically measured for insulation resistance. This is done in a temperature-humidity chamber or in an oven with a bell jar containing a saturated salt solution to provide the controlled humidity. When the humidity exceeds 65 to 70 percent, several molecular layers of water exist on the surface. This film of water can dissolve conductive ions, enhancing the rate of corrosion and degradation.

In the ANSI J-STD-004 document, the SIR test is performed at $85°C$/85 percent RH. The IPC-B-24 coupon (Fig. 46.4) with

0.4-mm lines and 0.5-mm spaces (Fig. 46.4) is used for this test. A 45- to 50-V polarizing bias and a −100-V (reversed bias) test voltage is used. Test data are taken at 24, 96, and 168 h, but the 96- and 168-h readings must pass the minimum SIR level of 100 MW.

46.10   ELECTROCHEMICAL MIGRATION

The SIR test just described includes a bias voltage during the test period. Thus, the test is also an electrochemical migration test. Electrochemical migration is the movement of an ionic species under the influence of a DC voltage. It can lead to electrical failure if (1) a short occurs due to surface dendrites (Fig. 46.5) or (2) an open circuit condition occurs due to depletion of the anode (Fig. 46.6). In the presence of moisture, metal ions may form at the anode and migrate toward the cathode, where they plate out, forming dendrites. The possible chemical reactions that can occur at the anode are:

$$Cu = Cu^{+n} + ne^-$$

$$Pb = Pb^{++} + 2e^-$$

$$Sn = Sn^{+n} + ne^-$$

$$H_2O = 2H^+ + \tfrac{1}{2}O_2 + 2e^-$$

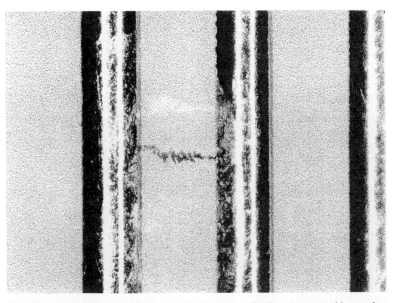

FIGURE 46.5   A lead dendrite bridges the 0.5-mm spacing between two solder conductor lines. This dendrite, which formed under a 10-V bias under aging conditions of 85°C and 85 percent RH, did not burn out because the circuit was designed to remove the bias voltage as soon as bridging occurred.

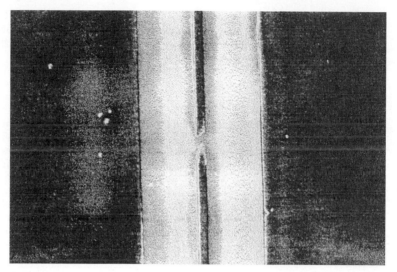

FIGURE 46.6   The 75-μm copper anode that runs down the center of this photo shows a gap due to corrosion caused by the soldering flux residues, resulting in an open circuit.

The following reactions can occur at the cathode:

$$Cu^{+n} + ne^- = Cu$$

$$Pb^{++} + 2e^- = Pb$$

$$Sn^{+n} + ne^- = Sn$$

$$2H_2O + O_2 + 4e^- = 4OH^-$$

$$2H_2O + 2e^- = 2OH^- + H_2$$

There are a number of factors that affect electrochemical migration. These include humidity, the nature of substrate and metallization, the presence of contamination, temperature, and voltage gradient.

At low humidity (20 percent) the substrate has a monolayer of water molecules bonded to the surface. Electrochemical migration cannot occur under these conditions because the water molecules are hydrogen bonded to the surface and are immobile. As the humidity increases, more layers of water molecules are added. The number of monolayers of water on the surface depends upon the nature of the substrate (epoxy, polyimide, ceramic, etc.) and the metallization. Yan[14] studied the conductivity of a thin film of moisture on alumina substrates as a function of relative humidity (RH). The data show that for films of thickness less than three monolayers the surface conductivity is 2 orders of magnitude below that of bulk water. Conductivity increases asymptotically with increased numbers of monolayers. Above 20 monolayers, the conductivity reached an equilibrium value. Cerofolini and Rovere[15] reported that metal ions are "scarcely mobile" on the first monolayer but become very mobile after several layers of water are present.

The presence of contaminants on the surface has a strong effect on moisture uptake and surface conductivity. Localized contaminants (such a flux residues) that are hygroscopic can lower the dew point and create water droplets. The dendrite in Fig. 46.5 is an example of this. Ionic contaminants enhance electrochemical migration, but the amount of enhancement depends on a number of factors including: (1) the solubility of the ion; (2) the mobility of the ion; (3) the effect of pH on solubility; (4) the reactivity of the ion; (5) temperature; and (6) relative humidity. The solubility of the ion and its mobility in water are important because conductivity depends upon both of these factors. As shown earlier, water reacts at the anode to create hydrogen ions, an acid medium, while hydroxide ions are created at the cathode, making it basic. Some ions are soluble in acid but not in base and may precipitate out before reaching the cathode.

Most contaminants from processing do not exist in isolation from one another. Rather, there may be several ionic species present, some of which will interact, creating new ionic species. A simple example of this is the reaction of moisture with copper. Normally the formation of the Cu^{++} ion is favored. However, in the presence of chloride, Cu^+ ions are favored and complexes such as $(CuCl_2)^-$ are formed.[16] Temperature is another important factor to consider. An increase in temperature can increase ion solubility, mobility, and reactivity.

Electrochemical corrosion becomes a more serious concern with today's fine-pitched circuitry because this potential failure mode increases as circuit spacing decreases. This is due to the increase in electric field, which is inversely proportional to the spacing between the conductors:

$$E = \frac{V}{d}$$

where E is the electric field
        V is the voltage
        d is the spacing between the conductors

A more subtle type of electrochemical migration is conductive anodic filament (CAF) formation, first reported in 1976 by researchers at Bell Labs.[17] This electrochemical failure mode of electronic substrates involves the growth of a copper-containing filament subsurface along the epoxy-glass interface, from anode to cathode (Fig. 46.7). A model developed by these Bell Labs researchers[18] in the late 1970s[19] details the mechanism by which CAF formation and growth occurs. The first step is a physical degradation of the glass-epoxy bond. Moisture absorption then occurs under high-humidity conditions. This creates an aqueous medium along the separated glass-epoxy interface that provides an electrochemical pathway and facilitates the transport of corrosion products.

Despite the projected lifetime reduction due to CAF, field failures were not identified in the 1980s. Recently, however, field failures of critical equipment have been reported.[20] Factors that affect this failure mode are substrate choice, conductor configuration, voltage gradient, and storage and use environment. Certain soldering fluxes[21] and HASL fluids, high humidity either in the storage or the use environment, and high voltage gradient enhance this failure mechanism. A recent study[22] indicates that the higher reflow temperatures needed for lead-free soldering will result in significantly higher incidence of CAF in the future.

46.11 SUMMARY

This chapter reviews the various types of soldering fluxes, their constituents, and their role in the soldering process. It discusses cleaning materials and makes recommendations on which cleaning agents are appropriate for which flux type. Methods for characterizing fluxes are described and failure mechanisms related to electrochemical migration are presented.

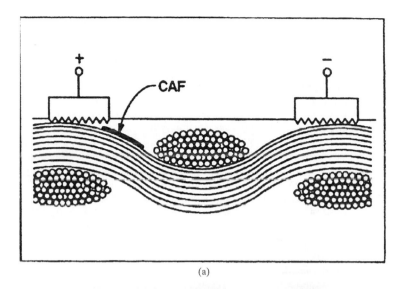

(a)

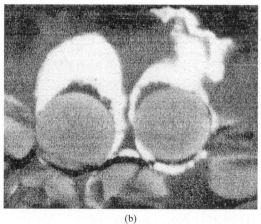

(b)

FIGURE 46.7 (*a*) Diagram of conductive anodic filament (CAF) formation in which a copper salt grows from the anode along the epoxy-glass interface. (*b*) SEM micrograph using backlighting to enhance the copper salt, showing the CAF growing at the epoxy-glass interface.

REFERENCES

1. D. A. Fisher et al., "Relative Effects on Stratospheric Ozone of Halogenated Methanes and Ethanes of Social and Industrial Interest," *Scientific Assessment of Ozone: 1989,* vol. II, World Meteorological Organization Global Ozone Research and Monitoring Project—Report no. 20, pp. 301–377.

2. Institute for Interconnecting and Packaging Electronic Circuits, *Terms and Definitions for Interconnecting and Packaging Electronic Circuits,* (ANSI/IPC-T-50), IPC, 2215 Sanders Rd., Northbrook, IL 60062-6135.

3. MIL-F-14256, "Flux, Soldering, Liquid (Rosin Base)."

4. F. M. Zado, "Effects of Non-Ionic Water Soluble Flux Residues," *Western Electric Engineer,* vol. 27, no. 1, 1983, p. 40.

5. J. Brous, "Electrochemical Migration and Flux Residues: Causes and Detection," *Proceedings of NEPCON West,* February 1992, pp. 386–393.

6. J. P. Mitchell and T. L. Welsher, "Conductive Anodic Filament Growth in Printed Circuit Materials," *Printed Circuit World Convention II,* Munich, Germany, published as IPC Technical Report WC-2A-5, June 1981.

7. L. A. Guth, "Low Solids Flux Technology for Solder Assembly of Circuit Packs," *Circuits Manufacturing,* vol. 29, no. 2, pp. 59–63.

8. F. R. Cala and A. E. Winston, *Handbook of Aqueous Cleaning Technology for Electronic Assemblies,* Electrochemical Publications, Isle of Man, UK, 1996.

9. Institute for Interconnecting and Packaging Electronic Circuits, *Post Solder Solvent Cleaning Guidelines* (IPC-SC-60), IPC, 2215 Sanders Rd., Northbrook, IL 60062-6135.

10. Institute for Interconnecting and Packaging Electronic Circuits, *Post Solder Semiaqueous Cleaning Guidelines* (IPC-SA-62), IPC, 2215 Sanders Rd., Northbrook, IL 60062-6135.

11. Institute for Interconnecting and Packaging Electronic Circuits, *Post Solder Aqueous Cleaning Guidelines* (IPC-AC-62A), IPC, 2215 Sanders Rd., Northbrook, IL 60062-6135.

12. Institute for Interconnecting and Packaging Electronic Circuits, *Guidelines for Cleaning Printed Boards and Assemblies* (IPC-CH-60), IPC, 2215 Sanders Rd., Northbrook, IL 60062-6135.

13. R. Breunsbach, "New Developments in Simplified, Low Cost, Semi-Aqueous Emulsion Cleaning Technology." *Proceedings of NEPCON West,* 1992, pp. 1217–1225.

14. B. D. Yan et al., "Water Adsorption and Surface Conductivity Measurements on Aluminum Substrates," *36th Electronic Components Conference,* 1986.

15. G. F. Cerofolini and C. Rovere, "The Role of Water Vapor in the Corrosion of Microelectronic Circuits," *Thin Solid Films,* vol. 47, 1977, p. 83–94.

16. H. H. Uhlig, *Corrosion and Corrosion Control,* 3d ed., Wiley, New York, 1971, pp. 327–340.

17. P. J. Boddy, R. H. Delaney, J. N. Lahti, and E. F. Landry, "Accelerated Life Testing in Flexible Printed Circuits," *14th Annual Proceedings of Reliability Physics,* 1976, pp. 108–117.

18. D. J. Lando, J. P. Mitchell, and T. L. Welsher, "Conductive Anodic Filaments in Reinforced Polymeric Dielectrics: Formation and Prevention," *17th Annual Proceedings of Reliability Physics,* 1979, pp. 51–63.

19. J. N. Lahti, R. H. Delaney, and J. N. Hines, "The Characteristic Wearout Process in Epoxy-Glass Printed Circuits for High Density Electronic Packaging," *17th Annual Proceedings of Reliability Physics,* 1979, pp. 39–43.

20. W. J. Ready, L. J. Turbini, S. R. Stock, and B. A. Smith, "Conductive Anodic Filament Enhancement in the Presence of a Polyglycol-Containing Flux," *34th Annual Proceedings of Reliability Physics,* 1996, pp. 267–273.

21. J. A. Jachim, G. F. Freeman, and L. J. Turbini, "Use of Surface Insulation Resistance and Contact Angle Measurements to Characterize the Interaction of Three Water Soluble Fluxes with FR-4 Substrates," *IEEE Transactions on CPMT, Part B. Advanced Packaging,* vol. 20, no. 4, November 1997, pp. 443–451.

22. L. J. Turbini, W. R. Bent, and W. J. Ready, "Impact of Higher Melting Lead-Free Solders on the Reliability of Printed Wiring Assemblies," *Journal of Surface Mount Technology,* vol. 13, no. 4, October 2000, pp. 10–14.

CHAPTER 47
PRESS-FIT CONNECTIONS

Gary M. Freedman
*Compaq Computer Corp. Alpha High Performance Systems,
Marlboro, Massachusetts*

47.1 INTRODUCTION

Electronic connectors are available in three types of configurations:

1. Solder tail (through-hole)
2. Surface-mount
3. Press-fit

The press-fit connector has been in use since the 1970s. It is also called *press pin* (see Fig. 47.1), *compliant pin*, and by a number of trade names. Once exclusively the realm of passive backplanes, more recently press-fit connectors have gained in popularity and are commonly incorporated on complex mother boards adjacent to active components and on associated daughter cards. Just as circuit boards have become more complex (thicker; higher layer count with densely routed circuit traces and high component count), so too have press-fit connectors. They are available with high pin density (pin pitch down to 2 mm) with pin count in the thousands per connector and even stripline shielding for enhanced high-speed signal performance. Press-fit connectors can be applied to either or both sides of a printed circuit assembly (PCA) and they are repairable. The press pin process is easier and more reliable than soldering and does not subject the PCA to additional thermal or chemical processes, an advantage for reliability especially advantageous for dense circuitry.

Connector bodies may or may not be compatible with the thermal rigors of reflow or wave, so press-fit connectors are typically applied after all mass soldering of the PCA is completed. Further, press pin connectors are not generally soldered into place (as through wave soldering), since rework, which is discussed later, would be rendered impractical.

47.1.1 Advantages of Press-fit Systems

There is a resurgence in press-fit connector popularity driven by the increased board complexity and density of PCAs and the appearance of backplanes populated with active components. Some advantages are as follows:

- Press-fit connectors are typically used on very thick boards that would be difficult or impossible to wave-solder. Therefore, complex assemblies can be limited to the one or two thermal cycle exposures incurred during surface mounting—a plus for reliability. This is important

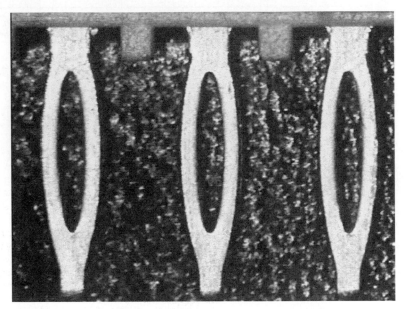

FIGURE 47.1 Photograph of compliant press-fit leads prior to force fit.

for dense printed circuit assemblies where the extra thermal step of wave soldering poses a threat to the solder joint integrity of previously surface mounted components (see re-reflow, Chap. 43, Sec. 43.7.1) or the overall mechanical integrity of the assembly.

- Using press-fit connectors is a means of lead (Pb) abatement because no solder is needed for press-fit installation. Given the climate of increasing importance of ecological responsibility and the potential for lead abatement legislation in several countries, use of press-fit components will likely increase.

47.1.2 Press-Fit System Catagories

The press-fit process is simple in theory and generally so in practice, relying on oversized connector leads forced into in plated through-holes (PTHs) in the PWB. There are two main categories of press-fit systems:

1. One system uses a relatively rigid lead that is designed to slightly deform the board's plated through-hole upon insertion. This is a variation of the proverbial square pin in a round hole.
2. The most prevalent system, the collapsible pin (Fig. 47.1), folds in upon itself axially in a controlled manner when pressure is applied during the insertion process, as depicted in Fig. 47.2. The plated through-hole barrel is only lightly skived and minimally deformed, with the resultant forces from the press-fit lead sufficient for reliable pin-to-barrel contact.

47.1.3 Press-Fit Design Issues

Press-fit connector leads are designed to work within a tight range of PTH barrel sizes. As they are pressed into place, the contact design must balance several factors.

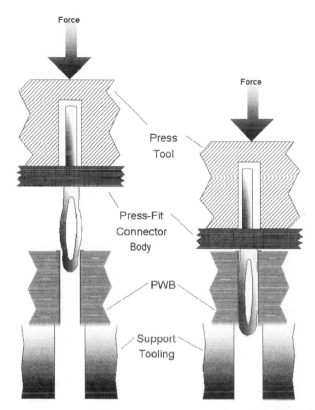

FIGURE 47.2 As the press tool is driven by the connector press ram, the compliant section of the pin collapses inward upon itself and is forced into intimate contact with the plated through-hole barrel of the board.

47.1.3.1 *Connector Issues*

- Compliant pin radial force exerted against the PWB plated through-hole barrel wall must be maximized to reliably anchor the connector into the PWB.
- Pin-to-barrel radial force must be high enough to form a gas-tight seal for long-term electrical and environmental contact reliability.
- The pin must be strong enough axially to prevent buckling during the pressing operation.
- Pin radial forces must be optimized so as not to damage the PTH barrel.

47.1.3.2 *Load and Insertion Issues.* The load necessary to apply a press-fit connector will depend on a variety of factors, including:

- Lead design
- Lead material
- Plated through-hole surface finish/material
- Plating thicknesses
- Lead size
- Plated through-hole size
- Number of leads per connector

Since pressing force generally ranges from some tens of grams to kilograms per lead, the mechanical advantage of a machine press is required for press-fit connector application.

47.1.3.3 Materials Issues.   For purposes of affordability, mechanical integrity, electrical properties, and materials conservation, all electrical contacts, regardless of assembly methodology, are made from inexpensive base metals and plated with a minimum thickness of a more precious or practical material. Because most materials oxidize readily upon exposure to the atmosphere, gold or other noble metals are often used to prevent or retard oxidation of SMT or PTH component leads or PWB solder lands and plated though-hole barrels. However, most circuit boards and component leads are covered with less noble materials such as tin or tin/lead and instead rely upon a fluxing agent to remove naturally occurring oxides prior to and during the soldering process. In a solder joint, the lead and pad interface materials are wetted by and sealed in solder. The solder gives the interconnection mechanical rigidity and seals the contact surfaces from the environment.

In the case of the press-fit connector lead, there is neither a physical wetting of material nor metallic encapsulation to protect the interconnection; additionally, no chemical fluxing agents are used. Instead, press-fit connectors are anchored solely by the mechanical interference between the connector lead and the PTH barrel of the PWB. Oxides on the surface of the press-fit pin and mating PTH barrel are broken loose by the high frictional forces associated with the press-fit assembly operation, so nobility of materials or material shelf life is less a consideration than in soldering. The press-fit process imparts a gas-tight contact between the connector pin and the plated through-hole barrel wall. The notion of a gas-tight seal as the product of proper press pinning is crucial to press-fit reliability. If an electrical contact is to be reliable, the interface between the two mating surfaces must remain chemically and mechanically stable. The gas-tight, smeared, metal-to-metal contact attained during the press-fit operation mitigates oxidation of either contact and prevents fretting corrosion, which is a common failure mechanism for mechanically mated contacts subjected to vibration. Press-fit interconnection is considered at least as reliable as a soldered connection.

47.2 BOARD REQUIREMENTS

Because the basic purpose of the press-fit connector is to mate with a printed wiring board, the characteristics of the board are critical to the success of the total system.

47.2.1 Surface Finish

The surface finish of the PWB's PTH can have a significant effect on the force required to seat a press pin connector. Solder-coated barrels provide lubrication, which eases the compliant pin into the PTH. When this surface finish is a product of the hot-air solder leveling (HASL) process, care must be taken by the PWB manufacturer to avoid excess solder buildup at the hole openings or within the hole. HASL finishes have been seen to cause solder slivering as the connector plows its way through the solder (see Fig. 47.3). The sliver may break free of its source and cause an electrical short elsewhere in the system.

Fusible coatings such as tin or tin/lead may also form a meniscus within the plated through-hole as a result of the finish application or subsequent SMT or wave-soldering thermal cycles. The meniscus may alter hole diameter enough to cause overly high press force requirements or may increase propensity for solder sliver formation. High-aspect-ratio holes (PWB thickness:plated through-hole diameter) may tend to have irregular plating where the thickness closest to the board's surface layer is within acceptable thickness limits while plating thickness in the middle of the PTH may have unusually thin deposits. This is caused by plating bath stagnation and exhaustion (a zone of poor electrolyte circulation within the

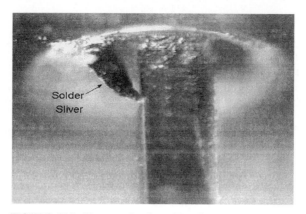

FIGURE 47.3 Photograph of a solder sliver pushed out of a HASL-coated PTH.

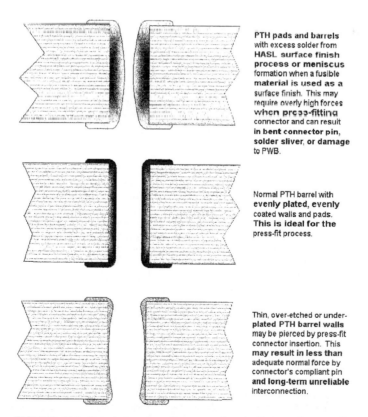

PTH pads and barrels with excess solder from HASL surface finish process or meniscus formation when a fusible material is used as a surface finish. This may require overly high forces when press-fitting connector and can result in bent connector pin, solder sliver, or damage to PWB.

Normal PTH barrel with evenly plated, evenly coated walls and pads. This is ideal for the press-fit process.

Thin, over-etched or under-plated PTH barrel walls may be pierced by press-fit connector insertion. This may result in less than adequate normal force by connector's compliant pin and long-term unreliable interconnection.

FIGURE 47.4 Barrel shape can have a profound effect on the press-fit process.

high-aspect-ratio barrel). These irregularities (See Fig. 47.4) can have a triply detrimental effect on the press pinning operation:

- If the walls are excessively thick, press force may exceed the mechanical strength of the pin and the pin may fold over.
- If the force is exceedingly high for the board structure, it may damage the PWB, causing broken barrel interconnections, damage to closely routed internal traces, or even barrel shearing and ejection.
- Normal force from the connector's compliant section may be enough to tear through an overly thin barrel wall.

47.2.2 Bare Copper

OSP-copper finish thickness can generally be controlled fairly well except within the high-aspect-ratio hole as for any other surface finish. But, owing to its lack of intrinsic lubrication, copper can be notably difficult to press into as compared to nickel/gold, tin/lead, and silver surface finishes.

47.3 EQUIPMENT BASICS

There are several types of mechanical presses used for forcing press-fit connectors into PWBs. The lever-activated arbor press is the most primitive but will suffice for simple, low-lead-count, coarse-pitch connectors. One step up is the pneumatically driven press. Although it is easier to operate than the lever-activated arbor press, it does little for process control or assembly repeatability. Higher on the evolutionary scale is the pneumatic-hydraulic combination (the so-called air-over-oil press), which offers a somewhat more controllable process. High-precision pressing cycles are mandated by the latest press pin connectors with their delicate molded body features, fragile electrical shielding, and fine pin pitch, so it is advisable to use a machine that is computer controlled. Aside from force and speed precision for process reproducibility, press cycle data can be stored along with process recipes for ram force, speed of compression, and component location if the press is equipped with programmable axes. Data logging is critical for statistical process control and is useful for press cycle root cause defect analysis or machine troubleshooting. The state-of-the-art connector press is electromechanically actuated, relying on a motor-driven coarse-pitch lead screw with tachometer speed control, z-axis positional encoders, and load cell feedback for an accurate pressing cycle (see Fig. 47.5).

Some commercially available machines are automated to the point that they are able to shuttle boards into the press, select the connector-appropriate pressing tool and supporting tool anvil, rotate the tools to the proper orientation, and press multiple connectors sequentially on a single board or multiple nested boards. There are several ways to control the press-fit process. Methods and merits of each will be discussed shortly, but first a more detailed review of the mechanics of the press-fit process is in order.

47.4 PRESSING CYCLE

Initially, the press tool and support plate are chosen for the connector. The operator manually places the connector into the appropriate plated through-holes in the PWB. The ram and tool are aligned to the connector to be pressed. When press-fit connector leads are forced into PWB plated through-holes during a pressing cycle, there is a characteristic fingerprint of force vs. vertical distance traveled for each connector and PWB type. The various events and slopes that comprise the fingerprint are shown in Fig. 47.6.

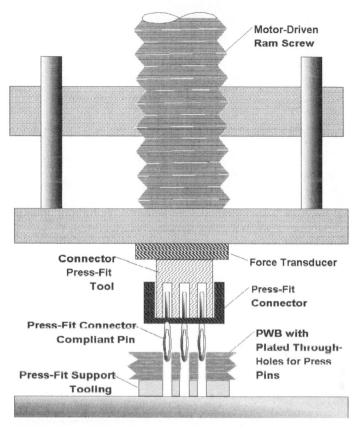

FIGURE 47.5 Electromechanical connector press schematic.

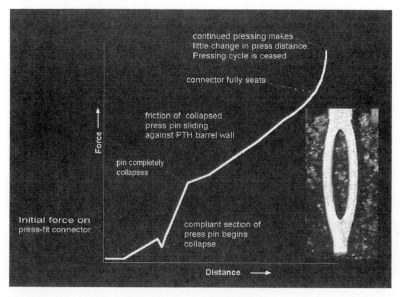

FIGURE 47.6 Force-vs.-distance plot of a press-fit cycle.

As pressing progresses, connector leads are forced into the plated through-holes and the compliant sections of its pins begin to deform elastically, then plastically. Continued resistance is encountered due to the friction of the collapsed pins sliding along the barrels of the plated through-holes. Another inflection in the plot can be noticed where the connector fully seats. Further force makes little or no change in pressed height and the press cycle ceases.

47.5 PRESS ROUTINES

For best results, a method that allows real-time measurement and control of the pressing process is preferred. There are four most commonly used pressing options:

1. Uncontrolled pressing
2. Press to height
3. Press to force
4. Press to gradient

Of these, only the last three take advantage of force sensing, distance sensing, or real-time feedback and control of the connector press. Connector complexity and equipment availability will dictate press-fit assembly methodology.

47.5.1 Uncontrolled Pressing

Initially, uncontrolled pressing was the most prevalent technique, but it has lost ground to the more sophisticated pressing methods described later. It commonly relies on an arbor press and operator muscle to force the press-fit connector into the PWB. There is neither force sensing nor speed control. It is the least reliable way to press connectors into PWBs. With more delicate connectors, if too high a ram speed is used, the connector pins may buckle. Although this type of pressing is discouraged, it may be adequate for low-lead-count, coarse-pitch connectors in noncritical assemblies. As with any press routine, a suitable support fixture (see Fig. 47.7) must be used beneath the board to keep the assembly from breaking or flexing. The fixture should incorporate relieved areas to accept protrusion of press pins below the bottom side of the board and clearances for bottom-side components, as shown Fig. 47.7.

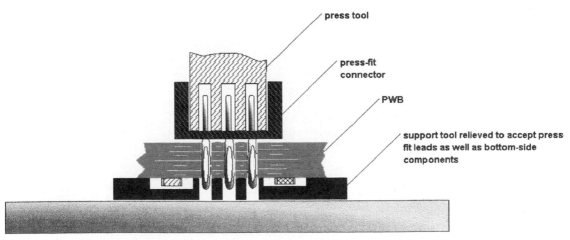

FIGURE 47.7 Supporting fixture is critical for good press-fit application. It must have clearances for press pin protrusion and any bottom-side components.

47.5.2   Press to Height

As previously described, connector presses have improved greatly. Some are capable of precision pressing, which includes means of precisely determining ram height above a datum such as top surface of the PWB (see Fig. 47.8). If board-to-board thickness is consistent, then the connector can be forced into the board to a predetermined standoff height as required by the connector specification. Just as connector pin dimensions vary from connector to connector, so do plated through-holes (PTHs) in the PWB. Pressing to height ensures that the connector will be seated the same way each time regardless of the pin or hole condition and resultant force required to seat the connector. Unfortunately, variations in PWB thickness or variability in molded cross section of the connector body can impart imprecision and negate some of the benefits of this approach.

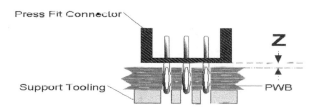

FIGURE 47.8   Press to height allows forcing the connector to a predetermined standoff distance *Z* from the surface of the PWB.

47.5.3   Press to Force

This method relies upon the intelligence of an instrumented connector press to sense that a predetermined force limit has been reached, at which time the pressing cycle is terminated. The force limit can be roughly set to coincide with the average force required per compliant pin (available from the press-fit connector manufacturer's specification) multiplied by the number of pins per connector. Application of upper and lower tolerances of specified force per pin coupled with knowledge from empirical trials will help refine the acceptable press force envelope for a given set of materials. This technique is highly sensitive to the material conditions of the compliant pins and the plated through-hole diameters.

47.5.4   Press to Gradient

As previously stated, the slope or gradient of the final portion of the force-vs.-distance plot can be used to trigger the end of the press cycle and retract the ram. It is a useful routine ensuring that regardless of pin size and hole size, board or connector variations, the connector will not be overpressed. This has been found to be a useful routine for reproducible pressing when a very steep slope is used and refined empirically. Gradients in the range of ≥75 percent are most reliable.

47.6   REWORK FOR PRESS-FIT CONNECTORS

As a result of the press cycle, the compliant pins of a press-fit connector are plastically deformed. Therefore, once used, the same connector cannot be removed and reinserted in a circuit board. However, most press-fit connectors are designed to be reworked, i.e., repaired or replaced. In some cases, individual leads/contacts can be replaced. In other cases, banks of leads/contacts can be substituted. Some press-fit connectors require complete removal of the damaged connector for replacement with a new one. There are many different press-fit connectors on the market and each has its own manufacturer recommended repair strategy.

47.6.1   Replacement Cycles

Press-fit assemblies are generally designed to accommodate three connector replacement cycles. The three cycles are predicated on two things:

1. The normal force exerted by the press pin's compliant section on the PWB's plated through-hole barrel surface

2. The quality and condition of the PWB's plated through-hole barrel surface after subsequent press-fit connector removals and replacements

If the radial force of the compliant section of a press pin is excessive, it might cut through the barrel and even damage surrounding innerlayer interconnects or traces. Instead, the radial force is optimized to permit a gas-tight seal without excessive deformation or deep skiving of the plated through-hole barrel. Once a connector or connector pin is replaced, the new connector/pin must remake the gas-tight seal for reliable contact. Each time that a new press pin is inserted into a plated through-hole, there is an opportunity for the barrel to deform. The more insertions into the barrel, the deeper the deformation. It is for that reason that the industry standard design goal for compliant pin connectors is generally targeted at three replacement cycles. Beyond three there is the possibility that the barrel will be thinned and damaged, impairing press pin contact reliability.

47.6.2   Rework Tools

Press pin rework tools vary greatly. In some cases they are designed to remove and replace a single pin. In other cases the tool may be designed to help dislodge an entire connector from a circuit board. Some tools are designed to remove the molded connector body only. Then individual pins are pulled one by one in preparation for replacement. Sometimes multiple press pin leads are gang-molded into discrete wafers and the wafers are ganged and held by a clip to form a single press-fit connector. In this case, individual wafers can sometimes be removed and replaced rather than having to replace single leads or the entire connector. Repair tools and repair strategies are generally available from connector manufacturer.

During rework, support the board such that it is not flexed during the repair operation to avoid damage to the board, components, or solder joints. Verify circuit integrity by electrical testing before and after repair to ensure that it is not impaired as a consequence of the repair cycle. If a connector is replaced because of pin stubbing on the surface of the circuit board upon the original installation, ensure that the surface and underlying traces of the circuit board have not been damaged by the errant lead.

47.7   PWB DESIGN/BOARD PROCUREMENT TIPS

47.7.1   Design Tips

- To avoid process problems and damage to adjacent components, allow sufficient clearance in the PCA design to permit room for the pressing tool and any associated rework tools.

- When possible, use press pin contacts slightly longer than the board thickness so that connector pins will protrude on the bottom side of the board after the pressing cycle. Pin protrusion will allow for easy inspection of the finished assembly. A missing pin will be evidence of pin stubbing or buckling.

- Provide enough clearance for any press pin connector rework tools.

- Follow press-fit connector manufacturer's recommendation for board finished hole size, tolerances, and acceptable or tested surface finishes. Cross section sample boards with high aspect ratio (depth:diameter) holes to qualify the board vendor. Ensure that wall thickness of plated through-holes is uniform throughout their depth and meets connector specifications.

47.7.2 Press-Fit Process Tips

There are several things that can be done to ensure adequacy of the press-fit operation in manufacturing:

- Prior to insertion into the board, check press-fit connector pins for straightness and integrity. An inexpensive go/no-go gauge (see Fig. 47.9) can be machined to check that connector pins are not bent. It should be constructed so that it neither compresses the compliant section of the press-fit pin nor damages its surface coating. This is especially helpful on fine-pitch press-fit connectors (2 mm centers) and is better than relying on visual inspection for pin field integrity.

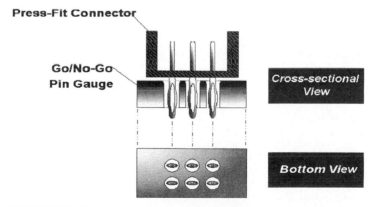

FIGURE 47.9   Gauge for checking connector pin position prior to pressing. This is best used for high-lead-count, fine-pitch press-fit connectors.

- Keep holes in circuit board clear of extraneous solder, flux, or other contaminants from surface mounting, wave soldering, or hand soldering processes. Use of a wave pallet or Kapton® tape to shield press-fit locations is recommended. Take care to inspect first articles to ensure that Kapton® tape residue does not foul press-fit holes. Liquid spot masks should be avoided in the area of press-fit holes as they may contaminate pin and barrel contact surfaces.

- Use pressing tools recommended by connector manufacturer to avoid damage to the connector or the PCA.

- Adequately support the PCA during the pressing operation so that connectors can be fully homed and the board is not overflexed. Excessive board flexure may cause solder joint cracking, broken PWB circuit traces, or board delamination. This is especially important for the characteristically inflexible ball-based solder joints of area array devices such as BGAs, CCGAs, etc.

- Best support will be localized directly under the connector to be pressed. Ensure that the support includes clearance for back-side pin protrusion; otherwise pins will stub and bend on the supporting anvil.

- Support tooling design should carry appropriate relief features for bottom-side components.

- Check connector after pressing for even bottom-side pin protrusion, top-side connector body seating height, damage to connector body, mating pin positions (within the connector as opposed to within the board), and damage or movement of electrical shields if connector is so equipped.

- Check the PCA for surface damage in the area of the pressing operation. A misplaced or damaged press tool or board support can damage components or overly compress the laminate, causing innerlayer trace damage.

- In the case of hot-air solder-leveled (HASL) boards, check the bottom side of the board for solder sliver formation. Brush slivers out if required and inspect. Loose slivers may cause electrical shorting on that board or other boards in the enclosure. Work with PWB vendor to ensure that drilled and plated holes are sized appropriately for the press-fit process and that excess solder from the hot-air solder-leveling process does not adversely alter the finished hole size or shape.

- In the case of high-friction surface finishes, such as OSP-copper, forces applied during the press-fit process may be so high as to cause damage to the PTH barrel or adjacent inner- or outerlayer circuit traces. Check board visually and by electrical testing as part of a first article inspection.

- Check for excessive tool wear or damage. A damaged tool can ruin a connector or PWB.

- Apply press-fit connectors after all SMT and wave-soldering operations.

- Do not try to solder press pin connectors in place. Press-fit connector bodies may not be reflow compatible. Soldering a press-fit connector will also make rework difficult or impossible.

QUALITY CONTROL
AND RELIABILITY

CHAPTER 48
ACCEPTABILITY OF FABRICATED BOARDS

A. D. Andrade (Retired)
Sandia National Laboratories, Livermore, California

48.1 INTRODUCTION

There has never been a "standard" printed wiring board (PWB) produced for multiple customers and multiple products. Each board represents an application-specific design. As component technologies change, the capabilities of and expectations for PWB fabrication also change and evolve to accommodate the associated interconnection requirements. In addition, with the increased reliance on contract fabrication and assembly, there are added complications of distance, language, and often culture to take into consideration when defining product acceptability elements of contracts. To be successful in this environment, it is critical to have clear methods of communication in the design, fabrication, and acceptability of PWBs. This can be done by establishing and following some basic rules.

This chapter gives guidelines for developing a clear understanding, at all levels of fabrication, of acceptability criteria. Detailed standards for specific purposes are available from a variety of sources. However, those presented here rely on IPC-A-600. While acknowledging that these documents are revised routinely and frequently, it is important to use the most current. However, it is critical for the contract to be definitive as to which standard and revision is being used.

48.2 DEVELOPING ACCEPTABILITY CRITERIA

Some elements of the user-manufacturer agreement that must be defined at the beginning of the design to fabrication cycle are as follows.

48.2.1 Basic Rules

The first step in developing a clear understanding of what constitutes acceptable performance is to agree on basic concepts. These include, in order of precedence:

1. Use of both English and metric dimensions to prevent converting of measurement dimensions. PBs can be designed on one continent, fabricated on another, and assembled on a third.

2. Use of only proper definitions as stated in IPC-T-50. This includes all communications from contract through design and acceptance.

3. Procurement documentation, "contract."

4. Master drawing reflecting the customer's detailed requirements.

5. Other documents to the extent specified by the customer.

6. The end item performance specification (e.g., IPC-6012) when invoked by the customer.

7. The acceptability document (e.g., IPC-A-600) indicating version to be used, when invoked by customer.

Note that before any transactions take place, a preliminary meeting (team meeting) should be held to discuss these basic steps and manufacturability of design to prevent misinterpretation.

48.2.2 Inspection—Is It Necessary?

The first printed wiring boards were fabricated out of inexpensive materials and were usually single- or double-sided, with the more expensive types having plated through-holes. During this era, defective boards found during assembly were usually thrown away. Printed wiring boards, however, have evolved into complex circuit components. They are often multilayer, rigid-flex printed circuit board (printed wiring and printed components), or a combination with several layers of circuitry and high-density surface pattern circuitry.

The cost of boards can be compared to that of a building structural foundation. Initial unit value is usually insignificant compared to total building structure costs; however, the value increases exponentially after the building construction is completed. The printed board's value increase is dependent on type, design complexity, number and type of components, and replacement component availability. This increased value, like the structural foundation, should be considered when making a decision on inspection.

48.3 DEVELOPING THE ACCEPTANCE CRITERIA AGREEMENT

Quality assurance operations are considered very costly; however, savings realized by reduced rework and improved customer relations greatly reduce the overall inspection cost. Customer relations can be greatly improved by meeting with the customer prior to printed board fabrication and reviewing acceptable and rejectable attributes.

48.3.1 Common Acceptance Standards

One method is to have both parties review and agree upon a common visual acceptance standard (e.g., IPC-A-600), and agree upon the target/nonconforming attributes by class designation. Jointly mark up two copies of the acceptable attributes agreed to. List the marked-up documents in the procurement contract.

48.3.2 Establishing a Team Environment

Quality assurance operations can be performed by the purchaser or contracted out partially or totally to an independent company. It is important to be clear about who is responsible for which elements, and to ensure that all other parties use the same documentation and methodology. For example, for printed wiring board fabrication technology capable of conductor widths and spacing less than 0.005 in (0.127 mm), reported down to 0.002 in (0.051 mm), visual inspection of surface patterns on pattern densities less than 0.005 in (0.127 mm) is questionable. Many companies are replacing surface visual inspection of high-density patterns with automated electrical or image scanning methods. Wherever the quality assurance operation is performed, however, the same tools and methods must be used upon subsequent inspection and acceptance. In addition, there must be access to a calibrated facility equipped with the required types of mechanical gauges, as well as microsectioning, chemical analysis, dimensional measuring, and electrical and environmental testing.

48.3.3 Meeting Quality Assurance Requirements

Quality assurance requirements are usually met by one of the following methods:

1. Inspection data submitted by the fabricator are reviewed for compliance with design requirements.
2. Inspection data submitted by the fabricator are reviewed and a sample lot inspection is performed.
3. A complete inspection is performed to all design requirements including destructive testing.

Where and how the inspections are performed is a question of economics, required test equipment, and availability of experienced personnel. Prior to making a decision, management should review all aspects of the question, such as equipment cost and maintenance, work volume, time to and from the quality assurance facility location, turnaround time, and qualified personnel availability.

48.4 USE OF TEST PATTERNS

Test patterns are useful mechanisms for performing destructive tests such as peel strength, flammability, thermal stress, and solderability testing as an indication of the integrity of the finished printed boards. For a detailed discussion of test patterns, with example patterns, used in process capability testing and process qualification, see Chap. 38.

The use of test patterns located outside the finished pattern contour area used to monitor all processes during routine fabrication and also for microsectioning can be a controversial subject. Some claims have been made that test patterns located outside the contour area usually have thicker plating than the actual printed board pattern and therefore are not representative of the actual pattern. Others debate that the cost savings of not destroying actual circuits overshadow any slight differences in plating thickness.

Multiple sets of test coupons are required on MIL-standard fabricated printed boards. These coupons provide lot-to-lot process control and indirectly indicate printed board quality fabricated on the panels. An additional test coupon is suggested for solderability testing of multilayer or rigid-flex printed boards. The average multilayer or rigid-flex board has one or more planes internally (ground or voltage type). Plated through-holes are usually connected to these planes through a thermal resistor–type interconnect to reduce thermal heat sink effects during machine soldering assembly, thus providing a uniform thermal soldering environment. Several different thermal resistor designs are available through the different CAD programs. However, the designs are not usually created for functional purposes but for ease in CAD programming. Therefore it is desirable to create a solderability test coupon that simulates the actual multilayer or rigid-flex board design with planes and thermal resistors. Solderability testing can then simulate actual solderability performance, indicating assembly machine soldering performance. If the fabrication panel allows placement of test patterns near the center of the panel, this would improve the acceptability confidence level.

The use of test patterns for the acceptance of printed boards is therefore an individual question and must be resolved for each design requirement and for the economic advantage of various inspection operations.

48.5   DETERMINATION OF ACCEPTABILITY

The question of acceptability must be answered for individual printed board design functions. Functionality through operational life should be the ultimate criterion for acceptance. A printed board that will see extreme environmental conditions should not be inspected according to the same requirements applied to a printed board for an inexpensive toy or radio. Some of the characteristics normally inspected come under the heading of workmanship. In the majority of designs those characteristics are cosmetic and pertain to how the printed board looks and not how it functions. In most cases the characteristics are inspected to establish a confidence level for integrity of the finished PWB and materials. The inspection results should be used to establish a fabrication quality level and not to scrap parts that will meet the functional criteria. Scrapping functional parts has a considerable impact on unit cost. However, the same workmanship characteristics can have an effect on the function of some designs. The effects of the different characteristics are discussed later in this chapter. The determination of printed boards that do not meet specified acceptance requirements but are functional should be made by a materials review board. The levels of acceptance have to be established by each company; they are dependent on the functional criteria to which the printed board will be subjected. The Institute for Interconnecting and Packaging Electronic Circuits (IPC) Association of Connecting Electronic Industries has established recommended guidelines for acceptance, based on end use, categorized as Classes 1, 2, and 3. The classes are defined as follows:

Class 1—General Electronic Products: includes consumer products and some computer and computer peripherals suitable for applications where cosmetic inperfections are not important and the major requirement is function of the completed printed board.

Class 2—Dedicated Service Electronic Products: includes communication equipment, sophisticated business machines, and instruments where high performance and extended life are required, and for which uninterrupted service is desired but is not critical. Certain cosmetic imperfections are allowed.

Class 3—High Reliability Electronic Products: includes equipment and products where continued performance or performance on demand is critical. Equipment downtime cannot be tolerated, and the equipment must function when required (e.g., life support systems, flight control systems). Printed boards in this class are suitable for applications where high levels of asistance are required and service is essential.

Each group is then subdivided into three acceptance categories: target, acceptable, and nonconforming. The acceptance guidelines in this chapter, however, utilize only two categories: target and nonconforming. Refer to IPC-A-600 for complete target and acceptability criteria.

48.6   THE MATERIALS REVIEW BOARD

The materials review board (MRB) is usually made up of one or more representatives from the departments of quality, production, and design. The purpose of the materials review board is to effect positive corrective action within a short time to eliminate the cause of recurring discrepancies and prevent recurrence of similar discrepancies. The board's responsibilities include:

1. Reviewing questionable printed boards or materials to determine compliance or noncompliance with quality and design requirements

2. Reviewing discrepant boards for effects on design functionality
3. Authorizing repair or rework of nonconforming materials when appropriate
4. Establishing responsibility and/or identifying causes for nonconformance
5. Authorizing scrappage of excessive quantities of materials

48.7 VISUAL INSPECTION

For a complete discussion of inspection processes and tools, see Chap. 49. For this chapter, however, it is considered that visual inspection is the inspection of characteristics that can be seen in detail with 1.75 (3 diopters) to 10× maximum. Note that large viewing screens are usually rated in diopters. One diopter is 0.25× magnification.

The viewing of printed board microsections for defects is the exception. Defects such as resin smear and plated through-hole quality require magnifications of 50× up to 500×, depending on the characteristic. Adequate magnification should be used to clearly define the attributes being inspected.

Note that the use of higher magnification for visual-type attributes can produce false results due to illumination and attribute contour effects. Visual inspection criteria are difficult to define grammatically, because individuals interpret words differently. An effective way to define visual inspection criteria is to use line illustrations and/or photographs. The IPC utilized this method in the publication "Acceptability of Printed Boards Manual," to "visually standardize the many individual interpretations to specifications on printed boards."

Another method is the use of an audiovisual projector. Slides of line illustrations and photographs depicting visual inspection preferred nonconforming criteria along with the related description in audio provide a reproducible inspection method. Note that pictorial standards and audiovisual aids are excellent methods for training inspectors, quality engineers, and design engineers.

Of the different types of inspection, surface visual inspection costs the least. Visual inspection by the unaided eye is usually performed on 100 percent of the printed boards or on a sample taken following an established sampling plan. MIL-STD-105, Sampling Procedures and Tables for Inspection by Attributes, is frequently used for this purpose. Inspection for visual defects, following a sampling plan, is done on the premise that the boards were 100 percent visually screened during the fabrication process. Defects that usually can be detected by visual inspection may be divided into three groups:

1. Surface defects
2. Base material defects
3. Other defects.

48.7.1 Surface Defects

Surface defects include dents, pits, scratches, surface roughness, voids, pinholes, inclusions, and markings. Dents, pits, scratches, and surface roughness usually fall in the class of workmanship. The defects, when minor, are normally considered to be cosmetic, and they usually have little or no effect on functionality. However, they can be detrimental to function in the edge board contact area (Fig. 48.1).

Voids in conductors, lands, and plated through-holes can be detrimental to function, depending on the degree of defect. Pinholes and inclusions are in the same category. Voids or pinholes, either of which reduce the effective conductor width, reduce current-carrying capacity and can affect other design electrical characteristics, e.g., inductance, impedance, etc. Voids

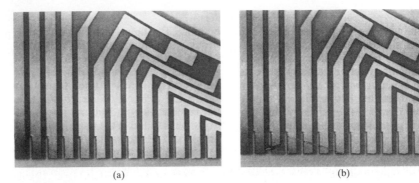

(a)   (b)

FIGURE 48.1   Edge board contact area. (*a*) Target: edge board contact area free of delamination, pits, pinholes, dents, nodules, and scratches. (*b*) Nonconforming: (1) scratch depth at the contact area exceeds the microinch surface roughness requirements; (2) delamination of one of the edge board contacts. *(Source: IPC.)*

in the hole wall plating area result in reduced conductivity, increased circuit resistance, and voids in plated through-hole solder fillets. Large voids in plated through-holes can result in hole barrel cracks during the assembly soldering operation. Soldering temperatures (approximately 500°F) cause *z*-axis expansion, thus stressing weaker plating areas (Fig. 48.2).

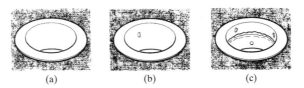

(a)   (b)   (c)

FIGURE 48.2   Voids in hole. (*a*) Target: no voids in hole. (*b*) Acceptable: no more than three voids in the hole; total void area does not exceed 10 percent of the hole wall area. (*c*) Nonconforming: voids exceed 5 percent of the total printed wiring board thickness; circumferential void present. *(Source: IPC.)*

Voids in lands are also detrimental to solderability. Pinholes or voids can undermine the top metal plate (Figs. 48.3 and 48.4). The degree of undermining depends on when during the fabrication process the defect occurred. The defects in this group are defined as follows:

1. *Dent:* A smooth depression in the conductive foil that does not significantly decrease foil thickness.
2. *Pit:* A depression in the conductive layer that does not penetrate entirely through it.
3. *Scratch:* A slight surface mark or cut.
4. *Void:* The absence of substances in a localized area.
5. *Inclusion:* A foreign particle, metallic or nonmetallic, in a conductive layer, plating, or base material. Inclusions in the conductive pattern, depending on degree and material, can affect plating adhesion. Metallic inclusions in the base material reduce the electrical insulation properties and are not normally acceptable if minimum spacing requirements are violated.
6. *Marking* (*legend*): A method of identifying printed boards with part number, revision letter, manufacturer's date code, etc. The condition is usually considered minor, but if there are different revisions to the same part number, markings that are missing or partially obscured could have an effect on functionality (Fig. 48.5).

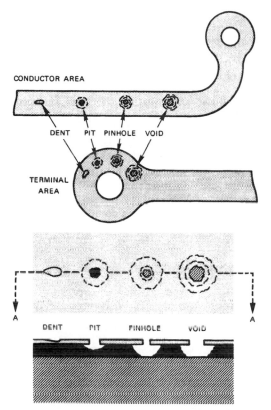

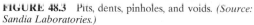

FIGURE 48.3   Pits, dents, pinholes, and voids. *(Source: Sandia Laboratories.)*

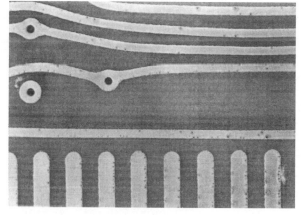

FIGURE 48.4   Severe pitting and pinholing. *(Source: IPC.)*

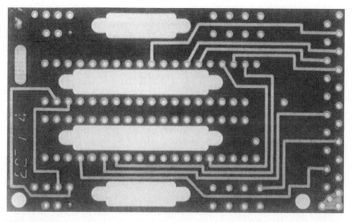

FIGURE 48.5   Part numbers partially obscured and missing. *(Source: Sandia Laboratories.)*

48.7.2 Base Material Effects/Defects

Visual inspection is also used to detect the following board material effects/defects: measling, crazing, blistering, lamination, weave texture, weave exposure, fiber exposure, and haloing (Fig. 48.6).

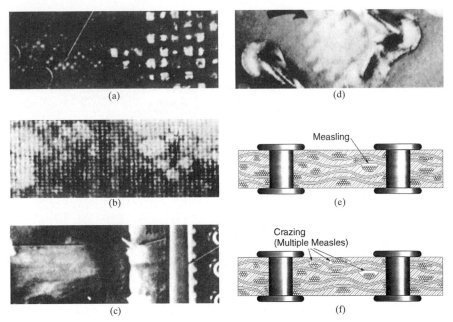

FIGURE 48.6 Base material defects: (*a*) blistering, (*b*) fiber exposure, (*c*) measling, (*d*) crazing (*Source: Sandia Laboratories*), (*e*) crazing, and (*f*) measling. (*Source: IPC.*)

These effects/defects have been a source of controversy as to what is good or what is bad. The IPC formed a special committee in 1971 to consider base material defects and to better define them with illustrations and photographs. The conditions are defined and discussed here.

1. *Measling:* an internal condition occurring in laminated base material in which the glass fibers are separated from the resin at the weave intersection. This condition manifests itself in the form of discrete white spots or "crosses" below the surface of the base material. A report compiled by the IPC, titled "Measles in Printed Wiring Boards," was released in November 1973. The report stated that "measles may be objectionable cosmetically, but their effect on functional characteristics of finished products are [*sic*], at worst, minimal and in most cases insignificant." The IPC Acceptability Subcommittee readdressed the subjects of measles and crazing in 1994 and verified the 1973 findings. As of 1994 the IPC has not obtained any data indicating that measles or crazing are detrimental to function. Therefore, starting with IPC-A-600 version E, all acceptance and/or rejection criteria were deleted. Measles and crazing were defined for information purposes only.

2. *Crazing:* an internal condition occurring in the laminate base material in which the glass fibers are separated from the resin at the weave intersections. This condition manifests itself in the form of connected white spots or crosses below the surface of the base material. The 1994 Acceptability Subcommittee examined the mechanics of crazing. It was determined that crazing looked like measling from the surface, due to light defraction, but was not interconnected and was actually separated at each fiber junction (knuckle) and was actually multiple measles. The weave pattern actually alternates from one side of the fiber bundle knuckle to the other. If a continuous condition exists (e.g., within a fiber bundle), it is termed *separation*. See Figs. 48.6(*e*) and 48.6(*f*). Note that, if the measles or crazing condition exists in high-voltage circuits, the acceptance of the product should be reconsidered.

Note that confusion exists with the term *crazing*. Whereas the printed wiring technology terms it an internal effect, most other technologies and the dictionary define crazing as a surface effect.

3. *Blistering:* a localized swelling and separation between any of the layers of a laminated base material, or between base material and conductive foil. It is a form of delamination.

4. *Delamination:* a separation between plies within the base material, or between the base material and the conductive foil, or both. Blistering and delamination are considered to be major defects, dependent on the degree (refer to IPC-A-600). Whenever a separation of any part of the board occurs, a reduction in insulation properties and adhesion occurs. The separation area could house entrapped moisture, processing solutions, contaminants, or electromigration and could contribute to corrosion and other detrimental effects in certain environments (Fig. 48.7).

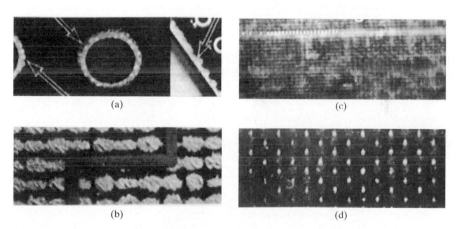

(a)

(c)

(b)

(d)

FIGURE 48.7   Base material defects: (*a*) haloing, (*b*) weave exposure, (*c*) delamination, (*d*) weave texture. *(Source: IPC.)*

There also is the possibility the delamination or blister area will increase to the point of complete board separation, normally during the assembly soldering operation. Last, but not least, is the question of solderability in plated through-holes. Entrapped moisture, when subjected to soldering temperatures, has been known to create steam that blows holes through the plated side walls, exposing the resin and glass of the plated through-holes and creating large voids in the solder fillet.

5. *Weave texture:* a surface condition of base material in which a weave pattern of glass cloth is apparent although the unbroken fibers of the woven cloth are completely covered with resin.

6. *Weave exposure:* a surface condition of base material in which the unbroken fibers of woven glass cloth are not completely covered by resin. Weave texture and weave exposure differ in the degree of defect (Fig. 48.7). A condition of weave texture after the board has been completely fabricated is considered a minor defect. However, if the condition materializes during processing, a judgment must be made concerning the possible attack of subsequent processing chemicals. Weave texture, usually caused by the lack of sufficient resin, can become weave exposure if processing chemicals attack the thin resin layer. Weave exposure is considered a major defect. The exposed glass fiber bundles allow wicking of moisture and entrapment of processing chemical residues.

7. *Fiber exposure:*   a condition in which reinforcing fibers within the base material are exposed in machined, abraded, or chemically attacked areas. (See also weave exposure.)

8. *Haloing:*   mechanically induced fracturing or delamination on or below the surface of the base material; it is usually exhibited by a light area around holes, other machined areas, or both.

48.7.3   Resin Smear

Resin smear is transferred from the base material onto the surface or edge of the conductive pattern normally caused by drilling. Excessive heat generated during drilling softens the resin in holes and smears it over the exposed internal copper areas. The condition creates an insulator between the internal land and subsequent plated through-holes, and the result is open circuits. The defect is removed by chemical cleaning. Inspection for resin smear is performed by viewing vertical and horizontal microsections of plated through-holes.

Chemical cleaning is the chemical process used in the manufacture of multilayer boards. Its purpose is to remove only resin from conductive surfaces exposed in the inside of the holes (Fig. 48.8). Sulfuric, chromic, and permanganate acids are some of the chemicals used in the chemical cleaning process; subsequent inspection indicates whether it has fulfilled its purpose.

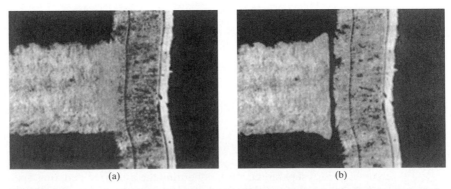

(a)   (b)

FIGURE 48.8   Smear. (*a*) Target: no evidence of resin smear between layer and plating in the hole. (*b*) Nonconforming: evidence of resin residue or resin between internal layer and plating in the hole. (*Source: IPC.*)

48.7.4   Registration, Layer to Layer: X-ray Method

The x-ray method provides a nondestructive way to inspect layer-to-layer registration of internal layers of multilayer boards. It utilizes an x-ray machine and, usually, Polaroid film. The multilayer board is x-rayed in a horizontal position. The x-ray photos are then examined for hole breakout of the internal lands The lack of an annular ring denotes severe misregistration (Fig. 48.9).

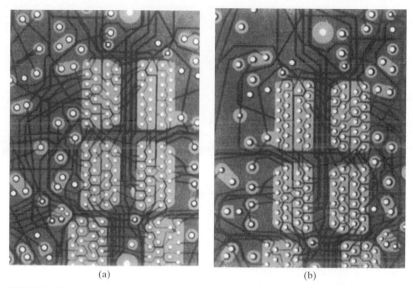

(a)	(b)

FIGURE 48.9 Layer to layer registration, x-ray method. (*a*) Target: all layers accurately registered. (*b*) Nonconforming: extreme misregistration; insufficient measurable annular ring exists on a segment of circumference. (*Source: IPC.*)

48.7.5 Plated Through-Holes: Roughness and Nodulation

Roughness is an irregularity in the side wall of a hole; nodulation is a small knot or an irregular, rounded lump. Roughness and/or nodulation creates one or more of the following conditions:

1. Reduced hole diameter
2. Impaired lead insertion
3. Impaired solder flow through the hole
4. Voids in solder fillet
5. Possible entrapment of contaminants
6. Highly stressed areas in the plating

Although roughness and nodulation are not desirable, they are allowable in small amounts. Specifications have a tendency to use simple, generalized statements such as "good uniform plating practice" when defining acceptability criteria. Such statements require that a judgment be made by the inspector as to what is acceptable or rejectable. The use of visual aids allows judgment to be made by different inspectors within a close degree of consistency. Figure 48.10 illustrates the acceptance criteria recommended by the IPC.

48.7.6 Eyelets

Metallic tubes, the ends of which can be bent outward and over to fasten them in place, are called eyelets. Eyelets are used to provide electrical connections with mechanical strength on printed boards. Acceptability of eyeleted printed boards is based on eyelet installation. Eyeleted boards should be inspected for the following conditions:

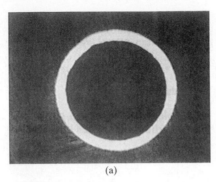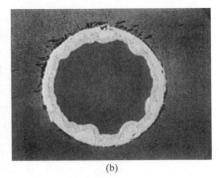

(a)   (b)

FIGURE 48.10   Plated through-hole roughness or nodulation. (*a*) Target: (1) plating is smooth and uniform throughout the hole; (2) there is no evidence of roughness or nodulation. (*b*) Non-conforming: (1) roughness or nodulation reduces plating thickness below minimum requirements; (2) roughness or nodulation reduced finished hole size below minimum requirements; (3) excessive roughness or nodulation permits outgassing of the hole when it is solder-dipped. *(Source: IPC.)*

1. Form of the flange and/or roll should be set in a uniform spread and be concentric with the hole.

2. Splits in the flange or roll should be permissible provided they do not enter the barrel and provided they allow proper wicking of solder through the eyelet and around the setting.

3. Eyelets should be set sufficiently tight that they cannot move.

4. Eyelets should be inspected for improper installation, deformations, etc.

5. A sample lot of eyeleted holes should be microsectioned to inspect for proper installation. IPC acceptance criteria for roll and funnel flange eyelets are shown in Figs. 48.11 and 48.12.

48.7.7   Base Material Edge Roughness

Base material edge roughness occurs on the printed board edge, cutouts, and non-plated through-hole edges (Fig. 48.13). It is classified as a workmanship condition, and it is created by dull cutting tools causing a tearing action instead of a clean cutting action.

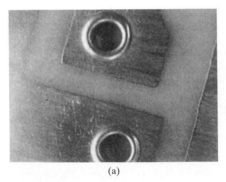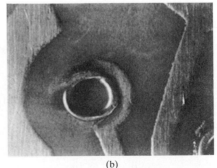

(a)   (b)

FIGURE 48.11   Roll flange eyelets. (*a*) Target: (1) eyelet set uniformly and concentric to hole; (2) strain or stress marks caused by rollover kept to a minimum. (*b*) Nonconforming: (1) eyelet flange uneven or crushed; (2) splits entering the barrel. *(Source: IPC.)*

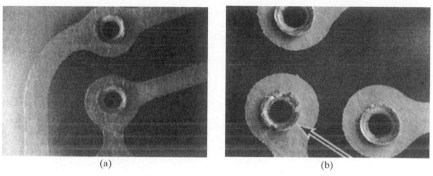

(a)   (b)

FIGURE 48.12   Funnel flange eyelets. (*a*) Target: (1) funnelet set uniformly and concentric with hole; (2) strain or stress marks caused by setting kept to a minimum. (*b*) Nonconforming: (1) funnelet periphery uneven or jagged; (2) splits enter into barrel. (*Source: IPC.*)

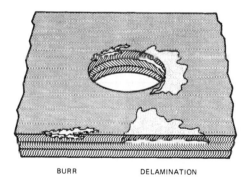

BURR   DELAMINATION

FIGURE 48.13   Base material edge roughness. (*Source: Sandia Laboratories.*)

48.7.8   Solder Mask

A solder mask is a coating material used to mask off or to protect selected areas of a pattern from the attachment of solder. A solder mask is also used as a conformal coating. Attributes inspected are registration, wrinkles, and delamination. Misregistration at land areas reduces or prevents adequate solder fillet formation. The minimum land area to obtain adequate solder fillets should be the minimum acceptance criteria. Wrinkles and delamination provide sites for moisture absorption and contaminants.

48.7.9   Visual Inspection

The board attributes subjected to nondestructive and destructive visual inspection are listed in Table 48.1.

TABLE 48.1 Visual Inspection Chart

Subject	Nondestructive	Destructive
Dents	X	
Pits	X	
Scratches	X	
Voids	X	X
Inclusions	X	
Markings	X	
Measling	X*	
Crazing	X*	
Blistering	X	
Delamination	X*	
Weave texture	X	
Weave exposure	X	
Haloing	X*	
Chemical cleaning		X*
Registration, layer to layer, x-ray method	X	
Plated through-hole roughness and nodulation	X*	
Eyelets	X	X
Base material edge roughness	X	X*
Solder mask	X	

* Verification of many of the internal effects visible from the surface, is done by microsection. However, microsection is a destructive method. Recent technology advances in x-ray inspection provide computer-enhanced laminography. This method allows inspection in thin increments (called slices); then the images are combined by computer graphics. This method is nondestructive; however, equipment costs are high. The method should be available at independent test laboratories.

48.8 DIMENSIONAL INSPECTION

Dimensional inspection is the measurement of the printed board attributes such that dimensional values are necessary to determine compliance with functional requirements. The methods of inspection vary, but the basic inspection equipment consists of gauges and measuring microscopes. More sophisticated equipment is available; it includes comparators, numerical control measuring equipment, coordinate measuring systems, microohmmeters, beta back-scatter gauges, eddy current hand probes, etc.

Dimensional inspection is usually performed on a sampling plan basis. One such plan, termed *acceptable quality level* (AQL), is specified as the maximum percent of defects that, for the purpose of sampling, can be considered statistically satisfactory for a given product. The following attributes fall into the category of dimensional inspection.

48.8.1 Annular Ring

The portion of conductive material completely surrounding a hole is called the annular ring (Fig. 48.14). The primary purpose of the annular ring is that of a flange surrounding a hole; it provides an area for the attachment of electronic component leads or wires. Annular ring width of 0.010 in (0.254 mm) is a standard requirement, but some specifications have allowed rings as small as 0.005 in (0.127 mm) or less. Figure 48.15 shows holes neatly centered in the land, holes on the extreme edge of the land, and holes breaking the edge.

The annular ring can also be used for determining registration between the pattern and the holes (Fig. 48.16). Some designers dimension a land to each datum on the master drawing. By

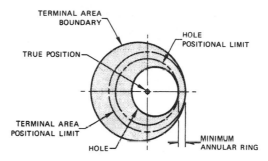

FIGURE 48.14 Annular ring. *(Source: Sandia Laboratories.)*

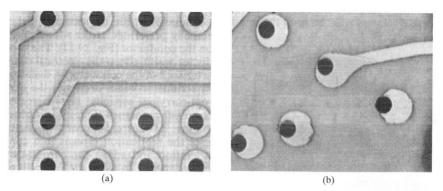

(a)   (b)

FIGURE 48.15 Land registration. (*a*) Target: holes neatly centered in the land. (*b*) Nonconforming: holes not centered in the land. *(Source: IPC.)*

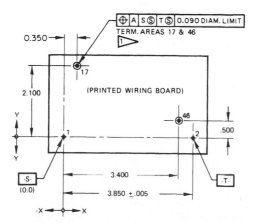

FIGURE 48.16 Land-to-datum registration. *(Source: Sandia Laboratories.)*

verifying the dimensions on the printed boards and then verifying that the minimum widths of the annular rings on all other lands are within the drawing requirements, the pattern is found to be in registration with the drilled hole location. Front-to-back registration is also inspected in this manner. Holes that extend beyond the land are sometimes nonconforming (see contract specification).

48.8.2 Conductor Width

The observable width of a conductor at any point chosen at random on the printed board, normally viewed from vertically above unless otherwise specified, is usually referred to as the overall conductor width (OCW). The conductor width affects the electrical characteristics of the conductor. A decrease in conductor width decreases the current-carrying capacity and increases electrical resistance. A significant difference in conductor width is realized by differences in fabrication process (panel vs. pattern plating). As an example, a nominal conductor width of 0.25 mm (0.010 in) fabricated with 1-oz copper cladding of 0.03 mm (0.0014 in) thickness and an additional 1 oz of copper electroplated on top of the cladding would have a cross-sectional area of 0.71 mm (0.028 in). Fabricating by the panel plating process, the 0.25-mm (0.010-in)-wide conductor would be reduced to a cross-sectional area of 0.54 mm (0.0214 in) (78 percent of nominal). If fabricated by the pattern plate process, the cross-sectional area would be 0.66 mm (0.0260 in) (93 percent of nominal). The fabrication process therefore has a significant effect on cross section reduction.

Most current-carrying capacity graphs take the fabrication process into consideration and adjust the current-carrying capacity to allow for a margin of safety. Although the conductor width definition is very basic, there are two different interpretations as to where on the conductor the measurement is performed. (1) The minimum conductor width (MCW) is measured at the minimum width of the conductor, and (2) the conductor width (OCW) is measured at the observable width, (Fig. 48.17). The minimum conductor width usually can only be measured on a conductor cross section, and it is a destructive inspection. Plating outgrowth during pattern plating can prevent the minimum width from being seen unless cross-sectioning is performed. The conductor width measurement (OCW) is always nondestructive and easily measured (see Fig. 48.18, dimension C). Note that new nondestructive equipment has recently been introduced to measure conductor resistance, which can be related to cross-section area.

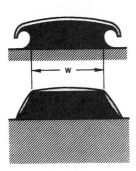

FIGURE 48.17 Minimum conductor width. *(Source: Sandia Laboratories.)*

The difference between the minimum conductor width (MCW) and the overall conductor width (OCW), which is measured from vertically above, can have an effect on the current-carrying capacity, inductance, and impedance. The difference can be significant on narrow conductors (0.13 mm [0.005 in] or less), especially those fabricated by the panel plating process. IPC-6012, Qualification and Performance Specification for Rigid Printed Boards, addresses conductor width requirements and allows added width reductions for edge roughness, nicks, pinholes, and scratches of 30 percent (Class 1) and 20 percent (Classes 2 and 3) from the minimum specified on the master drawing. The document states this reduction applies to the minimum conductor width called out on the master drawing. However, if no minimum is specified, the minimum allowable default width is 80 percent of the conductor pattern supplied on the procurement documentation. Therefore, question the minimum width requirement, if not stated on the master drawing, as a courtesy to the customer.

48.8.3   Conductor Spacing

Conductor spacing is the distance between adjacent edges (not centerlines) of isolated conductive patterns and a conductive layer. The spacing between conductors and/or lands is designed to allow sufficient insulation between circuits. A reduction in the spacing can cause electrical leakage or affect the capacitance. The cross-sectional width of conductors is usually nonuniform. Therefore, the spacing measurement is taken at the closest point between the conductors and/or land (see Fig. 48.18, dimension B).

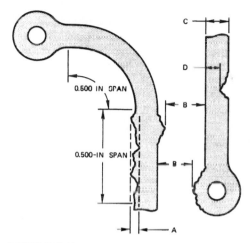

FIGURE 48.18   Edge definition. *(Source: Sandia Laboratories.)*

48.8.4   Edge Definition

The fidelity of reproduction of the pattern edge relative to the original master pattern is called edge definition. It falls in the cosmetic effective category and does not normally affect functionality. It can, however have an effect on high-voltage circuits: a corona discharge can be caused at the irregular conductor edges. Measurement of edge definition is performed by measuring the distance from the crest to the trough (see Fig. 48.18, dimension A).

A popular specification is 0.005 in (0.127 mm) from crest to trough. Isolated indentations that do not reduce the conductor width by more than 20 percent are usually allowed (see Fig. 48.18, dimension D). Also allowed are isolated projections that do not reduce the conductor spacing below specification requirements. As stated earlier, projections can produce corona discharge in high-voltage circuits. Figure 48.19 illustrates isolated indentation and projection and edge definition.

48.8.5   Hole Specifications

Hole size is the diameter of the finished plated through-hole or unplated hole. A plated through-hole is a hole in which electrical connection is made between internal or external conductive patterns, or both, by the deposition of metal on the wall of the hole. An unplated hole, termed an *unsupported hole*, is a hole containing no conductive material or any other type of reinforcement. The hole size measurement is performed to verify that the hole meets

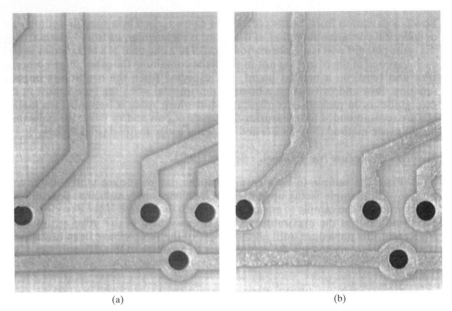

(a)   (b)

FIGURE 48.19   Edge definition. (*a*) Target: conductor edges are smooth and even within tolerance. (*b*) Nonconforming: conductor edges are poorly defined and outside tolerance. *(Source: IPC.)*

minimum and maximum drawing requirements. The size requirement is usually associated
with a fit requirement of a component lead, mounting hardware, etc., plus adequate clearance
for solder. Plated through-holes providing layer-to-layer interconnection, where no components are soldered into the hole, are called via holes. Via holes do not have a fit requirement
and therefore only the plating integrity is critical. Two basic methods are used to measure hole
size: (1) drill blank plug or suitable gauge (Fig. 48.20) and (2) optical. The latter method is uti-

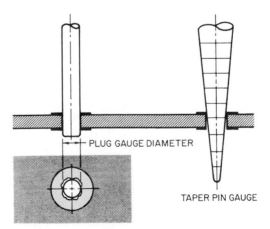

FIGURE 48.20   Hole-measuring methods. (*a*) Drill blank
plug method. (*b*) Kwik-Chek or taper pin method of measuring the hole diameter. *(Source: Sandia Laboratories.)*

lized when soft coatings over the copper are used. The optical method prevents deformation of the soft coatings within the hole. When Kwik-Chek or drill blank plug gauges are used, the inspector should acquire a soft touch to prevent damage to the hole. The Kwik-Chek or drill blank gauges should be cleaned prior to use to prevent solderability degradation. Plating nodules are sometimes present in the hole and restrict the penetration of the gauges. Forcing the gauges into the holes causes the nodules to be dislodged, resulting in voids in the plated through-hole sidewall.

48.8.6 Bow and Twist

Bow is the deviation from flatness of a board characterized by a roughly cylindrical or spherical curvature such that, if the board is rectangular, its four corners are in the same plane. Twist is the deformation parallel to a diagonal of a rectangular sheet such that one of the corners is not in the plane containing the other three corners. Bow and twist, on a printed board, are inspected when the conditions impair function. Bow and/or twist on a board are detrimental when the board must fit in card guides or in packaging configurations in which space is limited. Two methods are recommended for measuring the degree of bow and/or twist: the indicator height gauge method (see Fig. 48.21) and the feeler gauge method (see Fig. 48.22).

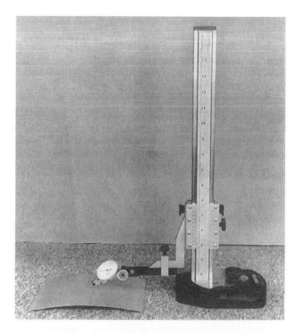

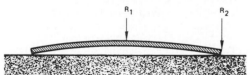

FIGURE 48.21 Indicator height gauge method. (*Source: Sandia Laboratories.*)

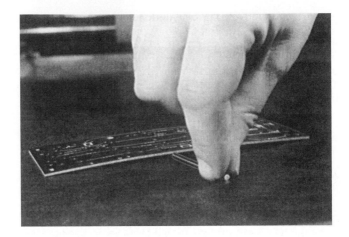

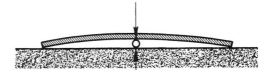

FIGURE 48.22 Feeler gauge method. (*Source: IPC.*)

48.8.6.1 Procedure 1 (Bow) (See Fig. 48.23)

1. Place the sample to be measured on the datum surface with the convex surface of the sample facing upward. For each edge, apply sufficient pressure on both corners of the sample to ensure contact with the surface. Take a reading with a dial indicator at the maximum vertical displacement of this edge, denoted as R_1. Repeat this procedure until all four edges of the sample have been measured. It may be necessary to turn the sample over to accomplish this. Identify the edge with the greatest deviation from datum. This is the edge to be measured in steps 2 and 3.

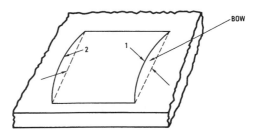

FIGURE 48.23 Bow. (*Source: IPC.*)

2. Take a reading with the dial indicator at the corner of the sample contacting the datum surface, or determine R_2 by measuring the thickness of the sample with a micrometer (denoted R_2 in Fig. 48.24).

3. Apply sufficient pressure so that the entire edge contacts the datum surface. Measure the length of the edge and denote as L.

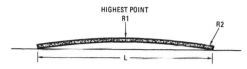

FIGURE 48.24   Bow measurement. *(Source: IPC.)*

4. Calculate bow for this edge as follows:

$$\text{Percent bow} = \frac{R_1 - R_2}{L} \times 100$$

The result of this calculation is the percent of bow.

Repeat the procedure for the other three edges and record the largest value percent of bow for the sample.

48.8.6.2   *Procedure 2 (Twist) (See Fig. 48.25)*

1. Place the sample to be measured on the datum surface with any three corners of the sample touching the surface. Apply sufficient pressure to ensure that three corners are in contact with the datum surface. Take a measurement from the datum surface to the lifted corner and record the reading. Repeat this procedure until all four corners of the sample have been measured. It may be necessary to turn the sample over to accomplish this. Identify the corner with the greatest deviation from datum. This is the corner to be measured in steps 2 and 3.

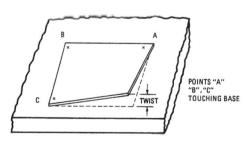

FIGURE 48.25   Twist. *(Source: IPC.)*

2. Place the sample to be measured on the datum surface with three corners touching the surface; insert suitable shims under the raised corner so that it is just supported. When the correct shim thickness is used, the three corners will be in contact with the datum surface without applying pressure to any corner.

3. Without exerting any pressure on the sample, take a reading with the dial indicator at the maximum vertical displacement, denoted R_1 in Fig. 48.26, and record the reading. Without disturbing the sample, take a reading with the dial indicator on the top surface of the sample at the edge contacting the datum surface or determine R_2 by measuring the thickness of the sample with a micrometer. Note that for fabricated boards, both readings must be made on base material.

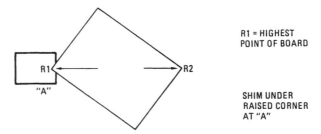

FIGURE 48.26   Twist measurement with shim. *(Source: IPC.)*

4. Measure the diagonal of the sample (for rectangular boards) and record the reading. For nonrectangular boards, measure from the corners exhibiting displacement diagonally to the point on the opposite end of the board.

Calculate as follows:

1. Deduct R_2 reading from R_1 reading and divide the result by 2.

$$\text{deviation} = \frac{R_1 - R_2}{2}$$

2. Divide the measured deviation (step 1) by the recorded length and multiply by 100. The result of this calculation is the percent of twist.

$$\text{percent twist} = \frac{R_1 - R_2}{\text{Length}} \times 100$$

The presentation is expressed in a percentage allowable bow and twist and may be applied to either metric or customary dimensions; it is an expression of percentage per unit of measure. Thus 1.5 percent allowable bow and twist translates to $0.015 \times$ inches or millimeters.

The maximum bow or twist allowed for the entire portion of the specimen or production board is 1.5 percent unless otherwise specified on the master drawing. The 1.5 percent allowable bow and twist is for standard printed wiring boards. Boards that will have surface-mounted devices (SMDs) attached may require a tighter tolerance for SMD interconnection reliability. Check with the purchaser before fabrication and question the subsequent assembly requirements.

48.8.7 Conductor Pattern Integrity

Several methods are used to determine conductor pattern integrity, i.e., the quality or state of being complete. Some methods include the use of comparison equipment and overlays. The use of positive or negative overlays made from the master pattern is an inexpensive and effective method. Overlaying the film on the finished printed board enables differences to be easily detected. Overlays can also be used to determine if conductor widths are within tolerance and annular ring and contours are within drawing requirements.

48.8.8 Contour Dimensions

Contour dimensional inspection verifies that the outside border dimensions are within the drawing requirements. Contour dimension requirements can be considered fit requirements.

Both undersized and oversized printed boards can affect functionality, depending upon the degree of requirement violation. Measuring methods vary from the use of a ruler or calipers to sophisticated numerical control equipment. The sophistication of the method is naturally dependent on the required dimensions and tolerances.

48.8.9 Plating Thickness

48.8.9.1 Nondestructive Methods.   The plating process is used in the fabrication of numerous printed board designs to produce plated through-holes and conductive pattern circuitry. When the plating process is utilized, a plating thickness requirement is usually specified on the drawing or in the accompanying specifications. Verification of the plating thickness on the pattern and in the holes is usually required to ensure circuit functionality. Two of the presently popular ways to measure plating thickness nondestructively are by the beta backscatter and micro-ohm methods.

Beta Backscatter Method.   This method utilizes an energy source (a radioisotope) and a detector (a Geiger-Müller tube). It functions on the beta radiation backscatter principle. It allows thickness measurements to be made within a short period of time, and it can be utilized on metallic, nonmetallic, magnetic, and nonmagnetic materials. The beta backscatter method can be used to measure plating thickness on the circuitry and in the plated through-holes during or after the board fabrication process and prior to or after the etching operation. It is useful only in measuring the top metal thickness, and it reads the thickness as an average.

Micro-ohm Method.   The micro-ohm method functions as a four-wire resistance bridge circuit that imposes a constant current across the area being measured and measures the voltage drop across the same area. The answer is given in micro-ohm units. In plated through-hole measurements, it treats the hole as a cylindrical resistor. The method is effective only after the circuitry has been etched. When two different metals are used on top of one another, only the lowest-resistance metal thickness is detected. Like the beta backscatter method, the micro-ohm method reads the thickness only as an average. The probe configuration is available in two different configurations: (1) a single segment, and (2) a multiple-segment type. Using the single probe, take several measurements (three or more) of each selected plated through-hole and average the measurements of each independent hole. The multiple segment probe takes several measurements of each plated through-hole electronically without requiring repositioning of the of the probe. The measurements of each independent hole are given as an average plating thickness. Check your equipment manual to determine if one or multiple readings are performed on each measurement operation.

48.8.9.2 Destructive, Microsection Method.   The verification of plating thickness on the surface and in the holes by the microsection method usually requires three measurements at three different locations on each plated hole side wall (Fig. 48.27). The results are reported either individually or as an average. Care must be exercised to select locations free from voids;

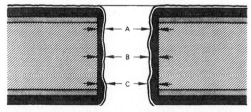

FIGURE 48.27   Verification of plating thickness in holes by use of vertical cross sections. *(Source: Sandia Laboratories.)*

they are usually 0.010 in (0.254 mm) apart. Examples of plating thickness are shown in Fig. 48.28. Typical plating thickness requirements are as follows:

Copper—0.001 in (0.0254 mm) minimum

Nickel—0.0005 to 0.0010 in (0.0127 to 0.0254 mm)

Gold—0.000050 to 0.000100 in (0.00127 to 0.00254 mm)

Tinlead—0.0003 in (0.00762 mm) minimum

Rhodium—0.000005 to 0.000020 in (0.000127 to 0.000508 mm)

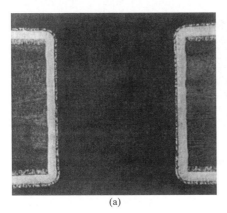

(a)   (b)

FIGURE 48.28 Examples of plating thickness. (*a*) Target: plating is uniform and meets thickness requirements. (*b*) Nonconforming: plating thickness less than requirements. (*Source: Sandia Laboratories.*)

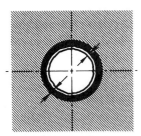

FIGURE 48.29 Verification of plating thickness in holes by use of horizontal cross sections. (*Source: Sandia Laboratories.*)

Polishing vertical plated through-hole cross sections to the mean (center) of the hole is critical. If the hole is polished less than or greater than the mean of the hole, a plating thickness error is introduced. Horizontal plated through-hole cross sections are recommended as references (Fig. 48.29). Although horizontal plated through-hole cross sections are usually more accurate for plating thickness measurements, they do not allow adequate inspection of other attributes such as voids, plating uniformity, adhesion, etchback, and nodules. Surface plating thickness measurements are taken on vertical cross sections of conductor areas. Microsection mounts are usually etched with an appropriate etchant to show grain boundaries between the copper-clad foil and copper plating. Copper plating surface measurements usually exclude the copper foil thickness unless otherwise specified.

48.8.9.3 *Method of Preparing Microsections (IPC Test Method 2.1.1).* Cut from the printed board or test coupon a specimen for vertical evaluation that contains at least three of the smallest holes adjacent to the edge. Cut additional specimens from the board or test coupon for horizontal evaluation. The test equipment and apparatus necessary for preparation of the specimen is as follows:

- Glass plate, 5 × 7 in (127 × 177.8 mm)
- Aluminum rings, 1¼ in (31.75 mm) ID

- Silicone release agent
- Room temperature curing potting material
- Wooden spatulas
- Plastic cups (at least 6 oz)
- Saw or shear
- Engraver
- 240-grit abrasive
- Double-coated tape
- Metallographic polishing tables
- 240-, 320-, 400-, and 600-grit disks
- Number 2 liquid alumina polish
- Polishing cloths
- Chemical etchants

For metallographic evaluation, the following equipment is necessary:

- Microscope and camera accessories
- Filar eyepieces or graduated reticle
- Engraver
- Photographic film or Polaroid film
- Filter lens

The procedure for preparing the specimen is as follows:

1. Clean glass plate and aluminum rings and dry thoroughly.
2. Apply strip of double-coated tape to plate to support specimen. Apply thin film of release agent to glass plate and ring.
3. Sand long edge of perpendicular specimen until edges of conductor pads appear and specimen will stand on edge on a flat surface. Use 240-grit abrasive. Note that equipment and procedures are available from equipment manufactures allowing multiple cross sections to be processed at the same time.
4. When secondary plating thickness is being measured, overplate specimen with a harder electroplated metal. Specimens may be overplated as per ASTM method E3-58T.
5. Measure inside diameter of plated through-holes prior to encapsulation.
6. Stand specimen on edge on double-coated tape in aluminum ring with plated through-hole edge down. For parallel specimens, delete tape and lay specimen flat on glass plate inside ring.
7. Mix potting material and pour to one side of the specimen until it flows through the holes. Support specimen in the vertical position if necessary. Continue pouring until ring is full. Avoid entrapment of air.
8. Allow specimen to cure at laboratory temperature. Accelerated curing at a higher temperature following manufacturer's instructions is permissible, provided cracking or deformation does not occur.
9. Identify specimen promptly by engraving.
10. If more than one specimen is potted in one ring, the specimens should be spaced apart to facilitate filling the holes. Specimens should be identified with a strip of paper marked with traceability information and molded into the mounting.

The grinding and polishing procedure is as follows:

1. Rough-grind the face of the specimen to the approximate center of the plated through-holes by using 240-, 320-, 400-, and 600-grit disks in that order.
2. Flush away all residue by using tap water at room temperature. Wash hands repeatably to avoid carrying over course grids.
3. Rotate specimen 360° about axis of wheel and opposite direction of rotation of wheel. Keep face of specimen flat on wheel.
4. Micropolish by using a nylon disk and alumina polish number 2 until specimen is smooth and free of sanding marks and a clear sharp image of plating lines is evident.
5. Rinse specimen thoroughly and, while it is still wet, chemically etch by using a cotton swab to highlight plating boundaries.
6. Lightly rub twice across specimen and immediately rinse in distilled water. Repeat if necessary.
7. Rinse thoroughly after etching to eliminate carryover of acids to microscope.
8. Dry specimen prior to viewing through microscope.

To make the microscope examination, proceed as follows:

1. Place specimen on microscope stage and adjust to get specimen centered with eyepiece.
2. Focus and adjust lighting for best viewing; then scan specimen.
3. With filar micrometer or graduated reticle, make three thickness measurements of each plating on both walls of the plated through-hole.

The following is the procedure for photomicrographing (composite photographs):

1. Locate specimen, place flat on microscope stage, and mount camera to microscope tube. Insert film pack.
2. Set lighting and exposure time prearranged settings.
3. Focus on one part of one wall and push camera plunger.
4. Immediately remove negative from camera and develop.
5. Take sufficient photographs to illustrate entire length of wall.
6. Repeat steps 1 to 5 and photograph opposite wall.
7. If defective junctions, voids, etc., appear, enlarged photographs may be made. Enlargements should be referenced to photographs.

48.8.10 Undercut (After Fabrication)

Undercut is the distance on one edge of a conductor measured parallel to the board surface from the outer edge of the conductor, excluding overplating and coatings, to the maximum point of indentation on the same edge. Measurement of the degree of undercut present is not usually required, because undercut is included in the minimum conductor width measurement. If, however, the undercut measurement is required separately, it is performed by first measuring the minimum conductor width and then subtracting that dimension from the design width of the conductor and dividing by 2 (Fig. 48.30). For example, given:

1. Design width of conductor dimension a
2. Minimum conductor width dimension b
3. Undercut dimension c

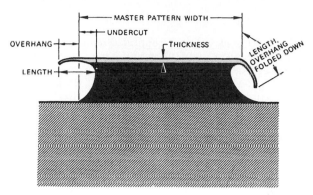

FIGURE 48.30 Undercut and outgrowth. *(Source: Sandia Laboratories.)*

Then

$$\frac{a-b}{2} = c$$

48.8.11 Outgrowth

Outgrowth is the increase in conductor width at one side of a conductor, caused by plating build-up, over that delineated on the production master. The measurement of the degree of outgrowth is required in some specifications. Excessive outgrowth can eventually lead to slivering (a thin metallic piece that breaks off of the conductive pattern). Outgrowth length dimensions of 1.5 times or less the outgrowth thickness, at the point where the outgrowth material extends beyond the main conductor configuration, are usually allowed. The degree of outgrowth is measured on a conductor cross section (Fig. 48.30).

An example of excessive outgrowth is depicted in Fig. 48.31. A method of determining if the outgrowth is prone to slivering utilizes an ultrasonic cleaner; the total printed board is suspended in water in an ultrasonic cleaner for 1 to 2 min. If the board is prone to slivering, metallic slivers will be evident along pattern edges (Fig. 48.32).

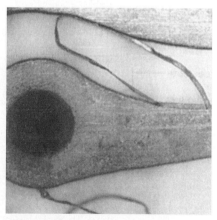

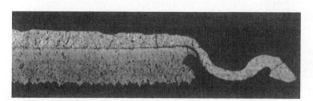

FIGURE 48.31 Outgrowth extension is approximately 10 times the thickness. *(Source: Sandia Laboratories.)*

FIGURE 48.32 Metallic slivers present after ultrasonic test. *(Source: Sandia Laboratories.)*

48.8.12 Etchback

Etchback is a process for the controlled removal of nonmetallic materials from side walls of holes to a specified depth. It is used to remove resin smear and to expose additional internal conductor surfaces. The degree of etchback is critical to function. Too much etchback creates excessively rough hole side walls and causes weak plated through-hole structures. Etchback requirements range from 0.0005 to 0.001 in (13 to 25 μm). The requirement is usually specified as a minimum and a maximum. The degree of etchback is usually measured by utilizing vertical plated through-hole cross sections of multilayer boards (Fig. 48.33). Typical etchback acceptance criteria recommended by the IPC are shown in Fig. 48.34.

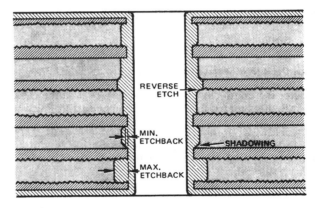

FIGURE 48.33 Etchback configurations. *(Source: IPC.)*

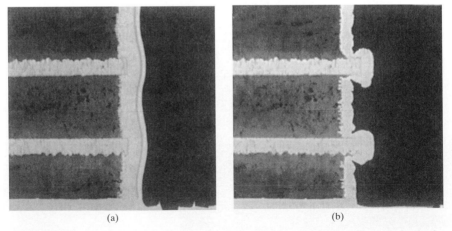

FIGURE 48.34 Etchback. (*a*) Target: uniform etchback of base laminate; uniform plating in the plated through-hole. (*b*) Nonconforming: nonuniform and excessive etchback of base laminate results in unacceptable nonuniform plating in the hole. *(Source: IPC.)*

48.8.13   Registration, Layer to Layer

Layer-to-layer registration refers to the degree of conformity of the position of a pattern, or portion thereof, with its intended position or with that of any other conductor layer of a board. It is required to ensure electrical connection between the plated through-holes and the internal layers. Misregistration of internal layers increases electrical resistance and decreases conductivity. Severe misregisteration creates an open circuit condition—a complete loss of continuity. Two popular methods of measuring layer-to-layer registration are the x-ray method, described earlier, and the microsection method, which is similar to the plating thickness inspection method. The microsection method consists of measuring each internal land area on a vertical plated through-hole cross section and determining the centerline. The maximum variation between centerlines is the maximum misregisteration (Fig. 48.35).

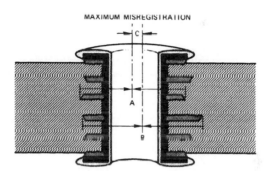

FIGURE 48.35   Registration, layer to layer—microsection method. *(Source: IPC.)*

48.8.14   Flush Conductor, Printed Boards

A flush conductor is a conductor whose outer surface is in the same plane as the surface of the insulating material adjacent to the conductor. Flush conductors are primarily used in rotating switches, commutators, and potentiometers. Common to those uses are brush or wiper and conductor pattern combinations. Making the conductors flush in these applications reduces wiper vibrations, wear, and intermittent or noisy signals. The height of the step allowed between the pattern and base material is dependent on the relative wiper speed, the materials used, and the degree of signal error and electrical noise tolerable in the circuit. A commonly accepted height allowance is as follows:

Up to 50 rpm—80 to 200 μin (2.032×10^{-4} to 5.08×10^{-3} mm)

51 to 125 rpm—better than 50 μin (1.27×10^{-4} mm)

126 to 500 rpm—better than 30 μin (0.762×10^{-4} mm)

Inspection of the degree of flushness is performed with height gauges or measuring microscopes (Fig. 48.36).

48.8.15   Summary

The dimensional board attributes subjected to nondestructive and destructive inspection are listed in Table 48.2.

(a)   (b)

FIGURE 48.36 Flush printed circuits. (*a*) Target: conductor is flush to the board surface. (*b*) Nonconforming: conductor is not flush to the board surface and does not meet the specified tolerance. (*Source: IPC.*)

TABLE 48.2 Dimensional Inspection Chart

Subject	Nondestructive	Destructive
Annular ring	x	
Pattern-to-hole registration	x	
Conductor width	x	x
Conductor spacing	x	
Edge definition	x	
Hole size	x	
Bow and twist	x	
Conductor pattern integrity	x	
Contour dimensions	x	
Plating thickness	x	x
Undercut		x
Outgrowth		x
Etchback		x
Layer-to-layer registration		x
Flush circuits	x	x

48.9 MECHANICAL INSPECTION

Mechanical inspection is applied to characteristics that can be verified with qualitative physical testing. Mechanical inspection methods can be either nondestructive or destructive. The following methods are used to verify printed board fabrication integrity. (Detailed procedures can be obtained from the IPC.)

48.9.1 Plating Adhesion

A common method of inspecting for plating adhesion is the tape test, which is described in detail in the IPC test method IPC-TM-650, 2.4.1, and military specification MIL-P-55110. The basic test method is shown in Fig. 48.37. Another method of determining plating adhesion is used during the preparation and viewing of the microsection mounts. If adhesion is poor, the

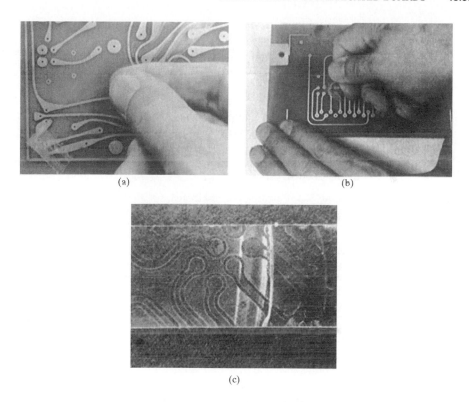

(a)   (b)

(c)

FIGURE 48.37   The tape test. (*a*) Step 1: place pressure sensitive tape across the circuits to be tested and press the tape onto the circuits. Eliminate all air bubbles with finger. (*b*) Step 2: lift tape on one end enough to get a grip. Pull tape off the printed board at approximately 90° to the board. Use a rapid pull. (*c*) Step 3: a clear tape is the preferred test result. (*Source: IPC.*)

layers of plating will separate during the microsection specimen preparation or will indicate lack of adhesion at plating boundaries during microsection mount viewing. Figure 48.38 indicates lack of adhesion at the different plating boundaries (copper foil to copper plate and copper plate to tinlead).

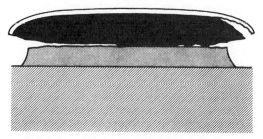

FIGURE 48.38   Plated metal adhesion. (*Source: Sandia Laboratories.*)

48.9.2 Solderability

Solderability inspection measures the ability of the printed pattern to be wetted by solder for the joining of components to the board. It involves the use of three terms:

1. *Wetting:* the formation of a relatively uniform, smooth, unbroken and adherent film of solder to a basis material

2. *Dewetting:* A condition that results when molten solder has coated a surface and recessed, leaving irregularly shaped mounds of solder separated by areas covered with a thin film of solder and leaving the basis metal not exposed

3. *Nonwetting:* the partial adherence of molten solder to a surface that it has contacted, leaving the basis metal exposed

Many methods for inspecting printed boards for solderability both quantitatively and qualitatively have been established. However, the most meaningful method is that used in the assembly soldering operation, hand soldering, wave soldering, drag soldering, etc. The IPC recently released an approved ANSI/J-STD-003 specification entitled "Solderability Tests for Printed Boards." This specification addresses the recommended test methods and subsequent storage that have no adverse effect on the solderability of those portions of the printed wiring board intended to be soldered. Test coupons or product are submitted to accelerated aging to a length of time defined by a coating durability requirement called out on the master drawing or procurement contract, followed by solderability testing. The coating durability requirements are:

Category 1: minimum coating durability, intended to be soldered within 30 days from time of manufacture and likely to experience minimum thermal exposure.

Category 2: average coating durability, intended for boards likely to experience storage of up to 6 months from the time of manufacture and moderate thermal or solder exposure.

Category 3: maximum coating durability, intended for boards that are likely to experience long storage (over 6 months from time of manufacture) and that may experience severe thermal or solder processing steps, etc. It should be recognized that there may be a cost premium or delivery delay associated with boards ordered to this durability level.

The IPC has adopted the criteria illustrated in Fig. 48.39 for printed board solderability acceptance.

48.9.3 Alloy Composition

Two popular alloys used in printed board manufacture are tinlead and tinnickel. The composition of the tinnickel bath is usually 65 percent tin and 35 percent nickel. During electroplating, the composition of the deposit remains nearly constant despite fluctuations in bath composition and operation conditions. However, the composition of the tinlead deposits can vary with bath composition and operation conditions, and will influence the melting temperature of the alloy and affect solderability. Specifications usually require the tin content to be between 50 and 70 percent. Other finishes such as electroless nickel, rhodium, or palladium are used for specific projects but must be called out in the master drawing. Methods available for analyzing the alloy composition on the plated printed board include wet analysis, atomic absorption, and the beta backscatter. The beta backscatter method is relatively new and is gaining in popularity because of the ease of obtaining the alloy composition nondestructively, but results are not as precise as with the other methods.

48.9.4 Thermal Stress Solder Float Test

Temperature-induced strain or straining force that is exerted upon the printed board and tends to stress or deform the board shape can be a serious problem during the soldering operation.

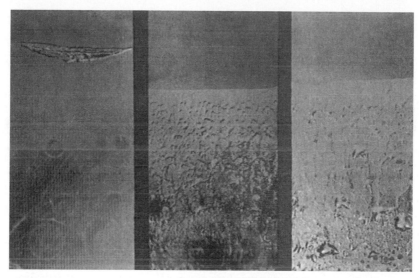

FIGURE 48.39 Solderability acceptance criteria are visible at left surface. (*a*) and (*b*) Conforming. (*c*) Nonconforming: surface shows severe dewetting and nonwetting. *(Source: IPC.)*

FIGURE 48.40 Continuity, thermal stress, moisture resistance test pattern. *(Source: IPC.)*

Thermal stress inspection is performed to predict the behavior of the printed boards after soldering. Plated through-hole degradation, separation of platings or conductors, and laminate delamination are detectable by the thermal stress test. The printed board specimen is (1) conditioned to reduce moisture, (2) placed in a dessicator on a ceramic plate to cool, (3) fluxed, (4) floated in a solder bath, and (5) placed on an insulator to cool. Visual inspection for defects is followed by cross-sectioning the plated through-holes and inspecting with magnification for integrity. Test patterns also are used for thermal stress inspection. (See Fig. 48.40 for a test pattern used in this and other connections.)

48.9.5 Peel Strength

Peel strength is the force per unit width required to peel the conductor or foil from the base material. It is usually associated with the acceptance testing of copper-clad laminate material upon receipt, but it is sometimes utilized to test the adhesion of the conductors to the finished board. Peel strength tests of conductors are usually performed after the specimens have been dip or flow soldered.

A calculation with width in millimeters will, of course, yield the same results. Peel strength tests of conductors are a good method of ensuring that the conductor-to-laminate adhesion is sufficient to withstand assembly soldering operations. Actual printed boards or test patterns can be utilized for the test. Figure 48.41 is an example of a typical peel strength test pattern.

FIGURE 48.41 Peel strength test pattern.

FIGURE 48.42 Typical bond strength test coupon. *(Source: IPC.)*

48.9.6 Bond Strength (Terminal Pull)

The bond strength test is a test of the plating adhesion to the laminate in the hole. Poor adhesion in the holes can occur due to the drilling and electroless copper deposition operations, the state of cleanliness in the hole being out of control, or the use of undercured laminate material. Whatever the cause, poor plating adhesion in the hole affects functionality. Lack of plating adhesion to the laminate in the hole is detectable during microsection analysis. A terminal pull test should be performed to substantiate the condition (see Fig. 48.42).

48.9.7 Cleanliness (Resistivity of Solvent Extract)

Printed boards are fabricated using wet chemistry and mechanical techniques. Some of the chemical baths contain metallic salts or are corrosive in nature. These conditions could affect the functionality of the boards, i.e., reduced electrical insulation resistance, corrosion of the metal pattern, etc. Another common problem is electromigration of metal between conductors. This condition is usually associated with low operating voltage (10 V or less) and requires three components: (1) moisture, (2) metallic contaminants, and (3) voltage. The cleanliness test measures the resistivity of the wash solution the board has been washed with. Commercial equipment is available to perform the cleanliness testing. Note that when a solder mask is required on printed wiring boards, the cleanliness test is performed on the uncoated boards before coating.

48.9.8 Mechanical Inspection Attributes

The mechanical board attributes subjected to nondestructive and destructive inspection are listed in Table 48.3.

TABLE 48.3 Mechanical Inspection Chart

Subject	Destructive	Nondestructive
Plating adhesion	X	X
Solderability		X
Alloy composition	X	X
Thermal stress (solder float test)		X
Peel strength, printed wiring conductor		X
Bond strength (Terminal pull)		X
Cleanliness	X	

48.10 ELECTRICAL INSPECTION

Electrical inspection is performed to verify circuit integrity after processing and also to substantiate that the electrical characteristics of the processed board meet design intent. Electrical inspection methods are either destructive or nondestructive. Nondestructive tests are usually performed on the actual printed boards. Destructive tests are performed either on printed circuit boards utilized for destructive acceptance testing or on test patterns fabricated outside the border of the board. Two popular nondestructive electrical tests are insulation resistance and continuity tests. These are usually performed on 100 percent of complex printed boards, especially multilayer ones. Care should be taken to prevent arcing as the probes approach the printed board pattern when the insulation resistance test is performed. An easily activated switch in series with one of the probes allows the probes to make contact with the pattern prior to current flow. To prevent probe impressions on soft metal coatings, apply the same coating on the probe tips.

48.10.1 Continuity

Continuity tests are performed on multilayer-type circuits to verify that the printed circuit pattern is continuous. They can be performed with an inexpensive multimeter or with more elaborate equipment such as a computer-enhanced bed-of-nails tester. This type of tester either can be preprogrammed with all the circuit nets or can use a known good multilayer board termed a golden board. The tester digitizes the golden board net circuits, then tests all subsequent boards to the digitized nets. In the printed circuits industry the continuity test is performed in one of two ways: (1) as a go/no-go test to verify that the pattern is continuous, or (2) to verify that the pattern is continuous as well as to verify the integrity of the measured pattern. The latter results are reported in electrical resistance values (ohms). The preferred method is to perform the continuity test on all printed boards submitted for acceptance. This is especially recommended for multilayer circuitry, the internal patterns and interconnections of which can not be inspected visually after fabrication. Test coupons are sometimes used for the continuity test. See Fig. 48.40 for a pattern used on a typical coupon.

48.10.2 Insulation Resistance (Circuit Shorts)

The purpose of this test, as stated in MIL-STD-202, is to measure the resistance offered by the insulation members of a printed board to an impressed direct voltage that tends to produce a leakage of current through or on the surface of those members. Low insulation resistances can disturb the operation of circuits intended to be isolated, by permitting the flow of large leakage currents and the formation of feedback loops. This test also reveals the presence of contaminants from processing residues.

FIGURE 48.43 Insulation resistance, dielectric withstanding volt test pattern. *(Source: IPC.)*

In printed circuitry the test is performed by opens either between conductors on the same layer or between two different layers. The bed of nails allows all noncommon nets to be tested against each other. The test is also utilized before and after thermal shock and temperature cycling tests. Test voltages of 40 to 500 V DC and minimum insulation resistances of 200 to 500 MΩ are popular. The insulation resistance test is performed on either the actual printed board or a test coupon fabricated on the same panel as the board (Fig. 48.43). When special conditions such as isolation, low atmospheric pressure, humidity, or immersion in water are required, they should be specified in the test method instruction.

48.10.3 Current Breakdown, Plated Through-Holes

The current breakdown test is used to determine if sufficient plating is present within the plated through-hole to withstand a relatively high current potential. The time and current selected determine if this test is destructive or nondestructive. IPC test method TM-650-2.5.3 recommends a current of 10 A for 30 s. The test is performed as follows:

1. Place a load resistor of predetermined value across the negative and positive terminals of a current-regulated power supply.
2. Adjust the supply for a current of 10 A or any other desired value.

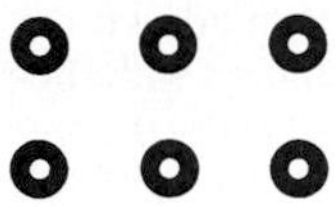

FIGURE 48.44 Typical current breakdown test pattern. *(Source: IPC.)*

3. Remove one end of the resistor from the positive supply terminal.

4. Connect the desired plated through-hole between the disconnected end of the resistor and the positive supply terminal.

5. Perform the test for the desired time.

The test is performed either on an actual printed board or on a test pattern (Fig. 48.44).

48.10.4 Dielectric Withstanding Voltage

This test is used to verify that the component part can operate safely at its rated voltage and withstand momentary overpotentials due to switching, surges, and similar phenomena. It also serves to determine whether insulating materials and spacings in the component part are adequate. It is thoroughly defined in MIL-STD-202, Electronic Components Method 301.

One of three different test voltages (500, 1000, 5000 V) is usually specified. The test is performed on either an actual board or a test coupon. Voltage is applied between mutually insulated portions of the specimen or between insulated portions and ground. The voltage is increased at a uniform rate until the specified value is reached. The voltage is held for 30 s at the specified value and then reduced at a uniform rate.

Visual examination of the part is performed during the test for evidence of flashover or breakdown between contacts. (See Fig. 48.43 for a typical test pattern.) The test can be either destructive or nondestructive, depending on the degree of over potential used.

48.10.5 Electrical Inspection Attributes

The electrical board attributes subjected to nondestructive and destructive inspection are listed in Table 48.4.

TABLE 48.4 Electrical Inspection Chart

Subject	Nondestructive	Destructive
Continuity	x	
Insulation resistance	x	
Current breakdown, plated through-holes	x	x
Dielectric withstanding voltage	x	x

48.11 ENVIRONMENTAL INSPECTION

Environmental inspection consists of performing specific tests to ensure the printed board will function under the influence of climatic and/or mechanical forces to which it will be subjected during use. Environmental tests are performed on preproduction printed boards to verify design adequacy. Specific tests are sometimes specified as part of the printed board acceptance procedure to expose a prospective failure situation. Specific environmental testing, as part of the acceptance procedure, is prevalent in high-reliability programs. Note that it is recommended that two cycles of component assembly soldering simulation preconditioning be performed prior to environmental testing. Test information will then relate to actual conditions the printed wiring board will be subjected to.

Environmental tests are performed either on actual printed boards or coupon test patterns. This section briefly reviews some to the popular environmental tests. Specific details on performing the tests are referenced at the end of the chapter in the section titled Related Test Methods.

48.11.1   Thermal Shock

The thermal shock test is particularly efficient in identifying (1) printed board designs with areas of high mechanical stress and (2) the resistance of the printed board to exposure of high- and low-temperature extremes. This test is required in IPC-6012 for product acceptance and is also used for printed wiring board supplier qualification inspection.

A thermal shock is induced on a printed board by exposure to severe and rapid differences in temperature. The test is usually performed by transferring the board from one temperature extreme (e.g., 125°C) to the other (e.g., –65°C) rapidly, usually within 2 min. A resistance difference in excess of 10 percent between the 1st and 100th cycles is considered a reject. Thermal shock effects on the board include cracking of plating in the holes and delamination. See Fig. 48.45 for a typical thermal shock test pattern (refer to IPC-TM-650, method 2.6.7.2, and MIL-STD-202, method 107C, for procedure). Note that continuous electrical monitoring during thermal shock cycling will display intermittent electrical circuits. Intermittent circuits usually occur at temperature transitions from cold to hot or hot to cold.

FIGURE 48.45 Typical thermal shock and moisture resistance test pattern. *(Source: IPC.)*

48.11.2   Moisture and Insulation Resistance

The moisture resistance test is an accelerated method of testing the printed board for deteriorative effects of high humidity and heat conditions typical of tropical environments. The test conditions are usually a relative humidity of 90 to 98 percent at a temperature of 25 to 65°C. After the required test cycles are completed, the board is subjected to insulation resistance testing. Test specimens should exhibit no blistering, measling, warp, or delamination after the moisture resistance test. See Figs. 48.44 and 48.45 for typical test patterns.

48.12   *SUMMARY*

To select the printed board attributes to be inspected in an individual program, proceed as follows:

1. Review the environment in which the printed board will operate and the life expectancy of the board.
2. Review the electrical and mechanical parameters associated with functionality.
3. Consider the total assembly unit cost and the importance of the printed board before determining if a quality assurance program is economical.
4. Consider both functionality and economics in quality assurance program selection. Functionality should prevail, however.

5. Design the quality assurance program for at least a 90 percent confidence level by using sampling plans when appropriate.

6. Select for inspection the attributes that will satisfy requirements 1 and 2.

7. Select test methods that verify the printed board's functionality and integrity. Test methods may require modification or creation to satisfy the quality assurance requirements.

48.13 TEST SPECIFICATIONS AND METHODS RELATED TO PRINTED BOARDS

Attributes	Method
Annular ring	IPC-TM-650,2.2.1
Base material edge roughness	IPC-TM-650,2.1.5
Blistering	IPC-TM-650,2.1.5
Chemical cleaning	IPC-TM-650,2.2.5 & 2.1.1
Conductor spacing	IPC-TM-650,2.2.2.
Conductor width	IPC-TM-650,2.2.2
Crazing	IPC-TM-650,2.1.5
Current breakdown, plated through-hole	IPC-TM-650,2.5.3
Dents	IPC-TM-650,2.1.5
Dielectric withstanding voltage	IPC-TM-650,2.5.7
	MIL-STD-202,301
Edge definition	IPC-TM-650,2.2.3
Etchback	IPC-TM-650,2.2.5
Eyelets	IPC-TM-650,2.1.5
Haloing	IPC-TM-650,2.1.5
Hole size	IPC-TM-650,2.2.6
Inclusions	IPC-TM-650,2.1.5
Insulation resistance	IPC-TM-650,2.5.9,
	2.5.10,2.5.11,
	MIL-STD-202,302
Layer-to-layer registration, destructive	IPC-TM-650,2.2.11
Markings	IPC-TM-650,2.1.5
Measling	IPC-TM-650,2.1.5
Microsections, methods of preparing	IPC-TM-650,2.1.1
Moisture resistance	IPC-TM-650,2.6.3
	MIL-STD-202,106C
Overhang	IPC-TM-650,2.2.9
Pinholes	IPC-TM-650,2.1.5
Pits	IPC-TM-650,2.1.5
Plating adhesion	IPC-TM-650,2.4.10
Plating thickness, destructive	IPC-TM-650,2.2.13

	ASTM-B-567-58
Plating thickness, nondestructive	IPC-TM-650, 2.2.13.1
	ASTM-B-567-72
Scratches	IPC-TM-650,2.1.5
Solderability	IPC-TM-650,2.4.12 and 2.4.14
Surface roughness	IPC-TM-650,2.1.5
Bond strength (terminal pull)	IPC-TM-650,2.4.20 and 2.4.21
Thermal shock	IPC-TM-650, 2.6.7 and
	MIL-STD-202,107C
Thermal stress	IPC-TM-650,2.6.8
Voids	IPC-TM-650,2.1.5
Warp and twist	IPC-TM-650,2.4.22
Weave exposure	IPC-TM-650,2.1.5
Weave texture	IPC-TM-650,2.1.5

48.14   GENERAL SPECIFICATIONS RELATED TO PRINTED BOARDS

The following printed board specifications are some of those used throughout the industry. The specifications usually cover both processing and acceptance requirements.

IPC specifications	
IPC-FC-250	Specifications for Double Sided Flexible Wiring with Interconnections
ANSI/IPC-D-275	Design Standard for Rigid Printed Boards and Rigid Printed Board Assemblies
IPC-6012	Qualification and Performance Specification for Rigid Printed Boards
IPC-D-300	Printed Circuit Board Dimensions and Tolerances
IPC-MC-324	Performance Specification for Metal Core Boards
IPC-D-325	Documentation Requirements for Metal Core Boards
IPC-D-326	Information Requirements for Manufacturing Electronic Assemblies
ANSI/IPC-A-600	Acceptability of Printed Wiring Boards
IPC-SS/QE-605	Printed Board Quality Evaluation Slide Set and Handbook
IPC-TM-650	Test Methods Manual
IPC-OI-645	Standard for Visual Optical Inspection Aids
IPC-SM-840	Qualification and Performance of Permanent Polymer Coatings (Solder Mask) for Printed Boards

Military specifications	
MIL-Q-9858	Quality Assurance
MIL-P-55110	Printed Wiring
MIL-P-55640	Multilayer Printed Wiring

MIL-P-50884	Flexible Printed Wiring
MIL-STD-105	Sampling Procedure and Tables for Inspection by Attributes
MIL-STD-202	Test Methods for Electronic and Electrical Component Parts
MIL-STD-810	Environmental Test Methods
MIL-STD-1495	Multilayer Printed Wiring Boards for Electronic Equipment

Other publications

ASTM-B-567-72	Measurement of Coating Thickness by the Beta Backscatter Principle
ASTM-A-226-58	Standard Method of Test for Local Thickness of Electrodeposited Coatings
EIA-RS-326	Solderability of Printed Wiring Boards
IEC No.326	Performance Specification for Single and Double Sided Printed Wiring Boards
UL 796	Printed Wiring Boards

REFERENCES

1. C. R. Draper, *The Production of Printed Circuits and Electronics Assemblies,* Robert Draper Ltd., London, 1969.
2. C. A. Harper, *Handbook of Electronic Packaging,* McGraw-Hill, New York, 1969.
3. F. W. Kear, "The Design and Manufacture of Printed Circuits," part 14, *PCM-PCE Magazine,* February 1970.
4. Institute of Interconnection and Packaging Electronic Circuits, IPC Design Guide, sec. 9, Quality Assurance.
5. J. A. Scarlett, *Printed Circuit Boards for Microelectronics,* Van Nostrand Reinhold, New York, 1970.
6. H. R. Shemilt, "Inspection of Printed Circuits," *Electronics Manufacturer Magazine,* vol. 15, 1971.
7. V. C. Barr, Sandia National Laboratories, Livermore, CA, private correspondence relating to tests of several thermal resistance design effects on solderability of plated through-holes.
8. T. A. Estes, Conductor Analysis Technologies, Inc., private correspondence on the use of equipment to determine cross-sectional area of conductors.

CHAPTER 49
ACCEPTABILITY OF PRINTED CIRCUIT BOARD ASSEMBLIES

Bruce Wooldridge
DSC Communications Inc., Plano, Texas

49.1 UNDERSTANDING CUSTOMER REQUIREMENTS

First and foremost in determining what acceptance criteria will be used to build or manufacture a printed circuit board assembly (PCBA) is to look at the contract between the supplier and the purchaser of the product. Always determine the needs and/or wants of the customer through contractual communications or, less formally, through talking to the customer if a contract is not detailed enough or a contract does not exist.

In some businesses, contracts are not the norms, such as for consumer electronics built for retail sale. In this case, the company must determine the level of quality desired based on company culture and goodwill, reputation for quality products, and the life cycle of the product. The company should decide to build product to an industry standard such as IPC-A-610, Acceptability of Electronic Assemblies, or develop their own workmanship manual that should closely match IPC-A-610 levels of quality. In any case, the workmanship manual must support the company's objectives to achieve customer satisfaction with their product.

Detailed specifications may be called out in a contract, such as in military specifications, or there may not be any documentation that identifies requirements. In the latter case, the recommendation is to revert to industry standards or an internal quality requirement such as a workmanship manual. Three common possibilities for acceptance criteria are discussed as follows.

49.1.1 Military, Telecommunications, and Consumer Specifications

Reference to these types of specifications are common in the applicable markets. No great amount of detail will be discussed; however, they are all worthy of some discussion.

49.1.1.1 Military Specifications. Generally, military specifications are not used in any market except for military products. There are a few exceptions to this, and the best example is MIL-STD-105, Sampling Procedures and Tables for Inspection by Attributes. This specification is used in many industries for multiple types of equipment. There are many military specifications that are applied to military electronics contracts. The ones with the most relevance to workmanship or acceptability criteria at the assembly level are MIL-STD-2000 and MIL-P-28809. There are many lower-level specifications for components and PCBs that can also be invoked. MIL-STD-2000 is titled Standard Requirements for Soldered Electrical and

Electronic Assemblies and MIL-P-28809 is titled Printed Wiring Assemblies. MIL-STD-2000 is used mainly for high-reliability requirements needed on military equipment and contains more stringent criteria than MIL-P-28809; therefore, this discussion focuses on MIL-STD-2000 criteria.

MIL-STD-2000 is a very detailed document and defines the defect types that must be reworked, repaired, or scrapped and cannot be used as is on government equipment. This, of course, takes away some flexibility from the manufacturer to evaluate a given defect type and make the decision to use the defect without rework or repair if it does not affect form, fit, or function.

MIL-STD-2000 also defines the assembly chemistry for a manufacturer in terms of solder composition, flux types, cleanliness, conformal coating, and solder mask. Any deviation from criteria specified in the document requires a qualification and procuring agency approval before use.

Solder connection characteristics for both plated-through-hole and surface-mount devices are defined by solder finish, physical attributes, fractures, voids, solder coverage, and wetting and filleting. The printed wiring board (PCB) requirements after assembly are detailed in terms of conductor finish and condition, conductor separation from the board, cleanliness of the assembly, PCB weave exposure, delamination, measles, haloing, and bow and twist. In addition, part markings must remain legible after assembly processing.

Also significant about MIL-STD-2000 requirements are the training of personnel and the subsequent certification that must be obtained in order to be authorized to work on product delivered under a MIL-STD-2000 contract. The specification also details methods for process control and defect reduction to be used in the assembly of PCBAs. One hundred percent inspection is required unless the following conditions are met, at which time a sample-based inspection may be utilized.

1. Training on utilization of process control and statistical methods is provided to personnel.

2. Quantitative evidence must be maintained that shows the process is in statistical control and is a capable process.

3. Sampling techniques must be statistically based and consistent with data collection requirements for maintaining process control.

4. Criteria for switching between sampling and 100 percent inspection must be defined. Sampling cannot be used if defect rates are above 2700 parts per million. If processes go out of control, 100 percent inspection must be instituted for the lot.

5. If defects are discovered in the sample, all hardware in that lot must be 100 percent inspected for other occurrences of the defect found.

49.1.1.2 Telecommunications-Bellcore TR-NWT-000078.[1] One of the most recognized requirements specifications in the telecommunications industry is Bell Communications Research (Bellcore) TR-NWT-000078, Generic Physical Design Requirements for Telecommunications Products and Equipment. This document was generated by Bellcore to be applied by their client companies for products to be used in the telecommunications network typical of a client company.

The specification is widely used by the Regional Bell Operating Companies (RBOC), and suppliers to the RBOCs are audited to the design and manufacturing requirements delineated within. Workmanship acceptability criteria are listed in some sections of the document; however, for the most part, TR-NWT-000078 does not show in any great detail the overall workmanship acceptability criteria for PCBAs.

49.1.1.3 Consumer Electronics. The key workmanship concerns for electronic assemblies for consumer products are the intended life cycle of the product and functionality of the unit. In many cases, the functionality can be achieved with less-than-acceptable workmanship attributes, but the product will lose some of its useful life since reliability of solder joints or mechanical strength may be affected. As stated earlier in this chapter, the company must determine the level of quality desired.

49.1.2   ANSI/J-STD and IPC-A-610 Industry Standards

The American National Standards Institute (ANSI) and the Institute for Interconnecting and Packaging Electronic Circuits (IPC) specifications are generally recognized as the standards for requirements and workmanship for PCB assemblies in many companies, both nationally and internationally. These specifications are gaining more recognition in the international arena and the trend to use them more is being seen.

49.1.2.1   ANSI/J-STD-001, Requirements for Soldered Electrical and Electronic Assemblies.   Another requirements document that is a little newer on the scene is the ANSI/J-STD-001, Requirements for Soldered Electrical and Electronic Assemblies,[2] which was first released in April 1992. This standard was a jointly developed document by the Electronics Industry Association (EIA) and the IPC.

ANSI/J-STD-001 classifies three levels of electronic assemblies based on end-item use. The three classifications were established to reflect differences in producibility, complexity, functional performance requirements, and verification frequency. The classes are as follows:

Class 1, General Electronic Products.   Includes consumer products, some computer and computer peripherals, and hardware suitable for applications where the major requirement is a function of the completed assembly.

Class 2, Dedicated Service Electronic Products.   Includes communications equipment, sophisticated business machines, and instruments where high performance and extended life are required, and for which uninterrupted service is desired but not critical. Typically the end-use environment would not cause failures.

Class 3, High-Performance Electronic Products.   Includes equipment for commercial and military products where continued performance or performance on demand is critical. Equipment downtime cannot be tolerated, end-use environment may be uncommonly harsh, and the equipment must function when required, such as in life-support systems and critical weapons systems.

ANSI/J-STD-001 addresses many of the same subjects as MIL-STD-2000,[3] but in most cases for Class 1 and 2 equipment, the requirements are less stringent. The Class 3 requirements in the majority of the subject areas are the same as MIL-STD-2000 requirements; however, there still remain a few areas that the military has not yet accepted. The IPC continues to try to get the joint military community to accept the ANSI/J-STD-001 specification as the military equipment standard and supersede all applicable military standards that currently exist.

49.1.2.2   ANSI/J-STD-002, Solderability Tests for Component Leads, Terminations, Lugs, Terminals and Wires.   The ANSI/J-STD-002,[4] Solderability Tests for Component Leads, Terminations, Lugs, Terminals and Wires, was released in April 1992 to complement the requirements of ANSI/J-STD-001. This standard prescribes the recommended test methods, defect definitions, acceptance criteria, and illustrations for assessing the solderability of electronic component leads, terminations, solid wire, stranded wire, lugs, and tabs.

Solderability evaluations are made to verify that the solderability of component leads and terminations meets the requirements established in ANSI/J-STD-002 and that subsequent storage has had no adverse effect on the ability to solder components to an interconnecting substrate. Determination of solderability can be made at the time of manufacture, at receipt of the components by the user, or just prior to assembly and soldering.

For a more detailed discussion of solderability issues refer to Sec. 49.5.

49.1.2.3 *ANSI/J-STD-003, Solderability Tests for Printed Boards.*   The ANSI/J-STD-003, Solderability Tests for Printed Boards,[5] was released in April 1992 to complement the requirements of ANSI/J-STD-001. This standard prescribes the recommended test methods, defect definitions and illustrations for assessing the solderability of printed board surface conductors, attachment lands, and plated-through-holes.

The solderability determination is made to verify that the PCB fabrication processes and subsequent storage have had no adverse effect on the solderability of those portions of the PCB intended to be soldered. This is determined by evaluation of the solderability specimen portion of a board or representative coupon which has been processed as part of the panel of boards and subsequently removed for testing per the method selected.

The objective of the solderability test methods described in ANSI/J-STD-003 is to determine the ability of printed board surface conductors, attachment lands, and plated-through-holes to wet easily with solder and to withstand the rigors of the PCB assembly processes.

For a more detailed discussion of solderability issues, refer to Sec. 49.5.

49.1.2.4 *IPC-A-610, Acceptability of Electronic Assemblies.*   As a companion document to ANSI/J-STD-001, a second document, IPC-A-610, Acceptability of Electronic Assemblies,[6] is used by many companies as the stand-alone workmanship standard for their products. The relationship of these two documents is that ANSI/J-STD-001 establishes the acceptability requirements for PCBA soldering, while IPC-A-610, when addressing soldering, is the complementary document. IPC-A-610 depicts the pictorial acceptability criteria for the requirements identified in the ANSI/J-STD-001 document. IPC-A-610 also addresses additional criteria which defines handling and mechanical workmanship requirements. A large percentage of the acceptability criteria defined in this chapter will be those criteria shown in the IPC-A-610.

The IPC-A-610 describes the acceptability criteria for producing quality soldering interconnections and assemblies. The methods used must produce a completed solder joint conforming to the acceptability requirements described in IPC-A-610. The IPC-A-610 document details the acceptance criteria for each class in three levels of quality: *target condition, acceptable,* and either *nonconforming defect* or *nonconforming process indicator.*

Target Condition.   This is a condition that is close to perfect and in the past has been labeled as preferred; however, it is a desirable condition and not always achievable and may not be necessary to ensure reliability of the assembly in its service environment.

Acceptable Condition.   This characteristic indicates a condition identified that, while not necessarily perfect, will maintain the integrity and reliability of the assembly in its service environment. Acceptable can be slightly better than the minimum end-product requirements to allow for shifts in the process.

Nonconforming Defect Condition.   A nonconforming defect is an identified condition that is insufficient to ensure the form, fit, or function of the assembly in its end-use environment. The manufacturer shall disposition (rework, repair, or scrap) the nonconforming product based on design, service, and customer requirements.

Nonconforming Process Indicator Condition.   Nonconforming process indicator is a condition which identifies a characteristic that does not affect the form, fit, or function of a product.

- Such condition *does* violate acceptance stated in IPC-A-610 or any other customer requirement.
- Such condition is a by-product of the process being out of control due to material, poor design, and operator- or machine-related causes.
- Such condition requires the manufacturer to get the process under control, identify causes, and take corrective action. The product affected will be dispositioned "Use as is."

If the manufacturer does not have a documented, customer-approved process control system, all nonconforming process indicators must be treated as nonconforming defects.

Nonconforming process indicators are characteristics that should be used to improve the process, because their occurrence signals a lack of good workmanship to the customer.

49.1.3 Workmanship Manuals

Many companies use a workmanship document of some sort. The IPC-A-610 is used frequently by manufacturing and quality personnel to determine acceptable quality levels of their product. There are companies that have developed very good workmanship manuals for their own use, but they also sell their manuals to any company that wishes to use them.

If a need exists to develop a unique workmanship manual, then this manual should be derived based on the contract requirements. It is desirable to use a manual already in existence since it can become very costly to develop your own.

49.2 HANDLING TO PROTECT THE PCBA

The handling of PCBs both before and after the soldering operation can be very important in terms of inducing possible damage to the board or contaminating the board such that subsequent operations are affected. There are three subjects that should be well thought out and controlled in an assembly operation for PCBAs: electrostatic discharge (ESD) protection, contamination prevention, and physical damage prevention.

49.2.1 ESD Protection

ESD is the rapid discharge of a voltage potential into an electronic assembly. ESD-sensitive components found on the assembly and the amount of current generated by the discharge will determine if electrostatic overstress or complete failure takes place.

Some electronic devices are more sensitive to electrostatic overstress (EOS) damage than others (Table 49.1). The degree of such sensitivity within a particular device is related directly to the manufacturing technology employed.

Where sensitive components are handled, protective measures must be taken to prevent component damage. Improper and careless handling accounts for a significant portion of ESD damage to components and assemblies. Before handling ESD-sensitive components, equipment should be carefully tested to ensure that it does not generate damage-causing spikes. The preferred workstation to be utilized in electronic assembly is shown in Fig. 49.1.

TABLE 49.1 Approximate EOS/ESD Damage Range for Some Components

Device type	Range of min. EOS/ESD susceptibility, V
VMOS	30–1800
MOSFET	100–200
GaAsFET	100–300
EPROM	100+
JFET	140–7000
SAW	150–500
Op-amp	190–2500
CMOS	250–3000
Schottky diodes	300–2500
Film resistors (thick, thin)	300–3000
Bipolar transistors	380–7800
ECL (PDC board level)	500–1500
SCR	680–1000
Schottky TTL	100–2500

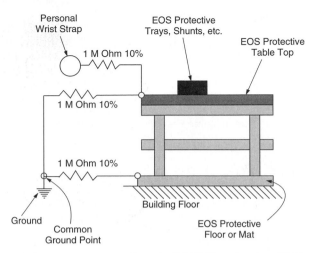

FIGURE 49.1 Target condition for EOS/ESD workstation. *(IPC.)*

Static charges are created when nonconductive materials are separated. Destructive static charges are often induced on nearby conductors, such as human skin, and discharged as sparks passing between conductors. This can happen when a PCBA is touched by a person having a static charge potential. The electronic assembly can be damaged as the discharge passes through the conductive pattern to a static-sensitive component. Static discharges may be too low to be felt by humans (less than 3500 V) and still damage ESD-sensitive components.

Sensitive components and assemblies must be enclosed in conductive, static-shielding bags, boxes, or wraps when not being worked on, unless otherwise protected. ESD-sensitive items must be removed from protective enclosures only at static dissipative or antistatic workstations. For ESD safety, a path-to-ground must be provided for static charges that would otherwise discharge on a device or board assembly. Provisions are made for grounding the worker's skin, preferably via a wrist strap or a heel strap, provided conductive flooring is used.

49.2.2 Contamination Prevention

The key to any contamination problem is to prevent it from happening. The cost of subsequent operations required to clean a product or rework a product are orders of magnitude more than any cost associated with preventing the contamination from occurring. These contaminants can cause soldering and solder mask or conformal coating problems. There are any number of possibilities of inducing contamination in an assembly environment, such as dirt, dust, machine oils, and process residues; however, in many cases, contamination is caused by the human body as salts and oils on the skin.

In a high percentage of assembly areas, good housekeeping will take care of contamination from the environment. It should be a common event on each shift to clean workstations, sweep floors, dust fixtures, empty trash, etc. This not only prevents contamination, but it also helps to keep personnel morale at a higher level. After all, we all desire to work in a nice, clean environment. It will also serve as a positive point if visitors or possible customers visit the assembly area.

To prevent contamination from the human body, every individual must be aware of the possibility of contaminating the PCBA. PCBAs should be touched only on the edges away from any edge connectors prior to the soldering operation (Fig. 49.2). Where a firm grip on the board is required due to any mechanical assembly procedures, gloves or finger cots should be worn. If the PCBA is to be conformal coated after the soldering operation, the handling of the

FIGURE 49.2 PCBA edge handling. *(IPC.)*

board is still very important to prevent contamination from fingerprints. In this case, gloves or finger cots should be worn until conformal coating is complete.

49.2.3 Physical Damage Prevention

Improper handling can damage components and assemblies. Typical defects associated with handling are cracked, chipped, or broken components, bent or broken terminals, scratched board surfaces, damaged traces or lands, fractured solder joints, and missing SMT components. Physical damage caused by handling can ruin assemblies and cause a high scrap rate of components or assemblies. Scrap is costly and must be avoided to have an efficient and high-quality operation.

Well-maintained handling equipment is also very important in preventing physical damage. One good example is conveyor systems when utilized. PCBAs can be caught in conveyors and damaged beyond rework or repair and, unless the area of operation is staffed, the conveyor system can damage multiple assemblies in a very short period of time.

49.3 PCBA HARDWARE ACCEPTABILITY CONSIDERATIONS

Most electronic assembly designs include a small percentage of mechanical assembly that uses hardware of various types to complete the assembly. Some of the more common component types and the acceptability criteria associated with each are discussed in the following sections.

49.3.1 Component Types

49.3.1.1 Threaded Fasteners.   Hardware stack-up for all threaded fasteners must be identified on engineering documentation. The stack-up can be critical, depending on the types of material used for both the hardware and the PCB (Fig. 49.3). Any missing hardware is a problem that must be corrected for obvious reasons. Any damage to hardware that prevents it from accomplishing what it was designed to do is unacceptable. A good example of this is screws or nuts that have been stripped, cross-threaded, or damaged to the point that a screw or nut driver is no longer able to tighten or loosen the part (Fig. 49.4).

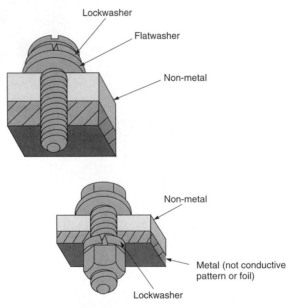

FIGURE 49.3   Hardware mounting for threaded fasteners. *(IPC.)*

FIGURE 49.4   PCBA hardware damage. *(IPC.)*

A minimum of 1½ threads shall extend beyond the threaded hardware unless the hardware could interfere with other components. The maximum extension of threaded hardware is 3 mm plus 1½ threads for hardware up to 25 mm long and 6.3 mm plus 1½ threads for hardware greater than 25 mm long.

Threaded fasteners shall be tight to the specified torque on engineering documentation. If torque is not specified on engineering documentation, a generic torque table should be available for use in the assembly environment. This is sometimes included in a workmanship manual.

49.3.1.2   Mounting Clips.   Uninsulated metallic components, clips, or holding devices must be insulated from underlying circuitry. Minimum electrical spacing between land and uninsulated component body must not be violated (Fig. 49.5).

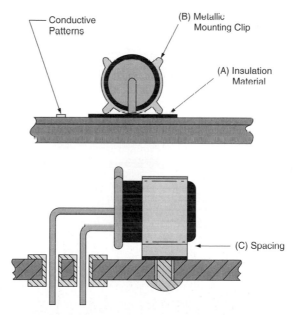

FIGURE 49.5   Component mounting clip insulation requirements. *(IPC.)*

The clip or holding device must make contact with the component sides on both ends of the component. The component must be mounted with its center of gravity within the confines of the clip or holding device. The end of the component may be flush or extend beyond the end of the clip or holding device if the center of gravity is within the confines of the clip or holding device (Fig. 49.6).

49.3.1.3   Heat Sinks.   Visual inspection should include hardware security, component or hardware damage, and correct sequence of assembly. Heat sinks must be mounted flush (in contact) with the surface to provide for adequate thermal conductivity. The component must be in contact with three quarters of the mounting surface to be considered flush (Fig. 49.7). It is unacceptable to have the heat sink mounted on the wrong side of the board, a bent or cracked heat sink, or missing fins on the heat sink. Any hardware mounts must be tight enough to prevent the component from being moved.

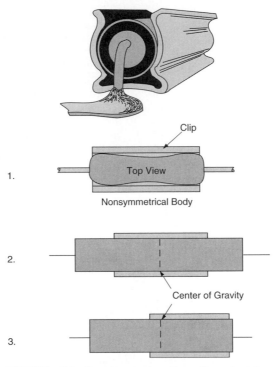

1. Top View

Clip

Nonsymmetrical Body

2.

Center of Gravity

3.

FIGURE 49.6 Component mounting clip orientation requirements. *(IPC.)*

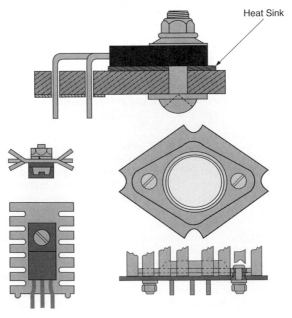

Heat Sink

FIGURE 49.7 Heat-sink acceptability requirements. *(IPC.)*

49.3.1.4 Terminals. Terminals that are to be soldered to a land may be mounted such that they can be turned by hand, but should be stable in the z axis. Terminals may be bent if the top edge does not extend beyond the base and no other mechanical damage such as fractures or breakage to the terminal or the solder joint have occurred (Fig. 49.8). Ordinarily the terminals utilized are turret, bifurcated, hook, and pierced or perforated terminals.

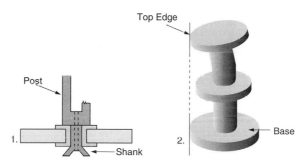

FIGURE 49.8   Terminal acceptability requirements. *(IPC.)*

49.3.1.5 Rivets and Funnels. Rivet and funnel barrels that extend above the substrate should be swaged or rolled to create an inverted cone, uniform in spread and concentric to the hole being mechanically fastened. The swaged or rolled flange should not be split, cracked, or otherwise damaged to the extent that mechanical strength is compromised or allows contaminating materials to be entrapped in the rivet or funnel (Fig. 49.9).

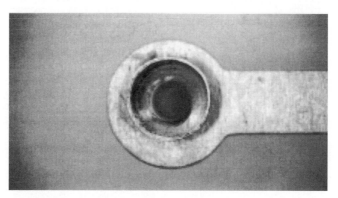

FIGURE 49.9   Rivet swage. *(IPC.)*

After swage or roll of the rivet or funnel, the rolled area should be free of circumferential splits or cracks. It may have a maximum of three radial splits or cracks, provided that the splits or cracks are separated by at least 90° and do not extend into the barrel of the rivet or funnel (Fig. 49.10).

The rivet mandrel cannot be pulled below the seating plane that is defined as the surface of the substrate being mechanically fastened (Fig. 49.11). As long as the mandrel was captured

FIGURE 49.10 Rivet swage split into barrel. *(IPC.)*

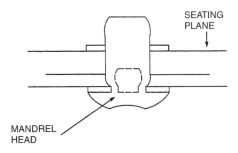

FIGURE 49.11 PCBA rivet—mandrel pulled below seating plane. *(DSC Communications.)*

long enough to form the rivet to secure the material, the mandrel need not be present; i.e., it may have fallen out of the rivet (Fig. 49.12). In all cases, the rivet or funnel must be mechanically secure in the z axis.

49.3.1.6 Rivet-Mounted Ejectors, Handles, and Connectors. Ejectors, handles, and connectors should not exhibit any cracks in the component material emanating from the roll pin or rivet that mounts the component to the PCBA. Roll pins should not protrude more than 0.015 in from the surface of the ejector, handle, or connector (Fig. 49.13). The major consideration for roll pins should be to ensure that any protrusion does not mechanically interfere with any other assembly. Damage to the part, PCB, or securing hardware is unacceptable.

Connector damage can also be the pins of the connector being pushed or bent out of specification. When connectors are mated there is the chance that the female portion of the connector pin will be pushed backwards and bent. This prevents adequate contact area between the male and female pins of the connector if it will mate at all. This circumstance is an unacceptable condition. There also exists the chance that the male connector pins can be bent. When the male pin is bent significantly, the mating of the female

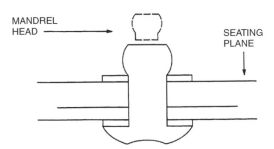

FIGURE 49.12 PCBA rivet—mandrel secured material correctly. *(DSC Communications.)*

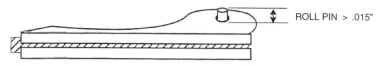

ROLL PIN > .015"

FIGURE 49.13   PCBA ejector roll—pin protrusion—nonconforming. *(DSC Communications.)*

and male connectors can cause mechanical damage to the connector casing due to the male pin being forced into the wrong female connector slot. In most cases, the damage is deformation, cracking, or breakage of the connector casing material, which are unacceptable.

When connector pins are installed (such as compliant pin, press fit), the pins must be straight within 50 percent of the pin thickness from the perpendicular. For Class 1 and 2 equipment, the PCB land may be lifted less than or equal to 75 percent of the annular ring width. Any land which is lifted across more than 75 percent of the annular ring or has been fractured is unacceptable (Fig. 49.14). For Class 3 equipment, no lifted or fractured lands are acceptable. For all classes, visibly twisted pins, damaged pins, or pins inserted such that there is a nonstandard pin height beyond a specified engineering tolerance is unacceptable.

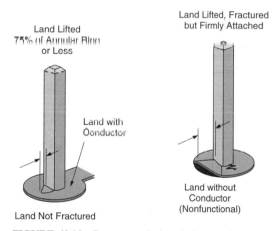

Land Lifted
75% of Annular Ring
or Less

Land with
Conductor

Land Not Fractured

Land Lifted, Fractured
but Firmly Attached

Land without
Conductor
(Nonfunctional)

FIGURE 49.14   Connector pin installation requirements. *(IPC.)*

49.3.1.7 Faceplates.   Faceplates must be clean and free of scratches or damage on the front surface. This is important since this is the surface that is commonly viewed by the customer and presents the cosmetic appeal to the product being purchased. An overexaggerated example of this is the purchase of an automobile. Few, if any, consumers would be willing to accept a new automobile with noticeable scratches in the painted surface.

A good rule of thumb for scratch criteria that some companies have adopted as a workmanship standard includes the following:

1. A scratch must be visible when viewed under the following conditions to be considered a defect:
   a. From a distance of 18 in
   b. No magnification utilized
   c. Normal room lighting used in the assembly area

2. A scratch must be visible from more than one angle to be considered a defect.

3. Metal or plastic faceplate scratches that exceed 0.125 in in length are considered a defect.

Surfaces exposed to frequent viewing should be free of blisters, runs, nicks, gouges, blemishes, or other abrasions that would detract from the general appearance of the finish. Faceplates must also be securely fastened to the PCB, i.e., tight enough to prevent physical movement.

49.3.1.8 Stiffeners.   Stiffeners are commonly used on PCBA designs that are large in dimension to prevent warpage of the PCBA before, during, or after assembly operations. If a stiffener has kept a PCBA from warping outside acceptability criteria, it has done its job; however, the stiffener must meet the following criteria to be acceptable.

1. Any marking or color coatings must be permanent. Loss of marking such that it is not legible, or fading or loss of color outside the color standard being used, is unacceptable.

2. The stiffener must be properly seated and mechanically fastened. If the soldering operation is used to mechanically fasten the stiffener to the PCBA, then good wetting to the stiffener such that the stiffener is mechanically sound is required.

3. It is unacceptable to have a loose stiffener on a PCBA; i.e., the stiffener is not securely fastened to the PCBA.

49.3.2 Electrical Clearance

Electrical clearance of hardware to components or traces carrying current must be controlled by the design engineer. In the past, IPC-A-610 required that 0.030 in minimum be maintained between hardware and any current-carrying material. This is no longer felt to be a requirement since lower levels of currents and voltages are frequently used in some PCBA designs today (Fig. 49.15).

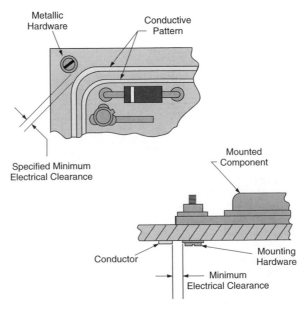

FIGURE 49.15   Electrical clearance—hardware to components. *(IPC.)*

It, therefore, becomes very important for the designer to identify the minimum electrical clearance required between hardware and current-carrying materials on the design documentation. For any given PCBA design, the electrical clearance should be verified by assemblers or inspectors to ensure that a shorting condition does not occur.

One example of a specification that has a definite requirement in this area is the Bellcore TR-NWT-000078 that requires a minimum of 0.005 in electrical clearance between uninsulated, noncommon conductive surfaces.

49.3.3 Physical Damage

Physical damage to hardware components in most cases refers to enough damage to a component to render it unusable in its application—for example, threads that are damaged to the point that mechanical fastening cannot be performed or a component that is fractured or broken such that it cannot perform the function for which it was intended.

Subjective judgments are common in this area since it is very much dependent on the application of the hardware in the given design, the environment in which the final product will be used, and life cycle expectations of the product.

49.4 COMPONENT INSTALLATION OR PLACEMENT REQUIREMENTS

Component installation or placement is the first step in the assembly of the PCBA. It may begin by preparation of the leads of a given package type in order to provide the correct lead protrusion, form the leads to fit the PCB plated-through-holes or pads, or put a bend in the leads which will serve as a standoff from the component to the PCB.

49.4.1 Plated-Through-Hole (PTH) Lead Installation

There are several requirements that apply to all PTH components. When using a polarized component, it must be oriented correctly. In all cases, the orientation must be correct or the board is unacceptable (Fig. 49.16), and when lead forming is required, the lead form must provide stress relief. Physical damage to the lead itself cannot exceed 10 percent of the diameter of the lead. Exposed basis metal as a result of lead deformation is a process indicator but does not render the board unacceptable.

49.4.1.1 *Axial Leaded Components.*   The target condition for axial components is that the entire body length of the component should be parallel and in contact with the board surface, provided the component dissipates less than 1 W of power. If the component dissipates more than 1 W of power, it must be mounted a minimum of 1.5 mm above the PCB surface to prevent burning or scorching the surface. The maximum space between the component and the PCB surface shall not violate the requirements for lead protrusion and cannot be greater than 3.0 mm for Class 1 and 2 and 0.7 mm for Class 3 equipment (Fig. 49.17).

Leads must extend from the component body at least one lead diameter (L) or thickness, but not less than 0.8 mm from the body or weld, before the start of the lead bend radius (Fig. 49.18).

No physical damage, such as chips or cracks to axial components, should be allowed; however, it is acceptable to allow minor damage. It becomes unacceptable if the insulating cover is damaged to the extent that the metallic element is exposed or the component shape is deformed (Fig. 49.19).

49.4.1.2 *Radial Leaded Components.*   The target condition for radial components is that the body is perpendicular to the PCB and the component base is parallel to the PCB. It is an

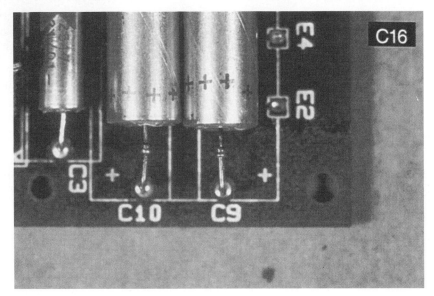

FIGURE 49.16   Component orientation—polarity. *(IPC.)*

FIGURE 49.17   Axial leaded component—above PCBA mounting. *(IPC.)*

acceptable condition to have the component tilt no more than 15° from the perpendicular. The space between the component base and PCB must be between 0.25 and 2.0 mm to be acceptable (Fig. 49.20). For components that have a coating meniscus, lack of visible clearance between the coating meniscus and the solder fillet is unacceptable but can be acceptable for Class 1 and 2 equipment under the following conditions:

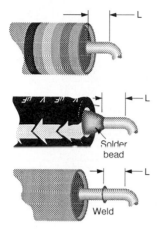

FIGURE 49.18 Lead extension from component body. *(IPC.)*

1. There is no risk of thermal damage to the component.
2. The component mass is less than 10 g.
3. The voltage is not greater than 240 V_{ac} RMS or 240 V_{dc} (Fig. 49.21).

Minor physical damage such as scratches, chips, or crazing to radial components is acceptable provided they do not expose the component substrate or active element and structural integrity is not compromised (Fig. 49.22).

49.4.2 Surface-Mount Technology (SMT) Placement

Coplanarity is very important in the placement of SMT components with leads. Occasionally, lead preparation to form the leads of SMT components is necessary to achieve the proper coplanarity for manual or automatic placement of components on the PCB. Most times, however, the components are purchased packaged with leads prepped and ready to be automatically placed by pick-and-place equipment. Another critical parameter for all surface mounted components is the accuracy of the placement of the component onto the pads of the PCB.

49.4.2.1 Chip Components. Side overhang is acceptable up to one-half (one-quarter for Class 3) the width of the component end cap or PCB pad. End-cap overhang is not acceptable for all classes. The end-cap solder joint width is acceptable with a minimum solder joint length of one-half (three-quarters for Class 3) the component end cap or PCB land, whichever is less. A side solder joint length is not required; however, a properly wetted fillet must be evident.

FIGURE 49.19 Damaged axial leaded component. *(IPC.)*

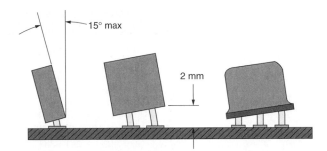

FIGURE 49.20 Radial leaded components—tilt and spacing. *(IPC.)*

FIGURE 49.21 Radial component coating meniscus in the PTH. *(IPC.)*

The maximum solder fillet height may overhang the land or extend onto the top of the end-cap metallization; however, the solder should not extend farther on the component body. The minimum solder fillet height must cover one-quarter the thickness, or height, of the component end cap. The component end cap must have overlap contact with the PCB pad. There is no minimum contact length specified (Fig. 49.23).

49.4.2.2 MELF or Cylindrical Components. Side overhang is acceptable up to one-quarter the diameter of the end cap. End-cap overhang is not acceptable for all classes. The end-cap solder joint width is acceptable with a minimum solder joint length of one-half the diameter of the component end cap. A side solder joint length must be a minimum of one-half (three-quarters for Class 3) the end-cap thickness as measured from the end of the component toward the center of the component. The maximum solder fillet height may overhang the land or extend onto the top of the end-cap metallization; however, the solder should not extend farther on the component body. The minimum solder fillet height must cover one-quarter the thickness, or height, of the component end cap for Class 3 equipment and for Classes 1 and 2,

FIGURE 49.22 Damaged radial leaded component. *(IPC.)*

the minimum solder fillet must exhibit proper wetting. The component end cap must have overlap contact with the PCB pad. There is no minimum contact length specified (Fig. 49.24).

49.4.2.3 Castellated Termination Leadless Chip Carrier Components.
Side overhang is acceptable up to one-half (one-quarter for Class 3) the castellation width. End overhang of the castellation is not acceptable for all classes. The castellation end solder joint width is acceptable with a minimum solder joint length of one-half (three-quarters for Class 3) the castellation width. A minimum side solder joint length is one-eighth the castellation solder fillet height. The maximum solder fillet height is not a specified parameter for this type

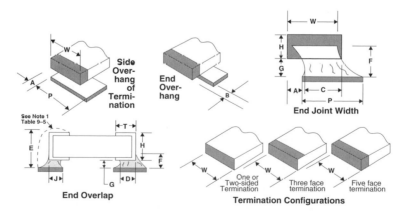

W = Width of Termination Area ● T = Length of Termination ● H = Height of Termination ● P = Width of Land

FIGURE 49.23 Chip components—placement and soldering requirements. *(IPC.)*

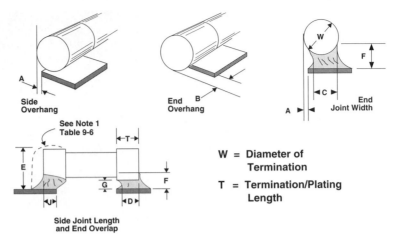

FIGURE 49.24 MELF or cylindrical components—placement and soldering requirements. *(IPC.)*

component. The minimum solder fillet height must cover one-quarter the castellation height (Fig. 49.25).

49.4.2.4 Gullwing Leaded Components. Side overhang is acceptable up to one-half (one-quarter for Class 3) the width of the component lead or 0.5 mm, whichever is less. Lead toe overhang is acceptable provided it does not violate minimum conductor spacing or solder joint fillet requirements for all classes. The end lead solder joint width is acceptable with a minimum solder joint length of one-half (three-quarters for Class 3) the component lead width. The lead side solder joint minimum length is one-half (three-quarters for Class 3) the component lead width. For maximum solder fillet height on *high*-profile devices such as QFPs

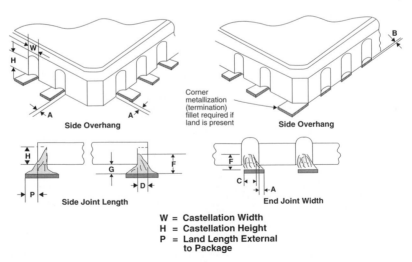

FIGURE 49.25 Castellated termination leadless chip carrier components—placement and soldering requirements. *(IPC.)*

and SOLs, the solder may extend to, but not touch, the package body or end seal. For maximum solder fillet height on *low*-profile devices such as SOICs and SOTs, the solder may extend to the package. The solder must not extend under the package in any case. The minimum solder fillet height must cover one-half the lead thickness (Fig. 49.26).

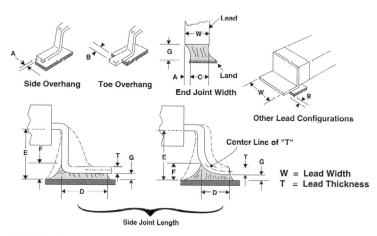

FIGURE 49.26   Gullwing leaded components   placement and soldering requirements. *(IPC.)*

The same criteria for acceptability of placement and soldering of gullwing components can be applied to round or coined (flattened) leaded components if utilized.

49.4.2.5 *J-Leaded Components.*   Side overhang is acceptable up to one-half (one-quarter for Class 3) the width of the component lead. Lead toe overhang is not specified for any class. The end lead solder joint width is acceptable with a minimum solder joint length of one-half (three-quarters for Class 3) the component lead width. A side solder joint length must be a minimum of 1½ times the lead width. The maximum solder fillet height is not specified; however, the solder fillet may not touch the component package body. The minimum heel solder fillet height must cover one-half the thickness of the component lead. The minimum solder thickness from PCB pad to component lead is not specified; however, there must be sufficient solder to form a properly wetted fillet (Fig. 49.27).

49.4.2.6 *Ball Grid Array (BGA) Components.*   BGA package components are relatively new types of components being used on PCBs. Although extremely desirable for the designers to use to get more function in a smaller space, this type of component package is virtually impossible to inspect visually. The prevailing thought on verifying compliance to workmanship standards is to verify the function through an in-circuit and/or functional test of the BGA circuit. If cost justification exists, due to volume or reliability requirements such as could be the case on Class 3 equipment, x-ray equipment can be used to verify the quality and integrity of the solder joints. This particular option is quite costly in terms of the upfront capital investment of the x-ray equipment but may pay for itself many times over in customer satisfaction.

49.4.3 Use of Adhesives

Adhesives can be used in surface-mount and PTH applications.

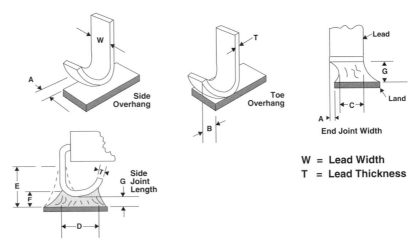

FIGURE 49.27   J-leaded components—placement and soldering requirements. *(IPC.)*

For SMT components, the most common use of adhesive is placing components onto the secondary side of the PCB, i.e., the side opposite the PTH components. In many process flows, the secondary-side SMT components are placed onto adhesive between PCB pads; the PCB is processed through an adhesive curing cycle; then the components are wave soldered onto the PCB along with the PTH components. The SMT components are acceptable as long as the adhesive has not contaminated the solder joint. If adhesive has contaminated the component solderable surface, lead or end cap, or the PCB pad such that an acceptable solder joint is not achieved, the PCBA is not acceptable.

Normally in PTH component application, adhesive is used to give large-profile and/or heavy components more mechanical stability. When used in this manner the following adhesive acceptance criteria apply:

- The adhesive must adhere to a *flush-mounted axial* component for 75 percent of the component length and 25 percent of its diameter on one side. The buildup of adhesive must not exceed 50 percent of the component diameter and adhesion to the mounting surface must be evident (Fig. 49.28).

- The adhesive must adhere to a *vertically mounted axial* component for 50 percent of the component length and 25 percent of its circumference, and adhesion to the mounting surface must be evident (Fig. 49.29).

- For components elevated from the PCB which weigh 7 g or more per lead, the component should be bonded to the mounting surface in at least four places and at least 20 percent of the total periphery of the component is bonded. The adhesion from component to mounting surface must be evident (Fig. 49.30).

49.5   COMPONENT AND PCB SOLDERABILITY REQUIREMENTS

Component solderability, which includes the solderability of the PCB, is probably the most important single characteristic to consider in building PCBAs. To be successful in soldering PCBAs, good wetting must be achieved. It is an ongoing effort to ensure that all PCBs and electrical components purchased and received from suppliers maintain good solderability.

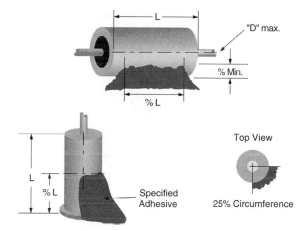

FIGURE 49.28 Adhesive bonding—axial components. *(IPC.)*

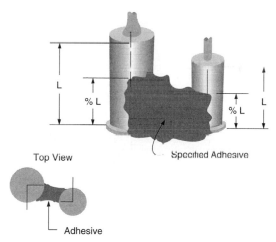

FIGURE 49.29 Adhesive bonding—multiple axial components. *(IPC.)*

The key to success in this all-important task is to have an excellent working relationship
with the suppliers and manufacturers of the components to ensure good, solderable components. Until partnership relationships are achieved with suppliers, there should be a sample of
each lot or batch of components received that are subjected to solderability testing. This testing should be conducted in accordance with ANSI/J-STD-002 and ANSI/J-STD-003. These
documents were written to specifically address solderability requirements for electronic
assembly components and PCBs.

Packaging and handling of components is important to maintain good solderability since
many contaminants that could affect solderability can be transferred to solderable surfaces as
a result of handling by equipment or personnel. Another possible problem that must be
avoided is aging of the components. As tin-lead component lead finishes age, they will oxidize,
and this oxidation will affect solderability of the component to the PCB. A rule of thumb used
by many companies is to suspect solderability on components 2 or more years old. In some

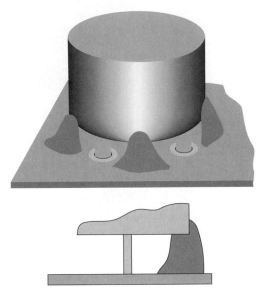

FIGURE 49.30   Adhesive bonding—elevated components >7 g per lead. *(IPC.)*

cases this can be tracked easily for components marked with a date code. Other components
too small to mark or that for some other reason are not marked with a date code cannot be
traced easily. Many times companies keep records on the date of receipt and base the age of
the component on this date. Although not completely accurate, it serves the purpose. If components are known to be over 2 years old, a new solderability sample should be pulled to
determine if the components are still usable. Many times, they are not usable due to oxidation
or some other contaminant the component has been exposed to over time.

49.6   SOLDER-RELATED DEFECTS

All solder joints should exhibit wetting as shown by a concave meniscus between the component and PCB being soldered. The outline of the component being soldered should be easily
determined. The most common solder defects produced by the soldering process are discussed in the following sections.

49.6.1   Plated-Through-Hole Solder Joint Minimum Acceptable Conditions

Table 49.2 shows the minimum acceptable plated-through-hole solder joint criteria.

49.6.2   Solder Balls or Solder Splash

Solder balls/splashes that violate minimum electrical design clearance, are not encapsulated
in a permanent coating, or attached to a metal surface are unacceptable. Solder balls/splashes
that are within 0.13 mm of lands or traces or exceed 0.13 mm in diameter are considered to

TABLE 49.2   Minimum Acceptable PTH Solder Joint Criteria

Criteria	Class 1	Class 2	Class 3
Circumferential wetting on the primary side, including lead and barrel	Not specified	180°	270°
Vertical fill of solder*†	Not specified	*†	*†
Circumferential fillet, and wetting on the secondary side	270°	270°	330°
% of original land area covered with wetted solder, primary side	0	0	0
% of original land area covered with wetted solder, secondary side	75	75	75

* A total maximum of 25 percent depression, including both solder source and destination sides, is permitted.
† A total maximum of 50 percent depression, including both solder source and destination sides, is permitted on PTHs connected to voltage or ground planes. Solder must extend 360° around the lead and 100 percent wet PTH barrel walls to lead on the secondary side (Fig. 49.31).

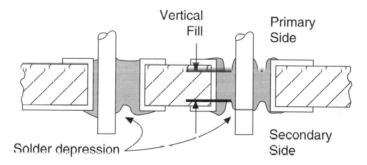

Minimum Acceptable for all Classes per Table 49.2

FIGURE 49.31   PTH solder fill requirements. *(IPC.)*

be process indicators. Also considered as process indicators are more than five solder balls/splashes of a diameter of 0.13 mm or less in a 600-mm² area (Fig. 49.32).

49.6.3   Dewetting and Nonwetting

Dewetting is a condition that results when molten solder coats a surface and then recedes to leave irregularly shaped mounds of solder that are separated by areas that are covered with a thin film of solder and with the basis metal not exposed (Fig. 49.33). *Nonwetting* is the partial adherence of molten solder to a surface that it has contacted, while the basis metal remains exposed (Fig. 49.34).

Dewetting and nonwetting of solder joints is generally caused by contaminants on either the component leads or on the PCB PTH or pads. A minimum of dewetting is allowed on solder joints assuming that the solder joint meets minimum requirements as defined in Table 49.2 and Sec. 49.4.2 by component package type, and good wetting is evident on the portion of the solder joint which does not display dewetting. Nonwetting is not acceptable since adequate wetting is not achieved and indicates a serious solderability problem on the component or the PCB.

FIGURE 49.32   Solder balls and solder splash requirements. *(IPC.)*

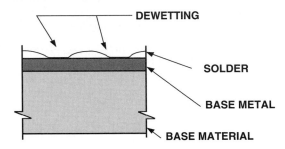

FIGURE 49.33   Solder dewetting. *(IPC.)*

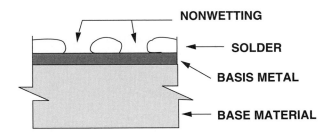

FIGURE 49.34   Solder nonwetting. *(IPC.)*

49.6.4   Missing and Insufficient Solder

Missing solder is an obvious unacceptable condition since solder provides electrical continuity and some measure of mechanical fastening of the component to the PCB.

Insufficient solder becomes an unacceptable condition for SMT components when minimum solder fillet requirements defined in Sec. 49.4.2 are not achieved. Insufficient solder becomes unacceptable for PTH components when the minimum requirements of Sec. 49.6.1 and Table 49.2 are not achieved.

49.6.5   Solder Webbing and Bridging

Solder bridged between electrically noncommon conductors creates a shorting condition and is an unacceptable condition. Solder webbing is a continuous film of solder that is parallel to, but not necessarily adhering to, a surface that should be free of solder. Solder webbing is also an unacceptable condition.

For components with leads, with the exception of criteria stated in Sec. 49.4.2.4, solder cannot come into contact with the component body or end seal. If solder contacts the component body or end seal it is an unacceptable.

49.6.6 Lead Protrusion Problems

Measurement of lead protrusion is defined as the distance from the top of the PCB land to the outermost part of the component lead, which can include any solder projection from the lead. Solder projections (icicles) are unacceptable if they violate lead protrusion maximum requirements or electrical clearance, or if they pose a safety hazard; otherwise, solder projections are an acceptable condition on both SMT or PTH components.

For single-sided PCBAs, lead or wire protrusion must be a minimum of 0.5 mm for all classes. For double-sided and multilayer PCBAs in all classes, the minimum lead protrusion is that the lead end be visible in the solder. The maximum lead protrusion for Class 1 is that there be *no danger of shorts* when the PCBA is used in its assembly application. For Class 2, the maximum lead protrusion is 2.5 mm and for Class 3 the maximum lead protrusion is 1.5 mm (Fig. 49.35).

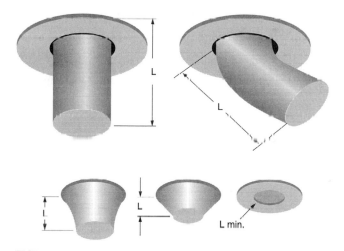

FIGURE 49.35   Lead protrusion requirements. *(IPC.)*

One notable exception to the lead protrusion requirement is for PCBs greater than 2.3 mm thick (nominal). For components with preestablished lead lengths—i.e., the leads are not cut to length by the board assembler—lead protrusion may not be visible and the PCBA is still considered to be acceptable. Some examples of components this may apply to are single/dual in-line packaged ICs (SIPs/DIPs), sockets, transformers/inductors, pin grid arrays (PGAs), and power transistors.

This relaxation of lead protrusion requirements is driven by the fact that some components are not offered in an array of different lead lengths by their manufacturers. When these types of components are used on PCBs of a thickness of 2.3 mm or greater, the lead will not protrude beyond the land by design. This exception should not be used as a license by the designer to select components on PCBAs which do not meet lead protrusion requirements. In many cases, components are available in longer lead lengths to accommodate the use of thick PCBs and should be utilized.

49.6.7 Voids, Pits, Blowholes, and Pinholes

Solder cavities (voids, pits, blowholes, pinholes, dewetting) are acceptable, provided the lead and land/pad are wetted and solder fillets meet requirements per Table 49.2. The areas adjacent to the solder cavity must be properly wetted and the bottom of the solder cavity must be visible with no basis metal exposed.

49.6.8 Disturbed or Fractured Solder Joints

Solder joints may be disturbed, i.e., rough, granular, or uneven in appearance, provided the wetting coverage criterion in Table 49.2 is met. Solder joints that have been fractured or cracked are unacceptable for all classes of equipment.

For solder joints in which leads have been trimmed after soldering, the solder joint cannot be damaged by the cutters due to physical shock. If lead cutting is required after solder, the solder joint shall be reflowed or visually inspected at 10× magnification to assure that the solder connection was not damaged as a result of the cutting operation. No fractures or cracks are allowed between the lead and solder (Fig. 49.36).

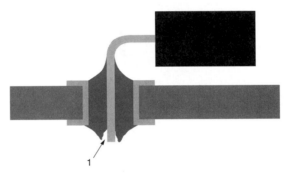

FIGURE 49.36   Solder joint—lead to fillet fracture. *(IPC.)*

49.6.9 Excess Solder

Excess solder conditions which produce a solder fillet that is slightly convex, or bulbous, and in which the lead is no longer visible are considered process indicator defects. This is an acceptable condition, assuming the condition is, indeed, excess solder and not the loss of lead protrusion due to component float, component tilt, or leads cut too short because of incorrect lead preparation. The exception for PCBs 2.3 mm or greater in thickness still applies (Fig. 49.37).

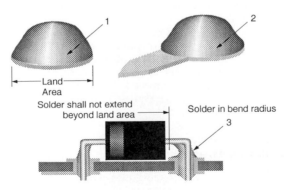

FIGURE 49.37   Excess solder requirements. *(IPC.)*

49.6.10   Solder Requirements for Vias

Plated-through-via-holes used only for interfacial connection do not need to be filled with solder if they were not exposed to a soldering process. This is usually achieved through temporary or permanent masks over the vias during the soldering process. Plated-through-holes or vias without leads when exposed to a soldering process should meet the acceptability requirements listed here.

1. The target condition is to have the holes completely filled with solder and the top of the lands show good wetting.
2. The minimum acceptable condition is the sides of the plated-through-hole are wetted with solder.
3. When solder has not wet the sides of the plated-through-hole, it will be considered a process indicator defect and product will not be rejected (Fig. 49.38).

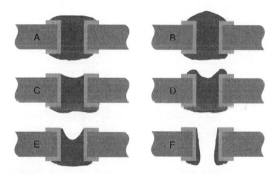

FIGURE 49.38   Via-hole solder fill requirements. *(IPC.)*

49.6.11   Soldering to Terminals

When soldering wire to terminals, the lead outline must be visible and good wetting between the wire and terminal must be evident to be acceptable. The wire insulation to terminal gap can be near zero if the insulation has not melted into the solder joint and a full 90° wrapped solder connection is evident. Slight melting of the insulation is acceptable. If the insulation gap is too large and allows potential shorting of wire to an electrically noncommon conductor, the joint is unacceptable. If the wire insulation is severely burned and the melt by-product intrudes into the solder joint, the joint is unacceptable (Fig. 49.39).

49.7   PCBA LAMINATE CONDITION, CLEANLINESS, AND MARKING REQUIREMENTS

49.7.1   Laminate Conditions

Laminate defect conditions may be caused by the laminator, PCB fabricator, or the assembly of the PCB. The major laminate conditions seen are measling, crazing, blistering, delamination, weave exposure, and haloing.

49.7.1.1   Measling and Crazing.   *Measling* is an internal condition occurring in laminated base material in which the glass fibers are separated from the resin at the weave intersections.

FIGURE 49.39   Soldering wire to a terminal. *(IPC.)*

This condition manifests itself in the form of white spots or crosses below the surface of the base material and is usually related to *thermally* induced stress.

Crazing is an internal condition occurring in laminated base material in which the glass fibers are separated from the resin at the weave intersections. This condition manifests itself in the form of connected white spots or crosses below the surface of the base material and is usually related to *mechanically* induced stress (Fig. 49.40).

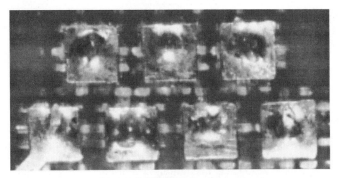

FIGURE 49.40   PCB crazing. *(IPC.)*

Measling or crazing that occurs as a result of an inherent weakness in the laminate is a warning of a potentially serious problem. If measling or crazing occur in the assembly process, they will usually not propagate further nor evolve into a more serious problem.

By the time a PCB enters an assembly process, the operator who observes measling or crazing cannot determine the source of the problem. This fact puts even more emphasis on getting high-quality PCBs from suppliers. This can be done through increased receiving inspection, source inspections, or a comprehensive qualification program and partnership with suppliers in which process control is utilized in the PCB fabrication and parametric data from the supplier is sent to the buyer to review for compliance to requirements. This kind of information can also be used to justify a dock-to-stock program if desired.

Evidence to date is that even boards with severe measles have functioned adequately over long periods of time and in harsh environments. In fact, the IPC has no data that shows a measled board has ever failed in the field in the absence of no other serious defects. The only acceptability criterion for measling and crazing is that the assembly be functional in its application.

49.7.1.2   Blistering and Delamination.   *Blistering* is a localized swelling and separation between any of the layers of the base material or between the material and the metal cladding. *Delamination* is a separation between any of the layers of the base material or between the base material and the metal cladding.

Blistering/delamination cannot exceed 50 percent (25 percent for Class 3) of the distance between plated-through-holes or subsurface conductors.

49.7.1.3   Weave Exposure.   *Weave exposure* is a surface condition of base material in which the unbroken fibers of woven glass cloth are not completely covered by resin. Weave exposure is acceptable if it does not reduce dielectric spacing as specified in engineering documentation between the weave exposure and a conductor. Class 3, however, does not accept any weave exposure.

49.6.1.4   Haloing and Edge Delamination.   *Haloing* is a condition existing in the base material in the form of a light area around holes or other machined areas on or below the surface of the base material. Haloing is acceptable if the penetration of the haloing or edge delamination does not reduce edge spacing more than 50 percent of the distance to the nearest conductor or 2.5 mm maximum, whichever is less.

49.7.2   PCBA Cleanliness

Board cleanliness is needed to ensure that contaminants are sufficiently removed which could affect functionality at present or in the future. Some contaminants can actually promote growth of undesirable substances on the PCBA which can cause shorting or corrosion which would affect the PCBA functional integrity.

No visible residue from cleanable or any activated fluxes is allowed. Class 1 equipment suppliers may not be required to remove cleanable residues if qualification testing is performed which demonstrates no need for cleaning the PCBAs. No-clean or low-residue flux residues may be allowed if the PCBA is not conformal coated. If conformal coating is used, these residues are not acceptable since they will, in most cases, detrimentally affect the conformal coating ability to bond to the components or the PCB being coated.

For processes utilizing corrosive fluxes, the solvent extract conductivity (SEC) cleanliness test must be performed. When SEC is performed, the surface contamination level must meet $1.5\ \mu g/cm^2$ or less of NaCl equivalent in order for the PCBA cleaning process to be acceptable. In the event of failure, immediate process corrections must be made before any additional product is assembled.

Fluxes are considered to be noncorrosive if they meet the following criteria:

1. Copper mirror test as defined by flux type L requirements of IPC-SF-818, General Requirements for Electronic Soldering Fluxes, flux-induced corrosion test.

2. Halides test as defined by flux type L, Class 3 requirements of IPC-SF-818, General Requirements for Electronic Soldering Fluxes, presence of halides in flux test.

3. Surface insulation resistance must meet a minimum of 2×10^4 MΩ per IPC-B-25.

4. Electromigration resistance requirements must be met. The test sample shall be examined at 10× magnification with no evidence of filament growth that reduces conductor spacing by more than 20 percent.

Particulate matter such as dirt, lint, dross, lead clippings, etc., are not acceptable on PCBAs. Metallic areas or hardware on the PCBAs may not exhibit any crystalline white deposits, colored residues, or rusty appearance.

49.7.3 PCBA Marking Acceptability

Marking provides both product identification and traceability. It aids in assembly, in-process control, and field repairs. The methods and materials used in marking must serve the intended purposes and must be readable, durable, and compatible with the manufacturing processes as well as the end use of the product.

Fabrication and assembly engineering drawings should be the controlling documents for the locations and types of markings on PCBAs. Marking on components and fabricated parts should withstand all tests, cleaning, and assembly processes to which the item is subjected and shall remain legible (capable of being read and understood). Acceptability of marking is based on whether it is legible. If a marking is legible and cannot be confused with another letter or number, it is acceptable. Components and fabricated parts do not have to be installed so that reference designators are visible after installation. Missing, incomplete, or illegible characters in markings is unacceptable.

49.8 PCBA COATINGS

It should be noted that not all PCBA designs use conformal coating; however, when used, it must meet the following acceptability criteria. All PCBAs that utilize top- or bottom-layer etch (trace) runs have a solder mask over them if a solder wave or static solder bath process is used to solder components onto the PCB. Without solder mask, the solder would bridge between many solder wettable surfaces causing an uncontrollable problem.

49.8.1 Conformal Coating

Conformal coating is an insulating protective covering that conforms to the configuration of the objects coated when it is applied to a completed PCBA. Conformal coatings should be homogeneous, transparent, and unpigmented. The conformal coating should be properly cured and not exhibit tackiness. Defects associated with conformal coating are limited to those identified in Table 49.3.

Conformal coating thickness requirements for three types of coatings are listed as follows.

1. Types ER (epoxy), UR (urethane), and AR (acrylic)   0.05 to 0.08mm
2. Type SR (silicone)   0.08 to 0.13mm
3. Type XY (para-xylene)   0.01 to 0.05mm

The thickness may be measured on a coupon that has been processed with the assembly. No conformal coating on the tip of leads is required.

TABLE 49.3   Limits of Conformal Coating Defects
(Percentage of PCBA surface area)

Conformal coat defect	Class 1	Class 2	Class 3
Voids and bubbles	10	10	5
Adhesion loss	10	5	5
Foreign material	5	5	2
Dewetting	10	5	5
Ripples	15	10	5
Fisheyes	15	10	5
Orange peel	15	10	5

49.8.2 Solder Mask

Solder mask is a film coating used to provide dielectric and mechanical shielding during and after soldering operations. Solder mask material may be applied as a liquid or a dry film.

Cracking of solder mask after the soldering and cleaning operation is acceptable for Class 1 and 2 equipment but unacceptable for Class 3 equipment. After assembly soldering and cleaning operations, wrinkling of solder mask over tin-lead (SMOTL) plated traces is acceptable provided the solder mask has not been lifted or degraded to the point of flaking, peeling, or loose solder mask on the PCBA. Flaking, peeling, or loose solder mask on the PCBA is unacceptable. Wrinkling of solder mask over bare copper (SMOBC) traces is not acceptable

After assembly of SMOTL boards, any unacceptable flaking, peeling, or loose solder mask on the PCBA may be removed, leaving solder mask which is adequately adhered to the PCBA. The removal of this solder mask will render an acceptable PCBA

49.9 SOLDERLESS WRAPPING OF WIRE TO POSTS (WIRE WRAP)

Many applications of wire wrap are still utilized in equipment design, and standards of acceptance should be used. Standards of acceptance in this area are shown in IPC-A-610 and in Bellcore TR-NWT-000078. The following is taken mainly from the Bellcore specification since it is slightly more detailed than the IPC specification.

49.9.1 Wrap Post

The wrap post cannot be bent or twisted before or after the wire is wrapped to the post. The wrap straightness will not exceed approximately 1 post diameter or thickness from its perpendicular position. The term "approximately" here implies that this is not a requirement that should be measured routinely and can be accepted based on visual observation and subjective judgment. After connection, the post must not be twisted more than 15° from its original position in order to be acceptable.

It should be noted that some wrap posts are used, standalone, as headers or test points and the question of tilt or post bend acceptability requirements continues to be raised. For this type of application, the tilt requirements should be loosened unless there are valid engineering concerns preventing it. Generally the tilt or bend will not exceed approximately 2 post diameters or thicknesses from its perpendicular position.

49.9.2 Wire Wrap Connection

Table 49.4 shows the number of wire wrap turns of insulated and uninsulated wire used on the wrapped connection. The connection is made using automatic or semiautomatic wire-wrapping devices. For this requirement, countable turns are those turns of bare wire or insulated wire in intimate contact with the corners of the terminals starting at the first contact with a terminal corner and ending at the last contact with a terminal corner (Fig. 49.41).

Maximum turns of bare and insulated wire are governed only by tooling configuration and space available on the ter-

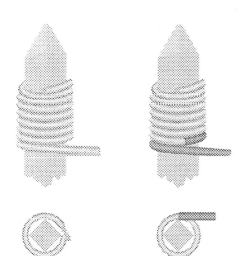

FIGURE 49.41 Wire wrap connection acceptability. *(IPC.)*

TABLE 49.4   Wire Wrap Turn Requirements

| Wire gauge | Terminal size, in^2 | Minimum number of turns | |
		Uninsulated wire	Insulated wire
#20 and #22	0.025–0.045 in	5	None
#24	0.025–0.045	6	None
#26	0.025	6	None
#26	0.045	7	None
#28 and #30	0.025	7	¾

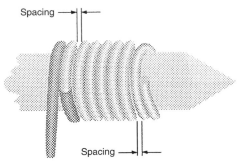

Spacing

Spacing

FIGURE 49.42   Single wire wrap spacing. *(IPC.)*

minal. Stripped ends of wires used for previous solderless wrapped connections shall not be reused. Electrical clearance between terminals as specified by engineering documentation must be maintained. Wire ends should in no case project to the extent that the required electrical clearance is compromised. In no case should the wire end project more than 0.125 in (1 wire diameter for Class 3) away from the terminal.

49.9.3   Single Wire Wrap Spacing

The wrapped conductors will be free of gaps (that is, each wrap will be in contact with the previous wrap) with turns not overlapping. The first insulated turn on a wrapped post cannot exceed a maximum of 0.050 in above the pin base of shoulder (wrappable surface of the pin). The first and last one-half-turns may have a space between turns, provided the space does not exceed 1 diameter of the uninsulated wire. Excluding the first and last one-half-turns, the wrapped conductors may have a single space between them, provided the opening does not exceed one-half the nominal diameter of uninsulated wire (Fig. 49.42).

49.9.4   Multiple Wire Wrap Spacing

Typically no more than three wires are wrapped to a single post. When more than one wrap is used on a single post the following requirements apply. The maximum spacing between consecutively wrapped wires is 2 uninsulated wire diameters with a preferred spacing of one-half the diameter of the uninsulated wire. The final wire wrap turn on a post must not extend to within 1 uninsulated wire diameter of the tapered portion of the wire wrap post's tip. The first insulated turn of a higher-level wire wrap may overlap the last turn of uninsulated wire on a lower-level wrap by a maximum of 1 turn (Fig. 49.43).

49.10   PCBA MODIFICATIONS

All modifications to PCBAs should be defined and detailed in approved engineering and/or methods documents. Jumper wires are considered as components and are defined by documentation for routing, termination, staking, and wire type.

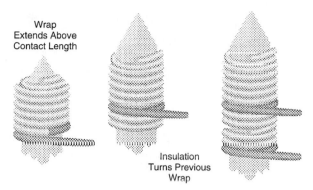

FIGURE 49.43   Multiple wire wrap requirements. *(IPC.)*

49.10.1   Cut Traces

Trace cuts should be at least 0.030 in in width as a minimum, with all loose material removed. Trace cuts should also be sealed with an approved sealant to prevent absorption of moisture. Care should be taken when removing etch from PCBs to prevent damaging the laminate material.

49.10.2   Lifted Pins

Lifted pins should be cut short enough to prevent the possibility of being shorted to the pad from which they were lifted should they be pushed back down. If the component hole from which the pad was lifted does not contain a jumper wire, it should be filled with solder.

49.10.3   Jumper Wires

Jumper wires may be used on all classes of equipment electronic assemblies and in thick-film hybrid technology. The wire may be terminated in plated-through-holes, on terminal stand-offs, on circuit lands, or on component leads. It should be noted here that Bellcore TR-NWT-000078 requires that wires be terminated only in plated-through-holes and that for Class 3 equipment a wire cannot be placed into the same plated-through-hole with a component lead.

Recommended jumper wire is solid insulated copper wire, tin-lead plated, 22 to 30 gauge with insulation capable of withstanding soldering temperatures. The insulation must have some resistance to abrasion and have a dielectric resistance equal to or better than the board insulation material. Jumper wire shall be insulated if greater than 1.0 in in length or it could possibly short between lands or component leads.

Jumper wires are to be routed in the shortest *x-y* route possible. Wire routing must be documented for each part number. Assemblies having the same part number must be routed in the same pattern (Fig. 49.44).

When jumper wires are used on the *primary side* of the PCBA, no wire is to pass over or under any components. The wire may pass over lands, provided the wire can be moved away from the land for component replacement. Care must be taken to avoid running wire near or in contact with heat sinks to prevent damage to the wire from excessive heat exposure (Fig. 49.45).

When jumper wires are used on the *secondary side* of the PCBA, the jumper wire should not pass through component footprints unless the layout of the assembly prohibits routing in

FIGURE 49.44   Jumper wire—*x-y* routing. *(IPC.)*

FIGURE 49.45   Jumper wire—routing over components. *(IPC.)*

other areas. If this condition occurs, it should be designated as a process indicator. The exception to this is for edge connectors on the PCBA. Jumper wires should not pass over test points or vias used as a test point.

Jumper wires should be staked to the base material with an approved adhesive. Uncured adhesive is unacceptable on the completed PCBA. The wire should be spot bonded along its route and must not be applied to lands, pads, or components. The interval of staking must be defined in the applicable engineering documentation but the wire must be staked at all changes in direction of the wire. Jumper wires must be staked or taut enough to prevent lifting the wire

FIGURE 49.46   Jumper wire—staking. *(IPC.)*

FIGURE 49.47   Jumper wire installation requirements. *(IPC.)*

above the height of adjacent components. No more than two
jumper wires may be stacked on a given route (Fig. 49.46).

When a jumper wire is attached to leads on the secondary
side of the PCBA or to axial components on the primary side
of the PCBA, it must form a full 180 to 360° loop around the
component lead. When a jumper wire is soldered to other
component package styles, the wire should be lap soldered to
the component lead.

Jumper wires may be installed into a plated-through-hole
with another component lead for Class 1 and 2 equipment;

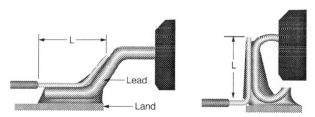

FIGURE 49.48   Jumper wire connection—leaded components. *(IPC.)*

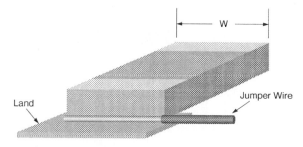

FIGURE 49.49   Jumper wire connection—chip components. *(IPC.)*

however, this is unacceptable for Class 3 equipment. Jumper wires may also be installed into via holes (Fig. 49.47).

For surface-mount components, the minimum length of the lap joint between the wire end and the lead or land must be the length L for leaded components and width W for leadless components (Figs. 49.48 and 49.49).

REFERENCES

1. TR-NWT-000078, *Generic Physical Design Requirements for Telecommunications Products and Equipment,* Bell Communications Research (Bellcore); Red Bank, N.J., issue no. 3, Dec. 1991.

2. ANSI/J-STD-001, *Requirements for Soldered Electrical and Electronic Assemblies,* Electronic Industries Association and Institute for Interconnecting and Packaging Electronic Circuits.

3. MIL-STD-2000, *Standard Requirements for Soldered Electrical and Electronic Assemblies.*

4. ANSI/J-STD-002, *Solderability Tests for Component Leads, Terminations, Lugs, Terminals and Wires,* Electronic Industries Association and Institute for Interconnecting and Packaging Electronic Circuits.

5. ANSI/J-STD-003, *Solderability Tests for Printed Boards,* Electronic Industries Association and Institute for Interconnecting and Packaging Electronic Circuits.

6. IPC-A-610, *Acceptability of Electronic Assemblies,* Institute for Interconnecting and Packaging Electronic Circuits.

CHAPTER 50
ASSEMBLY INSPECTION

Stig Oresjo
Agilent Technologies, Loveland, Colorado

Bruce Bolliger
Agilent Technologies, Singapore

50.1 INTRODUCTION

This chapter covers the various reasons why manufacturers inspect printed circuit assemblies, how they have implemented and enhanced visual inspection, what automated process inspection systems they are using, and how they have implemented these automated systems. The scope of this chapter includes only inspection of printed circuit assemblies during the assembly process, as typically shown in Fig. 50.1. Thus, it includes inspection of solder paste after the paste printing process step, components after the component placement process step, and solder joints after the solder curing process step. Not included, however, is incoming inspection of components and the bare printed circuit board. The focus of this chapter is on production use of inspection, not the collection of measurements during process development in an R&D environment.

50.1.1 Visual Inspection

Manufacturers of printed circuit assemblies have always visually inspected their boards at various points in the assembly process. There are a variety of reasons for visual inspection of assemblies, including eliminating cosmetic defects, detecting obvious process defects quickly, and meeting military specifications. But, with the advent and growth of surface-mount technology, visual inspection of printed circuit assemblies has also grown in importance and prevalence. SMT solder joints must carry a much bigger burden for mechanical or structural reliability than do plated through-hole solder joints. The pin-in-through-hole solder joint carries much of the mechanical burden, helping to keep the component attached to the printed circuit board. But with SMT solder joints, it is often only the solder that keeps the component attached to the board. In many cases, only visual inspection could judge the mechanical reliability of an SMT solder joint.

As SMT geometries have continued to shrink and solder joints have become denser on printed circuit boards, visual inspection has become more difficult, so visual inspection results have become less consistent and reliable. In addition, new types of components, such as pin grid arrays or ball grid arrays, completely hide their solder joints from view. Yet, as Fig. 50.2 indicates, achieving a high assembly process yield is more important as the number of solder joints per assembly increases.

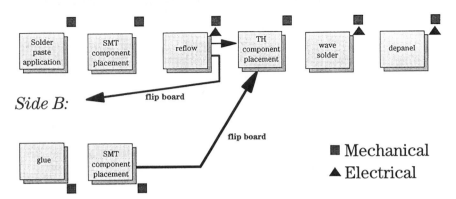

FIGURE 50.1 Generic manufacturing process for surface-mount technology printed circuit assemblies, including the possible locations within the process for inspection or testing of mechanical or structural attributes and electrical characteristics of an assembly.

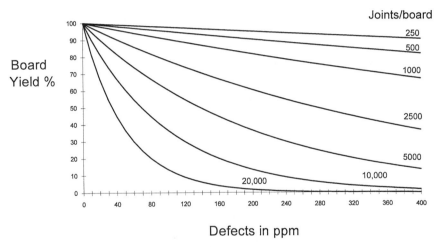

FIGURE 50.2 Printed circuit board yield decreases dramatically with increases in the number of solder joints per assembly if the defect rate per solder joint remains constant. For instance, at a 40-ppm defect rate the yield drops from 96 percent at 1,000 joints per assembly to 45 percent at 20,000 joints.

50.1.2 Automated Inspection

Inspection is an important source of process information without which high yields are very difficult to obtain. Consequently, manufacturers have employed the following range of techniques to either simply enhance visual inspection or fully automate inspection:

- Microscopes, mono- and stereoscopic, with magnification from 4× to 10×
- Real-time video images created using simple light, x-rays, thermal imaging, or acoustics
- Fully automated process test systems using light, laser, or x-ray imaging

The automation of inspection has evolved into systems that resemble automated test equipment used to make electrical measurements and find electrical defects. These automated process test systems acquire real-time images, process the images to find and measure features within the image, and make accept/reject decisions based on this image processing. Thus these automated systems remove the human—and human judgment—from the inspection process altogether.

50.2   REASONS FOR INSPECTION

Manufacturers inspect printed circuit assemblies during production for a variety of reasons. Most of these reasons fall into the following categories:

1. Improvement of process fault coverage
2. Improvement in ability to meet customer specifications
3. Detection of process defects as quickly after they occur as possible
4. Ability to decrease process defect rates through statistical process control (SPC)

50.2.1   Process Fault Coverage

The goal of higher fault coverage is the prevention of any defective printed circuit assemblies from reaching assembly of the board into its final product, whether final assembly is at the same site or at a separate customer site. Figure 50.3 shows a typical spectrum of SMT process defects. Some of these defects, such as misaligned solder joints and solder balls, can be discovered only by inspection, not by electrical, functional, or in-circuit testing. For instance, marginal solder joints are those that pass electrical tests just after assembly completion, but eventually fail because of inferior mechanical strength. Insufficient solder and components partially misaligned with printed circuit pads are the two most common causes of low mechanical strength. Voids in solder joints and cosmetic defects, such as solder balls away from solder joints and "cold" or dull solder, are other examples of defects that can be found only by some kind of inspection.

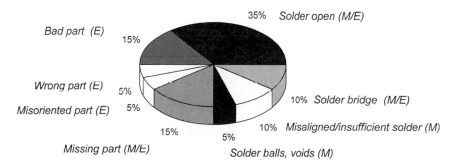

FIGURE 50.3 Typical process fault spectrum for surface-mount assemblies. Process faults marked E can be detected by electrical tests, those marked M can be detected by mechanical or structural inspection or testing, and those marked by E and M can be detected by either method.

50.2.2   Customer Specifications

Many printed circuit assembly workmanship specifications, including MIL-STD-2000A and IPC/EIA J-STD-001C, require inspection of solder joints to ensure conformance to these specifications. For instance, MIL-STD-2000A prescribes, for all defense and aerospace government contractors, the attributes of solder joints to be inspected and describes the conditions that result in nonconformance. Most commercial customers of contract manufacturers also have very specific solder joint and cosmetic specifications that require some type of inspection of printed circuit assemblies. Although process control is theoretically preferred over 100 percent inspection, practically speaking, a majority of printed circuit production lines include some kind of inspection to ensure conformance to specifications. At the very least, customer specifications covering cosmetics normally require visual inspection.

50.2.3 Quick Defect Detection and Correction

Quick detection of process shifts or defects can lower rework costs in several different ways.

50.2.3.1 Preventing Defects. If a particular process step has drifted out of its control limits, discovering this fact as soon as possible will prevent more defects from occurring. If the problem is found quickly enough, and corrected immediately, perhaps no actual defects will occur.

50.2.3.2 Lowering Rework Costs. Defects found earlier in the process are often easier to repair. For instance, if a defect in solder paste deposition is found before a component has been placed in the paste, it is fairly inexpensive to wipe the solder paste off and start over. If this same defect is found after the solder is reflowed, the solder joint itself will have to be touched up, a more difficult and expensive rework step. The same is true for repairing component placement defects before the solder is cured, particularly for missing or misaligned components.

50.2.3.3 Making Defect Diagnosis Easier. Finally, defects found earlier in the assembly process are often more easily diagnosed, shortening the overall repair time. An example is the inspection of solder joints. Inspection of solder joints detects defects specific to the solder joints and quickly determines the exact location and characteristic of the defect. Waiting until a later electrical test stage could make diagnosis more difficult because at that point there could be several causes of the defect, such as a defective component and defects in other connections, in addition to the actual defective solder joint.

50.2.4 Statistical Process Control

Statistical process control requires reliable data that can be analyzed either in real time or historically. Visual inspection collects defect data, such as the number of solder joint defects per assembly right after the solder curing process (either reflow or wave soldering). Some manual and automated inspection techniques also take quantitative measurements of key assembly parameters, such as solder paste volume or solder joint fillet height. To the extent that these data are repeatable, manufacturers use defect data or measurements to characterize the amount of process variation from assembly to assembly or from solder joint to solder joint. When the amount of variation starts to drift outside its normal range or outside its control limits, manufacturers can stop the assembly process until the process is adjusted to eliminate this drift. Historical analysis of the defect or measurement data also helps discover the cause of process variation. Eliminating the sources of process variation reduces the process defect rate, thus creating savings in rework costs and increasing product reliability.

50.3 VISUAL INSPECTION

Visual inspection is the visual comparison by a human of some attribute of the printed circuit assembly with specified standards that describe the acceptable range for that attribute. The inspector normally picks up the printed circuit assembly or places it under a microscope and carefully observes particular attributes, such as the condition or bend radius of component leads or the wetting of solder joints to a lead. Visual inspection always involves human judgment in comparing the attribute to its conditions of conformance to standards.

50.3.1   General Inspection Issues

As Fig. 50.1 indicates, visual inspection can occur after each of the several printed circuit assembly process steps. But visual inspection often has different purposes, depending on where in the assembly process it occurs. These purposes fall into the following three major categories:

1. Quick detection of an assembly process step that is not operating within its normal range
2. Detection of process defects as specified by the customer or industry or internal standards
3. Detection of cosmetic defects

Visual inspection after any process step other than reflow soldering and wave soldering usually has the purpose of quick detection of a process step out of its control limits. (Control limits may, in this case, be based just on operator judgment, not on any statistical analysis of process capability.) This type of visual inspection is usually done by a production line operator as only a small part of his or her normal responsibilities. The visual inspection is short—normally less than 1 min—and consists of a quick scan of the assembly for obvious defects. After solder paste printing, the operator visually inspects for pads with very little solder paste or paste bricks that consistently do not line up with the pads adequately. After component placement, the operator visually inspects for missing components, pins not through holes, components grossly misaligned with the appropriate pads, and a number of other improper conditions. After glue deposition, the operator inspects for pads without glue dots or with excess glue and for glue dots that are not aligned with the pads. If any of the defective conditions occurs more than is normal for that process step, the operator typically stops the equipment for that process step and readjusts the equipment or notifies production engineering or management.

50.3.2   Solder Joint Inspection Issues

Visual inspection after the reflow and wave-soldering process steps can also be just a quick scan for obvious defects to detect a process condition outside of control limits. In this case, the operator visually inspects for solder bridges, large solder balls or solder splashes, lifted leads, and a number of other improper conditions. Often, however, visual inspection after the soldering process steps is aimed at finding solder joints that do not meet specifications.

Visual inspection of solder joints against specification often covers 100 percent of the solder joints on an assembly. Inspecting samples of solder joints in conjunction with a documented process control system is also done, but 100 percent inspection is more common. So inspecting for this purpose can be a lengthy process, taking as long as a half hour for assemblies with 4000 solder joints being measured against military specifications. As a rule of thumb, this kind of visual inspection has a throughput of about five joints per second. Thus, inspection of solder joints against specifications normally requires visual inspectors dedicated to this function with no other responsibilities.

The visual inspector must be very familiar with the specifications of the attributes for each solder joint type. Each solder joint can have as many as eight different criteria for defects, and each assembly typically has more than six or so different solder joint types corresponding to different component packages. As an example, Fig. 50.4 shows the specification attributes for one component type (rectangular passive chips). Table 50.1 gives the corresponding conformance criteria for each attribute for this component type. More importantly, the visual inspector must be highly trained to make an accurate judgment of conditions on the borderline between good

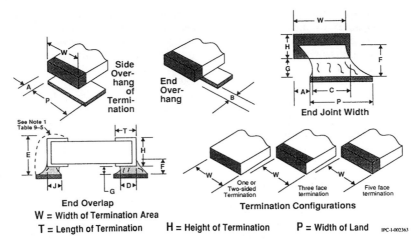

FIGURE 50.4 Inspection attributes for solder joints of surface-mounted rectangular passive chips.

TABLE 50.1 Dimensional Criteria—Rectangular or Square End Components (Dimensions in Millimeters)

Feature	Dim.	Class 1	Class 2	Class 3
Maximum side overhang	A	50% (W) or 50% (P), whichever is less; Note 1	50% (W) or 50% (P), whichever is less; Note 1	25% (W) or 25% (P), whichever is less; Note 1
Maximum end overhang	B	Not permitted	Not permitted	Not permitted
Minimum end joint width	C	50% (W) or 50% (P), whichever is less	50% (W) or 50% (P), whichever is less	75% (W) or 75% (P), whichever is less
Minimum side joint length	D	Note 3	Note 3	Note 3
Maximum fillet height	E	Note 4	Note 4	Note 4
Minimum fillet height	F	Note 3	Note 3	(G) + 25% (H) or (G) + 0.5 mm (0.02 in), whichever is less
Solder fillet thickness	G	Note 3	Note 3	Note 3
Height of termination	H	Note 2	Note 2	Note 2
Minimum end overlap	J	Required	Required	Required
Width of land	P	Note 2	Note 2	Note 2
Width of termination	W	Note 2	Note 2	Note 2

Note 1. Shall not violate minimum electrical clearance.
Note 2. Unspecified parameter or variable in size as determined by design.
Note 3. Properly wetted fillet shall be evident.
Note 4. The maximum fillet may overhang the land or extend onto the top of the end cap metallization; however, the solder shall not extend further onto the component body.

and bad. For instance, accurately determining whether a fillet height is one-quarter of the way up a component side that is only 0.05 mm high to begin with takes a lot of practice. Visual inspectors typically do not use any tools to help make these judgments. Rulers or calipers are very difficult or impossible to use to measure solder joint dimensions or thicknesses. Using reticules in microscopes in conjunction with coordinate-measuring machines is possible, but is usually much too time consuming to be done on a regular basis.

Visual inspection for cosmetic defects often occurs at the end of the assembly process, or even after all of the electrical tests on the assembly have occurred. The visual inspector looks for scratches, partial delaminations, solder splashes or solder balls away from solder joints, and any other condition that does not affect assembly performance but does make the assembly look like it may not be a high-quality product.

50.3.3   Standards for Visual Inspection

Many standards cover printed circuit board assemblies. Most major electronics manufacturers have their own internally developed workmanship standards. Several industry and military standards also exist. However, the joint industry standard IPC/EIA J-STD-001C, Requirements for Soldered Electrical and Electronic Assemblies, is the standard most often referenced for criteria defining reliable solder connections. This standard was jointly developed by the Institute for Interconnecting and Packaging Electronic Circuits (IPC) and the Electronic Industries Association (EIA). The IPC is headquartered in Lincolnwood, Illinois, and the EIN's engineering department is headquartered in Washington, DC. The IPC/EIA J-STD-001 was first approved by ANSI on July 1, 1992. Revision C was released in March 2000.

50.3.3.1   IPC/EIA J-STD-001 Requirements for Solder Joints.   The IPC/EIA J-STD-001C standard describes materials, methods, and verification criteria for producing quality solder connections on printed circuit assemblies. It covers criteria and methods for both pin-through-hole and surface-mount technology solder connections. It also reflects requirements for three different classes of end products.

Class 1, general electronic products: includes consumer products and some computers and computer peripherals

Class 2, dedicated service electronic products: includes communications equipment, critical business machines, and instruments where high performance is required and for which uninterrupted service is desirable

Class 3, high-performance electronic products: includes commercial and military equipment where continued performance or performance on demand is imperative

The Class 3 requirements of the IPC/EIA J-STD-001C, except in isolated instances, match the requirements found in the MIL-STD-2000A covering the same specification areas. Table 50.1, which is taken from the IPC/EIA J-STD-001 standard, shows specifications for one type of surface-mount solder joint for all three classes of end products.

50.3.3.2   Supplementary IPC Standards.   The IPC/EIA J-STD-001C standards can be augmented for use in printed circuit assembly by the following standards:

IPC-A-610C, Acceptability of Printed Board Assemblies: for overall workmanship requirements (released January 2000)

IPC-9191, General Guidelines for Implementation of Statistical Process Control (SPC): for establishing a process control plan (released November 1999)

50.3.4   Capabilities of Visual Inspection

Visual inspection serves a number of important purposes well, but it also has several important limitations.

50.3.4.1 Advantages. Visual inspection is still usually the only method of reliably detecting cosmetic defects. Visual inspection can be an accurate and cost-effective method of defect detection for nonsubjective defects, such as larger missing components, larger misoriented or reversed components, or solder bridges, which are clearly either defective or not. And when low volume or lack of technical resources prevents the use of automation or more sophisticated tools, visual inspection is one alternative that finds many of the more subjective defects, such as solder joints with insufficient solder, lifted leads, or poor wetting. Finally, visual inspection can quickly detect when a process step has drifted significantly out of its control limits.

50.3.4.2 Disadvantages. Visual inspection also has many limitations, including:

- Low rate of solder joint defect detection repeatability, particularly for fine-pitch SMT, which results in high false accepts or defect escapes and high false reject rates
- Low rate of component defect detection repeatability for smaller components such as 04-02 or 02-01 passive components
- Inability to see hidden solder joints in component types such as some connectors, pin grid arrays, and ball grid arrays
- Inability to collect quantitative measurements in addition to defect data

Repeatability Limitations. Several studies have documented the low repeatability rate of visual inspection of solder joints. One such study was conducted by AT&T at its Federal Systems Division.[1] This study showed that even the same inspector inspecting the same assembly twice had a defect call repeatability rate of only about 50 percent. Two different inspectors inspecting the same assembly had a defect call repeatability rate of only about 28 percent. This study did not include any very-fine-pitch SMT solder joints or 06-03 passive components, which are more difficult to visually inspect.

To alleviate this severe limitation somewhat, manufacturers have implemented the use of microscopes with a magnification level of 10×. Often, stereoscopic microscopes are used to provide visual inspectors a better three-dimensional view. Not as frequently, manufacturers have implemented light sources and cameras to capture real-time magnified video images of the assembly being inspected. The lighting and high-resolution video images can make it easier for visual inspectors to see the acceptance criteria for which they are searching. These enhancements do improve the repeatability rate of visual inspection. But the requirement for subjective human calls and the tedium of carefully inspecting thousands of connections per hour still results in repeatability rates much lower than desired. Low repeatability means many missed defects that escape to final assembly or to customers, and wasted and possibly damaging rework of good connections.

Hidden Solder Joints. Several component types used in printed circuit assemblies do not provide visual access to solder joints. These components include packages with bump arrays where the solder joints are distributed in a matrix under the entire component body, such as ball grid arrays. For these components, all of the joints except those on the edges of the array are completely hidden from view. Other packages, such as J-leaded components, where the connections are under the component body at the component edges, and 0.5-mm-pitch gull-wing components, where the solder joint heel is only 0.08 mm high and behind the component leads, also make visual inspection more difficult. Some manufacturers have attempted to solve this problem by using a penetrating imaging technique, such as x-ray, acoustics, or thermography, to acquire a live, magnified video image that the visual inspector can look at to make a defect call. However, the images resulting from these techniques are either less consistent or less well defined than a visual image, still requiring the inspector to make a difficult judgment call. Therefore, a low repeatability rate persists.

High-Complexity Boards. Many printed circuit boards today are entirely too complex to allow accurate visual inspection. Components as small as 04-02 or even 02-01 passive components are simply too small for accurate visual detection of missing parts and particularly misalignment defects. Add to this the fact that many printed circuit boards have hundreds of these component types per board, and the opportunity for the visual inspector to make a mistake becomes enormous. A further obstacle to accurate inspection are "families" of boards, where the basic layout of each board type is the same, but specific board types vary by purposely not mounting specific components. In these cases, a missing component can be completely acceptable even in places where the board layout may provide for a component to be mounted.

No Quantitative Measurements.   Accurate quantitative measurements of dimensions ranging from 0.05 to 0.5 mm are not possible with visual inspection. Quantitative measurements provide much more information about a process step, allowing much tighter process control and providing insight into the causes of process variation, without which process improvement is a hit-or-miss proposition. Some manufacturers have implemented semiautomatic measurement tools to allow operators to take quantitative measurements. Examples include optical-focusing microscopes and semiautomatic laser triangulation equipment. These tools do collect useful quantitative measurements, but are typically limited to sampling solder paste depositions. Sampling is required because the tools are very slow. Measurement of solder paste height, volume, and registration with pads is possible because of easy visual access to solder paste depositions as well as their simple rectangular shape.

Automated inspection techniques using automated process test equipment significantly overcome the repeatability and measurement limitations of visual inspection. It is for this reason that many manufacturers are implementing automated inspection.

50.4 AUTOMATED INSPECTION

The automation of inspection has evolved into automated process test systems that resemble automated test equipment used to make electrical measurements and find electrical defects. Automated process test systems generate images of the item to be inspected (normally solder paste, components, or solder joints), digitally analyze the image to locate and measure key features, and, based on these measurements, automatically decide whether a defect exists or not. Just like visual inspection, automated process test systems do not require physical contact with the printed circuit assembly to generate the desired images. Unlike visual inspection, automatic inspection removes human subjectivity from defect detection, thereby increasing repeatability rates typically by an order of magnitude. Many of the automated process test systems also provide accurate, repeatable, quantitative measurements that directly correspond to process parameters, thus providing the means for process control and improvement.[2]

50.4.1 Measurements by Automated Systems

Figures 50.5(*a*) through 50.5(*c*) show examples of measurements made by automated process test systems.

50.4.1.1 Solder Paste Measurements.   Typical solder paste measurements, shown in Fig. 50.5(*a*), are volume, area of pad covered, height, and misalignment with the pad. These quantitative measurements provide information about the paste viscosity, stencil registration, cleanliness, snap-off, and squeegee speed and pressure that can lead to improvement in the paste printing process.

50.4.1.2 Component Placement Measurements.   Typical component placement measurements, depicted in Fig. 50.5(*b*), include whether or not the component is missing, misaligned, or skewed. These measurements are usually attribute measurements, just separating good from bad. But quantitative dimensional measurement of the amount of misalignment, just after the placement process step, is also possible. Component placement measurements provide information about the rate of placement accuracy.

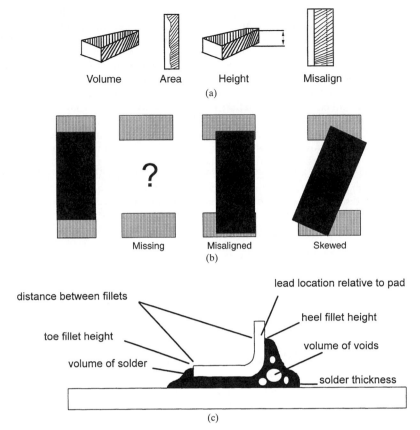

FIGURE 50.5 (*a*) Typical solder paste measurements made by automated process test systems, including the volume of paste deposition, area of the pad covered by paste, height of the paste deposition, and offset of the paste from the pad. (*b*) Typical component placement attributes inspected by automated process test systems, including missing component, components misaligned along or across the pads, and skewed components. (*c*) Typical solder joint measurements made by automated process test systems, including volume of solder, toe and heel fillet heights, distance between fillets, void volume, average solder thickness across the solder joint, and offset between solder joint and pad.

50.4.1.3   Solder Joint Measurements.   Solder joint measurements, such as fillet heights, average solder thickness across the pad, void volume, and pin-to-pad offsets, as shown in Fig. 50.5(*c*), provide information about the paste printing process, the component placement process, and the solder curing process steps. Attribute measurements, such as solder bridges, opens, or insufficient solder, are most common. But quantitative measurements of solder joints are also possible and provide greater insight into process parameters. For instance, the variation of heel fillet heights and void volume across an assembly provides insight into the oven temperature profile and spatial distribution. The variation of average solder thickness provides insight into squeegee speed and pressure and stencil cleanliness. And the pin-to-pad offsets measure the component placement accuracy.

50.4.2 Types of Automated Process Test Systems

Automated process test systems normally are dedicated to one type of measurement capability: solder paste, component placement, or solder joint. For example, systems for solder paste measurements do not normally also make component placement measurements. The cost of combining different measurement capabilities into one system would typically make that system prohibitively expensive. More importantly, to reduce manufacturing costs, manufacturers want to implement linear, sequential production lines where an assembly always flows in one direction and goes through each machine only once per assembly side. So automated process test systems fall into three major categories:

1. Those that make three-dimensional solder paste measurements
2. Those that make quantitative two-dimensional solder paste and/or component placement measurements
3. Those that make solder joint measurements and attribute component measurements

For all three types, the automated process test system compares the measurements taken against a specified conformance range to automatically accept or reject a solder paste brick, component placement, or solder joint as being within specification.

50.5 THREE-DIMENSIONAL SOLDER PASTE AUTOMATED PROCESS TEST SYSTEMS

50.5.1 Operating Principles

Automated process test systems designed for three-dimensional (3-D) solder paste measurements use structured light sources and cameras or digital LED sensors to generate "three-dimensional images" of solder paste depositions. The structured light is normally some variation of a sheet of light or a laser line or spot. As Fig. 50.6 shows, the structured light sweeps across a

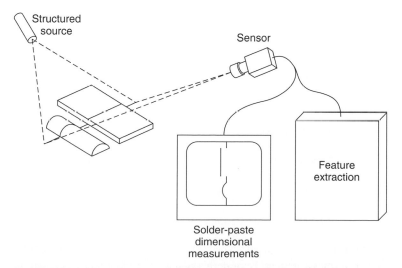

FIGURE 50.6 Schematic of automated process test system for solder paste measurement. The camera or LED sensor obtains images with discontinuities in the laser line scan; the image processing software finds these discontinuities, measures them, and calibrates them to real physical dimensions.

solder paste brick, creating discontinuities and other features in the image that are calibrated to actual physical dimensions, such as height and area.

50.5.1.1 View Magnification. To achieve adequate accuracy, only a small portion of the assembly, called a *view,* is imaged at a single time. Magnification increases and digital quantization error decreases as the size of the view decreases. Views normally vary in size from 10 to 25 mm in diameter. So the automated process test system may use more than 100 views to cover a typical assembly. The system moves the assembly or the image sensor on a positioning stage from view to view. The sum of move time plus image acquisition time normally makes up most of the total inspection time for a particular assembly.

50.5.1.2 System Throughput. The test speed for solder paste inspection systems varies anywhere from 2 to 15 cm²/s. The high end of this speed range can be fast enough to keep up with many automated printed circuit production lines. But for the slower systems, or even for the fastest systems on very high-volume production lines, manufacturers inspect a specific subset of paste depositions on every assembly coming down the production line.

50.5.2 Applications

Although for the fastest systems defects on all of the depositions per assembly can be detected, the speed of most systems does not allow defect detection to be their main purpose. Instead, manufacturers use these systems to monitor the process by tracking the key quantitative solder paste measurements, normally height and volume, against control limits. The systems then alarm any condition where control limits are exceeded or definite drifts are detected, allowing the manufacturer to shut down the production line until appropriate adjustments have been made. (The alarm generated by the system is normally an obvious flashing message on a computer monitor.) Because the paste printing process step causes a large percentage of all production defects, tight production control of this process step using quantitative measurements can significantly lower the process defect rate. In fact, automated solder paste inspection systems have been linked through software to paste printing systems, allowing not only process control, but also closed-loop feedback for semiautomatic adjustment of paste printing parameters based on the paste measurements.

50.5.3 Advantages and Disadvantages

Automated 3-D inspection of solder paste depositions has the following major advantages:

- Real-time process control of paste printing to lower defect rates and rework costs
- Quantitative measurements to help permanently eliminate the causes of paste defects
- Defect detection where rework is easiest, before component placement and solder cure

Automated inspection of solder paste depositions has the following limitations:

- It is often too slow to cover all paste depositions on an assembly.
- It does not measure or detect defects in component placement or solder curing.

50.6   *TWO-DIMENSIONAL COMPONENT PLACEMENT AUTOMATED PROCESS TEST SYSTEMS*

50.6.1   Operating Principles

Automated process test systems designed for two-dimensional (2-D) component placement and solder paste quantitative measurements typically use optical techniques consisting of multiangle light sources and CCD cameras to generate two-dimensional images. As Fig. 50.7 indicates, these systems extract specific features, such as edges of the components or solder paste deposits, in the images and then use these features to determine quantitative measurements of component and deposit misalignment or deposit area.

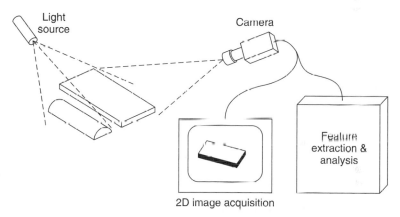

FIGURE 50.7   Schematic of automated process test systems for component placement defect detection. The camera sensor obtains images of component positions relative to the printed circuit board: the image processing software extracts features from the image and compares them to present position limits to flag a placement as defective.

Optical automated process test systems also image only a small portion, or *view*, of the assembly at a time. These systems normally can use somewhat bigger views than the 3-D solder paste systems because the features being extracted often do not require as much magnification as do measurements of solder paste depositions. However, inspection of small components, such as 04-02 and 02-01 passive components, or very-fine-pitch solder-paste deposits, can require the same level of magnification and therefore views as small as the 3-D systems. In general, 2-D systems are typically two to three times faster than 3-D systems and range in inspection speed between 10 and 20 cm^2/s. The prices of 2-D component and solder paste placement automated process test systems are typically somewhat less than those of the 3-D solder paste systems with the fastest inspection speed capability.

50.6.2   Applications

The majority of 2-D component placement systems are placed directly following the pick-and-place process step in the production line. These systems therefore only measure the amount of misalignment of components and are generally fast enough to cover all components on a printed circuit board and keep up with the production line cycle time. So manufacturers often use these systems to detect all misalignment defects as well as missing components. But the most important purpose of these systems is process control, just as it is for the 3-D systems. The intent is to discover process drift early enough to prevent defects by tracking the key quantitative misalignment measurements against control limits. The systems then alarm any condition where control limits are exceeded or process drifts are detected, allowing the manufacturer to shut down the production line until appropriate adjustments have been made. Some system suppliers are currently working with pick-and-place equipment suppliers to develop closed-loop feedback software links allowing semiautomatic adjustment of the pick-and-place equipment.

The automated inspection of solder paste deposits using these 2-D measurement systems is a relatively new development. Normally, inspection of only a small percentage of the solder paste deposits is combined with the component misalignment measurements. Component misalignment measurements cover the passive components, while the solder paste measurements cover deposits for ball grid array (BGA), chip-scale package (CSP), or fine-pitch quad flat pack (QFP) devices. Therefore, these systems are placed within production lines after the pick-and-place systems for passive devices but before the pick-and-place systems for the larger area array and leaded devices. These systems serve the same purpose as those only meant for component placement measurement, both detecting defects and monitoring measurements within control limits to discover process drift as early as possible.

Another recent variation of these 2-D systems is the addition of relatively slow 3-D measurement capability. With these systems, all solder paste deposits are measured for the standard 2-D measurements of misalignment and area. In addition, a small percentage (normally around 10 percent) of the solder paste deposits are also measured for height. Manufacturers place these systems within production lines right after solder paste printing; thus no component measurements are made. The advantage of these systems is the capability of 100 percent 2-D measurement coverage and good process control with a sample of 3-D measurements, at a lower price than for the fastest 3-D systems.

50.6.3 Advantages and Disadvantages

Automated inspection of 2-D component placement has the following major advantages:

- Real-time detection of systematic process defects where equipment is out of adjustment
- Detection of 2-D defects such as misalignment where rework is easier, before solder cure

Automated inspection of component placement has the following limitations:

- It does not measure or detect defects in solder curing, and normally does not detect defects in solder printing.

50.7 SOLDER JOINT AUTOMATED PROCESS TEST SYSTEMS

Solder joints have much more complex shapes than do solder paste depositions and components, so taking measurements of solder joints normally requires more complex imaging techniques than does measuring solder paste and components. Automated process test systems for solder joints have used a variety of imaging technologies, including optical and x-ray imaging, thermography, cooling profiles of laser-heated solder joints, and ultrasonic imaging. But three technologies have dominated in these systems:

1. Optical imaging using multiple light sources and cameras
2. Transmission x-ray imaging
3. Cross-sectional x-ray imaging

50.7.1 Optical Imaging Systems

50.7.1.1 Operating Principles. Optical automated process test systems for solder joints are similar to, but more complex than, those for component placement. (Note that these systems also detect component defects such as missing, misaligned, or misoriented components, but their complex design is required for solder joint inspection.) Multicolor, multiangled light sources are normally required to provide enough information in the images to allow sophisticated image processing algorithms to accurately detect the required features in the images. If the light source is not multicolored or multiangled, then many different cameras are usually used instead, each camera mounted at a different angle relative to the board being inspected. The multiple light sources and cameras create and detect shadows from various angles to detect features of solder joints of all types oriented in different directions on the printed circuit assembly. These optical systems often use higher magnification, particularly for the smallest passive components and fine-pitch components, and therefore capture a smaller portion of the assembly at any one time. The smaller views allow more image pixels per solder joint feature, allowing more accurate image processing and corresponding defect calls. The use of smaller views slows the throughput of these systems when compared to component placement inspection systems. However, in general these systems are faster than 3-D solder paste measurement systems. Their throughput typically falls in the range of 10 to 15 cm²/s.

The systems' image processing software uses sophisticated algorithms to extract specific features of solder joints, such as edges and areas of the solder joint in a specific range of angles relative to the board. Analysis of these extracted features then determines whether a defect exists or not. Defects detected include absence of solder, bridges, and grossly insufficient or excessive solder. These systems typically do not make quantitative measurements.

50.7.1.2 Application. These solder joint optical inspection systems generate only attribute data. For instance, these systems detect the existence of a solder bridge between two joints or the absence or presence of a toe fillet on solder joints. But they do not measure the height of the heel fillet or the amount of solder in the solder joint. These systems typically do not measure how far a component is misaligned from its proper placement, but instead simply determine whether or not the component is misaligned more than a predetermined amount. These attribute data are not as useful for process control, and thus manufacturers use these systems strictly to detect defects. Normally these systems do, however, alarm a condition where the same defect has occurred several times consecutively or within a specific number of assemblies, indicating that some part of the process needs adjustment.

Optical systems cannot inspect hidden solder joints, such as those for ball grid arrays, pin grid arrays, and in some cases J-leaded devices, and can have high false accept or reject rates for some component types, such as fine-pitch components at or below 0.5-mm pitch or SOT devices. If tall components are placed on the board very close to smaller components, optical systems can also have difficulty inspecting these smaller components.

50.7.1.3 Advantages and Disadvantages. Automated optical inspection of solder joints and component attributes has the following advantages:

- It eliminates visual inspection by automating solder joint defect detection, thereby also reducing unnecessary rework due to false reject calls.
- It reduces rework analysis time by pinpointing defects to the exact solder joint.
- It affords real-time process control of all three process steps—paste printing, component placement, and solder cure—to lower defect rates and rework costs.

Automated optical inspection of solder joints has the following limitations:

- Test throughput is not always fast enough to allow inspection of all solder joints within the manufacturing cycle time for the printed circuit assembly.

- A significant learning curve is required to become expert at developing solder joint tests with both low false accept and false reject rates.

50.7.2 Transmission X-ray Systems

50.7.2.1 Operating Principles. Transmission x-ray systems radiate x-rays from a point source perpendicularly through the printed circuit assembly being inspected, as depicted in Fig. 50.8. An x-ray detector picks up a varying amount of x-rays depending on the thickness of metals that the x-rays are penetrating and converts the x-rays to light photons for a camera to create a grayscale image. The x-ray source is filtered so that metals of only a certain density range, i.e., lead, tin, gold, and silver, will absorb the x-rays. The copper leads and frames of components sitting on top of solder joints do not absorb the x-rays and are therefore practically invisible to the x-ray detector. Thus, x-ray systems can easily see the entire solder joint, no matter what component material may be on top of the joint blocking its optical or visual access. The resulting x-ray image will be darker wherever the lead/tin solder is thicker in the solder joint. The image processing capability of the system then searches for features, such as the heel and toe fillets, the sides of the solder joint, and even voids internal to the joint based on grayscale readings of the solder joint x-ray image. The system then uses predetermined decision rules to compare the grayscale readings to acceptance criteria to automatically accept or reject a solder joint. For example, the system might compare the relative grayscale reading for the heel fillet region, the center of the solder joint, and the toe fillet region. The acceptance criteria might state that the heel fillet reading should be twice that of the center and that the toe fillet reading should be 50 percent higher than that of the center. If the actual readings do not meet these criteria, then the solder joint is reported as being defective.

The bottom of Fig. 50.8 shows an x-ray image of a gull-wing solder joint that shows the center of the joint as much darker than the heel fillet region. This solder joint is clearly defective,

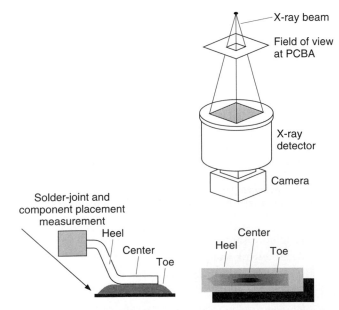

FIGURE 50.8 Schematic of transmission x-ray automated process test system for solder joint defect detection. The x-ray detector converts a varying amount of x-rays to light, based on how much various parts of the solder joint absorb: the camera converts the light photons to an image, which is then processed to find solder joint features and flag defects accordingly.

as the heel fillet region should always be darker and have a higher grayscale reading than the center of the joint, where the solder is thinnest for mechanically good solder joints. (The system's image processing capability is able to detect much more subtle changes in grayscales than can the human eye, allowing very accurate relative readings from one solder joint to the next.)

50.7.2.2 Application.   Transmission x-ray technology works well for single-sided surface-mount assemblies. These automated process test systems accurately detect solder joint defects such as opens, insufficient solder, excess solder, bridges, misalignment between pin and pad, and voids for most surface-mount solder joint types, including J-leads, gull-wings, passive chips, and small-outline transistors. These systems also detect missing components and reversed tantalum capacitors. Based on trends in grayscale reading, these systems also can accurately detect process drifts through real-time process control charting.

For double-sided assemblies, however, the transmission x-ray images of solder joints on the top side will overlap with the images of solder joints on the bottom side. The x-rays are absorbed by any solder in their path through the printed circuit assembly from the source to the detector. These overlapping images make accurate solder joint measurement impossible. Transmission x-ray imaging also cannot easily distinguish between the top, bottom, and barrel of plated through-hole (PTH) solder joints or the bottom and ball of ball grid array (BGA) solder joints. So transmission x-ray systems cannot be used for accurate measurement of and defect detection in solder joints on double-sided assemblies or for PTH and BGA solder joints.

50.7.3   Cross-Sectional X-ray Systems

50.7.3.1 Operating Principles.   Cross-sectional x-ray systems radiate x-rays at an acute angle from vertical through the printed circuit assembly being inspected. As Fig. 50.9 indicates, images from all around the particular view being inspected are added together or integrated to essentially create an x-ray focal plane in space. This focal plane creates a cross-sectional image, approximately 0.2 to 0.4 mm in thickness, right at the focal plane by blurring everything above and below the focal plane into the background, or *noise*, of the image. By moving the top side of an assembly into the focal plane, cross-sectional images of only the solder joints on the top side are created. By moving the bottom side of an assembly into the focal plane, cross-sectional images of only the solder joints on the bottom side are created. Separate images of top and bottom sides are always created, preventing any image overlap from the two sides.

50.7.3.2 Application.   Cross-sectional x-ray automated process test systems work well for all types of printed circuit assemblies, including single-sided and double-sided, surface-mount, through-hole, and mixed-technology assemblies. These systems accurately detect the same solder joint and component defects as do transmission x-ray systems, but, in addition, the cross-sectional x-ray systems accurately detect insufficient solder conditions for ball grid array and pin-through-hole solder joints.

Some cross-sectional x-ray automated process test systems go beyond just grayscale readings of specific solder joint features. By carefully calibrating grayscale readings to actual solder thickness, it is possible to generate real-world measurements, in physical units rather than grayscale numbers, of fillet heights, solder and void volume, and average solder thickness for the entire joint. Figure 50.10 shows an example of these calibrated measurements. This figure includes the actual cross-sectional x-ray image of tape automated bonded (TAB) solder joints. The profile shown at the top of the x-ray image is generated by the system in physical dimensional units by interpreting and calibrating the grayscale readings of pin 193 in the x-ray image. The table below the x-ray image includes example measurements for both pin 193 and pin 194.

Analysis of these physical thickness measurements of solder joints provides the information required for process characterization and improvement. For instance, variations in average solder thickness or volume for the solder joints across a single assembly or from assembly to assembly provide insight into the quality level of the paste printing process as well as sources of defects.

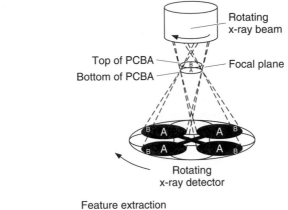

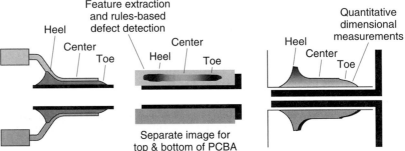

FIGURE 50.9 Schematic of cross-sectional x-ray automated process test system for solder joint measurement. Adding images around a circle from a rotating x-ray beam and detector creates a focal plane that captures just the solder joints of interest, nothing below or above. The image processing software then finds and measures solder joint features and flags defects accordingly.

50.7.4 Advantages and Disadvantages of X-ray Inspection

X-ray solder joint inspection systems can reach average inspection speeds of around 80 to 120 joints per second. X-ray solder joint inspection systems also have higher prices, typically about 50 to 100 percent more than the price of the optical solder joint systems with the fastest inspection speed capability.

Automated x-ray inspection of solder joints has the following major advantages:

- Its defect detection capability is extremely high.
- It eliminates visual inspection by automating solder joint defect detection, thereby also reducing unnecessary rework due to false reject calls.
- It reduces rework analysis time by pinpointing defects to the exact solder joint.
- It affords real-time process control of all three process steps—paste printing, component placement, and solder cure—to lower defect rates and rework costs.
- It provides quantitative measurements to help permanently eliminate the causes of defects from all three process steps.
- It reduces failures at final assembly and in the field due to defective hidden solder joints and marginal solder joints due to insufficient solder, misalignment, or excessive voids.
- It can also be used when lead-free solder becomes more common.

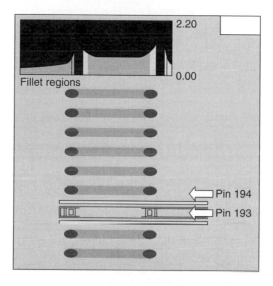

Good board (pin 6)

Reference designator	Inspection point	Thickness (in 0.001")
U1 pin 193	Pad	0.59
	Heel	1.18
	Center	0.69
	Toe	1.34
U1 pin 194	Pad	0.58
	Heel	1.20
	Center	0.68
	Toe	1.30

FIGURE 50.10 Cross-sectional x-ray image of tape automated bond (TAB) solder joints. Image processing software converts the grayscale readings of the image for pin 193 into the side profile of solder thickness shown above the image. The actual calibrated measurements of average solder thickness across the pad, heel fillet height, center thickness, and toe fillet height processed from the images of pins 193 and 194 are shown in the table below the x-ray image. These measurements indicate that both of these solder joints are good.

Automated inspection of solder joints has the following limitations:

- Test throughput is not always fast enough to inspect all solder joints within the manufacturing cycle time for the printed circuit assembly.
- A significant learning curve is required to become expert at developing solder joint tests with both low false accept and false reject rates.

50.8 IMPLEMENTATION OF AUTOMATED PROCESS TEST SYSTEMS

Successful implementation of automated process test systems into printed circuit assembly production lines requires a significant investment in training, process analysis, and system integration. The implementation can be a lengthy process that requires concerted effort by engineers or skilled technicians. Listed here are highlights of what several manufacturers have learned are key aspects of successfully implementing automated inspection systems.

1. *Assess requirements carefully.* Start by carefully assessing the requirements for automated process testing in the particular production environment into which the system will be integrated. Determine exactly what kinds of defects are most important for the inspection system to detect, what measurements will most help with process improvement, and what benefits will generate the quickest financial return on investment.[3] This assessment must consider the testing and measurement capability that already has been implemented as well as new requirements arising from future printed circuit assembly designs.

2. *Evaluate a select set of systems thoroughly.* Select a small number of automated process test systems to evaluate thoroughly and compare against the system requirements. The evaluation should include a benchmark using printed circuit assemblies from production to determine the system's capabilities to accurately detect the important defect types within the required false reject rate, repeatedly make the required measurements, and not exceed the required test time. Elements of cost of ownership should be well understood, including test development time, maintenance skills and cost, expected system downtime, and supplier maintenance services and prices.

3. *Plan for factory system interfaces.* Consider and plan carefully for interfaces to other factory systems. These systems include board-handling equipment, bar code-reading systems, CAD systems for automatic download of board layout and component package information, and quality data management systems for statistical process control (SPC) and historical quality tracking.

4. *Focus on SPC measurements.* Start with a focus on SPC measurements instead of defect detection. Until the process variation is reduced, most manufacturers will encounter either a false reject rate or a false accept rate that is higher than desired. Allowing one rate or the other to be too high while focusing on reducing the process variation first avoids time-consuming, unproductive tweaking of acceptance thresholds. Reducing process variation requires correlating measurements to the process parameters causing the variation and defects, and then properly adjusting these process parameters.[4]

5. *Define defects carefully.* With an understanding of the selected system's capability, carefully define the defects that must be detected for product quality and reliability. Many of the visual inspection criteria used in the past are not appropriate for automated inspection systems because the system takes objective and different measurements.

6. *Invest enough resources.* Do not underestimate the initial resource investment required to obtain optimum benefit from an automated process test system. The implementation plan should include dedicated technical support for the first 6 months of operation and test development. Developing a thorough understanding of the measurement results and correlating the data to process parameters is key to successful use of the system. Implementation should also address the fact that production personnel will have to be convinced of the accuracy of the system's test results before full benefit can be obtained from the system.

50.9 DESIGN IMPLICATIONS OF AUTOMATED PROCESS TEST SYSTEMS

Automated process test systems in general do not require many changes or limitations in the design of printed circuit assemblies. Because these systems use noncontact measurement techniques, fixturing requirements presents very few design limitations, for instance. However, the following requirements will facilitate automatic inspection if they are considered during printed circuit design.

50.9.1 Automated Board Handling Requirements

- Parallel edges of the assembly or panel that have adequate clearance (typically at least 3 mm), to allow board handling clamps or belts to grab the assembly.
- Alignment fiducials on three corners of the assembly or panel. (For fiducials to be useful for x-ray systems, they should be tinned with solder at least 12.5 μm in thickness.)
- Extra alignment fiducials near components with lead pitch of less than 0.5 mm.
- Bar code identification of assembly number and serial number at a predefined location on each printed circuit assembly.
- Adequate board rigidity without a fixture to prevent excessive board vibration during movement. Panels with prerouted breakaway boards or bare boards less than 30 mil in thickness present the biggest challenge.
- Component, heat sink, or daughter board height above or below the bare board that does not exceed the height clearance of the targeted automated process test system.

50.9.2 Test Development Ease of Use Requirements

- As few suppliers as possible (ideally one) for each component type. Variation in lead and component package dimensions from supplier to supplier for the same component forces longer and more difficult development of inspection routines for each printed circuit assembly type.
- Uniform pad shapes and sizes, particularly for each component package type. Variation in pad size and shape within a component package type forces longer and more difficult development of inspection routines for each printed circuit assembly type.
- Clearly visible solder joints for optical and structured light automated process test systems.
- No components opposite or under dense structures, such as transformers, large capacitors, or thick steel heat sinks, for transmission x-ray automated process test systems.
- No silk-screened outlines around components. Although these may be useful for visual inspection, they just confuse automated optical inspection systems.

REFERENCES

1. A. J. Donnel et al., "Visual Soldering Inspection Inconsistencies—Interpretation of MIL-SPEC Visual Acceptance Criteria," AT&T Bell Laboratories, 1988.
2. Michael Lancaster, "Six Sigma in Contract Manufacturing," *Proceedings of Surface Mount International Conference,* San Jose, CA, 1991.
3. Dennis L. Baird, "Using 3D X-ray Inspection for Process Improvements," *Proceedings of Nepcon West Conference,* Hughes Aircraft Company, Cahners Publishing, Des Plaines, IL, 1993.
4. Thilo Sack, "Implementation Strategy for an Automated X-ray Inspection Machine," *Proceedings of Nepcon West Conference,* IBM Corporation, Cahners Publishing, Des Plaines, IL, 1991.

CHAPTER 51
DESIGN FOR TESTING

Kenneth P. Parker
Agilent Technologies, Loveland, Colorado

51.1   INTRODUCTION

In the latter part of the 1970s, it became clear that forces of technology were causing an evolution in board complexity that was quickly outstripping testing technology. It is quite possible to design boards that essentially cannot be tested in an economical sense. This will doom projects and products to failure.

The author has personal experience here, having consulted in the mid-1970s with a Silicon Valley start-up company that had created one of the first dedicated word processors. This product was based on a customized processor that had been designed and sent into production with absolutely no thought about how it was to be produced in volume. There was no test strategy. When it was determined that volume production was impossible, the design team was brought onto the production floor in a last-ditch effort to produce shippable units. High-priced design engineers could only produce one or two units a day, and they were no longer available for developing the next product. The company soon failed, a casualty of untestable design. The marketplace never got a chance to determine whether their product idea was a winner, because their design process was a loser!

In the 1970s and into the 1980s, it was very common for a design department to be physically and organizationally removed from the test department. The nature of product life cycles dictates that by the time a test department starts ramping up test program development for a board, the design team is off on the next project and finds it distracting to go back and help the test engineers with testing problems. Thus, in those years, the designers were unknowing contributors to difficult test problems. But what is different today, with the emphasis on outsourcing and contract manufacturing? Now, a design may be outsourced to one contractor, board layout to a second, test development to a third, and actual production to a fourth, all scattered about the globe.

Even if boards can be tested, the question bears asking, "Is there something a designer can do that will make testing easier, cheaper, more thorough, etc.?" The answer is that there is a great deal a designer can do (or fail to do) that will affect the testability of a board. The technology that addresses this problem is called design for testability (DFT).

By the middle of the 1980s, testing became a bottleneck in product development. The situation ultimately became severe enough that attention was paid to the effects that design has on testing. A landmark survey on DFT technology by Williams and Parker[1] brought DFT out of common lore and into the design lexicon of the electronics industry. That paper is still remarkably current* nearly 20 years later. In it were coined the terms *ad hoc testability* and

---

* This survey does not cover the topic of standards-based testing (Sec. 51.5), because the testability standardization work promulgated by the IEEE began in the later 1980s.

structured testability. Before jumping into the DFT discussion, however, some definitions are crucial.

51.2   DEFINITIONS

These definitions are fully discussed in Chap. 52 and are repeated here in an abbreviated form for reference.

1. A *defect* is an unacceptable deviation from a norm.
2. A *fault* is a physical manifestation of a defect.
3. A *fault syndrome* is a collection of measured deviations from an expected good outcome.
4. A fault is *detected* when an operation with an expected outcome is conducted and this outcome is not observed.
5. A fault is *isolated* when an operation with an expected good outcome and one or more failing syndromes is conducted and the outcome matches a member of the set of failing syndromes.
6. A *test* is one or more experiments that are specifically constructed to detect (and possibly isolate) failures.
   - A *detection test* has an expected good outcome.
   - An *isolation test* has an enumeration of possible fault syndromes indexed to specific failures.

The technology of testing, as covered in Chap. 52, is highly influenced either positively or negatively by the design of boards being tested. If the preceding definitions are not clear, then the discussion of loaded board testing should be digested first.

51.3   AD HOC DESIGN FOR TESTABILITY

Ad hoc design for testability consists of a set of simple design rules of the form "Do this, don't do that," where "this" and "that" are often not motivated with reasons. For example, when designing a board with ICs that have preset or clear pins, a rule might read as follows:

"Tie unused preset and clear pins off through a 100-Ω resistor to a power rail; do not tie them directly."

The first-level reason for this is that a test engineer might want to access the preset/clear functions during testing, even though the designer did not use these functions. If these pins are tied through a resistor, a test engineer may still be able to manipulate them by applying a tester resource to them that can drive a signal in spite of the resistor. If these pins are tied directly to a power rail, the test engineer will never have that option. What might the difference be? Well, in a deeply sequential circuit, controlling the preset and clear functions might make the difference between a test that runs in milliseconds vs. hours. Clearly, hours of testing (per board) are impractical, so the bottom line may be the difference between a thorough test and one that lacks significant fault coverage, affecting quality.

The real reason for the various ad hoc DFT rules is that, to effectively and economically test a circuit, one must be able to control and observe the circuit's behavior. Most rules are related to controllability and/or observability of the circuit. The rule just cited is a controllability rule. Observability rules typically suggest ways one might be able to monitor signals that are deeply embedded in combinatorial circuitry, or that are activated only rarely by complex sequential events.

Ad hoc DFT is essentially the only way many products can have improved testability when those products are constructed with off-the-shelf merchant parts. Large, vertically integrated companies have the advantage of being able to customize testability into the heart of a design, including the very ICs themselves. Application-specific ICs (ASICs) allow more of this as well.

51.3.1 Physical Access

Testing is performed once the board to be tested is connected by some adaptor mechanism to the test system. This may be accomplished via the edge connector(s) of the board, where the tester is given the same access that the board gets in its end application. But far more common is "bed-of-nails" in-circuit test access, where the board to be tested is physically mounted on a platen and depressed into a field of precisely positioned spring-loaded probes ("nails") that contact hundreds or perhaps thousands of internal board nodes. (See Sec. 52.4.2 in Chap. 52.) This can be a challenging mechanical proposition to implement, particularly when high-volume, reliable manufacturing is the goal.

Board designers must consider physical attributes of their boards early in the design process. They have a size target and then often find that there are density issues that may require fine-line geometries and two-sided component mounting to solve. The fact that in-circuit bed-of-nails access may be needed for testing should also be considered very early. See Ref. 2 for an excellent discussion of how test target pads need to be provided, and how artifacts of the board layout (particularly vias) may be used to satisfy some of these needs. However, due to the density revolution, full nodal access, which has been the holy grail of in-circuit testing, is often impractical. This leads to the question of what access is most important when full access is impossible. The answer to this question comes from the domain of circuit design.

51.3.2 Logical Access

Sometimes it is impractical to gain physical access to all nodes of a circuit. For example, Fig. 51.1(*a*) shows an IC containing a large amount of complex logic, much of it deeply buried and effectively inaccessible from the I/O pins. Figure 51.1(*b*) shows the same IC with two additional gates added. The first is an exclusive OR gate that collects three buried signals, adds

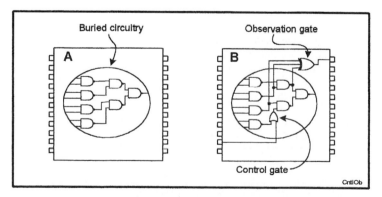

FIGURE 51.1   (*a*) An integrated circuit with deeply buried logic. (*b*) Same IC with controllability and observability logic added.

them together,* and brings the result to a spare I/O pin. This makes the states of those three signals much more observable. The second gate is an OR gate that allows us to insert a 1 from a spare input pin† into a deeply buried portion of the circuit. The assumption here is that the modified signal rarely reaches the 1 state in normal operation, so this extra controllability enhances our ability to manipulate the circuit for test purposes. Adding these extra gates can make it much easier to control and observe the performance of the IC. However, the downside is that we consumed a small amount of silicon area for the two gates, and, perhaps more importantly, we utilized two additional I/O pins. For many ICs, the extra cost of the pins could be a negative factor.

The choice of the optimal places to insert controllability and observability is not always clear, but rather is a trade-off of extra circuitry and pins vs. the difficulties that may be experienced in creating tests. There may be additional concerns as well with respect to circuit performance. For example, adding the controlling OR gate in Fig. 51.1 could add some signal propagation delay into the affected pathway, with detrimental results on system performance. When ad hoc circuit modifications are made for the sake of testability, it is likely that "seat-of-the-pants" decisions will be made rather than critical analysis. One person's decisions might be remarkably different than another's, with resulting great variability in testability improvements.

51.4 *STRUCTURED DESIGN FOR TESTABILITY*

Structured DFT was born inside companies that had vertical control over their designs, from custom ICs through systems. They also were well aware of their testing costs and realized that initial design decisions had a large impact on these downstream costs.

These companies studied the controllability and observability problems and instituted design rules into their design processes that, when followed, would guarantee that a circuit was testable. In the test department, where they also had complete control, they could utilize these added features with customized test development processes, gaining greatly enhanced levels of automation.

One of the earliest and most prominent structured DFT schemes was IBM's Level Sensitive Scan Design (LSSD), which was developed in the 1970s.[1] It is the precursor to what is called full internal scan technology now. In (greatly simplified) summary, LSSD design discipline requires every memory element (flip-flop or latch) to be constructed such that it obeys a testability protocol. This protocol allows two modes of operation: first, the normal operation of being a memory element in a design; and second, an operation for testing purposes, in which all memory elements can be connected into a serial shift register that can be loaded and unloaded by serial shifting. This makes every memory element a control point and an observation point within a circuit. No other memory elements are allowed in the design, e.g., no asynchronous feedback is allowed. This guarantees that circuitry between any control/observation point is combinatorial, not sequential (Fig. 51.2).

As might be imagined, these rules were looked upon by designers as restrictions of their creativity. Ed Eichelberger of IBM, a major proponent of LSSD, happens to stand about 5.5 feet tall. When asked in 1977 how designers received the LSSD rules, he quipped that, at the start, he was over 6 feet tall. Structured testability is not easy to implement. It requires commitment from the whole organization, starting with management.

The next piece of the puzzle was IBM's test generation software, which was able to automatically construct complete tests for combinatorial circuits (known as the D-algorithm and

---

* The addition is modulo-2 yielding a single bit. Any single-bit error delivered to the three inputs of the exclusive OR will cause the output to change from its expected state.

† This spare input pin should be held to a 0 value when the IC is performing its normal function. A pull-down resistor to ground could assure this, whereas during test, a tester signal could assert a 1 value when needed for testing.

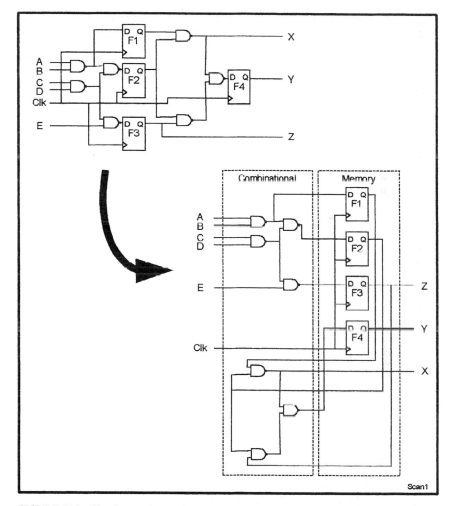

FIGURE 51.2   Circuitry made up of gates and flip-flops, redrawn to show that it can be represented as a bank of memory elements and a combinatorial circuit.

its derivatives). These tests could be shifted into the circuit and applied, and the results shifted out. This takes a lot of bookkeeping, but computers are good at that.

By using the LSSD discipline, IBM could verify that its designs would be completely testable, and those tests could be created by a computer program. Other companies such as Sperry Univac, Amdahl, Hitachi, etc., had similar proprietary structured approaches. Most smaller, nonintegrated companies were not able to participate in structured DFT—that is, until nonproprietary industry standards came into play.

51.5   STANDARDS-BASED TESTING

In the closing years of the 1980s, it became apparent that some sort of structured testability technology had to become accessible to the electronics industry at large. A small group of

European companies led by Philips formed the Joint European Test Action Group (JETAG) and began work on a testability standard. The effort quickly attracted the notice of North American companies, giving rise to the JTAG standard. As the proposal took shape, it was turned over to the IEEE, which ultimately produced IEEE Standard 1149.1-1990, Standard Test Access Port and Boundary-Scan Architecture,[3] in 1990.* Soon after 1149.1 came into being, a companion effort created IEEE Std 1149.4-1999, Standard for a Mixed-Signal Test Bus.[4] A complete coverage of these standards is beyond the scope of this chapter (see Ref. 5), so a brief summary is given here for an overview.

51.5.1   IEEE 1149.1, Boundary-Scan for Digital Circuits

The IEEE 1149.1 standard is a design discipline for digital ICs. It is a set of rules impressed primarily on the I/O structures of a device that allow two modes of operation, normal mode and test mode. In normal mode, the device performs its intended function. In test mode, the device obeys a protocol that has mandatory, optional, and customizable elements. The mandatory elements must exist, with the others being left as design options. The principle mandatory element of interest is a test mode dedicated to external test or EXTEST. When an 1149.1-compliant device is in EXTEST mode, its I/O pins are divorced from their normal operation and all internal functions of the device. Instead, the inputs become observation resources and the outputs become control resources[†] for test purposes. These resources are under control of the 1149.1 serial scan protocol. One can think of the I/O pins of the device being connected to shift register cells; states can be shifted in that will finally appear on all output pins (control) and the states of all input pins can be captured and shifted out (observe). This gives 1149.1-cognizant software a powerful tool for controlling and/or observing board-level node states. Figure 51.3 shows a simplified overview of the architecture.

Boundary register cells interposed between the IC pins and the internal logic surround the normal content of the IC called the mission logic. A small state machine called the test access port (TAP) is used to control the test functions. Four mandatory test pins (test clock [TCK], test mode select [TMS], test data in [TDI], and test data out [TDO])[‡] give standardized access to the test functions. All 1149.1 devices have a 1-bit BYPASS register used to bypass the (much longer) boundary register if it is not needed in a given testing activity. Figure 51.3 also shows an optional IDCODE register that can be shifted out to uniquely identify the IC, its manufacturer, and its revision.

It is intended that collections of ICs (called chains, as shown in Fig. 51.4) with 1149.1 be connected TDO-to-TDI so that they may form a long, shiftable register structure. The primary use for the 1149.1 EXTEST capability is to conduct board-level tests for shorts and opens. This is an example of how resources included in an IC design may be used to help with the testing problem at other levels in the manufacturing process.

Briefly, interconnections between ICs are tested as follows. Consider the circuitry in Fig. 51.5. Some circuit nodes (also known as *nets* or *traces*) are accessible with a bed of nails and some are not. (To avoid clutter, the TCK and TMS signals are not shown.) Boundary scan can be used to test all the nodes; those with nails can be tested by coordinating nails with boundary scan resources and those without nails are tested solely with boundary scan.

---

* An IEEE standard has as a suffix the year of its creation or last update. A standard must be updated and/or reaffirmed every 5 years. Up to two supplements to a standard may be issued within the 5-year cycle. Users of a standard should keep up to date with it.

† Bidirectional signal pins can both observe and control the nodes to which they are connected.

‡ An optional fifth pin called Test Reset (TRST*) is an asynchronous active low reset for the 1149.1 circuitry. Because any TAP can be reset by five clock pulses to TCK while TMS is held high, TRST* is not actually needed for resetting an 1149.1 device. It is often included as a fail-safe measure with a board-level pull-down resistor providing a constant reset to the TAP. Many 1149.1-compliant ICs do not include the TRST* pin because the extra pin required may be too costly.

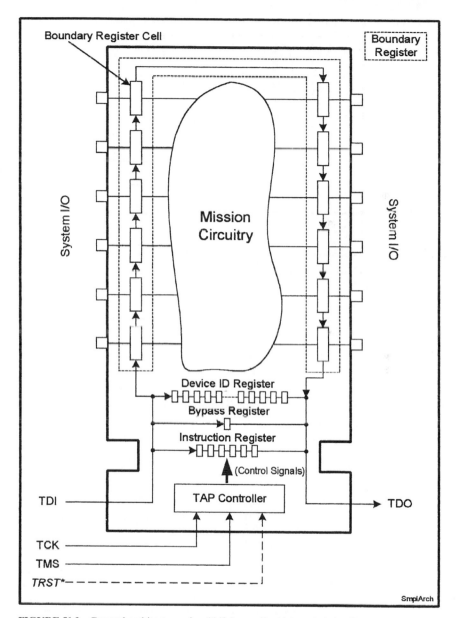

FIGURE 51.3   General architecture of an 1149.1-compliant integrated circuit.

As an example of this, boundary scan can be used to supply test patterns to nodes through a serialization process that a human would find laborious, but that is simple for a computer. When a set of patterns has been delivered (to control) and monitored (to observe) by the appropriate boundary register cells, we have uniquely identified each node with a "signature." A defect such as a short will cause two nodes to have deviant signatures, as shown in Fig. 51.6. Software can correlate observed deviations with the known boundary scan structure of each IC and the board netlist to yield a diagnostic message.

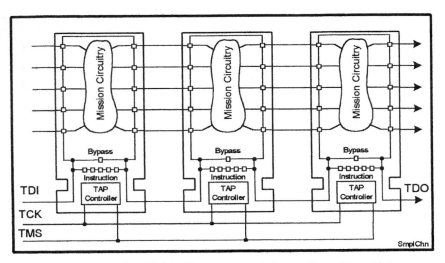

FIGURE 51.4 A collection (chain) of boundary scan devices can be used to test interconnections between ICs.

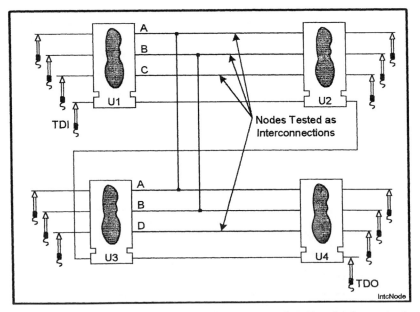

FIGURE 51.5 A set of boundary scan ICs with interconnections. Note that four nodes do not have bed-of-nail access.

The EXTEST capability can also be used during system testing to see if there are any system integration problems such as bad connections in backplanes and cabling. An IC designer may not see much attraction in EXTEST, but the 1149.1 standard offers other test modes that will allow a designer to access internal scan paths or built-in self-test functions. The 1149.1 standard's name has two parts, *Standard Test Access Port* being the first and crucially impor-

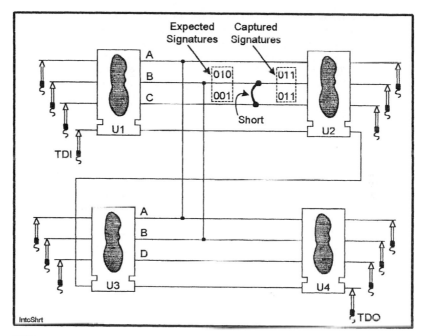

FIGURE 51.6 Interconnect testing drives unique patterns assigned to each node from drivers to receivers. In this case, a short is shown that creates a wire OR result.

tant one. It signifies that the standard anticipates being used as a standardized protocol for accessing any on-chip, board-level, or system-level testability scheme. In support of this, the standard is deliberately extensible, allowing clever designers to implement additional operational modes that can be used to solve unique testing problems.

The 1149.1 standard has proven itself to be quite useful and it has several contributions. First, it allows the creation of software that can automatically write tests for boards where in the past the same level of test effectiveness was nearly impossible to achieve, and then only with weeks or even months of skilled labor. It is not uncommon to see a boundary scan board test prepared in a single day that otherwise might have taken weeks. Second, 1149.1 ICs can "read" their input pins and scan out the result. This allows diagnostic software to pinpoint the location of open solder problems where in the past an IC might have been falsely indicted as faulty. Third, it allows tests to be performed on digital circuits without 100 percent accessibility to board nodes. With the trend toward miniaturization of components making it difficult to provide full nodal access, boundary scan is allowing the elimination of many access points. Of course, not all points may be eliminated, so one must understand which are still necessary. Finally, because many industry segments are affected by the test problem, a standard offers a way for everyone to benefit. One can find a large number of applications and tools available to solve testing problems that would not have been possible without a standard.

51.5.2 IEEE 1149.4, Boundary-Scan for Mixed-Signal Circuits

Boundary scan (1149.1) is a digital testability standard. However, there is also a trend in the superintegration of circuitry toward higher mixed-signal digital-analog content in our designs. The IEEE has developed a mixed-signal testability bus with a new standard 1149.4.[4-6] This standard is constructed as a superset of 1149.1 boundary scan, adding two additional analog test pins to the definition. The goal of the standard is to support opens and shorts testing of mixed-signal boards and to provide the capability of making analog value measurements of discrete analog components such as resistors, inductors, and capacitors without direct nodal accessibility, that is, a full bed of nails. (See Fig. 51.7.) It has been likened to in-circuit testing without a bed of nails, which is again not without caveats. (The elimination of test access points will still have to be done with thoughtful deliberation.)

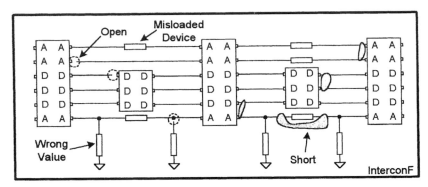

FIGURE 51.7   A mixed-signal circuit with some possible defects. IC pins marked A and D are analog and digital, respectively.

A mixed-signal device constructed with 1149.4 has the general architecture shown in Fig. 51.8. It is in many ways identical to an 1149.1 IC, but it has an additional analog test access port (ATAP) that is used to facilitate the control and observability of analog signals at device pins. The ATAP brings into the IC two additional analog signals that are used during testing.

With 1149.4, digital device pins are treated exactly as done with an 1149.1 boundary register. Analog pins have an augmented structure called an analog boundary module (ABM) as part of the boundary register. The ABM allows an analog pin to be tested for simple shorts and opens (this is called 1149.1 interconnect test emulation) as well as allowing the injection and/or observation of analog signals via the ATAP.

An ATE system can utilize the test resources in an 1149.4 IC as shown in Fig. 51.9. This requires the coordination of a digital test sequencer with analog test resources—in this case, a current source and voltmeter. Pathways from these resources and through the 1149.4 IC can be used to make measurements on discrete analog components on a board, even with no bed-of-nails access to the components.

The pathways needed for a measurement are provided by silicon switches in the ABM. Because switches implemented this way have significant nonlinearity and nonnegligible impedances, these must be accounted for in the measurement processes. Figure 51.10 shows how an analog device can be tested with two measurements. First, the tester's current source is connected such that current can flow along AT1 into the IC. There it flows on the AB1 bus inside to the ABM connected to pin 1 of the IC. The current flows through Z and then into pin 2 of the IC. Pin 2's ABM directs the current to ground, completing the current path. In Fig. 51.10(*a*), the ATE system's voltmeter is connected via the path AT2-AB2 to the ABM on pin 1, where it can observe the voltage at the top of Z. In Fig. 51.10(*b*), the voltmeter is switched to measure the voltage at pin 2, the bottom of Z. Subtracting the two voltage measurements gives the voltage drop across Z for a known current. Ohm's law gives us the value of Z, which we can check for the right value. This process works even with suboptimal silicon switches

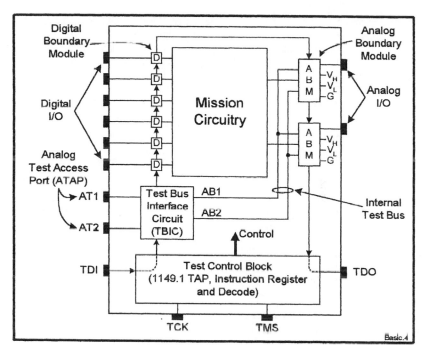

FIGURE 51.8   General architecture of an 1149.4-compliant IC.

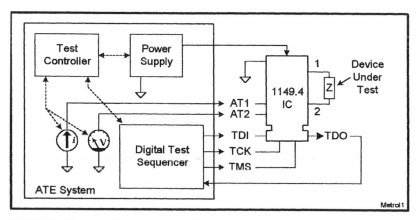

FIGURE 51.9   An ATE system can use 1149.4 resources to gain access to discrete analog components.

because we are using a current source to provide the known current and because the volt-meter pathway consumes a vanishingly small current and thus does not affect the accuracy of the measurement.

In the future, with these and other standards, test engineers may be able to do complex tests on superdense circuitry with far less nodal access than they were afforded in the past. This will be an enabling technology because without it the electronics industry may find it uneconomical to produce superdense designs except in very high-end applications.

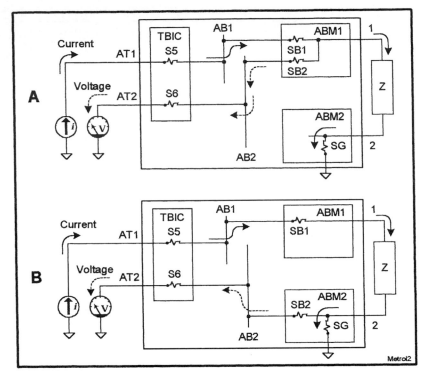

FIGURE 51.10 Two measurements (*a*) and (*b*) can be used to measure the voltage across Z for a known current.

REFERENCES

1. T. W. Williams and K. P. Parker, "Design for Testability—A Survey," *Proceedings of the IEEE*, vol. 71, no. 1, January 1983, pp. 98–112.

2. M. Bullock, "Designing SMT Boards for In-Circuit Testability," *Proceedings of the International Test Conference*, Washington DC, September 1987, pp. 606–613.

3. IEEE Standard 1149.1-1990, Standard Test Access Port and Boundary-Scan Architecture (includes IEEE Standard 1149.1a-1993), IEEE Inc., 345 E. 47th St., New York, NY 10017, USA.

4. IEEE Standard 1149.4-1999, Standard for a Mixed-Signal Test Bus, IEEE Inc., 345 E. 47th St., New York, NY 10017, USA.

5. K. P. Parker, *The Boundary-Scan Handbook: Analog and Digital*, 2d ed., Kluwer Academic Publishers, Norwell, MA, 1998.

6. K. P. Parker, J. E. McDermid, and S. Oresjo, "Structure and Metrology for an Analog Testability Bus," *Proceedings of the International Test Conference*, Baltimore, MD, October 1993, pp. 309–322.

7. R. W. Allen Jr. et al., "Ensuring Structural Testability of High-Density SMT Circuit Packs," *AT&T Technical Journal*, March/April 1994, pp. 56–65.

8. K. P. Parker, *Integrating Design and Test: Using CAE Tools for ATE Programming*, Computer Society Press of the IEEE, Washington, DC, 1987.

9. K. P. Parker, "The Impact of Boundary-Scan on Board Test," *IEEE Design and Test of Computers*, August 1989, pp. 18–30.

CHAPTER 52
LOADED BOARD TESTING

Kenneth P. Parker
Agilent Technologies, Loveland, Colorado

52.1 INTRODUCTION

Printed circuit boards, like everything else in the electronics industry, have been undergoing rapid technological evolution. This is only natural because everything, from the boards themselves, to the CAD systems that create them, to the components that populate them, to the assembly methodologies used to fabricate them, has been undergoing similar changes. These changes have common themes: greater functional density, better performance, improved reliability, and lower cost.

In the early 1990s, the move toward surface-mount technology (SMT) accelerated to the point that SMT designs became the rule. SMT supplanted, in large part, the familiar 100-mil centered through-hole package technology. The change came slowly, embraced by leading-edge applications that needed the density improvements that came with SMT. Many held back because they did not have a need for higher densities and could not justify the risk of putting new processes in place to manufacture with SMT. The process of perfecting SMT brought to light a surprising fact—it was more efficient once the necessary automation was put in place and perfected. There are now applications that use SMT not because of the density improvements it affords, but rather because of the efficiency it provides. This is confirmed by the fact that many new devices can no longer be obtained in the old-style through-hole packages.

Along with SMT came increases in lead pitch density. First there was 50-mil pitch; soon came 25-mil pitch, then came 15-mil pitch, and so on. Other technologies such as tape automated bonding (TAB), chip-on-board (COB), multichip modules (MCM), and ball grid arrays (BGAs) are gaining acceptance. The industry has for some time been in the midst of a packaging revolution. This revolution has applied to boards as well. The average board, by virtue of high-density interconnect (HDI), now has more layers, finer lines and spaces, buried vias, devices mounted on both sides, and so forth. The net result is that loaded printed circuit boards are becoming incredibly densely packed with highly sophisticated components. We are in fact experiencing a density revolution. Figure 52.1 shows a range of common electronic devices superimposed on a penny to show scale.

Testing has been impacted by all these changes. If perfect components were fed through perfect processes utilizing perfect machines run by perfect employees—testing would not be needed! Unfortunately, nearly perfect components are fed through processes that are subject to drift in many of the hundreds of variables that govern them, using machines that require careful calibration and preventive maintenance, operated by people who tire and err. For these reasons, testing is still an important part of loaded board manufacturing. However, as a result of the miniaturization brought on by the density revolution, our ability to gain the physical and

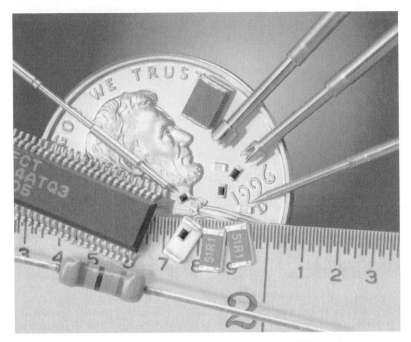

FIGURE 52.1   A US penny shows the size of some common PC board components, some older, some recent. The small black devices near Lincoln's bow tie are 04-02 (40×20 mils) surface-mount resistors. (Now 02-01 [20×10 mils] devices are being used.) Some 100-, 75-, and 50-mil test probes are shown for comparison. Note the lettering on the penny is done with 10-mil lines and the width of the circular part of the 9 is 35 mils, a common probe target size. *(Photograph courtesy of Agilent Technologies.)*

electrical access needed for testing purposes is increasingly hampered. Because access to a circuit is crucial for testing, accessibility difficulties make testing steadily more difficult to accomplish. On top of this, the electronics industry is expected to provide continued improvements in cost, reliability, and quality. Testing plays a crucial role in these improvements, as will be seen.

52.2   THE PROCESS OF TEST

In the earliest years of the electronics industry—indeed, before anyone may have called it an industry—there was no concept of test. A product was put together and shipped, because inherent in the putting together was a basic appreciation for how the product was supposed to look and behave. This crafting quickly gave way to mass production* of somewhat more complicated items by workers who were not themselves experts in the craft. Of course, today, workers with little or no fathoming of what it is they are producing turn out items of unbelievable complexity. If they had to know everything about the process, it would undoubtedly be too expensive to make the product.

---

* Perhaps the reader remembers the TV ads for Zenith televisions from about 25 years ago. They showed individual axial lead components being hand-wired onto lug strips and then soldered. The voice-over mentioned *"hand-crafted"* quality as if to imply that automated manufacturing (being used by Japanese manufacturers like Sony) was suspect.

It is also important to appreciate that boards are often themselves components of systems. Boards might have a respectable yield from a board test process of, say, 97 percent, which is to say, 3 of 100 still contain defects.* However, if 20 such boards are used in a system, the probability that the system will turn on is only 54 percent.† Because debugging a system is usually far costlier and more difficult than testing a board, there is much interest in higher post-test yields. This is why yields are being pushed into the 99 percent and higher levels.‡

Test has gone through three stages of evolution during this time. It has been used as a sorting process, then as a repair driver, and finally as a process monitor.

52.2.1 Test as a Sorting Process

Test can be used to sort boards into two piles, good and bad. Essentially, such testing provides only one bit of information about a board. This one bit of information provides little clue as to what the problem is or how to repair it. If all we intend to do with a bad board is discard it, then one bit of information may be enough. In some applications, for example making $2 digital watches by the million, it may be economically impossible to attempt to repair bad units. However, as will be seen soon, discarding bad boards may also lead to discarding of valuable information about the health of our manufacturing process.

52.2.2 Test as a Repair Driver

It is very often economically justifiable to repair a bad board, if it can be done quickly and with little skilled analysis needed. This is where modern test systems begin to come into their own. More bits of information are needed per bad board; a diagnostic test is needed that accurately resolves failures into defect reports that can be acted upon to effect repairs. Because a freshly repaired board will need retesting to ensure it is now defect free, one also would like to find as many defects as feasible in each pass across a tester to avoid a glut of work in progress.

52.2.3 Test as a Process Monitor

When repairing boards, valuable information about the health of our board manufacturing process is at our fingertips. Indeed, the repair process may have the ultimate defect resolution because faults detected by a tester may still not be perfectly correlated with defects (see Sec. 52.3). If one views test as a process monitor, one can gain valuable insight into what is happening upstream in many of the subprocesses that come together to produce boards. For example, one may see solder opens as a chronic problem. Further examination may show that they are happening in one area of a board more than in any other. That may lead to the examination of the solder application process, which shows that a solder screen squeegee has an uneven amount of solder paste on it. The root cause of this uneven distribution of paste can then be sought and corrected.

---

* It is important to differentiate test yield from defect coverage of a test. The yield of the manufacturing process before test might be (say) 50 percent, and the yield after test may be 97 percent. This may be accomplished with a test that detects (covers) 90 percent of the important defects.

† From basic probability theory, $(0.97)^{20} = 0.54$ is the chance that 20 independently manufactured boards with a probability of 0.97 of being fault free will all be fault free.

‡ Measuring yields in percentages is telling. In the IC industry, yields are measured in parts per million. One ppm corresponds to 99.9999 percent, and 250 ppm, a good IC yield, equals 99.975 percent yield. Will the board industry ever get into these ranges? Will it make sense to try?

It is quite important for the tester to do a good job of resolving faults into defects. Using an example of solder opens on digital devices, digital in-circuit test (detailed in Sec. 52.5.2) alone may indict devices as failing when it is really more of an open solder problem. The repair technician may notice this. However, it may well be overlooked because replacing the device "fixes" the problem. This pollutes the information about our process, causing one to call the IC vendor with complaints when one should be examining the solder process. If tests are focused more closely on defects than on faults, then one may be able to better trust tester-derived process information. This requires test engineers to be students of their manufacturing process and available test technologies so that test techniques are kept in balance with the defects that are most important.

52.3   DEFINITIONS

Testing is a word often used in a vacuum. What are tests actually looking for? The answer to this question has a huge impact on how to go about testing effectively. It makes little sense to test meticulously for problems with low likelihood of occurring, just as it makes little sense to inadequately test for likely problems. However, many of the terms used in the testing industry are overloaded such that surprisingly little agreement on their definitions may be found. Worse, this confusion is often not realized, which complicates communications between process people, test people, and equipment vendors.

52.3.1   Defects, Faults, and Tests

52.3.1.1   Defects.   A *defect* is an unacceptable deviation from a norm. As an example, a bond wire could be missing in an IC. This defect may in turn be due to a problem (the "root cause") in a wire-bonding machine, such as it misfeeding wire. Other deviations may not be considered defects, but rather acceptable variations. For example, the plastic used to encapsulate a component may have significant variations in color that are still considered acceptable.

Defects require remedial action of some sort. Most often, defects lead to a repair operation that removes the defect and restores the product to acceptable status. In some cases, a defect may not be economically repairable and thus causes scrap. It is also possible that a defect will be overlooked and thus be delivered within a product to an end customer. In some cases, such as the manufacture of $2 wristwatches, some amount of defective product may be deemed acceptable. However, for many products, defects perceived by end customers can have grave economic consequences and are strenuously avoided.

52.3.1.2   Faults.   A *fault* is a physical manifestation of a defect. (The word *fault* is often used synonymously with *failure.*) Thus a fault reveals the presence of a defect. A single defect may cause several faults (i.e., have several different manifestations), and a single fault could be the manifestation of several different defects.

A defect shows up as a fault. For example, the aforementioned missing bond wire on an input to a logic gate may cause it to see a permanent logic 1 rather than a time-varying signal. Several other defects can produce this same fault behavior. For example, missing solder between the input pin and the board, an electrostatically damaged input buffer within the IC, or a broken printed circuit trace between the upstream driver and the input will all exhibit the same faulty behavior. Similarly, one defect may cause several fault manifestations. For example, an open solder joint on an input pin (particularly a reset input) may cause an IC to show incorrect results, and these incorrect results vary in time.

An observed fault is not always a reliable pointer to a defect. For example, if an IC loaded onto a board has defective solder on one input pin causing an open circuit, this may appear to the IC to be a permanent, stuck-at-1 fault on that input. This faulty behavior may not be readily apparent because the effect of the erroneous logic 1 must propagate through the internal workings of the IC before its effects (improper output behavior) are seen.* When this faulty output behavior is finally observed, it can be a highly challenging task to relate this observed behavior to the input stuck-at fault caused by the defective solder. This is illustrated in Fig. 52.2.

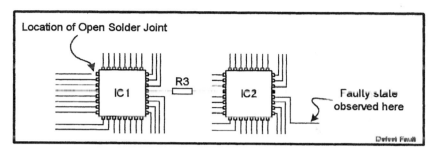

FIGURE 52.2   A defective solder joint on an input pin to IC1 causes a fault on an output of IC2. Note that IC1, R3, and IC2 are suspect, as well as six solder joints, two IC input buffers, two IC output buffers, and four IC bond wires.

52.3.1.3 *Fault Syndromes.* A *fault syndrome* is a collection of measured deviations from an expected good outcome. For example, a digital device may produce an incorrect response on several of its outputs when its inputs are stimulated in a certain sequence. The incorrect pins, their states at the point where they were incorrect, and the point in time where each failure was noted form a fault syndrome.

52.3.1.4 *Fault Detection.* A fault is *detected* when an operation with an expected outcome is conducted and this outcome is not observed. For example, when you turn on the power to a personal computer, you expect to see it boot up. If it doesn't, then you have detected a failure of some kind. There could be a plethora of reasons for this result, but you have little information to act upon. In other words, the underlying defect may not be easily resolvable.

52.3.1.5 *Fault Isolation.* A fault is *isolated* when an operation with an expected good outcome and one or more failing syndromes is conducted and the outcome belongs to the set of failing syndromes. For example, as your personal computer boots up, it may do a series of self-test operations. If any of these fail, a diagnostic message may appear on the screen that specifically identifies a failure. Again, the underlying defect may not be easily resolvable, but there is more information to act upon than you had when the fault was simply detected.

52.3.1.6 *Tests.* A *test* is one or more experiments that are specifically constructed to detect (and possibly isolate) failures. A *detection test* has an expected good outcome. An *isolation test* has an enumeration of possible fault syndromes indexed to specific failures. Thus, when a test fails, it provides us with information: a single bit of detection information, or a fault syndrome that may match an enumerated outcome, pointing to a fault.

A well-constructed detection test will detect a large number of potential failures. A well-constructed isolation test will accurately resolve a failure from a large list of potential failures. Note that an isolation test may encounter a failure that it was not constructed to isolate. If it fails (that is, if it detects the fault), it may not produce a syndrome that matches one from its

* See Sec. 52.6 for a discussion of how a new testing methodology may pinpoint a defect that conventional methodologies have difficulty isolating.

list, or it may produce a match that is erroneous. In either case, it has been reduced to a detection test.

Remember that an isolation test may point to a fault, but the fault may not be a good indicator of the actual defect that is present. For this reason, it is important to construct tests that target expected faults and that accurately resolve underlying defects. When defects are resolved correctly, it is then much easier to find and correct the causal problems.

There are, in general, three categories of faults being tested: performance faults, manufacturing defects, and specification failures.

52.3.2 Performance Faults

A performance fault is a fault in the performance of a system that occurs due to a mismatch of important parameters among the system's components. This mismatch is the defect. As a common example, the path delay seen by a digital signal as it passes through several components may exceed the intended design value, causing a malfunction. No single component in the path is defective, but the cumulative contributions of several components cause the performance fault. The repair for this defect is to replace one or more components in the path with new components specifically selected to give the proper delay.

There are several problems with testing for performance faults. First, the test developer must know about the circuit design in great detail. Second, it is difficult to set up a test that can resolve faults into specific defects (for example, the mismatch of parameters in several components). Third, it is difficult to avoid being confused by unanticipated defects that produce behaviors similar to the defects of interest.

Solving these problems implies great knowledge and understanding of a board design. Indeed, in some instances in a carefully designed board, the designer may have specific knowledge of some critical parameter that has to be precisely managed and can alert those responsible for test. However, much of the testing for performance faults carried out in the past was not done with this knowledge, but rather because of a lack of this knowledge. In the past, tools that could help control the key parameters of a design were unavailable, or designers were using components that were incompletely specified, or perhaps they were too trusting of their "seat-of-the-pants" instincts. It was expected that performance testing would give adequate coverage of any problems that might occur. In effect, performance testing was used to validate a design after the fact.

The expectation that performance testing will somehow protect products from the effects of poor design is now obsolete. It amounts to wildly shooting in the dark against a well-hidden, stealthy enemy force of unknown size and distribution. Given the ever increasing complexities of boards, it is simply not reasonable to expect that test engineers could stumble at random across effective tests for all possible design problems, and certainly not within the lifetime of the design. With the increase in effectiveness of design tools, designers should no longer be relying on test for design validation. Testing for performance faults will still be important, but it must be used in its proper role—to verify that critical parameters identified in the design process are properly controlled.

52.3.3 Manufacturing Defects

A manufacturing defect is a defect resulting from a problem in the manufacturing process. Manufacturing defects tend to be fairly gross in nature. Table 52.1 gives a list of potential manufacturing defects.

Manufacturing defects are the result of the havoc inherent in manufacturing processes. These defects result in faults that may be very easy to detect and to correlate with their root cause(s), but this is a function of the test approach. Some manufacturing defects can still be difficult to detect and resolve, as the example given of a solder open on a device input pin shows (Fig. 52.2).

TABLE 52.1   Examples of Manufacturing Defects Seen at the Board Level, Sources
of the Problems and Causes

Defect	Source(s)	Cause(s)
Shorts between solder pins	Wave/reflow soldering	Too much solder, solder screen defect, pin misregistration, bent pins
Solder open	Solder application, wave/reflow soldering	Too little solder, solder screen defects, tombstoning, bent pins
Missing component	Placement, soldering	Shock, abrasion, too little glue
Wrong component	Placement setup, inventory, handling	Handling error, mismarked packages, operator error, wrong specifications
Misoriented component	Placement setup	Handling error, operator error
Dead component	Placement, soldering	Dead on arrival, handling damage, electrostatic damage

52.3.4   Specification Failures

Specification failures are similar to performance faults. Performance specifications are checked
against requirements for the full range of operating conditions expected, such as temperature,
humidity, vibration, electronic noise, etc. Specification test is often a regulatory or contractual
requirement. One may argue that these specifications are largely unnecessary because, if a circuit design is robust, makes use quality components, is accurately assembled, and is thoroughly
tested for manufacturing defects, one does not have to make it perform all of its functions to
know that it works. This is the "proof by construction" argument, used by successful manufacturers of kitchen matches and bomb detonators. Full specification test may also be impossible in
practical terms because there may be too many combinations of circuit functions vs. operating
ranges. If only a subset of combinations is to be checked, this begs the question of which subset.

If specification test is required nonetheless, such testing is usually carried out with custom-
tailored test equipment that can simulate the range of operational environments of interest.
Such testing may be quite time consuming, and the cost of the supporting test equipment may
range from trivial to hyperexpensive. At one extreme, a manufacturer of I/O cards for personal computers may simply plug each one into a computer and see if it performs a simple
loopback test. At the other extreme, a manufacturer of guidance computers for missiles may
require a missile shot on a test range, full telemetry, and support from the Air Force and Navy.
The bottom line is that you should seriously question the motivations for and expectations of
testing for specification faults.

52.4   TESTING APPROACHES

We have seen a spectrum of faults, performance faults, manufacturing defects, and specification faults. Each has evolved a test technology.

52.4.1   Testing Boards for Performance Faults

Performance testers are often known as edge connector functional testers. They connect to the
edge connectors of the board under test with mating connectors that are then adapted to the
tester resources by a customized fixture called an adapter. In most cases, no other connections
are made by the test fixturing to any internal nodes of the circuit. In essence, the board is tested
in an environment that resembles its application environment to some degree.

Some performance testers have guided probe capability.[1,2] A guided probe is a manually positioned test probe with a measurement (and sometimes stimulus) capability and supporting software. It is used to temporarily gain access to internal nodes of a circuit, one at a time, where observations of internal circuit behavior during a test can be made and processed by software to provide enhanced fault resolution.

How does this software know what the nodes of a circuit are supposed to be doing during a test? One could enter all these data by hand if one knew the design of a circuit well, but this is impossible for any circuit that consists of more than a handful of gates. The most popular way to get these data is by logic simulation of the circuit.*[2] Good circuit simulation is often used by designers to prove the circuit behaves they way they expect it to under various input conditions (input vectors) they are interested in. Fault simulation is used to study how a circuit behaves when that circuit is perturbed by a modeled fault. An example of a modeled fault is a stuck-at-0 on a gate output, meaning no matter what the gate's inputs are, the output is 0. The single stuck-at model is the most prevalent fault model in existence. It assumes that only one stuck-at (either 0 or 1) may exist in a circuit at any one time. A fault simulator can be used to predict, for a given modeled fault, how the circuit will behave when stimulated with input (test) vectors. If the modeled fault causes one or more observation points (like circuit output pins) to deviate from normal behavior, we can declare the fault detected by that vector. In the common case of sequential circuits, it is more accurate to say all vectors detected the fault up to and including the one that caused the observable deviation.

Another technology used by some performance testers is the fault dictionary.[2] Fault dictionaries are prepared with fault simulators. A fault dictionary is a three-dimensional data structure of Boolean true/false bits (see Fig. 52.3). The first dimension is an enumeration of all test vectors. The second dimension is a list of modeled faults that are to be simulated. The third dimension is an enumeration of the circuit's output pins. A given bit in the fault dictionary is true if the corresponding output pin fails on the corresponding vector for a given fault.

---

* Logic simulation implies the treatment of the digital nature of the circuit, typically omitting any interaction with any analog portions of the circuit. Analog simulators do exist, but are implemented in a very different technology from digital simulators. Mixed-signal simulation is not anywhere near as viable as digital simulation, which presents a problem for test engineers.

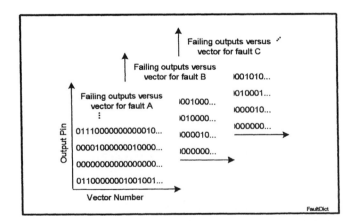

FIGURE 52.3   A fault dictionary. For each "page" corresponding to a modeled fault, bits set to 1 indicate that a given output pin fails for a given vector number. When testing is performed, the failing output(s) and failing vector number are noted and then each page is searched at those coordinates to look up a matching fault. (Note that more than one fault may produce a match.)

For a given vector and fault, the output pins that are expected to fail can be looked up in this dictionary. In a testing scenario, we take a failing test vector and list of failed output pins, and attempt to look up a fault (or faults) that match the failed pin behavior for that vector. Fault dictionaries work well when actual defects that cause a circuit to fail closely match the fault model. When defects occur that vary from the modeled faults, then a dictionary lookup may come up empty (no matching fault), or matches may be found but point to the wrong faults, or so many matches may be found that it is impractical to examine them all to see which (if any) are the actual problem. Dictionary generation is computationally intensive, dictionaries may consume huge amounts of storage space,* and dictionary lookups may not work. This technology was invented along with small-scale integrated bipolar technologies such as early TTL or ECL. The advent of large-scale integration and CMOS technology† has made fault dictionary technology nearly useless in board testing practice.

Performance testers are intended to emulate the environment the board would encounter in its native application. This means the tester itself needs to be carefully customized to supply this environment, or it must be a general-purpose tester with a great degree of flexibility. Flexibility brings cost, so commercial functional testers tend to be among the most expensive. Programming such machines is a complex task, due both to the flexibility of the tester and to the nature of the test requirements. It is difficult at best to get automated support for functional test programming. Typically, such programming takes extreme patience and a high level of skill—again, two more costs. To compound this, any last-minute design changes to the board may cause expensive and time-consuming test modifications or invalidate the tests altogether.

52.4.2 Testing Boards for Manufacturing Defects

Manufacturing defects can be detected by a functional test, but because functional testers are expensive and difficult to program, other techniques have arisen to test for them. These testers have one or more of the following advantages:

1. They are much easier to program, often requiring less than 10 percent of the time of functional test development.

2. Automatic test program generators that analyze the circuit design database do most of the work and relieve the skill required of the programmer.

3. Their programs are much less sensitive to design changes because they take a divide-and-conquer approach; the effects of design changes are localized, requiring that only a portion of the test be reprogrammed.

4. They offer much better defect resolution for many defects because localized portions of the circuit are being analyzed, reducing diagnostic complexity.

The principal technology is the in-circuit tester. This tester utilizes a high degree of nodal access—that is, connection to printed circuit traces—to perform its work. Nodal access is provided by a unique fixture called the bed of nails. (See Fig. 52.4.) This fixture is made up of a platen that supports the board under test. The platen is drilled with holes below each target point for nodal access. In the holes are spring-loaded "nails" that contact the target points on

---

* For example, say a circuit has 100,000 gates. The number of stuck-at faults to be simulated may easily be 200,000. A set of test vectors for these may easily exceed 500,000. If the circuit has a fairly normal complement of 250 outputs, then a fault dictionary for this example would consume the product of faults times vectors times outputs (2.5×10^{13} bits, or about 3.125 terabytes, give or take a gigabyte). Fault simulation times lasting months, even on very fast computers, have been reported. Of course, larger circuits will need more!

† CMOS VLSI circuits have several defect modes that are inconvenient to model in popular fault simulators, and thus fault models are more likely not to match defects. Two of the offending defect modes are bridging faults (intermetal shorts) and metallization opens that may actually introduce capacitive memory into a circuit. These are difficult to represent with traditional fault-modeling techniques.

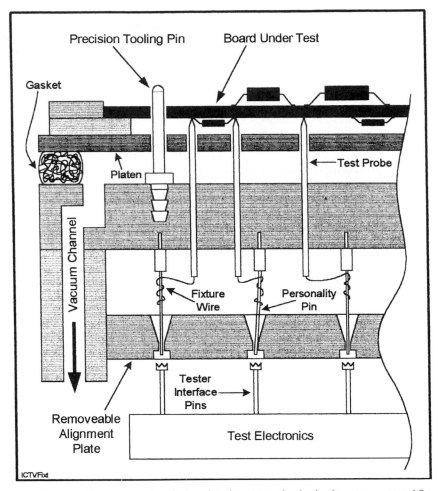

FIGURE 52.4 Cutaway drawing of a board resting on top of an in-circuit vacuum-actuated fixture, the bed of nails. The interface pins (the mechanical interface to the ATE pin electronics) are placed very close to reduce path lengths.

the board. (Special clamshell fixtures allow nail access to both sides of a board.) The platen with board forms the movable top layer of a vacuum chamber. When vacuum is actuated, the platen with board moves down, depressing the nail springs and causing nail contact with board nodes. (Sometimes mechanical actuation is used in place of or to assist vacuum.) The nails are wired (typically with wire-wrap technology) to the stimulus/measurement resources of the tester. These resources contain mechanical relays that allow connection of various tester functions to a given nail.

Once a tester has access to the nodes of a board, it can perform in-circuit (also called in situ) tests. The idea is to test components as if they are standing alone, while they in fact are part of a board circuit. The actual electrical processes used in in-circuit testing are explained in Sec. 52.5.

52.4.3 Testing Boards for Specification Faults

There is not much to add here because often, testing for specification defects is very similar to testing for performance faults. However, if for (say) contractual reasons, you are required to show your boards working as specified, you may be required to construct "live" situations where they are performing in a realistic system setting. Again, the extreme case of a missile shot on a test range could be involved. The question must be, is there an easier way?

52.5   IN-CIRCUIT TEST TECHNIQUES

The problem of in-circuit testing can be subdivided into two main categories, analog in-circuit testing and digital in-circuit testing.

52.5.1 Analog In-Circuit Test

Analog in-circuit test addresses testing for shorts in the printed wire circuitry; analog compo-nents, passive devices such as resistors, inductors, and capacitors; and simple semiconductor components such as diodes and transistors.* Analog in-circuit testing is conducted without applying power to a board; that is, it is an unpowered test methodology.

Shorts (unwanted connections between nodes) are tested for first because subsequent test-ing will profit from the assumption that shorts are not a factor, and subsequent testing may need to apply power to the board. Unpowered shorts testing may be accomplished by apply-ing a small DC voltage[†] to a node while all others are connected to ground. If current flow is observed that is below a computed threshold, then the node cannot be shorted to any of the grounded nodes. If the current flow is above the programmed threshold, there may be an unexpected path to at least one of the grounded nodes. The destination node(s) can be deter-mined by linearly searching the grounded nodes for the current flow (this is slow) or by using half splitting techniques (which are fast[‡]) to determine the other node(s) the current is flow-ing to. When the algorithm has finished stimulating all nodes sequentially, it can declare which nodes are shorted and use x-y position data to show where to look for the problems. Typically, shorts are repaired before continuing with other tests because they may confuse the resolu-tion of defects and may cause physical damage when the power is later turned on.

Unpowered tests on analog components such as resistors, capacitors, inductors, etc., are performed next. Again, low stimulus voltages keep semiconductor junctions turned off. AC voltages are applied and phase-shifted currents are measured in order to deduce the values of reactive components (capacitors, inductors). For a simple measurement of a lone[§] impedance $R,$ apply a voltage to one terminal of the component (through the bed of nails) and connect a

---

*   Complex analog devices such as analog or mixed-signal ICs are not very amenable to analog in-circuit testing because they require power to be applied to the board. Simple diodes and single transistors can be tested by in-circuit stimulus that essentially examines the characteristics of their semiconductor junctions.

    [†] The reason for using a small stimulus voltage (typically less than 0.2 V) is to prevent current flow through semi-conductor junctions that may exist between nodes. This voltage will not turn on a junction. These junctions may exist in parasitic form within ICs and are not always documented.

    [‡] A half-splitting technique (also called a binary chop) is a fundamental algorithm of computer science. It works by successively considering half of a set of items (grounded nodes that are receiving current in this case) while removing the other half from consideration. It recursively divides sets repeatedly until only one item is left to consider. It has a com-plexity related to the logarithm (base 2) of the size of the original set of items. By comparison, a linear process has a com-plexity linearly related to the size of the original set.

    [§] This component may not be alone in a physical sense, but because low stimulus voltages are used, other connected devices such as ICs may be electrically quiescent so that the component is electrically alone.

current-measuring device (conceptually, an ammeter with zero impedance to ground) to the other component terminal. The current flow observed due to the known stimulus voltage is related to R by Ohm's law ($R = V * I$). See Fig. 52.5(b).

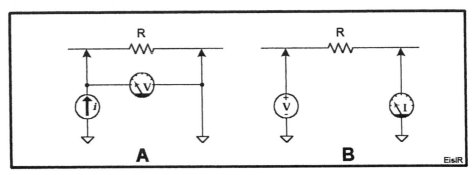

FIGURE 52.5 An analog component on a loaded board can be accessed by two in-circuit probes that connect its terminals with tester resources. There are two approaches: force a known current and measure a voltage (a) and force a known voltage and measure a current (b).

However, some discrete analog components may be connected to one another in ways that prevent simple measurement of the current flowing through the component due to parallel pathways that sidetrack some of the current. In Fig. 52.6, when applying a voltage to one terminal of R, a current also flows in a parallel path through R_a and R_b. The ammeter does not measure the true current through R because current from the parallel path also gets to the ammeter.

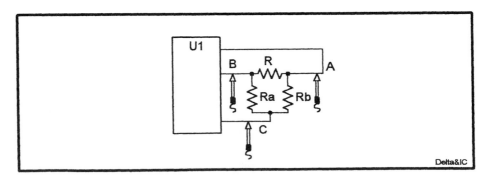

FIGURE 52.6 Three analog devices interconnected with an integrated circuit (U1).

The parallel path problem can be solved with a process called guarding. Guarding is accomplished by using a third nail to connect ground to the node marked C as shown in Fig. 52.7. When C is grounded, all the sidetracked current goes to ground because there is no voltage drop across R_b to attract current to or from the ammeter. All the current seen by the ammeter comes through R, so again Ohm's law gives the value of R. This is a classic three-wire measurement. It turns out that, for general component topologies, multiple guards (ground connections) may be required, but these are still considered three-wire measurements. In some cases where enhanced accuracies are needed or where there are extreme ratios of component

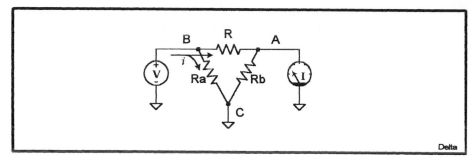

FIGURE 52.7   Equivalent circuit for Fig. 52.6 when making a guarded measurement.

values in a network, additional sense wires are used to eliminate errors that are due to small voltage drops in fixture nail, wire, relay, and trace contributions. (See Fig. 52.8.) See Ref. 3 for a discussion of enhanced measurement accuracies.

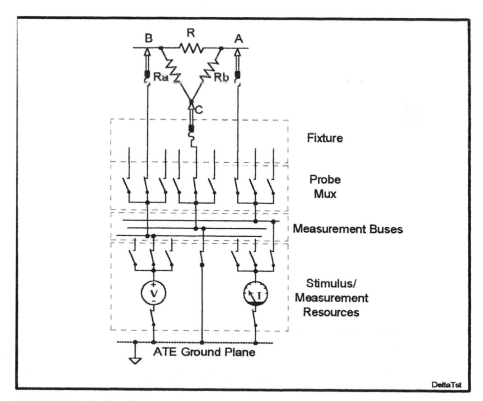

FIGURE 52.8   Routing of tester resources to the circuit to be tested.

52.5.2   Digital In-Circuit Test

Digital in-circuit test focuses on the digital components residing on a board, and requires that power be applied to activate the digital logic contained within the ICs. Just as analog components can be tested without removing them from a board, digital components can be tested the same way. The key technology is backdriving. When applying digital stimuli to a digital device's inputs (via the bed of nails), a tester's driver must overcome the voltage levels that connected upstream devices are producing. This is done by equipping the digital in-circuit tester with powerful, low-impedance drivers that can backdrive upstream drivers with enough current to create the desired signal voltage in spite of their state. With tester receivers connected by nails to the device's outputs, the tester can monitor these outputs for expected responses to the stimuli. (See Fig. 52.9.)

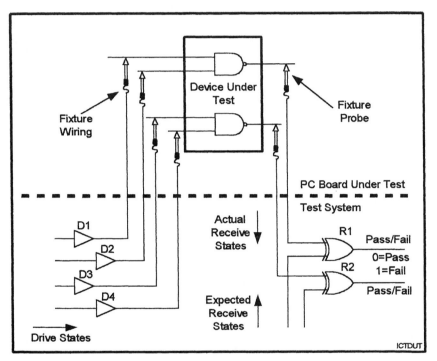

FIGURE 52.9   In-circuit digital test setup with full nodal access. The device under test is most often connected to other devices.

A problem with backdriving is that the tester drivers are abusing the upstream drivers in other devices* while testing the device of interest. Studies[4] have shown that this is a legitimate concern and that damage to upstream devices may occur. For example, overdriven silicon junctions or device bond wires may heat up enough to be damaged. Sometimes this damage may occur surprisingly fast (within milliseconds). This problem can be solved by careful application of tests with an eye toward their duration. If backdrive testing can be done quickly, and/or with appropriate cooling intervals, then damage can be successfully avoided.[5] This has allowed digital in-circuit testing to become the dominant testing technology.

A great advantage of digital in-circuit testing is that it is performed directly on the inputs and outputs of a targeted device. If the device should fail testing, this is seen directly, rather

---

* This abuse can be mitigated by "conditioning" upstream drivers. One method is to disable the upstream drivers by devoting additional tester drivers to driving states on disabling inputs (when they exist) such as output enable pins. Of course, the problem may simply be moved upstream, because the disabling inputs may also be connected to upstream drivers that are now being overdriven in place of the original drivers. This conditioning process may be carried out to several levels, if desired.

than having its identity masked by interactions with other devices. This is a major differentiation from functional testing. Faults typically can be resolved to two categories of defects: failed ICs or solder opens on I/O/power pins.

Another major differentiator is the ease—indeed, automation—of test programming that is possible with digital in-circuit testing. Tests for ICs can be prepared as if the ICs were standing alone,* stored in a library, and recalled from the library when needed. Modern digital in-circuit testers may have library tests for tens of thousands of devices. For custom, one-of-a-kind ICs for which a library test may not exist (e.g., ASICs), it is still substantially easier to create a test for just the one device than it is for a collection of ICs.

52.5.3 Manufacturing Defect Analyzer (MDA)

A manufacturing defect analyzer (MDA) is essentially a very low-end analog in-circuit tester. One way it maintains low equipment cost is by not having power supplies to power up a board. Another cost savings comes by having only rudimentary programming and operating software. Some amount of test accuracy and yield must be traded for these savings.

52.5.4 General-Purpose In-Circuit Tester

The workhorse of the electronics industry is the general-purpose in-circuit tester that merges support for analog and digital in-circuit tests. An example of a widely used system is shown in Fig. 52.10. It contains power supplies for powering boards and often contains sophisticated

---

* This assumes that the IC does not have any topological constraints on its I/O pins, such as having an input pin connected directly to ground or an output pin fed back to an input pin. In such cases, the prepared test may be incompatible with these constraints and may only be used after incompatible segments of the test are deleted, reducing fault coverage.

FIGURE 52.10 Example of a commercial in-circuit test head and operator terminal, with a printed circuit board mounted in the testing position on top of the bed-of-nails fixture.

analog in-circuit programming tools and extensive libraries of digital tests. The typical test and repair flow for this tester is shown in Fig. 52.11. Note the early exit to repair for boards that fail shorts testing. This avoids applying power to boards that contain shorts because these may present hazards to the board and human operator, plus they also confuse the diagnosis of faults later in testing.

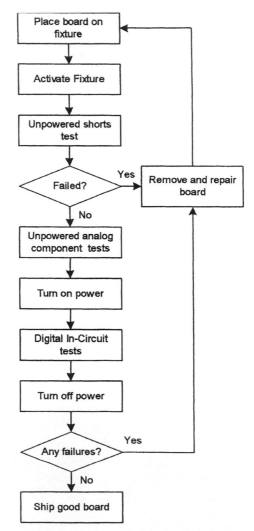

FIGURE 52.11   Typical test and repair flow for an in-circuit tester.

52.5.5   Combinational Tester

In situations where a manufacturing line has a variety of technologies in production, a need may exist for both functional and in-circuit testing. Thus, hybrid functional/in-circuit testers (commonly called combinational testers) exist. These machines give test engineers a full complement of tools to address testing problems. They also allow "one-stop" testing, where manufacturing defects and functional performance faults can be detected at one site in the production flow.

A combinational tester utilizes a bed of nails to perform in-circuit analog and digital tests. It may also utilize edge connector access to perform functional tests. In some cases these approaches are hybridized by constructing a two-stage bed-of-nails fixture that has two lengths of nails used in the bed, and a platen that has two stations of depression during operation. The first station is full depression, which brings all nails into contact with the circuit board for standard in-circuit access. The second station is partial depression, where only the longer nails still contact the board, perhaps at the board edge, for functional testing. The removal of the shorter nails removes the electrical loading they present to internal circuit nodes and allows the board to operate in a more natural environment.

52.6 ALTERNATIVES TO CONVENTIONAL ELECTRICAL TESTS

Alternative tests use radically different approaches to specifically address the resolution of defects. These may be needed to address blind spots in traditional electrical test methodologies. Two specific examples of defects that are difficult to resolve by conventional tests are given here.

First, consider a board with a large number of bypass capacitors. All of these capacitors are connected between power and ground so their parallel capacitances are summed. Using an analog in-circuit test, it is possible to test for this summed capacitance. However, if one capacitor is missing or tombstoned, the tester most likely will not notice because the resulting decrement in capacitance will likely fall well within the summed tolerances of the summed capacitances. The performance of the board at higher operating frequencies may be adversely affected, however, due to a loss of noise immunity. This could be detected (possibly) by a performance test. However, if this performance test failed, it could be very difficult to resolve which capacitor was missing. One non-traditional way to test for this problem is to use human visual inspection to look for missing capacitors. This may not find solder opens, however. Another approach would be to use an automated x-ray laminographic tester to check for the existence and solder integrity of each capacitor. Even the orientation of polarized capacitors can be verified. (See Chap. 50 for a discussion of x-ray inspection.)

Second, consider again the open solder problem on a digital device input. This defect may manifest itself as improper device behavior, but it would be wrong to replace the device (which would also fix the solder open) because the device is not at fault. Digital in-circuit testing has trouble resolving solder opens on input pins from bad devices.* Final resolution can be obtained either by visual inspection or by using a handheld probe to see if board signals reach IC legs, but this is becoming increasingly difficult as packaging dimensions continue to shrink (e.g., TAB) and newer attachment technologies (e.g., ball grid arrays) are used that prohibit visual inspection or probing altogether.

An alternative approach to the open solder problem uses a capacitive coupling technique to look for opens. The technique exploits the fact that many ICs have a leadframe that forms the conductive path from the legs of the device to the die bond wire pads. Using the bed of nails, all but one node attached to the IC can be grounded (this is an unpowered technique) and a small AC signal can be applied to the node that remains. An insulated metal plate pressed against the top of the IC forms the top plate of a capacitor and the stimulated IC leg and lead frame conductor form the bottom plate. (See Fig. 52.12.)

---

* Ignore for the moment the possibility of using boundary scan to solve this problem, which is discussed in Chap. 51.

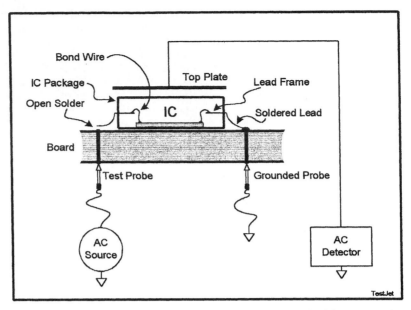

FIGURE 52.12   Capacitive leadframe testing used to find open solder joints.

Figure 52.13 shows an equivalent circuit for capacitive opens test for a properly soldered IC lead and for an open solder condition. The capacitance C1 may be on the order of 100 fF (100×10^{-15} F), which is small enough to require sophisticated detection electronics to measure in the face of environmental noise. Now, if the IC leg is soldered to the stimulated board node, the correct capacitance will be measured. If the solder joint is open, then a second small capacitance C2 now exists in series with the first. This reduces the measured capacitance by a factor of 2 to 10. Capacitive leadframe test allows testing of complex ICs for solder opens without knowing what the ICs actually do and without applying power. The technique requires no complicated programming, and gives accurate resolution of solder defects. The technique has been extended to allow testing of solder integrity for connectors and switches. A fixture for capacitive leadframe testing is shown Fig. 52.14.

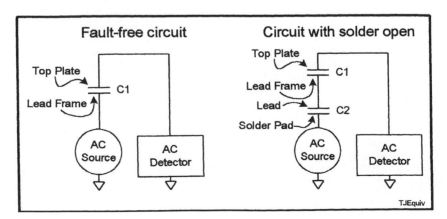

FIGURE 52.13   Equivalent circuits for fault-free and open solder conditions.

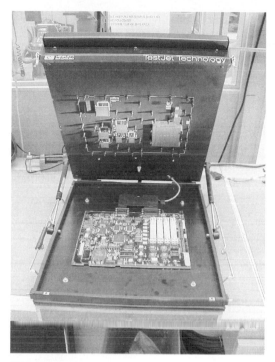

FIGURE 52.14 A test fixture with board on a bed of nails, with clamshell top fixture containing capacitive sensor plates. Notice eight closely spaced sensor plates (right) that test for opens in eight (white) connectors on the board.

52.7 TESTER COMPARISON

Table 52.2 summarizes the types of tester for comparative costs and capabilities. A majority of manufacturers value good diagnostic resolution and fault coverage, so it is not surprising that in-circuit and combinational testers are in widespread use. Next in prevalence are the MDA testers, which are typically used where coverage and diagnostic resolution may be sacrificed

TABLE 52.2 Costs and Capabilities of Various Testers

Tester type	Typical cost ($)	Programming time	Diagnostic resolution	Fault cover	Comments
MDA	10^4–10^5	1–2 days	Fair	Poor	No digital coverage; requires known good board for programming.
In-Circuit	10^5–10^6	5–10 days	Best	Good	Fixturing is a major portion of preparation time.
Combo	10^5–10^6	10–30 days	Best	Best	Functional test programming is a major portion of preparation time.
Functional	10^5–10^6	1–4 months	Fair	Fair	Very high skill required for test preparation and interpreting results.
Specification	10^3–10^8	Weeks to years	Poor	?	Very high skill required for test preparation and interpreting results.

for ease and speed of programming, usually in low-cost, high-volume products. Functional and specification testers are becoming rare, and are often only justified by the existence of contractual or regulatory requirements.

REFERENCES

1. W. A. Groves, "Rapid Digital Fault Isolation with FASTRACE," *Hewlett Packard Journal,* vol. 30, no. 3, March 1979.

2. K. P. Parker, *Integrating Design and Test: Using CAE Tools for ATE Programming,* Computer Society Press of the IEEE, Washington, DC, 1987.

3. D. T. Crook, "Analog In-Circuit Component Measurements: Problems and Solutions," *Hewlett Packard Journal,* vol. 30, no. 3, March 1979.

4. G. S. Bushanam et al., "Measuring Thermal Rises Due to Digital Device Overdriving," *Proceedings of the International Test Conference,* Philadelphia, PA, October 1984, pp. 400–407.

5. V. R. Harwood, "Safeguarding Devices Against Stress Caused by In-Circuit Testing," *Hewlett Packard Journal,* vol. 35, no. 10, October 1984.

CHAPTER 53
RELIABILITY OF PRINTED CIRCUIT ASSEMBLIES

Judith Glazer*

*Hewlett-Packard Company, Electronic Assembly
Development Center, Palo Alto, California*

This chapter describes the response of functional printed circuit board assemblies (PCAs) to environmental stresses—that is, their reliability in service—and the influence of design, materials, and manufacturing decisions on this behavior. A variety of stresses may be present in the service environment of the assembly. *Thermal* stresses come from fluctuations in the ambient temperature in the service environment of the assembly or from power dissipation of high-power devices mounted on the printed circuit board (PCB). There are also thermal stresses associated with assembly and rework. *Mechanical* stresses may be due to bending and flexing of the assembly during later assembly steps or in service, mechanical shock during transportation or use, or mechanical vibration, for example, from cooling fans. *Chemical* sources of environmental stresses include atmospheric moisture, corrosive gases (for example, smog or industrial process gases), and residual chemically active contaminants from the assembly processes (for example, from flux). These environmental stresses may act singly or in concert with one another and the electrical potential differences that exist when the assembly is functioning to cause electrical failures in the printed circuit assembly (PCA). This chapter will focus on the reliability of the PCB and the interconnects to it. Reliability of the electrical components themselves is beyond the scope of this chapter (see Fig. 53.1).

By defining reliability as the response of functioning assemblies to environmental stresses, we have excluded the large class of production defects that are detected in the testing processes immediately after manufacturing or that will cause the assembly to be nonfunctioning from the outset. This chapter will focus on the delayed effects of manufacturing defects and the wear-out mechanisms of properly manufactured product.

The remainder of this chapter is organized into six major sections:

53.1 Fundamentals of Reliability

53.2 Failure Mechanisms of PCBs and Their Interconnects

53.3 Influence of Design on Reliability

53.4 Impact of PCB Fabrication and Assembly on Reliability

---

* Significant portions of this chapter are drawn from T. A. Yager, "Reliability," chap. 30, *Printed Circuits Handbook*, 3d ed. (Coombs, ed.), 1988.

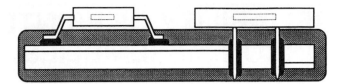

FIGURE 53.1   Schematic drawing of a printed circuit board assembly. This chapter focuses on the reliability of the printed circuit board and the interconnect between the printed circuit board and the components (shaded area in the drawing).

    53.5   Influence of Materials Selection on Reliability

    53.6   Burn-in, Acceptance Testing, and Accelerated Reliability Testing

Where applicable, each section covers printed circuit boards, printed circuit board assemblies, and components and their packages, in turn. Section 53.2 is the core of the chapter; it covers the fundamental of the failure mechanisms and is the assumed underlying basis of the subsequent sections. The breadth of this chapter, the complexity of the failure mechanisms involved, and the rapid evolution of the field mean that this chapter can provide only a brief overview of important topics in PCB and PCA reliability, many of which are the subject of books in their own right. The reader is encouraged to refer to the references and suggestions for further reading given at the end of the chapter before attempting quantitative reliability predictions.

53.1 FUNDAMENTALS OF RELIABILITY

53.1.1 Definitions

The reliability of a component or system can be defined as the *probability* that a functioning product at time zero will function *in the desired service environment* for a *specified amount of time.* Without these three parameters, the question "Is *x* reliable?" cannot be answered yes or no. Since reliability describes the probability that the product is still functioning, it is related to the cumulative number of failures. Mathematically, the reliability of an object at time *t* can be stated as

$$R(t) = 1 - F(t)$$

where $R(t)$ is the reliability at time *t* (i.e., the proportion of parts still functioning) and $F(t)$ is the fraction of the parts or systems that have failed at time *t*. Time may be measured in calendar units or some other measure of service time such as on/off cycles or thermal or mechanical vibration cycles. The unit of time that makes sense depends on the failure mechanism. When several failure modes are present, it is often helpful to think in terms of several time scales.

    A plot of the failure rate of a product as a function of time typically takes the shape of a "bathtub" curve (see Fig. 53.2). This curve illustrates the three phases that occur during the lifespan of a product from a reliability perspective. In the first, infant mortality phase, there is an initially high but rapidly declining failure rate caused by infant mortality. Infant mortality is typically caused by manufacturing defects that went undetected during inspection and testing and lead to rapid failure in service. Burn-in can be used to remove these units before shipment. The second phase, the normal operating life of the product, is characterized by a period of stable, relatively low failure rates.

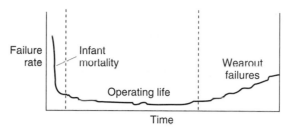

FIGURE 53.2 Classic bathtub reliability curve showing the three stages during the life of a product from a reliability perspective: infant mortality, steady-state, and wear-out.

During the operating life, failures occur apparently randomly and the failure rate r is roughly constant with time. An exponential life distribution is often assumed to describe the behavior in this region. In that case,

$$r = \left(\frac{N_t}{N_o} \right)\left(\frac{1}{\Delta t} \right)$$

and

$$R(t) = e^{-rt} - e^{\frac{-t}{\text{MTBF}}}$$

where N_t = number of failures in time interval Δt
N_o = number of samples at the beginning of the interval
MTBF = mean time between failures

During the third phase, the wear-out period, the failure rate increases gradually due to wear-out phenomena until 100 percent of the units have failed. For some systems, the second steady-state region may not exist; for solder joints, the wear-out region may extend over most of the life of the assembly. Understanding wear-out phenomena, which manifest themselves in properly manufactured parts after a period of service, and predicting when they will significantly affect the failure rate, are the primary focuses of this chapter.

Most wear-out phenomena can be characterized by cumulative failure distributions governed by either the Weibull or the log-normal distribution. Weibull distributions have been successfully used to describe solder joint and plated-through-hole fatigue distributions, while log-normal distributions are generally associated with electrochemical failure mechanisms. While these distributions may be quite narrow in some cases, their use should serve as a reminder that even with nominally identical samples, failures will be statistically distributed over time.

A practical use of fitting a distribution to reliability data is to extrapolate to smaller failure rates or other environmental conditions. To simplify the equations, the expressions in the text refer to the mean life of the relevant portion of the assembly. If the constants that define the failure distribution are known, the time to reach a smaller proportion of failures may be readily calculated. For example, for failure modes that are described by a Weibull distribution, the time t to reach $x\%$ failures is given by:

$$t(x\%) = t(50\%) \left[\frac{\ln(1 - 0.01x)}{\ln (0.5)} \right]^{1/\beta}$$

where β is the Weibull shape parameter, usually between 2 and 4 for solder joint failures.

53.1.2 Reliability Testing

Almost every reliability test program must solve the problem of determining whether an object is reliable in a calendar time period that is much shorter than the expected use period. Obviously, one cannot spend 3 to 5 years testing a personal computer that will be marketed for an even shorter time span or 20 years testing a military system. Depending on the failure mechanism, there are two approaches, which may be combined: (1) accelerate the frequency of the occurrence that causes failure and test the ability of the object to survive the expected number of events, or (2) increase the severity so that fewer occurrences are needed. Drop tests that simulate shock during transportation are an example of the first approach. Since the time between drops does not affect the amount of damage caused, a lifetime of drops can be conducted in rapid succession. However, the effect of temperature and humidity on corrosion over the lifetime of the product can be tested only by increasing the temperature, the humidity, or the concentration of contaminants, or some combination of these. The difficulty is ensuring that the test reproduces and/or correlates to the failure mechanism in service.

To use this data for making true reliability predictions—that is, *the probability of failures at a given time under given conditions*—testing must be continued until enough parts fail that a life distribution can be estimated. Unfortunately, this process can be time consuming and qualification tests are often substituted. Qualification test protocols specify a *maximum number of failures that may be observed in a specified period* in a sample of specified size. If few or no failures occur, a qualification test provides almost no information about the failure distribution; for example, the probability of failure during the next time interval is unknown. This limitation of qualification testing is minimized when the life distribution for properly manufactured samples is already known or can be estimated based on experience with similar designs. Many of the "reliability" tests described in Sec. 53.6 are actually qualification tests.

Many reliability or qualification testing schedules follow neither of these schemes. Instead, they test the ability of the product to survive a sequence of tests under extremely severe conditions for a short time or small number of exposures. Again, this type of testing may be adequate when it is supported by long experience with both the product type and its use environment; however, it is risky because it is not based on ensuring that probable failure modes will not occur in the life of the product. When new technologies or geometries are introduced, the old tests may not always be conservative. By the same token, irrelevant failure modes that would not occur in service may be introduced by the harsh test conditions.

53.2 FAILURE MECHANISMS OF PCBs AND THEIR INTERCONNECTS

This section will discuss the most important failure mechanisms of PCBs and the interconnects between PCBs and the components mounted on them. The discussion of PCB failure mechanisms will be more detailed since interconnect failures have been described far more extensively elsewhere. Whatever the environmental stress or the material response, these failures ultimately manifest themselves in terms of the functionality of the assembly, first as a change in electrical resistance between two points and then as electrical shorts or opens.

53.2.1 PCB Failure Mechanisms

PCB failure mechanisms fall into three groups: thermally induced failures, of which plated-through-holes are the most important example; mechanical failures; and chemical failure mechanisms, of which dendritic growth is the most important example.

53.2.1.1 *Thermally Driven Failure Mechanisms.*

PCBs are exposed to thermal stresses in a variety of situations. These may be either prolonged exposure to an elevated temperature or isolated or repeated temperature cycles. These temperature cycles can cause various PCB failures. The most important sources of thermal stress are:

- *Thermal shocks and thermal cycles during PCB manufacturing.* Thermal shocks are usually defined as temperature ramps faster than ~30°C/s, but include any ramp fast enough that temperature differentials play an important role. Examples include solder mask cure and hot-air solder leveling.

- *Thermal shocks and cycles during printed circuit assembly.* Examples are glue cure, solder reflow, wave soldering, and rework using a soldering iron, hot air, or molten solder pot.

- *Ambient thermal cycles in service.* Examples are going from inside to outside temperatures or ground to upper atmospheric temperatures, and elevation in box temperature due to heat dissipation from functioning electronic components.

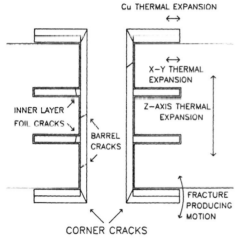

FIGURE 53.3 Schematic diagram of a plated-through-hole in a cross section of a four-layer printed circuit board showing common failure locations under thermal stress.

The primary PCB failure mechanisms accelerated by these thermal stresses are plated-through-hole cracking and delamination of the laminate.

Plated-Through-Hole Failures Due to Thermal Shocks or Cycling. Plated-through-holes (PTHs) are the most vulnerable features on PCBs to damage from thermal cycling and the most frequent cause of printed circuit board failures in service. PTHs include holes for through-hole (TH) components and vias that make electrical connections between layers. Figure 53.3 shows the common failure locations. Most organic resin-matrix substrate materials are highly anisotropic, with a much higher CTE above the glass transition temperature T_g in the through-thickness (z) direction than in the plane of the woven matrix cloth (the x-y plane of the board). Since above T_g the CTE climbs sharply, aggressive thermal cycles can result in large strains in the z direction and, consequently, on the PTHs (see Fig. 53.4[1]). The PTH acts like a rivet, which resists this expansion, but the Cu barrel is stressed and may crack, causing electrical failure. Figure 53.4 also illustrates the increasing strain on the barrel associated with a high temperature excursion. Failure may occur in a single cycle or may take place by initiation and growth of a fatigue crack over the course of a number of cycles. For high-aspect-ratio through-holes subject to repeated thermal shocks from room temperature to solder reflow temperatures (220 to 250°C) during board fabrication (e.g., hot-air solder leveling) and assembly (reflow, wave soldering, rework), it is not unheard of to encounter failures after 10 or fewer of these thermal cycles.

On a physical level, the number of thermal cycles to failure is affected by the strain imposed on the Cu in each cycle and the fatigue resistance of the copper. These factors are in turn controlled by a number of environmental, material, and manufacturing parameters. Low-cycle metal fatigue, in which most of the strain is plastic strain, can be treated approximately with the Coffin-Manson relation:

$$N_f \propto \frac{1}{2} \left(\frac{\varepsilon_f}{\Delta \varepsilon} \right)^m$$

where N_f = number of cycles to failure
  $\Delta \varepsilon$ = strain
  ε_f = strain ductility factor, which correlates closely with tensile ductility
  m = constant near 2.

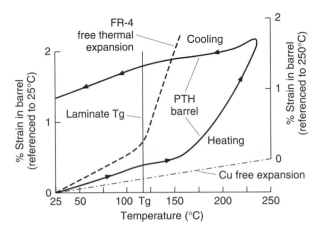

FIGURE 53.4 Strain vs. temperature for FR-4 (epoxy-glass), copper, and a PTH barrel in an FR-4 board during a single thermal cycle from 25 to 250 to 25°C. While the thermal expansion of the individual materials is fully reversible, much of the strain in the Cu PTH barrel is plastic, so most of the strain is not reversed during cooling. Note that the rate of thermal expansion of the FR-4 increases sharply at T_g. Results from Ref. 1.

This relation will significantly underestimate life for high cycle fatigue which can occur after repeated thermal cycling in service. The strain $\Delta\varepsilon$ can be estimated by finite element modeling or analytically. If no other data are available, ε_f for electroplated Cu can be approximated as 0.3.

The number of cycles to failure can be increased by increasing $\varepsilon_f/\Delta\varepsilon$, primarily by decreasing $\Delta\varepsilon$ by:

- Decreasing or eliminating thermal shocks by preheating the board before hot-air leveling, wave soldering, rework with a solder pot, etc.
- Decreasing the size of the thermal cycle (see Fig. 53.5[2]). Decreasing the size of the thermal cycle is the single most effective measure for increasing the life of the PTH, especially if the thermal cycle exceeds T_g.
- Decreasing the free thermal expansion of the laminate over the thermal cycle. The free thermal expansion can be reduced primarily by choosing a laminate material with a higher T_g, but also by choosing a laminate material (e.g., with Aramid fibers) with a low CTE below T_g (see Fig. 53.6).
- Decreasing the PTH aspect ratio (usually quoted as board thickness divided by finished hole size) by decreasing the board thickness or increasing the hole diameter (see Fig. 53.7). Aspect ratios tend to be higher in boards with eight or more layers because of their thickness and via density; aspect ratios greater than 3:1 require good-quality plating and aspect ratios higher than 5:1 are not recommended, in part because of the difficulty of achieving adequate plating thickness in the center of the barrel.
- Increasing the Cu plating thickness (see Fig. 53.8[2]). Increasing the plating thickness also increases the distance a fatigue crack must propagate to cause an electrical failure.
- Using Ni plating over the Cu (see Sec. 53.5.1.3 for more discussion).

The ratio $\varepsilon_f/\Delta\varepsilon$ can be increased by:

- Increasing the Cu ductility (increases ε_f) and yield strength (decreases ε). Cu strength and ductility are often inversely related, so these two factors must be balanced against one

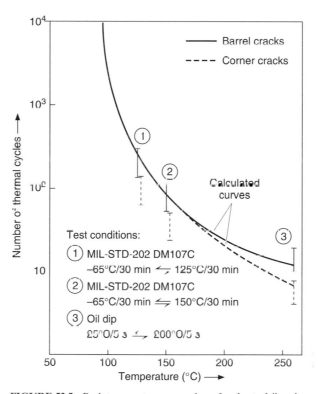

FIGURE 53.5 Peak temperature vs. number of cycles to failure by PTH barrel (solid bars) or corner (dashed bars) cracking in three different tests. Calculated lines are shown for comparison. Results are for pyrophosphate copper and FR-4. Other parameters are total strain energy required to cause fracture, 50 J/cm³; hole radius, 0.45 mm; distance from hole center to free end, 0.8 mm; plating thickness, 0.02 mm; distance from hole center to pad edge, 0.8 mm; board thickness, 2 mm. Results from Ref. 2.

another. However, the strength-ductility relationship can be altered by the choice of plating bath and plating conditions.

The number of cycles to failure can be dramatically decreased by defects in the hole wall or Cu plating in the hole or PTH knee that act as stress concentrations (increasing the local stresses and strains) and/or facilitate crack initiation. Because of the importance of this failure mode, it has been extensively studied experimentally and with analytical modeling techniques and more quantitative models are available.[3,4,5]

Laminate and Cu/Laminate Adhesion Degradation.   When a PCB is exposed to elevated temperatures for long periods of time, the adhesion between the Cu and the laminate and the flexural strength of the laminate itself will gradually degrade. Discoloration is usually an early symptom.

Several standards tests are used to compare the thermal resistance of different laminate materials. Cu adhesion is measured using a peel test.[6] Adhesion at elevated temperatures or after elevated temperature exposure gives some insight into the ability of the material to withstand rework and other high-temperature processes. Flexural strength stability is compared by measuring the times at 200°C before the flexural strength decreases to 50 percent of

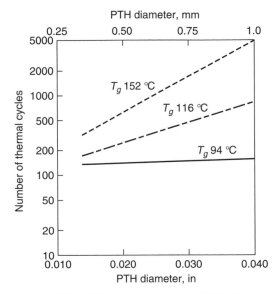

FIGURE 53.6   Effects of substrate T_g and PTH diameter on mean number of cycles to failure. The thermal cycle was 2-h cycle with extremes at −62 and +125°C. Multilayer printed circuit board thickness 0.10 in (2.5 mm); Cu in unfilled PTH is 1.2 mil (30 μm) thick. Results from Ref. 33.

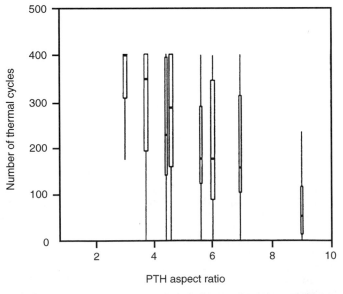

FIGURE 53.7   Cycles to failure vs. PTH aspect ratio for −65 to +125°C thermal shock cycles. Various hole diameters, board thicknesses, and board constructions. *(After Ref. 3.)*

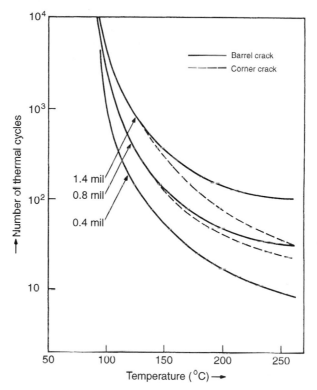

FIGURE 53.8 Effect of PTH plating thickness on number of thermal cycles to failure for thermal cycles with the indicated peak temperatures. For acid sulfate copper and FR-4 boards. Other hole parameters are the same as Fig. 53.5. *(After Ref. 2.)*

its original value. The quality of the bond between the resin and reinforcement is compared by measuring the time it takes for copper-clad laminate to blister during a solder float test at 290°C.[7]

53.2.1.2 Mechanically Induced Failures. PCBs may be mechanically loaded by test fixtures or processing equipment, when PCAs are loaded into card cages or fixtured into place with brackets, or when the assembly experiences mechanical shock or vibration in use. In general, once the PCA has been assembled, the interconnects to the components are the weak links in mechanical loading situations, not the boards themselves.

53.2.1.3 Electrochemical Failure Modes. The primary function of a printed circuit board is to provide electrical connections with the desired, stable, low-impedance and high-impedance insulation between them. A high surface insulation resistance (SIR) value is usually assumed by the circuit designer. Exposure to humidity, especially when ionic contaminants are present, is a common cause of insulation resistance failures that is accelerated by elevated temperatures and electrical bias. The impedance often decreases slowly over a long period of time. If the SIR value falls below the designed level, there will be cross talk between circuit elements that should be isolated, and the circuit may not function properly. Insulation resistance deterioration is particularly harmful for analog measurement circuits. If these circuits are used to measure low-voltage, high-impedance sources, changes in circuit impedance can result in the deterioration of instrument performance. Medical products that use sensors attached to a patient also pose special concerns, because deterioration of insulation resistance has the potential to cause electrical shock. For general applications, the surface resistivity is usually specified to exceed 10^8 $\Omega/\square$, but for these specialized applications higher values may be required. Electrochemical failures are usually accelerated by temperature, humidity, and applied bias.

High humidity is a significant cause of reliability problems because many corrosion mechanisms require water to operate. A humid environment is an excellent source of water, even when it is not condensing. Polymers commonly used in PCBs are hygroscopic; that is, they absorb moisture readily from the environment. This phenomenon is reversible; the moisture can be driven out of the PCB by baking it. The amount of moisture absorbed and the time to reach equilibrium with a humid environment depend on the laminate material, its thickness, the type of solder mask or other surface coating, and the conductor pattern.

The moisture absorbed by the PCB and ionic contaminants on or in the PCB play a role in a number of failure modes. Because the permittivity of water is much higher than that of most laminate materials, the increased water content can significantly affect the dielectric constant of the laminate, and thus may affect the electrical functionality of the board by increasing the capacitive coupling between traces. Absorbed and adsorbed water can lower SIR values, especially in the presence of ionizable contaminants (often from flux residues) and DC bias. The introduction of no-clean assembly processes has significantly increased the importance of measuring SIR values because contaminants left on the PCB remain there after assembly. Industrial pollutants can also be a source of ions that accelerate corrosion. In addition, typical industrial pollutants such as NO_2 and SO_2 can damage many materials used in PCAs, particularly elastomers and polymers. Some important mechanisms that cause failures due to low insulation resistance include dendritic growth and metal migration, galvanic corrosion, and conductive anodic filament growth. Whiskering can also cause electrical shorts, but neither electrical bias nor moisture is required.

Conductive Contaminant Bridging. Bridging of circuits by conductive salts may occur if plating, etching, or flux residues remain on the board. These ionic residues are good conductors of electricity in a moist environment. They tend to migrate across both metallic and insulating surfaces to form shorts. Corrosive byproducts, such as chlorides and sulfides formed in industrial environments, are chemically similar and can also cause shorting. An example of this type of failure is shown in Fig. 53.9.[8]

Dendritic Growth. Dendritic growth occurs by the electrolytic transfer of metal from one conductor to another; consequently, it is also termed *electrolytic metal migration.* It is also referred to as *electromigration,* although it should not be confused with the process that occurs in aluminum conductors in integrated circuits, which has a different mechanism. An example of a dendritic growth failure is shown in Fig. 53.10. Dendrites form on surfaces (including the interior surfaces of cavities) when the following conditions are met:

- Continuous liquid water film, a few molecules or more in thickness
- Exposed metals, especially Sn, Pb, Ag or Cu, that can be oxidized at the anode
- Low-current dc electrical bias

It is significantly accelerated by the presence of hydrolyzable ionic contaminants (for example, halides and acids from flux residues or extracted from polymers). Delaminations or voids that promote the accumulation of moisture or contaminants can promote dendritic growth. Conductive anodic filament growth (discussed later) is a special case of dendritic growth. Time to failure is inversely proportional to spacing squared and voltage. The failure mechanisms in accelerated tests have been reviewed.[9]

Dendritic growths usually form from cathode to anode. Metallic ions formed by dissolution at the anode are transported along a conductive path and reduced and deposited at the

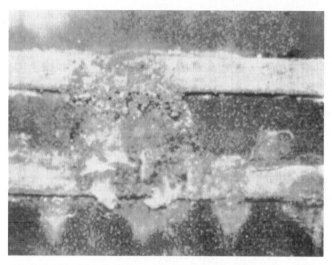

FIGURE 53.9 Migration of corrosion products across the surface of FR-4 bridging two conductors. From Ref. 11. *(From IPC-TR-476. Courtesy of Department of Defense.)*

FIGURE 53.10 Transmitted light micrograph through a PCB that failed in the field. Dendritic growth has formed at the interface of a UV-cured screened solder mask and the FR-4 surface.

cathode. The dendrite resembles a tree, since it consists of a stalk with branches. When the growth touches the other conductor there is an abrupt rise in current, which sometimes destroys the dendrite but may also cause an electrical circuit to temporarily malfunction or damage a device.

It has been proposed that the absorption of moisture produces an electrochemical cell. The following electrode reactions for Cu are an example:

At the anode: $Cu \rightarrow Cu^{n+} + ne^{-}$

$H_2O \rightarrow \frac{1}{2}O_2 + 2H^+ + 2e^{-}$

At the cathode: $H_2O + e^{-} \rightarrow \frac{1}{2}H_2 + OH^{-}$

where the majority of the leakage is due to the electrolysis of water. Copper metal is dissolved at the anode and migrates to the cathode, where it is no longer soluble. The dendrite that forms follows the resulting pH gradient.[10] The voltage difference between the cathode and anode also affects the rate of dendrite growth. When the cathode and anode are the same metal (e.g., Cu), the voltage difference is determined primarily by the applied bias, although the access of moisture and air also has an effect. Corrosion can be accelerated in a crevice because an oxygen concentration differential between the anode and the cathode develops. When the metals are dissimilar, galvanic corrosion may occur without a bias voltage.

If there is an applied bias, dendritic growth will occur almost instantly if the cathode and anode are under water. A simple laboratory experiment can prove the point. A 6-V bias across two conductors is sufficient to induce rapid growth (readily observable with a low-power microscope) even when distilled or deionized water bridges the conductors, although growth will occur faster with tap water.[11]

Galvanic Corrosion.   Galvanic corrosion occurs between dissimilar metals because they have differing affinities for electrons (i.e., they are more or less electronegative). Galvanic series have been compiled for many common metals and alloys (see Table 53.1). Metals near the top of the series (noble metals) do not corrode; those near the bottom corrode easily. When these metals are near each other, the more noble metal becomes the cathode, the less noble the anode. Moisture is required to couple the two metals electrically. Applied bias is generally not required, but may accelerate the reaction if the polarity is correct. When the anode is very small compared to the cathode, its corrosion can be very rapid. Conversely, if the anode is much larger than the cathode, corrosion is unlikely to be serious, particularly if the difference in electronegativity is small.

Conductive Anodic Filament Growth.   Conductive anodic filament growth (CAF) causes electrical shorts when a metal that dissolves anodically is redeposited at the interface between the glass (or other) fibers and the resin matrix of a printed circuit board. Conductive anodic filament growth is promoted by delamination at the glass-polymer interface, which may in turn be promoted by various environmental stresses including high temperatures (greater than about 260°C for FR-4) and thermal cycling. Shorts seem to occur most rapidly when a single fiber bundle connects two pads. Once delamination has occurred, the metal migration that causes shorts to occur is promoted by increasing temperature, relative humidity, and applied voltage. Small conductor spacings also significantly decrease times to failure.[12] In multilayer boards, failures occur faster on outerlayers than innerlayers because the surface layer absorbs moisture more readily. By the same reasoning, solder mask and conformal coating both increase the time-to-failure because they slow the absorption of moisture from the atmosphere into the board.

Whiskers.   Whiskers are faceted filament-like structures that grow spontaneously on the surface of a plated metal and can cause shorts between closely spaced conductors (see Fig. 53.11). Whiskering can be differentiated from other causes of shorts such as dendritic growth, because neither an electrical field nor moisture is required for whiskers to form. Whiskering is a particular problem with pure tin. The whiskers grow in response to internal stresses in the plating or external loads. Sn whiskers are commonly 50 µm long and 1 to 2 µm in diameter.

TABLE 53.1   Standard Electromotive Force Potential (Reductions Potentials) for Elements Commonly Found in Electronic Assemblies

	Reaction	Standard potential (Volts vs. standard hydrogen electrode)
Noble	$Au^{3+} + 3e^- = Au$	+1.498
	$Cl_2 + 2e^- = 2Cl^-$	+1.358
	$O_2 + 4H^+ + 4e^- = 2H_2O$ (pH 0)	+1.229
	$Pt^{3+} + 3e^- = Pt$	+1.2
	$Ag^+ + e^- = Ag$	+0.799
	$Fe^{3+} + e^- = Fe^{2+}$	+0.771
	$O_2 + 2H_2O + 4e^- = 4OH^-$ (pH 14)	+0.401
	$Cu^{2+} + 2e^- = Cu$	+0.337
	$Sn^{4+} + 2e^- = Sn^{2+}$	+0.15
	$2H^+ + 2e^- = H_2$	0.000
	$Pb^{2+} + 2e^- = Pb$	−0.126
	$Sn^{2+} + 2e^- = Sn$	0.136
	$Ni^{2+} + 2e^- = Ni$	−0.250
	$Fe^{2+} + 2e^- = Fe$	−0.440
	$Cr^{3+} + 3e^- = Cr$	0.744
	$2H_2O + 2e^- = H_2 + 2OH^-$	−0.828
	$Na^+ + e^- = Na$	−2.714
Active	$K^+ + e^- = K$	−2.925

Source: A. J. deBethune and N. S. Loud, *Standard Electrode Potentials and Temperature Coefficients at 25C,* Clifford A. Hampel, Skokie, Ill., 1964.

Once started, they may grow as fast as 1 mm per month. The tendency toward whisker growth is influenced by a variety of factors including plating conditions and the characteristics of the substrate. Growth may be inhibited by a Cu or Ni barrier layer beneath the tin coating. Pb seems to suppress whisker growth; eutectic Sn Pb solder is considered almost immune. Whiskers do not cause the corrosion resistance or solderability of the tin coating to deteriorate, so tin may be used as a temporary finish. To avoid whiskering, plated pure Sn should not be used on closely spaced conductors that could short during service, such as connector terminations or component leads.[13,14]

53.2.2   Interconnect Failure Mechanisms

53.2.2.1   Thermally Driven Failure Mechanisms

Thermal Fatigue of Solder Joints.   Thermal fatigue of solder joints has been extensively researched in the last decade. The mechanism of fatigue, accelerated testing methods, and methods for predicting life have all been described at length, although there is still much controversy about many of the details.[15–17] These references also illustrate how modern finite element methods can be used to model the strains in the solder under both operating and accelerated test conditions. This section briefly reviews some of the important principles underlying solder joint thermal fatigue.

The focus of the discussion is on surface-mount solder joints, which have been extensively researched; however, many of the same principles also apply to through-hole solder joints. Through-hole joints are generally less prone to solder joint fatigue failures, so long as the through-hole barrel is full of solder. Ideally, complete fillets should be observed on both surfaces of the board. Reviews of this topic can be found in Refs. 18 and 19.

Thermal fatigue in solder joints occurs because of the thermal expansion (CTE) mismatch between the PCB and the component interconnected by the solder joint (see Fig. 53.12). The

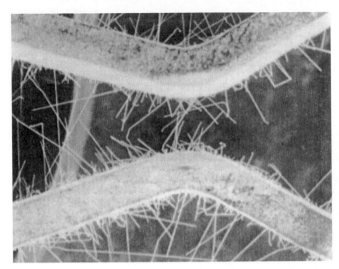

FIGURE 53.11   Tin whisker growth on a tin-plated surface. From Ref. 11. *(From IPC-TR-476. Courtesy of Burndy Corporation.)*

imposed thermal cycle ΔT results in an imposed cyclic strain $\Delta\varepsilon$ of the solder joint, which is generally the weakest part of the system. The relationship is simple under the assumptions that the part and substrate are rigid, the solder joints are relatively small, and that homogeneous shear deformation caused by the global CTE mismatch predominates:

$$\Delta\varepsilon = \frac{(\Delta T)(\Delta\alpha)l}{h} \tag{53.1}$$

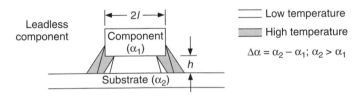

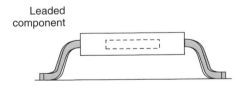

FIGURE 53.12   Schematic illustration of strains imposed on solder joints to leadless and leaded surface-mount components during a thermal cycle. Although the relative displacement of the substrate and component body is the same, the strain on the joint is reduced in the leaded case by the deflection of the lead.

where $\Delta\alpha$ = difference in thermal expansion coefficients of component and substrate
         l = distance between center of component and joint
         h = height of solder joint

If the component has leads or if the substrate is flexible, there will be some compliance in the system that will reduce the strain imposed on the solder joints. The local mismatch between the solder and the component lead or the pad or via metallization on the substrate can also contribute to the strains imposed on the solder.

Like plated-through-holes (PTHs), solder joints fail by a low-cycle-fatigue mechanism which can be crudely approximated by the Coffin-Manson relation

$$N_f = \frac{1}{2}\left(\frac{\varepsilon_f}{\Delta\varepsilon}\right)^m \tag{53.2}$$

where, again, N_f is the number of cycles to failure, ε_f is the fatigue ductility, and m is an empirical constant near 2. However, unlike the PTH case, the number of cycles to failure also depends on the frequency at which the cycles are imposed and the hold time at each temperature extreme. The reason for this dependence is that, for solders, the primary deformation mechanism causing thermal fatigue failures is creep.

The phenomenon of creep and its connection to fatigue are fundamental to understanding thermal fatigue of solder. Creep is time-dependent deformation that occurs gradually in response to a fixed imposed stress or displacement (see Fig. 53.13). Creep occurs by a variety of thermally activated processes. These processes play an important role only when the temperature exceeds half the melting temperature (in degrees Kelvin) of the material and, even then, the rate of deformation increases strongly with increasing temperature. For electronic solders, even room temperature is well above half the melting temperature; consequently, creep is the most important deformation mechanism of solder. When a displacement is first imposed, the strain is a combination of elastic and plastic strain. The elastic deformation is reversible and damages the microstructure relatively little, while the plastic deformation is permanent and contributes more significantly to the initiation and propagation of fatigue cracks in the solder (see Fig. 53.14). Given time, the creep process relieves some or all of the elastic stress through further permanent deformation. This additional deformation does further microstructural damage and increases the amount of plastic strain imposed when the thermal cycle is reversed. Because there is less time for creep to occur, rapid thermal cycles are less damaging than slow cycles or cycles with long hold times at the temperature extremes, a fact which is important in designing accelerated reliability tests as well as in service. The importance of creep makes the fatigue behavior of solder different from structural metals, such as copper, aluminum, or steel.

In summary, the effects of the thermal cycling profile on solder joint thermal fatigue life are as follows:

• *Temperature extremes:* Decreasing the size of the thermal excursion is the single most effective way to increase the life of the solder joints. Since creep occurs more rapidly at higher temperatures, decreasing the peak temperature of the thermal cycle further decreases the amount of creep deformation that occurs during the hold at high temperature.

• *Frequency:* The thermal fatigue damage per cycle is greater at lower cycling frequencies because there is more time for creep to occur, increasing the amount of permanent deformation. (Recall that most of the damage is

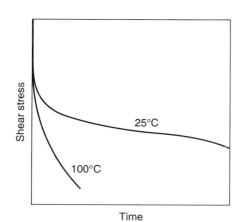

FIGURE 53.13 Typical behavior of solder in response to a constant applied displacement (for example, due to thermal expansion). The initial stress is relaxed over time as the solder elongates.

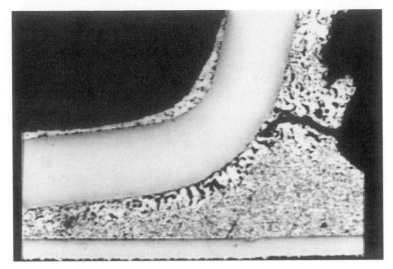

FIGURE 53.14 Thermal fatigue failure in eutectic Sn-Pb solder for a TSOP component. *(Photo courtesy K. Gratalo.)*

caused by the plastic deformation that occurs during each cycle, not by the cyclic stressing of the joint.)

- *Hold time:* As long as the stress on the solder joint remains nonzero, the thermal fatigue damage per cycle is increased if the hold time is lengthened, again because there is more time for creep to occur. Once the stress relaxation process has gone to completion, no further damage occurs and increasing the hold time further has no effect.

- *Thermal shock:* If the thermal cycle is extremely rapid, the components of the PCA may not be at the same temperature; consequently, the imposed strains may be larger or smaller than at slower rates.

Although the designer may be able to influence the peak temperature through cooling schemes, the thermal cycling profile and the frequency of thermal cycles in service are largely fixed by the application.

Solder joint fatigue life may be increased by decreasing the strain $\Delta\varepsilon$ imposed on the solder joint by:

- Choosing a package with a compliant attachment scheme. In this case, part of the strain is taken up by deflection of the lead, reducing the amount of strain in the solder. For these packages, the joint life can be further extended by decreasing the lead stiffness and increasing the joint area.

- Decreasing $\Delta\alpha$, the difference in the package and substrate thermal expansion coefficients, by carefully selecting the CTE of the package and substrate (see Secs. 53.5.3.1 and 53.5.1.1, respectively).

- Decreasing the size of the package, decreasing Δl.

- Increasing the height h of the solder joint.

Solder joint fatigue life may also be increased by:

- Decreasing the local CTE mismatch that occurs at the interfaces between the solder and the component lead and substrate metallization. While the substrate metallization is usually

Cu, which is moderately well matched to solder (17 vs. 25 ppm/°C), the leads may be made of a low-expansion metal such as Alloy 42 (~5 ppm/°C) or Kovar, as well as of Cu.

- Decreasing the mean stress imposed on the solder joint (for example, due to residual stress after assembly).

- Increasing c_f or decreasing the creep rate of the solder by controlling the solder joint microstructure or by selecting an alternative solder. (Finer microstructures, which may be achieved by faster reflow cooling rates, have significantly greater fatigue lives because they are resistant to fatigue crack initiation and propagation. Unfortunately, solder microstructures coarsen over time, even at room temperature. Some solders, such as Sn-4Ag and 50In-50Pb, seem to have significantly improved fatigue lives over eutectic Sn-Pb; however, their higher reflow temperatures are not necessarily compatible with FR-4. See Sec. 53.5.2.2.)

Thermal Shock.   Thermal shocks (>30°C/s) cause failures because differential heating or cooling rates introduce large additional stresses into the assembly versus thermal cycling. Under thermal cycling conditions, it is a good assumption that all components of the assembly are at approximately the same temperature. (High-power components may be an exception.) Under thermal shock conditions, different portions of the assembly are temporarily at different temperatures because their heating or cooling rates are not the same. These temperature transients are caused by differences in thermal mass and thermal conductivity across the assembly; they are caused by component selection and placement decisions and by differences in the physical properties of the materials used in the assembly. The temperature differences across the assembly and any resulting warpage can enhance the stresses normally imposed during temperature changes due to differences in CTE. Thermal shocks can cause reliability problems, such as solder joint failures in overload and crazing in conformal coatings leading to corrosion failures, as well as a range of component failures. Because of the differential thermal stresses that may be induced, thermal shock can cause failures that do not occur during slower thermal cycling between the same temperature extremes. On the other hand, rapid thermal shock cycles actually cause less solder joint fatigue than slower thermal cycles; because little creep occurs, more cycles are needed to cause solder fatigue failure.

53.2.2.2 Mechanically Induced Failures.

PCAs can also fail in response to externally imposed mechanical stresses, for example, due to mechanical shock or vibration in shipping or in use. These failures can be divided into two categories: overload failures and mechanical fatigue failures, which are caused by mechanical shocks and vibration, respectively. Susceptibility to mechanical failures is closely linked to design of the PCA and the housing in which it is installed. The design determines the resonant frequency of the board, which in turn determines its response to external mechanical stresses. Cantilevers with low natural frequencies, such as an edge-mounted PCB with an unanchored large mass in the center, are particularly prone to failure. Depending on connector design and mounting scheme, solder joints to surface-mount connectors can also be vulnerable, particularly if there are many connector insertion cycles. See Ref. 20 for a more detailed discussion of design methodologies for mechanical durability.

Overload and Shock Failures of Solder Joints.   When a printed circuit assembly is flexed, jolted, or otherwise stressed, solder joint failures may occur. In general, solder is the weakest material in the assembly; however, when it connects a compliant structure such as a leaded component to the board, the lead flexes and the solder joint is not placed under much stress. Solder joints to leadless components will see large stresses since the board can bend and the components themselves are usually rigid. These stresses can occur if the assembly is mechanically shocked, for example, if the unit is dropped, or during further assembly processes if the PCB is bent through a significant radius. The primary method of eliminating this failure mode is through package selection; however, other factors also play a role, including PCB design, process control during printed circuit board fabrication and assembly, and the shear strength, tensile strength, and ductility of the solder. Solder joints are particularly prone to failure in tension because the brittle intermetallics at the interface between the solder and the substrate are stressed. Joints with thick intermetallic layers are more susceptible.

Mechanical (Vibration) Fatigue.   Vibration (a common source is an improperly mounted fan) can cause solder joint fatigue by repeatedly stressing the interconnects. Metal fatigue can occur even when this stress is well below the level that causes permanent deformation (that is, the yield stress). As for thermal fatigue of solder joints, the number of cycles to failure in mechanical vibration fatigue can be described by the Coffin-Manson relation. However, in contrast to thermal fatigue, failures usually occur after very large numbers of small, high-frequency cycles in which most of the strain in the solder is elastic ($\varepsilon = \sigma/E$, where σ is the stress and E is the elastic modulus of the solder). Consequently, creep does not play an important role in vibration fatigue. Although the damage per cycle is minimal, the number of cycles can be extremely high; they are often imposed at 50 or 60 Hz. Over time, a crack can nucleate, and subsequent cycles serve to propagate this crack. Again, the risk is much higher for joints to large leadless parts, since there is no compliant structure to take up part of the stress. The amount of damage to the solder joint depends on the imposed strain in each cycle, which depends largely on whether the excitation frequency is close to the natural frequency of the board. The mass of the component (including any heat sink) also plays an important role.

53.2.2.3   *Electrochemically Induced Failures.*   The electrochemical failure mechanisms accelerated by temperature, humidity, and electrical bias that were described in Sec. 53.2.1.3 for printed circuit boards also apply to the remainder of the PCA. The solder used for the interconnects and the metal component terminations and lead frame finishes can also be involved in the reactions. The large number of dissimilar metals increases the complexity of the situation and the possibility of galvanic corrosion in a humid environment. In addition, contaminants introduced during printed circuit assembly such as flux residues can contribute to the failures.

53.2.3   Components

Although the failure mechanisms of electronic components in general have been described in detail elsewhere[21–24] and are beyond the scope of this chapter, there are a few failure mechanisms that are specifically associated with electronic assembly. In addition, component de-rating for high-temperature service should be evaluated if the unit will be exposed to a severe service environment. Component failures due to thermal shock, exceeding the maximum allowable component temperatures, and plastic package cracking can occur during reflow or wave soldering. These assembly-related failure mechanisms are described briefly as follows.

53.2.3.1   *Thermal Shock.*   Multilayer ceramic capacitors may crack if exposed to thermal transients exceeding 4°C/s. These cracks are usually invisible, but may be the sites for dendritic growth in service, when the assembly is exposed to moisture under applied bias. Capacitors with high values and larger thicknesses are most susceptible. These failures can be avoided by following the manufacturer's requirements for maximum temperature excursion and rate of temperature change.

53.2.3.2   *Overtemperatures.*   Many components, including connectors, inductors, electrolytic capacitors, and crystals, cannot survive the SMT reflow process, although most will survive wave soldering. Problems can include melting of internal soldered connections, melting or softening of polymeric capacitor dielectrics, and expansion of elastomeric materials. These failures can be prevented by carefully following the manufacturer's recommendations for maximum processing temperature.[25]

53.2.3.3   *Molding Compound Delamination in Plastic SMT Packages.*   Plastic-packaged integrated circuits are generally transfer molded using a filled epoxy-based compound. The plastic can absorb moisture, which tends to accumulate at interfaces within the package, such as the die attach paddle. Subsequent heating can cause the moisture to vaporize, causing delamination at the interface, eventually leading to package failure. This delamination phenomenon is also termed package cracking or "popcorning." The newer thin SMT components (e.g., TSOPs and TQFPs) are more susceptible because the distance the moisture must diffuse through the plastic to reach internal interfaces is shorter. Delamination when these components are exposed to high temperature may be prevented by ensuring that the components are dry through dry storage and/or baking.[23]

53.3 INFLUENCE OF DESIGN ON RELIABILITY

Design has a major influence on the reliability of any product. The implications of the demands of the application and expected service environment on product design should be considered as early as possible, since they can influence a wide range of decisions, including integrated circuit partitioning, package and substrate selection (which will impose specific design rules and electrical performance characteristics), component layout and box design, and heat sinking and cooling. IPC Standard D-279, Design Guidelines for Reliable Surface Mount Technology Printed Board Guidelines, is a good place to start in considering these issues. Section 53.2 has already described how design can promote or hinder certain failure mechanisms. Section 53.4 discusses the influence of materials, which are selected during the design process, on PCB and interconnect failures. This section highlights the importance of good thermal and mechanical design.

The size of the thermal cycles imposed on the PCA during power on/off cycles can have an overwhelming influence on the reliability of integrated circuits, solder joints, and plated-though-holes, especially if the external service environment is not particularly severe; consequently, good thermal design is critical to reliability. The thermal cycles imposed on the assembly can come from joule heating from high-power components and from ambient heating. Integrated circuit reliability depends on maintaining low enough junction temperatures, usually below 85 to 110°C, depending on the IC technology. During continuous operation, solder joint temperatures should be kept below about 90°C to avoid the extensive intermetallic growth and grain coarsening that occur during long-term exposure to higher temperatures. As described under Sec. 53.2.1.1 and 53.2.2.1, the size and number of thermal excursions directly affect the fatigue life of both solder joints and plated-though-holes. Component spacings, orientations, air velocities, and enhancements such as thermally enhanced packages, heat sinks, and fans can all have major effects on the thermal cycle experienced by the assembly. PCBs can also be enhanced with metal cores to improve heat dissipation.

As mentioned in the earlier descriptions of specific failure modes, package selection and via and PTH specification can have a major effect on reliability. Although small holes can be desirable from a design density perspective, use of the smallest holes (aspect ratios of 5:1 or greater) should be minimized to minimize the risk of PTH failures. This is especially true if the design includes large through-hole parts which are likely to be reworked frequently (for example, due to test escapes). Similarly, some package styles are more susceptible to solder joint fatigue than others. The effect of integration on reliability can depend on the difference in package styles for the options under consideration. Integration can have a positive effect by reducing the total number of connections that could fail. On the other hand, if it requires a large package with a large CTE mismatch to the substrate, integration may reduce the reliability of the assembly.

The effect of externally imposed mechanical shock and vibration on PCA reliability is largely determined by design factors, although substrate and package selection also play a role. Component placement and PCA mounting in the box determine the natural frequency of the board and, consequently, the extent to which the board deflects. High-mass packages, often due to large heat sinks, are particularly susceptible, especially if there is a large lever arm.

53.4 IMPACT OF PCB FABRICATION AND ASSEMBLY ON RELIABILITY

53.4.1 Effect of PCB Fabrication Processes

53.4.1.1 *Laminate and Lamination.*   Delamination in PCBs may occur between the laminate materials or between the laminate material and the Cu foil. One cause of delamination is defective laminate material. Defects such as incomplete bonding at the resin/fiber interface can result in delamination due to formation of voids at these interfaces. Other common causes of delamination are excessive lamination pressure and/or temperature, contamination at interfaces, heavily oxidized copper foil surfaces, and lack of oxide treatment to enhance adhesion between copper innerlayers and prepreg. Debonding increases the risk of conductive anodic filament growth because it provides a place for moisture to accumulate. It can also result in increased stresses on the plated-through-holes (PTHs) during thermal cycling.

Laminate voids and resin recessions are separations of the laminate material from the copper conductor that may occur during multilayer PCB lamination. Most acceptability specifications prohibit voids larger than 0.076 mm (0.003 in); however, smaller voids are not generally considered to be detrimental to reliability. Some of the causes of laminate voids are entrapped air during lamination, improper flow of resin, and improper epoxy cure, perhaps due to improper lamination pressure and/or temperature, inappropriate heating rate, or too little prepreg.

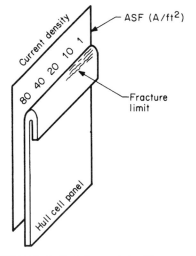

FIGURE 53.15 Ladwig panel ductility test. *(After Ref. 26.)*

53.4.1.2 *Cu Foil.*   The major cause of innerlayer foil cracks seems to be poor ductility of the Cu. Poor foil ductility can have a more significant effect on PTH reliability than such well-known culprits as insufficient plating thickness and excessive etchback. A minimum of 8 percent elongation is required for 1-oz foil to eliminate this problem. Foils plated in a Hull cell can be easily evaluated for room-temperature ductility using a 180° bend test. This technique is illustrated in Fig. 53.15, in which the sample panel is bent flat parallel to the axis along which current is varied.[26] Fractures occur at current densities producing low-ductility copper. This test can also be used to evaluate the influence of bath chemistry on ductility or as a bath monitor. Poor copper ductility can be correlated to the microstructure observed in metallographic cross sections.[27]

53.4.1.3 *Drilling and Desmear.*   Poor drilling and desmear (etchback) can cause PTH failures by providing stress concentrations that cause fatigue cracks to initiate. They can also cause voids and cracks at the interface to the copper plating, which can trap chemicals during plating and then contribute to conductive anodic filament growth. The following paragraphs describe the effects of poor desmear and some drilling defects which can cause poor plating, such as resin smear, rough walls, loose fibers, and burrs.

Resin smear can cause weak connections between plated-through-holes and innerlayer copper that fail under environmental stress. There is always some resin smear, which is removed by the desmear (etchback) process. If the desmear process is not effective, or if the resin smear is excessive, poor interconnection to the innerlayers can result. Possible causes of excessive smear are a dull drill or the wrong feed rate or drill speed, all of which can cause increased drill heating, resulting in more smear.

Similar errors in drilling setup can cause rough hole walls, loose fibers, or burrs. These defects are not serious in and of themselves, but can lead to rough plating or copper nodules, which introduce stress concentrations. Rough walls are typically associated with an incorrect feed rate or drill speed, or insufficiently cured material. Loose fibers may be caused by incorrect drilling parameters or improper cleaning. Burrs are usually associated with too fast a drill feed or a dull drill.

Poor drill registration can also decrease reliability of innerlayer via connections or the soldered connection to through-hole components. Poor registration can cause breakout on innerlayers; i.e., the drill hole may fall outside the pad on the innerlayer it is intended to connect to. Breakout increases the probability of PTH barrel failures. Breakout on outerlayers means that the solder fillet for a through-hole component will be partially missing, resulting in decreased reliability for some critical components.

Whether caused by excessive resin smear or not, poor etchback can result in a weak connection between the plating in the hole and the innerlayer copper. Etchback (see Fig. 53.16) removes laminate resin and woven glass in the hole so that the internal copper projects slightly into the hole, permitting the plating to make contact with the innerlayer foil on three sides. This strength is important to prevent cracking at the interface under thermal shock conditions. A review of innerlayer cracking in or around the electroless copper at the junction between the innerlayer foil and the electroplated Cu in the hole suggests that *negative* etchback, in which the electroplated Cu projects into the laminate, may also give good results. Zero etchback, when innerlayer foil is flush with the hole wall, is the most dangerous case because the bond line between the foil and the plated copper is located at the point of maximum stress.[28] Causes of insufficient etchback include improper lamination and curing, hardened epoxy smear, a depleted smear removal bath, or a host of process control issues, including improper bath temperature, agitation, or time exposure.

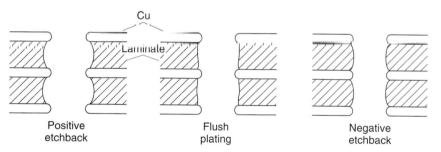

FIGURE 53.16   Schematic illustrations of positive, flush, and negative etchback. *(After M. W. Gray.[28])*

53.4.1.4 Plating.   Defects from the plating process can be responsible for a variety of PTH reliability problems; in addition, as previously described, problems in earlier process steps such as drilling and desmear often show up as plating defects.

Uniform coverage of the hole with electroless Cu is critical to the strength of the through-hole and the adhesion of the metallization to the laminate. Oxidation of the innerlayer copper prior to electroless Cu plating is one source of poor plating adhesion. Poor control of bath composition can have the same effect.

The adhesion of the electroplated Cu to the electroless Cu and the ductility of the electroplated Cu also strongly affect PTH reliability. If the adhesion between the layers is poor, this interface may be the weak point where failure initiates when the PTH is subjected to thermal stresses. Causes can include tarnished electroless copper that is insufficiently microetched, burning the electroless copper with too much current in electrolytic plating, and film contamination in the electrolytic copper.[28] Susceptibility to innerlayer cracks may be identified by looking for cracks in microsections after a solder float test. The fatigue life of the copper is directly related to its ductility. Plating process parameters and plating additives can strongly affect the plating ductility. For example, Mayer and Barbieri[29] found that good thermal shock resistance of electrodeposits of acid copper depended on proper concentrations of three additives:

1. A leveling agent, to smooth over surface imperfections. (Without the leveling agent imperfections are reproduced in the deposit.)
2. A ductility-promoting agent, which functions to produce the equiaxed grain structure.
3. A carrying agent, to guide the other two components to create the equiaxed structure. (Striations occurred with insufficient carrying agent.)

Additive levels below a certain threshold make the bath more susceptible to impurity effects. For example, iron contamination of 100 mg/L without the recommended concentration of ductility-promoting agents was found to produce a columnar grain structure at the hole corners. Similarly, organic contaminants such as photoresist can produce laminar deposits.

Insufficient plating thickness in the barrel also directly reduces PTH reliability because the stress and, consequently, the strain in the Cu are increased. Overall insufficient plating thickness can be caused by a depleted bath or insufficient plating time, among other things. Insufficient plating thickness in individual holes can also occur as a result of nonuniformities in plating current caused by nonuniform copper feature density. It is particularly difficult to obtain adequate plating thickness in the center of high-aspect-ratio PTHs; good process control is important for aspect ratios greater than 3:1. Good coverage is difficult to obtain for aspect ratios greater than 5:1 by electroplating.

What constitutes "sufficient" plating thickness in PTHs is a subject of some controversy. Specifications range from 0.5 to 1 mil (12 to 25 μm) Cu thickness in the barrel. There are at least two reasons why there is no one right specification. First, different applications provide different levels of thermal stress and demand different levels of reliability. Second, design factors such as the aspect ratio of the plated holes determine the susceptibility of the PTHs to thermal fatigue. The IPC recommends an average minimum copper-plating thickness of 0.5 mil for consumer products (Class 1) and 1.0 mil for general industrial and high-reliability applications (Classes 2 and 3).

Poor coverage at the PTH knee can significantly accelerate PTH failures because it means the plating is thin at a point of high stress. It can be caused by excessive concentration of the organic leveling agents added to an electroplating bath.

53.4.1.5 *Solder Mask Application.*

If it is properly applied, solder mask plays an important role in reducing the possibility of insulation resistance failures on PCBs. The solder mask protects the substrate from moisture and contaminants, which would otherwise promote shorting under electrical bias. The ability of the solder mask to perform this function depends on good conformity and adhesion of the solder mask to a clean, dry substrate. If solder mask conformity or adhesion is poor, moisture and other contaminants may accumulate at crevices or delaminations between the solder mask and the substrate. Substrate cleanliness is particularly critical, because in addition to causing poor solder mask adhesion, it can also provide the ionic species needed for rapid electromigration. When the laminate material absorbs moisture readily (e.g., polyimide, aramids), baking before reflow may be required to prevent solder mask delamination (as well as reinforcement/resin delamination). Other causes of poor adhesion or conformity include moisture on the board when the solder mask is applied, improper solder mask lamination or coating parameters, and improper solder mask cure parameters. Incomplete solder mask cure can create local soft pockets, which are common sites for delamination or contaminant entrapment. Solder mask over solder should also be avoided because delamination caused by reflow of the solder may allow contaminants to be entrapped.

53.4.2   Effects of Printed Circuit Assembly Processes

53.4.2.1 Stencil Printing and Component Placement.   Stencil printing and component placement generally do not cause reliability problems; however, poor stencil design or stencil printing or placement process control can create issues with solder volume and component cracking. Very low solder volume can result in weak solder joints that fail rapidly in thermal fatigue or by overload. In some cases, excessive solder volume can also accelerate solder joint fatigue failures because the compliance of the lead is reduced. Assuming that the stencil was designed and manufactured correctly, low solder volumes are usually due to small or missing paste bricks caused by a clogged stencil aperture, a stencil that needs cleaning, or improper stencil-printing parameters. Paste bridging can cause low solder volumes on some joints and high volumes on others because one joint may rob solder from another. Paste bridging may be caused by improper stencil design or stencil-printing parameters or by excessive force when placing chip carriers (e.g., PLCCs) and quad flat packs (e.g., PQFPs). Excessive placement force can also cause component cracking, particularly for small leadless ceramic components.

53.4.2.2 Reflow.   The reflow process attaches SMT and some TH components to the PCB by melting solder paste to form solder joints using an oven with a controlled thermal profile (see Fig. 53.17) and, in some cases, a controlled atmosphere (often N_2). Reliability problems that can arise due to improper reflow parameters can be grouped into three categories: damaged components, poor solder joints, and, for no-clean assemblies, cleanliness issues.

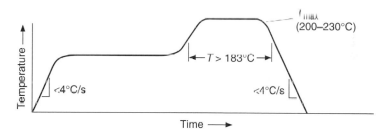

FIGURE 53.17   Schematic reflow profile for eutectic Sn-Pb solder paste, FR-4 substrate and typical surface-mount components illustrating key features from a reliability perspective.

Component Damage.   The reflow process is responsible for most of the assembly-process-related component failures described in Sec. 53.2.3. These failures include molding compound delamination in plastic packages that have absorbed moisture (popcorning) and component failures due to overheating or thermal shock caused by excessive heating or cooling rates. All of these problems are preventable with good procedures and process controls.

Package cracking can be prevented by storing components in the unopened dry bags in which they are shipped and by baking the moisture out of components that have been exposed to ambient conditions for too long. The manufacturer's recommendations regarding bake-out conditions and maximum exposure times before reflow should be followed, but a good general guideline is that packages that have been exposed to the atmosphere for more than 8 h should be baked to a moisture content below 0.1 percent by weight immediately prior to use; a bake of 125°C for 24 h is usually safe, although shorter times may be acceptable. Note that the same concerns apply for rework and second-side reflow of double-sided boards; for example, if the boards are stored for several days between reflow steps, bake-out before the second reflow step may be required.

Component failures due to overheating or thermal shock can be prevented by monitoring the reflow temperature profile in several locations on the PCB to ensure that it meets the manufacturer's specifications for temperature-sensitive components. Measuring the board profile is important because temperatures on the board can differ significantly from oven panel temperatures and from the ambient temperature in each oven zone. Temperatures may also vary significantly across the board if there are large differences in the thermal mass of the components or in component density. Areas of the assembly that are devoid of components are particularly sensitive to overheating, which can damage the laminate as well as any small components in the area. Temperature variations across the assembly tend to be much smaller for ovens with predominantly convection heating than for ovens with predominantly infrared heating. A poor reflow profile can also cause a variety of other problems; some of the others that also affect reliability will be mentioned here.

Poor Solder Joints.   A sound solder joint wets both the component termination and the substrate well, does not contain large or numerous voids, and does not have excessively thick intermetallic layers at the interfaces. When using solder paste, the reflow profile is the dominant factor in achieving these goals. Good wetting requires solderable incoming materials, but it also requires a reflow profile that gives the flux sufficient time to act in the right temperature range. In addition, the profile should ensure that all parts of the board are at least 15°C over the melting temperature of the solder for at least several seconds. "Cold," improperly formed solder joints can occur if the solder does not fully melt or if oxidation prevents the solder balls in the paste from melting together. The latter problem can be caused by an improper reflow profile or the wrong reflow atmosphere. Voiding is generally caused by a reflow profile that does not permit enough time for the solvents in the paste to boil off before the solder melts. All of these problems can be avoided by ensuring that the reflow profile of the board and reflow atmosphere (e.g., O_2 level) correspond to the manufacturer's recommendations for the solder paste.

Excessively long reflow times (time above the solder liquidus) can cause thick intermetallic layers to form at the interface between the solder and the component termination or substrate. Formation of an intermetallic layer at the solder interface indicates good metallurgical bonding, but thick intermetallic layers are undesirable because intermetallics are brittle and prone to fracture, especially if the joint is stressed in tension rather than shear. Because solder joint fatigue takes place in the solder rather than in the intermetallics or at the solder/intermetallic interface, the basic mechanism is unaffected. Nonetheless long reflow times and the accompanying thick intermetallic layers should be avoided. Cross-sectioning can be used to judge the extent of intermetallic growth; as long as the intermetallic layer thickness is relatively small compared to the joint thickness, reliability should not be adversely affected.[30] (Note however that minimizing reflow time is still a good thing; the reliability of all the components on the board is adversely affected by time at elevated temperature, both during processing and in service. Unfortunately, in developing a reflow profile, there is often a tradeoff between reflow time and peak temperature.)

Cleanliness Issues.   An improper reflow profile can also cause solder balling and increase the amount of flux residue remaining on the board after reflow. The reliability concerns associated with these process issues are discussed in Sec. 53.4.2.4. Solder balls can be caused by a combination of improper paste storage or handling, incompatibility between the flux and reflow atmosphere and a reflow profile that does not conform to the manufacturer's specification.

53.4.2.3 Wave-Soldering Process.
Improper wave-soldering practice can cause reliability problems. The root cause is generally thermal shock, overheating of the top side of the board, or contamination of the solder bath.

Component Cracking.   Ceramic components such as resistors and capacitors will crack under thermal shock conditions. When they are located on the bottom of the board they can be heated rapidly by the solder wave. Prevention is relatively simple; the assembly must be preheated before it hits the solder wave. A temperature difference between the component and the solder wave of less than 100°C is recommended; a typical preheat temperature is 150°C.

Hot Cracking.   Hot cracking, also known as partial melting, can cause previously sound solder joints to fail during the wave-soldering process. A typical mixed TH/SMT assembly is manufactured by assembling the surface-mount components to the top side, inserting the through-hole components, and wave soldering these components to the board from the bottom side. The first step of the wave-soldering process usually involves preheating the entire board. During the wave-soldering process, the SMT joints on the top side will be further heated due to conduction of heat through the board, particularly if there are many vias. If these solder joints reach the melting temperature of the solder (usually 183°C), the joints will begin to melt. If the joints melt completely, the assembly may be intact after reflow; however, if they only begin to melt, the surface tension of the solder is insufficient to prevent cracks from forming between the portions that are still solid. This type of failure is often detected as an intermittent in the field, since in-circuit test fixtures may bring the two halves of the joint into mechanical contact, causing the joint to appear electrically good.

Solder Bath Contamination.   Solder bath contaminant levels should be regularly monitored and limited to levels found in IPC-S-815. Many metals found on component terminations will dissolve into molten eutectic Sn-Pb solder. High Cu concentration is a relatively common occurrence that is associated with a rough solder surface and causes poor solderability. High Au concentrations can embrittle solder joints (see Sec. 53.5.1.3 for a discussion of this phenomenon).

53.4.2.4 Cleaning and Cleanliness.

Improper handling procedures and improper selection and application of solder paste and wave-solder fluxes and their associated cleaning processes can cause ionic residues to be left on the board that result in low surface insulation resistance. Low SIR values can cause failures in and of themselves for some sensitive circuits and in other cases set up the conditions for further corrosion that eventually result in short circuits. Sodium and potassium ions and halide ions are the most commonly quoted culprits for these failures. The major source of sodium and potassium ions is handling, i.e., fingerprints. The primary sources of halide ions are soldering fluxes.

The elimination of chlorofluorocarbons (CFCs) mandated by the Montreal Protocol has caused most SMT manufacturers to switch to water cleaning or a no-clean process. Water cleaning has been used by most printed circuit board manufacturers for some time, but outgoing cleanliness was not carefully monitored since the boards were cleaned again after assembly. Both the no-clean and water-clean assembly approaches must meet certain criteria to provide reliable assemblies.

In a no-clean assembly process, there is no cleaning step after SMT or TH assembly. The finished assembly has whatever contaminants were present on the incoming board and components, plus any additional contaminants added during the assembly process. These contaminants are generally flux residues, both from the solder paste and the flux applied for wave soldering, although adhesives and fingerprints are other potential sources. A no-clean flux should have a low solids content so that it leaves little residue and be free of ionic contaminants such as halides that promote corrosion. Use of a flux that contains halides will result in low SIR readings and may result in shorting due to corrosion, particularly if the assembly is exposed to a humid environment. However the incoming components and boards are cleaned, it is important that they are also free of halides when they arrive for assembly. Although SIR testing provides the best correlation with reliability, an ionic contamination test may be used for statistical process control. The measurement method may be found in MIL-P-28809.

Solder balls may also be a problem on no-clean assemblies. Solder balls are formed during reflow of solder paste when some solder is left behind when the solder melts and beads up and by spattering during wave soldering. These solder balls are usually washed off by solvent or water cleaning; however, in a no-clean process they remain on the board. Solder balls can cause shorts by bridging the pads of small capacitors or resistors or the leads of fine-pitch quad flat packs.

In a water-clean assembly process, the assemblies are cleaned with jets of deionized or saponified water after SMT and TH assembly. This process will work only if the flux residues and other contaminants are sufficiently soluble in either water or saponified water. It also depends on good access to the residues; consequently, a minimum component standoff that permits cleaning is required if it is possible for flux to get underneath the component body during assembly. It is almost as important that the board be thoroughly dried because water is an excellent medium for galvanic corrosion. Proper drying can be quite difficult even with substantial air flow since water has a much lower vapor pressure than CFCs do. If the component standoff is low, capillary action holds water in the small gap. If water cleaning is done in midprocess (e.g., before a reflow or wave-soldering step), plastic components may absorb moisture; in this case, the board must be baked out to prevent package cracking in subsequent high-temperature processes (see Secs. 53.2.3 and 53.4.2.2).

The rework process should not be overlooked in planning a flux and cleaning strategy. Compared to the automated processes that proceed it, it is typical to use a more aggressive flux and more of it to do rework. Use of a halide-free flux or proper cleaning after rework is essential to prevent cleanliness-related reliability problems.

Finally, the cleaning process itself can damage the PCA. Ultrasonic cleaning can damage components with internal wire bonds or die attach. It has also been observed to cause fatigue cracking of solder joints to LEDs and SOT-23s when the energy density was high because these components have terminations that are mechanically resonant near the generator frequencies. Solvent cleaning can attack the polymers used in solder masks, PCBs, conformal coatings, and components. D-limonene (terpene)-based solvents should be tested carefully for compatibility with exposed plastics and metals.

53.4.2.5 Electrical Test and Depanel.

The electrical testing and depanel processes can impose large mechanical stresses on the PCB and its components. In-circuit electrical test utilizes a bed of nails or two beds in a clamshell arrangement to contact each electrical node on the board. The probes must contact the board with sufficient force to make good electrical contact. If the board is not properly fixtured or if the loading in a clamshell fixture is unbalanced, the resulting deflections can cause solder joint or component cracking. These cracks may cause electrical failures immediately or after some period of service. Depaneling, the process of separating individual images from a larger panel, is done by a variety of methods. The associated mechanical deflections or vibration can cause component cracking or solder joint fatigue.

53.4.2.6 Rework.

Rework, whether repair of open or shorted solder joints or replacement of defective components, has a significant negative effect on component reliability. If there were not enough other incentives for low process defect rates, the effect on product reliability would be enough. Reworking the quality in does not bring the board to the quality level that would have been reached if the boards were built right the first time. Some of the ways rework processes can adversely affect reliability are described here.

Thermal Shock During Rework.   Thermal shock to components is a concern during rework as it is during reflow. The maximum heating or cooling rate is driven by the requirements for ceramic capacitors and should not exceed 4°C/s.[31]

Rework of large through-hole components, such as pin grid arrays (PGAs) and large connectors, poses special problems. If it is improperly done it can result in PTH failures. Because the damage during these large thermal cycles is cumulative, the number of rework operations at a given site should be monitored and limited to a safe number. The number of cycles that will cause a fatigue crack to initiate in the copper in the barrel and propagate to failure depends on the aspect ratio of the PTH, the type and thickness of plating in the hole, the substrate material, etc.

Due to the large number of joints that must be melted at once and the large thermal mass of the components, rework of large TH components is often done with a solder pot. The thermal shock caused when the molten solder hits the board can cause PTH cracking due to z-axis expansion. A preheat step (to about 100°C for FR-4) helps to reduce the damage. The time the board is in contact with the solder fountain should also be minimized since dissolution of the copper plating inside the PTH occurs during this time. Thinning the plating in the PTH tends to increase the strain in the Cu during thermal cycling, further accelerating failure. If the total time for part removal and replacement is kept under 25 s, little dissolution is measured.[32] Weakening of the PTH by copper dissolution during PGA rework can be essentially eliminated by using NiAu plating. Although the thin Au coating that protects the Ni dissolves almost instantly during soldering, Ni dissolves quite slowly and effectively prevents thinning of the PTH metallization.

Damage to Adjacent Components.   Rework can also damage components adjacent to the one being repaired or replaced. The hot cracking phenomenon during wave soldering can also occur in solder joints near the rework site if they reach the melting temperature of the solder. At slightly lower temperatures, rapid intermetallic growth can occur. Temperature-sensitive components can also be damaged. To prevent these problems, localized heating and shielding should be used and the temperature of adjacent components monitored. The generally recommended maximum temperature is 150°C. There are wide variations in the amount of heating of adjacent components among different types of rework equipment and between process protocols.[33]

Other Rework Concerns.   Rework can cause a host of moisture-related problems including measling and package cracking. Both of these problems can be prevented by baking the PCA beforehand to drive out moisture and by minimizing the peak temperature and time at elevated temperature during rework. Rework temperatures also weaken the adhesive bond between the copper conductors on the PCB and the laminate material; use of force to remove components when the solder is not completely molten can cause the pad to lift off the board. The latter can be a particular problem when using a soldering iron.[21,31]

53.5   INFLUENCE OF MATERIALS SELECTION ON RELIABILITY

53.5.1   PCB

53.5.1.1   Substrate.   Difunctional FR-4 is the workhorse material for high-reliability PCBs because its moderate z-axis expansion and moisture uptake characteristics are available at relatively low cost. Alternative substrate materials (see Table 53.2) are generally selected for more favorable properties in one or more of the following three areas: thermal performance, including maximum operating temperature and glass transition temperature; thermal expansion coefficient; and electrical properties, such as dielectric constant. Thermal performance characteristics and thermal expansion coefficient can have a significant effect on PCB and solder joint reliability. Other characteristics of these materials, such as moisture absorption, can also affect reliability.

PTH reliability can be improved by selecting a laminate with a lower z-axis CTE or a higher T_g. The damage caused to the PTH during a thermal cycle depends on the total z-axis expansion during the temperature change. Since the CTE is much lower below T_g than above,

TABLE 53.2   Physical Properties of Some Printed Circuit Board Laminate Materials

Material	CTE, x,y ppm/°C	CTE, z ppm/°C	T_g, °C
Epoxy glass (FR-4, G-10)	14–18	180	125–135
Modified epoxy glass (polyfunctional FR-4)	14–16	170	140–150
Epoxy aramid	6–8	66	125
Polyimide quartz	6–12	35	188–250

Source: *After IPC-D-279.*

the PTH strain can be reduced by increasing the T_g so that more or all of the cycle is below T_g (see Fig. 53.18a). Figure 53.18b shows that the increase in life can be quite significant. The strain imposed on the PTH can also be reduced by decreasing the CTE at temperatures below T_g, but the effect on total z-axis expansion is much smaller.

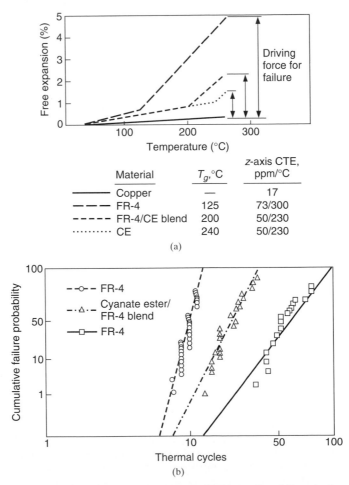

Material	$T_g,°C$	z-axis CTE, ppm/°C
—— Copper	—	17
— — FR-4	125	73/300
- - - - FR-4/CE blend	200	50/230
········ CE	240	50/230

(a)

(b)

FIGURE 53.18 (a) Effect of differences in CTE below T_g and T_g on the free z-axis expansion of FR-4 (epoxy-glass), cyanate ester (cyanate ester-glass) and a cyanate ester/epoxy blend. T_g and CTE for each material below/above T_g are indicated. Cyanate ester is abbreviated CE in the figure. Cu is shown for comparison. (b) Weibull plot of PTH failures for these substrates during thermal shock cycling between 25 and 260°C. PTHs are 0.029-in diameter on 0.100-in grid on 0.125-in-thick laminate. *(After Fehrer and Haddick.[4])*

A variety of specialty resins with increased T_g is available, albeit at higher prices. Modified FR-4 materials with higher functionality offer the best combination of improved T_g at a reasonable price. Further improvements in T_g and other characteristics can be obtained with bismaleimide triazine (BT), GETEK, cyanate ester, and polyimide, but at greater price penalties.

Interconnect failures due to thermal fatigue of solder joints can be reduced by closely matching the x-y plane thermal expansion properties of the substrate to at-risk components. Large leadless ceramic components that are used because of their hermeticity pose a particular risk. Possible approaches include altering the laminate reinforcement material, adding constraining metal cores or planes, and switching to a ceramic substrate. The first two approaches are discussed here. A more extensive discussion of these options can be found in Ref. 33.

A lower x-y plane thermal expansion coefficient laminate can be obtained by replacing the continuous-filament E-glass used in most FR-4 PCBs with an alternative material. The CTE decreases as the fraction of silica dioxide (SiO_2) decreases and the level of quartz (as well as the cost) increases in the progression E-glass, S-glass, D-glass, and finally quartz, which has a CTE about one-tenth of E-glass. Aramid (Kevlar) actually has a negative CTE, but it is available in only a few glass styles. Some of the disadvantages of aramid fibers are higher z-axis expansion and higher moisture absorption relative to glass fibers that can result in decreased susceptibility to PTH failures and corrosion-related insulation resistance failures, respectively. Aramid fiber is also used to make nonwoven paper fabric which has a lower modulus, but also a much smoother surface because there is no weave pattern. This form has better dimensional stability and reduced microcracking during thermal cycling.

Low-thermal-expansion metal cores or planes can also lower the overall substrate CTE because they constrain the expansion of the polymer material they are laminated to (see Fig. 53.19). Copper-Invar-copper (CIC) is the most widely used material for constraining metal cores (also termed polymer-on-metal or POM construction), followed by copper-molybdenum-copper (CMC). The PCB and core are bonded with a rigid adhesive, usually in a balanced construction to minimize warping. Other special processing is also required. The CTE of the assembly can be estimated using a simple model for composite structures most often written as

$$\text{CTE(overall)} = \frac{\Sigma E\alpha t}{\Sigma E t}$$

where E, α, and t are the elastic modulus, CTE, and thickness, respectively, of the various layers. A more sophisticated model can be found in Ref. 34.

An example of the low overall CTE that can be obtained using a CIC core is shown in Fig. 53.20. Unfortunately, the constrained x-y axis expansion results in increased z-axis expansion that can reduce PTH reliability to dangerously low levels, especially in an environment in which the full mil-spec thermal cycle of −55 to +125°C is imposed. Consequently, use of polyimide is recommended with CIC cores. Because of its high T_g and low CTE below T_g, polyimide imposes much lower strains on the PTH for a given thermal cycle than other dielectrics.

Constructions utilizing constraining low-CTE metal planes usually use CIC layers in place of ground and power planes in a standard multilayer board. The same PTH reliability concerns that hold for CIC core boards apply to these PCBs as well. PTH reliability can be improved by using polyimide resin and by using CuNiAu or CuNiSn metallization in the PTHs. These substrates are easier to manufacture than metal core boards because standard PCB fabrication techniques can be used for the most part.

Resin material can affect fiber/resin delamination, one of the prerequisites for conductive anodic filament growth. Measling occurs at about 260°C for FR-4, but may occur at lower temperatures for boards with more hygroscopic resins.

53.5.1.2 Solder Mask.

The three major types of solder mask—liquid screen-printed, dry film, and liquid photoimageable (LPI)—come with different benefits and concerns from a reliability perspective. The solder mask material should be selected for its compatibility with the heat and solvent characteristics of the assembly process, its capability to provide good conformity over surface features on the PCB, and its ability to tent vias if required. Since many of these characteristics are product-specific, only a few general guidelines can be provided here. Where tenting of vias is required to keep solder, moisture, or flux from wicking up

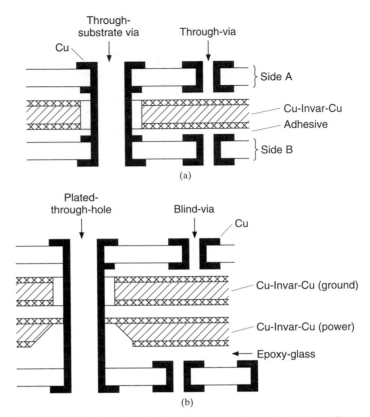

FIGURE 53.19 Examples of alternative construction using Cu-Invar-Cu to achieve low-*x-y* CTE: (*a*) metal core (sides A and B may be multilayer boards) and (*b*) metal plane constructions. (*After F. Gray.*[34])

under components, dry film solder mask should be used. However, excessively thick solder mask, particularly dry film over closely spaced traces, can result in crevices. If the solder mask cannot flow enough to adhere closely to the board, the resulting crevices can entrap contaminants such as flux that can accelerate corrosion later. LPI solder mask provides excellent coverage, resolution, and alignment to other features, but it generally cannot be used to tent vias. IPC-SM-840 defines the performance and qualification requirements for solder mask.

53.5.1.3 Metal Finish. The metal finish on the SMT and TH pads can have an impact on PTH reliability and on the reliability of the solder joints made to these pads. Common metal finishes for solder-mask-over-bare-copper (SMOBC) boards include hot-air solder leveling (HASL or HAL), organic-coated copper (OCC), and electroless NiAu. Galvanically plated CuNiAu and CuNiSn made by another processing route are also available. These finishes provide a solderable finish for later printed circuit assembly. The pros and cons of the various finishes are discussed in turn.

Of the common metal finishes, HASL is the only one which can directly reduce reliability of the board. In a typical HASL process, the board receives a severe thermal shock when it is dunked into a bath of molten eutectic Sn-Pb solder. The PTHs can survive only a certain number of solder shocks without failure; this process uses up one of these thermal cycles before the board leaves the fabricator.

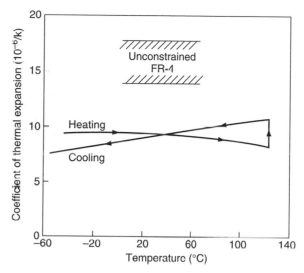

FIGURE 53.20 Example of the low overall CTE that can be achieved for a metal core construction similar to Fig. 53.19a. Two 0.055-in (1.4-mm)-thick multilayer FR-4 boards were bonded to a 0.085-in (2.2-mm)-thick copper-Invar-copper core. Data shown are for the third thermal cycle. (*After F. Gray.*)

Organic-coated copper provides a consistent, flat, solderable metal finish. Exposed copper after printed circuit assembly has been a persistent reliability concern, because it is generally not permitted on HASL boards. While exposed Cu on HASL boards is associated with poor solderability, which may be due to contaminants that were not removed before the HASL process, there is little evidence that exposed Cu on a properly processed OCC board causes reliability problems. Surface insulation resistance (SIR) testing shows that OCC boards have comparable or better performance than HASL boards in high-temperature, high-humidity storage tests.

CuNiAu boards fabricated either with the NiAu as the Cu etch resist or by the SMOBC process followed by electrolessly plating Ni and Au can confer improved PTH reliability. There are two mechanisms for the observed improvement: the enhanced rivet effect provided by the Ni and the elimination of Cu dissolution during solder shocks such as wave soldering or PGA rework. For high-aspect-ratio holes, electroless Ni confers an additional benefit because the plating thickness in the barrel is more consistent than for conventional electroplating.

In the simple picture of PTH failure shown in Fig. 53.3, the comparatively low CTE metal-plated PTH acts as a rivet that resists the z-axis expansion of the PCB. Because Ni has a higher elastic modulus than Cu, it strains less under the stress imposed by the expanding PCB. Consequently, adding Ni plating lowers the strain imposed on the Cu and lessens the amount of fatigue damage. In this model, the Ni protects the Cu, increasing PTH life.

The ability of the Cu to withstand the forces imposed on it by thermal expansion of the PCB is also dependent on the thickness of the Cu in the PTH. Unfortunately, in the SMOBC process all subsequent steps after pattern or panel plating reduce the Cu thickness from the plated amount. Nickel plating is resistant to the etches and developers used in later processing steps, so it protects the underlying copper from thinning due to dissolution. The HASL process and rework of large through-hole connectors or pin grid arrays (PGAs) can have particularly negative effects. Cu dissolves rapidly into molten eutectic Sn-Pb solder. During the HASL process or component removal and replacement with a solder fountain, large amounts of Cu can dissolve from the knee of the PTH. Nickel barrier plating minimizes this effect because Ni dissolves far more slowly in eutectic Sn-Pb solder than Cu does.

Use of Au plating can embrittle the eutectic Sn-Pb solder most commonly used in electronic assembly. Gold plating of various thicknesses is used for a variety of reasons, including as a solderability preservative over nickel plating, for connector contacts, and to provide wire-bondable pads. Reliability problems can arise because Au has a high solubility in eutectic Sn-Pb solders at reflow temperatures and dissolves extremely rapidly. In most cases, the Au finish on a PCB or component termination will be completely dissolved into the solder. In a wave-soldering process, the Au will be washed into the bath, requiring monitoring and bath changes to maintain the Au concentration at a low level that does not affect the process. However, in a reflow process, this Au remains in the finished solder joint. To avoid embrittlement of the solder by the $AuSn_4$ and $AuSn_2$ intermetallics that can form, the Au concentration should be kept below a critical level that most authors set at 3 to 5 percent by weight.[35,36]

For most components used today, a nominal 5-μin (0.1-μm) thickness Au flash to preserve solderability is harmless. However, if thicker Au plating (e.g., for connector contacts or wire bonding) is used, if the component lead pitch is less than 0.5 mm, or if the component lead terminations are also Au-plated, care should be taken to ensure that the Au concentration remains below the 3 to 5 percent by weight limit. For some applications in which use of thick Au is unavoidable, 50In-50Pb solder, in which Au dissolves very slowly, has been used to get around this problem.[37] Selective thick Au plating is another option.

The Au concentration in the finished reflowed solder joint can be estimated using the following equation:

$$\text{Wt. \% Au} = \frac{(\text{Au volume})}{[(\text{Au volume}) + (\text{solder volume})(\rho_{solder/Au})]}$$

where $\rho_{solder/Au}$ is the ratio of the density of the solder to the density of Au (0.4552 for 63Sn-37Pb solder). If the Au is less than 1 μm thick, it is usually valid to assume that all of the Au on the joined surfaces has dissolved into the joint. The solder volume should include any solder plated on either the component or board termination as well as that applied by stencil printing. It is common to specify only a minimum Au plating thickness; it is important to use a representative value to calculate the expected Au content in the solder joint.

53.5.2   Interconnect Material

53.5.2.1   Eutectic Sn-Pb Solder.   Eutectic Sn-Pb solder, 63Sn-37Pb, and near-eutectic Sn-Pb solders, including 60Sn-40Pb and 62Sn-36Pb-2Ag, are used in the overwhelming majority of soldered electronic assemblies. From a reliability perspective, the most important characteristics of these solders are their susceptibility to creep and fatigue because ambient temperatures are so close to the metal temperature of the solder, their ability to dissolve common termination metals rapidly and in large amounts, and their tendency to form thick intermetallic layers with termination metals.

Although solder joint thermal fatigue is a major source of PCA field failures, the industry has used the same solder alloy for several decades. At present, there is no generally agreed-on alternative to eutectic Sn-Pb solder that has improved fatigue resistance as well as the favorable processing characteristics of eutectic Sn-Pb solder. However, there has been a tremendous surge in research into alternative solders, especially Pb-free solders, in the last decade and one can expect improved alloys in the future. There is some evidence that solders containing 2% Ag have improved properties in thermal cycling to high temperatures.

Many common termination metals dissolve rapidly in eutectic Sn-Pb solder, including Ag, Au, and Cu.[36] The dissolved metals can alter the properties of the solder. Reliability can also be impacted if the termination metal is dissolved entirely, the most notable case being Ag (or AgPd with less than 33% Pd) terminations on ceramic resistors and capacitors. If the entire termination thickness is dissolved, the solder will dewet from the ceramic part, leaving an open joint if the entire termination is dissolved, or a substantially weakened one if it is dissolved only in some areas. Use of 63Sn-36Pb-2Ag substantially reduces this problem, since the presence of Ag in the solder reduces the dissolution rate of the Ag forming the termination.

Finally, eutectic and near-eutectic Sn-Pb solders form intermetallics with the termination metals that influence the properties of the finished joints. On the most common termination metals, Cu and Ni, continuous intermetallic layers form: Cu_3Sn and Cu_6Sn_5 on copper, and Ni_3Sn_4, Ni_3Sn_2, and Ni_3Sn on nickel. These intermetallic layers are composed of compounds that are hard and brittle in comparison to both the solder and the termination metals. Although there have been few systematic comparisons made, it is general wisdom that when the intermetallic layers are very thick, solder joint reliability is reduced. While solder joint thermal fatigue mostly involves cracking in the solder, the intermetallics can affect their ability to withstand mechanical stresses, particularly in tension. The Ni-Sn intermetallics are particularly brittle. In every case, an effort should be made to minimize the total time the solder is molten during processing and to minimize the time above about 150°C once the solder joint is formed.

53.5.2.2   Other Solders.
A number of other solders are used in specialized applications, including 80Sn-20Pb for lead finishes, 50In-50Pb for solder on thick Au, high-Pb solders such as 95Pb-5Sn and 97Pb-3Sn for flip-chip assemblies (usually on ceramic substrates), and low-temperature solders such as 58Bi-42Sn and 52In-48In, where hierarchical soldering is desirable. Further information can be found in Refs. 23, 24, and 38.

53.5.2.3   Conductive Adhesives.
Electrically conductive adhesives are used today for specialized applications such as connections to LCD displays and attachment of small resistors and capacitors. These materials consist of conductive particles, usually silver flakes or carbon, suspended in a polymer matrix, most commonly epoxy. The electrical resistance of the contact to the PCB tends to be unstable over time, so these materials are not suitable for applications requiring a constant, low-resistance contact. The primary failure mechanism is moisture migration through the epoxy to the interface, resulting in oxidation of the contact metal. Adhesion strength is also a reliability concern. New materials suitable for a broader range of applications are under development. Further information can be found in Ref. 39.

53.5.3   Components

Components and their packages influence many of the field failures of electronic assemblies. Packages are primarily designed and selected for their ability to protect the electronic components inside; for example, ceramic packages may be selected over plastic ones for their greater hermeticity. This section will discuss the ways in which package selection can influence solder joint and cleanliness-related failures.

53.5.3.1   Package Selection to Minimize Solder Joint Thermal and Mechanical Failures.
Minimizing solder joint thermal and mechanical fatigue failures means minimizing the global and local mismatches in the system and introducing compliance that minimizes the stresses and strains transmitted to the solder joints. Figure 53.21 illustrates the important features of these systems. The following describes how component parameters can influence the incidence of solder joint failures.

Surface-Mount Technology vs. Through-Hole Components.   Although there is little compliance in the system, the reliability of through-hole solder joints generally exceeds that of surface-mount joints in thermal fatigue (assuming good solder fillets are present in both cases), because the loading geometry makes it difficult for a crack to propagate far enough to cause an electrical failure. However, the PTHs themselves may be susceptible to failure if they are exposed to even a few thermal cycles well above T_g.

Plastic vs. Ceramic Package.   The global mismatch between the component body and substrate is minimized for most printed circuit boards if the package is plastic rather than ceramic. Most electronic ceramics have CTEs in the neighborhood of 4 to 10 ppm. Since the printed circuit board CTE is 14 to 18 ppm in-plane below T_g, the match to plastic packages which usually have average CTEs of 20 to 25 is better. The overall CTE of a plastic package can be significantly below that of the plastic if the die is large compared to the total package body. For example, TSOP components can have overall CTEs as low as 5.5 ppm. It is also worth recalling that component-level reliability must be considered; plastic packages suffer from other disadvantages versus ceramic packages, such as moisture absorption.

Leaded vs. Leadless Surface-Mount Components.   Leadless surface-mount components with peripheral solder joints (e.g., leadless ceramic chip carriers, LCCCs) are more susceptible to solder joint failures due to thermal and mechanical stresses than leaded components because there is no compliance in the system (see Fig. 53.21). A compliant lead can take up relative displacement between the component body and the substrate during mechanical or thermal stressing. In doing so, it minimizes the stress and strain imposed on the solder joint, thus reducing the likelihood of failures. Large leadless components should be avoided whenever possible. If they must be used, the substrate must have as close a CTE mismatch as possible and be protected from mechanical stresses. A conformal coating should be considered.

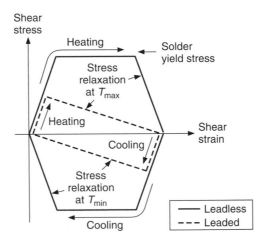

FIGURE 53.21   Schematic illustration of the strain in the solder during a thermal cycle for leaded and leadless surface-mount components. *(After W. Engelmaier)* (IPC-TP-797, Surface Mount Solder Joint Long-term Reliability: Design, Testing, Prediction.)

Ball grid arrays (BGAs) are a new style of leadless SMT component with an areal array of solder joints. The reliability of these components has come under intensive study. Plastic BGAs are less susceptible to solder joint fatigue failures than ceramic BGAs because the laminate and plastic body match the CTE of the PCB much better. At this time, it seems that there will be size and power limitations to ensure solder joint reliability.

Lead Compliance.   As described, leadless components cause more reliability problems than leaded components because the entire displacement is imposed on the solder joint; however, there are also large differences in compliancy among leaded surface-mount components (see Fig. 53.22). Body height plays an important role because it determines the length of the

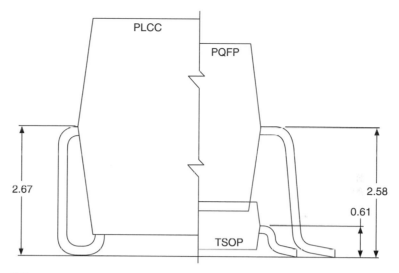

FIGURE 53.22 Schematic illustration of different surface-mount component lead types with widely differing compliance: (*left*) J-lead, (*right*) gullwing leads.

compliant beam. Other important lead characteristics that affect compliance are lead shape (e.g., J-lead vs. gullwing) and lead thickness (stiffness is proportional to thickness cubed).

Lead frame material also plays a role in determining solder joint life, although it is not as important as lead frame geometry. Common lead frame materials are Cu and Alloy 42 (Fe-42Ni); Alloy 42 has a better CTE match to silicon (and a greater mismatch to the solder), but it is much stiffer than Cu (see Table 53.3)

Thin small-outline packages (TSOPs), which are becoming increasingly common memory packages, pose the greatest solder joint reliability risks of any packages commonly used today. These packages generally have Alloy 42 lead frames and very low standoff from the PCB, resulting in a very stiff lead that transfers most of the relative displacement between the component and the substrate to the solder joint. The situation is exaggerated for this component because the overall CTE of the package is quite low. While adequate solder joint reliability can be achieved with TSOPs in many situations, some vendors have opted to encapsulate the solder joints with a filled epoxy to better distribute the stresses.[40]

53.5.3.2 Component Selection for Cleanliness. If fluid cleaning is used to remove flux residues after assembly, then a minimum component standoff (distance between the component body and the seating plane of the leads) is critical to ensuring proper cleaning and drying, and, therefore, resistance to corrosion and moisture-related failures. Industry standards for low standoff components permit the components to have 0–0.25 mm standoff from the board. These low standoffs permit corrosive residues and the cleaning fluid to be trapped

TABLE 53.3 CTE and Elastic Modulus at Room Temperature of Some Important Packaging Materials

	Cu	Alloy 42 (Fe-42Ni)	63Sn-37Pb Solder	Si
CTE, ppm/°C	17	5	25	3.5
E, GPa	130	145	~35	113

under the component. Poor fluid access can mean that flux residues are not removed. Drying is almost as important; it was not a major issue with highly volatile CFCs, but when the cleaning fluid is water, leaving it behind can promote moisture-driven failure mechanisms. Components which meet the standards for high standoff have a minimum standoff of 0.20 or 0.25 mm, which should be sufficient for cleaning and drying with water and most other cleaning fluids in use today. Using a no-clean process essentially eliminates these concerns about component standoff; however, an added concern is that any contamination of the external surface will remain. SIR testing should be conducted to ensure that the flux residues and other contaminants left on the board are not harmful.

53.5.3.3 *Component Termination Selection for Joint Integrity*   Surface-mount component termination finishes are generally tin-lead or tin, although other finishes are occasionally used. Good solderability is the foundation needed for forming a strong solder joint. Incoming cleanliness is another requirement that should go without saying, but has taken on increased importance with the advent of no-clean processes. With copper leads, formation of excessive quantities of Cu_3Sn and Cu_6Sn_5 intermetallics can occur if reflow times and temperatures are excessive. With pure tin, whisker growth can be a concern. When Au is used as a metal finish, the Au can be expected to dissolve rapidly into the solder joint. To avoid embrittling the solder, the finished joint should contain less than 3 to 5 weight percent Au (see Sec. 53.5.1.3).

Ceramic and ferrite components such as multilayer ceramic capacitors, chip resistors, and chip inductors are generally terminated with a fired-on silver or silver palladium paste. Because silver dissolves easily into molten Sn-Pb solder, a Ni/Sn or Ni/Au overplate is recommended.

53.5.4 Conformal Coatings

Conformal coatings are used for PCAs that require exceptional resistance to moisture, solvents, or abrasion. A wide variety of polymers are used to perform this function, including phenolic, silicone and urethane lacquers and silicone rubber, polystyrene, epoxy, and paraxylylene coatings. Epoxy and polyurethane-based coatings are the most commonly used. If conformal coatings are used in conjunction with a solder mask, they must be chemically compatible.[21,23]

Conformal coatings work by keeping contaminants away from the circuitry and preventing moisture from accumulating on the surface of the assembly. Since all conformal coatings are permeable to moisture, interfacial adhesion is essential to their function. Contaminants on the board that reduce the adhesion of the coating or trap moisture can cause the coating to fail, as can thermal stresses. When contaminants trap moisture, the coating will bubble up (vessicate), providing gaps where corrosion can occur. An ionograph is not always the right tool to detect the harmful level of ionic contamination; cleaning with both polar and nonpolar solvents before application is recommended.

A conformal coating that is not well matched to the use environment can actually promote new failure mechanisms that would not be found in the uncoated assembly. If it fills the gap beneath the component, the coating may place additional stresses on solder joints in thermal cycling by reducing or eliminating the compliance in the leads. It may also generate excessive stresses on the components if the service temperature drops below the T_g of the coating. Some coatings are not stable in hot, humid conditions.

53.6 BURN-IN, ACCEPTANCE TESTING, AND ACCELERATED RELIABILITY TESTING

This section reviews environmental stress testing procedures used to identify unacceptable parts or to estimate part life. These tests can be categorized by their intended goals: 100 percent screening to eliminate early failures (burn-in), acceptance testing on a sampling basis, and life distribution estimation (accelerated reliability estimation). Burn-in, more generally known as environmental stress screening (ESS), is an important subject in its own right that has been extensively written about; only its goals are described here. Acceptance or qualification testing and accelerated reliability testing are discussed together in sections covering testing of PCBs and PCAs under various environmental stresses.

Burn-in, also known as *environmental stress screening* (ESS), is used to eliminate early failures when infant mortality due to latent defects is a problem by exposing all parts to worst-case, but realistic, conditions. The conditions should not be too severe because burn-in decreases the useful life of the parts. On the other hand, in addition to increasing the reliability of the parts that are shipped, burn-in provides rapid feedback on process defects that can cause field failures. The most common use of burn-in in electronics is for integrated circuits, particularly memory chips, designed into leading-edge integrated circuit manufacturing processes.

Accelerated reliability tests are designed to cause failures that would occur by wear-out sometime during the service life of the part and provide the data for estimating the *life distribution* of the part. Estimating the life distribution requires that the test be continued until a large percentage of the parts fail. *Qualification tests* may be conducted under similar or even more severe conditions, but they are essentially pass/fail tests that are terminated after a specified period of time. Because there are few if any failures in a successful qualification, little new reliability information can be gained from this type of test. These tests should not be used routinely on all parts because they will greatly shorten the life of the parts.

Unfortunately, there is no standard suite of reliability tests in the industry, nor is there likely to be in the near future. There are several reasons for this. First, there are numerous service environments. The IPC has identified seven major application segments for electronic assemblies. Second, within these categories, the environment experienced by the assembly may differ from the external service environment, depending on product-specific design parameters, such as power dissipation and cooling efficiency, which influence the temperature and humidity in the vicinity of the assembly. Third, the desired product life and acceptable failure rates vary widely among applications and manufacturers. Finally, but of equal importance, as technology evolves, the tests must evolve, too. Use of reliability tests that once made sense but now are either unduly conservative or do not bring out potential failure modes of new designs has a high price in overdesigned assemblies that are larger or more costly than necessary and field failures that could have been predicted and prevented. This section describes some commonly used tests and methodologies for designing tests for new technologies or new applications.

As stated at the outset, reliability can be defined only after the service environment has been identified, and the acceptable failure rate over a specified service life is specified. If this environment proves unacceptable, the design can be modified by improving cooling, package hermeticity, cleaning, etc. If it is unclear whether the PCA will achieve the designed reliability goals, accelerated reliability tests should be designed to estimate the life distribution.

53.6.1 Design of Accelerated Reliability Tests

There are seven steps in accelerated reliability test design.

1. *Identify the service environment and the acceptable failure rate over a specified service life.*
2. *Identify actual environment of the PCA (modified service environment).* The service environment should be translated into the ambient environment actually experienced by the PCA. For example, the temperature experienced by the PCA is influenced by both power dissipation and cooling. The mechanical environment is influenced by shock-absorbing material, resonances, and so on.
3. *Identify probable failure modes* (e.g., solder joint fatigue, conductive anodic filament growth). Accelerated reliability tests are based on the premise that the frequency and/or severity of the environmental exposure can be increased to accelerate the incidence of the failure that

occurs in service in a known way, i.e., that the data can be used to predict the life distribution in the in-service PCA environment. This assumption makes sense only if the same failure modes occur in the test as in real life. It cannot be overemphasized that the accelerated tests must be designed around the real failure modes. Probable failure modes may be identified from past service experience, the literature, or preliminary testing or analysis.

4. *For each failure mode, construct an acceleration model.*   An acceleration model that allows test data to be interpreted in terms of the expected service environment is crucial to life distribution estimation. It is also extremely helpful in designing good tests, so ideally the acceleration model should be developed before the accelerated reliability tests are carried out. Equation (53.2) for solder joint reliability plus Eq. (53.1) for strain in solder joints to rigid components is an example of an acceleration model. It predicts that increasing strain will decrease the number of cycles to failure in a specific way. Within a certain temperature range, increasing the temperature cycling range is a way of increasing the strain.

In general, the acceleration model should be based on the rate-controlling step in the failure process. In some cases, the rate will be determined by an Arrhenius type equation; for example, if diffusion is the rate-controlling process:

$$D = D_o \exp\left(\frac{-E_a}{kT}\right) \quad \text{and} \quad x \propto \sqrt{Dt} \tag{53.3}$$

$$t_2 = \left(\frac{D_1}{D_2}\right)t_1 = t_1 \exp\left(\frac{-E_a}{k}\left[\frac{1}{T_1} - \frac{1}{T_2}\right]\right)$$

where D = diffusion rate
D_o = diffusion constant
E_a = activation energy for the process
k = Boltzmann constant

and T_1 and T_2 and t_1 and t_2 are two temperatures and corresponding equivalent diffusion times

Note that even when temperature is an important factor, *an Arrhenius relationship may not exist;* in the preceding thermal cycling example, the failure rate is roughly proportional to $(\Delta T)^2$. Some acceleration models will be explored in the following sections.

The limits of applicability of an acceleration model are as important as the model itself. Increasing or decreasing the temperature too much may promote new failure modes that would not occur in service or invalidate the quantitative acceleration relationship. For example, if the temperature is elevated above the T_g of the board, the z-axis CTE increases sharply and the modulus decreases, which may actually lessen the strains imposed on solder joints, but may also promote PTH failures.

Finite element modeling (FEM) can be invaluable in developing and/or applying acceleration models for thermal and mechanical tests. Two-dimensional nonlinear modeling capability will usually be required in order to get meaningful results. Models can be constructed to estimate the stresses and strains in the material (e.g., the Cu in a PTH barrel or the solder in a surface-mount or through-hole joint) under operating conditions as well as under test conditions. These estimates will be far more accurate than the simple models provided in this overview because they can account for the interactions between materials in a complex structure and both elastic and plastic deformation.

5. *Design tests based on the acceleration models and accepted sampling procedures.*   Using the acceleration model and the service environment and life, select test conditions and test times that simulate the life of the product in a much shorter period of time. The sample size must be large enough that it is possible to determine whether the reliability goal (acceptable number of failures over the service life) has been met.[41] Ideally, the life distribution in the accelerated test should be determined, even when the test period must be extended to do so.

6. *Analyze failures to confirm failure mode predictions.*   Since an accelerated test is based on the assumption that a particular failure mode in the accelerated test is the same one that occurs in service, it is important to confirm by failure analysis that this assumption is valid. If the failure mode in the accelerated test is different from the one expected, several possibilities should be considered. (1) The accelerated test is introducing a new failure mode different from the one that will occur in service. Usually this means that the acceleration of one parameter (e.g., frequency, temperature, humidity) was too severe. (2) The initial determination of the dominant failure mode was incorrect. In this case, to understand the significance of the test results, a new acceleration model must be developed for this failure mode. The new failure mode may be promoted more or less effectively by the test conditions than the mode originally assumed. (3) There may be several failure modes. In this case, the two failure distributions should be considered separately, so that life predictions will be meaningful. The difficulty in determining which of the above scenarios holds is that for genuinely new technologies or service environments, the failure mode in service may not be known. In these situations, it is desirable to conduct a parallel test with less aggressive acceleration for comparison.

7. *Determine life distribution from accelerated life distribution.*   The accelerated life distribution should be determined by fitting the data with the appropriate statistical distribution, such as the Weibull or log-normal distribution. The life distribution in service can be determined by transforming the time axis of the life distribution using the acceleration model. This predicted life distribution in service can then be used to estimate the number of failures in the specified service life.

The following discussion of testing for some specific failures will provide examples of this methodology.

53.6.2 Printed Circuit Board Reliability Tests

53.6.2.1 Thermal.   PTH failures are the predominant source of PCB failures in service and predicting them is the primary goal of PCB testing at elevated temperatures. PTH reliability testing should simulate the thermal excursions of a PTH throughout its life. Generally, the most severe thermal cycles are experienced during assembly and rework.

Two basic types of tests are conducted: thermal stress or solder float tests, and thermal cycling tests. Both of these tests are intended to be accelerated tests for the PTH, not for the laminate; the thermal stress test, in particular, is expected to severely degrade the laminate. The delamination test is similar to a solder float test, but is conducted at a lower temperature specified by the laminate manufacturer; typically, a different fluid is required.

The most commonly accepted thermal stress test is MIL-P-55110 (also found in IPC-TM-650). Following baking at 120 to 150°C (250 to 300°F), the specimens are immersed in an RMA flux and floated in a eutectic (or near-eutectic) Sn-Pb solder bath at 288°C (550°F) for 10 s. Other investigators use a bath at 260°C. Following the test, the samples are cross-sectioned and the PTHs are examined for cracks. This is a severe test that ensures that the sample will survive a single wave-soldering or solder pot rework cycle.

Most thermal cycling tests for PCBs cycle the PCB repeatedly over a wide temperature range; many are actually thermal shock tests using liquid-liquid cycling. The results of five accelerated tests with different temperature extremes, ramp rates, and dwell times have been compared by the IPC, which also provides a simplified analytical model to estimate PTH life.[3] The results of all tests suggest the same approaches for maximizing PTH reliability, but they do not all correlate well quantitatively. Two of the most common tests are (1) oven cycling from −65 to +125°C, and (2) thermal shock cycling between oil or fluidized sand baths at +25 to 260°C. Figure 53.23 shows a suitable test coupon that contains 3000 PTHs interconnected in series, several PTH sizes, and varying annular ring sizes. The PTHs can be monitored during the testing. Figure 53.18*b* shows the type of data that can be collected in this type of test.

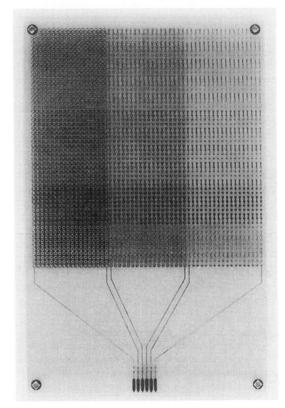

FIGURE 53.23 PTH reliability test coupon. This coupon contains three sets of 1000 PTHs interconnected in series on four layers. Each set is a different hole size. The pad size is also varied. Similar designs are available from the IPC.

53.6.2.2 Mechanical. Printed circuit boards are rarely subjected to mechanical tests that could cause electrical failures; however, adhesion of both Cu and solder mask to the laminate is critical and is often tested. Loss of solder mask adhesion can provide a place for corrodants and moisture to accumulate, which can be the cause of electrical failures when the board is exposed to temperature and humidity.

Adhesion is commonly tested using the peel test described in IPC-TM-650, Method 2.4.28. The simplest version of this test is conducted by scribing the adherent and dividing it into small squares. If the Cu or solder mask pulls off with a piece of tape with strong adhesive, the adhesion is inadequate. More quantitative tests that measure the actual peel strength are performed primarily by laminate and solder mask suppliers.

53.6.2.3 Temperature, Humidity, Bias. These tests are designed to promote corrosion on the PCB surface and conductive anodic filament growth, either of which can cause insulation resistance failures.

Surface insulation tests utilize two interleaved Cu combs with an imposed dc bias across the combs. These combs may be designed into existing boards or a coupon such as the IPC-B-25 test board shown in Fig. 53.24 may be used. The measured resistance (ohms) from the comb pattern can be converted to surface resistivity (ohms per square) by multiplying the measured

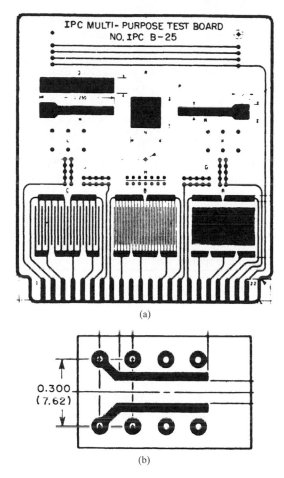

FIGURE 53.24 Test coupons used to check moisture, insulation, and metal migration resistance: (*a*) the IPC-B-25 test board, used to qualify the process; (*b*) The Y coupon, designed to be incorporated into production boards for statistical process control. From IPC-SM-840.

resistance by the square count of the pattern. The square count is determined geometrically by measuring the total length of the parallel traces between the anode and cathode and dividing by the separation distance. Special precautions are needed to make accurate measurements of insulation resistance.[42] Measurements of resistance above 10^{12} are very difficult and require careful shielding. Measurements of resistance below 10^{12} can be conducted in most laboratory environments if certain precautions are taken.

The actual tests are usually conducted at elevated temperature and humidity with an applied dc bias. A test for moisture and insulation resistance of bare printed circuit boards is included in IPC-SM-840A. The severity of the test depends on the intended use environment; for typical commercial products (Class 2), the test is conducted at 50°C, 90% RH, and 100 V_{dc} bias for 7 days. The minimum insulation resistance requirement is 10^8 Ω. The military test procedure for moisture and insulation resistance is specified in MIL-P-55110.[43] The moisture resistance test should be conducted in accordance with MIL-STD-202, Method 106, with applied polarization voltage ($100\,V_{dc}$) and Method 402, Test condition A.[44] IPC-SM-840A also includes a test for electromigration resistance. The test is conducted at 85°C/90% RH at a 10 V_{dc} bias with a limiting current of 1 mA for 7 days. A significant change in current constitutes a failure. The samples are also microscopically inspected for evidence of electrolytic metal migration. A common test for dendritic growth due to flux residues is 85/85/1000 h at $-20\,V_{dc}$ bias. These tests are empirically based; however, several investigators have attempted to develop acceleration factors for these and similar tests.[45,46]

53.6.3 Printed Circuit Assembly Reliability Tests

53.6.3.1 Thermal. Most thermal cycling of PCAs is intended to accelerate solder joint thermal fatigue failures. In spite of the existence of an IPC standard, there is no standard accelerated test today that is suitable for all component and substrate combinations and all service environments. There are several acceleration models in the literature, each of which seems to fit the data well in at least some situations. All are based on a combination of empirical observations and fundamental arguments under simplifying assumptions. This topic remains a subject of active research since in some cases the predictions are significantly different. There is also a move to replace thermal cycling tests with mechanical cycling tests, which could be conducted in a shorter period of time; however, these tests are even further from standardization. Finally, for some components which dissipate a significant amount of power (usually 1 W or more), cycling the ambient temperature (which heats from the outside) may give quite different results from power cycling (in which heating occurs from the inside); for example, the failure location may shift from corner joints (which see the largest displacements) toward solder joints located near the chip (because they are hotter). Consequently, while thermal cycling will be adequate for most ASICs, memory chips, etc., power cycling should be considered for microprocessors, particularly those that dissipate more than a few watts.

Thermal shock testing is commonly used to test components, but it is not necessarily a substitute for thermal cycling. Because the temperature ramp is extremely rapid and the dwell at the extremes is generally short, there is little time for creep; consequently, the number of cycles to failure is increased. Furthermore, the rapid temperature change can induce differential thermal stresses that may be larger than those experienced during thermal cycling. These stresses can induce early failures, particularly if the failure is not in the solder.

There are some principles for designing thermal cycling tests to accelerate solder fatigue that seem to be generally agreed on. The following guidelines apply to gradual temperature cycling due to ambient heating inside a unit (e.g., due to power dissipation). If the unit will be subjected to extreme temperatures or thermal shock in service, these generalizations may not apply. A sample cycling protocol is shown in Fig. 53.25.

- The maximum test temperature should be below the T_g of the printed circuit board, for FR-4 below about 110°C. At T_g, the CTE of the board increases rapidly, but many other properties also change; for example, the elastic modulus of the board decreases. To avoid approaching the melting temperature of the solder and changing the mechanism of solder creep, the maximum temperature should also be kept below about $0.9T_m$, where T_m is the melting temperature of the solder in Kelvin. For eutectic Sn-Pb solder, T_m is 137°C, well above T_g. But for printed circuit board materials with high T_g values or low-melting-temperature solders, this restriction may take precedence. Using a peak temperature above these limits results in unpredictable acceleration.

- The minimum temperature should be high enough that creep is still the primary deformation mechanism of the solder, that is, at least $0.5\,T_m$, or -45°C for eutectic Sn-Pb solder. Many investigators prefer a higher minimum temperature (-20 or 0°C) to ensure that creep occurs fast enough to relieve the imposed shear stress during the allowed dwell time. Using too low a minimum temperature may seem to increase the acceleration factor (increased ΔT) while actually decreasing it (decreased $\Delta\varepsilon$), resulting in an overly optimistic life prediction.

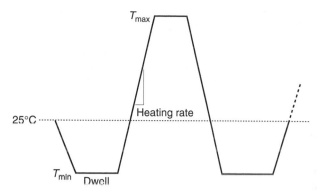

FIGURE 53.25 Schematic thermal cycling profile for testing solder joint thermal fatigue.

- The rate of temperature cycling should not exceed 20°C/min and the dwell time at the temperature extremes should be at least 5 min. The purpose of controlling the cycling speed is to minimize thermal shock and the stresses associated with differential heating or cooling. The dwell time at the temperature extremes is an absolute minimum needed to permit creep to occur. A longer dwell time is recommended, particularly at the minimum temperature.

As described here, there are several acceleration factors in the literature that may be applied to make life estimations from the accelerated test data that may be obtained using a thermal profile that fits the preceding criteria. One of the simplest expressions is due to Norris and Landzberg:

$$\frac{N_{op}}{N_{test}} - \left(\frac{v_{op}}{v_{test}}\right)^{1/3} \left(\frac{\Delta T_{test}}{\Delta T_{op}}\right)^2 \left(\frac{\phi_{test}}{\phi_{op}}\right)$$

where N_{op} and N_{test} = life under operating and accelerated test conditions, respectively

v = cycling frequency

ϕ = mean temperature[47]

Another widely used expression may be found in IPC-SM-785. This expression tries to account for the effect of power cycling and the dwell time in the thermal profile; it also makes it possible to make predictions for one component or solder joint geometry based on data for another similar one. In its most simplified form, the acceleration factor for tests meeting the preceding criteria for FR-4 and eutectic Sn-Pb solder joints may be *approximated* as

$$\frac{N_{op}}{N_{test}} = \frac{\Delta T_{op}^{2.4}}{\Delta T_{test}} \qquad \text{for leadless surface-mount attachments}$$

$$\frac{N_{op}}{N_{test}} = \frac{\Delta T_{op}^{4}}{\Delta T_{test}} \qquad \text{for compliant-leaded surface-mount attachments}$$

assuming the test assemblies are nearly identical to the assemblies that will be put into service.

Mechanical fatigue cycling is increasingly being used as a quick way to induce solder joint failures. The goal is to *simulate* the thermal fatigue failure process in a much shorter test. The validity of this approach is still being investigated; although the imposed strain in each cycle is intended to be the same, the mechanical test eliminates thermomechanical effects (including creep) because the cycles are about two orders of magnitude faster. Nonetheless, mechanical cycling can certainly provide useful comparisons between different designs or package styles. The tests are usually conducted at a constant temperature with fixturing that puts the solder joints in shear when bending or tensile displacements are applied.

53.6.3.2 Mechanical.   Mechanical vibration and shock can cause solder joint failures, particularly for large, rigid components or components with large, heavy heat sinks. Mechanical shock tests are usually modeled after drops that may occur during transportation or use. The test drops are generally quite severe, but few in number, since the system is not expected to be subjected to repeated drops in service. One common test uses a maximum acceleration of about 600 g, a maximum velocity of about 300 in/s, and a shock pulse of about 2.5 ms duration. The test setup is shown schematically in Fig. 53.26.

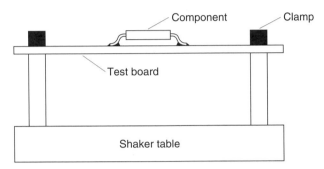

FIGURE 53.26   Schematic of setup used to test the resistance of printed circuit assemblies to out-of-plane mechanical vibration and shock.

On the other hand, the PCA may be exposed to millions of mechanical vibration cycles during its lifetime.[48] Depending on the application, both in-plane and out-of-plane vibration may play an important role. The damage done by these cycles depends primarily on whether the cycling frequency is near a natural frequency of the board, where large deflections can take place. For in-plane vibration, random vibration over a wide excitation frequency range is performed at a constant power spectral density. Most surface-mount components have high natural frequencies, and solder joint failures are rarely observed. For out-of-plane vibration, the following procedure is recommended:

- Design and fixture the test board so that each specimen is clamped at both ends and contains a single component at the center (see Fig. 53.26).
- Impose the vibration using a shaker table with sinusoidal excitation.
- Find the natural frequency of the sample by sweeping the frequency at a low amplitude (to prevent inadvertent damage before the start of the test). This natural frequency is near, but not the same as, the natural frequency at the larger amplitude that will be used for testing.
- Conduct the test by sweeping the frequency in a narrow range around the natural frequency previously determined. The amplitude can be set to correspond to a certain power spectral density or to achieve a desired deflection of the PCB.

53.6.3.3 Temperature, Humidity, Bias.   The primary purpose of these tests is to identify surface insulation resistance (SIR) degradation due to corrosive materials left on the board from the assembly process or due to galvanic couples set up in the assembly process. The usual test procedure is to use SIR comb patterns on the PCB like those described in Sec. 53.6.1.3 and to subject the assembly to 85°C/85% relative humidity/−20V$_{dc}$ for 1000 h. The bias voltage is dependent on the test device or test vehicle chosen.

One acceleration model that is applied is the modified Eyring model, which was developed for moisture-induced corrosion in plastic packages:

$$t_{50\%} = \left[A \exp\left(\frac{E_a}{kT} \right) \right]\left[\exp\left(\frac{C}{H_r} \right) \right]\left[D\exp\left(\frac{-V}{B} \right) \right] \tag{53.4}$$

where $t_{50\%}$ = time at which 50 percent of parts have failed
$A, B, C,$ and D = empirical constants
E_a = thermal activation energy
k = Boltzmann constant
T = temperature in degrees Kelvin
H_r = relative humidity
V = reverse-biasing voltage[41,46]

The time to failure is also dependent on the concentration of ionic contaminants. The default industrial ionic contamination limit comes from MIL-STD-28809A; it is the equivalent of 3.1 µg/cm² of NaCl. For small plastic packages, enough data have been collected to show that an empirical acceleration factor for temperature and relative humidity AF_{T,H_r} applies:

$$AF_{T,H_r} = 2^{(T+H_r)_{test} - (T+H_r)_{service}} \tag{53.5}$$

where AF = acceleration factor
T = temperature in °C
H_r = relative humidity in percent[46,49]

The rule of thumb for the effect of reverse bias is very device-specific; the following relationship was established for 20-V Schottky diodes in SOT-23 packages:[46]

$$AF_V = 7700 \exp\left(\frac{-V}{12.32} \right)$$

When both acceleration factors apply, the total acceleration factor is:

$$AF_{total} = (AF_{T,H_r})(AF_V)$$

53.7   SUMMARY

Reliability of electronic assemblies is a complex subject. This chapter has touched on only one aspect of the problem: understanding the primary failure mechanisms of printed circuit boards and the interconnects between these boards and the electronic components mounted on them. This approach provides the basis for analyzing the impact of design and materials choices and manufacturing processes on printed circuit assembly reliability. It also provides the foundation for developing accelerated testing schemes to determine reliability. It is hoped that the fundamental approach will enable the reader to apply this methodology to new problems not yet addressed in mainstream literature.

REFERENCES

1. M. A. Oien, "Methods for Evaluating Plated-Through-Hole Reliability," *14th Annual Proceedings of IEEE Reliability Physics,* Las Vegas, Nev., April 20–22, 1976.
2. K. Kurosawa, Y. Takeda, K. Takagi, and H. Kawamata, "Investigation of Reliability Behavior of Plated-Through-Hole Multilayer Printed Wiring Boards," IPC-TP-385, IPC, Evanston, Ill., 1981.

3. IPC-TR-579, "Round Robin Reliability Evaluation of Small Diameter Plated Through Holes in Printed Wiring Boards," Institute of Interconnecting and Packaging Electronic Circuits, Lincoln-wood, Ill. The IPC recently initiated a second round robin study, the results of which should be available in about 1996.

4. F. Fehrer and G. Haddick, "Thermo-mechanical Processing and Repairability Observations for FR-4, Cyanate Ester and Cyanate Ester/Epoxy Blend PCB Substrates," *Circuit World,* vol. 19, no. 2, 1993, pp. 39–44.

5. D. B. Barker and A. Dasgupta, "Thermal Stress Issues in Plated-Through-Hole Reliability," in *Thermal Stress and Strain in Microelectronics Packaging,* J. H. Lau (ed.), Van Nostrand Reinhold, 1993, pp. 648–683.

6. IPC-TM-650, Method 2.4.8.

7. L. D. Olson, "Resins and Reinforcements," in *ASM Electronic Materials Handbook, Vol. 1: Packaging,* ASM International, Materials Park, Ohio, 1989, pp. 534–537.

8. D. W. Rice, "Corrosion in the Electronics Industry," *Corrosion/85,* paper no. 323, National Association of Corrosion Engineers, Houston, 1985.

9. J. J. Steppan, J. A. Roth, L. C. Hall, D. A. Jeannotte, and S. P. Carbone, "A Review of Corrosion Failure Mechanisms during Accelerated Tests: Electrolytic Metal Migration," *J. Electrochemical Soc.,* vol. 134, 1987, pp. 175–190.

10. D. J. Lando, J. P. Mitchell, and T. L. Welsher, "Conductive Anodic Filaments in Reinforced Polymeric Dieletrics: Formation and Prevention," *17th Annual Proceedings of IEEE Reliability Physics Symposium,* San Francisco, April 24–26, 1979, pp. 51–63.

11. "How to Avoid Metallic Growth Problems on Electronic Hardware," IPC-TR-476, Sept. 1977.

12. B. Rudra, M. Pecht, and D. Jennings, "Assessing Time-of-Failure Due to Conductive Filament Formation in Multi-Layer Organic Laminates," *IEEE Trans. CPMT-Part B.,* vol. 17, August 1994, pp. 269–276.

13. J. W. Price, *Tin and Tin Alloy Plating,* Electrochemical Publications, Ayr, Scotland, 1983.

14. D. R. Gabe, "Whisker Growth on Tin Electrodeposits," *Trans. Institute of Metal Finishing,* vol. 65, 1987, p. 115.

15. C. Lea, *A Scientific Guide to Surface Mount Technology,* Electrochemical Publications, Ayr, Scotland, 1988.

16. J. H. Lau (ed), *Solder Joint Reliability: Theory and Applications,* Van Nostrand Reinhold, New York, 1991.

17. D. R. Frear, S. N. Burchett, H. S. Morgan and J. H. Lau, eds., *Mechanics of Solder Alloy Interconnects,* Van Nostrand Reinhold, New York, 1994.

18. J. H. Lau (ed.), *Thermal Stress and Strain in Microelectronics,* Van Nostrand Reinhold, New York, 1993.

19. S. Burchett, "Applications—Through-Hole," in *The Mechanics of Solder Alloy Interconnects, op. cit.,* pp. 336–360.

20. E. Suhir and Y.-C. Lee, "Thermal, Mechanical, and Environmental Durability Design Methodologies," in *ASM Electronic Materials Handbook, Vol. 1: Packaging, op cit.*

21. IPC-D-279, "Design Guidelines for Reliable Surface Mount Technology Printed Board Assemblies," to be published.

22. L. T. Manzione, *Plastic Packaging of Microelectronic Devices,* Van Nostrand Reinhold, New York, 1990.

23. *ASM Electronic Materials Handbook, Vol. 1: Packaging, op. cit.*

24. R. R. Tummala and E. J. Rymaszewski (eds.), *Microelectronic Packaging Handbook,* Van Nostrand Reinhold, New York, 1989.

25. For a more comprehensive list of components that may be at risk, see IPC-D-279, "Design Guidelines for Reliable Surface Mount Technology Printed Board Assemblies," App. C, to be issued.

26. L. Zakraysek, R. Clark, and H. Ladwig, "Microcracking in Electrolytic Copper," *Proceedings of Printed Circuit World Convention III,* Washington, D.C., May 22–25, 1984.

27. G. T. Paul, "Cracked Innerlayer Foil in High Density Multilayer Printed Wiring Boards," *Proceedings of Printed Circuit Fabrication West Coast Technical Seminar,* San Jose, Aug. 29–31, 1983.

28. M. W. Gray, "Inner Layer or Post Cracking on Multilayer Printed Circuit Boards," *Circuit World,* vol. 15, no. 2, pp. 22–29, 1989.

29. L. Mayer and S. Barbieri, "Characteristics of Acid Copper Sulfate Deposits for Printed Wiring Board Applications," *Plating and Surface Finishing,* March 1981, pp. 46–49.

30. J. L. Marshall, L. A. Foster, and J. A. Sees, "Interfaces and Intermetallics," in *The Mechanics of Solder Alloy Interconnects,* D. R. Frear, H. Morgan, S. Burchett, and J. Lau (eds.), *op. cit.*, pp. 42–86.

31. M. Economou, "Rework System Selection," *SMT,* February 1994, pp. 60–66.

32. J. Lau, S. Leung, R. Subrahmanyan, D. Rice, S. Erasmus, and C. Y. Li, "Effects of Rework on the Reliability of Pin Grid Array Interconnects," *Circuit World,* vol. 17, no. 4, pp. 5–10, 1991.

33. F. L. Gray, "Thermal Expansion Properties," in *ASM Electronic Materials Handbook, Vol. 1: Packaging, op. cit.*

34. P. M. Hall, "Thermal Expansivity and Thermal Stress in Multilayered Structures," in *Thermal Stress and Strain in Microelectronics Packaging,* J. H. Lau (ed), *op. cit.,* pp. 78–94.

35. J. Glazer, P. A. Kramer, and J. W. Morris, Jr., "The Effect of Gold on the Reliability of Fine Pitch Surface Mount Solder Joints," *Circuit World,* vol. 18, 1992, pp. 41–46.

36. G. Humpston and D. M. Jacobson, *Principles of Soldering and Brazing,* ASM International, Materials Park, Ohio, 1993, Chap. 3.

37. F. G. Yost, "Soldering to Gold Films," *Gold Bulletin,* vol. 10, 1977, pp. 94–100.

38. J. Glazer, "Metallurgy of Low Temperature Lead-free Solders: A Literature Review," *International Materials Review,* vol. 40, 1995, pp. 65–93.

39. H. L. Hvims, *Adhesives as Solder Replacement for SMT,* vols. I and II, Danish Electronics, Light and Acoutics, Hoersholm, Denmark, January 1994.

40. A. Emerick, J. Ellerson, J. McCreary, R. Noreika, C. Woychik, and P. Viswanadham, "Enhancement of TSOP Solder Joint Reliability Using Encapsulation," *Proceedings of the 43d Electronic Components and Technology Conference,* IEEE-CHMT, 1993, pp. 187–192.

41. P. Tobias and D. Trindade, *Applied Reliability,* Van Nostrand Reinhold, New York, 1986.

42. ASTM D 257-78, Standard Test Methods for DC Resistance or Conductance of Insulating Materials.

43. Military specification for Printed Wiring Boards.

44. Military standard test method for Electronic and Electrical Component Parts.

45. P. J. Boddy, R. H. Delaney, J. N. Lahti, E. F. Landry, and R. C. Restrick, "Accelerated Life Testing of Flexible Printed Circuits: Part I and II," *14th Annual Proceedings of the IEEE Reliability Physics Symposium,* Las Vegas, Nev., April 20–22, 1976.

46. K.-L. B. Wun, M. Ostrander, and J. Baker, "How Clean is Clean in PCA," *Proceedings of Surface Mount International,* San Jose, Calif., 1991, pp. 408–418.

47. K. C. Norris and A. H. Landzberg, *IBM Journal of Research and Development,* vol. 13, 1969, p. 266.

48. For example, see J. H. Lau, "Surface Mount Solder Joints Under Thermal, Mechanical, and Vibration Conditions," in *The Mechanics of Solder Alloy Interconnects,* D. R. Frear, H. Morgan, S. Burchett, and J. Lau (eds.), *op. cit.,* pp. 361–415.

49. E. B. Hakim, "Acceleration Factors for Plastic Encapsulated Semiconductors," *Solid State Technology,* Dec. 1991, pp. 108–109.

FURTHER READING

ASM Electronic Materials Handbook, Vol. 1: "Packaging," ASM International, Materials Park, Ohio, 1989.

Frear, D. R., H. Morgan, S. Burchett, and J. Lau, *The Mechanics of Solder Alloy Interconnects,* Van Nostrand Reinhold, New York, 1994.

IPC-A-600D, Acceptability of Printed Boards, August 1989.

IPC-A-610A, Acceptability of Electronic Assemblies, Feb. 1990.

IPC-D-279, Design Guidelines for Reliable Surface Mount Technology Printed Board Assemblies, to be issued.

IPC-SM-785, Guidelines for Accelerated Reliability Testing of Surface Mount Solder Attachments, Nov. 1992.

de Kluizenaar, E. E., "Reliability of Soldered Joints: A Description of the State of the Art," *Soldering and Surface Mount Technology,* Part I: no. 4, pp. 27–38; Part II: no. 5, pp. 56–66; Part III: no. 6, pp. 18–27, 1990.

Lau, J. H. (ed.), *Solder Joint Reliability,* Van Nostrand Reinhold, New York, 1991.

Lau, J. H. (ed.), *Thermal Stress and Strain in Microelectronics Packaging,* Van Nostrand Reinhold, New York, 1993.

Lea, C., *A Scientific Guide to Surface Mount Technology,* Electrochemical Publications, Ayr, Scotland, 1988.

Tummala, R. and E. J. Rymaszewski (ed.), *Microelectronic Packaging Handbook,* Van Nostrand Reinhold, New York, 1989.

CHAPTER 54
COMPONENT-TO-PWB RELIABILITY

Mark Brillhart
Cisco Systems, San Jose, California

54.1 INTRODUCTION TO PACKAGE RELIABILITY

The printed wiring board (PWB) designer is faced with a myriad of packaged device types, field use environments, connectors, PWB materials, cost considerations, and real estate restrictions that all must be optimized without degrading system reliability. The radical increases in package-to-board input/output (I/O), reductions in pitch, and overall rise in surface-mount component density on PWBs pose tremendous challenges with respect to reliability and create complex design trade-offs that must be considered while maintaining product reliability at required levels. Time to market and aggressive product life cycle pressures also require PWB designers to consider the impact that alternate designs and material sets may have on reliability in a rapid fashion.

54.1.1 Packaging Challenges

The reliability of many types of leaded components (see Figs. 54.1 and 54.2) have been assessed in great detail.[1] While these devices are by no means perfectly reliable at the first and second levels (definitions of first- and second-level interconnects can be found in Fig. 54.3), they have historically posed a lesser risk than grid array packaged components, such as ceramic ball grid arrays (CBGAs), plastic ball grid arrays (PBGAs), ceramic solder column carriers (CSCCs), flip-chip ball grid arrays (FCBGAs), and chip-scale packages (CSPs). Figures 54.4 through 54.6 show various grid array packaged devices. Many of these types of packages are either increasing in body size, decreasing in pitch, or both, to a point where the standard reliability envelope is being pushed to extreme levels. Additionally, most grid array packaged devices are high in cost (in many instances the grid array packaged devices are the most costly component on a printed circuit assembly) and thus cannot be deployed in a redundant design. Lack of redundancy creates critical path single points of failure (SPOFs). Failure of one of these devices can lead to catastrophic system-level failures. The proliferation of high-I/O, small-pitch grid array packaged devices throughout the electronics industry impacts almost all PWB engineers because any design has a high probability of containing one or more of these types of components.

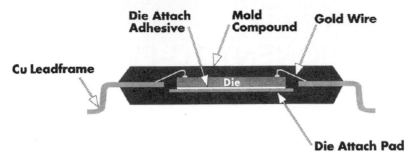

FIGURE 54.1 Diagram of a quad flat pack. Note the copper (Cu) leads that are attached by gold wire bonds to the die. This is the primary (or first-level interconnect) between the die and the package. The second-level interconnect will be created when the Cu leadframe is soldered to a motherboard via some type of surface-mount attach process. *(Courtesy of Amkor Technology, Chandler, Arizona.)*

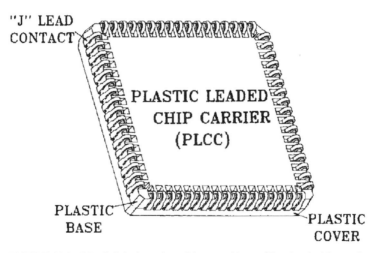

FIGURE 54.2 J-leaded device oriented bottom side up. *(Reprinted with permission from J. Lau,* Ball Grid Array Technology, *McGraw-Hill, New York, 1995, p. 20.)*

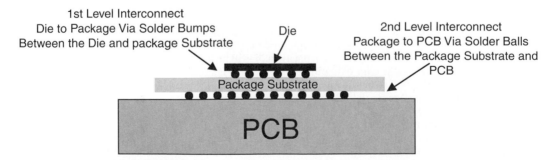

FIGURE 54.3 First- and second-level interconnects defined, employing a flip-chip assembly as an example. The first-level interconnect is the primary connection between the silicon die and the package substrate. In this example it is created by the solder bumps between the die and the package. The second-level interconnect is the "next" level of connection between the package substrate and the PWB. In this example the second-level interconnect is created when the solder balls on the bottom side of the package are attached via SMT to the PWB.

FIGURE 54.4 Typical plastic ball grid array packages, various body sizes and ball arrays. All of these packages employ a wire-bond first level interconnect (to connect the die to the package substrate) and solder balls as a second-level interconnect (to attach the package substrate to the PWB). *(Courtesy of Amkor Technology, Chandler, Arizona.)*

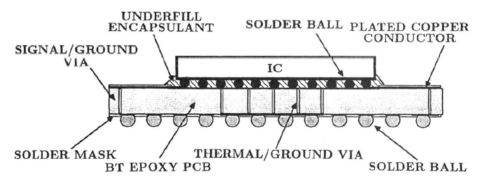

FIGURE 54.5 Cross section of a flip-chip PBGA. Note the silicon die (IC) attached to the package substrate (BT EPOXY PWB) with solder balls. The underfill encapsulant protects the solder balls from thermal fatigue. The thermal ground vias allow power to be dissipated from the IC through the package substrate and into the PWB. Solder mask is employed to define the conductive pads on the bottom of the package substrate that the solder balls are attached to. This creates a solder mask-defined pad (SMD). *(Reprinted with permission from J. Lau,* Ball Grid Array Technology, *McGraw-Hill, New York, 1995, p. 33.*

54.1.2 Overview

Designers need tools to assess experimental reliability data, transform laboratory results into actual field loading conditions, and rapidly assess the reliability of different PWB layouts, material sets, and package types. Engineers must also have the capabilities to address complex loading conditions such as mini- and power cycles and how they impact interconnect perfor-

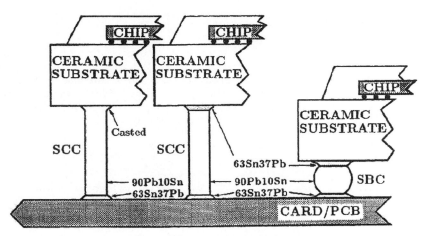

FIGURE 54.6 Cross section of a ceramic solder column carrier and ceramic ball grid array (CBGA). Solder column carriers (SCC) and a solder ball carrier (SBC) are shown. The SBC is often referred to as a CBGA package due the fact that it is a ceramic package substrate that employs solder balls to create a second-level interconnect between the package substrate and the PWB. The solder column carrier employs a column of high-lead solder (which is either casted onto the ceramic package substrate or attached by a layer of eutectic 63Sn/37Pb solder paste) to create a second-level interconnect between the ceramic package substrate and the PWB. Typical ceramic ball grid array packages cannot achieve required reliability when body sizes exceed 32 mm per side. Ceramic packages having larger body sizes either employ SCC-type technology or connectors to create a second-level interconnect. (*Reprinted with permission from J. Lau,* Ball Grid Array Technology, *McGraw-Hill, New York, 1995, p. 27.*)

mance. This chapter focuses on reliability of the PWB-to-package interconnect (second-level interconnect) of grid array devices. The chapter is broken up into the following sections:

- Section 54.2, "Variables that Impact Reliability," addresses many of the common variables impacting reliability that designers encounter when developing a new PWB. Practical examples and best practices are included where applicable.

- Section 54.3, "Experimental Tools for Estimating Solder Joint Life," contains a discussion of many common experimental tools currently employed throughout the industry to assess interconnect reliability. Types of data generated and methods for interpretation are presented, along with guidelines for assessing the quality of experimental results obtained via different types of reliability assessment tools.

- Section 54.4, "Rapid Assessment Tools," presents analytical and numerical rapid reliability assessment tools. A methodology for rapid assessment is discussed, along with applicability of the tools and guidelines for when to or when not to employ a specific approach to assess interconnect reliability.

- Section 54.5, "Power and Minicycles," covers the topic of minicycles vs. power cycles and how both loading modes impact interconnect reliability. Methods for dealing with multiple loading modes are also discussed.

- Section 54.7, "Practical Examples," discusses experimental tools, analytical techniques, and numerical approaches for reliability assessments.

54.2   VARIABLES THAT IMPACT RELIABILITY

Electronic packages are composed of complex materials that are assembled via intricate processes and then subjected to a wide range of service environments. It is crucial to note that both the die packaging processes and surface-mount technologies can impart stresses to and create residual stresses in the final packaged assembly that impact interconnect reliability. Assessing the reliability of package devices requires that all variables that can hinder or enhance reliability be considered.

54.2.1   Actual Product Environment

Printed circuit assemblies are deployed in a wide range of environments. Packages can be subjected to extreme temperatures and humidity. Automotive applications, for example, can see temperature ranges from −55 to well over 95°C[2] along with up to 100 percent humidity. The rate at which temperature and humidity changes occur can be quite severe under the hood, causing components to transition from extreme cold to heat in a matter of minutes. This creates a very difficult environment to model, as the rate of occurrence of thermal and humidity cycles must be thoroughly understood. Severe shock and vibration can also be imparted to assemblies in this type of service environment, which adds to the complexity of reliable assessment. Great care must be taken when assessing these types of aggressive environments and their impact on packaging reliability. A thorough experimental analysis that employs thermal couples (to assess temperature conditions) and accelerometers (to assess shock and vibration loading conditions) should be employed to determine the actual end use environmental conditions. These stressful environments are the primary driver for automotive applications employing extreme protection measures such as full encapsulation/potting for an electronic device placed under the hood.

Computer room environments represent the other extreme, with temperatures and humidity controlled to very tight tolerances. While servers, switches, hubs, routers, and other devices are housed in a controlled environment, their extremely large packages, coupled with power and minicycles (discussed in a subsequent section), create a set of loading conditions that are different from those found in automotive applications but equally challenging. Worst-case loading conditions for a variety of end use environments can be found in Table 54.1.[2] Note the broad variation in temperature extremes, frequency of thermal cycles, and expected service life.

TABLE 54.1   Examples of Worst-Case Environments for Different Use Categories[2]

Use Category	T_{min} (°C)	T_{max} (°C)	t_D (hrs)	Cycles per year	Typical years of service	Approximated acceptable failure risk (%)
Consumer	0	+60	35	12	1–3	1
Computers	+15	+60	2	1460	5	0.1
Telecom	−40	+85	12	365	7–20	0.01
Commercial aircraft	−55	+95	12	365	20	0.001
Industrial and automotive passenger compartment	−55	95	12	20–185	10	0.1
Military ground and ship	−55	+95	12	100–265	10	0.1
Space	−55	+95	1–12	365–8760	5–30	0.001
Military avionics	−55	+95	1–2	365	10	0.01
Automotive under hood	−55	+125	1–2	40–1000	5	0.1

T_{min}, minimum temperature; T_{max}, maximum temperature; t_D, dwell time at the operating temperature.

Prior to initiating any reliability assessment, it is crucial to understand the thermal, humidity, and mechanical loading conditions that a packaged device will be exposed to throughout its expected lifetime. Temperature maximums and minimums and frequencies and rates at which the component cycles between extremes must be thoroughly understood. Complex thermal cycles (such as an outdoor product experiencing daily temperature changes coupled with a more local effect such as under-the-hood heating and cooling due to engine usage) must also be considered. Section 54.5 provides methodologies for assessing those types of loading conditions. Humidity and vibration loading conditions must also be evaluated as part of an overall package qualification process.

54.2.2 Design Choices

Although environmental impact has a large effect on interconnect reliability, there are some basic design choices that can have a positive (or negative) impact on solder joint reliability. The high-density revolution has forced PWB designers to place increasingly complex surface-mount devices in smaller and smaller spaces. This PWB density increase (where density refers to the number of components per unit area of PWB) may force designers to consider mirror BGA placement (see Figs. 54.7 and 54.8) or to place multiple devices on one side of a PWB. Both of these design challenges are discussed in the following two sections.

FIGURE 54.7 Double-sided BGA in mirror configuration. Note that the PCA is an exact mirror about the center line of the PWB. In some instances the top- and bottom-side packages share common vias.

FIGURE 54.8 Double-sided BGA in quasi-mirror configuration. Note that although the packages are not right over each other, there is some overlap when looking from the top down or from the bottom up.

54.2.2.1 Double-Sided (Mirror BGA). Many printed circuit board designers are faced with a board space dilemma. The density revolution has forced designers to place as many devices as physically possible on all surfaces of a PWB. The beginning of the density revolution occurred years ago when designers began implementing double-sided technologies to allow for the placement of surface-mount components on both sides of a PWB. The continued push toward higher device density has resulted in placement of BGA components in mirror and quasi-mirror configurations (see Figs. 54.7 and 54.8).

The reduction in package reliability when employing a mirrored double-sided configuration has been experimentally assessed.[3,4] Test vehicles consisting of daisy-chained packaged dies mounted on experimental motherboards were subject to thermal cycle loading conditions while resistance was continually monitored throughout testing. Packaged daisy-chained dies were also mounted in nonmirror configurations to act as controls. Figures 54.9 and 54.10

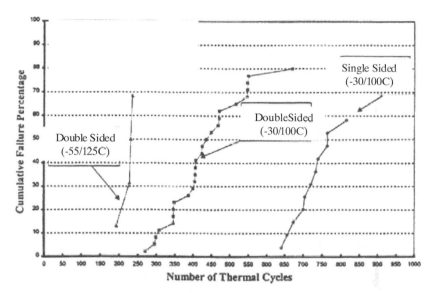

FIGURE 54.9 Reliability data for mirror and single-sided BGAs. Note the reduction in reliability when switching from single- to double-sided (mirror) configurations (when comparing the single-sided −30/100°C curve to the double-sided −30/100°C curve). Also, it can be seen that a decrease in the minimum temperature from −30 to −55°C results in an additional reduction in interconnect reliability. (*Reprinted with permission from* Proceedings of The Chip Scale Packaging Symposium, SMTA International, *September 1999, p. 76.*)

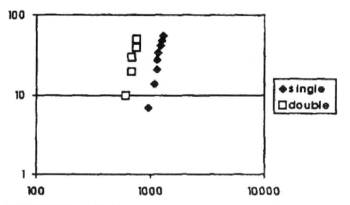

FIGURE 54.10 Reliability data for mirror BGAs. Note the reduction in reliability when the double-sided configuration is employed The solid diamonds indicate failure data for single-sided chip-scale packag (CSP) devices; the hollow squares express failure data for identical CSP devices that are mounted in mirror configuration. (*Reprinted with permission from* SMTA International, SMTA Proceedings, *September 1999, p. 243.*)

show failure rates for CSP packages in single-sided and double-sided mirror positions.[3,4] Note the almost 50 percent reduction in fatigue performance when a device is placed in a mirror configuration.

Due to the dramatic decrease in assembly reliability when attempting mirrored configurations, it is recommended that these types of configurations be avoided whenever possible. When it is impossible to avoid a mirror configuration, a quasi-mirror configuration (see Fig. 54.8) with no common through-hole vias is recommended. It is also essential to assess the impact on reliability. A standard rule of 50 percent reduction in fatigue life when employing a mirror configuration may be taken as a first-pass approximation. It is also recommended that a thermal cycle test vehicle be built and cycled to ensure that reliability degradation will still keep service life at an acceptable level when either a mirror or quasi-mirror configuration is considered.

54.2.2.2   High Density Due to Multiple Packages in a Small Area.   A second source of reliability degradation due to density increases can be observed when a designer attempts to place many BGA packages in a small area on a thin PWB (see Fig. 54.11). The effective coefficient of thermal expansion (CTE) of the packaged die or module tends to be much lower than that of the actual PWB. During reflow, the PWB and packages are in a stress-free state. Upon cooling, however, the packaged die contracts at a reduced rate (due to the lower effective CTE) than the underside of the PWB, causing a convex-type curvature to be imparted to the PWB.

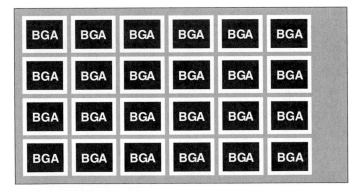

FIGURE 54.11   High-density placement of BGAs on top side of a PWB. This diagram illustrates a scenario where multiple BGA devices are all placed on one side of a board.

Warpage Issues.   The impacts of warpage on a PWB assembly are multiple. First, warpage must typically be overcome when a PWB is placed in a card cage. Card cage placement of warped boards results in a rapid change in the loading of the solder joints that may lead to microcrack formation or, in severe instances, global solder joint fracture. Severe PCA warpage may create a card that physically cannot be placed in a card cage, resulting in a nonmanufacturable product. Warped boards should be clamped in a controlled fashion to reduce the stress imparted to solder joints upon installation in a card cage.

Die and Package Stress Issues.   A second issue associated with board warpage is die and package stress. The package experiences an amount of stress as it constrains the PWB from contracting during cooling after reflow. If this stress is severe enough, then the silicon may crack, resulting in a failure of the ASIC. Also, severe stress may create multiple types of package delamination and/or package cracking. The extreme danger here is that a device could pass full first-level qualification (first-level qualification typically consists of thermal cycle, humidity and temperature exposure, extreme-temperature storage, and shock and vibration experiments performed on packaged devices not attached to a motherboard) yet suffer from die cracking due to stresses imparted from a warped PWB. This is one of the primary reasons that many packaged ASIC suppliers and consumers are requesting that second-level qualification test vehicles (where second level refers to the interconnect between the package and the motherboard, a second-level qualification is always conducted on devices attached to a motherboard) include daisy chains at the first-level interconnect. Inclusion of first-level joints or wire bonds allows for the impact of second-level attach on first-level reliability to be assessed simultaneously with second-level reliability during a second-level qualification.

Back-Side Component Issues. A third challenge related to warped PWBs concerns the back-side components in double-sided designs. During cooldown after reflow, the top of the PWB is restrained from contraction due to the smaller effective CTE of the packaged device. This results in the bottom side of the PWB overcontracting during warpage. The state of stress on bottom solder joints due to warpage may be severe enough to fail solder joints or crack the die in bottom-side components. Additionally, the back-side components may be subject to residual stresses imparted after the second reflow (when the top-side components are attached via SMT) that may impact the long-term reliability of both the devices and their second-level interconnects.

Solder Joint Geometry Issues. A final issue that must be addressed in cases of severe warpage is the shape of the postassembly solder joints. Joints can become excessively deformed, creating a dangerous state where the cross section is too small to withstand fatigue. Joints can also become compressed and in some instances can contact adjacent joints, causing shorts.

54.2.3 Package Design Parameters

One of the key components that impacts second-level interconnect reliability is the package itself. Although the primary function of the package is to pass signals, power, and ground from the silicon device to the PWB, the package also provides protection for the ASIC and enables a more robust interconnect than direct chip attach. Package design parameters can have a large impact on the reliability of the solder joints. Critical design parameters are discussed in the following text.

54.2.3.1 Overall Dimensions (L × W × H).

Package size continues to grow as I/O density increases to meet the demands of ever increasing application-specific integrated circuit (ASIC) functionality. As ASIC complexity increases, memory is embedded in devices, and silicon feature sizes decrease, I/O count continues to be driven upward. Package lengths and widths have surpassed 50 mm. Package layer counts for both ceramic and laminate (plastic) material sets continue to increase at a rapid rate.

54.2.3.2 Distance from Neutral Point.

Increases in overall length and width tend to coincide with increases in the distance from the center of the package to the outer ball (see Fig. 54.12). This dimension is referred to as the distance from the neutral point (DNP). As DNP increases, the amount of stress and subsequent damage inflicted on solder joints during thermal cycles encountered in field environments increases. Thus packages with larger DNPs (with all other material and design parameters equivalent) will have lower interconnect reliability. This is crucial when assessing second-level qualification data.

Most packaged ASIC suppliers qualify their products by thermal cycle experiments. The resistance in each solder joint is typically monitored in real time by incorporating a daisy-chain package and some type of data acquisition system. The end results of these experiments are failure distributions typically presented as Weibull plots (for a discussion of Weibull plots, see Ref. 5) that show percentage of failure as a function of time (Fig. 54.13). Note that if a package with a larger DNP than one that was qualified is considered, then there could be severe decreases in

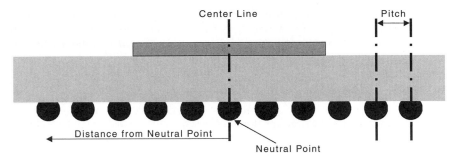

FIGURE 54.12 Distance from neutral point (DNP) for a BGA-type package. The DNP increases when moving away from the center (Center Line) of the package. Pitch is also indicated in this diagram. Pitch is defined as the center-to-center distance between two adjacent interconnects (in this example, between two adjacent solder balls).

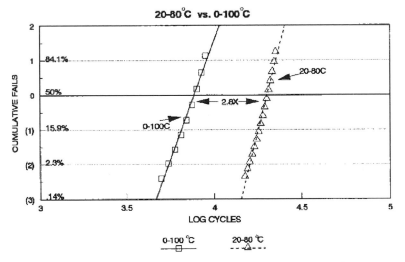

FIGURE 54.13 Weibull plots for same CCGA package cycled under two different loading conditions, cross section of flip-chip PBGA. The hollow triangles and dashed line are employed to indicate failures and associated failure distribution of a CCGA package cycled from 20 to 80°C (the least severe case presented in this figure). The hollow squares and solid line indicate distribution of failures and associated distribution for an identical package cycled from 0 to 100°C. Notice that an increase in the severity of the thermal cycle loading conditions (from 20–80 to 0–100°C) creates an associated decrease in fatigue life. (*Reprinted with permission from J. Lau,* Ball Grid Array Technology, *McGraw-Hill, New York, 1995, p. 163.*)

reliability and a full assessment should be conducted. However, if a package with a smaller DNP (10 to 20 percent smaller) than the package qualified (with all other material and design parameters being equal) is considered, then the proposed package should be at least as reliable as the qualified package.

54.2.3.3 Ball Pitch and Ball Size. Ball pitch is a package design variable that has an impact on reliability due to secondary results of changes. Ball pitch is the center-to-center distance between two adjacent solder balls (see Fig. 54.12). In theory one could change ball pitch and keep the same solder joint size. In practical applications, decreases in ball pitch also require a decrease in solder ball and pad size to physically accommodate the decrease in distance between adjacent balls. This decrease in ball size with decreasing pitch results in the formation of smaller balls with smaller stand-off between the package and the motherboard. Smaller solder balls and decreased stand-off have been shown to decrease solder joint reliability (see Fig. 54.13). The dashed line in Fig. 54.14 represents finite element analysis results for a 35-mil ball. The solid line represents finite element analysis results for a 30-mil ball. Note that the 35-mil ball is 1.3 times more reliable than the 30-mil ball. Thus decreases in pitch can result in decreases in reliability due to the subsequent decreases in solder ball size necessary to accommodate adjacent balls and the resulting reduced stand-off heights due to smaller solder balls.

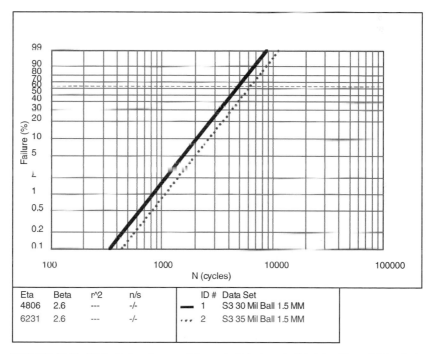

Eta	Beta	r^2	n/s		ID #	Data Set
4806	2.6	---	-/-	—	1	S3 30 Mil Ball 1.5 MM
6231	2.6	---	-/-	...	2	S3 35 Mil Ball 1.5 MM

FIGURE 54.14 Weibull plots for two different ball size/stand-off heights. The solid line indicates the FEA predicted failure distribution for an interconnect based on a 30-mil stand-off height. The dashed line indicates the FEA predicted failure distribution for an interconnect based on a 35-mil stand-off height. Note that a reduction in stand-off height correlates with a decrease in fatigue life.

The PWB designer is, therefore, faced with a dilemma when considering a reduction in ball pitch. Not only is there a risk of reduced assembly reliability, but there is also a new design challenge. A direct consequence of decreased pitch is added complexity with respect to ability to route all traces out from the package to the PWB. Designers are sometimes forced to add additional layers to PWBs in order to accommodate reduced-pitch parts. This layer increase results in increased price and, in some instances, drastic reductions in capacity due to the increased complexity associated with fabricating high-layer-count PWBs.

54.2.3.4 Ball Array (Full or Depopulated). Ball array is another variable that has an indirect impact on fatigue life. There is a coupled effect between the location of the edge of the packaged die, the solder ball array, and the state of stress in the solder joints. The silicon die has a much lower CTE than laminate substrates. Thus the die can create a state of overconstraint, producing excessive stress in the solder joints under and near its edge. This overconstraint tends to be worst in joints near the edge of the die. This effect is only a factor in laminate packages.

Figure 54.15 contains schematics of one configuration where a full array is employed and the edge of the die falls over the solder joints. Figure 54.16 illustrates an alternative design where a region near the perimeter of the die is depopulated of all solder balls (the depopulated configuration). Many laminate package suppliers offer a depopulated option to enhance reliability by not placing solder joints in the worst-case regions under the perimeter of the die. Additionally, many package designers further reduce risk of silicon-induced joint failure by only placing redundant power and ground or nonfunctional thermal balls under the core. If redundant balls or thermal balls do crack, the impact on device performance is negligible.

Die Perimeter

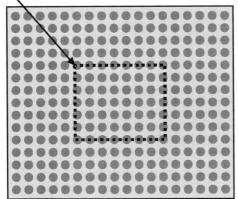

FIGURE 54.15 Full array with solder joints under the die perimeter. Solder ball locations are indicated by the array of circles. The die perimeter is indicated by the black dashed rectangular outline. Note that the die edge falls directly over solder joints.

Die Perimeter

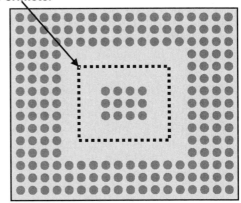

FIGURE 54.16 Depopulated array with no joints under the perimeter of the die. Solder ball locations are indicated by the array of circles. The die perimeter is indicated by the black dashed rectangular outline. Note that the die edge falls in an area of depopulated solder balls.

54.2.3.5 *Heat Spreader/Stiffener Ring.* A cross-sectional schematic of a flip-chip assembly with a heat slug and stiffener ring is shown in Fig. 54.17. The primary reason for employing a ring and slug is for thermal management. A secondary benefit is that the stiffener ring and heat slug reduce the potential for out-of-plane warpage to occur during reflow and thermal cycling. An additional benefit of incorporating a heat slug and stiffener ring into a laminate package design is improved handlability due to increased robustness of the packaged die.

54.2.4 Substrate Material

The choice of substrate material can have a large impact on solder joint reliability. Some substrates require employing solder ball alternatives (such as ceramic solder column carriers, see Fig. 54.6) or even the elimination of a soldered solution due to interconnect reliability considerations. A wealth of information regarding ceramic and laminate substrates can be found in the literature.[6-8] This section focuses on the impact of those materials on solder joint reliability.

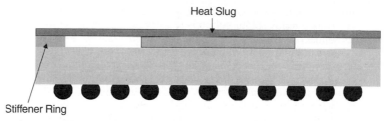

FIGURE 54.17   Flip-chip BGA with stiffener ring and heat slug. The stiffener ring provides added rigidity to the package and also provides mechanical support for the heat sink.

The two key material properties of a substrate that impact interconnect reliability are modulus of elasticity and coefficient of thermal expansion. Modulus of elasticity is defined as the ratio of stress to strain and can be used to describe a material's ability to resist deformation when subjected to load. Coefficient of thermal expansion relates the change in length of material when heated or cooled and can be used to describe a material's deformation when subjected to temperature changes. Reference 9 contains detailed discussions of modulus of elasticity and coefficient of thermal expansion. A tremendous amount of material property data can be found in the literature.[6-8] Figure 54.18 contains a sketch of a simplified solder joint interconnect. Note that the driving force for solder joint fatigue during field use thermal cycling is the coefficient of thermal expansion mismatch between the assembled package and the substrate. Additionally, the modulus of elasticity of the substrate can impact the fatigue life of solder joints by increasing (or decreasing) the driving force imparted to the interconnect.

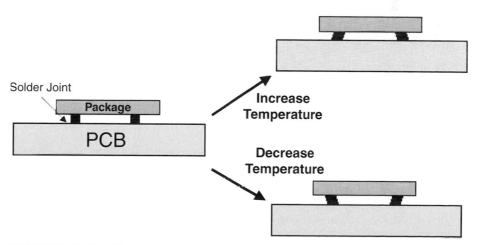

FIGURE 54.18   Simplified CTE missmatch effects. As the temperature increases, the PWB expands at a higher rate than the package due to the package having a lower effective CTE. As the temperature decreases, the opposite effect is observed.

Ceramic materials provide excellent CTE match between the die and package, which reduces the risk of first-level solder bump failure in flip-chip applications. Unfortunately, this superb CTE match between the ceramic and the silicon die creates a large CTE mismatch between the ceramic package and the PWB. This CTE mismatch creates severe stress on solder balls (see Fig. 54.18). This CTE mismatch, combined with DNP issues (previously discussed), limits ceramic BGA body sizes to approximately 32×32 mm. Ceramics with larger body sizes typically employ solder columns (Fig. 54.6) or compression-type sockets for second-level interconnect.

The CTEs of laminate packages tend to match the CTEs of PWBs fairly well. This CTE match allows for larger package body sizes to be employed when using a laminate-based BGA than with a ceramic package. It is crucial to note that many other variables independent of second-level reliability must be optimized when selecting a package. Power requirements, signal integrity, first-level reliability (interconnect between the die and package), and cost must all be balanced when selecting the correct package for a specific ASIC.

54.3 EXPERIMENTAL TOOLS FOR ESTIMATING SOLDER JOINT LIFE

PWB designers are also responsible for the reliability of the printed circuit assembly (PCA). Acceleration transforms, finite element analysis, estimated field loading conditions, and experimental data can be combined to assess the expected service life of any interconnect. The experimental data play a crucial role in reliability assessments. Poor data or data obtained under incorrect experimental conditions can lead to grossly inaccurate reliability assessments. "Garbage in equals garbage out" is nowhere more true than in interconnect reliability assessments.

A myriad of experimental tools have been developed to rapidly assess second-level reliability. These include thermal cycling, thermal shock, air-to-air cycling, liquid-to-liquid cycling, mechanical bending, mechanical deflection, and hyper-Peltier cooled thermal cycling. It is crucial for the PWB designer to understand the techniques employed, the applicability of the results, and quality of the data when assessing the reliability of any package that will be placed on his or her PWB. Many techniques fail to consider ability to map results into end use conditions. Although rapid assessments are desirable, their benefit diminishes if the failure modes are fictitious and not indicative of expected field failures.

54.3.1 Thermal Cycle

Thermal cycle experiments have been the current packaging industry standard for assessing second-level interconnect reliability. These types of tests tend to produce solder joint fatigue and solder joint–pad interfacial fractures that are typically seen in field failures. The fact that thermal cycle tests produce the same physical failures allows for detailed acceleration transformations and finite element–based life assessments to be made employing thermal cycle data as input.

Thermal cycle experiments are typically performed in single-chamber ovens. Dual-chamber systems are sometimes employed, but single-chamber equipment tends to dominate the test space. The main drawback of a dual-chamber oven is the lack of control over the thermal loading profile when transferring from one chamber (or thermal zone) to the other. Regardless of single vs. dual chamber, temperature is cycled from a maximum (T_{max}) to a minimum (T_{min}) value. Temperature is ramped from one temperature extreme to another at a controlled rate (referred to as the ramp rate). Temperature is also held at the maximum and minimum values for a predetermined time known as the dwell time. Table 54.2 contains a list of typical experimental parameters for second-level reliability qualifications. Figure 54.19 contains a plot of temperature vs. time along with explanations of the key experimental parameters.

Data acquisition systems and/or event detectors (commonly referred to as glitch detectors) are employed to monitor resistance and/or scan for intermittent opens in real time. One of the early techniques for assessing solder joint integrity during thermal cycling was to periodically remove parts from the test chamber and measure resistance manually at room tem-

TABLE 54.2 Typical Thermal Cycle Test Parameters

Parameter	Value
Maximum temperature (T_{max})	100°C
Minimum temperature (T_{min})	0°C
Ramp rates from T_{min} to T_{max} and T_{max} to T_{min}	10°C/min
Dwell time at T_{max} and T_{min}	10 min

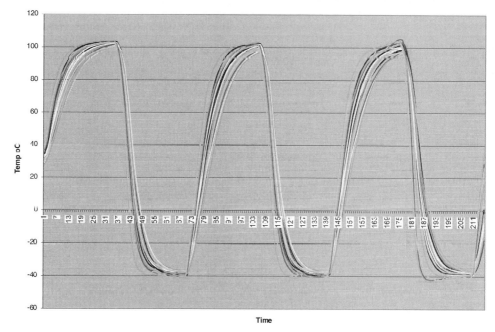

FIGURE 54.19 Typical thermal cycle profile. In this plot $T_{max} = 100°C$, $T_{min} = -40°C$ and the cyclic frequency is 60 min.

perature. The major risk associated with this technique is that opens are only observed at temperature extremes when the deformations due to CTE mismatch are largest. As the package cools to room temperature, most of the CTE-based deformations that pull cracked surfaces of solder joints are gone and there is a potential for cracked surfaces to be in enough physical contact to provide acceptable continuity. In some cases the cycle number at which opens are observed at temperature extremes is much shorter than the cycle at which the same opens are observable at room temperature.

54.3.2 Thermal Shock

Thermal shock experiments continuously cycle parts from a high to a low temperature in a very rapid fashion. These types of experiments are typically used as screening tools or to run parallel experiments on a known good part and a proposed new component to assess the relative performance of each part. Thermal shock experiments are quite rapid, with second-level test durations on the order of days rather than the months associated with thermal cycle test times.

The main drawback to thermal shock experiments lies in the fact that these types of tests tend to induce failure modes not observed in the field. Additionally, even if solder joint fatigue can be induced, it is very difficult to map the high ramp rates associated with thermal shock loading conditions to the more gradual thermal loading conditions that occur in the field. This test technique is limited to relative comparisons, as it is very difficult to map thermal shock results to end use environments.

54.3.3 Mechanical Shock and Vibration

Mechanical shock and vibration experiments are employed to simulate shipping shock and vibration. Very few methodologies exist for mapping shock and vibration test results into end use loading conditions. It is very difficult to relate shock and or vibration loading conditions to crack growth in solder joints. Thus shock and vibration are typically used as screening tools, with minimum survival requirements set for each package type.

54.3.4 Ball Shear/Pull

Ball shear testers are employed to quantify the strength of the interface between a solder ball and its package prior to second-level attach. Ball shear experiments can also be employed to qualitatively assess the solder ball–package pad interface as well. This is accomplished by assessing the failure mode observed during a ball shear experiment.

The emergence of electroless nickel immersion gold as a package metallization has created an opportunity for employing ball shear techniques as a quality control tool. One can run a series of ball shear experiments on a number of packages and examine the failure modes. If a large number of the failures occur at the solder-pad interface, then there is a strong probability that the electroless nickel immersion Au metallization has issues and will result in premature package failure. This procedure is usually employed on a per-batch basis.

54.4 RAPID ASSESSMENT TOOLS

PWB designers are often required to provide a reliability assessment of the card that they are designing. Experimental data supplied by vendors are obtained under ideal thermal loading conditions that do not simulate the actual end use thermal environment that the packaged device will experience. Designers must employ numerical and/or analytical tools (rapid assessment tools) to estimate the field life of the interconnects employed in their designs.

Rapid assessment tools assist in the evaluation of long-term reliability of solder joints. These tools are analytical in nature. They either transform accelerated experimental test results (such as thermal cycles) into expected field loading conditions, predict life based on techniques validated with existing experimental data, or both. The following text covers two common rapid assessment tools: Norris-Landsberg and finite element analysis.

54.4.1 Norris-Landsberg Acceleration Transformation

PWB designers typically receive a failure distribution plot (see Fig. 54.13) as part of a typical second-level package qualification report. This failure distribution plot is typically obtained in a thermal cycle oven under controlled temperature profiles that include the maximum and minimum temperatures (T_{max} and T_{min}, respectively) of the thermal cycle along with the frequency (f) of the experimental thermal cycle. The designer must then relate these experimental data obtained under ideal conditions to a different thermal cycle condition (or conditions) encountered in the product end use environment. Although relative comparisons of two packages thermally cycled under identical conditions can be made in a ranking sense, this type of analysis yields no information regarding the actual lifetime of the solder joints under field loading conditions. This section focuses on the mapping of a laboratory thermal cycle to an end use thermal profile. Multiple end use profiles are covered in a subsequent section of this chapter.

A tremendous number of researchers have explored the reliability of Sn/Pb solder alloys.[8] A wealth of data, material properties, and proposed constitutive relationships have also been captured in the literature.[6–8]

The fatigue behavior of many metals (including Pb/Sn solder alloys) can be described by the Coffin-Manson relationship (Eq. [54.1]),[6] which relates the fatigue life of a metal to the plastic strain the material experiences during loading. Plastic strain is the amount of irreversible (or permanent) deformation that occurs during a loading cycle.

$$\varepsilon = \varepsilon_e + \varepsilon_p = A N_f^{-c} + B N_f^{-m} \tag{54.1}$$

where ε = total strain per cycle
ε_p = plastic portion of the cyclic strain
ε_e = plastic portion of the cyclic strain
N_f = number of cycles to failure

and A, B, c, and m are constants.

Most solder joints fail in the low cycle fatigue region (less than 10,000 cycles to failure),[6] where strain is primarily plastic. When one does not consider the elastic component of strain, Eq. (54.1) can be simplified and rearranged to yield a relationship between the number of cycles to failure and the plastic strain (see Eq. [54.2]).[6]

$$N_f = (A/\varepsilon_p)^{1/m} \tag{54.2}$$

where ε_e = plastic portion of the cyclic strain
N_f = number of cycles to failure

and A and m are constants. This can be further simplified to the most common form of the Coffin-Manson relationship shown in Eq. (54.3).[6]

$$N_f \propto \varepsilon_p^c \tag{54.3}$$

where ε_p = plastic portion of the cyclic strain
N_f = number of cycles to failure

and c is a constant.

When any metal is cycled at a temperature above 50 percent of its melting temperature (homologous temperature), fatigue life is influenced by the strain rates, dwell times, and maximum temperature.[6] This is due to the occurrence of creep, along with plastic and elastic deformation, within these temperature regimes.[6] Norris and Landsberg have proposed a modified version of the Coffin-Mansion relationship that considers the effects of frequency (f) and maximum temperature (T_{max}) on solder joint fatigue life (Eq. [54.4]).[6,7] This relationship is based on detailed physical and empirical analysis of the relationships between temperature and frequency on the thermal fatigue behavior of Pb/Sn solder alloys employed in interconnects.[7]

$$N_f = (A/\varepsilon_p)^{1.9} f^{1/3} \exp\left(\frac{0.123}{k T_{max}}\right) \tag{54.4}$$

where ε_p = plastic portion of the cyclic strain
N_f = number of cycles to failure
k = Boltzmann's constant (eV)

$$f = \text{cyclic frequency}$$
$$T_{max} = \text{maximum temperature of the thermal cycle (K)}$$

and A is a constant.

One of the challenges that must be addressed when attempting to employ Eq. (54.4) directly to assess fatigue life is the determination of the plastic strain and the value of the constant A. It is crucial to remember that the goal of this analysis is to develop an acceleration transform (or acceleration factor) that can be employed to use known fatigue life data obtained under controlled laboratory conditions to estimate the number of cycles to failure under field loading conditions. The acceleration factor AF is defined as the ratio of the number of cycles to fail in the field N_{fF} to the number of cycles to fail in the lab N_{fL} (see Eq. 54.5).

$$AF = \frac{N_{fF}}{N_{fL}} \tag{54.5}$$

where AF = acceleration factor
N_{fF} = number of cycles to failure in the field
N_{fF} = number of cycles to failure in the lab

Substituting Eq. (54.5) for lab and field conditions into Eq. (54.5), one obtains

$$AF = \frac{N_{fF}}{N_{fL}} = \frac{(A/\varepsilon_{pF})^{1.9} f_F^{1/3} \exp\left(\dfrac{0.123}{kT_{maxF}}\right)}{(A/\varepsilon_{pL})^{1.9} f_L^{1/3} \exp\left(\dfrac{0.123}{kT_{maxL}}\right)} \tag{54.6}$$

where AF = acceleration factor
N_{fF} = number of cycles to failure in the field
N_{fL} = number of cycles to failure in the lab
ε_{pF} = plastic portion of the cyclic strain in the field
ε_{pL} = plastic portion of the cyclic strain in the lab
f_F = cyclic frequency
f_L = cyclic frequency
T_{maxF} = maximum temperature of the thermal cycle in the field (K)
T_{maxL} = maximum temperature of the thermal cycle in the lab (K)
k = Boltzmann's constant (eV)

where the subscript L indicates lab conditions and F indicates field conditions.

Equation (54.6) can be algebraically manipulated to yield Eq. (54.7). Note that the max subscripts on the maximum temperatures will be dropped to simplify the notation.

$$AF = \left[\frac{\varepsilon_{pL}}{\varepsilon_{pF}}\right]^{1.9} \left[\frac{f_F}{f_L}\right]^{1/3} \exp\left[1414\left(\frac{1}{T_F} - \frac{1}{T_L}\right)\right] \tag{54.7}$$

where AF = acceleration factor
ε_{pF} = plastic portion of the cyclic strain in the field
ε_{pL} = plastic portion of the cyclic strain in the lab
f_F = cyclic frequency in the field
f_L = cyclic frequency in the lab
T_F = maximum temperature of the thermal cycle in the field (K)
T_L = maximum temperature of the thermal cycle in the lab (K)

When considering package interconnects of the same geometry and material set, the strain term ε_p in Eq. (54.4) can be reduced to the change in temperature.[7] This is a key feature that enables the development of an acceleration transform. The goal of this analysis is to develop a transform that allows one to map laboratory thermal cycle loading conditions to field load-ing conditions for the same package. Thus it is valid to assume that the strain term can be reduced to the change in temperature, as shown in Eq. (54.8).

$$\varepsilon_p \propto \Delta T \tag{54.8}$$

where ε_p = plastic portion of the cyclic strain
ΔT = difference between maximum and minimum temperature of the thermal cycle

Substitution of Eq. (54.8) (including L and F subscripts for lab and field, respectively) yields the final form of the acceleration factor relationship, which can be seen in Eq. (54.9)[7]

$$AF = \left[\frac{\Delta T_L}{\Delta T_F} \right]^{1.9} \left[\frac{f_F}{f_L} \right]^{1/3} \exp\left[1414 \left(\frac{1}{T_F} - \frac{1}{T_L} \right) \right] \tag{54.9}$$

where AF = acceleration factor
f_F = cyclic frequency in the field
f_L = cyclic frequency in the lab
T_F = maximum temperature of the thermal cycle in the field (K)
T_L = maximum temperature of the thermal cycle in the lab (K)
ΔT_F = difference between maximum and minimum temperature of the thermal cycle in the field
ΔT_L = difference between maximum and minimum temperature of the thermal cycle in the lab

Figure 54.19 contains a plot of experimentally obtained failure rate data for a CCGA pack-age cycled under two different thermal loading conditions (20 to 80°C and 0 to 100°C). Figure 54.20 contains experimental failure data for a CBGA cycled under the same two conditions. The predicted acceleration factor was calculated to be 3.2 employing Eq. (54.9). This pre-

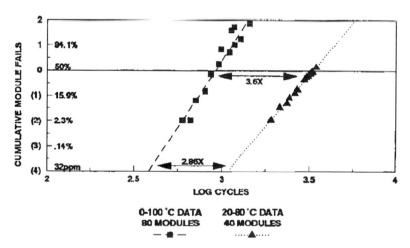

FIGURE 54.20 Weibull plots for CBGA package cycled under two different loading con-ditions. Solid triangles and dotted line are employed to indicate failures and associated fail-ure distribution of a CBGA package cycled from 20 to 80°C (the least severe case presented in this figure). Solid squares and dashed line indicate the failures and associated distribution for the identical package cycled from 0 to 100°C. Notice that an increase in the severity of the thermal cycle loading conditions (from 20–80°C to 0–100°C) creates an associated de-crease in fatigue life. (*Reprinted with permission from J. Lau,* Ball Grid Array Technology, *McGraw-Hill, New York, 1995, p. 162.*)

dicted value correlates quite well with both the CCGA data (acceleration factor of 2.8) and the CBGA data (acceleration factor ranges from 2.9 to 3.6). Note that this approach tends to be overly conservative when mapping large lab ΔT into small field ΔT (in the case of minicycles, to be discussed in a subsequent section).

54.4.2   Finite Element Analysis

Finite element analysis (FEA) is a powerful numerical tool that has been used to assess the reliability of solder joints. FEA can be employed to:

- Estimate the fatigue life of a new package
- Compare reliabilities of different packaging options
- Assess impact on reliability due to package
  - Material changes
  - Design changes
- Assess impact on reliability due to PWB
  - Material changes
  - Design changes
- Determine changes in warpage due to package density and location

FEA lifetime assessments are typically broken down into four phases:

- Phase 1: global model
- Phase 2: local model
- Phase 3: single-joint lifetime assessment
- Phase 4: statistical-based package reliability estimate

During phase 1, a global model of the package is built. A global model of one-eighth of a BGA package can be found in Fig. 54.21. Note that only one-eighth of the package is modeled.

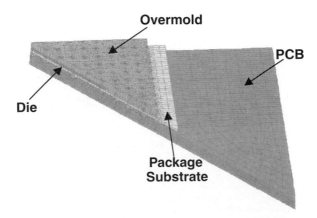

FIGURE 54.21   Global FEA model of one-eighth of a BGA. Note the PWB, package substrate, die, and overmold are all created by a series of elements.

This is due to the fact that the package is octant symmetric, and thus symmetry boundary conditions can be employed to reduce the model size, which in turn increases the speed at which a solution can be obtained.

The global model is then subject to a unit temperature rise, the worst-case joint is determined, and the displacements of the worst-case joints are calculated. Worst-case joint determination is typically based on a maximum stress or strain in the solder joints. It is essential to obtain the displacements of the worst-case joint because those displacements can now be incorporated in a local model that will simulate multiple thermal cycles on a worst-case joint.

The global model is only subject to a unit temperature rise to reduce the computational run time. An additional assumption employed in most global models to reduce run time is to assume all materials behave in a linear elastic, temperature-independent fashion. Because the goal of the global model is to determine the worst-case joint and the corresponding displacement, this approach is typically valid. Reference 10 contains a series of experimental cases and corresponding FEA predictions that validate the global/local approach and the linear elastic, temperature-independent unit temperature rise technique.

A local model of the worst-case joint is created and subjected to simultaneous temperature cycle and temperature-scaled displacement loading conditions obtained from the global model. The displacement field of the worst-case joint obtained in the global model when subjected to a unit temperature rise is scaled to match the expected displacement when the local model is subjected to actual thermal cycles. The advantage of this approach is that the local model can incorporate time and temperature material dependence because a reduced number of elements are employed when compared to the global model.

A local model of a solder joint can be found in Fig. 54.22. The local model is subjected to multiple thermal cycles and a final damage parameter is extracted. Plastic work and plastic strain are two of the most common local damage parameters employed in solder joint lifetime estimations. The type of damage parameter selected is usually dictated by the experimental data available for a specific joint type (BGA, CCGA) that correlate damage with fatigue life. A methodology that employs plastic work and a correlation between plastic work and crack initiation and propagation can be found in references 7 and 10.

The local model damage parameter is then employed to determine the fatigue life of a given joint. Techniques for assessing lifetime based on a damage parameter range from employing plastic strain (Coffin-Manson relationship, Eq. [54.2]) to complex considerations of crack initiation and crack growth rates correlated with plastic work.[7,10] Regardless of the technique, the output of this phase is an estimate of the reliability of one joint.

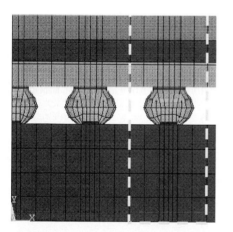

FIGURE 54.22 Close-up of three solder joints. The local model is created from one of the solder joints and its associated PWB, package substrate, die, and overmold. In this figure the rightmost ball and associated material for the local model are indicated by the area inside the dashed rectangle.

The final phase considers the reliability of all of the solder joints when determining the reliability of the package. Techniques employed here range from simply assuming that all joints are of equal reliability to performing a series reliability calculation to simultaneously solve an assumed distribution of failures (such as a Weibull distribution) considering each joint. It is also common to weigh the reliability of joints other than the worst-case joint by considering the displacement of all joints with respect to the displacement of the worst-case joint.[10] This allows for the impact of DNP to be incorporated in the potential distribution of failures among joints at different locations.

54.5 *POWER AND MINICYCLES*

54.5.1 Overview

Thermal cycle solder joint reliability experiments are performed in a laboratory, executing one specific thermal cycle profile that is repeated until the end of the test (Fig. 54.19). Section 54.4.1 presented techniques for mapping one laboratory thermal cycle into one field thermal condition. While this is an extremely useful tool, it is not sufficient in many circumstances to estimate field lifetimes based on thermal cycle qualification data.

Consider a processor in a server. The server may only be powered down and back up one time per month (a conservative estimate). Power-downs and -ups are referred to as power cycles. A reliability estimate could be made assuming that one thermal cycle occurred per month from the operating temperature within the region of the package to ambient. This would yield a tremendously high lifetime estimate that would unfortunately be misleading. It is important to consider what happens to the packaged device during typical usage of the server. The device will probably increase in temperature during usage and decrease when not in use. These smaller temperature rises can occur at multiple times throughout a day. These types of thermal conditions are referred to as minicycles and can have a dramatic impact on the long-term reliability of an interconnect. Section 54.5.2 presents a technique for incorporating mini- and power cycles in a lifetime assessment.

54.5.2 Miner's Rule

Miner's rule is employed when attempting to assess fatigue life under multiple fatigue loading conditions. It is commonly used when addressing the combined power and minicycle thermal loading conditions described in the previous section. Miner's rule is based on the premise that fatigue damage accumulates over each cycle. When a critical quantity of damage has been imparted on a system, the system will fail.

Miner's rule states that the sum of the ratios of the actual number of cycles to fail of a given fatigue loading condition to the number of cycles to fail if only that fatigue loading condition occurs is equal to 1. Equation (54.10) contains a general expression of Miner's rule.[6]

$$\Sigma \frac{N_{ai}}{N_{oi}} = 1.0 \qquad (54.10)$$

where N_{ai} = actual number of specific fatigue cycles that occur
N_{oi} = number of cycles to fail if only that specific fatigue cycle occurred

Miner's rule can be expanded and rearranged for power and minicycle considerations. Equation (54.11) contains that expression.

$$\frac{N_{poweractual}}{N_{poweronly}} + \frac{N_{miniactual}}{N_{minionly}} = 1.0 \qquad (54.11)$$

where $N_{poweractual}$ = actual number of power cycles that occur before failure
$N_{poweronly}$ = number of power cycles to fail if only power cycles occur
$N_{miniactual}$ = actual number of minicycles that occur before failure
$N_{minionly}$ = number of minicycles to fail if only minicycles occur

The Norris-Landsberg acceleration transformation can be combined with Miner's rule to estimate the fatigue life of a package. Note that the Norris-Landsberg method tends to be conservative when assessing minicycles whose changes in temperature are much smaller than the experimental temperature change applied when qualifying the part.

54.6   ACCELERATED STRESS TESTING

Accelerated stress testing (AST) techniques typically combine thermal cycle loading conditions with random vibration applied to printed circuit assemblies or, in some instances, to entire assembled systems. Loading states are usually stepped in severity until a failure occurs. The goal of AST is to induce a series of failures (with different failure modes) and allow designers to increase the robustness of each failure if deemed necessary. AST techniques do not predict expected field life of parts, interconnects, printed circuit assemblies, or systems. These techniques also can induce failure modes that may never occur under actual end use environmental loading conditions. Engineering judgment must be exercised when assessing if an AST failure mode needs to be addressed or if the field loading condition. A detailed discussion of AST techniques is beyond the scope of this section. It is recommended that the reader consult the literature for detailed information on this topic.[11]

54.7   PRACTICAL EXAMPLES

54.7.1   Comparing the Reliability of Two Packages Qualified Under Different Thermal Loading Conditions

Given:
Thermal cycle qualification data and test conditions for packages A and B (Table 54.3)

TABLE 54.3   Package A and Package B Qualification Data

	Package A	Package B
T_{max}	110°C	100°C
T_{min}	−10°C	0°C
Frequency	1 cycle/h	1 cycle/h
N_{50} (number of cycles to 50% failure)	2291	3575

Solution:
It is necessary to decide which thermal profile will be kept as a constant and which one will be mapped. This problem can be solved both ways. Either method requires utilizing Eqs. (54.5) and (54.9) to transform the number of cycles to failure. For clarity, let condition A refer to the thermal conditions that package A was qualified under and condition B refer to the qualification test conditions that package B was exposed to.

54.7.1.1   Map Package A Lifetime into Package B Thermal Loading Conditions.   Equation (54.5) can be modified to read

$$AF = \frac{N_{fAA}}{N_{fAB}} \qquad (54.12)$$

where AF = acceleration factor
N_{fAA} = number of cycles to failure in package A cycled from −10 to 125°C (condition A)
N_{fAB} = number of cycles to failure in package A cycled from 0 to 100°C (condition B)

Equation (54.9) can be rearranged to yield

$$AF = \left[\frac{\Delta T_B}{\Delta T_A}\right]^{1.9} \left[\frac{f_A}{f_B}\right]^{1/3} \exp\left[1414\left(\frac{1}{T_A} - \frac{1}{T_B}\right)\right] \tag{54.13}$$

where AF = acceleration factor
        F_A = cyclic frequency for condition A (1 cycle/h)
        F_B = cyclic frequency for condition B (1 cycle/h)
        T_A = maximum temperature for condition A (110°C, 383 K)
        T_B = maximum temperature for condition B (100°C, 373 K)
     ΔT_A = difference between maximum and minimum temperature of condition A
          (120)
     ΔT_B = difference between maximum and minimum temperature of condition B
          (100)

Substitution of the values in Table 54.3 into Eq. (54.13) yields an acceleration factor of 0.641. Package A would be expected to survive 3574 cycles ($N_{fAB} = N_{fAA}/AF = 2291/0.641$) if subjected to condition B thermal loading conditions. The packages appear to be comparable.

54.7.2 Assessing Expected Field Life of a New Package Based on Thermal Cycle Data

Given:
Thermal cycle qualification data for a new ball grid array package (use package B in Table 54.3); estimates of field loading conditions (Table 54.4)

TABLE 54.4 Field Conditions for Example

	Power cycle	Minicycle
T_{max}	65	80
T_{min}	25	65
ΔT	40	15
Frequency	1 cycle per 30 days, 30-min duration	4 cycles per day, 60-min durations

Solution:
The process for solving this problem is as follows. The experimentally obtained qualification data (presented in Table 54.3) must be mapped into number of power cycles to failure and number of minicycles to failure. These estimates assume that only power cycles occur and that only minicycles occur (that is the underlying assumption when mapping the experimental data to the end use power and minicycle conditions). After mapping the experimental data into the field conditions, Miner's rule is employed to assess the number of cycles to fail based on the simultaneous occurrence of both power and minicycles throughout the package life time.

When employing Eq. (54.9) to map power and minicycles, it is important to consider the implications of how the frequency term is interpreted. The Norris-Landsberg approach assumes that the frequency corresponds to the duration of the cycle, not how often that cycle occurs.

Equation (54.9) can be employed to determine the acceleration factors. Equations (54.14) and (54.15) contain the expressions for power and minicycles, respectively.

$$AF = \left[\frac{\Delta T_L}{\Delta T_P}\right]^{1.9} \left[\frac{f_P}{f_L}\right]^{1/3} \exp\left[1414\left(\frac{1}{T_P} - \frac{1}{T_L}\right)\right] \tag{54.14}$$

where AF = acceleration factor
        f_P = cyclic frequency for a power cycle (2 cycles/h)
        f_L = cyclic frequency in the lab (1 cycle/h)
        T_P = maximum temperature of a power cycle (65°C, 338 K)
        T_t = maximum temperature of the thermal cycle in the lab (100°C, 373 K)
    ΔT_P = difference between maximum and minimum temperature of a power cycle
          (40)
    ΔT_L = difference between maximum and minimum temperature of the thermal
          cycle in the lab (100)

$$AF = \left[\frac{\Delta T_L}{\Delta T_M}\right]^{1.9} \left[\frac{f_M}{f_L}\right]^{1/3} \exp\left[1414\left(\frac{1}{T_M} - \frac{1}{T_L}\right)\right] \tag{54.15}$$

where AF = acceleration factor
        F_M = cyclic frequency for a minicycle (1 cycle/h)
        f_L = cyclic frequency in the lab (1 cycle/h)
        T_M = maximum temperature of a minicycle (80°C, 353 K)
        T_L = maximum temperature of the thermal cycle in the lab (100°C, 373 K)
    ΔT_M = difference between maximum and minimum temperature of a minicycle (15)
    ΔT_L = difference between maximum and minimum temperature of the thermal cycle
          in the lab (100)

Substitution of the data in Tables 54.3 and 54.4 yields a power cycle acceleration factor of 10.6 and a minicycle acceleration factor of 45.6.

Equation (54.5) can be rearranged to yield Eq. (54.16)

$$N_{fi} = AF_i * N_{fL} \tag{54.16}$$

where N_{fi} = number of cycles to fail under mini- or power cycle (i = mini or i = power)
      N_{fL} = number of cycles to fail in the lab
      AF_i = acceleration factor for mini- or power cycle (i = mini or i = power)

Substitution of the lab number of cycles to fail from Table 54.3 into the Eq. (54.16) for mini- and power cycles yields 37,895 power cycles to fail (assuming only power cycles occur) and 163,020 minicycles to fail (assuming only minicycles occur).

Miner's rule (Eq. [54.11]) can now be applied to determine the actual number of cycles to fail. It is important to note that at this stage the number of minicycles to fail if only minicycles occur and the actual number of power cycles to fail if only power cycles occur are known. The actual number of mini- and power cycles to fail remain unknown; thus Eq. (54.11) is only one equation while two unknowns still exist.

It is important to consider the number of minicycles that occur per power cycle. In this case there is one power cycle per 30 days and four minicycles per day. This equates to 120 minicycles per power cycle. Equation (54.11) can now be written as

$$\frac{N_{poweractual}}{N_{poweronly}} + \frac{K * N_{poweractual}}{N_{minionly}} = 1.0 \tag{54.17}$$

where $N_{poweractual}$ = actual number of power cycles that occur before failure
      $N_{poweronly}$ = number of power cycles to fail if only power cycles occur
      $N_{miniactual}$ = actual number of minicycles that occur before failure
          K = number of minicycles that occur per power cycle

Equation (54.17) can be solved for the number of power cycles to fail, which yields a value of 1311 power cycles or 107.8 years. It is important to employ a confidence factor (safety factor). Taking 2 as a standard safety factor, the final fatigue life is estimated at 53.9 years.

REFERENCES

1. J. H. Lau (ed.), *Solder Joint Reliability: Theory and Applications,* Van Nostrand Reinhold, New York, 1991.

2. IPC-SM-785, Guidelines for Accelerated Reliability Testing of Surface Mount Solder Attachments, 1992.

3. E. Blackshear, "Advanced CSP and Fine Pitch Ball Grid Array Assembly Reliability," *SMTA 1999 Proceedings,* pp. 243–248.

4. N. P. Kim, K. Selk, B. Bjorndahl, R. Ghaffarian, J. K. Bonner, and S. Barr, "CSP Assembly Reliability: Commercial and Harsh Environments," *Proceedings of the Chip Scale Packaging Symposium,* SMTA International, 1999, pp. 69–76.

5. P. A. Tobias and D. C. Trindade, *Applied Reliability,* 2d ed., Van Nostrand Reinhold, New York, 1995.

6. R. R. Tummala, E. J. Rymaszewski, and A. G. Klopfenstein (eds.), *Microelectronics Packaging Handbook,* Chapman & Hall, New York, 1997.

7. J. H. Lau (ed.), *Ball Grid Array Technology,* McGraw-Hill, New York, 1995.

8. J. H. Lau (ed.), *Thermal Stress and Strain in Microelectronics Packaging,* Van Nostrand Reinhold, New York, 1993.

9. J. M. Gere and S. P. Timoshenko, *Mechanics of Materials,* 2d ed., PWS Publishers, California, 1984.

10. J. C. Riebling and M. V. Brillhart, "FEA Reliability Assessment Methodology Investigation to Improve Prediction Accuracy," *SMTA International,* 2000.

11. Accelerated Stress Testing, Thermotron Industries, 1998.

ENVIRONMENTAL ISSUES AND WASTE TREATMENT

CHAPTER 55
PROCESS WASTE MINIMIZATION AND TREATMENT*

Joyce M. Avery
Avery Environmental Services, Saratoga, California

Peter G. Moleux, P.E
Peter Moleux and Associates, Newton Centre, Massachusetts

55.1 INTRODUCTION

In the past, manufacturers of printed circuit boards have relied on end-of-pipe treatment and disposal for hazardous wastes generated in the fabrication process. These technologies are no longer optimal strategies for managing waste for two reasons. First, the potential liabilities involved with the handling and disposal of waste have increased and will continue to increase, and second, waste disposal costs have gone up significantly due to restrictions placed on land disposal. As a result, the industry is faced with the challenge of finding alternative methods for managing hazardous waste. This chapter presents a brief overview of some of the alternatives available to address this challenge, as well as a summary of some of the issues involved in implementation.

55.2 REGULATORY COMPLIANCE

Fabricators of printed circuit boards today are faced with a complex set of environmental requirements. In the United States there are three basic environmental statutes impacting the fabrication and assembly of printed circuit boards.

- Clean Water Act
- Clean Air Act
- Resource Conservation and Recovery Act (RCRA)

---

* This chapter is reprinted from the 4th edition. The basic issues, regulations, and processes are considered accurately stated and relevant to the 5th edition. For specific actions, however, it is recommended that waste treatment engineering and legal professionals be consulted to ensure that latest government expectations are understood at all levels of jurisdiction and that appropriate technology is applied to the resolution of specific issues.

55.2.1 Clean Water Act

The goals of the Clean Water Act are to "restore and maintain the chemical, physical, and biological integrity of the nation's waters." To accomplish these goals, discharges of industrial wastewater are subject to pretreatment requirements of federal, state, or municipal regulations. Industrial waste discharges are typically directed to a sewage treatment plant. Most sewage treatment plants use bacteria to biodegrade the organic matter present in the waste stream. Toxic materials such as copper, nickel, and lead from industrial discharges can pose a problem in two ways. These materials end up in the sludge from the sewage treatment process and can lead to disposal problems. Secondly, in high concentrations they can kill the bacteria in the treatment process, resulting in significant pollution of the receiving water. As a result, fabricators of printed circuit boards are required to pretreat their wastewater to specified levels prior to discharge to the sanitary sewer. The stringency of the requirements is ultimately determined by the use of the receiving water, as even minute amounts of toxics have been shown to have a negative impact on the aquatic environment. While the federal Clean Water Act specifies minimum pretreatment standards for fabricators of printed circuit boards, in most cases, state and local requirements may be more stringent. See Table 55.1 for an example of pretreatment requirements.

TABLE 55.1 Typical Pretreatment Requirements

Parameter	Limit, mg/l
pH	6.5–9.0
Copper	1.0
Nickel	0.5
Chromium	1.0
Silver	0.05
Cadmium	0.07
Zinc	0.5
Lead	0.2
Mercury	0.05
Aluminum	1.0
Selenium	0.2
Iron	2.0
Manganese	2.0
Tin	5.0
Cyanide	0.01
Phenol	0.05

55.2.2 Clean Air Act

The Clean Air Act established National Ambient Air Quality Standards (NAAQS) to achieve two goals:

1. Improve air quality in areas which fail to meet the standards.
2. Prevent significant deterioration of the air quality in clean air areas.

The states are responsible for achieving these standards by setting emission limitations and establishing timetables for compliance by sources. Printed circuit fabrication and assembly involves several processes that have an impact on air quality. Drilling, routing, sawing, and sanding create dust or airborne particulates. The plating process creates acid fumes and the etching process can generate ammonia if an ammoniacal etchant is used. Volatile organic compounds (VOCs) and lead particulates from the assembly process can pose a potential air pollution problem as well.

The technologies available for control of air pollutants include the following:

1. Electrostatic precipitators, baghouses (a type of dust collector with fabric bags mounted on frames), and cyclone separators are available to control airborne particulates.

2. Wet scrubbers containing a packed bed to provide surface area and water sprays are utilized for removing acid fumes. Addition of an acidic feed to the scrubbing liquid allows for the removal of ammonia. Addition of a caustic feed improves the scrubbing efficiency for other materials. Wet scrubbers are often used to prevent entrance of fumes back into the building from fresh air intakes.

3. Activated carbon filtration systems are utilized for removing chlorinated solvents and volatile organic compounds (VOCs). These can be regenerated on- or off-site.

55.2.3 Resource Conservation and Recovery Act

The Resource Conservation and Recovery Act (RCRA) goals are to protect human health and the environment by reducing or eliminating the generation of hazardous waste. To achieve this, the regulation mandates a system for managing hazardous wastes from cradle to grave. Anyone who generates, stores, treats, transports, or disposes of hazardous waste is subject to this regulation. The specific definitions of hazardous waste are spelled out in the federal statutes (40CFR Parts 260-280). Typically, however, the states have more stringent definitions. The states are responsible for the implementation and enforcement of RCRA. Wastes captured under this regulation in a printed circuit board fabrication facility include aqueous solutions with metals, acid or alkaline solutions with metals, sludges containing metals, etc. As a result, fabricators must comply with the following requirements.

1. Obtain an EPA identification number and apply for permits to generate, treat, store, or dispose of hazardous waste as appropriate.

2. Use appropriate containers for storage and disposal and approved manifests and labels for shipping.

3. Comply with technical requirements for on-site treatment of hazardous waste, including tank integrity standards, labeling, and secondary containment.

4. Keep records as appropriate for reporting to regulatory agencies.

A critical part of the regulation is to reduce and, where possible, eliminate the generation of hazardous waste. Waste minimization was specifically mandated in the 1984 Hazardous and Solid Wastes Amendments to the Resource Conservation and Recovery Act. This has had an enormous impact on the way waste is handled by printed circuit board facilities. Prevention of pollution has become the overriding goal in design with recycle and reuse technologies implemented only where pollution prevention is not feasible for technical and/or economic reasons. Chemical treatment of wastes should be utilized only where no other options exist.

55.3 MAJOR SOURCES AND AMOUNTS OF WASTEWATER IN A PRINTED CIRCUIT BOARD FABRICATION FACILITY

55.3.1 Major Sources of Waste

The following table indicates the major operations in the production of printed circuit boards with the waste streams they generate.

Source	Waste stream	Composition
1. Cleaning and surface preparation	Spent acid/alkaline baths Waste rinse waters	Metals, acids, alkalis
2. Electroless plating and deposition	Spent electroless copper bath Waste rinse water	Acids, palladium, complexed metals, chelating agents, formaldehyde
3. Pattern printing and masking	Spent stripper and developer solutions Waste rinse waters	Vinyl polymers, chlorinated solvents, organic solvents, alkalis
4. Electroplating	Spent plating bath Waste rinse waters	Metals, cyanide, sulfate
5. Etching	Spent etchant Waste rinse water	Ammonia, chromium, copper, iron, acids
6. Assembly	Aqueous and semiaqueous wastewaters	Lead, organics

55.3.2   Typical Amounts of Waste Materials

It is important to evaluate the waste generated in printed circuit board manufacturing in terms of the relative amounts of copper wasted from different processes in order to prioritize efforts at waste minimization. A waste audit should be performed to determine the types and amounts of waste generated. Figures 55.1 and 55.2 present two comparisons of copper-bearing wastes. The data presented are from two different printed circuit board fabrication facilities.

In Fig. 55.1, approximately 93 percent of the total amount of copper discharged was from the innerlayer and outerlayer etching process. The amount of copper discharged from the microetch baths (using sodium persulfate) was approximately the same as the copper contained in the rinses following acid copper and the microetch baths. That exact relationship is not always valid for every printed circuit board factory.

Figure 55.2 presents the results of an environmental audit at another printed circuit board facility in grams per hour of waste copper, excluding the large amounts of waste from I/L and O/L etching. The results presented in Fig. 55.2 closely resemble a "common" PCB factory.

The facility used to prepare Fig. 55.2 produced up to 1500 completed multilayer panels (18 in × 24 in) per 20-h day of production.

55.4   WASTE MINIMIZATION

55.4.1   Definitions

Waste minimization includes the procedures, operations, and equipment required to minimize the amount of waste produced. It includes anything that reduces the load on hazardous waste treatment or disposal facilities by reducing the quantity or toxicity of hazardous waste. Waste minimization should be approached as a hierarchy of options. The first step is *pollution prevention* or source reduction.

55.4.1.1   Pollution Prevention.   Pollution prevention is the use of materials, processes, or practices that reduce or eliminate the creation of pollutants or wastes. Pollution prevention is often cost effective because it may reduce raw material losses; reduce reliance on expensive end-of-pipe treatment technologies and disposal practices; conserve energy, water, chemicals, and other inputs; and reduce the potential liability associated with waste generation. Pollution prevention opportunities are intended to

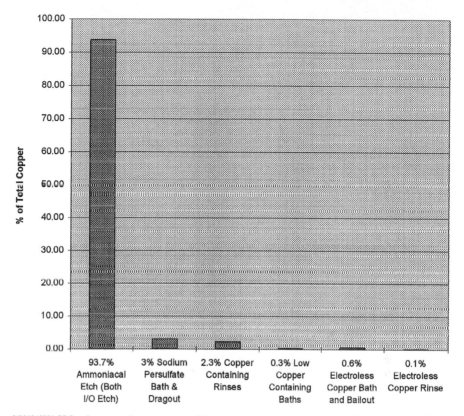

FIGURE 55.1 Amount of copper wasted for various waste streams expressed as a percent of total copper discharge.

1. Reduce the rinse water flow rate and volume.
2. Extend process bath life and reduce the dragout of a concentrated liquid (containing a contaminant such as copper).
3. Substitute materials where possible to eliminate or reduce the hazard of a particular waste.

55.4.1.2 *Recycling and Recovery.*   The next step in the hierarchy of waste minimization options is *recycling* or *recovery* of any waste that cannot be reduced or eliminated at the source. It includes reuse of waste in the process or recovery of metals from a waste before disposal. Recycling and recovery can occur both on- or off-site.

55.4.1.3 *Alternative Treatment.*   The last step in the waste minimization hierarchy is *alternative treatment.* Alternative treatments are selected to minimize the volume or hazard associated with a particular waste.

The benefits for implementing a waste minimization program can include savings in equipment and operating cost, recovery of natural resources, and significant reduction in risk of the liabilities associated with the disposal of hazardous waste. One measure of the effectiveness of a waste minimization project is the project's effect on the organization's cash flow. These projects should pay for themselves through reduced waste management and raw material costs.

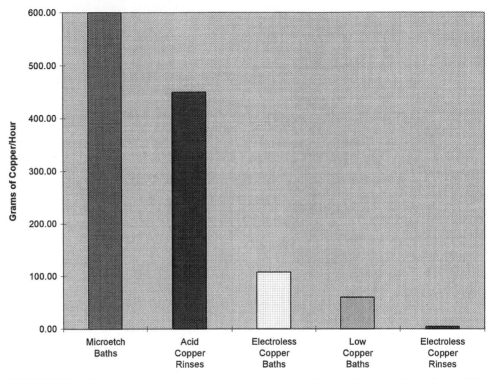

FIGURE 55.2 Major sources of copper expressed in grams per hour discharged. Inner- and outerlayer etching is excluded from this figure since it is so large in relation to the other contributors.

55.5 POLLUTION PREVENTION TECHNIQUES

Some common techniques available for pollution prevention in a printed circuit board fabrication facility are listed as follows. While this list is not all-inclusive, it provides an overview of the types of things that are important to consider. A brief description of each option is presented in the next section.

Rinse Water Reduction
1. Particulate filtration on deburr and panel-scrubbing operations with total or partial rinse water recycling
2. Etcher and conveyorized equipment design modifications
3. Immersion-type counterflow rinses
4. Alternating side spray rinses
5. Flexible orifice–type flow restrictors
6. DI and soft water for rinsing

Extend Bath Life
1. Filtration
2. Acid purification systems

3. In-tank electrolytic permanganate regeneration

4. Proper rack design and maintenance

5. Dragout recovery tanks

6. Monitor solution activity

Dragout Reduction

1. Etcher design

2. Automation

3. Rack design

4. DI and soft water for rinsing

Material Substitution

1. Acid tin as an etch resist

2. Eliminate thiourea from tin/lead stripping baths

3. Permanganate desmear

4. Direct metallization to eliminate electroless deposition and the use of formaldehyde

5. Nonaqueous waste resists

6. Use of nonproprietary chemistry to avoid chelating agents

55.5.1   Rinse Water Reduction

Most of the waste generated in the manufacture of printed circuit boards is from cleaning, plating, stripping, and etching. This section describes some of the techniques available for reducing the volume of rinse water used.

55.5.1.1   Particulate Filtration on Deburr and Panel-Scrubbing Operations.
Deburrers are used to remove stubs of copper formed after the drilling of holes in double-sided and multilayer panels before they enter the copper deposition process where copper is deposited within the holes recently drilled in the panel. Scrubbers are used to remove oxides from printed circuit laminates, clean the surface prior to a surface coating to provide better adhesion, and remove residuals after etching or stripping. In deburring and board scrubbing, particulate materials are added to the water and are removed by various methods based on size and the weight of the copper particle such that the wash water becomes suitable for up to 100 percent recycle. The types of filtration available for this operation are cloth, sand, centrifugation, and gravity settling with filtration.

55.5.1.2   Etcher and Conveyorized Equipment Design Modifications.
Use of recirculating rinse modules in the etcher and other conveyorized equipment will decrease the required flow rate of rinse water for that process step by about 50 percent without requiring significantly more floor space, compared to single-station spray rinse chambers (without recirculating rinses). In this application, fresh water is used for the final top and bottom nozzles in a rinse module. This water is collected in a sump located below the rinsing compartment. A pump recirculates this water through the first set of top and bottom nozzles (instead of using fresh water). As more fresh water enters the sump, the excess water overflows through a pipe fitting to drain. See Fig. 55.3.

Conveyorized equipment can also be used for, at least, innerlayer and outerlayer photoresist stripping and innerlayer and outerlayer photoresist developing, deburring, and panel scrubbing. Similar techniques to reduce water flow can be applied to these operations.

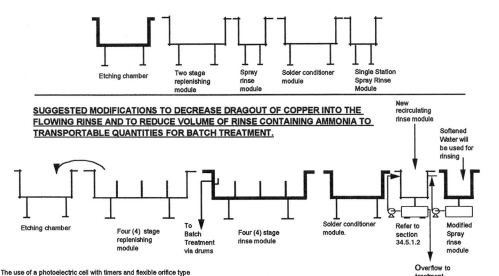

FIGURE 55.3   Suggested modifications to inner/outerlayer etching equipment.

55.5.1.3   *Immersion-Type Counterflow Rinses.*   Counterflow rinsing, employing several single-stage rinse tanks in series, is one of the most powerful waste reduction and water management techniques for innerlayer processing and in the electroless copper process. In counterflow rinsing, the plated part, after exiting the process bath, moves through several rinse contact stages, while water flows from stage to stage in the opposite direction. Over time, the first rinse reaches a steady-state concentration of process dragout contaminants that is lower than the process solution. The second rinse (away from the process bath) stabilizes at even a lower concentration. This enables less water to be used to produce the same cleanliness compared to a single-station rinse tank. The higher the number of rinse stations connected in series, the lower the rinse rate needed for adequate removal of the process solution from the panel. A reduction in rinse water use for a given process step of up to 90 percent over what would be used without counterflow rinsing can be achieved with this technique.

A multistage counterflow rinse system allows greater contact time between the panels and the rinse water, greater diffusion of the process chemicals into the rinse water, and more rinse water to come into contact with each panel. The disadvantage of multistage rinsing is that more steps are required and additional equipment and work space is required.

55.5.1.4   *Alternating Side Spray Rinses.*   Alternating side spray rinses can be used in place of or with immersion rinses in acid copper or solder plating lines. With this technique, rinse water pulsates on each side of a panel at different times so that penetration through holes can be achieved. The reduction in water consumption using an alternating side spray rinse for any given process step is about 85 to 90 percent of that consumed using single-stage immersion-type rinse. Another advantage of spray rinsing is that it will eliminate the need to dump the entire contents of an immersion-type rinse tank, for cleaning purposes, on a periodic basis.

55.5.1.5   *Flexible Orifice–Type Flow Restrictors.* Flow restrictors are available with flexible diaphragms and are rated for a specific flow rate in a specific pipe size. Typically a 2- to 3-gpm flow restrictor is recommended for a two-stage counterflow rinse tank. As long as a minimum water pressure exists in the water inlet line (about 20 lb/in² is considered minimum), the water flow into a rinse tank should not vary, although the water pressure will most likely vary. Flow restrictors are not adjustable after they are installed. By providing pipe union fittings both up and downstream from the flow restrictor fitting, the flow rate can be changed if necessary.

55.5.2   Extended Bath Life

Although process bath solutions are discarded infrequently, when spent or contaminated, they must be replaced. The resulting waste contains high concentrations of hazardous constituents. Bath life extension is thus an important means of minimizing waste in a printed circuit board manufacturing facility.

55.5.2.1   *Filtration.* Baths should be continuously filtered to remove impurities. Most manufacturers utilize spiral-wound cylindrical polypropylene-type filters with nominal pore-size ratings. Some manufacturers have decreased or actually eliminated the need to dump selected process baths, such as the brown oxide bath, by using a more expensive absolute-type mechanical filter.

The filter, in the brown oxide case, removes black copper metallic particles and other solids. Proper filtration will eliminate the cloudiness that may presently build up in the bath. A side benefit for filtering the brown oxide bath is the increased mechanical agitation in the bath that may improve the coating uniformity.

The filtration requirements for every bath should be reviewed with an objective to clean the bath and to decrease or eliminate its dumping frequency.

55.5.2.2   *Acid Purification Systems.* In the circuit board industry, mineral and semiaqueous acid solutions are used for activating metal and plastic surfaces and for the etching, cleaning, and stripping of racks and defective parts. These solutions become contaminated and lose strength (lowering the acid concentration) as a function of usage due to the increasing concentration of contaminants such as metals and the continuous dilution of the acid concentration. These less active acidic solutions are customarily disposed of by periodically dumping the bath at once, or slowly withdrawing a portion of, the bath.

From a process point of view, if the concentration of contaminants can be maintained low and the acid concentration maintained at the desired level, an improvement in performance of the acid solution will result (in other words, better quality control is provided using continuous bath purification).

Finding an economic method to continuously reuse 90 to 95 percent of an acid will (1) reduce incoming chemical consumption (for the factory), (2) reduce the cost to process parts (fewer rejects), (3) provide better quality control, and (4) reduce the chemicals (for example, caustic soda) required to neutralize the spent acid before discharge. Eliminating the need to handle spent acids will decrease the exposure of hazardous materials to employees and will reduce the wastewater treatment burden. This type of bath purification system may eliminate the need to dump selected concentrated wastes.

One technology for acid purification uses diffusion dialysis. Diffusion dialysis units utilize a membrane which allows anions to pass (i.e., Cl^- in the case of hydrochloric acid) while metal cations remain behind. This is the dialysis part of the process. At the same time, water molecules diffuse through the membrane in the other direction.

Each recovery unit produces two streams: a reclaimed acid stream, which is high in acid content and low in metal content, and a waste stream which contains a high concentration of metals and a low concentration of acid. The only required utilities are soft water and single-phase power.

The following commonly used acids found in the printed circuit industry can be purified using this technology:

Hydrochloric	Fluoboric
Hydrofluoric	Methane sulfonic
Nitric	Ferric chloride-hydrochloric acid
Nitric-hydrofluoric	Sulfuric acid

Methane sulfonic acid (MSA) is a relatively expensive acid used to strip (remove) solder (tin/lead) and copper metals. This acid is also used to condition metal surfaces prior to fluoborate free-solder electroplating when applying a final etch resist of solder onto multilayer printed circuit boards. By continuously recirculating the acid bath through a diffusion dialysis unit, a 95 percent reduction of chemical purchases (required by the process steps indicated in the first paragraph of this section) may result, producing less than a six-month return on the initial capital investment (ROI). Typically, 80 to 90 percent of the acid is recovered with 70 to 90 percent of the metals removed in an equal volume of dilute acid.

Another system uses only water in its process. The heart of this product is a shallow bed of special adsorption resin. This resin has the unique ability to adsorb strong mineral acids yet exclude metal salts. This is not an ion exchange process since only water is used, in a counter-current fashion, to desorb the acid from the resin.

35.5.2.3 *In-tank Electrolytic Permanganate Regeneration.*   Because permanganate is a powerful oxidizing agent only in basic conditions, desmear solutions usually contain 5 to 10% caustic soda. However, the alkalinity at high temperatures causes the permanganate (MnO_4^{-1}) ion to break down to manganate. The manganate reacts with water to form manganese dioxide (MnO_2), eventually forming a black sludge of MnO_2 which settles to the bottom of the desmear tank. This requires that the bath be dumped.

Oxidizing electrolytic regeneration systems are available to inhibit this reaction. A typical electrolytic regeneration system employs an in-tank probe composed of a copper rod as the cathode which is connected to a power supply. When current is applied, oxygen is generated in the permanganate solution which functions as the regenerate for the manganate. The efficiency of the system depends on the rate of manganate buildup, the amount of amperage generated, and the power supply to the in-tank regenerator. These probes attach to the top rim of the desmear tank and are always operating (as the permanganate ions will break down even when the bath is not active). In most cases, the use of this equipment has eliminated the need to dump this bath.

55.5.2.4 *Proper Rack Design and Maintenance.*   By using plastic-coated racks, only the tips and contactors (which electrically and physically connect and support the panel to the carrying rack), have exposed conductive metal and must be stripped. If one of these is removable, then the amount of metal to be chemically or electronically removed from the rack is also reduced. This should lengthen the useful life of the stripping bath and further reduce the cost for hauling away or on-site treatment (if nitric acid is used) of the spent rack stripper. Use of racks with disposable contactors eliminates the need to strip the racks and could eliminate the use of nitric acid.

55.5.2.5 *Dragout Recovery Tanks.*   The dragout tank's primary purpose is to collect most of the solution dragout before the panel enters the rinse tank. The dragout (recovery) tank is a rinse tank that operates without a continuous flow of feed water. It is positioned between a process bath and its rinse tank. The dragout tank can be either an immersion application (for the electroless and innerlayer operations) or a spray type application (on conveyorized equipment or for pattern- or panel-plating operations). After the panels have been removed from the process bath, they are immersed or sprayed in the dragout tank before they are rinsed in another tank. The concentrated rinse water can be collected separately from its following rinse.

Eventually, the concentration of the tank solution will increase to a point such that it provides an opportunity to

1. Recover metals.
2. Recycle a process bath since it is the same chemistry.

The disadvantage is that this opportunity may require additional production floor space.

A drip pan (also called a drain board) is one of the simplest methods for dragout recovery. The drip pan is located either under the rack while in transit or between tanks, and will capture drips of process solution from racks and panels as these are transferred between tanks. Drip pans not only save chemicals and reduce rinse water requirements, they also improve housekeeping by keeping the floor dry.

55.5.2.6   *Monitor Solution Activity.* Bath life can be extended by monitoring the bath activity and replenishing stabilizers and reagents. This will reduce dump frequencies and result in a reduction in waste.

55.5.3   Dragout Reduction

Dragout reduction is desirable as a waste minimization step because it results in the saving of process chemicals as well as a reduction in sludge generation.

55.5.3.1   *Etcher Design.* Etching machines can be the single largest source of copper waste in the discharge from a PCB facility. The amount of copper discharged and the rinse flow rate from that machine are a function of the machine design. An older etching machine often contains a single-stage etchant replenishing module positioned between the etching chamber and its continuously flowing single or multiple-stage water rinse chamber. Fresh etchant is fed to the replenishing module to wash the panels and that etchant (now containing copper washed off of the freshly etched panels), then flows (in a direction opposite to the direction of the panel movement) into the etching chamber. The continuous water rinse can contain from 100 to 500 mg/l or ppm of copper. Companies that use this type of equipment, with single-station rinse module, normally have floor space restrictions in their production area. See Fig. 55.3.

By way of comparison, adding a second stage replenisher station will reduce the range of copper concentration in the following rinse from 50 to 300 mg/l. Using a four-stage replenisher module and combination recirculating and then a single-stage rinse modules, will produce a rinse containing from less than 1.0 up to 2.0 mg/l of copper.

55.5.3.2   *Automation.* Computerized process control systems can be used for panel handling and process bath monitoring to prevent unexpected decomposition of a process bath, controlled rinse flow, and uniform panel withdrawal from each process bath. Since these systems require a significant capital expense for initial installation, typically only large printed circuit board companies will find this to be a cost-effective alternative.

55.5.3.3   *Rack Design.* Some plating rack designs allow the panels to be tilted to one side and the panel mounted at an angle (relative to the horizontal) to allow better solution drainage off of the panel. This will decrease the dragout.

55.5.3.4   *DI and Soft Water for Rinsing.* Natural contaminants found in water used for production purposes can contribute to the volume of sludge produced in waste treatment because they precipitate as carbonates and phosphates. Another advantage is that pretreated rinse water is cleaner; therefore, the normal diffusion of contaminants from the panel into the rinse water occurs more quickly. As a result, water requirements for each rinse may be reduced since rinse efficiency is improved.

55.5.4 Material Substitution

Material substitution is a waste minimization technique that should result in a reduced hazard associated with a particular waste or a simplied means of waste treatment. It is important that all process impacts are thoroughly understood prior to instituting any material changes.

55.5.4.1 Acid Tin as Etch Resist. Acid solder baths are often used as an etch resist before final etching of an assembled multilayer printed circuit board. The lead and fluoride present in this bath must then be removed in the waste treatment process. Acid tin can be used as the etch resist in place of solder under certain circumstances. An advantage of this bath is that it does not contain lead ions or fluorides. Use of this bath will eliminate these materials from the discharge.

35.5.4.2 Eliminate Thiourea. Thiourea is a known carcinogen which has been used as a chelating agent in Sn/Pb stripping baths. A peroxide-based bath can be substituted when the etch resist is acid tin, thus eliminating this concern.

35.5.4.3 Permanganate Desmear. To reduce the amount and toxicity of waste, most printed circuit board factories in the United States use the alkaline permanganate process instead of chromic acid for desmear. This technique has gained favor due to the wastewater treatment equipment requirements to reduce the chromic acid and the control problems (especially in humid environments) of 96% sulfuric acid. The plasma desmear process produces the least amount of hazardous waste, but that technique is used primarily for small batch operations and not high-volume continuous operations.

35.5.4.4 Direct Metallization. Alternatives to the electroless copper process have emerged to avoid the use of formaldehyde, a toxic and carcinogenic material. The use of formaldehyde has recently (as of June 1993) required extensive in-plant air monitoring and will require employers to have their affected employees periodically checked by a medical doctor. This is an additional unwanted expense for each PCB factory owner.

One alternative to the electroless process utilizes conductive polymers. Polymer thick film (PTF) technology is a method of screening polymer conductors, resistors, dielectrics, and protective coatings on a substrate or printed wire board to create basic circuitry or interconnections. The use of screened-on conductive films in place of the formaldehyde-containing electroless copper process is gaining acceptance in Asia (more so than in the United States at this time).

Other technologies are being applied commercially which use palladium or graphite to provide conductive pathways in the holes of multilayer printed circuit boards.

55.5.4.5 Nonaqueous Waste Resists. One of the most exciting waste minimization techniques currently being tested is the elimination of the need for rinsing from both photoresist developing and stripping operations. This new process may eliminate spent developer wastes and will eliminate the spent resist stripper bath dump.

The photoresist, being demonstrated at Circuit Center (Dayton, Ohio) and at E. G. & G. Mound, is applied in a production process as a thin film from a liquid solution onto a circuit board and air dried. When exposed to a high-intensity ultraviolet light source, these photoresists will decompose and vaporize as inert gas. There appears to be no rinsing required for these operations. In fact, these operations, as we currently know them, will be eliminated, thus reducing the cost of waste treatment. The existing 30-month demonstration project is a cooperative effort between the University of Dayton Research Institute, the Air Force, the U.S. Environmental Protection Agency, and the Edison Materials Technology Center (EMTEC).

Depending on the actual chemistry used, the use of this new process may reduce the hazardous air emissions from these processes commonly produced with conventional semiaqueous and aqueous chemistries containing butyl cellosolve or butyl carbitol.

55.5.4.6 Use of Nonproprietary Chemistry to Avoid Chelating Agents. Chelating agents are molecules that form a charged complex with a metal ion such as copper. They are used to enhance solubility and keep the metal ions in solution. Typical chelating agents used in printed circuit board fabrication include ferrocyanide, ethylenediaminetetraacetic acid (EDTA), phosphates, and ammonia. Chelated baths are intended to enhance etching, cleaning, and electroless plating but make waste treatment much more difficult because the metals are tightly bound in the complex, which inhibits precipitation. Often ferrous sulfate must be added to wastewaters to break the chelators prior to precipitation of metal hydroxides. The iron is precipitated as well as other metals, thus increasing the amount of sludge produced. Use of nonchelated chemistry where possible will eliminate this problem. If some chelating chemistry is required in the process, proper waste segregation should be implemented to minimize the volume of any chelated waste streams.

55.6 RECYCLING AND RECOVERY TECHNIQUES

55.6.1 Copper Sulfate Crystallization

55.6.1.1 Theory of Operation. The sulfuric acid/hydrogen peroxide etching reaction is:

$$Cu + H_2O_2 + H_2SO_4 \longrightarrow CuSO_4 + 2H_2O \tag{55.1}$$

Systems employing crystallizers are designed to

1. Lower the solution temperature to lower the solubility of copper.
2. Then remove copper in the form of copper sulfate pentahydrate crystals ($CuSO_4 - 5H_2O$).
3. Then reheat the bath (to prevent copper from clogging the return pipe line if the bath temperature is lowered further) before transferring the bath back to its working tank.

An advantage, from a waste treatment point of view, is that the bath is recycled and not routinely dumped. Other microetchants must be dumped when the copper concentration increases. The most significant advantage, from a process point of view, is that a low copper concentration can be maintained and the printed circuit board factory will minimize its purchase of microetchant chemicals.

A disadvantage to this process is that, at the time of this writing, there are no outlets available to receive and recycle the crystals, so the only alternative for manufacturers using crystallizers is to dissolve the copper sulfate crystals in spent dilute (10%) sulfuric acid. After the copper has dissolved, this material can be pumped to another holding tank for recirculation through an electroplating cell for reclaim of the copper.

55.6.2 Rinse Water Recycling

Rinse waters can be treated and reused in the process. This can be accomplished in two ways. Rinse water recycling can be accomplished with a point-of-source system. In this system, the flow from selected rinses is recirculated through cation and anion exchange columns and then returned to the point of origin. It may be necessary to add a carbon filtration step for rinses containing organics.

The second method would involve a central system. Selected rinses could be run through an ion exchange system for copper removal. Following this step, these rinses would pass through a general resin for cation removal, a general resin for anion removal, systems for organic removal such as activated carbon, ozonation or UV peroxide, and final filtration. In some cases, reverse osmosis could be substituted for the general ion exchange steps. Given the complexity of this treatment process, it would be cost effective only for large printed circuit board facilities with high water use rates.

55.6.3 Copper Recovery via Electrowinning

55.6.3.1 Theory of Operation. Electrowinning is the reduction of copper ions to solid metallic copper at a cathode. The following are the chemical reactions governing copper recovery in an acidic solution. The reduction reactions that occur at the *cathode* are:

$$Cu^{+2} + 2e^- \longrightarrow Cu \text{ (metal)} \tag{55.2}$$

$$2H^+ + 2e^- \longrightarrow H_2 \text{ (gas)} \tag{55.3}$$

The reaction that occurs at the *anode* is:

$$H_2O \longrightarrow \tfrac{1}{2}(O_2) + 2H^+ + 2e^- \tag{55.4}$$

The first reaction describes the principal objective of electroplating (the reduction of copper to its solid form on the cathode). As electroplating proceeds, the waste becomes more acidic, as noted in Eq. (55.4).

The speed of these reactions will be inhibited if oxidizing chemicals are present. Typical oxidizing agents include peroxides and persulfates, which are normal components of microetching baths. If these oxidizers are not chemically reduced in the solution entering an electroplating module, additional reaction time must be anticipated to reduce the oxidizers electrically. Following that, the reduction of copper will occur. For this reason, a chemical reduction step (addition of sodium bisulfite ($NaHSO_3$) or sodium metabisulfite ($Na_2S_2O_5$)) is recommended before the waste is recirculated through an electroplating module.

55.6.3.2 Central Systems. Copper can be recovered in a central system from the following sources:

1. Sulfuric acid regeneration of selective cation columns
2. Spent microetch baths
3. Dragout tanks
4. Copper sulfate electroplating bath bailout
5. Dissolved copper sulfate crystals from a crystallizer

The typical concentration of the preceding mixture (or any component of it) may range from 5 to 30 g of copper per liter. A central electrowinning system is shown in Fig. 55.4.

55.6.3.3 Parallel Plate Electrowinning Systems. The purpose of parallel plating, in a central system, is to recover copper as a 99.9 percent pure metal sheet (for resale) and to reduce the copper concentration in the liquid being recirculated to 1.0 g/l, or less. The efficiency of the reaction drops significantly below 1.0 g of copper per liter, as observed by the generation of heat in the bath by the electroplating process.

Many factors affect the ability of an electrolytic cell to recover copper. Important design parameters center around improving the mass transport. These include solution agitation and cathode agitation. Air agitation of the waste within the electroplating cell is used to increase the efficiency of the cell. Control of the air inlet rate, bubble size, current density and distribution are all critical to maintaining high efficiency in each parallel plate cell. Particles (so-called *dendrites*) or fines of metal can be dislodged and accumulate in dead spots with the plating cell tank. Electrical shorting out of the anode to cathode can occur when particles accumulate, creating the possibility of isolated locations where burning of the cell may occur.

55.6.3.4 High-Surface-Area Electrowinning Systems (HSA). HSA systems can be used to recover copper from a variety of solutions, including electroless copper concentrates and rinses. The use of high-surface-area cathodes improves the mass transport characteristics over flat plate cathodes. The HSA cathodes reduce electrode polarization potential and improve

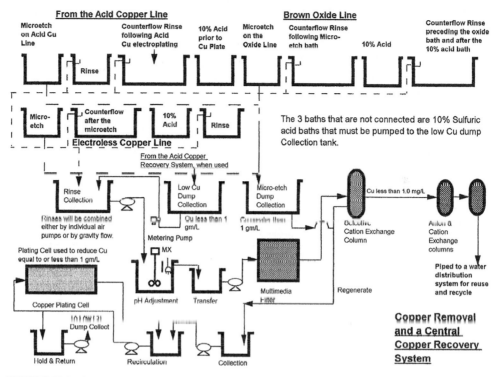

FIGURE 55.4 Central copper recovery system utilizing ion exchange for copper removal and electrowinning for copper recovery.

ion diffusion potential, allowing the copper to deposit rapidly on an HSA cathode from wastes with both high and low copper concentrations. HSA cathodes promote solution agitation by creating numerous bonding sites for copper. Expanded mesh carbon fibers or catalyzed foam are typical materials of construction in HSA cathodes. Some HSA systems require the purchase of new cathodes, while the cathodes in some systems can be regenerated for a period of time. To regenerate these cathodes they must be stripped of copper metal.

Stripping the cathode takes place in a second plating tank. The cathodes from the first tank must be manually removed and placed into the anodic position within the stripping tank. The copper is then recovered as a metal by electroplating the copper onto either a stainless steel or a copper laminate in a copper pyrophosphate electrolyte. The *reusable cathodes* are then available to be repositioned back to the first cell.

HSA systems can reduce the copper levels to low levels in an incoming waste stream, but it is important to note that copper is not recovered unless the stripping step takes place in the second plating tank.

55.6.3.5 *Point-of-Source Systems.*

Dragout tanks can be located between a process bath (for example, a copper-plating bath in the pattern-plating line) and its following rinse tank. Typically, dragout tanks are positioned *only* adjacent to the process tanks containing the most difficult wastes. These include the etching baths (if not conveyorized), the microetch baths (which produce the most copper), and the copper electroplating (usually sulfate-based) baths.

The contents of each dragout bath can be recirculated through its own plating cell for copper recovery, then back to the dragout tank. In this way, the concentration of copper is maintained at a low level. The plating cell, for this application, could be either a parallel plate or an HSA system.

If dedicated electroplating cells are not used to recover copper directly from a dragout bath, the dragout bath can periodically be discharged to a dilute (or concentrated) copper collection tank for recovery through an ion exchange system (or a central electroplating system, respectively).

55.7 ALTERNATIVE TREATMENTS

55.7.1 Selective Ion Exchange

Ion exchange is a process in which ions, which are held by electrostatic forces to charged functional groups on the surface of the ion exchange resin, are exchanged for ions of similar charge in a solution in which a resin is immersed. Ion exchange is classified as an adsorption process because the exchange occurs on the surface of a solid (a resin bead), and the exchanging ion must undergo a phase transfer from the solution phase to the surface of the solid. A single-pass selective cation exchange column can reduce the copper concentration in the influent stream to less than 1.0 mg/l. If two cation exchange columns are in series, as indicated in Fig. 55.4, the expected copper effluent concentration should be less than 0.5 mg/l. Organics present in the incoming waste can foul the resin and render it unusable; therefore, a carbon filtration step occurs ahead of the ion exchange column. Wastes with significant levels of organic material are not treated with this technology.

55.7.1.1 Theory of Operation.   Copper ions are normally present in their divalent (+2) state. The removal mechanism of divalent copper from an aqueous waste by ion exchange resins is:

$$Cu^{+2} + 2R\text{-}H \Longrightarrow 2H^+ + R_2\text{-}Cu \qquad (55.5)$$

where R = a cation exchange resin. Here the copper is "attached" to the resin beads.

Ion exchange equipment suppliers will usually furnish a water meter that totalizes the quantity of waste processed through the column. After the system has been calibrated by the equipment supplier, the water meter will activate an alarm after a specified volume of waste has been processed. An alarm will sound if the system requires manual regeneration. If the system is supplied with automatic regeneration, the water meter will deactivate a column while regeneration takes place. Where multiple columns are included within a system, valves must be adjusted to accommodate regeneration. This will be accomplished by manual or automated (pneumatic-, electric-, or hydraulic-actuated) techniques.

55.7.1.2 Regeneration.   To recover the copper, the cation resins are first rinsed with water and then washed with 5 to 15% by volume sulfuric acid. This proceeds by the following reaction:

$$R_2\text{-}Cu + 2H^{+1} \Longrightarrow Cu^{+2} + 2R\text{-}H \qquad (55.6)$$

where R = a cation exchange resin.

Copper is displaced from the resin by a proton (or hydrogen atom). The resulting sulfuric acid solution containing copper can be directed to an electroplating system to recover copper from the acidic solution. The remaining resin is washed with water to remove the residual sulfuric acid.

55.7.1.3 Spent Baths.   Certain spent baths can be bled into the ion exchange system. These typically include the copper sulfate electroplating dragout, acid cleaners, predips, microetch and rinses, rinses following cupric chloride and ammoniacal etchants, and copper waste from electrowinning after reduction to 1.0 ppm or less.

55.7.2 Removal of Copper from the Electroless Copper Bath

55.7.2.1 Theory of Operation.   It is well known that copper metal can be deposited onto metal (usually copper or palladium) within a PCB hole. Dissolved copper is automatically reduced (using formaldehyde as its reducing agent) to a solid metal and deposits itself onto the palladium inside of the hole while the panel is immersed in the electroless copper bath. The chemical reaction of copper deposition may be represented by

$$Cu\,(EDTA)^{-2} + 2HCHO + 4OH^{-1} \longrightarrow$$
$$Cu^0 + H_2 + 2H_2O + 2CHOO^{-1} + EDTA^{-4} \tag{55.7}$$

This process can be utilized to remove copper from the electroless copper baths and bailout. Products on the market include canisters or modules which contain a proprietary spongelike material deposited with palladium and copper. The electroless copper solution passes through the canister where the copper is autocatalytically reduced, trapped by the spongelike media and filtered out of the waste stream. See Fig. 55.5

Typically, a minimum of two canisters are connected in series. As the copper concentration in the first canister approaches 1.0 mg/l, the sponge material is replaced by the material in the second canister. New sponge material is added to the second canister.

55.7.3 Sodium Borohydride Reduction

Although not as common or as inexpensive as the autocatalytic method, one of the simplest methods for the treatment of electroless copper is by the use of another strong reducing

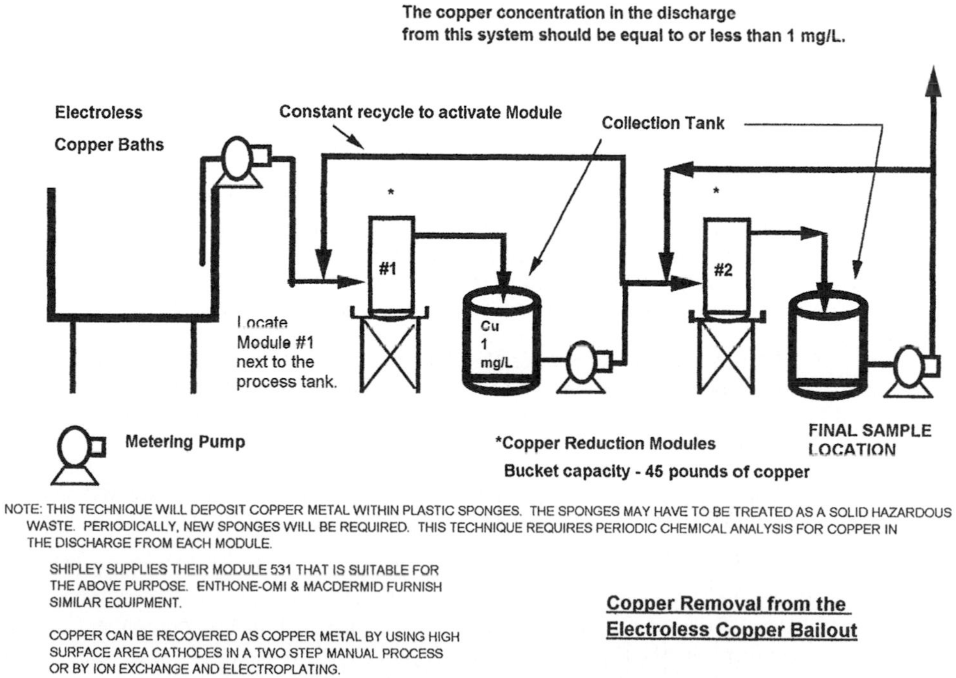

FIGURE 55.5 Copper removal from electroless copper bailout using copper reduction.

agent, sodium borohydride ($NaBH_4$). This chemical is commercially available as a solution of 40% caustic soda containing 12% by weight of sodium borohydride. The chemical reaction can be represented by

$$8MX + NaBH_4 + 2H_2O => 8M + NaBO_2 + 8HX \qquad (55.8)$$

where M is the cation with +1 valence and X is the anion.

The spent electroless bath has to be in an active state for this method to work efficiently. Adding sodium hydroxide (NaOH) and/or formaldehyde (HCHO) to the bath will shift the equilibrium in the bath to this reactive stage. When this stage is reached, a small addition of sodium borohydride will catalyze the reaction, and reduced copper will precipitate out as metallic fines.

A typical process for using sodium borohydride requires the use of an open-top batch tank that is located in a well-ventilated area. Since $NaBH_4$ can react to release hydrogen as a side reaction, any source of flames or sparks must be eliminated. A 0.5 to 1%, by volume, $NaBH_4$ solution is added slowly to the batch reaction tank that contains the spent electroless copper solution. After a few minutes, the solution will effervesce vigorously and fine, pink copper particles will begin to form and become visible to the naked eye. Small bubbles of flammable hydrogen gas will also evolve from the solution, and these must be rapidly removed by appropriate exhausts. Usually within about one hour, the solution will turn clear and copper will precipitate from the solution. These fines will eventually settle to the bottom of the batch tank and the copper fines should be removed as soon as the reaction has been completed. Otherwise the copper fines may reoxidize and dissolve back into solution. More $NaBH_4$ will then be necessary to remove the copper. After the copper fines have been removed, hydrogen peroxide (H_2O_2) can be added to the clear solution to oxidize any excess formaldehyde to the less toxic formic acid.

This process can be used for other metal-bearing waste streams besides electroless copper. However, it is typically not a cost-effective means of treating nonchelated waste.

55.7.4   Aqueous and Semiaqueous Photoresist Stripping Bath Treatment

A significant volume of waste produced from PCB factory is from the stripping of resists. Resist stripping machines can utilize on-line or off-line equipment to remove semisolid photoresists on a continuous basis from the operating bath. Removing the suspended solids within the solution provides a longer life and more efficient usage.

In any case, there will be times when the bath becomes spent and must be dumped. Waste from this process should be segregated for treatment so that the sludge produced can be classified as nonhazardous. In addition, good engineering practice dictates that spent photoresist stripping baths not enter a continuous precipitation system since organic photopolymers may not always settle in a clarifier. Under some conditions (at certain temperatures or with certain organics) stratification of the organic material in the clarifier can occur. As a result, the organic solids can float and any metal hydroxide particles nearby may become attached to the floating material.

The treatment procedure for this waste uses chemical precipitation (caused by lowering the pH) and filtration (to remove the photoresist and residual metals). The treated wastes can be slowly discharged into the existing final pH adjustment tank. The specific treatment procedure follows.

The untreated pH of this waste is approximately 12.0 pH units. Sulfuric acid is slowly added to lower the pH to about 9.0. After a stable pH is achieved, the pH will be lowered to about 6.0 by the addition of acidic proprietary chemicals available at the time of this writing. At this low pH, the dissolved resist will precipitate out of solution and the solids formed can be removed by pumping the waste through a filter press.

55.8 CHEMICAL TREATMENT SYSTEMS

55.8.1 Definition

Chemical treatment of wastes involves the addition of chemicals to precipitate metals out of solution. This results in a sludge which must be disposed of. The sludge is then typically sent to a reclaimer for recovery of the metals or licensed disposal site. The latter option is becoming increasingly difficult and expensive. It should be avoided due to the unknown risks associated with the landfill of hazardous wastes. Chemical treatment of wastes should be utilized only when pollution prevention or recycling is not technically or economically feasible.

55.8.2 Treatment Process

Nonchelated metal rinses and dumps are typically treated in a two-stage process. In the first stage, rough pH control is accomplished with the addition of caustic soda and sulfuric acid. In the second stage, the pH is elevated with caustic to precipitate the metals in the hydoxide form. Lime has been used in this step in the past, but its use has been discouraged since it increases the sludge volume significantly.

Alternatively, the metals can be precipitated as in the sulfide form with the addition of FeS at pH 9. This process is more difficult to control and presents a potential hazard due to the possible evolution of H_2S. However, the lower solubility of the metal sulfides provides an advantage where discharge limits are more stringent.

Chelated metal rinses and dumps such as those from the electroless copper process provide more challenge in chemical treatment since the chelating agents bind the metals and keep them in solution. This problem is often addressed by the addition of ferrous sulfate to break down the chelated metal complexes, allowing the metal to precipitate as the hydroxide. It is often necessary to add large quantities of this reagent such that the ratio of iron to copper is 8:1. The disadvantage of this method is that iron will precipitate out as well, thus significantly increasing the volume of sludge produced.

Regardless of the specific treatment chemistry used, there are common components to all chemical treatment systems.

55.8.3 Collection System

The collection system is the tank or group of tanks where wastes are collected prior to treatment. The collection tanks should be sized to provide a minimum residence time of 20 min. Segregation of wastes can improve the efficiency of the waste treatment system. For example, process bath dumps can be collected in separate tanks and metered into the rinse waste stream at a low flow rate. Also, chelated wastes can be collected and treated separately from nonchelated streams, thus minimizing the use of ferrous sulfate.

55.8.4 pH Adjust

This is typically a dual-tank system where the pH is elevated and metals are precipitated as either the hydroxide or the sulfide, as previously described. Each stage must be adequately sized to provide adequate retention time (usually a minimum of 30 min) to ensure the completion of the reaction. Each tank is equipped with a mixer, pH probe, and controller. In addition, pH recorders should be provided to effectively monitor system performance.

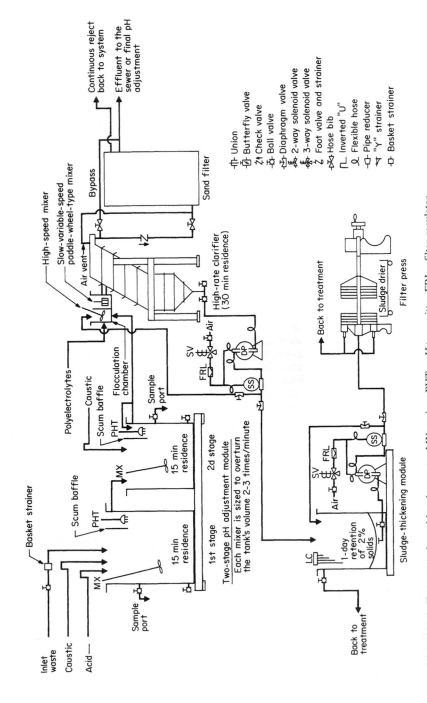

FIGURE 55.6 Treatment of metal-bearing wastes. MX = mixer; PHT = pH transmitter; FRL = filter, regulator, lubricator; SV = solenoid valve; SS = surge suppressor; DP = diaphragm pump; LC = level control. (*Courtesy of Baker Brothers/Systems.*)

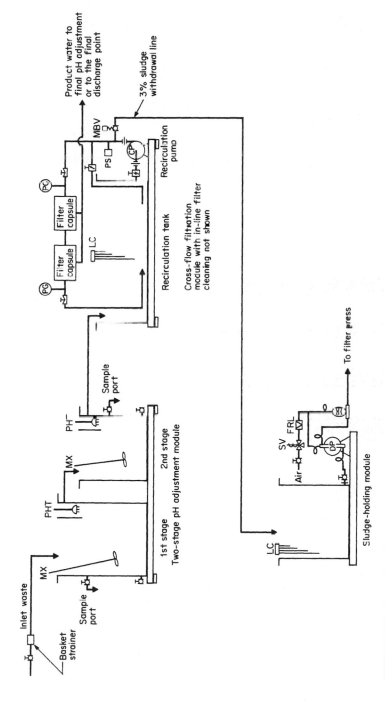

FIGURE 55.7 Cross-flow filtration. MX = mixer; PHT = pH transmitter; PG = pressure gauge; LC = level control; PS = pressure switch; MBV = motorized ball valve; CP = centrifugal pump; SV = solenoid valve; FRL = filter, regulator, lubricator; DP = diaphragm pump SS = surge suppressor. (*Courtesy of Baker Brothers/Systems.*)

TABLE 55.2 Advantages and Disadvantages of Various PCB Wastewater Treatment Alternatives

Categories of major wastes	Category of waste minimization or treatment	Advantages	Disadvantages
1. Copper-containing acidic rinses.	Selective rinse water recycling with copper recovery.	Recovered copper metal can be sold. Recycling will reduce the quantity of waste being discharged. Reduced sewer use charges.	Most expensive. Requires flow control. Requires thorough and periodic bath analysis. Must address spent process baths.
	Copper removal from preselected rinses using selective ion exchange with copper recovery by electroplating.	Recovered copper metal can be sold. Minimizes the quantity of copper discharged.	More expensive equipment than used for chemical treatment. Requires expensive selective cation resins. Must address spent baths.
	Chemical treatment to precipitate copper.	Least capital cost. Requires clarification with filtration and proper pretreatment chemistry.	Produces sludge. Increasing liability. Requires intensive effort for labor. Must address spent process baths.
2. Copper-containing acidic spent process baths.	Bath purification with partial or total reuse using crystallization, ion exchange, filtration, and/or diffusion dialysis.	Reduce or eliminate the need to dump a bath. Recovered copper can be sold.	Requires through and periodic bath analysis. Optimum application is copper removal from rinse waters.
	Bleed spent bath into rinses for copper removal with copper recovery.	Recovered copper can be sold. Minimizes quantity of copper discharged.	More expensive equipment than used for chemical treatment.
	Allow a higher contaminant concentration before dumping.	Less frequent need to dump a process bath.	Requires through and periodic bath analysis.
	Copper recovery from microetchants.	Most copper recovery using least floor space and cost.	Requires that these wastes be collected and processed separately.
	Chemical treatment to precipitate copper.	Used to concentrate copper into sludge.	Produces sludge. Increasing liability. Requires intensive effort for labor.
3. Ammonium chloride etchants and rinses.	Recycle 95% etchant and rinse water.	Reduces fresh chemical and hauling purchases. Recovered copper can be sold. Reduces waste quantity.	May be justified only when both inner and outer layers are etched with this chemistry. Ammonium sulfate etchant requires a different method for recycling.
	Pipe rinse into copper removal with copper recovery system.	Recovered copper can be sold.	More expensive equipment than used for chemical treatment. Must continue to haul spent etchant.

Waste source	Pollution prevention option	Advantages	Disadvantages
4. Electroless copper bath growth.	Process chemistry substitution using palladium-based and/or graphite-based alternatives.	Eliminates formaldehyde.	May produce excess organic wastes. May not be suitable for all applications.
	Copper recovery.	Reduces copper discharge.	Expensive compared to other methods.
	Autocatalytic copper removal onto a sponge.	Effective removal of copper metal complexed with EDTA. Possible copper recovery.	Produces a waste to be transported. Continuous activation, heat, and monitoring are required.
	Chemical treatment.	Least capital cost.	Produces sludge. Chelating agents are still present in the treated waste.
5. Develop/Resist Strip.	Use resists which do not emit wastes.	Eliminates hauling solids.	May not be universally applicable.
	Chemical treatment to reduce volume compared to hauling spent process baths off-site for disposal or dumping waste to the sewer.	Reduces quantity of this material to be hauled for off-site disposal.	Solids handling could be a problem. Expensive proprietary chemicals. Minimal organic load reduction in treated waste. Requires segregation.
6. Other wastes including tin, tin/lead, nickel, and gold.	Recover metal and chemistry from dragout tank.	Bath and metal recovery and recycling are possible.	Requires space to install dragout tanks. Requires more waste segregation and can be expensive as more recycling and recovery is required.
	Chemical treatment and membrane filtration.	Least capital cost.	Produces sludge. Increasing liability. Requires intensive effort for labor.
7. Printed circuit assembly cleaning wastes.	Use of water-soluble solder mask and flux.	Eliminates toxic chemicals. Allows rinse recycling.	Not universally applicable.
	Use of semiaqueous or saponified cleaners for RMA fluxes.	Eliminates use of CFCs.	Possible chemical treatment of bath VOC emission from semiaqueous baths.

55.8.5 Settling Process

Following the second-stage pH adjust, waste flows to the flocculation chamber in a clarifier where polyelectrolytes and recycled sludge are added and rapidly mixed. The waste then enters a second stage where a slow-speed mixer is used to enlarge the hydroxides to make settling more effective. Finally, the flocculated waste will flow down the clarifier. The precipitated solids settle to the bottom of the clarifier and the clarified liquid is discharged to a final pH adjust system or in some cases to a sand filter prior to final pH adjust if required to meet discharge standards. (See Fig. 55.6.)

55.8.6 Cross-Flow Microfiltration

Cross-flow microfiltration systems are used in place of a clarifier and sand filter where the maximum discharge limit for copper is below 1.0 mg/l. Treated wastes are collected in a recirculation tank and pumped at turbulent flow and 10 to 35 psig through a series of capsules containing tubular filters. The majority of the recirculated wastewater plus all the suspended solids returns to the recirculation tank. The suspended solids free water will pass through the side walls of the tubular filter. Sludge is withdrawn near the discharge of the recirculation pump. The 3% (total suspended solids) sludge is pumped to a sludge holding tank and then pumped to a filter press. (See Fig. 55.7.)

55.8.7 Sludge Thickening and Dewatering

The purpose of the sludge thickening tank is to increase the concentration of the 1 to 2% sludge from the clarifier to approximately 3%. This is accomplished by continuously decanting the water from the sludge thickening tank and directing it back to the rinse collection tank. The solids from the sludge thickening or holding tanks are then pumped to a filter press for dewatering to reduce sludge volume.

The total suspended solids in the sludge can be increased to about 35% with a filter press. In addition, sludge dryers are available which increase the solids concentration to about 70% by adding heat to the sludge. This reduces the volume of sludge to be hauled away.

55.9 ADVANTAGES AND DISADVANTAGES OF VARIOUS TREATMENT ALTERNATIVES

All of the techniques described in this chapter have advantages and disadvantages. None of them are appropriate in all situations. Each potential application requires specific analysis and a thorough understanding of the technical and economic issues involved prior to implementation. Table 55.2 presents a summary of the advantages and disadvantages of some of the technologies discussed in this chapter.

FLEXIBLE CIRCUITS

CHAPTER 56
FLEXIBLE CIRCUITS: APPLICATIONS AND MATERIALS

Dominique K. Numakura
DKN Research, Haverhill, Massachusetts

56.1   INTRODUCTION TO FLEXIBLE CIRCUITS

Flexible circuits are a form of printed wiring interconnect structure built on thin, flexible substrates. They are also bendable to complete 3-D (three dimensional) wiring that cannot be made by rigid circuit boards. Because of this flexibility, flexible circuits have many advantages compared to other wiring methods and they have many applications in electronic equipment that requires high-density wiring in a small space. As the samples in Fig. 56.1 show, most flexible circuits have cabling functions rather than mounting functions.

Tape automated bonding (TAB) has not been considered as a type of flexible circuit because of the slightly different manufacturing processes. However, the basic construction and materials of the final products are the same; therefore TAB is categorized as a type of flexible circuit in this book (see Fig. 56.2).

56.1.1   Advantages and Disadvantages of Flexible Circuits

A thin flexible circuit generates many supplemental advantages not available with other wiring methods (Table 56.1). On the other hand, flexible circuits also have many disadvantages due to unstable thin constructions. It is necessary to consider how to avoid the disadvantages when a flexible circuit is used in a packaging system. Otherwise overall process yield and productivity will be low, and the final cost will be relatively very high.

56.1.2   Economics of Flexible Circuits

The largest disadvantage of flexible circuits is that they cost more than rigid circuit boards of the same size. Usually, flexible circuits are larger than rigid circuit boards to allow 3-D wiring capability, and the total circuit cost increases but is often cheaper than using multiple rigid boards and associated connectors. That is, the cost of using flexible circuits must be compared to the total cost of rigid circuit boards in terms of:

- Connectors
- Wires
- Additional assembly costs

FIGURE 56.1 Various examples of flexible circuits.

FIGURE 56.2 TAB circuits. *(Courtesy of Shindo Denshi.)*

Because the total cost of a flexible circuit cannot be smaller than the cost of an equivalent rigid circuit board alone, designers should not use flexible circuits when the cost of a flexible circuit is compared with those of other wiring methods where the three-dimensional advantages are not needed, especially in the case of consumer applications that require low-cost wiring. Only when a flexible circuit is the solution for the wiring requirement should it be used (for example, when the requirement is for bendable wiring in a small space, light weight, long flex endurance, etc.). The critical point is whether the cost of flexible is smaller than the value of their performance.

TABLE 56.1 Advantages and Disadvantages of Flexible Circuits

Advantages	Disadvantages
Thin	Fragile
Light	Unstable dimensionally
Bendable	Low reliability—fatigue
Flexible	Difficult to design
Long flexing endurance	Complicated manufacturing process
3-D wiring	Difficult to handle
High-density wiring in a small space	Need special tools for assembling
Avoid connectors and soldering	Low yield in assembling
Avoid missed wiring	Difficult to rework
Low total cost, including all elements	High cost of circuit fabrication

Miniaturization of electronic products requires extremely lightweight wiring in a small space, and therefore there have been many opportunities for flexible circuits. Figure 56.3 shows an example of a flexible circuit using its advantages to be the best technical solution, and also the least expensive.

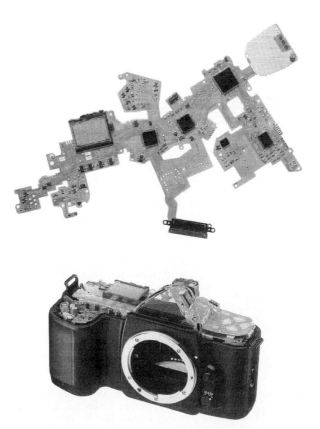

FIGURE 56.3 Main flexible circuit of a camera. *(Courtesy of Nikon.)*

56.2 APPLICATIONS OF FLEXIBLE CIRCUITS

As described, a flexible circuit should be adopted only when other wiring or packaging methods cannot satisfy the requirement. Table 56.2 shows how features of flexible circuits relate to appropriate applications. The table shows typical examples of flexible circuit applications in electronic products. Portable electronic products such as cameras, automobile instrument panels, and handheld calculators require reliable 3-D wiring in a small space. In some cameras, more than 10 flexible circuits are squeezed into a small space. Flexible circuits are used for the hinge wiring in many electronic products that have mechanical moving parts, such as notebook computers and palmtop computers. Many new electronic products, such as CD players and hard disk drives, small printers, and portable tape recorders, require a long-life dynamic flexible endurance. Multilayer rigid/flexible circuits have been effective in reducing wiring space and increasing wiring reliability in aerospace electronic equipment. Only flexible circuits could satisfy all these wiring requirements. Efforts to minimize the cost of flexible circuit wiring for consumer applications are ongoing. Often, the miniaturization of a consumer electronic product assumes a wiring system with multiple flexible circuits, especially if the equipment has mechanical moving portions.

TABLE 56.2 Features and Applications of Flexible Circuits

Features	Applications
High-density wiring in a small space	Camera, video camera, facsimile, ultrasound probe
Light weight	Calculator
3-D wiring	Automobile instrument panel, camera, notebook computer
Reliable wiring	Aerospace applications, industrial applications
Long flexing endurance	Printer, hard disk drive, floppy disk drive, CD drive, VCR

56.3 HIGH-DENSITY FLEXIBLE CIRCUITS

Serious progress in the density of flexible circuits was made during the second half of the 1990s owing to the extreme miniaturization of electronic products. These products incorporate a new technical concept in the design of the circuits. They have a much higher density than traditional flexible circuits, with dimensions for traces, spaces, and vias equivalent to those of rigid HDI boards. (Details are described in Chap. 57). They also have supplemental constructions on the circuit to allow higher termination capabilities. This type of construction is called HDI flexible circuits (i.e., flexible circuits for high-density interconnects).

56.3.1 Specifications

Table 56.3 and Fig. 56.4 show typical requirements for interconnection in the end product, and the preferred specification of circuit parameters for the use of high-density flexible circuits. Portable electronic products are exposed to more severe conditions than ever before. As a result, they need higher reliability for the circuits as listed in Table 56.3. In the case of portable telephones, they are often carried in the owner's pocket, where they may be exposed to a variety of conditions, including continuous vibration, mechanical shock, temperature, humidity, pocket lint, and even perspiration.

FIGURE 56.4 High-density double-sided flexible circuit with COF for mobile phone.

TABLE 56.3 Common New Requirements for High-Density Flexible Circuits

Requirements for end use	Preferred specification for flexible circuits
High-density wiring	Pitches smaller than 150 μm, Via holes smaller than 200 μm in diameter
Thin and light	Base 25 μm or thinner, conductor thinner than 12 μm
High-density SMT	Small opening on coverlay, tin or lead-free plating
High-density connection	High dimensional accuracy (~0.3-mm diameter)
Direct bonding	High-density flying lead, pitch ~ 100 μm
Chip-on-flex, flip-chip	High-density microbump array, pitch ~ 100 μm
High reliability for vibration	High bond strength of conductors
For mechanical shock	High bond strength between layers
For wide-range heat cycle	High heat resistance
For salty moistures	High insulation resistance

Other portable electronic products such as palmtop computers and notebook computers could be exposed to similar conditions.

56.3.2 Applications

Due to usage and assembly technologies, many applications require increased performance for high-density flexible circuits. The manufacturing processes for these circuits entail new requirements due to additional process conditions. Table 56.4 shows a list of applications and the associated advantages of using high-density flexible circuits for a given purpose.

TABLE 56.4 Major Applications of High-Density Flexible Circuits

Applications	Reasons to use high-density flexible circuits
Wireless suspension of disk drives	High-density wiring in a small space Flexible, thin, light, 3-D wiring Direct connection to MR heads
Interconnections of disk drive actuators	High-density wiring in a small space Flexible, thin, light, 3-D wiring Repeated reparability
Interposer of CSP	High-density wiring in a small space Thin, light, 3-D wiring
Flat-panel displays	High-density wiring in a small space Flexible, thin, light, 3-D wiring Chip-on-flex capability High-density connection
Ultrasound probes	High-density wiring in a small space Flexible, thin, light, 3-D wiring Direct connection to ultrasound device
Mobile phones	High-density wiring in a small space Flexible, thin, light, 3-D wiring SMT and COF capability
Inkjet printers	High-density wiring in a small space Flexible, thin, 3-D wiring Direct connection to printer head device

Without high-density flexible circuits, these applications could not achieve miniaturization or support valuable new functions. For example, specially developed high-density flexible circuits can provide highly reliable connections at a low cost for new MR head and giant MR technologies. Chip-scale package (CSP) technologies can be developed with high-density flexible circuits. Driver ICs for LCDs in the notebook PC are assembled with a set of TAB circuits. Due to the limited space within the camera bodies, small LCD monitors for video cameras and digital cameras are assembled using long-tailed high-density flexible circuits. Next-generation portable telephones will be wired by high-density flexible circuits.

56.4 MATERIALS FOR FLEXIBLE CIRCUITS

The basic difference between flexible circuits and rigid circuit boards is the thin and flexible materials used in the substrates of flexible circuits. Furthermore, because the flexible circuits have complicated constructions, materials other than copper-clad materials have been required to build a whole circuit.

56.4.1 Traditional Flexible Circuit Materials

Table 56.5 shows traditional major materials and typical examples of flexible circuits.

TABLE 56.5 Traditional Major Materials Used to Build Flexible Circuits

Application	Typical materials
Base substrates	Polyimide film (PI), polyester film (PET), polyester telephthalate Thin glass-epoxy, FEP, resin-coated paper
Conductors	ED copper foil, RA copper foil, stainless steel foil, aluminum foil, etc.
Copper-clad laminates (adhesive base)	Epoxy based, acrylic based, phenol based
Coverlay	Polyimide film (PI), polyester film (PET), flexible solder mask
Adhesive layer	Acrylic resin, epoxy resin, phenol resin, pressure-sensitive adhesives (PSAs)
Stiffener	Polyimide film (PI), polyester film (PET), glass-epoxy, metal boards, etc.

56.4.2 HDI-Oriented Flexible Circuit Materials

Since 1990, many new materials for flexible circuits have been developed in conjunction with design and manufacturing processes to satisfy new requirements in high-density interconnects (HDI) applications. These include:

- High-performance films
- Adhesiveless laminates
- Direct cast processes of liquid polyimide resins
- Photoimageable coverlay

See Table 56.6 for a list of new materials for HDI flexible circuits. There is a significant difference between traditional flexible circuits and HDI flexible circuits. For example, more than 80 percent of traditional flexible circuits were covered by adhesive-based copper-clad laminates that use the traditional polyimide films Kapton H™ or Apical AV™. Conversely, the majority of high-density flexible circuits used in large-volume applications utilize all new materials.

TABLE 56.6   Materials for High-Density Flexible Circuits

Application areas	New materials for high-density flexible circuits
Base dielectrics	Polyimide films (Kapton K, E, EN, KJ; Apical NP, HP; Upilex S) Liquid polyimide resin, PEN film, LCP films
Conductor materials	Ultrathin copper foils, sputtered copper, copper alloys Stainless steel foil
Copper-clad laminates	Adhesiveless laminates (cast type, sputtered/plated type, laminated type)
Coverlay	Photoimageable coverlay (PIC) (dry film type, liquid ink type)
Adhesive sheets	Hot-melt polyimide film

HDI applications require finer traces and microvia holes in severe manufacturing and application conditions. For manufacturing convenience, flexible circuit producers prefer thinner conductors and substrates. However, physical performances of thinner materials may not be optimal. Thinner materials impact both the performance and manufacturing yield of the final flexible circuit.

Dimensional stability of the materials in high-density flexible circuits in manufacturing is a key to good process yields. In order to produce reliable high-density flexible circuits, conflicts between materials, constructions, and manufacturing processes must be resolved. Measures to accomplish this objective have been undertaken. For example, new construction designs and manufacturing technologies now require use of the materials shown in Table 56.6.

56.5   SUBSTRATE MATERIAL PROPERTIES

Many kinds of materials have been developed as the major dielectric layers of flexible circuits; however, only polyimide films and polyester films have been the major core material for both substrate films and coverlay films with adhesive for traditional flexible circuits. The polyimide films have been the major substrate material for soldering applications because of their high temperature resistance and good balance of physical properties. Polyester films are not available for standard soldering, but they are a low-cost solution for large circuits such as those in automobile instrument panels and long cable circuits of printers.

56.5.1   Comparison of Substrate Materials

There have been examples of other materials, such as thin glass-epoxy, that have not had wide use. In order to satisfy the new concept of HDI flexible circuits, innovative uses of raw materials such as polyethylene naphthalate (PEN) film and liquid crystal polymer (LCP) film have been introduced. Comparisons are shown in Table 56.7.

56.5.2   Polyimide Film

Polyimide film continues to be the major substrate material for flexible circuits because only it can support high-temperature processes such as soldering and wire bonding. Both Kapton H film and Apical AV by Kaneka have been used in the flexible circuit industry for a long time. Both of these materials have a good property balance, and they are still the major substrates in traditional flexible circuits.

TABLE 56.7 Comparison of Substrate Materials

	Polyimide	Polyester	Thin glass-epoxy	PEN	LCP
Maximum operating temperature	>200°C	<70°C	~105°C	~90°C	>200°C
Standard thickness	12.5, 25, 50, 75, 125 μm	25, 50, 75, 100, 125, 188 μm	100, 150, 200 μm	(50 μm)	(50 μm)
Soldering	Applicable	Difficult	Applicable	Possible	Possible
Wire bonding	Possible	No	Difficult	Difficult	Possible
Color	Brown	Transparent	Transparent	Transparent	Milky
Moisture absorption	High	Low	Low	Low	Low
Dimensional stability	Acceptable	High	High	Good	High
Flexibility	High	High	Low	High	High
Cost	High	Low	Medium	Low	High

56.5.2.1 *Properties of Polyimide Films.* Table 56.8 shows major properties of typical poly-imide films, Kapton H and Apical AV. These have good mechanical properties and provide excellent performance as base substrate films and coverlay films in dynamic flexing applications. They have flame-retardant characteristics, and it is not difficult to achieve UL flame class 94-V-0 or 94-VTM-0. The largest disadvantage of polyimide film is its higher cost compared to other common plastic films such as PET.

There are several barriers to traditional polyimide films being the major substrate materials in HDI flexible circuits. Dimensional stability is the largest issue. Both Kapton H film and Apical AV by Kaneka have CTEs higher than 30 ppm, which is not acceptable for large-volume production processing of HDI flexible circuits. Relatively high moisture absorption is another major issue. High moisture levels in the films cause a lot of supplemental problems, mostly in increased temperature processes. Higher-performance polyimide films are required as the substrate material of HDI flexible circuits.

TABLE 56.8 Properties of Traditional Polyimide Films and New Polyimide Films

		Kapton H™	Apical AV™	Kapton E™	Apical NP™	Apical HP™	Upilex S™
Manufacturer		DuPont	Kaneka	DuPont	Kaneka	Kaneka	Ube
Thickness (μm)		12.5, 25, 50, 75, 125	12.5, 25, 50, 75, 125	12.5, 25, 50, 75	12.5, 25, 50, 75	12.5, 25, 50, 75	12.5, 25, 50, 75, 125
Tensile strength	MD	25.2	25.0	28.3	30.0	28.6	39.4
(kg/mm²)	TD	22.3	27.0	25.4	32.0	29.0	40.2
Elongation (%)	MD	85	119	16	82	40	22
	TD	83	114	32	73	38	21
Tensile modulus	MD	336.1	305.9	785.5	407.9	654.0	897.3
(kg/mm²)	TD	321.4	312.2	622.4	428.3	661.0	912.1
CTE (ppm) (100–200°C)	MD	27	35	3	17	12	14
	TD	31	31	12	13	11	15
CHE (ppm) (50°C,	MD	15	16	5	13	8	9
35–75% RH)	TD	16	15	8	12	6	10
Heat (%) shrinkage	MD	0.18	0.08	0.03	0.06	0.06	0.07
(200°C, 2 h)	TD	0.20	0.03	0.02	0.02	0.02	0.10
Water absorption (23°C, 24 h)		2.8	2.9	2.2	2.5	1.1	1.9
Alkaline resistance		17.4	25.9	5.5	22.9	4.2	~0
Chemical etching		Possible	Possible	Possible	Possible	Possible	Difficult

56.5.2.2   High-Performance Polyimide Film.   There have been several aggressive measures to commercialize high-performance polyimide films by various manufacturers. Upilex S by Ube Industry has excellent chemical stability, and it is a barrier for chemical processes such as bonding or chemical etching. Strong chemicals, such as hidrasine, are required to do the chemical etching. Upilex could be the major material as the substrate of TAB, which requires very high dimensional stability at high heat resistance. Newly developed polyimide films such as Kapton K, E, EN, etc., and Kaneka's Apical NP and HP, have higher dimensional stability and lower moisture absorption rates. It is possible to etch these films using mild chemistry. Basic properties of the materials are summarized in Table 56.8.

Hot-melt-type (thermoplastic) polyimide films have been developed as high-heat-resistant adhesive materials. TPI™ by Mitsui Chemical, Kapton KJ by DuPont, Upicel by Ube, and Pixeo by Kaneka are typical examples. These are coated on dimensionally stable polyimide films to ensure good physical performance.

56.5.2.3   Liquid Polyimide Resin (Photoimageable).   Many new HDI applications require special capabilities and properties on the substrates that are not provided by commercial polyimide films. In these cases, circuit manufacturers initiate production by manufacturing polyimide substrates. Sometimes manufacturers have to start from the synthesis of polyimide oligomers. Several liquid polyimide resins have been developed as the base materials of high-density flexible circuits. Some of these liquid polyimide resins can be photoimaged, and have been used in high volume as the dielectric layer and coverlay in the wireless suspension of hard disk drives. These liquid polyimide resins could be the major dielectric materials of special high-density flexible circuits that demand extremely high density. Compared to that of polyimide film, the cost of these materials is high. However, they broaden capability for nonstandard requirements such as ultrathin substrates. The properties are very dependent on the manufacturer.

56.5.3   Polyester Film

PET film (polyester film or polyethylene telephthalate) has been produced as a common film for general use for the following reasons.

- It can provide a low-cost solution.
- It has excellent mechanical performance at room temperature.
- It has low moisture absorption and its dimensional stability is excellent.

Unfortunately, PET film cannot keep its high performance at high temperatures, and it is available for only nonsoldering applications. (There are several soldering technologies that have been developed for PET-based flexible circuits, but they are very specialized.) Also, PET films are flammable and it is difficult to achieve UL flame class certification. There are not many available PET films that can satisfy the flammability requirements of UL-94. Typical properties are shown in Table 56.9.

56.5.4   Other Materials Used for Flexible Circuits

Thin glass-epoxy sheets less than 200 μm thick can be flexible and can be used as a base substrate of a flexible circuit. Their physical properties are basically the same as those of rigid circuit boards, and standard manufacturing processes and assembling processes for rigid circuit boards are available for these materials. Standard soldering processes are applicable. Glass-epoxy sheets do not have high flexibility, and they are not available for repeated flexing use. The basic properties of these materials are listed in Table 56.10.

Fluorized carbon polymer films have been applied as a low-loss substrate material for flexible circuits. However, they do not have good dimensional stability or adhesion characteris-

TABLE 56.9   Basic Properties of PET Films (Nonflammable Grade)

Test Items	Properties	
	MD	TD
Tensile strength (kgf/mm^2)	16	16
Elongation	110%	100%
Edge tear resistance (kgf/mm)	51	58
Folding endurance	37,000 times	39,000 times
Heat shrinkage	1.4%	−0.2%
CTE	30 ppm	
CHE	10 ppm	
Water absorption	0.7%	
Volume resistivity	$4 \times \exp 15$ Ω*cm	
Dielectric constant	3.0 for 1 MHz	
Dielectric loss	0.03 for 1 MHz	
Breakdown voltage	13.6 kV	
Light transmission	46%	
Chemical resistance	Stable in weak acid and alkaline solution	
	Stable in organic solvent	
Flammability	UL-94-VTM-0	

Data provided by Mitsubishi Plastic.

TABLE 56.10   Properties of Thin Glass/Epoxy Substrate

Items	Risho Industrial		Matsushita Electric
	MD	TD	(R-5766 series)
Tensile strength (kg/mm^2)	24.5	17.0	—
Elongation (%)	3.2	1.6	—
Thickness (μm)	120		50, 70, 100, 130, 150, 180, 200
Dimension change by heat (%)	−0.05		—
Dimension change by humidity (%)	0.03		—
Surface resistance (Ω)	$1 \times \exp 12$–13		$5.2 \times \exp 14$
Volume resistivity (Ω*cm)	$1 \times \exp 12$–13		
Dielectric constant	—		4.7
Dissipation factor (at 1 MHz)	—		0.015
Water absorption (%)	1.0		0.18
Breakdown voltage (kV/0.1 mm)	3.6		—
Bond strength (kg/cm)	—		1.39
Soldering resistance	—		260°C × 120 s
Flammability	—		UL-94-V-0

These materials are supplied as copper-clad laminates.

tics. High costs of the materials are the major reason that they cannot be standard in flexible circuits. In the last 20 years several heat-resistant films, such as polypalabalic acid film and polysulfon film, have been developed as the alternative substrate materials in flexible circuits instead of polyimide films. Unfortunately, there was no successful material from a business standpoint. Nowadays, polyethylene telenaphtalate (PEN) films and liquid crystal polymer (LCP) films have been considered as the new materials of flexible circuits.

56.6 CONDUCTOR MATERIALS

As flexible circuits have various mechanical stresses, their conductors also have requirements for more flexibility and toughness than traditional ED copper foils of standard rigid circuit boards. A rolled annealed (RA) copper foil is the solution for high flexing endurance. It is necessary for dynamic flexing use. RA copper foils are more expensive than ED copper, especially at smaller thicknesses. Therefore, high-ductility ED (HD-ED) copper foils have been developed that have a higher flexibility than ED copper foils and a lower cost than RA copper foils. The differences between the copper foils are shown in Table 56.11.

TABLE 56.11   Basic Properties of Copper Conductors

	ED foil	HD-ED foil	RA foil
Manufacturer	Furukawa Electric	Furukawa Electric	Japan Energy
Grade	STD foil	HD foil	RA
Thickness (μm)	9, 12, 18, 35, 70	18, 35	12, 18, 35, 70
Tensile strength (kg/mm^2)	TD 34	TD 32	MD 21.5
			TD 18.9
Elongation (%)	TD 9	TD 23	MD 12.8
			TD 9.5
Surface roughness (μm)	8	10	<3.5
Cost	Low	Middle	High

Data were measured for 35-μm foils

Due to the stringent requirements of fine-line etching, thinner copper foils have been developed with small surface profiles. RA copper foil 12 μm thick is a standard product. Low-profile 12-μm-thick ED copper foils have also been commercialized for fine-line etching. A special treatment has been developed to ensure reliable bond strength with each adhesive material. In the next 5 years, these materials will be standard and utilized in large volume. RA copper foils thinner than 10 μm are available; however, processing and high material costs need to be addressed before large-volume production can commence.

Sputtering and plating technologies have achieved conductors thinner than 5 μm with additive and semiadditive processes. Among others, these processes can generate copper, nickel, and gold conductors. In addition, they can produce fine traces at less than 10-μm pitch on flexible substrates. Plating baths and chemicals have been developed for these technologies.

Special metal and alloy foils other than copper ones have been developed for specific applications. Aluminum conductors have been developed as the low-cost solution. They have been applied for volume production of calculators, antennas, etc. But they could not be universal conductor materials of standard flexible circuits because of difficulty in soldering and the special chemistries needed for etching. The wireless suspension of disk drives has consumed a large volume of special stainless steel and copper alloy foils because of the special mechanical performance of the circuits. A high-resolution printer head has also utilized thin tungsten foils as the conductor of thermal printer head circuits. Nickel/chromium alloy foils have been developed as the conductor materials of flexible heater circuits.

56.7 COPPER-CLAD LAMINATES

The majority of flexible circuit manufacturers start the process with copper-clad materials. The properties of these materials depend on the capabilities of laminate manufacturers, even though the same base films and copper foils are used. The basic properties of each laminate material must be considered carefully.

56.7.1 Adhesive-Based Laminates

Historically, copper-clad laminates with acrylic or epoxy adhesives have been the major materials for flexible circuits. A special resin grade or special additives have been developed by each manufacturer to ensure reliable flexibility and bond strength. Other adhesive materials have been developed; however, they have not become standard adhesive materials in flexible circuits. The adhesive-based copper-clad laminates still represent more than 80 percent of the traditional flexible circuit market. The major properties of the materials are shown in Table 56.12.

TABLE 56.12 Basic Properties of Adhesive-Based Copper-Clad Laminates (Polyimide Base)

Items	Properties		
Manufacturer	DuPont	Nikkan	Shin-Etsu
Grade	Pyralux LF™	Nikaflex™	RAR Series™
Adhesive layer	Acrylic	Epoxy	Epoxy
Peel strength (kgf/ccm)	1.4	1.3	1.8
Dimensional stability MD	−0.08	−0.09	−0.09
(%) TD	−0.07	+0.03	−0.04
Flexing endurance MD	N/A	3200	2650
(MIT, 2.0R) TD		2950	2850
Insulation resistance (Ω)	$1.0 \times \exp 11$	$2.5 \times \exp 13$	$1.0 \times \exp 13$
Surface resistivity (Ω)	$1.0 \times \exp 13$	$2.7 \times \exp 14$	$1.0 \times \exp 14$
Volume resistivity (Ωcm)	$1.0 \times \exp 14$	$2.0 \times \exp 16$	$1.0 \times \exp 16$
Soldering resistance	288°C × 5 min	280°C × 10 s	280°C × 10 s
Flammability	No	UL-94-VTM-0	UL-94-VTM-0

The manufacturing process is illustrated in Fig. 56.5. Mostly, these films are processed in roll form. The surfaces of polyester films and polyimide films undergo special processes such as sandblasting and plasma treatment to achieve a reliable bond strength. A specially blended

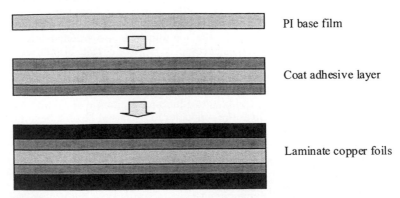

PI base film

Coat adhesive layer

Laminate copper foils

FIGURE 56.5 Manufacturing process for adhesive-type laminate, showing a polyimide-based film combined with an adhesive layer coat and laminated to copper foils.

adhesive resin is coated and dried on the film. Then a copper foil is laminated continuously under appropriate temperature and pressure. The surfaces of the copper foils receive a specific treatment according to the requirements of each laminate's manufacturer. The same process is repeated for double-sided copper-clad laminates. (Some laminate manufacturers, however, have developed simplified manufacturing processes that can make the double-sided laminate in one single process to reduce manufacturing cost.) A well-conditioned aging process is important to achieving reliable bond strength and flexible characteristics of the laminates as the raw material of flexible circuits.

All adhesive resins have lower heat resistances than polyimide films, and are the bottleneck for high-temperature processing of flexible circuits (e.g., lead-free soldering and wire bonding). The flame-retardant properties of the laminates depend on the composition of the adhesive materials used by each manufacturer. Usually, a normal flame-retardant component in an adhesive resin has a negative effect on bonding. Several adhesive resins contain halogens as the flame-retardant components, and will be eliminated for the ecological concerns. (For a discussion of this issue, see Chap. 6). As soldering temperatures increase, heat-resistant adhesives or adhesiveless laminates systems are required.

56.7.2 Adhesiveless-Based Laminates

Several laminates without adhesive layers have been developed as the advanced materials for next-generation flexible circuits. Lamination technology using epoxy resin or acrylic resin has been almost eliminated from HDI flexible circuits even though it uses new high-performance polyimide films as the substrates. Three types of adhesiveless copper-clad laminates have been developed (see Fig. 56.6):

- Cast
- Sputtering/plating
- Lamination

Each has different manufacturing processes and advantages as the major materials of HDI flexible circuits. Table 56.13 shows basic performances of typical adhesiveless laminate materials. Presently, there is no perfect solution that satisfies all requirements including cost. Suitable adhesiveless laminate materials should be chosen according to specifications of the final flexible circuits and convenience of manufacturing processes.

56.7.2.1 *Cast-Type Adhesiveless Laminates.*   The cast-type adhesiveless laminates have a good cost-performance balance. They have a high bond strength between substrate layers and conductor foils. They can also use special conductors such as copper alloy, nickel, stainless steel, etc. In addition, they have good flexibility to generate very thin substrate layers. They can have a wide range of copper thicknesses because of the process. There is no difficulty in making 70- or 105-μm copper laminates. The cast-type laminate has a relatively broader availability than the other adhesiveless laminates because it has existed longer. The process can produce both single-sided and double-sided laminates. The polyimide oligomer is coated on a metal foil, directly controlling mechanical stress on the materials. Air-floating conveyors have been employed to reduce the stress. A thermal treatment process is the key to ensuring good physical performance of the laminates. Generally, two or more polyimide layers are coated on the metal foil to ensure both good dimensional stability and high bond strength.

56.7.2.2 *Plating-Type Adhesiveless Laminates.*   The plating-type adhesiveless laminates can generate the thin conductors that are required for high-density circuits. These laminates provide stable conductor layers thinner than 10 μm at low cost. Technically, it is very possible to have 0.1- to 1.0-μm-thick conductors. Although the plating-type laminates provide more

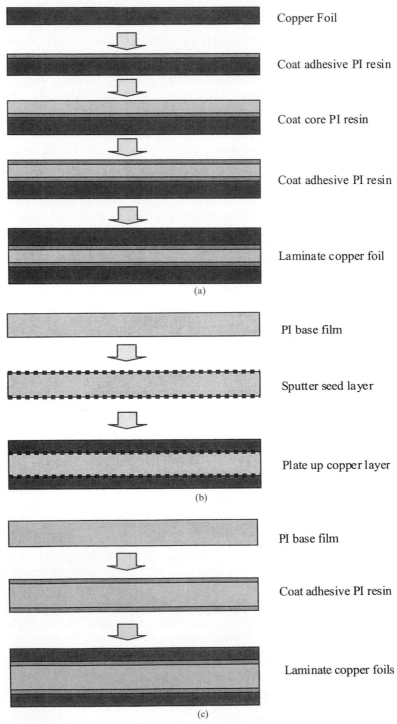

FIGURE 56.6 (*a*) Manufacturing process of adhesiveless laminate using the cast process. (*b*) Manufacturing process of adhesiveless laminate using plating process. (*c*) Manufacturing process of adhesiveless laminate using lamination process.

TABLE 56.13 Comparison of Adhesiveless Copper-Clad Laminates

Items	Sputtering/plating process	Cast process	Lamination process
Manufacturer	Gould Electronics	Nippon Steel Chemical	DuPont
Grade	GouldFlex™	Espanex™	Pyralux AP™
Choice of dielectric	Wide	Small	Small
Base etchability	Fair	Difficult	Difficult
Dielectric thickness (μm)	12.5–125	12.5–75	12.5–150
Choice of conductors	Small	Wide	Wide
Conductor thickness (μm)	~35	12–70	12–70
Bond strength	1.2	1.5	1.4
Double-sided	Available	Available	Available
Flexing endurance	N/A	180 (MIT 0.8R)	N/A
Dimensional change after etching	MD: −0.05%	MD: −0.02%	MD: −0.05%
	TD: 0.03%	TD: −0.02%	TD: −0.05%
Insulation resistance (Ω)	>1 × exp 10	>1 × exp 13	>1 × exp 14
Volume resistivity (Ωcm)	>1 × exp 14	>1 × exp 15	>1 × exp 17
Flammability	UL-94-VTM-0	UL-94-VTM-0	UL-94-VTM-0
Roll clad	Standard	Standard	Not available

substrate material choices, their basic physical properties depend heavily on substrates. The majority of manufacturers introduced the sputtering process to generate a seed layer for good bond strength between the substrate and the conductor. Sometimes seed layers may cause problems in the etching process because they cannot be etched.

There are small choices for the conductor materials. It is not optimal to make thick conductors, because this requires a longer plating process. Micropinholes in the conductor layer are the biggest issue with this material. Several manufacturers have introduced additional chemical processes to eliminate these pinholes.

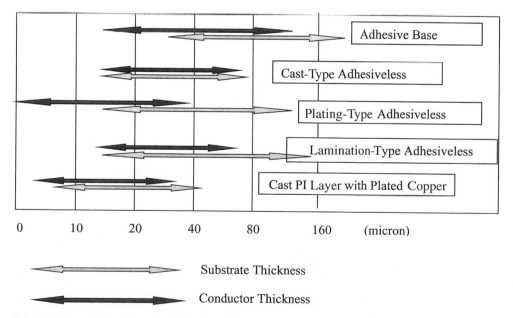

FIGURE 56.7 Thickness ranges of the laminates.

56.7.2.3 Lamination-Type Adhesiveless Laminates. Lamination-type adhesiveless laminates have advantages in small-volume production. Special polyimide films that have hot-melt polyimide resin on the surface of dimensionally stable polyimide film are available commercially. If a manufacturer has high-temperature heat press equipment, it is not difficult to produce the laminates. Lamination-type adhesiveless laminates allow many choices of conductor materials. This material has a variety of uses in multilayer rigid/flexible aerospace applications and in wireless suspension of disk drives due to the special combination of conductor materials with stainless steel and copper alloy foils. The significant investment required to produce roll cladding has limited its availability. The thickness ranges of substrates and conductors for each type material are illustrated in Fig. 56.7.

56.8 COVERLAY MATERIALS

Coverlay is one of the major differences between flexible circuits and rigid circuit boards. In addition to the solder mask used for assembly, coverlay is the mechanical protector for the fragile conductors on flexible circuits. Film-based coverlay and flexible solder mask have been standard materials for traditional flexible circuits. Several types of photoimageable coverlay materials have been developed to satisfy the severe requirements of HDI flexible circuits:

- Film coverlay
- Screen-printable coverlay (flexible solder mask)
- Dry film–type photoimageable coverlay
- Liquid-base photoimageable coverlay

56.8.1 Film Coverlay

Usually, the same films are chosen for the coverlay films as are used for substrates. The films are coated with semicured acrylic or epoxy resin, which is used as an adhesive layer. The surfaces of the adhesive layers are covered with release films when the materials are delivered to circuit manufacturers.

The process of manufacturing film coverlay is very complicated, which makes automation difficult and increases the cost. The manual registration process and the use of unstable film material are major issues when considering film coverlay for HDI flexible due to small hole capability and lower dimensional accuracy. Basic properties of the typical materials are shown in Table 56.14.

TABLE 56.14 Basic Properties of Film Coverlay (Polyimide Film Base)

Manufacturer	DuPont	Nikkan Industry
Grade	Pyralux LF	CISV
Adhesive	Acrylic	Epoxy
Lamination temperature	180°C	160°C
Storage condition	Room temperature	Refrigerator
Bond strength (kg/cm)	1.3	1.0
Surface resistivity (Ω)	$1.0 \times \exp 13$	$3.0 \times \exp 14$
Volume resistivity (Ω*cm)	$1.0 \times \exp 14$	$1.2 \times \exp 16$
Soldering resistance	288°C × 5 min	280°C × 10 s
Flammability	No	UL-94-V-0

56.8.2 Screen-Printable Coverlay Ink (Flexible Solder Mask)

Screen printing of the flexible solder mask is a possible low-cost solution, but it does not satisfy the requirements of HDI flexible circuits because of the poor resolution. Also, it does not provide good mechanical performance for dynamic flexing. Typical property examples are shown in Table 56.15.

TABLE 56.15 Basic Properties of Screen-Printable Coverlay

Manufacturer	Taiyo Ink	Nippon Polytech
Grade	S-222	NPR-5
Base resin	Epoxy	Epoxy
Color	Green	Green
Hardness	5H	H
Peel strength	100/100 (crosshatch)	100/100 (crosshatch)
Flexibility	N/A	0.5 mm diameter
Soldering resistance	260°C × 20 s	260°C × 20 s
Insulation resistance (Ω)	>1 × exp 13	>1 × exp 12
Flammability	UL-94-V-0	UL-94-V-0

56.8.3 Dry Film–Type Photoimageable Coverlay

From a processing standpoint, there are two types of PIC materials: dry film and liquid. The dry film type is easily processed if manufactured with vacuum laminators. However, the cost per unit area is higher for the dry film type than for liquid materials.

Dry film–type and liquid ink–type photoimageable coverlay materials have been developed with different resin matrixes to satisfy different technical requirements of HDI flexible circuits. A comparison of the materials is given in Table 56.16. Properties of the materials are shown in Tables 56.17 and 56.18.

TABLE 56.16 Comparison of Photoimageable Coverlay Materials

	Dry film type		Liquid ink type	
	Acrylic base/epoxy base	Polyimide base	Epoxy base	Polyimide base
Applicator	Vacuum laminator	Roll laminator	Screen print Spray coat Curtain coat	Screen print Spray coat Roll coat
Thickness range (μm)	25–50	25–50	10–25	10–20
Minimum opening (μm)	70	70	70	50
Flexibility	*	Good	*	Good
Heat resistance	Acceptable	Excellent	Acceptable	Good
Electrical properties	*	Good	*	*
Chemical properties	Acceptable	Good	Good	Excellent
Handling	Easy	Difficult	Fair	Difficult
Technical hardle	Low	High	High	Very high
Storage condition	Refrigerator	Room temperature	Refrigerator	Freezer
Quick turn	Easy	Easy	Slow	Slow
Material cost	Medium	High	Low	High
Process cost	Medium	High	Low	Low

* Depends on manufacturer.

TABLE 56.17 Properties of Liquid/Epoxy-Type Photoimageable Coverlay

Manufacturer	PolyTech	Coates	Asahi	Peters
Grade	NPR-80	Image Flex™	Focus Coat™	Elpemer™
Pencil hardness	5H	H, HB	4H	6H
Bond strength on copper	100/100	100/100	100/100	100/100
Flexibility	MIT test 0.5 mm R >500 times	1-mm mandrel: pass	180° bend: pass	3-mm mandrel: pass
Insulation resistance (Ω)	$2 \times \exp 12$ (IPC)	$5 \times \exp 11$ (IPC)	$6 \times \exp 13$ (JIS)	$2 \times \exp 14$ (VDE)
Volume resistance (Ωcm)	—	—	$5 \times \exp 14$	$1 \times \exp 15$
Dielectric strength	28 kV/mm	91 kV/mm	(2.1 kV/0.16 mm)	88 kV/mm
Soldering resistance	260°C × 10 s	260°C × 10 s	260°C × 1 min	288°C × 20 s
Flammability	UL-94-V-0	UL-94-VTM-0	N/A	UL-94-V-0
Resistance in Ni/Au plating	Pass	Pass	Pass	Pass
Exposure intensity (mJ/cm²)	150	400	300	150
Developer	Sodium carbonate	Sodium carbonate	Sodium carbonate	Sodium carbonate

56.8.3.1 Acrylic or Epoxy/Dry Film Type. Acrylic or epoxy/dry film types were the first materials developed to act as the flexible photoimageable coverlay for HDI flexible circuits. They have the same product concept as the photoimageable solder mask of rigid circuit boards. The same vacuum laminators and imaging equipment are available for these materials. The conditioning of the process is not difficult. They are good for small-volume production because of the flexible manufacturing process. They may need some more improvements in electrical performance and chemical resistance to be used in the general application of high-

TABLE 56.18 Properties of Different Photoimageable Coverlay Materials

	DuPont	NSCC	Nitto Denko	Toray
Material	Puralux PC™ Acrylic/dry film	Espanex SFP™ Polyimide/dry film	JR-3000 Polyimide/liquid ink	Photoneece™ Polyimide/liquid ink
Pencil hardness	3H	—	—	—
Tensile strength (MPa)	N/A	226	120	148
Elongation	>55%	—	11%	36%
CTE (ppm/K)	130	23	35	16.1
Bond strength on copper	100/100	0.8 kg/cm	100/100	100/100
Flexibility	MIT 0.38R 100 times	MIT 0.38R 1200 times	N/A	N/A
Surface resistance (Ω)	$>1.0 \times \exp 12$	$>1.0 \times \exp 13$	—	$>1.0 \times \exp 16$
Volume resistance (Ω*cm)	$3.4 \times \exp 16$	$1.0 \times \exp 16$	$5 \times \exp 15$	$>1.0 \times \exp 16$
Dielectric strength (kV/mm)	>80	—	240	>300
Dielectric constant	3.5–3.6	3.5	3.3	3.2
Loss tangent	0.03	0.007	0.6	0.002
Soldering resistance	260°C × 10 s	350°C (JIS)	>300°C	>300°C
Flammability	UL-94	UL-94-VTM-0	Self-distinguishable	Self-distinguishable
Resistance in Ni/Au plating	Pass	Pass	Pass	Pass
Exposure intensity (mJ/cm²)	200	400	500	150
Developer	Sodium carbonate	Lactic acid solution	Alkaline solution	Special chemical

The materials were developed assuming different applications; therefore measured data are not exactly equivalent.

density flexible circuits. The present version of the materials, however, gives good balanced performances for both high-density soldering and flexing endurance.

56.8.3.2   *Polyimide/Dry Film Type.*   The Espanex™ SFP series by Nippon Steel Chemical is the only example of this combination. Its physical performance is excellent. When applied to polyimide-base adhesiveless laminates, it provides very high heat resistance and good dimensional stability. Also, it has high electrical properties. It has been applied in high-density wiring of interconnection in the head suspension of disk drives. The major issues with this material are the complicated pattern generation process and high material cost. It requires multiple chemical processes using special solutions.

56.8.4   Liquid-Base Photoimageable Coverlay

Eventually, photoimageable coverlay could be the practical solution for HDI flexible circuits in large-volume production. It is similar to the solder mask of rigid circuit boards; however, the materials for flexible circuits must have high flexibility. All of the materials developed as solder masks for rigid boards are not available for flexible circuits, even though dynamic flexing is not required. The liquid PIC requires special coating equipment such as a screen printer or sprayer. It can, however, provide a low-cost coverlay for large-volume production.

56.8.4.1   *Epoxy/Liquid Ink.*   This material combination has been providing the best cost-performance combination for large-volume application. There have been a lot of manufacturers who have tried to develop a good photoimageable coverlay with this combination. Several successful examples are shown in Table 56.17. They provide good performance balances in mechanical, electrical, and chemical properties. A specialty is that these materials have high flexibility without serious curling during the thermal processing compared to other materials. These are available by screen printing, spray coating, and other coating methods. There are UL-certified grades available for commercial use.

56.8.4.2   *Polyimide/Liquid Ink.*   This material concept was generated as the reliable thin dielectric layer of flexible circuits for both substrates and coverlay. It has high reliability in wide thickness ranges. A 10-μm-thick layer could have sufficient performance to be a coverlay of a high-density flexible circuit. The major issues are short pot life and high cost. The ink materials must be kept at a temperature lower than 0°C. The materials require severe process conditions, especially curing. They require a temperature higher than 300°C. The cost of materials is 5 to 10 times higher than for epoxy-based materials.

56.9   STIFFENER MATERIALS

All kinds of sheet or board material could be used as stiffener materials for the flexible circuits; however, there are several common materials. Typical materials are listed in Table 56.19.

TABLE 56.19   Comparison of Stiffener Materials

	Paper/phenol	Glass-epoxy	Polyimide	Polyester	Metal
Thickness range (μm)	500–2500	100–2000	25–125	25–250	Very wide
Soldering	Possible	Possible	Possible	No	Possible
Adhesion by thermoset resin	No	Possible	Possible	No	Possible
Cost	Low	High	Medium	Low	Medium

56.10 ADHESIVE MATERIALS

There are several different types of adhesive materials that can be used in flexible circuits depending on the purposes. Table 56.20 shows comparison of the materials.

TABLE 56.20 Comparison of Adhesive Materials for Flexible Circuits

	Thermosetting epoxy and acrylic resin	Hot-melt-type polyimide resin	Pressure-sensitive adhesives (PSAs)
Reliability	High	High	Acceptable
Bond strength	High	High	Acceptable
Creep	Small	Small	Large
Soldering	OK	OK	OK
Application process	Complicated	Complicated	Simple
Process temperature	160–180°C	>330°C	Room temperature
Material cost	Low	High	Low
Total cost	High	Very high	Low

CHAPTER 57
DESIGN OF FLEXIBLE CIRCUITS

Dominique K. Numakura
DKN Research, Haverhill, Massachusetts

57.1 INTRODUCTION

Because of the broad wiring capabilities, it is possible to cover all 3-D wiring of an electronic product by one flex circuit or by a multilayer rigid/flex circuit without connectors and extra cables. However, the shape of the flex circuit may be very complicated and the size may be large. As a result, the final cost could be extremely high. If so, the use of flexible circuits is not an acceptable solution, especially for large-volume consumer applications. It may be better to divide the wiring into two or more parts to make each circuit simpler. A combination of simple circuits could be less expensive than one flex circuit, even though it mandates several connections between different circuits. A circuit designer should also consider the manufacturing convenience of flex circuits. There are serious differences among rigid circuit boards. A small design modification can sometimes reduce the total cost significantly.

57.2 DESIGN PROCEDURE

To achieve the best cost-performance combination for a wiring design with flex circuits, an appropriate design procedure is recommended, as illustrated in Fig. 57.1. The steps in the process are as follows:

1. A design concept of an electronic product is determined according to the basic specifications for the final product.
2. A logical circuit design and electronic circuit design are formulated.
3. A design of the equipment housing is made according to the final product concept. (This should sometimes be done independently in the case of consumer applications, especially for portable electronic products.)
4. The circuit design is formulated. Appropriate circuit selection is conducted depending on total conditions such as allowed space, sizes of components, operating temperature, possible assembling process, required reliability, etc.
5. A flex circuit or a combination with a flex circuit is considered as the solution for a small wiring space.

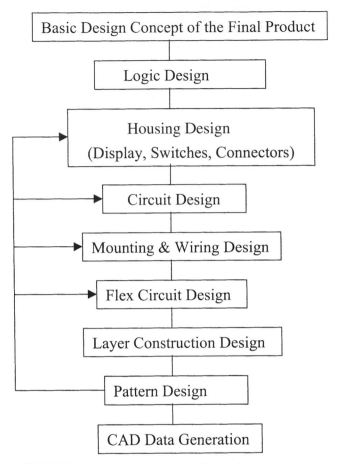

FIGURE 57.1   Design procedure for flexible circuits.

6. If the first circuit design does not satisfy the original specifications, the previous design steps are repeated to solve the conflicts.
7. A suitable value analysis is done to minimize the total wiring cost.

57.3 TYPES OF FLEXIBLE CIRCUITS

Flex circuits allow more variety in their basic structures and supplemental structures than rigid circuit boards, as shown in Table 57.1. An appropriate combination of a basic structure and supplemental structures should be chosen according to the requirements of the application.

57.3.1 High-Density Flexible Circuits

There are significant differences in design, materials, and manufacturing between traditional flexible circuits and new high-density flexible circuits developed for high-density interconnect

TABLE 57.1   Construction Types of Flex Circuits

Classification	Structure types
Basic structure	• Single-sided flexible circuits • Double-sided flexible circuits with or without through-holes • Multilayer flexible circuits with through-holes • Multilayer rigid/flexible circuits with through-holes • Blind via holes, inner via holes • Flying-lead structure
Supplemental structure	• Coverlay structure • Stiffener structure • Dimple structure • Microbump structure

(HDI). Figure 57.2 shows one way of defining the areas of high-density flexible circuits by technical hurdles of manufacturing for both trace densities and via hole sizes. Advanced etching technology is needed to give the high-density flexible circuits fine traces. Also needed are new microvia hole generation technologies other than traditional mechanical drilling, such as laser, plasma etching, and chemical etching.

Ultra-high-density flexible circuits have been developed for special applications. They have extremely fine circuits smaller than 20 μm in pitch and microvia hole connections smaller than 25 μm in diameter. Product concepts for ultra-high-density flexible circuits are very different from those for traditional flexible circuits in construction and manufacturing technologies.

New high-density flexible circuits have supplemental structures such as flying leads and microbump arrays to complete high-density terminations. As volume production is limited, the designers should consider the manufacturers' process capabilities seriously.

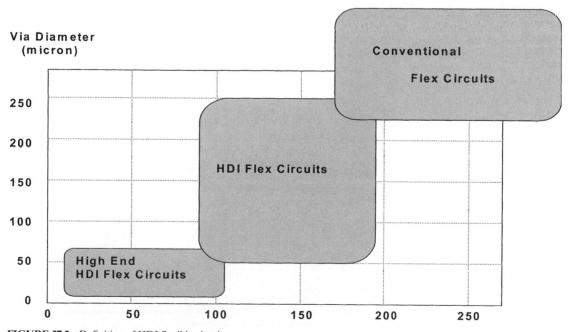

FIGURE 57.2   Definition of HDI flexible circuits.

57.3.2 Single-Sided Circuits

A bare single-sided flex circuit has a construction very similar to those of the single-sided rigid circuit boards, except for the thickness of the substrate (Fig. 57.3). As introduced in Chap. 56, there are several choices of substrate (see Table 57.2). A 25-μm-thick polyimide film substrate is the most common material for consumer applications, with large volume due to its low cost. A thicker film for higher reliability or a thinner film for higher flexibility may be chosen. The thickness of the adhesive layers must be counted when total thickness is considered. Possible fine-line capability is illustrated for each copper thickness in Fig. 57.4. Currently 18- and 35-μm copper foils are standard. Copper foils 12 μm thick are becoming the new standard to enable finer pattern etching. Thinner copper foils are available in sputtered/plated adhesiveless laminate materials. The fine trace etching capability depends very much on the manufacturer, especially the exposure process and etching process.

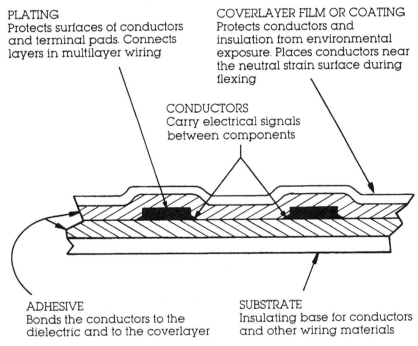

PLATING
Protects surfaces of conductors and terminal pads. Connects layers in multilayer wiring

COVERLAYER FILM OR COATING
Protects conductors and insulation from environmental exposure. Places conductors near the neutral strain surface during flexing

CONDUCTORS
Carry electrical signals between components

ADHESIVE
Bonds the conductors to the dielectric and to the coverlayer

SUBSTRATE
Insulating base for conductors and other wiring materials

FIGURE 57.3 Cross section of typical double-sided flexible circuit.

TABLE 57.2 Thickness of Base Substrates

	Base film (μm)	Adhesive (μm)
Polyester base/adhesive type	25, 38, 50, 75, 100, 125, 188	25–50
Polyimide base/adhesiveless type	12.5, 25, 38, 50, 75, 125	25–50
Polyimide base/adhesiveless type	12.5, 25, 30, 35, 40, 50	No
Glass-epoxy	100, 200	~0

Adhesive layer thickness is doubled for double-sided laminates.

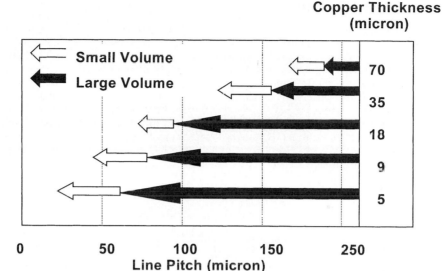

FIGURE 57.4   Fine-line capabilities. Potential line pitch is noted for copper thickness alternatives, along with whether a given thickness lends itself to high-volume production or requires too high a level of technical interaction.

57.3.3   Coverlay System of Flexible Circuits

The coverlay system of flexible circuits is one of the big differences from rigid printed circuit boards, as shown in Fig. 57.3. A coverlay has to have more functions than the solder mask of rigid boards. It should have not only solder dam capability, but also mechanical protection capability for the fragile conductors of flexible circuits. Also it is required to have high toughness to survive long flex endurance and optimize the major advantage of flexible circuits. Generally, the same film material laminated with appropriate adhesive as base substrate is selected as the coverlay material. However, the manufacturing process for this is complicated, and the final cost cannot be low. Flexible screen-printable inks are a low-cost solution. However, they do not have good mechanical properties and cannot be applied for dynamic flexing usage. A combination of polyimide-based screen-printable coverlay on adhesiveless copper-clad laminate provides the smallest total thickness with very long flexing endurance life. A comparison of cross sections is shown in Fig. 57.5.

High-density flexible circuits that have been developed since 1990 need a small opening with high accuracy on the coverlay. Because traditional coverlay technologies have no capabilities to meet these requirements, new technologies have been required. Many new technologies and new materials have been developed to satisfy the new requirements of high-density flexible circuits. The capabilities of the major technologies are shown in Table 57.3.

57.3.4   Surface Treatment Alternatives

Appropriate surface treatments for the bare conductors should be chosen according to termination technologies. Several example are listed in Table 57.4. Details are explained in Chap. 59.

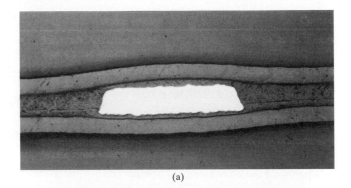

(a)

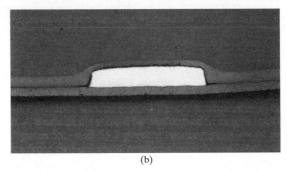

(b)

FIGURE 57.5   Cross section of a single-sided flexible circuit showing (*a*) adhesive-based laminate and film overlay; (*b*) adhesiveless laminate with liquid coverlay.

57.3.5   Double-Sided Circuits with Through-Holes

A double-sided flex circuit can have the same through-hole structures as rigid circuit boards, as shown in Fig. 57.6(*a*). However, there are several critical differences in microvia holes on flex circuits. Most of the differences come from a thin-film base substrate such as polyimide film. Electrically plated copper conductors are much more brittle than rolled annealed copper foils; therefore, thin copper plating is preferred to maintain the flexibility of the circuits. Generally, sufficient through-hole reliability is established with 15-μm-thick copper plating. Copper plating should be eliminated from the dynamic flexing area by using a suitable masking method.

Adhesiveless copper-clad laminates that have simple and uniform substrate layers and smooth copper surfaces have become popular in high-density flex circuits. Because of thin

TABLE 57.3   Varieties of Coverlay

	Material selection	Thickness (μm)	Dimensional accuracy (minimum opening)	Reliability (flexing endurance)
Traditional film coverlay	PI, PET	30–100	±0.3 mm (800 μm)	High (long)
Film coverlay + laser drilling	PI, PET	30–100	±50 μm (50 μm)	High (long)
Screen print liquid ink	Epoxy, PI	10–20	±0.3 mm (600 μm)	Acceptable (short)
Photoimageable (dry film type)	Epoxy, PI, acrylic	25–50	±50 μm (80 μm)	Acceptable (short)
Photoimageable (liquid ink type)	Epoxy, PI	10–20	±50 μm (80 μm)	Acceptable (short)

TABLE 57.4 Surface Treatments and Termination Technologies

Surface treatment	Termination technologies
Solder and lead-free solder	Soldering, FFC connector
Hot-air leveling	Soldering
OSP (preflux)	Soldering
Hard Ni/Au plating	Connector, pressure contact
Soft Ni/Au plating	Wire bonding, direct bonding
Tin	Soldering

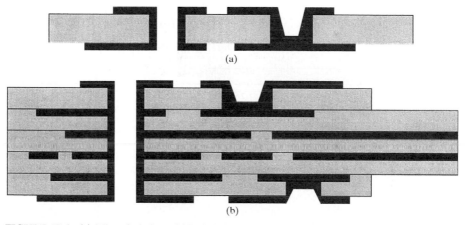

FIGURE 57.6 (*a*) Microvia hole and blind via hole for double-sided flexible circuit. (*b*) Multilayer rigid/flex circuit boards.

film materials, there are several choices in small hole generation and copper plating processes.

Flex circuits can contain both through-holes and blind via holes for double-sided circuits. Microsize blind via holes on double-sided flex circuits confer the ability to make SMT assembly density much higher.

The microvia hole capability depends on the process used by the manufacturer; however, general ideas are introduced in Chap. 58. Certainly the smaller via holes are available for the thinner substrates. Using the latest technology, such as excimer laser drilling, microvia holes smaller than 40 μm are available for 25-μm-thick adhesiveless polyimide base substrates. Microvia hole capabilities for polyimide substrates are shown in Fig. 57.7.

57.3.6 Multilayer Rigid/Flex

A simple multilayer flexible circuit does not make sense other than for its small thickness. These circuits will not have any more flexibility. Therefore, rigid/flexible constructions that contain both rigid parts and flexible parts have been developed. Typical shapes are illustrated in Fig. 57.6(*b*). This is not a simple combination of flexible circuits and rigid circuits. Generally, there are common flexible layers that are separated in the flexible parts. A flexible layer can have only

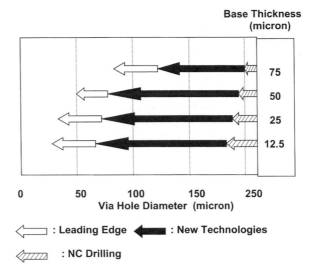

FIGURE 57.7 Microvia capabilities shown by base thickness and hole diameter. Also shown is the range of technologies from HDI to mechanical drilling required to create the vias.

one or two conductor layers to increase flexibility. The rigid parts are sandwiched by cap layers that consist of rigid circuit materials such as glass-epoxy or glass-polyimide. The constructions of rigid parts are very similar to those of rigid multilayer circuits except for the thin polyimide dielectric layer inside.

Through-hole and blind via hole constructions similar to those of multilayer rigid circuit boards are available using coated copper foil or photovia processes for multilayer flexible and rigid/flexible circuits. However, there are more varieties from which to choose in both materials and manufacturing processes. Basically, the same assembly technologies are available on the rigid parts. The latest portable electronic equipment, such as palmtop computers, requires high-density SMT assembling on the rigid parts.

There are no limits to the number of flex layers. More than 25 layers are built into one rigid/flex circuit for aerospace applications. However, the circuits require very complicated manufacturing processes, more so than the standard rigid multilayer circuits, and therefore their costs are much higher than costs for rigid boards of the same size. Design should be considered seriously prior to manufacturing to reduce the total cost based on the capability of the manufacturer. A higher circuit density can reduce the layer counts and the total manufacturing cost.

57.3.7 Flying-Lead Construction

Unsupported conductors, also known as flying leads or flying fingers, are not a new idea in the construction of flexible printed circuitry. The biggest advantage of flying-lead technology is its extremely high heat resistance and thermal conductivity when compared to standard conductors insulated by dielectric substrate materials. Based on the lack of insulation, flying leads can be treated as a bare conductor material similar to bare wire. There are many high-density flexible circuit applications that require flying-lead terminations. Details are discussed in Chap. 60.

Consideration must be given to specific surface treatments required on the flying leads to make each of the various application terminations. Currently, tin/lead plating is the most common treatment for various soldering processes; however, it is expected to be replaced by lead-free technologies. A soft gold plating with nickel under-plating is the standard surface treatment for direct wire-bonding processes. The purity and the thickness of the plated gold, along with the bond strength of the plated layer, are important factors in bond performance.

Copper thickness is a key consideration when developing a new circuit design involving flying leads. Although it is easier to manufacture high-density circuits utilizing thinner copper, the resultant flying lead may be extremely fragile and will not have the mechanical strength of the thicker copper circuits (see Fig. 57.8). Manufacturing yield plays an important role in the overall circuit costs, and thinner copper foils typically result in more damage during processing. It is important to understand this correlation as poor circuit design can effect manufacturing yields. Because the flying leads are basically bare conductors, suitable surface treatments should be conducted according to termination technologies.

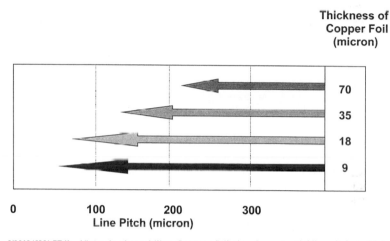

FIGURE 57.8 Flying lead capability of copper foil, showing potential line pitch vs. thickness of the copper foil.

57.3.8 Microbump Arrays

Many kinds of microbump arrays have been developed for high-density flexible circuits other than solder ball grid arrays. Recent technologies allow for a wide variety of shapes, materials, sizes, and pitches. An appropriate design should be chosen according to the requirements of the terminations. Figure 57.9 shows an example. Details are described in Chaps. 59 and 60.

57.4 CIRCUIT DESIGNS FOR FLEXIBILITY

One of the major features of flexible circuits is their long flexing endurance. A suitable layer construction and materials should be chosen. A basic idea is illustrated in Fig. 57.10. For best results, the conductor layer should be the center of the layer construction. It should have symmetrical layer constructions for both sides of the conductor layer. A rolled annealed copper foil should be used. But there is no exact relation between roll direction and flexing endurance; the exact performance should be evaluated for each material. It is another principle that the thinner the layer, the better. An 18-μm-thick copper foil has a longer flexing life than a 35-μm-thick copper foil. A 12.5-μm-thick polyimide film provides a longer flexing life than a 25-μm-thick

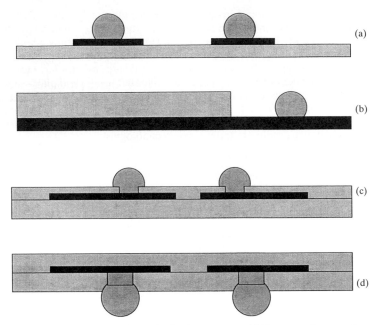

FIGURE 57.9 Constructions of microbump arrays. (*a*) Bumps on bare conductors. (*b*) Bumps on flying leads. (*c*) Bumps on cover lay. (*d*) Bumps through base substrate.

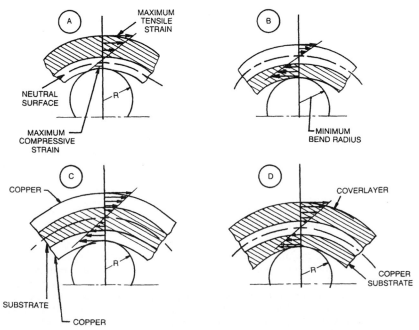

FIGURE 57.10 Bending of flexible circuitry showing (*a*) and (*b*) single-sided circuit without coverlay, (*c*) double-sided circuit, and (*d*) single-sided circuit with coverlay.

polyimide film. Adhesive layers should be as thin as possible for dynamic flexing. But thin adhesive layers do not have good bond strength and good encapsulation for traces, and therefore a suitable thickness should be determined based on experimental data. Generally, polyimide films have a longer life than polyester films. RA copper foils have a longer flexing life than ED copper foils.

Many kinds of bending modes on flexible circuits could be practical, but in IPC-TM-650 it is recommended to evaluate the flexing life using the movement shown in Fig. 57.11. The larger flexing radius provides a longer flexing life, as shown in Fig. 57.12. There are several ideas for reducing the risks of flexing failures in actual cases for electronic products; Fig. 57.13 shows examples.

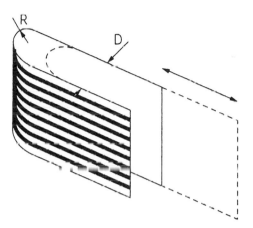

FIGURE 57.11 Typical dynamic flexing mode of flexible circuit specified by IPC-TM-650.

Basically, double-sided flexible circuits are not available for dynamic flexing. The flexibility depends on the configuration of conductors in both sides of the base substrate. Figure 57.14 shows examples. Bending the circuit with thin copper circuits on the outside of the flexible circuits may cause them to break, because the mechanical stresses are concentrated on the small traces.

A dynamic flexing area of a double-sided flexible circuit should have only one conductor layer. Coverlay on the other side of conductor should also be eliminated to enable symmetrical layer constructions. Copper plating for the through-holes of double-sided flexible circuits should be eliminated from dynamic flexing areas using a plating mask.

57.5 ELECTRICAL DESIGN OF THE CIRCUITS

There is no significant difference between the electrical performance of flexible circuits and that of rigid circuit boards. There are small differences in conductivity and insulation resistance depending on the materials. Also, there are small differences in dielectric constant and tangent delta. Therefore, similar calculations can be applied to determine electrical performances for a flexible circuit. However, flexible circuits have a relatively long parallel circuit for cabling capability, and impedance should be managed carefully, especially for high-speed circuits with high frequencies.

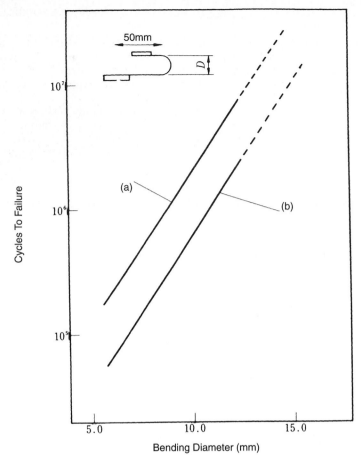

FIGURE 57.12 Flexing radius and life as specified by IPC-TM-650 test method. Bending Diameter (mm) (*a*) RA copper foil. (*b*) ED coppper foil.

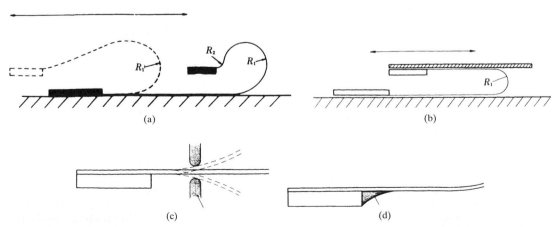

FIGURE 57.13 Design of dynamic flexing part of flexible circuits. (*a*) Unacceptable technique; (*b*) guide board; (*c*) loose guide; (*d*) silicone rubber.

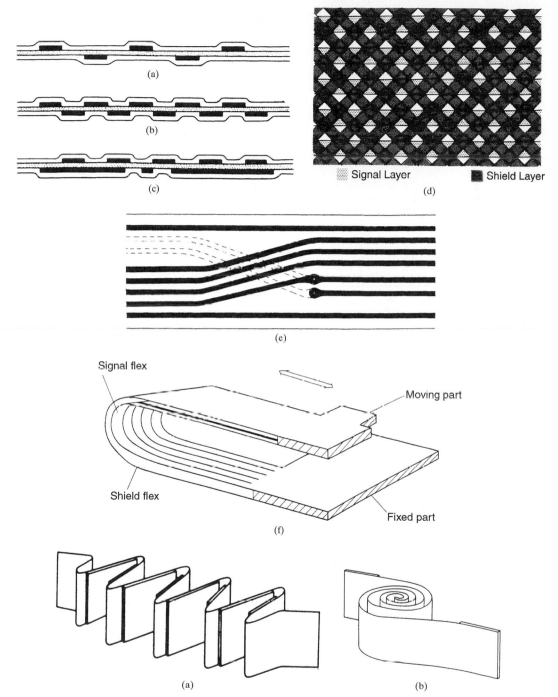

FIGURE 57.14 Circuit configuration of double-sided flexible circuits at flexing parts. (*a*) Preferred; (*b*) acceptable; (*c*) unacceptable; (*d*) meshed shield layer; (*e*) crossing traces; (*f*) separated shield layer for dynamic flexing; (*g*) fan-folded; (*h*) coiled.

57.6   *CIRCUIT DESIGNS FOR HIGHER RELIABILITY*

Because of the thin and fragile materials used, flexible circuits have lower mechanical reliability than rigid circuit boards. They have a low conductor bond strength and low base substrate tear strength. Nevertheless, most of the flexible circuits are subjected to more mechanical stresses due to movement. Therefore, special care is required in the circuit design to gain higher circuit reliability.

There are several ways to make the reliability higher. Figures 57.15 and 57.16 show several common ideas for increasing the reliability of flexible circuits by modifying conductor patterns. The trace patterns should be smooth slopes between different widths. The pad size should be as large as possible. Figure 57.17 shows an example applied for automobile instrument panels. The conductors are designed to be as wide as possible, and the spaces between conductors are kept in the safety range. A suitable coverlay opening could increase the reliability, as shown in Fig. 57.18. The amount of squeeze-out of coverlay adhesives should be minimized, however. The edge areas of stiffener boards are dangerous places in which mechanical stresses are concentrated. Figure 57.19 shows ways of reducing these risks.

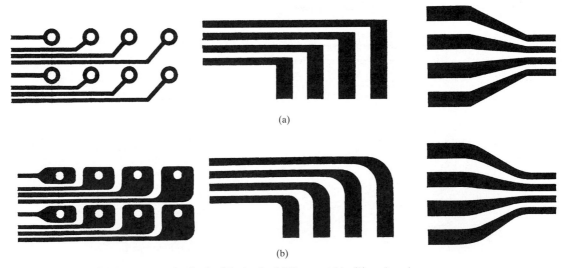

(a)

(b)

FIGURE 57.15   Reliable pattern design for flexible circuits. (*a*) Unacceptable; (*b*) preferred.

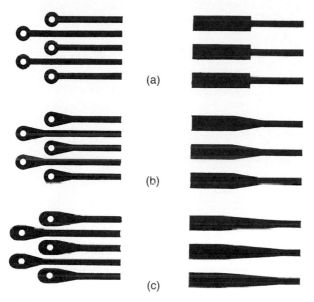

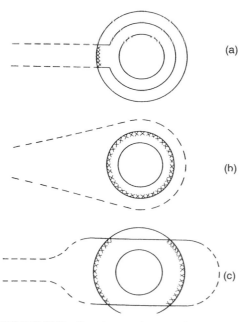

FIGURE 57.16   Reliable pattern design for flexible circuits. (*a*) Unacceptable; (*b*) acceptable; (*c*) preferred.

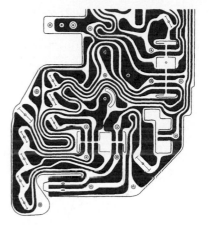

FIGURE 57.17   Conductor pattern designed for the instrument panel of an automobile.

FIGURE 57.18   Coverlay opening for reliable soldering pads. (*a*) Unacceptable; (*b*) acceptable; (*c*) preferred.

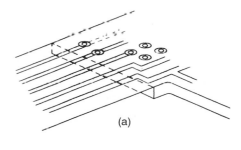

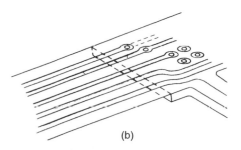

FIGURE 57.19   Circuit design at stiffener edge. (*a*) Unacceptable; (*b*) preferred.

CHAPTER 58
MANUFACTURING OF FLEXIBLE CIRCUITS

Dominique K. Numakura
DKN Research, Haverhill, Massachusetts

58.1 INTRODUCTION

Basically, it is possible to produce single- and double-sided flexible circuits and multilayer rigid/flexible circuits if there is a set of manufacturing facilities for multilayer rigid circuit boards. However, it is not easy to achieve high productivity with high yield for volume production of flexible circuits using these facilities. There are many major and minor differences. Specific equipment and process conditions are required for volume production of flexible circuits.

Table 58.1 shows the specialties for processing the flexible circuit materials. Most of the issues are caused by complicated structures and thin, fragile materials. The coverlay structures and the stiffener constructions require supplemental manufacturing processes. The varieties of outer shapes make material yields and productivity lower. Thin, fragile materials can easily incur serious mechanical damage due to rough handling. Thin materials can undergo significant dimensional changes in the manufacturing processes. They cause a pattern shift between the processes and make process yields lower. High-level automation has basic barriers, discussed later, when applied to the manufacturing process of flexible circuits due to unstable materials. Many processes must be conducted by manual methods, which are labor intensive. Process elements special to flexible circuits are required to achieve high manufacturing yield. Also, an appropriate total process design is required to achieve high productivity with high yield.

A roll-to-roll (RTR) manufacturing system is a high-productivity solution for volume production of flexible circuits. However, it is available for only the early steps of the long manufacturing processes. Also, many limitations apply.

58.2 SPECIAL ISSUES WITH HIGH-DENSITY FLEXIBLE CIRCUITS

High-density flexible circuits, developed recently, have special constructions and, therefore require supplemental processes. In addition, special technologies are required to generate finer circuits. Major items are listed in Table 58.2. Details of each process are described in each section.

TABLE 58.1 List of Specialties for Processing Flexible Circuits

Areas	Issues for processing flexible materials
Special constructions	Coverlay, stiffeners, etc. Complicated shapes
Materials	Thin, low stiffness Low dimensional stability Fragile for mechanical forces
Automation	Difficult for automation RTR available only for beginning

TABLE 58.2 New Manufacturing Technologies for High-Density Flexible Circuits

Processes	New technologies
Raw materials	Stable polyimide film (~20 ppm CTE) Thin copper foil (9, 12 μm) Adhesiveless copper-clad laminates
Microvia hole drilling	Laser (excimer, UV:YAG, carbon dioxide) Microhole punching High-accuracy x-y table Plasma etching, chemical etching
Copper plating	Direct plating Seed layer sputtering
Surface cleaning	New cleaning reagent
Etching resist	Thin dry film (~15 μm) Liquid resist
Exposure	Collimated light source, auto aligner Laser direct imaging
Etching	High-resolution etchant
Coverlay	Laser drilling Photoimageable coverlay (dry film type and liquid type)
Surface finishing	Electroless plating for Ni, Au, solder High-speed plating
Flying leads	Laser/chemical etching Chemical etching
Guide holes Inspection	Automatic high-accuracy punching machine Universal AOI Noncontact electrical tests
Roll-to-roll system	Low-tension EPC system Automatic aligner with CCD camera

58.3 BASIC PROCESS ELEMENTS

Figure 58.1 shows a standard manufacturing process flow for double-sided flexible circuits with through-holes and stiffener boards. All manufacturing processes for single-sided flexible circuits are included in these processes. The first half of the process (through pattern etching) has few choices. However, the second half has many options, depending on the design of the construction. The second half consumes more labor and its manufacturing cost is much higher

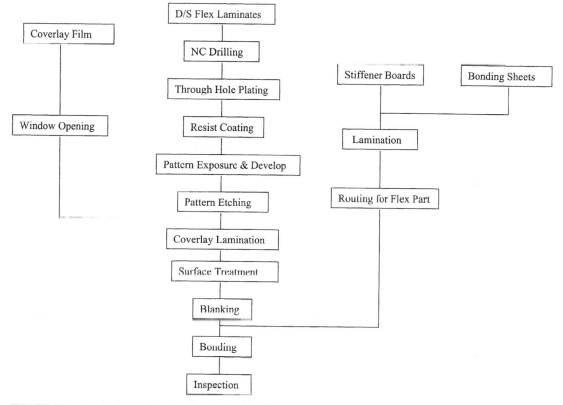

FIGURE 58.1 Standard manufacturing flow for double-sided flexible circuits.

than that of the first half. An appropriate consideration of process design should be conducted to increase the productivity and reduce the total cost. Good process design can reduce the total cost significantly, especially for complicated constructions. Typical ideas are introduced in the following sections.

58.3.1 Preparation of Materials

Most flexible circuit materials are supplied in rolls. The first process of manufacturing is to cut the rolled materials into workable sizes. It is possible to cut with a manual sharing tool; however, an automatic machine is recommended to get an exact size without handling damage.

Cleaning processes for copper foil surfaces for single-sided circuits should be conducted prior to cutting if roll-to-roll processing is available. This could reduce handling damage and increase the manufacturing yield significantly.

58.3.2 Direct Cast Processes

For special substrate constructions such as ultrathin substrates or ultra-high-density flexible circuits, a direct casting process of polyimide resin has been developed. (See Fig. 58.2 for a typical flowchart of the manufacturing process for direct die casting.) A photosensitive polyimide

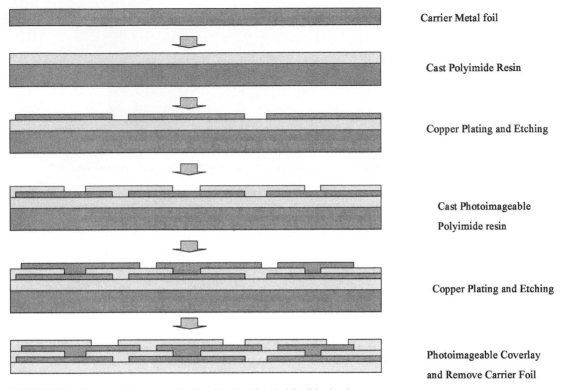

Carrier Metal foil

Cast Polyimide Resin

Copper Plating and Etching

Cast Photoimageable
Polyimide resin

Copper Plating and Etching

Photoimageable Coverlay
and Remove Carrier Foil

FIGURE 58.2   Direct casting process for fine-line double-sided flexible circuits.

oligomer is sometimes coated on a carrier stainless foil. Then an electroless and an electrocopper layer are plated as the thin conductor layer. A special surface treatment on the polyimide is required for good bond strength. The thin copper layer is then etched to generate fine traces. The process can be repeated to build multilayer construction with via connections. Finally, the carrier tape is removed for flexibility. Sometimes the carrier tape is partially etched and kept as a stiffener.

58.3.3 Mechanical Creation of Through-Holes

There may be several different processes for each via hole construction in the case of flexible circuits. Suitable processes, including material combination, should be designed according to final requirements (i.e., construction, circuit density, reliability, manufacturing volume, available equipment, urgency, etc.)

Figure 58.3 shows standard manufacturing processes for microvia holes for double-sided flexible circuits. A mechanical numerically controlled (NC) drilling process on double-sided flexible circuits can be done similarly to that used for rigid double-sided printed circuit boards. More laminates can be stacked up in the setup because of the thin materials. However, appropriate backing boards and drilling conditions should be determined to achieve good hole quality.

A punching process with a suitable die set is available for the thin, flexible copper laminate. It can provide a low-cost solution, but it needs a large-volume production to compensate for the high-cost die. A small-diameter punching process controlled by an NC system is avail-

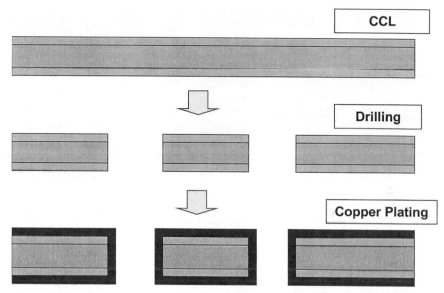

FIGURE 58.3 Standard manufacturing process for microvia holes in double-sided flexible circuits, starting with copper-clad laminate (CCL). The hole is created by drilling and then copper-plated.

able for thin, flexible copper-clad laminates instead of using the NC drilling process. These mechanical hole-generation processes cannot make blind via holes on double-sided flex circuits.

58.3.4 Through-Hole Plating

Through-hole plating processes similar to those used in rigid board fabrication are available for thin, flexible circuits. Both standard electroless copper-plating processes and direct copper-plating processes have been established for the seed layer in the holes. A thinner plated copper conductor is recommended to maintain the flexibility of the circuits. Selective copper plating with suitable masking is required for dynamic flexing circuits. Precise thickness control of the through-hole copper is required, because the electrically plated copper does not deposit uniformly on the surface of the original copper foil if the raw material sheet is not fixed exactly on a fixture. Design of fixtures and electrodes in the plating process is critical to achieve uniform copper thickness, as a thin, flexible sheet flops by agitation in the plating bath. As a result, it is not easy to achieve very uniform copper thickness for large panels. Nonuniform plated copper will cause lower etching yield and reduce through-hole reliability.

As an adhesiveless copper-clad laminate does not have low-heat-resistant layers such as epoxy or acrylic resin, it provides excellent through-shapes without desmearing processes. Figure 58.4 shows a comparison of a hole plated with an adhesive-based laminate and a hole plated with adhesiveless laminate.

58.3.5 Microvia Hole Processes

Microvia hole generation could be one of the critical items in building high-density flexible circuits. Figure 58.5 shows standard manufacturing processes for microvia holes, including blind via holes for double-sided flexible circuits. Several new technologies have been developed for microhole generation on thin polyimide dielectric layers, as shown in Table 58.3.

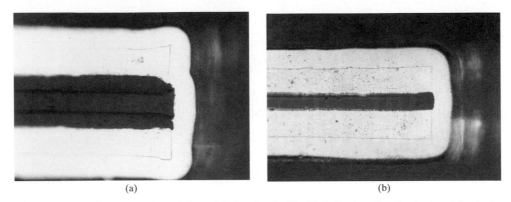

(a)   (b)

FIGURE 58.4 (*a*) Cross section of through-hole of a double-sided circuit with adhesive-based laminate. (*b*) Cross section of through-hole of a double-sided circuit with adhesiveless base laminate.

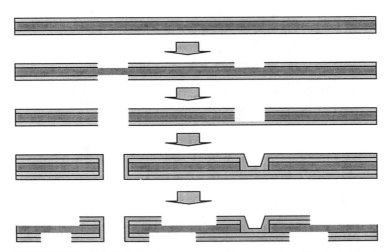

FIGURE 58.5 Basic high-density circuit generation process by subtractive method.

TABLE 58.3 Comparison of Technologies for Hole Generation on Dielectric Layers

	Drilling	Punching	Laser drilling	Plasma etching	Chemical etching	Photo polymer process
Hole diameter (μm)	~50	~70	~10	~70	~70	~70
TH for D/S	Yes	Yes	Yes	Yes	Yes	Yes
TH for multilayer	Yes	No	Possible	No	No	No
Blind via hole	No	No	Yes	Yes	Yes	Yes
Material choice	Wide	Wide	Wide	Fair	Small	Small
Technical flexibility	Small	Small	High	Fair	Small	Small
Chemical waste	No	No	Small	Small	Serious	Fair
RTR availability	No	Possible	Possible	Difficult	Possible	Possible

58.3.5.1   Mechanical Microvia Hole Creation.   The newly developed NC drilling machine can generate microvia holes smaller than 100 μm in diameter using special drill bits. Technically, the newest NC drilling system can generate 50-μm-diameter holes on 50-μm-thick flexible materials. However, it can drill fewer sheets in a stack than larger holes with slower speeds, and its productivity becomes smaller with smaller hole sizes. An NC microhole punching system can also generate microholes smaller than 80 μm in diameter on 50-μm-thick flexible materials. It has similar productivity issues to the NC drilling machines; however, multipunch die systems can increase productivity significantly.

58.3.5.2   Laser Microvia Hole Creation.   Several laser systems have been developed to generate microvia holes for flexible circuits. The excimer laser has a wide capability to generate small holes on most organic substrates. It can generate holes 10 μm in diameter on 25-μm-thick polyimide film, as shown in Fig. 58.6(a). The largest issue with the excimer laser is its slow speed. Appropriate process design is necessary to achieve a good productivity. The excimer is capable of drilling through copper foil, but this is much slower than for substrates alone. The UV:YAG laser is another choice for generating microvia holes smaller than 50 μm in diameter on flexible materials. It has a higher productivity rate than the excimer laser. It can drill through both copper foil and flexible substrates, as seen in Fig. 58.6(b). An issue with the UV:YAG laser is that it takes a long time to generate large holes. The carbon dioxide laser

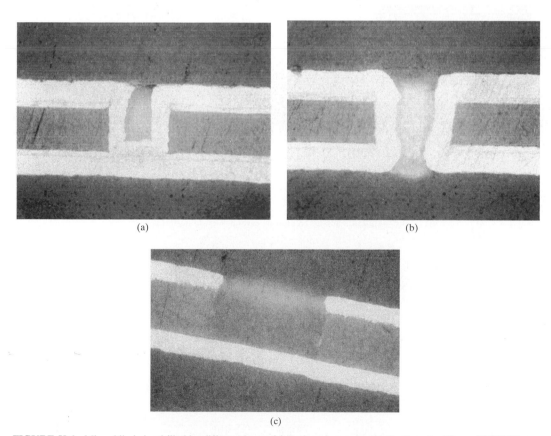

(a)   (b)

(c)

FIGURE 58.6   Micro-blind vias drilled by different lasers. (a) Excimer laser, plated hole diameter 25 μm. (b) YAG laser, plated hole diameter μm. (c) Carbon dioxide laser, plated hole diameter 100 μm. *(Source: Photo Machining.)*

has greater productivity than the excimer laser or the UV:YAG laser when used to generate via holes larger than 70 μm in diameter (see Figs. 58.6[c] and 58.7); however, it cannot drill through copper foil. As a result, copper foil etching is required to create a drilling mask before the laser operation. Ideally, it is possible to use a reel-to-reel manufacturing system for laser processes. Comparisons of the technical capabilities of these microvia processes are shown in Table 58.4.

58.3.5.3 Plasma and Chemical Microvia Hole Creation.

Plasma etching and chemical etching are the unique processes used to generate microvia holes on flexible circuit materials (see Figs. 58.8 and 58.9). Both can generate microvia holes smaller than 100 μm in diameter on 50-μm-thick polyimide substrates. Plasma processes can etch all kinds of organic materials, but standard alkaline etching can etch only specific polyimide materials such as Kapton™ or Apical™. Special chemicals are needed to etch Upilex™-type polyimide substrates, which have high dimensional stability. Processing costs of plasma processes and chemical processes do not depend on the number of holes or sizes in a unit area. They have a remarkably lower cost than laser processes if there are many holes in the same area. On the other hand, they have several disadvantages. Generally, there are many parameters that affect process capabil-

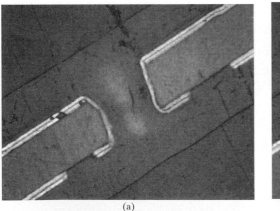 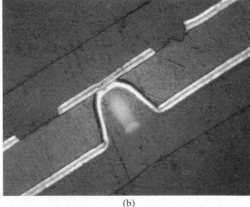

(a)   (b)

FIGURE 58.7 Microvia holes 50 μm in diameter created by carbon dioxide laser: (a) through-hole: (b) blind via hole (Source: Photo Machining.)

TABLE 58.4 Technical Capabilities of Microvia Hole Technologies for Double-Sided Flex Circuits

	Excimer	UV:YAG	Carbon dioxide	Plasma etch	Chemical etch
Hole size (μm)					
For small volume	~10	~15	~50	~70	~50
For large volume	25–150	25–75	75–250	~100	~75
Hole quality	Excellent	Good	Fair	Fair	Fair
Additional cleaning	Necessary	Necessary	Necessary	No	No
Technical hurdles	Low	Low	Fair	High	High
Cost	High	High	Fair	Low	Low

For only specific polyimide resins.

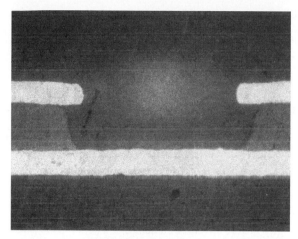

FIGURE 58.8   Microvia generated by plasma etching. The hole diameter is 200 μm.

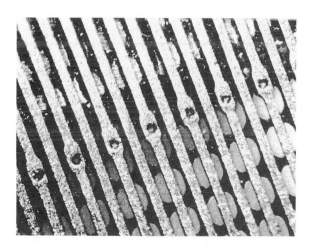

FIGURE 58.9   Microvia holes made by chemical etching. The hole diameter is 40 μm. *(Source: Asahi Fine Technology.)*

ities. It takes time to create suitable process conditions that provide reliable via holes with high productivity. Waste treatment is another issue, especially in the case of wet chemical etching. Chemical etching is a convenient technology to apply to a reel-to-reel manufacturing system. It takes a large investment to introduce a reel-to-reel system for plasma etching system. Capabilities to process materials of each technology are shown in Table 58.5.

Figure 58.10 shows examples of the cost comparison of the technologies for micro-through-holes on 50-μm polyimide substrates. Figure 58.10(*a*) shows cost comparison of micro-size NC drilling vs. the excimer laser and UV:YAG laser. NC drilling cost is inversely proportional to hole size. A conventional NC drilling process has small advantages over laser processes for holes smaller than 200 μm in diameter. Figure 58.10(*b*) shows a cost comparison of different laser systems. The UV:YAG laser is cost advantageous in the 25- to 75-μm hole diameter range.

TABLE 58.5   Material Availability

	Punching and drilling	Excimer laser	UV:YAG laser	Carbon dioxide laser	Plasma etching	Chemical etching
Copper	Yes	Slow	Slow	No	No	Yes
Epoxy	Yes	Yes	Yes	Yes	Yes	Difficult
FR-4	Yes	Slow	Slow	Yes	No	No
Kapton/Apical	Yes	Yes	Yes	Yes	Yes	Yes
Upilex	Yes	Yes	Yes	Yes	Yes	Difficult
Espanex	Yes	Yes	Yes	Yes	Yes	Difficult
NeoFlex	Yes	Yes	Yes	Yes	Yes	Difficult

However, its cost increases exponentially with hole size, especially for holes larger than 75 µm in diameter. The cost of the excimer laser process is more stable. It is proportional to hole sizes in the 20- to 200-µm diameter range. Carbon dioxide lasers show extremely lower drilling cost than the excimer laser and UV:YAG laser for the holes larger than 75 µm in diameter. It is very possible to generate more than 10,000 holes 100 µm in diameter in 1 min with the galvanometer, a Diamond Carbon Dioxide Laser System™ that has higher productivity than a carbon dioxide TEA laser system. Figure 58.10(*c*) shows a drilling costs analysis of a plasma etching system and a chemical etching system compared to a laser system or an NC drilling system. A unique cost feature of the etching procedure is that it does not depend on the number of holes or shapes, but on area sizes. Therefore, there are big cost advantages for circuits that have a large number of holes in a small area.

58.3.6   Surface Cleaning

Only a mild chemical cleaning process should be applied for thin flexible materials to eliminate serious damages. Soft scrubbing or soft brushing must be used when a mechanical process is necessary to remove tough stains or oxidation on the copper surface. A severe mechanical brushing will produce mechanical stress on the thin materials and will cause the loss of dimensional uniformity. A soft brushing process on a belt conveyor has been developed to minimize mechanical damage to flexible materials.

58.3.7   Resist Coating

The same etching resist inks used for rigid circuit boards and screen printers are available for flexible circuits. Also, the same dry films and laminators could be used for flexible circuits. However, flexible dry films should be selected to have a high etching yield. A thin dry film is recommended, especially for fine patterns. Dry films 15 or 20 µm thick can generate 30- to 40-µm lines/spaces on 18-µm-thick copper foils with a high yield. A liquid photoresist should be used for very-high-density traces with pitches smaller than 50 µm.

58.3.8   Pattern Generation

Similar exposure equipment is available for flexible circuits. A collimated light source machine is recommended for high-density traces with pitches smaller than 100 µm to achieve a high yield. An appropriate matching tool for flexible materials should be developed to achieve good pat-

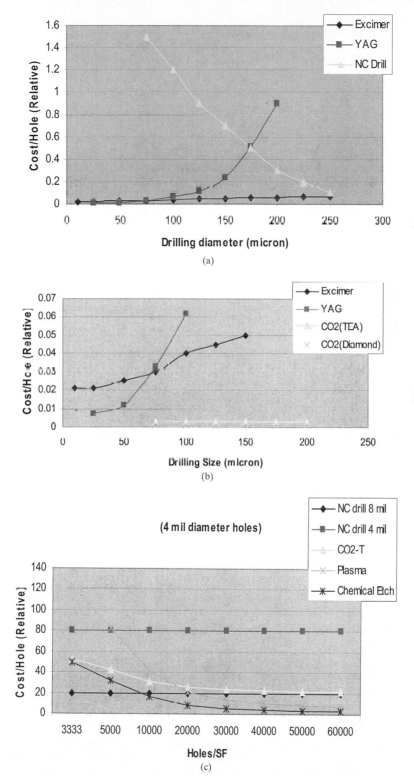

FIGURE 58.10 Relative cost comparisons by hole creation technology. (*a*) Comparison of NC drilling cost to Excimer and YAG lasers. (*b*) Comparison of alternative laser processes. (*c*) Comparison of NC, laser, plasma, and chemical etch.

tern alignment with high productivity. Laser direct imaging equipment is also available for prototypes of 100-μm-pitch circuits.

58.3.9 Etching

Ferric chloride or cupric chloride solution is recommended for use as the etchant for flexible circuits, especially for high-density circuits. An alkaline etchant has a higher process speed, but it is not good on fine circuits because of unstable etching rates. As an alkaline solution makes some chemical attacks on polyimide surfaces, it should not be used for severe requirements.

A special conveyor system should be designed for the wet processing of flexible circuits. Standard equipment designed for rigid boards will create many wrinkles and dents on the surface of copper foils. Higher-density conveyor wheels and smooth surfaces are required to reduce damage to the thin, fragile materials (Fig. 58.11).

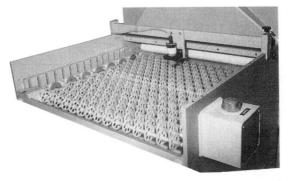

FIGURE 58.11   Special wheel conveyor system for thin, flexible materials showing high density and smooth wheels. *(Source: Camellia.)*

58.4   *NEW PROCESSES FOR FINE TRACES*

The generation of fine lines is the most basic process in building high-density flexible circuits. Three basic manufacturing concepts have been developed:

1. Advanced subtractive
2. Semiadditive
3. Fully additive

These methods have many variations to cover different construction concepts and specifications. Innovative technologies have been developed for each unit process, such as the collimated light source exposure machine and the laser direct imaging system.

58.4.1 Subtractive Process

The subtractive process is the most common technology for generating copper-based circuitry on flexible film substrates. It is basically the same method used for rigid boards, but may be automated for use with a roll-to-roll process, as discussed later.

Subtractive technology with traditional materials cannot produce high-density traces with pitches smaller than 150 μm. However, many new technologies have been introduced for each key item, as shown in Table 58.6. The most advanced subtractive process can generate 30-μm line/space traces on 18-μm copper foil using roll-to-roll manufacturing systems. It is possible to produce 20-μm-pitch traces on double-sided flexible circuits with 5-μm-thick copper.

TABLE 58.6 New Technologies Needed to Generate Fine-Line Traces on Flexible Substrates

Processes	New technologies
Surface treatment	New soft etching chemicals (organic)
	Sputtering of seed layer
Resist coating	High-resolution liquid etching resist and plating resist (spray coater, dipping coater, etc.)
	Thin dry film (~15 μm)
	High-sensitivity dry film with high resolution (10 mJ/cm²)
	Wet laminator
Exposure	Collimated light source with glass mask (~15-μm line/space)
	Autoalignment system for double-sided circuits (with accurate dimension compensation system, ~5 μm)
	Roll-to-roll conveyor system with low tension
	Laser direct imaging system (30–40 μm line/space)
Etching	High-resolution etchant (cupric chloride, ferric chloride)

58.4.2 Semiadditive Processes

Two semiadditive processes—sputtering and wet chemical electroless plating—were developed to generate fine-line traces on specific polyimide substrate materials. An advantage of these processes is that the conductor layers can have varying thicknesses. Both processes create suitable electroless plating processes to achieve reliable bond strength on the substrates. The sputtering process can generate thin conductive layers as the reliable seed layers. The wet chemical electroless plating process produces low-cost seed layers. The latest technology makes 15-μm lines/spaces available on a 6-μm-thick copper conductor layer. Figure 58.12 shows the process flow for the semiadditive method.

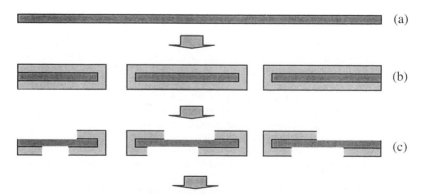

FIGURE 58.12 Basic high-density circuit generation process flow for semiadditive method. (*a*) Plain film substrate. (*b*) Drilling via holes and plating of copper layer. (*c*) Etching of copper layer.

58.4.3   Fully Additive Processes

There are several fully additive processes that can produce extremely fine traces on flexible substrate materials with a high aspect ratio. The basic concept for via holes is different from that for other technologies, however. A suitable combination of substrate materials, seeding processes, plating mask resists, and plating processes must be arranged. The most advanced process can generate 5-μm line/space gold traces on thin polyimide substrate. Figure 58.13(*a*) shows a process flow for the fully additive method, while Fig. 58.13(*b*) shows a 5-μm line/space flexible circuit made with a fully additive process.

58.4.4   Process Comparison

A comparison of the technologies is shown in Table 58.6. The advanced subtractive process can provide the best cost-performance solution when building a high-density circuit at pitches larger than 50 μm. Semiadditive and fully additive processes can be used to generate fine-line traces with pitches smaller than 40 μm.

58.5   COVERLAY PROCESSES

The coverlay construction of flexible circuits demands another special process. It is very different from the solder mask used for rigid circuit boards. Furthermore, recent high-density flexible circuits need new coverlay technologies. Table 58.7 summarizes the comparison of traditional coverlay technologies and new technologies. Traditional film coverlay has an excellent balance of physical properties and can provide high reliability, especially for long dynamic flexible endurance. Unfortunately, it requires very complicated, labor-intensive processes, and it is difficult to introduce automatic manufacturing systems. Another big issue is that it cannot make small openings with high dimensional accuracy because of the manual handling of thin film materials. Fine openings smaller than 200 μm in diameter with high dimensional accuracy better than 100 μm positioning have been required for flexible circuits that need high-density interconnect (HDI), such as SMT of 06-03 components. Screen printing of flexible ink can provide a low-cost solution for volume production. However, it cannot provide a better solution for small openings with high dimensional accuracy.

58.5.1   Film Coverlay

A typical coverlay manufacturing process is shown in Fig. 58.14. Generally the same film materials are chosen for the substrate layer and the coverlay film. However, the epoxy-based coverlay films have a short shelf life and must be kept in a refrigerator. A B-stage adhesive such as epoxy or acrylic resin is coated on one side of the film covered with a release sheet. Openings for access windows are made by NC drilling or punching with a release sheet. The surface of traces must be cleaned before coverlay lamination takes place. It is difficult to introduce an automation system for the lamination process because the materials are very unstable after window openings are made. Manual alignment is the solution for this process unless the circuit needs an accurate registration higher than 0.5 mm. After the coverlay film is temporarily fixed on the circuit, the adhesive layer is cured in a heat press or autoclave. A suitable curing condition should be chosen carefully based on the circuit design and material conditions. Epoxy resins change flow properties significantly under the pressure, even though they are kept at a low temperature; therefore, detailed conditioning should be performed. A vacuum press or a vacuum autoclave helps to make good lamination, eliminating voids or air traps beside the traces.

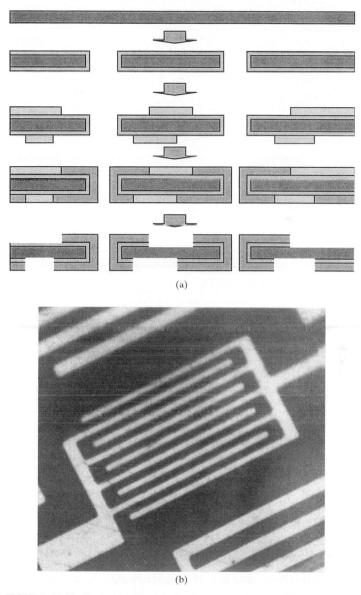

(a)

(b)

FIGURE 58.13 Basic high-density circuit generation process flow for fully additive method. (*a*) Starting with unclad substrate, holes are created and a thin layer of copper is deposited, followed by a resist to stop deposition in unwanted areas, then a further build-up of copper, and, finally, the removal of the resist and underlying copper to form the finished circuit. (*b*) Example of fully additive circuit showing 5-μm lines and spaces. (*Source: Dynamic Research.*)

TABLE 58.7 Coverlay Process Options

	Accuracy (minimum opening)	Reliability (flexing endurance)	Material selection	Equipment/ tooling	Technical hurdles	Cost
Traditional film coverlay	Low (800 μm)	High (long)	PI, PET	NC drill, heat press	Need long experience	High
Film coverlay + laser drilling	High (50 μm)	High (long)	PI, PET	Heat press, laser	Low	High
Screen print liquid ink	Low (600 μm)	Acceptable (short)	Epoxy, PI	Screen printer	Medium	Low
Photoimageable, dry film type	High (80 μm)	Acceptable (short)	Epoxy, PI, Acrylic	Laminator, exposure, developer	Medium	Medium
Photoimageable, liquid type	High (80 μm)	Acceptable (short)	Epoxy, PI	Coater, exposure, developer	Relatively high	Low

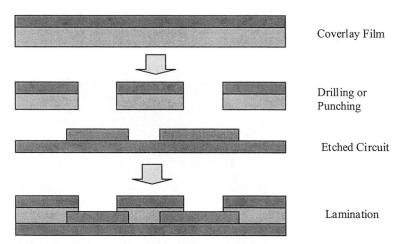

Coverlay Film

Drilling or Punching

Etched Circuit

Lamination

FIGURE 58.14 Traditional film coverlay process.

58.5.2 Screen-Printing Coverlay

Basically, the same process and the same screen printer are available for applying liquid coverlay materials as a solder mask on rigid circuit boards. The only difference in the processes is the handling of the unstable flexible materials. Any vacuum used to fix the material on the printing table should be set to a low value to make the circuit flat without causing damage. A strong vacuum makes many small spots on the coverlay. The post-baking should be conducted with a suitable carrier that can keep the material flat to avoid supplemental damages (Fig. 58.15).

58.5.3 Photoimageable Coverlay

As photoimageable solder mask systems for rigid circuit boards were developed in the 1980s, similar ideas were tried for flexible circuits. Ideally, fine openings and high dimensional accuracy were expected. Unfortunately, the materials for rigid circuit boards were too brittle for flexible circuits, and therefore new materials were required. DuPont developed a dry film–type

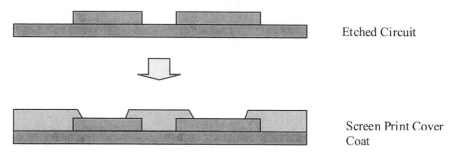

FIGURE 58.15 Typical screen-printing process using an ink cover coat.

photoimageable coverlay (Pyralux PC™) and Nippon PolyTech developed a liquid ink–type photoimageable coverlay (NPR-80) in the early 1990s. These were practically the first materials available for flexible circuits. They had volume applications in automobiles, disk drives, cameras, etc. There were several problems with early versions, but there have been remarkable improvements, and application areas have expanded. More material manufacturers have been commercializing different materials, and therefore there are more choices nowadays. Reliability performances depend very much on materials and manufacturers. Material selection should be conducted carefully.

There are significant differences in the manufacturing processes between the dry film type and the liquid ink type (see Fig. 58.16). The dry film type materials are laminated by a vacuum laminator to ensure good encapsulation of conductors. The same laminators designed for dry film type solder mask are available. Roll-to-roll vacuum laminators have been developed, but have not gained wide acceptance. Some materials are applicable by a simple heat roll laminator, but supplemental processes are required for good encapsulation. UV exposure with traditional equipment and development generates openings on the plain photoimageable layers. Usually, additional thermal curing processes are required to complete the chemical cross-linking reaction in the molecules.

There are several choices for application processes of liquid ink photoimageable coverlay. Screen printing could be an easy method for most circuit manufacturers because they have been using the processes and machines for other purposes for a long time. Newly introduced spray process or curtain coating could be a low-cost solution for volume production. Drying is necessary before pattern generation. There is no big difference in UV exposure and development process other than the detailed conditions and chemicals from dry film–type materials. Postbaking processes for curing could be more important for good physical performances. Figure 58.17 shows a finished photoimageable coverlay with a hole diameter of 100 μm.

58.5.4 Laser-Drilling on Film Coverlay

Excimer laser drilling of traditional film coverlay can provide a perfect solution for both physical performances and fine openings with high dimensional accuracy. A traditional film coverlay is laminated on the etched traces without prepunching or drilling. (See Fig. 58.18 for a manufacturing flowchart for laser drilling on film coverlay.) The latest excimer laser system can make sharp 50-μm openings on film coverlay without serious changes of physical properties (Fig. 58.19). It can keep the good mechanical, electrical, and chemical properties of the film coverlay, and the circuits are available for dynamic flexing. The largest issue with this technology is the low productivity and high processing cost. The processing time and cost are proportional to the total size of the opening areas; therefore it could be expensive when there are high-density SMTs. Several new laser systems such as the YAG laser and carbon dioxide laser have been developed to achieve a higher processing speed and higher productivity. A carbon dioxide laser

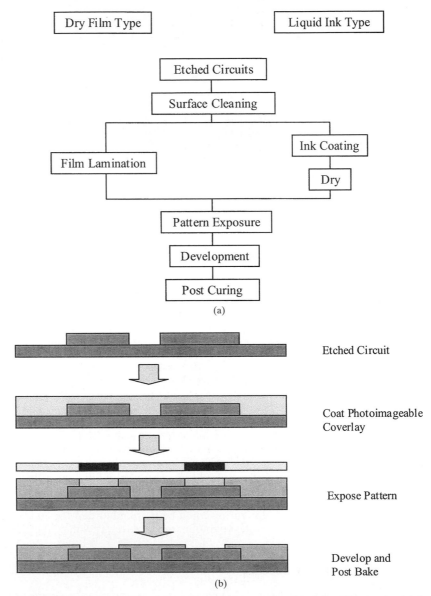

FIGURE 58.16 An example of a manufacturing process for photoimageable coverlay. (*a*) A schematic flowchart of a generic process using either dry film or liquid resist. (*b*) A graphic flowchart, starting with the etched circuit, showing the manufacturing process for a photoimageable coverlay.

with a galvano mirror has a speed one order higher than that of an excimer laser, even though engraving quality is worse than that of the excimer laser (Fig. 58.20). It cannot remove organic materials perfectly from copper surfaces, and it requires a supplemental cleaning process such as plasma etching or chemical etching.

A laser process does not require supplemental tooling such as punching die or photomask. Also, its initial conditioning is significantly simpler than that for other technologies. There-

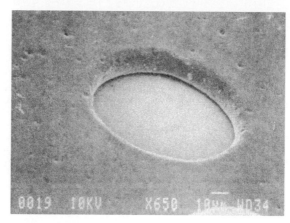

FIGURE 58.17 Small opening with photoimageable coverlay. Hole diameter is 100 μm. *(Source: Nippon Polytech.)*

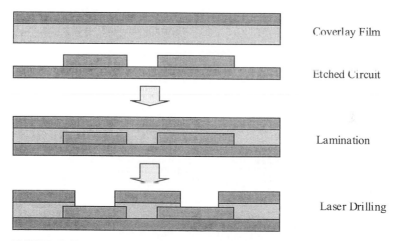

Coverlay Film

Etched Circuit

Lamination

Laser Drilling

FIGURE 58.18 Flowchart of process for laser drilling of film coverlay.

FIGURE 58.19 Small coverlay opening drilled by excimer laser. Hole diameter is 100 μm. *(Source: Shinozaki.)*

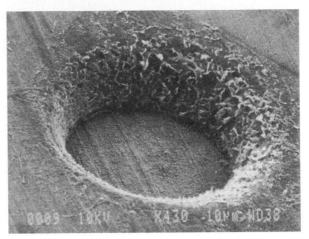

FIGURE 58.20 Small coverlay opening drilled by carbon dioxide laser. Hole diameter is 200 μm. *(Source: Shinozaki.)*

fore, a laser process could be very valuable for prototype jobs and small-volume quick-turnaround jobs.

58.6 SURFACE TREATMENT

Electro- and electroless plating of tin, nickel, gold, and solder are the common surface treatments for flexible circuits. Solder plating by a hot-air leveler could be a low-cost solution. However, it entails severe high-temperature process conditions, and uniformity of the solder is not acceptable for several applications.

Several new surface finishing technologies have been developed to satisfy the requirements of new termination technologies. A soft nickel/gold plating will be the major process for the wire-bonding processes. Tin- and lead-free solder plating will be an alternative treatment to traditional solder plating. Organic preflux will be a low-cost solution for next-generation soldering. Microbump arrays will be constructed by advanced processes. Optimized combinations of plating processes can build many kinds of bump shapes at a reasonable cost (see Table 58.8).

TABLE 58.8 Surface Treatments of Flexible Circuits

	Materials	Process conditions
Electroplating	Ni, Au, Sn, solder, etc.	Wet process
Electroless plating	Ni, Au, Sn, solder, etc.	Wet process
Hot-air leveling	Solder	>260°C
Solder roll coating	Solder	>260°C
OSP	Organic molecules	Wet process

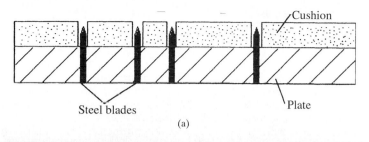

(a)

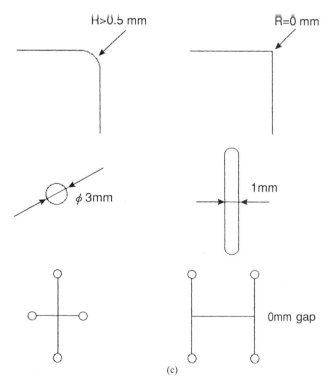

(b)

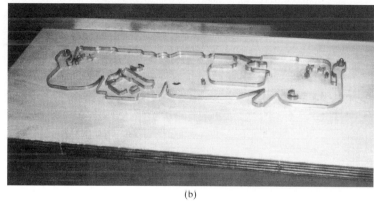

(c)

FIGURE 58.21 Steel rule die for blanking flexible circuits. (*a*) A cross section of the configured die. (*b*) A sample finished die. (*c*) Limitations on shapes that can be designed into a steel rule die.

58.7 *BLANKING*

A punching press similar to those used in the processing of rigid boards could be used for the blanking of flex circuits. Smaller tons are required to cut thin materials. Steel rule dies are the specialty for flex circuits (see Fig. 58.20). They can be prepared in a few days, and their costs are much smaller than those of hard tools. Their weight is much smaller than that of hard tools and operators can manage them by hand. A typical cross section is shown in Fig. 58.21(*a*), a finished steel rule die is shown in Fig. 58.21(*b*), and some limitations on the configurations that can be used are illustrated in Fig. 58.21(*c*). Steel rule dies have less dimensional accuracy, lower productivity, and a shorter life than hard tools, and they need maintenance more frequently. They are good at quick-turnaround jobs and small- to medium-volume productions. Table 58.9 shows a comparison.

TABLE 58.9   Comparison of Punching Dies

Items	Steel rule dies	Hard tool dies
Accuracy	~0.2 mm for whole	~0.01 mm for whole
Minimum hole size	3 mm diameter	0.4 mm diameter
Minimum slit width	Zero	0.5 mm
Minimum corner radius	0.5 mm	0.1 mm
Delivery time	2–4 days	More than 3 weeks
Weight	Light, manage by hand	Heavy, need a lift
Tooling cost	Low	High
Processing cost	High	Low

A hard punching die is recommended for use for high dimensional requirements and large-volume production. Zero clearance between the male and female dies is required for thin materials. Special technology and periodical maintenance are needed when zero-clearance punching die sets are used.

Combinations of steel rule dies and hard tool dies are the trend for achieving high productivity and high dimensional accuracy at a low cost. Quick-change die systems are the new trend for high productivity for small-volume productions. An NC routing process is also available for the blanking of flex circuits. However, special backing boards and conditions are required to process unstable flexible materials.

It is very important to have accurate punching guide holes to create good perimeters for flex circuits. Unfortunately, a standard NC drilling machine does not provide accurate positions for each hole in a work sheet because of distortions during the previous manufacturing processes. An appropriate compensation for dimensional changes is necessary for each guide hole. Several punching machines that have self-alignment systems that use CCD cameras are available to ensure accurate positioning with guide holes. The latest machine has a level of accuracy better than 50 μm and high productivity based on automatic x-y table.

Very different ideas are required to achieve a severe perimeter that allows tight spaces from conductors (smaller than 50 μm). A laser router that has self-alignment capability could be a possible solution. However, the processing speed is slow and expensive. A chemical etching process of the base could be a low-cost solution for volume production; however, the manufacturing process is very different from a standard process and a special process design is required. Its applicable construction and materials are limited, and suitable design changes are recommended.

58.8 STIFFENER PROCESSES

It is difficult to apply full automation to the process of making stiffener boards because of the many varieties. Most of the processes are conducted manually and are labor intensive. Sometimes stiffener boards account for a big percentage of the total manufacturing cost.

An adhesive film is laminated on a rigid board with a release sheet first. The boards are then routed by punching or by NC router. It is a very simple process if the adhesive material is pressure-sensitive adhesives (PSAs). Each piece of stiffener is placed on the flex circuits with appropriate pressure, mostly by hand. The process is more complicated when a thermoset-type adhesive material is required. A temperature higher than 160°C with a pressure higher than 20 kg/cm² is required for more than 30 min. A similar heat press used for multilayer circuit boards or film coverlay is necessary. Dummy boards should be prepared to make the pressure uniform. A heat press with a vacuum chamber or autoclave could provide uniform adhesion for irregularly shaped stiffener boards.

58.9 PACKAGING

Many ideas have been introduced for packaging fragile HDI flex circuits. The packaging should provide easy handling capability, including the processing of the assembling side. A reeled tape modified from TAB technology is a valuable solution for volume production. A sheet carrier is another possible solution for medium- and large-scale production. Reusable cartridge or container designed for each circuit is recommended for volume production.

58.10 ROLL-TO-ROLL MANUFACTURING

Roll-to-roll (RTR) manufacturing is a specialty developed for rolled materials of flex circuits. It can reduce manual handling and defects caused by human errors; therefore, when running effectively, it will increase productivity significantly and reduce manufacturing cost. On the other hand, it requires a large investment and advanced engineering capabilities. It is not appropriate for small-volume production, and can create a lot of scrap when not set up correctly. Table 58.10 shows advantages and disadvantages of RTR manufacturing systems for flex circuits.

TABLE 58.10 Advantages and Disadvantages of RTR Systems

Advantages	Disadvantages
• High productivity	• Small flexibility for small volume
• High process yields	• Much scrap caused by small mistakes
• Low manufacturing cost	• Large investment cost
	• Limited for quick turnaround
	• Low flexibility for design changes

Table 58.11 shows available manufacturing processes and typical examples. Most RTR systems are developed for single-sided flex circuits. There have been significant technical barriers to applying the RTR system for double-sided flex circuit with through-holes. The NC drilling

TABLE 58.11 Availability of RTR Processes

Process	Technical difficulty
Surface cleaning	Standard machines available
Via hole generation	Possible, but difficult
Via hole plating	Big investment
Resist coating	Standard machines available
Pattern generation	Standard machines available
Developing	Standard machines available
Etching/strip resist	Standard machines available
Coverlay	Possible, but difficult
Surface treatments	Standard machines available
Blanking	Possible, but difficult
Inspection	Possible, but difficult
Stiffener boards	Difficult

process is the bottleneck in introduction of the RTR system. A punching process, which is basically a continuous-flow process element, is an alternative technology to NC drilling of the RTR system to be used for double-sided flex circuits. However, it requires an expensive punching die and is not flexible for a design change.

The new via hole technologies such as laser drilling and chemical etching are an innovative solution for generating small holes in rolled materials. A blind-via hole design is preferred for these RTR manufacturing processes.

58.11 DIMENSION CONTROL

Due to increasing density requirements and inherently unstable film substrate characteristics, dimensional control has become a key issue for both circuit fabricators and their end customers. In order to achieve such densities while maintaining high production yields, careful dimensional consideration in both the design and manufacturing processes is necessary. An exact concept for dimension control of the whole manufacturing processes is required for high-density flex circuits. There are many factors that impact the dimensional stability of high-density flex circuits. These factors can be categorized as follows:

- Materials
- Circuit design
- Manufacturing processes

58.11.1 Materials

Substrate material selection may be the most important item to consider for dimensional control of high-density flex circuits. (For a more detailed discussion of flexible circuit materials, see Chap. 56.)

58.11.1.1 Polyimide. Kapton H™ and Apical AV™ are standard substrate materials for traditional flexible circuits. Unfortunately, these films have high moisture absorption rates, which create excessive dimensional instability, making them unsuitable for high-density flexible circuits. Several newer polyimide films such as Upilex S™, Kapton E™, and Apical NP™ and HP™ have been commercialized to satisfy the requirements for increased dimensional stability. A comparison of these base materials is shown in Table 56.8. Although these new polyimide materials have lower moisture absorption and CTEs than traditional polyimide films, their dimensional change is still not zero. In addition, process variations of 0.02 to 0.05 percent are common from batch to batch during film production.

Generally, material manufacturers fabricate polyimide film with a wide web. Special conditioning is necessary to ensure uniform properties for the whole web. Unfortunately, there are usually deviations from one side to another. Because the master roll is slit in halves or thirds according to the customer's convenience, each roll has different mechanical properties depending on its location in the master roll. If the master roll is slit into three rolls, a roll in the middle may have more uniform properties than the others. If the master is slit into two rolls, relatively wide edges will be wasted.

58.11.1.2 Copper-Clad Laminate. Most manufacturers of flexible circuits initiate their processes with copper-clad laminates. During production of these laminates, mechanical stresses are generated (by heat and pressure) that can become locked into the material. These stresses are partially released during the circuit manufacturer's etching operation and a certain amount of dimensional reversion takes place. In actuality, these stresses (or dimensional stability characteristics) differ in the machine direction (MD) vs. the transverse direction (TD). It is therefore important to review all the mechanical stresses in a laminated material and to consider some form of stress relieving prior to circuit fabrication.

IPC-TM-650 recommends a method of testing dimensional stability in a 9 × 10-in rectangle for both MD and TD. One standard-size test sheet is too small to represent the whole web. Several samplings are required to evaluate the whole web. Yaw direction changes and local changes of dimensions that cannot be measured by the IPC test method are sometimes critical for high density flexible circuits. Detailed investigation should be conducted.

58.11.1.3 Coverlayer Selection. Coverlayer selection should be considered the second important material factor for controlling dimensional stability. Generally this layer is adhered to the circuitry using heat and pressure, which can induce additional dimensional distortion. Careful consideration should be taken if effective dimensional stability is to be achieved.

58.11.2 Circuit Design

As discussed earlier, most laminate materials undergo mechanical stresses, which generate nonuniform dimensional changes during the etching process. After etching, an etched area with substrate only has different physical properties from a nonetched area with substrate and copper foil. During manufacturing, the laminate materials generate different dimensional changes. A uniform circuit pattern design is necessary to control dimensional stability. A dummy copper pattern in an empty area can help to maintain optimal dimensional stability.

58.11.3 Critical Processes for Dimension Control

The critical manufacturing processes for dimensional control of flexible circuits are shown in Tables 58.12 and 58.13. In general, a high temperature or a wet chemical process can alter the dimensions of flexible circuits. Increased mechanical tensions can also impact dimensional control. The major manufacturing steps for standard double-sided flexible circuits with throughholes are listed here. Each step should be considered carefully to determine its contribution to the overall distortion.

1. *NC drilling:* Although NC drilling does not significantly impact the final dimension, certain scale factors based on assumed movements during the following operations may be factored into the drill pattern.
2. *Through-hole plating:* Because the electric field in a through-hole plating bath is generally nonuniform, the copper thickness becomes uneven. An extremely thickly plated copper con-

TABLE 58.12   Key Items for Dimension Control in Flex Circuit Manufacturing

Categories	Key items for dimension control
Materials	Base film, copper, coverlay, bonding material
Design	Layer constructions, circuit density, circuit balance
Manufacturing process	Wet processes, heat processes, mechanically stressed processes

TABLE 58.13   Critical Processes for Dimensional Control

Process	Category	Importance
NC drilling	Mechanical	Medium
TH plating	Wet and heat	High
Cleaning	Wet and heat	Medium
DF lamination and print	Mechanical	High
Development	Wet and heat	High
Etching and stripping of resist	Wet	High
Coverlay, film lamination	Heat and pressure	High
Coverlay, screen printing	Heat	High
Coverlay, photoimageable	Heat	High
Lamination for multilayer	Heat and pressure	High
Plasma etching	Heat	Medium
HASL	High heat	High
Electro- and electroless plating	Wet and heat	Medium
Baking	Heat	High
RTR process	Mechanical	Medium

Heat means temperatures higher than 80°C.

ductor will create mechanical stresses. During the etching process these stresses are released, generating some dimensional changes. In an extreme case, a copper-plated sheet cannot lie flat. An appropriate set of work fixtures and plating electrodes should be designed to create a uniform plating thickness.

3. *Dry film lamination and pattern printing:* Due to the relatively low temperature and lack of wet processing (moisture), dry film lamination and pattern printing have little effect on dimensional control. Pattern imaging is, however, an opportunity to apply appropriate dimensional corrections for the following processes.

4. *Etching and stripping of resists:* The most critical wet process in flexible circuit manufacturing is etching. Dimensional changes can occur when the by-products of the etching process are combined with the various materials. It is important to identify how to correct the various dimensional alterations, particularly as additional factors are incorporated into the process.

5. *Coverlay processes:* There are several coverlay processes that have been developed with varying process conditions for different concept materials. Film coverlay undergoes a manufacturing process with a temperature greater than 160°C and a pressure over 20 kg/cm². Cover coat ink withstands a screen-printing process with a drying temperature higher than 130°C. A newly developed photoimageable coverlay must be baked at over 150°C. The majority of the heat and pressure processes impact dimensional stability significantly. Additional factors, such as press pad heating, make dimensional changes even more complicated. Conditioning should be reviewed in detail for each construction.

6. *Lamination pressing for multilayers:* Like the coverlay process, lamination pressing involves high temperature and pressure. Due to the number of layers, however, dimensional changes are more complicated because each level needs to be considered.

CHAPTER 59
TERMINATION OF FLEXIBLE CIRCUITS

Dominique K. Numakura
DKN Research, Haverhill, Massachusetts

59.1 INTRODUCTION

Major termination technologies for flexible circuits are listed in Table 59.1. Most termination technologies developed for rigid printed circuit boards are available for flex circuits with small modifications. SMT components can also be assembled through the same processes used for rigid circuit boards, such as standard pick-and-place and reflow soldering without significant conditioning. Most of the card-edge connectors designed for rigid circuit boards are also available for flex circuits with the addition of suitable stiffener boards. In addition, because of the unique wiring capabilities and special construction of flexible circuits, many specialized termination technologies have been developed for them. These include solder fusing, ACF connections, flying-lead direct soldering, dimple contacts, and FFC connectors.

To accommodate the miniaturization of electronic products, many termination technologies have also been developed for high-density interconnect (HDI) flexible circuits. These technologies are important in the assembly of various electronic components in small spaces, such as portable telephones, miniature LCD displays, keypad switches, antennas, connectors, microphones, speakers, battery cases, etc. It is common for several different termination processes to be used in each electronic product. Examples of applications include wireless suspension of disk drives and chip-scale packaging (CSP). These technologies use high-density flexible circuits as the interposers. Multiple special termination capabilities are required for these interposer flexible circuits.

All of the termination technologies of flexible circuits have both advantages and disadvantages. No technology can satisfy all requirements at the same time. In general, flexible circuits have numerous wiring options, and therefore a thorough review of the technologies and the wiring requirements is required to design the appropriate termination.

59.2 SELECTION OF TERMINATION TECHNOLOGIES

In order to determine the optimal cost-performance combination, the requirements from the assembly side and the capability of the termination technologies need to be determined. Table 59.2 lists the critical items that are required to design a suitable termination.

TABLE 59.1   Termination Technologies for High-Density Flexible Circuits

Rigid PWB based	Flexible circuit based
• SMT	• Solder fusing
• COB (COF)	• ACM (ACF)
• Wire bonding	• Direct bonding of flying leads
• Conductive adhesive	• Dimple contact
• Card-edge connectors	• Microbump contacts
	• FFC connectors

TABLE 59.2   Critical Items in the Design of High-Density Interconnects

Critical Item	Description
Objectives	What kinds of devices?
Termination density	Points per cm or cm^2
	Number of connections in the allowed space
Permanent or nonpermanent	Repeated connections or not?
Circumstances used	Temperature, humidity, etc.
Condition processed	Temperature
Reliability	Total performance
Cost	Material and processing

59.2.1 Termination Objectives

There are several types of electronic components that can be attached to flexible circuits. Table 59.3 categorizes these from the termination standpoint, assuming the use of high-density flexible circuits.

Because of the nature of their use, there are a limited number of locations where imaging devices such as LCDs and CCDs can be placed in electronic products. As a result, there are fewer choices for wiring materials other than flexible circuits. Newer imaging devices have numerous pixels in a limited area, which requires more connections with smaller pitches. In addition, it is often important to place the driver IC close to the device. Therefore, a flexible circuit is required for both wiring and COF. A large LCD, typically used in a notebook computer, often needs several TAB circuits to create the wiring between a rigid board and the LCD de-

TABLE 59.3   Categories of Electronic Components

Category	Components
SMT devices	Discrete chip components, QFP, CSP, MCM, bare IC chips
Imaging devices	Large FPD, small FPD, CCD, linear sensors, 2-D sensors, printers
Wiring devices	Rigid PWB, ceramic-based circuits, flex circuits, cable, wires, connectors
Specialty devices	Keyboards, switches, printer heads, antennas, actuators, sensors, microphones, speakers, chimes

vice. A flat-panel display can be wired by a single high-density flexible circuit, which reduces the cost of the entire package.

It is common for a flexible circuit to be connected to other circuit boards. However, an increased number of traces and a higher density of the connections is then required. The placement of specialty components such as keyboards, switches, antennas, actuators, printer heads, sensors, microphones, speakers, and buzzers in small electronic applications is determined automatically by the basic design of the end products. Although these components do not require numerous wires, they do need high-density connections due to the extreme miniaturization of portable electronic products.

59.2.2 Nonpermanent or Permanent Termination

The permanence of a termination is determined by connection repeatability. For example, if connections are changed regularly, a nonpermanent termination should be utilized. Due to the number of reconnections, inkjet printers and electrical test probes should have nonpermanent terminations. Conversely, if a connection is to remain relatively constant, a permanent termination method should be employed. Termination method capabilities are profiled in Table 59.4.

TABLE 59.4 Capabilities of Termination Methods

	Permanent or nonpermanent	Termination density
SMT soldering	Permanent	350-µm pitch (2-D)
SMT wire bonding	Permanent	150-µm pitch (4 directions)
Solder fusing	Permanent	150-µm pitch
ACM connection	Permanent	50-µm pitch (100 contacts in 1 mm²)
Direct bonding	Permanent	50-µm pitch (1- or 2-D)
Pressure contacts	Nonpermanent	20 contacts in 1 cm² (1- or 2-D)
FFC connectors	Nonpermanent	300-µm pitch

59.2.3 Termination Density

In order to design a suitable termination solution, the density and the number of terminations need to be considered. An overview of each method's termination density is profiled in Table 59.4. Due to higher connection reliability, a contact array can produce more connections than a contact line in the same space with a larger pad pitch. Numerous wires must fit together in a limited space.

59.2.4 Wiring of Imaging Devices

A high-density flex circuit could be the best solution for the wiring of a small imaging device. In general, these devices have hundreds of pads that need to be connected with COF for the driver IC close to the device. The flex circuit usually requires a long tail to be connected to the other device. Only a high-density flex circuit can satisfy all the requirements as a single circuit. There are several choices in the design of the constructions of the COF area and the bonding area. An adhesiveless clad laminate having a stable substrate with polyimide resin is required to create a high connection yield for a high-resolution device. An ACM connection is a low-risk solution because of the low processing temperature. A direct bonding process with fine flying leads can provide reliable connections for COF and wiring with the imaging device. Due to increased temperature, it requires a higher dimension control.

59.3 PERMANENT CONNECTIONS

If there is no functional reason to remove the connection during normal operation, a permanent connection is recommended because of the lower costs. Examples of this type of connection include:

- Soldering, general
- Through-hole leaded components
- General surface-mount technology components
- Solder fusing
- Wire bonding
- Direct bonding flip-chip

59.3.1 Soldering

Soldering is still the most popular and common termination technology for flexible circuits. All solder and termination materials and equipment are available with suitable tooling guides. If adhesiveless laminates are used as the major materials, high temperatures can be utilized with lead-free soldering. In the case of SMT assembling, the flexible circuit should be fixed on a carrier frame. This allows it to be processed by the same mounting and soldering equipment as used for rigid boards.

59.3.2 Through-Hole Leaded Components

Often, through-hole leaded electronic components are necessary for specific applications. Processes similar to those used for single-sided rigid boards are used to attach them to the flexible material. However, because through-hole leaded components require clinching of the leads and are heavier than SMT components, special fixtures and control are required when placing them on a fragile flexible circuit. This is to minimize mechanical stresses, especially when components are connected by soldering, as shown in Figs. 59.1(*a*) and 59.1(*b*).

59.3.3 SMT General Issues

Most SMT technologies can be used for flexible circuits without significant alteration (Fig. 59.2). In general, the same circuit design rules can be applied to SMT for a high-density flexible circuit. A suitable thin stiffener sheet such as 125-μm-thick polyimide film should be bonded to support SMT components. Adhesiveless copper-clad laminates need to be used as the raw material to withstand the high temperatures of typical lead-free soldering processes. A flexible photoimageable solder mask needs to be applied to cover the fine traces. A special dynamic flexible-grade photoimageable coverlay should be used if the flexible circuit has a portion of repeated flexing. A stiffener of suitable thickness needs to be bonded on the back side of an SMT area of a thin, single-sided flexible circuit. A thicker substrate layer should be used for double-sided assembly.

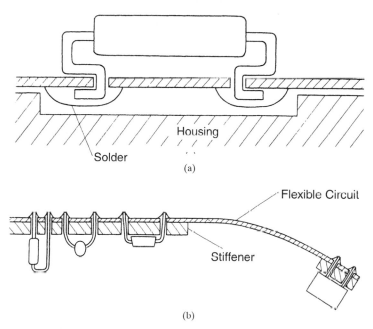

Housing

Solder

(a)

Flexible Circuit

Stiffener

(b)

FIGURE 59.1 Examples of direct soldering of flexible circuits. (*a*) without stiffener; (*b*) with stiffener boards.

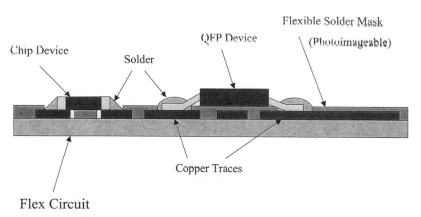

Chip Device

Solder

QFP Device

Flexible Solder Mask

(Photoimageable)

Copper Traces

Flex Circuit

FIGURE 59.2 SMT soldering of high-density flexible circuits.

An appropriately designed stiffener board can increase assembly productivity without a big cost increase. A larger stiffener board and adhesive sheets are punched beforehand for the preparation of the assembling processes. The stiffener board has extra frame space that contains appropriate guide holes for the convenience of part mounting and soldering. The flexible circuits fixed on the stiffener boards can be managed the same as rigid printed circuit boards in the assembling processes. A standard mounting machine places small SMT components without additional tooling. In addition, the same high-temperature soldering equipment used for reflow soldering in an SMT line can be used here. In-circuit tests are conducted with an aluminum board. As the final process of the assembly, unnecessary parts of the stiffener boards are cut off, and the whole circuit is folded to make a part of the module casing. An appropriate seal is made with the other casing parts to protect the module from the outside stress.

59.3.4 Solder Fusing

Although solder fusing (Figs. 59.3 and 59.4) is not a new technology for making connections for flex circuits, it is still a low-cost and reliable solution for multitrace connections. There has also

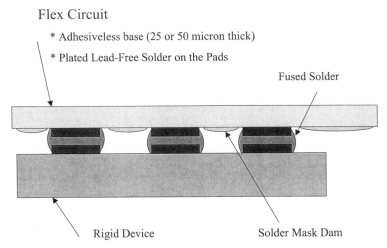

FIGURE 59.3 Flexible circuit termination by solder fusing.

FIGURE 59.4 Flexible circuits connected to a rigid circuit board by solder fusing.

been remarkable progress in fine-pitch capability. The newest applicator, coupled with well-controlled solder plating on fine traces of flex circuits, can provide a 150-μm-pitch connection with the other circuit pads. It is also possible to produce 30 connections in a 5-mm square, but high fusing temperatures with lead-free soldering are required. Because traditional adhesive materials such as epoxy or acrylic resin cannot survive the process, adhesiveless laminates need to be used as the major materials.

59.3.5 Wire Bonding

A wire-bonding process similar to COB has been applied for high-density flexible circuits without serious changes. Due to the high heat resistance of new adhesiveless laminates, standard bonding materials and equipment for flexible circuits can be used (Fig. 59.5). The wire-bonding process could be a standard assembly technology for the chip-on-flex (COF) system. Numerous driver ICs for LCD devices have been assembled by these methods.

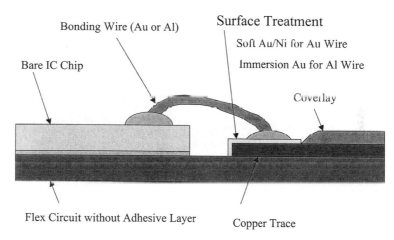

FIGURE 59.5   Wire bonding on an HDI flexible circuit.

59.3.6 Direct Bonding

Although flying-lead construction is not a new technology, fine-line traces created direct bonding. When a flying lead has a fine line smaller than 100 μm wide with nickel/gold plating on it, it can be processed by the same process as the wire-bonding process. This can produce high-density, reliable connections between a flexible circuit and another circuit or device (Figs. 59.6 and 59.7). The basic idea was developed with tape automated bonding (TAB), but flexible circuits expand the capabilities and applicable areas.

59.3.7 Flip-Chip

Flip-chip mounting of bare IC chips has been common in consumer portable products. The same technologies as used in rigid circuit boards are available for flexible circuits. Adhesiveless copper-clad laminate should be used when high-temperature processes are applied. Cast-type adhesiveless laminates could provide high reliability because of high bond strength.

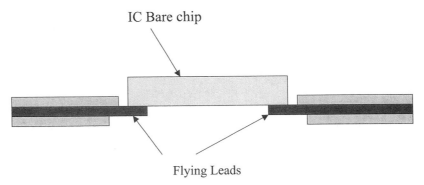

FIGURE 59.6 Direct bonding of flying leads.

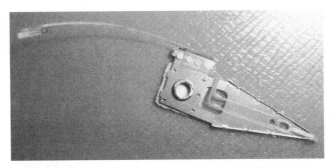

FIGURE 59.7 Termination of head suspension with flying leads. *(Source: K R Precision.)*

59.4 SEMIPERMANENT CONNECTIONS

Semipermanent connections have been required for rework, replacement of components and modules, and repeated connections. Realistic assembly processes for most electronic products are not perfect, and therefore semipermanent terminations are required for rework and replacement of electronic components or modules in usual assembly processes to increase final manufacturing yields. Also, temporary connections are required for testing and adjustment of electronic products and modules. These often need 5 to 10 reconnections at least, with 100 reconnections required in an extreme case. Basic data on typical semipermanent termination technologies are summarized in Tables 59.5 through 59.7.

TABLE 59.5 Semipermanent Interconnects with Flexible Circuits

	Objectives	Application examples
Pressure contacts	Flex-rigid boards	Facsimile
Anisotropic conductive rubber	Flex-flex or rigid boards	Flat panel displays
Dimple contacts	Flex-rigid parts (flex-flex)	Inkjet printers
Microbump array	Flex-rigid boards and parts	Board testers

TABLE 59.6 Performance of Semipermanent Technologies

	Connection density	Repeatability	Trace density of flex circuits
Pressure contacts	150-μm pitch	~20	150-μm pitch
Anisotropic conductive rubber	100-μm pitch, 200 contacts per cm^2	~1000	100 μm pitch

59.4.1 Pressure Contact Termination

The basic idea of the technology is very simple and it is not new. Pad patterns of the same pitch plated with gold are generated on a flexible circuit and opposite circuits, and they are attached to each other with uniform pressure using a suitable rubber strip (Fig. 59.8). There is no high-temperature process to complete the connection; polyester base materials are available. The maximum connection number and density depend on the capability of dimension control and uniformity of the circuits. It is possible to make 200 connections with 150-μm pitch because of the all-room-temperature processes. Neither special application equipment nor special constructions on flexible circuits are required except simple tooling, and therefore it could be a low-cost solution. Unfortunately, connection and disconnection take time, and this is not a suitable technology for multiple repeated connections.

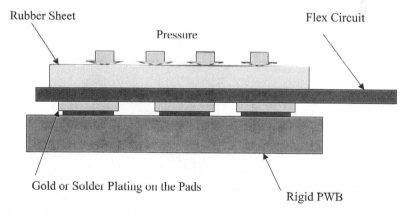

Rubber Sheet

Pressure

Flex Circuit

Gold or Solder Plating on the Pads

Rigid PWB

FIGURE 59.8 Pressure contact of flexible circuits.

TABLE 59.7 Items to Be Considered in Semipermanent Termination Technologies

	Supplemental parts	Applicators	Remarks
Pressure contacts	Rubber strip and fixture	Simple tooling	Good dimensional control required
Anisotropic conductive rubber	Fixture and ACR strip	Simple tooling	Not available for high currents High contact resistance High noise level

59.4.2 Anisotropic Conductive Material

Newly developed anisotropic conductive rubber has generated several new termination technologies for high-density flexible circuits (Figs. 59.9 and 59.10). The latest material has 100-μm pitch resolution in one direction. Another type of sheet rubber can have more than 200 contacts in 1 cm². These types of rubber can make more varieties for traditional pressure contacts. A dis-

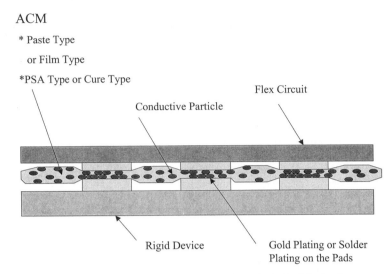

FIGURE 59.9   Ansiotropic conductive material (ACM) of a flexible circuit.

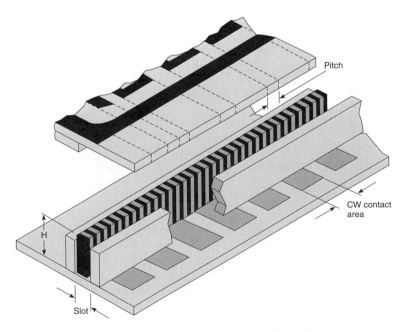

FIGURE 59.10   Pressure contacts with ansiotropic conductive rubber.

advantage of the technology is high contact resistance, which makes it unavailable for large-current circuits. Also, it generally has a relatively high noise level. Table 59.8 gives the basic properties.

TABLE 59.8 Basic Properties of Anisotropic Conductive Materials

Manufacturer	Toshiba Chemical	3M
Grade	XAP	
Type	Paste	Film
Gel time	15 s	
Contact resistance	<0.0015 Ω	
Current capacity	>2 A	
Insulation resistance	>1 × exp13	
Breakdown voltage	>500 V	
Peel strength	>1 kg/cm	

59.5 NONPERMANENT CONNECTIONS

Many repeated connections are required for replacement of disposable components such as inkjet printer cartridges and biological sensor devices. These require more than 1000 repeated connections for a flexible circuit cable. More than 1 million repeated reliable contacts on flexible circuits are required for switching circuits and test probe circuits. Flat-panel switches and electrical test probe fixtures for microcircuits are typical examples. As flex circuits have high-density cabling capabilities other than the mounting capabilities of electronic components, there have been serious requirements for high-density nonpermanent termination.

There have been two kinds of approaches to realizing high-density nonpermanent flex circuits. The first is improvement in connection densities of traditional termination technologies. Crimp tabs, pressure contacts, key contacts, and FFC connectors could be categorized in this group. Another approach is a new concept. Anisotropic conductive rubber, dimple contacts, and bump contacts are typical examples. The technologies are compared in Tables 59.9 through 59.11.

59.5.1 General Connectors

Most of the general connectors designed for other wiring technologies are applicable without serious changes. (Fig. 59.11 shows a set of alternative designs.)

TABLE 59.9 Nonpermanent Interconnects with Flexible Circuits

	Objectives	Application examples
Crimp tabs	Flex-flex	Tach panel switches
Key pad contacts	Flex internal	Keyboards
FFC connectors	Flex-flex or rigid boards	Computer peripherals
Area array connectors	Flex-rigid boards	Industrial equipment
Dimple contacts	Flex-rigid parts (flex-flex)	Inkjet printers
Microbump arrays	Flex-rigid boards and parts	Board testers

TABLE 59.10 Performance of Nonpermanent Technologies

	Connection density	Repeatability	Trace density of flex circuits
Crimp tabs	1.27-mm pitch	~100	1.27-mm pitch
Keypad contacts	10 in 1 cm^2	10 million	200-μm pitch
FFC connectors	0.3-mm pitch	~100	150-μm pitch
Area array connectors	100 contacts per cm^2	~500	100-μm pitch
Dimple contacts	20 contacts per cm^2	~1000	0.5-mm pitch (150-μm pitch)
Microbump arrays	400 contacts per cm^2	~1 million	60-μm pitch

59.5.1.1 Card-Edge Connectors. All kinds of connectors designed for rigid circuit boards are available for flexible circuits. They can be mounted and soldered on flexible circuits supported with stiffener boards of suitable thickness. Card-edge connectors designed for rigid circuit boards are also available. A flexible circuit with a suitable stiffener board on the bottom can be inserted into the connector. However, the newest FFC connectors designed specifically for flexible circuits have a much higher density in a smaller housing and should be replaced by FFC connectors for large volume.

59.5.1.2 Area Array Connectors. Higher density than 150-μm pitch on double-sided flexible circuits with through-holes is required for high-pin-count connections. All kinds of connectors designed for rigid printed circuit boards are available for flexible circuits with some modification on flexible construction. A newly designed array connector for high-density flexible circuits can make 300 connections in a 10 × 15-mm area (Fig. 59.11). Special constructions are not necessary, other than 0.5-mm-pitch array soldering. However, traces with pitches smaller than 150 μm are required for high pin counts. They can survive 1000 repeated contacts.

59.5.1.3 Crimp Tabs. Crimp tabs were originally developed for connections of flat conductor cables; however, they are also applied for flexible circuits because of the similar construction. Female tabs are firmly stapled through dielectric layers mechanically by an applicator (Figs. 59.12 through 59.14). Male tabs are stapled on the other cable. Both are covered with plastic housings. Because there is no soldering process, polyester-based materials are available. The standard pitch was 2.54 mm for a long time. A 1.25-mm system was introduced recently, doubling the connection density. The connectors can make reliable repeated connections more than 100 times. Total cost per connection is relatively high.

TABLE 59.11 Items to Be Considered for Nonpermanent Termination Technologies

	Supplemental parts	Applicators	Remarks
Crimp tabs	Tab and housing	Applicator	Low connection density.
Keypad contacts	Conductive keypads or a sheet	No	Density depends on mechanical key sizes.
FFC connectors	ZIF connectors	SMT soldering	Soldering on the other circuit.
Area array connectors	Male and female connectors	Special soldering	High-density traces are required on flex circuits.
Dimple contacts	(Housing)	Simple tooling	Suitable forming process shall be developed.
Microbump arrays	(Housing)	Special fixtures	Suitable processes to build microbumps must be developed.

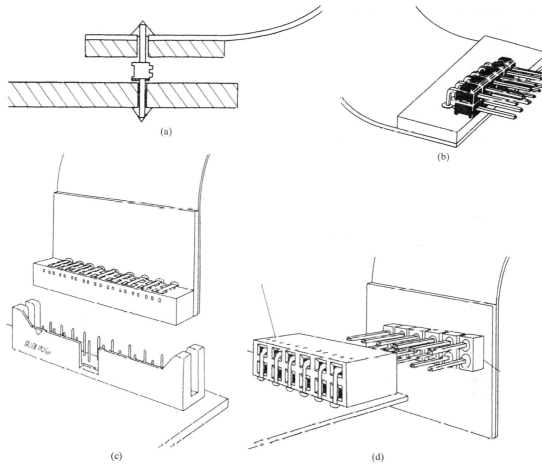

(a)

(b)

(c)

(d)

FIGURE 59.11 Examples of soldering lead–type connectors on flexible circuits.

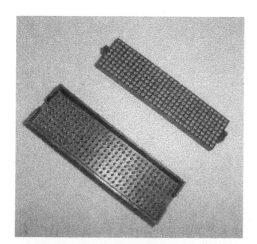

FIGURE 59.12 High-pin-count area array connectors available for high-density flexible circuits.

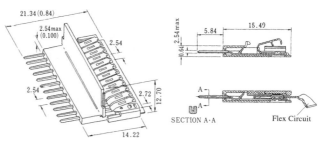

FIGURE 59.13 Flying connector (male) for flex circuits (2.54-mm pitch).

FIGURE 59.14 Clamp-type connector for flexible circuits.

59.5.2 FFC Connectors for Flexible Circuits

Originally, FFC connectors were developed for laminated cable with flat conductors to connect to another circuit board. A similar construction of flexible circuits was developed to attach the connector. At this time, the standard pitch of the connectors was 2.54 mm. Due to increased trace densities of flexible circuits, small-pitch FFC connectors are now required. The newest connector can make 0.3-mm-pitch connections for 30 traces. Zero in force (ZIF) mechanisms were introduced for easier handling. Due to limited height requirements for thin board assembly, a 1.2-mm-high connector from the board surface is now available.

59.5.3 Bump Array Contacts and Dimple Array Contacts

59.5.3.1 Dimple Contacts. Dimpled pads of flexible circuits can be a low-cost solution for high-density connections of flexible circuits. One hundred contacts in 12.5 mm^2 between flexible circuits and the other circuit boards are possible. Appropriately shaped dimples are formed for a gold-plated pad by pressure and temperature. Dimpled pads with hard gold plating have enough life for several hundred connections/disconnections (Figs. 59.15). The dimples are supported with bumped rubber to create long reliable contacts. The pitches of the dimples cannot be smaller than 1.5 mm, but high-density termination can be realized with the area dimple array.

59.5.3.2 Microbump Array. Because of thin substrates, various constructions of microbump arrays have been developed. A plating combination of nickel and hard gold on a copper bump

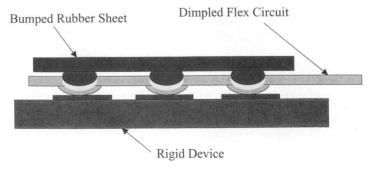

FIGURE 59.15 Dimple contacts for flexible circuits.

provides reliable contact performances. A microbump array built on flexible circuits by a plating process enables high-density contacts with the other circuit boards. The process has a range of wide flexibility to generate various shapes of microbumps. It is possible to build more than 1000 microbumps in 10 mm² if 30-μm line/space circuits are available. A mushroom-head copper bump plated with hard nickel and hard gold has enough life for millions of contacts (see Fig. 59.16).

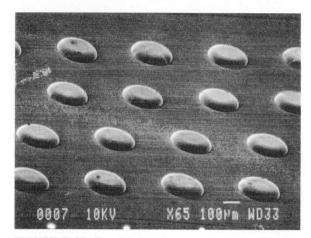

FIGURE 59.16   Flat disk-shaped microbump array on flexible circuit for area contacts. Contact metal is copper/nickel gold at 0.5-mm pitch. *(Source: Asahi Fine Technology.)*

59.5.3.3 *Keypad Contacts.*

There are several different constructions for keypads on flexible circuits. Thick film circuits on polyester film could be a low-cost solution with 1 million reliable contacts. However, polyimide-based circuits are necessary with soldering of components. The size of the mechanical key is the bottleneck in creating a high-density switch array. Ten switches are possible in 1 cm².

59.6   HIGH-DENSITY FLEXIBLE CIRCUIT TERMINATION

Table 59.12 shows new termination technologies used for high-density flexible circuits.

TABLE 59.12   New Termination Technologies for High-Density Flexible Circuits

Termination technology	Termination density (μm pitch)	Repeatability	Requirement for flex circuits
Lead-free reflow	300	~2	Higher heat resistance (260°C for 30 s)
Lead-free hot bar reflow	150	~2	High heat resistance (400°C for 10 s) High dimensional accuracy (~0.02%)
Wire bonding	150	1	High heat resistance (450°C for 3 s)
Direct bonding	50	1	Flying-lead construction (50-μm pitch)
High-density BGA for flip-chip	200	1	Microbump array (200-μm pitch)
Anisotropic conductive materials	60	~5	High dimensional accuracy (~0.01%)
Small-pitch ZIF connectors	300	100	High dimensional accuracy (~0.05%) uniform tin plating
Contact probe array	200	~1000	Microbump array hard gold plating on nickel

CHAPTER 60
SPECIAL CONSTRUCTIONS OF FLEXIBLE CIRCUITS

Dominique K. Numakura
DKN Research, Haverhill, Massachusetts

60.1 INTRODUCTION

Flexible circuits are used in many situations that require complex constructions and supplemental structures designed for the application. This chapter introduces the major constructions in use, along with their manufacturing processes. A combination of rigid and flexible circuits can be called *rigid/flex* or *rigid/flexible*. The term *rigid/flexible circuits* is used throughout this chapter.

60.2 MULTILAYER RIGID/FLEXIBLE CIRCUITS

Figure 60.1 shows a standard manufacturing flow for multilayer rigid/flexible circuits. It can be seen that this is basically a combination of rigid multilayer and flexible circuit processes. However, the successful accomplishment of this combination requires a high level of skill in both areas, and it is important to have a clear understanding of the capabilities and limitations of the chosen fabricator before designing this type of circuit into a product.

60.2.1 Basic Constructions

There are many different concepts in the construction design of multilayer rigid/flexible circuits. Figure 60.2 shows the basic construction of a multilayer rigid/flexible circuit, both from a plane view and in cross section. Figure 60.3 shows an example of this construction. More than 30 layers are built for extreme cases, mostly for aerospace applications. Because of the requirement for high reliability, fine pattern designs cannot be used. In addition, leaded components are specified instead of SMT components. The design usually calls for relatively large lines/spaces and large-diameter through-holes with thick copper plating. There are several types, designated the folding type (Fig. 60.4), the flying tail type (Fig. 60.5), the bookbinder type, etc., according to their shapes.

On the other hand, consumer applications such as palmtop computers have been requiring high-density rigid/flexible circuits at very a low cost. They are specifying traces finer than 100 μm and blind via holes smaller than 150 μm in diameter with high-density SMT. Many tech-

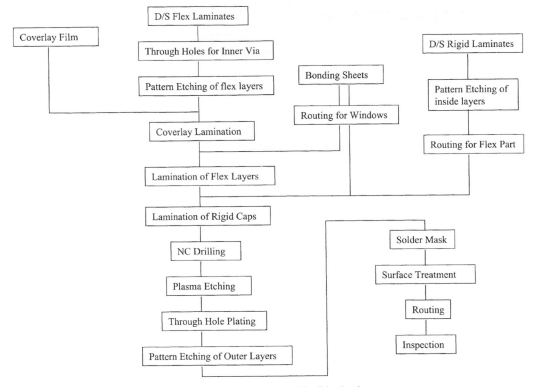

FIGURE 60.1 Standard manufacturing flow for multilayer rigid/flexible circuits.

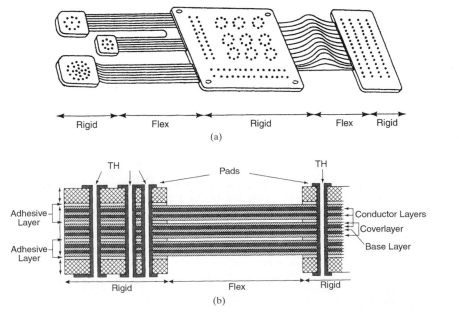

FIGURE 60.2 Basic construction of multilayer rigid/flexible circuits. (*a*) Plane view. (*b*) Cross section.

FIGURE 60.3 Example of simple multilayer rigid/flexible circuit. *(Source: Toshiba.)*

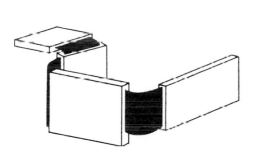

FIGURE 60.4 Folding-type rigid/flexible circuit.

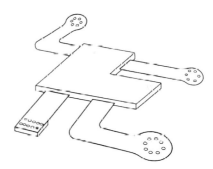

FIGURE 60.5 Flying tail–type rigid/flexible circuit.

nologies have been developed to satisfy these requirements. The use of a build-up process technology could be one of the solutions.

60.2.2 Materials

Several supplemental materials are necessary to build a multilayer rigid/flexible circuit, as shown in Table 60.1. It should be noted that there has been significant technical progress in the development of these high-performance materials.

The materials must have high heat resistance and high dimensional stability to survive during several high-temperature processes. Thicker polyimide films (50 μm thick or thicker) are recommended because the basic dielectric materials are required to have good stability in the manufacturing processes. An acrylic adhesive system in copper-clad laminates, as well as coverlay film and bonding sheets that have higher heat resistance than epoxy systems, are also recommended for use in manufacturing. Adhesiveless copper-clad laminates, made by casting or lamination processes, usually perform better in high-temperature processes. They are also valuable because they reduce the total thickness of the finished board. Hot melt–type polyimide adhesive systems, including coverlay films and bonding sheets, have been developed to

TABLE 60.1   Materials for Multilayer Rigid/Flexible Circuits

Materials required	Traditional materials	High-performance materials
Flexible substrates	Traditional polyimide films (Kapton H™, Apical AV™)	New polyimide films (Kapton E™, Apical HP™, Upilex S™)
Copper-clad laminates (double-sided)	Polyimide film substrate Acrylic adhesive (epoxy adhesive)	Polyimide film substrate Adhesiveless laminate (cast materials or laminated materials)
Coverlay	Traditional polyimide films coated with acrylic or epoxy adhesives	New polyimide films coated with hot-melt polyimide adhesives
Bonding sheets	Acrylic resin films Epoxy resin films Polyimide films coated with acrylic adhesives on both sides	New polyimide films coated with hot-melt poyimide resin on both sides
Rigid substrates	Glass/epoxy boards	Glass–BT resin boards Glass-polyimide boards

have very high reliability and smaller thicknesses. They can reduce smear level in drilled holes significantly. But they must be processed at temperatures higher than 300°C, and therefore require special facilities and conditioning.

60.2.3   Manufacturing Process Flow

There are many varieties of manufacturing processes for multilayer rigid/flexible circuits because of their complicated structures. Figures 60.6 and 60.7 show a typical layer construction using the standard manufacturing process for multilayer rigid/flexible circuits shown in Fig. 60.1. As seen in Fig. 60.1, the process starts with flexible double-sided copper-clad laminates. A conductive pattern is generated by a normal etching process. Through-holes are made prior to etching when inner via holes are required. All traces are covered with plain coverlay

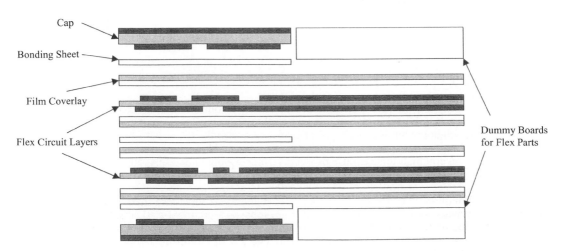

FIGURE 60.6   Material construction for multilayer rigid/flexible circuits.

FIGURE 60.7 Example of typical multilayer rigid/flexible circuit with via holes. *(Source: Toshiba.)*

films, which have no openings. The multiple covered flexible circuits are bonded with bonding sheets, which are already routed to make openings for nonbonding flexible areas.

Caps (the outside layers of rigid parts) are made with rigid double-sided copper-clad laminates. Traces are printed and etched only on the inside surfaces of the laminates as the first process. Flexible parts of the boards are removed through NC routing or punching. Both flexible layers and caps are bonded with bonding sheets that have been routed for flexible parts. Appropriate dummy boards are prepared for the flexible part during the bonding process. A vacuum press is preferred for good bonding quality. An autoclave system is recommended for complicated circuit constructions, because the autoclave can create uniform pressure on the entire circuit. Suitable baking should be done prior to lamination or bonding processes.

As the laminated board can be handled in the same way as multilayer rigid boards, the same through-hole processes are available, except desmearing, which is dependent on the combination of materials used (see later). The drilling process is conducted after enough baking in basically the same way as for rigid boards. A plasma etching process should be applied to remove smears of acrylic resins in the holes. The etchback conditions should be determined carefully. The same through-hole plating process is available; however, detailed conditions must be determined based on reliability test data.

The rest of the processes are very similar to those used for multilayer rigid boards. Outerlayer etching, coverlay (solder mask), and surface treatment are conducted in the same way. When the dummy boards are removed after routing, the circuit becomes a rigid/flexible circuit.

60.2.4 Through-Hole Process

When inner via holes are required in flexible layers, the same through-hole process should be conducted prior to etching of innerlayers as for rigid board innerlayers. Figure 60.7 shows an example of inner via holes in flexible parts of a multilayer rigid/flexible circuit, and Fig. 60.8 shows a cross section of a multilayer rigid/flexible circuit plated through-hole. The through-hole drilling can be accomplished by the same machine as used for rigid multilayer boards.

FIGURE 60.8   Cross section of a through-hole in a multilayer rigid/flexible circuit.

However, the drilling conditions should be determined carefully for each material's construction. Baking conditions prior to drilling could be critical in creating reliable through-holes.

Desmearing of the rigid/flexible circuits is different from that of standard rigid multilayer boards. The acrylic adhesive materials will undergo serious swelling during standard permanganate desmearing processes, lowering through-hole reliability. A plasma etching process is recommended to remove the smears of acrylic adhesive materials in the drilled holes. Use of a polyimide adhesive system can reduce the smear level significantly, but this also requires a high-temperature process in special press equipment.

60.2.5   Build-up Process and Blind Via Holes

Similar build-up processes with blind via holes are used with multilayer rigid and rigid/flex circuits; however, instead of the traditional manufacturing process used for multilayer rigid/ flex circuits, more layers are built on the rigid part with resin-coated foils. Figure 60.9 shows a cross section of a blind microvia hole in a multilayer rigid/flexible circuit made by a build-up process with copper foil precoated with epoxy resin. A single-sided adhesiveless copper-clad laminate with bonding sheet or prepreg sheet could be an alternative material to resin-coated foil. Figure 60.10 shows a micro–blind via hole in a multilayer rigid/flexible cir-

FIGURE 60.9   Micro–blind via hole on a multilayer rigid/flexible circuit made by a build-up process using resin-coated foil. The hole diameter is 150 μm. *(Source: Photo Machining.)*

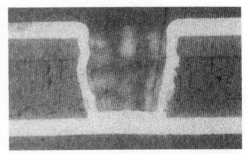

FIGURE 60.10   Micro–blind via hole on a multilayer rigid/flexible circuit made by a build-up process using adhesiveless laminate prepreg. *(Source: Photo Machining.)*

cuit, made by a build-up process using an adhesiveless laminate prepreg and having a hole diameter of 200 μm.

60.2.6 Bookbinder Construction

A multilayer rigid/flexible circuit that has multiple conductor layers in its flexible parts will not bend effectively if all the layers are of the same length. Figure 60.11 shows a bookbinder construction that is recommended for this application. Special tooling and equipment are required, however, and the productivity is comparatively extremely low. As a result, this process is not used for consumer or other applications that require low cost.

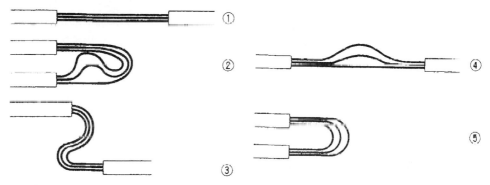

FIGURE 60.11   Design of bookbinder construction of multilayer rigid/flexible circuit.

60.3 FLYING-LEAD CONSTRUCTION

New HDI flexible circuit applications have arisen involving flying-lead construction. Applications such as wireless suspension of hard disk drives, interposers for chip-scale packaging, and ultrasound probes depend on the reliability of this high-density interconnect. In competitive markets such as these, it is imperative that flying-lead applications offer low cost and high volume. Traditional manufacturing processes used to create flying leads, such as prepunching or predrilling, are not conducive to the quality levels and manufacturing yields required to achieve low cost, and alternative processes are needed, as described later.

60.3.1 Basic Design

A flying-lead structure is composed of a single layer of high-density-pitch copper conductors accessed from both sides of the substrate. Generally, the substrate side of a conductor on a circuit board is not electrically accessible because the conductive foil is typically adhered to the substrate board in sheet form. In the case of a flex circuit, however, the substrate can be removed by various methods. By removing the coverlayer in the same area, dual access is created.

Figure 60.12 indicates two basic structures of dual-access conductors, in both the single-sided and double-sided types of flexible circuits. Basically, a coverlayer is applied to the side

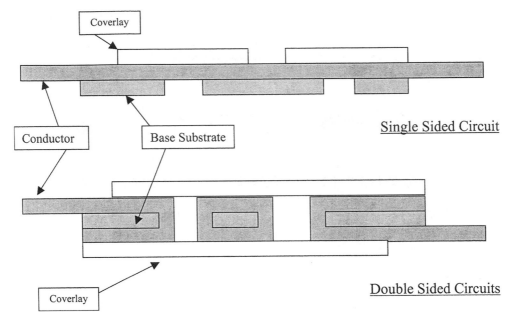

FIGURE 60.12 Basic constructions of flying leads for flexible circuits.

of the conductive layer opposite the substrate. The coverlayer can be processed easily to create the top-side exposure. The flying lead is created when the base substrate is also removed in the same area.

60.3.2 MANUFACTURING PROCESSES FOR FLYING LEADS

The fragility of flying leads, caused by the lack of dielectric support, is probably the largest disadvantage in the manufacturing process. Proper circuit design and layout is the key to the manufacturing process yield and subsequent cost and physical performance of the circuit. The most conventional manufacturing process for generating a flying-lead construction involves a prepunching process similar to the device hole process of tape automated bonding (TAB). Figure 60.13 shows this process schematically. There is no proprietary technology or special tooling required for manufacturing the prepunched window in the base substrate. This process, however, offers the manufacturer very little flexibility and is fairly labor intensive, which makes it difficult to maintain high productivity rates.

The handling of unstable punched films, along with thin, fragile copper foils, makes it difficult to achieve the high manufacturing yields required to support the cost model associated with high-density flexible circuits. These unstable film substrates also make it difficult to maintain dimensional accuracy between the fingers and the window, a key element for HDI flex. Adhesive flow or squeeze-out from the edge of the punched film along the conductor can also be an issue when dealing with small window exposures. Openings of less than 1.0 mm are difficult to create with high dimensional accuracy and yield.

Alternative technologies have been developed to generate high-density flying-lead designs with higher accuracy. Laser ablation, plasma etching, and chemical etching are just a few of the possibilities and are illustrated schematically in Fig. 60.14. A comparison of small opening ca-

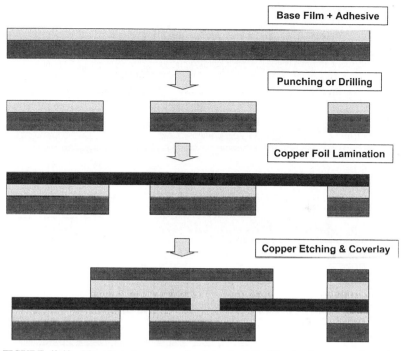

FIGURE 60.13 Manufacturing process for flying-lead flexible circuit construction using prepunching method.

pabilities of each of the processes is illustrated in Fig. 60.15, while a comparison of the technical capabilities of the processes is shown in Table 60.2. With these manufacturing technologies, many kinds of flying-lead constructions became possible.

60.3.3 Laser Processing

The excimer laser process has the capability to generate small openings with clearly defined edges on high-density flexible circuits. The narrow slit openings can be made smaller than 50 μm wide. The excimer process is available for all types of flexible circuit materials with high accuracy and fine openings. A carbon residue, typically removed by a suitable wet cleaning process, is found around the edges of the openings. One of the biggest disadvantages of excimer laser system is its slow speed, which accounts for the high processing cost when dealing with larger openings.

The UV:YAG laser process offers a higher productivity rate than the excimer laser for very small openings. It has the ability to cut both copper foil and organic substrate materials with good quality, which makes it suitable for cutting small openings on a wide range of high-density flexible circuits. Unfortunately, the productivity rate becomes extremely slow when cutting larger openings. Actually, it becomes impractical from a cost standpoint to utilize a UV:YAG laser system for openings larger than 100 μm².

Carbon dioxide TEA lasers, as well as diamond carbon dioxide lasers, have a much higher productivity rate when producing openings wider than 70 μm. The definition of the opening and quality of the edges, however, is somewhat less desirable than for other laser systems. In order to facilitate surface plating, a suitable cleaning process such as plasma is required to re-

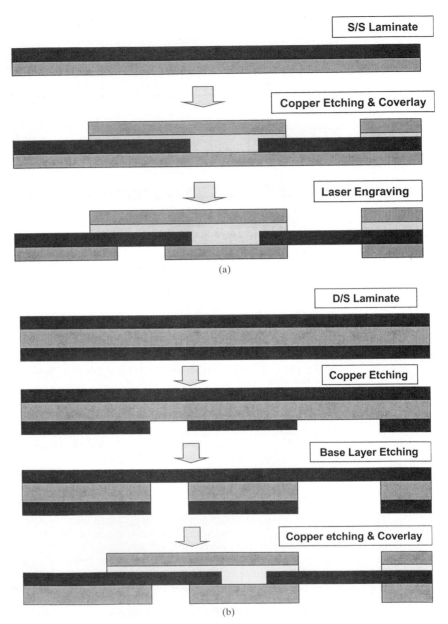

FIGURE 60.14 Manufacturing process for flying-lead flexible circuit construction using (*a*) laser ablation and (*b*) plasma etching or chemical etching.

move the thin residue remaining on the copper surface. A copper mask process helps to improve both quality and speed of the carbon dioxide laser process.

The excimer laser process, like the UV:YAG laser system, minimizes the potential for thermal damage on very thin flying leads. The carbon dioxide laser generates a remarkable amount of heat. The beam power density of the laser must be controlled carefully in order to process 18-μm copper with high-density flying leads narrower than 100 μm wide.

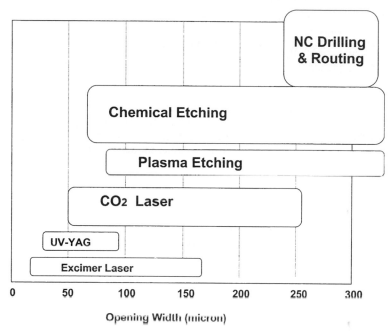

FIGURE 60.15 Comparison of opening sizes of flying-lead flexible circuit manufacturing processes.

TABLE 60.2 Comparison of Flying-Lead Manufacturing Technologies

	Prepunching (NC routing)	Laser ablation	Plasma etching	Chemical etching
Minimum slit width (μm)	800	50	100	70
Material availability	Fair	Wide	Wide	Limited
Design flexibility	Limited	Wide	Limited	Limited
Capability for high-density flying leads	Limited	Wide	Fair	Fair
Technical hurdles	Low	Fair	High	High
Registration	Difficult	Fair	Fair	Fair
Damage on flying leads	Serious	Small	Fair	Fair
Process cost	High	High	Low	Low
Investment for large volume	Small	High	High	Medium

60.3.4 Plasma Etching and Chemical Etching

Plasma and chemical etching processes are other suitable choices for generating high-density flying-lead structures. Although their capabilities are somewhat less than those of laser processing, these processes can provide a lower-cost model of high-volume processing. A copper foil mask is required in order to define the exposure area, and specific process conditions are required to remove organic substrates effectively. These processes are quite effective for single-sided substrates. Double-sided substrates can be processed in this manner given very specific materials and construction.

A big advantage of these processes is the unlimited number of holes or openings that can be processed simultaneously. Performance value is high, while manufacturing costs remain fixed when larger sizes or higher numbers of holes are processed. The plasma process is capable of etching all kinds of organic materials; however, it must be performed in a vacuum chamber. Capital investment is relatively large for high-volume production. Another issue involves uniform quality over a large working area/panel size in a vacuum chamber of limited size. Depending on the process conditions, plasma etching creates a wall slope of 30 to 60° at the edge of the openings, and it is difficult to make holes smaller than 100 μm in diameter on a 50-μm-thick polyimide layer.

Chemical etching also offers a low-cost process for creating access openings in high volume on polyimide-based high-density flexible circuits. This process offers the capability of generating openings smaller than 100 μm in diameter or slit width on 25-μm substrates, while providing a very small slope on the wall definition of the opening. The issue in this process is finding the appropriate chemicals and creating the suitable conditions to etch the various flexible substrates.

Sodium hydroxide and potassium hydroxide are popular chemicals for etching Kapton-type polyimide materials. Dangerous chemistries and tightly controlled process conditions are required to etch dimensionally stable polyimide substrates such as Upilex™. Serious consideration must be given to choosing the appropriate circuit design, materials, and manufacturing processes. On the other hand, chemical processing such as this requires no specialized capital equipment, and it is extremely suitable for high-volume roll-to-roll manufacturing systems. Typical high-density flying leads produced by new processes are shown in Fig. 60.16.

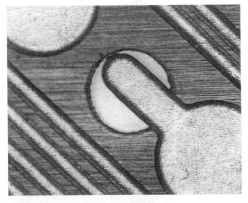

FIGURE 60.16   An example of a tab-shaped flying-lead flexible circuit construction. *(Source: Asahi Fine Technology.)*

60.3.5 Economic Comparison

Each technology has its own cost structure, which is dependent on the conditions. Figure 60.17 shows a comparison from an economic study of the various technologies. This study was conducted on 25-μm-thick polyimide substrate with a 300-mm² working size. The opening size is assumed to be 0.1×3 mm, with flying leads at a 150-μm pitch processed on 18-μm-thick copper.

Plasma and chemical etching processes have little dependency on the size or numbers of openings because they are based on batch processing of the materials in sheets. In contrast, the relative costs of the excimer and carbon dioxide laser process are proportional to the size of the area being processed as well as the opening sizes. This is primarily based on the small size

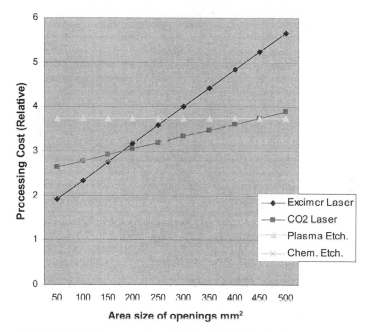

FIGURE 60.17   Cost comparison of flying-lead flexible circuit manufacturing processes.

of the laser beam as compared to the size of the opening, and on the need to overlap with each pass of the laser in order to remove the substrate in larger areas. The carbon dioxide laser, however, has higher manufacturing costs than the excimer laser process because of the additional chemical processes required to remove the organic residue left on the copper surface, but it has a lower associated manufacturing cost for larger opening sizes due to faster ablation speeds.

60.4   TAPE AUTOMATED BONDING

TAB has been developed as the special wiring material for IC chips and liquid crystal displays (LCDs). Although it was developed separately from standard flexible circuits, they have similar constructions and they use similar materials, and can be categorized as one kind of flexible circuits.

6.4.1   Basic Concepts

Basic construction of TAB is the same as that of the flying-lead structure of flex circuits except for the web size used in manufacturing and the sprocket holes made for reel-to-reel conveyor systems which give TAB an appearance similar to that of cinema film. A 35-mm-wide web is the most common standard width for TAB. Webs 70 and 105 mm wide are available for larger circuits. Because of the small working areas, TAB is good at producing fine traces with flying leads. The reeled circuits are convenient for automation of circuit manufacturing and termination. A short strip circuit made by standard manufacturing processes for flex circuits is also called a TAB circuit because it has the same functions. (Examples of TAB are shown in Fig. 60.18.)

FIGURE 60.18 Examples of TAB high-density flexible circuits.

60.4.2 Manufacturing Processes

The manufacturing flow of single-sided TAB is the same as that of the traditional prepunch manufacturing process of flying-lead construction. Basically, all manufacturing processes of TAB are designed for automatic reel-to-reel equipment. Because of the narrower material width, they have high capability for fine-line generation with good dimensional accuracy. The chemical etching process is also available to generate device holes. The technology can generate 40-μm-pitch traces as the flying leads.

60.5 MICROBUMP ARRAYS

A variety of shapes and materials for microbump arrays on flexible circuits have been developed to accommodate specific applications.

60.5.1 Basic Design and Applications

There are several choices of places to build microbumps on flexible circuits, as described in Chap. 57 (see Sec. 57.3.8). The same processes used to build ball grid arrays on rigid circuit boards are appropriate for flexible circuits with flexible photoimageable solder mask. It is possible to build microbumps on bare conductors using plating masks. Many microbumps are built on fine flying leads without solder masks. A valuable construction of a flexible circuit is microbumps built on the other side of the circuit through the dielectric layer. This can provide a reliable microbump array because of the stable construction.

Many different shapes have been developed for microbump arrays. The fusing solder process can make a spherical bump shape easily. Electrical plating of copper, nickel, gold, and combinations of these elements can provide many varieties of bump shapes. Changing the combination ratios and other factors can generate flat disks, flat domes, straight columns, etc. (Fig. 60.19). The best combination can provide 100- to 150-μm-pitch microbump arrays on 50-μm-thick dielectric layers.

Most of the microbump array constructions of flexible circuits have been developed for high-density termination (e.g., flip-chip, CSP, test probes, etc.). A suitable design shape and materials should be chosen according to the requirements of the termination.

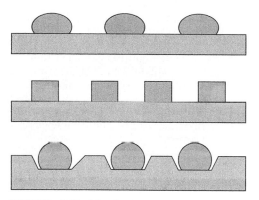

FIGURE 60.19 Microbump arrays on flexible circuits: examples of bump shapes.

60.5.2 Manufacturing

There are two kinds of manufacturing processes for the microbump array on flex circuits:

1. The same technology as for ball grid arrays may be used. Solder paste is dispensed on the pads of the circuits with a liquid dispenser or screen printer. The solder paste will form spherical shapes for a reflow process. The process is available for relatively large solder balls.

2. Electroplating processes are used for small solder balls and nonspherical microbumps.

Suitable combinations of different metal plating can provide wide varieties of microbumps. There are several factors that determine the shape and size of the bumps. The thickness of the dielectric layer and the size of the opening are the major factors in determining the bump shapes. The bump shapes will be more spherical when the aspect ratio of the opening is large. The bump shapes will be like flat discs when the aspect ratio is small. An appropriate plating mask will help to build straight, high bumps.

60.6 THICK-FILM CONDUCTOR FLEX CIRCUITS

A thick-film circuit is a type of additive manufacturing process. If an organic resin matrix is used, the conductor can be flexible, and this process has been applied in flexible circuits.

60.6.1 Materials

A silver paste conductor with an acrylic resin matrix is the most popular conductor material as the major thick-film conductor material for flexible circuits. It can provide very flexible conductor layers via a simple screen-printing process. Copper-based and carbon-based paste materials have been developed as the low-cost materials, but their conductivity is very low and unstable, and therefore their applicable areas are limited.

60.6.2 Applications

A common application for thick-film conductors is as shield layers for high-speed flexible circuit cable. If a shielding layer of both outerlayers is built with copper foils as parts of a multilayer flexible circuit, the whole circuit will lose flexibility. However, a flexible circuit cable built with shielding layers of silver paste thick-film conductors can maintain good flexibility. A silver paste thick-film flexible circuit built on polyester film can provide the lowest-cost solution for large circuits. Unfortunately, the conductivity of the traces is much lower than for copper foil conductors, and this material is not available for power circuits or signal layers of high-speed circuits. Standard soldering is not available because of the acrylic-based matrix. The largest application of thick-film-based flexible circuits is in membrane switches, which need large circuit spaces but do not carry large currents. The majority of membrane switches are covered by silver paste thick-film flexible circuits. Major switching circuits of keyboard switches are also covered by the same constructions.

60.6.3 Manufacturing Process

It is a very simple process to build a thick-film flexible circuit. There are two steps:

1. Screen printing on a flexible film followed by baking
2. Blanking.

This still requires that the process conditions be controlled carefully. The basic properties of these circuits change significantly depending on these conditions. A prebaking step is recommended for polyester film substrate to eliminate thermal shrinkage during baking for printed conductor paste. It is not difficult to apply the roll-to-roll manufacturing system for large volume because of the simplicity of the processes.

CHAPTER 61
QUALITY ASSURANCE OF FLEXIBLE CIRCUITS

Dominique K. Numakura
DKN Research, Haverhill, Massachusetts

61.1 INTRODUCTION

Because of the different design concepts and materials, there are several significant differences in quality assurance for flex circuits vs. that for typical rigid circuit boards. In addition, special test methods are required to guarantee the finished quality of high-density flex circuits, and new inspection technologies have been developed for this purpose.

61.2 BASIC CONCEPTS IN FLEXIBLE CIRCUIT QUALITY ASSURANCE

The basic concept of quality assurance for flexible circuits is the same as that for rigid boards in the sense that any critical defect must be eliminated before shipping. This includes failures due to opens and shorts; dimensional degradation of pad arrays; serious defects on conductors, substrates, and coverlay, etc. These defects need to be inspected for each circuit whether it is flexible or rigid. Technically, however, there are several differences in quality assurance for flexible circuits because of their additional functions, such as dynamic flexing capability.

A common issue is damage incurred during the manufacturing process because of the fragile materials. Even though the fine traces are inspected by a high-resolution automatic optical inspection (AOI) system during processing, it is possible that subsequent processes may cause new damage to the inspected circuits. This is a special issue for flying leads, which have no mechanical support under the traces. Therefore, a final inspection is necessary at the end of the manufacturing process.

The basic quality assurance concept for high-density flexible circuits is the same as for other circuit types. However, a difference for high-density flexible circuits is the acceptable defect size, which is one order smaller than that for traditional flexible circuits, necessitating higher inspection capabilities. Furthermore, new high-density flexible circuits contain additional structures such as flying leads and microbump arrays that require additional inspection capabilities for reliable termination. Exact 3-D accuracy and uniform surface conditions are required. Dimensional allowances are smaller than ±0.3 percent, and sometimes ±2-μm accuracy is required. This is true not only for trace pitches, but also for linewidths and spaces. Three-dimensional measurements are required to guarantee the quality of microbump arrays. Usually, it takes time to measure exact dimensions for each circuit. This is one of the major costs of the circuits when the specification is very tight compared to the process capability.

There have been several basic technical issues regarding quality assurance for high-density flex circuits, as listed in Table 61.1. Because of the fine circuits and the severe acceptable level of the defects, low-magnification scopes and low-resolution AOI do not work effectively. Magnification of 20× to 50× is required for visual inspections of fine circuits, and higher resolutions are required for AOI systems. Low contrast between copper and polyimide base creates another difficulty for traditional AOI equipment. It is not easy for common AOI systems to distinguish the copper traces from the circuits on the other side of a double-sided circuit built on adhesiveless substrate, because the base layer is very transparent and the sensor detects both sides of the circuits as one.

TABLE 61.1   Key Inspection Issues for High-Density Flex Circuits

Item	Issues
Finer traces	Resolution of AOI
Micro–blind via holes	Low contrast for AOI Resolution of AOI
Unstable materials	Dimensional accuracy
Photoimageable coverlay	Low contrast for AOI
Fragile traces	Damage during processes Damage from test probe pins
Fragile flying leads	Damage after AOI
Microbump arrays	3-D constructions
Cosmetic defects	Serious failures caused by small defects

61.3 AUTOMATIC OPTICAL INSPECTION SYSTEMS

AOI is the most valuable inspection equipment for the detection of dimensional defects in the traces and spaces of the printed circuits, especially for high-density traces, as well as for detection of cosmetic defects on the surfaces. Further sensitivities are required for AOI systems to detect small defects on coverlay, flying leads, microbumps, narrow space perimeters, etc. These have low contrast, different brightness, different defect shapes, etc. Traditional AOI systems are usually not appropriate for inspection in these circumstances.

Traditional AOI systems were designed for rigid circuit boards, and they were good at inspection of only the copper trace qualities. They could work for 50-μm-line/space fine traces, but their inspection speed could be very slow. There has been remarkable progress in resolution and productivity in the newer AOI systems; as a result, systems can detect 1.5-μm defects on 15-μm traces. Also, the systems have other capabilities to detect various defects on polyimide films and coverlay films. A careful review of AOI capabilities, compared to the specified design requirements, should be done prior to committing to particular equipment.

61.4 DIMENSIONAL MEASUREMENTS

It seems to be an impossibility to require accurate dimensions on an unstable, thin, flexible material. However, high-density interconnects in particular demand very high accuracy in the termination area of the circuits, even though the dimensional allowance is smaller than the dimensional stability of flexible substrate materials. Therefore, high-accuracy 2-D or 3-D dimension measurement technologies are required.

There has been continuous progress in dimension-measuring equipment. The newest 3-D measurement equipment has a resolution higher than 0.2 µm and accuracy higher than 1 µm, with statistical data in a 200-mm square 100 mm high. It has automatic measuring capabilities to detect each measuring point, producing statistical data. It can provide standard deviation data for each point with C_{pk} value.

Usually, there are several critical points that require exact dimensional accuracy in a flexible circuit. They are guide marks and termination areas, and customers ask exact tolerance for each dimension. The stable dimensions, such as punched perimeters processed by blanking dies, can be represented with the first article. However, unstable dimensions such as linewidth and pad pitch should be checked for each circuit. A statistical method is available. The measurement can be conducted by sampling when C_{pk} is larger than 1.3, but 100 percent measurement should be done when C_{pk} is smaller than 1.0. See MIL-STD-414 for help in reducing the number of samples.

61.5   ELECTRICAL TESTS

Electrical tests of high-density flex circuits could be one of the largest issues. Electrical open/short performance must be guaranteed as the minimum function of flex circuits. A traditional contact electrical testing method presents two basic barriers in inspection of high density flex circuits. The first is a geometric limit of probe pins. The costs of test fixtures are in inverse proportion to pitches of pin arrays, and 150-µm-pitch arrays could be the physical limit of pin probes. The second problem is physical damage caused by probe pins on fine traces, especially on fragile flying leads. A bus bar trace for a plating process is a problem in detecting shorts in the circuits with the traditional pin contact electrical test method. A new additional electrical test method is required.

A noncontact electrical test system can detect all open/short failures for 50-µm-pitch fine traces without pin contacts. It does not make any dents or cause damage on the fine traces. It works for fragile, fine flying leads because it is a noncontact inspection system. The basic principle of the method is illustrated in Fig. 61.1. An electric field sensor or a magnetic field sensor prepared for each circuit will detect all opens and shorts in branched circuits. Another big advantage of the noncontact test system is that it works for general circuits that have a common bus bar for electrical plating. The circuits must be isolated to conduct the electrical short test when traditional contact test methods are being used.

A noncontact electrical test system can inspect more than 5000 traces in a few seconds, eliminating physical damage to the circuits. It is also possible to use it in reel-to-reel processes. On the other hand, it still requires electrical contact on the other side of the fine traces, and therefore appropriate extra leads are required for pin or conductive rubber contacts.

61.6   INSPECTION SEQUENCE

Table 61.2 shows major quality assurance items for the final high-density flexible circuit products. It is difficult to check all items after each manufacturing process. Therefore, an appropriate inspection sequence in the total manufacturing flow should be designed.

On the other hand, it is a basic concept that failures should be eliminated as early as possible to ensure a high process yield, from a quality control standpoint, especially for long manufacturing processes. Figure 61.2 shows the general manufacturing flow of high-density flexible circuits, and a recommended inspection procedure.

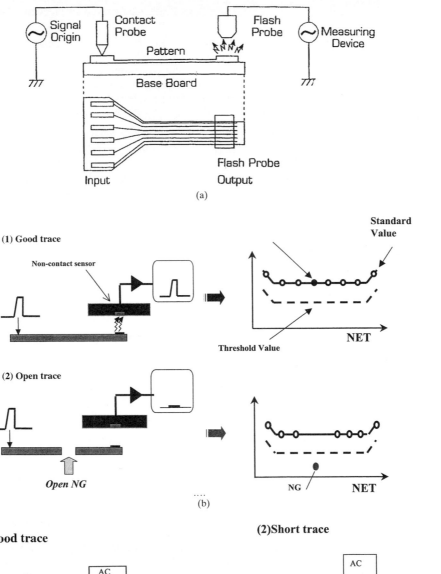

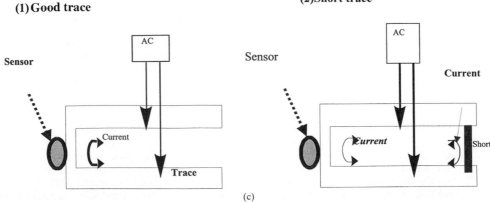

FIGURE 61.1 Noncontact electrical testing of flexible circuits. (*a*) General principle of noncontact electrical testing, showing contact only at the input. (*b*) Principle of noncontact testing for open circuits, where NG means no go (rejected). (*c*) Principle of noncontact testing for short circuits, using a bus bar circuit. (*Source: OHT.*)

TABLE 61.2 Quality Assurance Items for High-Density Flex Circuits

Item	Requirements
Physical properties	Sampling tests of raw materials Sampling tests of coupons
Opens/shorts	100% electrical inspection
Reliability of microvia holes	Cross sections obtained by sampling Daisy-chain test by coupon
Dimensional accuracy	Sampling by MIL-414 or 100%
Trace quality	100% inspection
Coverlay opening	Visual inspection, 100% or by sampling
Flying leads	Sampling test for dimensions 100% for straightness
Cosmetic defects	100% visual inspection, audit by MIL-105

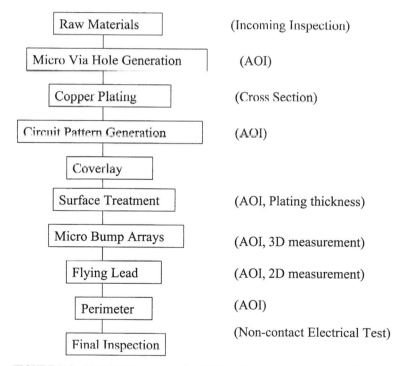

FIGURE 61.2 Manufacturing process for HDI flexible circuits and key inspection steps.

61.7 RAW MATERIALS

Basic properties of the raw materials should be guaranteed by material vendors; however, it is still necessary to conduct incoming inspections by circuit manufacturers. There is less experience with the new materials developed for high-density flex circuits than with traditional materials. Physical properties and cosmetic defects should be inspected by sampling.

Micro-size defects larger than 25 μm in diameter, such as pinholes and dents, will cause serious failures in 60-μm-pitch traces. Material vendors and circuit manufacturers should come to an agreement about acceptable levels, as well as about AQL and sampling.

61.8 FLEXIBLE CIRCUIT FEATURE INSPECTION

Flexible circuit features that require special consideration include:

- Micro-size holes
- Trace quality
- Coverlay and surface treatments

61.8.1 Micro-Size Hole Inspection

Several new technologies have been developed to generate blind holes smaller than 100 μm. These include lasers, plasma etching, and chemical etching. Unfortunately, the reliability of micro-hole generation processes has not been established exactly, and a suitable inspection system should be used. Traditional hole checkers do not work on blind holes. An AOI system is sensitive to the low contrast of copper foil surfaces.

There is no perfect nondestructive test method for evaluating the reliability of microvia holes. It is not easy to create suitable conditions for uniform plating of copper in very small holes. As a result, cross-sectional inspections should be done periodically. Also, a daisy-chain coupon arranged beside the circuits in the work should be tested periodically.

61.8.2 Trace Quality

As there is no perfect circuit generation process for fine traces, a high-resolution AOI system is required for high-density flex circuits. All serious defects larger than one-fourth of conductor widths or spaces must be detected by AOI. This means the AOI system must have a resolution higher than 5 μm for 25-μm traces. An appropriate sensitivity should be set in the AOI system to distinguish the copper traces from the polyimide base film and the traces on the other side of the film. It makes sense to apply AOI before etching to avoid a material loss. An algorithm of PC-Micro II™ can reduce the inspection time by applying high-resolution capabilities focusing on specific areas.

61.8.3 Coverlay and Surface Treatments

Coverlays and surface treatments are very difficult for current AOI systems to inspect because of the small optical contrast between coverlay materials and conductors or substrate materials. It may be better to apply AOI after a surface treatment, which will increase the contrast. Gold plating and tin plating create a clear contrast, and a small defect can be detected by AOI. The thickness and quality of surface treatments should be inspected immediately after the processes.

61.8.4 Microbump Arrays

Both cosmetic inspections and dimensional measurements are required for microbump arrays. Also, new ideas for inspecting 3-D constructions are needed. The uniformity of the bumps and cosmetic defects could be inspected by a suitable AOI system. The 3-D size of the bump and array pitches should be measured by a high-accuracy 3-D measuring system.

61.8.5 Flying Leads

The appropriate places for inspection of flying leads depend on the manufacturing process of the circuits. Flying leads should be inspected just after they are generated by copper etching or by engraving of the substrate. Both AOI inspection and dimensional measurement are required. Usually, flying leads have a severe dimensional allowance for both linewidth and pitches for reliable termination.

61.8.6 Electrical Testing (Opens/Shorts)

The electrical function is the basic and most important performance of the circuit. Unfortunately, the reliability of manufacturing processes for high-density flex circuits has not been established perfectly. There are many possibilities to cause further damage to the traces after the etching process and the AOI process, and therefore a 100 percent electrical test is necessary at the end of manufacturing to guarantee the circuit's performance. This is similar in objective to rigid bare board testing, and Chaps. 37 to 40 provide a detailed discussion of these issues. Certainly, final electrical tests must not cause any additional damage to the fragile, fine traces. The best solution is a noncontact electrical test. An appropriate design of noncontact sensors combining flash probe, flash shock, common flash, and surface probe of OHT can reduce the total inspection cost. A suitable circuit pattern should be arranged at the beginning. The inspection should be conducted before blanking for handling convenience when final circuits are very tiny.

61.8.7 Physical Properties (Final Properties)

The properties of raw materials do not represent final properties of the circuits. They undergo change during the manufacturing process, and therefore final properties should be measured at the end of the manufacturing process. Generally, most of electrical, mechanical, and chemical properties are different from those of raw materials. Critical properties such as flexing endurance, insulation resistance, and heat resistance should be measured carefully.

IPC-TM-650 indicates a flexing endurance test method based on a flexing mode as shown in Fig. 61.3. Unfortunately, the test method takes a very long time (sometimes more than 1 month) to give a final result by conductor failure. JIS-C5016 indicates an alternative test method called the MIT test, as shown in Fig. 61.4. This method does not require more than 1 h for one test.

61.8.8 Cosmetic Quality

There is no perfect automatic inspection system that can detect all cosmetic defects. Visual inspection at high magnification is the most reliable inspection method of eliminating all kinds of defects in high-density flex circuits. A stereomicroscope with high magnification helps the inspection procedure, but inspectors need training with exact criteria sample books. A final audit to confirm the level of the inspections is also recommended. MIL-STD-105 could help to reduce number of samples for the audit.

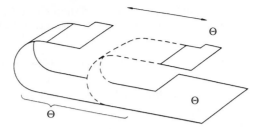

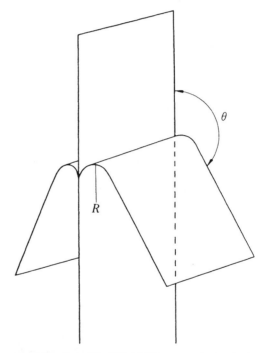

FIGURE 61.3 IPC test method for flexing endurance.

FIGURE 61.4 JIS-C5016 MIT test mode for flexing endurance. R is the radius of the bend during the test; θ is the angle from vertical.

61.9 STANDARDS AND SPECIFICATIONS FOR FLEXIBLE CIRCUITS

Most of the major standards associations have generated new standards or upgraded existing ones for flexible circuits. Major flexible circuit standards and relating standards are listed in the following sections.

61.9.1 IEC

IEC 249-2-15: flexible copper-clad polyimide film of defined flammability grade

IEC 326-7: specification for single- and double-sided flexible printed boards without through-hole connections

IEC 326-8: specification for single- and double-sided flexible printed boards with through-hole connections

IEC 326-9: specification for multilayer flexible printed boards with through-hole connections

IEC 326-10: specification for flex/rigid double-sided printed boards with through-hole connections

IEC 326-11: specification for flex/rigid multilayer printed boards with through-hole connections

61.9.2   IPC

IPC-FC-231C: Flexible Bare Dielectrics for Use in Flexible Printed Wiring

IPC-FC-232C: Specification for Adhesive Coated Dielectric Films for Use as Cover Sheets for Flexible Printed Wiring

IPC-FC-234: Pressure Sensitive Adhesives Assembly Guidelines for Single Sided and Double Sided Printed Circuits

IPC-FC-241C: Flexible Metal-Clad Dielectrics for Use in Fabrication of Flexible Printed Wiring

IPC-RF-245: Performance Specification for Rigid-Flex Printed Boards

IPC-D-249: Design Standard for Flexible Single and Double-Sided Printed Boards

IPC-FC-250A: Specification for Single and Double-Sided Flexible Printed Wiring

IPC FA 251: Guidelines for Assembly of Single and Double-Sided Flex Circuits

IPC-6013A: Qualification and Performance Specification for Flexible Printed Boards

61.9.3   JIS

JIS-C 5016: test methods for flexible printed wiring boards

JIS-C 5017: flexible printed wiring boards, single-sided and double-sided

JIS-C 6471: test methods for copper-clad laminates of flexible printed wiring boards

JIS-C 6472: copper-clad laminates for flexible printed wiring boards (polyester film, polyimide film)

61.9.4   JPCA

JPCA-BM 01: copper-clad laminates of flexible printed circuits, polyester films, and polyimide films

JPCA-FC 01: single-sided flexible printed circuits

JPCA-FC 02: double-sided flexible printed circuits

JPCA-FC 03: visual defect criteria for flexible printed circuits

61.9.5   MIL

MIL-STD-2118: Flexible and Rigid-Flex Printed-Wiring for Electronic Equipment

MIL-C-28809: Circuit Card Assemblies, Rigid, Flexible and Rigid-Flex

MIL-P-50884C: Printed Wiring, Flexible and Rigid-Flex

61.9.6   UL

UL796F: Flexible Materials Interconnect Constructions

APPENDIX
SUMMARY OF KEY COMPONENT, MATERIAL, PROCESS, AND DESIGN STANDARDS

David W. Bergman
IPC, Northbrook, Illinois

The following documents center on surface-mount technology. These documents have been developed by standards organizations in the United States and internationally. The letters in each document designation indicate the organization that has responsibility for the document:

EIA: Represents documents prepared by the Electronic Components, Assemblies, Equipment and Supplies Association (ECA) of the Electronic Industries Alliance.

JEDEC: Represents documents of the Solid State Technology Association of the Electronic Industries Alliance.

IPC: Represents documents prepared by the IPC—Association Connecting Electronics Industries.

MIL: Represents documents prepared by the military.

DoD: Represents documents prepared by the Department of Defense.

J-STD: Represents joint industry standards (more than one association).

SURFACE MOUNT COUNCIL PUBLICATIONS

SMC-TR-001: An Introduction to Tape Automated Bonding Fine Pitch Technology

SMC-WP-001: Soldering Capability

SMC-WP-002: An Assessment of the Use of Lead in Electronic Assemblies

SMC-WP-003: Chip Mounting Technology

SMC-WP-004: Design for Excellence

SMC-WP-005: PWB Surface Finishes

COMPONENTS, GENERAL

JESD-625A: Requirements for Handling Electrostatic-Discharge-Sensitive (ESDS) Devices

EIA-186-E: Standard Test Methods for Passive Electronic Component Parts—General Instructions and Index

EIA-481-2-A: Embossed Carrier Taping of Surface Mount Components for Automatic Handling (16, 24, 32, 44, and 56 mm)

EIA/IS-47: Contact Termination Finish Standard for Surface Mount Devices

EIA-PDP-100: Registered and Standard Mechanical Outlines for Electronic Parts

EIA-JEP-95: JEDEC Registered and Standard Outlines for Solid State and Related Products

EIA-JESD30B: Descriptive Designation System for Semiconductor Device Packages

IPC-9501: Component Qualification for the Assembly Process

IPC-9502: PWB Assembly Soldering Process Guideline for Electronic Components

IPC-9503: Moisture Sensitivity Classification for Non-IC Components

IPC-9504: Assembly Process Simulation for Evaluation of Non-IC (Preconditioning Non-IC) Components

JEDEC Pub 124: Guidelines for the Packing, Handling, and Repacking of Moisture-Sensitive Components

PASSIVE COMPONENTS

Capacitors

EIA-198-E: Ceramic Dielectric Capacitors Classes I, II, III, and IV

EIA-469-C: Standard Test Method for Destructive Physical Analysis of High Reliability Ceramic Monolithic Capacitors

EIA-479A: Film-Paper, Film Dielectric Capacitors for 50/60 Hz Voltage Doubler Power Supplies

EIA-535: Series of Detail Specifications on Fixed Tantalum Capacitors

EIA-595: Visual and Mechanical Inspection Multilayer Ceramic Chip Capacitors

EIA-CB-11: Guidelines for the Surface Mounting of Multilayer Ceramic Chip Capacitors

EIA-535: Series on Fixed Tantalum Capacitors

EIA/IS-35: Two-Pin Dual In-Line Capacitors

EIA/IS-36: Chip Capacitors, Multi-Layer (Ceramic Dielectric)

EIA/IS-37: Multiple Layer High Voltage Capacitors (Radial Lead Chip Capacitors)

EIA/IS-717: Surface Mount Tantalum Capacitor Qualification Specification

IEC-384-3: Sectional Specification, Tantalum Chip Capacitors

IEC-384-10: Sectional Specification, Fixed Multilayer Ceramic Chip Capacitors

IECQ Draft: Blank Detail Specification, Fixed Multilayer Ceramic Chip Capacitors

IECQ-PQC-31: Sectional Specification, Fixed Tantalum Chip Capacitors with Solid Electrolyte

IECQ-PQC-32: Blank Detail Specification, Fixed Tantalum Chip Capacitor

Resistors

> EIA-575: Resistors, Rectangular, Surface Mount, General Purpose
>
> EIA-576: Resistors, Rectangular, Surface Mount, Precision
>
> EIA/IS-34: Leaded Surface Mount Resistor Networks, Fixed Film

ACTIVE COMPONENTS

> EIA-JEP-95: JEDEC Registered and Standard Outlines for Solid State and Related Products
>
> EIA-JESD11: Chip Carrier Pinouts Standardized for CMOS 4000, HC, and HCT Series of Logic Circuits
>
> EIA-JESD21-C: Configurations for Solid State Memories
>
> EIA-JESD22-B: Test Methods and Procedures for Solid State Devices Used in Transportation/Automotive Applications (series format—consists of over 16 different test procedure documents)
>
> EIA-JESD-26A: General Specification for Plastic Encapsulated Microcircuits for Use in Rugged Applications
>
> EIA-JESD30: Descriptive Designation System for Semiconductor Device Packages
>
> IPC/JEDEC J-STD-020A: Moisture/Reflow Sensitivity Classification for Plastic Integrated Circuits Surface Mount Devices
>
> IPC/JEDEC-033: Standard for Handling, Packing, Shipping and Use of Moisture Reflow Sensitive Surface Mount Devices
>
> JEDEC Pub 124: Guidelines for the Packing, Handling, and Repacking of Moisture-Sensitive Components

ELECTROMECHANICAL COMPONENTS

Connectors

> EIA-364-C: Electrical Connector/Socket Test Procedures Including Environmental Classifications
>
> EIA/IS-47: Contact Termination Finish Standard for Surface Mount Devices
>
> IPC-CI-408: Design and Application Guidelines for the Use of Solderless Surface Mount Connectors

Sockets

> EIA-5400000: Generic Specification for Sockets for Use in Electronic Equipment
>
> EIA-540H000: Sectional Specification for Burn-In Sockets Used with Ball Grid Array Devices for Use in Electronic Equipment

Switches

> IECQ-PQC-41: Detail Specification, Dual-in-Line Switch, Surface Mountable, Slide Actuated US0003

EIA-448-23: Surface Mountable Switches, Qualification Test

EIA-5200000-C: Generic Specification for Special-Use Electromechanical Switches of Certified Quality

Printed Boards

IPC-6011: General Performance Requirements for Printed Boards

IPC-6012: Performance Specification for Rigid Printed Boards

IPC-6013: Performance Specification for Flexible Printed Boards

IPC-6018: Microwave End Product Board Inspection and Test

IPC-6105: Performance Specification for Organic Multichip Module Structures (MCM-L)

MIL-PRF-31032: Printed Circuit Board/Printed Wiring Board Manufacturing, General Specification

MIL-P-50884: Military Specification for Printed Wiring, Flexible, and Rigid Flex

Materials

IPC-4101: Laminate/Prepreg Materials Standard for Printed Boards

IPC-CF-148: Resin Coated Metal for Multilayer Printed Boards

IPC-3406: Guidelines for Electrically Conductive Surface Mount Adhesive

IPC-3408: General Requirements for Anisotropically Conductive Adhesive Films

IPC-3407: General Requirements for Isotropically Conductive Adhesives

IPC-CF-152A: Metallic Foil Specification for Copper/Invar/Copper (CIC) for Printed Wiring and Other Related Applications

IPC-CC-830A: Qualification and Performance of Electrical Insulation Compounds for Printed Board Assemblies

IPC-SM-840C: Qualification and Performance of Permanent Solder Mask for Printed Boards

J-STD-004: Requirements for Soldering Fluxes

J-STD-005: General Requirements and Test Methods for Electronic Grade Solder Paste

J-STD-006: General Requirements and Test Methods for Soft Solder Alloys and Fluxed and Non Fluxed Solid Solders for Electronic Soldering Applications

IPC-STD(P116): Qualification and Performance of Flip Chip Underfill Materials

DESIGN ACTIVITIES

IPC-D-279: Reliability Design Guidelines for Surface Mount Technology Printed Board Assemblies

IPC-D-316: Design Guide for Microwave Circuit Boards Utilizing Soft Substrates

IPC-D-317: Design Standard for Electronic Packaging Utilizing High Speed Techniques

IPC-C-406: Design and Application Guidelines for Surface Mount Connectors

IPC-SM-782A: Surface Mount Land Patterns (Configuration and Design Rules)

IPC-H-855: Hybrid Microcircuit Design Guide

IPC-D-859: Design Standard for Multilayer Hybrid Circuits

IPC-2221: Generic Standard on PWB Design

IPC-2222: Sectional Standard on Rigid PWB Design

IPC-2223: Sectional Design Standard for Flexible Printed Boards

IPC-2225: Design Standard for Organic Multichip Modules (MCM-L) and MCM-L Assemblies

IPC-(P107): Design Standard for Advanced IC Interconnection Mounting Structures

IPC-(P111): Design Standard for Flip Chip/Chip Scale Assembly Configurations

IPC-STD(P201): Mechanical Outline Standard for Ball Grid Arrays and Other High Density Technology

IPC (P207): Design Standard for High Density Array or Peripheral Leaded Component Mounting Structures

IPC-(P211): Design Standard for Assembling Array or Peripheral Leaded (BGA, QFP, etc.) Components for Printed Boards and Other Interconnecting Structures

J-STD-026: Semiconductor Design Standard for Flip Chip Applications

J-STD-(P102): Mechanical Outline Standard for Flip Chip or Chip Scale Configurations

MIL-STD-2118: Design Standard for Flexible Printed Wiring

IPC-2615: Printed Board Dimensions and Tolerances

COMPONENT MOUNTING

EIA-CB-11: Guidelines for the Surface Mounting of Multilayer Ceramic Chip Capacitors

IPC CM-770D: Guidelines for Printed Board Component Mounting

IPC-SM-784: Guidelines for Direct Chip Attachment

IPC-SM-780: Electronic Component Packaging and Interconnection with Emphasis on Surface Mounting

IPC-MC-790: Guidelines for Multichip Module Technology Utilization

J-STD-012: Implementation of Flip Chip and Chip Scale Technology

J-STD-013: Implementation of Ball Grid Array and Other High Density Technology

IPC/JPCA-2315: Design Guide for High Density Interconnects (HDI) and Microvias

IPC-4104: Specification for High Density Interconnect (HDI) and Microvia Materials

IPC-6016: Qualification and Performance Specification for High Density Interconnect (HDI) Layers or Boards

IPC-7095:

SOLDERING AND SOLDERABILITY

EIA/IS-46: Test Procedure for Resistance to Soldering (Vapor Phase Technique) for Surface Mount Devices

EIA-448-19: Method 19: Test Standard for Electromechanical Components, Environmental Effects of Machine Soldering Using a Vapor Phase System

EIA-638: Surface Mount Solderability Test

EIA-534: Applications Guide, Soldering and Solderability Maintenance of Leaded Electronic Components

IPC-TR-460A: Trouble Shooting Checklist for Wave Soldering Printed Wiring Boards

IPC-TR-462: Solderability Evaluation of Printed Boards with Protective Coatings Over Long-Term Storage

JESD22: B102C Solderability Test Methods

IPC-TR-464: Accelerated Aging for Solderability Evaluations

J-STD-001: Requirements for Soldered Electrical and Electronic Assemblies

J-STD-002: Solderability Tests for Component Leads, Terminations, Lugs, Terminals and Wires

J-STD-003: Solderability Tests of Printed Boards

IPC-S-816: Troubleshooting for Surface Mount Soldering

IPC-AJ-820: Assembly and Joining Handbook

QUALITY ASSESSMENT

EIA-469-C: Standard Test Method for Destructive Physical Analysis of High Reliability Ceramic Monolithic Capacitors

IPC-A-600E: Acceptability of Printed Boards

IPC-A-610: Acceptability of Printed Board Assemblies

MIL-STD-883: Methods and Procedures for Microelectronics

J-STD-028: Performance Standard for Flip Chip/Chip Scale Bumps

IPC-STD-(P203): Performance Standard for Ball Grid Array Bumps and Columns

IPC-A-20/21: Stencil Pattern for Solder Paste Slump Test

IPC-A-24: Flux/Board Interaction Board

IPC-A-36: CFC Cleaning Alternatives Artwork

IPC-A-38: Fine Line Round Robin Test Pattern

IPC-A-48: Surface Mount Airforce Mantech Artwork

IPC/JEDEC J-STD-035: Acoustic Microscopy for Non-Hermetic Encapsulated Electronic Components

SURFACE MOUNT PROCESS

IPC-SC-60: Post Solder Solvent Cleaning Handbook

IPC-SA-61: Post Solder Semi Aqueous Cleaning Handbook

IPC-AC-62: Post Solder Aqueous Cleaning Handbook

RELIABILITY

IPC-SM-785: Guidelines for Accelerated Surface Mount Attachment Reliability Testing

IPC-D-279: Design Guidelines for Reliable Surface Mount Technology Printed Board Assemblies

NUMERICAL CONTROL STANDARDS

IPC-2511: Generic Requirements for Implementation of Product Manufacturing Description Data and Transfer Methodology

IPC-2512: Sectional Requirements for Implementation of Administrative Methods for Manufacturing Data Description

IPC-2513: Sectional Requirements for Implementation of Drawing Methods for Manufacturing Data Description

IPC-2514: Sectional Requirements for Implementation of Printed Board Manufacturing Data Description

IPC-2515: Sectional Requirements for Implementation of Bare Board Product Electrical Testing Data Description

IPC-2516: Sectional Requirements for Implementation of Assembled Board Product Manufacturing Data Description

IPC-2517: Sectional Requirements for Implementation of Assembly Circuit Testing Data Description

IPC-2518: Sectional Requirements for Implementation of Bill of Material Product Data Description

EIA-227-A: One-Inch Perforated Tape

EIA-267-C: Axis and Motion Nomenclature for Numerically Controlled Machines

EIA-274-D: Interchangeable Variable Block Data Contouring, Format for Positioning and Contouring/Positioning Numerically Controlled Machines

EIA-358-C: ANSI X3.4 American National Standard Code for Information Interchange for Numerical Machine Control Perforated Tape

EIA-408: Interface Between Numerical Control Equipment on Data Terminal Equipment Employing Parallel Binary Data Interchange

EIA-431: Electrical Interface Between Numerical Control and Machine Tools

EIA-441: Operator Interface Function of Numerical Controls

EIA-484-A: Electrical and Mechanical Interface Characteristics and Line Control Protocol Using Communication Control Characters for Serial Data Link between a Direct Numerical Control System and Numerical Control Equipment Employing Asynchronous Full Duplex Transmission

EIA-494-B: 32 Bit Binary CL (BCL) and 7 Bit ASCII CL (ACL) Exchange Input Format for Numerically Controlled Machines

TEST METHODS

JESD22-B105-B: Test Method B105-B Lead Integrity

EIA-638: Surface Mount Solderability Test

JESD22-B108: Coplanarity Test for Surface Mount Semiconductor Devices

IPC-TM-650: Test Methods Manual

J-STD-(P104): Test Methods for Flip Chip of Chip Scale Products

JEDEC Standard 22 Series, Test Methods

REPAIR

IPC-7711: Rework of Electronic Assemblies

IPC-7721: Repair and Modification of Printed Boards

TERMS AND DEFINITIONS

IPC-T-50F: Terms and Definitions for Interconnecting and Packaging Electronic Circuits

HOW TO OBTAIN THESE DOCUMENTS

Following are the addresses of the IPC and EIA, as well as other sources for documents shown in this SMT listing.

IPC—Association Connecting Electronics Industries
2215 Sanders Road, #250
Northbrook, IL 60062-6135

Electronic Industries Alliance (EIA)
2500 Wilson Boulevard
Arlington, VA 22201-3834

Global Engineering Documents
2805 McGaw Avenue
Irvine, CA 92713
Phone: 1-800-854-7179

Military documents are available from:

Standardization Documents
Order Desk, Building 4D
700 Robbins Avenue
Philadelphia, PA 19111-5094

Central office of the IEC:

International Electrotechnical Commission (IEC)
3 Rue de Varembe
1211 Geneva 20, Switzerland

IEC documents are available from:

American National Standards Institute (ANSI)
11 West 42nd Street
New York, NY 10036

GLOSSARY*

ACCELERATOR: A chemical that is used to speed up a reaction or cure, as cobalt naphthenate is used to accelerate the reaction of certain polyester resins. It is often used along with a catalyst, hardener, or curing agent. The term "accelerator" is often used interchangeably with the term "promoter."

ACCURACY: The ability to place the hole at the targeted location.

ADDITIVE PROCESS: A process for obtaining conductive patterns by the selective deposition of conductive material on an unclad base material.

ADHESIVE: Broadly, any substance used in promoting and maintaining a bond between two materials.

AGING: The change in properties of a material with time under specific conditions.

AMBIENT TEMPERATURE: The temperature of the cooling medium, such as gas or liquid, which comes into contact with the heated parts of an apparatus (or the normal temperature of the surrounding environment).

ANNULAR RING: The circular strip of conductive material that completely surrounds a hole.

ARC RESISTANCE: The time required for an arc to establish a conductive path in a material.

ARTWORK MASTER: An accurately scaled configuration used to produce the production master.

BACKUP MATERIAL: A material placed on the bottom of a laminate stack in which the drill terminates its drilling stroke.

BASE MATERIAL: The insulating material upon which the printed wiring pattern may be formed.

BASE MATERIAL THICKNESS: The thickness of the base material excluding metal foil cladding or material deposited on the surface.

BLIND VIA: Conductive surface hole that connects an outerlayer with an innerlayer of a multilayer PWB without penetrating the entire board.

BLISTERING: Localized swelling and separation between any of the layers of the base laminate or between the laminate and the metal cladding.

BONDING LAYER: An adhesive layer used in bonding other discrete layers during lamination.

BOND STRENGTH: The force per unit area required to separate two adjacent layers by a force perpendicular to the board surface; usually refers to the interface between copper and base material.

BOW: A laminate defect in which deviation from planarity results in a smooth arc.

B STAGE: An intermediate stage in the curing of a thermosetting resin. In it a resin can be heated and caused to flow, thereby allowing final curing in the desired shape.

B-STAGE LOT: The product from a single mix of B-stage ingredients.

B-STAGE RESIN: A resin in an intermediate stage of a thermosetting reaction. The material softens when heated and swells when in contact with certain liquids, but it may not entirely fuse or dissolve.

---

*Some terms may not be included in the glossary as they are treated in detail in the text. Please also see the index and subject chapters.

BURIED VIA: Conductive surface hole that connects one innerlayer to another innerlayer of a multilayer PWB without having a direct connection to either the top or bottom surface layer.

BURR: A ridge left on the outside copper surfaces after drilling.

CAPACITANCE: The property of a system of conductors and dielectrics which permits the storage of electricity when potential difference exists between the conductors.

CAPACITIVE COUPLING: The electrical interaction between two conductors caused by the capacitance between the conductors.

CARBIDE: Tungsten carbide, formula WC. The hard, refractory material forming the drill bits used in PWB drillings.

CATALYST: A chemical that causes or speeds up the cure of a resin but does not become a chemical part of the final product.

CERAMIC LEADED CHIP CARRIER (CLCC): A chip carrier made from ceramic (usually a 90–96% alumina or beryllia base) and with compliant leads for terminations.

CHIP CARRIER (CC): An integrated circuit package, usually square, with a chip cavity in the center; its connections are usually on all four sides. (See *leaded chip carrier* and *leadless chip carrier.*)

CHIP LOAD (CL): The movement of the drill downward per revolution; usually given in mils (thousandths of an inch) per revolution.

CHLORINATED HYDROCARBON: An organic compound having chlorine atoms in its chemical structure. Trichloroethylene, methyl chloroform, and methylene chloride are chlorinated hydrocarbons.

CIRCUIT: The interconnection of a number of electrical devices in one or more closed paths to perform a desired electrical or electronic function.

CLAD: A condition of the base material, to which a relatively thin layer or sheet of metal foil (cladding) has been bonded on one or both of its sides. The result is called a metal-clad base material.

CNC: Computer numerically controlled. Refers to a machine with a computer which stores the numerical information about location, drill size, and machine parameters, regulating the machine to carry out that information.

COAT: To cover with a finishing, protecting, or enclosing layer of any compound.

COLD FLOW: The continuing dimensional change that follows initial instantaneous deformation in a nonrigid material under static load. Also called creep.

COLLIMATION: The degree of parallelism of light rays from a given source. A light source with good collimation produces parallel light rays, whereas a poor light source produces divergent, nonparallel light rays.

COMPONENT HOLE: A hole used for the attachment and electrical connection of a component termination, including pin or wire, to the printed board.

COMPONENT SIDE: The side of the printed board on which most of the components will be mounted.

COMPOUND: A combination of elements in a stable molecular arrangement.

CONDUCTIVE FOIL: The conductive material that covers one side or both sides of the base material and is intended for forming the conductive pattern.

CONDUCTIVE PATTERN: The configuration or design of the electrically conductive material on the base material.

CONDUCTOR LAYER 1: The first layer having a conductive pattern, of a multilayer board, on or adjacent to the component side of the board.

CONDUCTOR SPACING: The distance between adjacent edges (not centerline to centerline) of conductors on a single layer of a printed board.

CONDUCTOR THICKNESS: The thickness of the copper conductor exclusive of coatings or other metals.

CONDUCTOR WIDTH: The width of the conductor viewed from vertically above, i.e., perpendicularly to the printed board.

CONFORMAL COATING: An insulating protective coating which conforms to the configuration of the object coated and is applied on the completed printed board assembly.

CONNECTOR AREA: The portion of the printed board that is used for providing external (input–output) electrical connections.

CONTACT BONDING ADHESIVE: An adhesive (particularly of the nonvulcanizing natural rubber type) that bonds to itself on contact, although solvent evaporation has left it dry to the touch.

CONTROLLED IMPEDANCE: The matching of substrate material properties with trace dimensions and locations to create a specific electric impedance as seen by a signal on the trace.

COPOLYMER: See *polymer.*

CORE MATERIAL: The fully cured inner-layer segments, with circuiting on one or both sides, that form the multilayer circuit.

CORNER MARKS: The marks at the corners of printed board artwork, the inside edges of which usually locate the borders and establish the contour of the board.

COUPON: One of the patterns of the quality conformance test circuitry area. (See *test coupon.*)

CRAZING: A base material condition in which connected white spots or crosses appear on or below the surface of the base material. They are due to the separation of fibers in the glass cloth and connecting weave intersections.

CROSS-LINKING: The forming of chemical links between reactive atoms in the molecular chain of a plastic. It is cross-linking in the thermosetting resins that makes the resins infusible.

CROSS TALK: Undesirable electrical interference caused by the coupling of energy between signal paths.

CRYSTALLINE MELTING POINT: The temperature at which the crystalline structure in a material is broken down.

CTE: Coefficient of thermal expansion. The measure of the amount a material changes in any axis per degree of temperature change.

CURE: To change the physical properties of a material (usually from a liquid to a solid) by chemical reaction or by the action of heat and catalysts, alone or in combination, with or without pressure.

CURING AGENT: See *hardener.*

CURING TEMPERATURE: The temperature at which a material is subjected to curing.

CURING TIME: In the molding of thermosetting plastics, the time in which the material is properly cured.

CURRENT-CARRYING CAPACITY: Maximum current which can be carried continuously without causing objectionable degradation of electrical or mechanical properties of the printed board.

DATUM REFERENCE: A defined point, line, or plane used to locate the pattern or layer of a printed board for manufacturing and/or inspection purposes.

DEBRIS: A mechanically bonded deposit of copper to substrate hole surfaces.

DEBRIS PACK: Debris deposited in cavities or voids in the resin.

DEFINITION: The fidelity of reproduction of the printed board conductive pattern relative to the production master.

DELAMINATION: A separation between any of the layers of the base laminate or between the laminate and the metal cladding originating from or extending to the edges of a hole or edge of the board.

DIELECTRIC CONSTANT: The property of a dielectric which determines the electrostatic energy stored per unit volume for a unit potential gradient.

DIELECTRIC LOSS: Electric energy transformed into heat in a dielectric subjected to a changing electric field.

DIELECTRIC LOSS ANGLE: The difference between 90° and the dielectric phase angle. Also called the dielectric phase difference.

DIELECTRIC LOSS FACTOR: The product of dielectric constant and the tangent of dielectric loss angle for a material.

DIELECTRIC PHASE ANGLE: The angular difference in phase between the sinusoidal alternating potential difference applied to a dielectric and the component of the resulting alternating current having the same period as the potential difference.

DIELECTRIC POWER FACTOR: The cosine of the dielectric phase angle (or sine of the dielectric loss angle).

DIELECTRIC STRENGTH: The voltage that an insulating material can withstand before breakdown occurs, usually expressed as a voltage gradient (such as volts per mil).

DIMENSIONAL STABILITY: Freedom from distortion by such factors as temperature changes, humidity changes, age, handling, and stress.

DIRECT IMAGING: The exposure of photoresist material with a laser without the use of positive or negative phototool.

DISSIPATION FACTOR: The tangent of the loss angle of the insulating material. Also called loss tangent or approximate power factor.

DRILL FACET: The surface formed by the primary and secondary relief angles of a drill tip.

DRILL WANDER: The sum of accuracy and precision deviations from the targeted location of the hole.

DUMMY: A cathode with a large area used in a low-current-density pulsating operation for the removal of metallic impurities from solution. The process is called "dummying."

DWELL POINT: The bottom of the drilling stroke before the drill bit ascends.

EDGE-BOARD CONTACTS: A series of contacts printed on or near an edge of a printed board and intended for mating with a one-part edge connector.

EDX: Energy dispersive x-ray fluorescent spectrometer.

ELASTOMER: A material which at room temperature stretches under low stress to at least twice its length but snaps back to its original length upon release of the stress. Rubber is a natural elastomer.

ELECTRIC STRENGTH: The maximum potential gradient that a material can withstand without rupture. It is a function of the thickness of the material and the method and conditions of test. Also called dielectric strength or disruptive gradient.

ELECTROLESS PLATING: The controlled autocatalytic reduction of a metal ion on certain catalytic surfaces.

ELEMENT: A substance composed entirely of atoms of the same atomic number, e.g., aluminum or copper.

EMULSION SIDE: The side of the film or glass on which the photographic image is present.

ENTRY MATERIAL: A material placed on top of a laminate stack.

EPOXY SMEAR: Epoxy resin which has been deposited on edges of copper in holes during drilling either as a uniform coating or as scattered patches. It is undesirable because it can electrically isolate the conductive layers from the plated-through-hole interconnections.

ETCHBACK: The controlled removal of all the components of the base material by a chemical process acting on the sidewalls of plated-through holes to expose additional internal conductor areas.

ETCH FACTOR: The ratio of the depth of etch to lateral etch.

EXOTHERM: A characteristic curve which shows heat of reaction of a resin during cure (temperature) vs. time. The peak exotherm is the maximum temperature on the curve.

EXOTHERMIC REACTION: A chemical reaction in which heat is given off.

FIBER EXPOSURE: A condition in which glass cloth fibers are exposed on machined or abraded areas.

FILLER: A material, usually inert, added to a plastic to reduce cost or modify physical properties.

FILM ADHESIVE: A thin layer of dried adhesive. Also, a class of adhesives provided in dry-film form with or without reinforcing fabric and cured by heat and pressure.

FLEXURAL MODULUS: The ratio, within the elastic limit, of stress to corresponding strain. It is calculated by drawing a tangent to the steepest initial straight-line portion of the load-deformation curve and using the equation $E_B = L^3 m/4bd^3$, where E_B is the modulus, L is the span (in inches), m is the slope of the tangent, b is the width of beam tested, and d is the depth of the beam.

FLEXURAL STRENGTH: The strength of a material subjected to bending. It is expressed as the tensile stress of the outermost fibers of a bent test sample at the instant of failure.

FLUOROCARBON: An organic compound having fluorine atoms in its chemical structure, an inclusion that usually lends stability to plastics. Teflon* is a fluorocarbon.

---

*Trademark of E. I. du Pont de Nemours & Company.

GEL: The soft, rubbery mass that is formed as a thermosetting resin goes from a fluid to an infusible solid. It is an intermediate state in a curing reaction, and a stage in which the resin is mechanically very weak.

GEL POINT: The point at which gelation begins.

GLASS TRANSITION POINT: The temperature at which a material loses properties and becomes a semiliquid.

GLASS TRANSITION TEMPERATURE: The temperature at which epoxy, for example, softens and begins to expand independently of the glass fabric expansion rate.

GLUE-LINE THICKNESS: Thickness of the fully dried adhesive layer.

GRID: An orthogonal network of two sets of parallel lines for positioning features on a printed board.

GROUND PLANE: A conducting surface used as a common reference point for circuit returns, shielding, or heat sinking.

GULL WING LEAD: A surface mounted device lead which flares outward from the device body.

HALOING: A light area around holes or other machined areas on or below the surface of the base laminate.

HARDENER: A chemical added to a thermosetting resin for the purpose of causing curing or hardening. A hardener, such as an amine or acid anhydride for an epoxy resin, is a part of the chemical reaction and a part of the chemical composition of the cured resin. The terms "hardener" and "curing agent" are used interchangeably.

HEAT-DISTORTION POINT: The temperature at which a standard test bar (ASTM D 648) deflects 0.010 in under a stated load of either 66 or 264 psi.

HEAT SEALING: A method of joining plastic films by simultaneous application of heat and pressure to areas in contact. The heat may be supplied conductively or dielectrically

HIGH DENSITY INTERCONNECT (HDI): Ultrafine-geometry multilayer PWB constructed with conductive surface microvia connections between layers. (Microvia is usually defined as a hole with a diameter less than 0.006 in.) These boards also usually include buried and/or blind vias and are made by sequential build-up lamination.

HOLE PULL STRENGTH: The force, in pounds, necessary to rupture a plated-through hole or its surface terminal pads when loaded or pulled in the direction of the axis of the hole. The pull is usually applied to a wire soldered in the hole, and the rate of pull is given in inches per minute.

HOOK: A geometric drill bit defect of the cutting edges.

HOT-MELT ADHESIVE: A thermoplastic adhesive compound, usually solid at room temperature, which is heated to fluid state for application.

HYDROCARBON: An organic compound containing only carbon and hydrogen atoms in its chemical structure.

HYDROLYSIS: The chemical decomposition of a substance involving the addition of water.

HYGROSCOPIC: Tending to absorb moisture.

I-LEAD: A surface mounted device lead which is formed such that the end of the lead contacts the board land pattern at a 90° angle. Also called a butt joint.

IMPREGNATE: To force resin into every interstice of a part, as of a cloth for laminating.

INHIBITOR: A chemical that is added to a resin to slow down the curing reaction and is normally added to prolong the storage life of a thermosetting resin.

INORGANIC CHEMICALS: Chemicals whose molecular structures are based on other than carbon atoms.

INSULATION RESISTANCE: The electrical resistance of the insulating material between any pair of contacts, conductors, or grounding devices in various combinations.

INTERNAL LAYER: A conductive pattern contained entirely within a multilayer board.

IPC: Institute for Interconnecting and Packaging Electronic Circuits. A leading printed wiring industry association that develops and distributes standards as well as other information of value to printed wiring designers, users, suppliers, and fabricators.

IR: Infrared heating for solder-reflow operation.

J-LEAD: A surface mounted device lead which is formed into a "J" pattern folding under the device body.

JUMPER: An electrical connection between two points on a printed board added after the printed wiring is fabricated.

LAMINATE: The plastic material, usually reinforced by glass or paper, that supports the copper cladding from which circuit traces are created.

LAMINATE VOID: Absence of epoxy resin in any cross-sectional area which should normally contain epoxy resin.

LAND: See *terminal area.*

LANDLESS HOLE: A plated-through hole without a terminal area.

LASER PHOTOPLOTTER (laser photogenerator, or LPG): A device that exposes photosensitive material, usually a silver halide or diazo material, subsequently used as the master for creating the circuit image in production.

LAYBACK: A geometric drill bit defect of the cutting edges.

LAYER-TO-LAYER SPACING: The thickness of dielectric material between adjacent layers of conductive circuitry.

LAY-UP: The process of registering and stacking layers of a multilayer board in preparation for the laminating cycle.

LCCC: Leadless ceramic chip carrier.

LEADED CHIP CARRIER: A chip carrier (either plastic or ceramic) with compliant leads from terminations.

LEADLESS CHIP CARRIER: A chip carrier (usually ceramic) with integral metallized terminations and no compliant external leads.

LEGEND: A format of lettering or symbols on the printed board, e.g., part number, component locations, or patterns.

LOOSE FIBERS: Supporting fibers in the substrate of the laminate which are not held in place by surrounding resin.

MAJOR WEAVE DIRECTION: The continuous-length direction of a roll of woven glass fabric.

MARGIN RELIEF: The area of a drill bit next to the cutting edge is removed so that it does not rub against the hole as the drill revolves.

MASTER DRAWING: A document that shows the dimensional limits or grid locations applicable to any or all parts of a printed wiring or printed circuit base. It includes the arrangement of conductive or nonconductive patterns or elements; size, type, and location of holes; and any other information necessary to characterize the complete fabricated product.

MEASLING: Discrete white spots or crosses below the surface of the base laminate that reflect a separation of fibers in the glass cloth at the weave intersection.

MICROSTRIP: A type of transmission line configuration which consists of a conductor over a parallel ground plane separated by a dielectric.

MICROVIA: Usually defined as a conductive hole with a diameter of 0.006 in or less that connects layers of a multilayer PWB. Often used to refer to any small-geometry connecting hole the creation of which is beyond the practical capabilities of traditional mechanical drilling processes.

MINOR WEAVE DIRECTION: The width direction of a roll of woven glass fabric.

MIXED ASSEMBLY: A printed wiring assembly that combines through-hole components and surface mounted components on the same board.

MODULUS OF ELASTICITY: The ratio of stress to strain in a material that is elastically deformed.

MOISTURE RESISTANCE: The ability of a material not to absorb moisture either from air or when immersed in water.

MOUNTING HOLE: A hole used for the mechanical mounting of a printed board or for the mechanical attachment of components to a printed board.

MULTILAYER BOARD: A product consisting of layers of electrical conductors separated from each other by insulating supports and fabricated into a solid mass. Interlayer connections are used to establish continuity between various conductor patterns.

MULTIPLE-IMAGE PRODUCTION MASTER: A production master used to produce two or more products simultaneously.

NAILHEADING: A flared condition of internal conductors.

NC: Numerically controlled. Usually refers to a machine tool, in this case a drilling machine. The most basic type is one in which a mechanical guide locates the positions of the holes. NC machines are usually controlled by punched tape.

NEMA STANDARDS: Property values adopted as standard by the National Electrical Manufacturers Association.

NOBLE ELEMENTS: Elements that either do not oxidize or oxidize with difficulty; examples are gold and platinum.

OILCANNING: The movement of entry material in the z direction during drilling in concert with the movement of the pressure foot.

ORGANIC: Composed of matter originating in plant or animal life or composed of chemicals of hydrocarbon origin, either natural or synthetic.

PAD: See *terminal area*.

PADS ONLY: A multilayer construction with all circuit traces on inner layers and the component terminal area only on the surface of the board. This construction adds two layers but may avoid the need for a subsequent solder resist, and since inner layers usually are easier to form, this construction may lead to higher overall yields.

pH: A measure of the acid or alkaline condition of a solution. A pH of 7 is neutral (distilled water); pH values below 7 represent increasing acidity as they go toward 0; and pH values above 7 represent increasing alkalinity as they go toward the maximum value of 14.

PHOTOGRAPHIC REDUCTION DIMENSION: The dimension (e.g., line or distance between two specified points) on the artwork master to indicate the extent to which the artwork master is to be photographically reduced. The value of the dimension refers to the 1:1 scale and must be specified.

PHOTOMASTER: An accurately scaled copy of the artwork master used in the photofabrication cycle to facilitate photoprocessing steps.

PHOTOPOLYMER: A polymer that changes characteristics when exposed to light of a given frequency.

PINHOLES: Small imperfections which penetrate entirely through the conductor.

PINK RING: The appearance of a halo of copper around the hole of a multilayer PWB.

PITS: Small imperfections which do not penetrate entirely through the printed circuit.

PLASTICIZER: Material added to resins to make them softer and more flexible when cured.

PLASTIC LEADED CHIP CARRIER (PLCC): A chip carrier packaged in plastic, usually terminating in compliant leads (originally "J" style) on all four sides.

PLATED-THROUGH HOLE: A hole in which electrical connection is made between printed wiring board layers with conductive patterns by the deposition of metal on the wall of the hole. (See *PTH*.)

PLATING VOID: The area of absence of a specific metal from a specific cross-sectional area: (1) When the plated-through hole is viewed as cross-sectioned through the vertical plane, it is a product of the average thickness of the plated metal times the thickness of the board itself as measured from the outermost surfaces of the base copper on external layers. (2) When the plated-through hole is viewed as cross-sectioned through the horizontal plane (annular method), it is the difference between the area of the hole and the area of the outside diameter of the through-hole plating.

PLOWING: Furrows in the hole wall due to drilling.

POLYMER: A high-molecular-weight compound made up of repeated small chemical units. For practical purposes, a polymer is a plastic. The small chemical unit is called a mer, and when the polymer or mer is cross-linked between different chemical units (e.g., styrene-polyester), the polymer is called a copolymer. A monomer is any single chemical from which the mer or polymer or copolymer is formed.

POLYMERIZE: To unite chemically two or more monomers or polymers of the same kind to form a molecule with higher molecular weight.

POTLIFE: The time during which a liquid resin remains workable as a liquid after catalysts, curing agents, promoter, etc., are added. It is roughly equivalent to gel time.

POWER FACTOR: The cosine of the angle between the applied voltage and the resulting current.

PRECISION: The ability to repeatedly place the hole at any location.

PREPRODUCTION TEST BOARD: A test board (as detailed in IPC-ML-950) the purpose of which is to determine whether, prior to the production of finished boards, the contractor has the capability of producing a multilayer board satisfactorily.

PRESS PLATEN: The flat heated surface of the lamination press used to transmit heat and pressure to lamination fixtures and into the lay-up.

PRESSURE FOOT: The tubelike device on the drilling machine that descends to the top surface of the stack, holding it firmly down, before the drill descends through the center of the pressure foot. The vacuum system of the drilling machine separates through the pressure foot to remove chips and dust formed in drilling.

PRINTED WIRING ASSEMBLY DRAWING: A document that shows the printed wiring base, the separately manufactured components which are to be added to the base, and any other information necessary to describe the joining of the parts to perform a specific function.

PRINTED WIRING LAYOUT: A sketch that depicts the printed wiring substrate, the physical size and location of electronic and mechanical components, and the routing of conductors that interconnect the electronic parts in sufficient detail to allow for the preparation of documentation and artwork.

PRODUCTION MASTER: A 1:1 scale pattern used to produce one or more printed wiring or printed circuit products within the accuracy specified on the master drawing.

PROMOTER: A chemical, itself a feeble catalyst, that greatly increases the activity of a given catalyst.

PTH: Plated-through holes. Also refers to the technology that uses the plated-through hole as its foundation.

QUADPACK: Generic term for surface mount technology packages with leads on all four sides. Commonly used to describe chip carrier-like devices with gull wing leads.

QUALITY CONFORMANCE CIRCUITRY AREA: A test board made as an integral part of the multilayer printed board panel on which electrical and environmental tests may be made for evaluation without destroying the basic board.

RAW MATERIAL PANEL SIZE: A standard panel size related to machine capacities, raw material sheet sizes, final product size, and other factors.

REFRACTIVE INDEX: The ratio of the velocity of light in a vacuum to the velocity in a substance. Also, the ratio of the sine of the angle of incidence to the sine of the angle of refraction.

REGISTER MARK: A mark used to establish the relative position of one or more printed wiring patterns, or portions thereof, with respect to desired locations on the opposite side of the board.

REGISTRATION: The relative position of one or more printed wiring patterns, or portions thereof, with respect to desired locations on a printed wiring base or to another pattern on the opposite side of the base.

RELATIVE HUMIDITY: The ratio of the quantity of water vapor present in the air to the quantity which would saturate the air at the given temperature.

REPAIR: The correction of a printed wiring defect after the completion of board fabrication to render the board as functionally good as a perfect board.

RESIN: High-molecular-weight organic material with no sharp melting point. For current purposes, the term "resin," "polymer," and "plastic" can be used interchangeably.

RESIST: A protective coating (ink, paint, metallic plating, etc.) used to shield desired portions of the printed conductive pattern from the action of etchant, solder, or plating.

RESISTIVITY: The ability of a material to resist passage of electric current through its bulk or on a surface.

RIFLING: Spiral groove or ridge in the substrate due to drilling.

RIGID/FLEX: A PWB construction combining flexible circuits and rigid multilayer PWBs, usually either to provide a built-in connection or to make a three-dimensional form that includes components.

ROCKWELL HARDNESS NUMBER: A number derived from the net increase in depth of an impression as the load on a perpetrator is increased from a fixed minimum load to as higher load and then returned to minimum load.

ROUGHNESS: Irregular, coarse, uneven hole wall on copper or substrate due to drilling.

SCHEMATIC DIAGRAM: A drawing which shows, by means of graphic symbols, the electrical interconnections and functions of a specific circuit arrangement.

SEM: Scanning electron microscope.

SEQUENTIAL BUILD-UP: A process for making multilayer PWBs in which already-finished multilayers are laminated together to form a higher-layer-count final board, or in which additional outerlayers are added to finished multilayer PWBs.

SHADOWING: Etchback to maximum limit without removal of dielectric material from conductors.

SHORE HARDNESS: A procedure for determining the indentation hardness of a material by means of a durometer.

SINGLE-IMAGE PRODUCTION MASTER: A production master used to produce individual products.

SINGLE-IN-LINE PACKAGE (SIP): Component package system with one line of connectors, usually spaced 0.100 in apart.

SMC: Surface mounted component. Component with terminations designed for mounting flush to printed wiring board.

SMD: Surface mounted device. Any component or hardware element designed to be mounted to a printed wiring board without penetrating the board.

SMEAR: Fused deposit left on copper or substrate from excessive drilling heat.

SMOBC: Solder mask over bare copper. A method of fabricating a printed wiring board which results in the final metallization being copper with no other protective metal; but the non soldered areas are coated by a solder resist, exposing only the component terminal areas. This eliminates tin-lead under the solder mask.

SMT: Surface mount technology. Defines the entire body of processes and components which create printed wiring assemblies without components with leads that pierce the board.

SOJ: SOIC package with J-leads rather than gull wing leads.

SOIC: Small-outline integrated circuit. A plastic package resembling a small dual-in-line package (DIP) with gull wing leads on two sides for surface mounting.

SOT: Small outline transistor. A package for surface-mounting transistors.

SPECIFIC HEAT: The ratio of the thermal capacity of a material to that of water at 15°C.

SPINDLE RUNOUT: The measure of the wobble present as the drilling machine spindle rotates 360°.

STORAGE LIFE: The period of time during which a liquid resin or adhesive can be stored and remain suitable for use. Also called shelf life.

STRAIN: The deformation resulting from a stress. It is measured by the ratio of the change to the total value of the dimension in which the change occurred.

STRESS: The force producing or tending to produce deformation in a body. It is measured by the force applied per unit area.

SUBSTRATE: A material on whose surface an adhesive substance is spread for bonding or coating. Also, any material which provides a supporting surface for other materials used to support printed wiring patterns.

SURFACE RESISTIVITY: The resistance of a material between two opposite sides of a unit square of its surface. It may vary widely with the conditions of measurement.

SURFACE SPEED: The linear velocity of a point on the circumference of a drill. Given in units of surface feet per minute—sfm.

TERMINAL AREA: A portion of a conductive pattern usually, but not exclusively, used for the connection and/or attachment of components.

TEST COUPON: A sample or test pattern usually made as an integral part of the printed board, on which electrical, environmental, and microsectioning tests may be made to evaluate board design or process control without destroying the basic board.

TETRA-ETCH*: A nonpyrophoric (will not ignite when exposed to moisture) proprietary etchant.

TETROFUNCTIONAL: Describes an epoxy system for laminates that has four cross-linked bonds rather than two and results in a higher glass transition temperature, or T_g.

T_g: Glass transition temperature. The temperature at which laminate mechanical properties change significantly.

THERMAL CONDUCTIVITY: The ability of a material to conduct heat; the physical constant for the quantity of heat that passes through a unit cube of a material in a unit of time when the difference in temperatures of two faces is 1°C.

THERMOPLASTIC: A classification of resin that can be readily softened and resoftened by repeated heating.

THERMOSETTING: A classification of resin which cures by chemical reaction when tested and, when cured, cannot be resoftened by heating.

THIEF: An auxiliary cathode so placed as to divert to itself some current from portions of the work which would otherwise receive too high a current density.

THIXOTROPIC: Said of materials that are gel-like at rest but fluid when agitated.

THROUGH-HOLE TECHNOLOGY: Traditional printed wiring fabrication where components are mounted in holes that pierce the board.

THROWING POWER: The improvement of the coating (usually metal) distribution ratio over the primary current distribution ratio on an electrode (usually a cathode). Of a solution, a measure of the degree of uniformity with which metal is deposited on an irregularly shaped cathode. The term may also be used for anodic processes for which the definition is analogous.

TWIST: A laminate defect in which deviation from planarity results in a twisted arc.

UNDERCUT: The reduction of the cross section of a metal foil conductor caused by the etchant removing metal from under the edge of the resist.

VAPOR PHASE: The solder-reflow process that uses a vaporized solvent as the source for heating the solder beyond its melting point, creating the component-to-board solder joint.

VIA: A metallized connecting hole that provides a conductive path from one layer in a printed wiring board to another. (1) *Buried via*—connects one inner layer to another inner layer without penetrating the surface. (2) *Blind via*—connects the surface layer of a printed wiring board to an internal layer without going all the way through the other surface layer.

VOID: A cavity left in the substrate.

VOLUME RESISTIVITY: The electrical resistance between opposite faces of a 1-cm cube of insulating materials, commonly expressed in ohm-centimeters. The recommended test is ASTM D 256 51T. Also called the specific insulation.

VULCANIZATION: A chemical reaction in which the physical properties of an elastomer are changed by causing the elastomer to react with sulfur or some other cross-linking agent.

WATER ABSORPTION: The ratio of the weight of water absorbed by a material to the weight of the dry material.

WEAVE EXPOSURE: A condition in which the unbroken woven glass cloth is not uniformly covered by resin.

WEAVE TEXTURE: A surface condition in which the unbroken fibers are completely covered with resin but exhibit the definite weave pattern of the glass cloth.

WETTING: Ability to adhere to a surface immediately upon contact.

WICKING: Migration of copper salts into the glass fibers of the insulating material.

WORKING LIFE: The period of time during which a liquid resin or adhesive, after mixing with catalyst, solvent, or other compounding ingredients, remains usable. (See *potlife*.)

---

*Trademark of W. L. Gore and Associates, Inc.

INDEX

ABOUT THE EDITOR

Clyde F. Coombs, Jr., is one of the best-selling editors in professional publishing. He developed and edited four editions of *Printed Circuits Handbook,* three editions of *Electronic Intrument Handbook,* and *Communications and Network Test and Measurement Handbook,* three of McGraw-Hill's top-selling technical handbooks. An experienced electronics engineer and manager, Mr. Coombs is currently semiretired from Hewlett-Packard. He lives in Los Altos, California.